OXYGEN TRANSPORT TO TISSUE XXI

ADVANCES IN EXPERIMENTAL MEDICINE AND BIOLOGY

Recent Volumes in this Series

Volume 462
ADVANCES IN BLADDER RESEARCH
Edited by Laurence S. Baskin and Simon W. Hayward

Volume 463
ENZYMOLOGY AND MOLECULAR BIOLOGY OF CARBONYL METABOLISM 7
Edited by Henry Weiner, Edmund Maser, David W. Crabb, and Ronald Lindahl

Volume 464
CHEMICALS VIA HIGHER PLANT BIOENGINEERING
Edited by Fereidoon Shahidi, Paul Kolodziejczyk, John R. Whitaker, Agustin Lopez Munguia, and Glenn Fuller

Volume 465
CANCER GENE THERAPY: Past Achievements and Future Challenges
Edited by Nagy A. Habib

Volume 466
CURRENT VIEWS OF FATTY ACID OXIDATION AND KETOGENESIS: From Organelles to Point Mutations
Edited by P. A. Quant and S. Eaton

Volume 467
TRYPTOPHAN, SEROTONIN, AND MELATONIN: Basic Aspects and Applications
Edited by Gerald Heuther, Walter Kochen, Thomas J. Simat, and Hans Steinhart

Volume 468
THE FUNCTIONAL ROLE OF GLIAL CELLS IN HEALTH AND DISEASE: Dialogue between Glia and Neurons
Edited by Rebecca Matsas and M. Tsacopoulos

Volume 469
EICOSANOIDS AND OTHER BIOACTIVE LIPIDS IN CANCER, INFLAMMATION, AND RADIATION INJURY, 4
Edited by Kenneth V. Honn, Lawrence J. Marnett, and Santosh Nigam

Volume 470
COLON CANCER PREVENTION: Dietary Modulation of Cellular and Molecular Mechanisms
Edited by AICR

Volume 471
OXYGEN TRANSPORT TO TISSUE XXI
Edited by Andras Eke and David T. Delpy

OXYGEN TRANSPORT TO TISSUE XXI

Edited by

Andras Eke
Semmelweis University of Medicine
Budapest, Hungary

and

David T. Delpy
University College London
London, England

KLUWER ACADEMIC / PLENUM PUBLISHERS
New York, Boston, Dordrecht, London, Moscow

Library of Congress Cataloging-in-Publication Data

Oxygen transport to tissue XXI / edited by Andras Eke and David T. Delpy.
p. cm. -- (Advances in experimental medicine andbiology ; v. 471)
"Proceedings of the 26th Annual Meeting of the International Society on Oxygen Transport to Tissue, held August 23-28, 1998, in Budapest, Hungary"--T.p. verso.
Includes bibliographical references and index.
ISBN 0-306-46281-8. -- ISBN 0-306-46281-8
1. Tissue respiration Congresses. 2. Oxygen--Physiological transport Congresses. I. Eke, Andras. II. Delpy, David T. III. Title: Oxygen transport to tissue 21. IV. Title: Oxygen transport to tissue twenty-one. V. Series.
[DNLM: 1. Oxygen Consumption--physiology Congresses. 2. Biological Transport--physiology Congresses. 3. Oxygen--metabolism Congresses. W1 AD559 v.471 1999]
QP121.O887 1999
572'.47--dc21
DNLM/DLC
for Library of Congress 99-40549
CIP

ISOTT '98

Proceedings of the 26th Annual Meeting of the International Society of Oxygen Transport to Tissue, held August 23–28, 1998, in Budapest, Hungary

ISSN 0065-2598

ISBN 0-306-46281-8

233 Spring Street, New York, New York 10013

10 9 8 7 6 5 4 3 2 1

A C.I.P. record for this book is available from the Library of Congress

Printed in the United States of America

Dr. Britton Chance

The editors wish to dedicate this volume in honor of Dr. Britton Chance and to commemorate his 85th birthday. Dr. Britton Chance played a key role in the early days of the International Society on Oxygen Transport to Tissue in the 1970s, and since then has been an active member whose scientific participation is greatly valued at the annual meetings of this society.

Dr. Chance has been an important figure in the development of modern biochemistry and biophysics. On considering his scientific achievements to date, the diversity of his fundamental and original contributions to many areas of science, including mathematics, electronics, biochemistry and clinical medicine is overwhelming. These contributions have been documented in an unprecedented number of publications describing studies from the molecular up to the whole-body level. His work has furthermore provided stimulus for other scientists and laboratories to continue and expand his studies, often leading to the development of new technological approaches. For example, his invention of the dual-wavelength spectrophotometer made the precise measurement of enzyme action in isolated organelles and tissues possible, leading subsequently to a series of key developments in biophysics, biochemistry and molecular biology. The later invention of a time-sharing spectrophotometer expanded knowledge of the function of myoglobin and cytochrome and enhanced the understanding of the effects of cardiac ischemia, while at the same time providing a new tool for the study of cardiac structure and function in vivo. Dr. Chance's combination of electro-optical and electromagnetic techniques have additionally played a major role in the development of NMR spectroscopy for biochemical studies in humans.

During his outstanding academic career at the University of Pennsylvania, where he was nominated professor of Biophysics and Physical Biochemistry, Dr. Chance has received numerous awards and honors and is a member of many renowned scientific societies. His strong and continuing commitment to research combined with his ability to transfer concepts and techniques from one discipline to another make him an outstanding scientist of our era.

Peter Vaupel

INTERNATIONAL SOCIETY ON OXYGEN TRANSPORT TO TISSUE

1998

Officers

President:	Andras Eke, Hungary
Past President:	Antal G. Hudetz, USA
President-elect:	Harold M. Swartz, USA
Secretary:	David F. Wilson, USA
Treasurer:	Peter E. Keipert, USA
Chairman of the Knisely Award Committee:	Duane F. Bruley, USA

Executive Committee

Sanja Batra, USA
David Benaron, USA
Karlfried Groebe, Germany
David Harrison, United Kingdom
Luis J.C. Hoofd, The Netherlands
Josef Moravec, France
Paul Okunief, USA
Karel Rakusan, Canada

Archivist

Ian S. Longmuir, USA

Scientific Program Committee

ISOTT

James B. Bassingthwaighte, USA
David T. Delpy, United Kingdom
Marco Ferrari, Italy
David K. Harrison, United Kingdom
Karel Rakusan, Canada
Peter W. Vaupel, Germany
David F. Wilson, USA

Local

Andras Eke
Janos Hamar
Alexander Juhasz-Nagy
Emil Monos
Zoltan Nagy
Istvan Nyary

Local Organising Team

Andras Eke (President, ISOTT)
Peter Herman
Zita Kasza
Judit Spacsek

SPONSORS

ISOTT is a scientific society and as such is most grateful for the generosity of the following sponsors who helped ISOTT'98 to meet the expectations of the members of our society and the scientitists attending our meeting. However, it does not endorse any of their products or services. This applies also to any material which may be distributed in mailings or at its meetings.

Sponsors

- Alliance Pharmaceuticals
- Hemox Therapeutics
- Radiometer Copenhagen
- Berlex Biosciences

Exhibitors

- Richter Gedeon
- Hamamatsu Deutschland
- Oxford Optronix

REVIEWERS

Barnikol, WKR, Germany
Blasko, Gy, Hungary
Coremans, JMCC, The Netherlands
Delpy, DT, United Kingdom
Dunn, JF, USA
Eke, A, Hungary
Erdmann, W, The Netherlands
Faithful, NS, USA
Giovannini, I, Italy
Groebe, K, Germany
Gronow, G, Germany
Hamar, J, Hungary
Harrison, DK, United Kingdom
Juhasz-Nagy, A, Hungary
Keipert, P, USA
LaManna, JC, USA
Pittman, RN, USA
Popel, AS, USA
Rakusan, K, Canada
Severinghaus, JW, USA
Thews, O, Germany
Tomita, M, Japan
van Beek, JHGM, The Netherlands
Vaupel, P, Germany
Wilson, DF, USA

TECHNICAL EDITOR
Visser-Isles, L, The Netherlands

PREFACE

The International Society on Oxygen Transport to Tissue (ISOTT) held its 26th annual meeting from August 23–26, 1998, and met for the second time in Budapest. As captured in the design of the ISOTT'98 logo, the venue of the conference was the Budapest Hilton in the heart of the historic Castle District in Buda, across from Hungary's coronation church—the Matthias Church—and the fairy-tale-like Fisherman's Bastion; a special place with a historical touch situated atop the Castle Hill, from where participants enjoyed the spectacular panoramic views of Budapest exquisitely laid out on both sides of the Danube.

In preparation for ISOTT'98, major emphasis was given to the application of informatics in collecting, presenting, and disseminating scientific and other information associated with the meeting. Electronic submission of the abstracts made it possible to publish the illustrated Abstracts on the ISOTT'98 Web Site well before the meeting. Following the meeting, an ISOTT'98 CD was published as a digital, searchable record of ISOTT'98.

The scientific program was designed with the active participation of prospective attendees, in that priority and momentum was given to the topics selected by those visiting the Web Site. Poster and oral presentations were considered as equivalently effective formats, reflected by the fact that posters were on display throughout the meeting.

The scientific program was comprised of 114 presentations by participants from Europe, North America, Asia, and Australia. It started with the Physiome Symposium emphasizing the need to integrate the available knowledge base into models, and this was followed by systematic overviews of different aspects of the transport of oxygen from air to the consuming cells of the body. Keynote presentations by leading workers in the field helped to maintain the focus on current issues and also to put new developments into a broader perspective. In most cases these presentations lead up to a session of oral presentations and poster discussions.

The Award Ceremony, during the Society Dinner given at the Gundel, was the highlight of the social program. We congratulate Dr. Martin Wolf, the recipient of the Melvin H. Knisely Award, and Dr. Nathan A. Davies, who received the Dietrich W. Lübbers Award from the Society.

The organizers hope that those who attended ISOTT'98 enjoyed its scientific presentations, the lively scientific debates, meeting old and making new friends in the ISOTT "family".

This volume is the XXIth in the ISOTT Series. The editors believe you will find it a valuable and up-to-date source of scientific knowledge in the field of oxygen transport to tissue. All manuscripts were reviewed for their scientific merits and editorial acceptability. The editors gratefully acknowledge the work of the scientific reviewers and our technical editor, Ms. Laraine Visser-Isles. Manuscripts were rejected or revised, when necessary, so that the high standard of the ISOTT Series is maintained. To capture all typographical errors was not possible without an unnecessary delay in producing this volume. Please accept our apologies for any you may find.

It is our privilege to dedicate this volume to Dr. Britton Chance on the occasion of his 85th birthday. The editors invited Dr. Peter Vaupel to pay tribute to Dr. Chance's fundamental, multifaceted achievements in science that have and continue to inspire generations of scientists around the world.

Andras Eke
David T. Delpy

CONTENTS

Section 6. Methods and Instrumentation

PROPERTIES OF A NOVEL SUPEROXIDE DISMUTASE-HEMOGLOBIN CONJUGATE OF POSSIBLE VALUE IN BLOOD SUBSTITUTE FORMULATION

Peter J. Anderson and Jean H. O'Connor

Department of Biochemistry
University of Ottawa
Ottawa, K1H 8M5, Ontario, Canada

INTRODUCTION

Because of concerns about blood-borne diseases, there is a continuing interest in the development of a temporary blood replacement for alleviating the harmful effects of extensive blood loss. It has long been recognized that to develop a carrier capable of delivering oxygen to tissues based on hemoglobin, the chemical modification of hemoglobin is necessary. Unmodified hemoglobin has too short a retention time in circulation and too high an oxygen affinity to be useful in treating hemorrhagic shock. Therefore cross-linking of hemoglobin subunits or molecules has received considerable attention as a basis for the formulation of a safe efficacious blood substitute. Both native stroma free hemoglobin and chemically modified forms of hemoglobin have been shown to facilitate the formation of potentially harmful forms of oxygen, although the extent of this varies with the form of the protein (Alayash, 1995). The conditions under which administration of a hemoglobin-based oxygen carrier would be advised are those which appear to be associated with tissue injury accompanying reperfusion. It is thought that tissue damage occurs following treatment of hemorrhagic shock because of a burst of formation of toxic forms of oxygen accompanying oxygen resupply to tissues (Granger and Korthius, 1995; Biro et al., 1995). The inclusion of features which would be expected to alleviate or prevent the generation of harmful forms of oxygen would therefore seem valuable in developing therapeutic formulations. The following work describes a novel chemical modification which can be used to make a modified form of hemoglobin containing an enzyme which functions physiologically to prevent the accumulation of

Oxygen Transport to Tissue XXI, edited by Eke and Delpy.
Kluwer Academic / Plenum Publishers, New York, 1999.

harmful forms of oxygen. Superoxide dismutase when covalently joined to hemoglobin has been shown to have an extended lifetime in circulation. The modified enzyme when injected into rats has a half-life in circulation in an active form which might be expected to be of benefit with respect to minimizing reperfusion injury following resuscitation with a hemoglobin-based blood substitute.

METHODS

Proteins

Superoxide dismutase (SOD) from bovine erythrocytes was obtained from Calbiochem. Human hemoglobin was prepared from outdated blood obtained from the Canadian Red Cross as previously described (Biro et al., 1991). It was stored frozen at –80°C at a concentration of 1.2 μM and transferred to the appropriate buffer by dialysis immediately prior to use. There was no evidence of significant formation of methemoglobin under these conditions.

Cross-Linking

Polyglutamate (average molecular weight, 10,000 Daltons) and the carboxyl group activating carbodiimide 1-ethyl-3-(3-dimethylaminopropyl)-carbodiimide were obtained from Sigma. N-hydroxysulfosuccinimide was obtained from Pierce Chemical Co. Solutions of these reagents at a concentration of 75 mM of each were allowed to react for 15 minutes on ice in Hepes buffered saline, pH 7.4 (HBS). Polymerizations were initiated by mixing this reagent 2:1 (vol:vol) with superoxide dismutase to a final concentration of 0.175 μM with or without hemoglobin (0.35 μM final concentration). Reactions were allowed to proceed for 1 hour on ice and then terminated by addition of glycine to a final concentration of 200 mM. Reagents were then removed by dialysis against HBS.

Characterization of Protein Products

Proteins were separated on the basis of molecular weights by size exclusion chromatography on a 0.5 × 30 cm column of Superdex 200 HR (Pharmacia). The column was eluted at a flow rate of 0.6 mL/min with HBS using a Bio-Rad UV detector set at 280 nm to monitor eluted protein. Fractions were collected at minute intervals and subsequently further characterized using SDS-PAGE (Bio-Rad, 4–15% gradient minigels) and Coomassie Brilliant Blue staining to detect protein.

Detection of SOD Activity

The enzyme was measured in solution by following the reduction of cytochrome c by superoxide using the xanthine-xanthine oxidase system to generate superoxide (Flohé and Ötting, 1984). The rate of reduction is a function of the concentration of SOD present. SOD competes for the superoxide and decreases the rate of reduction of cytochrome c. Cytochrome c reduction was followed by monitoring the absorbance change at 550 nm using a Beckman DU-65 spectrophotometer. This assay could not be used in the presence of high concentrations of hemoglobin due to absorbance of light of this wavelength by heme. SOD was detected in the presence of hemoglobin following protein separation by polyacrylamide gel electrophoresis in the absence of sodium dodecyl sulfate using 7.5% acrylamide gels. Reduced riboflavin when reoxidized in air generates superoxide which converts colorless nitroblue tetrazolium to insoluble blue formazan. The presence of SOD scavenges the superoxide and prevents the color change allowing the detection of SOD on the gel (Beauchamp and Fridovich, 1991).

Radiolabelling of Protein Solutions

The protein content of pooled high molecular weight protein fractions eluted from the Superdex was determined using the BCA protein assay (Pierce Chemical Co.). After adjusting protein concentrations to 5 mg/mL with HBS buffer, 100 μL samples of polymerized hemoglobin, polymerized SOD, polymerized SOD-hemoglobin and unpolymerized SOD and hemoglobin were radiolabelled by reductive methylation (Jentoft and Dearborn, 1983). To the protein solutions, 5 μL aliquots of ^{3}H formaldehyde (New England Nuclear) of a specific radioactivity of 58.75 mCi/mM, 2 mCi/mL in 10% methanol and 20 μL aliquots of 0.2 M sodium cyanoborohydride (Sigma) were added. After 10 minutes the labelled protein was purified using a Pharmacia HiTrap desalting column eluted with HBS at a flow rate of 1 mL/min. and stored at −20°C prior to use. Assays of the SOD solutions indicated that the radiolabelling did not inhibit the enzyme activity.

Animal Studies

These were carried out on male rats ranging between 250 and 300 grams in weight using a protocol approved by the University of Ottawa Animal Care Committee. The rats were euthanized at the end of the experiment. Protein solutions were infused through the tail vein of rats anaesthetized with halothane as a single 1 mL bolus. Blood samples (1 mL) were collected at 5, 20, 120, and 240 minutes into HBS buffer containing 10 mM EDTA and red blood cells were removed by centrifugation. Radioactivity in serum was monitored by scintillation counting of duplicate aliquots and SOD activity was measured on aliquots after concentration of serum by lyophilization.

RESULTS

Figure 1 shows elution profiles obtained for SOD polymerized by cross-linking with polyglutamic acid in the presence and absence of hemoglobin when separated by size exclusion chromatography. In the case of the SOD polymerized on its own, aliquots of samples collected at one minute intervals were assayed for SOD activity using the inhibition of cytochrome c reduction by superoxide to monitor SOD activity. The activity profile exactly parallelled the absorbance profile indicating that the chemical modification of the protein by the cross-linker did not inhibit the enzyme activity (data not shown). Figure 1 (a) indicates that, under the conditions used, less than 50% of the SOD polymerized on its own was cross-linked and that the major proportion of the cross-linked SOD had an elution time consistent with the molecular weight of a SOD dimer. When hemoglobin was present during the polymerization, Figure 1 (b), two thirds of the absorbance had an elution time consistent with a molecular weight greater than that of hemoglobin tetramer or SOD dimer (64 kD).

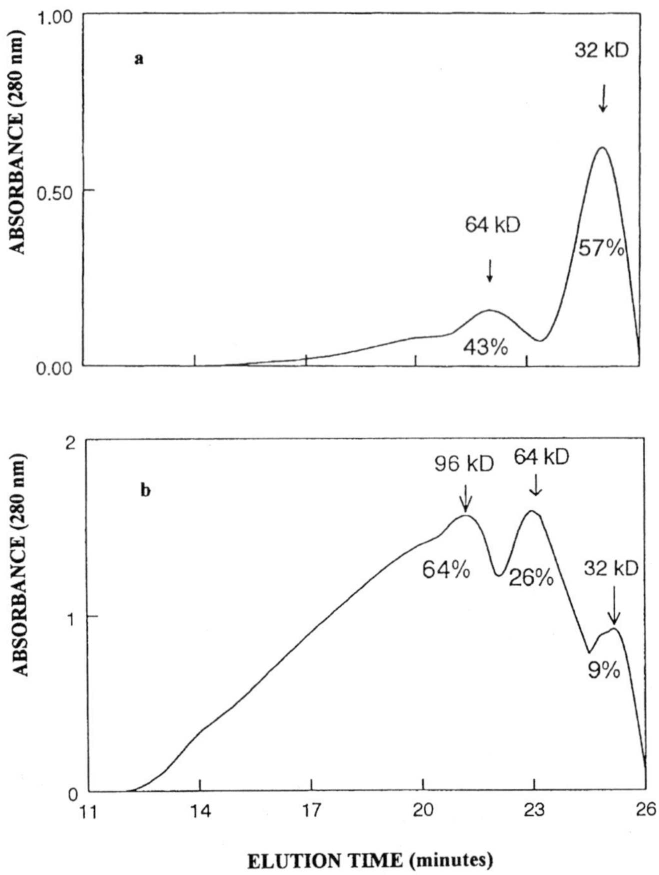

Figure 1. Elution profiles of polyglutamate cross-linked SOD (a) and of SOD cross-linked with polyglutamate in the presence of hemoglobin (b). Separation conditions are described in the text.

The cross-linked proteins were further examined by polyacrylamide gel electrophoresis. When sodium dodecyl sulfate is present during electrophoresis, proteins are separated on the basis of the size of covalently linked polypeptide chains. Aliquots of fractions eluted from the size exclusion chromatography column taken at minute intervals from 14 to 27 minutes were examined. Staining with Coomassie blue indicated that protein with a mobility corresponding to a molecular weight greater than that of the 97 kD marker protein was present in fractions from the fourteenth to twentieth minute when fractions either polymerized SOD or SOD polymerized in the presence of hemoglobin were analysed. There was considerably more high molecular weight protein in the fractions from the polymerized SOD hemoglobin separation.

To allow the detection of enzyme activity after separation, aliquots of column fractions were separated on polyacrylamide gels in the absence of sodium dodecyl sulfate. Under these conditions protein mobility is a function of both size and net charge. However, the larger the protein, the lower the expected mobility. Figure 2 shows the results obtained when aliquots of the fractions obtained from the size exclusion separation of SOD polymerized in the presence of hemoglobin were further separated by poly-

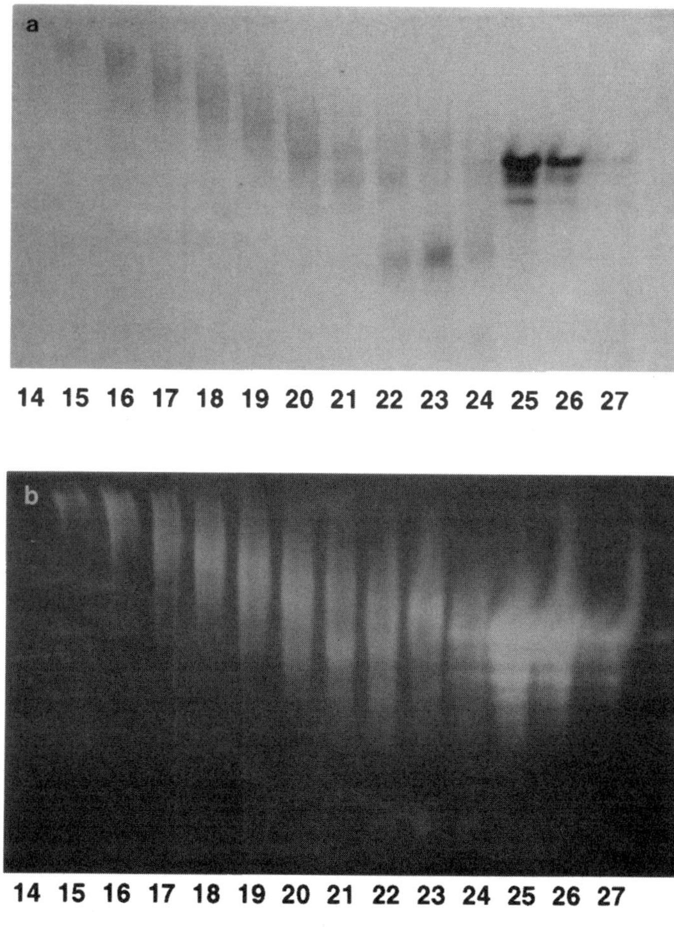

Figure 2. Patterns obtained by SDS-PAGE of column fractions. Aliquots of fractions eluted from the size exclusion column at the elution times indicated were separated and protein located by Coomassie blue staining (a) or by using a nitro blue tetrazolium assay to detect SOD activity.

acrylamide gel electrophoresis in the absence of detergent. Two identical gels were run. One was stained with Coomassie blue, Figure 2 (a), and the other, Figure 2 (b), was analysed using nitro blue tetrazolium to locate SOD activity. It is apparent that the protein stain and the SOD activity test detect material of identical mobilities indicating the presence of catalytically active high molecular weight forms of the enzyme in fractions 14 through 20. When SOD polymerized in the absence of hemoglobin was similarly analysed, high molecular weight SOD activity was also detected, but there was considerably less such material than was obtained when the polymerized hemoglobin SOD fractions were analysed.

On the basis of these results, fractions eluting from 14 to 20 minutes from the size exclusion column were pooled and concentrated by lyophilization and subsequently radiolabelled by reductive methylation as described, prior to injection into animals. Figure 3 shows the results obtained when the radioactivity in serum was determined as a function

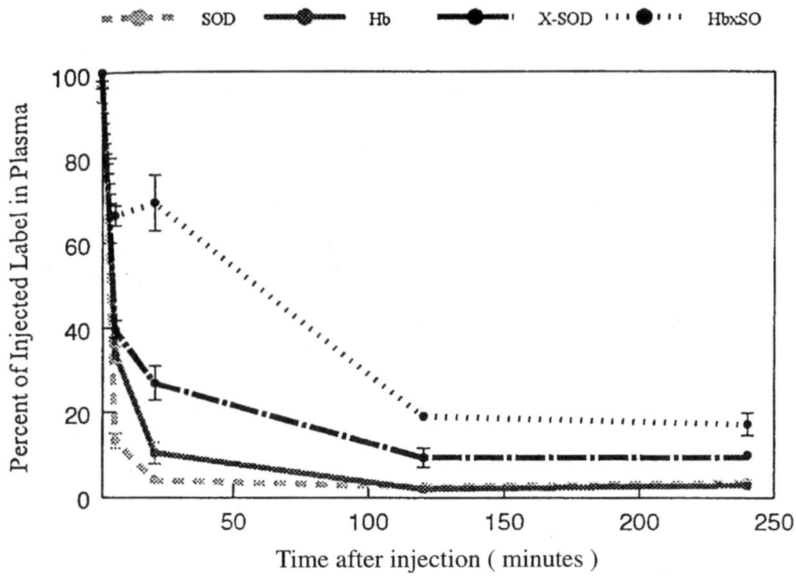

Figure 3. Time course of the decay of radiolabel from the serum of rats after injection with labelled protein. SOD, radiolabelled superoxide dismutase; Hb, radiolabelled hemoglobin; X-SOD, high molecular weight radiolabelled cross-linked SOD isolated by size exclusion chromatography; HbxSO, high molecular weight SOD cross-linked in the presence of hemoglobin isolated by size exclusion chromatography.

of time after injection. Unpolymerized radiolabelled SOD and unpolymerized radiolabelled hemoglobin were tested as well as SOD polymerized with polyglutamic acid on its own or in the presence of hemoglobin. Total radioactivity in plasma was calculated assuming that the blood volume of rats was 7% of body weight.

It can be seen that unpolymerized forms of SOD and of hemoglobin disappear from the circulation very rapidly. Polymerization of SOD increases the time that SOD remains in circulation. A two-fold higher percentage of the injected radioactivity remained in circulation for 4 hours when polymerized SOD was injected as remained after 20 minutes when unpolymerized SOD was injected. In the case of the SOD polymerized in the presence of hemoglobin, more than 4 times the percentage remaining 20 minutes after the injection of unpolymerized SOD remained after 4 hours. It is apparent that polymerization increases time in circulation based on measurement of radiolabel disappearance. In complex systems such as serum, the inhibition of the rate of reduction of cytochrome c can only give an estimate of the apparent SOD content (Flohé and Ötting, 1984). To determine whether injected polymerized SOD remained in active form in the circulation inhibition of reduction was compared using equivalent aliquots of serum of rats injected with hemoglobin and rats injected with SOD in polymerized and unpolymerized forms. Only in the rats injected with the SOD polymerized in the presence of hemoglobin was there significantly more SOD activity per μL of serum in samples taken four hours after injection. It was found that this serum contained 25 +/– 12% more SOD activity than the serum samples taken from rats with the other treatments at this time. This was in reasonable agreement with the prolonged retention of polymerized SOD estimated on the basis of disappearance of injected radioactivity.

DISCUSSION

The present work explores the potential of a novel method of cross-linking as a means of producing forms of the enzyme SOD which could have value in hemoglobin-based blood substitute formulation. It is clear that activated polyglutamic acid can be used to produce high molecular weight forms of SOD. The chemical modification which forms the cross-link does not inhibit the activity of the enzyme. The inclusion of hemoglobin during cross-linking results in the production of much more high molecular weight material than when SOD is cross-linked on its own (Figure 1). It would therefore seem that the copolymerization of SOD and hemoglobin may be a useful approach to the development of a hemoglobin based oxygen carrier to treat hemorrhagic shock.

To be relevant to blood substitute development, high molecular weight forms of SOD need to be catalytically active. This was shown to be the case, both when SOD was polymerized on its own or when it was cross-linked to hemoglobin. The nitro blue tetrazolium assay after protein separation on native polyacrylamide gels proved to be particularly useful in measuring SOD activity in the presence of hemoglobin. This is demonstrated in Figure 2. The SOD activity on the gel as indicated by inhibition of nitro blue tetrazolium has the same distribution as the total protein as indicated by Coomassie staining. This indicates that SOD cross-links to hemoglobin and that this cross-linking does not alter the ability of the enzyme to scavenge superoxide.

A SOD product of potential use in blood substitute formulation needs to have an extended lifetime in circulation. As has been previously shown (Mao et al., 1993), native SOD when injected is very rapidly cleared from circulation. Close to 90% of injected labelled unpolymerized SOD (Figure 3) is cleared by the first time point and after 20 minutes less than 5% remains. It is apparent that cross-linking SOD to hemoglobin greatly extends the time in circulation. Close to 20% of injected label remains in circulation after 4 hours in a form that shows increased apparent SOD activity in serum compared to serum from rats not treated with this form of the enzyme.

Recently the effects of a form of hemoglobin polymerized by glutaraldehyde in the presence of catalase and superoxide dismutase on secondary hemoprotein-associated free radical events have been described (D'Agnillo and Chang, 1998). Although this protein was shown to possess antioxidant properties, the protein conjugate was not characterized and the activities of its enzyme components were not directly determined. The cross-linking procedure based on glutaraldehyde used has been shown to produce a high oxygen affinity form of hemoglobin unlikely to be of use as an oxygen delivering blood replacement (Guillochon et al., 1985) and facilitates potentially deleterious autooxidation (Alayash, 1995). It would therefore seem useful to explore other methods of cross-linking to incorporate properties designed to alleviate potentially harmful properties of hemoglobin polymers which could lead to oxygen toxicity when used to replace blood in the treatment of shock due to hemorrhage. On the basis of the present study polyglutamate would appear to offer a potentially valuable method of producing a high molecular weight form of hemoglobin with active SOD incorporated as a superoxide scavenger. Presumably catalase could also be included in this formulation but it is a high molecular weight protein and has a much longer circulation time in unmodified form than SOD (Mao et al., 1993). The cross-linking to hemoglobin of SOD by polyglutamic acid involves amino group modifications. It has previously been shown that such alterations of hemoglobin result in a form of the protein which has oxygen binding properties which could facilitate its use as an oxygen delivering blood replacement (Anderson et al., 1996).

It is concluded that the novel cross-linking procedure described may be of value in the development of a hemoglobin based blood substitute.

ACKNOWLEDGMENT

This work was supported by a grant from the Bayer/Canadian Red Cross Research and Development Fund.

REFERENCES

Alayash, AI (1995) Effects of intra- and intermolecular cross-linking on the free radical reactions of bovine hemoglobins, Free Radic. Biol. Med. 18:295–301.

Anderson, PJ, Vaccani, JP, and Biro, G (1996) The role of amino groups in hemoglobin oxygen binding, Adv. Exp. Med. Bio. 388:611–615.

Beauchamp, C and Fridovich, I (1971) Superoxide dismutase: improved assay applicable to acrylamide gels, Anal. Biochem. 44:276–287.

Biro, G, Anderson, PJ, Curtis, S, and Cain, S (1991) Stroma-free hemoglobin: its presence in plasma does not improve oxygen supply to resting hindlimb vascular bed of hemodiluted dogs, Can. J. Physiol. Pharmacol. 69:1656–1662.

Biro, G, Ou, C, Ryan-Macfarlane, C, and Anderson, PJ (1995) Oxyradical generation after resuscitation of hemorrhagic shock with blood or stroma free hemoglobin, Artif. Cells Blood Substit. Immob. Biotechnol. 23:631–645.

D'Agnillo, F and Chang, T (1998) Absence of hemoprotein-associated free radical events following oxidant challenge of cross-linked hemoglobin-superoxide dismutase catalase, Free Radic. Biol. Med. 24:906–912.

Flohé, L and Ötting, F (1984) Superoxide dismutase assays, Methods in Enzymology 105:93–104.

Granger, D and Korthius, R (1995) Physiologic mechanisms of post ischemic tissue injury, Ann. Rev. Physiol. 57:311–332.

Guillochon, D, Vijayakshmi, M, Thiam-Sow, A, and Thomas, D (1985) Effects of glutaraldehyde on hemoglobin: functional aspects and Mossbauer parameters, Biochem. Cell Biol. 64:29–37.

Jentoft, N and Dearborn, D (1983) Protein labelling by reductive alkylation, Methods in Enzymology 91:570–579.

Mao, GD, Panakkezhum, D, Lopaschuk, G, and Poznansky, MJ (1993) Superoxide dismutase (SOD)-catalase conjugates, J. Biol. Chem. 268:416–420.

REACTIONS OF CROSS-LINKED METHAEMOGLOBINS WITH HYDROGEN PEROXIDE

Jackie Dunne,[1] Dimitri A. Svistunenko,[1] Abdu I. Alayash,[2]
Michael T. Wilson,[1] and Chris E. Cooper[1]

[1]Department of Biological Sciences
University of Essex, Colchester, CO4 3SQ
United Kingdom
[2]Food and Drug Administration
Bethesda, Maryland 20892

1. INTRODUCTION

The development of haemoglobin based oxygen carriers is intended to avoid the problems which can arise from the use of blood. Blood can only be stored for a short period of time and blood groups must be matched prior to use. In addition, the risk of the transfer of infectious agents such as viruses has become a major factor in the development of cell free haemoglobin solutions.

Haemoglobin free in the blood has an increased oxygen affinity due to the absence of 2,3 bisphosphoglycerate, and because of its decreased concentration the haemoglobin molecule can dissociate into dimers. Cross-linking of the haemoglobin molecule between subunits has produced haemoglobins which have both a stable structure and an oxygen affinity comparable to that of cellular haemoglobin. Such cross-links have been successfully introduced between both α and β subunits using several reagents (Benesch, 1994; Chatterjee et al., 1986). Further modification by polymerisation and conjugation of the haemoglobin has also been achieved (MacDonald and Pepper, 1994; Xue and Wong, 1994). Haemoglobin can form ferryl haem from the ferrous state in the presence of hydrogen peroxide (Guilivi and Davies, 1990). Methaemoglobin (ferric) will also react with hydrogen peroxide to produce ferryl haem and globin based radicals (McArthur and Davies, 1993; Patel et al., 1996) (eq. 1). This latter process can happen *in vivo* when oxyhaemoglobin autoxidises to methaemoglobin and reacts with naturally occurring hydrogen peroxide or lipid peroxides.

Oxygen Transport to Tissue XXI, edited by Eke and Delpy.
Kluwer Academic / Plenum Publishers, New York, 1999.

$$Fe^{3+} + R + H_2O_2 \rightarrow Fe^{4+} = O^{2-} + H_2O + R^{\cdot +} \quad \text{(eq. 1)}$$

The reaction with hydrogen peroxide also results in the formation of a covalent link between the globin and the haem group for myoglobin and haemoglobin (Catalano et al., 1989; Kelder et al., 1991). Both ferryl haem and globin radicals can react further with other substrates. Oxidation of different substrates by these reactive haemoglobin species is considered to be part of the mechanisms by which tissue damage can occur.

The use of cell free haemoglobin may lead to an increase in the formation of ferryl haem and globin radicals and hence tissue damage. This is because inside the blood cell the amount of methaemoglobin is kept relatively low as it is reduced by the enzyme methaemoglobin reductase. Haemoglobin free in the blood, however, has no such protection leading to an increase in methaemoglobin levels. Haemoglobins which are cross-linked between subunits have also been shown to have increased autoxidation rates compared to non cross-linked haemoglobin (Benesch, 1994; Yang and Olsen, 1989).

An increase in cell free methaemoglobin levels could lead to a greater formation of both ferryl haem and globin radicals due to the reaction with hydrogen peroxide. However, as yet there have been no systematic studies of the reaction products of cross-linked methaemoglobins with hydrogen peroxide. The reactions of the met oxidation state of two cross-linked haemoglobins under consideration as blood substitutes were therefore compared to that of non cross-linked haemoglobin. The haemoglobins used were α DBBF haemoglobin (which is cross-linked between lys 99 residues of the α subunits by bis(dibromosalicyl) fumarate) and PHP haemoglobin (which is cross-linked between the lys 82 residue of one β subunit and the N terminal of the other and then conjugated with polyoxyethylene). As well as being studied as a blood substitute PHP haemoglobin is also undergoing current clinical trials as a nitric oxide scavenger.

2. METHODS

Oxyhaemoglobin was oxidised to methaemoglobin by the addition of excess potassium ferricyanide. Ferricyanide was removed by passing the solution down a Sephadex G25 column equilibrated with 50 mM sodium phosphate buffer containing 10 μM diethylenetriamine pentaacetic acid (DTPA) at pH 7.4. Hydrogen Peroxide (100 μM) was added to methaemoglobin (50 μM) in 50 mM sodium phosphate buffer containing 10 μM DTPA pH 7.4 at 37°C. Optical spectra of the reaction mixture were taken at 1 minute intervals on a Cary 5E spectrophotometer.

After one hour the samples were acidified by the addition of HCl. The free haem was extracted three times into equal volumes of butanone, and the haem retained by the apoprotein measured optically.

EPR samples were prepared by the addition of hydrogen peroxide and methaemoglobin as above, samples of the reaction mixture were removed and frozen at specific time points. The spectra were measured on a Bruker EMX EPR spectrometer with an Oxford Instruments liquid helium flow cryostat.

3. RESULTS

3.1. Optical Spectroscopy

The addition of hydrogen peroxide to methaemoglobin resulted in a loss in the absorbance of the met spectrum at 630 nm with the formation of peaks at 545 nm and

580 nm which have been identified as characteristic of ferryl haemoglobin (Guilivi and Davies, 1990).

For both non cross-linked and DBBF haemoglobins the maximum absorbance change resulting from ferryl formation occurred two minutes after peroxide addition (Figure 1). PHP haemoglobin showed the absorbance changes characteristic of ferryl formation, but this change was considerably less than that of the other haemoglobins being only 15% of the expected value.

Extraction of the haem group into butanone showed that, following hydrogen peroxide treatment, in both non cross-linked and DBBF haemoglobins some 10% of the haem was retained within the globin as a result of the formation of a covalent link between the haem group and the globin. For PHP haemoglobin covalent linking was 2% after reaction with hydrogen peroxide. The degree of this covalent linking in control methaemoglobin which had not been reacted with hydrogen peroxide was 1%.

3.2. EPR Spectroscopy

EPR spectra of the methaemoglobins after the hydrogen peroxide addition show a loss of the g = 6 signal of the high spin ferric haem group as the methaemoglobin is oxidised to ferryl (results not shown). This loss of the g = 6 feature in the EPR spectrum of PHP is much less and is consistent with its lower ferryl formation. The reaction of non cross-linked haemoglobin and hydrogen peroxide gives a spectrum after 30s which shows

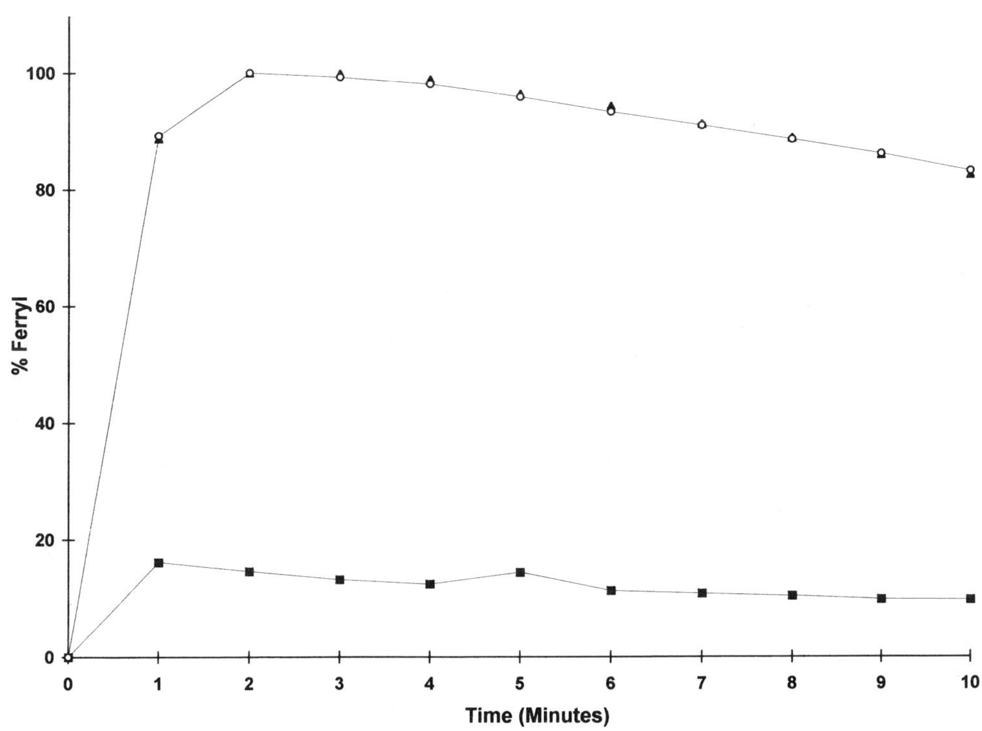

Figure 1. Extent of ferryl formation of cross-linked haemoglobins. Ferryl Formation of Non Cross-linked (□), DBBF (▲) and PHP (■) haemoglobins following the addition of hydrogen peroxide (100 μM) to methaemoglobin (50 μM) in phosphate buffer (50 mM) pH 7.4 at 37°C. The results for non cross-linked haemoglobin and DBBF haemoglobin overlap to such an extent that the individual points are difficult to see.

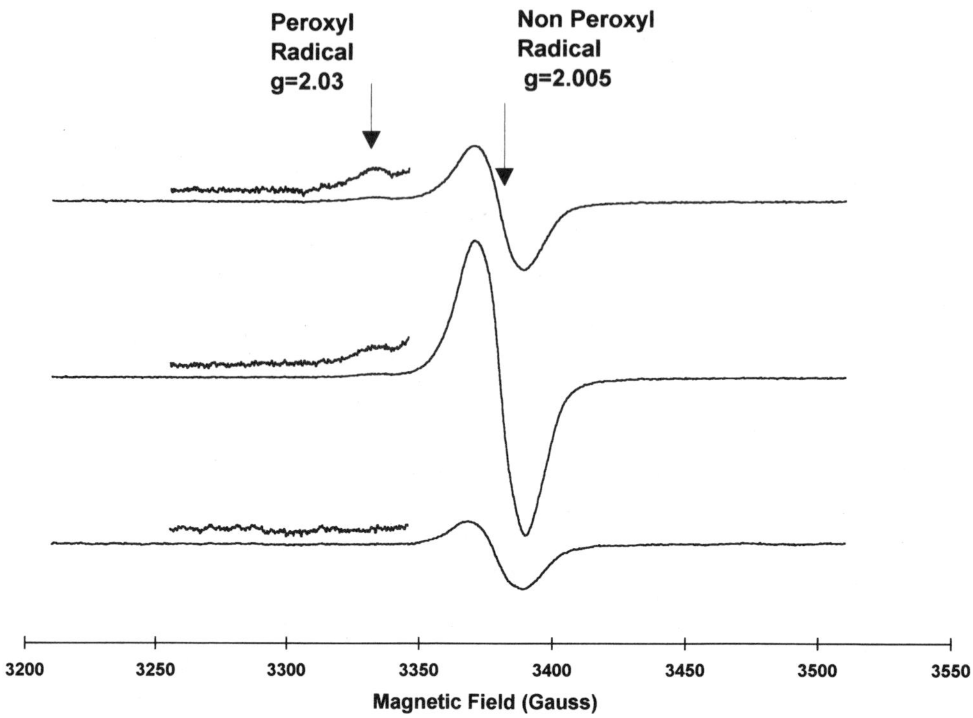

Figure 2. EPR of globin based free radicals in cross-linked haemoglobins. EPR spectra of non cross-linked (top), DBBF (middle) and PHP (bottom) haemoglobins 30s after addition of Hydrogen Peroxide. Experimental conditions were, Microwave power P = 3.18 mW, Frequency $\nu = 9.48$ GHz, Temperature T = 10 K, Gain = 4×10^4, Modulation Amplitude = 3 G, Time constant $\tau = 82$ ms, Number of scans = 1, Sweep Time = 84 s.

two radical signals (Figure 2, top spectrum). These are a peroxyl radical at g = 2.03 and the non peroxyl radical at g = 2.005 (Svistunenko, Patel and Wilson, 1996). DBBF shows both of these radicals (Figure 2, middle). Although both spectra show similar amounts of the peroxyl radical after 30s, the amount of the non peroxyl radical is greater in DBBF compared to the non cross-linked haemoglobin. PHP haemoglobin (Figure 2 bottom) 30s after the addition of hydrogen peroxide has also formed the non peroxyl radical, but in a much smaller amount. There was no peroxyl radical detected in this spectrum.

Spectra of the DBBF sample measured at a lower microwave power (Figure 3) show evidence for a third radical signal in the g = 2.005 region. This is most clearly seen following subtraction of the spectrum 30s post peroxide addition, from that at 90s. A free radical signal with a triplet hyperfine structure is seen to decrease in concentration (Figure 3 bottom), as indicated by the inverted spectrum. At the same time as this new radical is decreasing in concentration we see a rise in the peroxyl radical signal at g = 2.03.

4. DISCUSSION

Oxidation of methaemoglobin by hydrogen peroxide gives ferryl haem (McArthur and Davies, 1993; Patel et al., 1996). Studies of ferrous DBBF haemoglobin have shown it has an increased tendency to form ferryl haem compared to ferrous non cross-linked

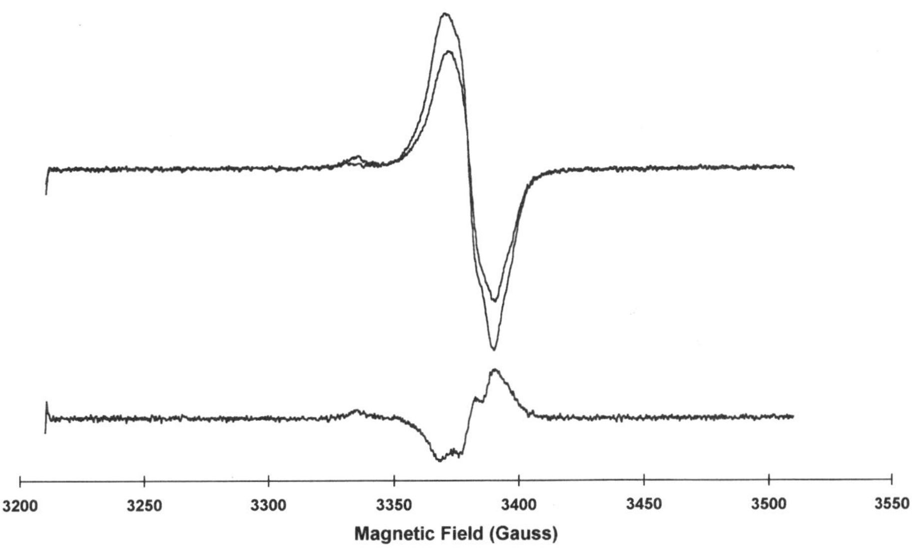

Figure 3. New radical species bound to DBBF haemoglobin. EPR spectra of DBBF haemoglobin at 30s and 90s after Hydrogen Peroxide addition (Top) the spectrum at 30s is the larger of the two spectrum. Subtraction of 30s spectrum from 90s spectrum (bottom). Experimental conditions were, Microwave power P = 50 μW, Frequency $\nu = 9.48$ GHz, Temperature T = 10 K, Gain = 4×10^4, Modulation Amplitude = 3 G, Time constant $\tau = 82$ ms, Number of scans = 1, Sweep Time = 84 s.

haemoglobin (Cashon and Alayash, 1995). There is also an increased formation of haem globin covalent links (Osawa et al., 1993). However, met DBBF showed a similar formation of both ferryl haem and covalent links to the non cross-linked haemoglobin. Despite this similarity the formation of globin radicals by this haemoglobin was clearly different from that of the ordinary non cross-linked haemoglobin.

The reactions of non cross-linked haemoglobin with hydrogen peroxide produced globin radicals. Both the previously identified peroxyl and non peroxyl radicals were observed in the EPR spectra during the reaction. These radicals were also observed DBBF haemoglobin, together with a third species. Studies of the kinetics of the reaction of non cross-linked haemoglobin with hydrogen peroxide indicated that the non peroxyl radical is not the precursor to the peroxyl radical (Svistunenko et al., 1996). This, it was postulated, was formed from another globin radical which subsequently reacted with oxygen. The continuing formation of a peroxyl radical between 30s and 90s seen in the DBBF spectra while the signal assigned to the third radical declines, indicates that this third species may be converted into the peroxyl radical. Introduction of cross-linking between the α subunits apparently decreases the rate of formation of peroxyl radicals, allowing the precursor to be seen.

What is the nature of this new radical? The globin molecule offers a number of possible sites where the radical may be harboured. Studies of both myoglobin and haemoglobin have suggested radicals can form on several residues including tyrosine (Davies, 1991; McArthur and Davies, 1993; Tschirret-Guth and Ortiz de Montellano, 1996) and tryptophan (DeGray et al., 1997; Gunther et al., 1995). It has also been proposed that the radical can delocalise within the globin molecule (Guilivi and Cadenas, 1998). The new radical we have seen in DBBF has a clear hyperfine structure. This structure suggests the radical signal is being split by a nearby nucleus. The 3 line structure suggests coupling with two equivalent protons (possible on a tyrosine residue) although nitrogen coupling cannot be completely ruled out. The identification of this new radical site should provide further information on the mechanisms by which free radicals are formed in haemoglobin.

The optical spectra of PHP indicates that ferryl haemoglobin is formed by the reaction with hydrogen peroxide. The formation of a small amount of free radical also indicates the reaction is occurring. The small amount of covalent linking detected is consistent with the smaller amounts of ferryl and radicals observed. The structure of the PHP molecule would appear to be limiting, but not preventing, the oxidation of the methaemoglobin. The lack of detectable peroxyl radicals may indicate they are not being formed by this haemoglobin. It is more likely that a small amount of peroxyl radicals have been formed and have decayed to an undetectable level at 30s.

For DBBF and PHP haemoglobins the modification of the molecule alters the reactivity of the met oxidation state towards hydrogen peroxide. The different modifications have resulted here in both increased and decreased extents of radical formation. Although the tendency of DBBF haemoglobin to become oxidised to its ferryl form is similar to native met haemoglobin A, there is an increase in the amount of free radical detected. Together with an increased autoxidation rate (Yang and Olsen, 1989) this may indicate that *in vivo* there is a greater potential for tissue damage. For PHP haemoglobin the formation of relatively small amounts of both ferryl haem and globin radicals and hence lower potential for tissue damage may favour its use in vivo.

ACKNOWLEDGMENTS

We are grateful to Neil Barnard for technical assistance and we acknowledge financial support from the Wellcome Trust (CEC, MTW and DAS) and the BBSRC (JD).

REFERENCES

Benesch, RE (1994) Bispyridoxal polyphosphates as Specific Intramolecular Cross-linking agents for Haemoglobin. Methods Enzymol. 231:267–280.

Cashon, RE and Alayash, AI (1995) Reaction of Human HbA(O) and Two Cross-Linked Derivatives with Hydrogen Peroxide—Differential Behaviour of the Ferryl Intermediate. Arch. Biochem. Biophys. 316:461–469.

Catalano, CE, Choe, YS, and Ortiz de Montellano, PR (1989) Reactions of the Protein Radical in Peroxide treated Myoglobin, Formation of a Haem Protein Cross-link. J. Biol. Chem. 264:10534–10541.

Chatterjee, R, Welty, EV, Walder, RY, Pruitt, SL, Rogers, PH, Arnone, A, and Walder, JA (1986) Isolation and Characterisation of a new Hb Derivative Cross-Linked between the alpha chains (lys99alpha1–lys99alpha2) J. Biol. Chem. 261:9929–9937.

Davies, MJ (1991) Identification of a Globin Free Radical in Equine Myoglobin treated with Peroxides. Biochim. Biophys. Acta 1077:86–96.

DeGray, JA, Gunther, MR, TschirretGuth, R, Ortiz deMontellano, PR, and Mason, RP (1997) Peroxidation of a Specific Tryptophan of Metmyoglobin by Hydrogen Peroxide. J. Biol. Chem. 272:2359–2362.

Guilivi, C and Cadenas, E (1998) Heme Protein radicals; Formation, fate, and Biological consequences. Free Rad. Biol. Med. 24:269–279.

Guilivi, C and Davies, KJA (1990) A Novel Antioxidant role for Hemoglobin, The comproportionation of ferryl hemoglobin with oxyhemoglobin. J. Biol. Chem. 265:19453–19460.

Gunther, MR, Kelman, DJ, Corbett, JT, and Mason, RP (1995) Self Peroxidation of Metmyoglobin results in formation of an Oxygen reactive Tryptophan centred Radical. J. Biol. Chem. 270:16075–16081.

Kelder, PP, Fischer, MJE, de Mol, NJ, and Jansson, LHM (1991) Oxidation of chlorpromazine by methaemoglobin in the presence of hydrogen peroxide. Formation of a chlorpromazine radical cation and its covalent binding to methaemoglobin. Arch. Biochem. Biophys. 284:313–319.

MacDonald, SL and Pepper, DS (1994) Haemoglobin Polymerisation. Methods Enzymol. 231:287–308.

McArthur, KM and Davies, MJ (1993) Detection and reactions of the globin radical in haemoglobin. Biochim. et Biophys. Acta 1202:173–181.

Osawa, Y JF D, Meyer, CA, and Alayash, AI (1993) Differential Susceptibilities of the Prosthetic Heme of Hemoglobin based Blood Substitutes: Implications in the Design of Safer Agents. Biochem. Pharmacol. 46:2299–2305.

Patel, RP, Svistunenko, DA, Darley-Usmar, VM, Symons, MCR, and Wilson, MT (1996). Redox Cycling of Human Methaemoglobin by H2O2 Yields Persistent Ferryl Iron and Protein Based Radicals. Free Rad. Res. 25:117–123.

Svistunenko, DA, Patel, RP, and Wilson, MT (1996) An EPR Investigation of Human Methaemoglobin Oxidation by Hydrogen Peroxide: Methods to Quantify all Paramagnetic Species Observed in the Reaction. Free Rad. Res. 24:269–280.

Tschirret-Guth, RA and Ortiz de Montellano, PR (1996) Protein radicals in myoglobin dimerization and myoglobin catalyzed styrene epoxidation. Arch. Biochem. Biophys. 335:93–101.

Xue, H and Wong, TF (1994) Preparation of Conjugated Haemoglobins. Methods Enzymol. 231:308–322.

Yang, T and Olsen, KW (1989) The Effect of Crosslinking by Bis(3,5 Dibromosalicyl) Fumarate on the autoxidation of Hemoglobin. Biochem. Biophys. Res. Comm. 163:733–738.

A MODEL FOR EVALUATION OF ARTIFICIAL OXYGEN CARRIERS REGARDING CIRCULATION, RESPIRATION AND METABOLISM IN ANESTHETIZED SPONTANEOUSLY BREATHING GUINEA PIGS

Mustafa Özbek,* Ulrike Domack, and Wolfgang K. R. Barnikol

Bereich Klinische Physiologie
Private Universität Witten/Herdecke und SanguiBioTech AG
D-58455 Witten/Ruhr, Germany

SUMMARY

The evaluation of artificial oxygen carriers requires experiments with suitable animals. Many investigators do this with the classical laboratory animal, the rat, but it has a quite different oxygen pressure of half saturation (p50 = 36 mmHg) from that of humans (26 mmHg). It was demonstrated that induced changes of the p50 value in animals provokes substantial changes in important cardiovascular parameters. Therefore, we decided to develop a guinea pig model for evaluation of artificial oxygen carriers, because it has a p50 of about 24 mmHg that is very similar to that of man. We found an anesthesia combination using fentanyl/droperidol/urethane to be very suitable for the narcosis, because important cardiovascular and respiratory parameters remain normal. Our model allows assessment of arterial pressure, oxygen uptake, carbon dioxide release, cardiac output, total peripheral resistance, shock index, blood lactate level, and, in particular, it allows to differentiate the oxygen transport in blood. We evaluated two hemoglobin-based oxygen carriers: bovine and porcine hemoglobin polymerized with glutaraldehyde in isoncotic solution (n = 3). Control experiments (n = 4) were done with an isotonic albumin solution. The protocol comprised a so-called exchange phase (I) with different degrees of hemodilution and a so-called time phase (II), an observation

* This publication contains parts of the thesis of M. Özbek.

Oxygen Transport to Tissue XXI, edited by Eke and Delpy.
Kluwer Academic / Plenum Publishers, New York, 1999.

period with a hematocrit of 10% over about 3 hours. In the control group substantial changes in mean arterial pressure, total peripheral resistance, shock index, oxygen uptake and of blood lactate level were seen. All these effects were prevented by the artificial oxygen carriers tested. The carriers proved to be very effective, as small quantities in blood effectively restored these parameters, presumably via synergetic effects. Moreover, the experiments clearly revealed the limitation of hemodilution, at least in guinea pigs: below a hematocrit of 20% all parameters mentioned above changed significantly. The animal model presented proved to be appropriate for the evaluation of artificial oxygen carriers.

1. INTRODUCTION AND PROBLEM

The evaluation of artificial oxygen carriers for humans necessarily requires animal experiments. Thus many investigators—including us—established models using the classical laboratory small animal the rat (Waschke et al., 1993; Pötzschke et al., 1994; Gulati et al., 1994; Abassi et al., 1997). But we had to recognize that the rat is an inadequate animal for this purpose because of its high oxygen pressure of half saturation of blood (p50) being about 36 mmHg; in contrast, man has a value of about 26 mmHg. Studies have shown that experimental variation in p50 value causes substantial cardiovascular changes in vivo (Soulard et al., 1983; Liard et al., 1993; Kunert et al., 1996).

Therefore we tried to apply our experimental procedure in the rat model (Pötzschke et al., 1994) to the guinea pig which has a p50 value of about 24 mmHg. However, the classical narcotics for the rat, such as pentobarbital sodium and urethane alone decrease the normal arterial blood pressure in guinea pigs (normally about 50 mmHg) and/or substantially suppresses spontaneous respiration. For a neuroleptanesthesia, we found a combination of fentanyl/droperidol/urethane to be suitable for our purpose: with this anesthesia, important parameters of respiration, circulation and metabolism remained within normal range (Amdur et al., 1958; Brown et al., 1989).

Here we report on the evaluation of two different hemoglobin-based oxygen carriers (HBOC) employing the new model: porcine and bovine hemoglobin, both crosslinked with glutaraldehyde.

2. METHODS AND MATERIALS

2.1. Anesthesia and Surgical Preparation

Albino guinea pigs (Charles-River/D) of either sex with body weight (bw) from 750 to 1000 g were used. The animal were fasted 16–18 hours before the experiment. For anesthesia, a mixture (in $mg.kg^{-1}$) of fentanyl 0.2, droperidol 10 and urethane 400 was dissolved in isotonic solution and given intraperitoneally (i.p.). To maintain anesthesia, which was controlled by examining the pedal pain reflex, a supplemental dose of 20–30% of the initial amount was given i.p. The rectal temperature was kept between 36–37°C using a heated table.

A tracheostomy was performed using a cannula with an outer diameter (od) of 3.2 and an inner diameter (id) of 2.4 mm. The animals were always breathing spontaneously. The right atrium or the right ventricle was catheterized through the right vena jugularis externa using a polyethylene tube (od: 1.2, id: 0.8 mm) to obtain mixed venous blood and for injection of exchange solutions. The right arteria femoralis was cannulated with a polyethylene tube (od: 0.4, id: 0.2 mm) for continuous recording of arterial blood pressure. The left arteria femoralis was also prepared with a similar tube and used for blood sampling and for injections. A Ringer solution, 1–2.5 ml/hour, was administered through the left femoral artery or the external jugular vena to maintain the electrolyte-water balance of the animal. The whole preparation lasted about 75 minutes. The first measurements began after a further 60 minutes at steady state.

2.2. Measurements and Calculations

The mean arterial blood pressure (MAP) and heart frequency (HF) were obtained from a Gould SP1400 pressure monitor. The shock index (SI) is calculated as:

$$SI = HF/MAP$$

The tracheal tube was connected to the microvalve system with a very small dead space that enables to measure the respiratory frequency (RF) and the ventilation (Vt) and to sample the expired gas (Barnikol et al., 1994). For expiratory gas flow measurement a Fleisch pneumotachograph (Hugo Sachs, Freiburg/D) and an amplifier from Ganshorn Electronic (Niederlauer/D) were used. The signal was integrated to assess ventilation (Vt) using an EG & G Multiplier/Divider Model 193 (München/D) and choosing a time constant of 10 sec. The volume fractions of respiratory gas were analyzed with a Capnomac II AGM-123 (Hoyer, Bremen/D), the inspired gas was always air. Carbon dioxide release ($V'CO_2$) and oxygen uptake ($V'O_2$) were calculated from the differences between the fractions of inspired and expired gases by multiplication with Vt regarding standard conditions. The respiratory quotient (RQ) is calculated as:

$$RQ = V'CO_2/V'O_2$$

Total O_2 content in arterial and mixed venous blood (CaO_2 and CvO_2) were measured using a Lex-O_2-Con device from Lexington Instruments (USA). The cardiac output (CO) was calculated according to Fick's principle:

$$CO = V'O_2/(CaO_2 - CvO_2)$$

Total peripheral resistance (TPR) was calculated as the ratio between the MAP and CO assuming the mean arterial pressure as the driving pressure:

$$TPR = MAP/CO$$

The partial pressures of oxygen and carbon dioxide (pO_2 and pCO_2) and the pH value in arterial and venous blood were measured at 37°C with a blood gas analyzer from AVL Omni (Bad Homburg/D).

Total hemoglobin content (ct tHb) was determined photometrically by the hemoglobin cyanide method. The hematocrit (Hct) was measured by a micro method. The initially measured values for hemoglobin and hematocrit are abbreviated as 'ct tHb_0' and 'Hct_0' respectively. For a given experiment where the animal has received HBOC solution, the following calculations were performed. Briefly, to differentiate the content of native hemoglobin (ct nHb) from the content of HBOC (ct HBOC) in the same guinea pig blood, we have used hematocrit value which is measured actually (Hct_A):

$$ct\ nHb = Hct_A \cdot ct\ tHb_0/Hct_0$$

Thus the actual content of HBOC is given as:

$$ct\ HBOC = ct\ tHb - ct\ nHb$$

The dissolved oxygen concentration was calculated as the product of the average oxygen solubility coefficient of blood and the partial oxygen pressure. The oxygen content of the guinea pig hemoglobin is the product of ct nHb, Hüfner's number (O_2 bounding capacity of 1 g Hb is 1.34 ml $O_{2\ STPD}$) and the oxygen saturation of nHb based on the oxygen binding curve of guinea pig blood regarding the actual pH value and partial gas pressures of blood. The content of oxygen bound by HBOC results from the difference between the total oxygen content and the sum of dissolved oxygen content and the content of oxygen bound by nHb. The oxygen uptake of the different oxygen carriers within the blood was calculated as the product of the corresponding arterio-venous difference of oxygen content and cardiac output. This allows a differentiation of the oxygen transport. Details of the above calculations are given elsewhere (Pötzschke et al., 1994; Dinkelmann, 1997). The oxygen binding properties of the hemoglobins were determined with the photometric thin layer method (Barnikol et al., 1978).

The Lact MPR 1 test method is performed for measurement of the lactate content in arterial blood to monitor indirectly and globally oxygen deficiency; for blood glucose the Accutrend test was used, both were purchased from Boehringer (Mannheim/D).

2.3. Exchange Solutions

The control solution contained 5.6 g human albumin in 100 ml and the following crystalloids (in mmol/L) Na 147.0, Cl 156.0, K 4.0, Ca 2.5.

Stroma-free porcine and bovine hemoglobin were polymerized with glutaraldehyde, the aldimines were reduced with sodium cyanoborhydride. The average molecular weight of the polymers was adjusted by ultra filtration. Before application we adjusted the oncotic pressure of the HBOC solutions to that of the blood of guinea pig by adding 2.4 g/dl human albumin (Immuno/D). The hemoglobin content of both preparations was 8.0 g/dl, solved in the crystalloid solution given above. Details for preparation and characterization of the HBOC polymers are given elsewhere (Domack, 1997).

2.4. Experimental Protocol

The experiments comprised a control group of 4 animals (A, n = 4) receiving only albumin (control) solution and the test group of 3 animals (B, n = 3) receiving control solution and HBOC solution, respectively. Each experiment consisted of two phases (I and II), which are displayed with the aid of hemoglobin content in Figure 1 as a typical example. In phase I (so-called exchange phase), isovolemic blood exchanges were performed in several steps with 1–4 ml solution. The hematocrit reflects the degree of the hemodilution. In both groups the Hct was reduced to about 20% (about 6 g Hb/dl, see Figure 1). From this state of exchange the control animals received further albumin solution, but the test animals received HBOC solution (see arrow in Figure 1) until in both cases a hematocrit of about 10% was reached. The hemoglobin content of blood of the control animals decreased to about 3 g/dl, whereas in the test animals it was maintained at about 6 g/dl.

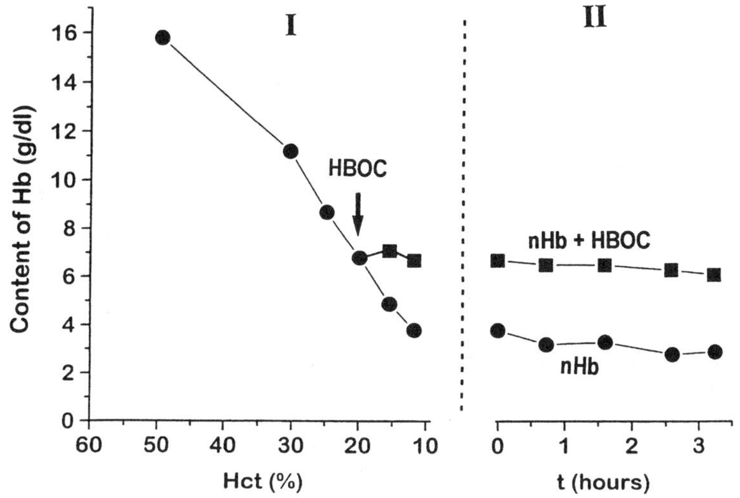

Figure 1. Illustration of the two phases showing the hemoglobin (Hb) content of blood, phase I: exchange phase, II: time phase, nHb: native hemoglobin of the animal, HBOC: hemoglobin-based oxygen carrier, Hct: hematocrit, ●: control, ■: test.

Phase II is an evaluation period in both groups with time (time phase) at a hematocrit of 10%. During both phases at certain hematocrit values and times, respectively, all the above-mentioned parameters were assessed. An animal experiment, excluding the preparation time, lasted about 7 hours.

3. RESULTS

In Table 1 important properties of HBOC solutions are given. Figure 2 shows the cardiovascular effects in the control (A) and test (B) groups. The upper part shows changes in mean arterial pressure (MAP). Both groups exhibit a decrease in MAP down to a hematocrit of 20%. During further hemodilution to a Hct of 10% in the control group (A) the MAP declined to one third of the initial value. In contrast, the test group (B) shows a recovery of MAP which is almost maintained also in the following time phase (II) whereas in the control group (A) the pressure remains low at about 20 mmHg. The middle part of Figure 2 shows the effects on relative total peripheral resistance (TPR).

Table 1. Properties of the two applied hemoglobin based artificial oxygen carriers (HBOC), both are dissolved in Ringer's solution containing (in mmol/L) Na 147.0, Cl 156.0, K 4.0, Ca 2.5

	Content of HBOC (g/dl)	Content of H.Alb. (g/dl)	Molecular weight (Dalton)	p50[1] (mmHg)	Proton Bohr factor[1] [Δ log p50/Δ pH]	Hill index [n50]
Porcine HBOC	8.0	2.4	350,000	17.4	−0.17	2.45
Bovine HBOC	8.0	2.4	250,000	25.6	−0.22	2.20

H.Alb.: Human Albumin, p50: oxygen partial pressure (pO_2) at 50% of O_2 saturation, $n50 = \Delta \log(SO_2/[1 - SO_2])/\Delta \log(pO_2)$ ($40\% < SO_2 < 60\%$, SO_2: oxygen saturation of hemoglobin, pO_2: oxygen partial pressure). Conditions 1: pCO_2 (partial pressure of carbon dioxide) = 40 mmHg, temperature = 37°C, pH = 7.40.

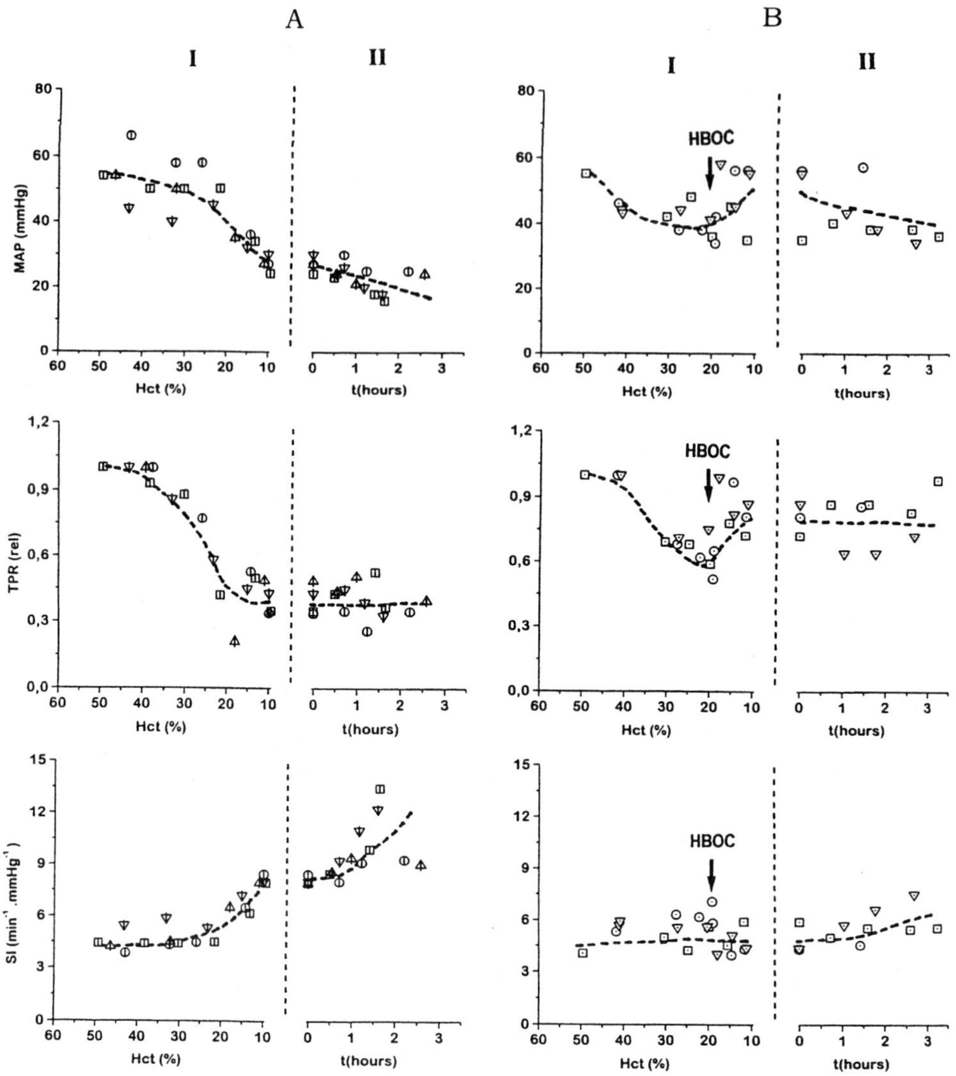

Figure 2. Changes of main cardiovascular parameters in phases I and II, A: control group, B: test group, MAP: mean arterial pressure, TPR (rel): relative total peripheral resistance, SI: shock index, Hct: hematocrit, each symbol represents one animal.

Clearly, a hemodilution down to a Hct of 20% decreases TPR in both groups. With application of HBOC solution in the test group, the TPR is restored to nearly 80% of its initial value parallel to MAP. In contrast, the TPR of the control group is further reduced. During phase II the recovery of TPR in group B is maintained. The lower part of Figure 2 shows the effects on the shock index (SI). This quantity did not change in both groups down to a hematocrit of 20%. But additional hemodilution to a Hct of 10% caused a doubling of the quantity in phase I and a triplication at the end of phase II in group A. In contrast, in group B the HBOC solution totally prevented the increase of SI in both phases.

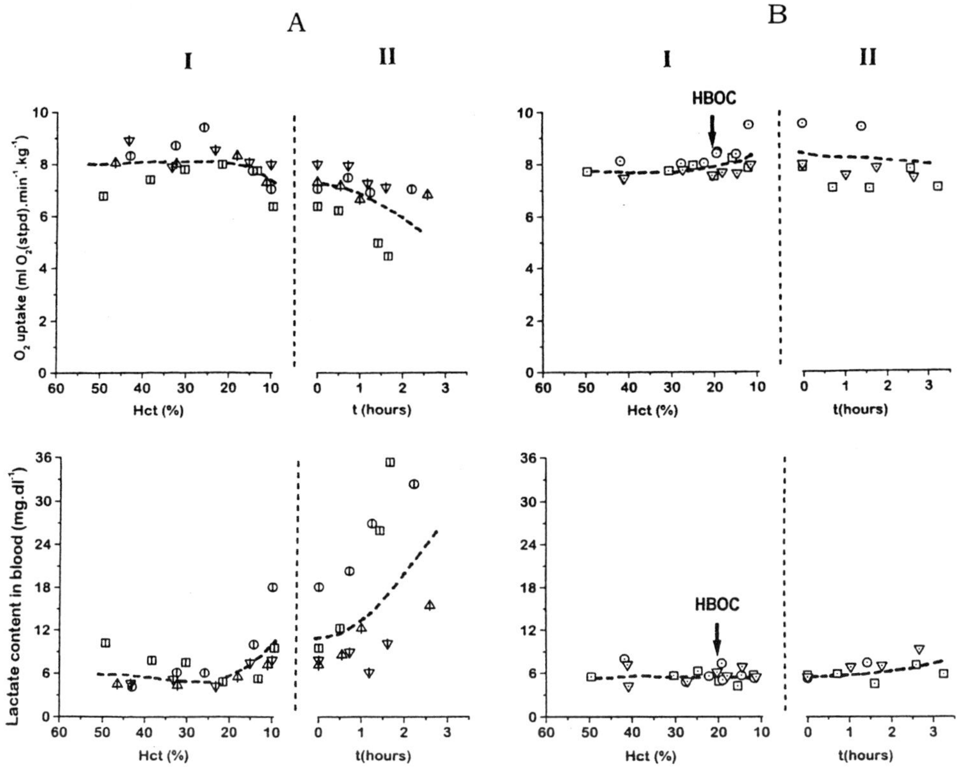

Figure 3. Effects on respiratory and metabolic parameters in phases I and II, A: control group, B: test group, Hct: hematocrit, each symbol represents one animal.

Figure 3 shows respiratory and metabolic effects. In the upper part the values of oxygen uptake are seen. In both groups oxygen uptake remains constant during phase I down to a Hct of 20%. Further hemodilution led to a significant decrease in oxygen uptake and carbon dioxide release ($p < 0.05$) during time phase in the control group (the latter is not shown here). The decrease of O_2 uptake begins already in phase I by lowering Hct of 20%. This is not seen in the test group. The ventilation of the animal ($n = 7$) was determined to be between 145 and 215 ml min^{-1} kg^{-1} at the beginning of phase I in both groups. A slight increase was seen only in phase II in both groups. The lower part of Figure 3 shows changes in blood lactate level. It was kept constant down to a Hct of 20% in both groups. However, it increased drastically together with a decrease in oxygen uptake (see Figure 3) by further hemodilution with human albumin in group A up to the end of the experiment, especially in phase II, while the added HBOC in group B totally prevented the increase of the lactate level, also in phase II. No significant changes in blood glucose level were seen.

Figure 4 shows a typical example of the analysis for the differentiation of the oxygen transport. Here all three forms of oxygen transport are displayed in case of an HBOC experiment: by native hemoglobin of blood (1), by HBOC (2), and by dissolved oxygen (3). The calculations of the contributions for oxygen transport are explained above briefly. These were given in detail by (Pötzschke et al., 1994). It is seen that the third transport form does not change significantly during the whole experiment, being

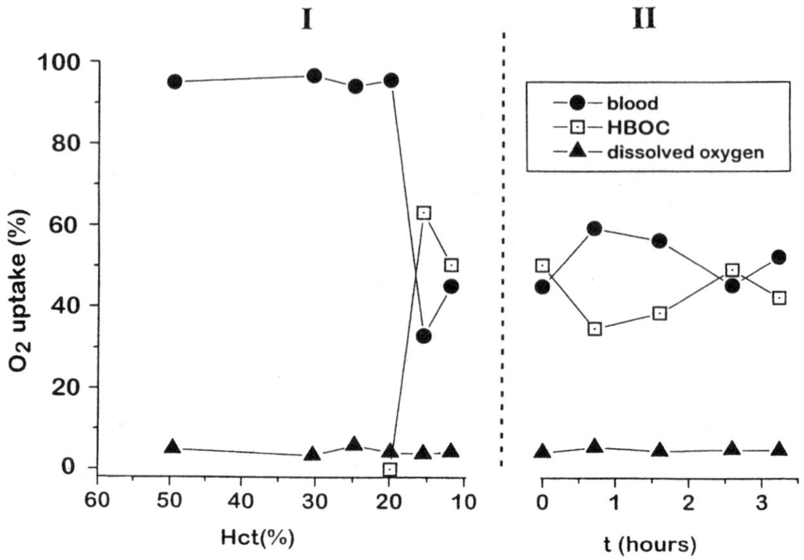

Figure 4. Representative example (porcine HBOC) for differentiation of oxygen transport. The relative contributions of the three different forms of oxygen transport in phases I and II.

about 5%; down to a Hct of 20% the main part of oxygen transport (95%) is necessarily affected by the native hemoglobin (blood) of animal. But immediately after dosing the artificial oxygen carrier (HBOC) takes over nearly half of the total transport. This remains the case also for phase II.

4. DISCUSSION AND CONCLUSIONS

In this study we present a new guinea pig model and an experimental protocol for evaluation of artificial oxygen carriers. The animal model allows the determination of important parameters of circulation, respiration and metabolism in anaesthetized spontaneously breathing guinea pigs. We have chosen this model, because the p50 of guinea pig blood (22–26 mmHg) is definitely closer to the p50 of human blood (26 mmHg) than to the p50 of the blood of the rat (34–36 mmHg). Thus, if the p50 value of HBOC is close to that of guinea pig or man, cardiovascular effects and any other effects of HBOC may be due to the efficiency of HBOC but not due to the change of the oxygen affinity.

It was a problem to find a suitable anesthesia procedure. Although some anesthesia methods for guinea pigs were recommended previously, there was a lack of data about some main physiological parameters of respiratory and cardiovascular function during narcosis (Evans, 1979; Brown et al., 1989). After many trials, a modified neuroleptanesthesia procedure was found to be more suitable than others (Evans, 1979; Brown et al., 1989) for our purpose.

In all seven animals used, we observed an inter individual variation of cardiac output between 160 and 350 ml $min^{-1} kg^{-1}$ whereas the MAP of all animals was normal. Consequently, the TPR also varies greatly. Therefore, we displayed the TPR effects in relative form. Such a variation of CO was surprising and not seen in rat (Pötzschke et al., 1994).

With our animal model, employing a two-phase procedure (exchange and time phase) we have clearly revealed the limits of hemodilution, at least in guinea pig. By hemodilution with an isoncotic albumin solution a critical hemodilution level was reached with a Hct value of 20%. Below this Hct, depressive effects on important cardiovascular, respiratory and metabolic functions could be demonstrated. Some parameters such as mean arterial pressure, total peripheral resistance, shock index and lactate level changed more extensively than others.

Both artificial oxygen carriers, at low level of only 3 g/dl, appear to provide sufficient oxygen supply for the animals in a critical situation, restoring and maintaining the normal functional parameters for hours. Especially in case of TPR, it is astonishing that such a small content of HBOC could restore this parameter in spite of further dilution, which leads to a decrease of blood viscosity. This 'over compensation' can only be achieved by a biological reaction. A nitric oxide (NO) scavenging effect due to our solution should not be the reason because there was no monomer hemoglobin in our HBOC, so that extravasation is not probable. We rather attribute this effect of vasoconstriction to an augmented oxygen supply in peripheral tissues. This assumption can be supported by (Rohlfs et al., 1998), who have found no correlation between NO scavenging reaction of Hb solutions and systemic vasopressor effects. A novel mechanism reported by (Vollrath et al., 1998) may be an alternative explanation of vasoconstriction due to hemoglobin.

The enormous efficacy of acellular HBOC solution was also observed during the exchange experiment: HBOC at a concentration of only 0.5 g/dl brought the arterial pressure back to the normal level when reduced beforehand by hemodilution. Application of an equal volume of albumin solution in the same situation shows no such effect. The high efficacy of HBOC is also seen by the differentiated analysis of oxygen transport: in spite of minor content, HBOC takes over 50% of the whole O_2 transport in blood.

These results may be explained with the aid of in vitro experiments reported by (Vandegriff and Winslow, 1995): it was found that thin layers of acellular HBOC solution release oxygen much quicker than suspension of red blood cells, both with the same hemoglobin content, even so with the increased oxygen affinity of HBOC. The same is true for artificial capillaries with 25 μm diameter (Page et al., 1998a) and it is even better in capillaries of 10 μm diameter (Page et al., 1998b). Also in native capillaries acellular HBOC solutions should release oxygen better than whole blood. Moreover, it was found that a relatively small replacement of red blood cell hemoglobin with HBOC causes a significant increase in oxygen release from the artificial capillary; in half to half mixtures of HBOC solution with erythrocyte suspension, the oxygen release is indistinguishable from that of pure HBOC solution (Page et al., 1998b). This means that HBOC transports oxygen not only better on its own, but it also enhances oxygen release from red blood cells exhibiting a synergistic effect, presumably by the mechanism of facilitated diffusion of oxygen in hemoglobin solution (Wittenberg 1966).

The model proves to be appropriate for the evaluation of artificial carriers. In order to further differentiate the efficacy of oxygen transport of the two HBOC types investigated, we must perform more experiments.

ACKNOWLEDGMENTS

We thank Mrs. Stephanie Hammon for the careful preparation and characterization of the two types of HBOC and Mrs. Karola Krause and Mrs. Birgit Wintzek for their skilful assistance during the animal experiments.

REFERENCES

Abassi Z, Kotob S, Pieruzzi F, Abouassali M, Keiser HR, Fratantoni JC, and Alayash AI: Effects of polymerization on the hypertensive action of diaspirin cross-linked hemoglobin in rats. J. Lab. Clin. Med. 129:603–610, 1997.

Amdur MO and Mead J: Mechanics of respiration in unanesthetized guinea pigs. Am. J. Physiol. 192:364–368, 1958.

Barnikol WKR, Döhring W, and Wahler W: Eine verbesserte Modifikation der Mikromethode nach Niesel and Thews (1961) zur Messung von O_2-Hb-Bindungskurven in Vollblut und konzentrierten Hb-Lösungen. Respiration 36:86–95, 1978.

Barnikol WKR, Hiller B, and Guth S: Ein Mikro-Atemstromventil für spontan atmende narkotisierte Kleintiere. Biomed. Technik 39:57–62, 1994.

Brown JN, Thorne PR, and Nuttall AL: Blood pressure and other physiological responses in awake and anesthetized guinea pigs. Laboratory Animal Science 39(2):142–148, 1989.

Dinkelmann S: Präparation und in vitro Charakterisierung eines künstlichen Sauerstoffträgers auf der Basis von Schweinehämoglobinpolymeren und seine Evaluierung am Kleintier, Thesis, Mainz/D, 1997.

Domack U: Entwicklung und in vivo-Evaluation eines künstlichen Sauerstoffträgers auf der Basis von Rinderhämoglobin, Thesis, Mainz/D, 1997.

Evans EF: Neuroleptanesthesia for the guinea pig. An ideal anesthetic procedure for long-term physiological studies of cochlea. Arch. Otolaryngol. 105:185–186, 1979.

Gulati A, Sharma AC, and Burhop KE: Effect of stroma-free hemoglobin and cross-linked hemoglobin on the regional circulation and systemic hemodynamics. Life Sciences 55(10):827–837, 1994.

Kunert MP, Liard JF, and Abraham DJ: RSR-13 an allosteric effector of hemoglobin, increases systemic and iliac vascular resistance in rats. Am. J. Physiol. 271(Heart Circ. Physiol. 40):H602–H613, 1996.

Liard JF and Kunert MP: Hemodynamic changes induced by low oxygen affinity in dogs. Am. J. Physiol. 264(Regulatory Integrative Comp. Physiol. 33):R396–R401, 1993.

Page TC, Light WR, McKay CB, and Hellums JD: Oxygen transport by erythrocyte/hemoglobin solution mixtures in an in vitro capillary as a model of hemoglobin-based oxygen carrier performance. Microvasc. Res. 55(1):54–64, 1998a.

Page TC, Light WR, and Hellums JD: Experimental determination of intraluminal oxygen transport by hemoglobin based oxygen carriers in 10 μm capillaries. 26th Annual Meeting of the International Society on Oxygen Transport to Tissue, Budapest, 1998b.

Pötzschke H, Guth S, and Barnikol WKR: Divinyl sulfone cross-linked hyperpolymeric human haemoglobin as an artificial oxygen carrier in anaesthetized spontaneously breathing rats. Advances in Experimental Medicine and Biology 345:205–214, 1994.

Rohlfs RJ, Bruner E, Chiu A, Gonzales A, Gonzales ML, Magde D, Magde MD, Vandegriff KD, and Winslow RM: Arterial blood pressure responses to cell-free hemoglobin solutions and the reaction with nitric oxide. J. Biol. Chem. 273(20):12128–12134, 1998.

Soulard CD, Teisseire BP, Teisseire LJ, and Herigault RA: Consequences of an acute increase in P50 in anaesthetized guinea pigs. Respiration Physiology 51:21–30, 1983.

Vandegriff KD, and Winslow RM: A Theoretical analysis of oxygen transport: A new strategy for the design of hemoglobin-based red cell substitutes, pp 143–154 (chapter 10). In Blood Substitutes. Physiological Basis of Efficacy, Edited by Winslow RM, Vandegriff KD, and Intaglietta M, Birkhäuser Boston, 1995.

Vollrath B, Cook D, Megyesi J, Findlay JM, and Ohkuma H: Novel mechanism by which hemoglobin induces constriction of cerebral arteries. European Journal of Pharmacology 361:311–319, 1998.

Waschke K, Krieter H, Albrecht DM, van Ackern K, and Kuschinsky W: Modifiziertes Hämoglobin als Blutersatzmittel im Rattenmodell. Anaesthesist 42:90–92, 1993.

Wittenberg JB: The molecular mechanism of hemoglobin-facilitated oxygen diffusion. J. Biol. Chem. 241:104–114, 1966.

4

REDOX STATES OF CEREBRAL TISSUES OF RATS SUBSTITUTED BY THE LIPOSOME-ENCAPSULATED HEMOGLOBIN

Jun Sakanoue,[1] Mamoru Tamura,[1] Kunihiko Nakai,[2] Ichiro Sakuma,[3] and Akira Kitabatake[3]

[1]Division of Biophysics
Research Institute for Electronic Science
Hokkaido University, Sapporo 060-0812, Japan
[2]Environmental Health Sciences
Tohoku University School of Medicine
Sendai 980-8575, Japan
[3]Department of Cardiovascular Medicine
Hokkaido University, School of Medicine
Sapporo 060-0815, Japan

INTRODUCTION

Many kinds of artificial blood that have the ability to transport oxygen into tissues have been proposed. Hemoglobin vesicles, in particular, have advantages over other ones because of their longer half-lives and lower viscosity in circulation.[10,11,12] Neo Red Cells (NRC) is the liposome-encapsulated human hemoglobin, and the $P_{50}O_2$ of NRC is changeable in the range of 13–60 Torr in accord with the variation in the addition of inositolhexaphosphate (IHP) as an allosteric effector.[8]

Near-infrared spectrometry (NIRS) has been developed extensively for non-invasive optical monitoring in clinical medicine. Redox states of hemoglobin, myoglobin and cytochrome oxidase, a terminal enzyme in the respiratory chain are measurable by this technique noninvasively in cerebral tissues.[1,2,6,7] However, little is known about the actual behavior of redox states of cerebral tissues when substituted by artificial blood, except for experiments using fluorocarbon.[3]

The purpose of this study was to determine whether oxygen could be carried into tissues to adequately maintain the physiology of rats under conditions in which 80% of

Oxygen Transport to Tissue XXI, edited by Eke and Delpy.
Kluwer Academic / Plenum Publishers, New York, 1999.

whole blood was substituted for by NRC. Using NIRS, we estimated the oxygenation of hemoglobin and oxidation of cytochrome oxidase in cerebral tissues.

MATERIALS AND METHODS

Materials

Liposome-encapsulated hemoglobin containing 6%(w/w) human hemoglobin (Neo Red Cells, NRC) was synthesized according to the standard procedure[8] of TELMO corporation (Tokyo, Japan). Two NRC with different $P_{50}O_2$ values were employed (Figure 1). The $P_{50}O_2$ of the first type was adjusted to 60 Torr by the addition of 3 mM IHP (NRC60), and that of the second one was 14 Torr without the addition of IHP (NRC14).

Animal preparation

Male Wistar rats (body weight: 230–260 g) were anesthetized with pentobarbital (50 mg/kg, i.p.), and paralyzed with pancuronium bromide (0.2 mg/kg, i.v.). They were tracheotomized and ventilated mechanically. Femoral arteries and veins were cannulated, and blood gases and hemoglobin content were measured frequently with an ABL 30 blood gas analyzer (Radiometer, Copenhagen, Denmark). The electroencephalogram (EEG) was recorded at the occipital region during blood substitution. The blood substitution was carried out by transfusion from a femoral vein (0.75–1.0 ml/min) and drawing from a femoral artery. The rats were ventilated with 100% O_2 during the substitution. In the experiments on blood substitution, 1.5%(w/w) bovine serum albumin was added to NRC for osmoregulation, and both NRC and the body temperature of the rats were maintained at 37 degree C. After completing the substitution, the O_2–N_2 ratio was changed by stepwise decreases of fractional inspired oxygen (FiO_2), and finally rats were killed by 100% N_2.

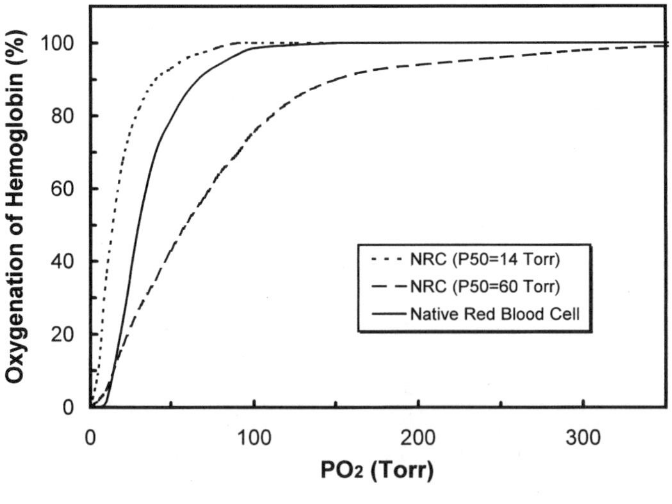

Figure 1. Oxygen-binding curves for two kinds of liposome-encapsulated hemoglobin and native red blood cells.

Measurements of NIRS

For the measurements of redox states of hemoglobin and cytochrome oxidase, the four-wavelength method[4,5,7] was employed. For simultaneous measurements of oxy-Hb, deoxy-Hb, and cytochrome oxidase, a spectrophotometer with a rotating disc having four band-pass filters (Unisoku, Osaka, Japan) was used. The light source contained four wavelengths (700, 730, 750, and 805 nm), and was guided onto the calvaria of rats by the use of an optical fiber. Light transmitted through the cerebral tissues was detected in the mouth and guided to a photomultiplier. The changes in oxy-Hb, deoxy-Hb, total-Hb, and reduced cytochrome oxidase were calculated by the following formulas.

$$[oxy\text{-}Hb] = -0.254A_{700} - 1.59A_{730} + 1.844A_{750},$$
$$[deoxy\text{-}Hb] = 0.763A_{700} - 1.479A_{730} + 0.716A_{750},$$
$$[total\text{-}Hb] = 0.509A_{700} + 0.111A_{730} + 2.560A_{750},\ and$$
$$[cytochrome\ oxidase] = -1.676A_{700} + 0.7A_{730} + 1.976A_{750} - A_{805}.$$

In this paper, each concentration expresses a percentage value calculated from the maximum changes caused by aerobic (100% O_2) to anaerobic (100% N_2) transition.

RESULTS

Changes in Cerebral Oxygenation Caused by the Blood Substitution

Figure 2 shows typical changes in oxygenation states of hemoglobin and cytochrome oxidase during the blood substitution by NRC60 and following stepwise decreases of FiO_2. Oxy-Hb and total-Hb decreased with the substitution, and finally reached a plateau at 60% of initial values. On the other hand, oxidation of cytochrome oxidase was at a constant value of 100% during the substitution. After the substitution, reduction of hemoglobin started with the value of 30% FiO_2, and half of the hemoglobin was reduced with 15% FiO_2. Finally, hemoglobin was completely reduced at 0% O_2 (100% N_2). In contrast to hemoglobin, cytochrome oxidase was not reduced with the decreases of FiO_2 from 100% to 15%, and started to be reduced when the level of FiO_2 fell below 10% O_2, where 70% of hemoglobin was reduced. Finally, cytochrome oxidase was also reduced completely with 100% N_2. The blood pressure was elevated transiently at the beginning of the substitution by the volume effect (10 min), but was almost constant during the blood substitution and decreasing of FiO_2. The EEG amplitude recorded during the blood substitution did not change until 80% of whole blood was substituted for by both NRC60 and NRC14 (data not shown).

Effects of $P_{50}O_2$ of NRC on Oxygenation States in Cerebral Tissues

Figure 3(A) shows the oxygenation states of hemoglobin plotted against the oxygen partial pressures in venous blood (PvO_2) measured by gas analysis. We previously measured the changes in PO_2 values of blood in femoral vein and cervical vein with decreasing in FiO_2, and the differences between them were confirmed to be less than 15% (data not shown). In consequence, the PO_2 values in femoral vein were employed for plotting in Figure 3 instead of those in cervical vein. In the case of NRC60, reduction of hemoglobin started below the PvO_2 value of 110 Torr, and half of the hemoglobin was reduced

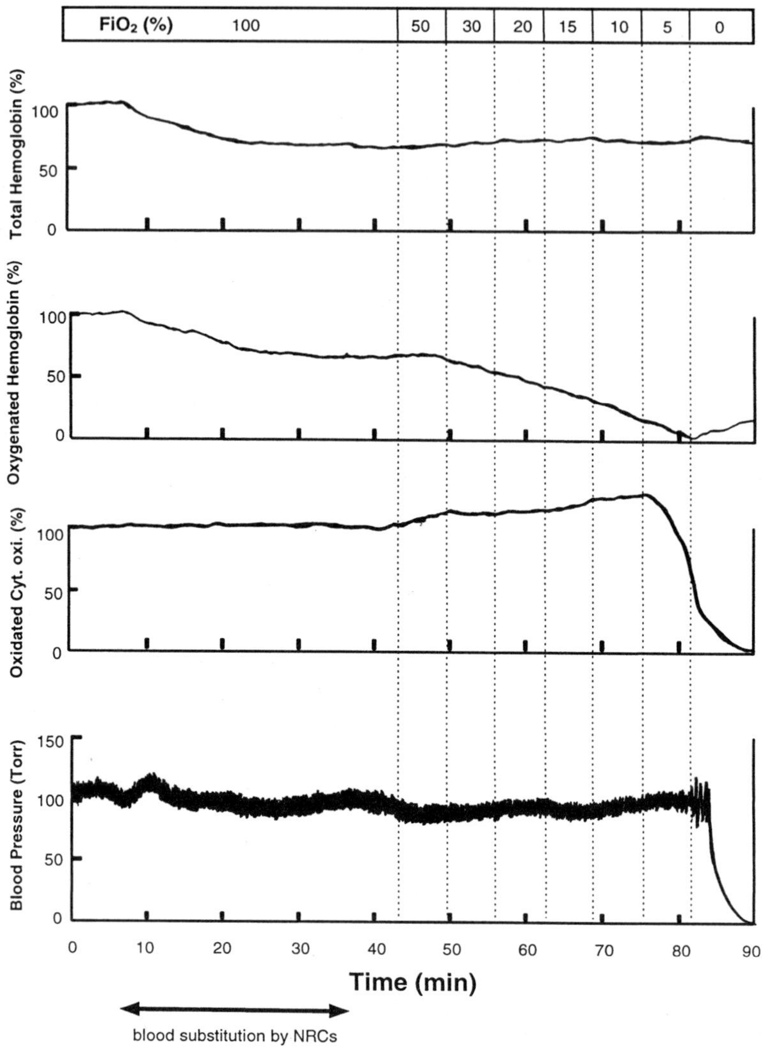

Figure 2. Typical changes of oxygenation states in cerebral tissue and blood pressure in a rat with blood substituted by liposome-encapsulated hemoglobin. After the 80% substitution with NRC60, fractional inspired oxygen (FiO_2) was controlled by stepwise decreases from 100% to 0%. Oxygenated hemoglobin and oxidized cytochrome oxidase are expressed as percentage values calculated from the 100% value.

at the PvO_2 value of 50 Torr. In the case of NRC14, reduction of hemoglobin started with the PvO_2 value of 35 Torr, and half of the hemoglobin was reduced at the value of 17 Torr. In consequence, the plots of NRC60 were largely right-shifted compared with the plots of NRC14. Figure 3(B) shows the oxidation of cytochrome oxidase plotted against the values of PvO_2. In the plots of NRC60, reduction of cytochrome oxidase started with the PvO_2 value of 30 Torr, and half of the cytochrome oxidase was reduced at 10 Torr. In the plots of NRC14, cytochrome oxidase was reduced quickly below 5 Torr.

Figure 4 shows the relationship of oxygenation states between hemoglobin and cytochrome oxidase plotted using the values in Figure 3. Both trends in plots for NRC60 and NRC14 were almost identical. The reduction of cytochrome oxidase started when

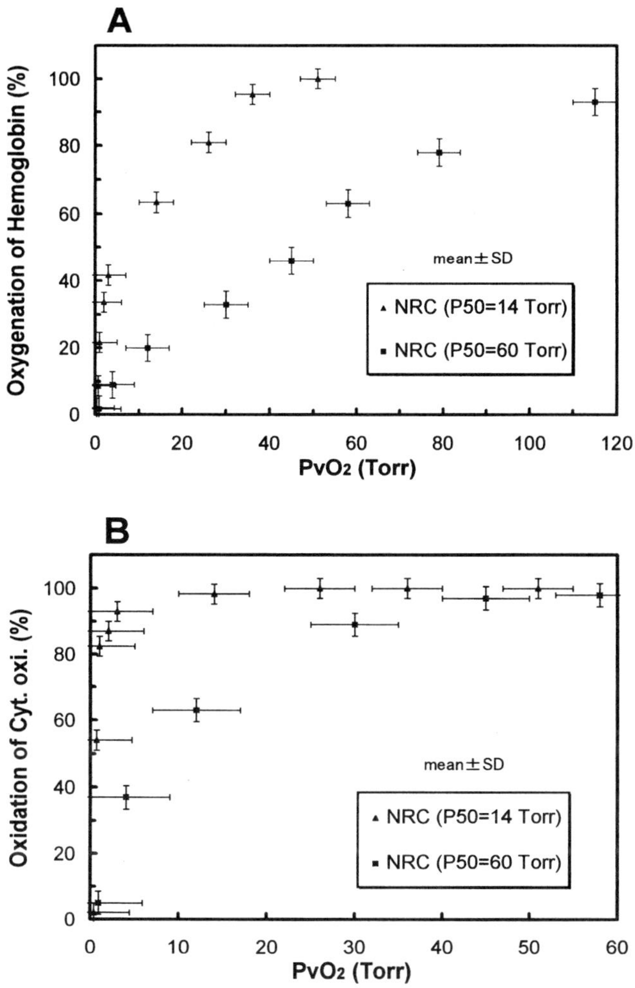

Figure 3. Relationship between oxygenation states in cerebral tissues and oxygen partial pressure in a vein. PvO_2 was measured from a small amount of blood drawn from the femoral vein. Oxygenated hemoglobin (A), and oxidated cytochrome oxidase (B) are presented as percentage values calculated manually from Figure 2.

the ratio of oxy-Hb was below 45%, and reduction was almost parallel to the hemoglobin reduction.

DISCUSSION

NRC contain about 6%(w/w) of human hemoglobin, almost one-half that of native red cells.[8,11] In consequence, the total content of hemoglobin in rats with NRC substitution is around 60% of the initial one. The decreases in total-Hb and oxy-Hb observed in the process of blood substitution in Figure 2 were attributed to this loss of hemoglobin. Takahashi[11] reported that injection of NRC induced hypertension in rabbits. However, we could remove the elevation in blood pressure by careful control of infusion and by withdrawing blood except for the initial volume effect.

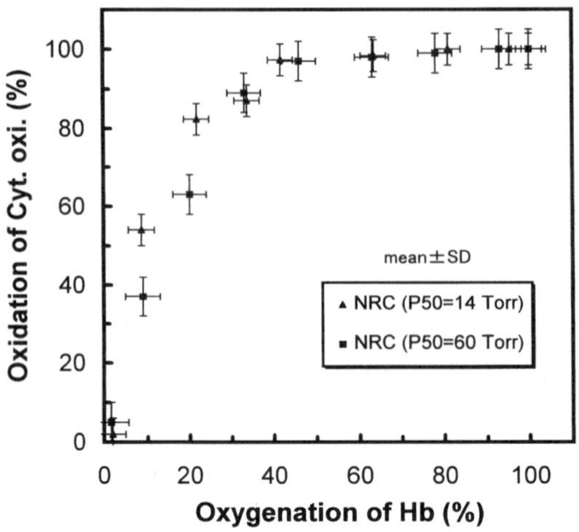

Figure 4. Relative oxidation states of cytochrome oxidase with respect to relative oxygenation states of hemoglobin. Average values (n = 4) were plotted using the percentage values in Figure 3.

In our studies, a decrease of EEG amplitude induced by hypoxia was not observed above 10% FiO_2 (data not shown). The sustained EEG amplitude provided evidence that the function of brain was maintained after the substitution by both NRC14 and NRC60.

Figure 3(A) demonstrates that, with decreases in F_iO_2, hemoglobin in the rats with NRC60 substitution was reduced more quickly than with NRC14 substitution. The changes in hemoglobin oxygenation in Figure 2 came from the hemoglobin reduction in the veins, because the volume percentage of veins among blood vessels is in excess of 70% in cerebral tissues. The difference of oxygenation ratio seen in Figure 3(A) is explained by the difference of oxygen-binding constants between NRC60 and NRC14 (Figure 1). The lower the binding constant, the quicker the reduction caused by decreases PvO_2. Figure 3(B) demonstrates that cytochrome oxidase can sustain its oxidation state in the rats with NRC14 substitution more effectively than that with NRC60 below PvO_2 value of 20 Torr, although NRC14 has a higher binding constant to oxygen than NRC60. This observation seems strange, but is accounted for by the difference of oxygen content between arteries and veins. In hypoxic hypoxia condition, such as 10% FiO_2, PaO_2 values were around 50 Torr in both substituted rats (data not shown), while PvO_2 values for NRC14 and NRC60 substituted rats were 5 Torr and 30 Torr, respectively. From Figure 1, total amount of oxygen delivered to tissues by NRC14 is about 50%, almost threefold larger than that of NRC60, and cytochrome oxidase can highly sustain its oxidation level even in the hypoxic condition.

In Figure 4, plots of oxidation of cytochrome oxidase for both NRC14 and NRC60 show an identical predilection to be reduced by hypoxia. The physiological condition of the rats was maintained by total amount of oxygen delivered by liposome-encapsulated hemoglobin, and did not depend on P_{50} values of NRC. This result is reasonable, because the binding constant of cytochrome oxidase is larger than that of hemoglobin by two orders of magnitude.

CONCLUSION

In the future, studies will examine the oxygen-transportation ability of liposome-encapsulated hemoglobin depending on $P_{50}O_2$ in various pathologic models in detail. For instance, NRC14 might be effective in hypoxic hypoxia condition. Alternatively, blood transfusion with NRC60 can be effective treatment for hemorrhagic shock or ischemia damage.

REFERENCES

1. Cooper CE and Springett R (1997) Measurement of cytochrome oxidase and mitochondrial energetics by near-infrared spectroscopy. *Phil. Trans. R. Soc. Lond.* B 352:669–676.
2. Delpy DT and Cope M (1997) Quantification in tissue near-infrared spectroscopy. *Phil. Trans. R. Soc. Lond.* B 352:649–659.
3. Ferrari M, Williams MA, Wilson DA, Thakor NV, Traystman RJ, and Hanley DF (1995) Cat brain cytochrome-c oxidase redox changes induced by hypoxia after blood-fluorocarbon exchange transfusion. *Am. J. Physiol.* 269 (*Heart Circ. Physiol.* 38):H417–424.
4. Hazeki O and Tamura M (1988) Quantitative analysis of hemoglobin oxygenation state of rat brain in situ by near-infrared spectrophotometry. *J. Appl. Physiol.* 64(2):796–802.
5. Hazeki O and Tamura M (1989) Near infrared quadruple wavelength spectrophotometry of the rat head. *Adv. Exp. Med. Biol.* 248:63–69.
6. Hoshi Y and Tamura M (1993) Dynamic multichannel near-infrared optical imaging of human brain activity. *J. Appl. Physiol.* 75(4):1842–1846.
7. Hoshi Y, Hazeki O, Kakihana Y, and Tamura M (1997) Redox behavior of cytochrome oxidase in the rat brain measured by near-infrared spectroscopy. *J. Appl. Physiol.* 83(6):1842–1848.
8. Ogata Y, Goto H, Sakaguchi K, Suzuki M, Ohsaki K, Suzuki K, Saniabadi AR, Kamitani T, and Takahashi A (1994) Characteristics of Neo Red Cells, their function and safety: in vivo studies. *Art. Cells, Blood Subs., and Immob., Biotech.* 22(3):875–881.
9. Rudolph AS, Sulpizio A, Hieble P, Macdonald V, Chavez M, and Feuerstein G (1997) Liposome encapsulation attenuates hemoglobin-induced vasoconstriction in rabbit arterial segments. *J. Appl. Physiol.* 82(6):1826–1835.
10. Rudolph AS (1988) The freeze-dried preservation of liposome encapsulated hemoglobin: a potential blood substitute. *Cryobiology* 25:277–284.
11. Takahashi A (1995) Characterization of Neo Red Cells (NRCs), their function and safety in vivo tests. *Art. Cells, Blood Subs., and Immob., Biotech.* 23(3):347–354.
12. Usuba A, Motoki R, Ogata Y, Suzuki K, and Kamitani T (1995) Effect and safety of liposome-encapsulated hemoglobin Neo Red Cells (NRCs) as a perfusate for total cardiopulmonary bypass. *Art. Cells, Blood Subs., and Immob., Biotech.* 23(3):337–346.

5

CORTICAL BLOOD OXYGENATION CHANGES IN THE LEFT AND RIGHT OCCIPITAL AREA INDUCED BY SELECTIVE VISUAL STIMULI IN HUMANS

Willy N. J. M. Colier,[1] Valentina Quaresima,[2] Rüdiger Wenzel,[3] Marco C. van der Sluijs,[1] Berend Oeseburg,[1] Marco Ferrari,[2] and Arno Villringer[3]

[1]Department of Physiology
University of Nijmegen
P.O. Box 9101, 6500 HB Nijmegen
The Netherlands
[2]Dipartimento di Scienze e Tecnologie Biomediche
Università L'Aquila, 67100 L'Aquila
Italy
[3]Department of Neurology
Charité, Humboldt-University Berlin, 10117 Berlin
Germany

1. INTRODUCTION

In the last few years near infrared spectroscopy (NIRS) has become a more widespread tool to map the human brain function, in particular the motor cortex areas (Gratton et al., 1995; Hirth et al., 1996; Obrig et al., 1996; Colier et al., 1997) and visual cortex areas (Kato et al., 1993; Hoshi et al., 1993; Meek et al., 1995; Wenzel et al., 1996; Gratton et al., 1998). In these studies it was shown that NIRS is able to detect changes in cerebral oxygenation due to visual stimulation in the occipital cortex. Most subjects respond to a visual stimulus with an increase in the oxyhemoglobin concentration ([O_2Hb]) and a decrease in the deoxy-hemoglobin concentration ([HHb]) of smaller magnitude than the [O_2Hb] change (Wenzel et al., 1996).

All the previous studies reported only the response in either left or right hemisphere. The aim of this study was to localize simultaneously in both hemispheres the human

Oxygen Transport to Tissue XXI, edited by Eke and Delpy.
Kluwer Academic / Plenum Publishers, New York, 1999.

visual cortex oxygenation response to selective visual stimuli. For this purpose a multichannel approach was used using highly sensitive NIRS instrumentation.

2. METHODS

2.1. Near Infrared Spectroscopy

The theory of NIRS has been described extensively (Delpy et al., 1997). Briefly, changes in optical density in the near infrared region at different wavelengths are converted into *changes* in concentration of [O_2Hb] and [HHb]. Quantification of concentration changes is possible by introducing the differential path length factor (Delpy et al., 1988). This factor accounts for the light scattering in the tissue. Two pulsed continuous wave NIRS instruments, the OXYMON, were used for this study. The Department of Physiology and the Instrumentation Department of the University of Nijmegen built this instrument. A detailed description of the instrument and comparison with two other NIRS instruments has been reported (van der Sluijs et al., 1998). The instrument uses 3 wavelengths (775, 848 and 901 nm). The sampling rate is 10 Hz. The light is transmitted and collected by 2 m fiber optic bundles (optodes) with an active diameter of 4 mm.

2.2. Experiment

Six healthy volunteers participated in the study (2 male, 4 female; age 21–42 years) after giving their written informed consent. The local Ethics Committee of the Humboldt-University Berlin approved the experimental protocol. The volunteers were sitting comfortably on a chair in a quiet and dark room. Before the start of the experiment the volunteers were familiarized with the procedures and the paradigms. Two receiving and the two transmitting optodes were horizontally positioned over the right and left occipital regions at the level of the calcarine sulcus. The transmitting optodes were placed 1 cm to left and right of the midline. The receiving optodes were placed 4.2 cm laterally to the corresponding transmitting optodes. The optodes were inserted into a special holder with fixed angle and distance and fixed to the head using an elastic strap. Before the beginning of the protocol, the oxygenation response was checked using a full-field green and white reversing checkerboard (Figure 1a). If no oxygenation change could be detected the optode holder was moved vertically (less than 1 cm) until a response was found.

2.3. Protocol

The stimulus was a green and white pattern-reversal left or right hemi-field checkerboard reversing at the rate of 8 Hz (Figure 1b). Each cycle consisted of either a left or right hemi-field stimulus lasting 10 s followed by a 20 s rest period. Twelve cycles were randomized in order to obtain six left and six right stimuli. A 60 s baseline period preceded the twelve cycles.

2.4. Data Analysis

Data sets for each volunteer were analyzed individually. A concentration baseline value was determined during the 10 s prior to the stimulus. A differential path length factor of 6.0 was used to estimate the concentration changes. The heart rate signal was

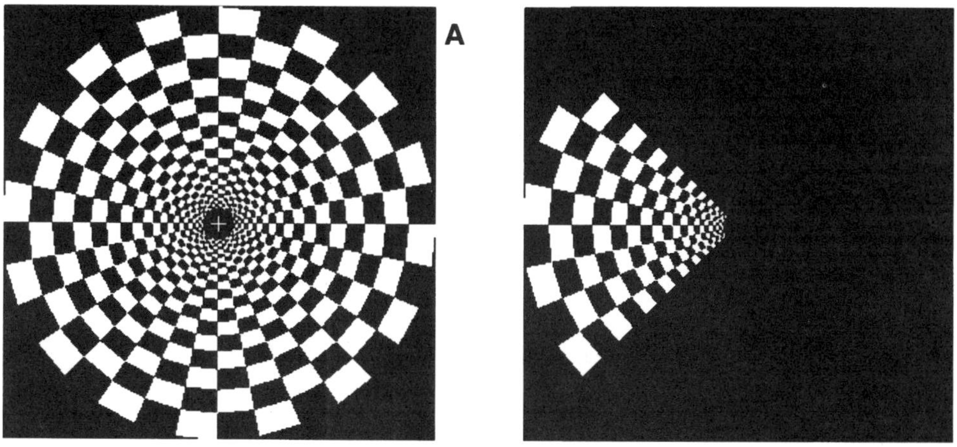

Figure 1. The stimulus to localize the visual area is a full-field green and white reversing (8 Hz) checkerboard is shown in panel A. Panel B shows the left hemi-field stimulus.

smoothed with a 25-point Savitzky-Golay filter (Savitzky et al., 1964). A band-pass filter was used to remove slow variations of oxygenation changes. Left hemisphere oxygenation changes in response to either six left or right hemi-field stimuli were averaged for every individual. The same procedure was used for the right hemisphere oxygenation changes.

3. RESULTS AND DISCUSSION

Figure 2 shows the response to a full-field stimulus used to localize both left and right visual areas. As previously described (Wenzel et al., 1996) visual stimulation provoked an increase of [O_2Hb] accompanied by a smaller decrease in [HHb].

Figure 3 reports the response of both hemispheres to 12 randomized hemi-field stimuli in two subjects (A and B). The oxygenation changes reported in panel A show the capability of NIRS to detect the localized changes: left hemi-field stimuli induce a decrease of [HHb] in the right visual area and no decrease in the left visual area and *viceversa*. As expected the amplitude of the responses is smaller than those obtained using the full-field stimuli. More difficult is the detection of the oxygenation changes in the tracings reported in panel B. This can be due to: 1) smaller activated area; 2) smaller oxygenation response; 3) lower penetration of NIRS light due to the skull thickness; 4) anatomical variations of the visual area. Although there is a lower signal-to-noise ratio in panel B, it is possible to detect localized changes by averaging the responses to six left and right stimuli. This is shown in subject 5 of Figure 4.

Both [O_2Hb] and [HHb] changes were time averaged over all six left and right hemi-field stimuli for all individuals. Figure 4 reports the oxygenation changes of [O_2Hb] and [HHb] in the left and right hemispheres simultaneously in response to six right hemi-field visual stimuli in each individual volunteer. From these averaged data it is evident that no significant response can be detected in the hemisphere stimulated with the ipsilateral hemi-field stimuli.

In subjects 4 to 6 in Figure 4 a periodical signal with a frequency of approximately 0.1 Hz can be observed. The origin of this signal is probably vasomotion. The periodical

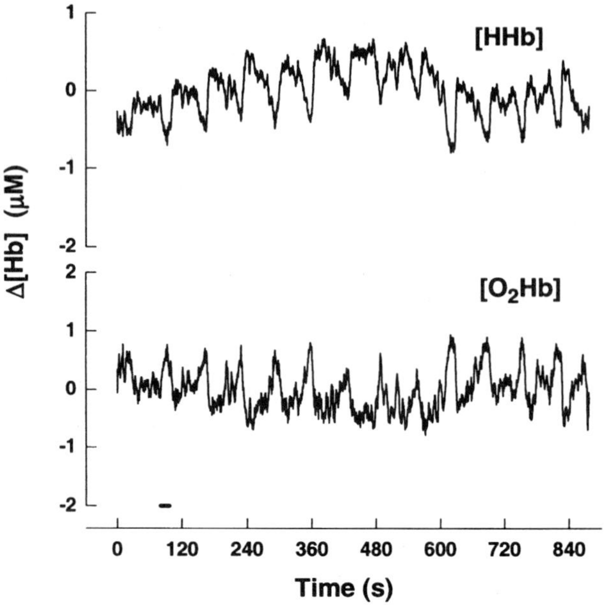

Figure 2. Visual hemisphere oxyhemoglobin [O_2Hb] and deoxy-hemoglobin [HHb] changes (μM) in response to 12 full-field stimuli in a single volunteer. The stimulus was a green and white reversing (8 Hz) full-field checkerboard with a 16 s stimulus duration. The period between stimuli was 50 s. The small horizontal bar indicates the duration of the first stimulus.

signal is not averaged out of the final signal because of the chosen time intervals of the stimulus and inter-stimulus periods: 10 and 20 s, respectively. This phenomenon is also described by others (Biswal et al., 1997, Elwell et al., 1998). A solution can be found in choosing different, irregular, time intervals or short stimulus durations.

4. CONCLUSION

In this study we have shown that it is possible to localize simultaneously in both hemispheres of the occipital cortex oxygenation changes in response to a hemi-field paradigm using a novel multi-channel NIRS approach. Our results confirm previous findings obtained using several brain imaging techniques. We conclude that multi-channel NIRS offers the advantages to measure directly not only concentration changes of [HHb] as measured by functional magnetic resonance imaging, but also of [O_2Hb] at the bedside with low cost instrumentation with potential usefulness to investigate the pathophysiology of vision.

ACKNOWLEDGMENTS

W.N.J.M. Colier, V. Quaresima and R. Wenzel equally contributed to this work. This research has been supported in part by EU Contract BMH4-CT96.1658, the Dutch Technology Foundation and "Programma Vigoni 1997–98".

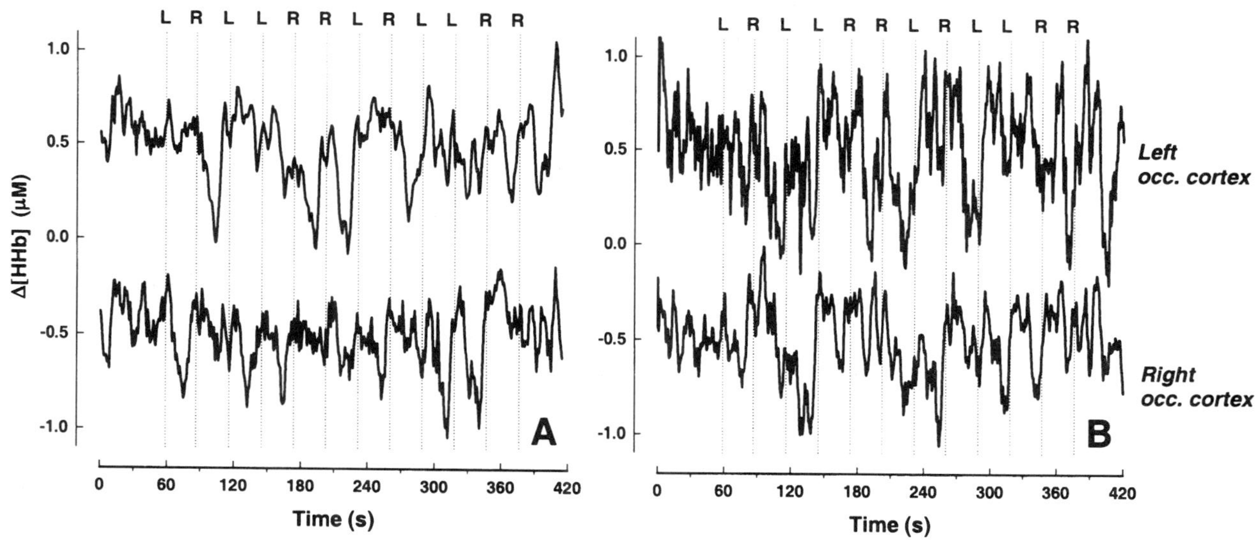

Figure 3. A) Right and left visual cortex deoxy-hemoglobin [HHb] changes in μM in response to randomized left and right hemi-field stimuli in a single volunteer. The stimulus was a green and white reversing (8 Hz) hemi-field checkerboard with a 10 s duration. B) Right and left visual cortex oxygenation changes in response to hemi-field visual stimuli in a single volunteer, showing smaller and less clear reactions. The upper tracings of both panels refer to the left hemisphere. The lower panels refer to the right hemisphere. Vertical dotted lines indicate the start of either a left (L) or right (R) hemi-field stimulus.

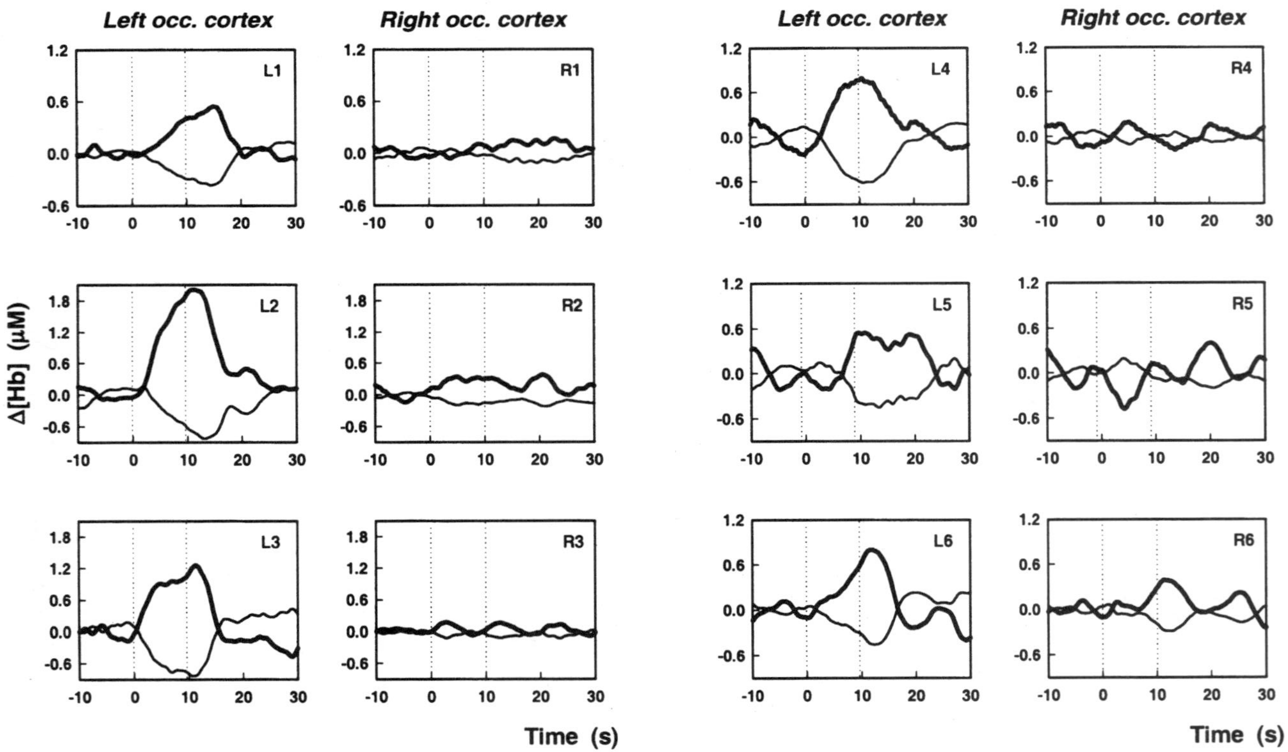

Figure 4. Right and left hemispheres oxygenation changes in response to right hemi-field visual stimuli for all six subjects. Both oxy- and deoxy-hemoglobin concentration changes ([O_2Hb] and [HHb], respectively) in μM were time averaged over all six right hemi-field stimuli. Left and right panels refer to the left and right visual cortex areas, respectively. The vertical lines indicate the stimulus duration (10 s).

REFERENCES

Biswal B, Hudetz AG, Zerrin Yetkin F, Haughton VM, and Hyde JS (1997) Hypercapnia reversibly suppresses low-frequency fluctuations in the human motor cortex during rest using echo-planar MRI. J Cereb Blood Flow Metab 17:301–308.

Colier WNJM, Quaresima V, Barattelli G, Cavallari P, Sluijs M van der, and Ferrari M (1997) Detailed evidence of cerebral hemoglobin oxygenation changes in response to motor cortical activation revealed by a continuous wave spectrophotometer with 10 Hz temporal resolution. SPIE 2979:390–396.

Delpy DT and Cope M (1997) Quantitation in tissue near-infrared spectroscopy. Phil Trans R Soc Lond B 261:649–659.

Delpy DT, Cope M, Zee P van der, Arridge S, Wray S, and Wyatt J (1988) Estimation of optical pathlength through tissue from direct time of flight measurements. Phys Med Biol 33:1433–1442.

Elwell CE, Springett R, Hillman E, and Delpy DT (1998) Oscillations in cerebral haemodynamics: implications for functional activation studies. In: Abstract book ISOTT 1998, Budapest, p. 47.

Gratton G, Fabiani M, Friedman D, Franceschini MA, Fantini S, Corballis P, and Gratton E (1995) Rapid changes of optical parameters in the human brain during a tapping task. J Cogn Neurosc 7:446–456.

Gratton G, Fabiani M, Goodman-Wood MR, and Desoto MC (1998) Memory-driven processing in human medial occipital cortex: An event-related optical signal (EROS) study. Psychophysiology 35:348–351.

Hirth C, Obrig H, Villringer K, Thiel A, Bernarding J, Mühlnickel W, Flor H, Dirnagl U, and Villringer A (1996) Non-invasive functional mapping of the human motor cortex using near-infrared spectroscopy. NeuroReport 7:1977–1981.

Hoshi Y and Tamura M (1993) Dynamic multichannel near-infrared optical imaging of brain activity. J Appl Physiol 75:1–5.

Kato T, Kamei A, Takashima S, and Ozaki T (1993) Human visual cortical function during photic stimulation monitoring by means of near-infrared spectroscopy. J Cereb Blood Flow Metab 13:516–520.

Meek JH, Elwell CE, Khan MJ, Romaya J, Wyatt JS, Delpy DT, and Zeki S (1995) Regional changes in cerebral haemodynamics as a result of a visual stimulus measured by near infrared spectroscopy. Proc R Soc Lond B261:351–356.

Obrig H, Hirth C, Junge-Hülsing JG, Döge C, Wolf T, Dirnagl U, and Villringer A (1996) Cerebral oxygenation changes in response to motor stimulation. J Appl Physiol 81:1174–1183.

Savitzky A and Golay MJE (1964) Smoothing and differentation of data by simplified least squares procedures. Anal Chem 36:1627–1639.

Sluijs MC van der, Colier WNJM, Houston RJF, and Oeseburg B (1998) A new and highly sensitive continuous wave near infrared spectrophotometer with multiple detectors. SPIE 3194:63–72.

Wenzel R, Obrig H, Ruben J, Villringer K, Thiel A, Bernarding J, Dirnagl U, and Villringer A (1996) Cerebral blood oxygenation changes induced by visual simulation changes. J Biomed Opt 1:399–404.

MEASUREMENT OF ARTERIAL, VENOUS, AND INTERSTITIAL pO_2 DURING ACUTE HYPOXIA IN RAT BRAIN USING A TIME-RESOLVED LUMINESCENCE-BASED OXYGEN SENSOR

J. F. Dunn, C. I. Nwaigwe, and M. Roche

Biomedical NMR Research Center, Hanover, New Hampshire USA

ABSTRACT

This is the first publication using a fiber optic "optode" and a luminescence based pO_2 detection method for assessing neural tissue oxygenation. The system was used to simultaneously monitor pO_2 in tissue (P_tO_2) and venous blood (P_vO_2) during normoxia, hyperoxia and hypoxia. P_aO_2 was varied by changing inspired oxygen (FiO_2) from 0.3 to 0.13. Tissue and arterial pO_2 were measured in 5 rats while the simultaneous venous measurements were undertaken in 3 animals. The P_tO_2 was 29 ± 10 at an arterial pO_2 of 116 ± 10 (mean $\pm$ SE, n = 5). The P_vO_2 was consistently higher than P_tO_2 although P_vO_2 approached P_tO_2 as PaO_2 declined to 50 mmHg and was lower than tissue pO_2 during the complete hypoxic period in one animal. These data indicate that brain venous pO_2 is not representative of brain tissue pO_2 and support published models predicting that in brain P_vO_2 is higher than P_tO_2.

1. INTRODUCTION

Oxygen tension in brain may relate to the metabolic homeostasis in brain such that there may be predictable values which, under specific conditions, would correlate with impairment of metabolic processes or with cell death. Venous oxygen tension (P_vO_2) has

Address for correspondence: Jeff F. Dunn, Dartmouth Medical School 7786 Vail Hanover, NH 03755. Phone: 603-650-1677; Fax: 603-650-1717; email: jeff.dunn@dartmouth.edu

Oxygen Transport to Tissue XXI, edited by Eke and Delpy.
Kluwer Academic / Plenum Publishers, New York, 1999.

been used as an indicator of mean brain tissue pO_2 (mean P_tO_2), but it is not clear if this is a reliable determinant of brain tissue oxygenation. Models in muscle have shown that under normoxic control conditions the mixed venous pO_2 does approximate mean tissue pO_2 (Tenney, 1974), but this may not be the case for brain (Bruley et al., 1973; Metzger, 1976; Bruley, 1980; Sharan et al., 1989). We undertook to quantify these parameters in a time-course study on rat brain.

Real-time, continuous assessment of P_tO_2 and P_vO_2 requires monitoring venous blood which flows specifically from the brain and not from peripheral tissues. Venous outflow has been measured by cannulating a retroglenoid vein (Nilsson and Siesjo, 1983), but even though there was minimal extra cerebral contamination, small amounts of non-cerebral flow could significantly change venous pO_2. The same problem exists with the internal jugular vein. It has been shown that hemoglobin saturation in the internal jugular was higher than in the sagittal sinus in ventilated piglets (Rudinsky and Meadow, 1991). As a result, we concluded that venous pO_2 measurements must be made within the sagittal sinus in order to be assured of measuring cerebral venous drainage.

The pO_2 measurements were obtained by monitoring luminescence in a ruthinium based compound fixed to the end of a fiber optic "optode". The advantages are significant for an implanted oxygen sensor. The calibration is very stable and the precision of the result increases with declining pO_2. In addition, the fiber optic positioning system allows one to cannulate vessels. The method has been described in part by Collingridge et al. (Collingridge et al., 1977), who worked on implanted tumors and the system is now commercially available. This is the first report of the use of this instrument for any region other than a tumor.

2. METHODS

Male Sprague-Dawley rats (300–350 g), were purchased from Charles River laboratories, kept on a 12 hr diurnal cycle and fed *ad libidum*. Animals were initially anesthetized with 2% isoflurane in air through a nose cone until a trachiostomy was performed. Subsequently, the animals were volume ventilated with a respiratory rate of 60–80/min., a minute ventilation of 120 ml/min., and a maintenance level of 1.5% isoflurane.

The femoral artery was cannulated with PE 50 tubing for blood pressure recording and sampling of arterial blood for gas analysis with a Ciba Corning model 288 gas analyzer. The femoral vein was cannulated for infusion of pancuronium (0.1 mg/kg per hr). The animals were placed on a pediatric warming pad with a circulating water system for temperature regulation. Rectal temperatures were monitored and maintained at 36.5–37.5°C.

Approximately 10 mm of skull was exposed and two small holes were carefully made through the bone using the tip of a 23G needle. The holes were 5 mm posterior to the bregma and 4 mm lateral to the midline. A thermocouple was advanced through one hole and a fiber optic "optode" (diameter approximately 250–300 µm) through the other to a depth of approximately 5 mm from the surface of the skull. This placed the tips into the region of the optic tracks or the thalamus. This was confirmed by subsequent sectioning of the formalin fixed brains. A third hole was made on the midline suture, just anterior to the lambda and an optode inserted for assessment of sagittal sinus pO_2.

The tissue and sagittal sinus oxygen measurements were made with an Oxylite + Oxydata system (Oxford Optronics, Oxford, UK). The fiber optic tip was coated with a silicone rubber polymer containing a ruthenium luminophor. Excitation at 470 nm was achieved with a photodiode. The rate of signal change was monitored electronically over 3 time "bins" and the data used to calculate pO_2 based on a modified Stern-Volmer equation. The optodes were calibrated by Oxford Optronics at 5 oxygen tensions (from 0 to 76 mmHg) at 37°C. Since oxygen quenches the signal, lower pO_2 values have less noise and are more precise. Data were digitized using the OxyData system and transferred to a PC for analysis using the software "PowerLab" (ADI Instruments, Castle Hill Australia).

After surgery, the animals were exposed to inspired gasses with varying proportions of oxygen in the order of 0.30, 0.25, 0.20, 0.13, 0.30 for approximately 20–30 minutes each. Blood samples (approximately 100 µl) were taken at the end of each period, before changing the gas mixtures. A total of 3 animals were included in this complete protocol. In addition, 2 more animals had P_tO_2 and P_aO_2 samples taken without the P_vO_2 measurements.

3. RESULTS

Cerebral P_tO_2 was 29 ± 6 (mean ± SE, n = 5 animals) when P_aO_2 was 116 ± 10 (mean ± SE, n = 5) and the FiO_2 was 0.25. At an FiO_2 of 0.13, P_tO_2 was 13 ± 8 and P_aO_2 was 48 ± 3 (mean ± SE, n = 5). The tissue and venous data were obtained from the average of the 3 minutes before withdrawal of blood for arterial analysis (36 data points).

Figure 1 shows time-course traces of P_tO_2 and P_vO_2. For most of the experimental times, the P_tO_2 was less than the P_vO_2. Reductions of FiO_2 to 0.20 have little influence on P_tO_2 or P_vO_2 in "B" and "C". Reduction of FiO_2 to 0.13 caused significant reductions in both P_tO_2 and P_vO_2 in all cases. In two animals (A and B), large spikes in PvO_2 were seen after the change to an FiO_2 of 0.13. These correspond to times when the N_2 or O_2 were being further adjusted to achieve the proper FiO_2. In "A", a slow oscillation was observed in P_vO_2 throughout the experiment.

After the FiO_2 was changed from 0.13 to 0.30, the P_tO_2 recovered to the mean value (or higher) observed at the previous period when the FiO_2 was 0.30. There was an overshoot of P_tO_2 in "A" and "B" and overshoot of P_vO_2 in "B" upon returning to an FiO_2 of 0.30 from 0.13.

The trace was continued after KCl euthanasia (under isoflurane anesthesia) in Figure 1C. Both P_tO_2 and P_vO_2 show steep declines with the P_tO_2 approaching an asymptote at a higher pO_2 than the P_vO_2 (which approximated 0 mmHg).

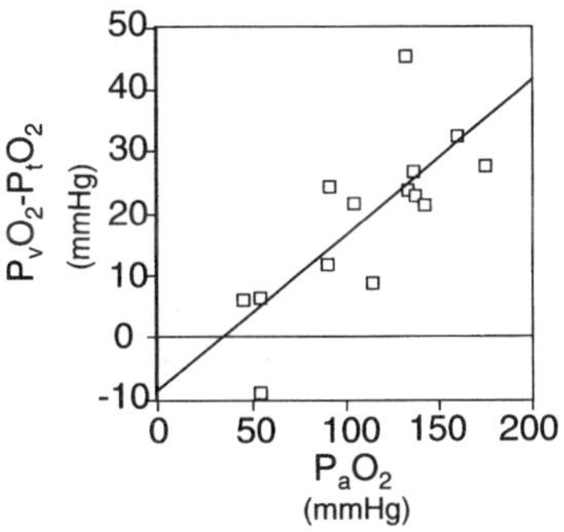

Figure 1. Representative time-course data of sagittal sinus pO_2 (P_vO_2) and interstitial pO_2 (P_tO_2) during time-course studies where inspired oxygen tension (FiO_2) was changed to induce acute hyperoxia and hypoxia. Vertical bars indicate the times when FiO_2 was changed. The initial FiO_2 was 0.30 except in "B" where it was 0.20. The rapid decline in pO_2 in 1C indicates when the rat was euthanized with KCl.

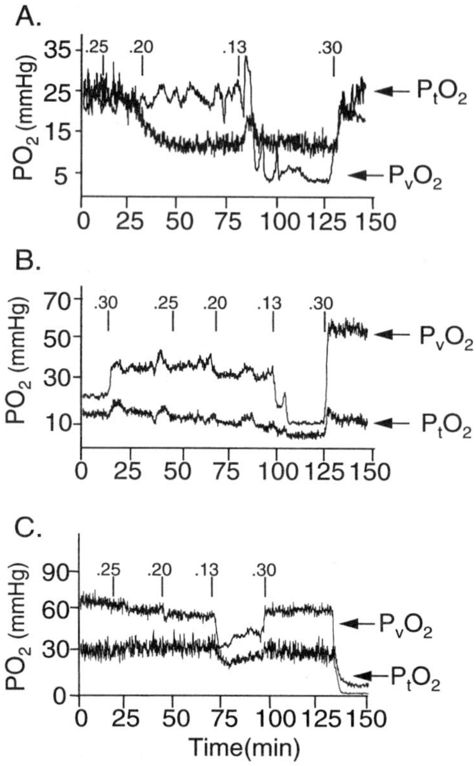

Figure 2. Sagittal sinus pO_2 (P_vO_2)—interstitial pO_2 (P_tO_2) vs. P_aO_2 in a rat during hyperoxia, normoxia and hypoxia. All points except one are above the line of unity. The correlation is significant with $y = 0.251 \times -8.8$ ($r^2 = 0.58$, $n = 14$).

Figure 2 shows a plot of the difference between venous and tissue pO_2 at different values of PaO_2. All but one of the data points show venous pO_2 being higher than tissue pO_2. There is a strong trend for venous pO_2 to approach tissue pO_2 at lower values of arterial oxygen tension.

4. DISCUSSION

An understanding of the regulation of oxygen delivery to the brain is necessary for understanding the mechanisms which might be used to improve delivery in disease states.

Due to the structure of the vascular system, it is difficult to continuously monitor cerebral P_vO_2 in small animals. Implanted electrodes have been the only method available without more complex surgery (Nilsson and Siesjo, 1983). The recent availability of a fiber optic based measuring system makes such measurements in animals (and potentially people) much simpler as it involves cannulating the vessel with a flexible fiber optic.

The measuring technique has been previously described and applied to studies of tumors (Collingridge et al., 1977). The optodes are supplied pre-calibrated for both temperature and pO_2. This has the potential advantage of providing a standard calibration curve for all optodes. However we have found that the calibrations are not all identical at this time, and further calibration may be required. To ensure that differing calibrations did not negate our main conclusions, the two optodes used in Figure 1C were placed in a 37°C saline bath after the experiment and gas bubbled through at varying oxygen tensions. Above 20 mmHg, the two were identical. When recording less than 10 mmHg, the two deviated such that the optode used in the venous sample recorded approximately 3 mmHg less than the tissue. In this subject, the error would have minimized the differences between the arterial and venous samples since the P_vO_2 was higher than P_tO_2. As a result, although absolute values may have deviated by a few mmHg, the observation that P_vO_2 was higher than P_tO_2 would still hold.

The mean P_tO_2 (29 mmHg at an FiO_2 of 0.25) falls within the range of values reported using microelectrodes. Microelectrode studies have been used for decades for measurement of P_tO_2 in brain. When very small tip diameters (1–5 μm) are used, large ranges in P_tO_2 and very high spatial heterogeneity are reported. The P_tO_2 may vary from 0 to more than 70 mmHg over sub-millimeter distances in the cortex of an anesthetized animal, but median pO_2 values fall in the range of approximately 10–40 mmHg depending on experimental conditions (Lubbers, 1968; Bicher et al., 1971; Leniger-Follert and Lubbers, 1976; Metzger et al., 1980; Ivanov et al., 1982; Fennema et al., 1989; Gaab et al., 1990). Using electron paramagnetic resonance oximetry, which reports oxygenation on the surface of an implanted crystal of lithium phthalocyanine, cerebral P_tO_2 was reported to be 24 mmHg in rats under isoflurane anesthesia (Liu et al., 1995).

It is expected that at high P_aO_2 (when Hb saturation is higher than 95%) large changes in P_aO_2 would have little or no effect on P_tO_2 or P_vO_2. This is because of the small oxygen carrying capacity of the serum. There is a negligible change in PvO_2 and P_tO_2 as the FiO_2 was changed from 0.30 to 0.20 (Figure 1), whereas there was significant decline in both when the FiO_2 declined to 0.13. This is largely a result of the shape of the oxy-hemoglobin dissociation curve and the relative oxygen carrying capacity of serum. At high FiO_2 values (and so high arterial saturations), a change in either FiO_2 or P_aO_2 will have only a small influence on either P_vO_2 or P_tO_2. However, when P_aO_2 declines through the steep region of the oxyHb dissociation curve, changes in P_aO_2 cause large changes in both P_vO_2 and P_tO_2. This result is consistent with modeling of brain oxygen tensions (Sharan et al., 1989).

Some specific details emerge from the data in Figure 1. Firstly, the response time for the pO_2 measurement is fast relative to the time-scale of the experiment. This can be seen in the data at the end of the experiment in Figure 1C when the animal was euthanized with KCl while under anesthesia. The P_vO_2 declined within seconds. This shows that the combined times for diffusion to the optode and into the rubber tip are small relative to the time-course of this experiment. Secondly, it is possible for blood sampling to cause such a significant hemodynamic insult as to influence both P_vO_2 and P_tO_2. This can be seen Figure 1B as the small, periodic bursts in pO_2 before the 0.25, 0.20 and 0.13 changes. In this one animal, blood withdrawal from the femoral artery correlated with transient increases in pO_2.

The main goal of this paper was to determine the relative pO_2 in the tissue and venous drainage of the brain. Figure 2 shows that P_vO_2 is usually higher than P_tO_2 over a wide range of values for P_aO_2. Although the scatter is large, there is only one data-point where P_vO_2 was less than P_tO_2, and that occurred during hypoxia. A linear correlation between P_vO_2-P_tO_2 and P_aO_2 exists, indicating that P_vO_2 approaches P_tO_2 at low arterial oxygen tensions.

Brain pO_2 has been modeled by a variety of groups (Bruley et al., 1973; Metzger, 1976; Bruley, 1980; Sharan et al., 1989). Sharan et al. incorporated much of this information when they modeled brain vascular oxygen tensions based on data from a sheep (Sharan et al., 1989). They began with a Krogh cylinder model but incorporated counter current exchange, the range of vessel sizes in the exchange units, hematocrit, and the oxyhemoglobin dissociation curve. Their data indicate that P_vO_2 should be in the range of 40 mmHg when PaO_2 is 80 mmHg or higher and that tissue pO_2 should be in the range of 30 mmHg. This is consistent with our results.

In summary, we have used a novel fiber-optic based, luminescence system to measure P_tO_2 and P_vO_2 during normoxia, hypoxia and hyperoxia in rat brain. The normoxic values are within the range of previously published values. Brain venous pO_2 is higher than tissue pO_2, but begins to approximate tissue pO_2 as arterial pO_2 declines.

ACKNOWLEDGMENTS

This work was supported by NIH RO1 CA67431.

REFERENCES

Bicher HI, Bruley D, Knisely MH, and Reneau DD (1971) Effect of microcirculation changes on brain tissue oxygenation. J Physiol 217(3):689–707.

Bruley DF (1980) Probabilistic solutions and models: oxygen transport in the brain microcirculation. In: JF G, AP, ed. Mathematics of microcirculation phenomena. New York: Raven Press, 133–158.

Bruley DF, Reneau DD, Bicher HI, and Knisely MH (1973) Theoretical studies of brain tissue oxygenation considering the deoxygenation rate of the red cell. Biblio Anatom 11:507–514.

Collingridge DR, Young WK, Vojnovic B, Wardman P, Lynch EM, Hill SA, and Chaplin DJ (1977) Measurement of tumor oxygenation: a comparison between polarographic needle electrodes and a time-resolved luminescence based optical sensor. Rad Res 147:329–334.

Fennema M, Wessel JN, Faithfull NS, and Erdmann W (1989) Tissue oxygen tension in the cerebral cortex of the rabbit. Adv Exp Med and Biol 248:451–460.

Gaab MR, Poch B, and Heller V (1990) Oxygen tension, oxygen metabolism, and microcirculation in vasogenic brain edema. Adv Neurol 52:247–256.

Ivanov KP, Derry AN, Vovenko EP, Samoilov MO, and Semionov DG (1982) Direct measurements of oxygen tension at the surface of arterioles, capillaries and venules of the cerebral cortex. Pflugers Arch 393:118–120.

Leniger-Follert E and Lubbers DW (1976) Behavior of microflow and local pO_2 of the brain cortex during and after direct electrical stimulation. A contribution to the problem of metabolic regulation of microcirculation in the brain. Pflugers Arch 366(1):39–44.

Liu KJ, Bacic G, Hoopes PJ, Jiang J, Du H, Ou LC, Dunn JF, and Swartz HM (1995) Assessment of cerebral pO_2 by EPR oximetry in rodents: effects of anesthesia, ischemia and breathing gas. Brain Res 685:91–98.

Lubbers DW (1968) The oxygen pressure field of the brain and its signficance for the normal and critical oxygen supply of the brain. In: Lubbers DW, Luft UC, and Witzleb GT, ed. Oxygen Transport in blood and tissue. Theime, Stuttgart:124–139.

Metzger H (1976) The influence of space-distributed parameters on the calculation of substrate and gas exchange in microvascular units. Math Biosci 30:31–45.

Metzger H, Heuber-Metzger S, Steinacker A, and Struber J (1980) Staining pO_2 measurement sites in the rat brain cortex and quantitative morphometry of the surrounding capillaries. Pflugers Arch 388(1):21–27.

Nilsson B and Siesjo BK (1983) A venous outflow method for measurement of rapid changes of the cerebral blood flow and oxygen consumption in the rat. Stroke 14(5):797–802.

Rudinsky BF and Meadow WL (1991) Internal jugular venous oxygen saturation does not reflect sagittal sinus oxygen saturation in piglets. Biol Neonate 59(5):322–328.

Sharan M, Jones MD Jr, Koehler RC, Traystman RJ, and Popel AS (1989) A compartmental model for oxygen transport in brain microcirculation. Annals of Biomed Eng 17(1):13–38.

Tenney SM (1974) A theoretial analysis of the relationship between venous blood and mean tissue oxygen pressures. Respir Physiol 20:283–296.

7

FRACTAL ANALYSIS OF SPONTANEOUS FLUCTUATIONS IN HUMAN CEREBRAL HEMOGLOBIN CONTENT AND ITS OXYGENATION LEVEL RECORDED BY NIRS

Andras Eke and Péter Hermán

Experimental Research Department
II. Institute of Physiology
Semmelweis University of Medicine
P.O. Box 446, 1448 Budapest
Hungary

1. INTRODUCTION

The NIRS technology provides a non-invasive tool to monitor hemodynamics from the human brain cortex (Jöbsis 1977; Wyatt et al., 1986). Our interest recently has turned toward studies utilizing this technology in understanding the complexity of hemodynamics in the human brain as captured in high resolution hemoglobin time series. In our definition a time series or signal is complex when it is difficult to recognize a pattern such as it is being stable or periodic in its structuring. A complex signal cannot be adequately characterized by descriptive statistical measures such as its mean, standard deviation, etc., or by its frequency spectrum because by doing so other relevant information that is present in these signals are discarded or not revealed. Hence, adequate models, and methods need to be identified for such signals to be characterized in their entirety. Similar to our previous study on red blood cell flux time series acquired from the brain cortex of the rat (Eke et al., 1997) we have applied the fractal model as implemented in a combination of the power spectral density (PSD) and the scaled windowed variance (SWV) methods in the analysis of human cerebrocortical hemoglobin signals in an attempt to assess their temporal pattern.

2. METHODS

2.1. Near Infrared Spectroscopy (NIRS) of the Brain

The Hamamatsu NIRO 500 near infrared spectrophotometer (Hamamatsu Photonics, Herrsching, Germany) was used to record total (HbT), oxygenated (HbO_2) and reduced (Hb) hemoglobin content from the human brain cortex (Figure 1). An interoptode distance of 4 cm was used and the probe was secured just below the frontal hairline in a midline position. Measurements were made in 5 healthy male volunteers of 24–28 years of age with their informed consent. The subjects were comfortably seated in an armchair and were allowed to read at their leisure.

2.2. Creating Time Series from a Continuous NIRS Record

Time series with 32,678 data points were created by reading the data from the NIRO 500 unit's serial port representing samples of the spectrophotometer's analog record corresponding to relative concentration in cerebrocortical hemoglobin @2 Hz/12 bits for 285 minutes (Figure 1).

2.3. Fractal Analysis of Hemoglobin Time Series

The PSD method was used to characterize the signals by their frequency spectrum. The overall performance of this method was optimized by the following signal conditioning steps: windowing, end-matching (Fougere, 1985) and the exclusion of the high-

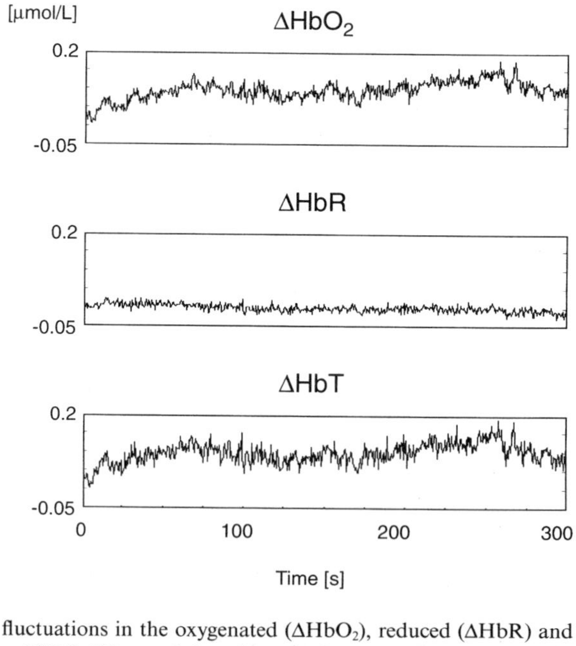

Figure 1. Spontaneous fluctuations in the oxygenated (ΔHbO_2), reduced (ΔHbR) and total (ΔHbT) hemoglobin content detected by a NIRO 500 near infrared spectrofotometer from the frontal brain cortex in man. All traces show changes relative to the control level. The first 300 seconds from a record of 285 minutes is shown.

frequency range from the analysis. This method looks at the frequency distribution of power amplitude ($|A(f)|^2$) within the signal and yields the power spectral density plot such as the one shown in Figure 2. As seen in the upper linear plot, there is no single frequency with high power amplitude emerging from the spectrum. There is, however, a continuum of frequencies present in the signal with power increasing as frequency decreases. At the bottom, the PSD-plot is shown in a log-log fashion. If the power structure can be linearly fitted on this plot by the following model, $|A|^2 \propto c\,(1/f^{-s})$ (where c is a constant determined by the data, and s is the slope of the fit), it suggests that there is a fractal relationship between the frequency elements and their power present within at least 2 decades of temporal scales (frequencies, i.e.) (Shepers et al., 1992). This, so-called 1/f pattern, hence is referred to as scale-independent, or self-affine. The spectral index of the signal is obtained as $\beta = -s$. Random signals in the whole range of β varying from -1 to 3 can be classified according to their spectral slope or index by a dichotomous model. Signals with $-1 < \beta < 1$ are stationary and called fractional Gaussian noises (fGn's), those with $1 < \beta < 3$ are non-stationary with stationary increments and are called fractional Brownian motions (fBm's). The overall measure of the fractal correlation between the sampled signal elemets, H, the Hurst coefficient can be derived from β as $H_{fGn} = (\beta + 1)/2$ or $H_{fBm} = (\beta - 1)/2$. Note, that signals being considered "random" by the general use of the term (that is uncorrelated) are merely special cases of $\beta = 0$ (random noise) and $\beta = 2$ (random walk) both yielding $H = 0.5$. For the rest of the signals some degree of correlation is present.

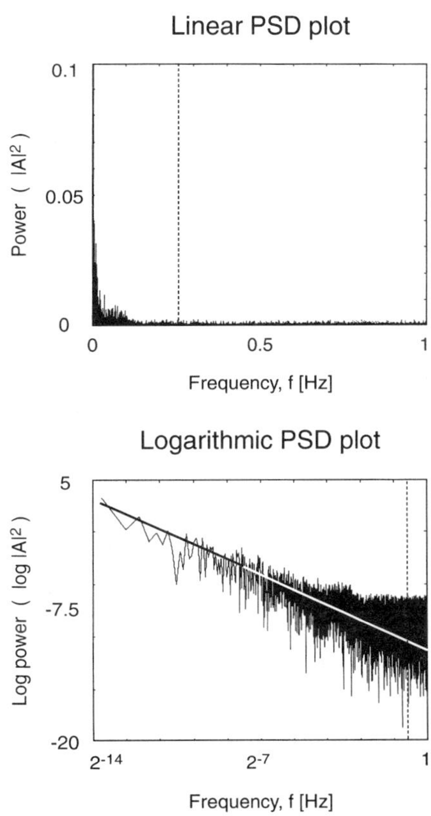

Figure 2. Power distribution in a hemoglobin concentration time series within the 2^{-14} and 1 Hz frequency range. Dotted line indicates the high frequency range of power estimates that needs to be excluded when the spectral index (β) is to be determined by fitting the power slope across the low frequency estimates (shown in the lower panel). With this modification the Fourier method, denoted as H_{PSD}, yields a reliable estimate of β and from it the Hurst coefficient. An exponential decay in the spectral estimates towards the high frequencies is characteristic to a fractal time series in which the fractal pattern is captured by $0 < H_{PSD} < 1$. For 1-dimensional time series $H = 2 - D$, where D is the fractal dimension.

Classification based on the fGn/fBm model is critical when attempts are to be made to obtain a reliable value for H, because certain fractal analytical methods can be applied to either fGn or fBm signals, some like the spectral method to both signal categories. Improper selection of a fractal tool for a given signal class results in serious error in calculating H. H varies between 0 and 1. H close to 0 shows a noisy, while H close to 1 a smooth appearance.

Based on extensive numerical analysis of several fractal tools, in addition to the PSD method we identified the bridge detrended scaled windowed variance method (bdSWV) as a highly reliable estimator of H for fBm signals (Cannon et al., 1997). This method looks at a descriptive statistical parameter of the signal, the standard deviation SD, within varying window size, τ. H can be obtained as the slope of $\log[SD(\tau)/SD(\tau_0)]$, where $SD(\tau_0)$ is the standard deviation in a reference window. The bdSWV method is a variant of the standard SWV method, which improves its precision by applying linear detrending of the signal between the first and last points in each window. The advantage of the spectral method is that it is applicable to both types of signals, while bdSWV is good only for motion signals. Hence, the signals need to be classified if analysis by bdSWV is to be done, or if H is to be obtained by the PSD method.

The flow chart followed in this study in arriving at H values for hemoglobin signals recorded from the brain cortex of humans is shown in Figure 3. Sampling the raw signal created the time series that were rendered to fractal analysis. First, PSD plots were generated and β was estimated. H_{PSD} was calculated from β and later compared with H obtained by bdSWV. Signal class was determined based on the value of β. If class proved fBm, bdSWV method could be directly applied to calculate H, if it proved fGn, then first the signal needed to be summed in order to turn it to fBm and an estimate of its H subsequently obtained by the bdSWV method. H's obtained by PSD and bdSWV were eventually compared. Their similarity gives confidence that the fractal correlation captured by these values is indeed present in the signal.

3. RESULTS

In statistically representative number of trials ($n = 100$) signals with known values of H were generated (Davies and Harte, 1987) and evaluated for H by the PSD and bdSWV methods. The mean square error (MSE) was calculated to determine the overall degree of bias and variance in the estimates. For both methods, MSE was low (for PSD 0.0079, for bdSWV 0.0017), thus these methods were regarded reliable in the fractal analysis of temporal signals, provided the length of the time series was sufficiently long (Bassingthwaighte et al., 1994; Cannon et al., 1997). We have found an acceptable compromise between the need for long signal length on one hand, and tolerable recording time on the other by setting the signal length to $N = 32{,}678$.

Hemoglobin signals recorded by NIRS in resting conditions did show spontaneous fluctuations (Figure 1). These fluctuations were present irrespective whether or not the subject was reading.

All HbT and HbO_2-signals turned out to be firmly of fBm types, whereas the reduced Hb-signals were split because of their borderline signal character as judged by the dichotomous fGn/fBm model. Hence, the mean H value for the reduced Hb-signal group is not reported. Spontaneous fluctuations were more apparent in the total and oxygenated traces (Figure 1). The similarity of H values obtained by the two methods,

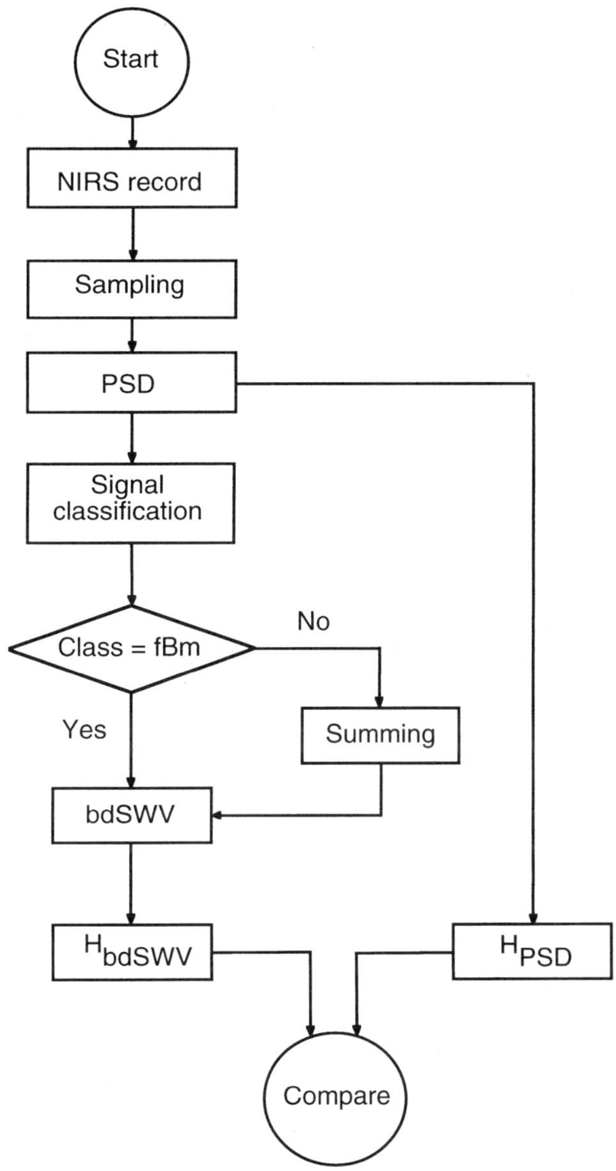

Figure 3. Flowchart of the fractal analysis of hemoglobin concentration time series created from NIRS records of the human brain cortex by sampling the output of the NIRO 500 spectrophotometer. Power spectral density (PSD) and bridge detrended scaled windowed variance methods (bdSWV) were used to estimate the Hurst coefficient (H_{PSD}, H_{bdSWV}) of the time series. Signal classification according to a dichotomous model is a critical step in the analysis and is shown in a diamond shaped block.

$H_{PSD,HbT} = 0.13 \pm 0.08$ vs. $H_{bdSWV,HbT} = 0.16 \pm 0.06$ and $H_{PSD,HbO2} = 0.23 \pm 0.06$ vs. $H_{bdSWV,HbO2} = 0.22 \pm 0.07$, indicates that this range of H represents the presence of a fractal correlation in these signals. H values found in this study were similar to those previously found, though for much smaller temporal scales, in the rat brain cortex for red blood cell flux time series (Eke et al., 1997).

4. DISCUSSION

The Hamamatsu NIRO 500 spectrophotometer updates its serial data port at a maximum frequency of 2 Hz. Reliable estimate of H with a precision of 0.1 required data sets of at least 32,678 long (Cannon et al., 1997), hence data collection had to last for 285 minutes for each subject. In preliminary experiments we tested our subjects' ability to cope with the length of data collection, during which period they were restricted to sit in an armchair. It was noted, that the task of continuously sitting for this length of time tended to make the subjects sleepy and prompted reactions to stay alert. In order to assure that the level of alertness of our subjects stay steady at the level observed when beginning the session, our subjects were allowed to read their university course books at their leisure. We regard a potential shift of alertness resulting from reading less detrimental to the steady state conditions of cerebral neuronal activity and associated hemodynamics, than slipping into a completely different state of mind manifested in drowsiness or falling asleep.

Human cerebral hemoglobin content recorded by NIRS shows spontaneous fluctuations at rest. This fluctuation can certainly be ascribed to vasomotion along the flowpath and within the monitored region due to elements present there. Possibly due to the great number of vasomotive segments of the monitored vasculature in play, no single frequency dominates the power spectrum of these signals. These signals are complex in that no simple pattern (oscillations, i.e.) can be recognized in their temporal dynamics. In this study the fractal model demonstrated the presence of statistical self-similarity or 1/f fractal pattern in these sufficiently long, high-definition physiological signals. There are other methods, like chaos analysis (Griffith, 1995), to investigate the nature of complex dynamic behavior in biological systems. The spectral method has been traditionally used in the temporal analysis of biological signals, typically to identify solitary frequencies in the signal spectrum to be used to characterize simple patterns, like oscillations. In contrast, we used the PSD method in the analysis of overall signal structuring utilizing a wide range of frequencies present in the signal. The fractal model as it is formulated by the spectral approach looks at multiple scales defined by the frequencies of the spectrum and by fitting for its power slope estimates the scale-independent temporal correlation structure captured by H.

The actual mechanisms of how these complex signals are made up is currently under investigation.

ACKNOWLEDGMENTS

This study was supported in part by OTKA Grants I/3 2040, T 016953 and NIH Grant TW00442.

REFERENCES

Bassingthwaighte JB and Raymond GM (1994) Evaluating rescaled range analysis for time series. Ann Biomed Eng 22:432–444.

Cannon MJ, Percival DB, Caccia DC, Raymond GM, and Bassingthwaighte JB (1997) Evaluating scaled windowed variance methods for estimating the Hurst coefficient of time series. Physica A 241:606–626.

Davies and Harte (1987) Test for Hurst Effect. Biometrika 74:95–101.

Eke A, Herman P, Bassingthwaighte JB, Raymond GM, Balla I, and Ikrenyi C (1997) Temporal fluctuations in regional red blood cell flux in the rat brain cortex in a fractal process. Adv Exp Med Biol 428:703–709.

Fougere PF (1985) On the accuracy of spectrum analysis of red noise processes using maximum entropy and periodogram methods: Simulation studies and application to geographical data. J Geograph Res 90(A5):4355–4366.

Griffith TM and Edwards DH (1995) Complexity of chaotic vasomotion is insensitive to flow and pressure but can be regulated by external control. Am J Physiol 269(38):H656–668.

Jöbsis FF (1977) Noninvasive infrared monitoring of cerebral and myocardial oxygen sufficiency and circulatory parameters. Science 198:1264–1267.

Schepers HE, van Beek JHGM, and Bassingthwaighte JB (1992) Four methods to estimate the fractal dimension from self-affine signals. IEEE Eng Med Biol 11:57–64 and 71.

Voss RF (1988) Fractals in nature: From characterization to simulation. In: The Science of Fractal Images. (Peitgen HO, and Saupe D, Eds) Springer Verlag, New-York, Berlin, Heildelberg, London, Paris, Tokyo, pp. 21–70.

Wyatt JS, Cope M, Delpy DT, Edwards AD, Wray S, and Reynolds EOR (1986) Quantification of cerebral oxygenation and haemodynamics in sick newborn infants by near infrared spectrophotometry. Lancet 2(8515):1063–1066.

8

OSCILLATIONS IN CEREBRAL HAEMODYNAMICS

Implications for Functional Activation Studies

C. E. Elwell, R. Springett, E. Hillman, and D. T. Delpy

Department of Medical Physics and Bioengineering
University College London
London, United Kingdom

1. INTRODUCTION

The blood oxygenation level dependent (BOLD) contrast signal in functional magnetic resonance imaging (fMRI) provides highly spatially resolved deoxygenation maps of the activated cortex (Turner 1995). In common with many other activation measurement techniques, (including optical measurements) fMRI relies upon multiple repetition of the stimulus followed by signal averaging, baseline subtraction and extensive thresholding techniques to provide difference images of activation.

One of the roles of this type of data manipulation is to account for non activation related changes in the measured cerebral parameters. It is well known that the "resting" brain is in a constant state of haemodynamic redistribution (Halsey & McFarland 1974, Dóra & Kovách 1980) and hence may not be considered to provide a static baseline for activation studies. An important aspect of the status of the resting brain is the phenomena of vasomotion—the name given to rhythmical oscillatory behavior of capillary blood flow of different frequencies caused by the contraction and relaxation of the precapillary vessels (Berne & Levy 1997). The phenomenon of vasomotion has been known for some time and was in fact first described in 1854 (Schniff 1854). Several studies have been designed to elucidate its origin however many of the techniques employed have the disadvantage of requiring exposed cortex preparations and anaesthesia. An example of one such study on the exposed cortex of the rat has revealed slow (typically 0.1 Hz) oscillations in optical attenuation signal, which have been ascribed to the phenomenon of vasomotion (Mayhew et al., 1996). Laser doppler studies have also demonstrated the presence of "spontaneous" oscillations in flow (Hudetz et al., 1992). The advent of non

Oxygen Transport to Tissue XXI, edited by Eke and Delpy.
Kluwer Academic / Plenum Publishers, New York, 1999.

invasive optical techniques has provided new opportunities for investigating the resting brain. Near infrared spectroscopy (NIRS) has found widespread use as a monitor of tissue oxygenation and in the past decade has been applied specifically to measurements of cerebral activation related changes in haemodynamics (Villringer et al., 1993, Meek et al., 1994).

The purpose of this study was (i) to use near infrared spectroscopy to investigate the presence and characteristics of vasomotion type signals in the intact resting adult human brain and (ii) to assess the implications of these signals upon the design of functional activation studies.

2. METHODS

2.1. Experimental Methods

The subjects for the study comprised 10 healthy adult volunteers, of median (range) age 31 (25–43) years. Light from a stabilized tungsten halogen light source was filtered with a 610 nm long pass filter and transmitted to the head of the subject using a fibre optic bundle. Transmitted light was collected with a second fiber bundle and focused onto the entry slits of a spectrograph (270 M, Instruments SA). Spectra with a resolution of 0.27 nm/pixel were collected between 650 and 980 nm on a cooled CCD detector (Wright Instruments, Enfield, London). The exposure time was set to allow a sampling rate of 19 Hz. The optodes were placed 3 cm apart over the occipital cortex, 1 cm above and to the right of the inion. Nasal airflow was measured using a pair of thermistors and the subject's pulse waveform was recorded using a pulse oximeter (Model 520, Novametrix, USA). These data were collected and stored simultaneously with the spectral information. Measurements were taken over a period of seven minutes in each subject during which they were awake in a darkened room with no auditory or visual stimuli. In all cases a support was used to prevent head movement.

2.2. Data Analysis

Changes in concentration of oxy (ΔHbO_2) and deoxyhaemoglobin (ΔHb) were obtained by fitting changes in the attenuation spectra between 780 and 900 nm to in-vitro absorption spectra of HbO_2 and Hb which had been corrected for wavelength dependence of pathlength (Wray et al., 1988, Essenpreis et al., 1993). Changes in total haemoglobin concentration (ΔHbT) were calculated from the sum of the ΔHbO_2 and ΔHb signals. Throughout this paper all concentration changes will be expressed in units of μmolar.cm.

Labview (National Instruments, Texas, USA) was used to provide a power spectrum of the raw, unaveraged ΔHbO_2, ΔHb and ΔHbT data. A power spectrum of both the nasal airflow and heart pulse waveform data was also performed to calculate the respiratory rate and heart rate for each subject.

To determine the implications of the presence of relatively slow oscillations in cerebral oxygenation and haemodynamics on BOLD fMRI data, the NIRS data was averaged to mimic a typical echo planar imaging (EPI) BOLD data acquisition sequence. This was achieved by averaging 110 ms of data every 4 seconds where 110 ms represents the echo time (TE) and 4 seconds repetition time (TR) of an EPI sequence.

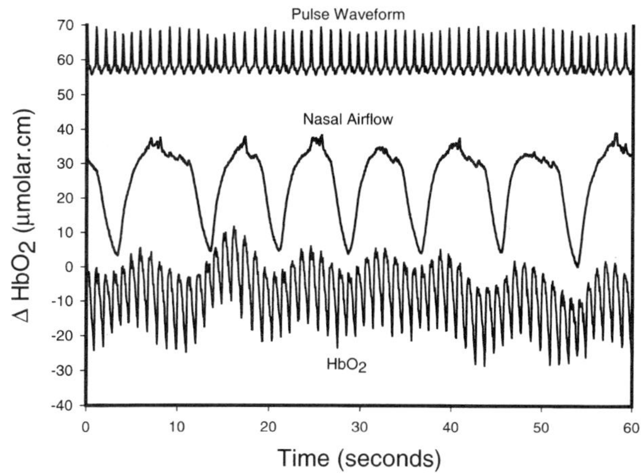

Figure 1. An example of the HbO_2, respiratory and cardiac data collected from one subject.

3. RESULTS

Figure 1 shows an example of the nasal airflow, pulse waveform and ΔHbO_2 data collected from one subject clearly demonstrating the presence of both respiratory and heart pulse related oscillations in the cerebral HbO_2 signal. Figure 2 shows a longer section of data in which a mean filter (rank 15) has been used to filter out the heart pulse effects and to reveal a slower oscillatory signal.

Figure 3 summarises the results of frequency analysis of the ΔHbO_2 signal for all subjects. A persistent slow oscillation was seen in all subjects at a range of frequencies associated with vasomotion (mean ± SD = 0.082 ± 0.016 Hz) which could be clearly separated from those related to cardiac and respiratory changes.

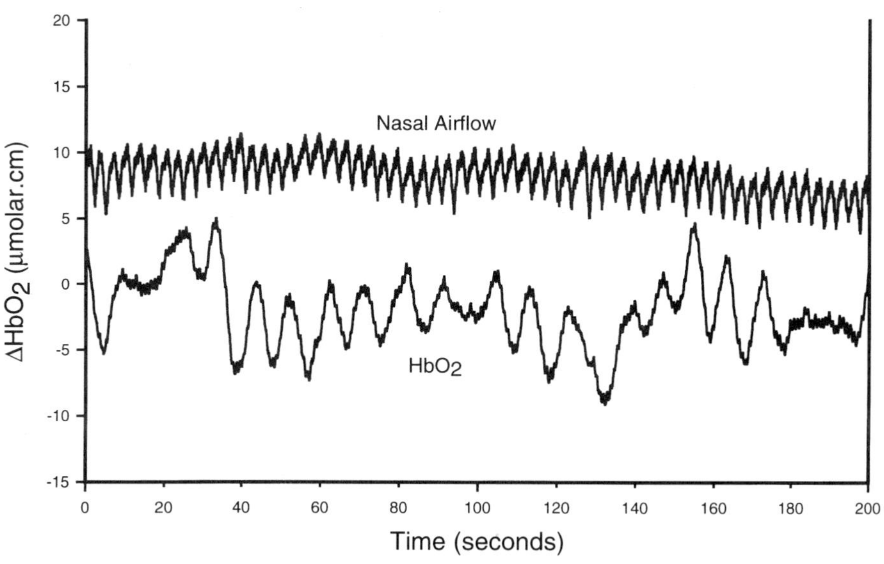

Figure 2. Evidence of the slow oscillations in HbO_2 signal in a single subject.

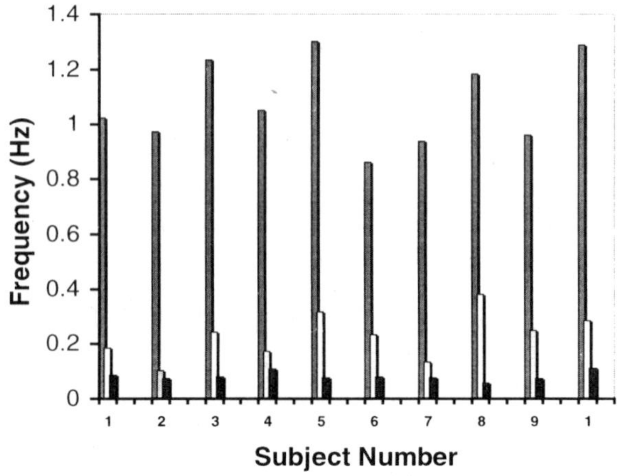

	Frequency (Hz)	Period (s)
heart rate	**1.08** ± 0.16	**0.9** ± 0.15
breathing rate	**0.22** ± 0.07	**4.5** ± 0.81
vasomotion	**0.082** ± 0.016	**12.2** ± 2.3

Figure 3. A histogram of the frequency components of the HbO_2 signal for each subject. The table documents the mean ± SD frequency and period for each of the three oscillations.

The cardiac, respiratory and vasomotion related oscillations were evident on each of the ΔHbO_2, ΔHb and ΔHbT signals. The relative magnitudes of the amplitudes of the three frequency components in each signal are shown in Figure 4.

It was clear that in all subjects there were a range of vasomotion frequencies. Figure 5 shows the power spectrum of the ΔHbO_2 data for a single subject illustrating the cardiac, respiratory and vasomotion peaks. Viewed on an expanded frequency scale (Figure 6) the range of vasomotion frequencies can be seen.

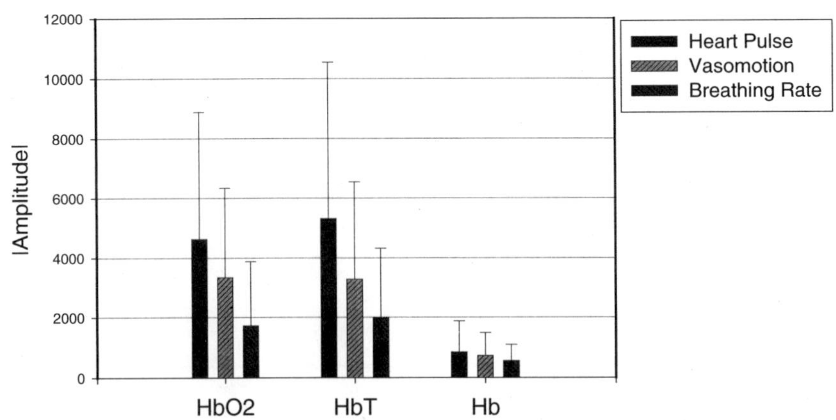

Figure 4. The relative amplitudes of the frequency components of the HbO_2, HbT and Hb data for all subjects.

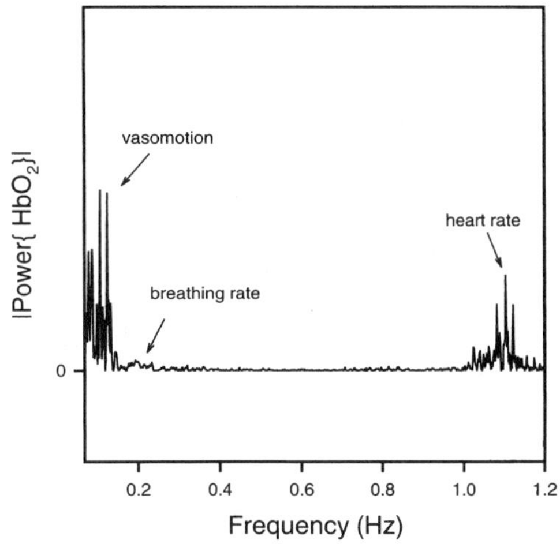

Figure 5. Power spectrum of the HbO_2 signal for a single subject.

Following the fMRI averaging, vasomotion oscillations were still visible in the NIRS data at a mean ± SD frequency of 0.075 ± 0.012 Hz in the ΔHbO_2 and ΔHbT signals and 0.082 ± 0.017 in the ΔHb signal. Figure 7 demonstrates the effect of fMRI averaging on the ΔHbT data for a single subject. The power spectrum of the HbO_2, Hb and HbT signal for this subject, after fMRI averaging, is shown in Figure 8.

4. DISCUSSION

This study has shown that in the intact, resting adult brain, cardiac and respiratory related oscillations in cerebral haemodynamics are clearly visible. In addition, oscillations in cerebral HbO_2 associated with vasomotion have also been confirmed. These oscillations, which appear to be predominantly arterial, have periods between 9 and 17 seconds.

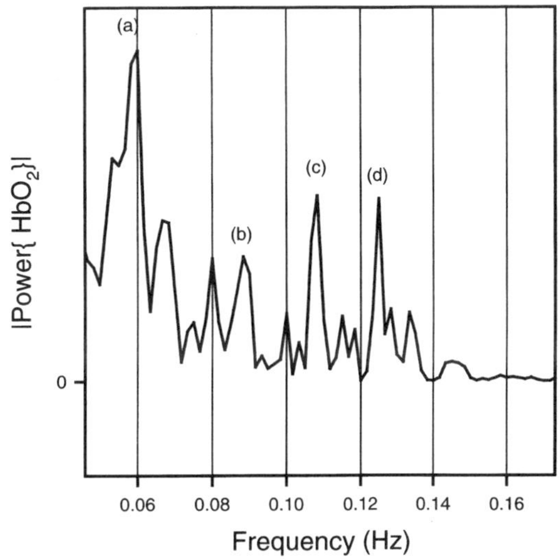

Figure 6. Expanded plot of Figure 5 demonstrating the distribution of vasomotion frequencies. The corresponding periods of the peaks shown are (a) 16.7 s, (b) 11.4 s, (c) 9.3 s and (d) 8 s.

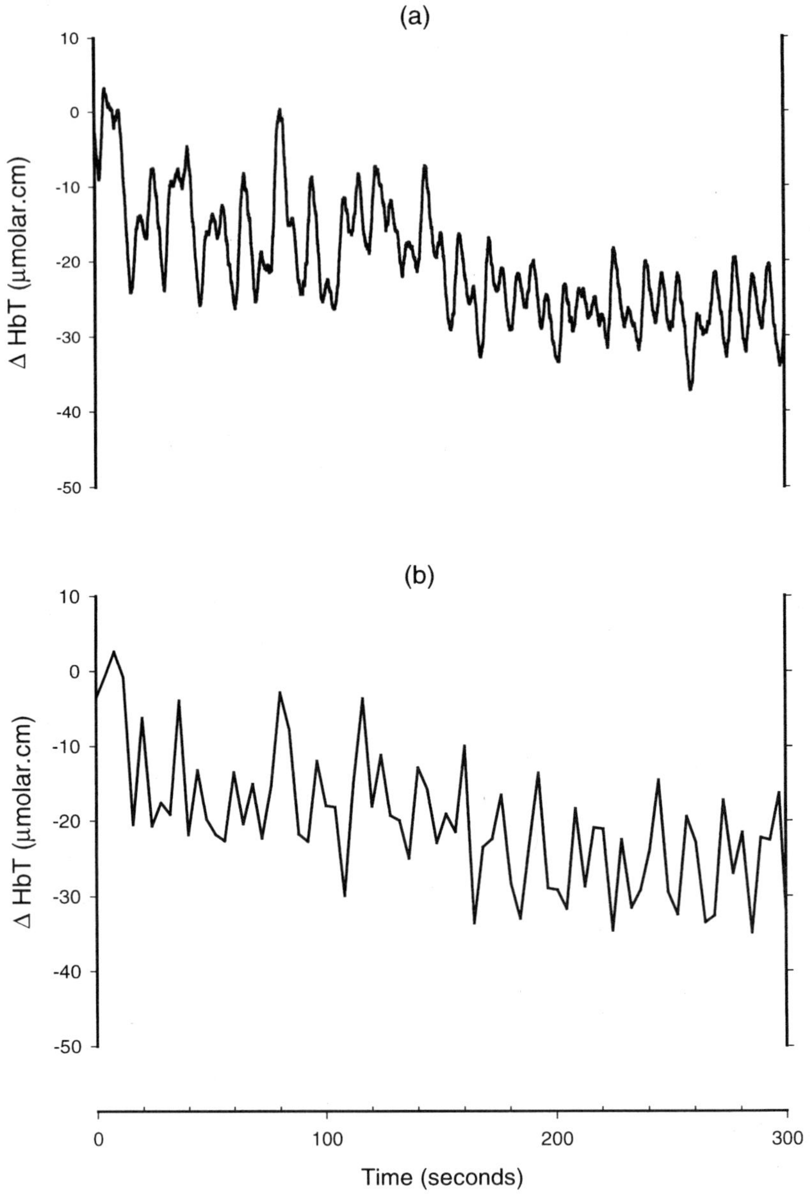

Figure 7. The ΔHbT signal (a) before and (b) after the simulated fMRI averaging assuming TE = 110 ms and TR = 4 s. The slow vasomotion oscillations are still clearly visible after averaging.

Since it was first described, vasomotion has been demonstrated in a variety of species including bats, frogs and guinea pigs and in a number of organs including skin, skeletal muscle, intestine and the brain. It is believed to originate from the spontaneous activity of the vascular smooth muscle that exhibits patterns of continuous contraction and dilation (Funk & Intaglietta, 1983). For some time research in this field was hindered by the invasive nature of available techniques for the continuous monitoring of haemodynamics. This, coupled with the sensitivity of any observations to the use of anaesthet-

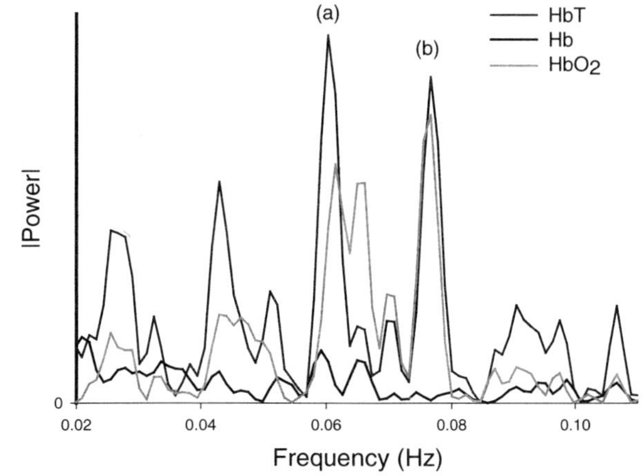

Figure 8. The power spectra of the HbO_2, Hb and HbT signals of a single subject after simulated fMRI averaging. The labelled peaks correspond to periods of (a) 16.7 s and (b) 13 s.

ics, prevented truly *spontaneous* haemodynamic events to be routinely monitored. NIRS provides the opportunity for the continuous, non invasive assessment of quantitative cerebral haemodynamics and oxygenation in the intact adult brain. Data from other groups using related technology for exposed cortex preparations indicate the presence of oscillations similar to those described here. Vern et al. (Vern et al., 1998) used visible reflectance spectroscopy to measure local light intensity changes in the exposed cortex of the cat during stages of waking and sleep. Oscillations were seen in the 590 nm signal (related to blood volume) and 603–590 nm signal (related to cytochrome aa_3) in all behavioral states at a mean ± SEM frequency of 0.14 ± 0.001 Hz which did not differ significantly between sleep states. Hoshi et al. (Hoshi & Tamura 1997), using interoptode spacings between 2 and 3 cm over the temporal, occipital and frontal regions, describe fluctuations in cerebral Hb and HbO_2 in resting adult subjects with frequencies ranging from 0.003–0.03 Hz, although no likely explanation for these oscillations is discussed.

It is important to note that most *in vitro* studies have investigated vasomotion effects at a very local level (<2 mm^2 or even single capillary loops (Hudetz et al., 1992)). In NIRS studies however, a much larger elliptical region approximately 4 cm long, 2–3 cm wide and 1–3 mm deep is interrogated (Okada et al., 1997, Firbank et al., 1998), yet significant effects are still seen over this larger volume. Further investigation is required to determine whether the relatively "macroscopic" events recorded with NIRS are likely to have the same origin as those observed in the exposed cortex.

The paramagnetic tracer responsible for the BOLD contrast signal is deoxyhaemoglobin, although the exact contribution to the BOLD signal of changes in cerebral blood volume and flow is still under debate (Buxton et al., 1998). Our data suggests that the vasomotion effects are primarily seen in the ΔHbO_2 signal, and are less prominent in the ΔHb signal. However, as demonstrated in Figure 8, there are significant vasomotion oscillations present in the ΔHbT (or cerebral blood volume) signal, the magnitude of which may well impact on BOLD measurements.

In most functional activation studies, the low signal/noise ratio makes necessary the repetition of stimulus presentation (alternating with the "rest" condition) over a number of cycles. Given the typical time constant of the haemodynamic response (6–10 s) (Bandetti et al., 1995), stimulation protocols are very often based around repetition periods between 10 and 20 s. Using a combination of experimental and theoretical methods, Hutton et al. (1998) determined an optimal inter-stimulus interval for visual stimuli of 18 s. The possible aliasing artifacts which could be incurred by running activation protocols at these frequencies in the presence of the described vasomotion signals are obvious. In addition, the possibility of entraining the vasomotion signal to the stimulus must also be taken into account. In fMRI the problem is compounded by the relatively poor temporal resolution of the technique where standard acquisition sequences result in an effective sampling rate of approximately 0.25 Hz. As shown by the data presented in this paper, vasomotion related oscillations can still be seen at this sampling rate and may not be adequately dealt with by averaging and thresholding techniques.

Furthermore the magnitude of these resting oscillations can be equivalent to, or even exceed those of the haemodynamic changes observed during functional activation. In an identical experimental setup to the one described here, (data not presented) NIRS measurements were recorded over the occipital cortex of a subject during a visual stimulus activation study. The stimulus comprised of a black and white checkerboard pattern (reversing at 8 Hz) for 18 seconds alternating with a blank screen for 36 seconds. During activation the mean increase in HbO_2 was 8 µmolar.cm compared to mean changes during vasomotion in the resting brain of approximately 10 µmolar.cm.

Recent NIRS studies using systems with adequate signal to noise to observe single evoked responses have demonstrated that the magnitude of the evoked response itself varies with each stimulus (Colier et al., 1997). Since this characteristic is lost in repeated averaging techniques, further work with this type of instrumentation may be the key to providing more information about the effects of the slow oscillations in cerebral oxygenation haemodynamics on functional activation studies.

ACKNOWLEDGMENTS

The author is grateful to colleagues at the Wellcome Department of Cognitive Neurology, Queens Square, London for valuable discussions. This work was funded by the United Kingdom Medical Research Council and Hamamatsu Photonics.

REFERENCES

Bandetti PA, Wong EC, Binder JR, Rao SM, Jesmanowicz A, Aaron EA, Lowry TF, Forster HV, Hinks RS, and Hyde JS (1995) Diffusion and Perfusion: Magnetic Resonance Imaging, Raven Press, New York, 335–349.

Berne RM and Levy MN (1997) Cardiovascular Physiology (7th Edition), Mosby Year Book, St. Louis, p. 152.

Buxton RB, Wong EC, and Frank LR (1998) Dynamics of blood flow and oxygenation changes during brain activation: The balloon model. Magn. Res. Med. 39:855–864.

Colier WNJM, Quaresima V, Barattelli G, Cavallari P, van der Sluijs M, and Ferrari M (1997) Detailed evidence of cerebral oxygenation changes in response to motor cortical activation revealed by a continuous wave spectrophotometer with 10 Hz temporal resolution. Proc. SPIE 2979:390–396.

Dóra E and Kovách AGB (1980) Metabolic and vascular volume oscillations in the cat cerebral cortex. Acta Physiol Acad Sci Hung 57:261–275.

Essenpreis M, Elwell CE, Cope M, van der Zee P, Arridge SR, and Delpy DT (1993) Spectral dependence of temporal point spread functions in human tissues. Appl. Opt. 32(4):418–425.

Firbank M, Okada E, and Delpy DT (1998) A theoretical study of the signal contribution of regions of the adult head to near infrared spectroscopy studies of visual evoked responses. Neuroimage 8: 69–78.

Funk W and Intaglietta M (1983) Spontaneous arteriolar vasomotion In: Vasomotion and quantitative capillaroscopy. (Messmer K, Hammersen F, Eds) Karger, Basel, pp. 66–82.

Halsey JHJ and McFarland S (1974) Oxygen cycles and metabolic autoregulation. Stroke 5:219–225.

Hoshi Y and Tamura M (1997) Fluctuations in the cerebral oxygenation state during the resting period in functional mapping studies of the human brain. Med. Biol. Eng. & Comp. 35:328–330.

Hudetz AG, Roman RJ, and Harder DR (1992) Spontaneous flow oscillations in the cerebral cortex during acute changes in mean arterial pressure. J. Cereb. Blood Flow & Metab. 12:491–499.

Hutton C, Howesman A, Joesphs O, Friston K, and Turner R (1998) The effect of inter-stimulus on signal response in fMRI. Proc ISMRM:S1430.

Mayhew JE, Askew S, Zheng Y, Porrill J, Westby GW, Redgrave P, Rector D, and Harper RM (1996) Cerebral vasomotion: A 0.1 Hz oscillation in reflected light imaging of neural activity. Neuroimage 4:183–193.

Meek JH, Elwell CE, Khan AJ, Romaya J, Wyatt JS, Delpy DT, and Zeki S (1994) Changes in occipital haemodynamics due to a visual stimulus detected by near infrared spectroscopy. J. Physiol. 479.P:.

Okada E, Firbank M, Schweiger M, Arridge SR, Cope M, and Delpy DT (1997) Theoretical and experimental investigation of near infrared light propogation in a model of the adult head. Appl. Opt. 36:21–31.

Schniff (1854) Ein accessorisches Arterienherz beim Kaninchen. Arch. Physiol. Heilk (Leipzig) XIII:523–527.

Turner R (1995) Functional mapping of the human brain with magnetic resonance imaging. Seminars in The Neurosciences 7:179–194.

Vern BA, Schette WH, Leheta B, Juel VC, and Radulovacki M (1998) Low frequency oscillations of cortical oxidative metabolism in waking and sleep. J. Cereb. Blood Flow & Metab. 8,2:215–226.

Villringer A, Planck J, Hock C, Schleinkofer L, and Dirnagl U (1993) Near infrared spectroscopy (NIRS)—a new tool to study hemodynamic changes during activation of brain function in human adults. Neuroscience Letters 154:101–104.

Wray S, Cope M, Delpy DT, Wyatt JS, and Reynolds EOR (1988) Characterisation of the near infrared absorption spectra of cytochrome aa_3 and haemoglobin for the non invasive monitoring of cerebral oxygenation. Biochim. Biophys. Acta 933:184–192.

9

ON THE OXYGENATION OF HEMOGLOBIN IN THE HUMAN BRAIN

Albert Gjedde, Peter Høst Poulsen, and Leif Østergaard

PET Center, Aarhus University Hospitals
Aarhus 8000, Denmark

ABSTRACT

We test three hypotheses arising from a model of oxygen delivery to brain tissue. The hypotheses claim that mitochondrial oxygen is negligible in brain tissue such that oxygen consumption depends solely on the mean capillary oxygen tension for a given capillary density; that capillary density is adjusted to satisfy the average steady-state oxygen requirement; and that sudden changes of brain function are subserved by changes of blood flow which adjust the mean capillary oxygen tension in the required direction. The results of the tests did not reject the two former hypotheses and only partly rejected the latter: Sudden changes of blood flow are not always accompanied by increases of oxygen consumption. When they are, changes of apparent capillary density (physiological recruitment) can occur.

INTRODUCTION

Oxygen extraction fraction declines with higher oxygen consumption, thus establishing a higher average capillary partial pressure of oxygen (Widén, 1991) and more highly oxygenated hemoglobin in human brain. Measurements of blood-oxygenation-level-dependent magnetic resonance signal changes due to blood-oxygenation-level dependent (BOLD) contrast further show that brain activity generally leads to increases of venous hemoglobin oxygenation, as evidenced by the decline of signal losses due to the paramagnetic deoxyhemoglobin (Ogawa et al., 1990; Kwong et al., 1992).

As an explanation of these findings, it has been suggested that oxygen delivery to brain could be limited by low diffusibility to the extent that oxygen in brain tissue is much lower than in the brain vasculature (Gjedde et al., 1991; Kassissia et al., 1995). If this were the case, the delivery of oxygen would depend greatly on the average partial pres-

Oxygen Transport to Tissue XXI, edited by Eke and Delpy.
Kluwer Academic / Plenum Publishers, New York, 1999.

sure of oxygen in capillaries and on the tissue conductivity of oxygen which in turn would depend on the microvascular surface area and the distance between capillaries (Krogh, 1919; Gjedde et al., 1991). The mechanism underlying the regulation of oxygen delivery is important to the issues of physiologic and pathologic re- or uncoupling of flow and metabolism during changes of brain activity (Gjedde, 1996, 1997).

This negligible-tissue-oxygen model is one extreme of a range of oxygen distribution models, the other extreme of which is the Krogh venous equilibration model. The two extremes are not mutually exclusive, as tissue oxygen might rise significantly during conditions of increased blood flow if the oxygen demand is not proportionately increased (e.g., in hypercapnia or anesthesia). If too low to influence oxygen delivery, the oxygen tension in the tissue nonetheless must be exactly adjusted to the oxygen tension required for cytochrome oxidase to mediate oxygen reduction to water. When the oxygen demand rises, the capillary oxygen tension must therefore rise to maintain the necessary tissue oxygen tension. In contrast, when blood flow or the average capillary partial pressure of oxygen rises for reasons other than increased oxygen demand, the consequent increase of tissue oxygen tension must in turn elevate the venous oxygen tension, to the extent permitted by the diffusion limitation. It is likely that homeostatic mechanisms allow neurons to alternate between several such metabolic states, as dictated by external circumstance, when states like hypercapnia or anesthesia are encountered naturally.

Crone (1963) described the simple diffusion of substances subject to low permeability of the blood-brain barrier. This description is valid only when the accumulation of the substance in the tissue is so low that little or no substance returns to the vasculature in a single transit (no backflux). Buxton and Frank (1997) recently applied this description to the blood-brain transfer of oxygen from blood to brain tissue. However, the inherent permeability of the blood-brain barrier to oxygen is so great that no diffusion limitation is likely to be imposed by the cerebral capillary endothelium alone.

Although the binding of oxygen to hemoglobin significantly reduces the availability of freely diffusible oxygen (Gjedde et al., 1991), the tension of oxygen at the abluminal side of the cerebral microvasculature must be sufficient to drive the oxygen from the capillaries to the site of metabolism. Because the partial pressure gradient in the tissue is not likely to be less than the gradient in the endothelial wall of the capillaries, the greater decline of oxygen tension must occur extravascularly rather than in the endothelial wall. Further, the anatomical arrangement of the brain microvasculature is not orderly (i.e., axially symmetric) unlike that of muscle capillaries; the brain capillaries are intertwined so as to make almost any geometric model artefactual, and the position of mitochondria in neurons and glial cells varies among different brain regions. The details of a barrier to oxygen delivery are therefore in doubt, and the simplest one-dimensional description is as likely to be accurate as any more elaborate model.

It is the purpose of this presentation to test predictions of a simple one-dimensional model of oxygen diffusion to brain tissue. The model was previously presented by Gjedde (1997). These predictions were tested by comparison to published measurements of oxygen consumption and blood flow in the awake human brain (Gjedde et al., 1990; Widén, 1991; Kuwabara et al., 1992; Gjedde and Kuwabara, 1993; Vafaee et al., 1998, 1999).

HYPOTHESES

1. Under steady-state circumstances, the brain tissue oxygen tension is negligible at the end of the diffusion path, i.e., in or near mitochondria. This hypothesis was tested by calculating the brain mitochondrial oxygen tension, using a formula relating the tissue oxygen tension to measurements of oxygen diffusibility (L), net oxygen extraction fraction (E), and the rate of oxygen consumption (J) in human brain,

$$\overline{P}^{\,mit}_{O_2} = P_{50}\sqrt[h]{\frac{2}{E}-1} - \frac{\overline{J}}{L} \tag{1}$$

where $\overline{J}$ is the net steady-state oxygen consumption rate, L the tissue oxygen conductivity (in units of μmol hg^{-1} mmHg^{-1} min^{-1}), P_{50} the half-saturation tension of the hemoglobin oxygen dissociation curve, $\overline{P}^{\,mit}_{O_2}$ the mitochondrial oxygen tension (i.e., the oxygen tension at the sites of oxygen consumption), and h the Hill-coefficient of the oxygen dissociation curve. To determine L, a formula was derived for the relationship between oxygen diffusibility and the density of capillaries.

2. When the mitochondrial oxygen tension is negligible, brain oxygen consumption in the steady-state must depend on the mean capillary oxygen tension for a given capillary density. The average capillary oxygen tension in turn must depend on the density of capillaries for a given blood flow. We claim that the capillary density is adjusted to satisfy the average steady-state oxygen requirement of each brain region, according to the equation,

$$\overline{J} = L_{max}P_{50}\sqrt[h]{\frac{2}{E}-1} - J_{50} \tag{2}$$

where L_{max} is the maximum oxygen diffusibility attainable in the tissue, and J_{50} the steady-state oxygen consumption rate at which an oxygen diffusibility of half the attainable maximum is required.

3. When the mitochondrial oxygen tension is negligible, transient changes of brain oxygen consumption must be subserved by acute changes of cerebral blood flow for a given density of capillaries. We claim that blood flow rises to establish a higher average capillary oxygen tension. We tested the hypothesis by determining the relationship between oxygen consumption, cerebral blood flow, and mitochondrial oxygen tension according to the equation,

$$J = L\,P_{50}\left(\sqrt[h]{\frac{2}{E}-1} - \eta\sqrt[h]{\frac{1}{E}-1}\right) \tag{3}$$

where η is an index of mitochondrial oxygen tension, relative to the mean venous oxygen tension. At steady-state, Hypothesis 1 claims that η is nil. Hypothesis 3 predicts that oxygen consumption can only rise acutely if capillary density or blood flow rises and extraction falls. If extraction falls more than needed to sustain a given increase of oxygen consumption, η will rise above zero.

THEORY

Steady-State Mitochondrial Oxygen Tension (Hypothesis 1)

Oxygen Diffusibility as Function of Capillary Density. The one-dimensional model of oxygen diffusion is the steady-state solution to Fick's first law,

$$J = \beta\frac{D\,A_v}{\overline{x}}\Delta P \tag{4}$$

where J is the quantity of oxygen transported per unit tissue volume, β the oxygen solubility coefficient, D the oxygen diffusion coefficient in brain tissue, A_v the capillary surface area per unit tissue volume, $\bar{x}$ the average length of the diffusion path, and ΔP the partial pressure difference. The product $\beta\ D$ is Krogh's permeability or diffusion constant. The term $D\ A_v/\bar{x}$ is Crone's permeability-surface area product PS. The coefficients were combined to form a single oxygen conductivity coefficient,

$$L = \beta \frac{DA_v}{\bar{x}} \tag{5}$$

Gjedde et al. (1990) derived an equation for the density of parallel exchange vessels per unit tissue volume. In the case of randomly organized exchange vessels in brain, this equation is more accurately the following (Weibel et al., 1981),

$$N_v = \frac{A_v}{4\pi r} \tag{6}$$

Given a simple hexagonal arrangement of oxygen exchange vessels (capillaries and other microvessels from which oxygen can escape) around the sites of oxygen consumption, the average diffusion path of oxygen can be approximated by (Weibel et al., 1981),

$$\bar{x} = \frac{1}{2.5\sqrt{N_v}} \tag{7}$$

from which it follows that

$$\frac{A_v}{\bar{x}} = 10\pi r (N_v)^{\frac{3}{2}} \tag{8}$$

and hence that

$$L = 10\pi r \beta D (N_v)^{\frac{3}{2}} \tag{9}$$

which establishes the oxygen diffusibility from known or measurable quantities.

Oxygen Partial Pressure Difference as Function of Oxygen Extraction. The oxygen partial pressure difference ΔP, i.e., the difference between the average capillary oxygen tension and the mitochondrial oxygen tension, is therefore the ratio between the oxygen consumption rate and the tissue conductivity of oxygen,

$$\Delta P = \bar{P}_{O_2}^{\text{cap}} - \bar{P}_{O_2}^{\text{mit}} = \frac{J}{L} \tag{10}$$

where the oxygen consumption can be described also as the product of the oxygen concentration, the net extraction fraction, and the blood flow,

$$J = EFC_{O_2} \tag{11}$$

where E is the average net extraction fraction, F the blood flow, and C_{O_2} the arterial oxygen concentration. The extraction fraction defines an average capillary oxygen saturation, assuming even clearance of oxygen from all portions of the capillary bed,

$$\frac{E}{2} = 1 - \bar{S}_{O_2}^{\text{cap}} \tag{12}$$

which in turn is a function of the average capillary oxygen tension expressed by the Hill equation for the average capillary oxygen saturation,

$$\overline{S}^{\,\mathrm{cap}}_{O_2} = \frac{1}{1+\left[\dfrac{P_{50}}{\overline{P}^{\,\mathrm{cap}}_{O_2}}\right]^h} \tag{13}$$

Combining the equations above, the average mitochondrial tension is given by equation (1),

$$\overline{P}^{\,\mathrm{mit}}_{O_2} = P_{50}\sqrt[h]{\frac{2}{E}-1} - \frac{J}{10\pi r\beta D(N_v)^{\frac{3}{2}}} \tag{14}$$

Steady-State Regional Flow-Metabolism Relationship (Hypothesis 2)

The hypothesis claims that the average or steady-state level of functional activation defines a baseline from which the brief functional excursions occur. The hypothesis further claims that these excursions occur according to equation (1), each region having an oxygen conductance commensurate with the demands of its average degree of functional activation. Assuming that capillary density and oxygen diffusibility are higher in a tissue in which oxygen consumption is permanently elevated, the maximum oxygen conductance achievable in brain tissue is defined as follows,

$$L_{\mathrm{max}} = L + \Delta L \tag{15}$$

where ΔL is the quantity by which the current conductance (L) would increase to reach the maximum conductance. This quantity in turn defines a metabolic rate,

$$J_{50} = \Delta L \overline{P}^{\,\mathrm{cap}}_{O_2} \tag{16}$$

from which it follows that

$$\overline{J} = L_{\mathrm{max}}\left(P_{50}\sqrt[h]{\frac{2}{E}-1}\right) - J_{50} \tag{17}$$

which, when plotted for several regions, is a line of slope L_{max} and ordinate intercept J_{50} if L_{max} and J_{50} are common to these regions, as contended in Hypothesis 2. Then, J_{50} is the characteristic oxygen consumption rate at which a regional oxygen conductivity (L) of half the maximum value is required. The abscissa intercept must be P_{50} for values of E approaching unity.

Transient Changes of Flow-Metabolism Relationship (Hypothesis 3)

The question of the blood flow needed for sudden neuronal activation concerns the magnitude of $\overline{P}^{\,\mathrm{mit}}_{O_2}$ during the earliest stages of functional activation of brain tissue; will $\overline{P}^{\,\mathrm{mit}}_{O_2}$ approach and eventually reach $\overline{P}^{\,\mathrm{ven}}_{O_2}$ under some physiological conditions? If $\overline{P}^{\,\mathrm{mit}}_{O_2}$ is defined as a fraction of the average venous oxygen tension, i.e., as $\eta\overline{P}^{\,\mathrm{ven}}_{O_2}$, equation (14) changes to,

$$\overline{J} = L\,P_{50}\left(\sqrt[h]{\frac{2}{E}-1} - \eta\sqrt[h]{\frac{1}{E}-1}\right) \tag{18}$$

which is the fundamental equation underlying the hypothesis. Paradoxically, the need for a higher oxygen tension leads to oxygenation rather than deoxygenation of hemoglobin in brain. The decline of the deoxyhemoglobin can be calculated from measurements of net extraction, according to the equation,

$$\overline{M}_{Hb-dO_2} = \frac{E}{2} V_c\, C_{hb} + E\, V_v\, C_{hb} \tag{19}$$

which by inclusion of equation (18) modifies to,

$$\overline{M}_{Hb-dO_2} \approx \frac{H\, C_{\mathrm{hb}}^{\mathrm{ercs}} (V_c + 2V_v)}{1 + \left(\dfrac{J}{10\pi r \beta D N_v^{1.5} P_{50} (1 - 0.67\eta)} \right)^h} \tag{20}$$

where M_{hb-dO_2} is the total amount of deoxyhemoglobin in the brain vasculature, V_c is the volume of capillary blood, V_v is the volume of venous blood, the product $L\ P_{50}$ is J_{50}, H is the hematocrit, and $C_{\mathrm{hb}}^{\mathrm{ercs}}$ is erythrocyte hemoglobin concentration. Thus, although the blood-oxygenation-level-dependent (BOLD) contrast is a function of the tissue content of deoxygenated hemoglobin, numerous factors influence this quantity, including hematocrit, vascular volumes, oxygen consumption, recruitment of capillaries, and the additional factors which influence the magnitude of P_{50} and h. Although metabolism (J) is one of the factors, and blood flow is not, the effect of blood flow is felt through the tissue oxygen index (η) which changes in the direction of blood flow when metabolism remains unchanged.

RESULTS

Capillary Density and Mitochondrial Oxygen Tension (Hypothesis 1)

Correlative measurements of cerebral blood flow, oxygen consumption and capillary density are rare. Gjedde et al. (1990), Kuwabara et al. (1992), and Gjedde and Kuwabara (1993) used glucose transport measured with [^{18}F]fluorodeoxyglucose to determine the densities of capillaries in human cerebral cortex shown in Table 1. The densities given in Table 1 were recalculated from the original measurements, using the formula for randomly oriented (rather than parallel capillaries) given by Weibel et al. (1981) [equation 6 above]. The product βD is known as Krogh's permeability or diffusion constant and has a value of 1 pmol cm^{-1} min^{-1} mmHg^{-1} for oxygen in an average tissue much like brain tissue (Altman and Dittmer, 1971, Table 14). The radius of microvessels (r in equation 14) was chosen as 2 µm. Therefore, the product $10\ \pi\ r\ \beta\ D$ has the value of 0.0065 pmol min^{-1} mmHg^{-1}, to be multiplied by $N_v^{3/2}$ to obtain L. Equation (7) yielded an average oxygen diffusion path of 21 µm. The average value of L, the oxygen diffusibility or conductivity, for human cerebral cortical gray matter was found to be 0.044 µmol cm^{-3} min^{-1} mmHg^{-1}, as calculated by means of equation (9).

Inserting the values into equation (14), we calculated the oxygen tension at the far end of the average diffusion path to be insignificantly different from zero, as shown in Table 1. The average tissue P_{O_2} would therefore be expected to be around 20 mmHg.

Table 1. Steady-state capillary density and oxygen extraction measurements

Reference	Region	CMR_{O_2} µmol cm^{-3} min^{-1}	E_{O_2} ratio	N_v mm^{-2}	L^* µmol cm^{-3} min^{-1} mmHg^{-1}	$P_{O_2}^{mit}$ mmHg
Gjedde et al. 1990	Cerebral Ctx	1.8	0.47	462	0.065	11
		1.9	0.46	360	0.044	−3
Kuwabara et al. 1992	Somatosensory Ctx	1.6	0.42	289	0.032	−9
Gjedde & Kuwabara 1993	Cerebral Ctx	1.4	0.43	294	0.033	−2
Means		1.7	0.44	351	0.044	−1
±SEM		0.1	0.01	40	0.008	4

* $10\ \pi\ r\ \beta\ D = 0.0065$ pmol min^{-1} mmHg^{-1} (Weibel, 1984).

Regional Steady-State Flow-Metabolism Relation in Human Brain (Hypothesis 2)

Table 1 revealed a trend towards an inverse relationship between the average steady-state oxygen consumption and the calculated extraction fraction, suggesting that a higher average oxygen consumption is associated with a combination of higher mean capillary oxygen tension and higher oxygen conductivity, the latter established by a higher capillary density in highly oxygen-consuming regions. A priori, the association of higher capillary oxygen tension with higher capillary density seems paradoxical but Hypothesis 2 claims that the density is maintained by this higher tension or the higher blood flow, as expressed in equation (2).

To test the present mathematical treatment of the oxygen diffusion limitation model, equation (2) was applied to measurements of oxygen consumption in different regions of normal human brain, as presented by Widén (1991) and illustrated in Figures 1 and 2. Figure 1 shows the relationship between regional values of oxygen consumption and oxygen extraction fraction averaged from values reported by Widén (1991) and Figure 2 shows the regression of equation (2) to this data, for C_{O_2} 9 mM and P_{50} 26 mmHg. The results of the regression are shown in Table 2.

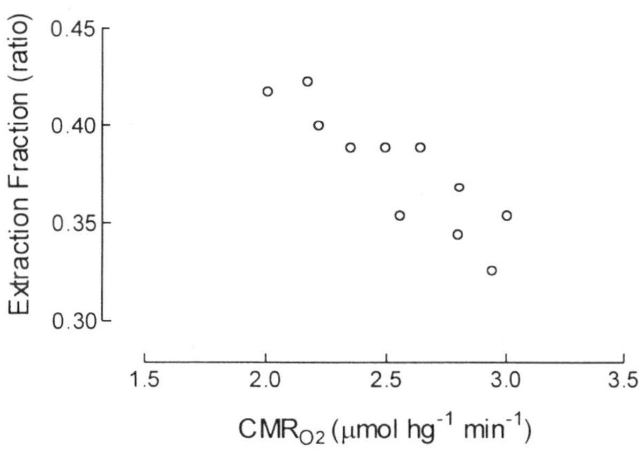

Figure 1. Relation between oxygen extraction and oxygen consumption in different regions of the human brain. Note the decline of the fractions. Each point lumps numerous pixels. Recalculated after data presented by Widén (1991).

Table 2. Regression of equation (2) to data from Widén (1991)

Parameter	Estimate	Unit
Maximum L (L_{max})	0.18 ± 0.03	$\mu mol\, cm^{-3}\, min^{-1}\, mmHg^{-1}$
CMR_{O_2} At Half-Maximum L (J_{50})	5.2 ± 1.4	$\mu mol\, cm^{-3}\, min^{-1}$
Abscissa Intercept ($\approx P_{50}$)	29	mmHg

The results can be interpreted in two ways: Either the regions of the human brain examined by Widén (1991) have identical oxygen conductivities (i.e., the magnitude of the slope) and the same high end-path (mitochondrial) oxygen tension (i.e., the magnitude of the ordinate intercept relative to the magnitude of the slope), close to the value of P_{50}, or the regional oxygen conductivities vary positively with the average metabolic demands of the neurons in each of these regions. The estimate of negligible mitochondrial oxygen tension favors the latter interpretation. According to this interpretation, the regional oxygen conductivity reaches a steady-state value predicted below when an average net oxygen consumption has been maintained for some time.

The steady-state conductivity can be predicted by modifying equation (15),

$$L = L_{max} - \frac{J_{50}}{P_{O_2}^{cap}} \tag{21}$$

according to which,

$$L = L_{max}\left(1 - \frac{P_{50}}{\overline{P}_{O_2}^{cap}}\right) \tag{22}$$

which by insertion of equation (9) for the relationship between N_v and L becomes

$$N_v = \left[\frac{L_{max}}{10\pi r\beta D}\left(1 - \frac{P_{50}}{P_{O_2}^{cap}}\right)\right]^{\frac{2}{3}} \tag{23}$$

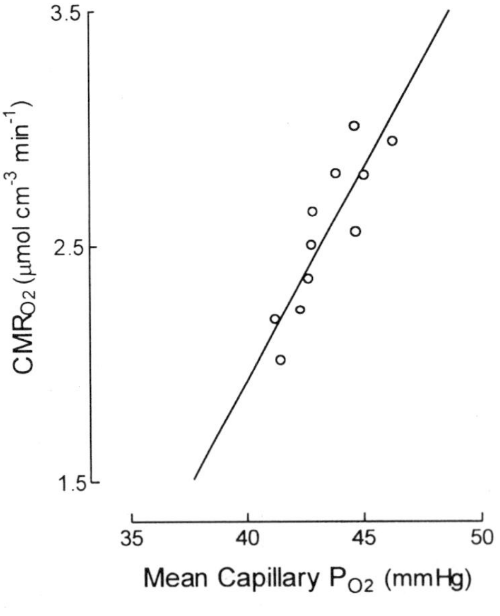

Figure 2. Relationship between mean capillary oxygen tension and regional oxygen consumption in data presented of *Figure 1*. The parameters of the regression line are listed in *Table 2*.

which shows that the capillary density is a hyperbolic function of the capillary oxygen tension, such that the capillary density increases when the average capillary oxygen tension remains elevated for some time. Substituting the term N_{max}, the maximum attainable capillary density in cerebral cortex, for $(L_{max}/[10\ \pi\ r\ \beta\ D])^{2/3}$, we obtain,

$$N_v = N_{max}\left(1 - \frac{P_{50}}{P_{O_2}^{cap}}\right)^{\frac{2}{3}} \tag{24}$$

such that the steady-state regional oxygen consumption is given by,

$$\bar{J} = 10\pi r\beta D(N_{max})^{\frac{3}{2}}\left(P_{O_2}^{cap} - P_{50}\right) \tag{25}$$

and the oxygen consumption corresponding to the half-maximum oxygen conductivity is given by,

$$J_{50} = 10\pi r\beta D(N_{max})^{\frac{3}{2}} P_{50} \tag{26}$$

where N_{max} is the maximum attainable capillary density in human cerebral cortex, equal to $940\,mm^{-2}$.

Transient Changes of Flow-Metabolism Relationship (Hypothesis 3)

The observation of negligible end-path tissue means that η is nil in equation (18). In the absence of recruitment (i.e., no change of N_v), sudden increases of J can only be accomplished by lowering E, the extraction fraction, which in turn requires an increase of blood flow which is larger than the needed increase of oxygen consumption. Figure 3 is a graph of equation (18) for $\eta = 0$, showing the oxygen extraction and oxygen consumption as functions of cerebral blood flow.

When η is nil, it follows that the $J/(LP_{50})$ ratio must approach 3/2 when the net oxygen extraction fraction is close to 0.50. Assuming that 20% of the vessel volume contains fully oxygenated blood (arteries and arterioles) and 70% contains venous blood (in venules and veins) in brain, equation (20) predicts that 36% of the hemoglobin in brain is deoxygenated under normal circumstances.

Observations of the relationship between blood flow, capillary density, and oxygen metabolism during transient and less transient changes of human cortical function are

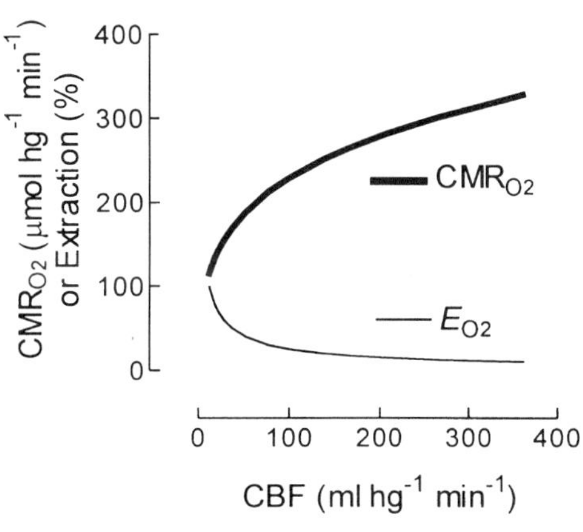

Figure 3. Flow-metabolism couple during transient changes of oxygen metabolism in single brain region, according to equation (18) for $\eta = 0$ (zero mitochondrial oxygen tension).

Table 3. Steady-state capillary density and oxygen extraction measurements

Reference	Condition	CMR_{O_2}	E_{O_2}	N_v	L	$P_{O_2}^{mit}$
		μmol cm^{-3} min^{-1}	ratio	mm^{-2}	μmol cm^{-3} min^{-1} $mmHg^{-1}$	mmHg
Baseline (Table 1)	Cerebral Ctx	1.7	0.44	351	0.044	0
Kuwabara et al. 1992	Vibratory Stimulus	1.6	0.33	340	0.041	6
Gjedde et al. 1990	Ischemia (36 hours)	0.6	0.56	257	0.027	12
	Ischemia (8 days)	0.5	0.18	202	0.019	30

summarized in Table 3. The findings show that some forms of transient activation (vibratory stimulus) are not accompanied by significant increases of oxygen consumption or capillary density, despite the increase of blood flow elicited by the neuronal stimulation. In that case, mitochondrial oxygen tension is expected to increase, as shown in the table.

The table also shows that subacute ischemia (within 36 hours) is accompanied by a decline of capillary density, undoubtedly caused by the blocking of parts of the microvascular bed. However, the blood flow has declined less than expected from the decline of oxygen consumption, leading to a substantial increase of average mitochondrial oxygen tension. This development is further accentuated in subchronic ischemia (after 8 days) when the blood flow has returned towards normal, despite a further decline of oxygen consumption, leading to considerably increased tissue oxygen.

The increased tissue oxygen can be expressed as an elevation of the venous equilibration index η by means of equation (18). Values of η were calculated from the numbers listed in Table 3 according to this equation and illustrated in Figure 4. The graph shows that the venous equilibration index approaches unity in subchronic ischemia.

Equation (20) yields an estimate of the amount of deoxygenated hemoglobin in the tissue from determinations of the oxygen extraction fraction. The deoxygenated hemo-

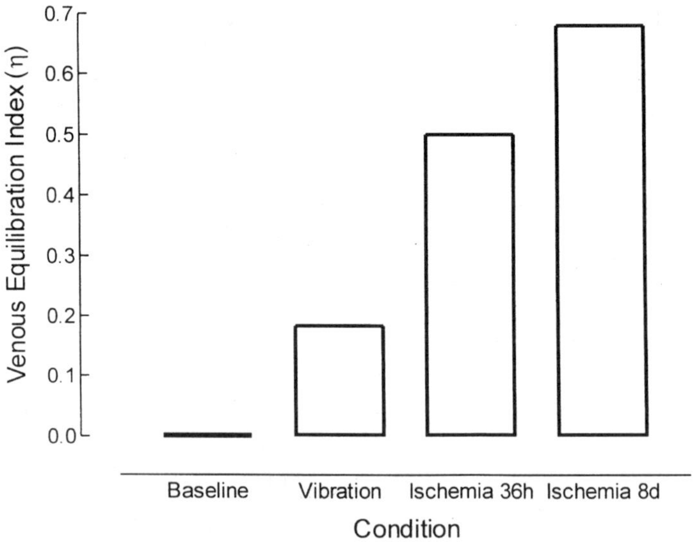

Figure 4. Increase of mitochondrial oxygen tension in physiological and pathological conditions, including functional activation and ischemia. Note approach of mitochondrial oxygen tension towards full venous equilibration. Values of η were calculated from numbers listed in *Table 3*.

Table 4. Stimulation of visual cortex (Vafaee et al., 1999)

Frequency	CBF ml cm^{-3} min^{-1}	CMR_{O_2} μmol cm^{-3} min^{-1}	E_{O_2} (ratio)	$P_{O_2}^{cap}$ mmHg	L μmol cm^{-3} min^{-1} mmHg^{-1}	η (ratio)
Baseline	0.59	2.8	0.43	41	0.068	0.0
1 Hz	0.74	3.1	0.38	43	0.071	0.0
4 Hz	0.75	3.2	0.38	43	0.074	0.0
8 Hz	0.77	2.9	0.34	45	0.074	0.167

globin is important because it is the origin of the so-called BOLD (blood-oxygenation level dependent) contrast detected by magnetic resonance imaging during transient changes of blood flow. Equation (20) predicts that deoxygenation must fall to 30% for a 10% rise of the brain oxygen delivery when the venous equilibration index (η) remains zero. According to the model, a 10% increase of oxygen the consumption at the same tissue oxygen level requires a 35% increase of blood flow. A 10% increase of net oxygen consumption raises the quantities of oxygenated hemoglobin in capillaries, venules and veins.

However, equation (20) revealed that other factors also influence the quantity of oxygenated hemoglobin in the upward or downward directions. Paradoxically, the deoxygenation is actually reduced (and hence the oxygenation increased) by increase of oxygen consumption, increase of hematocrit, derecruitment of capillaries, arteriolar dilatation, venular constriction, alkalosis, 2,3-DPG-decline, or temperature fall. In contrast, the deoxygenation is increased by hematocrit decline, decrease of oxygen consumption, recruitment, arteriolar constriction, venular dilatation, acidosis, 2,3-DPG-increase, and temperature increase. The deoxygenation signal is also affected by the magnitude of η; an increase further augments the BOLD contrast.

To answer the question whether neuronal activations occur in which 1) mitochondrial oxygen does not rise above zero, or 2) recruitment obviates, in part or in whole, the need for an excessive increase of blood flow, Vafaee et al. (1999) measured blood flow and oxygen during stimulation of visual cortex with a checkerboard stimulus reversing at several different frequencies. The results are cited in Table 4. Blood flow and oxygen consumption both rose at the frequencies 1 and 4 Hz but the blood flow increase alone was not sufficient to account for the increase of oxygen consumption, necessitating the calculation of a minimum increase of the oxygen conductivity to explain the measured increase of the oxygen consumption.

Assuming no changes of any variables other than those listed in Table 4, the quantity of deoxyhemoglobin at each of the frequencies was calculated according to equation (20) and shown in Figure 5. The decline of the deoxyhemoglobin, underlying the BOLD contrast change, was greatest (11%) at the frequency at which the increase of the oxygen consumption was the smallest (3%).

DISCUSSION

Two models of blood-brain oxygen transfer have been advanced to account for the BOLD contrast phenomenon in functional activation studies of brain tissue. One model

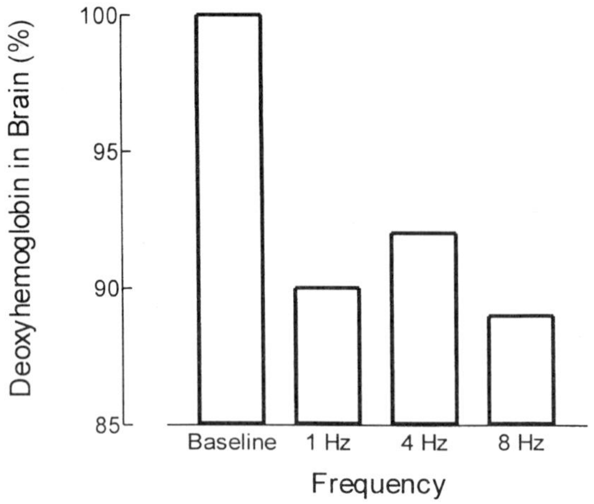

Figure 5. Oxygenation of hemoglobin in brain. Amount of deoxyhemoglobin in brain as function of neuronal activation during stimulation of visual cortex with reversing checkerboard stimulus. Amount was calculated from data shown in *Table 4* by means of equation (20). Decline of deoxyhemoglobin is source of BOLD (blood-oxygenation level dependent) signal.

is the capillary oxygen tension model considered in this presentation (Gjedde, 1997). This model has two extremes, complete diffusion limitation at one end and complete venous equilibration at the other.

A somewhat different model has been presented by Buxton and Frank (1997) who treated oxygen delivery to brain tissue as a special case of the Crone model of blood-brain transfer (Crone, 1963). Their treatment led to an equation for the relationship between the oxygen extraction fraction and blood flow which is similar in shape to the one predicted by the capillary oxygen tension model for complete diffusion limitation. Buxton and Frank (1997) described oxygen extraction from the capillary as an exponentially declining function identical to the Crone-Renkin equation for the unidirectional clearance (Gjedde, 1995, equations 8 and 23),

$$J = C_{O_2} F(1 - e^{-PA_v/(\alpha F)}) \tag{27}$$

where P is the diffusional permeability of oxygen equal to the ratio $D/\bar{x}$, and α is the ratio between the whole-blood oxygen content and physically dissolved oxygen in plasma water established by the hemoglobin oxygen dissociation curve. The consequence of this treatment is the same as for flow tracers: The higher the flow, the higher the unidirectional clearance is but the relationship is non-linear; at low flow rates the clearance equals the flow rate, at higher flow rates it approaches an upper limit equal to the PA/α ratio, equal to 0.30 ml cm^{-3} min^{-1}, corresponding to a maximum oxygen delivery rate of less than 3 µmol cm^{-3} min^{-1}. Thus, the main difference between the two models is the upper limit to oxygen consumption inherent in the model of Buxton and Frank (1997).

Hypothesis 1: Average Steady-State Tissue Oxygen Tension

The examination of several sets of data listed in Table 1 did not reject the hypothesis that the mitochondrial oxygen tension is near zero in brain regions under steady-state conditions. Calculation of the mean mitochondrial oxygen tension yielded values so low that the reserve capacity must be regarded as non-existing. In this case, oxygen consumption can only rise if capillaries and mitochondria could move closer together (recruitment) or when the oxygen pressure head in the capillaries increases.

Several arguments support the calculation that low mitochondrial oxygen may be the rule at steady-state, rather than the venous equilibration which requires that oxygen molecules move as rapidly across the tissue as along the capillaries. Kassissia et al. (1995) showed that the first-pass extraction of oxygen in brain capillaries is lower (and the first-pass transmission or throughput therefore higher) than expected of a freely diffusible gas. This finding suggests that the net extraction of oxygen is similar to the first-pass extraction, i.e., between 40% and 50%.

The low oxygen tension at the far end of the diffusion path (mitochondrial oxygen tension) is not at variance with the many measurements of a mean tissue oxygen tension well above zero. The oxygen profile through the tissue is established by the diffusion of oxygen from the capillaries along the path of average length 21 μm. It is therefore expected that measurements of the average tissue oxygen are evenly distributed across this profile.

The negligible oxygen tension at the far end of the diffusion path would be expected to cause problems in situations of low ambient oxygen pressure, experienced at high altitude and in airplanes. According to the present theory, only an increase of blood flow could compensate for the low ambient oxygen pressure. However, the particular shape of the hemoglobin oxygen dissociation curve makes the necessary increase of blood flow minimal at arterial oxygen pressures down to half of the normal value, as shown in Figure 6. A doubling of blood flow is not required until the arterial oxygen tension drops below one quarter of the normal value. This curve is in excellent agreement with actual measurements of blood flow at different ambient oxygen pressures (e.g., Kozniewska et al., 1987).

Hypothesis 2: Regional Steady-State Flow-Metabolism Relation

The examination of a single set of data (Widén, 1991) did not reject the hypothesis that capillary density is adjusted to limit the otherwise necessary increase of mean capillary oxygen tension in highly metabolizing tissues. However, it is not known how the capillary density is regulated in the steady-state. Some studies show that an inverse relationship exists between the steady-state extraction fraction on one hand and the capillary density and steady-state blood flow rate on the other in different brain regions (Widén 1991; Poulsen et al., 1997). Higher steady-state metabolic rates must therefore be coupled to higher average capillary oxygen tensions. In other words, the higher capillary density of regions of higher metabolic rate does not alone establish a greater oxygen gradient; an increased capillary oxygen tension contributes to the gradient.

Whether this is always so, and why it should be so (as opposed to having an even greater variation of capillary densities commensurate with the same extraction fraction in all regions), is not known; perhaps the increased oxygen tension or blood flow in the microvessels is the factor which maintains the greater capillary density, as regulated by a

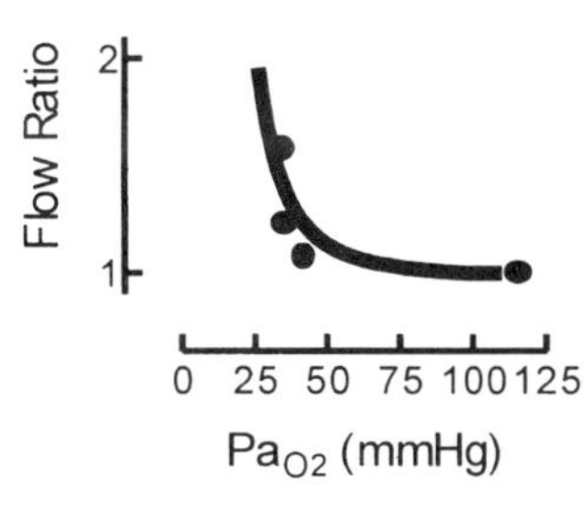

Figure 6. Calculated effect of hypoxemia on blood flow increase necessary to preserve mean capillary oxygen tension. Note diminutive change required at low degrees of hypoxemia, due to shape of hemoglobin oxygen dissociation curve. Circles represent observations made by Kozniewska et al. (1987) in rat.

signal from needy cells. Blood flow therefore plays an intermediary role in the signal from cells to circulation. Thus, in cases of transiently altered neuronal work, associated with a use for more oxygen, blood flow waxes and wanes to establish the necessary oxygen gradient. To do so, the blood flow must increase more, and decline less, than the oxygen demand, as shown in Figure 3.

Accumulating evidence supports the prediction that the oxygen consumption transiently must wax and wane in parallel with the altered blood flow, to settle eventually at the steady-state level associated with the readjusted capillary density, blood flow, and capillary oxygen pressure. If increases of oxygen consumption persist, the cited findings suggest that the shift towards a higher blood flow and a higher capillary oxygen tension is accompanied by a commensurate readjustment of capillary density. This readjustment, in turn, eliminates the signal for the elevated blood flow and is accompanied by the partial normalization of the blood flow. In situations of reduced ambient oxygen pressure, the findings predict a signal from needy cells leads to the blood flow increase shown in Figure 3 and that the moderately elevated blood flow eventually leads to microvascular proliferation if the pressure reduction persists.

Hypothesis 3: Transient Flow-Metabolism Coupling

The examination of several sets of data (Tables 3 and 4) and abundant reports in the literature did not reject the hypothesis that blood flow rises more than oxygen consumption during functional activations. It is puzzling that the oxygen consumption sometimes doesn't rise at all (vibrotactile stimulations), suggesting that additional factors prevent the neurons from using the additional oxygen, or obviate the need to do so. The increased blood flow supplies other compounds in greater amounts, including glucose and amino acids, and removes other substances at greater speed, including lactate and pyruvate.

The cited findings suggest that a signal from insufficiently oxygenated tissue elicits a blood flow increase, also in cases in which the involved neurons may be incapable of utilizing the oxygen. This incapacity may be physiological, as in some cases of somatosensory stimulation, or pathological, as in cases of ischemia or profound hypoxemia. Table 3 provides a curious analogy between functionally activated and pathologically incapacitated neurons, in both cases revealing the inability of the affected neurons to fully consume the oxygen provided. In ischemia, the apparent elevation of tissue oxygen may be the result of an inhomogeneity of blood flow reduction, with still functioning neurons "calling" for more oxygen which other neurons cannot use.

REFERENCES

Altman PL and Dittmer DS (1971) *Biological Handbooks: Respiration and Circulation*, Federation of American Societies for Experimental Biology, Bethesda.

Buxton R and Frank R (1997) A model for the coupling between cerebral blood flow and oxygen consumption during neuronal stimulation. *J Cereb Blood Flow Metab* **17**:64–72.

Crone C (1963) The permeability of capillaries in various organs as determined by use of the "indicator diffusion method". *Acta Physiol Scand* **58**:292–305.

Gjedde A, Kuwabara H, and Hakim A (1990) Reduction of functional capillary density in human brain after stroke. *J Cereb Blood Flow Metab* **10**:317–316.

Gjedde A, Ohta S, Kuwabara H, and Meyer E (1991) Is oxygen diffusion limiting for blood-brain transfer of oxygen? In: *Brain Work and Mental Activity*, Lassen NA, Ingvar DH, Raichle MME, Friberg L, eds., Alfred Benzon Symposium **31**: pp. 177–184.

Gjedde A and Kuwabara H (1993) Absent recruitment of capillaries in brain tissue recovering from stroke. *Acta Neurochir* **[Suppl] 57**:35–40.

Gjedde A (1995) Compartmental analysis. In: *Principles of Nuclear Medicine*, 2nd edition, eds Wagner HNJr, Szabo Z, Buchanan JW. Saunders, Philadelphia, pp. 451–461.

Gjedde A (1996) PET criteria of cerebral tissue viability in ischemia. *Acta Neurol Scand* **[Suppl] 166**:3–5.

Gjedde A (1997) The relation between brain function and cerebral blood flow and metabolism. In: *Cerebrovascular Disease*, Batjer HH, ed., Lippincott-Raven, New York, pp. 23–40.

Kassissia IG, Goresky CA, Rose CP, Schwab AJ, Simard A, Huet PM, and Bach GG (1995) Tracer oxygen distribution is barrier-limited in the cerebral microcirculation. *Circ Res* **77**:1201–1211.

Kozniewska E, Weller L, Höper, Harrison DK, and Kessler M (1987) Cerebrocortical microcirculation in different stages of hypoxic hypoxia. *J Cereb Blood Flow Metab* **7**:464–470.

Krogh A (1919) The number and distribution of capillaries in muscles with calculations of the oxygen pressure head necessary for supplying the tissue. *J Physiol (London)* **52**:409–415.

Kuwabara H, Ohta S, Brust P, Meyer E, and Gjedde A (1992) Density of perfused capillaries in living human brain during functional activation. *Progr Brain Res* **91**:209–215.

Kwong KK, Belliveau JW, Chesler DA, Goldberg IE, Weisskoff RM, Poncelet BP, Kennedy DN, Hoppel BE, Cohen MS, and Turner R (1992) Dynamic magnetic resonance imaging of human brain activity during primary sensory stimulation. *Proc Natl Acad Sci USA* **89**:5675–5679.

Ogawa S, Lee TM, Kay AR, and Tank DW (1990) Brain magnetic resonance imaging with contrast dependent on blood oxygenation. *Proc Natl Acad Sci USA* **87**:9868–9872.

Poulsen PH, Smith DF, Østergaard L, Danielsen EH, Gee A, Hansen SB, Astrup J, and Gjedde A (1997) In vivo estimation of cerebral blood flow, oxygen consumption and glucose metabolism in the pig by [O-15]water injection, [O-15]oxygen inhalation, and dual injections of [F-18]fluorodeoxyglucose. *J Neurosci Meth* **77**:199–209.

Vafaee MS, Marrett S, Meyer E, Evans AC, and Gjedde A (1998) Increased oxygen consumption in human visual cortex: Response to visual stimulation. *Acta Neurol Scand* **98**:85–89.

Vafaee MS and Gjedde A (1999) Frequency-dependent changes in cerebral metabolic rate of oxygen during activation of human visual cortex. *J Cereb Blood Flow Metab* **19**: in press.

Weibel ER, Taylor CR, Gehr P, Hoppeler H, Mathieu O, and Maloiy GMO (1981) Design of the mammalian respiratory system. IX. Functional and structural limits for oxygen flow. *Respiration Physiology* **44**:151–164.

Weibel ER (1984) *The Pathway for Oxygen. Structure and Function in the Mammalian Respiratory System*, Harvard University Press, Cambridge.

Widén L (1991) How shall we measure brain work. In: *Brain Work and Mental activity*, Lassen NA, Ingvar DH, Raichle MME, Friberg L, eds., Alfred Benzon Symposium **31**: pp. 127–139.

10

REDUCED DILATOR AND CONSTRICTOR RESPONSE OF THE MIDDLE CEREBRAL ARTERY AFTER COLD LESION OF THE RAT PARIETAL CORTEX

C. Görlach,[1] Z. Benyó,[2] and M. Wahl[1]

[1]Department of Physiology
University of Munich, Pettenkoferstr. 12, D-80336 München
Germany
[2]Experimental Research Department—2nd Institute of Physiology
Semmelweis University of Medicine
Üllői út 78/A, H-1082 Budapest
Hungary

1. INTRODUCTION

Cold injury to the exposed brain through the intact dura was used as the model for brain injury in the present study (Klatzo et al., 1958). The histological features of the cold lesion in the rat resemble those of human cortical contusion. Although the freezing injury is an artificial insult of the brain compared with the complexity of traumatic damage in humans, the model is highly standardized and yields reproducible focal disruption of the blood brain barrier (BBB) (Stoffel et al., 1997). The resulting brain edema leads to secondary brain damage, thus increasing the necrosis due to the primary insult (Eriskat et al., 1994). In this respect, this experimental approach mimics the pathophysiology of mild human traumatic head injury (Klatzo et al., 1958, Wahl et al., 1988). In this model of brain injury in the parietal cortex we tested in vitro the vasoreactivity of middle cerebral artery (MCA) and its distal branches, a main artery which supplies the lesioned area with blood. We investigated the vascular response to endothelin (ET)-1. This compound has been suggested as a possible mediator of secondary brain damage due to its high potency and extremely long-lasting action (Feger et al., 1994). Furthermore, the cerebral tissue concentration of ET-1 is increased after brain injury or ischemia (Barone et al., 1994; Yamada et al., 1995). Because of the ET-NO network (Ehrenreich and Schilling, 1995) we also examined whether the NO system was func-

Oxygen Transport to Tissue XXI, edited by Eke and Delpy.
Kluwer Academic / Plenum Publishers, New York, 1999.

tionally altered in the MCA, with a focus on the ET-3-induced dilation. Since we found that neurotrauma reduces the vasomotor responses to ETs we also examined vasoreactivity in response to some other stimuli.

2. MATERIALS AND METHODS

2.1. Animal Preparation and Experimental Protocol

Three groups of anaesthetized male rats were investigated: untreated controls, sham-operated and cold-lesioned animals. In the lesioned group, a cold lesion was applied to the left cortex. A copper cylinder (Ø 5 mm), precooled to −78°C by acetone/dry ice, was impressed 1 mm onto the intact dura for 60 s. After suturing the skin, the wound was treated with Xylocain spray. The animals were killed 4, 24, or 48 h after the cold lesion.

ET-1 concentration was determined in EDTA plasma samples from shams and lesioned animals (after 24 h or 48 h) with a commercial ELISA test kit.

Ring segments of the MCA and its branches were prepared for measurement of isometric force as described by Feger et al. (1994). The ring segments were mounted on two L-shaped prongs of tungsten wire (50 μm diameter), one of which was mounted rigidly to the bath and the second connected to a force transducer. The transducers were mounted on a microdrive to set the resting tension at 0.8–2.3 mN (mean 1.31 ± 0.02 mN) in MCA and its branches.

A reference contraction was obtained in all segments by exchanging the bath solution for a 124 mM K^+-Krebs' solution (NaCl replaced by KCl). The functional state of the endothelium was tested by cumulative application of bradykinin (BK) (10^{-8}–10^{-5} M) following precontraction with 10^{-4} M uridine triphosphate (UTP). To test the dilator response independent of the endothelium, sodium nitroprusside (SNP) (10^{-3} M) was applied.

The contractile action of ET-1 ($10^{-12} - 3 \times 10^{-7}$ M) was studied under resting tension conditions. To study whether the blood vessels are under tone from endogenously released endothelins, the specific ET_A receptor antagonist BQ 123 ($10^{-12} - 3 \times 10^{-5}$ M) was applied cumulatively immediately after the accommodation period.

The relaxant action of ET-3 (10^{-12}–10^{-8} M) was studied by cumulative application following precontraction of the MCA with prostaglandin $F_{2\alpha}$ ($PGF_{2\alpha}$) (10^{-5} M).

3. RESULTS

3.1. Influence of the Parietal Cortical Lesion on the Contractile Effect of ET-1, ET-3, 124 mM K^+ and $PGF_{2\alpha}$

Administration of the selective ET_A-receptor antagonist BQ123 ($10^{-12} - 3 \times 10^{-5}$ M) to MCA and its branches from sham-operated or lesioned animals did not induce any vasomotor effects, indicating that the tone of the isolated vessels is not influenced by endogenously released ETs acting at ET_A-receptors. Application of ET-1 (10^{-12} – 3×10^{-7} M) resulted in a dose-dependent contraction under resting conditions in MCA ring segments of shams. The -log of agonist concentration with half maximal effect (pD_2) was 8.36 ± 0.12 and maximal effect (E_{max}) 2.41 ± 0.15 mN at 10^{-7} M (Figure 1). In ipsilateral segments obtained 24 and 48 h after cold lesion the effect of ET-1 was significantly

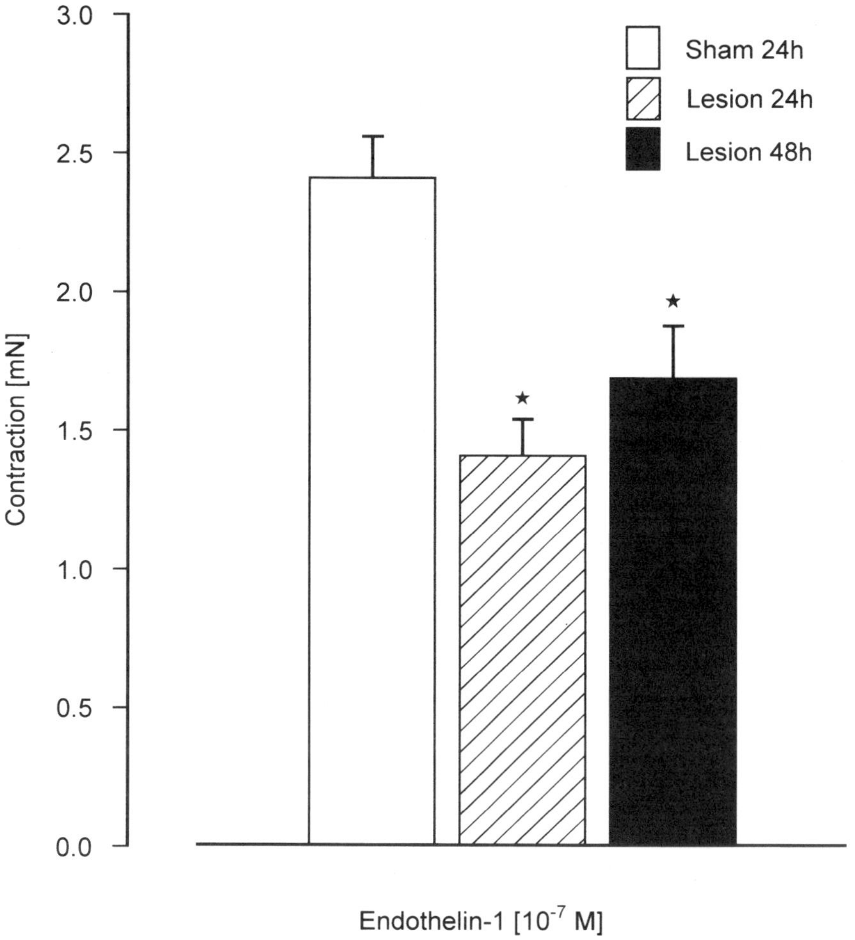

Figure 1. Maximal contractile effect of 10^{-7} M endothelin-1 obtained in shams (24 h) or in animals 24 and 48 h after cold lesion in middle cerebral artery. $*p < 0.05$ versus sham. Data were taken from Görlach et al. (1998b).

reduced in the concentration range $10^{-8} - 3 \times 10^{-7}$ M. The E_{max} was 1.41 ± 0.13 mN 24 h and 1.69 ± 0.19 mN at 48 h after cold lesion at 10^{-7} M, respectively (Figure 1). Similarly, the contractile effect of 10^{-7} M ET-3 was also reduced 24 h after lesion but without reaching a level of significance ($p \leq 0.05$, Tukey-HSD). Moreover, cold lesion in the parietal cortex changed the response to a variety of contractile agents. This is shown by a significantly reduced effect due to 124 mM K^+ or 10^{-5} M $PGF_{2\alpha}$.

3.2. Influence of the Parietal Cortical Cold Lesion on the Dilating Response to ET-3, BK and SNP

After precontraction with $PGF_{2\alpha}$, ET-3 relaxed the MCA at concentrations of $10^{-12} - 3 \times 10^{-8}$ M with a pD_2 of 9.20 ± 0.19 and an E_{max} of 0.42 ± 0.07 mN at 10^{-8} M in sham-operated animals. The dilation was reduced 24 and 48 h after cold lesion by 19 and 62% at 10^{-8} M, respectively. Because vasodilation was very small and the number of observations limited, the reduction of the vasodilating effect of ET-3 after cold lesion compared with shams did not reach significance. However, the trend, although not significant, is in agreement with previous findings using the same model and basilar artery (BA) in which significant reductions of the ET-3 induced dilation were observed (Görlach et al., 1998a). BK was applied to test whether the hyporeactivity to ET-3 resulted from a lack of ability of MCA and its branches to release NO. BK has been shown to elicit relaxation of the rat MCA via release of NO from the vascular endothelium (Görlach and Wahl, 1996). In UTP (10^{-4} M) precontracted sham segments BK elicited a concentration dependent (10^{-8}–10^{-5}) dilation with a pD_2 of 7.08 ± 0.10 and an E_{max} of 0.65 ± 0.06 mN at 10^{-6} M (Figure 2). The BK-induced dilation in ring segments of MCA was reduced after cold lesion. There was a significant difference at 10^{-6} M between the values obtained 24 h after lesion compared with shams (Figure 2). This demonstrates a weakened NO-cGMP system after cold lesion. A further indication of an impaired guanylate cyclase/cGMP system in smooth muscle cells is the fact that the dilating effect of SNP in MCA was reduced at all times, after cold lesion.

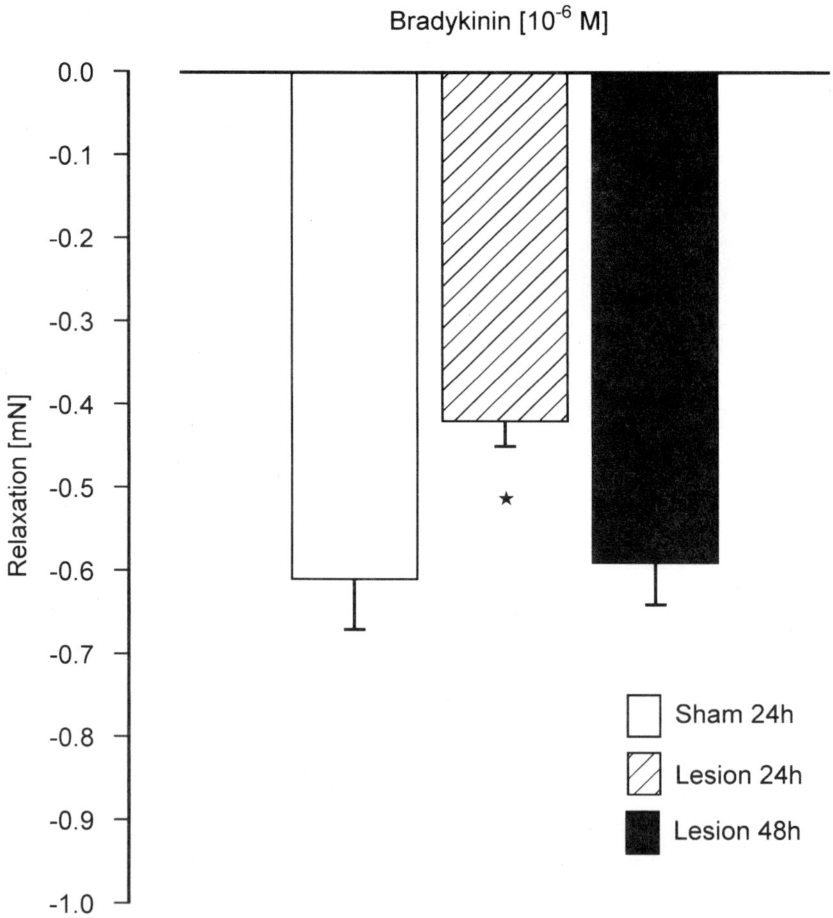

Figure 2. Maximal dilating effect of 10^{-6} M bradykinin obtained in shams (24 h) or in animals 24 and 48 h after cold lesion in middle cerebral artery. *$p < 0.05$ versus sham. Data were taken from Görlach et al. (1998b).

3.3. Determination of Plasma ET-1 Concentration

Shams after 24 and 48 h had an ET-1 concentration of 0.67 ± 0.05 and 0.74 ± 0.10 pmol/l, respectively. The ET-1 concentration was significantly increased by 81% 24 h after cold lesion to 1.25 ± 0.06 pmol/l but 48 h after the cold lesion it was only increased by 8% (not significantly).

4. DISCUSSION

In the present study we demonstrated, for the first time, an unspecific reduction of vascular reactivity in MCA and its large branches after cold lesion. The vessels showed a significantly diminished contraction to ET-1 which was 59% 24 h and 70% 48 h after cold lesion. This contraction is mediated via stimulation of ET_A receptors located on the smooth muscle (Feger et al., 1994). Activation of the ET_A receptor may open Ca^{2+} channels or may activate G-protein triggered phospholipase C and proteinkinase C (Görlach et al., 1998c). The finding of a reduced contractile response to ET-1 after cold lesion conflicts with a paper of Armstead (1996) who used the cranial window and the model of a fluid percussion injury in the newborn pig. In that study the small dilation normally observed in response to 10^{-12} ET-1 was converted into constriction after injury. The constrictor effect of higher concentrations of ET-1 were even more pronounced after damage.

In our model of a cold lesion in the parietal cortex, the tone of the MCA and its branches apparently was not influenced by endogenous ET-1. This is evident from the fact that BQ-123, a specific ET_A antagonist, had no effect per se in the isolated MCA after cold lesion and in sham-operated animals. The data showing no effect of an ET_A antagonist are also in apparent contradiction to the study of Armstead (1996), in which pretreatment with BQ-123 diminished the constriction of pial arteries. Similarly, in the cat model of a focal cerebral ischemia, Patel et al. (1996) have found that perivascular application of ET antagonists dilates pial arteries. The apparent discrepancy between these studies and the present one may be explained by the different time of measurement, experimental design, and species. For instance, we obtained the isolated MCA without pharmacologically induced tone 24 h after cold lesion from the rat. In contrast, Armstead (1996) used pretreatment of BQ 123 in an open cranial window in the pig and Patel et al. (1996) applied the drugs 30–210 min after MCA occlusion perivascularly to individual arteries of the cat.

We found not only a reduced contractile response of MCA and its branches to ET-1 but also to other contractile agents, i.e. 124 mM K^+, and 10^{-5} M $PGF_{2\alpha}$. The reduced effect to K^+ and $PGF_{2\alpha}$ imply a more or less general reduction of the vasomotor response to contractile agents. The present study also demonstrates a diminished response of MCA to several dilating stimuli after cold lesion. The maximal degree of relaxation due to ET-3 was less in MCA than in BA (Görlach et al., 1998a) which may be explained by a lower density of ET_B receptors. The relaxation was reduced by 19 and 62% at 24 and 48 h after cold lesion, respectively, at a concentration of 10^{-8} M. Previous work from our laboratory has demonstrated that the vasodilator effect of ET-3 is mediated by endothelial ET_B-receptors and the release of NO (Schilling et al., 1995). We believe these endothelial ET-receptors act as ET_{B1} receptors. The reduction of vasodilation in response to ET-3 after cold lesion was partly due to a diminished release of NO from the endothelium. This was evident from the reduced vasorelaxant response to BK, which is mediated via endothelial B_2 receptors and the release of NO (Görlach and Wahl, 1996). Furthermore, the activity of guanylate cyclase in the smooth muscle cells is obviously reduced since the effects of SNP in MCA were diminished significantly.

The functional lack of ET_B-receptors has been demonstrated also recently in BA, using the same brain injury model (Görlach et al., 1998a), when we found in this artery, remote from the lesion, a reduced effect of ET-3 after 24h and a nearly abolished action 48h after cold lesion. In contrast, the ET-1 induced contraction was not diminished. The ET_B receptor mRNA expression in the BA was not altered 48h after the neurotrauma, at a time when the functional impairment had reached its maximum, making the possibility of down-regulation of the ET_B-receptor at the transcriptional level very unlikely. In agreement with the loss of ET_B-mediated vasodilation in BA 24h and 48h after cortical cold lesion, however, we found a delayed reduction of ET_B-receptor immunoreactivity in BA, and in brain stem intraparenchymal vessels, that coincides with the functional alterations in BA reactivity to ET-3.

Development of tachyphylaxis may explain the delayed impairment of vasodilation to ET-3 after the cortical lesion. The tachyphylaxis of ET_B receptors might result from an increased concentration of ETs. This would mean that endogenous ETs are released during or after cold lesion and their action is weakened. Indeed, plasma ET-1 concentration was increased by 81% 24h after cold lesion. In addition tissue ET-1 was enhanced 1–5 days after cold lesion in the rat (Yamada et al., 1995) and high plasma levels of ETs are found in humans after stroke or subarachnoid hemorrhage (Ehrenreich et al., 1992; Ziv et al., 1992). Furthermore, we have found that tachyphylaxis develops to the dilating component of the ET-3 response in the BA of normal animals during repeated application of cumulative concentration effect curves. Tachyphylaxis was also found in situ using microapplication of the same dose of the specific ET_B agonist BQ 3020 to feline pial arteries. Moreover, a loss of ET_B-receptor mediated vasodilation also occurs after focal ischemia (Touzani et al., 1997).

Besides the development of tachyphylaxis to ETs after cold lesion there may be another reason for a more general reduction of vasoreactivity to several other contractile (124 mM K^+, $PGF_{2\alpha}$) and relaxing (BK, SNP) stimuli. During cold lesion spreading depressions originate in the penumbra of the lesion (Obrenovitch, 1995). These spreading depressions have been found to induce an unspecific reduction of vasoreactivity to constrictor and dilating compounds (Wahl et al., 1987).

ACKNOWLEDGMENT

This work was supported by grants of BMBF, DAAD-MÖB and by the von Humboldt Foundation. The authors are grateful to Mrs. E. Held for expert secretarial help and to Mrs. J. Specker for excellent technical assistance.

REFERENCES

Armstead WM (1996) Role of endothelin in pial artery vasoconstriction and altered responses to vasopressin after brain injury. J Neurosurg 85:901–907.

Barone FC, Globus MY-T, Price WJ, White RF, Storer BL, Feuerstein GZ, Busto R, and Ohlstein EH (1994) Endothelin levels increase in rat focal and global ischemia. J Cereb Blood Flow Metab 14:337–342.

Ehrenreich H, Lange M, Near KA, Anneser F, Schoeller LAC, Schmid R, Winkler PA, Kehrl JH, Schmiedek P, and Goebel F-D (1992) Long term monitoring of immunoreactive endothelin-1 and endothelin-3 in ventricular cerebrospinal fluid, plasma and 24-h urine of patients with subarachnoid hemorrhage. Res Exp Med 192:257–268.

Ehrenreich H and Schilling L (1995) New developments in the understanding of cerebral vasoregulation and vasospasm: the endothelin-nitric oxide network. Cleveland Clin J Med 62:105–116.

Eriskat J, Schürer L, Kempski, Q, and Baethmann A (1994) Growth kinetics of a primary tissue necrosis from a focal lesion. Acta Neurochir Suppl. 60:425–427.

Feger GI, Schilling L, Ehrenreich H, and Wahl M (1994) Endothelin-induced contraction and relaxation of rat isolated basilar artery: Effect of BQ-123. J Cereb Blood Flow Metab 14:845–852.

Görlach C and Wahl M (1996) Bradykinin dilates rat middle cerebral artery and its large branches via endothelial B_2 receptors and release of nitric oxide. Peptides 17:1373–1378.

Görlach C, Sirén A-L, Knerlich F, Feger G, Fricke A, Barth M, Schilling L, Ehrenreich H, and Wahl M (1998a) Delayed loss of ET-B-receptor mediated vasorelaxation after cold lesion of the rat parietal cortex. J Cereb Blood Flow Metab 18:1357–1364.

Görlach C, Benyó Z, and Wahl M (1998b) Reduced reactivity of the middle cerebral artery and its large branches after cold lesion. J Neurotrauma 15:1067–1075.

Görlach C, Benyó Z, and Wahl M (1998c) Endothelin-1-induced contraction in cerebral vessels is mediated by activation of the phospholipase C/proteinkinase C cascade. Kidney Int 54, Suppl. S224–S225.

Klatzo I, Piraux A, and Laskowski EJ (1958) The relationship between edema, blood-brain barrier and tissue elements in local injury. J Neuropathol Exp Neurol 17:548–564.

Obrenovitch TP (1995) The ischaemic penumbra: twenty years on. Cerebrovasc Brain Metab Rev 7:297–323.

Patel TR, Galbraith S, McAuley MA, and McCulloch J (1996) Endothelin-mediated vascular tone following focal cerebral ischaemia in the cat. J Cereb Blood Flow Metab 16:679–687.

Schilling L, Feger GI, Ehrenreich H, and Wahl M (1995) Endothelin-3-induced relaxation of isolated rat basilar artery is mediated by and endothelial ETB-Type endothelin receptor. J Cereb Blood Flow Metab 15:699–705.

Stoffel M, Berger S, Staub F, Eriskat J, Jacob K, and Baethmann A (1997) The effect of dietary α-tocopherol on the experimental vasogenic brain edema. J Neurotrauma 14:339–348.

Touzani O, Galbraith S, Siegl P, and McCulloch J (1997) Endothelin-B receptors in cerebral resistance arterioles and their functional significance after focal cerebral ischemia in cats. J Cereb Blood Flow Metab 17:1157–1165.

Wahl M, Lauritzen M, and Schilling, L (1987) Change of cerebrovascular reactivity after cortical spreading depression in cats and rats. Brain Res 411:72–80.

Wahl M, Unterberg A, Baethmann A, and Schilling L (1988) Mediators of blood-brain barrier dysfunction and formation of vasogenic brain oedema. J Cereb Blood Flow Metab 8:621–634.

Yamada G, Hama H, Kasuya Y, Masaki T, and Goto K (1995) Possible resources of endothelin-1 in damaged rat brain. J Cardiovasc Pharmacol 26:S486–S490.

Ziv I, Fleminger G, Djaldettl R, Achiron A, Melamed E, and Sokolovsky M (1992) Increased plasma endothelin-1 in acute ischemic stroke. Stroke 23:1014–1016.

11

LOSS OF ET_B-RECEPTOR-MEDIATED RELAXATION IN BASILAR ARTERY AFTER COLD LESION OF THE RAT PARIETAL CORTEX

C. Görlach,[1] A.-L. Sirén,[3] F. Knerlich,[3] G. Feger,[2] A. Fricke,[3] L. Schilling,[4] H. Ehrenreich,[2,3] and M. Wahl[2]

[1]Department of Physiology, University of Munich
Pettenkoferstr. 12, D-80336 München
Germany
[2]Department of Neurology and Psychiatry
University of Göttingen
[3]MPI Experimental Medicine Göttingen
[4]Department of Neurosurgery
Medical Faculty Mannheim
University of Heidelberg

1. INTRODUCTION

The endothelins (ETs) can exert both constrictor and dilator effects. These opposite actions have been attributed to the existence of different ET receptors. The ET_A receptor has high affinity for ET-1 and ET-2 but low affinity for ET-3, while the ET_B-receptor is non-selective in this respect. Stimulation of ET_A receptors, which are located on vascular smooth muscle, causes vasoconstriction (Feger et al., 1994). This vasomotor reaction is mediated by activation of phospholipase C and protein kinase C (Görlach et al., 1998a). ET_B receptors of arteries mediate vasodilation which is coupled to NO-production in endothelial cells (Schilling et al., 1995).

ET-1 applied to the adventitial side of cerebral arteries evokes cerebral ischemia and neuronal damage (Sharkey et al., 1993). Furthermore, the cerebral tissue concentration of ET-1 is increased after brain injury or ischemia (Barone et al., 1994; Yamada et al., 1995). We investigated the effects of ETs in rat basilar artery (BA) since ET-1 might be a mediator of secondary brain damage due to its high potency and extremely long-

Oxygen Transport to Tissue XXI, edited by Eke and Delpy.
Kluwer Academic / Plenum Publishers, New York, 1999.

lasting action. Because of the ET-NO network (Ehrenreich and Schilling, 1995) we also examined, whether the NO system was functionally altered in the BA, with a focus on the ET-3-induced dilation. Since we found that neurotrauma reduces the vasodilator response to ET-3 in BA, we investigated by reverse transcriptase—polymerase chain reaction (RT-PCR) and immunohistochemistry whether the expression of the ET_B-receptor mRNA or protein is changed after cold lesion.

2. MATERIAL AND METHODS

2.1. Animal Preparation

Three groups of anaesthetized rats were investigated: untreated controls, sham-operated and cold-lesioned animals. In the lesioned group, a cold lesion was induced to the left cortex. For cold lesion, a copper cylinder (∅ 5 mm), precooled to −78°C by acetone/dry ice, was impressed 1 mm onto the intact dura for 60 s. The wound was treated with Xylocain spray. The animals were killed 16, 24, 40 or 48 h after the cold lesion.

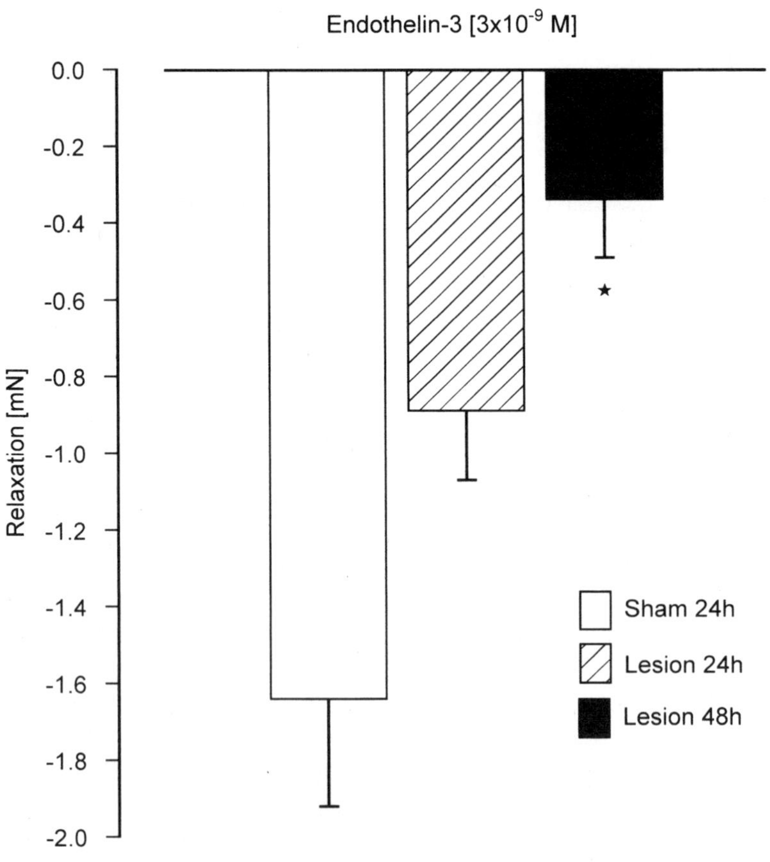

Figure 1. Dilating effect of ET-3 (3×10^{-9} M) obtained in BA segments after 24 or 48 h in sham-operated or cold lesioned basilar artery in rats. * $p < 0.05$ versus sham. Data were taken from Görlach et al. (1998b).

Ring segments of basilar artery of 20 rats were prepared for measurement of isometric force as described by Feger et al. (1994). The ring segments were mounted on two L-shaped stainless steel wires (70 μm diameter), one of which was mounted rigidly to the bath and the second connected to a force transducer. The transducers were mounted on a microdrive to set the resting tension at 2.5–3.5 mN in BA and at 6–7.5 mN in aorta.

A reference contraction was obtained in all segments by exchanging the bath solution for a 124 mM K^+-Krebs' solution (NaCl replaced by KCl). The functional state of the endothelium was tested by cumulative application of acetylcholine (ACh, 10^{-8}–10^{-4} M) following precontraction with 10^{-6} M 5-hydroxytryptamine (5-HT) in BA or 10^{-5} M 5-HT in aorta. To test the dilator response independent of the endothelium, sodium nitroprusside (SNP, 10^{-3} M) was applied.

The relaxant action of ET-3 (10^{-12}–10^{-8} M) was studied by cumulative application to BA following precontraction with prostaglandin $F_{2\alpha}$ ($PGF_{2\alpha}$, 3×10^{-6} or 10^{-5} M). In some segments of aorta the dilator action of the ET_B-receptor agonist IRL 1620 (10^{-8} M) was investigated. The contractile action of ET-1 (10^{-12}–10^{-6} M) and the specific ET_A receptor antagonist BQ123 (10^{-12}–3×10^{-5}) were studied in BA under resting tension conditions.

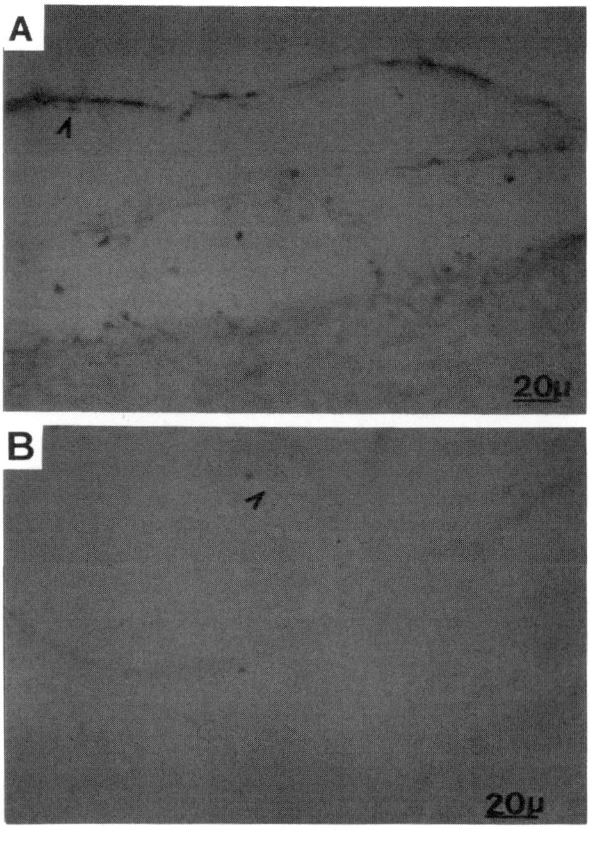

Figure 2. Expression of immunoreactivity for ET_B-receptors in BA of sham-operated rats (A) and in lesioned rats (B) 48 h after cold injury. Data were taken from Görlach et al. (1998b).

2.2. RNA Isolation and PCR

For each sample (48 h sham operated and lesioned animals), basilar arteries from 6 rats were pooled for RNA isolation by the CsCl-gradient method (Sambrook et al., 1989) and used for reverse transcriptase–polymerase chain reaction (RT-PCR) with lamin as housekeeping gene. The expression of ET-1, ET_A- and ET_B-receptors was determined.

2.3. Immunohistochemistry

Immunohistochemistry of the ET_B-receptors was performed as described by Sairanen et al. (1997).

3. RESULTS

3.1. Influence of the Parietal Cortical Cold Lesion on ET-3 Induced Vasomotor Response

After precontraction with $PGF_{2\alpha}$, ET-3 relaxed the BA at concentrations of 10^{-10} to 10^{-8} M with a pD_2 of 9.06 ± 0.31 and an E_{max} of 1.64 ± 1.0 mN at 3×10^{-9} M in sham-operated animals. At higher concentrations (5×10^{-8} and 10^{-7} M) ET-3 induced contractions with an E_{max} of 2.01 ± 1.49 mN in sham-operated animals that was statistically not different from the lesioned animals. The dilation was reduced 24 h after cold lesion by 33%, and 48 h after cold lesion by 73% at 3×10^{-9} M. ACh was applied to test whether this resulted from a lack of ability of BA to release NO. The ACh (10^{-8}–10^{-4} M)-induced dilation in ring segments of BA (E_{max} 2.56 ± 0.27 mN, pD_2 6.65 ± 0.09) was not impaired at any time after cold lesion. This demonstrates an intact NO-cGMP system after cold lesion. A further indication of an intact guanylate cyclase/cGMP system is the fact that the dilating effect of SNP in BA was not altered at any time after cold lesion.

To test whether the loss of vasodilation via the ET_B-receptor after cold lesion was specific for BA we investigated the dilating effect of IRL 1620, a specific ET_B-receptor agonist in ring segments of rat thoracic aorta. In $PGF_{2\alpha}$-precontracted rat thoracic aorta ring segments of sham-treated animals IRL1620 induced a relaxation of 2.91 ± 1.92 mN. Forty eight hours after cold lesion, IRL 1620 (10^{-8} M) elicited only a weak relaxation of 1.41 ± 1.27 mN in $PGF_{2\alpha}$-precontracted rat thoracic aorta which was statistically significant ($p < 0.05$) from shams.

3.2. Influence of Parietal Cortical Cold Lesion on ET-1 Induced Vasoconstriction

Administration of the selective ET_A-receptor antagonist BQ123 (10^{-12}–3×10^{-5} M) to BA from control or lesioned animals did not induce any vasomotor effects, indicating that the tone of the vessels is not influenced by endogenously released ETs acting on ET_A-receptors. Application of ET-1 (10^{-12}–10^{-6} M) resulted in a dose-dependent contraction under resting conditions in BA ring segments of untreated controls. The pD_2 was 8.03 ± 0.22 and E_{max} was 6.35 ± 0.70 mN. In segments obtained 16 and 40 h after cold lesion, ET-1 elicited the same contraction as in controls. Moreover, cold lesion in the parietal cortex does not change the response to a variety of contractile agents. This is evident from the same effect due to 124 mM K^+, 10^{-6} M 5-HT or 10^{-5} M $PGF_{2\alpha}$ as in segments of sham-operated animals.

3.3. Influence of Parietal Cortical Cold Lesion on Expression of ET_B Receptors in the BA

To explore further the impairment of the ET_B-receptor-mediated vasodilation after neurotrauma, RNA was isolated from the BA and analysed by RT-PCR. The mRNAs for ET-1, ET_A and ET_B are constitutively expressed in the BA of both sham-operated and lesioned animals, indicating that the alterations in ET_B-receptor expression were more likely to occur at the posttranscriptional level. Indeed, in the sham animals, immunoreactivity for ET_B-receptor was found in the BA and in its intraparenchymal blood vessels of the brain stem, while 24 and 48 h after cortical cold lesion ET_B immunoreactivity in the BA and in the brain stem intraparenchymal vasculature was reduced.

4. DISCUSSION

The present study demonstrates, for the first time, a delayed loss of the BA to ET_B-receptor stimulation following a standardized neurotrauma. The functional lack of ET_B-receptors could further be demonstrated morphologically by reduced antibody staining of the ET_B-receptor protein. The loss of ET_B-receptors results in the disappearance of the vasorelaxing effect of ET-3. This reduction of vasodilation after cold lesion is apparently specific for ET_B-receptors. Previous work from our laboratory has demonstrated that the vasodilator effect of ET-3 or the specific agonist IRL 1620 is mediated by endothelial ET_B-receptors and the release of NO (Schilling et al., 1995). The reduction of vasodilation found after cold lesion was not due to diminished possible release of NO from the endothelium nor to a reduced activity of guanylate cyclase in the smooth muscle cells. This was also evident from the unaltered vasomotor response to ACh and SNP.

Development of tachyphylaxis may explain the delayed impairment of vasodilation to ET-3 or IRL 1620 in the BA and aorta, two large blood vessels remote from the cortical lesion. The tachyphylaxis of ET_B receptors might result from an increased expression of ETs. This would mean that endogenous ETs are released during or after cold lesion and their action is weakened. Tissue ET-1 was increased 1–5 days after cold lesion in the rat (Yamada et al., 1995), and high plasma levels of ETs were found in humans after stroke or subarachnoid hemorrhage (Ziv et al., 1992; Ehrenreich et al., 1992) and in rats after cold lesion (Görlach et al., 1998c). Moreover, we have found that tachyphylaxis develops to the dilating component of the ET-3 response in the BA of normal animals during repeated application of cumulative CEC curve. Tachyphylaxis was also found *in situ* using microapplication of the same dose of the specific ET_B agonist BQ 3020 to feline pial arteries. Furthermore, a loss of ET_B-receptor mediated vasodilation also occurs after focal ischemia (Touzani et al., 1997). The ET_B-receptor mRNA expression in the BA was not altered 48 h after the neurotrauma, at a timepoint when the functional impairment had reached its maximum, rendering the possibility of a down-regulation of the ET_B-receptor at the transcriptional level very unlikely. In agreement with the loss of ET_B-mediated vasodilation in BA 24 h and 48 h after cortical cold lesion, however, we found a delayed reduction of ET_B-receptor immunoreactivity in BA and in brain stem intraparenchymal vessels that coincides with the functional alterations in BA reactivitity to ET-3.

In the model of a cold lesion in the parietal cortex, the tone of the BA was apparently not influenced by a component of the endogenous endothelin system. This is evident from the fact that BQ 123, a specific ET_A antagonist, had no effect per se in the isolated BA after cold lesion and under control conditions. Similarly, BQ 788, a specific ET_B antagonist, has no effect per se in the isolated BA under control conditions, a finding which was also obtained by Touzani et al. (1977) in feline pial arteries. Since the ET_B receptors were down-regulated after cold lesion their contribution to the resting tone of BA could not be investigated. The data showing no effect of an ET_A antagonist are in apparent contradiction to a study of Armstead, 1996. This author used the model of a fluid percussion injury in the newborn pig and found that local application of BQ 123 diminished the constriction of pial arteries. Similarly, in the cat model of a focal cerebral ischemia, Patel et al. (1996) found that perivascular application of ET antagonists dilates pial arteries. The apparent discrepancy between these studies and the present one may be explained by the different timepoint of measurement, experimental design, species, size and location of the artery under investigation: the BA, as a large cerebral artery, is remote from the primary cold lesion in the parietal cortex. We found not only an unaltered contractile response of the BA to ET-1, ET-3 but also to other contractile agents, i.e. 124 mM K^+ and 5-HT, in lesioned and sham-operated animals. In addition, a normal dilating response to ACh or SNP demonstrate an intact vasomotor responsiveness of this vessel except ET_B-receptor stimulation.

ACKNOWLEDGMENT

This work was supported by BMBF. The authors are grateful to Mrs. E. Held for expert secretarial help and to Mrs. J. Specker for excellent technical assistance.

REFERENCES

Armstead WM (1996) Role of endothelin in pial artery vasoconstriction and altered responses to vasopressin after brain injury. J Neurosurg 85:901–907.

Barone FC, Globus MY-T, Price WJ, White RF, Storer BL, Feuerstein GZ, Busto R, and Ohlstein EH (1994) Endothelin levels increase in rat focal and global ischemia. J Cereb Blood Flow Metab 14:337–342.

Ehrenreich H, Lange M, Near KA, Anneser F, Schoeller LAC, Schmid R, Winkler PA, Kehrl JH, Schmiedek P, and Goebel F-D (1992) Longterm monitoring of immunoreactive endothelin-1 and endothelin-3 in ventricular cerebrospinal fluid, plasma, and 24 h-urine of patients with subarachnoid hemorrhage. Res Exp Med 192:257–268.

Ehrenreich H and Schilling L (1995) New developments in the understanding of cerebral vaso-regulation and vasospasm: the endothelin-nitric oxide network. Cleveland Clin J Med 62:105–116.

Feger GI, Schilling L, Ehrenreich H, and Wahl M (1994) Endothelin-induced contraction and relaxation of rat isolated basilar artery: Effect of BQ-123. J Cereb Blood Flow Metab 14:845–852.

Görlach C, Benyó Z, and Wahl M (1998a) Endothelin-1 induced contraction in cerebral vessels mediated by PLC/PKC cascade. Kidney International 54 (Supp. 67):00–00.

Görlach C, Sirén A-L, Knerlich F, Feger G, Fricke A, Barth M, Schilling L, Ehrenreich H, and Wahl M (1998b) Delayed loss of ET_B-receptor-mediated vasorelaxation after cold lesion of the rat parietal cortex. J Cereb Blood Flow Metab (in press).

Görlach C, Benyó Z, and Wahl M (1998c) Reduced reactivity of the middle cerebral artery and its large branches after cold lesion. J Neurotrauma (in press).

Patel TR, Galbraith S, McAuley MA, and McCulloch J (1996) Endothelin-mediated vascular tone following focal cerebral ischaemia in the cat. J Cereb Blood Flow Metab 16:679–687.

Sairanen TR, Lindsberg PJ, Brenner M, and Sirén A-L (1997) Global forebrain ischemia results in differential cellular expression of interleukin-1β (IL-1β) and its receptor at mRNA and protein level. J Cereb Blood Flow Metab 17:1107–1120.
Sambrook J, Fritsch EF, and Maniatis J (1989) Molecular Cloning: A Laboratory Manual. 2nd ed. Cold Spring Harbor Laboratory Press.
Schilling L, Feger GI, Ehrenreich H, and Wahl M (1995) Endothelin-3-induced relaxation of isolated rat basilar artery is mediated by an endothelial ET_B-Type endothelin receptor. J Cereb Blood Flow Metab 15:699–705.
Sharkey J, Ritchie IM, and Kelly PAT (1993) Perivascular microapplication of endothelin-1: a new model of focal cerebral ischaemia in the rat. J Cereb Blood Flow Metab 13:865–871.
Touzani O, Galbraith S, Siegl P, and McCulloch J (1997) Endothelin-B receptors in cerebral resistance arterioles and their functional significance after focal cerebral ischemia in cats. J Cereb Blood Flow Metab 17:1157–1165.
Yamada G, Hama H, Kasuya Y, Masaki T, and Goto K (1995) Possible sources of endothelin-1 in damaged rat brain. J Cardiovasc Pharmacol 26:S486–S490.
Ziv I, Fleminger G, Djaldetti R, Achiron A, Melamed E, and Sokolovsky M (1992) Increased plasma endothelin-1 in acute ischemic stroke. Stroke 23:1014–1016.

12

REGULATION OF CEREBRAL OXYGEN DELIVERY

Fahmeed Hyder,[1] Robert G. Shulman,[2] and Douglas L. Rothman[1]

[1]Department of Diagnostic Radiology
[2]Department of Molecular Biophysics and Biochemistry
Yale University, New Haven, Connecticut

1. INTRODUCTION

Under normal conditions, it is generally accepted that for the adult mammalian cortex, almost all the energy required for cerebral ATP generation is supplied by oxidation of glucose through the tri-carboxylic acid cycle (Siesjö, 1978), and cerebral metabolic rates of oxygen and glucose use (i.e., CMR_{O2} and CMR_{glc}, respectively) are regionally modified to fulfil metabolic needs through regulation of cerebral blood flow (CBF) and volume (CBV) (Roy and Sherrington, 1890). However, some positron emission tomography (PET) results have reported that during brain activation of awake humans CBF increases by a greater fraction than CMR_{O2} (Fox and Raichle, 1986; Fox et al., 1988). In the last few decades, ever since *in vivo* cortical measurements could be made of CBF and/or CMR_{O2}, the mechanisms for the proportionality of changes between CBF and CMR_{O2} due to physiological perturbations have remained intangible, and the measured stoichiometries have varied reflecting a variety of experimental methods and/or conditions. However, regionally tight relationship between CMR_{O2} and CBF has been observed at rest with a ratio of about ~1 : 1 (Kety et al., 1947–8; Raichle et al., 1976).

In a recent publication, Buxton and Frank (1997) have presented a model in which they extended the Renkin-Crone model of solute delivery (Crone, 1963; Renkin, 1959) to show that the fractional increases of CBF and CMR_{O2} during cerebral activation, as in the PET data of Raichle and coworkers (Fox and Raichle, 1986; Fox et al., 1988), actually represents a mechanistic limitation in the ability to enhance oxygen delivery. The two main assumptions in their model are that cerebral tissue oxygen tension is low such that the capillary-tissue oxygen partial pressure (pO_2) gradient is determined primarily by the capillary oxygen tension, and oxygen delivery can only be increased through perfusion.

Oxygen Transport to Tissue XXI, edited by Eke and Delpy.
Kluwer Academic / Plenum Publishers, New York, 1999.

Their assumption that there is no change in the effective diffusivity of oxygen (D) from blood to tissue with increased perfusion, indicates that the capillary bed has little flexibility with respect to changes in perfusion. This model predicts a highly nonlinear relationship between changes in CBF and CMR_{O2}, because a large increase in CBF is required to achieve a small increase in the capillary pO_2. Although the model is able to fit the PET data of Raichle and coworkers where a ratio of about 0.2:1 was reported for relative changes in CMR_{O2} and CBF (Fox and Raichle, 1986; Fox et al., 1988), reported changes in CBF and CMR_{O2} from additional PET activation studies of awake humans (Roland et al., 1987; Seitz and Roland, 1992) are not allowed by the Buxton and Frank model. A further limitation of their model is that the predicted highly non-linear relationship between CBF and CMR_{O2} does not agree with the close to linear relationships between CBF and CMR_{O2} measured across ranges of graded anaesthesia in rats and humans (Hyder et al., 1998a).

Here we present an extended model of cerebral oxygen delivery where certain characteristics are derived from a number of recent models (Bassingthwaighte, 1974; Buxton and Frank, 1997; Crone, 1963; Gjedde, 1997; Hyder et al., 1998a; Pries et al., 1996; Renkin, 1959; Siesjö, 1978; Walley, 1996). The present model relaxes one of the constraints in the model of Buxton and Frank by allowing for changes in effective diffusivity, D, of the capillary bed with changes in CBF. By allowing for change in D, which can be manifested via alterations in capillary oxygen tension, hematocrit, and/or blood volume, oxygen delivery is allowed to be more responsive to perfusion. It is important to point out that the model of Buxton and Frank is the limiting case of the present model in which the ratio of $(\Delta D/D)/(\Delta CBF/CBF)$ is equal to zero. We present *in vivo* data of CBF and CMR_{O2} from rat and human cortex under various levels of activity to test the predictions of the current model. The ability of the current model to fit the wide range of data indicates that the diffusivity properties of the capillary bed plays an important role in regulating cerebral oxygen delivery *in vivo*.

2. THE MODEL

Since the equilibrium between deoxy- and oxyhemoglobin (Hb and HbO_2) can be shifted by altering the blood oxygenation, Y, the oxyhemoglobin dissociation curve describes the oxygen carrying property of blood,

$$Y = C_{Hb(O2)n}/(C_{Hb(O2)n} + C_{Hb}) \tag{1}$$

where $C_{Hb(O2)n}$ and C_{Hb} are the contents of oxy- and deoxyhemoglobin, respectively, in a particular volume of blood, and the oxyhemoglobin concentration, $C_{Hb(O2)n}$, is equal to the sum of deoxyhemoglobin, C_{Hb}, and oxygen, C_{O2}.

$$C_{Hb(O2)n} = C_{Hb} + nC_{O2} \tag{2}$$

Oxygen is transported by blood in two discrete components, hemoglobin and plasma, and because the former is the main oxygen carrier in blood it also has the higher oxygen affinity. Since oxygen extraction into the tissue from an infinitesimally thin element of blood occurs all through capillary transit, the changing temporal profile of the total oxygen content of blood (C_T) can be described in relation to the oxygen content in the plasma (C_P) by

$$dC_T/d\tau = -kC_p \tag{3}$$

where the C_T is equivalent to $C_{Hb(O2)n}$, C_P is equivalent to C_{O2}, and the constant k is determined by the spatial gradients of oxygen tension across the volume element, and is the first-order rate constant of oxygen loss from the capillary (Buxton and Frank, 1997; Crone, 1963; Gjedde, 1991; 1997; Hyder et al., 1998a; Renkin, 1959). The main assumption in equation [3] is that while oxyhemoglobin is the main carrier of oxygen in blood, once oxygen is released into the plasma it is very rapidly transported across the capillary endothelium (Clark et al., 1985) and almost completely metabolized (Kassissia et al., 1995). If it is assumed that through an elapsed time of $\Delta\tau$ the ratio of transient oxygen content values in plasma and blood is constant, i.e., $r = C_P/C_T$, then it can be shown that

$$C_T(\tau + \Delta\tau) = C_P(\tau)\exp(-kr\Delta\tau) \tag{4}$$

which is equivalent to

$$C_T(T_c) = C_P(0)\exp(-[kr]_{net}\,T_c) \tag{5}$$

where $\Delta\tau$ is the mth equivalent fraction of the capillary transit time, T_c, and $[kr]_{net}$ is the net kr product for the whole transit time, T_c. Since r is determined by the relationship between hemoglobin fractional oxygenation and the average capillary oxygen tension, from equations [1] and [2],

$$r = (1/n)(2 - 1/Y) \tag{6}$$

Owing to the cooperativity of oxygen binding, r will vary with τ in this current model, which is in contrast to the model of Buxton and Frank. From the use of permeability surface area product in the Renkin-Crone relationship and assumption that the oxygen tension is negligible throughout the brain extracellular and intracellular space in the Buxton and Frank model, the resistivity can be localized to the capillary endothelium in both cases. Although equation [5] is similar to the Renkin-Crone relationship and the Buxton and Frank model, in the current model the resistance of oxygen transport is localized to a range of physiological episodes which occur in various locations and/or phases of oxygen transport, which together are termed effective diffusivity for oxygen (see equation [10]), rather than one particular site and/or period (Hyder et al., 1998a). Oxygen extraction fraction around the capillary may be calculated from (Crone, 1963; Renkin, 1959; Siesjö, 1978)

$$OEF_c = (C_T(0) - C_T(T_c))/C_T(0) \tag{7}$$

where $C_T(0) > C_T(T_c)$, and given equation [5] OEF_c is also equivalent to

$$OEF_c = 1 - \exp(-[kr]_{net}\,T_c) \tag{8}$$

Since at steady-state, perfusion is constant at all points along the capillary so that T_c may be calculated from the relationship (Stewart, 1894; Meier and Zierler, 1954)

$$T_c = CBV_c/CBF_c \tag{9}$$

which by substitution leads to

$$OEF_c = 1 - \exp(-D_c/CBF_c) \tag{10}$$

where D_c is defined as the effective diffusivity for oxygen from the capillary to the mitochondria which is the point of efficient consumption (Clark et al., 1985; Kassissia et al., 1995). Therefore, the total oxygen extraction per capillary can be depicted by the summation of the effect of an infinitesimally thin element of blood flowing down a capillary during transit as it fails to maintain oxygen, and as a consequence of which the flowing blood goes through changing spatial gradients of oxygen tension between the traversing volume element and the abluminal side of the capillary endothelium. The changing values of C_T and C_P in the traversing volume reflects the cooperative binding of oxygen by hemoglobin in whole blood within the capillary (see equation [6]). The macroscopic picture is derived through averaging across a group of identical capillaries: effective diffusivity of the capillary bed (D), transit time of the capillary bed (T), perfusion in the capillary bed (CBF), oxygen extraction fraction of the tissue (OEF), and oxygen consumption in the tissue (CMR_{O2}) such that:

$$OEF = 1 - \exp(-D/CBF) \tag{11}$$

Equation [11] is similar to previous analytical expressions of a microvascular tissue unit which is composed of a mass of tissue that is irrigated by a collection of similar capillaries that are uniformly separated (Bassingthwaighte, 1974; Buxton and Frank, 1997; Crone, 1963; Gjedde, 1991; 1997; Renkin, 1959; Walley, 1996), although the interpretation of the diffusivity constant D is somewhat different here (Hyder et al., 1998a). It has been shown previously that for a distribution of transit times an expression similar to equation [11] is valid provided the distribution is reasonably symmetrical and peaked about the average value (Pries et al., 1996; Walley, 1996). An alternate expression for OEF determined from Fick's equation is given by (Siesjö, 1978)

$$OEF = CMR_{O2}/(CBF\,C_a) \tag{12}$$

where C_a is the average capillary arterial oxygen concentration in the bed. The unidirectional flow of oxygen molecules across the endothelium is very efficient (Kassissia et al., 1995; Lenigert-Follert, 1975; Metzger and Heuber, 1977) and this process is driven by the average capillary oxygen tension. Combination of equations [11] and [12] shows that

$$CMR_{O2} = (CBF\,C_a)(1 - \exp(-D/CBF)) \tag{13}$$

Equation [13] shows that oxygen utilization is linked to perfusion via the effective diffusivity of the capillary bed. By rearrangement of equation [12] the relationship between fractional changes in CMR_{O2}, CBF, and OEF may be expressed as

$$\Delta CMR_{O2}/CMR_{O2} = (1 + \Delta CBF/CBF)(1 + \Delta OEF/OEF) - 1 \tag{14}$$

where in equation [14], and below, the terms without and with the "Δ" indicates basal and relative differences, respectively, due to a physiologic perturbation. In the present analysis, the effective diffusivity for oxygen within the capillary bed is assumed to be coupled to perfusion according to a parameter Ω which we define as

$$\Omega = (\Delta D/D)/(\Delta CBF/CBF) \tag{15}$$

The term Ω is a measure of the coupling between the changes in effective diffusivity of the capillary bed for oxygen delivery. If the value of Ω is constant over the range of CBF changes then equation [15] may be substituted into equation [14] to yield

$$\Delta CMR_{O2}/CMR_{O2} = \alpha[1 - \exp\{-(D/CBF)(\beta/\alpha)\}]/OEF - 1 \tag{16}$$

where $\alpha = (1 + \Delta CBF/CBF)$, $\beta = (1 + (\Delta CBF/CBF)\Omega)$, and $\Delta OEF/OEF$ is given by

$$\Delta OEF/OEF = [1 - \exp\{-(D/CBF)(\beta/\alpha)\}]/OEF - 1 \tag{17}$$

Provided that $\Delta CBF/CBF$, Ω, and the basal value of OEF are known, the values of $\Delta CMR_{O2}/CMR_{O2}$ and $\Delta OEF/OEF$ may be calculated using the above relationships. It is also useful to define a parameter which measures the coupling between changes in oxidative metabolism and perfusion which we define as

$$\eta = (\Delta CMR_{O2}/CMR_{O2})/(\Delta CBF/CBF) \tag{18}$$

For meaningful comparison of η values obtained at various levels of activity, it is convenient to compare with the awake state. However, different stimuli affect CBF and CMR_{O2} in an arbitrary nature from a particular resting condition (Siesjö, 1978), and therefore, comparison of η values obtained at various levels of activity may not reflect an exact view of oxygen delivery within the autoregulation range. To remove the unknown effect of physiological perturbations on CBF and/or CMR_{O2}, η in equation [18] can be normalized by a unitless constant

$$\xi = [(CMR_{O2}\text{ (alter state)}/CBF\text{(altered state)}]/[CMR_{O2}\text{ (awake)}/CBF\text{(awake)}] \tag{19}$$

where by

$$\Psi = \eta/\xi \tag{20}$$

Depending on Ω and the basal value of OEF, the calculated value of Ψ may either be close to constant over the physiological range of changes in perfusion or highly nonlinear as proposed by Buxton and Frank allowing a wide range of data to be fitted. As such the model does not validate any particular reported value of Ψ, only reliable experiments can do that.

3. ASSUMPTIONS OF THE MODEL

1 If $\Delta\tau$ is the *m*th equivalent fraction of the capillary transit time, T_c, then within that time the rate of oxygen loss from an infinitesimally thin element of blood flowing through the capillary can be described by a first-order rate constant k, during which time the ratio C_p/C_T $(=r)$ is constant.

2 Oxygen delivery can be influenced by the both CBF and D of the capillary bed (see equation [13]), since it has been demonstrated that the physiological capacity for oxygen diffusivity of the capillary bed can be altered by local capillary

oxygen tension, hematocrit, and/or blood volume (Ducharme et al., 1991; Federspiel and Sarelius, 1984; Moss et al., 1976).

3 The assumption that oxygen extraction is directly proportional to the capillary oxygen tension (see equation [7]) is supported by the reported low cerebral oxygen tension values (Lenigert-Follert, 1975; Metzger and Heuber, 1977), and by a recent tracer study which showed efficient diffusivity and metabolism of oxygen (Kassissia et al., 1995). In the current model, the low cerebral oxygen tension is maintained over the range of autoregulation by modulation of both CBF and Ω (see equation [17]).

4 An ensemble of capillary transit-times, lengths, or volumes, which can be a consequence of topological and/or geometric heterogeneity of the capillary network in the microvascular tissue unit (Pries et al., 1996), gives rise to a distribution of Ω values about a mean. A symmetrical distribution of Ω values about a mean would produce equivalent observations (Pries et al., 1996).

5 Other assumptions are similar to those defined and stated previously (Bassingthwaighte, 1974; Buxton and Frank, 1997; Crone, 1963; Gjedde, 1991; 1997; Hyder et al., 1998a; Pries et al., 1996; Renkin, 1959; Siesjö, 1978; Walley et al., 1996). In particular, [a] the brain is assumed to be well perfused by plasma and hemoglobin, but the capillary bed has the capacity to change its local oxygen capacity (Ducharme et al., 1991; Federspiel and Sarelius, 1984; Homer et al., 1981; Moss et al., 1976) by altering the number of plasmatic capillaries (Abounader et al., 1995; Tajima et al., 1992; Vetterlein et al., 1990; Villringer et al., 1994) or intra-capillary stacking of erythrocytes (Federspiel and Sarelius, 1984; Homer et al., 1981; Moss et al., 1976; Stewart, 1894; Villringer et al., 1994); [b] oxygen is assumed to be carried in the blood by plasma and hemoglobin and the exchange between these pools is extremely fast such that the oxygen saturation curve represents the equilibrium of the exchange process (Buxton and Frank, 1997; Gjedde, 1991; 1997); [c] the assumption of topological and geometric homogeneity of the capillary bed (Crone, 1963; Gjedde, 1997; Pries et al., 1996; Renkin, 1959; Walley et al., 1996) reflects symmetric topology and geometry of metabolism in tissue and is supported by mitochondrial aggregation around capillaries (Krogh, 1919; Kreuzer, 1982; Walley, 1996); [d] the assumption of the plasma pool of oxygen being well mixed (Buxton and Frank, 1997; Crone, 1963; Gjedde, 1997; Pries et al., 1996) suggests an enhancement of oxygen transfer by an element of moving blood and indicates a steady-state picture of oxygen extraction as this element transits (Bassingthwaighte, 1974; Buxton and Frank, 1997).

4. DATA AND SIMULATIONS

All human data and most rat data were obtained from the literature. Absolute values of CMR_{O2} and CBF are shown in Figures 1(A) and (B) for the rat and human cortices, respectively, for various levels of activity. In Figures 1(A), the rat data for the conditions of graded anesthesia (●), awake and activation (□), and seizure (▲) were obtained from the literature (Ginsberg et al., 1987; Hägerdal et al., 1975; Meldrum and Nilsson, 1976; Nilsson and Siesjö, 1975; 1976, Ueki et al., 1988), and our laboratory by nuclear magnetic resonance methods (Hyder et al., 1996; 1997; 1998b). In Figure 1(B), the human data for the conditions of graded anesthesia (●), awake and activation (□),

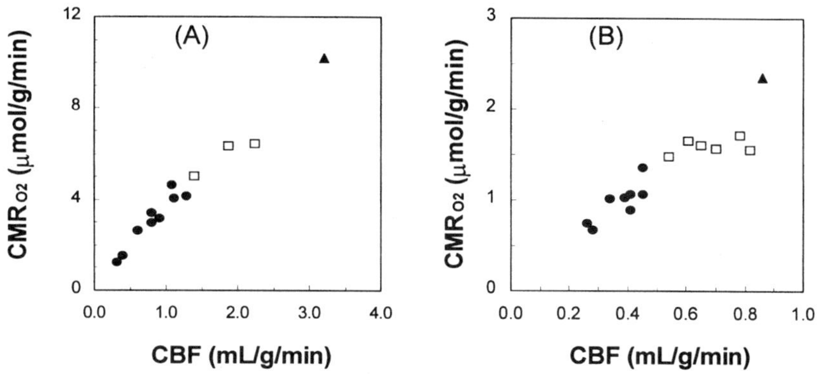

Figure 1. Absolute values of CMR_{O2} and CBF for various levels of activity were obtained from the literature for the cerebral cortex of the rat (A) and human (B). In (A), the graded anesthesia data (●) were obtained from Hägerdal et al. (1975), Meldrum and Nilsson (1976), Nilsson and Siesjö (1975; 1976), Ueki et al. (1988), and Hyder et al. (1996; 1997; 1998b); the awake and activation data (□) were obtained from Nilsson and Siesjö (1975), Ginsberg et al. (1987), Ueki et al. (1988), and Hyder et al. (1996; 1997); and the seizure data (▲) were obtained from Meldrum and Nilsson (1976). For the data from Ginsberg et al. (1987), CMR_{O2} was calculated from CMR_{glc} based on the stoichiometry described in Hyder et al. (1996; 1997). In (B), the graded anesthesia data (●) were obtained from Pierce et al. (1962) and Smith and Wollman (1972); the awake and activation data (□) were obtained from Davis et al. (1998), Fox and Raichle (1986), Fox et al. (1988), Kety et al. (1947–8), Roland et al. (1987), and Seitz and Roland (1992); and the seizure data (▲) were obtained from Siesjö (1978). The *in vivo* data were used to determine values of Ψ (see equation [20]) and the mean values of Ψ were 0.91 ± 0.47 ($n = 18$) and 0.94 ± 0.56 ($n = 20$), respectively.

1986; Fox et al., 1988; Kety et al., 1947–8; Pierce et al., 1962; Roland et al., 1987; Seitz and Roland, 1992; Smith and Wollman, 1972). All simulations were carried out in MATLAB (Natick, MA). The *in vivo* data were used to determine values of Ψ (see equation [20]). The mean values of Ψ were 0.91 ± 0.47 ($n = 18$) and 0.94 ± 0.56 ($n = 20$), respectively. For each data point, a Ω value was calculated with its respective Ψ value, using equation [21], at basal OEF of 0.3,

$$\Omega = \left(1 + (\Delta CBF/CBF)^{-1}\right)(\vartheta/\rho) - (\Delta CBF/CBF)^{-1} \tag{21}$$

where $\vartheta = \ln[1 - OEF(\gamma/\alpha)]$, $\rho = \ln[1 - OEF]$, $\alpha = (1 + \Delta CBF/CBF)$, and $\gamma = (1 + (\Delta CBF/CBF)\Psi)$. A mean Ω value was obtained that would accommodate the wide range of activity covered by the *in vivo* data. The point of death from the awake state, where $(\Delta CMR_{O2}/CMR_{O2})/(\Delta CBF/CBF)$ or η is equal to -1, was included (by five times the weighting as any other point) for the determination of a mean value of Ω such that the point of break from autoregulation could be visualized by the model's predictions. In each case, the mean value of Ω was used in equation [16] to determine the model's predictions of $\Delta CMR_{O2}/CMR_{O2}$ as a function of $\Delta CBF/CBF$. The fits were then compared to the *in vivo* data. For comparison to the extreme case, fits were also created with $\Omega = 0$, as in the case of the Buxton and Frank model.

5. RESULTS

Mean values of Ω for the rat and human data that were used for the fits are 0.64 and 0.69, respectively. The *in vivo* rat and human data of $\Delta CMR_{O2}/CMR_{O2}$ vs.

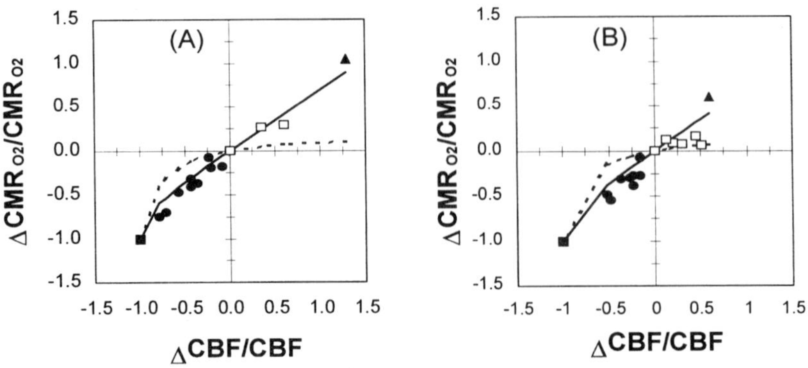

Figure 2. From the data shown in Fig. 1 for rat and human cerebral cortex on CMR_{O2} and CBF coupling, mean values of 0.64 and 0.69 were calculated for Ω, respectively, for the entire range of activity levels shown. The *in vivo* rat (A) and human (B) data of $\Delta CMR_{O2}/CMR_{O2}$ vs. $\Delta CBF/CBF$, and the model predictions of $1 > \Omega > 0$ (———) and $\Omega = 0$ (- - - - -) are shown in Figs. 2(A) and 2(B), respectively. In both cases, the awake, non-anesthetized, resting state was used as the control condition for the other data points, and the symbols ●, □, ▲, and ■ represent the conditions of graded anesthesia, awake and activation, seizure, and death, respectively. The fits to the *in vivo* data are clearly better when $1 > \Omega > 0$ (———), showing how the flexibility in the current model allows these well established results to be explained. In contrast, the Buxton and Frank approach of $\Omega = 0$ (- - - - -) cannot be extended below the awake level.

$\Delta CBF/CBF$, and the model predictions of $\Omega > 0$ (——) and $\Omega = 0$ (- - - - -) are shown in Figures 2(A) and 2(B), respectively. The fits to the *in vivo* data are clearly better when $\Omega > 0$, showing how the flexibility in the current model allows these well established results to be explained. In contrast, the Buxton and Frank approach cannot be extended below the awake level.

6. DISCUSSION

Several but not all (Davis et al., 1998; Roland et al., 1987; Seitz and Roland, 1992) human brain functional data have found that during brain activation $\Delta CBF/CBF \gg \Delta CMR_{O2}/CMR_{O2}$ (Fox and Raichle, 1986; Fox et al., 1988). However, the finding of $\Delta CBF/CBF \gg \Delta CMR_{O2}/CMR_{O2}$ during a functional challenge of the awake brain had been interpreted as an uncoupling between perfusion and oxidative metabolism (Fox and Raichle, 1986; Fox et al., 1988; Raichle, 1987). Recently, Buxton and Frank (1997) proposed that the greater fractional increase in CBF than in CMR_{O2} during activation is a consequence of oxygen delivery being proportional to the pO_2 gradient between the capillary and the tissue. The Buxton and Frank model predicts a highly non-linear relationship with low proportionality ratios between changes in CMR_{O2} and CBF at low basal OEF values due to the limited ability of perfusion to increase the capillary-tissue pO_2 gradient. Also, Gjedde (1997) has presented an alternative model in which oxygen delivery is limited by the maximum partial pressure difference achievable between the capillary plasma and the oxygen consuming mitochondria. Although no specific barrier to oxygen diffusion is required in the Gjedde model for a low starting OEF, the ability of CBF to increase oxygen delivery to the tissue is similarly limited. Here we extend this approach by considering the effect of a changing effective diffusivity of the capillary bed for oxygen. As signified by a physiologic parameter Ω (see equation [15]), for a fixed ratio between changes in effective diffusivity of the capillary bed and perfusion, a close to constant proportionality is maintained between changes in CBF and CMR_{O2} throughout the physiologic activity range (Siesjö, 1978). The current model provides better fits to the majority of *in vivo* rat and human data of changes in CBF and CMR_{O2} than does the model of Buxton and Frank, which is the limiting case when $\Omega = 0$.

As equation [16] shows, an increased Ω elevates the ratio between $\Delta CMR_{O2}/CMR_{O2}$ and $\Delta CBF/CBF$ (Hyder et al., 1998a). A significant assumption in the current model is that changes in perfusion and effective diffusivity of the capillary bed are tightly regulated in order to maintain a low cerebral oxygen tension, and as a consequence of which, oxygen delivery is primarily determined by the capillary oxygen tension. The assumption that capillary-tissue pO_2 gradient is proportional to the capillary oxygen tension is quite likely because very low cerebral oxygen tensions have been observed *in vivo* (Lenigert-Follert, 1975; Metzger and Heuber, 1977). Furthermore, most of the oxygen which efficiently enters the tissue (Clark et al., 1985) is almost completely metabolized (Kassissia et al., 1995). Even if the cerebral oxygen tension were not negligible, provided that it was maintained at a constant and lower value than the capillary oxygen tension, similar trends would be observed. In this alternative case, changes in oxygen delivery would still be determined predominantly by changes in the capillary oxygen tension and the coupled effective diffusivity for oxygen of the capillary bed (Crone, 1963; Renkin, 1959).

Although the *in vivo* data were obtained from large volumes of tissue, a microscopic characterization should be able to predict the macroscopic observations. The microscopic picture of this model is the ability of the capillary bed to increase effective diffusivity for oxygen, which can be manifested by increase in capillary oxygen tension, hematocrit, and/or blood volume (Ducharme et al., 1991; Federspiel and Sarelius, 1984; Homer et al., 1981; Moss et al., 1976) during increased brain activity. Increased CBV in the activated cortex has been observed (Fox and Raichle, 1986; Fox et al., 1988; Roland et al., 1987; Seitz and Roland, 1992), and this finding supports the Roy and Sherrington (1890) proposal that changes in CBF and CBV regulate changes in CMR_{O2} and CMR_{glc}. The swelling of capillary diameter during altered brain function (Atkinson et al., 1990; Cox et al., 1993; Narain et al., 1995; Romaines et al., 1993; Villringer et al., 1993; 1994) variation in the number of plasmatic and hematic capillaries (Abounader et al., 1995; Tajima et al., 1992; Vetterlein et al., 1990; Villringer, 1994) and/or intra-capillary stacking of erythrocytes (Federspiel and Sarelius, 1984; Homer et al., 1981; Moss et al., 1976; Stewart, 1894; Villringer et al., 1994) can contribute to the changing in capillary oxygen tension, hematocrit, and/or blood volume. All these findings support the view that changes in the number of plasmatic vs. hematic capillaries or intra-capillary stacking of erythrocytes can modulate oxygen delivery and demand. Since local changes in capillary oxygen tension, hematocrit, and/or blood volume, would modify the effective diffusivity for oxygen of the capillary bed, it is not necessary to assume that there is absolutely no capillary readjustment within a microvascular tissue unit, as did Buxton and Frank (1997). While the term capillary readjustment generally reflects, to a certain extent, increases in the blood volume of the capillary bed, the term presented here suggests a modifying capillary bed with respect to capillary oxygen tension, hematocrit, and/or blood volume. Owing to the generally better fits obtained by the current model to *in vivo* data, compared with the case of no increase in effective diffusivity of the capillary bed, suggests that either one or all mechanisms may be operational in increasing this parameter *in vivo*.

7. CONCLUSION

Recent models of cerebral oxygen delivery (Buxton and Frank, 1997; Gjedde, 1997) have suggested that at low basal values of OEF oxygen delivery is limited by the inefficiency of perfusion to raise capillary oxygen tension. The current model weakens this limitation by allowing changes in the capillary effective diffusivity to oxygen, D, with changes in CBF. The change in D with respect to CBF, signified by Ω, allows oxygen delivery to be more sensitive to changes in perfusion. An appropriate value of Ω asserts a close to ~1 : 1 ratio between changes in CBF and CMR_{O2} throughout a wide range of activity. The model of Buxton and Frank in which $\Omega = 0$ is the limiting case of the current model. This model shows that when $\Omega > 0$, the limitation imposed by oxygen delivery on increases in CMR_{O2} predicted by the Buxton and Frank model does not hold within the physiological range. In contrast to the Buxton and Frank model, where the data could only fit data where $\Omega = 0$, the current model can fit a wide range of data so that it makes no claims to establishing the relation between changes in CBF and CMR_{O2}, only good data can do that. The present model is able to fit all available *in vivo* data, and depicts the transition of hemodynamic and metabolic events for a large range of physiological cases. This characteristic of the model signifies that the oxygen diffusivity properties of the capillary bed, which is adjusted with respect to perfusion, plays a crucial part in regulating cerebral oxygen delivery *in vivo*.

ACKNOWLEDGMENTS

This work was supported by NIH grants DK-27121 (RGS), and NS-32126 (DLR) and NSF grant DBI-9730892 (FH). FH would like to acknowledge Drs. Kevin L. Behar, Richard P. Kennan, Edward J. Novotny, and Ognen A.C. Petroff for support and advice.

REFERENCES

Abounader R, Vodel J, and Kuschinsky W (1995) Patterns of capillary plasma perfusion in brains of conscious rats during normocapnia and hypercapnia. *Circ Res* 76:120–126.

Atkinson JLD, Anderson RE, and Sundt Jr. TM (1990) The effect of carbon dioxide on the diameter of brain capillaries. *Brain Res* 517:333–340.

Bassingthwaighte JB (1974) A concurrent flow model for extraction during transcapillary passage. *Circ Res* 35:483–503.

Buxton RB and Frank LR (1997) A model of the coupling between cerebral blood flow and oxygen metabolism during neural stimulation. *J Cereb Blood Flow Metab* 17:64–72.

Clark Jr. A, Federspiel WJ, Clark PAA, and Cokelet GR (1985) Oxygen delivery from red cells *Biophys. J* 47:171–181.

Cox SB, Woolsey TA, and Romaines CM (1993) Localized dynamic changes in cortical blood flow with whisker stimulation corresponds to matched vascular and neuronal architecture of rat barrels. *J Cereb Blood Flow Metab* 13:899–913.

Crone C (1963) The permeability of capillaries in various organs as determined by the "indicator diffusion" method. *Acta Physiol Scand* 58:292–305.

Davis TL, Kwong KK, Weisskoff RM, and Rosen BR (1998) Calibrated functional MRI: Mapping the dynamics of oxidative metabolism. *Proc Natl Acad Sci USA* 95:1834–1839.

Ducharme R, Kapadia P, and Dowden J (1991) A mathematical model of the flow of blood cells in fine capillaries. *J Biomech* 24:299–306.

Federspiel WJ and Sarelius IH (1984) An examination of the contribution of red cell spacing to the uniformity of oxygen flux at the capillary wall. *Mircovasc Res* 27:273–285.

Fox PT and Raichle ME (1986) Focal physiological uncoupling of cerebral blood flow and oxidative metabolism during somatosensory stimulation in human subjects. *Proc Natl Acad Sci USA* 83:1140–1144.

Fox PT, Raichle ME, Mintun MA, and Dence C (1988) Nonoxidative glucose consumption during focal physiologic neural activity. *Science* 241:462–464.

Ginsberg MD, Dietrich WD, and Busto R (1987) Coupled forebrain increases of local cerebral glucose utilization and blood flow during physiologic stimulation of a somatosensory pathway in the rat: Demonstration by double-label autoradiography. *Neurology* 37:11–19.

Gjedde A (1997) The relation between brain function and cerebral blood flow and metabolism. In: *Cerebrovascular Disease* (Batjer HH, Ed) Lippincott-Raven, Philadelphia, pp. 23–40.

Gjedde A (1991) Is oxygen diffusion limiting for blood-brain transfer of oxygen? In: *Brain Work and Mental Activity, Alfred Benzon Symposium 31* (Lassen NA, Ingvar DH, Raichle ME, Friberg L, Eds) Munksgaard, Copenhagen, pp. 177–184.

Hägerdal M, Hamp JR, Nillson L, and Siesjö BK (1975) The effect of induced hypothermia upon oxygen consumption in rat brain. *J Neurochem* 24:311–316.

Homer LD, Weathersby PK, and Kiesow LA (1981) Oxygen gradients between red blood cells in the microcirculation. *Mircovasc Res* 22:308–232.

Hyder F, Chase JR, Behar KL, Mason GF, Siddeek M, Rothman DL, and Shulman RG (1996) Increased tricarboxylic acid cycle flux in rat brain during forepaw stimulation detected with ^{1}H-[^{13}C] NMR. *Proc Natl Acad Sci USA* 93:7612–7617.

Hyder F, Rothman DL, Mason GF, Rangarajan A, Behar KL, and Shulman RG (1997) Oxidative glucose metabolism in rat brain during single forepaw stimulation: a spatially localized ^{1}H[^{13}C] NMR Study. *J Cereb Blood Flow Metab* 17:1040–1047.

Hyder F, Shulman RG, and Rothman DL (1998a) A model for the regulation of cerebral oxygen delivery. *J Appl Physiol* 85:554–564.

Hyder F, Kennan RP, Sibson NR, Mason GF, Behar KL, Rothman DL, and Shulman RG (1998b) Cerebral oxygen delivery *in vivo*: NMR measurements of CBF and CMR_{O2} at different levels of brain activity. *Soc Magn Reson Med Abstr* p. 1160.

Kassissia IG, Goresky CA, Rose CP, Schwab AJ, Simard A, and Huet PM (1995) Tracer oxygen distribution is barrier-limited in cerebral microcirculation. *Circ Res* 77:1201–1211.

Kety SS, Woodford RB, Harmel MH, Freyhan FA, Appel KE, and Schmidt CF (1947–8) Cerebral blood flow and metabolism in schizophrenia. The effect of barbiturate semi-narcosis, insulin coma and electroshock. *Amer J Psychiat* 104:765–770.

Kreuzer F (1982) Oxygen supply to tissues: the Krogh model and its assumptions. *Experimentia* 38:1415–1426.

Krogh A (1919) The number and the distribution of capillaries in muscles with the calculation of the oxygen pressure head necessary for supplying the tissue. *J Physiol (Lond)* 52:409–41.

Lenigert-Follert E (1975) Direct determination of local oxygen consumption of the brain cortex in vivo. *Plügers Arch* 372:175–179.

Meier P and Zierler KL (1954) On the Theory of the indicator-dilution method for measurement of blood flow and volume. *J Appl Physiol* 6:731–744.

Meldrum BS and Nilsson B (1976) Cerebral blood flow and metabolic rate early and late in prolonged seizures induced in rats by bicuculline. *Brain* 99:523–542.

Metzger H and Heuber S (1977) Local oxygen tension and spike activity of the cerebral grey matter of the rat and its response to short intervals of O_2 deficiency or CO_2 excess. *Plügers Arch* 370:201–209.

Moss GS, Dewoskin R, Rosen AL, Levine H, and Palani CK (1976) Transport of oxygen and carbon dioxide by hemoglobin-saline solution in the red-cell-free primate. *Surg Gynecol Obstet* 142:357–362.

Narain SM, Esfahani P, Blood AJ, Sikkens L, and Toga AW (1995) Functional increases in cerebral blood volume over somatosensory cortex. *J Cereb Blood Flow Metab* 15:754–765.

Nilsson B and Siesjö BK (1975) The effect of phenobarbitone anaesthesia on blood flow and oxygen consumption in the rat brain. *Acta Annaesthesiol Scand Suppl* 57:18–24, 1975.

Nilsson B and Siesjö BK (1976) A method of determining blood flow and oxygen consumption in rat brain. *Acta Physiol Scand* 17:273–282.

Pierce Jr EC, Lambersten CL, Deutsch S, Chase PA, Linde HW, Dripps RD, and Price HL (1962) Cerebral circulation and metabolism during thiopental anaesthesia and hyperventilation in man. *J Clin Invest* 41:1664–1671.

Pries AR, Secomb TW, and Gaehtgens P (1996) Relationship between structural and hemodynamic heterogeneity in microvascular networks. *Am J Physiol* 270:H545–H553.

Raichle ME, Grubb RL, Gado MH, Eichling JO, and Ter-Pogossian MM (1976) Correlation between regional cerebral blood flow and oxidative metabolism. *Arch Neurol* 33:523–526.

Raichle ME (1987) Circulatory and metabolic correlates of brain function in normal humans. In: *Handbook of Physiology. The Nervous System. Higher Functions of the Brain.* Bethesda, MD: AM Physiol Soc, sect 1, vol V, pt. 2, chapt 16, pp. 633–674.

Renkin EM (1959) Transport of potassium-42 from blood to tissue in isolated mammalian skeletal muscle. *Am J Physiol* 197:1205–1210.

Roland PE, Eriksson L, Atone-Elander S, and Widen L (1987) Does mental activity change the oxidative metabolism of the brain? *J Neurosci* 7:2373–2389.

Romaines CM, Woolsey TA, Blocher NC, Wang D-B, and Robinson OF (1993) Blood flow in single surface arterioles and venules on the mouse somatosensory cortex measured with videomicroscopy, flourescent dextrans, non-occluding flourescent beads, and computer-assisted image analysis. *J Cereb Blood Flow Metab* 13:359–271.

Roy CS and Sherrington CS (1890) On the regulation of the blood supply of the rat brain. *J Physiol (Lond)* 11:85–108.

Seitz RJ and Roland PE (1992) Vibratory stimulation increases and decreases the regional cerebral blood flow and oxidative metabolism: a positron emission tomography (PET) study. *Acta Neurol Scand* 86:60–67.

Siesjö BK (1978) *Brain Energy Metabolism.* Wiley, New York.

Smith AL and Wollman H (1972) Cerebral blood flow and metabolism. *Anesthesiology* 36:378–400.

Stewart GN (1894) Researches on the circulation time in organs and on the influences which affect it: parts I–III. *J Physiol (Lond)* 15:1–27.

Tajima A, Nakata H, Lin S-Z, Acuff V, and Fenstermacher J (1992) Differences and similarities in albumin and red blood cell flows through microvessels. *Am J Physiol* 262:H1515–H1524.

Ueki M, Linn F, and Hossmann K-A (1988) Functional activation of cerebral blood flow and metabolism after global ischemia of rat brain. *J Cereb Blood Flow Metab* 8:486–494.

Vetterlein F, Demmerle B, Bardosi A, Göbel U, and Schmidt G (1990) Determination of capillary perfusion pattern in rat brain by timed plasma labelling. *Am J Physiol* 258:H80–H84.

Villringer A, Planck J, Hock C, Schleinkofer L, and Dirnagl U (1993) A new tool to study hemodynamic changes during activation of brain function in human adults. *Neurosci Lett* 154:101–104.

Villringer A, Them A, Lindauer U, Einhäul K, and Dirnagl U (1994) Capillary perfusion in rat brain cortex. *Circ Res* 75:55–62.

Walley KR (1996) Heterogeneity of oxygen delivery impairs oxygen extraction by peripheral tissues: theory. *J Appl Physiol* 81:885–894.

13

A LINK BETWEEN VASOMOTION AND SPONTANEOUS OSCILLATIONS OF OXYGEN IN RAT BRAIN

D. M. Karkan, C. van Breemen, P. L. Skarsgard, G. J. Lagaud, and D. F. Hardwick

Department of Pharmacology and Therapeutics and Department of Pathology
The University of British Columbia

Early recordings with oxygen electrodes show that under steady state conditions, tissue oxygen tension (Po_2) undergoes spontaneous fluctuations (Clark, 1958; Cooper et al., 1966; Hardwick, 1996; Misrahy et al., 1962). These oscillations differ between organs and species but a general pattern of 1–10 c/min with an amplitude of 10–30% of the mean Po_2 has been observed (Morita-Tsuzuki et al., 1992; Hasibeder et al., 1994; Braun et al., 1992; Eickhoff and Jacobsen, 1982; Towell et al., 1976; Manil et al., 1984). Recordings of oscillations in flowmotion by other methods such as laser Doppler and magnetic resonance imaging (MRI) (Beinder et al., 1994; Morita-Tsuzuki et al., 1992; Hudetz et al., 1992; Biswal and Hudetz, 1996; Ren et al., 1993; Biswal et al., 1995, 1996) appear to exhibit similar frequencies of oscillation as the Po_2.

While the exact mechanism of the regulation of these oscillations remains unclear, it has been hypothesized that arteriolar vasomotion is an important factor in formation of these fluctuations (Tsai et al., 1993; Braun et al., 1992; Hasibeder et al., 1994). Studies by intravital microscopy suggest that order three arterioles exhibit significant vasomotion at a frequency of the order of 0.1 Hz (Bertuglia et al., 1996). This frequency is similar to fluctuations in local blood flow recorded by laser doppler flowmeter. However, no simultaneous recording of Po_2 oscillation by an oxygen electrode and vasomotion by intravital microscopy has been conducted to link vasomotion and Po_2 oscillations. Therefore, although vasomotion has been assumed as the main cause of spontaneous changes in flowmotion, its secondary effects on tissue oxygen tension remain unknown. Detailed experiments to explore this issue have been difficult to conduct (Silver, 1978).

The ongoing debate about the role of vasomotion in tissue oxygenation and its mechanism of action emphasizes the significance of analyzing the factors involved in the generation of these oscillations. In the present study, we sought to provide further evi-

Oxygen Transport to Tissue XXI, edited by Eke and Delpy.
Kluwer Academic / Plenum Publishers, New York, 1999.

dence to support the direct vascular control of Po_2 fluctuations. The main purpose of this report was to link Po_2 fluctuations *in vivo* with spontaneous vasomotion of pressurized isolated arterioles *in vitro*. We chose the rat brain as we could record oxygen tension continuously without extensive anesthesia. Parallel to these experiments, isolated arterioles dissected from the measurement sites were examined *in vitro* under pressurised conditions. The frequency of spontaneous oscillations and the effects of nifidipine on their formation were compared between *in vitro* and *in vivo* preparations.

MATERIAL AND METHODS

Animals and Surgery

Animal experiments were approved by the University of British Columbia animal care unit. Wistar rats (250–300 g) were anesthetized with pentobarbitone (60 mg/kg). After reflecting the cranial soft tissue a hole was made in the frontal skull with a dental drill (Pronto Dental Equipment, OH) to access the cerebral cortex. Animals were placed in the supine position in a temperature-regulated (37°C) tubular restrainer (Harvard Instruments, MASS) 30 minutes before the measurements. In some experiments a catheter (Baxter Healthcare, IL, CA) was placed in the jugular artery. This catheter was inserted in the artery and connected to a blood pressure transducer. Anesthesia was maintained at a low dose of 0.2 mg/kg and no ventilation was necessary.

Po_2 Measurements

The animals recovered from anesthesia 30 minutes before oxygen tension measurements. We determined the values of oxygen tension using a polarographic O_2-sensitive gold needle electrode (Eppendorf, Hamburg, Germany). Probe diameter was 350 µm, response time 2 seconds, and sensitivity: 6 ± 3.0 PA. Calibrations were performed in 0.9% saline solution equilibrated with calibration gases immediately before and after the measurements. Gold electrode was introduced, and recordings were made at a depth of 2–4 mm from the surface of the brain tissue.

Pressurised Arterioles

Middle cerebral and small branches of the femoral artery were obtained from adult male Sprague-Dawley rats (200–250 g). The animals were anaesthetized with intraperitoneal injection of sodium pentobarbital (Somnotol, 30 mg/kg) and heparin (Hepalean, 100 IU/kg), and killed by decapitation. The brain was removed and immersed in cold oxygenated physiologic saline solution (PSS). A segment of artery (0.6–1.0 mm long) was carefully dissected from surrounding connective tissue and transferred to the experimental chamber of a pressure myograph filled with oxygenated PSS at 20°C. One vessel was taken from each experimental animal. The proximal aspect of the artery was fed onto a glass microcannula (tip diameter ~60 µm), and tied with a single strand (20 µm) of braided 4-O nylon suture. After flushing the artery with PSS to remove intraluminal blood, the distal aspect of the vessel was similarly cannulated and tied. The intraluminal pressure was gradually increased to 60 mm Hg with an electronic pressure servo system (Living Systems, Burlington, VT), the bath temperature was slowly increased to 37°C, and the vessel was then equilibrated for 60 minutes. During this equilibration period, vessels spontaneously and reliably developed myogenic tone, with significantly reduced luminal diameters. Once attained, myogenic tone and vessel diameter remained stable unless perturbed by changes in transmural pressure or the addition of vasoactive compounds. The PSS entered the experimental chamber at a flow rate of 20 ml/min, passing through an external reservoir that was bubbled with 95% O_2–5% CO_2. A heating pump connected to a heat exchanger maintained the PSS at 37°C. Buffer temperature and pH were monitored by microprobes situated in the tissue bath: pH was maintained at 7.40 ± 0.04 U by adjustment of the gassing rate. The arteriograph containing a pressurized artery was placed on the stage of an inverted microscope with a monochrome video camera attached to a viewing tube. Arterial dimensions were measured using a video system that provides automatic continuous digital measurements of luminal diameter and wall thickness. The information is updated every 17 ms, and the precision of the diameter measurements is within 1% (Nilsson and Sjoblom, 1984).

Drug Treatment

Effect of nifedipine (0.1 mg/kg) iv was examined on tissue oxygen tension. Recordings of PO_2 were made 2 minutes after the injection. Effect of the same agent (0.01 μM–1 μM) was used to determine the dependence of spontaneous vasomotion on plasmalemmal Ca^{2+} influx.

Statistical Analysis

Statistical differences between data sets in different experimental conditions were analyzed by Student's t-test, multiple regression, and analysis of variance (ANOVA). All data obtained through the Po_2 electrode were processed by Fourier statistical analysis software (Coulbourn Instruments, PA.). In the case of drug administration, effects of changes in blood pressure were compensated by measuring the dependence of these oscillations on blood pressure.

RESULTS

Fluctuations of EEG recordings during Po_2 measurements did not correlate with those obtained by oxygen electrode and did not remain constant under control conditions. Overall, no correlation among rhythmic changes in systemic measurements and in vivo Po_2 oscillations, was observed.

In a group of 8 animals, the oxygen tension in brain was 43 ± 5.7 mm Hg. The oxygen tension was similar in all animals if anesthesia was not too deep and if 30 minutes were allowed for recovery from surgery. Figure 1a shows an example of the fluctuations in the brain. These fluctuations were complex exhibiting superimposed smaller oscillations but their amplitude was usually relatively constant. The oscillations had a main frequency of 10 c/min and amplitude of 15 ± 5.7% of mean value. A secondary frequency of 24 c/min was observed. Figure 1b shows the relative spectrum analysis of Po_2 values recorded in the cortex.

Small arterioles identified near the tip of the oxygen electrode were dissected and examined *in vitro*. The average diameter of isolated pressurized arterioles was measured at 20 mm Hg. Cerebral arteries had an average diameter of 126 ± 6 μm (n = 5). During the initial 1/2 hour stabilization period, at the pressure range of 60–120 mm Hg, arteri-

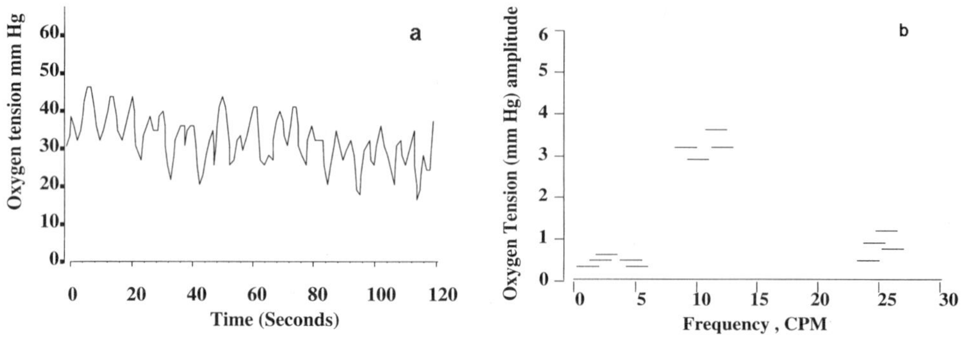

Figure 1. A representative recording of oxygen tension by Eppendorf oxygen electrode at 4 mm depth in rat brain cortex. a) Oscillations of oxygen tension recorded during 2 minutes. b) Fast Fourier transform spectrum shows a main frequency of 12 c/min with amplitude 20 ± 5.7% of mean value.

oles exhibited spontaneous vasomotion. The frequency of these oscillations matched those observed in vivo. Figure 2 shows a representative tracing of spontaneous oscillations in diameter of pressurized brain arterioles.

Injection of Ca^{2+} antagonist (nifidipine) *in vivo* decreased the frequency of fluctuations by 60.3 ± 2% and it also caused vasodilation and attenuation of spontaneous vasomotion amplitude in isolated pressurized cerebral arteries (Figure 3).

DISCUSSION

Frequency and amplitude of oscillations observed in our study correlate with previous observations of spontaneous changes in oxygen tension in brain (Misrahy et al., 1962). However, the frequency of Po_2 oscillations do not match the fluctuations recorded by other methods such as laser-Doppler flowmetry (Bertuglia et al., 1996; Biswal and Hudetz, 1996; Linden et al., 1995; Morita-Tsuzuki et al., 1992).

Records of vasomotion and changes in blood flow as measured by laser-Doppler

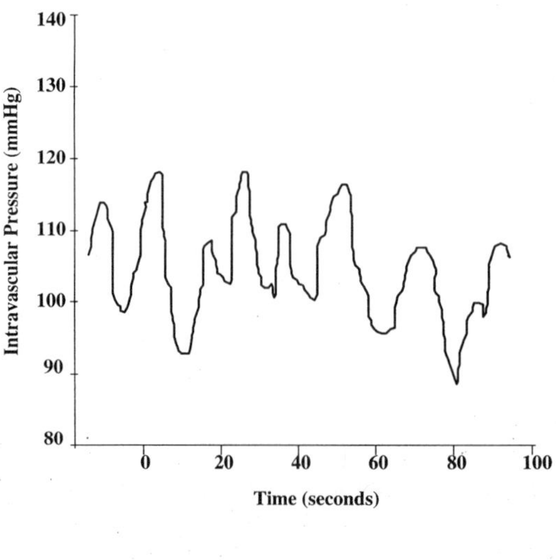

Figure 2. A representative recording of spontaneous vasomotion of middle cerebral arteries of adult male rats. The proximal aspect of the artery was fed onto a glass microcannula (tip diameter ~60 µm), and tied with a single strand (20 µm) of braided 4-O nylon suture, the distal aspect of the vessel was similarly cannulated and tied. The intraluminal pressure was gradually increased to 60 mm Hg, the bath temperature was 37°C, and the vessel was then equilibrated for 60 minutes.

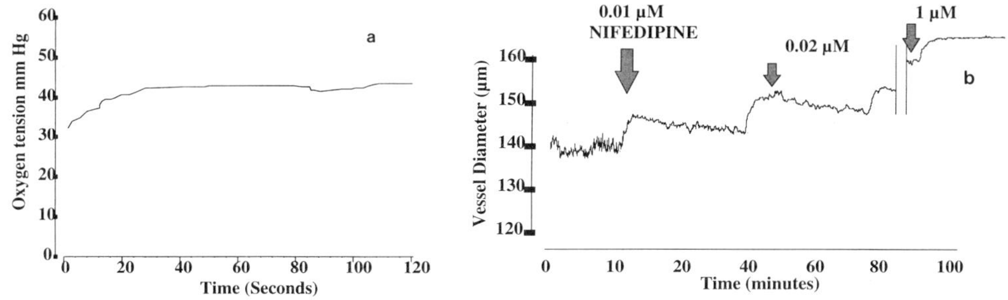

Figure 3. Effect of calcium antagonist on spontaneous oscillations in brain. a) Po_2 oscillations in brain was abolished 2 minutes after the injection of 1 mg/kg nifidipne. b) Spontaneous vasomotion in isolated pressurized arterioles was abolished by 0.01 µM–1 µM addition of nifidipine.

flowmetry appear to be more rhythmic than the Po_2. This difference may depend on the sampling volume used by each method. Because the sampling volume of laser-Doppler flowmeter are not well defined, a detailed comparison between the number of microvessels in each sampling volume is not possible.

Tissue oxygen tension as measured in our experiments was limited to the area assessed by the tip of the Po_2 electrode. It is known that this electrode consumes a large quantity of oxygen (Brizel et al., 1994). Therefore, the Po_2 measured by this electrode may not be accurate. However, measurements with larger (Misrahy et al., 1962) or smaller (Braun et al., 1992; Manil et al., 1984; Biswal et al., 1997) electrodes show similar frequency of fluctuations in Po_2.

The gold electrode used in our study caused minimum vascular disruption, as shown by histological studies which failed to show signs of bleeding at the measurement sites or incursion channels. The zone of measurement thus produces a mix data from the extracellular space and immediately adjacent extravascular O_2 (Vaupel, 1995). It is interesting to note that although previous investigators have used smaller (Braun et al., 1992) or larger (Misrahy et al., 1962) electrodes, the frequency and amplitude of O_2 oscillations reported in those studies correlate well with ours.

The main finding in the present study is a correlation between in vivo measurements of oxygen and the in vitro spontaneous vasomotion. Our results show that pressurized arterioles undergo spontaneous vasomotion at the same frequency as Po_2 oscillations in vivo. A correlation between frequency of Po_2 oscillations and individual arteriolar vasomotion rate has not been reported previously. However, reports on frequency of vasomotion of order 3 arterioles observed via intravital microscopy are similar to our recordings (Colantuoni et al., 1984). This resemblance provides further support for direct vasculature control of Po_2 oscillations.

In our study, Ca^{2+} antagonists abolished these fluctuations in vivo as well as in vitro. As shown previously, spontaneous activity of smooth muscle cells can be blocked by omission of Ca^{2+} in the superfusate (Harder et al., 1987; Hayashida et al., 1987). We also observed that isolated arterioles did not undergo vasomotion at pressures above or below the physiological range, supporting the concept that a vascular pacemaker acts as a local oscillator that can be affected by intraluminal pressure in a graded fashion (Bouskela, 1989).

In conclusion, our study supports the hypothesis that fluctuations in extracellular Po_2 are secondary to arteriolar vasomotion. Our study is also consistent with the hypothesis of pacemaker control of vasomotion. Correlation between in vivo and in vitro results in this study suggests that arteriolar vasomotion can affect several aspects of mass transport including oxygen transport. Further experimental work is required to determine whether and under what circumstances consequences of vasomotion become essential to other tissue dynamics and cellular homeostasis.

ACKNOWLEDGMENTS

We would like to thank Dr. A. I. Minchinton at B.C. Cancer Research Center for his generous assistance with the polarographic electrode. This study was supported by Medical Research Council of Canada No B1421.

REFERENCES

Beinder, E, Trojan, A, Bucher, HU Huch, A, and Huch, R (1994) Biology of the Neonate. 65:7–15.

Bertuglia, S, Colantuoni, A, Arnold, M, and Witte, H (1996) Microvascular Research. 52:235–244.

Biswal, B, DeYoe, AE, and Hyde, JS (1996) Magnetic Resonance in Medicine. 35:107–113.

Biswal, B, Hudetz, AG, Yetkin, FZ, Haughton, VM, and Hyde, JS (1997) Journal of Cerebral Blood Flow & Metabolism. 17:301–308.

Biswal, B, Yetkin, FZ, Haughton, VM, and Hyde, JS (1995) Magnetic Resonance in Medicine. 34:537–541.

Biswal, BB and Hudetz, AG (1996) Microvascular Research. 52:1–12.

Bouskela, E (1989) Microvascular Research. 37:339–351.

Braun, RD, Linsenmeier, RA, and Yancey, CM (1992) Microvascular Research. 44:73–84.

Brizel, DM, Rosner, GL, Harrelson, J, Prosnitz, LR, and Dewhirst, MW (1994) International Journal of Radiation Oncology, Biology, Physics. 30:635–642.

Clark, LC, Misrahy, G, Phyllis Fox (1958) Jour. Physiol. 13:85–91.

Colantuoni, A, Bertuglia, S, and Intaglietta, M (1984) American Journal of Physiology. 246:H508–H17.

Cooper, R, Crow, W, Walter, G, and Winter, AL (1966) Brain Research. 3:174–190.

Eickhoff, JH and Jacobsen, E (1982) Biotelemetry & Patient Monitoring. 9:175–184.

Harder, DR, Gilbert, R, and Lombard, JH (1987) American Journal of Physiology. 253:F778–F781.

Hardwick, DF (1996) Pathology International. 46:652.

Hasibeder, W, Germann, R, Sparr, H, Haisjackl, M, Friesenecker, B, Luz, G, Pernthaler, H, Pfaller, K, Maurer, H, and Ennemoser, O. (1994) American Journal of Physiology. 266:G978–G986.

Hayashida, N, Okui, K, and Fukuda, Y (1987) Japanese Journal of Physiology. 37:183–196.

Hudetz, AG, Roman, RJ, and Harder, DR (1992) Journal of Cerebral Blood Flow & Metabolism. 12:491–499.

Linden, M, Sirsjo, A, Lindbom, L, Nilsson, G, and Gidlof, A (1995) American Journal of Physiology. 269: H1496–H1500.

Manil, J, Bourgain, RH, Van Waeyenberge, M, Colin, F, Blockeel, E, De Mey, B, Coremans, J, and Paternoster, R (1984) Advances in Experimental Medicine & Biology. 169:231–239.

Misrahy, GA, Hardwick, DF, and Brooks, CJ (1962) American Journal of Physiology. 202:225–231.

Morita-Tsuzuki, Y, Bouskela, E, and Hardebo, JE (1992) Acta Physiologica Scandinavica. 146:431–439.

Nilsson, H and Sjoblom, N (1984) Journal of Hypertension—Supplement. 2:S103–S105.

Ren, TY, Nuttall, AL, and Miller, JM (1993) Acta Oto-Laryngologica. 113:609–614.

Silver, IA (1978). "Cellular microenvironment in relation to local blood flow, in: "cerebral vascular smooth muscle and its control", Ciba Foundation Symposium 56. Elsvier/expectra Medica/North, Amesterdam.

Towell, ME, Lysak, I, Layne, EC, and Bessman, SP (1976) Journal of Applied Physiology. 41:245–250.

Tsai, AG, Intaglietta, M, Braun, RD, Linsenmeier, RA, and Yancey, CM (1993) International Journal of Microcirculation: Clinical & Experimental. 12:75–88.

Vaupel, P (1995). In "Function analyse biologische system", Vol. 24, Plenum, Frankfort, pp. 211–218.

14

EVALUATION OF APOLIPOPROTEIN E GENOTYPES ON CEREBRAL BLOOD FLOW AND METABOLISM FOLLOWING TRAUMATIC BRAIN INJURY

Mary E. Kerr,[1] Marilyn Kraus,[2] Don Marion,[3] and Ilyas Kamboh[1]

[1]University of Pittsburgh
[2]Western Psychiatric Institute and Clinic
[3]University of Pittsburgh Medical Center
360 Victoria Building
3500 Victoria Street
Pittsburgh, Pennsylvania 15261

1. INTRODUCTION

For nearly 15 years, the link between a positive history of head injury and an increased risk for Alzheimer disease has been recognized (Heyman, Wilkinson, et al., 1984; Mortimer and French Huton, 1985). More recently, the Apolipoprotein E gene (APOE) has been identified as a major genetic factor predisposing one to Alzheimer's disease. It is this association between head injury and Alzheimer disease that initiated the hypothesis that one's genetic code, specficially, one's APOE allele, may influence the rehabilitation process following TBI.

Mayeux and colleagues (Mayeux, Ottman, et al., 1995; Mayeux, Ottman, et al., 1993), using a case-control design, found evidence that both a positive history of head injury and APOE4 allele increased one's risk for AD. They found a ten-fold increase in the odds ratio for Alzheimer disease (odds ratio = 10.5; 95%, c.i. 1.3–87.8) when both a history of head injury and the APOE4 genotype were present. These findings provide preliminary support for the notion of a synergistic response between an environment assault (i.e. head trauma) and the presence of APOE4.

Oxygen Transport to Tissue XXI, edited by Eke and Delpy.
Kluwer Academic / Plenum Publishers, New York, 1999.

APOE has also been linked to Aβ protein deposits and neurofibril tangles following head injury (Nicoll, Roberts, et al., 1996; Roberts, Allsop, et al., 1990). Evidence of a potential association between AD and head injury was demonstrated when amyloid deposits, traditionally found in the cerebral cortex of patients with AD, were also found in the patients with TBI. (Geddes, Vowles, et al., 1996; Roberts, Allsop, et al., 1990). Roberts and colleagues, (Roberts, Allsop, et al., 1990; Roberts, Gentleman, et al., 1991; Benveniste, Kwon, Sampson et al., 1994) in a series of studies detected extensive beta amyloid protein immunoreactive plaques in boxers as well as in patients with TBI. In approximately 30% of patients with fatal TBI, they found amyloid beta protein deposits. The majority of these patients (78.3%) carried an APOE4 allele (Graham, Gentleman, et al., 1996; Zumarelli, Nicoll, et al., 1996).

Nicoll's work also supported the presence of amyloid beta protein deposits following TBI. They reported that diffuse amyloid protein deposits were found in 45% of the patients with an APOE4 allele versus 10% of the patients without an APOE4 allele within the first two weeks following a severe head injury. He posited one explanation for this relationship: that the APOE4 genotype may influence the overexpression of brain's metabolic processes of amyloid beta as part of the acute phase response following TBI.

Yanker and others (Yanker and Mesulam, 1991; Goetting and Preston, 1990) also suggested that amyloid deposition contributes to neuronal degeneration by destroying the cell membrane and cytoskeleton and forming neuritic plaques. This genetic influence on amyloid plaque deposition and neural regeneration could confound or influence the cell's metabolism after an acute injury and has the potential to impact on short and long term outcomes. Evidence for this association was also demonstrated when neurofilimentary tangles, traditionally found in the cerebral cortex of patients with AD, were also found in the cortex of boxers who had no evidence of Alzheimer symptoms or pathology (Roberts, et al., 1990; Geddes, et al., 1996; Nicoll, et al., 1996). Nicoll et al. (1996) provided the first evidence that the presence of specific APOE genotypes has a gene dose effect on the response and recovery following head injury. Nicoll and colleagues examined the relationship between the deposition of Aβ and genotype in 90 nonsurvivors following a TBI. They found diffuse deposits of Aβ plaques in the cerebral cortex of 26% of the patients (n = 23). Individuals who were homozygous for APOE2 (2/2) had no deposits of Aβ while individuals who were homozygous for APOE4 (4/4) all had deposits of Aβ. When compared with the APOE4 allele frequency, they found a gene dose effect with Aβ deposition increasing as the number of APOE4 alleles increased. In the presence of the APOE4 allele, deposition of Aβ was more diffuse which may contribute to neurodegeneration following head injury. They also found that the APOE2 allele was associated with the lack of Aβ deposition and suggest that the APOE2 allele may have a neuroprotective effect.

Extending this association of APOE4 alleles to outcomes, Sorbi and others found that patients with APOE4 allele, in combination with a coma and/or intracerebral hemorrhage, were more likely to die (Sorbi, Macmias, et al., 1995; Poirer, Baccichet, et al., 1993). These studies provide early evidence that there may be a synergistic effect between the poor recovery following a head trauma and the presence of APOE4.

This study extends this association between one's APOE genotype and outcomes to examine the role of APOE genotype on factors that impact on outcomes of severe traumatic brain injury (TBI) with a Glasgow Coma Score of ≤8, specifically, cerebral blood flow and metabolism, during the acute phase following injury.

2. METHODS

The APOE genotypes were determined in 15 adults by genonmic DNA amplification and polymerase chain reaction and HhaI digestion. The patients were categorized based on the presence of the ApoE 4 allele.

2.1. ^{133}Xe CBF, avD_{O2} and CMR_{O2}

^{133}Xe CBF measurements were made using a mobile Cerebrograph (Novo 10a, Copenhagen, Denmark) at the patients' bedside in the intensive care unit (Obrist, Langfitt, et al., 1984). Ten detectors were placed symmetrically around the head (five on each hemisphere). The arterial and intracranial pressure was continuously monitored from an indwelling arterial and intraventricular catheter connected to transducers and a bedside monitor prior to and following the procedure. The peripheral arterial oxygen saturation was monitored continuously with a pulse oximeter. The end-tidal carbon dioxide was monitored and recorded on a chart recorder during the procedure.

The CBF was measured immediately following injection of a 25–35 mCi of ^{133}Xe intravenous bolus and 10 cc saline flush. After correction for recirculation and background activity, a two compartmental analysis of the ^{133}Xe clearance curves based on a modified Kety equation determined the regional, hemispheric and global CBF (Obrist, Thompson, et al., 1975) CBF was measured daily within the first five days after injury. The CBF values were corrected to a $PaCO_2$ of 35 mmHg assuming a 3% change in the global CBF per torr change in PaCO2.

The difference in the arterial and jugular venous O2 contents (avD_{O2}) was measured by inserting a catheter into the right jugular bulb and collecting simultaneous samples of jugular venous and radial arterial blood during the CBF study. Cerebral oxygen metabolism (CMRO2) was calculated as the product of the avjO2 and CBF based on the Fick equation.

2.2. APOE Genotyping

APOE genotyping was conducted using the polymerase chain reaction (PCR) method as decribed by Kamboh et al. (1995). DNA was isolated using a QIAGEN kit (Chatsworth, CA) and amplified using a forward (E1) and reverse primer (E2) through 40 cycles of denaturation (5 min at 95°C), annealing (1 min at 57°C) and extension (2 min at 72°C). The amplified DNA was digested with the restriction enzyme Hha I (BioLabs) and separated onto a polyacrylamide gel in 1 × Tris-borate buffer. Following staining with ethidium bromide solution, the APOE genotypes were typed by visualization under ultraviolet light.

2.3. Neurobehavioral Outcomes

The Disability Rating Scale (DRS) and the Glasgow Outcome Scale (GOS) were evaluated at 3 months post injury. On site neuropsychiatric evaluations were conducted at the patient's home or institution under the guidance of a neuropsychiatrist who was blinded to the patient's APOE genotype.

3. ANALYSIS

Statistical analyses were performed using a standard statistical software program, SPSS for Windows (SPSS 7.5 for Windows). Statistical analysis included descriptive statistics, and independent t-tests to describe the sample and to detect any medical or demostraphic differences between the two groups. Repeated measures analysis of variance were used to examined differences across in the CBF, avD_{O2} and CMR_{O2} during the first five days after injury between the two groups. Independent t-tests and chi-square analyses were used to detect significant differences in the neuropsychiatric outcomes by genotype.

4. RESULTS (Table 1, Figures 1–4)

Of the 15 subjects, 5 were females and 10 were males. There were 11 individuals genotyped with APOE 3/3 allele pairs and 4 subjects were genotyped with APOE 3/4 allele pairs. The average age was 41.3 (14.8) years. There was no significant differences in the age, best Glasgow Coma Score or type of head injury as indicated by the a cranial computed tomographic (CT) scan. Although the CT scans indicated that patients with APOE 3/4 allele pairs experienced a trend towards a greater amount of midline shift than those with APOE 3/3 allele pairs.

The global CBF during the first 24 hours for patients without an APOE4 allele was an average of 32.5 ml/100 g/minute (SD = 11.8) as compared to patients with an APOE4 allele with an average of 23.0 ml/100 g/minute (SD = 2.7). By 48 hours, there was no dif-

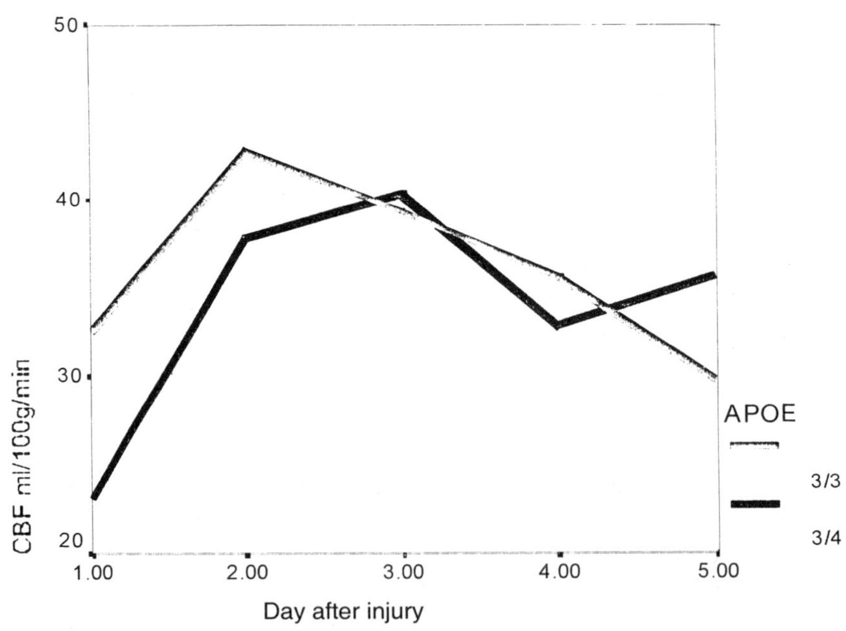

Figure 1. Comparison of CBF between patients with and without APOE4 alleles.

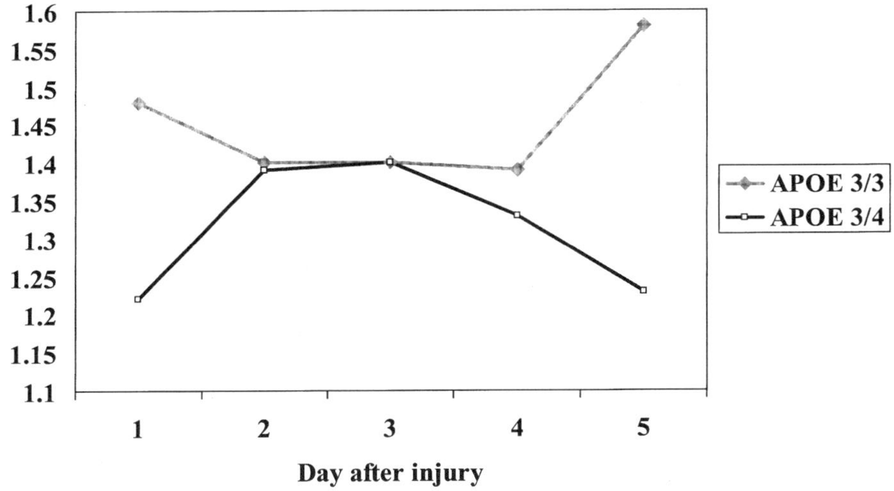

Figure 2. Comparison of CMRO2 between patients with and without APOE4 alleles.

ference in the CBF between the two groups. There was no significant difference in the CMR_{O2}. Three of the 4 patients with APOE4 alleles (75%) died as compared to 1 of the 11 patients without APOE4 alleles (9%). All 4 patients (100%) with APOE4 alleles had a poor outcome (defined as a GOS of 1 and 2) as compared to 4 of the 11 patients (36%)

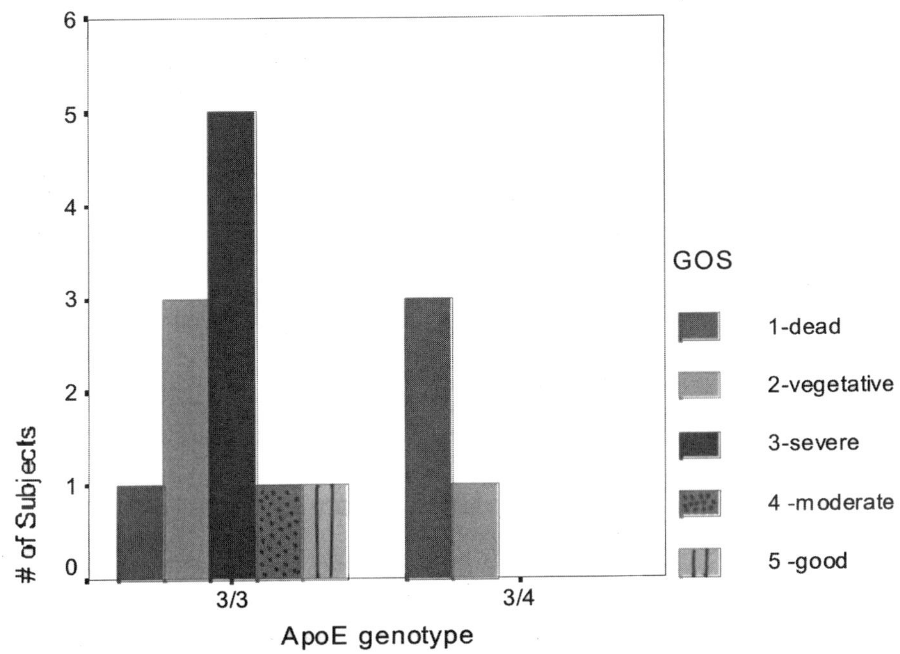

Figure 3. Comparison of the Glasgow Outcome Scale between patients with and without APOE4 alleles.

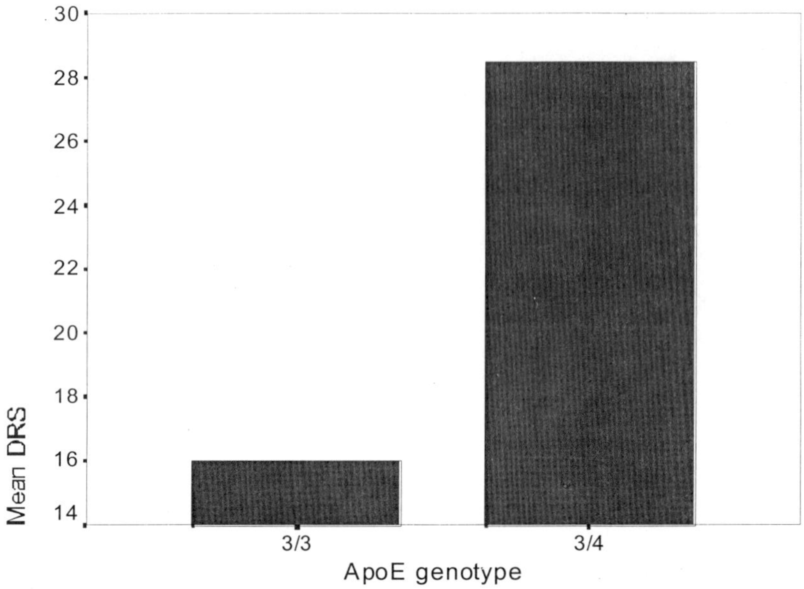

Figure 4. Comparison of the Disability Rating Scale between patients with and without APOE4 alleles.

Table 1. Comparison of demographics between patients with and without APOE4 alleles

Group Statistics

	APOE	N	Mean	Std. Deviation	Std. Error Mean
INITIAL_ INITIAL_GCS	33.00	11	5.00	1.26	.38
	34.00	4	4.50	1.29	.65
BEST_EYE	33.00	11	1.00	.00[a]	.00
	34.00	4	1.00	.00[a]	.00
BEST_VER BEST_VERBAL	33.00	11	1.00	.00[a]	.00
	34.00	4	1.00	.00[a]	.00
BEST_MOT BEST_MOTOR	33.00	11	2.45	1.29	.39
	34.00	4	2.50	1.29	.65
RPUPSZ R_PUPIL_SIZE	33.00	11	4.09	1.30	.39
	34.00	4	4.25	1.50	.75
LPUPSZ L_PUPIL_SIZE	33.00	11	3.64	1.50	.45
	34.00	4	3.75	1.50	.75
SAP INITIAL_BP_SYSTOLIC	33.00	9	136.89	43.44	14.48
	34.00	4	141.00	24.74	12.37
HR INITIAL_PULSE_RATE	33.00	9	90.22	22.13	7.38
	34.00	4	101.75	27.43	13.71

[a] t cannot be computed because the standard deviations of both groups are 0.

without APOE4 alleles. The mean DRS for patients with APOE4 alleles at 3 months was 28.3 (SD = 3.0) as compared to the mean DRS for patients without APOE4 alleles (M = 16.0) (SD = 8.6).

5. CONCLUSIONS

It is known that low CBF within the first 24 hours after TBI is related to morbidity. This study showed that there was a trend toward lower CBF in patients with APOE4 alleles without a difference in the CMRO2. There was, however, a difference in the outcomes, with the patients genotyped with an APOE4 allele having poorer outcomes. This study suggests that an association may exist between areas of low CBF immediately after head injury and APOE4 alleles ultimately impacting on long term outcomes. Further work in this area is warranted.

A potential genetic link between APOE and recovery following a TBI opens new areas for investigation. Of particular interest is, how the structural and functional effects of APOE genotypes affect the response and repair following injury. Understanding the allele specific interactions may provide the basis for future medical and pharmacologic interventions and treatment following a head injury. It raises tantalizing questions of whether response to treatment and nursing are genotype specific. For example, can more aggressive treatments be developed for those individuals whose neurons do not repair as readily as those individuals with neuroprotective genes? A better understanding of factors, such as APOE, that modify pathology in disease or injury to the brain will be critical to developing interventions to maximize recovery and improve functional outcome for these patients.

ACKNOWLEDGMENTS

We would like to thank W. Obrist, PhD, S. DeKosky, PhD, M. Butters, PhD, J. DeFeo, and A. Puccino, RN for their collaboration in the collection of data for this project.

REFERENCES

Benveniste EN, Kwon WJ, Sampson J, Pandya K, and Tang LP. Differential modulation of astrocyte cytokine gene expression by TGF-beta. *Journal of Immunology.* 1994;153(11):5210–21.

Geddes JF, Vowles GH, Robinson SF, and Sutcliffe JC. Neurofibrillary tangles, but not Alzheimer-type pathology in a young boxer. *Neuropathology Applied Neurobiology.* 1996;22:12–16.

Goetting MG and Preston G. Jugular bulb catheterization: Experience with 123 patients. *Critical Care Medicine.* 1990;18:1220–1223.

Graham DI, Gentleman SM, and Nicoll JA, et al. Altered beta-APP metabolism after head injury and its relationship to the aetiology of Alzheimer's disease. *Acta Neurochirurgica—Supplementum.* 1996;66:96–102.

Heyman A, Wilkinson WE, and Stafford JA, et al. Alzheimer's disease: a study of epidemiological aspects. *Annals of Neurology.* 1984;15:335–341.

Kamboh, MI, Aston, CE, and Hamman RF. The relationship of APOE polymorphism and cholesterol levels in normoglycemic and diabetic subjects in a biethnic population from the San Luis Valley, Colorado. Athersclerosis 1995;112:145–159.

Mayeux R, Ottman R, Maestre E, et al. Synergistic effects of traumatic head injury and apolipoprotein-eta4 in patients with Alzeheimer's disease. *Neurology.* 1995;45:555–557.

Mayeux R, Ottman R, and Tang M, et al. Genetic susceptibility and head injury as risk factors for Alzheimer's Disease among community-dwelling elderly persons and their first-degree relatives Annals of Neurology. *Annals of Neurology.* 1993;33:494–501.

Nicoll JR, Roberts GW, and Graham DI. Amyloid β-protein, APOE genotype and head injury. *Annals New York Academy of Sciences.* 1996:271–275.

Obrist WD, Langfitt TW, and Jaggi JL. Cerebral blood flow and metabolism in comatose patients with acute head injury. Relationship to intracranial pressure. *J Neurosurgery.* 1984;61:241–251.

Obrist WD, Thompson HK, Wang HS, and Wilkinson WE. Regional cerebral blood flow estimated by xenon inhalation. *Stroke.* 1975;6:245–256.

Poirer J, Baccichet A, Dea D, and Gauthier S. Cholesterol synthesis and lipoprotein reuptake during synaptic remodeling in hippocampus in adult rats. *Neuroscience.* 1993;55:81–90.

Roberts GW, Allsop D, and Bruton C. The occult aftermath of boxing. *Journal of Neurology, Neurosurgery and Psychiatry.* 1990;53:373–378.

Roberts GW, Gentleman SM, Lynch A, and Graham DI. β4 amyloid protein deposition in brain after head trauma. *Lancet.* 1991;338(1422):1422.

Sorbi S, Macmias B, and Piacentini S. ApoE as a prognostic factor for post-traumatic coma. *Nature Medicine.* 1995;1:852.

Yanker BA and Mesulam MM. Beta amyloid and the pathogenesis of Alzeheimer's disease. *New England Journal of Medicine.* 1991;26:1849–1855.

Zumarelli E, Nicoll JA, and Graham DI. Presenilin-1 polymorphism and amyloid beta-protein deposition in fatal head injury. *Neuroreport.* 1996;8(1):45–48.

CEREBRAL TISSUE OXYGEN SATURATION (SrO_2) FROM NEAR INFRARED SPECTROSCOPY (NIRS) MEASUREMENTS FOLLOWING 90°-HEAD-UP TILT

A. Martinez-Coll,[1] M. K. Morgan,[2] P. G. Cooper,[1] H. T. Nguyen,[3] and S. N. Hunyor[1]

[1]Cooperative Research Centre for Cardiac Technology
Block 4, Level 3 and
[2]Department of Neurosurgery
Royal North Shore Hospital
St. Leonards, NSW 2065 Australia
[3]University of Technology, Sydney
PO Box 123, Broadway, NSW 2007
Australia

1. INTRODUCTION

Changes in posture from supine to erect (head up) elicit changes in the hydrostatic pressure gradient acting on the systemic cerebral circulation. The initial effect of a head-up tilt (HUT) is a decrease in mean arterial pressure and a concomitant temporary decrease in blood flow (Patterson and Warren, 1952) which in turn triggers the appropriate myogenic mechanisms to maintain flow. At the venous end, the same changes in posture result in decreased blood volume.

Alterations in the concentrations of oxy- (HbO_2) and deoxy-hemoglobin (Hb) that result from tilting may be assumed to originate from a volume change in venous blood only, and can be used to calculate venous oxygen saturation (SvO_2) non-invasively using near infrared spectroscopy (NIRS) (Skov et al., 1993). Furthermore, regional cerebrovascular oxygen saturation (SrO_2), representative of an aggregate of arterial, venous, and capillary levels may be estimated using HUT to induce a change in cerebral venous blood volume which can be measured by NIRS.

NIRS is a non-invasive technique that allows monitoring of changes in oxygen saturation of haemoglobin and the redox state of cytochrome c oxidase. Like other forms

Oxygen Transport to Tissue XXI, edited by Eke and Delpy.
Kluwer Academic / Plenum Publishers, New York, 1999.

of oximetry, NIRS relies on the application of a modified Beer-Lambert law for highly scattering media such as biological tissues (Tamura et al., 1989). Visible light penetrates tissue very poorly, and it was not until 1977 that Jobsis explored the possibility of using infrared light in the range of 700–1000 nm to obtain information about "deep tissue" such as the brain (Brazy, 1991).

Several authors have utilised this technology to monitor cerebral hemodynamics and oxygenation in both neonates (Brazy et al., 1985; Cope and Delpy, 1988; Wyatt et al., 1990) and adults (Elwell et al., 1994; Owen-Reece et al., 1996; Pollard et al., 1996). Hemoglobin oxygen saturation can be determined from the ratio of HbO_2 to Hb by solving for chromophore content ratios, and allowing the variable path length to drop out of the calculation (Hoffman and Lubbers, 1985). Hemoglobin oxygen saturation is a clinically meaningful parameter, and in this case represents a measure of haemoglobin in the entire brain vasculature (McCormick et al., 1991).

The rate of oxygen uptake in brain is dependent on cerebral blood flow (CBF) and the difference between arterial and venous oxygen content (CaO_2-CvO_2) or arterial venous oxygen difference ($avDO_2$), written as:

$$\begin{aligned}\text{Oxygen uptake (or CMRO}_2) &= \text{CBF} \times (\text{CaO}_2 - \text{CvO}_2)\\ &= \text{CBF} \times (\text{avDO}_2)\end{aligned}$$

In most clinical conditions, the cerebral metabolic rate of oxygen consumption ($CMRO_2$) is thought to remain constant, while changes in $avDO_2$ provide valuable information regarding the oxygen supply-demand relationship. Oxygen contents in arterial and/or venous blood can be calculated from:

$$\text{oxygen content} = (\text{SO}_2 \times 1.34 \times [\text{Hbt}]) + (\text{PO}_2 \times .0003)$$

The second term on the right-side of the equation represents oxygen dissolved in plasma, which under relatively normal circumstances is small compared to that bound to hemoglobin and may, therefore, be ingored. SO_2 is the oxygen saturation (for either arterial or venous blood), 1.34 is the oxygen carrying capacity of hemoglobin, and [Hbt] is the normal hemoglobin concentration in blood (15 g/dL). From this relationship, arterial and venous oxygen saturations can be used to calculate $avDO_2$, and therefore, assess the adequacy of oxygen supply to the brain.

The aim of this study was to determine the proportional contribution of the cerebral venous vascular compartment following 90°-HUT on healthy volunteers as measured by near infrared spectroscopy (NIRS), and to calculate SrO_2 based on NIRS-derived SvO_2. This study describes a methodology to non-invasively "quantify" the cerebral venous blood volume contribution to the NIRS blood volume (BV) signal in each subject; allowing optimization of the calculated SrO_2. An estimate of SrO_2 coupled with arterial SO_2 from pulse oximetry can be used to provide a simple, continuous indicator of the oxygen supply-demand relationship.

2. MATERIAL AND METHODS

2.1. Subject Groups

Eleven healthy volunteers (6 males mean age 28.3, range 17–34 years and 5 females mean age 25, range 19–30 years) without a previous history of either cardiovascular or cerebrovascular insufficiency were recruited to the study following informed consent and approval from the Royal North Shore Hospital Human Research Ethics Committee.

2.2. Protocol

Each subject was placed in a supine position on a tilt table. The NIRS probe was positioned on the forehead along the midline of the eye, and secured under a rubber swimming cap. A 5-min rest period in the supine position was allowed before the table was tilted upright (90°) within 3 seconds. The upright position was maintained for a minimum of 15 seconds, after which the subject was returned to the horizontal position. The NIRS signal was allowed to stabilize and the procedure was repeated 3 times for each subject.

2.3. Near Infrared Spectroscopy (NIRS)

A prototype 30-mW variable gain, dual wavelength LED-NIRS instrument with a 4 cm optode separation was used. NIRS-measured signals at 769 nm and 850 nm were acquired throughout the procedure and stored for offline analysis. Venous blood volume (BV) contribution to the NIRS signal was calculated as change in the sum of the two signals at λ_{769} and λ_{850} from supine to erect or Δ BV $= \Delta[\text{Hbt}] \equiv \Delta\ [(\lambda_{850} + \lambda_{769})_{(\text{supine})} - (\lambda_{850} + \lambda_{769})_{(\text{head up})}]\ /\ [(\lambda_{850} + \lambda_{769})_{(\text{supine})}]$. Venous oxygen saturation (SvO_2) was estimated as

$$\begin{aligned} SvO_2 &= \Delta HbO_2/\Delta Hbt \text{ (or } \Delta\lambda_{850}/\Delta(\lambda_{850}+\lambda_{769}) \text{)} \\ &= \Delta[HbO2]_{\text{supine-head up}}/\Delta[Hbt]_{\text{supine-head up}} \\ &= \Delta\left[\lambda_{850(\text{supine-head up})}/(\lambda_{850}+\lambda_{769})_{(\text{supine-headup})}\right] \end{aligned} \tag{1}$$

while cerebral regional oxygen saturation (SrO_2) is calculated as

$$\begin{aligned} SrO_2 &= SvO_2 \times (\text{venous BV}) + SaO_2 \times (1 - \text{venous BV}) \\ &= \text{NIRS } SvO_2 \times (\text{NIRS\% venous}) + SaO_2 \times (1 - \text{NIRS\% venous}) \end{aligned} \tag{2}$$

SaO_2 was assumed at 98%.

2.4. Statistical Analysis

Means +/– SD were calculated for measurements of venous BV contribution, NIRS-derived SvO_2 as calculated from eq. 1, and SrO_2 as calculated from eq. 2 in all subjects (3 measurements in each subject; n = 33 in total). Means +/– SD were also obtained for SvO_2 and SrO_2 in both the supine and head up positions. Statistically significant differences between SvO_2 and SrO_2 were determined with paired t-tests. A value of $p = <0.05$ was considered significant.

3. RESULTS

The venous BV change from supine to head up (mean +/– SD) was 69.41 +/– 11.68%; while SvO_2, calculated from eq. 1 above, was 51.91 +/– 1.32%. Substitution of the venous BV and SvO_2 into eq. 2 yields an SrO_2 of 66.04 +/– 5.17%. Results are summarized in Table 1. Figure 1 (A) shows the spread of the acquired venous blood volume contribution data; while in (B) SvO_2 and SrO_2 are plotted together.

Table 1. Individual and group values for cerebral venous BV, SvO_2, and SrO_2.

Subject	Venous BV	SvO_2 (from Eq. 1)	SrO_2 (from Eq. 2)
1	80.81	51.68	60.58
2	60.57	52.09	70.19
3	60.30	52.85	70.77
4	70.81	51.40	65.00
5	73.64	55.00	66.33
6	63.4	51.05	68.24
7	81.42	51.66	60.27
8	82.22	51.61	59.86
9	76.47	51.20	62.21
10	43.96	50	76.89
11	70.06	52.46	66.09
Group values	69.41 +/-11.68 %	51.91 +/-1.32 %	66.04 +/-5.17 %

Means of three measurements in each subject are presented, as well as the mean +/– SD for all measurements (n=33).

SvO_2 was also calculated using HbO_2 and Hb in the supine and head-up positions by themselves (their ratio in either position: supine or head-up). For example, in the supine position, $SvO_{2(supine)} = HbO_{2(supine)}/Hbt_{(supine)}$ or $\lambda_{850(supine)}/(\lambda_{850} + \lambda_{769})_{supine}$, and similarly, for the head-up position. Although there was a statistically significant difference between SvO_2 as in eq.1 and $SvO_{2(supine)}$ ($p = <.0001$), when these values were substituted into eq. 2 for calculation of SrO_2, there was no difference between the two ($SrO_{2(eq.2)}$ = 66.04 +/– 5.17 while $SrO_{2(supine)}$ = 67.2 +/– 5.09; $p = .4715$). The same analysis was carried out for the "head-up"

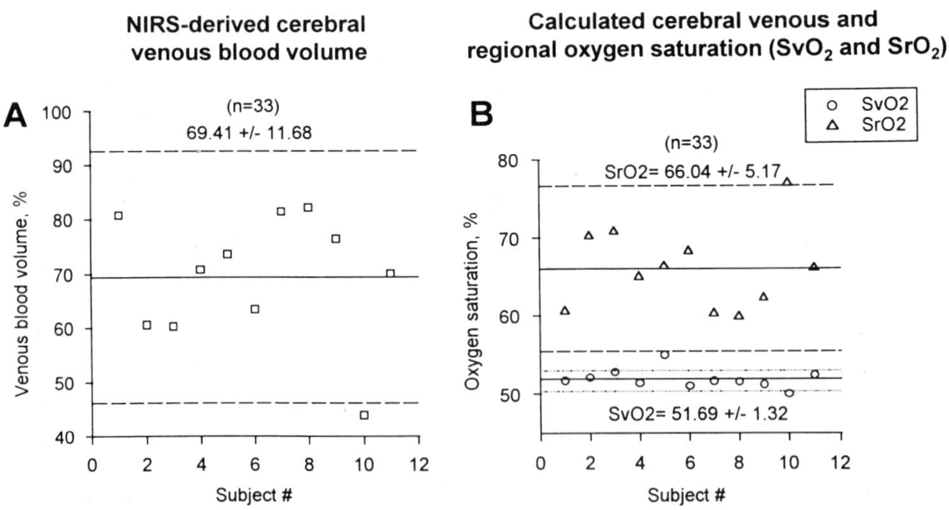

Figure 1. Panel **(A)** shows venous BV = 69.41 +/– 11.68%; while panel **(B)** shows the calculated venous oxygen saturation, $SvO_{2(Eq.1)}$ = 51.91 +/– 1.32% and regional oxygen saturation, $SrO_{2(Eq.2)}$ = 66.04 +/– 5.17%. Lines are means +/– 95% confidence intervals (1.96σ).

position; however, a statistically significant difference was found between the two ($SrO_{2(eq.2)}$ = 66.04 +/– 5.17 while $SrO_{2(head\text{-}up)}$ = 71.7 +/– 4.81; p = .0016).

4. DISCUSSION

We have described a method for calculating the percent cerebral venous contribution to the NIRS BV signal non-invasively. By tilting the subjects from supine to head up (90°), the change in NIRS-measured blood volume can be attributed to the venous compartment exclusively. The average cerebral venous blood volume contribution correlates well with the 75% suggested in the literature (Mchedlishvili, 1986).

Although venous oxygen saturation (SvO_2) estimates from NIRS data in our study are lower than those obtained by Skov et al in term neonates, the values are similar. They found cerebral venous oxygen saturation values of 30–80% (mean of 61%) following 15° head-down tilt; with half of the infants studied showing significantly higher SvO_2 (67.3% in the asphyxiated group compared to 53.44% in the preterm group) and CBF values (Skov et al., 1993). The normal arterial-jugular venous oxygen difference ($ajDO_2$) in adults is 7 ml O2/100 ml blood, corresponding to an SvO_2 of 54–75% (Dearden 1991); while in children $ajDO_2$ is 5 ml O_2/100 ml blood (Kennedy and Sokoloff, 1957), resulting in higher SvO_2 values. In a recent study Yoxall and colleagues compared measurements of cerebral venous oxyhemoglobin saturation determined by NIRS with direct invasive measurements of SvO_2 by co-oximetry. They found a mean SvO_2 of 60% in children, with a mean difference between the co-oximeter and the NIRS of 1.5% (Yoxall et al., 1995). The lower mean NIRS-derived SvO_2 for adults found in our study is expected since in adults not only is the $ajDO_2$ wider, but also there is a greater degree of NIRS signal attenuation.

Our results of SrO_2 (66.04% +/– 5.17%) also compare well with those of McCormick and collaborators who found good correlation between spectroscopically measured SrO_2 (63.86 +/– 3.43%) and estimated SrO_2 calculated from SvO_2 values measured invasively by co-oximetry in adults (n = 68; r = 0.74) (McCormick et al., 1991). Even though the SvO_2 levels in our study are in the lower end of normal, we believe that the calculated SrO_2 in adults is a better indicator of tissue oxygen saturation from NIRS data. Furthermore, our findings that SvO_2 determined as the ratio of HbO_2/Hbt in the supine position can be used to calculate an SrO_2 that is not different from that calculated through changes following HUT, suggest that an estimate of SrO_2 can be obtained continuously in the supine position without patient manipulation.

The mean percent cerebral venous contribution to the NIRS BV signal (three tilts) in subject #10 was significantly lower than the rest of the subjects (43.96 compared to 71.96); resulting in the highest SrO_2 values, despite a similar SvO_2 to the rest of the group. Although the data from subject #10 does impact on the variance of the various parameters, it does not significantly affect the overall statistics of the study. The observed differences in venous BV in this subject may be explained by a diminished sensitivity to changes in hydrostatic pressure resulting in unlikely values for the distribution of cerebral blood volume in the venous vasculature.

Jugular bulb venous oxygen saturation measuremensts are routinely used to assess cerebral hemodynamics in neurointensive care settings (for references see Feldman et al., 1997). On average two-thirds of the blood from a single internal carotid artery is drained through the ispilateral internal jugular vein, with an approximate 2.7% extracerebral contamination of jugular venous blood (Shenkin et al., 1948). Sampling of blood from the jugular bulb provides a measure of global cerebral oxygenation, while SrO_2 measured by NIRS provides an average regional saturation across all vascular beds. Interestingly enough in a study comparing carbon dioxide-induced changes in jugular venous oxygen saturation with changes in cerebral near infrared spectroscopy, the investigators found close correlation between changes in HbO_2 measured by the two approaches (Tateishi et al., 1995). Results from these studies, as well as from our own, suggest that a NIRS-derived SrO_2 may provide similar information to other clinically established, yet invasive techniques such as jugular bulb oximetry to assess adequate cerebral oxygenation.

4.1. Limitations and Assumptions of the Technique

The NIRS-derived cerebral venous blood volume is a function of interoptode spacing, the wavelengths used, and individual variations in head geometry which would all influence the photon attenuation due to both absorption and scattering. Wider optode separations may optimize the NIRS-measured changes in blood volume originating from deep tissue by minimizing the amount of superficial tissue contribution; however, at inter-optode distances of ≥4 cm, Delpy and collaborators consider that NIRS attenuation changes originate from the adult brain (Delpy, 1997). Another possible source of error in estimation of cerebral venous blood volume through a tilting protocol is due to changes in geometry from movement of the brain inside the skull, either away from or towards the optodes, altering pathlength.

We have shown, however, that SrO_2 in the supine position alone is not statistically different from $SrO_{2(Eq.2)}$ which is determined by tilting. Since our measured cerebral venous BV contribution is similar to the value quoted in the literature, we suggest that 75% can be used to avoid subject/patient manipulation. Continous estimates of the above mentioned parameters may be used to assess the adequacy of oxygen supply to the brain.

REFERENCES

Brazy JE (1991) Near-infrared spectroscopy. Clin. Perinatol. 18(3):519–534.

Brazy JE et al. (1985) Noninvasive monitoring of cerebral oxygenation in preterm infants: preliminary observations. Pediatrics 75(2):217–225.

Cope M and Delpy DT (1988) System for long-term measurement of cerebral blood and tissue oxygenation on newborn infants by near infra-red transillumination. Med. & Biol. Eng. & Comput. 26:289–294.

Dearden NM (1991) Jugular bulb venous oxygen saturation in the management of severe head injury. Curr Opin. Anesthesiol. 4:279–286.

Delpy DT and Martinez-Coll A (1997) personal communication.

Elwell CE et al. (1994) Quantification of adult cerebral hemodynamics by near-infrared spectroscopy. J. Appl. Physiol. 77(6):2753–2760.

Felman Z and Robertson CS (1997) Monitoring of cerebral hemodynamics with jugular bulb catheters. Crit. Care Clin. 13(1):51–77.

Hoffmann J and Lübbers DW (1985) Estimation of concentration ratios and the redox states of the cytochromes from noisy reflection spectra using multicomponent analysis methods. Adv. Exp. Med. Biol. 191:119–124.

Kennedy C and Sokoloff L (1957) An adaptation of a nitrous oxide method to study the cerebral circulation in children: Normal values for cerebral blood flow and cerebral metabolic rate in childhood. J. Clin. Invest. 36:1130–1137.

McCormick PW et al. (1991) Regional cerebrovascular oxygen saturation measured by optical spectroscopy in humans. Stroke 22:596–602

Mchedlishvili GI (1986) Arterial behaviour and blood circulation in the brain. (Bevan JA, Ed.) Plenum Press, New York, p.57.

Owen-Reece H et al. (1996) Use of near infrared spectroscopy to estimate cerebral blood flow in conscious and anaesthetised adult subjects. Br. J. Anaesth. 76:43–48.

Patterson JL and Warren JV (1952) Mechanisms of adjustments in the cerebral circulation upon assumption of the upright position. J. Clin. Invest. 31:653.

Pollard V et al. (1996) Validation in volunteers of a near infrared spectroscope for monitoring brain oxygenation in vivo. Anesth. Analg. 82:269–277.

Shenkin GA et al. (1948) Dynamic anatomy of the cerebral circulation. Arch. Neurol. Psychiatry 60:240–252.

Skov L et al. (1993) Estimation of cerebral venous saturation in newborn infants by near infrared spectroscopy. Pediatr. Res. 33(1):52–55.

Tamura M et al. (1989) In vivo study of tissue oxygen metabolism using optical and nuclear magnetic resonance spectroscopies. Annu. Rev. Physiol 51:813–834.

Tateishi A et al. (1995) Qualitative comparison of carbon dioxide-induced change in cerebral near-infrared spectroscopy versus jugular venous oxygen saturation in adults with acute brain disease. Crit. Care Med. 23:1734–1738.

Wyatt JS et al. (1990) Quantitation of cerebral blood volume in human infants by near-infrared spectroscopy. J. Appl. Physiol. 68(3):1086–1091.

Yoxall CW et al. (1995) Measurement of cerebral venous oxyhemoglobin saturation in children by near-infrared spectroscopy and partial jugular venous occlusion. Pediatr. Res. 38(3):319–323.

16

OPTICAL MONITORING OF NADH REDOX STATE AND BLOOD FLOW AS INDICATORS OF BRAIN ENERGY BALANCE

A. Mayevsky, S. Meilin, T. Manor, N. Zarchin, and J. Sonn

Department of Life Sciences
Bar-Ilan University
Ramat-Gan 52900, Israel

1. INTRODUCTION

Real-time evaluation of brain vitality *in situ* could be done by monitoring different parameters, which are complementary to each other. During the last 40 years the following four minimally invasive techniques were developed and applied to monitor the brain *in situ*:

(1) Cerebral Blood Flow (CBF) using laser Doppler flowmetry (Stern et al., 1977; Dirnagl et al., 1989).
(2) Hemoglobin oxygenation or saturation (HbO_2) by dual wavelength reflectometry (Rampil et al., 1992) or spectral analysis (Frank et al., 1989).
(3) Brain average oxygenation using oxygen electrodes (Mayevsky et al., 1980).
(4) Mitochondrial redox state by monitoring of NADH fluorescence using surface fluorometry (Chance et al., 1962; Jobsis et al., 1971b; Mayevsky and Chance, 1982).

In order to be able to evaluate the oxygen balance of the brain *in situ* it is important to monitor more than 1 parameter simultaneously in real-time. Various combinations of the above mentioned parameters were made in the past and have been reported (Mayevsky et al., 1980, 1991, 1992, 1996, Rampil et al., 1992). We have developed and used a combined two optical technique, namely, CBF and NADH redox state, into a single bundle of optical fibers thus enabling to correlate the two optical parameters that could be monitored together (Mayevsky et al., 1991). The pioneering work published by Chance and Williams in 1955 and later on in 1962 by Chance et al. established the foun-

Oxygen Transport to Tissue XXI, edited by Eke and Delpy.
Kluwer Academic / Plenum Publishers, New York, 1999.

dation for the close to 700 publications appeared since then on NADH fluorescence as an optical probe for mitochondrial function evaluation. Mayevsky and Chance published in 1982, for the first time, the multichannel fiber optic based surface fluorometer used for the simultaneous measurement of NADH redox state in 4 different body organs (brain, liver, kidney and testis) or 4 various locations on the same brain surface. As of today, 15 years later, we were able to extend our understanding of tissue O_2 balance by simultaneous monitoring the correlation between brain blood flow and mitochondrial NADH redox state in real time. In our previous work we combined the monitoring of the oxygenation of hemoglobin and NADH (Mayevsky et al., 1992; Rampil et al., 1992) or tissue pO_2 and NADH (Mayevsky et al., 1980) but it seems to us that the best combination is to monitor CBF and NADH redox state simultaneously (Mayevsky et al., 1991). This approach provides information about the two most important parameters that represent the regulated processes in brain energy state, namely, brain blood flow and mitochondrial function or redox state.

In their original publication, in 1955, Chance and Williams defined few metabolic states for isolated mitochondria as shown in Figure 2. They found that in the resting state—4, the NADH was almost completely reduced (99%) while the activation of the mitochondria by addition of ADP led to the active state—3 and NADH became more oxidized (53%). The use of surface fluorometry *in vivo* began in the late 50's and early 60's by Chance and his collaborators (Chance and Legallias, 1959; Chance and Thorell, 1959; Chance et al., 1962) and later on was changed to fiber optic surface fluorometry/reflectometry (Mayevsky and Chance, 1973, 1982; Mayevsky, 1984). According to the results accumulated since then (while monitoring NADH redox state) it seems that the resting brain is somewhere in the midline between state 4 and 3 (Figure 2). The same is probably true for other organs as shown for liver, kidney and testis (Chance et al., 1962). We can conclude that the NADH redox state is changing in a continuous way from minimal to maximal levels in the brain *in vivo* exposed to various perturbations.

Since brain blood flow is a regulated process as also in various organs, it was hypothesized that simultaneous monitoring of brain blood flow and NADH redox state will provide significant knowledge on the energy state of the tissue. In order to evaluate in real time both mitochondrial NADH redox state and brain blood flow we combined the optical fibers attached to the two instruments into one common bundle (Mayevsky et al., 1991; Mayevsky, 1992). Using this approach we were able to correlate between changes in CBF and mitochondrial redox state in the brain exposed to various pathological states in experimental animals (Mayevsky et al., 1991, 1995) and in clinical research (Mayevsky et al., 1996).

2. METHODS

In the present study we used a combined monitoring system by which CBF and NADH was measured simultaneously. The hypothesis to our study was that the multiparametric monitoring system will give us a new tool to differentiate between brain vitality exposed to the various pathological states.

The fiber optic bundle contained fibers that were connected to the two instruments used. As shown in Figure 1 the fluorometer and the laser Doppler flowmeter (LDF) used different light sources. The 366 nm light was filtered from a 100 watt Hg lamp (Mayevsky and Chance, 1982) while CBF was measured at 632.8 nm originated from HeNe laser (Mayevsky et al., 1995). The light was transmitted to the brain via a combined bundle

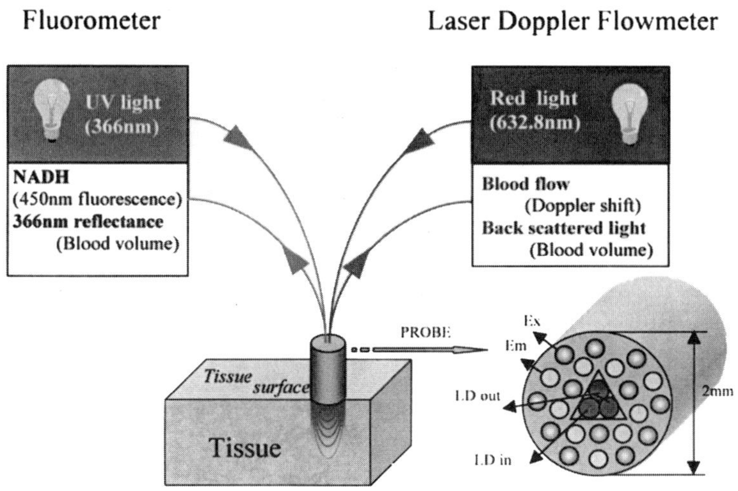

Figure 1. Multiparametric monitoring system by which a combination between cerebral blood flow, volume and mitochondrial redox state was assessed *in vivo*. The simultaneous measurements were done using two separate instruments. The combined probe contains two types of optical fibers: the Laser Doppler probe is comprised of one excitation fiber and two collecting fibers. For the NADH measurement, a randomly mixed group of excitation (Ex.) and emission (Em.) fibers was implemented.

of optical fibers as shown in Figure 1. We used the original probe of Perimed Inc. (Sweden) for the LDF while the NADH fibers were constructed in our laboratory (Optical Fibers Inc). Due to the different wavelengths, the depth of light penetration was different and therefore the volume of tissue sampled by the two instruments was different. The meaning of this is that the CBF measured deeper structures as compared to the NADH. Each of the used instruments provided also an indirect parameter, tissue reflectance, correlated to blood volume in the tissue sampled by the probe. Each of the instruments measured the intensity of the back scattered light which inversely correlated to the blood volume in the tissue (Jobsis et al., 1971a; Mayevsky, 1984a).

2.1. Local Cerebral Blood Flow

On-line measurement of CBF was done using the laser Doppler flowmeter (LDF) technique. The LDF apparently measures relative changes which are significantly correlated to the relative changes in CBF measured by other quantitative approaches (for review see (Wadhwani and Rapoport, 1990)). The change in the total back-scattered light is an indirect indicator for the blood volume in the examined tissue volume, but it cannot be calibrated due to the nature of the instrument.

2.2. NADH Redox State Fluorometry

For NADH monitoring from the surface of the brain, 366 nm excitation light was passed from a fluorometer to the brain via a bundle of quartz optical fibers. The emitted light (450 nm), together with the reflected light at the excitation wavelength, was transferred to the fluorometer via another bundle of fibers. Changes in the reflected light correlate to changes in tissue blood volume and facilitate the correction of hemodynamic artifacts appearing in the NADH measurement. The changes in the fluorescence and reflectance signals were calculated relative to the calibrated signals under normoxic conditions (Mayevsky, 1984a).

3. RESULTS

The right side of Figure 2 presents in a schematic way the results obtained in the brain of experimental animals exposed to various perturbations. The correlation between NADH redox state and CBF vary between various pathological states. The comparison point is the resting level of NADH and CBF measured in the normoxic brain after the stabilization period following the operation of the animal. Under deep anesthesia, ischemia or death, the NADH increased toward state 4 and 5 simultaneously with the decrease in CBF. It is assumed that the primary effect was on CBF (a decrease) while the NADH response (an increase) was due to a decrease in O_2 delivery to the brain. When the brain was activated by seizures or by a spreading depression wave (SD) the demand for more energy (Mayevsky, 1983a, 1983b; Mayevsky and Weiss, 1991) led to an oxidation of NADH (decrease) simultaneously with an increase in CBF to compensate for the extra O_2 needed. Under two other pathological conditions the NADH and CBF were also correlated. Exposing the brain to hypoxia led to an increase in NADH while CBF was increased (Mayevsky, 1992). Under the hyperbaric oxygenation (HBO) the NADH became more oxidized (decrease) while the CBF decreased (Mayevsky, 1984b; Miller et al., 1987) due to vasoconstriction response.

The significant value of using two-parameter monitoring approach (Mayevsky et al., 1998) as compared to the classical single parameter device is shown in Figure 3. The starting point of data accumulation is after placing the probe on the tissue and stabilization of the signals. The initial values of both CBF (cerebral blood flow) and NADH received an arbitrary value of 100 and serves as a reference point to the entire

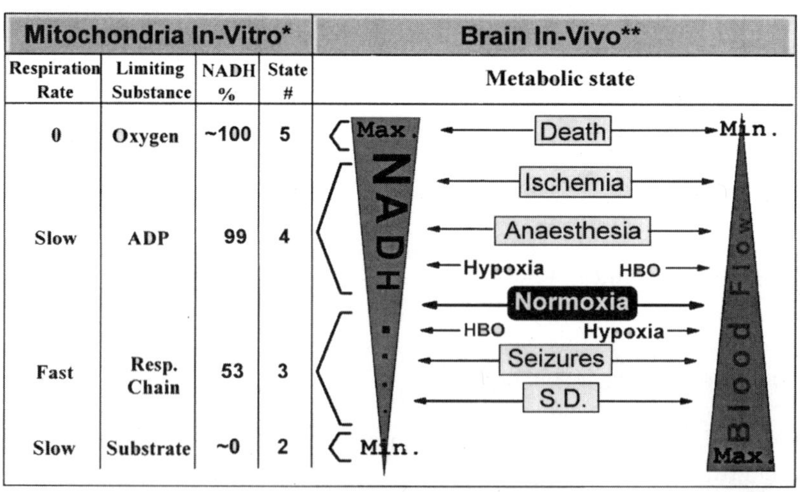

Figure 2. Comparison between the classical definition of Chance & Williams regarding the metabolic state of isolated mitochondria and the metabolic state of the *in vivo* brain (*according to Chance & Williams 1955; **according to Mayevsky 1984; HBO—Hyperbaric; S.D.—Spreading).

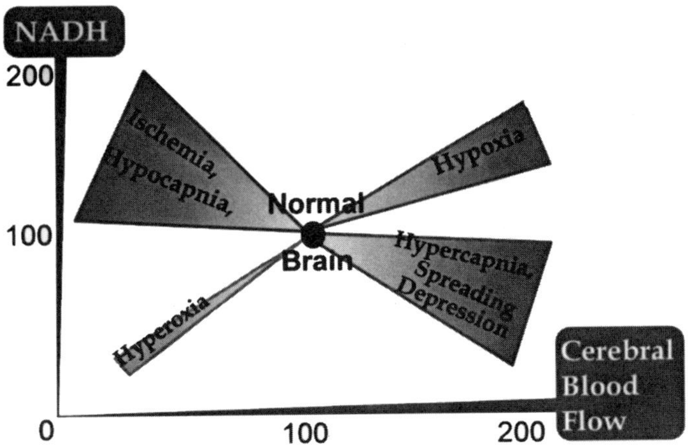

Figure 3. Schematic presentation of the correlation between cerebral blood flow (CBF) and the mitochondrial NADH redox state measured *in vivo.* Since it is impossible to use calibrated units for both CBF and NADH the scales used, covering the range between minimal and maximal monitored values as compared to the starting normoxic value defined as 100.

monitoring period. The aim of Figure 3 is to demonstrate the significant value and practicality of simultaneous monitoring of the two parameters in real-time. This schematic presentation is based on published and unpublished experimental data monitored in the brain.

The classically known CO_2 effect (hypo and hypercapnia) on the microcirculation was detected by a decrease and an increase in CBF. The NADH redox state may remain steady or may increase or decrease respectively according to the availability of O_2. In the activated brain (Spreading Depression) the NADH became more oxidized (Mayevsky and Chance, 1982; Mayevsky, 1992) while a significant increase in CBF was recorded (Mayevsky et al., 1991; Mayevsky and Weiss, 1991). Hypoxia and ischemia in the brain led to an increase in NADH while CBF was increased or decreased respectively (Mayevsky et al., 1991; Mayevsky, 1992). Under hyperoxic conditions (Torbati et al., 1979; Mayevsky, 1984b) the clear vasoconstrictive effect was noticed as a decrease in flow while NADH became more oxidized (decrease). The results presented in Figure 3 show that monitoring of one parameter either CBF or NADH will not provide a meaningful tool for determination of the physiological state of the tested tissue or organ under *in vivo* conditions.

4. DISCUSSION

The trend of minimally invasive brain monitoring is expanding for the last few years. Consequently new instruments based on various techniques are introduced to answer this demand in experimental as well as clinical situations. The main obstacle is obtaining information about the key processes taking place in the investigated organ.

When dealing with the brain, we have learned that in order to study the vitality of tissue, it is crucial to monitor as many parameters as possible. We suggest that for evaluating brain vitality, at least the two parameters described in the present study should be monitored. For better and deeper understanding of brain energy balance, we have

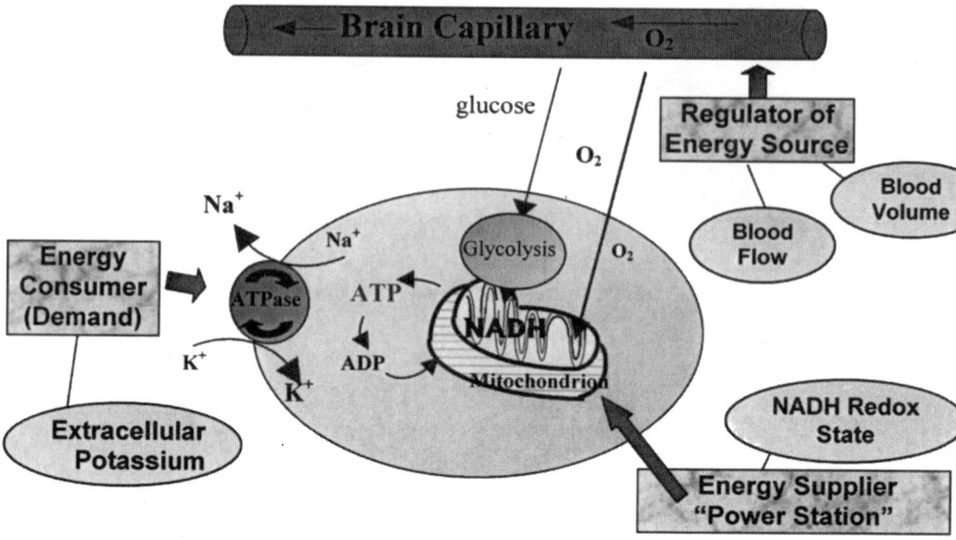

Figure 4. The interrelation between energy supply and demand (energy balance) in the brain.

monitored a third parameter, the extracellular K^+, as shown in Figure 4. However, since monitoring of K^+_e is much more complicated and time demanding, we feel that only when conditions are in favor (experimental or non urgent procedures), this third parameter should be monitored. Otherwise, the "simple" optical probe is sufficient for real-time monitoring of the brain energy state.

As shown in Figure 4, coupling between the main energy demanding processes (Na-K-ATPase) and the mitochondrial function as an energy supplier, is evident by the consumed ATP molecules. Any change in the activity of the pump will affect the rate of oxidative phosphorylation processes. In cases where the pump is activated (Epilepsy or Spreading Depression), oxygen consumption will increase (Mayevsky and Wiess, 1991) and NADH level will decrease (more of it will be oxidized). This elevated energy demand will affect the energy resources namely, blood flow to regulate the increased O_2 demand.

Any decrease in O_2 delivery to the brain (i.e. Stroke) will inhibit mitochondrial functions and end up in lower ATP production. The pump, having less fuel, will slow down and extracellular levels of K^+ will increase (Depolarization).

As shown in Figure 2 and 3, the relationship between CBF and mitochondrial functions (NADH) could be quantified and a "vitality index" could be calculated.

It is important to note that, activation of the brain (Epilepsy or Spreading Depression) under partial ischemic conditions, will end up in tissue deterioration. That is due to the limitation in O_2 delivery, needed for restoration of the normal ionic homeostasis. The ability of the brain to cope with such activation depends upon the level of ischemia and duration of the activated state.

ACKNOWLEDGMENTS

The study was supported by a grant from the committee for Research and Prevention in Occupational Safety and Health, Ministry of Labor and Social Affairs, Israel, and the Research Committee of Bar-Ilan University, Israel.

REFERENCES

Chance B, Cohen P, Jobsis F, and Schoener B (1962) Intracellular oxidation-reduction states *in vivo*. Science 137:499–508.

Chance B and Legallias V (1959) Differential microfluorometer for the localization of reduced pyridine nucleotide in living cells. The Review of Scientific Instruments 30:732–735.

Chance B, Legallias V, and Schoener B (1962) Metabolically linked changes in fluorescence emission spectra of cortex of rat brain, kidney and adrenal gland. Nature 195:1073–1075.

Chance B and Thorell B (1959) Localization and kinetics of reduced pyridine nucleotide in living cells by microfluorometry. J. Biol. Chem. 234:3044–3050.

Chance B and Williams GR (1955) Respiratory enzymes in oxydative phosphorylation. I. Kinetics of oxygen utilization. J. Biol. Chem. 217:383–393.

Dirnagl U, Kaplan B, Jacewicz M, and Pulsinelli W (1989) Continuous measurement of cerebral cortical blood flow by Laser Doppler Flowmetry with a rat stroke model. J. CBF and Metab. 9:589–596.

Frank KH, Kessler M, Appelbaum K, and Dummler W (1989) The Erlangen micro-lightguide spectrophotometer EMPHO I. Phys. Med. Biol. 34:1883–1900.

Haberl RL, Heizer ML, Marmarou A, and Ellis EF (1989) Laser doppler assessment of brain microcirculation: Effect of systemic alterations. Am. J. Physiol. 256:H1247–H1251.

Jobsis FF, O'Conner MJ, Rosenthal M, and VanBuren JM (1971a) Fluorometric monitoring of metabolic activity in the intact cerebral cortex. In: Anonymous Neurophysiology studied in man. Excerpta Medica, Amsterdam, pp 18–26.

Jobsis FF, O'Connor M, Vitale A, and Vreman H (1971b) Intracellular redox changes in functioning cerebral cortex. I. Metabolic effects of epileptiform activity. Neurophysiology 34:735–749.

Mayevsky A (1983a) Metabolic, ionic and electrical responses to experimental epilepsy in the awake rat. In: Baldy M, Moulinier DH, Ingvar DH, Meldrum BS (eds.) Proceedings First International Congress of Cerebral Blood Flow, Metabolism & Epilepsy. John Libbey, pp. 263–270.

Mayevsky A (1983b) Multiparameter monitoring of the awake brain under hyperbaric oxygenation. J. Appl. Physiol. 54:740–748.

Mayevsky A (1984a) Brain NADH redox state monitored *in vivo* by fiber optic surface fluorometry. Brain Res. Rev. 7:49–68.

Mayevsky A (1984b) Brain oxygen toxicity. Invited Review. In: Bachrach AJ, Matzen MM (eds.) Underwater Physiology. 8th Symposium Undersea Medical Society, Bethesda, Maryland, pp. 69–89.

Mayevsky A (1992) Cerebral blood flow and brain mitochondrial redox state responses to various perturbations in gerbils. In: Erdmann W (ed.) Oxygen Transport to Tissue XIV. Plenum Press, pp. 707–716.

Mayevsky A and Chance B (1973) A new long-term method for the measurement of NADH fluorescence in intact rat brain with implanted cannula. Adv. Exp. Med. Biol. 37A:239–244. New York Plenum Press.

Mayevsky A and Chance B (1982) Intracellular oxidation reduction state measured *in situ* by a multichannel fiber-optic-surface fluorometer. Science 217:537–540.

Mayevsky A, Doron A, Manor T, Meilin S, Zarchin N, and Ouaknine GE (1996) Cortical spreading depression recorded from the human brain using a multiparametric monitoring system. Brain Res. 740:268–274.

Mayevsky A, Flamm ES, Pennie W, and Chance B (1991) A fiber optic based multiprobe system for intraoperative monitoring of brain functions. SPIE Proc. 1431:303–313.

Mayevsky A, Frank K, Muck M, Nioka S, Kessler M, and Chance B (1992) Multiparametric evaluation of brain functions in the Mongolian gerbil *in vivo*. J. Basic Clin. Physiol. Pharmacol. 3:323–342.

Mayevsky A, Lebourdais S, and Chance B (1980) The interrelation between brain PO_2 and NADH oxidation-reduction state in the gerbil. J. Neurosci. Res. 5:173–182.

Mayevsky A, Meilin A, Rogatsky GG, Zarchin N, and Thom SR (1995) Multiparametric monitoring of the awake brain exposed to carbon monoxide. J. Appl. Physiol. 78:1188–1196.

Mayevsky A, Meilin S, Manor T, Ornstein E, Zarchin N, and Sonn J (1998) Multiparametric monitoring of brain oxygen balance under experimental and clinical conditions. Neurol. Res. 20:S76–S80.

Mayevsky A and Weiss HR (1991) Cerebral blood flow and oxygen consumption in cortical spreading depression. J. CBF and Metab. 11:829–836.

Miller WC, Suich DM, and Unger KM (1987) Hyperbaric environment for chronic hypoxemia. J. Hyperbaric Med. 2:211–214.

Rampil IJ, Litt L, and Mayevsky A (1992) Correlated, simultaneous, multiple-wavelength optical monitoring *in vivo* of localized cerebrocortical NADH and brain microvessel hemoglobin oxygen saturation. J. Clin. Monit. 8:216–225.

Stern MD, Lappe DL, Bowen PD, Chimosky JE, Holoway GA, and Keiser HR (1977) Continuous measurement of tissue blood flow by Laser Doppler spectroscopy. Am. J. Physiol. 232:H441–H448.

Torbati D, Parolla D, and Lavy S (1979) Organ blood flow, cardiac output, arterial blood pressure, and vascular resistance in rats exposed to various oxygen pressures. Aviat. Space Environ. Med. 50:256–263.

Wadhwani KC and Rapoport SI (1990) Blood flow in the central and peripheral nervous systems. In: Shephard AP, Oberg PA (eds.) Laser-Doppler Blood Flowmetry. Kluwer Academic Pub., Boston, pp. 265–304.

17

METABOLIC AND HEMODYNAMIC OSCILLATIONS MONITORED OPTICALLY IN THE BRAIN EXPOSED TO VARIOUS PATHOLOGICAL STATES

S. Meilin, A. Mendelman, J. Sonn, T. Manor, N. Zarchin, and A. Mayevsky

Department of Life Sciences, Bar-Ilan University, Ramat-Gan 52900, Israel

1. INTRODUCTION

Slow (<1 Hz) oscillations of cerebral blood flow (CBF) and oxidative metabolism have been previously reported in several animal species under different physiological/pathological conditions (Vern et al., 1998, Mayevsky and Ziv 1991, Hudetz et al., 1992, Deyoe et al., 1995). Spontaneous oscillations in cerebral circulation have been demonstrated as periodic variations of blood volume, oxygen availability, NAD^+/NADH oxidative state (Dora and Kovach 1981), cytochrome aa3 redox state (Jobsis 1978), ischemia and anesthesia.

In a previous study (Mayevsky et al., 1984) cerebral metabolic oscillations namely, NADH redox state, pO2 and Fp, were obtained in carotid occluded rats.

Reports of occasional dissociation of focal slow oscillations of cortical blood flow from the oscillations of either the cortical pO2 or the intramitochondrial redox state (Cooper et al., 1966, Dora et al., 1980, Vern et al., 1988) suggest the existence of primary metabolic, rather than (or in addition to) hemodynamic oscillators.

It is not yet known whether this phenomenon is related to normal cerebral function, or to a local response to pathological situations such as reduction in CBF.

In this study we summarize the data accumulated in our laboratory during the past few years showing metabolic and hemodynamic oscillations during various pathophysiological situations. The aim of this study was to characterize spontaneous oscillations of blood flow and mitochondrial NADH redox state in the cerebral cortex of rats, which underwent different treatments. Also, oscillation in CBF and NADH recorded from the human brain are presented.

Oxygen Transport to Tissue XXI, edited by Eke and Delpy.
Kluwer Academic / Plenum Publishers, New York, 1999.

2. METHODS

2.1. Local Cerebral Blood Flow

To measure in real time the CBF from the same cortical area as the NADH redox state we used the laser Doppler Flowmetry (LDF) technique. The LDF measure relative flow changes. Those changes were found to be correlated to CBF measured by 2 quantitative methods (Wadhwani and Rapoport 1996). The principle of the LDF is to utilize the Doppler shift namely; the frequency change that light undergoes when reflected by moving red blood cells. To quantitate and normalize CBF values we defined the reading value after death as 0 CBF. The value 100% was defined as CBF read on LDF scale during the control period.

2.2. NADH Redox State Fluorometry

The principle of NADH monitoring from the surface of the brain is that excitation light (366 nm) is passed from the fluorometer to the brain via a bundle of optical fibers made of quartz. The emitted light (450 nm), together with the reflected light at the excitation wavelength, is transferred to the fluorometer via another bundle of fibers. The changes in the reflected light (R) were correlated to the changes in tissue blood volume and also serve to correct for hemodynamic artifacts appearing in the NADH measurement. More details on this technique have been published by our group (Mayevsky and Chance 1982, Mayevsky 1984).

The LDF and NADH redox state are 2 parameters out of a large number of parameters monitored by the Multiprobe Assembly (MPA) system (Mayevsky et al., 1995, Meilin et al., 1996). Since oscillations occurred only in CBF and NADH redox state the details regarding other parameters monitored by the MPA are not given.

2.3. Animal Preparation

The animals were anesthetized by an intraperitoneal (IP) injection (0.3 ml/100 g) of Equithesin (eachml contains: pentobarbital, 9.72 mg; chloral hydrate, 42.51 mg; magnesium sulfate, 21.25 mg; propylene glycol, 44.34% w/v; alcohol, 11.5%; and water). After a midline incision of the skin a 6 mm hole was drilled in the parietal bone and the dura mater was gently removed. The MPA was held by a micromanipulator and was lowered to make gentle contact with the brain. A reference electrode was introduced into the neck area. Two stainless steel screws located in the skull were used to fixate the plexiglass holder of the MPA to the skull using dental acrylic cement.

2.4. Experiment Protocol

At least 90 minutes after surgery the animals underwent one of the various protocols:

Alcohol administration: The rats were injected IV with 30% alcohol, 3 ml/200 gr during 30 minutes.

CO exposure: The rats were placed in an airtight Plexiglas cage. The rats were exposed to 40 min of 1000 ppm CO followed by 20 min of 3000 ppm of CO (Mayevsky et al., 1995).

N-Arg administration: The rats were injected with 50 mg/kg of N^{ω}-nitro-L-arginine (N-Arg) IP (Meilin et al., 1995).

Ischemia: 24 h prior to brain surgery the common carotid arteries were isolated and occluded by 4-0 silk thread ligatures. 24 h later after, the brain surgery was completed and the MPA was placed. The animal was left to recover until fully awake.

Ischemia + Anesthesia: 24 h prior to brain surgery the common carotid arteries were isolated and occluded by 4-0 silk thread ligatures. During the monitoring period the animal was anesthetized by additional IP injections of 0.3–0.5 ml equithesin every 20–30 minutes.

Trauma: A special modified MPA system was used for human brain monitoring in the Intensive-Care Unit as published previously (Mayevsky et al., 1996). The patient was hospitalized due to comatose state induced by severe head injury (GCS = 8). Preparation of the patient for brain monitoring was performed in the neurosurgical operating theater under general anesthesia. The patient was prepared for monitoring as follows:

1. A 15 mm diameter hole was carefully drilled in the frontal bone area selected according to CT scan. The hole was drilled in a "normal" looking section of the brain.
2. After cleaning the exposed area of bone residues, the dura mater was cut to enable direct contact between the MPA probes and the brain.
3. The MPA holder was screwed into the skull to a predetermined depth to avoid any undue pressure on the brain.
4. The operated area was closed by suturing the skin around the MPA holder (covered until the beginning of the monitoring period).
5. The patient was transferred to the intensive care unit and after stabilization and connection to the routine patient monitoring system the sterilized MPA was introduced into the holder (for contact with the brain) and fastened to it in order to ensure measurements for up to 72 h.

3. RESULTS

Continuous monitoring of the brain exposed to those pathological states reveal that spontaneous hemodynamic and metabolic oscillations were developed and recorded under the various treatments. The oscillations were recorded in the following parameters: CBF, reflected light (correlated to CBV) and mitochondrial NADH redox state. Figure 1(A–E) presents records taken from one animal of each of the treated groups. It demonstrates typical changes in CBF reflected light and NADH during the oscillation period. The range of amplitude in CBF oscillation peaks is 6%–30% and NADH oscillations range is 5–10%. The oscillation frequency was not significantly different among the various treatments and was found to be in the range of 3.6–5.5 waves/min (Table 1). Although NADH signal follows the CBF, a clear negative correlation was found between

Table 1. Frequency (wave/min) of the oscillations obtained at various treatments

Treatment	Ischemia	Anesthesia + ischemia	Alcohol	N-Arg	3000 ppm CO	Trauma (human)
Frequency (wave/min)	3.9 ± 0.4	4.1 ± 0.42	3.7 ± 0.8	3.7 ± 0.17	5.5 ± 0.62	2.5

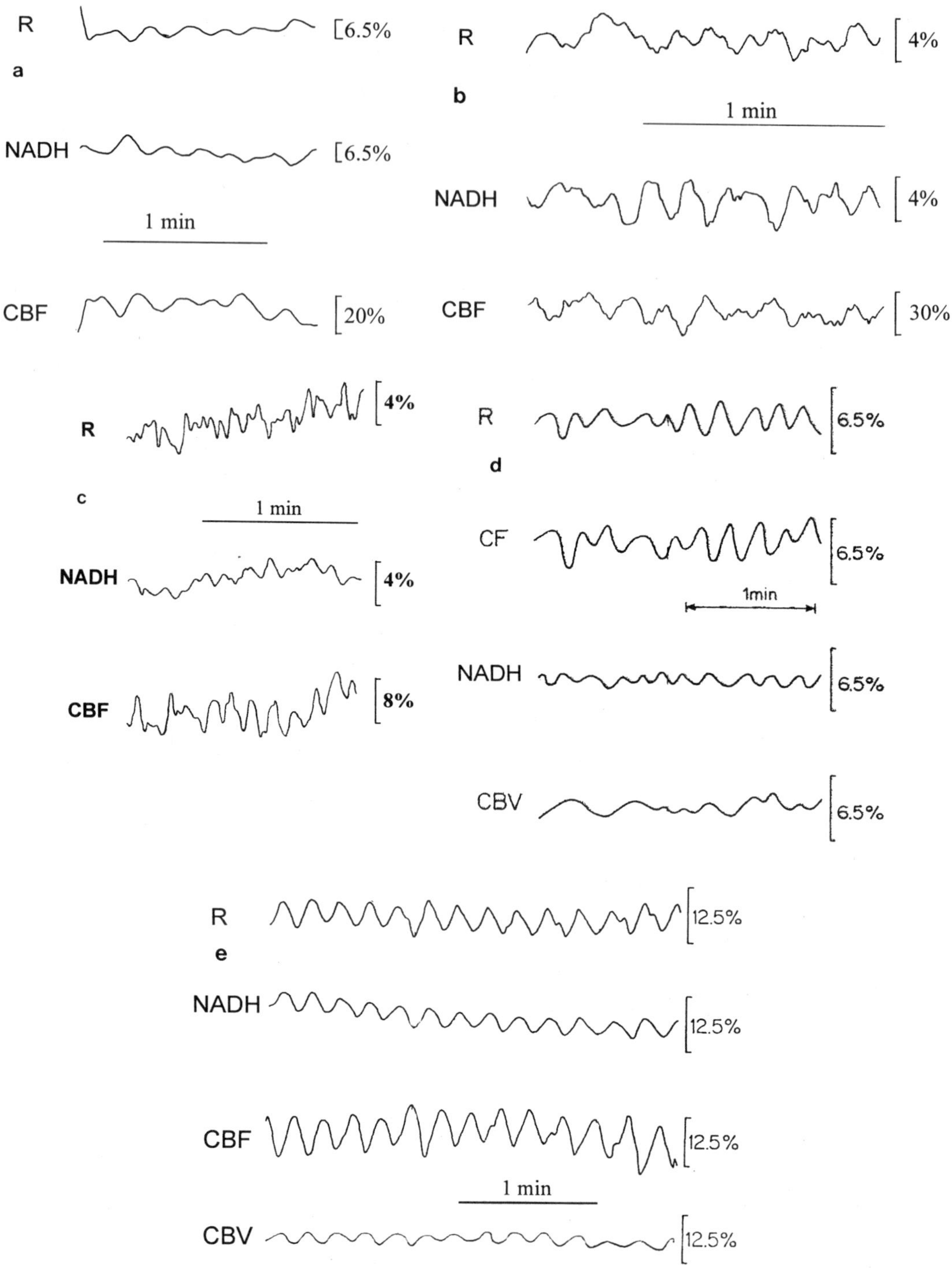

Figure 1. (A–E): Oscillations recorded from different animals under various treatments:
1A-Anesthetized animal monitored 24 h after 2 carotids occlusion procedure.
1B-Awake animal monitored after 24 h after 2 carotids occlusion procedure.
1C-Animal receiving alcohol IV.
1D-Oscillations monitored after N-Arg administration.
1E-Animal exposed to 3000ppm CO.
R-reflectance, NADH- mitochondrial NADH redox state, CBF- cerebral blood flow, CBV- cerebral blood volume.

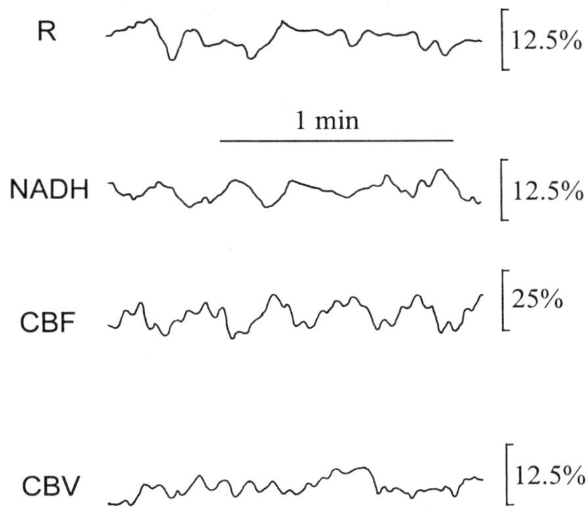

Figure 2. Oscillation recorded from the human cortex after trauma.

the 2 parameters in all treatments ($r = (-0.5) - (-0.8)$, $p < 0.01$). Any increase or decrease in CBF during the oscillation wave was followed by an increase in the mitochondrial redox state oscillation wave.

Although the frequency of the oscillation waves was found to be slower (2.5 waves/min) than those obtained in animals, the oscillations monitored from the injured patient were very similar to those recorded from the rat's cortex (Figure 2).

4. DISCUSSION

In this study we observed similar oscillations in terms of frequency, correlation and wave characterization under various treatments. The significant correlation between the hemodynamic and metabolic oscillations suggests that the trigger to this phenomenon could be the autoregulatory response to a minor lack of cellular metabolic energy. These findings are in agreement with a previous study carried out in our laboratory which demonstrates Ischemia-related oscillations of CBV and NADH redox state (Mayevsky et al., 1984). We observed cortical hemodynamic and metabolic oscillations only in pathological states related to either a reduction in substrate (such as CBF reduction, oxygen depletion) or to a reduction in tissue O_2 extraction.

In our previous study (Mayevsky and Ziv, 1991) we showed that pO2 was oscillated in the same frequency as the NADH as compared to the CBF measured in the present study. In an original work of Dora and Kovach (1981) the oscillations were induced by either blood pressure fluctuations, bleeding or by deep anesthesia. In the present study we were also able to detect more pathophysiological conditions that initiated the oscillations and also to correlate the NADH response to the microcirculatory oscillations (CBF). In our previous studies, the initiation factor of the oscillations was partial ischemia, suggesting that a systemic factor was involved. According to the present study it seem that various sources could initiate the oscillations. Not all of the factors are linked to a decrease in blood flow to the brain. The unique effect of carbon monoxide (CO) on the brain was an increase in CBF (Mayevsky et al., 1995). The fact that oscillations were induced under CO suggests that the inducing factor is located along the line of microcirculation to the mitochondrial activity.

Therefore we conclude that the source of the oscillations is a change in brain regulation of microcirculation and the pathological state developed. It is also clear that the oscillator is not central but rather a local one affected by various perturbations of CO levels, depth of anesthesia as well as events such as Trauma.

ACKNOWLEDGMENT

Supported by a grant from the committee for Research and Prevention in Occupational Safety and Health, Ministry of Labor and Social Affairs, Israel and by the Chief Scientist's Office Ministry of Health, Israel.

REFERENCES

Cooper R, Crow HJ, Greywalter W, and Winter AL (1966) Regional control of cerebral vascular reactivity and oxygen supply in man. Brain Res. 3:174–179.

Deyoe EA, Trusk TC, and Wong-Riley MT (1995) Activity correlates of cytochrome oxidase-defined compartments in granular and supragranular layers of primary visual cortex of the macaque monkey. Vis. Neurosci. 12:629–639.

Dora E and Kovach AGB (1981) Metabolic and vascular volume oscillations in the cat cortex, Acta Physiol. Acad. Sci. Hung. 53:261–275.

Hudetz AG, Roman RJ, and Harder DR (1992) Spontaneous flow oscillations in the cerebral cortex during acute changes in mean arterial pressure. J. Cereb Blood Flow Metab 12:491–499.

Jobsis FF (1978) In: Cerebral Vascular Smooth Muscle and its Control. Ciba Foundation Symposium 56 (New series), Elsevier North-Holland, Amsterdam, p. 46.

Mayevsky A (1984) Brain NADH redox state monitored *in vivo* by fiber optic surface fluorometry. Brain Res. Rev. 7:49–68.

Mayevsky A and Chance B (1982) Intracellular oxidation reduction state measured *in situ* by a multichannel fiber-optic-surface fluorometer. Science 217:537–540.

Mayevsky A, Doron A, Manor T, Meilin S, Zarchin N, and Ouaknine GE (1996) Cortical spreading depression recorded from the human brain using a multiparametric monitoring system. Brain Res. 740:268–274.

Mayevsky A, Meilin A, Rogatsky GG, Zarchin N, and Thom SR (1995) Multiparametric monitoring of the awake brain exposed to carbon monoxide. J. Appl. Physiol. 78:1188–1196.

Mayevsky A and Ziv I (1991) Oscillations of cortical oxidative metabolism and microcirculation in the ischemic brain. Neurol. Res. 13:39–47.

Mayevsky A, Ziv I, Ventura V, and Levy Y (1984) Multiparametric oscillations in the rat brain *in vivo*. In: Anonymous Oscillations in Physiological Systems: Dynamics and Control. The Institute of Measurement and Control, London, pp. 205–210.

Meilin S, Rogatsky GG, Thom SR, Zarchin N, Guggenheimer-Furman E, and Mayevsky A (1996) Effects of carbon monoxide exposure on the brain may be mediated by nitric oxide. J. Appl. Physiol. 81:1078–1083.

Meilin S, Zarchin N, Mayevsky A, and Shapira S (1995) Multiparametric responses to cortical spreading depression under nitric oxide synthesis inhibition. In: Weissman BA, Alon N, Shapira S (eds) Biochemical Pharmacological and Clinical Aspects of Nitric Oxide. Plenum Press, New York, pp. 195–204.

Vern BA, Schuette WH, Leheta B, Juel VC, and Radulovacki M (1988) Low-frequency oscillations of cortical oxidative metabolism in waking and sleep. J Cereb Blood Flow Metab 8:215–226.

Wadhwani KC and Rapoport SI (1996) Blood flow in the central and peripheral nervous systems. In: Shephard AP, Oberg PA (eds) Laser Doppler Blood Flowmetry. Kluwer Academic, Boston, pp. 265–304.

18

ALTERED LIGAND BINDING OF D_1 AND D_2 DOPAMINE RECEPTORS IN RESPONSE TO HYPOXIA AND POSTHYPOXIC REOXYGENATION

Stephanie J. Murphy, Thomas Liu, Michaela Windfuhr, Dekun Song, David F. Wilson, and Anna Pastuszko

Departments of Biochemistry and Biophysics and of Pediatrics
University of Pennsylvania
School of Medicine
Philadelphia, Pennsylvania 19104

INTRODUCTION

The physiological effects of dopamine, a major striatal neurotransmitter, are mediated by several distinct receptor subtypes. On the basis of structural, pharmacological, functional, and distributional similarities, these subtypes fall into one of two groups, the D_1-like receptor family and the D_2-like receptor family. The D_1-like receptors are linked to stimulation of adenylate cyclase via G_s proteins whereas the D_2-like receptors inhibit adenylate cyclase through a $G_{i/o}$ protein.[1] The D_1-like receptors are also associated with inositol phosphate production and calcium mobilization while the D_2-like receptors stimulate potassium efflux and indirectly inhibit inositol phosphate production. Thus, changes in the properties of these dopaminergic receptors could have a major impact on cellular metabolism. Several studies have examined the effect of ischemia-hypoxia on ligand binding properties of the striatal dopaminergic receptors, particularly on adult animals, and the authors suggested that the changes in the D_1 and/or D_2 receptors play a key role in the pathogenesis of damage to the striatum induced by ischemia. For example, Bonfenati et al.[4] proposed that reduction in the number of D_1 receptors plays a key role in the pathogenesis of ischemic striatal damage. The purpose of the present study was to

---

Address correspondence to: Dr. Anna Pastuszko, Department of Pediatrics and Biochemistry Biophysics, University of Pennsylvania, Philadelphia, Pennsylvania 19104, Tel. 215-898-6382; FAX 215-898-4217.

Oxygen Transport to Tissue XXI, edited by Eke and Delpy.
Kluwer Academic / Plenum Publishers, New York, 1999.

determine whether arterial hypoxia is associated with alterations in the affinities (K_d) and numbers (B_{max}) of the dopaminergic receptor subtypes, D_1 and D_2, in the striatum of newborn piglets.

MATERIALS AND METHODS

Animal Preparation

Newborn piglets, age 2–4 days, were used for this study. Anesthesia was induced with halothane (Halocarbon Laboratories, Augusta, SC; 4% mixed with 96% oxygen), and 1.5% lidocaine-HCl (Abbott Laboratories, North Chicago, IL) was used as a local anesthetic. The halothane was reduced to 0.6–0.8% after tracheotomy and Tubocurine-HCl was used to induce respiratory paralysis (Apothecon, Bristol-Meyers Squibb, Princton, NJ; 1.5 mg/kg), and the femoral artery and femoral vein were cannulated. Once the vessels were catheterized, halothane was withdrawn entirely and Fentanyl-citrate (Elkins-Linn, Inc., Cherry Hill, NJ; 30 μg/kg) was injected intravenously at approximately one hour intervals throughout the experiments. The animals were paralyzed with tubocurarine and mechanically ventilated with a mixture of oxygen and nitrous oxide (in control conditions with 21–22% oxygen and 78–79% N_2O). Following a control period, hypoxia was induced by lowering the fraction of inspired O_2 (FiO_2) to 7% for 1 h. At the end of the hypoxic period, the FiO_2 was raised to 21% for a 2 h period of reoxygenation. Sham-operated animals, ventilated with 21% FiO_2 for 3 h, served as controls. All animal use procedures were in strict accordance with the NIH Guide for the Care and Use of Laboratory Animals and were approved by the local Animal Care Committee.

Cortical Oxygen Pressure

The oxygen pressure in the microvasculature of the cortex was measured by the oxygen-dependent quenching of phosphorescence as described previously.[19,25] The oxygen probe (Pd-meso-tetra (4-carboxyphenyl) porphine, Porphyrin Products, Logan, Utah) was injected intravenously (5–8 mg/kg body weight) as a solution of 10 mg/ml probe in 60 mg/ml bovine serum albumin in saline, pH 7.4. Light from an EG&G flashlamp is filtered (an interference filter with a 535 nm center wavelength and a half-bandwidth of 30 nm) and conducted to the sample through a bifurcated light guide with randomized fibers, and focused on the tissue (approximately 5 mm diameter area) by a lens system on the end of the fiber. The emitted light was collected by the lens system on the end of the light guide and conducted to the photomultiplier through the other branch of the light guide. The instrument settings are such that the oxygen pressures given are those for the average of the microcirculation of tissue.

D_1 and D_2 Receptor Binding Assays

At the end of each study, the anesthetized animal was decapitated, the brain rapidly removed and the striata dissected out. The tissues were frozen and stored at −60°C until needed for assay. The striata were homogenized (Teflon/glass) at 4°C in isolation medium (50 mM Tris-HCl buffer, pH 7.4) diluted 1:50 (wt/vol) and centrifuged at 13,000 × *g* for 30 min in a Sorvall centrifuge. The supernatant was discarded, and the pellet washed twice as described above. The final pellet was resuspended in 2 mls of resuspension buffer (50 mM Tris-HCl, pH 7.4; 120 mM NaCl; 5 mM KCl; 1 mM $MgCl_2$; 1 mM $CaCl_2$; 10 μM paragyline; 0.01% ascorbic acid). The protein was then measured. Striatal membrane suspensions were then diluted to achieve a final concentration of 4 mg/ml for D_1 receptor binding assays and 2 mg/ml for D_2 receptor binding assays.

D_1 and D_2 receptor binding assays were carried out using modifications of the method described by Schettini et al.[20] Briefly, for D_1 receptor binding assays, [^{3}H]SCH23390 (87 Ci/mmol) used concentrations ranging from 0.1 nM to 3.0 nM. Nonspecific binding was obtained using 1 μM cold SCH23390. For D_2 receptors, [^{3}H]spiperone (99 Ci/mmol) binding was performed in the presence of 50 nM ritanserin to block 5-hydroxytryptamine$_2$ receptors, with concentrations of [^{3}H]spiperone ranging from 0.1 nM to 1.5 nM. Nonspecific binding was calculated using 10 μM(–)-sulpiride.

Samples containing 200 μg of protein and 50 μg of protein of striatal homogenate for D_1 and D_2 receptor binding assays, respectively were pre-incubated for 10 min at 37°C. The measurements were started by adding an aliquot of the radioactive ligand to the membrane suspension and incubating for 30 min at 37°C. The incubations were stopped by adding 4 mls of ice cold resuspension buffer. Reaction mixtures were then rapidly vacuum filtered through glass fiber filters (Schleicher and Schuell GF 50) and washed six times with approximately 4 mls of cold resuspension buffer. Filters were air-dried and the radioactivity was measured in a Searle Delta 300 liquid scintillation counter using an ICN aqueous scintillation fluid (EcoLite™).

Protein was measured by the method of Lowry et al.[14] with bovine serum albumin as the standard.

Statistical Evaluation

All physiologic data and binding parameters for the D_1 and D_2 receptors are expressed as means ± SEM. Significance of differences between mean values were assessed by repeated measures analysis of variance (ANOVA) and *post-hoc* analysis (Bonferroni test). Differences were considered to be statistically significant at $p < 0.05$.

RESULTS

Physiological Parameters of the Piglets during Hypoxia and Reoxygenation

The effects of one hour of arterial hypoxia and posthypoxic reoxygenation on the physiological parameters of newborn piglets are shown in Table 1. Hypoxia caused significant decrease in arterial blood pH from control values of 7.44 ± 0.01 to 7.18 ± 0.03 ($p < 0.001$). During reoxygenation, the pH increased to 7.36 ± 0.02, a value not significantly different from the control. The P_aO_2 decreased during hypoxia from 115.2 ± 2.1 Torr to 22.5 ± 0.05 Torr ($p < 0.001$). After two hours of reoxygenation the P_aO_2 had increased to 100.5 ± 3.1 Torr. The P_aCO_2 decreased from control levels of 37.7 ± 0.6 Torr to 32.0 ± 0.7 Torr during hypoxia ($p < 0.001$) and remained on this level during the reoxygenation period. MABP decreased significantly during the period of hypoxia from 89 ± 2 Torr to 57 ± 1 Torr ($p < 0.001$). Following reoxygenation, the blood pressure increased to 85 ± 2 Torr, a value comparable to the control value.

Table 1. The physiological parameters of newborn piglets during hypoxia and reoxygenation

Experimental Conditions	Blood pH	P_aO_2 (Torr)	P_aCO_2 (Torr)	MABP (Torr)
Control	7.44 ± 0.01	115.2 ± 2.1	37.7 ± 0.6	89 ± 2
Hypoxia	7.18 ± 0.03[c]	22.5 ± 0.5[c]	32.0 ± 0.7[c]	57 ± 1[c]
Reoxygenation				
60 min	7.11 ± 0.04[c]	112.5 ± 3.8	33.4 ± 1.5[a]	75 ± 2[c]
120 min	7.36 ± 0.02	100.5 ± 3.1[c]	32.1 ± 0.9[b]	85 ± 2

The values are the means ± SEM for seven experiments. [a]$p < 0.05$, [b]$p < 0.01$, [c]$p < 0.001$ for difference from control as determined by one way analysis of variance (ANOVA) with *post-hoc* Bonferroni test.

Effect of Hypoxia on Cortical Oxygen Pressure in Newborn Piglets

Hypoxia caused a progressive decrease in cortical oxygen pressure from normoxic values of 45.7 ± 1.7 Torr to 6.9 ± 1.1 Torr ($p < 0.001$) (Figure 1). During the first 10 minutes following reoxygenation the cortical oxygen pressure increased to 54.5 ± 5.9 Torr, a value not significantly different from control, and remained at this level throughout the remainder of the reoxygenation period.

The K_d and B_{max} of striatal D_1 and D_2 receptors during hypoxia and reoxygenation.

The effects of normoxia (control), hypoxia, and posthypoxic reoxygenation on the D_1 and D_2 receptors are shown in Tables 2 and 3, respectively. The values for K_d and B_{max} for the D_1 receptors under control conditions were 1.47 ± 0.07 nM and 27.9 ±

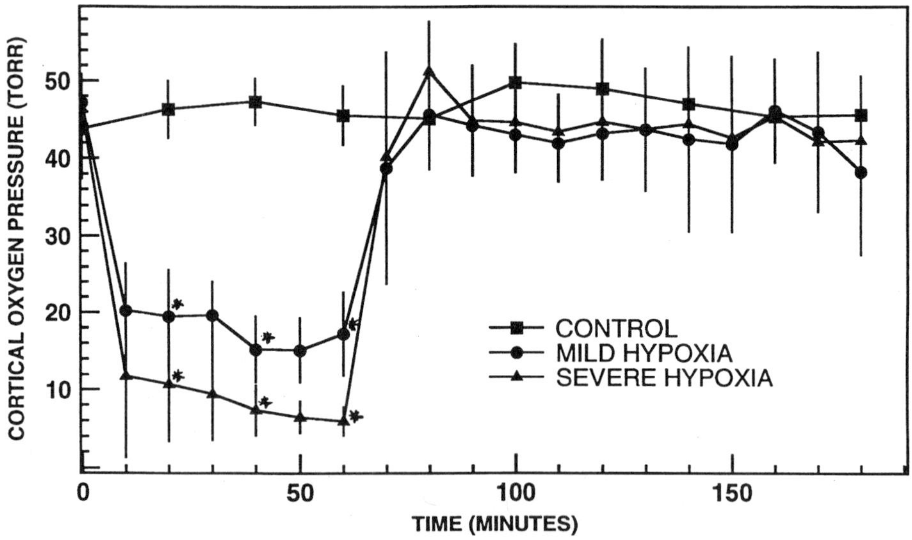

Figure 1. Changes in cortical oxygen in newborn piglets during hypoxia and posthypoxic reoxygenation. The cortical oxygen pressure was measured by oxygen-dependent quenching of phosphorescence as described in Materials and Methods. Hypoxia was initiated by decreasing the FiO_2 from 22% (control) to 7%–8% for 60 min and then returning the FiO_2 to control for a 2 h reoxygenation period. Results are given as means ± SEM values for 7 experiments. $^{**}p < 0.01$; $^{*}p < 0.001$, for significant difference compared with baseline values as determined by one way ANOVA with *post-hoc* Bonferroni test.

Table 2. Effects of normoxia, hypoxia, and posthypoxic reoxygenation on the kinetic parameters of D_1 receptors using the ligand, SCH23390

	K_d (nM)	B_{max} (fmoles/mg protein)
Control	1.47 ± 0.07	27.9 ± 5.2
Hypoxia	1.28 ± 0.05	22.8 ± 4.3
Reoxygenation	2.71 ± 0.08[a,b]	25.1 ± 6.7

Values are the means ± SEM for ten experiments. [a]$p < 0.01$ for difference from control and [b]$p < 0.01$ for difference from hypoxia as determined by one way analysis of variance (ANOVA) with *post-hoc* Bonferroni test.

5.2 fmoles/mg protein, respectively. One hour of hypoxia did not result in statistically significant changes in K_d or B_{max} for the D_1 receptors. After two hours of reoxygenation, the Kd (2.71 ± 0.08 nM) for the D_1 receptors was significantly higher ($p < 0.01$) than the values for control and hypoxia while B_{max} was not significantly altered. Control values for the D_2 receptor were 0.31 ± 0.08 nM and 482.6 ± 43.7 fmoles/mg protein, respectively. The K_d for the D_2 receptors was significantly increased by the period of hypoxia and reoxygenation to 0.60 ± 0.07 nM and 0.49 ± 0.06 nM, respectively. Statistically significant decrease in the B_{max} (358.2 ± 31.1 fmoles/mg protein, $p < 0.01$) was observed after two hours of reoxygenation.

DISCUSSION

The present study shows that in newborn piglets hypoxia generated by breathing 7% oxygen for one hour results in alterations in ligand binding by the D_1 and D_2 receptor subtypes in striatum. In the control piglets, the K_d values for the D_1 receptor subtype for SCH23390 and for the D_2 receptor subtype for spiperone in the controls are slightly higher than those reported in the literature, most of which are for mature animals. The K_d values reported for SCH23390 ranged from 0.2 nM to 0.63 nM[8,15,24,27] and are in good agreement with those cited for the native striatal receptor.[21,22] For the D_2 receptor subtype, the reported K_d values for spiperone range from 0.03 to 0.07 nM.[11,23] The range in K_d values in the literature could be partially due to species and/or age differences. It has been reported that B_{max} for the striatal dopamine receptors decreased with age in a variety of species, including humans. Membranes prepared from the striatum of young rats are reported to exhibit more dopamine receptors and to have an approximately 50% reduction in the number of D_2 receptor subtype mRNA compared to those isolated from senescent rats.[18]

Table 3. Effects of normoxia, hypoxia, and posthypoxic reoxygenation on the kinetic parameters of D_2 receptors binding of spiperone

	K_d (nM)	B_{max} (fmoles/mg protein)
Control	0.31 ± 0.08	463.6 ± 63.7
Hypoxia	0.60 ± 0.07[a]	517.3 ± 41.9
Reoxygenation	0.49 ± 0.06[a]	399.2 ± 31.1[b]

Values are the means ± SEM for eight experiments. [a]$p < 0.01$ for difference from control; [b]$p < 0.01$ for difference from hypoxia as determined by one way analysis of variance (ANOVA) with *post-hoc* Bonferroni test.

In the present study, arterial hypoxia followed by 2 hours of normoxic reoxygenation altered the properties of dopaminergic receptors. There was an increase in K_d (decreased ligand affinity) for the D_1 receptor subtype after two hours reoxygenation whereas the K_d for the D_2 receptor subtype K_d increased during hypoxia and reoxygenation. The D_2 receptor B_{max} decreased again during the reoxygenation period. These changes in properties of dopamine receptors following hypoxia/ischemia are in agreement with other reports that hypoxia/ischemia can lead to changes in the affinity and the number of D_1 and D_2 receptor in the brain. Chang et al.[6] reported that during ischemia there was a decreased affinity of D_1 receptors for ligands and a significant reduction in the number of D_2 receptors, suggesting desensitization and down regulation of D_1 and D_2 receptor subtypes, respectively. The effects appear to result from the excessive increase in extracellular dopamine during ischemia rather than being a response to that increase. Dawson et al.[7] demonstrated significant reduction of D_1 receptor subtype binding sites within the caudate nucleus at 2 h after occlusion of the middle cerebral artery. Similarly, Araki et al.[3] reported a marked decrease in the D_1 receptor subtypes in the dorsolateral striatum following ischemia. Autoradiographic studies of Zouakia et al.[28] showed that in perinatal rats hypoxia-ischemia delayed maturation of D_2 receptors; these receptors decrease bilaterally after one week recovery. These authors did not observe any changes in D_1 receptors. Johnson et al.[12] reported decreases in striatal binding density for both receptor types in neonatal rats 10 days after unilateral hypoxic/ischemic brain injury as well as an increase in K_d for D_1 and a decrease in B_{max} for both D_1 and D_2. Utilizing the same model in adult rats, Kostic et al.[13] reported significant reduction only in the binding capacity of D_2 receptor subtypes 9 to 11 weeks after injury.

The observed changes in the properties of the dopaminergic receptors can have a major impact on brain function. Studies of the late effect of ischemia demonstrated a decrease in D_1- and D_2-like receptor binding sites associated with cellular necrosis in the caudate putamen[17] and the disappearance of D_1 receptors has been related to neuronal necrosis in the dorsolateral striatum.[4] The mechanisms underlying the changes in dopaminergic receptors during hypoxia remain poorly defined. One possible mechanism could be through the regulation of the expression of these receptors. It is generally agreed that dopaminergic enervation induces an increase in the expression of D_2 receptor subtype mRNA, suggesting transcriptional control for up-regulation of these receptors.[2,5,10,16,26] Possible regulation of D_1 receptor subtype mRNA is less clear, with some reports indicating that dopaminergic denervation or chronic treatment with D_1 receptor antagonist SCH23390 decreased while others indicate increased D_1 mRNA levels.

Filloux et al.[9] reported that in neonatal rat the effects of hypoxia-ischemia on dopamine receptor mRNA differed between D_1 and D_2 receptors. Changes in D_1 receptor mRNA transcripts closely paralleled alterations in receptor number. The authors demonstrated that D_1 receptors appear to recover more quickly than do D_2 receptors from hypoxic-ischemic insults. Recovery in D_1 receptor protein was associated with a parallel recovery in D_1 receptor message. In contrast, decrements in D_2 receptor binding remain permanently affected after injury despite return of the D_2 receptor message to normal levels.

In conclusion, in neonatal striatum arterial hypoxia and posthypoxic reoxygenation can produce long lasting changes in the D_1 and D_2 dopamine receptor subtypes. This may occur either through regulation of receptor expression or via covalent modification of the receptors. These changes would be expected to affect subsequent brain metabolism and dopamine related behavior, as well as normal brain development.

ACKNOWLEDGMENT

Supported by grant NS-31465 from the National Institutes of Health.

REFERENCES

1. Andersen, PH, Gingrich, JA, Bates, MD, Dearry, A, Falardeau, P, Senoges, SE, and Caron, MG 1990. Dopamine receptor subtypes: beyond the D1/D2 classification. Trends Pharmacol. Sci. 11:231–236.
2. Angulo, JA, Coirini, H, Ledoux, M, and Schumacher, M 1991. Regulation by dopaminergic neurotransmission of dopamine D2 mRNA and receptor levels in the striatum and nucleus accumbens of the rat. Mol. Brain. Res. 11:161–166.
3. Araki, T, Murakami, F, Kanai, Y, Kata, H, and Kogure, K 1992. Autoradiographic analysis of dopamine D1 receptors in the gerbil brain following transient cerebral ischemia. Gen. Pharmac. 23:1073–1078.
4. Benfenati, F, Merlo Pich, E, Grimaldi, R, Zoli, M, Fuxe, K, Toffano, G, and Agnati, LF 1989. Transient forebrain ischemia produces multiple deficit in dopamine D1 transmission in the lateral neostriatum of the rat. Brain Res. 498:376–380.
5. Bernard, V, Le Moine, C, and Bloch, B 1991. Striatal neurons express increased level of dopamine D2 receptor mRNA in response to haloperidol treatment: a quantitative in situ hybridization study. Neurosci. 45:117–126.
6. Chang, CJ, Ishii, H, Yamamoto, T, and Spatz, M 1993. Effects of cerebral ischemia on regional dopamine release and D1 and D2 receptors. J. Neurochem. 60:1483–1490.
7. Dawson, DA, Robinson, MJ, Macrae, IM, Reid, JL, and McCulloch, J 1992. Autoradiographic evaluation of forskolin and D1 dopamine receptor binding in rat model of focal cerebral ischaemia. Brain Res. 577:210–217.
8. Dearry, A, Gingrich, JA, Falardeau, P, Fremeau, RT Jr, Bates, MD, and Caron, MG 1990. Molecular cloning and expression of the gene for a human D1 dopamine receptor. Nature. 347:72–76.
9. Filloux, FM, Adair, J, and Narang, N 1996. The temporal evolution of striatal dopamine receptor binding and mRNA expression following hypoxia-ischemia in the neonatal rat. Dev. Brain Res. 94: 81–91.
10. Gerfen, CR, Engber, TM, Mahan, LC, Susel, Z, Chase, TN, Monsma, FJ Jr, and Sibley, DR 1990. D1 and D2 dopamine receptor-regulated gene expression of striatonigral and striatopallidal neurons. Science. 250:1429–1432.
11. Gingrich, JA and Caron, MG 1993. Recent advances in the molecular biology of dopamine receptors. Ann. Rev. Neurosci. 16:299–321.
12. Johnson, M, Hanson, GR, Gibb, JW, Adair, J, and Filloux, F 1994. Effect of neonatal hypoxia-ischemia on nigro-striatal dopamine receptors and on striatal neuropeptide Y, dynorphin A and substance P concentrations in rats. Dev. Brain Res. 83:109–118.
13. Kostic, VS, Przedborski, S, Jackson-Lewis, V, Cadet, JL, and Burke, RE 1991. Effect of unilateral perinatal hypoxic-ischemic brain injury on striatal dopamine uptake sites and D1 and D2 receptors in adult rats. Neurosci. Lett. 129:197–200.
14. Lowry, OH, Rosenbrough, NJ, Farr, AL, and Randall, RJ 1951. Protein measurements with the Folin phenol reagent. J. Biol. Chem. 193:265–275.
15. Monsma, FJ, Maham, LC, McVittie, LD, Gerfen, CR, and Sibley, DR 1990. Molecular cloning and expression of a D1 dopamine receptor linked to adenylyl cyclase activation. Proc. Natl. Acad. Sci. USA. 87:6723–6727.
16. Neve, KA, Neve, RL, Fidel, S, Jankowsky, A, and Higgins, GA 1991. Increased abundance of alternatively splice forms of D2 dopamine receptor mRNA after denervation. Proc. Natl. Acad. Sci. USA. 88:2802–2806.
17. Przedborski, S, Kostic, V, Jackson-Lewis, V, Cadat, JL, and Burke, RE 1991. Effect of unilateral perinatal hypoxic-ischemic brain injury in the rat on dopamine D1 and D2 receptors and uptake sites: a quantiative autoradiographic study. J. Neurochem. 57:1951–1961.
18. Roth, GS and Joseph, JA 1994. Age-related changes in transcriptional and posttranscriptional regulation of the dopaminergic system. Life Sci. 55:2013–2035.
19. Rumsey, WL, Vanderkooi, JM, and Wilson, DF 1988. Imaging of phosphorescence: a novel method for measuring oxygen distribution in perfused tissue. Science. 241:1649–1651.

20. Schettini, G, Ventra, C, Florio, T, Grimaldi, M, Meucci, O, and Marino, A 1992. Modulation by GTP of basal and agonist-stimulated striatal adenylate cyclase activity following chronic blockade of D_1 and D_2 dopamine receptors: involvement of G proteins in the development of receptor supersensitivity. J. Neurochem. 59:1667–1674.
21. Seeman, P 1980. Brain dopamine receptors. Pharmac. Rev. 32:229–313.
22. Seeman, P 1992. Dopamine receptor sequences: therapeutic levels of neuroleptics occupy D2 receptors, clozapine occupies D4. Neuropsychopharmacology. 7:261–284.
23. Senogles, SE, Speigel, AM, Padrell, E, Iyengar, R, and Caron, MG 1990. Specificity of receptor-G protein interactions: discrimination of Gi subtypes by the D2 dopamine receptor in a reconstituted system. J. Biol. Chem. 265:4507–4514.
24. Sunahara, RK, Niznik, HB, Weiner, DM, Stormann, TM, Brann, MR, Kennedy, JL, Gelernter, JE, Rozmahel, R, Yang, YL, Israel, Y, Seeman, P, and O'Dowd, BF 1990. Human dopamine D1 receptor encoded by an intronless gene on chromosome 5. Nature. 347:80–83.
25. Wilson, DF, Pastuszko, A, DiGiacomo, JE, Pawlowski, M, Schneiderman, R, and Delivoria-Papadopoulos, M 1992. Effect of hyperventilation on oxygenation of the brain cortex of newborn piglets. J. Appl. Physiol. 70:2691–2696.
26. Xu, S, Monsma, FJ Jr, Sibley, DR, and Creese, I 1991. Regulation of D1a and D2 dopamine receptor mRNA during ontogenesis, lesion and chronic antagonist treatment. Life Sci. 50:383–396.
27. Zhou, Q-Y, Grandy, DK, Thambi, L, Kushner, JA, Van Tol, HHM, Cone, R, Pribnow, D, Salon, J, Buzow, JR, and Civelli, O 1990. Cloning and expression of human and rat D1 dopamine receptors. Nature. 347:76–79.
28. Zouakia, A, Guilloteau, D, Zimmer, l, Besnard, J-C, and Chalon, S 1997. Evolution of dopamine receptors in the rat after neonatal hypoxia-ischemia: autoradiographic studies. Life Sci. 60:151–162.

REGIONAL EXPRESSION OF HEAT SHOCK PROTEIN 72 MRNA FOLLOWING MILD AND SEVERE HYPOXIA IN NEONATAL PIGLET BRAIN

Stephanie J. Murphy,[1] Dekun Song,[1] Frank A. Welsh,[2] David F. Wilson,[1] and Anna Pastuszko[2]

[1]Department of Biochemistry and Biophysics and [2]Division of Neurosurgery
Medical School, University of Pennsylvania
Philadelphia, Pennsyl vania 19104

ABSTRACT

The present study examined the effect of hypoxia on expression of 72-kDa heat shock protein (hsp72) mRNA in the newborn brain. The studies were carried out in anesthetized and mechanically ventilated newborn piglets, age 3–5 days. Hypoxic insult was induced by decreasing the fraction of inspired oxygen (FiO_2) from 21% to 6% or 10% for 1 h. Oxygen pressure in the microvasculature of the cortex (cortical pO_2) was measured by oxygen dependent quenching of the phosphorescence of phosphor dissolved in blood. Following the two hours of normoxic recovery, regional expression of the 72-kDa heat shock protein (hsp72) mRNA was determined using *in situ* hybridization and autoradiography. Two grades of hypoxia were studied. Mild hypoxia (cortical pO_2 = 10–30 mm Hg) induced the expression of hsp72 mRNA predominantly in the subcortical white matter. In individual animals of this group, the extent of expression varied from isolated regions to widespread involvement of the white matter. Severe hypoxia (cortical pO_2 = 3–10 mm Hg) induced the expression of hsp72 mRNA in both white and gray matter regions, with strong expression occurring in the cerebral cortex of individual

Abbreviations used: hsp72, 72-kDa heat-shock protein; F_iO_2, fraction of inspired O_2

Address correspondence to: Dr. Anna Pastuszko, Department of Biochemistry and Biophysics, University of Pennsylvania Philadelphia, Pennsylvania 19104. Telephone: 215-898-6382; Telefax: 215-898-4217.

Oxygen Transport to Tissue XXI, edited by Eke and Delpy.
Kluwer Academic / Plenum Publishers, New York, 1999.

animals. The present results indicate that immature white matter is more sensitive than gray matter to the hypoxia induced expression of hsp72 mRNA. Further, increased expression of hsp72 mRNA may be an indicator of a pathologic degree of hypoxic stress, and the observed increase may indicate that in the newborn brain the immature white matter is particularly sensitive to injury by hypoxia-ischemia and reperfusion.

INTRODUCTION

Hypoxia-ischemia is a common cause of brain damage in the neonate, resulting in mental retardation, neurologic dysfunction, or death. However, the pathophysiology of perinatal hypoxia-ischemia and the factors that determine regional vulnerability in developing brain are not fully understood. Recent studies have demonstrated that hypoxia-ischemia induces the 72-kDa heat shock protein (hsp72) in the brain of immature rats (Ferriero et al., 1990; Blumenfeld et al., 1992; Munell et al., 1994). Induction of hsp72 has been correlated with cellular injury in experimental models of ischemia, excitotoxicity, and trauma (Gonzalez et al., 1989; Sloviter and Lowenstein, 1992; Nowak, 1993; Sharp and Sagar, 1994). Thus expression of hsp72 may be a useful tool for identifying regions of the neonatal brain at risk of hypoxic injury. The objective of the present study was to determine the regional expression of hsp72 mRNA in a model of arterial hypoxia in newborn piglets.

MATERIAL AND METHODS

Animal Preparation

Newborn piglets, 1–4 days of age, averaging 2 kg in body weight, were employed in this study. General anesthesia was induced with a mixture of 4% halothane, balance O_2; local anesthesia was produced with 1.5% lidocaine-HCl. Following a tracheotomy, the percentage of halothane was reduced to 0.6% to 0.8%, and the femoral artery and vein were cannulated. The animals were administered Fentanyl-citrate (30 µg/kg/h, i.v.), and halothane was withdrawn from the ventilation mixture. The animals were immobilized with Tubocurarine-HCl (0.3 mg/kg, i.v.) and ventilated mechanically with a mixture of 22% O_2, balance N_2O. A rectal temperature probe was inserted to record core temperature. The head of the animal was placed in a Kopf stereotaxic frame and a longitudinal scalp incision was made along the midline. After exposing the skull, a burr hole 12 mm in diameter was made over the parietal region of one hemisphere. A small area of the dura was exposed for measurement of cortical pO_2 (see below) through the dura. MABP, heart rate and end-tidal CO_2 were monitored continuously. Arterial blood samples were taken every 10–20 min for measurement of pH, p_aCO_2 and p_aO_2.

Following a 15-min control period, arterial hypoxia was induced by lowering the fraction of inspired O_2 (FiO_2) to the range of 0.06–0.10. Two grades of hypoxia were produced. In the first group (mild hypoxia, n = 6), FiO_2 was lowered until the cortical pO_2 fell from normal values of 47 mm Hg to the range of 10–30 mm Hg. In the second group (severe hypoxia, n = 5), FiO_2 was lowered until the cortical pO_2 fell from normal values to the range of 3–10 mm Hg. In both groups, arterial hypoxia was maintained for 1 h, at which time the FiO_2 was raised to 0.22 for a 2-h recovery period. Sham-operated animals, ventilated with 22% O_2 for 3 h (n = 5), served as controls. All animal use procedures were in strict accordance with the NIH Guide for the Care and Use of Laboratory Animals and were approved by the local Animal Care Committee.

Cortical pO_2 Measurements

Cortical pO_2 was measured using an O_2 dependent phosphorescence quenching method described previously (Rumsey et al., 1988; Wilson et al., 1992). In brief, the O_2-sensing molecule, Pd-meso-tetra(4-carboxyphenyl) porphine (Porphyrin Products, Logan, UT), was administered (5–8 mg/kg, i.v.) in a 10 mg/ml solution containing 60 mg/ml bovine serum albumin in physiologic saline, buffered to pH 7.4. Light from an EG & G flashlamp was passed through an interference filter (535 nm center wavelength and a half-bandwidth of 30 nm) and conducted to the exposed parietal cortex through a bifurcated light guide with randomized fibers. The light was focused on a 3 mm diameter area of tissue. Light emitted from the cortex was collected by a lens system on the end of the light guide, filtered with a long pass filter with 50% transmission at 630 nm, and measured by a photomultiplier. The resulting signal was amplified, digitized with a 12 bit, 1 MHz A/D board, and analyzed using a microcomputer. The digitized decay curve was fitted to a single exponential decay to determine the lifetime, and pO_2 was calculated using the Stern-Volmer equation. The pO_2 recorded corresponds to the mean value in the microcirculation. Measurements were taken every 10 min.

In Situ Hybridization

At the end of the 2-h period of reoxygenation, the animals were sacrificed by i.v. administration of a saturated KCl solution, and the brain was removed from the cranium and frozen in a bed of finely powdered dry ice. *In situ* hybridization was carried out using modifications of previously described methods (Welsh et al., 1992). In brief, the brain was sectioned in a cryostat at –25°C at a thickness of 14 µm. Sections were mounted on subbed slides, dried, fixed in 4% paraformaldehyde, and rinsed in 2X SSC (saline/sodium citrate). Hsp72 mRNA was detected using a 30-mer oligodeoxynucleotide probe with the following sequence: 5′-C GAT CTC CTT CAT CTT GGT CAG CAC CAT GG-3′. This sequence was reported specific to the inducible member of the family of 70-kDa heat-shock proteins (Nowak et al., 1990; Miller et al., 1991). The oligonucleotide probe was labeled on the 3′-end with [^{35}S]deoxyATP using terminal transferase. The labeled probe was isolated and hybridized to tissue sections overnight at 37°C in 4X SSC. After rinsing for 1.5 h each in 1X SSC and 0.5X SSC at 37°C, the sections were dried and exposed to Kodak SB film for 1–2 weeks.

The extent of hsp72 mRNA expression in specific regions was quantitated using image analysis of the autoradiograms (Image 1.47, NIH). For each animal, a single section at the level of the striatum was analyzed. Areas of expression were determined separately in four brain regions: subcortical white matter, internal capsule, cerebral cortex, and caudate nucleus. Areas exhibiting at least a 2-fold increase in optical density compared to corresponding regions of sham-operated controls were included in the area of expression. In the majority of cases, the optical density ranged between 3 and 5-fold higher than that in control sections. However, in some cases, the optical density was elevated only 2–3 fold and these were designated "weak expression" to distinguish them from stronger expression (see Table 2).

Statistical Analysis

All physiologic data were expressed as means ± SD. Differences between mean values were assessed by the repeated measures analysis of variance (ANOVA) and *post-hoc* analysis (Bonferroni test) in order to adjust the significance level for multiple comparisons. Differences were considered to be statistically significant at $p < 0.05$.

RESULTS

Effect of Mild and Severe Hypoxia on the Physiologic Parameters in Newborn Piglets

The physiological parameters of newborn piglets during hypoxia and reoxygenation are presented in Table 1. Reduction of FiO_2 decreased P_aO_2 to 23–27 and 20–21 mm Hg in the mild and severe hypoxia groups, respectively. Reoxygenation caused P_aO_2 to return to control levels in both groups. During mild hypoxia, MABP decreased to a steady level of 57–59 mm Hg, but severe hypoxia caused a progressive decline of MABP to 38 mm Hg. Reoxygenation was accompanied by an increase of MABP to 70 mm Hg in both groups. There were no significant differences in P_aCO_2 between groups at any time. Arterial pH decreased during both mild and severe hypoxia, but returned to normal levels during reoxygenation. The core temperature remained at 37.8°C ± 1.0°C for the duration of the experiment.

Table 1. Effect of hypoxia and reoxygenation on physiological parameters of newborn piglets

Experimental group	Time point (min)	P_aO_2 (Torr)	MABP (Torr)	P_aCO_2 (Torr)	pH
Control	0	109 ± 6	92 ± 22	40 ± 5	7.46 ± 0.05
	20	152 ± 59	88 ± 11	40 ± 8	7.43 ± 0.10
	40	134 ± 22	91 ± 19	38 ± 9	7.47 ± 0.07
	60	114 ± 17	86 ± 8	34 ± 6	7.46 ± 0.04
	120	106 ± 16	87 ± 20	35 ± 4	7.43 ± 0.11
	180	114 ± 18	89 ± 15	35 ± 8	7.37 ± 0.10
Mild hypoxia					
Preinsult	0	120 ± 21	87 ± 8	42 ± 5	7.41 ± 0.08
Hypoxia	20	23 ± 5**	59 ± 10*	35 ± 9	7.38 ± 0.14
	40	22 ± 5**	57 ± 9*	38 ± 4	7.20 ± 0.23*
	60	27 ± 8**	59 ± 16*	34 ± 7	7.08 ± 0.26**
Reoxygenation	120	120 ± 4	68 ± 7	38 ± 8	7.31 ± 0.14
	180	113 ± 12	72 ± 15	33 ± 4	7.37 ± 0.06
Severe hypoxia					
Preinsult	0	137 ± 21	96 ± 9	36 ± 4	7.50 ± 0.03
Hypoxia	20	21 ± 4**	58 ± 7*	36 ± 5	7.29 ± 0.13
	40	20 ± 3**	50 ± 6**	31 ± 12	7.05 ± 0.23**
	60	21 ± 5**	38 ± 5**	35 ± 20	6.83 ± 0.23**
Reoxygenation	120	111 ± 15	68 ± 12	35 ± 5	7.26 ± 0.08
	180	110 ± 18	70 ± 15	32 ± 3	7.35 ± 0.11

The results are presented as the means ± SD for 5–6 experiments. $^{*}p < 0.05$, $^{**}p < 0.001$ significantly different from control values at each time point as determined by one way analysis of variance (ANOVA) with *post-hoc* Bonferroni test.

Cortical Oxygen Pressure in Newborn Piglets during Hypoxia

Reduction of FiO_2 decreased cortical pO_2 from normoxic values of 47 ± 5 mm Hg (mean ± SD) to 19 ± 6 mm Hg by 20 min and to 17 ± 5 mm Hg by 60 min in the mild hypoxia group (Figure 1). In the severe hypoxia group, cortical pO_2 decreased to 11 ± 7 mm Hg by 20 min and to 6 ± 2 mm Hg by 60 min. In both groups, reoxygenation caused prompt return of cortical pO_2 to normal levels. Cortical pO_2 in sham-operated animals remained in the normal range during the experimental procedure.

Expression of hsp72 mRNA in Brain of Newborn Piglets during Hypoxia

In sham-operated animals, expression of hsp72 mRNA was low in all brain regions except for the choroid plexus and ventricular lining in two animals (Table 2). Both mild and severe hypoxia caused increased expression of hsp-72 mRNA, but the topography and intensity of gene expression varied depending on the degree of hypoxic stress (Figure 2). During mild hypoxia, increased hsp-72 mRNA was seen primarily in choroid plexus/ventricular lining and in the white matter. As the degree of the hypoxic insult increased within the mildly hypoxic group, there was an increase in the area of white matter showing increased gene expression before other structures such as internal capsule, midbrain, cortex and caudate nucleus were affected (Table 2).

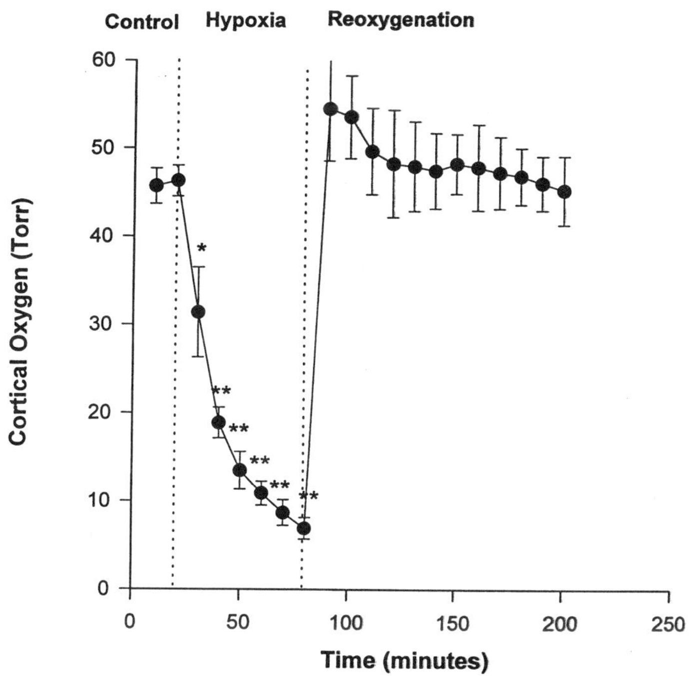

Figure 1. Effect of hypoxia and reoxygenation on cortical pO_2 in neonatal piglets. The fraction of inspired O_2 was lowered until cortical pO_2 fell into the range 10–30 mm Hg (mild hypoxia) or 3–10 mm Hg (severe hypoxia). After hypoxia for 1 h, the fraction of inspired O_2 was restored to normal (0.22) for 2 h. Control animals were sham-operated without hypoxia. Data are means ± SD for 5–6 animals in each group. Asterisks denote a significant difference ($p < 0.001$) from Control as determined by one way ANOVA with *post-hoc* Bonferroni test.

Table 2. Regional expression of hsp72 mRNA in neonatal piglet brain following mild and severe hypoxia

Experimental group	CP/VL	WM	IC	MB	CX	CN
Control						
1	0	0	0	0	0	0
2	0	0	0	0	0	0
3	0	0	0	0	0	0
4	+1	0	0	0	0	0
5	+2	0	0	0	0	0
Mild hypoxia						
1	0	0	0	0	0	0
2	+1	0	0	0	0	0
3	0	+2	0	0	0	0
4	0	+2	0	0	0	0
5	+2	+1	+1	+1	0	0
6	+2	+2	+2	+1	+1	+1
Severe hypoxia						
1	+2	+1	+1	0	0	0
2	+1	+2	0	0	+1	0
3	+2	+1	+1	+1	+1	0
4	+2	+2	+2	+2	+1	0
5	+2	+2	+2	+2	+2	+2

Grades of expression: 0 = no difference, +1 = mild to moderate increase in optical density, +2 = strong increase in optical density. CP = Choroid Plexus; VL = Ventricular Lining; WM = White Matter; IC = Internal Capsule; MB = Midbrain; CN = Caudate Nucleus; CX = Cortex.

Severe hypoxia induced the expression of hsp72 mRNA in both gray and white matter regions. Within the gray matter, the extent of expression in individual animals varied from widespread involvement of the cerebral cortex, caudate nucleus and deep gray matter regions to restricted patches in cerebral cortex and hypothalamus (Figure 2, Table 2). Within the white matter, strong expression of hsp72 mRNA was evident in the central areas, while expression was often absent in the tracts of white matter extending into the cortical gyri.

DISCUSSION

The present results demonstrate that arterial hypoxia induces regionally heterogeneous expression of hsp72 mRNA in the brain of neonatal piglets. Milder levels of hypoxia induced expression preferentially in the cerebral white matter, while severe hypoxia induced expression in both white matter and gray matter regions. The selective expression of hsp72 mRNA in white matter is similar to recent findings in the brain of 7-day-old rats, in which white matter was the only region to exhibit expression following the mildest levels of hypoxia-ischemia (Kobayashi and Welsh, 1995). The preferential white matter expression of hsp72 mRNA is not due to a limited capacity for expression in the immature gray matter, since severe hypoxia was capable of inducing robust expression of hsp72 in the gray matter. Rather, the selective white matter expression indicates that white matter is either more sensitive than gray matter to hypoxia or that the degree of hypoxia may be more severe in white matter.

Before further discussion of selective white matter expression of hsp72 mRNA, the

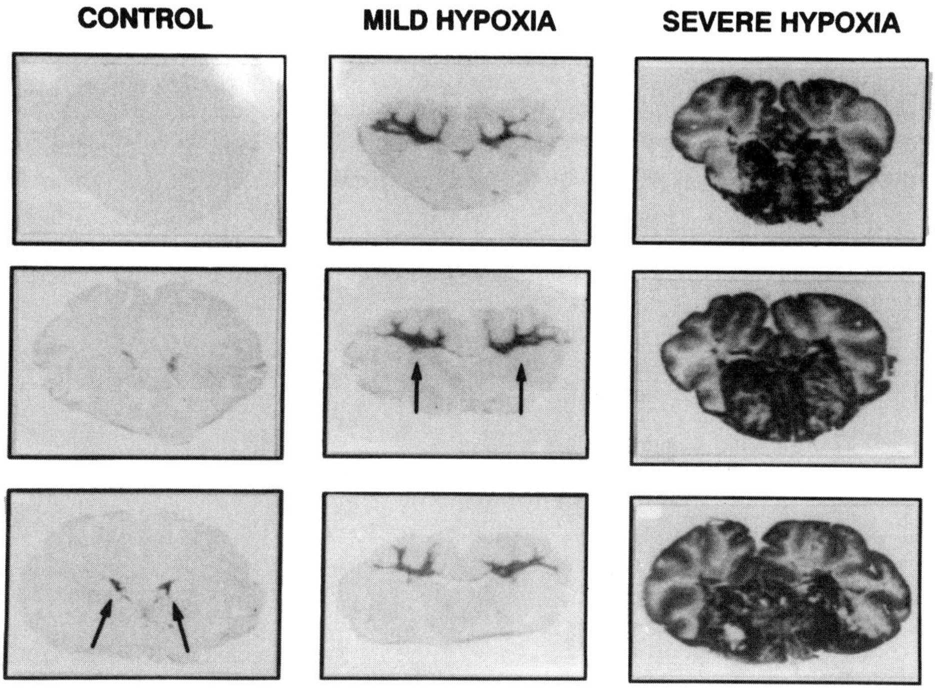

Figure 2. Regional expression of hsp72 mRNA following hypoxia in neonatal piglet brain. Autoradiograms from 3 brain sections in one animal from each group are presented. Increased optical density represents increased hybridization for hsp72 mRNA. In sham-operated controls, hsp72 mRNA was detected only in the choroid plexus and ventricular lining (control, arrows). Following mild hypoxia, expression of hsp72 mRNA occurred predominantly in the subcortical white matter (mild hypoxia, arrows). Severe hypoxia induced expression of hsp72 mRNA in both gray matter and white matter, although expression was low in the white matter extending into the cortical gyri (severe hypoxia).

biological significance of the expression should be considered. In previous studies, expression of hsp72 and its mRNA in unstressed brain was negligible (Vass et al., 1988; Nowak et al., 1990; Welsh et al., 1992). In sham-operated animals of the present study, expression of hsp72 mRNA was limited to the choroid plexus and ventricular lining, structures in which expression has been reported in control animals previously (Vass et al., 1988). Thus, the appearance of detectable amounts of hsp72 mRNA in the brain parenchyma following arterial hypoxia indicates that the degree of tissue hypoxia was sufficient to trigger expression of hsp72 mRNA. However, the precise molecular signals that couple tissue hypoxia to expression of hsp72 mRNA have not been identified. Induction of hsp72 by anoxia and other stresses is mediated, in part, by activation of a pre-existing protein factor (Wu et al., 1987; Wu et al., 1994). Activation of this factor *in vitro* was reported to require anoxic insults that cause prolonged depletion of cellular ATP (Benjamin et al., 1992). However, the degree of hypoxia *in vivo* required to activate this factor has not been defined. Nevertheless, previous studies indicate that potentially lethal degrees of cell stress are required to trigger expression of hsp72 (Gonzalez et al., 1989; Sloviter and Lowenstein, 1992; Sharp and Sagar, 1994). Thus, we conclude that selective white matter expression of hsp72 mRNA following mild arterial hypoxia indicates that the degree of hypoxic stress in white matter has the potential to cause long term disturbance in the function of this structure.

Previous studies of blood flow and metabolism also suggest that hypoxia may preferentially affect the white matter in neonatal brain (Duffy et al., 1982; Cavazzuti and Duffy, 1982). In newborn dogs, the hypoxic increase in cerebral blood flow was reported to be lower in subcortical white matter and corpus callosum than in gray matter structures (Cavazzuti and Duffy, 1982). In addition, hypoxia caused greater activation of glycolysis in white matter, presumably triggered by anaerobic alterations of regulatory metabolites (Duffy et al., 1982). Further, hypoxic alterations of tissue energy metabolites were more pronounced in subcortical white matter than in the adjacent cortex (Duffy et al., 1982). Thus, the limited increase in white matter blood flow during hypoxia may contribute to the activation of glycolysis, alteration of energy metabolites, and, as demonstrated in the present study, expression of hsp72 mRNA.

The regional variation of hsp72 mRNA expression within the brain of individual animals and between brain from different animals. First, within the white matter of individual animals, hsp72 mRNA was often expressed in the central portions of the white matter, but not in the peripheral extension of tracts into the cortical gyri. It is possible that the blood supply to the central white matter is inferior to that in peripheral white matter, due to the proximity of the latter to the more vascularized gray matter. Alternatively, the central white matter of neonatal animals is the focus of active myelination and, thus, may be more susceptible to hypoxia than other white matter regions (Rice et al., 1981). The variation in hsp72 mRNA expression between individual animals may have been caused by differences in cortical pO_2 during the hypoxic insult. However, the mean cortical pO_2 of individual animals during hypoxia did not correlate well with the resulting extent of expression. Further, the validity of the assumption that pO_2 measured in a single region of cortex accurately reflects the tissue pO_2 throughout the entire brain may be questioned, since regionally heterogeneous expression of hsp72 mRNA was evident in the brain of several animals.

In conclusion, there are probably several factors that contribute to the observed variations in hsp72 mRNA expression. The present results demonstrate that immature white matter is especially susceptible to hypoxia, at least with respect to expression of hsp72 mRNA. This selective white matter expression may be indicative of a greater degree of hypoxic stress and, thus, may help to explain the high pathologic vulnerability of the white matter in neonatal brain (Ranck and Windle, 1959; Banker, 1967; DeReuck et al., 1972; Brierley, 1976; Myers, 1977; Rice et al., 1981).

ACKNOWLEDGMENT

We thank Dr. Ellen G. Shaver and Ms. Valerie Harris for their technical assistance. This work was supported by NIH grants NS-29331 and NS-31465.

REFERENCES

Banker BQ (1967) The neuropathological effects of anoxia and hypoglycemia in the newborn. *Dev Med Child Neurol* 9:544–550.

Benjamin IJ, Kroger B, and Williams RS (1990) Activation of the heat shock transcription factor by hypoxia in mammalian cells. *Proc Natl Acad Sci USA* 87:6263–6267.

Blumenfeld KS, Welsh FA, Harris VA, and Pesenson MA (1992) Regional expression of c-fos and heat shock protein-70 mRNA following hypoxia-ischemia in immature rat brain. *J Cereb Blood Flow Metab* 12:987–995.

Brierley JB (1976) Cerebral hypoxia. In: *Greenfield's Neuropathology*, 3rd ed (Blackwood W, Corsellis JAN, eds), London, Edward Arnold, pp. 43–85.

Cavazzuti M and Duffy TE (1982) Regulation of local cerebral blood flow in normal and hypoxic newborn dogs. *Ann Neurol* 11:247–257.

DeReuck J, Chattha AS, and Richardson EP Jr (1972) Pathogenesis and evolution of periventricular leukomalacia in infancy. *Arch Neurol* 27:229–236.

Duffy TE, Cavazzuti M, Cruz NF, and Sokoloff L (1982) Local cerebral glucose metabolism in newborn dogs: effects of hypoxia and halothane anesthesia. *Ann Neurol* 11:233–246.

Ferriero DM, Soberano HQ, Simon RP, and Sharp FR (1990) Hypoxia-ischemia induces heat shock protein-like (hsp72) immunoreactivity in neonatal rat brain. *Dev Brain Res* 53:145–150.

Gonzalez MF, Shiraishi K, Hisanaga K, Sagar SM, Mandabach M, and Sharp FR (1989) Heat shock proteins as markers of neural injury. *Mol Brain Res* 6:93–100.

Kobayashi S and Welsh FA (1995) Regional alterations of ATP and heat-shock protein-72 kDa mRNA following hypoxia-ischemia in neonatal rat brain. *J Cereb Blood Flow Metab* 15:1047–1056.

Miller EK, Raese JD, and Morrison-Bogorad M (1991) Expression of heat shock protein 70 and heat shock cognate 70 messenger RNAs in rat cortex and cerebellum after heat shock or amphetamine treatment. *J Neurochem* 56:2060–2071.

Munell F, Burke RE, Bandele A, and Gubits RM (1994) Localization of c-*fos*, c-*jun*, and HSP70 mRNA expression in brain after neonatal hypoxia-ischemia. *Dev Brain Res* 77:111–121.

Myers RE (1977) Experimental models of perinatal brain damage: relevance to human pathology. In: *Intrauterine Asphyxia and the Developing Fetal Brain* (Gluck L, ed), Chicago, Year Book Press, pp. 37–97.

Nowak TS Jr (1993) Synthesis of heat shock/stress proteins during cellular injury. In: *Markers Of Neuronal Injury And Degeneration—Annals Of The New York Academy Of Sciences, Vol. 679* (Johannessen JN, ed), New York, The New York Academy Of Sciences, pp. 142–156.

Nowak TS Jr, Bond U, and Schlesinger MJ (1990) Heat shock mRNA levels in brain and other tissues after hyperthermia and transient ischemia. *J Neurochem* 54:451–458.

Ranck JB and Windle WF (1959) Brain damage in the monkey, *Macaca mulatta*, by asphyxia neonatorum. *Exp Neurol* 1:10–154.

Rice JE III, Vannucci RC, and Brierley JB (1981) The influence of immaturity on hypoxic-ischemic brain damage in the rat. *Ann Neurol* 9:131–141.

Rumsey WL, Vanderkooi JM, and Wilson DF (1988) Imaging of phosphorescence: a novel method for measuring oxygen distribution in perfused tissue. *Science* 241:1649–1651.

Sharp FR and Sagar SM (1994) Alterations in gene expression as an index of neuronal injury: heat shock and the immediate early gene response. *Neurotoxicity* 15(1):51–60.

Sloviter RS and Lowenstein DH (1992) Heat shock protein expression in vulnerable cells of the rat hippocampus as an indicator of excitation-induced neuronal stress. *J Neurosci* 12(8):3004–3009.

Vass K, Welch WJ, and Nowak TS Jr (1988) Localization of 7-kDa stress protein induction in gerbil brain after ischemia. *Acta Neuropathol* (Berl) 77:128–135.

Welsh FA, Moyer DJ, and Harris VA (1992) Regional expression of heat shock protein-70 mRNA and c-fos mRNA following focal ischemia in rat brain. *J Cereb Blood Flow Metab* 12:204–212.

Wilson DF, Pastuszko A, DiGiacomo JE, Pawlowski M, Schneiderman R, and Delivoria-Papadopoulos M (1992) Effect of hyperventilation on oxygenation of the brain cortex of newborn piglets. *J Appl Physiol* 70:2691–2696.

Wu C, Wilson S, Walker B, Dawid I, Paisley T, Zimarino V, and Ueda H (1987) Purification and properties of *Drosophila* heat shock activator protein. *Science* 238:1247–1253.

Wu C, Clos J, Giorgi G, Haroun RI, Kim S-J, Rabindran K, Westwood JT, Wisniewski J, and Yim G (1994) Structure and regulation of heat shock transcription factor. In: *The Biology of Heat Shock Proteins and Molecular Chaperones* (Morimoto RI, Tissieres A, and Georgopolous C, eds) Cold Spring Harbor Press, pp. 395–416.

20

EXCITATORY AND INHIBITORY PATHWAYS FOR ANOXIC PRECONDITIONING NEUROPROTECTION IN HIPPOCAMPAL SLICES

Miguel A. Pérez-Pinzón

Department of Neurology
University of Miami School of Medicine
Miami, Florida 33101

INTRODUCTION

The induction of tolerance against ischemia resulting from sublethal ischemic or anoxic insults has gained attention as a robust neuroprotective mechanism against conditions of stress such as anoxia/ischemia in heart and brain (Murry et al., 1986; Kato et al., 1992; Li et al., 1992; Liu et al., 1992; Walker et al., 1993; Pérez-Pinzón et al., 1997). There are different preconditioning paradigms both in heart and brain. For example, variations in preconditioning paradigms include, the number of preconditioning insults, types of preconditioning insults, and time between preconditioning and the test insults.

Most studies in brain suggest that preconditioning required several hours to develop. However, in recent studies we and others suggested that preconditioning with a rapid onset time course, similar to that in heart (within 1 h), can protect synaptic activity after anoxia in brain slices (Schurr et al., 1986; Schurr and Rigor, 1987; Pérez-Pinzón et al., 1996; Pérez-Pinzón and Born, 1999) and reduce histopathology after ischemia in intact brain (Pérez-Pinzón et al., 1997; Pérez-Pinzón et al., 1999b).

This suggested that mechanisms inducing this state of tolerance in heart and brain may be common to both organs. For example, it was found that: 1) activation of the adenosine A1 receptor induced ischemic/anoxic tolerance in both heart and brain; 2) whereas adenosine A1 receptor antagonists blocked preconditioning neuroprotection in both organs (Liu et al., 1991; Fitzgibbons et al., 1994; Gidday et al., 1995; Heurteaux et al., 1995; Pérez-Pinzón et al., 1996); and 3) activation of the K^+_{ATP} channel has been linked to preconditioning in both organs (Auchampach et al., 1991; Gross and Auchampach,

Oxygen Transport to Tissue XXI, edited by Eke and Delpy.
Kluwer Academic / Plenum Publishers, New York, 1999.

1992; Auchampach and Gross, 1993; Heurteaux et al., 1995; Riepe et al., 1997; Pérez-Pinzón and Born, 1999).

In the current study, we review additional pathways that may be common to both organs. The adenosine A1 receptor and the K^{+}_{ATP} channel due to their inhibitory properties in neurons fall within the inhibitory pathways of ischemic/anoxic preconditioning. We also investigated whether mild excitability as that promoted by an hyper-excitable agent or KCl-induced depolarization could also condition and thus protect brain slices from "test" anoxia, and whether extracellular calcium is involved in the neuroprotection afforded by preconditioning anoxia. We then proceed to discuss possible common pathways between the inhibitory and excitatory pathways.

MATERIALS AND METHODS

Slice Preparation

Description of slice preparation and synaptic activity measurements have been described in prior studies (Pérez-Pinzón et al., 1996; Pérez-Pinzón and Born, 1999).

Drugs

Adenosine and tolbutamide were purchased from SIGMA Chemical Company. Pinacidil, 2-chloroadenosine (2-CADO) and 8-cyclopentyl-1,3-dipropylxanthine (DPCPX) were purchased from Research Biochemical International. DPCPX was dissolved in a stock solution with 10% ethanol. The final ethanol concentration in ACSF was 0.5%. Statistical comparisons were made between groups using DPCPX and control slices superfused in ACSF containing 0.5% ethanol. In addition, no significant differences were found between results from control slices in normal ACSF and control slices in ACSF that contained a final concentration of 0.5% ethanol.

Study Groups and Experimental Paradigms

Slices were assigned to one of five groups. As in previous studies, for pharmacological emulation of preconditioning we used longer periods of AD for "test" ischemia, since more robust neuroprotection was found with these pharmacological agents than that found during anoxic preconditioning. Thus, for "test" anoxic insults, 1 min of AD was used for anoxic preconditioning experiments, and 2 min of AD for pharmacological emulation of preconditioning.

Group 1. Control group: slices were made anoxic for times sufficient to produce anoxic depolarization of 1 and 2 min duration.

Group 2. Preconditioning group: these slices underwent either one or three short anoxic insults (1 min of anoxia—conditioning insults) separated by 10 min. After 30 min of re-oxygenation following the third anoxic insult, anoxia was again induced ("test" anoxia) for a duration defined by an AD of 1 min (as in group 1).

Group 3. Chemical preconditioning group: slices were superfused with adenosine (100 μM), 2-chloroadenosine (10 μM), pinacidil (10 μM), DPCPX (10 μM) for 10 min, or

SD-depolarization was induced, by 200 mM KCl superfusion onto slices. After this time, the drug was washed out, or recovery from SD-induced depolarization was followed for 30 min. Then, "test" anoxia was induced and continued for 2 min of AD (as in group 1).

Group 4. Low calcium preconditioning: slices underwent three short periods of anoxia (as in Group 2). Five min prior to the first anoxic preconditioning insult, slices were superfused with a 0.5 mM $CaCl_2$ ACSF. Superfusion of the low calcium ACSF was terminated at the onset of reoxygenation after the third anoxic preconditioning insult. After 30 min of re-oxygenation, "test" anoxia was induced and AD was provoked. The duration of anoxia was prolonged to 1 min of AD (as in group 1 control).

Group 5. Slices were preconditioned (as in Group 2) but they were simultaneously superfused with an adenosine A1 receptor blocker (e.g., DPCPX) or a K^+_{ATP} channel blocker (tolbutamide). Superfusion of the blockers was started 1 min prior to "conditioning" ischemia (to allow for the solution to reach the slice at the time of anoxia), and terminated at the time of AD.

Statistical Analysis

Statistical significance among different groups was established with an ANOVA test. Post-hoc comparisons within groups were carried out with a Dunnett's test. Significance was accepted with $p < 0.05$. All data were expressed as mean ± SD.

RESULTS

Control slices had poor recovery of EP after anoxic insults. This poor recovery was inversely related to AD duration. When AD lasted for 1 min, recovery of EP was 39.2 ± 14.8% of control (n = 8) (Figure 1). When the anoxic insult was prolonged to a duration which produced AD for 2.0 min, evoked potential recovery was 16.3 ± 16.3% of control, (n = 6) (Figure 1) (Pérez-Pinzón and Born, 1999).

Three preconditioning insults significantly improved EP recovery following a subsequent "test" anoxic insult which produced AD for 1 minute (*p < 0.01). EP recovery was 147.2 ± 35.6% (n = 8) (Figure 1) (Pérez-Pinzón and Born, 1999). One preconditioning insult also significantly improved EP recovery to 163.0 ± 30.0% (***p < 0.001; n = 6) (Pérez-Pinzón et al., 1996).

Evoked potentials recovered from test anoxia when 100 µM adenosine, 10 µM 2-CADO were superfused onto slices 30 min prior to anoxia. Recovery of evoked potentials was 128 ± 23% (n = 6, *p < 0.05) and 88 ± 7% (n = 14, *p < 0.05) for adenosine and 2-CADO, respectively (Pérez-Pinzón et al., 1996).

Further support for the role of adenosine A1 receptors during preconditioning was defined when 5 µM DPCPX was superfused onto slices during "conditioning" episodes, the protection afforded by preconditioning was abolished. The recovery of evoked potentials in this group was 4.6 ± 1% (n = 8) (Pérez-Pinzón et al., 1996).

When 10 µM pinacidil, a K^+_{ATP} channel agonist, was superfused onto slices 30 min prior to anoxia, recovery of evoked potential amplitude was 128.1 ± 43.2% (n = 7, **p < 0.01), also indicating a recovery of electrical activity beyond that measured during baseline (Figure 1) (Pérez-Pinzón and Born, 1999). Additional support for the role of K^+_{ATP} channels during preconditioning was indicated when 5 µM of tolbutamide, a K^+_{ATP}

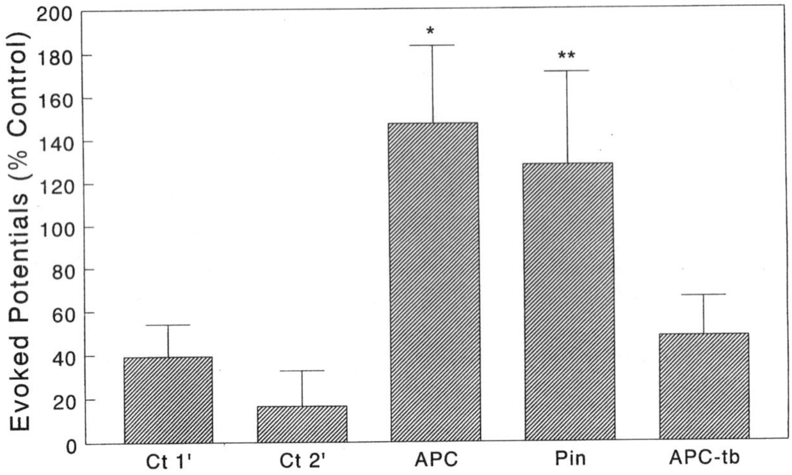

Figure 1. Average recovery of evoked field potentials in hippocampal slices 30 min after reoxygenation following "test" anoxia. Recovery is expressed as % of the pre-anoxia population spike amplitude. Significant recovery of evoked potential amplitudes were found in the anoxic preconditioning (APC) group and in the pinacidil (Pin) group (*$p < 0.05$ and **$p < 0.01$, respectively). Partial blockade of APC protection was obtained when slices where superfused with tolbutamide (APC-tb) during conditioning insults (adapted from Pérez-Pinzón and Born, 1999).

channel blocker, was superfused onto slices during "conditioning" episodes. The protection afforded by preconditioning was abolished in this instance (i.e., the recovery of evoked potentials in this group was 47.9 ± 17.6%; n = 11), which was not significantly different from control slices (Figure 1) (Pérez-Pinzón and Born, 1999).

Superfusion of slices with the excitatory agent DPCPX (10 μM) 30 min prior to anoxia, provoked an increase of evoked potentials to 108.6 ± 16.9% (n = 5) from baseline values. Slices pre-treated this way recovered evoked potential amplitudes following "test" anoxia to 117.8 ± 74.6% (n = 5, ***$p < 0.001$), also indicating a recovery of electrical activity (as compared to control—2 min of AD) (Pérez-Pinzón et al., 1999a).

When spreading depression was provoked by superfusing slices with 200 mM KCl 30 min prior to anoxia, a significant recovery of evoked potential amplitude following anoxia was observed (71.7 ± 30.1%; n = 5, **$p < 0.01$), (as compared to control—2 min of AD) (Pérez-Pinzón et al., 1999a).

When extracellular calcium concentration was lowered to 0.5 mM during APC, the recovery of electrical activity was partially blocked. The recovery of evoked potentials in this group was 80.6 ± 64.9%; (n = 8), which was not significantly different from control slices (1 min of AD) or the APC group (Pérez-Pinzón et al., 1999a).

DISCUSSION

The current study describes two different pathways involved in the induction of anoxic preconditioning neuroprotection. The first pathway appears to involve receptor/channels that when activated promote inhibition of electrical activity. Activation of the adenosine A1 receptor and the ATP-sensitive potassium channel conferred protection and promoted recovery of electrical activity in hippocampal slices following "test" anoxia (Pérez-Pinzón et al., 1996; Pérez-Pinzón and Born, 1999). The second pathway involves a transient increase in excitability or the induction of transient depolarization such as the one that occurs during spreading depression (Pérez-Pinzón et al., 1999a).

Inhibitory Pathway

We have previously shown that adenosine receptors may mediate APC in hippocampal slices. The adenosine A1 receptor antagonist DPCPX blocked APC while both adenosine and the A1 receptor agonist 2-CADO provided protection similar to that provided by APC (Pérez-Pinzón et al., 1996). Other reports also suggested the involvement of adenosine receptors in IPC protection in heart and brain.

For example, the nearly complete protection produced by IPC in the CA1 region of rat hippocampus 3 days after 6 min of global cerebral ischemia (Heurteaux et al., 1995) was abolished by DPCPX. In contrast, infusion of cyclopentyl adenosine (CPA) (1 mg/kg) (an adenosine A1 receptor agonist) 15 min prior to ischemia produced 70% protection of CA1 cells after 3 days of reperfusion. Adenosine was also implicated in protection of brain using a model of hypoxic preconditioning in neonates. In this study, the adenosine receptor blocker caffeine attenuated protection of neonatal brain afforded by hypoxic preconditioning (Fitzgibbons et al., 1994). Moreover, hypoxic preconditioning was also potentiated by blocking adenosine deaminase (which initiates adenosine catabolism) (Gidday et al., 1995).

Adenosine has been suggested to act as a "retaliatory" metabolite providing a negative feedback signal to enhance ATP production and inhibit ATP consumption (Belardinelli and Shryock, 1992; Rudolphi et al., 1992). This appears to be achieved by promoting increased glycogenolysis (Magistretti et al., 1986), increased glycolytic rate (Pasteur effect) (Wyatt et al., 1989; Gutierrez-Juarez et al., 1992), and increases in the adenylate pool (Whittingham et al., 1989). These effects are accompanied by inhibition of electrical activity (Pérez-Pinzón et al., 1996).

In brain, activation of the adenosine A1 receptor has been reported to open the K^+_{ATP} channel (Fellows et al., 1993). Thus, it has been suggested that the K^+_{ATP} channel may play a key role in the induction of APC. In a previous report we showed that this is likely the case since, blockade of this channel during "conditioning" insults inhibited neuroprotection, whereas transient superfusion of pinacidil, a K^+_{ATP} channel activator emulated APC (Pérez-Pinzón and Born, 1999). In agreement with these results, other studies have shown that K^+_{ATP} activation is required for neuroprotection in other preconditioning paradigms (ischemic preconditioning) (Heurteaux et al., 1995; Riepe et al., 1997).

Possible mechanisms by which the K^+_{ATP} channel protect during APC include that, a) this channel may counteract membrane depolarization that results from energy failure during anoxia (Ben-Ari et al., 1990; Jiang et al., 1992); and b) K^+_{ATP} channels have been shown to decrease neurotransmitter release in vitro and in vivo, as shown in substantia nigra and hippocampus (Amoroso et al., 1990).

Excitatory Pathways

We previously showed that both DPCPX and KCl-induced SD conditioned slices and promoted protection against a "test" anoxic insult (Pérez-Pinzón et al., 1999a). It is expected that during SD, or DPCPX induced depolarization or hyperexcitability, respectively, increases in cytosolic calcium occur. DPCPX has been shown to suppress an inhibitory purinergic tonus, which leads to sustained epileptiform activity in hippocampal slices (Alzheimer et al., 1989) and to cause increases in cytosolic calcium (Obrietan et al., 1995). SD in vitro is also associated with increases in cytosolic calcium in the CA1 regions of the hippocampus (Basarsky et al., 1998).

The protection afforded by SD-induced preconditioning in vivo has been demonstrated in previous studies and appears to be mediated in part by glutamate receptors, since ketamine (Taga et al., 1997) and MK-801 (Plumier et al., 1997) blocked this protection against cerebral ischemia. Blockade of NMDA receptors will limit Ca^{+2} entry into cells during depolarization-induced SD. These results are in agreement with results from our previous study, in which low $[Ca^{+2}]_o$ ACSF superfusion during APC partially blocked EP recovery after "test" anoxia (Pérez-Pinzón et al., 1999a).

A Putative Common Pathway

Figure 2 describes the possible biochemical common pathways by which anoxic or ischemic preconditioning promotes tolerance against such insults. At present, it is uncertain whether both pathways converge into a common pathway of neuroprotection. However, studies in heart suggest that protein kinase C might be that convergence point.

During anoxic preconditioning it is expected that several events will take place: 1) adenosine will be released and activation of the adenosine A1 receptor will occur. This in turn will activate the K^+_{ATP} channel. Activation of the adenosine A1 receptor will also promote PKC activation via production of diacylglycerol (DAG) and inositol trisphosphate; 2) also during anoxic preconditioning a mild accumulation of cytosolic calcium and free radical production will occur. Cytosolic calcium accumulation by itself will activate PKC, by activation of phospholipase C. Free radicals generated by mitochondrial inhibition during preconditioning insults will promote opening of the NMDA receptor and activation of phospholipase C. NMDA receptor activation during APC will induce

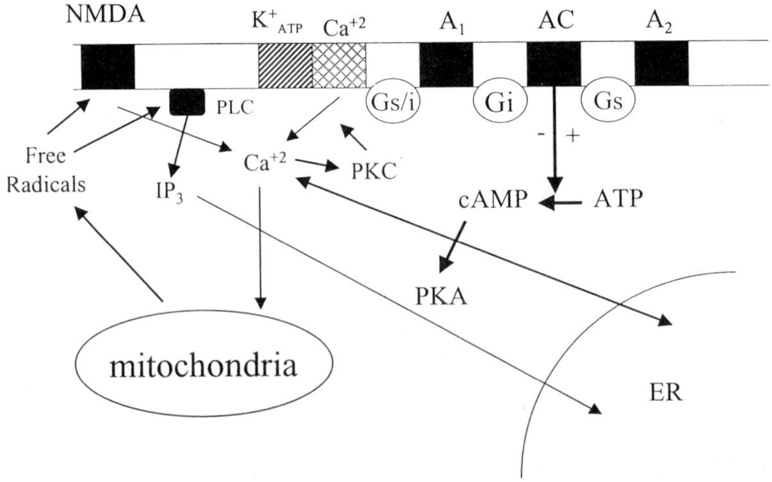

Figure 2. Schematic diagram describing putative convergent pathways for neuroprotection during anoxic preconditioning. N-methyl-D-aspartate (NMDA); protein kinase C (PKC); protein kinase A (PKA); phospholipase C (PLC); adenylate cyclase (AC); endoplasmic reticulum (ER).

cytosolic calcium accumulation, which in turn will activate phospholipase, which in turn will activate PKC by producing DAG and inositol trisphosphate. Thus, it seems logical that PKC might be that convergence point among different preconditioning paradigms presented in the current review.

The role of PKC during ischemic preconditioning in heart has been supported by numerous studies. Downey (1994) proposed the "translocation theory of ischemic preconditioning" attempting to define the role of PKC during preconditioning. This theory proposed that preconditioning insults would promote activation of the adenosine A1 receptor, which through the production of diacylglycerol (DAG) and inositol trisphosphate, would induce activation of PKC. During a "test" ischemic insult, PKC in membranes can be rapidly activated upon re-activation of A1 receptors. In this way, proteins responsible for protection against ischemia can be phosphorylated within the first few minutes of "test" ischemia in preconditioned myocardium. The translocation theory then states that the only difference between a preconditioned heart and a virgin heart is that the former has PKC translocated into its membranes.

In brain however, no evidence for such mechanism has been reported. In a previous study we showed that the role of PKC appears to be negligible in the anoxic preconditioning paradigm used (Pérez-Pinzón and Born, 1999). But, it is possible that activation of PKC may be protective in brain under conditions different to those shown in our previous study (Pérez-Pinzón and Born, 1999). It is also possible that concentrations for PKC emulation of APC in brain slices are different from those used to normally activate this protein in the CNS. Most importantly, definition of the role of different PKC isoforms is required to determine their role on preconditioning neuroprotection.

In conclusion, even though the mechanisms of anoxic preconditioning remain undefined, a more clear picture is emerging, which includes, activation of the adenosine A1 receptor, the opening of the K^{+}_{ATP} channel, mild accumulation of cytosolic calcium, and perhaps the activation of PKC. Additional studies are required to thoroughly define the role of the different isoforms of PKC, and to determine whether this type of preconditioning will promote prolonged protection after brain ischemia.

ACKNOWLEDGMENTS

The author thanks Dr. Myron Rosenthal for his helpful comments and criticisms of the manuscript. These studies were supported by PHS grant NS 34773.

REFERENCES

Alzheimer C, Sutor B, and Bruggencate G (1989) Transient and selective blockade of adenosine A1-receptors by 8-cyclopentyl-1,3-dipropylxanthine (DPCPX) causes sustained epileptiform activity in hippocampal CA3 neurons of guinea pigs. *Neurosci Lett* 99(1–2):107–112.

Amoroso S, Schmid-Antomarchi H, Fosset M, and Lazdunski M (1990) Glucose, sulfonylureas, and neurotransmitter release: role of ATP-sensitive K+ channels. *Science* 247:852–854.

Auchampach JA and Gross GJ (1993) Adenosine A1 receptors, K^{+}_{ATP} channels, and ischemic preconditioning in dogs. *Am J Physiol* 264(5 Pt 2):H1327–H1336.

Auchampach JA, Maruyama M, Cavero I, and Gross GJ (1991) The new K^+ channel opener Aprikalim (RP 52891) reduces experimental infarct size in dogs in the absence of hemodynamic changes. *J Pharmacol Exp Ther* 259(3):961–967.

Basarsky TA, Duffy SN, Andrew RD, and MacVicar BA (1998) Imaging spreding depression and associated intracellular calcium waves in brain slices. *J Neurosci* 18(18):7189–7199.

Belardinelli L and Shryock J (1992) Does adenosine function as a retaliatory metabolite in the heart? *News Physiol Sci* 7:52–56.

Ben-Ari Y, Krnjevic K, and Crepel V (1990) Activators of ATP-sensitive K^+ channels reduce anoxic depolarization in CA3 hippocampal neurons. *Neuroscience* 37(1):55–60.

Downey JM, Cohen MV, Ytrehus K, and Liu Y (1994) Cellular mechanisms in ischemic preconditioning: the role of adenosine and protein kinase C. *Ann NY Acad Sci* 723:82–98.

Fellows LK, Boutelle MG, and Fillenz M (1993) ATP-sensitive potassium channels and local energy demands in the rat hippocampus: an in vivo study. *J Neurochem* 61:949–954.

Fitzgibbons JC, Shah AR, Park TS, and Gidday JM (1994) Caffeine blocks the cerebroprotective actions of preconditioning in the hypoxic-ischemic neonate. *Soc Neurosci Abs* 20(1):618.

Gidday JM, Fitzgibbons JC, Maceren RG, Shah AR, Shah NR, and Park TS (1995) Inhibition of adenosine deaminase with deoxycoformycin (DCF) potentiates preconditioning cerebroprotection in a perinatal rat model of ischemic tolerance. *Soc Neurosci Abs* 21(1):513.

Gross GJ and Auchampach JA (1992) Blockade of ATP-sensitive potassium channels prevents myocardial preconditioning in dogs. *Circ Res* 70(2):223–233.

Gutierrez-Juarez R, Castrejon-Sosa M, Martinez-Valdez H, Blancas-Torres PG, Pina E, and Madrid-Marina V (1992) Activating effect of adenosine on rat erythrocyte glycolysis. *Int J Biochem* 24(3):433–436.

Heurteaux C, Lauritzen I, Widmann C, and Lazdunski M (1995) Essential role of adenosine, adenosine A1 receptors, and ATP-sensitive K^+ channels in cerebral ischemic preconditioning. *Proc Natl Acad Sci USA* 92(10):4666–4670.

Jiang C, Xia Y, and Haddad GG (1992) Role of ATP-sensitive K^+ channels during anoxia: major differences between rat (newborn and adult) and turtle neurons. *Journal of Physiology* 448:599–612.

Kato H, Araki T, Murase K, and Kogure K (1992) Induction of tolerance to ischemia: alterations in second-messenger systems in the gerbil hippocampus. *Brain Res Bull* 29(5):559–565.

Li YW, Whittaker P, and Kloner RA (1992) The transient nature of the effect of ischemic preconditioning on myocardial infarct size and ventricular arrhythmia. *Am Heart J* 123(2):346–353.

Liu GS, Thornton J, Van Winkle DM, Stanley AW, Olsson RA, and Downey JM (1991) Protection against infarction afforded by preconditioning is mediated by A1 adenosine receptors in rabbit heart. *Circulation* 84(1):350–356.

Liu Y, Kato H, Nakata N, and Kogure K (1992) Protection of rat hippocampus against ischemic neuronal damage by pretreatment with sublethal ischemia. *Brain Res* 586(1):121–124.

Magistretti PJ, Hof PR, and Martin JL (1986) Adenosine stimulates glycogenolysis in mouse cerebral cortex: a possible coupling mechanism between neuronal activity and energy metabolism. *J Neurosci* 6(9): 2558–2562.

Murry CE, Jennings RB, and Reimer KA (1986) Preconditioning with ischemia: a delay of lethal cell injury in ischemic myocardium. *Circulation* 74(5):1124–1136.

Obrietan K, Belousov A, Heller H, and Van den Pol AN (1995) Adenosine pre- and postsynaptic modulation of glutamate-dependent calcium activity in hypothalamic neurons. *J Neurophysiol* 74(5):2150–2162.

Pérez-Pinzón MA and Born JG (1999) Anoxic preconditioning in hippocampal slices: role of ATP-potassium channel and PKC. *Neuroscience* 89(2):453–459.

Pérez-Pinzón MA, Centeno J, and Born J (1999a) Calcium and increase excitability promote tolerance against anoxia in hippocampal slices. *Brain Res.* Submitted.

Pérez-Pinzón MA, Mumford PL, Rosenthal M, and Sick TJ (1996) Anoxic preconditioning in hippocampal slices: role of adenosine. *Neuroscience* 75:687–694.

Pérez-Pinzón MA, Vitro TM, Dietrich WD, and Sick TJ (1999b) The effect of rapid preconditioning on the microglial, astrocytic and neuronal consequences of global cerebral ischemia. *Acta Neuropathol*: In press.

Pérez-Pinzón MA, Xu GP, Dietrich WD, Rosenthal M, and Sick TJ (1997) Rapid preconditioning protects rats against ischemic neuronal damage after 3 but not 7 days of reperfusion following global cerebral ischemia. *J Cereb Blood Flow Metab* 17:175–182.

Plumier JC, David JC, Robertson HA, and Currie RW (1997) Cortical application of potassium chloride induces the low-molecular weight heat shock protein (Hsp27) in astrocytes. *J Cereb Blood Flow Metab* 17(7):781–790.

Riepe MW, Esclaire F, Kasischke K, Schreiber S, Nakase H, Kempski O, Ludolph AC, Dirnagl U, and Hugon J (1997) Increased hypoxic tolerance by chemical inhibition of oxidative phosphorylation: "Chemical preconditioning". *J Cereb Blood Flow Metab* 17(3):257–264.

Rudolphi KA, Schubert P, Parkinson FE, and Fredholm BB (1992) Adenosine and brain ischemia. *Cerebrovasc Brain Metab Rev* 4(4):346–369.

Schurr A, Reid KH, Tseng MT, West C, and Rigor BM (1986) Adaptation of adult brain tissue to anoxia and hypoxia in vitro. *Brain Res* 374(2):244–248.

Schurr A and Rigor BM (1987) The mechanism of neuronal resistance and adaptation to hypoxia. *Febs Lett* 224(1):4–8.

Taga K, Patel PM, Drummond JC, Cole DJ, and Kelly PJ (1997) Transient neuronal depolarization induces tolerance to subsequent forebrain ischemia in rats. *Anesthesiology* 87(4):918–925.

Walker DM, Walker JM, and Yellon DM (1993) Global myocardial ischemia protects the myocardium from subsequent regional ischemia. *Cardioscience* 4(4):263–266.

Whittingham TS, Warman E, Assaf H, Sick TJ, and LaManna JC (1989) Manipulating the intracellular environment of hippocampal slices: pH and high-energy phosphates. *J Neurosci Methods* 28(1–2):83–91.

Wyatt DA, Edmunds MC, Rubio R, Berne RM, Lasley RD, and Mentzer R, Jr (1989) Adenosine stimulates glycolytic flux in isolated perfused rat hearts by A1-adenosine receptors. *Am J Physiol* 257(6 Pt 2): H1952–H1957.

21

MECHANISM(S) OF MITOCHONDRIAL HYPEROXIDATION AFTER GLOBAL CEREBRAL ISCHEMIA

Miguel A. Pérez-Pinzón, Thomas J. Sick, and Myron Rosenthal

Department of Neurology
University of Miami School of Medicine
Miami, Florida 33101 USA

INTRODUCTION

In brain and other tissues, there has been a recent re-emphasis on the role of mitochondria in the cascade of events producing cell injury and death following anoxia/ischemia (e.g., Abe et al., 1995; Ankarcrona et al., 1995; Saris and Eriksson, 1995; Kluck et al., 1997; Yang et al., 1997).

For example, putative post-ischemic mitochondrial derangements in studies of cell injury mechanisms have been linked to production of reactive oxygen species and free radical-mediated injury (e.g. Kontos, 1989; Vlessis et al., 1990; Hall and Braughler, 1993), excitotoxic damage (Ankarcrona et al., 1995) and decreased energy production capability (Sims and Pulsinelli, 1987; Bogaert et al., 1994).

Important contributions toward understanding the mitochondrial changes produced by anoxia or ischemia have been derived from studying mitochondrial activity in intact, functioning tissues. This is possible with optical techniques, which non-invasively monitor reduction/oxidation (redox) activity of the electron carriers of the mitochondrial respiratory chain. During cerebral ischemia, for example, fluorescence (at the approx 460 nm wavelength characteristic of reduced nicotinamide adenine dinucleotide [NADH]) increased to the maximum level expected due to the inhibited transport of electrons in the respiratory chain. Light absorption at wavelengths characteristic of cytochromes (in particular, that of cytochrome a,a_3) also increased to a maximum indicating that the cytochromes have become fully reduced (see review in Rosenthal and Sick, 1988). Linked to these mitochondrial redox shifts, especially if the ischemic insult is prolonged beyond a short period, is the cascade of events associated with energy failure. These

Oxygen Transport to Tissue XXI, edited by Eke and Delpy.
Kluwer Academic / Plenum Publishers, New York, 1999.

include loss of ion homeostasis ("anoxic depolarization"), acidosis, release of neurotransmitters with putative excitotoxic consequences, accumulation of intracellular free calcium ion, fluid shifts to the intracellular compartment, shrinkage of the extracellular space with microvaculolation of the cytoplasm and ballooning and swelling of mitochondria.

Although much is yet to be learned about mitochondrial events during cerebral ischemia, the changes that occur to mitochondria during reperfusion/reoxygenation afterward are even less well understood. These changes include a decrease in fluorescence (at the NADH wavelengths) and decreases in light absorption at wavelengths that indicate absorption by cytochromes. Usually, these decreases continue to levels below pre-ischemic baselines suggesting that reperfusion is accompanied by the apparent reoxidation and then hyperoxidation of the mitochondrial respiratory chain components (e.g., Mayevsky et al., 1985; Rosenthal and Sick, 1988; Welsh et al., 1991; Pérez-Pinzón et al., 1997a,b; Pérez-Pinzón et al., 1998).

Defining the mechanism of the apparent post-ischemic mitochondrial hyperoxidation (PIMHo) may be central to understanding ischemic injury in brain and other tissues. Support for this concept is derived from correlations reported between PIMHo and subsequent tissue pathophysiology (e.g., Tanaka et al., 1986; Welsh et al., 1991; Rosenthal et al., 1995; Pérez-Pinzón et al., 1997a,b; Pérez-Pinzón et al., 1998). This concept is further supported by findings that increasing tissue oxygenation after ischemia increased PIMHo and decreased recovery of evoked potentials while decreasing tissue oxygenation had opposite effects (Feng et al., 1998). In addition, suppression of NADH hyperoxidation by antioxidants (ascorbate or glutathione) improved electrical recovery after anoxia in hippocampal slices (Pérez-Pinzón et al., 1997).

Mitochondrial hyperoxidation could result from several possible events in post-ischemic brain. An initial concern was that the optical signals of mitochondrial hyperoxidation may result not from intracellular events but from influences of large changes in hemoglobin volume and saturation. This possibility was discounted by findings that PIMHo amplitude was not linked to circulatory events (Rosenthal et al., 1995) and also that mitochondrial hyperoxidation occurred after anoxia in hippocampal slices which function independent of a vasculature (Pérez-Pinzón et al., 1997a,b; Pérez-Pinzón et al., 1998). These results confirmed that PIMHo is modulated by intracellular events (see Rosenthal et al., 1997).

Intracellular changes that could modulate the apparent mitochondrial hyperoxidation of electron carriers after ischemia include decreased substrate availability and disruption or loss of the electron carriers from the mitochondrial matrix. Consistent with either of these mechanisms were findings that antioxidants or lowered extracellular calcium ion activity limited mitochondrial hyperoxidation and improved electrical recovery after anoxia in hippocampal slices (Pérez-Pinzón et al., 1997a,b). These studies suggested that mitochondrial hyperoxidation is mediated, at least in part, by reactive oxygen species (ROS) and cytosolic calcium ion accumulation, which could impair key metabolic enzymes required in substrate production and/or could promote mitochondrial membrane disruption that could result in the translocation of the electron carriers of the respiratory chain.

The hypothesis suggested in this review is that mitochondrial membrane disruption underlies post-ischemic mitochondrial hyperoxidation (see Pérez-Pinzón et al., 1999). This was tested by determining whether levels of cytochromes in the cytosol were increased by global cerebral ischemia in intact rat brain and whether reducible cytochromes were decreased after anoxia in hippocampal slices.

RESULTS

Hippocampal Slices

Following 60 minutes of reoxygenation after anoxia in hippocampal slices, the intensity of fluorescence emission at wavelengths characteristic of NADH was decreased below baseline. This indicates that there was less reduced pyridine nucleotide than prior to the insult. When the apparent hyperoxidation of NADH was attenuated by ascorbate, evoked potential recovery was improved (Figure 1). These results confirm that the apparent mitochondrial hyperoxidation is modulated by intracellular events, which influence the redox activity of all members of the electron transport chain.

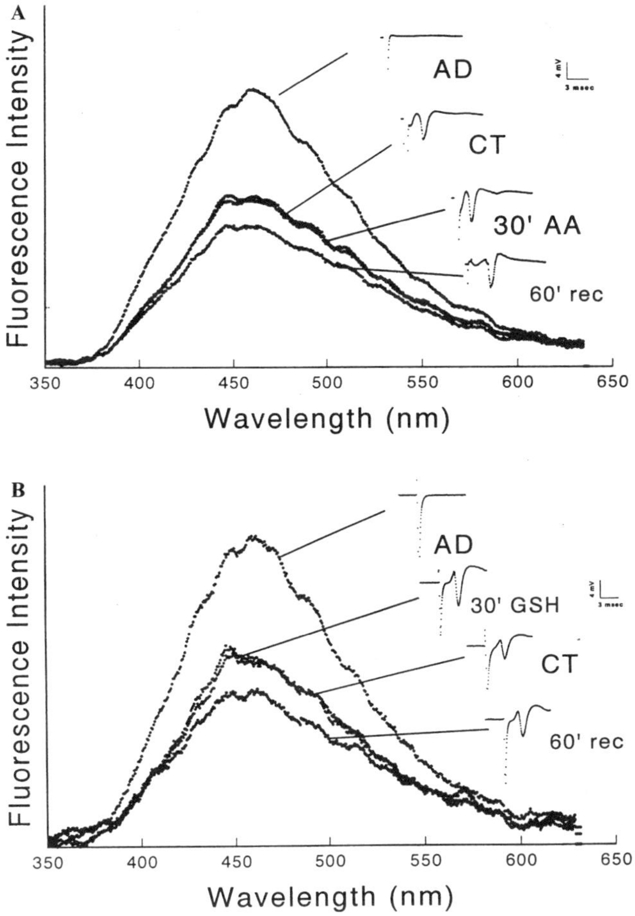

Figure 1. Examples of NADH fluorescence emission spectra and field potentials recorded from hippocampal slices: A) treated with 500 μ*M* ascorbic acid during normoxia, anoxia and 60 min following reoxygenation. Ascorbate (AA) promoted recovery of evoked potentials and lessened the post-anoxic hyperoxidation of NADH; and B) treated with 50 μ*M* glutathione (GSH) during normoxia, anoxia and 60 min following reoxygenation. Glutathione promoted recovery of evoked potentials and lessened the post-anoxic hyperoxidation of NADH (adapted from Pérez-Pinzón et al., 1997). Anoxic depolarization (AD); control or baseline values (CT); recovery (rec).

Comparisons were made of the amplitudes of the absorption peaks attributable to reduced cytochrome *c* in control slices and in "experimental" slices previously subjected to anoxia and reoxygenation. Amplitudes of these absorption peaks were maximized by subjecting both groups of slices to anoxia. Findings were that the shift toward reduction of cytochrome *c* produced by anoxia in experimental slices ($4.00 \pm 1.1 \times 10^{-3}$ log units; n = 9) was less than that in control slices ($5.08 \pm 0.9 \times 10^{-3}$ log units; n = 9). This 21% difference was significant to a level of $p < 0.05$ (Pérez-Pinzón et al., 1999).

These findings in hippocampal slices support the hypothesis that the level of reducible cytochrome *c* in mitochondria is decreased during the period of apparent mitochodnrial hyperoxidation following anoxia.

Intact Brain

If intramitochondrial cytochrome *c* is decreased during reoxygenation after anoxia/ischemia, it is reasonable to expect that cytosolic cytochrome *c* would be increased by a similar insult. This hypothesis was tested in intact rat brain following global cerebral ischemia. Absorption spectra obtained from the cytoslic fraction of post-ischemic rat neocortex were compared with spectra from control rats. In the post-ischemic rats (20 min ischemia followed by 30 min reperfusion), there was $5.44 \pm 1.8 \times 10^{-3}$ uM/mg of cytochrome *c* (n = 7) in the cytosolic fraction compared with $3.83 \pm 0.8 \times 10^{-3}$ uM/mg of protein of cytochrome *c* (n = 7) in sham operated rats. This increase of 42% in cytosolic cytochrome *c* is made more significant by findings that the brain spectra from either control or experimental rats did not exhibit absorption peaks characteristic of mitochondrial pigments other than cytochrome *c* (Pérez-Pinzón et al., 1999).

DISCUSSION

Present studies support the hypothesis that cytochrome *c* is lost from mitochondria to the cytosol during anoxia/ischemia or during reoxygenation/reperfusion afterward. This was not accompanied by loss of other electron carriers which is not surprising since cytochrome *c* is soluble, while the other electron carriers are bound to mitochondrial members or in the mitochondrial matrix and thus are more likely to remain associated with the membrane fraction.

Although the mechanism by which cytochrome *c* is released into the cytosol remains undefined, the fact of its release suggests that following conditions of energy failure, reperfusion may involve the mitochondrial transition pore and/or osmotic lysis of mitochondria. Evidence supporting the opening of the transition pore includes findings by Ouyang et al. (1997). These investigators showed that large molecules such as the mitochondrial aspartate aminotransferase (mAST) were released from mitochondria into the cytosol following transient focal cerebal ischemia. In related studies, Fiskum et al. (1997) showed that cytochrome *c* was released from isolated brain mitochondria in a calcium dose-dependent manner after 10 minutes of cardiac arrest in beagles. This observation is significant because calcium overload has been observed following anoxia/ischemia during the time corresponding to the putative release of cytochrome *c* into the cytosol. It is consistent with reports that the conductance of the permeability transition pore increased with pathological cytosolic Ca^{+2} accumulation and free radical production (Bernardi and Petronilli, 1996). Furthermore, the apparent hyperoxidation of NADH following anoxia in hippocampal slices increased dose dependently with increases in extracellular calcium ion activity (Pérez-Pinzón et al., 1997b).

There is evidence that antioxidants may limit changes in mitochondrial permeability transition pore (Bernardi and Petronilli, 1996; Chernyak and Bernardi, 1996). If true, this could provide a mechanistic explanation for findings that antioxidants decreased the apparent mitochondrial hyperoxidation and promoted recovery of electrical recovery following anoxia in hippocampal slices (Pérez-Pinzón et al., 1997a).

Disruption of the organizational structure of the mitochondrial matrix may also follow anoxia/ischemia. Such disruption could explain the apparent hyperoxidation of NADH and other electron carriers, which occurs despite the fact that they are not lost to the cytosol. Rather than the actual loss of these carriers, mitochondrial disruption could lead to an overall decrease in reducible electron carriers or in the capacity for electron transport along the chain of carriers. No matter the mechanism, such an effect could compromise energy production and likely underlie the apparent mitochondrial hyperoxidation.

Also unknown is whether changes in mitochondrial organization or pore conductance are the exclusive mechanisms of the apparent post-ischemic mitochondrial hyperoxidation. In fact, earlier studies had suggested that a condition analogous to the "substrate limitation" state (i.e. state 2) of substrate-starved, electron-limited mitochondria might be involved (e.g., Chance and Williams, 1956; Rosenthal et al., 1997). This is because inhibition of glycolysis by superfusion of iodoacetate (IAA) promoted oxidation of cytochrome a,a_3, increased tissue oxygen tension and slowed the post-stimulus restoration of ionic equilibrium in rat brain. Such changes were not seen during or after hypoxia.

The concept that a "state-2"-like condition exists in brain after anoxia or ischemia is compatible with the concept that these insults disrupt mitochondrial membrane and electron transport chain integrity. Although glycolytic activity may be sufficient to provide reducing equivalents to the respiratory chain after anoxia or ischemia, the capacity to accept these reducing equivalents may be limited. Therefore, a unifying hypothesis proposed here is that during the reoxygenation/reperfusion period after brain anoxia or ischemia, there is mitochondrial disruption, which limits electron transport and compromises mitochondrial function in a manner, which leads to pathophysiology. We believe that such mitochondrial disruption is indicated by the apparent mitochondrial hyperoxidation, which characterizes the post-anoxic/ischemic brain.

REFERENCES

Abe K, Aoki M, Kawagoe J, Yoshida T, Hattori A, Kogure K, and Itoyama Y (1995) Ischemic delayed neuronal death: A mitochondrial hypothesis. *Stroke* 26:1478–1489.

Ankarcrona M, Dypbukt JM, Bonfoco E, Zhivotovsky B, Orrenius S, Lipton SA, and Nicotera P (1995) Glutamate-induced neuronal death: A succession of necrosis or apoptosis depending on mitochondrial function. *Neuron* 15:961–973.

Bernardi P and Petronilli V (1996) The permeability transition pore as a mitochondrial calcium release channel: A critical appraisal. *J Bioenerg Biomembrane* 28(2):131–138.

Bogaert YE, Rosenthal RE, and Fiskum G (1994) Postischemic inhibition of cerebral cortex pyruvate dehydrogenase. *Free Radic Biol Med* 16(6):811–820.

Chance B and Williams G (1956) The respiratory chain and oxidative phosphorylation. *Adv Enzymol* 17:65–134.

Chernyak BV and Bernardi P (1996) The mitochondrial permeability transition pore is modulated by oxidative agents through both pyridine nucleotides and glutathione at two separate sites. *Eur J Biochem* 238(3):623–630.

Feng ZC, Sick TJ, and Rosenthal M (1998) Oxygen sensitivity of mitochondrial redox status and evoked potential recovery early during reperfusion in post-ischemic rat brain. *Resuscitation* 37:33–41.

Fiskum G, Auldrige ME, Murphy AN, and Rosenthal RE (1997) Cardiac arrest (CA) damages the ability of brain toaccumulate Ca^{+2} and potentiates release of cytochrome c. *Society for Neuroscience Abtracts* 23(1):547.

Hall ED and Braughler JM (1993) Free radicals in CNS injury. *Res Publ Assoc Res Nerv Ment Dis* 71:81–105.

Kluck RM, Bossywetzel E, Green DR, and Newmeyer DD (1997) The release of cytochrome c from mitochondria: A primary site for Bcl-2 regulation of apoptosis. *Science* 275(5303):1132–1136.

Kontos HA (1989) Oxygen radicals in CNS damage. *Chem Biol Interact* 72(3):229–255.

Mayevsky A, Friedli CM, and Reivich M (1985) Metabolic, ionic, and electrical responses of gerbil brain to ischemia. *Am J Physiol* 248(1 Pt 2):R99–R107.

Ouyang YB, Kuroda S, Kristian T, and Siesjoe BK (1997) Release of mitochondrial aspartate aminotransferase (mAST) following transient focal cerebral ischemia suggests the opening of a mitochondrial permeability transition pore. *Neurosci Res Commun* 20(3):167–173.

Pérez-Pinzón MA, Mumford PL, Rosenthal M, and Sick TJ (1997a) Antioxidants, mitochondrial hyperoxidation and electrical activity recovery after anoxia in hippocampal slices. *Brain Res* 754:163–170.

Pérez-Pinzón MA, Mumford PL, Carranza V, and Sick TJ (1997b) Calcium influx from the extracellular space promotes NADH hyperoxidation and electrical dysfunction after anoxia in hippocampal slices. *J Cereb Blood Flow Metabol* 18:215–221.

Pérez-Pinzón MA, Mumford PL, and Sick TJ (1998) Prolonged anoxic depolarization exacerbates NADH hyperoxidation and promotes poor electrical recovery after anoxia in hippocampal slices. *Brain Res* 786:165–170.

Pérez-Pinzón MA, Xu GP, Born J, Lorenzo J, Busto R, Rosenthal M, and Sick TJ (1999) Cytochrome c is released from mitochondria into the cytosol after cerebral anoxia or ischemia. *J Cereb Blood Flow and Metabol* 19(1):39–43.

Rosenthal M, Feng ZC, Raffin CN, Harrison M, and Sick TJ (1995) Mitochondrial hyperoxidation signals residual intracellular dysfunction after global ischemia in rat neocortex. *J Cereb Blood Flow Metab* 15(4):655–665.

Rosenthal M, Mumford PL, Sick TJ, and Pérez-Pinzón MA (1997) Mitochondrial hyperoxidation after cerebral anoxia/ischemia: Epiphenomenon or precursor of residual damage? *Adv Exp Med Biol* 428: 189–195.

Rosenthal M and Sick TJ (1988) Measurement of metabolic activity associated with ion shifts. *Neuromethods: The neuronal microenvironment*. New Jersey, The Humana Press Inc 187–245.

Saris N and Eriksson KO (1995) Mitochondrial dysfunction in ischaemia-reperfusion. *Acta Anaesthesiol Scand* 39(Suppl. 107):171–176.

Sims NR and Pulsinelli WA (1987) Altered mitochondrial respiration in selectively vulnerable brain subregions following transient forebrain ischemia in the rat. *Journal of Neurochemistry* 49(5):1367–1374.

Tanaka K, Dora E, Greenberg JH, and Reivich M (1986) Cerebral glucose metabolism during the recovery period after ischemia–its relationship to NADH-fluorescence, blood flow, EcoG and histology. *Stroke* 17(5):994–1004.

Vlessis AA, Widener LL, and Bartos D (1990) Effect of peroxide, sodium, and calcium on brain mitochondrial respiration in vitro: potential role in cerebral ischemia and reperfusion. *J Neurochem* 54(4):1412–1418.

Welsh FA, Marcy VR, and Sims RE (1991) NADH fluorescence and regional energy metabolites during focal ischemia and reperfusion of rat brain. *J Cereb Blood Flow Metab* 11(3):459–465.

Yang J, Liu XS, Bhalla K, Kim CN, Ibrado AM, Cai JY, Peng TI, Jones DP, and Wang XD (1997) Prevention of apoptosis by Bcl-2: Release of cytochrome c from mitochondria blocked. *Science* 275(5303): 1129–1132.

OXYGEN DEPENDENCY OF CEREBRAL Cu_A REDOX STATE DURING INCREASED OXYGEN CONSUMPTION PRODUCED BY INFUSION OF A MITOCHONDRIAL UNCOUPLER IN NEW-BORN PIGLETS

R. Springett,[1] J. Newman,[2] D. T. Delpy,[1] and M. Cope[1]

[1]Department of Medical Physics and Bioengineering and
[2]Department of Obstetrics, University College London
London WC1E 6JA, United Kingdom

1. INTRODUCTION

The mitochondrial uncoupler 2,4-dinitrophenol (DNP) is a membrane soluble weak acid that is capable of transporting protons across the proton-impermeable inner mitochondrial membrane (Skulachev, 1998). This results in a reduction of the electrochemical gradient across the membrane and stimulation of flux through the electron transport chain, and thus oxygen consumption, without increasing cytosolic ATP hydrolysis. Uncouplers are used extensively in studies on isolated mitochondria to collapse the proton gradient and stimulate respiration (Rumsey et al., 1990), and to study the effects of the proton gradient on the redox state of the components of the electron transport chain (Morgan and Wikström, 1991). In this study, it was not the intention to collapse the mitochondrial proton gradient but to use low doses of DNP in order to stimulate oxygen consumption.

Cytochrome oxidase is the terminal enzyme of the electron transport chain and reduces molecular oxygen to water using electrons donated from reduced cytochrome c. The Cu_A centre of cytochrome oxidase is close to the cytochrome c binding site on the cytosolic side of the inner mitochondrial membrane far from the binuclear centre at which molecular oxygen is reduced (Iwata et al., 1995; Tsukihara et al., 1995). In the oxidised state, the Cu_A centre has a broad absorption band centred on 840 nm and so, in theory, the Cu_A redox state can be monitored using near infra red (NIR) spectroscopy. However, attenuation changes due to the Cu_A centre are usually small compared to atten-

Oxygen Transport to Tissue XXI, edited by Eke and Delpy.
Kluwer Academic / Plenum Publishers, New York, 1999.

uation changes due to haemoglobin and therefore the Cu_A signal is sensitive to artefact. For this reason, we used a CCD-based NIR system and performed full spectral fitting which has been shown by modelling to be more robust in separating the Cu_A component than two, four or six wavelength algorithms (Matcher et al., 1995).

The aim of this study was to determine whether the Cu_A redox state became oxygen dependent when cerebral oxygen consumption ($CMRO_2$) was increased by infusing a solution of DNP into the internal carotid artery. Oxygen dependency at baseline and at 5 accumulated doses of DNP was determined by measuring arterial saturation (SaO_2) and Cu_A redox state during brief anoxic swings.

2. METHODS

Five piglets, born at term but less than 24 hours old, were sedated with Midazolam and then anaesthetised with isoflurane. A tracheotomy was performed and the piglets ventilated with an intermittent positive pressure ventilator using an oxygen/nitrogen gas mixture.

One common carotid artery was cannulated with a 3.5 French catheter and the catheter inserted 2–3 cm towards the brain into the internal carotid artery; the distal artery was tied off. A cannula was sited in the umbilical artery for measurement of mean arterial blood pressure and heart rate using a strain gauge pressure transducer and a second cannula sited in the umbilical vein for infusion of 10% glucose solution.

Blood samples were collected from the umbilical artery for gas (ABL505, Radiometer, Copenhagen, Denmark) and glucose/lactate (2300 STAT Plus, YSI Inc., Yellow Springs, Ohio, USA) analysis. Arterial saturation was monitored with a pulse oximeter (Model 8604FO, Nonin Medical Inc., Plymouth, MN, USA) positioned on the foot.

Rectal temperature was maintained at 38.5°C using a heated water mattress. A pilot study revealed large increases in rectal temperature during infusion of DNP and so the water temperature was reduced, typically from 45 to 25°C, during the course of the infusion.

The DNP solution was made by dissolving ≈90 mg of DNP (Sigma, Poole UK) in ≈20 ml physiological saline solution containing a small quantity (1 mmole) of sodium bicarbonate to give a solution of 20 mMolar DNP. This solution was further diluted with saline to give a final concentration of 10 mM DNP. The bicarbonate was necessary because the protonated form of DNP is only sparing soluble in water.

After surgery and prior to commencement of the study, there was a stabilisation period of at least an hour. The protocol consisted of six consecutive brief anoxic episodes equally spaced over a period of 2 hours. Each anoxic episode consisted of 5 minutes of baseline at an FiO_2 0.4, FiO_2 was then switched to 0.0 for 1.5 minutes and then returned to 0.4 for a further 13.5 minutes of recovery. The DNP solution was infused via the cannula sited in the carotid artery at a rate of 1.4 mg/Kg/hour beginning 10 minutes before the second anoxia.

Tissue attenuation spectra across the piglet head were simultaneously collected at 1024 wavelengths between 680 and 1000 nm every 5 seconds using a fibre-coupled tungsten halogen light source and a CCD spectrometer. Changes in deoxyhaemoglobin (Hb) and oxyhaemoglobin (HbO_2) concentration and Cu_A redox state were calculated by fitting chromophore spectra, corrected for the wavelength dependence of pathlength, to changes in attenuation between 780 and 900 nm (Wray et al., 1988; Essenpreis et al., 1993). A pathlength, calculated from the 840 nm water feature in the 2nd differential of baseline spectra and assuming an average water content of 85% was used to convert changes in HbO_2, Hb and Cu_A to units of μmoles/L. Total haemoglobin (HbT) is the sum of HbO_2 and Hb and is proportional to cerebral blood volume (CBV) at constant haematocrit.

All data are given as mean ± SD (n = 5 animals). Analysis of variance (ANOVA) was used to compare repeated measurements and a paired Student's *t*-test was to identify significant differences. The criterion for significance was $P < 0.05$.

To produce mean and SD error bars on time-course data, time zero for each anoxia was identified as the first point at which there was a substantial decrease in HbO_2 (typically 0.5 μmoles/L) and the data was then averaged on a point by point basis. Due to variations in ventilation parameters and gas mixing in the tubing, cerebral rexoygenation typically occurred at 70–80 seconds.

3. RESULTS

Typical root mean square (rms) noise values on the baseline were 0.2, 0.1 and 0.03 μmoles/L for the HbO_2, Hb and Cu_A signals respectively.

Physiological parameters prior to each anoxia are given in Table 1. All baseline parameters are in the normal range and, although pH is slightly alkalotic, this is normal for a new-born piglet. Infusion of DNP resulted in significant increases in $PaCO_2$ and decreases in arterial pH even though ventilation rate was increased during the infusion.

Figure 1 shows the accumulated dose response of changes in HbT, HbO_2, Hb and Cu_A in the baseline period prior to each anoxia. Filled circles represent data points at which there has been a significant ($p < 0.05$) change from baseline and the results from Figure 1 are reproduced in Table 2. The increase in HbT with increasing DNP dose is consistent with a rise in CBF; increases in CBF lead to increase pressure on the venous side, venous distension and an increase in CBV and thus HbT. Hb increases continuously to 11.4 ± 3.9 μmoles/L with increasing accumulated DNP dose although the change did not gain significance at the lowest dose. In contrast, HbO_2 is essentially constant except at the highest dose studied where there is a small, but not significant, increase of 2.6 ± 3.5 μmoles/L.

The results presented in the bottom panel of Figure 1 would suggest that there is a continuous oxidation in the Cu_A redox state but the changes at the lower doses are small compared to the system noise and so cannot be resolved. However, the changes gain sig-

Table 1. Physiological variables measured immediately before each anoxia

DNP (mg/Kg)	Baseline	0.23	0.7	1.17	1.63	2.1
$PaCO_2$ (KPa)	4.49 ± 0.77	4.76 ± 0.38*	4.76 ± 0.88**	6.01 ± 1.10**	5.71 ± 1.68	6.99 ± 0.86**
pH	7.51 ± 0.03	7.52 ± 0.01	7.53 ± 0.04	7.42 ± 0.05**	7.41 ± 0.10	7.32 ± 0.04**
SaO_2 (%)	98.6 ± 0.3	98.9 ± 0.8	98.8 ± 0.7	98.4 ± 0.5	98.2 ± 0.5	97.6 ± 0.8*
MAP (mmHg)	42.3 ± 6.6	46.3 ± 8.9	46.1 ± 7.7*	50.4 ± 7.6	46.5 ± 6.3	50.0 ± 9.0
HR (min^{-1})	163 ± 7	164 ± 7	172 ± 12	183 ± 15	200 ± 14**	209 ± 13**
T_R (°C)	38.5 ± 0.0	38.6 ± 0.1	38.7 ± 0.2	38.6 ± 0.2	38.8 ± 0.2	38.6 ± 0.3
Glucose (mM)	6.1 ± 1.3	6.4 ± 1.6	6.1 ± 1.4	6.0 ± 1.0	6.0 ± 1.4	6.4 ± 1.0
Lactate (mM)	2.73 ± 0.89	2.58 ± 0.17	2.49 ± 0.60	3.21 ± 0.82	3.19 ± 0.79	4.03 ± 1.17

The accumulated DNP dose is given in the column heading. $PaCO_2$ arterial CO_2 tension, pH arterial pH, SaO_2 arterial saturation, HR heart rate, T_R rectal temperature, glucose and lactate arterial glucose and lactate respectively. Values are mean ± SD (n = 5). * Significantly different from baseline ($p < 0.05$), ** significantly different from baseline ($p < 0.01$).

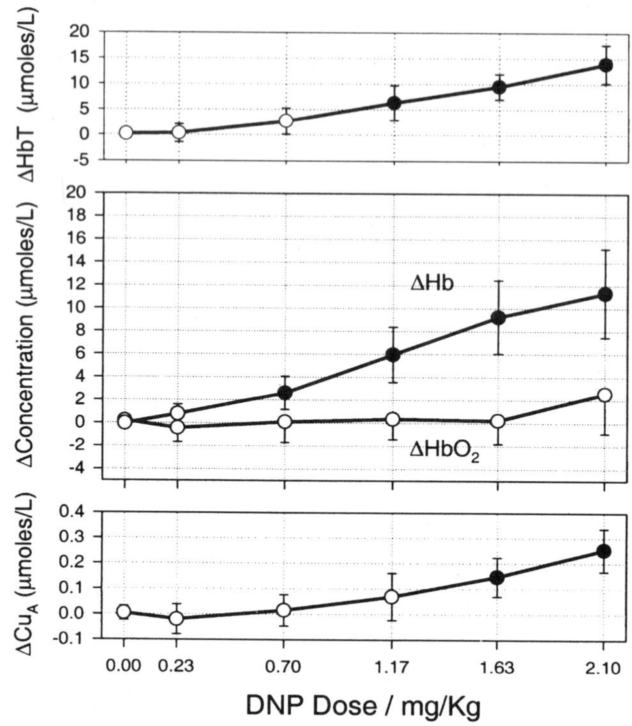

Figure 1. Changes in HbT, HbO_2, Hb and Cu_A in the baseline period prior to each anoxia. Filled circles represent data points at which there has been a significant changes ($p < 0.05$) from baseline.

nificance at an accumulated dose of 1.63 mg/Kg ($p < 0.01$) and the Cu_A centre attains a maximum oxidation of 0.25 ± 0.08 µmoles/L at the highest dose.

Figure 2 shows pooled data for changes in HbT, HbO_2, Hb and Cu_A for the baseline anoxic swing (left panels) and the fifth anoxic swing after an accumulated dose of 1.63 mg/Kg (right panels). The reduction in arterial saturation leads to a cerebral desaturation starting at time zero and resulting in a fall in HbO_2 and an increase in Hb, initially with little change in HbT. The reduction of oxygen tension at the mitochondrial level leads to a reduction in the Cu_A centre of cytochrome oxidase but there is a significant delay between the fall in HbO_2 and Cu_A in the baseline anoxia (left panels) whereas

Table 2. NIR variables measured immediately before each anoxia. The accumulated DNP dose is given in the column heading

DNP (mg/Kg)	Baseline	0.23	0.7	1.17	1.63	2.1
ΔHbO_2	0.2 ± 0.2	−0.5 ± 1.2	0.0 ± 1.8	0.4 ± 1.8	0.2 ± 2.0	2.6 ± 3.5
ΔHb	0.0 ± 0.0	0.8 ± 0.8	2.6 ± 1.5*	6.0 ± 2.4**	9.3 ± 3.2**	11.4 ± 3.9**
ΔHbT	0.2 ± 0.3	0.3 ± 1.8	2.7 ± 2.6	6.3 ± 3.4*	9.5 ± 2.5**	13.9 ± 3.8**
ΔCu_A	0.00 ± 0.03	−0.02 ± 0.06	0.02 ± 0.06	0.07 ± 0.09	0.15 ± 0.08**	0.25 ± 0.08**

All values, which are in units of µmoles/L, are mean ± SD (n = 5). * Significantly different from baseline ($p < 0.05$), ** significantly different from baseline ($p < 0.01$).

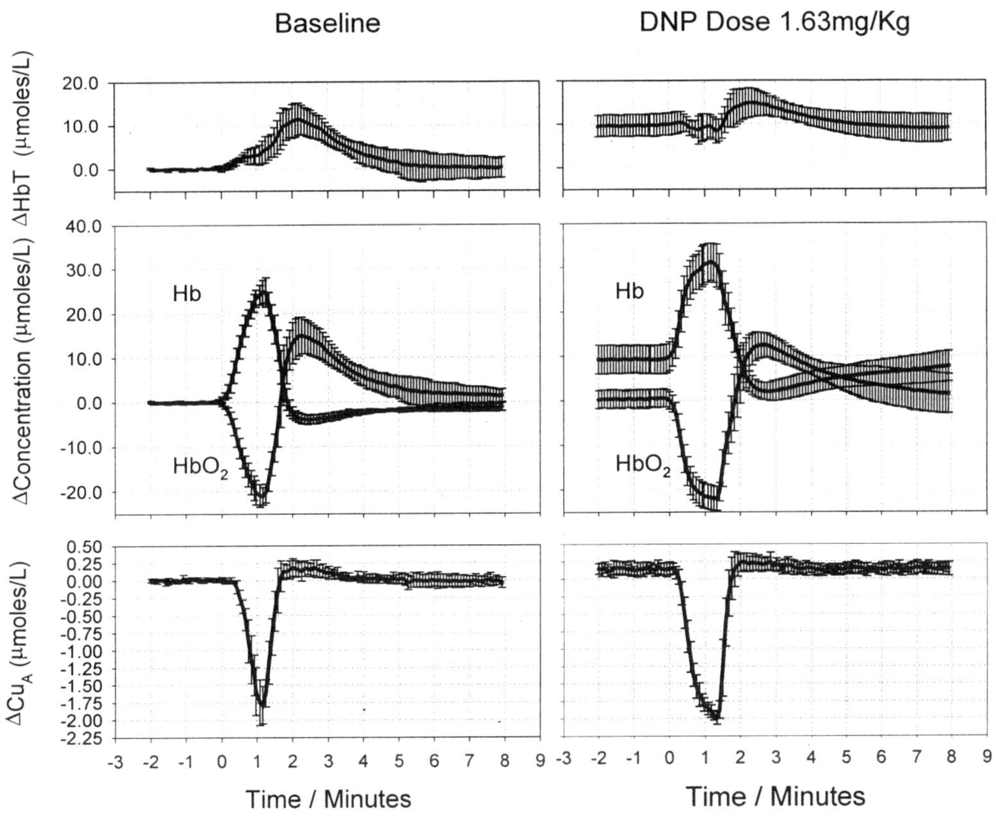

Figure 2. Changes in the NIR parameters during the first (left panels) and fifth (right panels) anoxic swing. Top panels: changes in HbT, Middle panels: changes in HbO_2 and Hb, Bottom panels changes in the Cu_A redox state. The beginning of anoxia, lasting 90s, is marked as time zero. Results are presented as mean ± SD (n = 5).

this delay is either very short or absent after an accumulated dose of 1.63 mg/Kg (right panels). The hypoxemia triggers an increase in CBF which leads to a hyperaemia as indicated by the increase in HbT. Reoxygenation begins at ≈80s with a simultaneous oxidation in the Cu_A signal and increase in HbO_2. Arterial saturation rapidly returns to normal levels but, at the elevated CBF triggered by the hypoxemia, there is an increase in HbO_2 and a decrease in Hb over baseline values. HbT reaches a maximum at approximately 2.5 minutes and then HbT, HbO_2 and Hb return to baseline values over the subsequent 6 minutes.

Figure 3 shows the change in Cu_A plotted against SaO_2 during the onset period of the anoxia (a) before infusion of DNP, (b) after an accumulated dose of 0.7 mg/Kg of DNP and (c) after an accumulated dose of 1.63 mg/Kg of DNP. The filled circles in the figure represent data points at which there has been a significant ($p < 0.05$) reduction in the Cu_A redox state from the baseline of the respective anoxia. As can be seen from the figure, the arterial saturation at the first significant reduction in the Cu_A centre depends on accumulated DNP dose and increases from 75.8 ± 5.0% in the baseline anoxia to 93.1 ± 4.5% with an accumulated dose of 1.63 mg/kg.

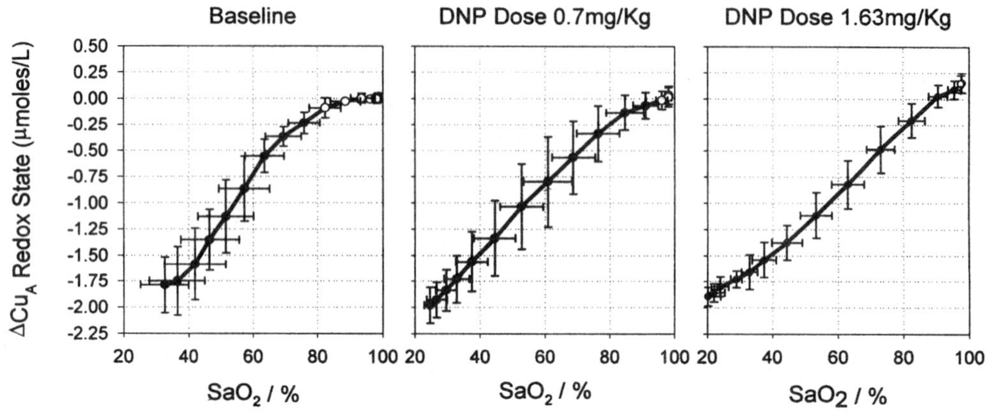

Figure 3. Mean ± SD (n = 5) changes in the Cu_A redox state plotted against SaO_2 during the onset of the first, third and fifth anoxia corresponding to an accumulated DNP dose of 0, 0.7 and 1.63 mg/Kg respectively. Solid circles indicate that the Cu_A redox state is significantly different ($p < 0.05$) from the anoxia baseline.

4. DISCUSSION

When DNP was infused into the common carotid artery of the neonatal piglet, an increase in heat generation and carbon dioxide production was observed which is consistent with mitochondrial uncoupling. However, it is unlikely that the DNP was completely absorbed in a single pass through the brain and it is probable that these effects are the result of systemic uncoupling. However, large increases in cerebral Hb were observed which could not be accounted for by the small decrease in arterial saturation or by the effects of the increase in $PaCO_2$ as there is usually a slight decrease in Hb during episodes of hypercapnia. It can therefore be concluded that the DNP produced a significant increase in cerebral oxygen consumption.

This study, in a global model, shows that Hb *always* increases with increasing oxygen consumption, even when CBF is augmented by increased $PaCO_2$. This is in contrast to functional activation where NIR studies have shown an increase in HbO_2 and a small decrease of Hb in the activated region (Kleinschmidt et al., 1996). These changes have been attributed to an increase in local CBF over and above the increase in local $CMRO_2$ (Villringer and Dirnagl, 1995).

At high oxygen tensions, the redox state of cytochrome c and also the Cu_A centre are expected to be independent of oxygen tension and determined by the metabolic state (Chance and Williams, 1956) and the activity of the dehydrogenases of the tricarboxylic acid cycle (Balaban, 1990). As oxygen tension is reduced towards zero, oxygen consumption becomes oxygen-limited and the components of the electron transport chain, including the Cu_A centre, are expected to become reduced. Although it is possible to determine a critical PO_2 for the electron transport chain in isolated mitochondria (Sugano et al., 1974), this cannot be simply related to cerebral oxygen delivery because, in-vivo, there is considerable heterogeneity in cerebral tissue PO_2 (PtO_2) (Lubbers et al., 1994). The PtO_2 at the majority of sites is much lower than arterial PO_2 which would suggest that substantial diffusion barriers exist between mitochondria and capillaries. However, large increases in tissue PtO_2 are observed when oxygen delivery is increased by increasing $PaCO_2$ (Metzger and Heuber, 1977) and thus there must be a balance between oxygen consumption and oxygen delivery.

As in previous studies (Cope et al., 1991), the results from the baseline anoxia swing show that oxygen delivery can be reduced with hypoxemia with initially no significant change in the Cu_A redox state. This would suggest that, at normoxia, cytochrome oxidase is operating at an oxygen tension well above the critical threshold. As oxygen consumption is increased, oxygen delivery must be either be increased to match demand or, due to the presence of diffusion barriers, mitochondrial PO_2 will fall and the tissue will become hypoxic. The results of this study show that the arterial saturation at which the Cu_A centre begins to reduce during the onset of anoxia rises with increasing accumulated dose of DNP and so, at normoxia, the cerebral mitochondria must be operating closer their critical threshold. At an accumulated dose of 1.63 mg/Kg the redox state of the Cu_A centre is essentially oxygen-dependent: any decrease in arterial saturation results in a significant reduction in the Cu_A centre. At the highest DNP dose, there was an additional increase in $PaCO_2$ which would have resulted in an increase in CBF and oxygen delivery. This increase in oxygen delivery was probably sufficient to prevent tissue hypoxia and a reduction in the Cu_A redox state.

Studies on isolated mitochondria have shown that, in coupled mitochondria, the Cu_A centre of cytochrome oxidase and cytochrome c are in redox equilibrium with the Cu_A centre being slightly more reduced due to its slightly more positive midpoint potential (Rich et al., 1988). When DNP is added in excess to isolated mitochondria, oxygen consumption is typically increased by a factor of 10 and large oxidations are observed in the Cu_A centre (Morgan and Wikström, 1991) and in cytochrome c (Wilson et al., 1988). However, in a previous in-vivo study (Kariman et al., 1986) in which the redox state of haem a, cytochrome c and cytochrome b were compared to ^{31}P NMR spectroscopy in an isolated rat head model, it was found that the components of the electron transport chain reduced and there was almost complete failure of phosphorous energetics when DNP was added to the perfusion medium. The authors explained their reduction as a result of relative hypoxia due to the high oxygen demand after addition of DNP. However, their single dose was over two orders of magnitude greater than the maximum does reported in this study ($\approx$300 mg/Kg assuming a rat head weight of 50 g compared to 2.1 mg/Kg in this study) which would lead to much greater uncoupling and increase in oxygen consumption than in our model. In addition, their rat heads were perfused with a perflurocarbon (FC43) which has a lower oxygen carrying capacity than haemoglobin and, in this study, the increase in arterial CO_2 tension would increase arterial PO_2 via the Bohr effect and augment any increase in CBF and therefore oxygen delivery.

Thus in our preparation, oxygen delivery, in comparison to oxygen consumption was sufficient to prevent hypoxia and so the small oxidations of the Cu_A centre, which are consistent with the results from isolated mitochondria if the proton gradient has been only partially reduced, can be observed. However, It should be noted that an oxidation of the Cu_A centre has been observed in the human neonate when $PaCO_2$ was increased (Edwards et al., 1991) and so it cannot be unequivocally concluded that uncoupling in-vivo oxidises the Cu_A centre of cytochrome oxidase.

ACKNOWLEDGMENTS

We are grateful to Hamamatsu Photonics KK for financial assistance.

REFERENCES

Balaban RS (1990) Regulation of phosphorylation in the mammalian cell. Am. J. Physiol. 258 (Cell Physiol. 27):C377–C389.

Chance B and Williams CM (1956) The respiratory chain and oxidative phosphorylation. Adv. Enzymol. 17:65–134.

Cope M, Van der Zee P, Essenpreis M, Arridge SR, and Delpy DT (1991) Data analysis methods for near infrared spectroscopy of tissue: problems in determining the relative aa_3 concentration. Proc. SPIE 1431:251–262.

Edwards AD, Brown GC, Cope M, Wyatt JS, McCormick DC, Roth SC, Delpy DT, Reynolds EOR (1991) Quantification of redox changes in human neonatal cerebral cytochrome oxidase. J. Appl. Physiol. 71:1907–1913.

Essenpries M, Elwell CE, Cope M, Van der Zee P, Arridge SR, and Delpy DT (1993) Spectral dependence of temporal point spread functions in human tissues. Appl. Opt. 32(4):418–425.

Iwata S, Ostermeier C, Ludwig B, and Michel H (1995) Structure at 2.8 Å resolution of cytochrome c oxidase from Paracoccus Denitrificans. Nature 376:660–669.

Kariman K, Chance B, Burkhart DS, and Bolinger LS (1986) Uncoupling effects of 2,4-Dintrophenol on electron transfer reactions and cell bioenergetics in rat brain in situ. Brain Res. 366:300–306.

Kleinschmidt A, Obrig H, Requardt M, Merboldt KD, Dirnagl U, Villringer A, and Frahm J (1996). Simultaneous recording of cerebral blood oxygenation changes during human brain activation by magnetic resonance imaging and near-infrared spectroscopy. J. Cereb. Blood Flow Metab. 16:817–826.

Lubbers DW, Baumgartl H, Zimelka W (1994) Heterogeneity and stability of local PO_2 distribution within the brain tissue. Adv. Exp. Med. Biol. 345:567–574.

Matcher SJ, Elwell CE, Cooper CE, Cope, M Delpy DT (1995) Performance comparison of several published tissue near-infrared spectroscopy algorithms. Anal. Biochem. 227:54–68.

Metzget H and Heuber S (1977) Local oxygen tension and spike activity of the cerebral grey matter of the rat and its response to short intervals of O2 deficiency or CO2 excess. Pflügers Arch. 370:201–209.

Morgan JE and Wikström M (1991) Steady-state redox behaviour of cytochrome c, cytochrome a, and Cu_A of cytochrome c oxidase in intact rat live mitochondria. Biochemistry 30:948–958.

Rich PR, West IC, and Mitchell P (1988) The location of Cu_A in mammalian cytochrome c oxidase. FEBS Lett. 233:25–30.

Rumsey WL, Schlosser C, Nuutinen EM, Robiolio M, and Wilson DF (1990) Cellular energetics and the oxygen dependence of respiration in cardiac myocytes isolated from the adult rat. J. Biol. Chem. 265:15392–15299.

Skulachev VP (1998) Uncoupling: new approaches to an old problem of bioenergetics. Biochim, Biophys, Acta 1363:100–124.

Sugano T, Oshino N, and Chance B (1974) Mitochondrial functions under hypoxic conditions. The steady states of cytochrome c reduction and of energy metabolism. Biochim. Biophys. Acta 347:340–358.

Tsukihara T, Aoyama H, Yamashita E, Tomizaki T, Yamaguchi H, Shinzawa-Itoh K, Nakashima R, Yaono R, and Yoshikawa S (1995) Structure of metal sites of oxidized bovine heart cytochrome c oxidase at 2.8 Å. Science 269:1069–1074.

Villringer-A and Dirnagl-U (1995) Coupling of brain activity and cerebral blood flow: basis of functional neuroimaging. Cerebrovasc-Brain-Metab-Rev. 1995 Fall; 7(3):240–276.

Wilson DF, Rumsey WL, Green TJ, Vanderkooi JM (1988) The oxygen dependence of mitochondrial oxidative phosphorylation measured by a new optical method for measuring oxygen concentration. J. Biol. Chem. 263:2712–2718.

Wray S, Cope M, Delpy DT, Wyatt JS, and Reynolds EOR (1988) Characterisation of the near infrared absorption spectra of cytochrome aa_3 and haemoglobin for the non invasive monitoring of cerebral oxygenation. Biochim. Biophys. Acta 933:184–192.

23

EFFECTS OF KETAMINE/XYLAZINE AND PENTOBARBITAL ANESTHESIA ON CEREBRAL TISSUE OXYGEN TENSION, BLOOD PRESSURE, AND ARTERIAL BLOOD GAS IN RATS

Satoshi Taie,[1] Steven B. Leichtweis,[2] Ke Jian Liu,[2] Minoru Miyake,[2] Oleg Grinberg,[2] Eugene Demidenko,[3] and Harold M. Swartz[2]

[1]Department of Anesthesiology and Emergency Medicine
Kagawa Medical School
1750-1, Miki-cho, Kita-gun, Kagawa 761,07
Japan
[2]EPR Center for Study of Viable Systems
Department of Diagnostic Radiology
Dartmouth Medical School
7785 Vail Building, Hanover, New Hampshire 03755
[3]Community and Family Medicine
Dartmouth Medical School, Box 7927
Hanover, New Hampshire 03755

1. INTRODUCTION

General anesthesia is needed in many animal experiments in order to carry the studies out in an effective and appropriate manner. The concentration of oxygen in tissues is often a critical variable in many of these studies. Anesthesia, however, can affect tissue oxygenation because it can alter many physiological parameters, especially cardiopulmonary and metabolic parameters. While these physiological parameters can be measured, the relationship between them and tissue oxygenation may not be straightforward and further, the effects of anesthetics on these parameters vary with the type of anesthetic, the route of administration, and the method of administration. Therefore, it seems very desirable to directly measure the effects of anesthesia on tissue oxygenation.

Oxygen Transport to Tissue XXI, edited by Eke and Delpy.
Kluwer Academic / Plenum Publishers, New York, 1999.

Electron paramagnetic resonance (EPR or completely equivalently, ESR) oximetry is a noninvasive technique which can measure pO_2 level in tissue repeatedly and non-invasively (Swartz and Walczak, 1993). Recently the development of new stable paramagnetic particles such as fusinite, India ink, and lithium phthalocyanine and the development of low frequency EPR instruments (1 GHz or lower) have enabled rapid progress to be made in the field of in vivo EPR oximetry (Vahidi et al., 1994; Swartz et al., 1994; Colacci et al., 1993, Liu et al., 1993).

Probably the most popular technique for anesthesia in laboratory studies of rodents is the bolus administration of injectable anesthetics. In this study using EPR oximetry, we investigated the effects of a bolus injection of two of the most commonly used and representative anesthetic regimes, ketamine/xylazine and pentobarbital, on cerebral pO_2, blood pressure, and arterial blood gases in rats. We obtained the data under the conditions that are often used in many experimental laboratories (use of spontaneously breathing animals with room air) in order to provide an understanding of the effects of anesthetics on cerebral pO_2 under typical experimental conditions in laboratory animals.

2. MATERIALS AND METHODS

2.1. Materials

Oxygen sensitive lithium phthalocyanine (LiPc) crystals were synthesized in our laboratory. The line width of LiPc is a linear function of pO_2 and independent of local metabolic processes, the presence of other paramagnetic species, and pH (5). LiPc crystals equilibrate with local tissue pO_2 within 30 sec, and the responses of the line width to changes in pO_2 are stable for at least 30 days in the brain (Liu et al., 1997). The high density of unpaired spins, combined with a narrow intrinsic line width of LiPc, allows measurements of pO_2 in the brain by one or more crystals with a diameter of ~ 0.2 mm.

2.2. Preparation of Animals

We used 35 male Sprague-Dawley rats weighing 350–450 g (Charles River Lab, Massachusetts). Several small LiPc crystals were implanted at a single site in each animal at least 5 days prior to the experiments under ketamine/xylazine anesthesia (50 mg/5 mg/kg. i.m.) in order to reduce possible variability due to trauma associated with the implantation. The LiPc crystals were placed directly using stereotactic techniques via a 25 G spinal needle into the cerebral cortex at a depth of 3 mm from the surface of the skull through 1 mm diameter drilled holes located 3.0 mm from the midline and 1 mm in front of the bregma.

The rats were divided into four groups; low dose ketamine/xylazine (50 mg/5 mg/kg i.m.; n = 12), high dose ketamine/xylazine (100 mg/10 mg/kg i.m.; n = 12), low dose pentobarbital (40 mg/kg i.p.; n = 6), high dose pentobarbital (80 mg/kg i.p.; n = 5). All drugs were diluted with physiological saline to a total volume of 0.5 ml. The low dose was based on reports in the literature and personal experience to obtain the minimum level of anesthesia that produced a lack of pain response against surgical stress (Wixon et al., 1987). The high dose was chosen to reflect a maximum amount recommended for use in rats.

After injection of the anesthetics the rats were allowed to breathe air spontaneously. Polyethylene catheters (PE-50, Clay-Adams, NJ) were inserted into the right femoral artery to monitor blood pressure. This procedure was accomplished within 15 min after injection of the anesthetics. The body temperature was maintained at 37.0 ± 0.5°C with a heating pad and monitored by a YSI-400 telethermometer throughout the experiment.

At the end of the experiment, the rats were dissected and it was confirmed that LiPc was implanted into a suitable site of the cerebral cortex. Gross and microscopic examination of the tissue around the implanted LiPc confirmed the absence of significant local perturbations from the LiPc.

2.3. EPR and Physiological Measurements

The spectra of LiPc were obtained using an EPR spectrometer constructed in our laboratory with a low frequency (1.2 GHz, "L-band") microwave bridge (Nilges et al., 1989). The rat was placed in the magnet and the head was positioned so that the brain was directly under the extended loop resonator which was adjusted to obtain the maximum signal from the LiPc in the cerebral cortex. Typical settings for the spectrometer were: incident microwave power, 10 mW; magnetic field center, 425 gauss; scan range, 1 gauss; modulation frequency, 27 kHz. Modulation amplitude was set at less than one-third of the EPR line width. Scan time was 2 min. 3–5 scans were usually averaged to achieve a better signal to noise ratio for the spectra.

After at least 10 min following the insertion of the catheter, the measurements of EPR spectra were started and continued until accurate measurements were not possible due to the movement of the animals as they began to recover from the anesthesia. The EPR line widths were converted to pO_2 using a calibration curve determined for each batch of crystals.

Blood pressure (mean arterial, MAP) was monitored continuously by a pressure transducer (Biopac Systems). Arterial blood (0.2–0.4 ml) was drawn into a 1 cc dry heparin coated syringe and blood gases were assessed immediately on a blood gas analyzer (Ciba-Corning 238 pH/Blood Gas Analyzer) at 30 min, 90 min, 150 min (high dose groups only) and 180 min (only high dose of ketamine/xylazine group) after injection of the anesthetics. After each sample was withdrawn, 0.5 ml of saline containing 0.006% heparin was infused into the catheters to prevent clotting; to avoid dehydration an additional 1.0 ml/hr of saline was infused.

2.4. Statistical Analysis

The data are displayed as means ± SE. Significant differences are noted in the tables. The analysis used an unpaired t-test with adjustments for missing values. The calculations were made with the statistical package S-Plus 4.5 (MathSoft, Seattle, 1997).

3. RESULTS

3.1. The Duration of Anesthesia

Table 1 summarizes the data on the usual duration of anesthesia for each group. The data from animals whose duration was not in these ranges were excluded from the further analyses. At the end point of EPR measurements the righting reflex was regained in all rats.

Table 1

Drug group	Drug dosages	Range of duration of anesthesia	# of rats in range/total number in group*
Ketamine/xylazine	50 mg/5 mg/kg	90–110 min	9/12
Ketamine/xylazine	100 mg/10 mg/kg	180–210 min	9/12
Pentobarbital	40 mg/kg	80–90 min	4/6
Pentobarbital	80 mg/kg	150–180 min	5/5

* Only data from animals whose duration of anesthesia was within the indicated range were used for further analyses

3.2. Arterial Blood Gas

Table 2 summarizes the values for the blood gases after injection of the anesthetics. There was no significant difference in any of the parameters between 30 min and 90 min for the low doses of the anesthetics and they were within the physiologically normal range. With the high dose of ketamine/xylazine and pentobarbital there were some significant changes over time for the PaO_2. The other parameters were within the normal range at all sampling points.

3.3. Mean Arterial Pressure

The time course of the mean arterial pressure during anesthesia is summarized in Figure 1. Within each dose group there were no statistically significant differences.

Table 2. Blood gas values during anesthesia with ketamine/xylazine and pentobarbital anesthesia. The values are means ± SE

		30 min	90 min
Low Dose			
pH	ketamine/xylazine	7.40 ± 0.02	7.47 ± 0.01
	pentobarbital	7.42 ± 0.01	7.47 ± 0.01
PaO_2	ketamine/xylazine	72.7 ± 7.5	79.0 ± 0.0
	pentobarbital	78.2 ± 2.5	90.4 ± 5.0
$PaCO_2$	ketamine/xylazine	39.1 ± 1.6	31.8 ± 3.3
	pentobarbital	40.4 ± 1.3	36.4 ± 2.1
BE	ketamine/xylazine	−0.3 ± 0.9	−0.5 ± 1.5
	pentobarbital	1.8 ± 0.7	2.6 ± 1.2

		30 min	90 min	150 min	180 min
High Dose					
pH	ketamine/xylazine	7.37 ± 0.01	7.39 ± 0.01	7.40 ± 0.02	7.43 ± 0.01
	pentobarbital	7.34 ± 0.04	7.38 ± 0.04	7.47 ± 0.01	
PaO_2	ketamine/xylazine	67.6 ± 6.0	68.8 ± 2.1#	77.7 ± 1.2	79.1 ± 3.7
	pentobarbital	53.3 ± 3.2*	82.0 ± 5.7	80.7 ± 5.5	
$PaCO_2$	ketamine/xylazine	43.8 ± 2.4	39.9 ± 0.9	38.7 ± 4.5	35.9 ± 2.1
	pentobarbital	50.0 ± 6.9	43.9 ± 6.5	35.7 ± 2.3	
BE	ketamine/xylazine	0.6 ± 0.7	−1.0 ± 0.7	−1.0 ± 1.2	−0.9 ± 0.7
	pentobarbital	0.0 ± 1.0	0.7 ± 1.5	2.0 ± 1.2	

# $p < 0.05$; vs 180 min * $p < 0.05$; vs 150 min

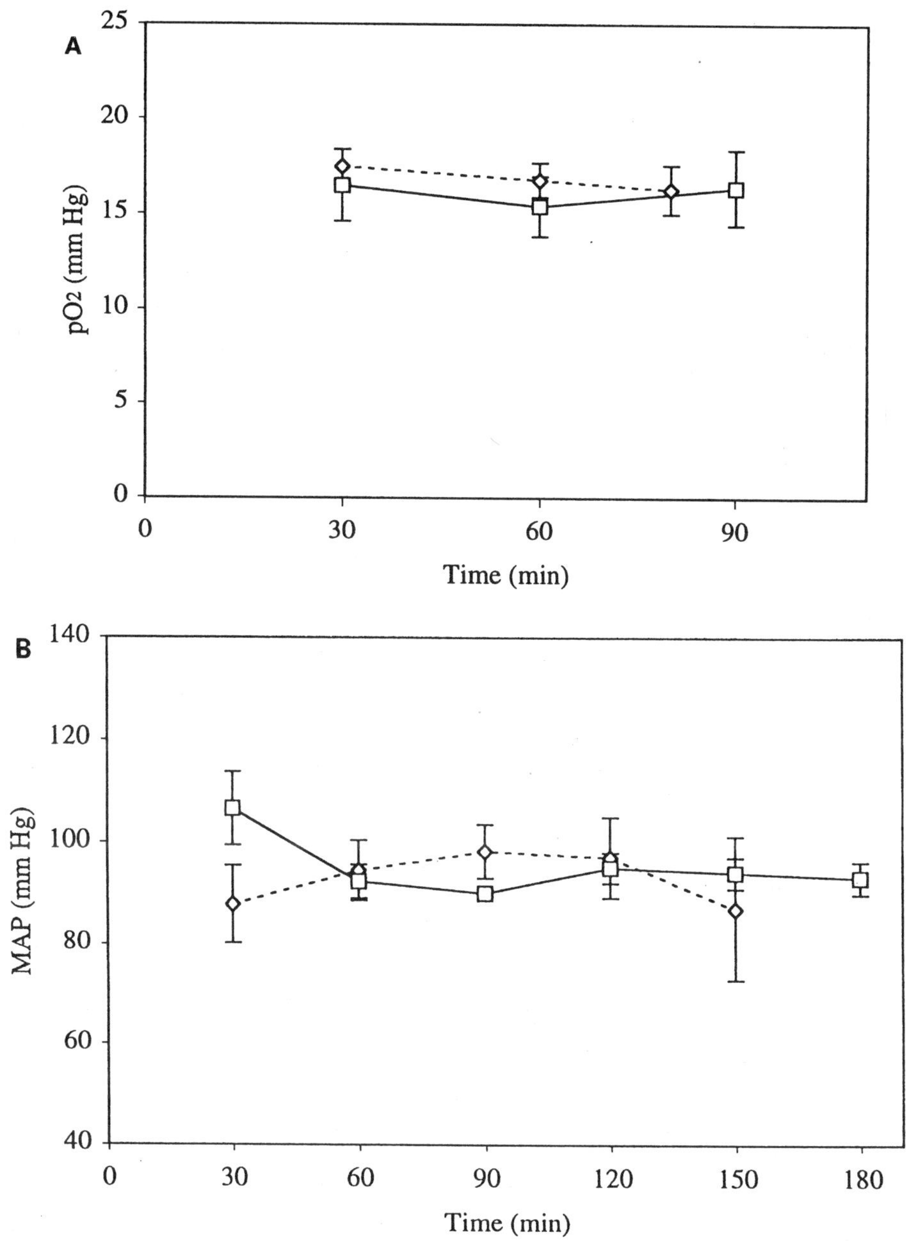

Figure 1. Time course of changes in mean arterial pressure (MAP) after injection of low (A) and high (B) doses of ketamine/xylazine □ and pentobarbital ◇. The values are means ± SE.

3.4. Cerebral pO_2 and the Relationship with MAP

The time courses of the cerebral pO_2 are summarized in Figure 2. There were no statistically significant differences between the two types of anesthetics for the various time points for the low dose anesthetics. In the high dose there were significant differ-

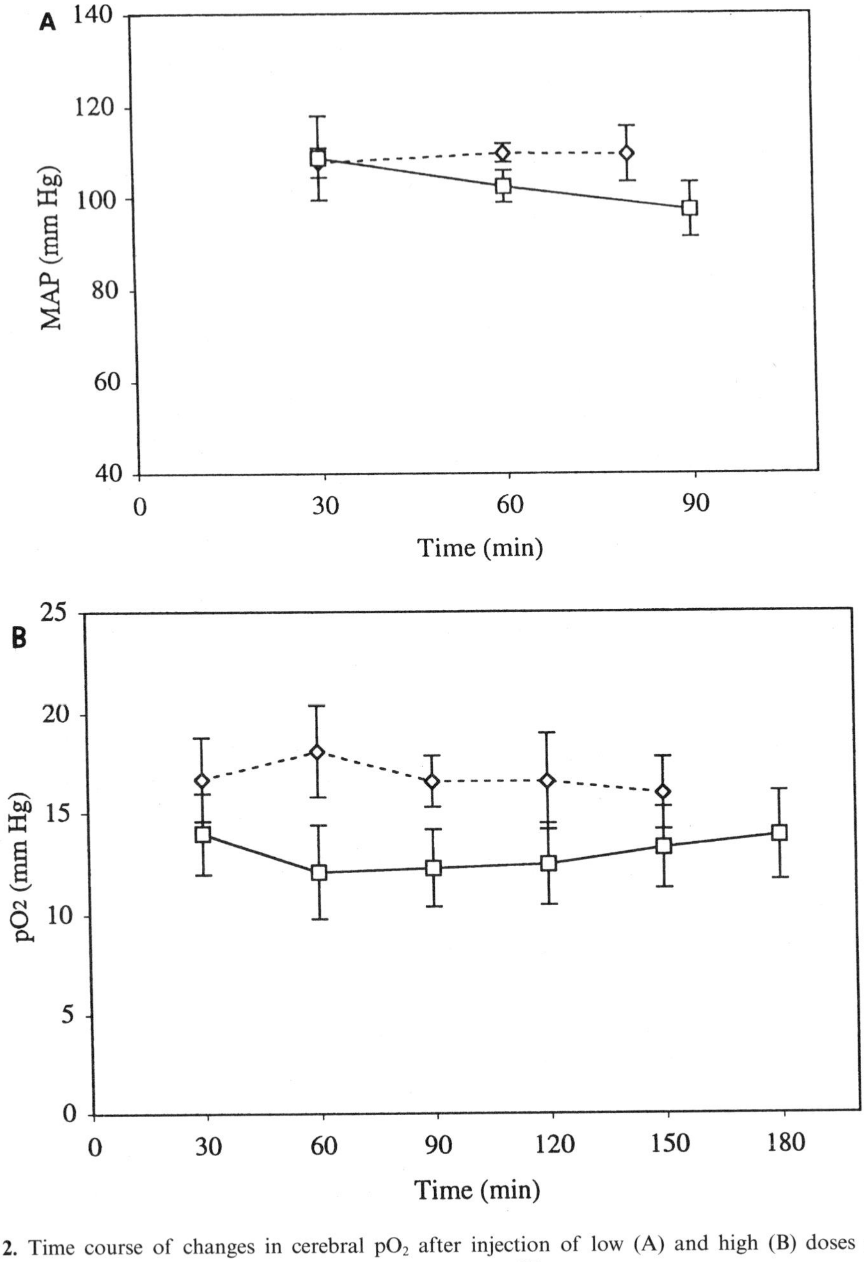

Figure 2. Time course of changes in cerebral pO_2 after injection of low (A) and high (B) doses of ketamine/xylazine □ and pentobarbital ◇. The values are means ± SE.

ences, as summarized in Table 3, with the values becoming significantly lower for ketamine/xylazine, compared to pentobarbital, for the 60 and 90 minute time points.

While the values of the cerebral pO_2 for each animal tended to be consistent throughout the period of adequate anesthesia, the values of the cerebral pO_2 varied within the groups, with the greatest variations observed in the animals who received the high dose of ketamine/xylazine. In order to determine if the variation in cerebral pO_2

Table 3. Statistical significance of cerebral pO_2 for two anesthesia agents at high doses

Time after induction (Min)	Ketamine/Xylazine		Pentobarbital		p-value for the difference
	Mean	SE	Mean	SE	
30	13.6	2.07	16.7	1.67	0.245
60	11.9	2.31	18.1	1.91	0.037-signif
90	12.1	1.32	17.4	1.57	0.010-signif
120	12.5	2.11	16.6	1.61	0.121
150	13.3	1.63	16.0	1.63	0.233

could be explained on the basis of variations in the MAP, the two types of data for these parameters for the high dose of ketamine/xylazine were compared. The results are summarized in Figure 3. There was no apparent correlation of the pO_2 and the blood pressure in individual animals or in the grouped data, indicating that these two parameters were not directly related.

4. DISCUSSION

This study was designed to determine the pO_2 in the brain under what often are "typical" laboratory conditions, i.e., using injectable anesthetics and spontaneous breathing of room air. In these experiments we found that the duration of anesthesia was consistent with the recovery time of the righting reflex, in keeping with our intention of reproducing typical conditions and making measurements during the time period when experiments are usually carried out (Wixon et al., 1987).

The cerebral pO_2 and cardiopulmonary parameters with the low dose of both anesthetics were relatively constant throughout the anesthetic periods. While we did not measure the cerebral pO_2 prior to anesthesia in these experiments, the values we obtained with even the low doses were lower than our previously published values for unanesthetized animals under similar conditions; for the latter the values consistently were more than 20 mm Hg (Liu et al., 1997; Liu et al., 1995). This occurred even though all of the parameters for the blood gases and mean arterial pressure values were within the physiologically normal range throughout the duration of anesthesia.

In contrast to the results with the low doses, the high doses of the anesthetics did result in significant differences in cerebral pO_2 at two time points (Table 3). It is possible that this occurred in part because of a decrease in cerebral oxygen consumption during pentobarbital anesthesia. This interpretation is consistent with a number of published studies on the effect of pentobarbital on cerebral oxygen consumption (Donegan et al., 1985; Albrecht et al., 1985; Laurent et al., 1982), while the response of cerebral blood flow to arterial CO_2 is substantially maintained during pentobarbital anesthesia (Wixon et al., 1987).

Regardless of their basis, the differences in cerebral pO_2 in the high dose groups were not explainable on the basis of changes in the MAP (Figure 3). This may be the most significant new finding from this study, because it very clearly demonstrates that it is not always possible to predict the actual cerebral pO_2 from the physiological measurements that traditionally are made.

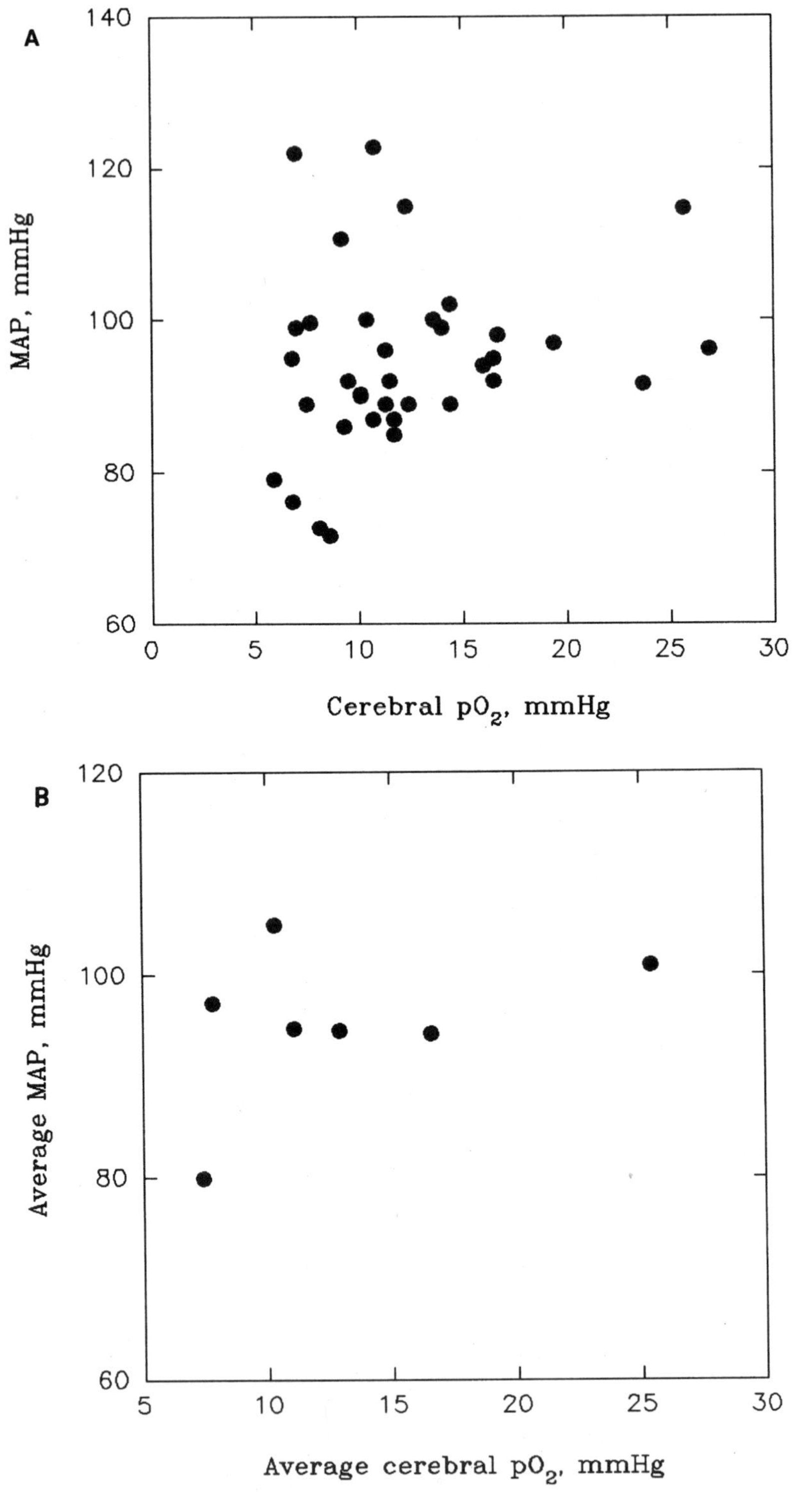

Figure 3. Relationship between MAP and cerebral pO_2 in seven rats receiving high dose ketamine/xylazine during period of anesthesia. Figure 3A shows individual values for each measurement; Figure 3B shows average values for each rat.

5. CONCLUSIONS

Several conclusions seem warranted from these studies:

1. The effects of the two injectable anesthetics on the cerebral pO_2 vary with the type and the dose of anesthetics and the time course after injection, and these are not necessarily consistent with the changes of the conventional cardiopulmonary parameters. Therefore, the effects of anesthetics on tissue pO_2 should be assessed directly for each anesthetic regime.
2. While in this study we did not directly measure the effects of the anesthetics on the pO_2 in other tissues, it seems reasonable to anticipate that these will not necessarily follow the pattern in the brain and therefore the pO_2 should be measured in all of the tissues of interest whenever possible.
3. The "typical" laboratory use of anesthesia in rodents seems suboptimal as indicated by the effects on cerebral pO_2.
4. These results confirm that in vivo EPR oximetry can be a useful tool for evaluating the effects of drugs on tissue pO_2.
5. These studies do not provide definitive information on what anesthetic procedures may be more suitable for studies in rats. We are in the process of completing studies under conditions that will enable to investigate the pertinent effects of various anesthetics under what may be more optimal conditions; our preliminary data suggest that the use of inhalation anesthetics and ventilation with higher concentrations of oxygen may be more optimal for many studies.

ACKNOWLEDGMENTS

This study was supported by NIH Grant PO1 GM51630 and used the facilities of the EPR Center for Viable Biological Studies, supported by NIH Grant P41 RR11602.

REFERENCES

Albrecht, RF, Cook, J, Hoffman, WE, Larscheid, P, Miletich, DJ, and Naughton, N, The interaction between benzodiazepine antagonists and barbiturate-induced cerebrovascular and cerebral metabolic depression, Neuropharmacol. 1985;24:957–963.

Colacci, S, Allecci, M, Gualtieri, G, Quaresimia, V, Urisini, CL, Ferrari, M, and Sotgiu, A, New experimental procedures for in vivo L-band and radio frequency EPR spectroscopy/imaging, J. Chem. Soc. Perkin Trans. 1993;2:2077–2082.

Donegan, JH, Traystman, RJ, Koehler, RC, Jones, MD. Jr, and Rogers, MC, Cerebral vascular, hypoxic, and autoregulatory responses during reduced brain metabolism, Am. J. Physiol. 249 (1985) 421–429.

Laurent, JP, Lawner, P, Simeone, FA, and Fink, E, Pentobarbital changes compartmental contribution to cerebral blood flow, J Neurosurg. 1982;56:504–510.

Liu, KJ, Gast, P, Moussavi, M, Norby, SW, Vahidi, N, Walczak, T, Wu, M, and Swartz, HM, Lithium phthalocyanine: A probe for electron paramagnetic resonance oximetry in viable biological system, Proc. Nat. Acad. Sci. USA, 1993;90:5438–5442.

Liu, KJ, Basic, G, Hoops, PJ, Jiang, J, Du, H, Ou, LC, Dunn, JF, and Swartz, HM, Assessment of cerebral pO_2 by EPR oximetry in rodents: effects of anesthesia, ischemia, and breathing gas, Brain Res. 1995; 685:91–98.

Liu, KJ, Hoopes, PJ, Rolett, EL, Dunn, JF, and HM. Swartz, HM, The effects of anesthesia on cerebral tissue oxygenation and cardiopulmonary parameters in rats, Adv. Expt. Biol. Med, 1997;411:33–39.

Nilges, MJ, Walczak, T, and Swartz, HM, 1 GHz in vivo ESR spectrometer operating with a surface probe, Phys. Med., 1989;5:195–201.

Swartz, HM, and Walczak, T, In vivo EPR: prospects for the 90's, Phys. Med., 1993;9:41–48.

Swartz, HM, Liu, KJ, Goda, F, and Walczak, T, India ink: A potential clinically applicable EPR oximetry probe, Magn. Reson. Med, 1994;31:229–232.

Vahidi, N, Clarkson, RB, Liu, KJ, Norby, SW, Wu, M, and Swartz, HM, In vivo and in vitro EPR oximetry with fusinite: a new coal-derived, particulate EPR probe, Magn. Reson. Med, 1994;31:139–146.

Wixon, SK, White, WJ, Hughes Jr, HC, Marshall, WK, and Lang, CM, The effects of Pentobarbital, Fentanyl-Droperidol, Ketamine/xylazine and Ketamine-diazepam on noxious stimulus perception in adult male rats, Lab. Anim. Sci. 1987;37:731–735.

Wixon, SK, White, WJ, Hughes, HC Jr, Lang, CM, and Marshall, WK, A comparison of Pentobarbital, Fentanyl-Droperidol, Ketamine-xylazine and Ketamine-diazepam in adult male rats, Lab. Anim. Sci. 1987;37:726–730.

Wixon, SK, White, WJ, Hughes, HC Jr, Lang, CM, and Marshall, WK, The effects of Pentobarbital, Fentanyl-Droperidol, Ketamine-xylazine and Ketamine-diazepam on arterial blood pH, blood gas, mean arterial blood pressure and heart rate in adult male rats, Lab. Anim. Sci. 1987;37:736–742.

CEREBRAL BLOOD VOLUME RESPONSES TO ACUTE $PaCO_2$ CHANGES IN HUMANS, ASSESSED WITH NEAR INFRARED SPECTROSCOPY

M. J. T. Van de Ven,[1] W. N. J. M. Colier,[2] B. T. P. Kersten,[1] B. Oeseburg,[2] and H. Folgering[1]

[1]Department of Pulmonology Dekkerswald
University of Nijmegen
PO Box 9001, 6560 GB Groesbeek
The Netherlands
[2]Department of Physiology
Faculty of Medical Sciences
University of Nijmegen
PO Box 9101, 6500 HB Nijmegen
The Netherlands

1. INTRODUCTION

Near Infrared Spectroscopy (NIRS) can be used to measure cerebral blood volume (CBV), and its reactivity to changes in $PaCO_2$. NIRS is a promising technique, which offers continuous measurements of CBV at bedside and is relatively cheap. Moreover, invasiveness is not necessary, which makes NIRS a very interesting tool for clinical use.

The assessment of CBV reactivity in response to $PaCO_2$ changes is not as common as the use of cerebral blood flow (CBF). However, there is a close relationship between CBV and CBF that has been extensively investigated by Grubb et al. (1974) and by Van Zijl et al. (1998). They showed a significant, nonlinear relationship between CBF and CBV ($r = 0.9$). Moreover, the use of CBV instead of CBF eliminates the problems related to the mean transit time (Shockley and LaManna, 1988).

The changes in CBF hemodynamics due to changes in arterial PCO_2 are well known (Kety and Schmidt, 1948; Grubb et al., 1974): hypocapnia results in vasoconstriction with

Oxygen Transport to Tissue XXI, edited by Eke and Delpy.
Kluwer Academic / Plenum Publishers, New York, 1999.

a decrease in CBV, whereas hypercapnia results in vasodilation with increased CBV (Edvinsson et al., 1993).

The optical technique of NIRS is based on the principle, that light in the NIR region is absorbed by chromophores of oxyhemoglobin [O_2Hb] and deoxyhemoglobin [HHb] (Jöbsis et al., 1977). Changes in light absorption are converted into concentration changes of these chromophores. By monitoring changes in [O_2Hb], it should be possible to measure and quantify CBV (Wyatt et al., 1990; Elwell et al., 1994).

In this study, we used NIRS to determine the cerebrovascular reactivity to acute hypo- and hypercapnia, expressed as CBV/$PaCO_2$, in healthy conscious humans. Secondly, we set out to establish reference values of $\Delta CBV/\Delta PaCO_2$ during hypercapnia and hypocapnia. These reference values of $\Delta CBV/\Delta PaCO_2$, measured in healthy subjects, will help to evaluate the response to CO_2 in patients with chronic obstructive pulmonary disease (COPD).

2. METHODS

2.1. Subjects

Fifteen healthy volunteers—5 male and 10 female, mean age 56 years (range 46 to 68 years)—participated in the study. None of them were on medication. At least two hours prior to the experiments, they had to abstain from caffeinated drinks and smoking. A short physical examination was done with a supplementary brief lung function test. All dynamic lung volumes (FEV_1) were within the normal range. All volunteers gave informed consent. The study was approved by the ethical committee of the Department of Pulmonology Dekkerswald, Groesbeek.

2.2. NIRS

We used a new NIRS system (OXYMON, Departments of Physiology and Instrumentation, University of Nijmegen, the Netherlands). This instrument allowed us to use a high sampling rate of 10 Hz. Furthermore, this instrument shows an improvement of 1 OD in sensitivity in contrast to other NIRS instruments (Van der Sluijs et al., 1997).

Changes in optical density (OD) in the near infrared region are measured at three different wavelengths: 905, 845 and 770 nm. A distance of 5.5 cm between transmitting and receiving fibers was used. This ensues deep enough light penetration into the brain (Harris et al., 1993). A pathlength factor of 6.0 was used (van der Zee et al., 1992). Data were stored on disk for off-line analysis and calculation of CBV.

2.3. CBV Measurements

Owing to light scattering in the tissue, only changes in concentration in the chromophores from an arbitrary zero can be determined. A modified Lambert-Beer law gives the relationship between the changes in concentration and absorption. The difference between [O_2Hb] and [HHb] is defined as the oxygenation-index [OI].

To quantify CBV, a slight desaturation is necessary. This was performed by lowering the FiO_2. The change in arterial saturation (SaO_2) was measured with a pulse oximeter (N200, Nellcor Puritan Bennet, USA), on the forehead, in the beat-to-beat mode.

When the change in SaO_2 was related to the change of OI, absolute values of CBV could be calculated. A detailed description of the analysis technique and quantification of CBV in adults using NIRS was described by Elwell et al. (1994), under the assumption that blood flow, blood volume and oxygen consumption remain constant during the desaturation procedure.

In formula, CBV is given by:

$$CBV(ml.100g^{-1}) = \frac{K * \Delta[OI]}{[Hb] * \Delta SaO_2}$$

where K is a constant, which accounts for molecular weight of hemoglobin, cerebral tissue density and the cerebral-to-large vessel hematocrit ratio. [Hb] represents the total hemoglobin concentration in blood (mmol/l), measured from a blood sample.

When stable readings of $[O_2Hb]$, [HHb] and SaO_2 at a lowered saturation level of 6–8% were seen, resaturation followed. Subsequently, a FiO_2 of 21% was given to maintain normoxia. Values of [OI] and SaO_2 at the applied desaturation level were obtained in a 20s time prior to resaturation. Values of [OI] and SaO_2 during normoxia were calculated from a 20s average. Steady state conditions are needed to obtain reproducible values of [OI] and SaO_2 at two levels. A steady state condition was predefined as a condition with stable values of $[O_2Hb]$, [HHb] and [tHb], SaO_2, mean arterial pressure (MAP), endtidal PCO_2 ($PetCO_2$), heart rate (HR) and respiratory rate (RR) for at least 20s. [tHb] is the sum of $[O_2Hb]$ and [HHb].

2.4. General Procedure

The subjects were in supine position during the whole experiment. Figure 1 shows the experimental setup. SaO_2 and HR were monitored with a pulse-oximeter, with the

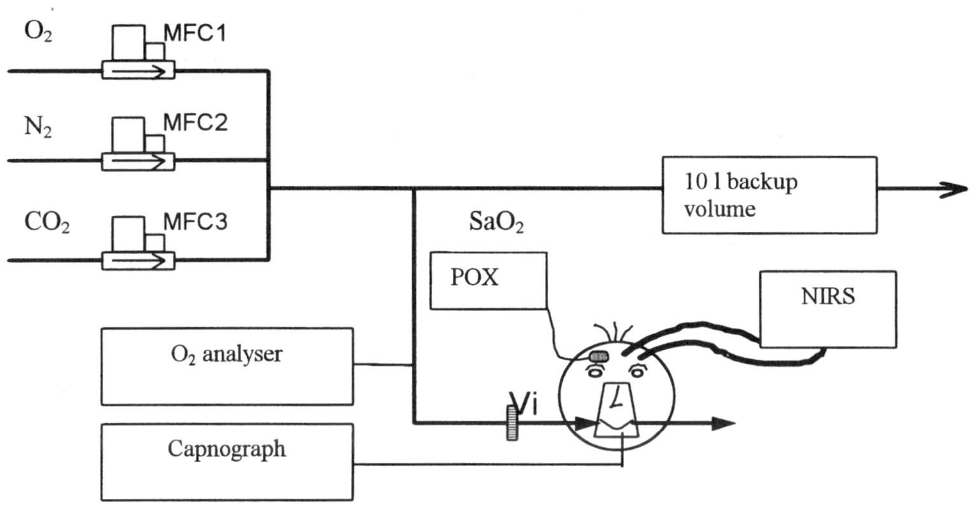

Figure 1. Experimental setup. MFC1–3: flow of oxygen (O_2), nitrogen (N_2), and carbon dioxide (CO_2), regulated with massflow controller. POX: pulse-oximeter; SaO_2: tensions of arterial saturation, measured in a beat-to-beat mode. NIRS: Near Infrared Spectroscopy. V_i: inspiratory minute ventilation (l/min), measured by pneumotachography.

sensor attached to the right-frontal forehead. The optodes of the NIRS were positioned on the left side of the forehead with the receiving probe 2 cm from the midline. The optodes were held in position by an elastic band around the head. Both optodes of the NIRS and the sensor of the pulse-oximeter, were not disconnected until the experiment was finished.

Arterial blood pressure was measured continuously from the middle finger of the right hand with a finger plethysmographic device (Finapres Ohmeda, USA). Blood pressure values were monitored as mean arterial blood pressure (MAP) from each step change in the protocol.

The subjects breathed through a facemask with separated valves for in- and expiratory gas mixture. Changes in inspiratory gas mixture of oxygen, nitrogen and carbon dioxide were induced by means of a computer controlled mass flow system (Bronckhorst, Hitec, The Netherlands). The FiO_2 was monitored continuously using an oxygen analyzer (OM-11, Beckman Inc., USA). Fast changes in inspiratory gas mixture could be performed; the aimed changes were reached within one breath.The expiratory part of the mask was connected to a combined capnograph/pulse-oximeter (N1000, Nellcor Puritan Bennet, USA) to monitor $PetCO_2$ (kPa), respiratory rate (min^{-1}) and inspiratory CO_2 ($PiCO_2$) (kPa). The analogue data were linked directly to the NIRS computer for real time display and storage simultaneously with the NIRS data.

Three different respiratory $PaCO_2$ conditions were measured. First, normocapnic $PetCO_{2,norm}$ baseline conditions were measured with a FiO_2 of 21%. Then, inspired CO_2 was given to induce hypercapnia. We aimed at a hypercapnic value of ($PetCO_{2,norm}$ +1) kPa. For that purpose, the $FiCO_2$ could vary between 3.5% and 6%. For hypocapnia, we predefined a $PetCO_2$ value of ($PetCO_{2,norm}$ −1) kPa. This condition could be performed by making the subject increase the respiratory rate and/or tidal volume. The subject had a visual feedback of his/her $PetCO_2$ on a meter. All three respiratory conditions were duplicated. When a stable respiratory $PetCO_2$ condition was reached, a slight desaturation was induced for calculation of CBV by lowering the FiO_2 to 11–13%. Arterialized capillary samples were taken during (each) stable CO_2 challenge, from a warmed fingertip, and analyzed within 2 minutes (Synthesis 25, Instrumentation Laboratory SpA, Italy). The Hb concentration, was determined on the same analyzer.

2.5. Statistical Analysis

Each subject was measured twice during three respiratory $PaCO_2$ conditions. Duplicate values of CBV and $PaCO_2$ were averaged afterwards.

During the whole experiment, values of $PetCO_2$, HR, RR and MAP were recorded continuously; they were expressed as means with standard deviation during each CO_2 challenge. A paired *t*-test was used to compare mean CBV and MAP values in normocapnic condition to mean CBV and MAP values during hypercapnic and hypocapnic conditions. The level of statistical significance was set at $p < 0.01$. Linear regression analysis was performed on the individual slopes of CBV/$PaCO_2$.

3. RESULTS

Table 1 shows the results of all measurements of CBV, $PaCO_2$, MAP, HR, $PetCO_2$ and RR, expressed as means with standard deviation. Compared to normocapnic CBV values, higher and lower values of CBV were seen in hypercapnia and hypocapnia, respec-

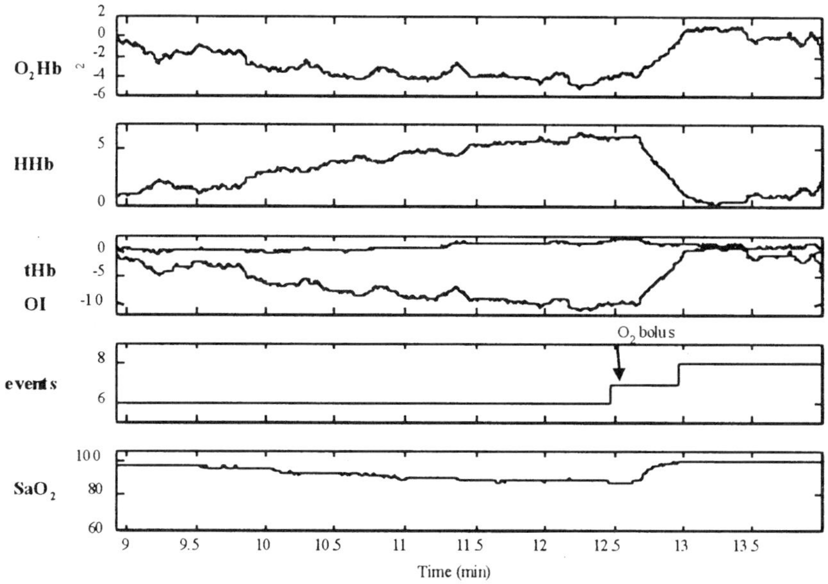

Figure 2. Outline of NIRS measurements ($[O_2Hb]$, [HHb], [OI], [tHb]) and measurements of the pulse oximeter (SaO_2) during slight desaturation, necessary for each CBV calculation. Events: moments to collect the data (which does not automatically correspond to applied changes in gasmixture).

tively ($p < 0.001$). The $PetCO_2$ in hypercapnia showed a value of ($PetCO_{2norm}$ + 1.2) kPa, with concomitant values of ($PaCO_{2norm}$ + 0.5) kPa. In hypocapnia, a ($PetCO_{2norm}$ − 1.1) kPa was achieved, with concomitant values of ($PaCO_{2norm}$ − 1.3) kPa. During each experiment, the same amount of desaturation was established. HR and RR remained stable during the three measured conditions. MAP increased significantly ($p < 0.001$) during both hyper- and hypocapnia. When values of systolic and diastolic arterial blood pressure are evaluated separately, mean systolic blood pressure alters more than mean diastolic blood pressure in all CO_2 conditions; mean sd in changes of systolic blood pressure was 10 mm Hg, whereas the mean sd in changes of diastolic blood pressure was 5 mm Hg.

Figure 3 shows the raw data of all subjects in three conditions; from each individual, 6 values of $PaCO_2$ with concomitant CBV values are plotted. Linear regression analysis was performed, resulting in the equation $CBV = 1.01 * PaCO_2 - 1.39$ ($r = 0.69$). (CBV: $ml.100\,g^{-1}$; $PaCO_2$: kPa).

We determined the individual $\Delta CBV/\Delta PaCO_2$ reactivity for the hypercapnic and for the hypocapnic response, taking the normocapnic situation as a starting point. Figure 4

Table 1. Age and mean values (with SD) for cerebral blood volume (CBV), capillary $PaCO_2$, change in saturation ($dSaO_2$), mean arterial pressure (MAP), heart rate (HR), end tidal CO_2 ($PetCO_2$) and respiratory rate (RR) during normocapnia, hypercapnia and hypocapnia

		Age (yrs)	CBV ($ml.100\,g^{-1}$) Mean (SD)	$PaCO_2$ (kPa) Mean (SD)	$dSaO_2$ (%) Mean (SD)	MAP (mmHg) Mean (SD)	HR (min^{-1}) Mean (SD)	$PetCO_2$ (kPa) Mean (SD)	RR (min^{-1}) Mean (SD)
n = 15	normocapnia	56 (7)	3.51 (0.71)	5.14 (0.36)	7.83 (2.29)	85 (4)	72 (10)	4.33 (0.42)	14 (3)
	hypercapnia		4.82 (1.12)	5.63 (0.29)	5.89 (1.75)	93 (6)	71 (10)	5.57 (0.46)	15 (4)
	hypocapnia		2.55 (0.72)	3.80 (0.24)	6.12 (0.85)	94 (5)	70 (8)	3.21 (0.14)	19 (4)

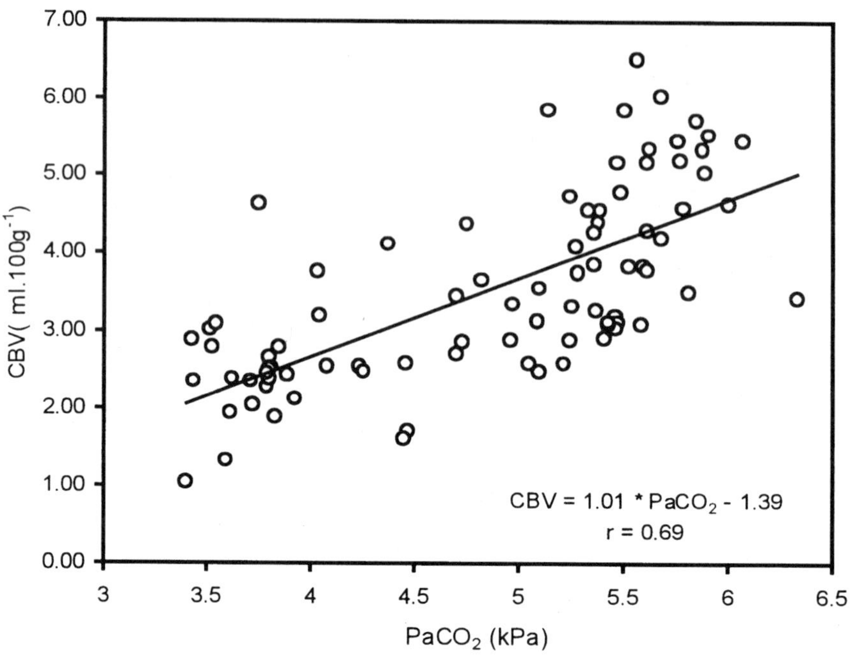

Figure 3. $CBV/PaCO_2$ reactivity from hypo- to hypercapnia: raw data of all subjects.

shows the results. Mean slopes in the hypercapnic and hypocapnic ranges were 2.60 (1.32–3.88) $ml.100\,g^{-1}.kPa^{-1}$ and 0.83 (0.64–1.02) $ml.100\,g^{-1}.kPa^{-1}$ (95% confidence interval), respectively.

4. DISCUSSION

We investigated cerebrovascular reactivity to acute induced hypercapnia and hypocapnia (hyperventilation), expressed ΔCBV over $\Delta PaCO_2$ ($ml.100\,g^{-1}.kPa^{-1}$) in healthy conscious volunteers with NIRS.

There was a difference between the $PetCO_2$ and $PaCO_2$. The $PetCO_2$ is always expected to be lower than the $PaCO_2$. We aimed at a higher level of hypercapnia, but this is difficult to induce in conscious humans due to buffering of CO_2 in the blood. However, despite the small increase in $PaCO_2$, a considerable increase in CBV was found.

During hypocapnia as well as during hypercapnia, the MAP rose in both cases with about 10 mm Hg. This rise, probably due to sympathetic activity, falls well within the range of cerebral autoregulation (Berne and Levy, 1998).

How does $\Delta CBV/\Delta PaCO_2$ from hypo- to hypercapnia compare to other studies? Brun et al. (1994) did a study on mechanically "ventilated preterm infants with arterial blood pressure and blood gases within the normal ranges in 24–48% oxygen supply". Using NIRS and [O_2Hb] as a tracer, the CBV/CO_2 reactivity was in the same range (0.89 $ml.100\,g^{-1}.kPa^{-1}$, 95% confidence interval = 0.63–1.26) as in the present study. Grubb et al. (1974) investigated the $CBV/PaCO_2$ reactivity in anaesthetized, intubated rhesus monkeys using a free diffusible (^{15}O-labeled water) and vascular tracer (^{15}O-labeled car-

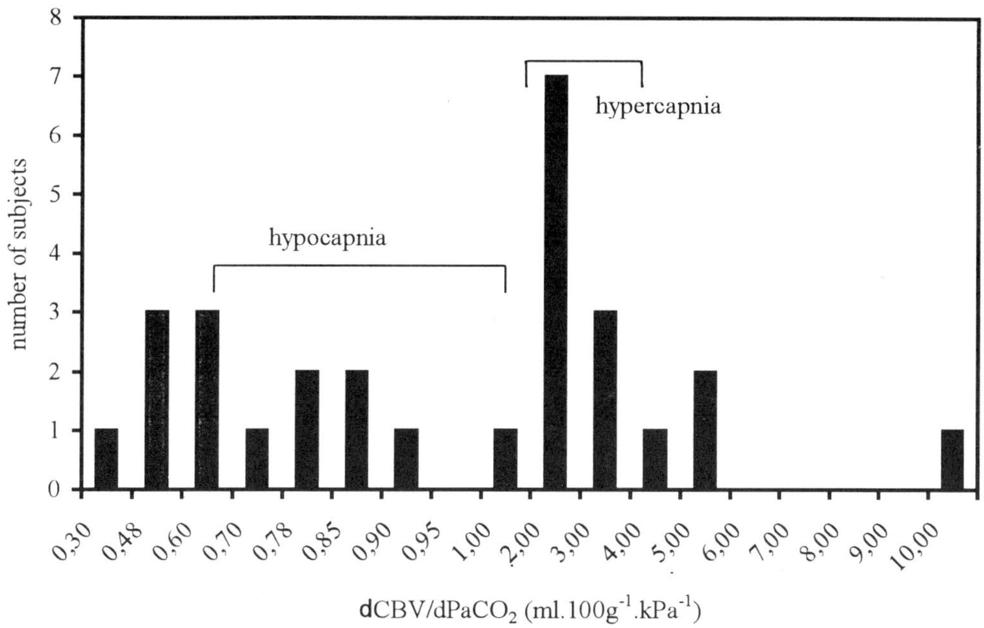

Figure 4. Individual $\Delta CBV/\Delta PaCO_2$ reactivity for the hypercapnic and hypocapnic response.

boxyhemoglobin) with detection by PET. They found a slope of $0.3\,ml.100\,g^{-1}.kPa^{-1}$. However, the study was done in animals, they used a different method and they measured during different circumstances (anaesthetized, mechanically ventilated).

Figure 4 shows the slopes of ΔCBV over $\Delta PaCO_2$ reactivity in the hypercapnic and hypocapnic region. Hypocapnia led to a much lower slope than hypercapnia. Hypocapnia results in lower responses of CBF and CBV than hypercapnia (Edvinsson et al., 1993). Interpretation of the CBV response to $PaCO_2$ must therefore be divided into a reactivity from normo- to hypercapnia and normo- to hypocapnia. A linear relationship was seen in both responses, but with a significantly different slope.

The use of CBV, instead of CBF, may be discussed. Using $[O_2Hb]$ as a tracer to measure CBF, NIRS cannot provide the standards to obtain reproducible values of CBF (Newton et al., 1997; van de Ven et al., 1998). Prior to this study, we determined the reproducibility for CBV measurements with NIRS, which was good (coefficient of variation (CV) ± 10%, Van de Ven et al., 1998), whereas the "reproducibility" for NIRS-CBF varied from (CV) 22 to 92%.

The absolute CBV values found for normocapnia ($3.51 \pm 0.71\,ml.100\,g^{-1}$) were compared to other CBV measurements in humans. They are in good agreement with other studies using NIRS (Elwell et al., 1994). However, other methods showed significant ($p < 0.01$) higher CBV values (Table 2). An explanation for this discrepancy was given by Owen-Reece et al. (1996). They quantified cerebral blood flow on the scalp and then on the dura after craniotomy in humans. Compared to the CBF values measured with fibres on the dura, they found an approximately threefold underestimate of CBF with fibres on the scalp. The role of extracerebral tissue on transmission of light through the cranial cavity seemed to be the main cause. Due to this underestimation, lowered values of CBV could be expected, when the fibres are placed on the scalp. The slightly higher

Table 2. Comparison of mean (normocapnic) CBV value to other CBV values.

Investigators	Number of subjects	Method	CBV (ml.100 g^{-1})
Van de Ven et al. (1998)	15	NIRS	3.51 ± 0.71
Elwell et al. (1994)	10	NIRS	2.85 ± 0.97
Leenders et al. (1990)	32	PET	4.3 ± 0.8*
Muizelaar et al. (1997)	10	CT	5.3 ± 0.4*

*, $p < 0.01$

CBV value found by Muizelaar et al. (1997) using Computed Tomography are based on independent measurements of CBF and transit time; he supposed an overestimation for CBF (and therefore for CBV) using Xenon/CT.

MRI did not report normal values of CBV in humans yet. However, there is a report studying CBV values measured by PET and MRI in pigs (Østergaard et al., 1998). PET CBV values were 2.5 times larger than absolute MRI CBV values. They explained this difference as a result of the (hypothetized) sensitivity of MRI to small vessels. This hypothesis is in agreement with the NIRS theory. Near infrared absorption changes reflect changes in the oxygenation of the microvasculature, i.e., tissue oygenation at the level of small blood vessels, capillaries, and intracellular sites of oxygen uptake (Mancini et al., 1994).

We conclude that CBV responses to acute $PaCO_2$ changes in humans can adequately be assessed by NIRS: hypercapnia results in an increase of CBV, whereas hypocapnia decreases CBV. Furthermore, the results of $\Delta CBV/\Delta PaCO_2$ reactivity are in agreement with other studies in human neonates using NIRS. Separated evaluation of CBV reactivity to hypo- and hypercapnia resulted in a linear relationship with significantly different slopes. The latter approach to evaluate CBV reactivity might be important for further analysis of pathological respiratory CO_2 conditions as in chronic obstructive pulmonary disease.

ACKNOWLEDGMENTS

We would like to thank Ing. MC van der Sluijs for his technical support and J. Evers for his assistance during the experiments.

This study is supported by the Dutch Astma Foundation (96.09).

REFERENCES

Berne RM and Levy MN (1998) Special circulations. In: Physiology. (Berne RM and Levy MN, Eds) Mosby, Inc. St. Louis, pp. 478–501.

Brun NC and Greisen G (1994) Cerebrovascular responses to carbon dioxide as detected by near-infrared spectrophotometry: comparison of three different measures. Ped Res 36(1):20–24.

Edvinsson L, MacKenzie ET, and McCulloch J (1993) Fundamental responses of the cerebral circulation. From: Cerebral blood flow and metabolism, Raven Press, New York, pp. 553–580.

Elwell C, Cope M, Edwards D, Wyatt J, Delpy D, and Reynolds EO (1994) Quantification of adult cerebral hemodynamics by near infrared spectroscopy. J Appl Physiol 77:2753–2760.

Grubb RL, Raichle ME, Eichling JO, and Ter-Pogossian MM (1974) The effects of changes in $PaCO_2$ on cerebral blood volume, blood flow, and vascular mean transit time. Stroke 5:630–639.

Harris DNF and Bailey SM (1993) Near infrared spectroscopy in adults. Anaesthesia 48:694–696.

Jöbsis FF (1977) Noninvasive infrared monitoring of cerebral and myocardial oxygen sufficiency and circulatory parameters. Science 198:1264–1267.

Kety SS and Schmidt CF (1948) The effects of altered arterial tensions of carbon dioxide and oxygen on cerebral blood flow and cerebral oxygen consumption of normal young men. J Clin Invest 27:484–492.

Leenders KL, Perani D, Lammertsma AA, Heather JD, Buckingham P, Healy MJR, Gibbs JM, Wise RJS, Hatazawa J, Herold S, Beaney RP, Brooks DJ, Spinks T, Rhodes C, Frackowiak RSJ, and Jones T (1990) Cerebral blood flow, blood volume and oxygen utilization; normal values and effect of age. Brain, 113:27–47.

Mancini DM, Bolinger L, Li H, Kendrick K, and Wilson JR (1994) Validation of near-infrared spectroscopy in man. J Appl Physiol 77(6):2740–2747.

Muizelaar JP, Fatouros PP, and Schröder ML (1997) A new method for quantitative regional cerebral blood volume measurements using computed tomography. Stroke 28(10):1998–2005.

Newton CRJC, Wilson DA, Gunnoe E, Wagner B, Cope M, and Traystman RJ (1997) Measurements of cerebral blood flow in dogs with near infrared spectroscopy in the reflectance mode is invalid. J Cerebr Blood Flow Metab 17:695–703.

Østergaard L, Smith DF, Verstergaard-Poulsen P, Hansen SB, Gee AD, Gjedde A, and Gyldensted C (1998) Absolute cerebral blood flow and blood volume measured by magnetic resonance imaging bolus tracking: Comparison with positron emission tomography values. J Cereb Blood Flow Metab 18(4):425–432.

Owen-Reece H, Elwell CE, Harkness W, Goldstone J, Delpy DT, Wyatt JS, and Smith M (1996) Use of near infrared spectroscopy to estimate cerebral blood flow in conscious and anaesthetized adult subjects. Br J Anaesth 76:43–48.

Shockley RP and LaManna JC (1988) Determination of rat cerebral cortical blood volume changes by capillary mean transit time analysis during hypoxia, hypercapnia and hyperventilation. Brain Res 454:170–178.

Van de Ven MJT, Colier WNJM, Walraven D, Oeseburg B, and Folgering H (1998) Cerebral blood flow in humans measured with near infrared spectroscopy is not reproducible. Adv Exp Med Biol (in press).

Van de Ven MJT, Colier WNJM, Walraven D, Oeseburg B, and Folgering H (1998) Reproducibility of cerebral blood volume measurements assessed with near infrared spectroscopy. Abstract ISOTT.

Van der Sluijs MC, Colier WNJM, Houston RJF, and Oeseburg B (1997) A new and highly sensitive continuous wave near infrared spectrophotometer with multiple detectors. SPIE 3194:63–72.

Van der Zee P, Cope M, Arridge SR, Essenpreis M, Potter LA, Edwards AD, Wyatt JS, McCormick DC, Roth SC, Reynolds EOR, and Delpy DT (1992) Experimentally measured optical pathlengths for the adult head, calf and forearm and the head of the newborn infant as a function of interoptode spacing. Adv Exp Med Biol 316:143–153.

Wyatt JS, Cope M, and Delpy DT (1990) Quantitation of cerebral blood volume in human infants by near infrared spectroscopy. J Appl Physiol 68(3):1086–1091.

Van Zijl PCM, Eleff SM, Ulatowski JA, Oja JME, Ulug AM, Traystman RJ, and Kauppinen RA (1998) Quantitative assessment of blood flow, blood volume and blood oxygenation effects in functional magnetic resonance imaging. Nat Med 4(2):159–167.

BRAIN OXYGENATION AFTER EXPERIMENTAL CLOSED HEAD INJURY

A NIRS Study

K. van Rossem,[1] S. Garcia-Martinez,[1] G. De Mulder,[2]
B. Van Deuren,[1] K. Engelborghs,[1] J. Van Reempts,[1] and M. Borgers[1]

[1]Department of Life Sciences–Neuropathology
Janssen Research Foundation
Beerse, Belgium
[2]Department of Neurosurgery
University of Antwerp, Belgium

1. INTRODUCTION

In blunt head trauma, primary brain injury takes the form of surface contusions and fractures, diffuse axonal injury and intracranial hemorrhage. Secondary damage may result from delayed pathological events including cerebral ischemia, brain edema and increased intracranial pressure (ICP). These events reinforce one another and may lead to a fatal disturbance of the balance between oxygen delivery and oxygen demand. Improvement or maintenance of this balance is a main issue in the treatment of severe head-injured patients. In this respect reduction of ICP is a key tool as it may improve brain perfusion and attenuate the pathologic cascade. Therapeutic interventions that adequately reduce ICP do not necessarily improve and may even impair tissue oxygenation (Unterberg et al., 1997a). This addresses the need for adequate monitoring of brain oxygenation during and after treatment.

In order to study the pathophysiology and pharmacological treatment of traumatic brain injury we have modified the closed head injury (CHI) rat model of Marmarou et al. (1994). This modified model expresses several features of severe head injury, including cell necrosis, diffuse axonal injury, brain edema and increased ICP (Engelborghs et al., 1998). The latter could be reduced with intravenous infusion of the osmotic diuretic mannitol (G. De Mulder, unpublished results), a therapy that is recommended in the management of traumatic intracranial hypertension in humans.

Near-infrared spectroscopy (NIRS) may provide a non-invasive tool for continuous monitoring of cerebral oxygenation (Jöbsis, 1977; Wyatt et al., 1986) and can be useful to identify impaired brain oxygenation after severe head trauma and evaluate effects of pharmacological interventions. In the present study we applied NIRS to study brain oxygenation and the effect of mannitol treatment after experimental CHI in the rat.

2. METHODS

2.1. Animal Preparation and Treatment

Animal housing and treatment complied with European Union directive for Animal Welfare #86/609.

Male Sprague-Dawley rats (Charles River, Sulzfeld, Germany) of 380–430 g were used for the experiments. Anesthesia was induced with 4% isoflurane in 30% O_2 and 70% N_2O during 5 min. Tracheal intubation was then performed and the animals were kept under 2% isoflurane in the same gas mixture during surgical preparation. Vascular catheters were inserted into the left femoral artery for arterial blood sampling and monitoring of mean arterial blood pressure (MABP) and into the left femoral vein for injection of solution.

Twenty min after induction of anesthesia rats were subjected to CHI using a weight-drop device (Engelborghs et al., 1998). They were positioned onto a horizontal platform mounted on springs to allow a vertical downward displacement upon impact. The platform was then placed under a vertical hollow Plexiglas column (inner diameter 20 mm, length 1.5 m) containing a steel cylinder of 400 g (diameter 19 mm) which can freely fall through the column from a designated height. The impact-site of the cylinder was covered with a silicon disk of 9 mm diameter and 2 mm thickness. The impact was leveled at the bregma of the skull with the inter-aural line as reference coordinate. Twenty min after induction of anesthesia the trachea tube was disconnected from the gas circuit and the cylinder was dropped from a height of 70 cm onto the unprotected head. After the impact the trachea tube was reconnected.

Immediately following CHI the rats were fixed into a stereotaxic apparatus (Model 900, David Kopf, Tujunga, CA, USA) and a temperature probe (1PT 100, Telerex nv, Belgium) was inserted in the rectum. The scalp, periost and parts of the temporal muscles were removed in order to expose the parietal and temporal bones. An emitting and receiving optic fiber (∅ 3 mm) were placed respectively onto the left and right temporal bones (frontal pole of the fibers at the level of bregma, angle of ±10° relative to both the coronal and sagittal plane). Optic gel (code 608, Cargille, New Jersey, USA) was applied between the fibers and the skull. The skull was then covered with black clay. The concentration of isoflurane was switched to 1.5% and maintained until the end of the experiment. Since large hemorrhages significantly increase light absorption and reduce the accuracy of the NIRS measurements, rats presenting a skull fracture or a visible massive intracranial bleeding after CHI were immediately excluded from the experiment and killed by exposure to 100% CO_2.

Body temperature was monitored continuously and maintained at 37°C with a control unit (model JSI0968, Janssen Scientific Instruments, Beerse, Belgium) connected to a heated pad. NIR chromophore changes and MABP were monitored continuously from 30 to 85 min after CHI. An arterial blood sample was taken at 40 min after CHI.

One group of animals was treated with mannitol. Treatment was given as a single i.v. infusion (1.3 g/kg) from 45 to 65 min after CHI. At the end of the measurement, the animals were killed by 100% N_2O breathing (terminal anoxia).

2.2. Near-Infrared Spectroscopy (NIRS)

The CCD spectrophotometer, NIRS software and algorithms have been developed at University College London (Cope et al., 1989; Matcher et al., 1995). Multiwavelength spectroscopy (spectrum 600–1000 nm) and second differential NIRS (Matcher and Cooper, 1994) were applied to monitor the optical pathlengths for three wavelengths (740 nm, 840 nm and 960 nm), the absolute concentration of deoxyhemoglobin (Hb) and changes in concentrations of oxyhemoglobin (HbO_2) and oxidized cytochrome oxidase (OxCytOx). Absolute [HbO_2] and [OxCytOx] were back-calculated by taking the minimum values obtained by terminal anoxia as zero. Brain hemoglobin saturation (HbSat) and the concentration of total hemoglobin (HbTot) could then be calculated: [HbTot] = [Hb] + [HbO_2]; HbSat = 100 × [HbO_2]/[HbTot]. Taking into account the possible interference of spectral contributors other than Cu_A to the cytochrome signal (Cooper et al., 1994), [OxCytOx] has to be regarded as an estimate of the absolute concentration of oxidized cytochrome oxidase.

2.3. Experimental Design and Data Analysis

Three groups of rats were studied according to a completely randomized experimental design: sham rats (n = 7); rats subjected to CHI without further treatment (CHI-control, n = 6); rats subjected to CHI and treated with mannitol (CHI-mannitol, n = 6).

Differences between groups were evaluated using the Wilcoxon-Mann-Whitney midrank sum test. Differences within groups were evaluated using the Wilcoxon matched-pairs signed-ranks test. Results shown in the text and tables are expressed as mean ± SD.

3. RESULTS

Physiological variables at 40 min after CHI are summarized in Table 1. No significant differences in systemic arterial pressure and arterial blood oxygenation were observed between sham and CHI animals. Table 2 summarizes the concentrations of brain hemoglobin and oxidized cytochrome oxidase at 40 min after CHI. HbSat was significantly reduced after CHI. In 4 CHI animals HbSat was below 30%. Two of them, both belonging to the CHI-mannitol group, showed values of [OxCytOx] below 4.5 μM.

Table 1. Physiological variables at 40 min after CHI

	Sham (n = 7)	CHI control (n = 6)	CHI mannitol (n = 6)
MABP (mmHg)	78.8 ± 15.8	79.1 ± 18.7	66.9 ± 13.0
Body temperature (°C)	36.8 ± 0.9	37.3 ± 0.2	36.7 ± 0.9
p_aCO_2 (mmHg)	40.4 ± 4.0	42.1 ± 4.9	43.5 ± 5.4
p_aO_2 (mmHg)	105.8 ± 14.9	110.3 ± 25.2	102.9 ± 12.6
pH	7.43 ± 0.03	7.43 ± 0.02	7.39 ± 0.04
SaO_2(%)	97.8 ± 1.1	97.8 ± 1.2	97.3 ± 0.8

Values are mean ± SD. No significant differences are observed between groups.

Table 2. Concentrations of brain hemoglobin and oxidized cytochrome oxidase, and the optical pathlengths for different wavelengths as assessed with NIRS at 40 min after CHI

	Sham (n = 7)	CHI control (n = 6)	CHI mannitol (n = 6)
[Hb] (mM)	18.9 ± 2.3	68.2 ± 55.9*	56.8 ± 45.4*
[HbO_2] (mM)	63.7 ± 7.6	41.0 ± 14.4*	35.0 ± 6.5*
[HbTot] (mM)	82.1 ± 9.4	108.7 ± 47.1	92.3 ± 44.0
HbSat (%)	77.0 ± 1.6	44.0 ± 22.0**	44.2 ± 16.8**
[OxCytOx] (mM)	6.1 ± 0.4	6.3 ± 0.5	5.4 ± 1.1
Path 740 nm (cm)	5.6 ± 0.8	5.0 ± 1.1	5.2 ± 0.9
Path 840 nm (cm)	5.1 ± 0.8	5.8 ± 0.5	5.6 ± 0.5
Path 960 nm (cm)	3.6 ± 0.4	3.9 ± 0.4	3.8 ± 0.2

Values are mean ± SD. * $P < 0.05$ vs. Sham; ** $P < 0.001$ vs. Sham

In all other animals [OxCytOx] fell within the range of 5.2 to 6.8 μM. No significant differences in [OxCytOx] were found between groups. Total hemoglobin was higher in some CHI animals but CHI groups did not differ significantly from shams.

The time course of the NIR chromophores in the three groups is represented in Figure 1. In sham animals [HbO_2], [Hb] and [OxCytOx] were reproducible and stable over time. HbSat gradually increased in CHI animals but remained significantly lower than in sham animals.

Mannitol treatment did not affect [HbTot] nor HbSat. In 5 out of 6 CHI-mannitol rats [OxCytOx] transiently increased during treatment, whereas in the control group the OxCytOx signal gradually decreased. The average change in [OxCytOx] during the treatment period, however, was not statistically different between the CHI-control and CHI-mannitol group (−0.21 ± 0.27 μM vs. 0.16 ± 0.44 μM, $p = 0.0656$). Mannitol did not induce significant changes of MABP.

No significant changes in optical pathlengths were observed in any group during the experiment.

4. DISCUSSION

This study clearly demonstrates that, in the acute phase after experimental CHI, cerebral oxygenation is impaired, and addresses the use of NIRS to identify such changes and monitor the effects of pharmacological treatment. NIRS relies on the relative transparency of biological tissue to light in the near-infrared region and the oxygen dependence of the absorption spectra of hemoglobin and cytochrome oxidase (Jöbsis, 1977). Conventional NIRS allows to monitor changes in concentrations of these chromophores by monitoring the changes in absorption of a limited number of wavelengths (Wyatt et al., 1986). With the aid of multiwavelength spectroscopy and second differential NIRS the optical pathlength and absolute concentration of deoxyhemoglobin can additionally be determined, assuming that the water concentration in the tissue is known and constant (Matcher & Cooper, 1994). The assumed water content (80%) corresponds to values found in normal adult Sprague-Dawley rats (Barzo et al., 1997; Unterberg et al., 1997). In the CHI rat model developed by Marmarou et al. (1994) a significant but small increase in brain water content (<2%) has been demonstrated (Barzo et al., 1997). In preliminary experiments we have found that in our model the water content was comparable in sham and CHI animals (unpublished results). Assuming the worst case, the error introduced

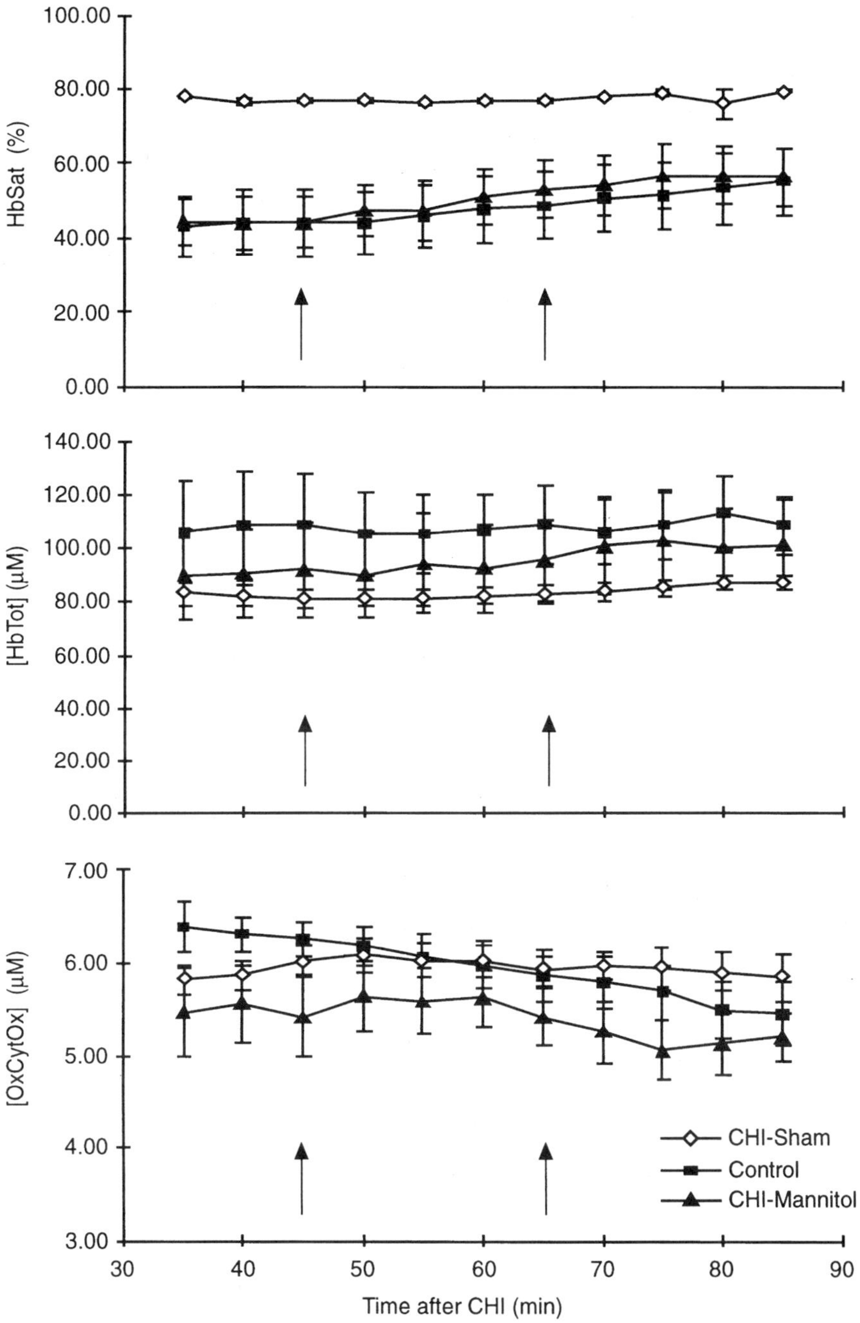

Figure 1. Time course of [HbTot], HbSat and [OxCytOx] in sham rats and rats subjected to closed head injury. Values are mean ± SE. Arrows indicate onset and end of mannitol treatment.

in the calculation of chromophore concentrations in brains with increased water content, remains very small. In addition, we did not observe significant changes in the optical pathlengths during the measurements, indicating that water content and tissue geometry did not change significantly.

In normal rats all NIRS chromophores were reproducible and stable in time. The estimated concentrations of oxidized cytochrome oxidase correspond with biochemically determined concentrations of total cytochrome oxidase in adult rats (Brown et al., 1991). In the acute phase after CHI total brain hemoglobin saturation with oxygen (HbSat) was consistently lower than in sham rats. This is in line with the observed increase of cerebral arteriovenous oxygen difference ($AVDO_2$) (Le Roux et al., 1997) and decreased tissue pO_2 (Dings et al., 1996) during the first hours after brain trauma in humans. Although additional studies are required to determine the underlying cause and evaluate a possible correlation with ICP, a reduction of cerebral blood flow (CBF) is the most likely explanation. In the acute phase after head trauma a reduced CBF is indeed often observed in patients (Bouma et al., 1992; Martin et al., 1997). Arterial vasoconstriction, impaired microcirculation, venous congestion and hemorrhages may all contribute to the decrease in HbSat.

In spite of a reduced HbSat, mitochondrial respiration, as reflected by the cytochrome oxidase signal, was still normal in most CHI animals. Low values of oxidized cytochrome oxidase were observed only when HbSat was below 30%, indicating that cytochrome oxidase becomes reduced only when hemoglobin is almost fully deoxygenated. This observation also demonstrates that measuring oxygenation at the cellular level in addition to monitoring of brain hemoglobin saturation may be essential and that NIRS provides a useful tool in this respect.

Intravenous injection of mannitol is a recommended therapy for intracranial hypertension in head-injured patients (Bullock et al., 1996) and also reduces ICP in our model (De Mulder, unpublished results). There is still controversy regarding the exact mechanisms by which mannitol exerts its beneficial effect. Delayed osmotic dehydration of the brain is a widely accepted mechanism of action. An immediate plasma-expanding effect that reduces the hematocrit and blood viscosity, increases cerebral oxygen delivery and results in a reflectory vasoconstriction with a reduction of the cerebral blood volume, is also proposed. Our results are not in line with the latter mechanism. Indeed, a vasoconstriction in addition to a reduced hematocrit would lead to a reduction of [HbTot] which was not observed in our experiments. HbSat was not influenced by mannitol treatment, whereas a tendency towards improved oxidative respiration, as indicated by the OxCytOx signal, was observed. A simultaneous increase of oxygen delivery and oxygen consumption can be hypothesized. Reactive oxygen species are known to inhibit the respiratory chain and protective effects of mannitol, e.g. reduction of ischemic damage, have been attributed to free radical scavenging (MacGovern et al., 1984) but this mechanism of action is not generally accepted (Gilbe et al., 1996).

In summary, the data presented in this study indicate that brain oxygenation is significantly impaired in the acute phase after experimental closed head injury and that treatment with mannitol tends to improve oxidative respiration without altering hemoglobin saturation. Efforts are ongoing to measure CBF, $AVDO_2$, ICP and NIR chromophores simultaneously in order to determine the underlying causes and to provide an adequate tool for the evaluation and improvement of pharmacological treatment after severe head injury.

REFERENCES

Barzo P, Marmarou A, Fatouros P, Ito J, and Corwin F (1997) MRI diffusion-weighted spectroscopy of reversible and irreversible ischemic injury following closed head injury. Acta Neurochir. [Suppl] 70:115–118.

Bouma GJ, Muizelaar JP, Stringer WA, Choi SC, Fatouros P, and Young HF (1992) Ultra-early evaluation of regional cerebral blood flow in severely head-injured patients using xenon-enhanced computerized tomography. J. Neurosurg. 77:360–368.

Brown GC, Crompton M, and Wray S (1991) Cytochrome oxidase content of rat brain during development. Biochim. Biophys. Acta 1057:273–275.

Bullock R, Chesnutt RM, Clifton G, Ghajar J, Marion DW, Narayan RK, Newell DW, Pitts LH, Rosner MJ, and Wilberger JE (1996) The use of mannitol in severe head injury. J. Neurotrauma 13(11):705–709.

Cooper CE, Matcher SJ, Wyatt JS, Cope M, Brown GC, Nemoto EM, and Delpy DT (1994) Near-infrared spectroscopy of the brain: relevance to cytochrome oxidase bioenergetics. Biochem. Soc. Trans. 22: 974–980.

Cope M, Delpy DT, Wray S, Wyatt JS, and Reynolds EOR (1989) A CCD spectrophotometer to quantitate the concentration of chromophores in living tissue utilising the absorption peak of water at 975 nm. Adv. Exp. Med. Biol. 247:33–40.

Dings J, Meixenberger J, Amschler J, Hamelbeck B, and Roosen K (1996) Brain tissue pO_2 in relation to cerebral perfusion pressure, TCD findings and TCD-CO_2-reactivity after severe head injury. Acta Neurochir. 138:425–434.

Engelborghs K, Verlooy J, Van Reempts J, Van Deuren B, Van de Ven M, and Borgers M (1998) Temporal changes in intracranial pressure in a modified experimental model of closed head injury. J. Neurosurg. in press.

Gilbe CE, Sage FJ, and Gutteridge JMC (1996) Mannitol: molecule magnifique or a case of radical misinterpretation? Free Rad. Res. 24(1):1–7.

Jöbsis FF (1977) Noninvasive, infrared monitoring of cerebral and myocardial oxygen sufficiency and circulatory parameters. Science 198:1264–1267.

Le Roux PL, Newell DW, Lam AM, Grady MS, and Winn HR (1997) Cerebral arteriovenous oxygen difference: a predictor of cerebral infarction and outcome in patients with severe head injury. J. Neurosurg. 87:1–8.

MacGovern GJ, Bolling SF, and Casale AS (1984) The mechanism of mannitol in reducing ischemic injury: hyper-osmolarity or hydroxyl scavenger. Circulation 70 [Suppl 1]:191–195.

Marmarou A, Montasser A, Foda MA, van den Brink W, Campbell J, Kita H, and Demetriadou K (1994) A new model of diffuse brain injury in rats. Part I: Pathophysiology and biomechanics. J Neurosurg. 80(2):291–300.

Martin NA, Patwardhan RV, Alexander MJ, Zane Africk C, Hong Lee J, Shalmon E, Hovda DA, and Becker DP (1997) Characterization of cerebral hemodynamic phases following severe head trauma: hypoperfusion, hyperemia, and vasospasm. J. Neurosurg. 87:9–19.

Matcher SJ and Cooper CE (1994) Absolute quantification of deoxyhaemoglobin concentration in tissue near infrared spectroscopy. Phys. Med. Biol. 39:1295–1312.

Matcher SJ, Elwell CE, Cooper CE, Cope M, and Delpy DT (1995) Performance comparison of several published tissue near-infrared spectroscopy algorithms. Anal. Biochem. 227:54–68.

Unterberg AW, Kiening KL, Härtl R, Bardt T, Sarrafzadeh AS, and Lanksch WR (1997a) Multimodal monitoring in patients with head injury: evaluation of the effects of treatment on cerebral oxygenation. J. Trauma 42(5):S32–S37.

Unterberg AW, Stroop R, Thomale UW, Kiening KL, Pauser S, and Vollmann W (1997b) Characterisation of brain edema following "controlled cortical impact injury" in rats. Acta Neurochir. [Suppl] 70:106–108.

Wyatt JS, Cope M, Delpy DT, Richardson CE, Edwards AD, Wray SC, and Reynolds EOR (1986) Quantification of cerebral oxygenation and haemodynamics in sick newborn infants by near infrared spectrofotometry. Lancet 2:1063–1066.

ADENOSINE IMPROVES CEREBRAL RECOVERY IN RAT AFTER CARDIAC ARREST AND RESUSCITATION

K. Xu, Y. Zhou, W. D. Lust, and J. C. LaManna

Departments of Neurology and Neurosurgery
Case Western Reserve University
School of Medicine
Cleveland, Ohio 44106

1. INTRODUCTION

Cardiac arrest and resuscitation produce global cerebral ischemia and reperfusion damage to the brain, which lead to high mortality and delayed neuronal death. Adenosine (ADO) has been suggested as an endogenous neuroprotective molecule, acting through multiple potential mechanisms (Sweeney, 1997). In this study we investigated the possible neuroprotective effects of adenosine on cerebral recovery following global ischemia induced by cardiac arrest and resuscitation.

2. METHODS

2.1. Animal Preparation

The experimental protocol employed by this study was approved by the Animal Care and Use Committee at our institution. Male Wistar rats (250–400 g) purchased from Charles River were housed in our animal resource facility for at least five days on a 12 hour day-night diurnal cycle. All the rats were deprived of food overnight (10–12 hours) before the experimentation in order to maintain a relatively constant plasma glucose concentration. Anesthesia was induced by halothane (5% halothane, 70% N_2O in O_2) and maintained with 2% halothane, 70% nitrous oxide in oxygen through a nasal cone. A Silastic catheter (0.64 mm ID, 1.19 mm OD; Dow Corning) was inserted through the external jugular vein into the right atrium. The ventral tail artery was cannulated

Oxygen Transport to Tissue XXI, edited by Eke and Delpy.
Kluwer Academic / Plenum Publishers, New York, 1999.

(Intramedic PE-50, 0.058 mmID, 0.965 mmOD) for the purpose of monitoring of systemic arterial blood pressure and to obtain samples for blood gas and plasma glucose determinations. The rats were allowed to recover from anesthesia for at least 1 hour while being restrained in plastic cages. Body temperature during this period and during ischemia and following recovery was maintained at 37°C by an infrared heat lamp (250 W, 45 cm above the body) regulated by feedback from a rectal temperature probe.

2.2. Induction of Total Cerebral Ischemia

Reversible total cerebral ischemia was achieved as previously reported (Crumrine & LaManna 1991). Cardiac arrest was induced in the conscious rat by the rapid sequential intra-atrial injection of d-tubocurare (0.6 mg) and ice-cold KCl solution (0.5 M; 0.12 ml/100 g of body weight). Resuscitated rats were orotreacheally intubated with a 14-gauge catheter(Angiocath) and the tracheal tube was attached to a rodent ventilator. Resuscitation was initiated 7 minutes after arrest. Ventilation (100% O_2, tidal volume of 10 cc/kg; respiratory rate of 80 breaths/min) was begun simultaneously with chest compression and administration of intravenous fluid (Plasmalyte 148; Baxter) at a rate of approximately 0.5 ml/min. Once a spontaneous heart beat returned, epinephrine (4–10 μg) was administered intravenously to bring the mean blood pressure above 80% of the pre-arrest value, at which point the rat was considered to be resuscitated. The duration of arrest was 12 minutes. After recovery, inspired oxygen content was decreased gradually to 21–30% O_2, balance N_2O. Ventilation was adjusted on the basis of blood gas analysis to normal ranges, until the rats regained spontaneous respiration. For the adenosine treated rats, infusion of adenosine was begun as soon as blood pressure was established. Infusion of adenosine (9-β-D-Ribofuranosyladenine; Sigma) was administered intravenously at a constant rate of 80 μg/kg/min for 60 or 90 minutes. For brain edema evaluation, rats were frozen in situ 1, 6, 24, and 48 hours (n = 5 for each group) after cardiac arrest and resuscitation. Non-arrested control rats who underwent all the surgical procedures except cardiac arrest were frozen in situ as above. The brains of frozen rats were stored at −80°C. For neuronal survival analysis, the rats were resuscitated and allowed to recover for 4 days if able. Mortality rates were determined. Rats that survived for 4 days and non-arrested controls were deeply anesthetized with halothane, perfused through the heart with about 50 ml 0.1 M PBS, and perfusion fixed with 4% paraformaldehyde. The brains were removed and embedded in paraffin for the purpose of hippocampal neuronal counts.

2.3. Tissue Water Content

The brains of frozen rats were removed in a glove box maintained at −30°C. Brain tissue samples (5–25 mg) from representative frozen brains were taken in a cryomicrotome (−30°C) by small scoop from cortical, hippocampal and brain stem regions. Tissue samples were quickly placed into tared, dried, capped Eppendorf tubes and re-weighed to determine tissue wet weights. The tissues were then dried to constant weight at 100°C to determine tissue dry weights. Water content (%) was calculated as (wet weight–dry weight)/wet weight × 100.

2.4. Hippocampal Cell Counts

Serial 5 μm coronal sections were made at the level of the anterior hippocampus from paraffin embedded brains and H&E stained. At the level of atlas plate 30 (Palkovits and Brownstein, 1988), the entire length of the pyramidal cell layer in the hippocampus was viewed under high power light microscopy (400×). Neurons with rounded cell bodies and clearly visible nucleoli were considered to have survived. The number of neurons surviving was evaluated in CA1, CA2 and CA3 regions of the hippocampus.

2.5. Statistical Methods

Statistical analyses were performed using SPSS v6.0 for Windows. Group comparisons were made by one-way ANOVA using the Tukey B method. Significance was considered at the $p < 0.05$ level.

3. RESULTS

3.1. Physiological Variables

As seen in Table 1, the physiological variables of the rats that recovered for 1–4 days were comparable between ADO-treated and untreated groups. Rats usually regained spontaneous respiration by 3 to 6 hours after recovery. Arterial oxygen tensions at 1 hour were higher than pre-arrest baseline in both groups because they were ventilated with a higher oxygen concentration (about 30%) for the first hour of the recovery period. Arterial pH recovered by 1 hour after arrest in most rats. Hypertension occurred right after resuscitation and was followed by a hypotension of about 30% lasting at least 1 hour. The post-ischemic blood pressure of the resuscitated rats usually returned to the pre-arrest value within 2–3 hours of recovery. Adenosine treatment, at this dose rate and duration of infusion, had no effect on blood pressure.

3.2. The Effect of Adenosine Treatment on Survival after Cardiac Arrest

As seen in Table 2 , ADO-treated rats had better recovery from cardiac arrest and resuscitation, especially those treated with adenosine for 90 minutes. In addition, these rats had fewer seizures and were able to walk and drink sooner than the untreated rats.

Table 1. Physiological variables for the long-term recovery (1–4 days) study (mean ± SD)

Variable	Untreated control (n = 22)	ADO-treated (n = 24)
Body weight (g)	313 ± 36	314 ± 38
MABP (mmHg)		
pre-	109 ± 7	107 ± 5
1 hr	79 ± 17	83 ± 16
Arterial pH		
pre-	7.41 ± 0.02	7.41 ± 0.03
1 hr	7.37 ± 0.06	7.40 ± 0.04
PaO_2 (mmHg)		
pre-	91 ± 8	94 ± 17
1 hr	122 ± 32	126 ± 27
$PaCO_2$ (mmHg)		
pre-	42 ± 2	43 ± 3
1 hr	38 ± 6	38 ± 4

Table 2. Survival rates after cardiac arrest and resuscitation

Experimental group	<1 day	1–3 days	4 days/total
Untreated	4	2	7/13
60 min ADO	3	3	8/14
90 min ADO	1	0	6/7

The rats treated for 90 minutes had a significantly better survival rate than the untreated group. At the end of the 4 days, 85.7% (6/7) in the 90 minute ADO-treated group survived, the other one died within first 24 hours after arrest. At the end point, the survival in the untreated group was 53.8% (7/13), the survival in 60 min ADO-treated group was similar to the untreated rats, 57.1% (8/14). Those rats that did not survive for 4 days suffered mostly from respiratory failure.

3.3. The Effect of Adenosine Treatment on Delayed Neuronal Death

The CA1 cell counts after 4 days of survival were 86 ± 5 (n = 6, mean ± sd) and 57 ± 23 (n = 8) for the 90 min and 60 min ADO-treated groups (Figure 1). These were significantly ($p < 0.05$, ANOVA) more than the CA1 cell counts (23 ± 7, n = 7) in the untreated group and similar to that of control, non-arrested rat brains (102 ± 11, n = 7). There were no differences under any conditions between cell counts in CA2 and CA3 regions.

3.4. The Effect of Adenosine on Brain Edema after Cardiac Arrest

Baseline water content (%) in the brainstem was 77.85 ± 0.74 (n = 5). In the untreated rats, after cardiac arrest, there was a trend toward increased water content in the brainstem that reached a level of statistical significance at 24 hours post-arrest, where

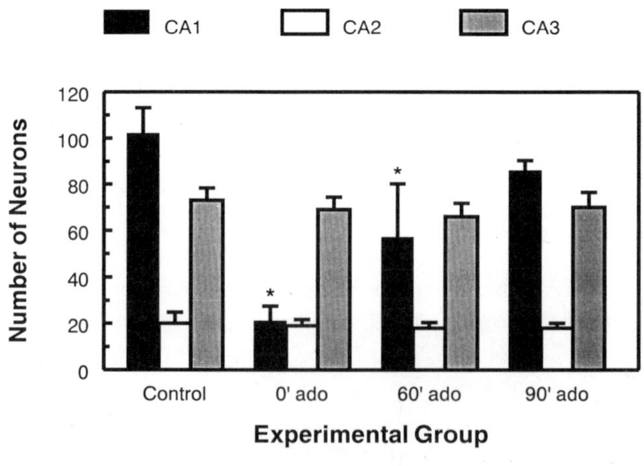

Figure 1. Neuronal cell counts (Mean ± st. dev.) from hippocampal CA1, CA2, and CA3 regions in untreated and adenosine treated rats. * = $p < 0.05$ from control.

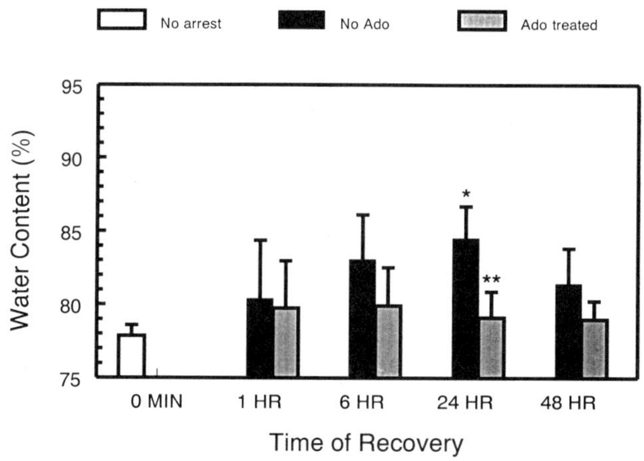

Figure 2. Wet weight/dry weight ratios (% water content, mean ± st. dev.) from brain stem samples of untreated and 90 min adenosine treated rats. * = $p < 0.05$ with respect to control, non-arrested value. ** = $p < 0.05$ with respect to the 24 hour, untreated group.

the water content was almost 10% higher than control. Rats treated with adenosine for 90 minutes after resuscitation showed no significant edema through the 2 days of measurement. Brain edema reached an earlier maximum in the cortex and hippocampus at about 6 hours in untreated rats, and 90 min ADO treatment suppressed the increased water content in these regions also.

4. DISCUSSION

Successful resuscitation from cardiac arrest results in reperfusion injury following global ischemia (Crumrine and LaManna, 1991; LaManna et al., 1995). Reperfusion injury manifests as increased post-resuscitation mortality and delayed neuronal degeneration. Adenosine has been proposed as an endogenous neuroprotective agent (Phillis, 1989), and suggested as a potential post-ischemic treatment approach (Marangos, 1990), because of its multiple interactions with the pathophysiological mechanisms activated during ischemia and reperfusion. During complete ischemia, energy failure is associated with a significant loss of adenylates (Crumrine and LaManna, 1991; Phillis et al., 1995; Newman et al., 1998), thus limiting the ability of the tissue to restore normal levels of ATP during recovery, despite a normal energy charge potential. It has been shown in hippocampal slice preparations in vitro, that adding adenosine to the bathing medium increased total adenylates and ATP content (Whittingham et al., 1989). Thus, it is possible that one of the mechanisms by which adenosine acts in ischemia is to replenish the adenylate pool. Administration of an adenosine analog, 2-chloroadenosine, was protective against ischemic cell loss in the rat hippocampus (Evans et al., 1987).

Alternatively, adenosine might be acting through 1) inhibition of glutamate neurons (Rudolphi et al., 1992; von Lubitz et al., 1995; Sweeney, 1997), 2) free radical scavenging (Rudolphi et al., 1992), 3) cerebral vasodilation (von Lubitz et al., 1995; Sweeney 1997), 4) inhibition of neutrophil adhesion (Cronstein et al., 1986; Rudolphi et al., 1992), and, 5) hypothermia (Jonzon et al., 1986).

Our results have shown that post-ischemic intravenous administration of adenosine following 12-min cardiac arrest and resuscitation in rats improved the survival rate and preserved hippocampal CA1 neurons after 4-day recovery. In addition, adenosine treatment was associated with a depressive effect on edema of cerebral cortex and brainstem during the first 2 days after cardiac arrest, when the rat is most vulnerable to brain failure.

In conclusion, the present findings indicate that post-ischemic adenosine treatment improves overall survival, and acts to preserve CA1 neurons, 4 days after cardiac arrest and resuscitation. Adenosine also appears to inhibit the development of brainstem and cerebral edema during the first 48 hours of reperfusion.

REFERENCES

Cronstein BN, Levin RI, Belanoff J, Weissmann G, and Hirschhorn R (1986) Adenosine: An endogenous inhibitor of neutrophil-mediated injury to endothelial cells. J Clin Invest 78:760–770.

Crumrine RC and LaManna JC (1991) Regional cerebral metabolites, blood flow, plasma volume and mean transit time in total cerebral ischemia in the rat. J Cereb Blood Flow Metab 11:272–282.

Evans MC, Swan JH, and Meldrum BS (1987) An adenosine analog, 2-chloroadenosine, protects against long term development of ischaemic cell loss in the rat hippocampus. Neurosci Lett 83:287–292.

Jonzon B, Berquist A, Li Y, and Fredholm BB (1986) Effects of adenosine and two stable adenosine analogues on blood pressure, heart rate and colonic temperature in the rat. Acta Physiol Scand 126:491–498.

LaManna JC, Griffith JK, Cordisco BR, Bell HE, Lin C-W, Pundik S, and Lust WD (1995) Rapid recovery of rat brain intracellular pH after cardiac arrest and resuscitation. Brain Res 687:175–181.

Marangos PJ (1990) Adenosinergic approaches to stroke therapeutics. Med Hypoth 32:45–49.

Newman GC, Hospod FE, Trowbridge SD, Motwani S, and Liu Y (1998) Restoring adenine nucleotides in a brain slice model of cerebral reperfusion. J Cereb Blood Flow Metab 18:675–685.

Palkovits M and Brownstein MJ (1988) Maps and Guide to Microdissection of the Rat Brain. Elsevier, New York, p. 233.

Phillis JW (1989) Adenosine in the control of the cerebral circulation. Cerebrovasc Brain Metab Rev 1:26–54.

Phillis JW, Perkins LM, Smith-Barbour M, and O'Regan MH (1995) Oxypurinol-enhanced postischemic recovery of the rat brain involves preservation of adenine nucleotides. J Neurochem 64:2177–2184.

Rudolphi KA, Schubert P, Parkinson FE, and Fredholm BB (1992) Neuroprotective role of adenosine in cerebral ischaemia. TIPS 13:439–445.

Sweeney MI (1997) Neuroprotective effects of adenosine in cerebral ischemia: window of opportunity. Neurosci Biobehav Rev 21:207–217.

von Lubitz DKJE, Lin RCS, and Jacobson KA (1995) Cerebral ischemia in gerbils: effects of acute and chronic treatment with adenosine A_{2A} receptor agonist and antagonist. Eur J Pharmacol 287:295–302.

Whittingham TS, Warman E, Assaf H, Sick TJ, and LaManna JC (1989) Manipulating the intracellular environment of hippocampal slices: pH and high energy phosphates. J Neurosci Meth 28:83–91.

HEMODYNAMIC, METABOLIC, IONIC, AND ELECTRICAL RESPONSES TO CORTICAL SPREADING DEPRESSION IN AGING RATS

N. Zarchin,[1] S. Meilin,[1] A. J. Rifkind,[2] and A. Mayevsky[1]

[1]Department of Life Science
Bar-Ilan University
Ramat Gan 52900, Israel
[2]LCMB Gerontology Research Center
NIH/NIA, Baltimore, Maryland

1. INTRODUCTION

Spreading depression (SD) is a phenomenon that has been described in the literature as depression of spontaneous electrical activity (Leao, 1944). It is accompanied by increased metabolic activity (Rosenthal and Somjen, 1973; Mayevsky and Chance, 1974; Hansen et al., 1980), changes in cerebral blood flow (Shibata et al., 1990; Piper et al., 1991) and fluctuations in ionic homeostasis (Hansen and Zeuthen, 1981; Mayevsky et al., 1982). Previous results from our laboratory indicate that the nature of the response to spreading depression is dependent on the ability of the brain to compensate for extra oxygen needed by increasing blood flow to the activated site (Mayevsky et al., 1982; Mayevsky and Weiss, 1991). It was also found that this ability of the brain is age-dependent, namely, adult brain showed "oxidation cycles" of the NADH in most conditions, while the young rat and the anesthetized gerbil brains exhibited "reduction cycle" of the NADH. On the other hand other studies demonstrated that aging acts intracellularly to limit the brain's capacity to increase oxidative phosphorylation to meet heightened ATP damage (Benzi et al., 1992). Furthermore, it is also already known that the metabolic functions of the brain as well as its blood supply inevitably decline with age (Meyer et al., 1993).

The aim of this study was to evaluate the function during cortical spreading depression of the aging brain using the multiparametric monitoring system (Mayevsky, 1992; Mayevsky et al., 1995).

Oxygen Transport to Tissue XXI, edited by Eke and Delpy.
Kluwer Academic / Plenum Publishers, New York, 1999.

2. METHODS

We used the multiparametric assembly (MPA) for the continuous simultaneous on-line measurement of hemodynamic, metabolic, ionic and electrical activities of the rat cerebral cortex (Mayevsky et al., 1995). The measurement included NADH redox state (using surface fluorometry), CBF (laser Doppler flowmetry), extracellular K^+, Ca^{2+} and H^+ concentrations and DC steady potential as well as ECoG (Figure 1).

The experiments were performed in 10 aging (25–30 months) Wistar rats and in 7 adult rats (2–3 months).

The rats were anesthetized by an intraperitoneal injection (0.3 ml/100 g) of Equithesin. The skull was exposed by a midline incision in the skin, and a 6 mm hole was drilled in the parietal bone. The dura mater was gently removed, and the MPA was located on the parietal cortex with a micromanipulator so that undue pressure on the brain was avoided. A 2 mm hole was drilled in the frontal bone ipsilateral to the MPA for a push-pull cannula enabling flushing the brain with KCl solution in order to initiate

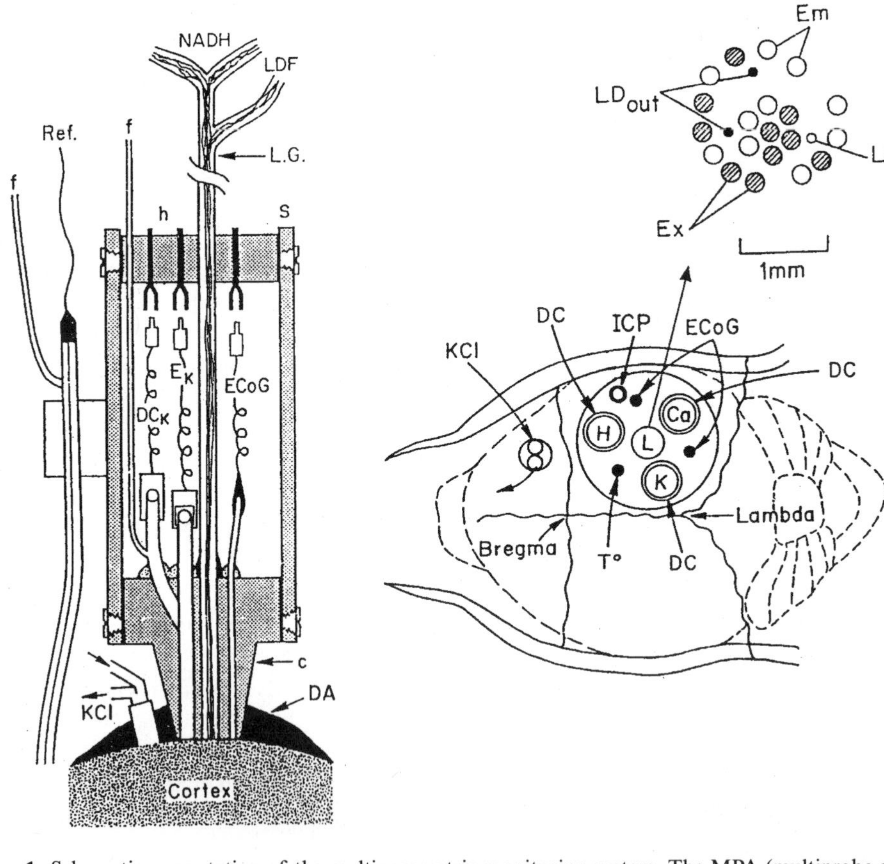

Figure 1. Schematic presentation of the multiparametric monitoring system. The MPA (multiprobe assembly) presented here contains 3 minielectrodes (K^+, Ca^{2+}, H^+) and is connected to the various amplifiers. The light guide (L) contains two sets of optical fibers, one for NADH measurement and other for laser Doppler flowmetry measurement of the cerebral blood flown (CBF). A thermistor for measuring brain temperature (T^0) as well as 2 electrodes for ECoG measurements (ECoG) are located between the other probes.

SD. The MPA cannula as well as 4 stainless steel screws and the push-pull cannula were cemented to the skull with dental acrylic cement.

3. RESULTS

Spreading depression (SD) was induced in 7 adult and 10 old rats. A typical response to cortical SD is demonstrated in Figure 2. This figure is a polygraph tracing taken from an adult animal after induction of SD by KCl solution. The typical SD is characterized by a transient increase in CBF followed by a decrease of NADH level (oxidation), electrical changes namely: ECoG suppression and DC negative shift and ion leak from the cells to the extracellular space (elevation in extracellular K^+ concentration, a decrease in extracellular Ca^{2+} concentration and bi-phasic change in pH).

Figures 3–6 are polygraph tracings of old rats' cortex demonstrating the various responses to cortical SD induction. The general response to SD induction demonstrated in Figure 3 is similar to the responses obtained in adult animal cortex (as was demonstrated previously in Figure 2). Although CBF and NADH responses are smaller than

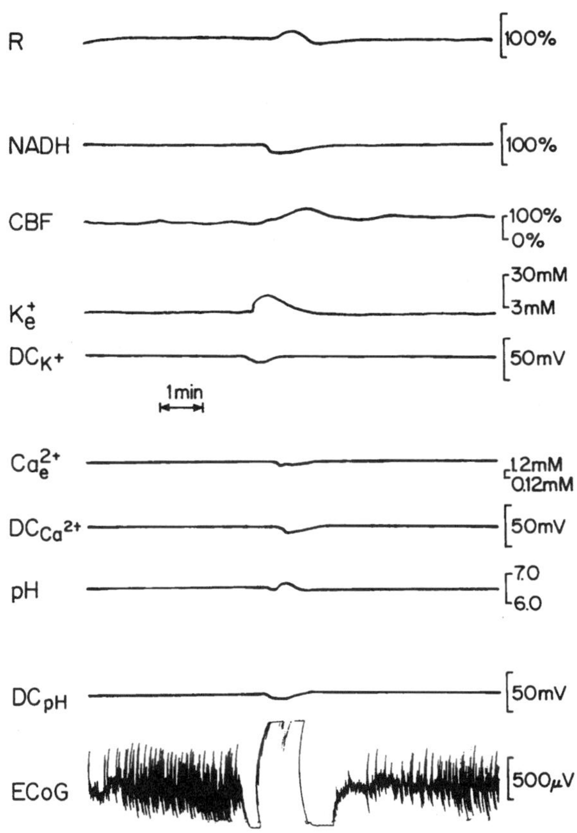

Figure 2. Polygraph tracing of the responses of the adult rat to cortical stimulation by KCl. R- reflectance; NADH- intramitochondrial NADH levels; CBF- cerebral blood flow; K_e^+, Ca_e^{2+}- extracellular potassium and calcium ions; pH- extracellular pH; $DC_{K}+$, $DC_{Ca^{2+}}$, DC_{pH}- direct current concentric the potassium ion, calcium ion and pH electrodes respectively; ECoG- electrocortigram.

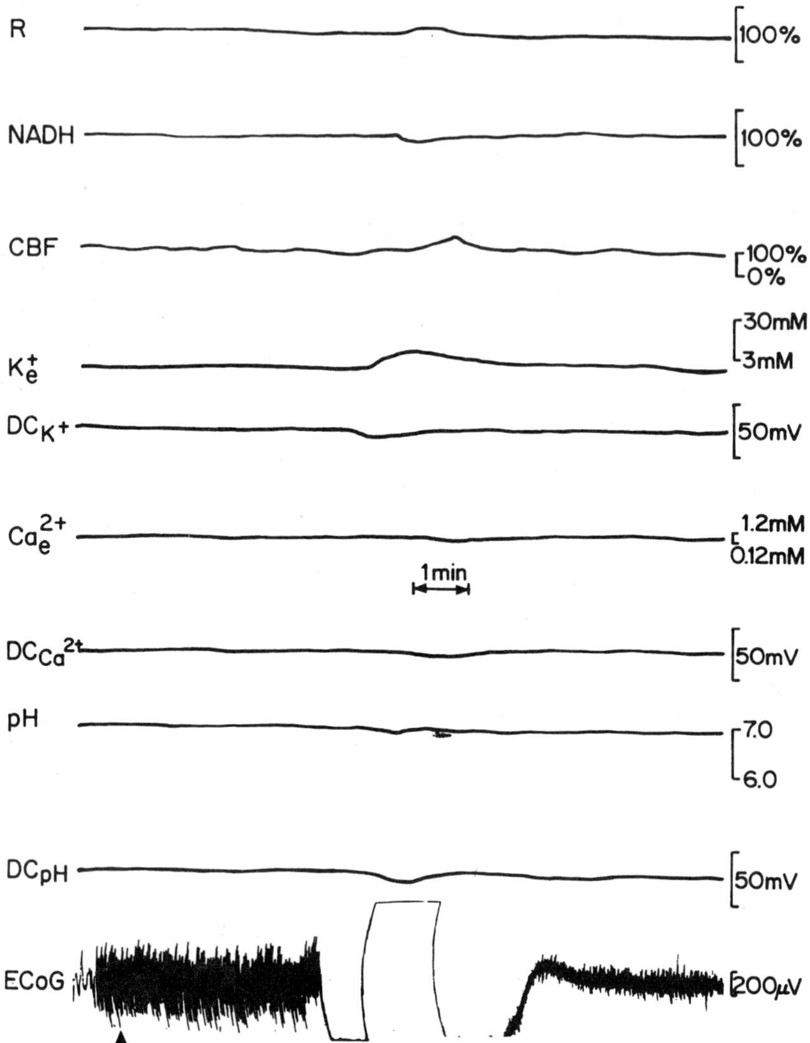

Figure 3. Polygraph tracing of an old rat cortex demonstrating typical normoxic response to KCl stimulation.

those monitored from the adult rat's cortex and the SD wave is longer, this SD response can be categorized as typical normoxic response.

In one old animal we observed "flipped SD" namely, as a result of flushing the cortex with KCl the CBF increased but immediately decreased and NADH increased. These changes were followed by a negative shift of the DC steady potential and by changes in the extracellular ions (elevation in K^+ level, reduction in Ca^{2+} level and bi-phasic change in pH) (Figure 4). Unfortunately the ECoG was not monitored in this experiment due to technical problems.

Another response of the old rat's cortex to KCl flush is demonstrated in Figure 5. In this animal the CBF increased but no metabolic change was observed as was recorded from the NADH signal. All other parameters responded normally, namely, ions leak, DC

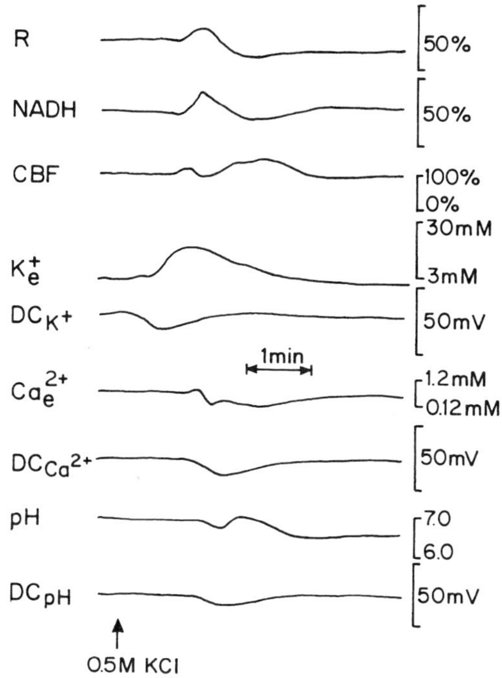

Figure 4. Polygraph tracing of an old rat cortex demonstrating "flipped" response to SD stimulation by KCl.

negative shift and ECoG suppression (extracellular Ca^{2+} concentration was not monitored in this experiment due to technical difficulties).

Figure 6 is a polygraph tracing of an old animal representing 5 out of 10 animals used in this study, which responded in the same way to SD induction by KCl. Those animals did not exhibit any hemodynamic, metabolic, or ionic responses. The only recorded change was the ECoG suppression.

4. DISCUSSION

In this study we observed various responses of the aged rat cortex during SD. Typical SD is characterized by a large efflux of potassium to the extracellular space leading to elevation in Na^+/K^+ ATPase pump activity. In order to supply the elevation in ATP demand the mitochondria are more activated and NADH oxidizes. As a result of these events, CBF increases and this facilitates oxygen supply. Negative shift of DC steady potential and a depression in ECoG occur (Mayevsky, 1992).

One out of 10 old rats exhibited such a typical response, suggesting that this animal's brain cortex was not influenced by the age in terms of tissue adaptation to an increase of energy demand.

However the remaining 9 animals exhibited different levels of brain tissue dysfunction. The first non-typical response to KCl flush was "flipped SD" namely, reduction of NADH and a decrease in CBF. This kind of response is well known as typical response of SD induced under limitation of oxygen delivery (such as ischemia) in adult rats (Mayevsky et al., 1982, 1996; Mayevsky, 1993). All the remaining animals exhibited different levels of brain tissue dysfunction under brain stimulation by SD. Only the

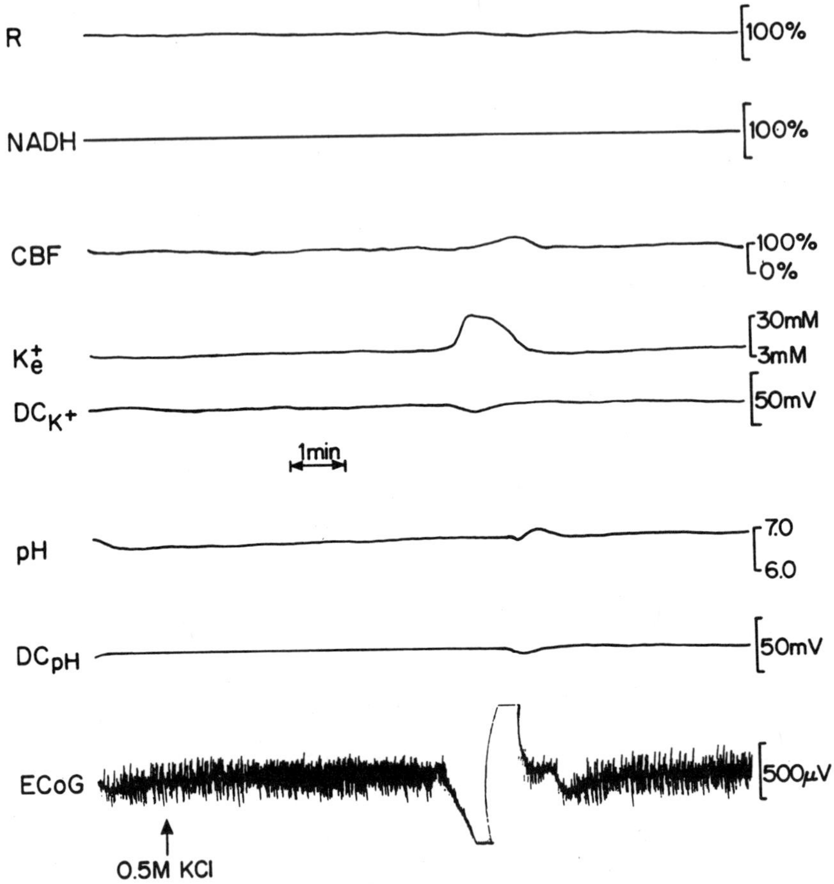

Figure 5. Polygraph tracing of an old rat cortex demonstrating cortex response to KCl stimulation. No changes were recorded in the NADH.

monitoring of the ECoG was the same. These results are similar to a large number of aging studies describing a decline in mitochondrial number and function (Ames et al., 1995; Takai et al., 1995; Fattoretti et al., 1996). A possible explanation for reduction in mitochondrial activity is an appreciable loss of cytochrome oxidase with age, such as reported in heart, skeletal and liver tissue (Wallace, 1992; Buerk and Nair, 1993).

Furthermore, documentation of morphological changes in the cerebral vasculature of the aging rats included thinning of the endothelium, increase in collagen and impaired reactivity to vasodilators stimuli which is thought to result from a decline in endothelial function (Davis et al., 1995). In addition, NO is thought to have a vasodilatation function during SD (Meilin et al., 1995). However, later studies measuring NO synthesis found an elevation in NO synthesis probably to compensate for the reduced sensitivity of the NO system in the aging process (Dorheim et al., 1994; Inada et al., 1996).

In 5 out of 10 old animals no ion (potassium, calcium and hydrogen) flux was observed.

Roberts et al. found age-related alteration in K^+ homeostasis in anoxic rats (Roberts et al., 1997). Early studies carried out by us demonstrated different kinds of

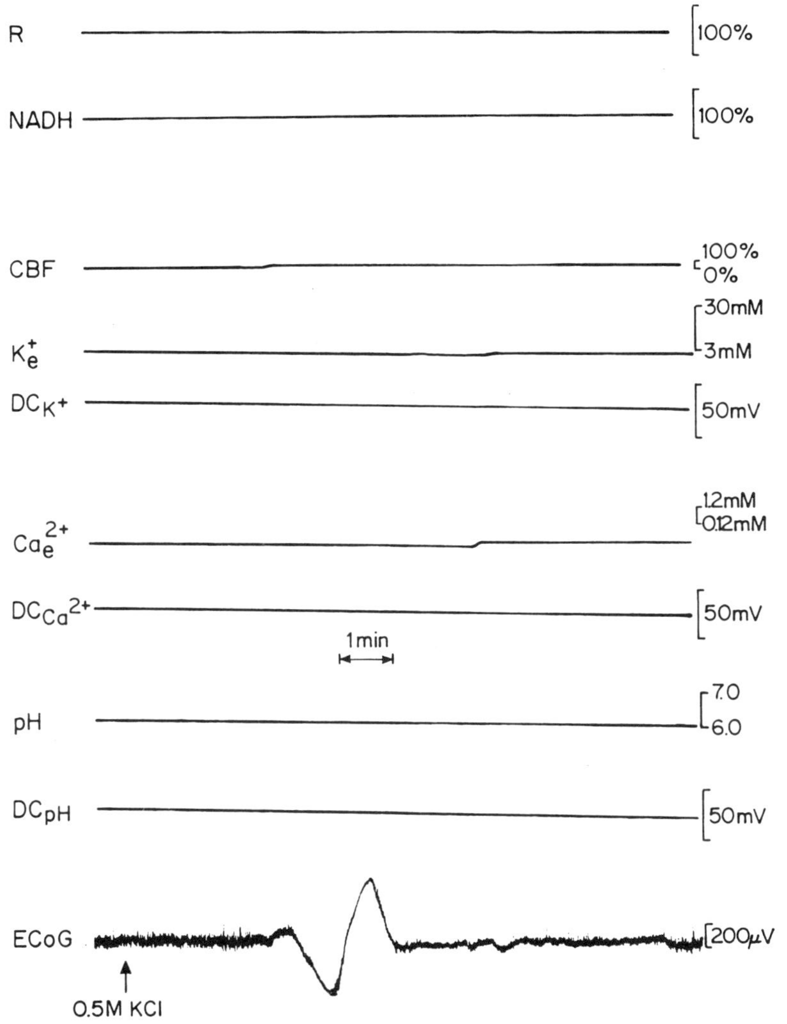

Figure 6. Polygraph tracing of an old rat cortex demonstrating cortex response to KCl stimulation. The only responding parameter was the ECoG.

age-dependent response to SD in young animals (Mayevsky et al., 1982). This study underlines the findings suggesting that SD is an age-dependent phenomenon.

In conclusion, our findings show that in old rats' brain the response to SD is in one of three conditions: 1. Normal brain- exhibiting typical response for SD stimulation. 2. Pathological condition type a- "flipped SD" as occurred in ischemic adult brain. 3. Pathological condition type b- metabolic, hemodynamic and ionic stability- suggesting dysfunction of the mitochondria with inability of blood vessels responding to the increasing oxygen demand.

Supported by the NIH/NIA, the Research Authority Bar-Ilan University and by the Health Science Research Center. Dept of Life Sciences Bar-Ilan University, Israel.

REFERENCES

Ames BN, Shigenaga MK, and Hagen TM (1995) Mitochondrial decay in aging. Biochim. Biophys. Acta 1271:165–170.

Benzi G, Pastoris O, Marzatico F, Villa RF, Dagani F, and Curti D (1992) The mitochondrial electron transfer alteration as a factor involved in the brain aging. Neurobiol. Aging 13:361–368.

Buerk DG and Nair P (1993) Pti_{O2} and $CMRo_2$ changes in cortex and hippocampus of aging gerbil brain. J. Appl. Physiol. 74:1723–1728.

Davis M, Mendelow AD, Perry RH, Chambers IR, and James OFW (1995) Experimental stroke and neuroprotection in the aging rat brain. Stroke 26:1072–1078.

Dorheim MA, Tracey WR, Pollock JS, and Grammas P (1994) Nitric oxide synthase activity is elevated in brain microvessels in Alzheimer's disease. Biochem. Biophys. Res. Commun. 205:659–665.

Fattoretti P, Bertoni-Freddari C, Caselli U, Paoloni R, and Meier-Ruge W (1996) Morphologic changes in cerebellar mitochondria during aging. Anal. Quant. Cytol. Histol. 18:205–208.

Hansen AJ, Quistorff B, and Gjedde A (1980) Relationship between local changes in cortical blood flow and extracellular K^+ during spreading depression. Acta Physiol. Scand. 109:1–6.

Hansen AJ and Zeuthen T (1981) Extracellular ion concentration during spreading depression and ischemia in the rat brain cortex. Acta Physiol. Scand. 113:437–445.

Inada K, Yokoi I, Kabuto H, Habu H, Mori A, and Ogawa N (1996) Age-related increase in nitric oxide synthase activity in senescence accelerated mouse brain and the effect of long-term administration of superoxide radical scavenger. Mech. Ageing Dev. 89:95–102.

Leao AAP (1944) Spreading depression of activity in cerebral cortex. J. Neurophysiol. 7:359–390.

Mayevsky A (1992) Cerebral blood flow and brain mitochondrial redox state responses to various perturbations in gerbils. In: Erdmann W (ed) Oxygen Transport to Tissue XIV. Plenum Press, pp. 707–716.

Mayevsky A (1993) Biochemical and physiological activities of the brain as *in vivo* markers of brain pathology. In: Bernstein EF, Callow AD, Nicolaides AN, Shifrin EG (eds) Cerebral, Revscularization. Med-Orion Pub., pp. 51–69.

Mayevsky A and Chance B (1974) Repetitive patterns of metabolic changes during cortical spreading depression of the awake rat. Brain Res. 65:529–533.

Mayevsky A, Doron A, Manor T, Meilin S, Zarchin N, and Ouaknine GE (1996) Cortical spreading depression recorded from the human brain using a multiparametric monitoring system. Brain Res. 740:268–274.

Mayevsky A, Meilin A, Rogatsky GG, Zarchin N, and Thom SR (1995) Multiparametric monitoring of the awake brain exposed to carbon monoxide. J. Appl. Physiol. 78:1188–1196.

Mayevsky A and Weiss HR (1991) Cerebral blood flow and oxygen consumption in cortical spreading depression. J. CBF and Metabol. 11:829–836.

Mayevsky A, Zarchin N, and Friedli CM (1982) Factors affecting the oxygen balance in the awake cerebral cortex exposed to spreading depression. Brain Res. 236:93–105.

Meilin S, Zarchin N, Mayevsky A, and Shapira S (1995) Multiparametric responses to cortical spreading depression under nitric oxide synthesis inhibition. In: Weissman BA, Alon N, Shapira S (eds) Biochemical Pharmacological and Clinical Aspects of Nitric Oxide. Plenum Press, New York, pp. 195–204.

Meyer JS, Terayama Y, and Takashima S (1993) Cerebral circulation in the elderly. Cerebrovasc. Brain Metab Rev. 5:122–146.

Piper RD, Lambert GA, and Duckworth JW (1991) Cortical blood flow changes during spreading depression in cats. Am. J. Physiol. 261:H96–H102.

Roberts EL, Chih CP, and Rosenthal M (1997) Age-related changes in brain metabolism and vulnerability to anoxia. Oxygen Transport to Tissue XVIII. Plenum Press, pp. 83–89.

Rosenthal M and Somjen G (1973) Spreading depression of cerebral cortex of cats. J. Neurophysiol. 36:739–749.

Shibata M, Leffler CW, and Busija DW (1990) Cerebral hemodynamics during cortical spreading depression in rabbits. Brain Res. 530:267–274.

Takai D, Inoue K, Shisa H, Kagawa Y, and Hayashi J (1995) Age-associated changes of mitochondrial translation and respiratory function in mouse brain. Biochem. Biophys. Res Commun. 217:668–674.

Wallace DC (1992) Mitochondrial genetics: a paradigm for aging and degenerative diseases? Science 256:628–632.

ARTERIOLAR AND CAPILLARY TISSUE SUPPLY AREAS IN MYOCARDIUM

Nicholas Cicutti,[1] Bindee Kuriya,[2] and Karel Rakusan[2]

[1]Department of Anaesthesia
[2]Department of Cellular and Molecular Medicine
University of Ottawa
Ottawa, Canada

1. INTRODUCTION

The principal role of the mammalian coronary microcirculation is to provide oxygen to the working heart. Terminal arteries and arterioles regulate coronary nutritive blood flow and the capillary net provides the structural basis for oxygen transport by the process of diffusion. Most of our previous research has focussed on the predominant component of the microvascular bed, namely, coronary capillaries. In this connection, the amount of tissue supplied by individual coronary capillaries and the variability of capillary spacing have been found to constitute two of the principal oxygen determinants (Turek et al., 1991). Both may be derived from measuring the size of individual capillary domains, defined as tissue cross-sectional areas closer to a given capillary than to any other. As the distribution of these domain areas is log-normal, the standard deviation of this distribution (SDlog) best serves as a heterogeneity index of capillary spacing. Our established histochemical method (Batra et al., 1989) has previously been used to provide color distinction within capillary regions: proximal capillary portions i.e., close to feeding arterioles having larger capillary domains stain blue, while those within distal capillary regions and with smaller domain areas stain red. Despite widespread acceptance of our technique, some reservation has nevertheless been put forward as to the validity of this approach in distinguishing unambiguously separate capillary regions.

Recent experimental and theoretical studies have proposed that arterioles can also supply neighbouring tissue with oxygen by direct diffusion in addition to serving as a conduit for convective blood flow (e.g., Popel et al., 1989; Ellsworth and Pittman, 1990; Schacterle et al., 1993).

Oxygen Transport to Tissue XXI, edited by Eke and Delpy.
Kluwer Academic / Plenum Publishers, New York, 1999.

The objectives of the present study were (1) to conduct an independent verification and determination of capillary domain areas and heterogeneity values in rat hearts subjected to altered hemodynamic status without resort to our dual staining methodology and (2) using our domain method, to investigate the potential contribution of small coronary arterioles as a diffusive oxygen source.

2. MATERIALS AND METHODS

2.1. Design

Animal care was in accordance with institutional guidelines. Male Lewis rats were obtained from Charles River Breeding Farms (Montreal, Canada). Potential changes in capillary and arteriolar domains due to altered hemodynamic load were investigated in three groups of hearts (n = 8/group) (1) hearts from normal control male 50-day-old rats (2) hearts from 50-day-old normal rats heterotopically transplanted into the abdominal cavity of their littermates; recipient animals killed 3 weeks later, and (3) hearts from rats 40 days subsequent to induction of pressure overload cardiac hypertrophy by abdominal aortic banding in 10-day-old neonatal animals (for experimental details see Rakusan et al., 1997). Briefly, heterotopic heart-lung isotransplantation was performed using a modification of a procedure by Kolar et al. (1995). Donor rats were anaesthetized with sodium pentobarbital (40 mg/kg i.p.), heparinized and bled. After opening the thoracic cavity, the heart was arrested with cold saline and placed on crushed ice. Afterwards, the aorta was cannulated and the heart was flushed with cardioplegic solution. The superior and inferior caval veins were ligated, the heart and lungs removed from the thoracic cavity and kept in cold saline. The aortic stump of the donor heart attached to the remaining 2 lower lobes of the right lung were anastomosed end-to-side with the abdominal aorta of the anaesthetized recipient. Subsequently, the transplant was perfused, via the coronaries, with recipient blood and started to beat spontaneously. The implanted hearts were used 21 days later for morphometric analysis.

The aortic constriction procedure was based on a protocol described previously (Rakusan and Campbell, 1991). Briefly, 10-day-old rats were anaesthetized with halothane and a midline abdominal incision was performed, with the intestinal wall reflected to one side in order to allow access to the abdominal aorta. A subdiaphragmatic suprarenal ligature was tied around the aorta with a 30-gauge hypodermic needle serving as the template. Forty days after the imposed constriction the hearts were collected for morphometry.

2.2. Morphometry

At the end of the experimental period (50 days), animals were weighed, killed with sodium pentobarbital overdose, and their hearts removed. The middle portion of the left ventricle was fixed in 1.5% glutaraldehyde, buffered to pH 7.4 with phosphate buffer and then used for morphometric analysis. Histological methods have been described previously (Rakusan and Campbell, 1991). Briefly, after fixation and dehydration in alcohol, the samples were embedded in historesin. Sections 1 µm thick from the midventricular free wall, were stained using Avallone's modification of Jones' silver methenamine method for staining basement membranes. Measurements were done by a single investigator in a blind fashion. From each heart five to seven projection drawings were obtained, each having an approximate area of 40,000 μm². With the aid of a digitizing tablet, the positions of capillaries both near and far from the smallest arterioles (diameter < 15 μm) were directly identified on cross-section. Using only the vessel diameter, it was occasionally difficult to distinguish the smallest arterioles from dilated capillaries or venules. While the arteriolar wall was always relatively thicker than the walls of the remaining vessels with layers of smooth muscle cells generally present, when there was some doubt we confirmed our observations at higher magnification. Subsequently, using our domain method (Rakusan and Turek, 1985; Turek et al., 1986) tissue supply domains and heterogeneity values for these three types of vessels (arterioles, proximal and distal capillaries) were measured. From the selected fields from each heart 8 to 12 arteriolar profiles (diameter < 15 μm) were located and arteriolar domain and heterogeneity values calculated. Proximal capillary profiles were defined based on the presence of immediately adjacent domains (apx. 6–10 proximal caps/arteriole). Distal capillary profiles were defined on the basis of vessels within individual fields not adjacent to, or overlapping with arteriolar domains (see Figure 1).

Results are presented as mean ± S.E.M. Initially, data were analyzed using a 2-way analysis of variance (ANOVA) with subsequent Bonferroni multiple comparisons for main effects. The potential significance of interaction effects were analyzed using least squares pairwise contrasts where appropriate.

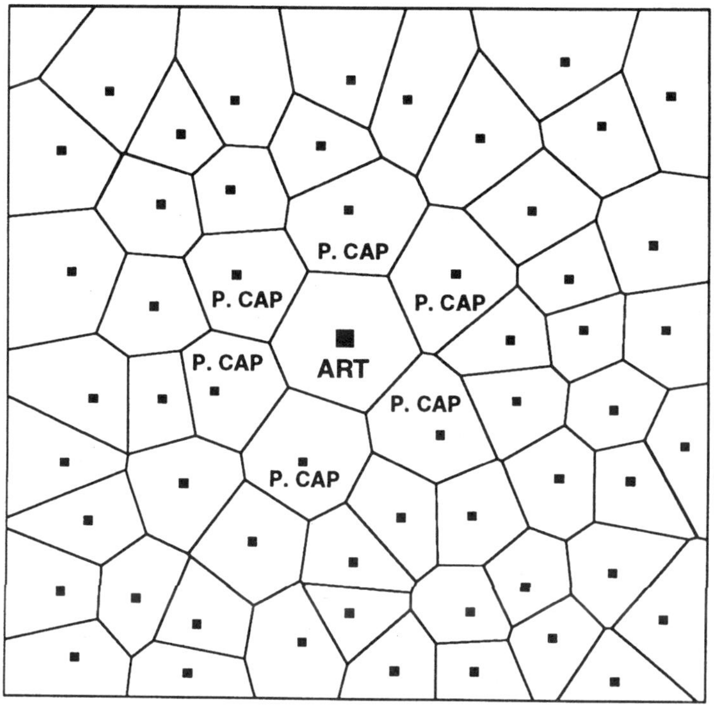

Figure 1. Schematic representation of theoretical tissue supply domain areas for arteriolar (ART), proximal capillary (P. CAP) and distal capillary vessels. Proximal capillary domains depicted as adjacent to arteriolar domain. All other domain areas depicted represent distal capillary domains.

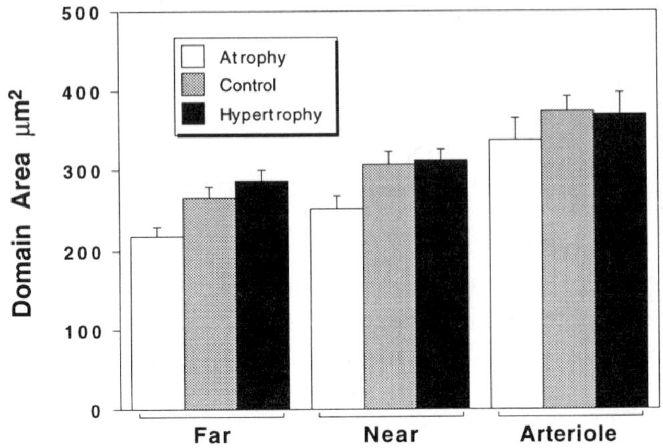

Figure 2. Theoretical tissue supply domain areas for smallest arterioles (Arteriole) Proximal capillaries (Near) and Distal capillaries (Far) in atrophic hearts (hollow bars), control hearts (grey bars) and hypertrophic hearts (black bars). Domain areas supplied by arterioles were larger than those for capillaries ($p < 0.01$), and not influenced by hemodynamic status. Capillaries nearest arterioles had significantly larger domain areas than capillaries further removed ($p < 0.05$). Capillary domains both far from and near arterioles were also significantly smaller in atrophic than in hypertrophic hearts. Control values were intermediate between experimental groups and not significantly different from them. Values depicted represent mean ± S.E.M.

3. RESULTS

3.1. Heart Weight Data

Three weeks after cardiac transplantation left ventricular mass decreased to approximately 50% of values at time of transplantation ie., relative to controls: 220 ± 6mg vs 444 ± 5mg; ($p < 0.001$). On the other hand, experimental aortic constriction lasting 40 days resulted in an 86% increase in left ventricular mass relative to controls: 825 ± 72mg vs 444 ± 5mg; ($p < 0.001$).

3.2. Morphometric Data

Overall, and irrespective of hemodynamic status, capillaries nearest arterioles had significantly larger domain areas than capillaries further removed from them: 290.5 ± 10.4 μm^2 vs 256 ± 10.0 μm^2 ($p < 0.05$). In all cases, the domain areas supplied by arterioles (361.5 ± 14.1 μm^2) were larger than those for capillaries ($p < 0.01$). Neither atrophy nor hypertrophy significantly influenced arteriolar domain size. Furthermore, domains both far from and near arterioles were significantly smaller in atrophic than in hypertrophic hearts: 217.1 ± 12.9 μm^2 vs 286.1 ± 15.9 μm^2 for domains far from arterioles, and 253.1 ± 16.0 μm^2 vs 311.6 ± 14.8 μm^2 for domains closer to arterioles ($p < 0.05$) see Figure 2. Values obtained from control hearts were intermediate between experimental groups and not significantly different from either.

In terms of the heterogeneity of vessel spacing, arterioles demonstrated the least variability based on SDlog values 0.060 ± 0.002 (arterioles) vs 0.068 ± 0.002 (cap. near) vs 0.072 ± 0.00 (capillaries far) ($p < 0.01$) see Figure 3. As with domain areas, heterogeneity values for arterioles obtained from either atrophic or hypertrophic hearts were

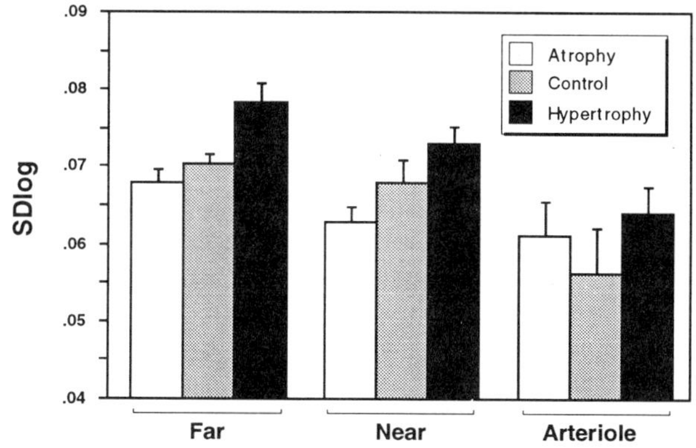

Figure 3. Heterogeneity of vessel spacing depicted as SDlog for smallest arterioles (Arteriole) Proximal capillaries (Near) and Distal capillaries (Far) in atrophic hearts (hollow bars), control hearts (grey bars) and hypertrophic hearts (black bars). Values for arterioles were significantly smaller than those for capillaries ($p < 0.01$), and not influenced by hemodynamic status. Capillary heterogeneity for vessels both far and near from arterioles was significantly smaller in atrophic than in hypertrophic hearts ($p < 0.05$). Values represent mean ± S.E.M.

not significantly different from controls. Overall, capillaries closest to arterioles were not significantly different from those further removed, or distal to the arterioles. However, when analyzed more specifically on the basis of altered cardiac mass, capillaries both far from and near arterioles depicted less heterogeneity in atrophic than in hypertrophic hearts: 0.068 ± 0.002 vs 0.078 ± 0.003 for SDlog values far from arterioles, and 0.062 ± 0.002 vs 0.073 ± 0.002 for SDlog values closer to arterioles ($p < 0.05$) see Figure 3.

4. DISCUSSION

Irrespective of the method used to visualize capillaries, difficulties still arise in applying reliable standards for the distinction of terminal arterioles, capillaries and collecting venules. Most researchers rely upon tissue cross sections to visualize the smooth muscle layer that is present in arterioles, as the method of differentiation. Others have simply chosen to classify vessels on the basis of their respective diameters.

The original observation by Lojda (1979) that capillaries had a non-uniform distribution of enzymes in the endothelium, which served to distinguish arteriolar and venular capillary regions served as the impetus for application of our dual staining methodology. The constituents of the incubating media and technical details of these methods have previously been described (for details see, Batra et al., 1989). This treatment stains the venous portion of capillaries red and the arterial portion of capillaries blue. Stained tissue sections are then available for detailed morphometry. Results derived from both cross and longitudinal sections are classified as a function of proximal and distal capillary regions. To the best of our knowledge, we were the first to apply this method to the quantitative analysis of coronary capillary geometry. Notwithstanding the widespread acceptance of our method, it should be noted that some doubt has been cast on the use of Alkaline Phosphatase (AP) and Dipeptidyl Peptidase IV (DPP IV) when used in concert, as unambiguous markers of arteriolar and venular capillary regions respectively. This may be due to the fact that the reason for a differential distribution of enzymes is not yet known. Although it has been suggested that DPP IV may be localized to venular regions because the efflux of small proteins occurs primarily on the distal side of the capillary bed (Lojda, 1979), such a limited explanation is not completely satisfactory. Therefore, we considered it important to verify the validity and reliability of our staining procedure by conducting an independent study using an alternate and widely used histochemical stain amenable to the methods of quantitative morphology. For this purpose we made use of the silver methenamine stain to help identify the smallest arterioles on cross section as well as the neighboring capillaries present in selected tissue fields. Our working hypothesis was predicated on the premise that capillaries in closest vicinity of arterioles (ie., with adjacent overlapping tissue supply domains) constituted proximal capillaries and that all other capillaries present in the same fields were distal or venular capillaries.

The preeminent finding from this study was that capillary domains in all cases were significantly larger for closely neighboring capillaries than those further removed. This was found in normal control hearts as well as those subjected to either experimental hypertrophy or atrophy. In this regard our results were remarkably similar to those obtained previously using our dual staining method. Although the absolute values for capillary domains reported here are somewhat lower (approximately 20%) than when using the double staining method this can be largely explained by the use fixed dehydrated tissue in the case of silver staining yielding somewhat higher capillary densities and correspondingly smaller domains than is the case with frozen tissue used for double staining. SDlog values as representing the heterogeneity of capillary spacing revealed significant differences, with higher values in hypertrophic hearts relative to atrophic hearts. These results were consistent for capillaries both near and far from arterioles. This increase in capillary dispersion may reflect a progressive impairment in tissue oxygenation in the hypertrophic heart model.

The classic model of Krogh presumed that all oxygen was delivered by convection from terminal arterioles and lost from capillaries by diffusion. Since then several authors have pointed out that in addition to serving as the primary controllers of convective blood and oxygen supply, under specific conditions arterioles may also be directly involved in tissue gas exchange. Various theoretical models have been presented (e.g. Popel et al., 1989; Hsu and Secomb, 1992). In addition, the theoretical models of Schacterle et al. (1991, 1993) suggested that significant diffusion of O_2 and CO_2 occurred in precapillary vessels and that this transport increased as vessel size decreased. The experimental work of Ellsworth and Pittman (1990) measured O_2 saturation at discrete locations along the lengths of individual capillaries in the hamster cheek pouch retractor muscle. Their results indicated that the amount of oxygen lost from individual capillaries could be markedly affected by the presence of arterioles that crossed the capillary pathway. These vessels could act either as a diffusive supply of oxygen for red blood cells within the capillary, or as an additional sink for the oxygen depending on the direction of the oxygen tension gradient. Kuo and Pittman (1990) estimated that approximately 10% of the oxygen that diffused across the arteriolar network was consumed by the surrounding tissue, with the remainder transported to capillaries and venules. Under conditions of hemoconcentration, the proportion of this precapillary oxygen loss could increase further.

The notion of an arteriolar contribution to diffusive oxygen supply in myocardium is also supported by our own results. The measured tissue supply domains for the smallest arterioles were consistently and significantly larger than for capillaries. These data would suggest that arterioles can supply neighboring tissue with oxygen by direct diffusion. This would also be compatible with the view that the high PO_2 and O_2 content of arterioles would permit larger diffusion distances for oxygen than would be the case along the capillary pathway where a progressive diminution in PO_2 from arteriolar to venular capillary regions occurs. Nevertheless, in our view the contribution of arterioles as a diffusive oxygen source is probably of secondary importance, particularly if one considers the numerical values for capillary density. There are approximately 600 capillary profiles per arteriole present on a given myocardial tissue cross-section (Rakusan et al., 1990; Rakusan and Wicker, 1990).

Heterogeneity values for arterioles were not affected appreciably by altering the hemodynamic load on the heart. This was consistent with what was found for domain areas. Presumably, since arterioles are operating for the most part at close to peak levels for PO_2 and O_2 content independent of hemodynamic status, there would appear to be less chance for structural and functional alterations in vessel heterogeneity to occur. Any such alteration would be more likely to occur down stream, in the capillary bed proper, where progressive and substantial reductions in PO_2 take place including localized variation in the actual numbers of perfused capillaries.

In conclusion, the present results consolidate and verify our previous findings concerning capillary domains based on colour distinction due to enzymatic localization. These data also support the notion of an arteriolar contribution to diffusive oxygen supply.

ACKNOWLEDGMENTS

Supported by a grant from the Heart and Stroke Foundation of Ontario.

REFERENCES

Batra S, Rakusan K, and Kuo C (1989) Spatial distribution of coronary capillaries: A-V segment staggering. In: Oxygen Transport to Tissue-XI. (Rakusan K, Biro GP, Goldstick TK, Turek Z Eds) Plenum, New York and London, pp. 241–247.

Ellsworth ML and Pittman RN (1990) Arterioles supply oxygen to capillaries by diffusion as well as by convection. Am J Physiol 258:H1240–H1243.

Hsu R and Secomb TW (1992) Analysis of oxygen exchange between arterioles and surrounding capillary-perfused tissue. J Biomech Eng 114:227–231.

Kolar F, MacNaughton C, Papousek F, Korecky B, and Rakusan K (1995) Changes in calcium handlings in atrophic heterotopically isotransplanted rat hearts. Basic Res Cardiol 90:475–481.

Kuo L and Pittman RN (1990) Influence of hemoconcentration on arteriolar oxygen transport in hamster striated muscle. Am J Physiol 259:H1694–H1702.

Lojda Z (1979) Studies on dipeptidyl(amino) peptidase IV(Glycyl-proline naphthlyamidase). Histochem 59:153–166.

Popel AS, Pittman RN, and Ellsworth ML (1989) Rate of oxygen loss from arterioles is an order of magnitude higher than expected. Am J Physiol 256:H921–H924.

Rakusan K and Turek Z (1985) Protamine inhibits capillary formation in growing rat hearts. Circ Res 57:393–398.

Rakusan K, Sarkar K, Turek Z, and Wicker P (1990) Mast cells in the rat heart during normal development and in cardiac hypertrophy. Circ Res 66:511–516.

Rakusan K and Wicker P (1990) Morphometry of the small arteries and arterioles in the rat heart: effects of chronic hypertension and exercise. Cardiovasc Res 24:278–284.

Rakusan K and Campbell SE (1991) Spatial relationship between cardiac mast cells and coronary capillaries in neonatal rats with cardiomegaly. Can J Physiol Pharmacol 69:1750–1753.

Rakusan K, Heron MI, Kolar F, and Korecky B (1997) Transplantation-induced atrophy of normal and hypertrophic rat hearts: effect on cardiac myocytes and capillaries. J Mol Cell Cardiol 29:1045–1054.

Schacterle RS, Adams JM, and Ribando RJ (1991) A theoretical model of gas transport between arterioles and tissue. Microvasc Res 41:210–228.

Schacterle RS, Ribando RJ, and Adams JM (1993) A model of brain arteriolar oxygen and carbon dioxide transport during anemia. J Cereb Blood Flow Metab 13:872–880.

Turek Z, Hoofd L, and Rakusan K (1986) Myocardial capillaries and tissue oxygenation. Can J Cardiol 2:98–103.

Turek Z, Rakusan K, Olders J, Hoofd L, and Kreuzer F (1991) Computed myocardial PO_2 histograms: effects of various geometrical and functional conditions. J Appl Physiol 70:1845–1853.

29

ROLE OF K^{+}_{ATP} CHANNELS IN THE METABOLIC ADAPTATION OF THE CORONARIES

L. Fazekas, P. Soós, R. Fazekas, V. Kékesi, and A. Juhász-Nagy

Department of Cardiovascular Surgery
Semmelweis University of Medicine
Városmajor u. 68., 1122 Budapest
Hungary

1. INTRODUCTION

In the coronary vascular bed multiple mechanisms interact with each other to control the blood supply of the heart muscle and oxygen transport to cardiac myocytes. Since the early works of Drury & Szent-Györgyi in 1929 and Berne in 1963 the most tempting hypothesis for this control is based on the metabolic theory. The intrinsic mechanism which matches coronary blood flow (CBF) to myocardial demand would be mediated by adenosine, a purine compound, released by cardiac myocytes in metabolic stress, which has a vasodilator effect on coronary vascular smooth muscle cells (VSMC). However, the natural P_1 purinergic antagonist methylxantines, while blocking the exogenous adenosine effects, fail to antagonize the metabolic autoregulatory responses i.e., the functional hyperemia in the same vascular bed. This is the greatest flaw of the adenosine hypothesis of coronary autoregulation. In the myocardial tissue the presence of inosine, the first breakdown product of adenosine, may be responsible for this difference because inosine, which has a vasodilator action of its own, strongly potentiates adenosine-induced vasodilation (Kékesi et al., 1986). Accordingly, inosine may function as a co-transmitter. If so, the coronary vasodilator action of inosine—which is also remarkably resistant to methylxantine blockade—should be blocked by the K^{+}_{ATP} channel antagonist sulfonylureas which effectively antagonize both adenosine vasodilation and reactive hyperemia.

In the last decade K^{+}_{ATP} channels were reported as regulating elements in coronary circulation (Noma, 1983; Daut et al., 1990). In this system the intracellular ATP concentration of the coronary VSMC would determine the blood supply of the myocardium. Thus, the feed-back loop would be closed on the subcellular level, omitting the most

Oxygen Transport to Tissue XXI, edited by Eke and Delpy.
Kluwer Academic / Plenum Publishers, New York, 1999.

attractive concept of the metabolic theory i.e., the feed-back on the tissue level from myocyte to VSMC.

In this study our first question was whether or not the vasodilator actions of adenosine and inosine are dependent on the activation of the ATP sensitive potassium channels; and whether or not these potassium channels alone play the key role in coronary adaptation or do they act as a coupling substance in the actions of the classical metabolic transmitters. To evaluate this possible unifying role of K^+_{ATP} channels, the metabolic adaptation of the coronaries was examined by the reactive hyperemia test (5–30 sec occlusion), and the coronary vascular response was measured by intracoronary infusion of the putative metabolic transmitters (adenosine 10–80 μg/min; inosine 1.25–10 mg/min) with or without the pre-treatment of the K^+_{ATP} channel antagonist glibenclamide (0.03–1.0 μmol/min).

2. METHODS

2.1. General Preparation

Adult mongrel dogs (n = 18) weighing 23.7 ± 1.4 kg were anesthetized with intravenous pentobarbital sodium (30 mg/kg). The animals were ventilated with room air and the chest was opened by transverse thoracotomy in the fifth intercostal space. The left anterior descending coronary artery (LAD) was dissected free and an ultrasonic flow probe [Transonic Systems Inc.] was placed around the vessel. A bent-tip butterfly needle (23-gauge) was inserted into the LAD for intracoronary (i.c.) administration of drugs. In all animals blood pressure (BP) was measured in the femoral artery and heart rate was calculated from BP tracing. All hemodynamic data were recorded on an IBM-compatible personal computer using DASYLab data acquisition software.

2.2. Experimental Setup

In the first set of experiments (n = 7) after hemodynamic stabilization reactive hyperemia tests (RH) were performed with 5, 10 and 30-sec occlusions and adenosine dose-response curves (adenosine infusion: 10, 20, 40 and 80 μg/min i.c. for ~3-min to steady state) were obtained with or without the pre-treatment of glibenclamide (0.03, 0.1, 0.3 and 1.0 μmol/min i.c.).

In the second set of experiments (n = 8) the dose-response curves of inosine (1.25, 2.5, 5.0 and 10.0 mg/min i.c. for ~3-min to steady state) were obtained with or without the same increasing doses of glibenclamide as in the first experimental setup.

In the time control series (n = 3) RH tests were performed with 5, 10 and 30-sec occlusion times and adenosine dose-response curves were obtained with the superinfusion of saline at infusion rates equivalent to those of glibenclamide in the first two sets of experiments.

2.3. Statistical Analysis

Results are expressed as means ± SEM. Analysis of variance for repeated measurements with Scheffe's contrasts were performed for multiple comparisons. Probability values less than 0.05 were considered significant.

3. RESULTS

As shown on the averaged reactive hyperemic response curves (Figure 1) increasing doses of glibenclamide decreased peak hyperemic flow (peak hyperemic flow after 30 sec occlusion; control vs. maximal glibenclamide effect: 113.6 ± 14.3 vs. 63.7 ± 9.7 ml/min, $P < 0.01$) as well as duration of the hyperemic responses (50% return of peak hyperemic flow after 30 sec occlusion; control vs. maximal glibenclamide effect: 35.3 ± 4.2 vs. 17.8 ± 2.1 sec, $P < 0.01$) which resulted in a significant decrease of the hyperemic excess flow (hyperemic excess flow after 30 sec occlusion; control vs. maximal glibenclamide effect: 60.5 ± 10.3 vs. 17.2 ± 2.8 ml, $P < 0.01$; Figure 2.A). In the time control series a slight decrease was also observed (Figure 2.B), which however did not prove significant.

In the unblocked state adenosine and inosine equally dilated the coronaries to 366 ± 68 and 310 ± 16 percent CBF (control = 100%), respectively. At its highest dose glibenclamide reduced these CBF maxima to 120 ± 22% and 117 ± 8% (Figure 3.A–B) while in the time control series the same blockade was not noticed (Figure 3.C).

Neither intracoronary administration of glibenclamide nor adenosine or inosine affected systemic blood pressure and heart rate (Table 1).

4. DISCUSSION

In the last decade K^+_{ATP} channels were reported as regulating elements in coronary circulation (Noma 1983; Daut et al., 1990). A fall in intracellular ATP concentration in

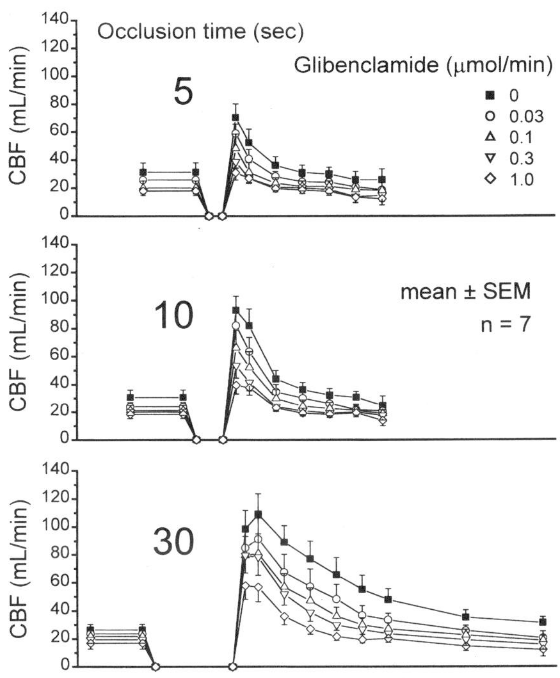

Figure 1. Effect of increasing doses of glibenclamide (0.03–1.0 μmol/min) on coronary reactive hyperemia after 5, 10 and 30 sec occlusion times. Data points are averaged from individual hyperemic response curves.

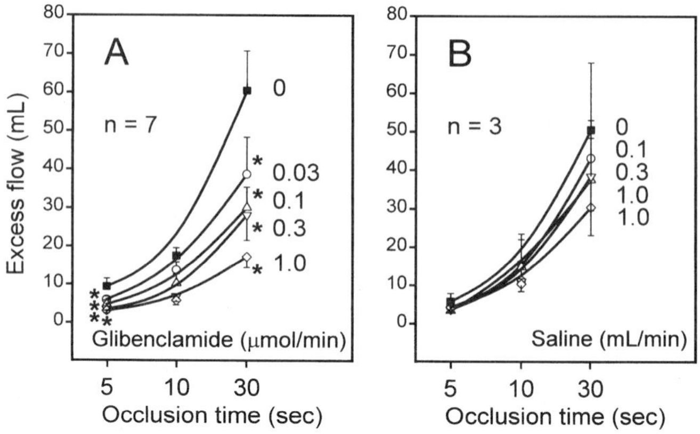

Figure 2. Blockade by increasing doses of glibenclamide (0.03–1.0 μmol/min) of coronary reactive hyperemia described by the hyperemic excess flow, A); and results of time control experiments with the same reactive hyperemia tests during saline infusion, B). All values are mean ± SEM. Data points between * symbols are significantly different from control values (glibenclamide 0) at $P < 0.05$ level.

coronary vascular smooth muscle cells would lead to the opening of K^+_{ATP} channels and by reducing the opening probability of voltage operated calcium channels it would result in dilation of coronary resistance vessels. In this regulating pattern the intracellular ATP concentration of the coronary VSMC determines the blood supply of the myocardium. However, it is not likely that even in extreme hypoxia it would sink to the μM level which is necessary for K^+_{ATP} channels to open (Stanfield, 1987). The alternative hypothesis that the VSMC ATP is compartmentalized (Paul, 1989) is still missing the most attractive concept of the metabolic hypothesis i.e., the nucleoside feed-back between the coronary VSMC and the supplied myocardium. Numerous papers have reported that sulphonylurea blockers of K^+_{ATP} ion channels, such as glibenclamide, antagonize several types of coronary vasodilator responses (Aversano et al., 1991, Komaru et al., 1991, Clayton et al., 1992, Narishige et al., 1994, Katsuda et al., 1995, Tada et al., 1997, Tanaka et al.,

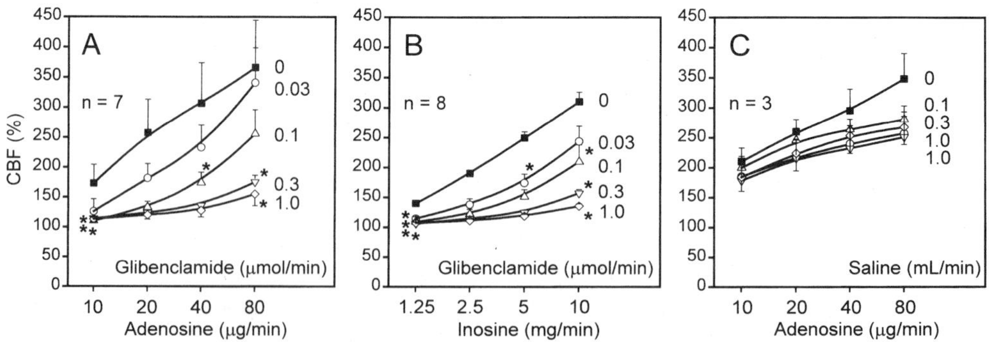

Figure 3. Blockade by glibenclamide (0.03–1.0 μmol/min) of coronary vasodilation induced by intracoronary doses of adenosine (10–80 μg/min), A); and inosine (1.25–10 mg/min), B). C) time control series with intracoronary administration of saline in infusion rates equivalent to those of glibenclamide in A) and B). All values are mean ± SEM. Data points between * symbols are significantly different from control values (glibenclamide 0) at $P < 0.05$ level.

Table 1. Effect of glibenclamide on systemic blood pressure and heart rate

	n = 7	Adenosine (μg/min)				n = 8	Inosine (mg/min)			
	Control	10	20	40	80	Control	1.25	2.5	5.0	10.0
A) Blood pressure responses to intracoronary glibenclamide in mmHg (mean ± SEM)										
Saline	108.7 ± 2.3	108.8 ± 3.4	108.0 ± 2.9	106.7 ± 2.8	106.4 ± 2.5	120.8 ± 5.9	117.9 ± 5.8	119.4 ± 5.2	120.1 ± 5.1	122.0 ± 4.6
Glibenclamide 0.03 μmol/min	102.4 ± 3.1	100.5 ± 3.2	102.7 ± 3.7	99.3 ± 3.1	100.8 ± 4.8	119.1 ± 4.3	120.8 ± 4.9	118.0 ± 3.8	118.0 ± 4.2	119.1 ± 4.3
Glibenclamide 0.1 μmol/min	96.1 ± 3.6	97.2 ± 3.7	97.1 ± 2.7	97.7 ± 4.0	98.5 ± 4.0	115.6 ± 4.7	117.6 ± 4.5	116.0 ± 4.3	117.6 ± 4.3	116.6 ± 4.9
Glibenclamide 0.3 μmol/min	98.9 ± 3.3	99.8 ± 3.2	100.0 ± 3.3	100.0 ± 3.2	101.2 ± 5.2	114.8 ± 3.9	116.6 ± 4.5	116.3 ± 4.3	118.9 ± 4.4	122.4 ± 6.0
Glibenclamide 1.0 μmol/min	97.9 ± 4.3	104.3 ± 4.7	100.7 ± 5.0	100.7 ± 5.5	95.8 ± 4.6	114.4 ± 4.8	117.6 ± 4.2	116.3 ± 4.5	118.4 ± 4.4	118.5 ± 4.6
B) Heart rate responses to intracoronary glibenclamide in beats/min (mean ± SEM)										
Saline	175.7 ± 7.2	173.3 ± 8.0	174.3 ± 7.2	172.9 ± 8.1	172.9 ± 8.1	158.5 ± 8.1	158.0 ± 8.4	157.8 ± 8.2	156.9 ± 8.0	156.9 ± 8.2
Glibenclamide 0.03 μmol/min	170.7 ± 10.5	165.8 ± 12.7	169.3 ± 11.3	169.3 ± 10.6	175.0 ± 10.6	157.3 ± 8.9	153.8 ± 7.7	156.0 ± 9.0	156.0 ± 9.0	156.0 ± 9.2
Glibenclamide 0.1 μmol/min	172.9 ± 9.6	170.0 ± 10.9	172.9 ± 9.6	171.4 ± 9.8	175.8 ± 10.4	155.8 ± 9.4	155.1 ± 8.9	154.4 ± 8.8	154.5 ± 8.4	154.6 ± 9.0
Glibenclamide 0.3 μmol/min	168.6 ± 9.7	165.0 ± 10.6	168.6 ± 9.7	170.0 ± 9.6	175.8 ± 9.0	151.9 ± 8.2	152.3 ± 8.5	151.6 ± 8.6	153.0 ± 8.5	152.9 ± 8.5
Glibenclamide 1.0 μmol/min	166.4 ± 10.6	165.8 ± 10.6	169.3 ± 9.0	169.3 ± 9.0	173.3 ± 9.5	149.6 ± 8.6	149.3 ± 8.6	147.0 ± 8.9	147.8 ± 9.2	148.3 ± 8.9

1997), however to our knowledge no detailed dose-response characteristics of the adenosine and inosine vasodilation are available with selective blockade of the cardiac VSMC K^{+}_{ATP} channels in the canine heart.

In this study we obtained detailed dose-response curves of the K^{+}_{ATP} channel antagonist glibenclamide and the possible metabolic coronary transmitters adenosine and inosine in *in situ* canine hearts. Parallel coronary responses were induced both by adenosine and inosine and these effects were equally dependent on the activation of the K^{+}_{ATP} channels, i.e. inosine was not resistant to glibenclamide as it was to the P_1 purinergic antagonist methylxantines. The selective K^{+}_{ATP} channel blockade was not associated with significant blood pressure and heart rate alterations, therefore the blocking action seemed to be specific. These results support the assumption that the K^{+}_{ATP} ion channels of the coronary smooth muscle cell membrane can be regarded the main substrate of dilator mechanism for both adenosine and inosine.

The selective blockade of the K^{+}_{ATP} channels significantly reduced reactive hyperemic coronary responses. This parallel and dose-dependent antagonism of both exogenous and endogenous coronary vasodilation supports the assumption of the unifying effect of coronary VSMC K^{+}_{ATP} channels. The specific mechanism by which adenosine and inosine are connected to K^{+}_{ATP} channels is not known. A possible coupling may be mediated from adenosine receptors to K^{+}_{ATP} channels via G proteins (Kirsch et al., 1990, Auchampach and Gross, 1993, Akatsuka et al., 1994). The role of G proteins and their possible connection to adenosine A_2 receptors in the adenosine and inosine vasodilatation needs further investigation.

In summary, we obtained full dose-response curves of adenosine and inosine during continuous K^{+}_{ATP} channel blockade by glibenclamide on the *in situ* canine heart. The coronary vasodilator action of inosine was found to be dependent on the activation K^{+}_{ATP} channels and we demonstrated that glibenclamide in a parallel and dose-dependent fashion antagonizes not only the exogenous nucleoside vasodilation but the endogenous, reactive hyperemic responses as well. From these results we may conclude that K^{+}_{ATP} channels can be regarded as the main substrate for adenosine and inosine vasodilation in the coronary vascular bed while inosine may function as a co-transmitter in the adenosine effect. The parallel and dose-dependent antagonism of both exogenous and endogenous coronary responses supports the assumption of these potassium channels having a key role in the metabolic adaptation of the coronaries not only on the subcellular, but via these nucleoside compounds, on the tissue level as well.

REFERENCES

Akatsuka Y, Egashira K, Katsuda Y, Narishige T, Ueno H, Shimokawa H, and Takeshita A (1994) ATP sensitive potassium channels are involved in adenosine A2 receptor mediated coronary vasodilatation in the dog. Cardiovasc Res 28:906–911.

Auchampach JA and Gross GJ (1993) Adenosine A1 receptors, KATP channels, and ischemic preconditioning in dogs. Am J Physiol 264:H1327–H1336.

Aversano T, Ouyang P, and Silverman H (1991) Blockade of the ATP-sensitive potassium channel modulates reactive hyperemia in the canine coronary circulation. Circ Res 69:618–622.

Berne RM (1963) Cardiac nucleotides in hypoxia: possible role in regulation of coronary blood flow. Am J Physiol 204:317–322.

Clayton FC, Hess TA, Smith MA, and Grover GJ (1992) Coronary reactive hyperemia and adenosine-induced vasodilation are mediated partially by a glyburide-sensitive mechanism. Pharmacology 44:92–100.

Daut J, Maier Rudolph W, von-Beckerath N, Mehrke G, Gunther K, and Goedel Meinen L (1990) Hypoxic dilation of coronary arteries is mediated by ATP-sensitive potassium channels. Science 247:1341–1344.

Drury AN and Szent-Györgyi A (1929) The physiological activity of adenine compounds with especial reference to their action upon the mammalian heart. J Physiol 68:213–237.

Katsuda Y, Egashira K, Ueno H, Akatsuka Y, Narishige T, Arai Y, Takayanagi T, Shimokawa H, and Takeshita A (1995) Glibenclamide, a selective inhibitor of ATP-sensitive K+ channels, attenuates metabolic coronary vasodilatation induced by pacing tachycardia in dogs. Circulation 92:511–517.

Kékesi V (1986) A hypothesis of metabolic adaptation in the coronaries: modulation by inosine of the adenine nucleoside-induced vascular response. In: Current Problems of Cardiovascular Surgery. (SzabóZ, Ed) Academic Press, Budapest, pp. 99–120.

Kirsch GE, Codina J, Birnbaumer L, and Brown AM (1990) Coupling of ATP-sensitive K^+ channels to A1 receptors by G proteins in rat ventricular myocytes. Am J Physiol 259:H820–H826.

Komaru T, Lamping KG, Eastham CL, and Dellsperger KC (1991) Role of ATP-sensitive potassium channels in coronary microvascular autoregulatory responses. Circ Res 69:1146–1151.

Narishige T, Egashira K, Akatsuka Y, Imamura Y, Takahashi T, Kasuya H, and Takeshita A (1994) Glibenclamide prevents coronary vasodilation induced by beta 1-adrenoceptor stimulation in dogs. Am J Physiol 266:H84–H92.

Noma-A (1983) ATP-regulated K^+ channels in cardiac muscle. Nature 305:147–148.

Paul RJ (1989) Smooth muscle energetics. Annu Rev Physiol 51:331–349.

Stanfield PR (1987) Nucleotides such as ATP may control the activity of ion channels. Trends Neurosci 10:335–339.

Tada H, Egashira K, Yamamoto M, Ueno H, Takemoto M, Shimokawa H, and Takeshita A (1997) Glibenclamide, a specific inhibitor of ATP-sensitive K^+ channels, inhibits coronary vasodilation induced by angiotensin II-receptor antagonists. J Cardiovasc Pharmacol 30:313–319.

Tanaka E, Mori H, Chujo M, Yamakawa A, Mohammed MU, Shinozaki Y, Tobita K, Sekka T, Ito K, and Nakazawa H (1997) Coronary vasoconstrictive effects of neuropeptide Y and their modulation by the ATP-sensitive potassium channel in anesthetized dogs. J Am Coll Cardiol 29:1380–1389.

30

UNEVEN FLOW DISTRIBUTION IN THE HEART INDUCED BY ENDOTHELIN

A. Juhász-Nagy,[1] V. Kékesi,[1] L. Fazekas,[1] B. Merkely,[1] and M. Tóth[2]

[1]Department of Cardiovascular Surgery
[2]First Department of Medicine
Semmelweis University of Medicine, Budapest
Hungary

1. INTRODUCTION

The safety of cardiac energy balance depends on the equilibrium and near uniformity of the intramyocardial tissue oxygen level ensured by the coronary supply. Since the myocardium is deprived of the sufficient capacity of anaerobiosis, the most important function of the smooth muscle in the coronary vessel wall is to adapt continuously to the changing O_2 demands of the heart muscle. At the same time, in contrast with expectations of conventional wisdom, the coronary vessels are well equipped with vasoconstrictor nervous elements: the effect of the local release of neurotransmitters, primarily norepinephrine, was found to be superimposed on the inherent, highly developed myogenic tone in the coronaries (Feigl, 1983). Traditionally, vasodilator modulation of this vascular tone is attributed (beside ß-adrenergic influences) to metabolic factors produced by the myocardial tissue.

Over the past 10–15 years, however, many new concepts have been developed involving substantial evidence of additional vasoactive factors that are released from the endothelial cells in response to a variety of physiologic stimuli. Among the most intriguing of these observations are the reports of the novel vasoconstrictor peptides endothelins (Yanagisawa et al., 1988). It has become increasingly clear that these endogenous agents are potent enough to elicit marked flow reductions in many vascular beds (including the coronaries) lasting for considerable periods of time (Rubanyi and Polokoff, 1994). However, the special conditions of endogenous compensatory mechanisms, enabling spastic reactions of such strength and duration to unfold are still obscure. The aims of the present investigations were, therefore, to clarify these issues by studying the action of

Oxygen Transport to Tissue XXI, edited by Eke and Delpy.
Kluwer Academic / Plenum Publishers, New York, 1999.

endothelin-1 (ET-1), the most potent of endothelins, on coronary blood flow (CBF) of the *in situ* dog heart.

2. METHODS

2.1. General Preparation

All experiments were done in accordance with the *Guide for Care and Use of Laboratory Animals* published by the United States National Institutes of Health (NIH publication No. 85-23, revised 1985). A total of 68 mongrel dogs of either sex weighing 14–27 kg was used. The animals were fasted overnight and anesthetized with pentobarbital sodium (30–35 mg/kg b.w., i.v.); subsequent small doses (15–30 mg) were given as needed. After tracheal intubation the animals were artificially ventilated with room air. One of the femoral arteries was cannulated for monitoring arterial blood pressure. Thoracotomy was performed in the fifth intercostal space(s) and the heart was suspended in a pericardial cradle. The left anterior descending (LAD) branch of the coronary artery was dissected free close to its origin and an electromagnetic flow probe (Statham SP 2201) of appropriate size was placed around the vessel. A bent-tip butterfly needle (23 gauge) was inserted into the LAD. The needle was kept open by continuous infusion (0.2 ml/min) of physiologic saline. In a small series of dogs (n = 6) the abdomen was opened in the midline, a suitable segment of the small intestine was exteriorized, and the supplying branch of the superior mesenteric artery, comparable in size to the LAD, was prepared for perfusion (v.i.).

Drugs were injected intracoronarially (i.c.). ET-1 (human, porcine, Sigma) was dissolved in physiologic saline. The peptide was administered as a bolus in a dose range of 1–1000 pmol or as an infusion (5–40 pmol/min).

2.2. Special Interventions

The animals were divided into different experimental series according to special requirements of the study.

In 8 dogs to ascertain basic dose-response relations ET-1 was injected in increasing doses into the LAD either with normal blood supply or after midpart occlusion of the vessel distal to its larger diagonal branches. After this intervention the blood flow in the vascular bed which remained open (termed residual flow) supplied the adjacent ischemic zone via collateral channels.

In 12 dogs a similar intervention was performed and infrared thermograms from the relevant exposed cardiac surface were serially recorded and analyzed with a special computer program according to the technique described earlier in detail (Papp et al., 1985). The sensitivity of the Thermovision equipment (AGA 750) was set to cover a range of 5°C. Evaluation of the thermograms was performed for three subunits: the non-ischemic and the ischemic LAD zones and the unaffected right ventricular region. Residual blood flow was also measured in the LAD.

In 6 dogs the LAD artery was cannulated and perfused from a carotid with the aid of a Masterflex roller pump at four different constant flow levels roughly corresponding to 30, 60, 100, and 135 percent values of the average normal levels measurable in the same vessel (resistographic measurements). For comparison, analogous measurements were performed in the mesentery of another 6 dogs.

The possible interference of ET-1 and exogenously provoked EDRF actions was studied in 13 dogs. Dose response relations were established for i.c. acetylcholine (0.11–1.0 μg/min) with simultaneous background infusions of subthreshold (5 and 20 pmol/min) and slightly constrictor (40 pmol/min) ET-1 doses. For comparison, lysin-vasopressin was infused in a dose (0.5 nmol/min) which elicits coronary vasoconstriction similar to the threshold dose of ET-1.

Action of ET-1 on metabolic vasodilator capacity of the coronaries was tested utilizing the reactive hyperemia (RH) test in two subseries of experiments. The magnitude of RH was characterized by the excess volume flow following 1 min LAD occlusion. ET was administered as a 1.0 nmol bolus. In the first subseries (n = 8) the hyperemic propensity was tested in the state of extreme vasoconstriction (CBF decrease >70%). In a second subseries (n = 15) the tests were performed at a more moderate CBF decrease (<30%), and immediately after LAD occlusions in 10 sec fractions blood samples were obtained from the coronary sinus for adenosine and inosine determinations. The animals in this subseries were treated with ET-1 alone (1.0 nmol) or with ET-1 during the background infusion of glibenclamide (1.0 μmol/min, RBI Inc.) i.e. during selective blockade of the membrane K_{ATP} channels.

Blood samples drawn from the coronary sinus (and simultaneously from the femoral artery) were immediately mixed with an ice-cold stop solution to inhibit adenosine and inosine formation and degradation. Adenine nucleoside contents in samples were determined by HPLC with UV detection as has been reported previously (Huszár et al., 1996).

2.3. Statistical Analysis

Analysis of variance for repeated measurements were performed for multiple comparisons. Results are expressed as mean ± SEM. Probability values less than 0.05 were considered significant.

3. RESULTS

To establish dose response relationships of the *in situ* dog heart i.c. ET-1 boluses (10^{-11}–10^{-9} mol) were injected into the LAD artery in the normal state and after midpoint occlusion of the same artery (Figure 1) ET-1 produced, without significant alterations in systemic blood pressure and heart rate, slowly developing intense vasoconstrictions in both states often preceded by a short-lived (<10 sec) slight dilator response. The 1.0 nmol dose decreased the regional blood supply to about one-tenth (9.2 ± 2.9%) of its initial value. Vasoconstrictor responses to ET-1 were never followed by compensatory (hyperemic) coronary vasodilations. More characteristically, as revealed by the time-course of the curves shown in Figures 1 and 2, by attaining a critical dose (>0.3 nmol), the vascular reaction turned into a fairly long-lasting vasospasm, i.e. the degree of the contractile response in the coronary smooth muscle exponentially "exploded".

Upon occlusion of the LAD, the directly affected ventricular region was supplied exclusively by collaterals. I.c. injection of ET-1 proximal to the site of occlusion elicited practically identical patterns of thermographically recorded cooling in the unaffected and the ischemic LAD zones, which suggests that the ET-induced vasoconstriction was not mitigated by the ischemic (vasorelaxant) negative feed back (Figure 2).

Direct resistographic measurements in the LAD area perfused artificially with con-

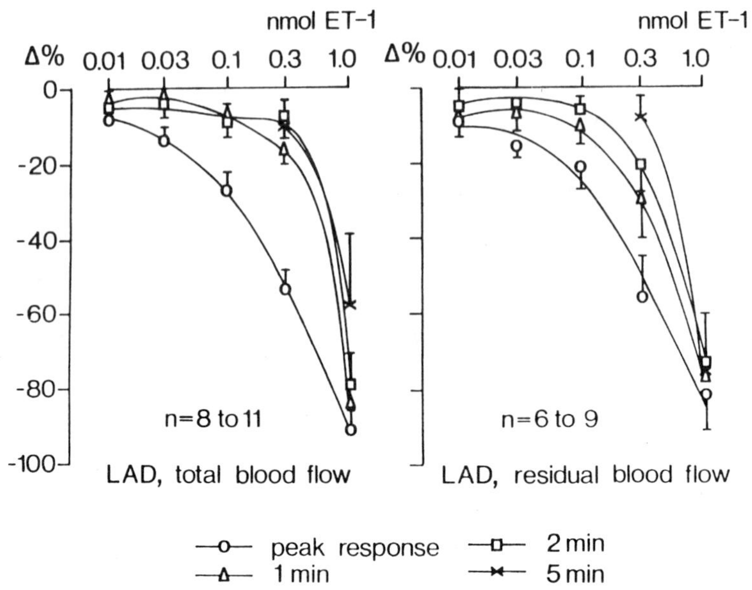

Figure 1. Dose response relations and time courses of ET-1-induced coronary blood flow decreases in the normal and the regionally ischemic dog heart. Mean values ± SEM.

stant flow at different CBF levels also indicated that the vasoconstrictor effect of ET-1 is well maintained in hypoperfusion, although its optimum lies in the normal and the mildly hyperperfused range (Figure 3). The coronary responses can be compared with those of the mesenteric vascular bed studied under identical conditions. Despite a basic qualitative similarity of these vascular patterns, the mesenteric ET-1 responses were

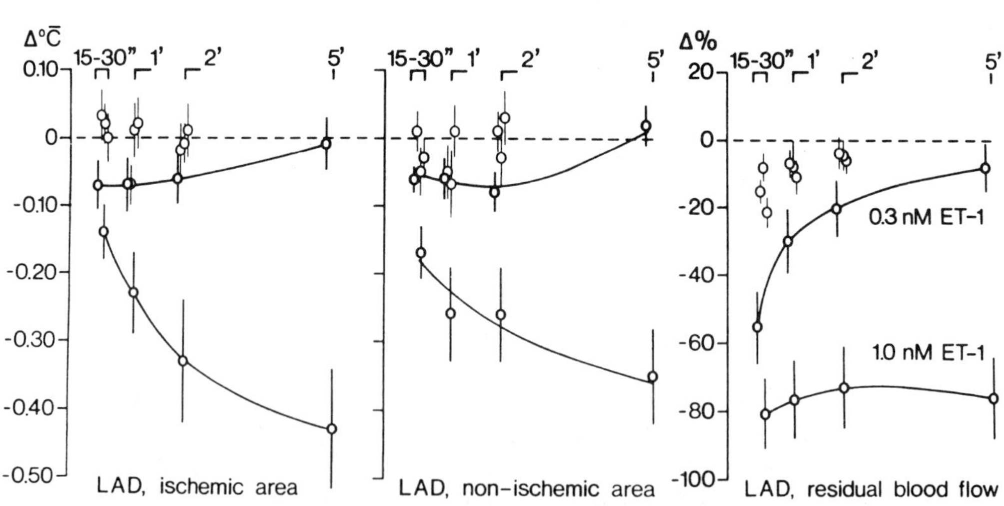

Figure 2. Time course of the ET-1 action in ischemic and normal left ventricular areas (thermography, left and middle panels) as compared to the hemodynamic response in the LAD artery (right panel). Unaffected right ventricular areas exhibited no change (not shown). Mean values ± SEM of 8-11 determinations. Effects elicited by smaller doses (0.03–0.1 nmol) are not connected with lines.

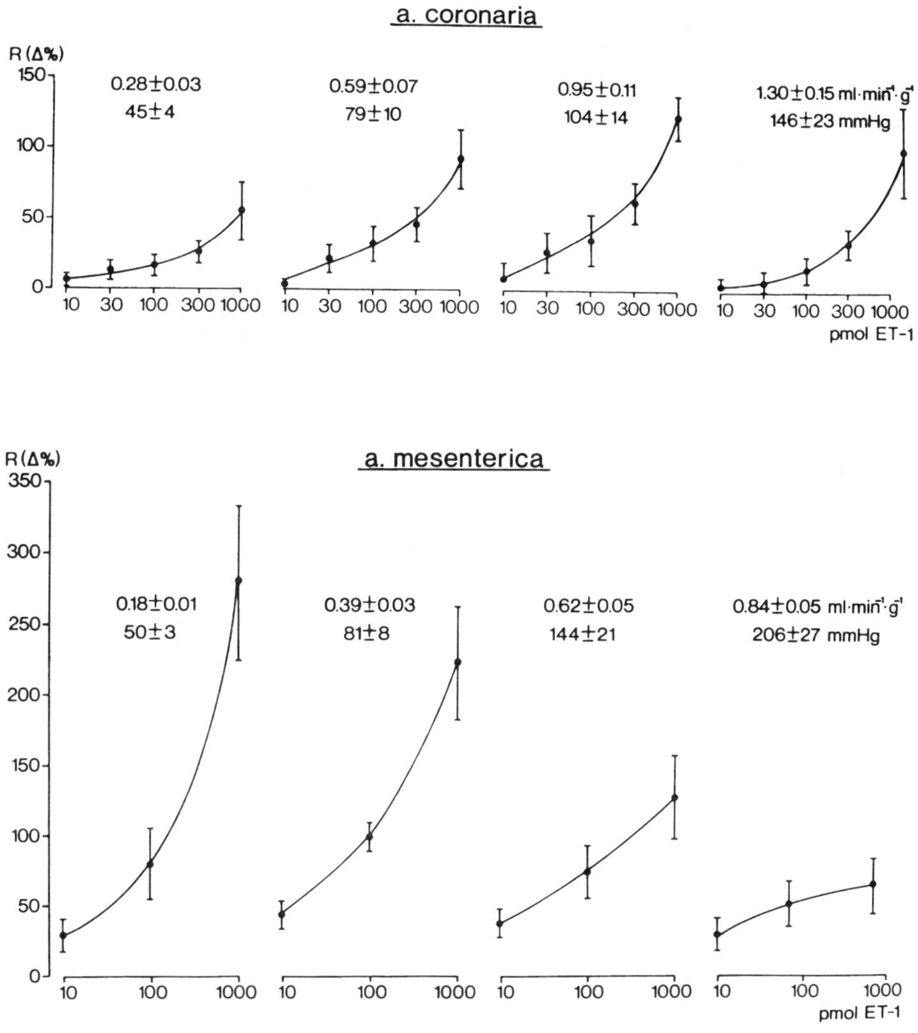

Figure 3. ET-1 induced increase of vascular resistance (R) in the artificially perfused coronary and mesenteric vascular bed. Mean values ± SEM of 6-6 determinations. Mean values of constant blood flows and the corresponding basal perfusion pressures are indicated.

found to be consequently augmented with decreasing basal flow, while the coronary vasoconstrictor effects were not. At the same time, in the mesenteric arteries higher perfusions (associated with an extremely increased basal distenting transmural pressure) revealed less pronounced ET-1 effects. In the heart such high perfusion pressure could not be attained experimentally without causing serious myocardial damage.

In order to examine whether the ET-1 action interfered with important nitric oxide-dependent coronary vasodilator mechanisms, the sensitivity to acetylcholine was tested during background ET-1 infusions. Even subthreshold peptide doses (5–20 pmol/min) significantly and dose-dependently decreased the acetylcholine effect, while the threshold background ET-1 dose drastically reduced coronary vasodilation (Figure 4). Background infusion of vasopressin with a comparable constrictor action failed to induce a vasodila-

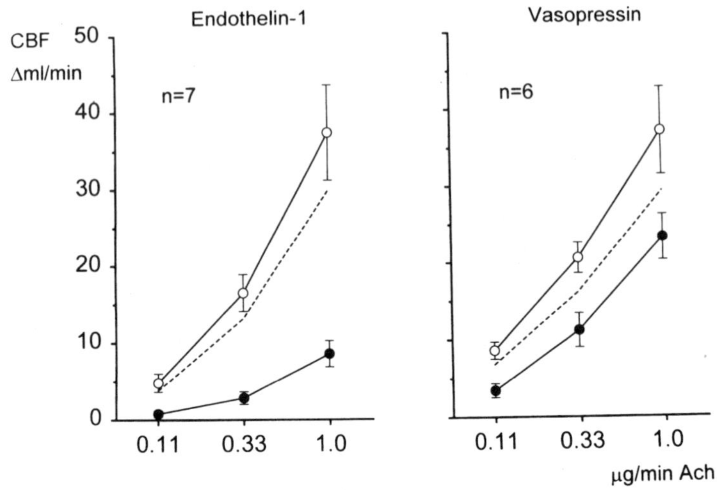

Figure 4. Effect of background ET-1 (40 pmol/min) and vasopressin (0.5 nmol/min) infusions on acetylcholine (Ach)-induced coronary flow increase. Open circles: control Ach effects. Dotted line indicates the calculated decrease of Ach effects that may be due to hemodynamic causes (vasoconstriction).

tor blockade of similar degree, suggesting that the underlying mechanism is independent, at least partly, of the hemodynamics of vasoconstriction.

The effect of ET-1 on coronary reactive hyperemia (RH) was studied in two consecutive experimental subseries. In the coronary bed under strong vasoconstrictor influence, when ET-1 drastically decreased basic CBF (31.7 ± 7.5 vs 13.3 ± 5.8 ml/min, $p < 0.001$, $n = 8$), reactive hyperemic flow excess was found to be reduced to one-fifth of the control (97.1 ± 17.6 vs 19.8 ± 9.3 ml, $p < 0.002$). This was associated with a more than twentyfold enhancement of the adenosine plasma concentration in the coronary sinus. (In these experiments there was no detailed measurement of the nucleoside output.)

Figure 5 depicts results dogs under a significant but less drastic coronary constrictor influence. ET-1 alone considerably reduced the RH response (81.7 ± 13.6 vs 43.4 ± 10.9 ml, $p < 0.001$) and significantly increased the fractionally determined adenine nucleoside release into the sinus. Pretreatment with the K_{ATP} channel blocker glibenclamide elicited a quantitatively similar degree of RH reduction (72.1 ± 11.7 vs 31.9 ± 5.5 ml, $p < 0.01$), and also increased nucleoside output via the sinus. On ET-1 administration during background glibenclamide infusion (i.e. when the blocker was simultaneously infused into the same artery), there was a tendency for further reduction of RH (to 21.8 ± 8.0 ml, n.s.). However, the nucleoside output from the heart was significantly augmented.

4. DISCUSSION

Endothelial cells release at least a dozen of major vasoactive factors which either relax or contract the vascular smooth muscle. The picture that is beginning to emerge suggests that the vascular tone is influenced simultaneously by two interacting regulatory systems: the network of vegetative (mainly sympathetic) nervous elements (counterbalanced by tissue metabolism), and the local paracrine system of endothelial cells which

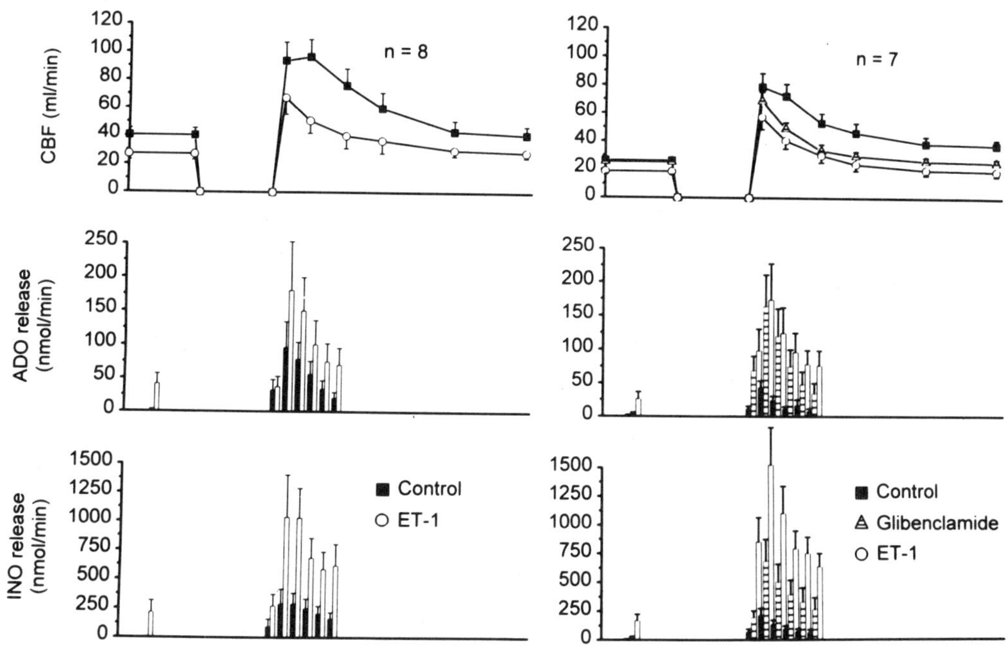

Figure 5. Effect of ET-1 alone (left) and of ET-1 administered during continuous glibenclamide infusion (right) on coronary blood flow (CBF), myocardial adenosine (ADO), and inosine (INO) release after coronary occlusion of 1 min duration.

transmits signals in response to chemical and physical changes occurring in the proximity of the vessels (Miller, 1991).

In agreement with a number of previous investigations (Rubanyi and Polokoff, 1994) the present results confirm the great sensitivity of the coronary vascular bed to ET-1, the most potent constrictor member of this peptide family. On i.c. administration, ET-1 yields a full spastic coronary dose response curve in a comparatively narrow range. Our present study indicates that the constrictor effect of ET-1 prevails under conditions that imply the involvement in the myocardium of compensatory (vasodilator) mechanisms. The main experimental proofs for this assumption are as follows: Long-lasting maintenance of the coronary spasm after nanomolar (higher) doses, i.e. after a non-linear "explosion" of ET-1 action, itself an ischemic stimulus (i); Extension of the vasoconstrictor effect to collateral-dependent regions (ii); Maintained (although somewhat blunted) constrictor potency of ET-1 in myocardial hypoperfusion (iii); Serious reduction of coronary reactive hyperemic capacity and the apparent antagonism with the putative mediator(s) of this phenomenon (iv).

These characteristic features of the ET-1 action make it a likely endogenous agent for triggering coronary vasospasm. In previous studies including ours (Aviado and Juhász-Nagy, 1981) various agents such as catecholamines or angiotensin II, have been postulated to be candidates for eliciting acute non-occlusive myocardial infarction. However, these agents, especially angiotensin II, proved to be singularly ineffective in producing long-lasting coronary constrictor effects (Kékesi et al., 1991). Regarding the capacity of angiotensin II. to induce abrupt signs of CBF inhomogeneity, its actions seem to comparable to those of ET-1. However, long-lasting angiotensin effects are limited as compared to ET-1 (Kékesi et al., 1991). At the same time the relation of catecholamines to compensatory adenosine effects is diametrically opposite to that of ET-1 (Dobson, 1983).

The unusually strong coronary constrictor action of ET-1 could tentatively be explained by a powerful inhibition of metabolic stimuli dilating the same vessels. This possibility was examined by measuring RH which is a reproducible index of metabolic autoregulation. It should be noted, however, that RH is not exclusively "chemically" mediated: the early increase in flow is partly due to myogenic responses elicited by reduced transmural pressure (Eikens and Wilcken, 1973). Moreover, transmitters other than adenine nucleosides may contribute to the flow adjustment (Feigl, 1983). Nevertheless, the decisive role played by adenine nucleosides makes these agents ideal markers of metabolic imbalance in flow control. The present studies proved for the first time the enhanced release of adenine nucleosides from the myocardium after ET-1 administration. Consequently, the RH-attenuating action of ET-1 can be ascribed to the decreased sensitivity of the coronaries and not to a blocked release of these transmitters. To reveal the role the breakdown products of adenine nucleotide metabolism in ET-induced attenuation of coronary adaptation, we used the specific K_{ATP} channel blocker glibenclamide which inhibits all types of actions related to these metabolites (see Fazekas et al., this volume). The results indicate a considerable overlap in the underlying mechanisms, suggesting that ET-1 and glibenclamide may share a common intracellular pathway in the inhibition of nucleoside-induced vasodilation. This conclusion is in agreement with recent evidence demonstrating that ET-1 blocks K_{ATP} channels (Miyoshi et al., 1992) probably via pertussis-toxin-sensitive G_i proteins (Watanuki et al., 1997).

Our findings may imply but do not prove the pathophysiologic action of ET-1 in the body. The assumption of this role is closely related to the effective endogenous ET-1 level which is reportedly very low in the plasma (Rubanyi and Polokoff, 1994). However, acting from within the coronary vessels is only one way in which the peptide can influence vascular tone: paracrine effects are probably more important and are poorly reflected in plasma levels. Recently, we have found that the pericardial fluid, which is a transsudate and thus may reflect the composition of interstitial fluid contains nearly two orders higher concentrations of ET-1 than plasma (Horkay et al., 1995 and 1997). Possibly this is one of the bases of ET-induced arrhythmogenesis, a pathophysiologic event analogous phenomenologically (Szokodi et al., 1998) with the peptide-induced inhomogeneity of flow: neither of them could be explained without supposing self-maintained and/or amplified local actions.

Figure 6 summarizes in a hypothetical scheme the mechanism of the complex regulatory action of ET-1 on CBF. Some elements of this block diagram are well known. It has repeatedly been shown, e.g., that a critical degree of CBF reduction stimulates ET-1 release from the heart (Nayler, 1991, Rubanyi and Polokoff, 1994). ET-induced relative insensitivity to metabolic dilator mechanisms may augment the consequences of this positive feed back action. Contrasted with these elements of instability, ET-1 enhances adenine nucleoside release, while adenosine may suppress myocardial ET-1 formation as was shown by Velasco et al. (1993). The final outcome apparently depends on the functional involvement and delicate balance of the above negative and positive feed back loops. The whole system clearly exhibits a sensitive dependence on initial conditions. It is difficult to define the ultimate physiologic concern of such a system. It was established that mixed systems (with both positive and negative feed back loops) are sources of chaotic behaviour (Glass et al., 1987). It is tempting to speculate that straightforward

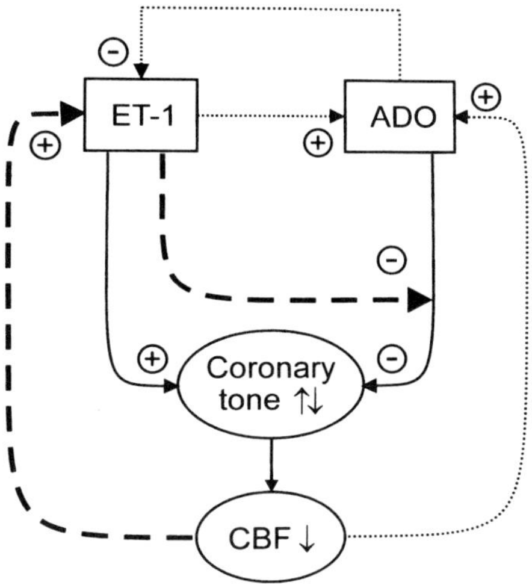

Figure 6. Hypothetical interaction of ET-1 and vasodilator adenine nucleosides (represented by adenosine, ADO) in the coronary circulation. Both *functionally* positive (thick dashed line) and negative (thin dotted line) feed back loops are present. Signs in small circles (+ and –) denote increasing and decreasing influences irrespective of their regulatory role. For further explanation see text.

homeostatic regulations have to be destabilized at times to produce new activity patterns in the body.

REFERENCES

Aviado DM and Juhász-Nagy A (1981) Alpha-adrenoceptors in ischemic canine heart blocked by phentolamine. Adv Physiol Sci 27:245–260.

Dobson JG Jr (1983) Mechanism of adenosine inhibition of catecholamine-induced responses in the heart. Circ Res 52:151–160.

Eikens E and Wilcken DEL (1973) Myocardial reactive hyperemia and coronary vascular reactivity in the dog. Circ Res 33:267–274.

Feigl EO (1983) Coronary physiology. Physiol Rev 63:1–205.

Glass L, Guevara MR, and Shrier A (1987) Universal bifurcations and the classification of cardiac arrhythmias. Ann NY Acad Sci 584:168–178.

Horkay F, Laine M, Szokodi I, Leppäluoto J, Vuolteenaho O, Ruskoaho H, Juhász-Nagy A, and Tóth M (1995) Human pericardial fluid contains the highest amount of endothelin of all mammalian fluids tested so far. J Cardiovasc Pharmacol 26(Suppl 3): S502–S504.

Horkay F, Szokodi I, Selmeci L, Merkely B, Kékesi V, Vuolteenaho O, Ruskoaho H, Juhász-Nagy A, and Tóth M (1997) Presence of immunoreactive endothelin-1 and atrial natriuretic peptide in human pericardial fluid. Life Sci 62:267–274.

Huszár É, Barát E, and Kollai M (1996) Isocratic high-performance liquid chromatographic determination of plasma adenosine. Chromatographia 42:318–322.

Kékesi V, Köngeter T, Tóth M, and Juhász-Nagy A (1991) Angiotensin II-induced coronary and myocardial actions in the normal and the regionally ischemic heart. J Mol Cell Cardiol 23:S19.

Miyoshi Y, Nakaya Y, Wakatsuki T, Nakaya S, Fujino K, Saito K, and Inoue I (1992) Endothelin blocks ATP-sensitive K^+ channels and depolarizes smooth muscle cells of porcine coronary artery. Circ Res 70:612–616.

Nayler WG (1991) Second Generation of Calcium Antagonists. Springer-Verlag, Berlin.

Papp L, Álló G, Szabó Z, and Juhász-Nagy A (1985) Natural history of acute regional myocardial ischaemia revealed by infrared thermography in the canine heart. Acta Morph Hung 33:123–142.

Rubanyi GM and Polokoff MA (1994) Endothelins: molecular biology, biochemistry, pharmacology, physiology, and pathophysiology. Pharmacol Rev 46:325–415.

Szokodi I, Horkay F, Merkely B, Solti F, Gellér L, Kiss P, Selmeci L, Kékesi V, Vuolteenaho O, Leppäluoto J, Ruskoaho H, Juhász-Nagy A, and Tóth M (1998) Intrapericardial infusion of endothelin-1 induces ventricular arrhythmias in dogs. Cardiovasc Res 38:356–364.

Velasco CE, Jackson EK, Morrow JA, Vitola JV, Inagami T, and Forman MB (1993) Intravenous adenosine suppresses cardiac release of endothelin after myocardial ischemia and reperfusion. Cardiovasc Res 27:121–128.

Watanuki M, Horie M, Tsuchiya K, Obayashi K, and Sasayama S (1997) Endothelin-1 inhibition of cardiac ATP-sensitive K^+ channels via pertussis-toxin-sensitive G-proteins. Cardiovasc Res 33:123–130.

Yanagisawa M, Kurihara H, Kimura S, Tomobe Y, Kobayashi M, Mitsui Y, Yazaki Y, Goto K, and Masaki T (1988) A novel potent vasoconstrictor peptide produced by vascular endothelial cells. Nature 322:411–415.

31

HEMODYNAMIC CHANGES CAUSED BY GLIBENCLAMIDE IN ISOLATED, WORKING, ERYTHROCYTE PERFUSED RAT HEART

Roger J. Legtenberg,[1] Ralph J. F. Houston,[1] Paul Smits,[2] and Berend Oeseburg[1]

[1]Department of Physiology 237
[2]Department of Pharmacology
Faculty of Medical Sciences
University of Nijmegen
Post box 9101, NL-6500 HB Nijmegen
The Netherlands

1. INTRODUCTION

During the past decades sulfonylurea (SU) derivatives, e.g. glibenclamide, have been extensively used in treatment of non-insulin dependent diabetes mellitus (NIDMM) (Groop, 1992). These SU derivatives exert their anti-diabetic effects by closing the so-called ATP-sensitive potassium (K_{ATP}) channels in the β-cell of the pancreas. By closing these channels membrane depolarization occurs followed by an increased Ca^{2+} influx and subsequent increased Ca^{2+} dependent exocytosis of the insulin-containing granules (Gerich, 1989). Because K_{ATP} channels not only exist in pancreatic β-cells but also in a variety of other tissues, such as vascular smooth muscle cells (Standen et al., 1989) and myocardial cells (Noma, 1983) there is reason to believe that SU derivatives may contribute to the higher mortality and morbidity by cardiovascular complications occurring in NIDDM.

The coronary blood flow (CBF) response to glibenclamide has been extensively studied in several species at baseline, during hypoxia and after ischemia. It has been reported that glibenclamide reduces basal CBF in anesthetized dogs (Samaha et al., 1992; Imamura et al., 1992) and in isolated rat hearts (Randall, 1995), so cardiovascular K_{ATP} channels are thought to directly influence basal coronary vascular tone. Hypoxic vasodilation is a common reaction to protect the heart from hypoxic damage and it seems that closing K_{ATP} channels (Daut et al., 1990) prevents this effect. Furthermore, closing

Oxygen Transport to Tissue XXI, edited by Eke and Delpy.
Kluwer Academic / Plenum Publishers, New York, 1999.

the K_{ATP} channel diminishes the post-ischemic vasodilation response (Aversano et al., 1991).

Although it is quite clear that the K_{ATP} channel contributes to coronary vascular tone, the effects of K_{ATP} channel blockers on cardiac functional recovery after a period of global ischemia are not distinct. Several groups suggest impaired cardiac function using the K_{ATP} channel blocker glibenclamide (Mitani, 1991; Auchampach et al., 1992), whereas others suggest improved cardiac function (Ali et al., 1993; Tosaki et al., 1994; Docherty et al., 1997). During myocardial ischemia, K_{ATP} channels open and hyperpolarisation of the cell membrane occurs. This abbreviates systole and decreases contractility. These actions are believed to protect the heart against ischemic damage (Nichols et al., 1991). When closing the K_{ATP} channel with e.g. glibenclamide this protective mechanism possibly is abolished (Mitani, 1992). However, glibenclamide may preserve high-energy phosphorus compounds during ischemia and serve in this way as protection (Tosaki et al., 1994).

Glibenclamide is frequently used in experiments to block the cardiovascular K_{ATP} channels and the concentrations are often higher than the therapeutic range used in NIDDM. Therefore the implications of the K_{ATP} blocking properties of glibenclamide for the treatment of NIDDM are very unclear, since so far none of the animal studies has been conducted using therapeutic concentrations of glibenclamide. Furthermore, the animal studies conducted did not have protein in their perfusate, while in the *in vivo* situation glibenclamide binds highly to protein. Furthermore in these animal studies the kinetics of endogenous adenosine (due to breakdown during ischemia) was not taken into account (uptake by erythrocytes *in vivo*) because of using a buffer solution as perfusate. Another disadvantage of using buffer solutions is the vasodilation occurring due to lower oxygen capacity, and therefore CBF effects are difficult to determine. We studied the hemodynamic changes induced by glibenclamide before and after ischemia in an isolated, working rat heart model. In contrast to other groups we added erythrocytes to the perfusate, as well as protein. Also we used more clinically relevant concentrations of glibenclamide to improve a possible extrapolation to the *in vivo* situation.

2. METHOD AND MATERIALS

2.1. Reagent and Experimental Groups

Glibenclamide was purchased from Sigma-Aldrich Chemie B.V., Zwijndrecht, The Netherlands. Four groups were studied: vehicle (n = 3), $0.25\,\mu mol.L^{-1}$ (n = 3), $1\,\mu mol.L^{-1}$ (n = 3) and $4\,\mu mol.L^{-1}$ glibenclamide (n = 3). The solutions were infused into the system by a syringe with a computer-controlled pump (1 : 10). The vehicle composition in the syringe was as follows: $1\,mmol.L^{-1}$ NaOH, 0.4% DMSO, 0.4% NaCl and 1.5% bovine serum albumin (fraction V, ICN Biomedicals B.V., Zoetermeer, The Netherlands). The glibenclamide concentrations were non-protein bound concentrations (measured by HPLC analysis).

2.2. Animal Model

The animal procedures described were approved by the Faculty Ethical Committee on Animal Research. A male Wistar rat was anesthetized with diethyl ether, the heart excised via thoractomy and placed in ice-cold buffer. The aorta was cannulated and

Langendorff perfusion with buffer was started within 8–10 min. after heart removal. Subsequently the left atrium was cannulated, pacing-wires were attached and after checking for leaks the system was switched to working configuration with erythrocyte suspension. Fluid columns were used to maintain the preload pressure at 2 kPa and the afterload pressure at 13 kPa. The buffer (95% O_2 and 5% CO_2) and erythrocyte suspension (18% O_2, 8% CO_2 and rest N_2) were equilibrated using membrane oxygenators. The composition of the modified Krebs-Henseleit buffer solution was as follows: NaCl 118, $CaCl_2$ 3.0, KCl 4.7, $NaHCO_3$ 25, $MgSO_4$ 1.2, KH_2PO_4 1.2, NaEDTA 0.5, Glucose 11.1 mmol.L^{-1}. The free calcium concentration in this composition was 1.6 mmol.L^{-1}. For the erythrocyte suspension albumin was added to the same buffer solution to 1.5% followed by adding erythrocytes to a hematocrit of 0.25. The temperature of the heart and perfusate was maintained throughout the experiment at normothermia of a rat (38°C). A more detailed description of the perfusion set-up has already been published (Olders et al., 1990).

2.3. Experimental Protocol and Statistical Analysis

First performance was assessed during 20 min. by changing the preload pressure from 2.0 to 1.3 and 2.7 kPa and holding the afterload pressure constant at 13 kPa. Secondly, 10 min. infusion of drug was performed, followed by 12 min. normothermic global ischemia. Thereafter reperfusion began during 20 min. with infusion of drug. Finally performance was assessed in the same way as the first time.

The results are presented as means ± standard deviations. Differences in CBF and cardiac functional recovery were statistically tested using one-way ANOVA followed by a Bonferroni comparison test. Statistical analysis was performed using Instat (version 3.00, Graphpad Software Inc.). It should be noted however that with a group size of three, in order to reach a power of 0.9 the expected difference must be around six times the standard deviation. A p-value lower than 0.05 was considered to be significant ($p < 0.05$ = *, $p < 0.01$ = **, $p < 0.001$ = ***).

3. RESULTS

Figure 1 shows the CBF response to vehicle, 0.25, 1 and 4 μmol.L^{-1} glibenclamide. The CBF significantly declined after first treatment with glibenclamide in a dose-dependent matter and after ischemia at the end of reperfusion CBF also declined dose-dependently.

In Figure 2 the cardiac functional recovery after treatment with vehicle and glibenclamide is shown. A significant improvement was seen with 1 and 4 μmol.L^{-1} glibenclamide.

4. DISCUSSION

4.1. Physiological Considerations

K_{ATP} channels seem to play an important role in regulating CBF, which is shown in our results by a dose-dependent decrease of CBF during normoxic situations. Similar observations have been seen at normoxia by Samaha et al. (1992) and Imamura et al.

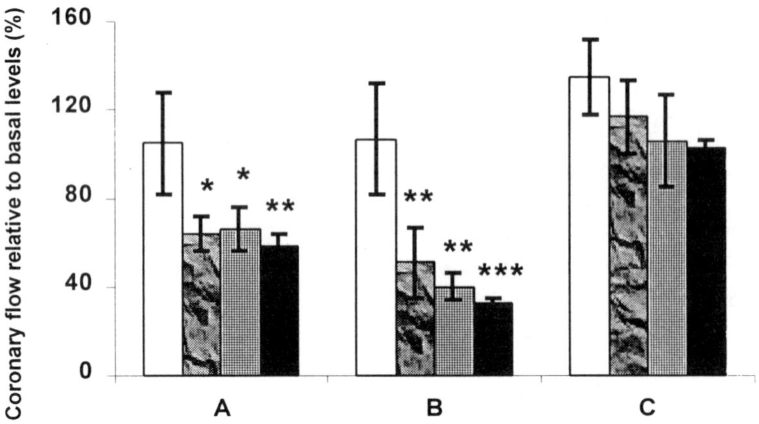

Figure 1. CBF levels relative to basal levels (measured during first performance assessment) after first treatment with glibenclamide (A), after reperfusion with treatment with glibenclamide (B) and during second performance assessment (C). Vehicle is represented by white bars, 0.25 μmol.L^{-1} glibenclamide by patterned bars, 1 μmol.L^{-1} glibenclamide by grey bars and 4 μmol.L^{-1} glibenclamide by black bars ($p < 0.05$ = *, $p < 0.01$ = **, $p < 0.001$ = ***, compared to control).

(1992) in canine models and in isolated rat hearts (Randall, 1995). However it should be noted that the former two investigations used anesthetized dogs and the last used buffer perfused Langendorff hearts whereas the present study was performed using an isolated working heart model with erythrocyte suspension. During reperfusion the K_{ATP} channels also play a key role as is indicated by a dose-dependent decrease in CBF. Even a trend is seen of a further decline in CBF during reperfusion compared to normoxic situations. This slight further decrease of CBF may be the result of a doubled administration time of glibenclamide (10 min. pre-ischemic, 20 min. post-ischemic), however the reactive hyperemia following ischemia also occurred during the 20 min. post-ischemic treatment

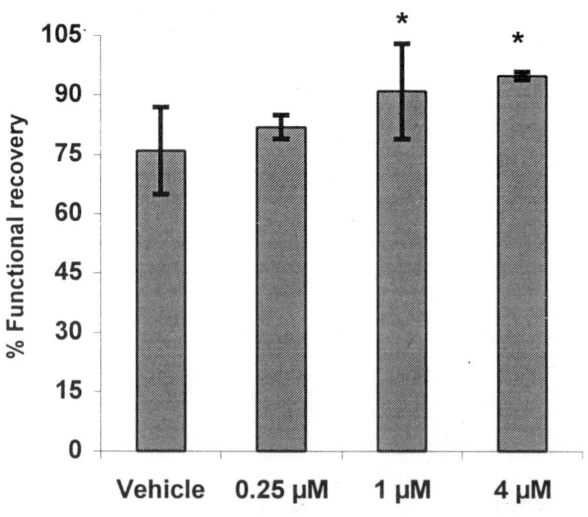

Figure 2. Cardiac functional recovery after treatment with vehicle, 0.25, 1 and 4 μmol.L^{-1} glibenclamide ($p < 0.05$ = *, compared to control).

with glibenclamide so it is not clear whether to ascribe the possible further decline in CBF to the ischemic period or to a time-dependent effect.

Our results indicate that glibenclamide slightly improves cardiac functional recovery after a period of ischemia. Similar findings were found by others (Ali et al., 1993; Tosaki et al., 1994) using also a working rat heart model but perfused with buffer solution and by Docherty et al. (1997) using a Langendorff buffer perfused rat heart model. However, the concentrations used by those groups were significantly higher than those used in this study. A possible explanation for the improved functional recovery may be the preservation of high-energy phosphorus compounds by glibenclamide (Tosaki et al., 1994). By preservation of in particular ATP the myocardium is less energy-depleted during ischemia and therefore it may recover better. Another explanation for the improved functional recovery may be the involvement of lactate in the activation of K_{ATP} channels in the myocardium (Docherty et al., 1997; Lederer and Nichols, 1989). Attenuation in lactate accumulation caused by glibenclamide clearly indicates an inhibitory effect on glycolysis and possibly glycogenolysis. This may be protective to the ischemic heart.

On the contrary other groups have suggested a worsened myocardial function using glibenclamide after an ischemic period (Mitani, 1991; Auchampach et al., 1992). It is believed that this effect is potentiated by depolarization followed by increased Ca^{2+} influx, thereby prolonging the ischemia-induced shortening of the action potential, increasing energy utilization and in this way being damaging to the heart.

Because the results shown in this study were pilot experiments, the strong trends are to be confirmed with more experiments.

4.2. Model Considerations

Many different models and protocols have been used to study effects of glibenclamide on cardiac hemodynamics. The primary model used to study hemodynamics before and after ischemia is an isolated heart model. These isolated hearts were frequently perfused with buffer solutions like Krebs-Henseleit or Tyrode. Due to the lower oxygen capacity of these buffer solutions the CBF in these models is much higher than in *in vivo* situations, actually near maximal vasodilation. Effects of glibenclamide on CBF in these models are therefore far more difficult to determine. Furthermore most models used a langendorff perfused heart instead of a working heart set-up. This can also be a source of difficulty in interpreting clinical relevance of results. To improve on these, we used an isolated working rat heart model. To the modified Krebs-Henseleit buffer used we added bovine erythrocytes and albumin. By using erythrocytes the CBF reflects more the *in vivo* value, so results are better interpretable. Albumin gives a better view of the pharmacological interaction in rat hearts of glibenclamide by means of drug-protein interaction and thus improving also the extrapolation to the *in vivo* situation in NIDDM patients.

4.3. Clinical Implications

It has been reported as early as in the 1970s that when treating NIDDM patients with tolbutamide (a K_{ATP} channel blocker) an increased mortality rate occurred (University Diabetes Group Program, 1970). Because of the K_{ATP} channel opening during ischemia, closure due to sulphonylurea derivative treatment is likely to further reduce mechanical function after an ischemic event as is shown by this clinical trial and as stated before by other groups (Mitani, 1991; Auchampach et al., 1992). The present study however suggests that there is a marked decrease in CBF after clinically relevant glibenclamide doses but this does not result in a more impaired myocardial function after a period of ischemia. Because we used a more clinically relevant model to study these effects it seems that it is not likely that there are very alarming side-effects of glibenclamide in the ischemic heart.

5. SUMMARY

Glibenclamide-induced cardiac hemodynamic changes before and after ischemia have been frequently studied. In general a Langendorff buffer perfused heart model was used to examine these effects. However these models used protein-free buffer perfusates. To improve clinical relevance and thereby enhance extrapolation to the *in vivo* condition we studied the effects of glibenclamide on cardiac hemodynamics using a working, erythrocyte perfused, rat heart model, where the perfusate was enriched with albumin. The results show a dose-dependent decline in CBF in normoxia and at the end of reperfusion (after an ischemic period) with glibenclamide treatment compared to control. Cardiac functional recovery improved with 1 and 4 μmol.L^{-1} glibenclamide concentrations. From this study it seems that there is a marked decrease in CBF but this did not result in impaired myocardial function after a period of ischemia, so it appears that there are no startling side effects of glibenclamide in the ischemic rat heart.

ACKNOWLEDGMENTS

The authors would like to thank Jos Evers and Maarten de Jong for their invaluable (bio-) technical assistance during the experiments. Yuen Tan is acknowledged for doing the HPLC analyses. Prof. dr. S.H. Skotnicki and Prof. dr. F.W.A. Verheugt are acknowledged for their valuable advice. This study was supported by Dutch Diabetes Foundation (grant 97–105).

REFERENCES

Ali K, Morimoto M, Fukaya Y, Furukawa Y, and Iida F (1993) Beneficial effects of the ATP sensitive K^+ channel opener cromakalim on recovery of cardiac function from cardioplegic arrest in isolated perfused rat hearts. Asia Pac J Pharmacol 8:173–179.

Auchampach JA, Maruyama M, Cavero I, and Gross GJ (1992) Pharmacological evidence for a role of ATP-dependent potassium channels in myocardial stunning. Circulation 86:311–319.

Aversano T, Ouyang P, and Silverman H (1991) Blockade of the ATP-sensitive potassium channel modulates reactive hyperemia in the canine coronary circulation. Circ Res 69:618–622.

Daut J, Rudolph WM, von Beckerath N, Meherke G, Gunther K, and Meinen LG (1990) Hypoxic vasodilation of coronary arteries is mediated by ATP-sensitive potassium channels. Science 247:1341–1344.

Docherty JC, Gunter HE, Kuzio B, Shoemaker L, Yang L, and Deslauriers R (1997) Effects of cromakalim and glibenclamide on myocardial high energy phosphates and intracellular pH during ischemia-reperfusion: ^{31}P NMR studies. J Mol Cell Cardiol 29:1665–1673.

Gerich JE (1989) Oral hypoglycemic agents. N Engl J Med 321:1231–1245.

Groop LC (1992) Sulfonylureas in NIDDM. Diabetes Care 15:737–754.

Imamura Y, Tomoike H, Narishige T, Takahashi T, Kasuya H, and Takeshita A (1992) Glibenclamide decreases basal coronary blood flow in anesthetized dogs. Am J Physiol 263:H399–H404.

Lederer WJ and Nichols CG (1989) Nucleotide modulation of the activity of rat heart ATP-sensitive K channels in isolated membrane patches. J Physiol (London) 66:913–931.

Mitani A, Kinoshita K, Fukamachi K, Sakamoto M, Kurisu K, Tsuruhara Y, Fukumura F, Nakashima A, and Tokunaga K (1991) Effects of glibenclamide and nicorandil on cardiac function during ischemia and reperfusion in isolated perfused rat hearts. Am J Physiol 261:H1864–H1871.

Nichols CG, Ripoll C, and Lederer WJ (1991) ATP-sensitive potassium channel modulation of the guinea pig ventricular action potential and contraction. Circ Res 68:280–287.

Noma A (1983) ATP-regulated K-channels in cardiac muscle. Nature 305:147–148.

Olders J, Boumans T, Evers J, and Turek Z (1990) An experimental set-up for the blood perfused working isolated rat heart. Adv Exp Med Biol 277:151–160.

Randall MD (1995) The involvement of ATP-sensitive potassium channels and adenosine in the regulation of coronary flow in the isolated perfused rat heart. Br J Pharmacol 166:3068–3074.

Samaha FF, Heineman FW, Ince C, Fleming J, and Balaban RS (1992) ATP-sensitive potassium channel is essential to maintain basal coronary vascular tone in vivo. Am J Physiol 262:C1220–C1227.

Standen NB, Quayle JM, Davies NW, Brayden JE, Huang Y, and Nelson MT (1989) Hyperpolarizing vasodilators activate ATP-sensitive K+-channels in arterial smooth muscle. Science 245:177–180.

Tosaki A and Hellegouarch A (1994) Adenosine triphosphate-sensitive potassium channel blocking agent ameliorates, but the opening agent aggravates, ischemia/reperfusion-induced injury. JACC 23:487–496.

University Group Diabetes Program (1970) A study of the effects of hypoglycemic agents on vascular complications in patients with adult-onset diabetes mellitus: II Mortality results. Diabetes 26:453–465.

HYPEROXIA AND HYPERCAPNIC ACIDOSIS DIFFERENTIALLY ALTER NUCLEAR FACTOR-κB ACTIVATION IN HUMAN PULMONARY ARTERY ENDOTHELIAL CELLS

Kei Takeshita,[1] Yukio Suzuki,[2] Kazumi Nishio,[2] Takuya Aoki,[2] Osamu Takeuchi,[3] Kyoko Toda,[3] Nagato Sato,[1] Katsuhiko Naoki,[1] Hiroyasu Kudo,[1] and Kazuhiro Yamaguchi[1]

[1]Department of Medicine, School of Medicine
Keio University, Tokyo
[2]Department of Internal Medicine
[3]Department of Biomedical Research
Kitasato Institute Hospital, Tokyo, Japan

1. INTRODUCTION

Although ventilator-associated pulmonary injury, such as oxygen toxicity or barotrauma, has given rise to critical problems for mechanically ventilated patients with serious lung diseases, mechanical ventilation is an essential procedure to maintain appropriate tissue oxygenation, pH, and P_{CO_2} in patients with respiratory distress (Dreyfuss and Saumon, 1998). In order to reduce the possibility of ventilator-associated pulmonary barotrauma, permissive hypercapnia has been applied as appropriate ventilatory support to acute respiratory distress syndrome (ARDS) (Feihl and Perret, 1994). In spite of the fact that there are some favorable prospective trials reporting that permissive hypercapnia improved mortality in patients with ARDS induced by heterogeneous background disorders (Hickling et al., 1994, Amato et al., 1998), efficacy and indication of this newly developed ventilator strategy are still controversial (Hudson, 1998). Permissive hypercapnia requires higher inspired oxygen concentration to maintain arterial oxygen tension than conventional mechanical ventilation mode and induces systemic hypercapnic acidosis for at least 48 hours (Carvahlo et al., 1997), thus arising the question whether hyperoxia and hypercapnic acidosis would exert a significant influence on sustained inflammation including ARDS.

Oxygen Transport to Tissue XXI, edited by Eke and Delpy.
Kluwer Academic / Plenum Publishers, New York, 1999.

Nuclear factor-κB (NF-κB) is a key transcription factor modulating adhesion molecule expression and cytokine production that play an important role in inflammation and immunity (Bares and Karon, 1997). NF-κB is also speculated to be a pivotal transcription factor mediating pulmonary inflammation such as oxygen toxicity, ARDS, and interstitial pneumonia (Rahman and MacNee, 1998). Although we previously reported that hyperoxia-induced lung injury was successfully protected by antioxidants (Aoki et al., 1996), which are also known to be suppressers of NF-κB (May and Ghosh, 1998), it is not elucidated how hyperoxia and/or hypercapnic acidosis alter NF-κB activity in the pulmonary vascular endothelium.

The present study was undertaken to determine the effects of hyperoxia and hypercapnic acidosis on NF-κB DNA binding activity in pulmonary artery endothelial cells *in vitro*.

2. METHODS

2.1. Endothelial Cell Culture

Human pulmonary artery endothelial cells (HPAEC; Kurabo, Osaka, Japan) were cultured in an endothelial cell growth medium (EGM-UV; Kurabo) supplemented with 10% fetal bovine serum (FBS), penicillin G (100 U/ml) and streptomycin (100 μg/ml; GIBCO, Grand Island, NY) at 37°C and 5% CO_2 in a humidified incubator. Cells used for experiments were from passages 8 to 12. Endothelial cells were grown on 25 cm^2 tissue culture flasks (Iwaki, Tokyo, Japan), then subcultured with 0.025% trypsin-0.05 mM EDTA (GIBCO). Endothelial cells were identified by their a pavement-like monolayer appearance, characteristic ultrastructure and positive immunofluorescent staining with a specific anti-human von Willebrand factor antibody (DAKO Japan, Tokyo, Japan). Cell viability exceeded 95%, as determined by the trypan-blue exclusion test.

2.2. Exposure of Endothelial Cell Monolayers to Hyperoxia

Confluent endothelial cell monolayers were exposed to either control gas (21% O_2, 5% CO_2, 1 atm) or hyperoxic gas (90% O_2, 5% CO_2, 1 atm) for up to 2 hours at 37°C in a humidified multi-gas incubator (APM-36; ASTEC, Fukuoka, Japan). The O_2 concentration in the incubator was monitored continuously and kept at 90 ± 0.5% during the exposure. P_{O_2} and P_{CO_2} in the culture medium were measured with a blood gas analyzer (model 178; Corning). Medium P_{O_2} was consistently higher than 580 Torr after a 15-minute exposure to a gas mixture containing 90% O_2. The culture medium pH did not differ between control and hyperoxic conditions. The trypan-blue exclusion tests showed over 85% endothelial cells were viable under hyperoxic conditions within 72 hours. Endothelial monolayers examined in terms of phase-contrast microscopy appeared to be morphologically normal at the time point of 72-hour exposure.

2.3. Exposure of Endothelial Cell Monolayers to Hypercapnic Acidosis

Endothelial cell monolayers cultured to confluence were exposed to the gas mixture with normocapnic gas composition (21% O_2, 5% CO_2, 1 atm) or that with hypercapnic composition (21% O_2, 10% CO_2, 1 atm) for up to 2 hours at 37°C in the same incubator used for hyperoxia experiments. The CO_2 concentration was monitored continuously and maintained at 10 ± 0.5% during the entire hypercapnic exposure. The medium was equilibrated with either normocapnic or hypercapnic gas 12 hours prior to conducting each experiment. The medium pH and P_{CO_2} under normocapnic and hypercapnic conditions were 7.24 ± 0.04, 45.2 ± 1.6 Torr and 6.91 ± 0.05, 76.4 ± 2.4 Torr, respectively, at the end of equilibration. The culture medium was replaced for the equilibrated medium just before experiments. After an additional 1 or 2-hour exposure to normocapnia or hypercapnia, pH and P_{CO_2} in the medium with HPAEC did not significantly differ from those before experiments. In some experiments, lipopolysaccharide (LPS; *E.coli* O55:B5, Sigma) was added to the medium at concentration of 1 µg/mL. The viability of endothelial cells under hypercapnic acidosis exceeded 90% within 48 hours as measured by the trypan-blue exclusion test. In phase-contrast microscopy, endothelial monolayers appeared morphologically normal at 48 hours of exposure to hypercapnic acidosis.

2.4. Nuclear Protein Extracts and Electrophoretic Mobility Shift Assay

Nuclear extracts were prepared from 10^6 cells by a rapid procedure (Andrews and Faller, 1991). Endothelial cell monolayers exposed to each condition were washed with ice-cold D-PBS and scraped with a cell scraper, then were collected and resuspended in 2 mL of cold D-PBS. Cells were pelleted for 10 seconds and resuspended in 400 µl cold buffer A (10 mM Hepes-KOH, pH 7.9, 10 mM NaCl, 1.5 mM MgC12, 0.5 mM dithiothreitol; DTT, 0.2 mM phenylmethanesulfonyl fluoride; PMSF) by flicking the tube. Cells were allowed to swell on ice for 10 minutes, then were vortexed for 10 seconds. Samples were centrifuged for 10 seconds, and the pellet was resuspended in 20 µl of cold buffer C (20 mM Hepes-KOH, pH 7.9, 420 mM NaCl, 1.5 mM MgC12, 0.2 mM EDTA, 0.5 mM DTT, 0.2 mM PMSF, 25% glycerol) and incubated on ice for 20 minutes. Cellular debris was removed by centrifugation for 5 minutes at 15,000 rpm and 4°C and the supernatant fraction (containing DNA binding proteins) was stored at −80°C. The protein concentration was determined by a spectrophotometer (Pharmacia, Uppsala, Sweden). The yield was 10–20 µg protein per 10^6 cells. Sequence of the double-stranded oligonucleotides used to detect the DNA binding activity of NF-κB was as follows (binding sites were underlined): NF-κB: 5′-GATCCAGA*GGGGACTTTCC*AAGAGG-3′ (Schreck et al., 1992). Oligonucleotides were annealed with the complementary strand to produce 5′-overhanging ends (*BamHI*), which enabled labeling by Klenow polymerase in the presence of dNTPs and with [^{32}P]dCTP (3000 Ci/mmol; Amersham, Tokyo, Japan) for the electrophoretic mobility shift assay. The labeled oligonucleotide was purified on push columns (Stratagene, La Jolla, CA). The typical binding reaction of 20 µl contained 10,000 cpm of ^{32}P-labeled double-stranded oligonucleotide, 10 µg of nuclear extraction in buffer C, 20 µg BSA, and 2 µg poly (dI-dC) (Sigma) in gel shift-binding buffer. Poly (dI-dC) was used to eliminate any nonspecific binding. Specific binding was demonstrated by competition with a 100–200 fold excess of unlabeled double-stranded oligonucleotide competitor. After incubation for 20 minutes at room temperature, samples were analyzed on 4% native acrylamide gels run at 170 V for 1.5 hours in 0.5XTBE buffer. Gels were then dried and visualized by autoradiography.

3. RESULTS

3.1. Effect of Hyperoxia on NF-κB Activity in HPAEC (Figure 1)

NF-κB DNA binding activity was not detected in the control condition. NF-κB DNA binding activity was induced by hyperoxic exposure for 1 hour and 2 hours in HPAEC.

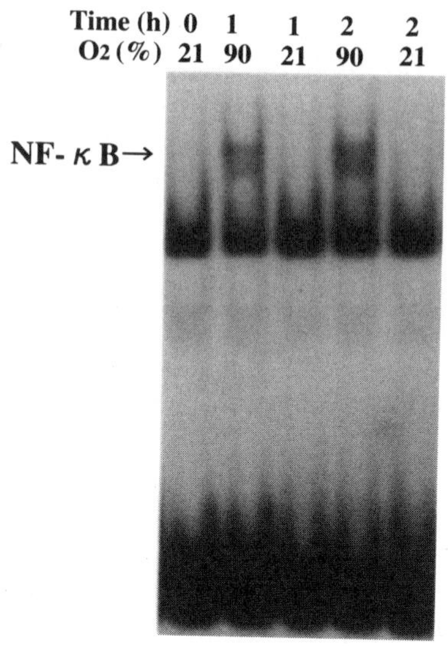

Figure 1. Effect of Hyperoxia on NF-κB Activation in HPAEC.

3.2. Effect of Hypercapnic Acidosis on NF-κB Activity in HPAEC (Figure 2)

Hypercapnic acidosis did not alter NF-κB DNA binding activity without LPS treatment. LPS elicited NF-κB activation at 1-hour and 2-hour in normocapnia as well as in hypercapnia. LPS-induced NF-κB activation was attenuated by hypercapnic acidosis at 1-hour and 2-hour in HPAEC.

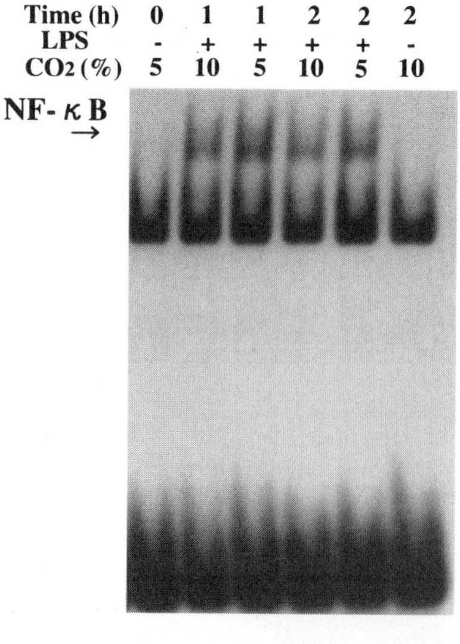

Figure 2. Effect of Hypercapnic Acidosis on NF-κB Activation in HPAEC.

4. DISCUSSION

4.1. Hyperoxia Activates NF-κB in HPAEC

NF-κB, heterodimetric protein consisting of 55-kDa (p55) and 65-kDa (p65) subunits, is present in cytoplasm complexed with its inhibiting protein IκB, which prevents NF-κB from translocation to the nucleus and binding to DNA (Baeuerle and Henkel, 1994; Baldwin Jr., 1996). Various inducers of NF-κB, including tumor necrosis factor-α (TNF-α), LPS, and interleukin-1, phosphorylate IκBα at Ser32 and Ser36 contained within the protein's N-terminal, resulting ubiquitination and degradation of phosphorylated IκBα. The degradation of IκB unmasks the nuclear localization sequences in the remaining NF-κB dimer, which in turn translocates to the nucleus and binds to specific promoter sites and activates gene transcription (Baeuerle and Baltimore, 1996; DiDonato et al., 1997). In the present study, we demonstrated that hyperoxia increased NF-κB DNA binding in HPAEC. Hyperoxia-induced reactive oxygen species production is believed to serve as the mediator activating NF-κB (Schreck et al., 1991). NF-κB was activated in this *in vitro* hyperoxia model after 2 hours of hyperoxic exposure. Shea et al. (1996) reported that NF-κB in pulmonary lymphocytes was activated by hyperoxic exposure for 24 hours in mice. The difference in onset of NF-κB activation between our *in vitro* model and the animal model may be related to the time required to overwhelm antioxidants such as superoxide dismutase and glutathione present in the medium or lungs. Since NF-κB is a key transcription factor influencing various inflammatory protein expressions, NF-κB may be a candidate for target to strategy of ventilator-associated hyperoxic lung injury.

4.2. Hypercapnic Acidosis Attenuates LPS-Induced NF-κB Activation in HPAEC

Permissive hypercapnia is a recently developed strategy for the medical management of ARDS with mechanical ventilation. In this ventilatory mode, priority is given to the prevention of pulmonary barotrauma or hyperinflation over the maintenance of normal alveolar ventilation and arterial CO_2 tension ($Pa{CO_2}$). Carvalho et al. (1997) reported that the minimal arterial pH and the maximal $Pa{CO_2}$ observed in the patients with permissive hypercapnia during the first week were 6.88 and 70 ± 18 Torr, respectively. In our experiment, medium pH after exposure to 10% CO_2 for 12h showed pH value of 6.91 ± 0.05 and $P{CO_2}$ of 76.4 ± 2.4 Torr. Hypercapnic acidosis *per se* did not affect NF-κB DNA binding activity, but attenuated LPS-induced NF-κB activation in HPAEC. Since LPS was known to activate NF-κB via several intracellular signal pathways including activation of protein kinases (Carter et al., 1998) and production of oxidants (Kurata et al., 1996), attenuation of NF-κB by hypercapnic acidosis may be associated with these signals. LPS-induced lung injury is a model of ARDS elicited by Gram-negative bacilli sepsis, that is at least partially mediated by endothelium-derived inflammatory proteins such as intercellular adhesion molecule-1 (Kumasaka et al., 1996). Hypercapnic acidosis is expected to attenuate LPS-induced expression of adhesion molecules and production of cytokines by attenuation of NF-κB activation in HPAEC. Our results suggested that permissive hypercapnia may have protective effect for ARDS induced by Gram-negative bacilli sepsis.

In conclusion, 1) administration of high concentration oxygen may augment NF-κB-mediated pulmonary inflammation, 2) hypercapnic acidosis elicited by permissive hypercapnia may have beneficial effects on pulmonary inflammation by attenuation of NF-κB activation.

REFERENCES

Amato MBP, Barbas CSV, Medeiros DM, Magaldi RB, Schettino GPP, Lorenzi-Filho G, Kairalla RA, Deheizelin D, Munoz C, Oliceira R, Takagaki TY, and Carvahlo CRR (1998) Effect of a protective-ventilation strategy in mortality in the acute respiratory distress syndrome. *N. Engl. J. Med.* 338: 347–354.

Andrews NC and Faller DV (1991) A rapid micropreparation technique for extraction of DNA-binding proteins from limiting numbers of mammalian cells. *Nucleic Acids Res.* 19:2499.

Aoki T, Suzuki Y, Suzuki K, Miyata A, Oyamada Y, Takasugi T, Mori M, Fujita H, and Yamaguchi K. (1996) Modulation of ICAM-1 expression by extracellular glutathione in hyperoxia-exposed human pulmonary artery endothelial cells. *Am. J. Respir. Cell. Mol. Biol.* 15:319–327.

Baeuerle PA and Henkel T (1994) Function and activation of NF-κB in the immune system. *Ann. Rev. Immunol.* 12:141–179.

Baeuerle PA and Baltimore D (1996) NF-κB: ten years after. *Cell.* 87:13–20.

Baldwin Jr, AS. (1996) The NF-κB and IκB proteins: new discoveries and insights. *Ann. Rev. Immunol.* 14:649–681.

Bares P and Karon M (1997) Nuclear Factor-κB—a pivotal transcription factor in chronic inflammatory diseases *N. Engl. J. Med.* 336:1066–1071.

Carter AB, Monick MM, and Hunningkake GW (1998) Lipopolysacharide-induced NF-κB activation and cytokine release in human alveolar macrophages is PKC-independent and TK- and PC-PLC-dependent. *Am. J. Respir. Cell. Mol. Biol.* 18:384–391.

Carvahlo CRR, Barbas CSV, Medeiros DM, Magaldi RB, Lorenzi-Filho G, Kairalla RA, Deheizelin D, Munoz C, Kaufmann M, Ferreria M, Takagak TY, and Amato MBP (1997) Temporal hemodynamic effects of permissive hypercapnia associated with ideal PEEP in ARDS. *Am. J. Respir. Crit. Care. Med.* 156:1458–1466.

DiDonato JA, Hayakawa M, Rothwarf DM, Zandi E, and Karin E (1997) A cytokine-responsive IκB kinase that activates the transcription factor NF-κB. *Nature.* 388:548–554.

Dreyfuss D and Saumon G (1998) Ventilator-induced lung injury. *Am. J. Respir. Crit. Care. Med.* 157:294–323.

Feihl F and Perret C. (1994) Permissive hypercapnia. *Am. J. Respir. Crit. Care. Med.* 150:1722–1737.

Hickling KG, Walsch J, Henderson S, and Jackson R (1994) Low mortality rate in adult respiratory distress syndrome using low-volume, pressure-limited ventilation with permissive hypercapnia: a prospective study. *Crit. Care. Med.* 22:1568–1578.

Hudson LD (1998) Protective ventilation for patients with acute respiratory distress syndrome. *N. Engl. J. Med.* 338:385–387.

Kumasaka T, Quinlan WM, Doyle NA, Condon TP, Sligh J, Takei F, Beaudet AL, Bennett CF, and Doerschuk CM (1996) Role of the intercellular adhesion molecule-1 in endotoxin-induced pneumonia evaluated using ICAM-1 antisense oligonuceotides, anti-ICAM-1 moinoclonal antibodies, and ICAM-1 mutant mice. *J. Clin. Invest.* 97:2362–2369.

Kurata S, Matsumoto M, Tsuji Y, and Nakajima H (1996) Lipopolysaccharide activates transcription of the heme oxygenase gene in mouse M1 cells through oxidative activation of nuclear factor κB. *Eur. J. Biochem.* 239:566–571.

May MJ and Ghosh S (1998) Signal transduction through NF-κB. *Immunology Today.* 19:80–88.

Rahman I and MacNee W (1998) Role of transcription factors in inflammatory lung diseases. *Thorax* 53:601–612.

Schreck R, Meier B, Männel DN, Dröge G, and Baeuerle PA (1992) Dithiocarbamates as potent inhibitors of nuclear factor κB activation in intact cells. *J. Exp. Med.* 175:1181–1194.

Schreck R, Rieber P, and Baeuerle PA (1991) Reactive oxygen intermediates as apparently widely used messengers in the activation of the NF-κB transcription factor and HIV-1. *EMBO J.* 10:2247–2258.

Shea LM, Beehler C, Schwartz M, Shenkar R, Tuder R, and Abraham E (1996) Hyperoxia activates NF-κB and increases TNF-α and IFN-γ gene expression in mouse pulmonary lymphocytes. *J. Immunol.* 157:3902–3908.

HETEROGENEITY OF LOCAL METABOLISM AND PERFUSION

Measuring Local Myocardial O_2 Consumption *in Situ* Using Frozen Samples

Johannes H. G. M. van Beek, Harald G. J. van Mil, David J. C. Alders, A. B. Johan Groeneveld, Antonie A. van Lambalgen, Frans J. J. de Kanter, Glenn J. Harrison, and Joli Bussemaker

Institute for Cardiovascular Research and Department of Chemistry
Vrije Universiteit
Van der Boechorststraat 7
1081 BT Amsterdam, The Netherlands

1. INTRODUCTION

Perfusion of the heart muscle is very heterogeneous, even in the healthy heart (Bassingthwaighte et al., 1989). The most important aspect is not the heterogeneity of perfusion or metabolism per se, but whether the blood delivers the oxygen necessary for metabolism. Whether perfusion is matched or not matched to aerobic metabolism is the question. Of course, a simple answer is that one must expect a good match in the healthy heart. Indeed, in healthy volunteers no myocardial lactate production is hidden under the net lactate extraction of the heart, not even during strenuous exercise (Kaijser and Berglund, 1992). However, it is predictable that mismatch will develop between perfusion and metabolism during ischemic disease and perhaps in other stressful situations such as cardiac hypertrophy.

1.1. Fractal Perfusion Heterogeneity in Animals and Humans

Perfusion heterogeneity was previously found in various animal species (Bassingthwaighte et al., 1989; Bussemaker et al., 1994a). Figure 1 shows an example of the spread of local blood flow in the porcine heart. The spread is much larger than expected based on measurement error and is relatively stable over time. Heterogeneity of

Oxygen Transport to Tissue XXI, edited by Eke and Delpy.
Kluwer Academic / Plenum Publishers, New York, 1999.

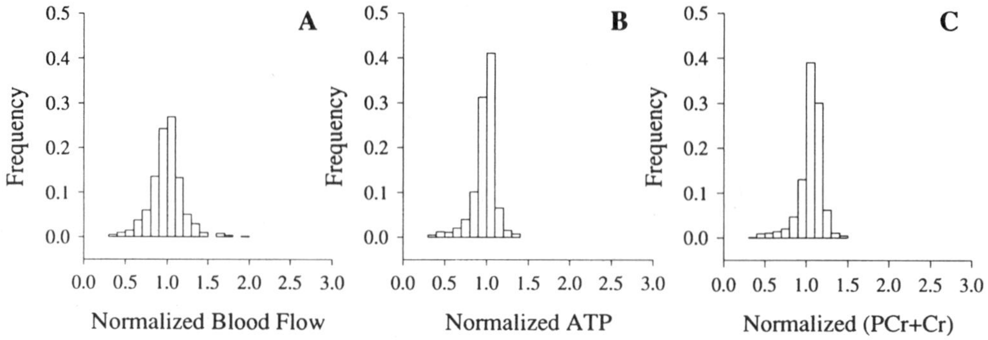

Figure 1. Histograms of the distribution of blood flow, ATP and total creatine content in porcine left ventricle (n = 7, 659 samples). Each ventricle was divided in 84–102 samples. The data were normalized to the median value before pooling. Mean left ventricular blood flow was 1.19 ± 0.21 ml min^{-1} g^{-1} wet weight (mean ± SD). Mean ATP content was 4.21 ± 0.24 μmol g^{-1}, the total creatine content (PCr + Cr) was 17.9 ± 1.0 μmol g^{-1}. Data replotted from Bussemaker et al. (1994b).

perfusion has also been assessed in the human heart. Local perfusion was measured using $[^{13}N]$-ammonia detected with positron emission tomography, PET (Meeder et al., 1997). The SD/mean of the left ventricular (LV) perfusion distribution was 0.17 in healthy human subjects, slightly smaller than in porcine LV where this value was 0.215 (Bussemaker et al., 1994b). The heterogeneity of perfusion was smaller in volunteers than in patients who were diagnosed to suffer from cardiac Syndrome X. These patients experience chest pain and show electrocardiographic signs of ischemia during exercise, but coronary arteriograms are normal.

The myocardial perfusion heterogeneity exhibits a self-similar fractal pattern (Bassingthwaighte et al., 1989). A simple fractal model is that flow is distributed heterogeneously in subsequent vascular branches: a fraction γ of total blood flow goes into one daughter vessel at a vascular bifurcation and the rest, fraction (1-γ), goes into the other daughter vessel (Van Beek et al., 1989). The blood flow is distributed unevenly at successive bifurcations, unless $\gamma = 0.5$. This fractal process generates a self-similar pattern of blood flow distribution at every scale. The histogram of the flow distribution calculated from this simple model gets broader if more generations of branching are taken into account, perfusing smaller regions. The distribution also gets broader when γ deviates farther from 0.5. The analysis of the PET data showed that healthy volunteers have values of γ declining from 0.485 to 0.47 for the first eight bifurcations of the blood vessel network model. However, the model analysis indicates that the blood flow distribution is significantly more asymmetrical in the third to seventh bifurcation for Syndrome X patients, with γ reaching a minimum of 0.46, indicating increased perfusion heterogeneity (Visser et al., 1998). This suggests that Syndrome X is due to deviations in microvascular blood flow distribution in small coronary vessels after three to seven vascular bifurcations.

1.2. Measurements of Local Oxygen Consumption

The matching between local perfusion and metabolism is of crucial importance. However, measurement of local O_2 consumption has been problematic. As a replacement, the relations between local flow and mitochondrial enzyme activity (Bussemaker et al.,

1994a) or between local flow and metabolic products (Bussemaker et al., 1994b, 1997) have been determined. Key metabolites of metabolism show broad distributions, clearly exceeding the spread caused by measurement error, see Figure 1. Blood flow is correlated significantly to local ATP content (Spearman rank correlation coefficient $R_s = 0.40$, n = 7 pigs of Figure 1), to local total creatine content ($R_s = 0.41$, Bussemaker et al., 1994b), and to local succinate dehydrogenase activity (Pearson r = 0.45, Bussemaker et al., 1994a).

Methods to measure local myocardial O_2 consumption in many regions simultaneously, by measuring ^{13}C enrichment via the Krebs cycle with NMR (Van Beek et al., 1996, 1998) and by PET (Bassingthwaighte et al., this volume) have recently been developed. The ^{13}C method is invasive but allows measurement in the animal laboratory with high spatial resolution, 2 mm, transmurally for the heart muscle. Here we discuss this method and demonstrate its feasibility in studies where subendocardial and subepicardial O_2 consumptions were derived from computer model optimization to measured ^{13}C enrichment of glutamate in the rabbit heart *in situ*.

1.3. Computer Models for Data Analysis

Methods to analyze perfusion heterogeneity and measure local aerobic metabolism depend strongly on computer model analysis. In the context of the development of the Physiome project (see Bassingthwaighte, this volume), we treat in the Discussion difficulties which may be encountered while setting up the computer analyses for this study. These difficulties are particularly due to the need to locally install computer analysis tools developed in other laboratories. We will discuss how the Physiome project (Bassingthwaighte, this volume) may help to improve this situation.

2. METHODS

2.1. Measurement of O_2 Consumption Using Frozen Tissue Samples and ^{13}C-NMR Spectroscopy

Oxygen consumption (VO_2) is proportional to the flux in the tricarboxylic acid (TCA) cycle. Usually the flux in the TCA cycle is estimated from NMR spectroscopy measurements of the gradual enrichment of glutamate with ^{13}C isotopes. This measurement is repeated every few minutes until a plateau is reached (Chance et al., 1983; Yu et al., 1995). However, perfusion is measured in many small regions simultaneously, and to study metabolism-perfusion matching VO_2 must therefore also be assessed in many regions simultaneously, which is not yet possible with *in vivo* NMR. Our strategy is to quick-freeze tissue samples after a well defined brief period of infusion of ^{13}C-enriched acetate which enters the TCA cycle. The ^{13}C NMR multiplets of glutamate, enriched via the TCA cycle, are then measured in extracts of the samples. We have shown that estimation of O_2 consumption is feasible from quick frozen tissue samples by computer simulation and by comparison with venous O_2 electrode measurements in isolated perfused rabbit hearts (Van Beek et al., 1996, 1997, in press).

^{13}C-isotope enters the TCA cycle via acetyl CoA and reaches the C4 of α-ketoglutarate in the first turn of the TCA cycle, from where it is transferred to the C4 of glutamate (G4) with high exchange flux, $J_{exchange}$, in both directions. Glutamate's C2 (G2, to which the amino-group is attached) and C3 (G3) are enriched in the next turn of the cycle. The ratio of the enrichment of the 2- or 3-carbon atom of glutamate to enrichment in the 4-carbon position depends uniquely on the turnover rate of the TCA cycle, as we derived from the data of E.M. Chance (Chance et al., 1983; Van Beek et al., in press). Even more information is obtained by also analyzing the G4, G3 and G2 multiplets of ^{13}C-glutamate which are caused by J-coupling between adjacent ^{13}C-isotopes (Malloy et al., 1990). With our computer models (see below) the distribution of ^{13}C in the TCA cycle during infusion of ^{13}C-enriched acetate can be calculated.

In the first simple version of the model two pools were discerned: the pool of acetyl CoA which is ^{13}C-enriched after acetate transport through the blood vessels, vessel wall permeation and activation via the enzyme acetyl CoA synthase. This enrichment is represented by 4 differential equations for the 4 possible ^{12}C/^{13}C combinations (isotopomers) in the two-carbon acetyl CoA molecule with time constant $\tau_{transport}$. The second pool represents all the TCA cycle intermediates plus glutamate and aspartate. The assumption is that glutamate is the dominant pool, and that the exchange between the TCA cycle intermediate α-ketoglutarate and glutamate is so fast that it is not limiting. Thus the second pool is represented by 32 differential equations for the 32 isotopomers of glutamate. This model has been described in detail previously (Van Beek et al., 1997; Van Beek et al., 1998).

Of course this model represents a simplification. There are many intermediary metabolites in the pathway and the exchange flux between α-ketoglutarate and glutamate, $J_{exchange}$, may limit the enrichment of glutamate to some extent (Yu et al., 1995). Therefore a more complex model has been developed, consisting of 132 differential equations. It consists of 6 pools: the acetyl CoA pool, the 6-carbon TCA cycle intermediates, α-ketoglutarate, the 4-carbon TCA cycle intermediates, glutamate and aspartate. The model is described in detail elsewhere (Van Beek et al., under review). The following model parameters apply to both models: $\tau_{transport}$, the TCA cycle flux, the fraction F_{c2} of acetyl CoA derived from the infused ^{13}C-enriched acetate, the glutamate pool size, the ratio of the anaplerotic flux to the TCA cycle flux, and the parameter h which indicates the fraction of ^{13}C going from G4 to G2, while (1-h) goes to G3 (Van Beek et al., in press). Parameter h reflects metabolite channeling in the TCA cycle, but is usually considered to be 0.5, indicating there is no channeling (Yu et al., 1995). For the 132 equations model extra parameters apply: the contents of the three TCA cycle pools and $J_{exchange}$, which has often been taken to be non-limiting (approaching infinity) in previous models. Because of the potential important effects of slow exchange between α-ketoglutarate and glutamate (Yu et al., 1995), the 132 differential equations model is applied here to investigate the influence of $J_{exchange}$.

The differential equations for both models were integrated and parameters optimized on the computer using the simulation interface SIMCON (developed by National Simulation Resource for Circulatory Mass Transport and Exchange, NIH, Seattle, USA).

2.2. Experimental Methods

In this study we apply the new ^{13}C method to the rabbit heart *in situ*. The left coronary main stem was cannulated and perfused from the carotid artery. Acetate (unenriched in ^{13}C) was infused into the coronary circulation. At $t = 0$ we switched to [2-^{13}C]-acetate (99% enriched). The left ventricular free wall was freeze-clamped at $t = 5$ min, freeze-dried and divided into subendocardial and subepicardial halves. The ^{13}C-multiplet areas of glutamate in the sample extracts were fitted with the TCA cycle model, yielding TCA cycle flux and acetyl CoA enrichment by non-linear least squares optimization. Details are given below.

The study was conducted on male New Zealand White rabbits (n = 6), weight ~3 kg. Anesthesia was induced with 0.1 mg/kg fentanyl citrate and 5 mg/kg fluanison (Hypnorm, Janssen Pharmaceutics) and 1.5 mg/kg midazolam intramuscularly, followed by a continuous infusion in the ear vein of fentanyl citrate (100 μg/kg/hour) and midazolam (1 mg/kg/hour). After tracheotomy artificial ventilation was started, giving 40% O_2/60% air at 2 L/min with 25 breaths/min, with positive end expiratory pressure 4 cm H_2O. Body temperature was measured rectally, and adjusted with a thermostatted pad to 38.5–39.0°C. The left femoral artery was cannulated to record arterial blood pressure and for blood sampling. The left carotid artery was cannulated to feed the extracorporeal circuit. The left jugular vein was cannulated for blood sampling and saline infusion. After sternotomy the pericardium was opened and a subclavian artery cannulated to enter a cannula into the aorta. The left atrium was cannulated to introduce a cannula into the left ventricle for pressure measurement.

Heparin was injected (1000 IU/kg bolus, followed by 400 IU/kg/hour). Ten minutes after the heparin injection the extracorporeal circuit was connected to the left carotid artery cannula, and the cannula connected to the circuit was led to the left coronary artery via the left subclavian artery.

After cannulation 30 min was allowed for stabilisation during infusion of unlabeled acetate in the extracorporeal circuit. Via a side branch 0.5 ml/min of Tyrode solution was added containing 40 mM acetate. Between t = 24 and 27 min radioactively labeled microspheres were injected into the extracorporeal circuit (1 batch of 0.2–0.4 × 10^5 spheres; 15 μm diam.; suspended in saline with 1 drop of 0.01% Tween during 30 sec, total volume 0.7 ml, flushed with 0.3 ml saline). From 5 sec before to 1 min 25 sec after injection, a reference sample was drawn from the extracorporeal circuit at 0.5 ml/min.

After 30 min, [2-^{13}C]-acetate was infused for exactly 5 min, replacing unenriched acetate. During this period blood samples were drawn to determine arterial (extracorporeal circuit) and venous (from the external jugular vein) blood gas values and hemoglobin content. Arterial pressure, LV pressure, coronary perfusion pressure and temperature were also recorded. One minute before the end of the experiment the dye Evans blue (1 ml over 1 min) was injected into the extracorporeal circuit to stain the area perfused from the extracorporeal circuit.

At exactly t = 5 min the LV free wall was quickly excised and Wollenberger clamped between two aluminum plates precooled to −80°C. The left ventricular sample was freeze-dried and cut into subendocardial and subepicardial halves. These two samples were extracted as described previously (Van Beek et al., in press) to measure the ^{13}C multiplets of glutamate and the deposition of radioactive microspheres. For details of the high resolution NMR spectroscopy see Van Beek et al. (in press). The NMR free induction decay signals were analyzed in the time domain using the computer program MRUI-VARPRO (Van den Bogaart, 1997). The contents of glutamate, citrate and aspartate were assayed in all tissue samples by standard biochemical methods, and were fixed in the mathematical models during parameter optimization.

In a separate preliminary experiment the hemoglobin oxygen saturation was measured in the coronary sinus of the rabbit heart *in situ*, using a Schwarzer IVH3 densitometer. After opening the chest and the pericardium a fiber optic catheter was introduced via the left rostral vena cava into the coronary sinus. This vein enters the coronary sinus rather than the right atrium, and does not exist in many other animal species. Blood flowing around the tip of catheter came from the left ventricular free wall. The heart was electrically paced. The IVH3 densitometer was calibrated against a Radiometer OSM3 hemoglobin saturation meter using a circuit with an artificial oxygenator. Results are given as mean ± SE.

2.3. Computer Simulations

The results of ^{13}C infusion experiments were simulated with the computer model to calculate the multiplet resonance areas after 5 min of infusion of ^{13}C-acetate. Then Gaussian noise was added to the calculated multiplets, with a standard deviation estimated from the time-domain fit using the Variable Projection method implemented in the MRUI-VARPRO program. Parameters of the model were then reestimated by fitting the model to the simulated data sets including the NMR measurement noise. By fixing one parameter to different values, the sensitivity of other parameters for the fixed parameter is obtained for simulated data sets and the *in situ* experimental data.

3. RESULTS

3.1. Simulation Results

Computer simulations had shown that the TCA cycle flux can be estimated if 7 or more distinguishable ^{13}C-NMR-multiplets of glutamate are measured from extracts of samples, after short infusion (5 min) of ^{13}C-enriched acetate (Van Beek et al., 1997, in press). The SD of the estimation of the TCA cycle flux is 0.75 μmol g^{-1} min^{-1} (n = 10 Monte Carlo simulations) with glutamate fixed (simulating cases where glutamate has been accurately assayed). The SD of the estimation is 1.25 μmol g^{-1} min^{-1} (n = 30 Monte Carlo simulations) if glutamate is estimated from the NMR intensities.

A simulated time course of the development of the multiplets during 5 min infusion of [2-^{13}C]-acetate is shown in Figure 2. Note that in Figure 2A the singlet of G2 and the doublet of G3 resulting from the ^{13}C-isotope in the neighboring G4 (G3D34) are almost the same because the labeling of acetyl CoA is 100% ($F_{c2} = 1$). The contribution of the doublet of G3 caused by G2 (G3D23) is negligible early after start of the infu-

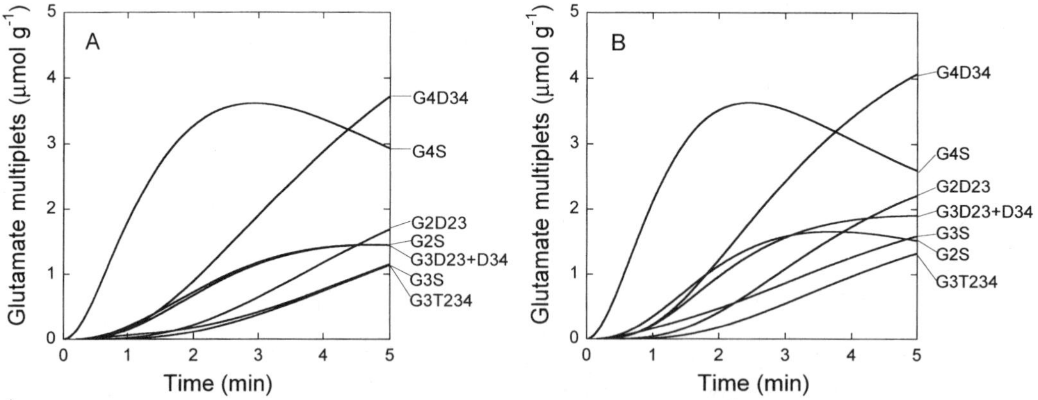

Figure 2. Simulation of the ^{13}C-enrichment of glutamate during 5 min infusion of [2-^{13}C]-acetate using the 36 differential equation model. All seven measurable multiplets are shown as a function of time. See text for meaning of the multiplet codes. J_{TCA} is 9.6 μmol g^{-1} dry weight min^{-1}, $\tau_{transport}$ = 40 sec. A. Acetyl CoA 100% enriched with ^{13}C ($F_{c2} = 1$). B. Acetyl CoA 85 % enriched ($F_{c2} = 0.85$).

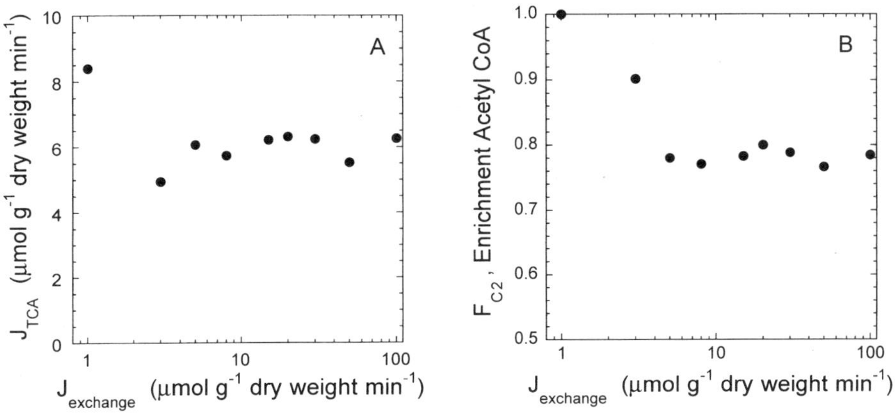

Figure 3. (A) The TCA cycle flux, J_{TCA}, and (B) the fractional enrichment of acetyl CoA, F_{c2}, estimated by parameter optimization, as a function of the fixed value of the exchange flux between α-ketoglutarate and glutamate, $J_{exchange}$. To this end the 132 differential equation model was used. The data set fitted is similar to the ones obtained in *in situ* experiments. The best fit to the measured data (least squares point) was found for $J_{exchange} = 8\,\mu mol\,g^{-1}$ dry weight min^{-1}.

sion because it only arises during the third turn of the TCA cycle. Please note that G3S not only comprises the singlet of G3 (which is very small in this case) but also the middle peak of the G3 triplet (G3T234), because the latter is indiscernable from the singlet. In the case where F_{c2} is lower than 1, G2S and G3D34 differ, see Figure 2B. Half of the ^{13}C goes to G2, half goes to G3, but the G3 does not pick up an adjacent ^{13}C isotope from acetyl CoA because the latter is not 100% ^{13}C-enriched.

The simulations show that the estimation of J_{TCA} and F_{c2} is rather insensitive to the exchange flux between α-ketoglutarate and glutamate provided this flux is >5 μmol g^{-1} dry weight min^{-1}, see Figure 3. We conclude that estimating J_{TCA} and F_{c2} can be done with the 36 differential equation model where the assumption is that $J_{exchange}$ is very high. The limited effect of $J_{exchange}$ on estimates of J_{TCA} and F_{c2} was corroborated in extensive simulations on other data sets (Van Beek et al., under review).

3.2. ^{13}C Method Applied to *in Situ* Heart

O_2 consumption is calculated from the flux in the TCA cycle using the equation (Van Beek et al., in press): $VO_2 = (3\text{-}F_{c2}) \cdot J_{TCA}$. It has already been shown in isolated hearts perfused with Tyrode's solution that a close correspondence exists between O_2 consumption measured with a Clark type electrode in coronary venous effluent samples and measured in frozen tissue samples with our new ^{13}C method using 5 min infusion of [2-^{13}C]-acetate (Van Beek et al., 1997, in press). In the isolated hearts there was significant correlation between O_2 consumption measured with the ^{13}C method and measured using samples for the blood gas machine (r = 0.86, n = 7). With a new double labeling protocol a high correspondence is found with O_2 consumption measured in the same region by sampling coronary venous effluent and measuring this in-line with a Clark-type oxygen electrode, r = 0.90, n = 26 (Van Beek et al., 1998, under review).

Here we apply the 5 min infusion method to the rabbit heart *in situ*. In two of the six experiments the protocol could be completed. The measured flow in the extracorporeal circuit was 3.8 ml/min in experiment 1 at mean arterial pressure 50 mmHg, and

Table 1. Parameter estimates obtained by fitting model I with 36 differential equations (3 parameters fitted) and model II with 132 differential equations (5 parameters fitted) to the measured NMR resonance areas for the left ventricular free wall[a]

Experiment	Transmural layer	J_{TCA}	J_{TCA}	F_{c2}	F_{c2}	VO_2	VO_2
		model I	model II	model I	model II	model I	model II
1	subendo	4.4 ± 0.6	4.7 ± 0.3	0.92 ± 0.08	0.83 ± 0.04	9.2 ± 1.2	10.2 ± 0.6
	subepi	4.0 ± 0.3	4.1 ± 0.3	1	0.99 ± 0.08	8.0 ± 0.6	8.3 ± 0.8
2	subendo	8.8 ± 1.4	8.8 ± 2.2	1	1	17.6 ± 2.8	17.7 ± 4.4
	subepi	5.9 ± 0.8	5.7 ± 0.3	1	1	11.9 ± 1.7	11.4 ± 0.6

[a] The biochemically measured glutamate content for each sample was fixed in the model. The results for four different samples are compared. The TCA cycle flux, J_{TCA}, and oxygen consumption, VO_2, are given in μmol g^{-1} dry weight min^{-1}, the enrichment fraction of acetyl CoA, F_{c2}, is dimensionless. The estimate ± standard error of the parameter are given (Chan et al., 1993), and for VO_2 the propagation of the standard error is calculated. For several cases F_{c2} reached the maximum constrained value, fraction 1, and the standard error was not calculated. Subendo, subendocardial layer. Subepi, subepicardial layer.

6.7 ml/min in experiment 2 at mean arterial pressure 75 mmHg. The acetate concentration of the infused blood was 5.3 and 3.4 mM, respectively. The hemoglobin content of the blood under our experimental conditions was 6.1 ± 0.2 mmol L^{-1} (n = 13 samples in 4 hearts, Radiometer OSM3). The results of parameter optimization are compared between the 36 and 132 differential equations model. Using a routine for solving ordinary nonlinear least square problems, SENSOP (Chan et al., 1993), parameters were estimated by fitting 7 glutamate multiplets. The results are compiled in Table 1. There is good correspondence between the two models. At systolic arterial pressure 115 mmHg and heart rate 260 min^{-1}, LV subendocardial and subepicardial MVO_2 were 17.6 ± 2.8 and 11.9 ± 1.7 μmol g^{-1} min^{-1}, for the 36 equations model. In this heart the local blood flows measured with radioactive microspheres were 4.93 and 2.84 ml g^{-1} dry weight min^{-1} in the subendocardial and subepicardial samples, respectively. At systolic pressure 60 mmHg and heart rate 210 min^{-1}, subendocardial and subepicardial MVO_2 were 9.2 ± 1.2 and 8.0 ± 0.6 μmol g^{-1} min^{-1}, respectively. $F_{exchange}$ was always above 10 μmol g^{-1} min^{-1} for the 132 equations model and therefore did not affect the value of J_{TCA}.

In one preliminary experiment the O_2 saturation of blood in the coronary sinus of the rabbit heart *in situ* was measured. At heart rate 240 min^{-1} the coronary sinus hemoglobin O_2 saturation was 48–49% while the arterial O_2 tension was kept at 150–200 mmHg. At higher electrically paced heart rate, 300 min^{-1}, the O_2 saturation declined to 41–42% although the systolic blood pressure and rate pressure product both decreased at the higher electrically paced heart rate. This shows the feasibility, in the future, of comparing VO_2 measured with the ^{13}C method in the LV free wall with measurements based on venous hemoglobin O_2 saturation for the same region. However, the ^{13}C method allows to measure local O_2 consumption in subregions of the LV free wall.

4. DISCUSSION

4.1. Measuring Local Oxygen Consumption Using Frozen Tissue Samples

By combining cryospectrophotometric values of hemoglobin saturation in local arterioles and venules with blood flow measured in adjacent tissue samples using radioactively labeled microspheres, the local O_2 consumption has been calculated (Weiss, 1979). The values for the rabbit LV free wall were 20.7 ± 6.8 and 15.4 ± 3.7 μmol g^{-1} dry weight

min^{-1} in subendocardium and subepicardium, respectively, at a heart rate of $230 \pm 9\,min^{-1}$, so that values we found with the ^{13}C method in rabbit heart *in situ* in experiment 2 are in the same range (see above). In experiment 1 with low arterial blood pressure the oxygen consumption was lower. The cryospectrophotometric method is proficient, but has, to the best of our knowledge, not been used in recent years.

In the open-chest rabbit the subendocardial VO_2 appears to be higher than the subepicardial VO_2 (this study, Weiss, 1979). This had also been found in isolated rabbit heart (Van Beek et al., in press). With our ^{13}C-method local O_2 consumption can be measured separately in subendocardial and subepicardial regions, 2–3 mm thick. The procedure yields absolute values of local oxygen consumption in 50–200 mg dry weight samples, which can be related to local blood flow measured with labeled microspheres. The ^{13}C method is useful for mapping the distribution of heterogeneous aerobic metabolism and its relation to local perfusion in health and disease. Given the generality of the underlying biochemical principles, the method is also applicable to other organs and animal species. The method to measure local oxygen consumption in many samples simultaneously requires merely access to state-of-the-art high-resolution NMR spectroscopy, available in many chemistry departments. A bottleneck for implementing the technique may be the local installation of the computer analysis software. Therefore we discuss this difficulty in the context of the Physiome (Bassingthwaighte, this volume).

4.2. Data Analysis and the Physiome

During the design of our computer models and the model analysis we were repeatedly confronted with delays during the local implementation of computer programs and interfaces stemming from incompatibilities between various computer platforms and file formats. For the time domain analysis of the NMR free induction decay, the program was available quickly and free of charge but the program had to be installed with a commercial computer package which required more random access memory than available. The analyses were finally run in batch mode without use of the graphical interface, which allowed larger data files to be analyzed, but is not easy to use.

Similarly, to run the ^{13}C isotope models with graphics and the capability to optimize parameters an advanced simulation interface was used and had to be installed locally. Installation of the package was not possible on several computer platforms locally available at that time and a computer compatible with the program was acquired. These difficulties with portability between computer platforms caused substantial delays in the development of the ^{13}C work reported above.

Perhaps these two examples illustrate difficulties often encountered in the transfer of computer programs and data. There is not a uniform format for operating systems, data file formats and programming languages. Even though it may be easy to transfer all sorts of files across computer networks, it is often a problem to run a certain advanced and complex program on the locally available computer. With the rapid pace of developments in computing, the problems of porting data, programs and graphical simulation interfaces become worse: technological and commercial developments make old data formats and computer programs obsolete, so that having a uniform standard becomes even more difficult to define and maintain.

A simple, partial and temporary solution is feasible for smaller programs. An executable program for the personal computer platform can probably be used in almost all institutions in the world, although there are limits to computational speed and sophistication of graphical presentation. Simple graphical capabilities may be built into the program.

An example is a fractal analysis program for the personal computer by Kleen et al. (1998). For ^{13}C isotope enrichment an instructive program was made by Malloy et al. (1990).

The accessibility of data and computer models would be improved a lot if these could be approached from all personal computer and workstation platforms via computer networks, just as a world wide web browser is used to view web pages. If providing the model depends on individual researchers, support would last only as long as the individual researcher is active in the field and financial support is available. It may be an idea, therefore, to implement the models on a central site that is similar in nature to a major biomedical library: securely funded so that one may rely on its continued existence. Once a model is accepted by this institute it could then be accessed at that central site. At this central site extensive expertise would be developed regarding the technology needed to make computer models available via computer networks. Of course, this would not mean that individuals and research groups could not make their models available, but it would provide a stable basis for models that are generally useful, and acceptance by this central institution could acquire a similar status as having a manuscript accepted by a major scientific journal. Likewise, such an institution could be the basis of a biomedical data warehouse with the goal of making useful biomedical data quickly accessible via computer networks. Such a scientific computer data and programs library for the computer network age could form an important core of the Physiome project.

5. SUMMARY

The feasibility of measuring local cardiac oxygen consumption in several regions simultaneously is shown. To this end ^{13}C-enriched acetate is infused briefly into the heart in the animal laboratory and the tissue is quick-frozen. The ^{13}C NMR spectra of extracts of the tissue samples must then be analyzed with computer models. A time consuming part of implementing the method may be to install the program on the locally available computer. It may be desirable that such analysis tools can be run directly via computer networks as part of the Physiome project.

ACKNOWLEDGMENTS

This study was supported by the Netherlands Heart Foundation (grant D94.016 and 96.127) and by an intramural grant from the Vrije Universiteit. The MRUI software package was kindly provided by Dr. A. van den Bogaart, Katholieke Universiteit Leuven, Belgium. It is currently funded by the European Union project TMR/Networks (ERB-FMRX-CT970160). We are very grateful to Drs. R. B. King and J. B. Bassingthwaighte for providing and modifying the SIMCON simulation interface (National Simulation Resource for Circulatory Mass Transport and Exchange, Center for Bioengineering, University of Washington, Seattle, NIH grant no. RR-01243). We thank Wim Gerrissen and Marcel Nederhoff for their assistance during the experiments, and Jan Paul Barends for computer support.

REFERENCES

Bassingthwaighte JB, King RB, and Roger SA (1989) Fractal nature of regional myocardial flow heterogeneity. Circ Res 65:578–590.

Bussemaker J, van Beek JHGM, Groeneveld ABJ, Hennekes M, Teerlink T, Thijs LG, and Westerhof N (1994a) Local mitochondrial enzyme activity correlates with myocardial blood flow at basal workloads. J Moll Cell Cardiol 26:1017–1028.

Bussemaker J, van Beek JHGM, Groeneveld ABJ, Hennekes M, Teerlink T, van Lambalgen AA, Thijs LG, and Westerhof N (1994b) Local blood flow correlates with energy metabolite content in the pig heart. In: The coupling of heterogeneous myocardial blood flow and energy metabolism. Bussemaker J (Ed) (ISBN 90-9007189-X). Vrije Universiteit, Amsterdam, pp. 13–29.

Bussemaker J, Groeneveld ABJ, Teerlink T, Hennekes M, Westerhof N, and Van Beek JHGM (1997) Low and high blood flow regions in the normal pig heart are equally vulnerable to ischaemia during partial coronary stenosis. Eur J Physiol (Pflügers Arch) 434:785–794.

Chan IS, Goldstein AA, and Bassingthwaighte JB (1993) SENSOP: A derivative-free solver for nonlinear least squares with sensitivity scaling. Ann Biomed Eng 21:621–631.

Chance EM, Seeholzer SH, Kobayashi K, and Williamson JR (1983) Mathematical analysis of isotope labeling in the citric acid cycle with applications to ^{13}C NMR studies in perfused rat hearts. J Biol Chem 258:13785–13794.

Kaijser L and Berglund B (1992) Myocardial lactate extraction and release at rest and during heavy exercise in healthy men. Acta Physiol Scand 144:39–45.

Kleen M, Habler O, Hutter J, Kemming G, Podtschaske A, Tiede M, Welte M, Keipert P, Batra S, Faithfull NS, Corso C, Zwissler B, and Messmer K (1998) Hemodilution and hyperoxia locally change distribution of regional pulmonary perfusion in dogs. Am J Physiol 274:H520–H528.

Malloy CR, Sherry AD, and Jeffrey FMH (1990) Analysis of tricarboxylic acid cycle of the heart using ^{13}C isotope isomers. Am J Physiol 259:H987–H995.

Meeder JG, Blanksma PK, van der Wall EE, Willemsen ATM, Pruim J, Anthonio RL, de Jong RM, Vaalburg W, and Lie KI (1997) Coronary vasomotion in patients with syndrome X: evaluation with positron emission tomography and parametric myocardial perfusion imaging. Eur J Nucl Med 24:530–537.

Van Beek JHGM (1997) Is local metabolism the basis of the fractal vascular structure in the heart? Int J Microcirc: Clin Exp 17:337–345.

Van Beek JHGM, Csont T, Bussemaker J, and Barends JPF (1996) Measuring local myocardial O_2 consumption in many samples. Ann Biomed Eng 24:S-19.

Van Beek JHGM, Csont T, De Kanter FJJ, and Bussemaker J (in press) Simple model analysis of ^{13}C NMR spectra to measure O_2 consumption using frozen tissue samples. Adv Exp Med Biol, in press.

Van Beek JHGM, De Kanter FJ, and Bussemaker J (1998) Measuring subendocardial versus subepicardial O_2 consumption in rabbit left ventricle using ^{13}C-NMR spectroscopy. J Moll Cell Cardiol 30:A61.

Van Beek JHGM, Van Mil HGJ, de Kanter FJJ, Alders DJC, King RB, and Bussemaker J (under review) A double label ^{13}C NMR method to measure local myocardial oxygen consumption using frozen tissue samples. Revision submitted.

Van den Bogaart A (1997) Quantitative data analysis of in vivo MRS data sets. Magn Res Chem 35:S146–S152.

Visser KR, Meeder JG, Van Beek JHGM, Van der Wall EE, Willemsen ATM, and Blanksma PK (1998) A mathematical model for the heterogeneity of myocardial perfusion using Nitrogen-13-Ammonia. J Nucl Med 39:1312–1319.

Weiss HR (1979) Regional oxygen consumption and supply in the rabbit heart—effect of nitroglycerin and propanolol. J Pharmacol Exp Ther 211:68–73.

Yu X, White LT, Doumen C, Damico LA, LaNoue KF, Alpert NM, and Lewandowski ED (1995) Kinetic analysis of dynamic ^{13}C NMR spectra: metabolic flux, regulation, and compartmentation in hearts. Biophys J 69:2090–2102.

34

ESTIMATION OF LACTATE THRESHOLD BY NEAR INFRARED SPECTROSCOPY

Caroline Angus, Debbie Welford, Martin Sellens, Steve Thompson, and Chris E. Cooper

Department of Biological Sciences
University of Essex
Colchester, Essex, CO4 3SQ, United Kingdom

1. INTRODUCTION

The anaerobic or lactate threshold (LT) is defined as the point at which blood lactate levels rise significantly above basal levels and correlates with athletic performance in endurance events (Weltman, 1995). Currently gas exchange measurements of whole body VO_2 and VCO_2 are the best non-invasive techniques of measuring the LT (Beaver et al., 1986). Near-infrared spectroscopy (NIRS) data has shown a correlation between muscle deoxygenation and blood lactate concentration (Rundell et al., 1997) and it has recently been suggested that NIRS can be used to determine the LT (Belardinelli et al., 1995; Bhambhani et al., 1997). NIRS provides a non-invasive method for the continuous measurement of changes in muscle haemo(myo)globin oxygenation and blood volume (Chance et al., 1992; Ferrari et al., 1997; Wilson et al., 1989). By identifying particular trends in the rate and extent of muscle deoxygenation these authors have shown correlations with gas exchange methods of measuring the LT.

However, the two NIRS methods used different criteria for measuring the LT. During an incremental cycling exercise there is at first an increase in muscle oxygenation, followed by a decrease to below the baseline resting level; the rate of the decrease increases as the work intensity increases. Belardinelli et al. (1995) used the point at which the muscle deoxygenation rate increased as a measure of the LT, whereas Bhambhani et al. (1997) used the point at which the muscle oxygenation trace returned to its pre-exercise level. The purpose of this study was to compare these two methods within a single study, using direct invasive blood lactate measurements of LT, rather than the indirect gas exchange methods used previously.

Oxygen Transport to Tissue XXI, edited by Eke and Delpy.
Kluwer Academic / Plenum Publishers, New York, 1999.

2. METHOD

The incremental cycle ergometer exercise protocol was the same as described by Bhambhani et al. (1997). The subjects were 7 male and 7 female students with a mean (SD) age of 22.5(4.4) years; they cycled at 60 rpm, with an increase in weight every 2 minutes such that the power increased by 30 W. The subjects continued until volitional fatigue. Dual wavelength NIRS measurements (Micro Run-Man-96, Philadelphia, PA) were taken, with the probe positioned on the right vastus lateralis muscle 150 mm from the knee and parallel to the major axis of the thigh. The distance between the light source and detection fibres was 3 cm. Blood samples for lactate measurements were taken from the finger at the end of every two-minute increment and analysed in a P-GM7 analyser (Analox Instruments Ltd, London). Gas exchange measurements of VO_2 VCO_2 were taken throughout the test using open circuit spirometry (Oxycon beta, Mijnhardt, Bunnik, The Netherlands).

The NIRS machine was calibrated (zeroed) whilst the subject was sitting on the cycle. NIRS data was collected at 1 s intervals. At the end of the test the NIRS data was replotted against time with a 20 s moving average. Changes in muscle oxygenation were determined essentially by subtracting the absorbance at 850 nm from that at 760 nm (Chance et al., 1992). These results were analysed using the two following methods to determine the LT of each subject.

1. The point at which the muscle oxygenation level returns to zero ($O_2 = 0$) (Bhambhani et al., 1995).
2. The point where there is an increase in the rate of deoxygenation (Belardinelli et al., 1995).

For comparison with the NIRS data the direct measurement of the lactate threshold was defined as the VO_2 at which the blood lactate had increased by 1 mM from its baseline level. The NIRS and blood lactate analyses were carried out independently of each other i.e. with the analyser unaware of the results obtained by the other methodology.

The two methods of detecting the LT from the NIRS data were then compared to the direct blood lactate method using linear regression to determine any positive correlation. If there was a positive correlation a Bland-Altman analysis (Bland et al., 1986) was used to determine if the methods were essentially identical.

3. RESULTS

Figure 1 shows the data averaged from 6 identical tests on a single subject (± SE on the blood lactate measurements). There was an initial rise in blood oxygenation during the warm-up phase. Deoxygenation of the muscle occurred approximately three minutes into the exercise—the $O_2 = 0$ point defined by Bhambhani et al. (1997). This was accompanied by a significant increase in blood volume. Throughout the exercise period the muscle blood volume rise correlated to the increase in global oxygen consumption, mirroring the known correlation between rises in muscle blood flow and muscle oxygen consumption. Near the end of the exercise (at about 12 min.) there was an increased rate of muscle deoxygenation—the deoxygenation rate change defined by Belardinelli et al. (1995). In this individual the deoxygenation rate change method (12 min.) is clearly more closely related than the $O_2 = 0$ method (3 min.) to the true LT point (10 min.).

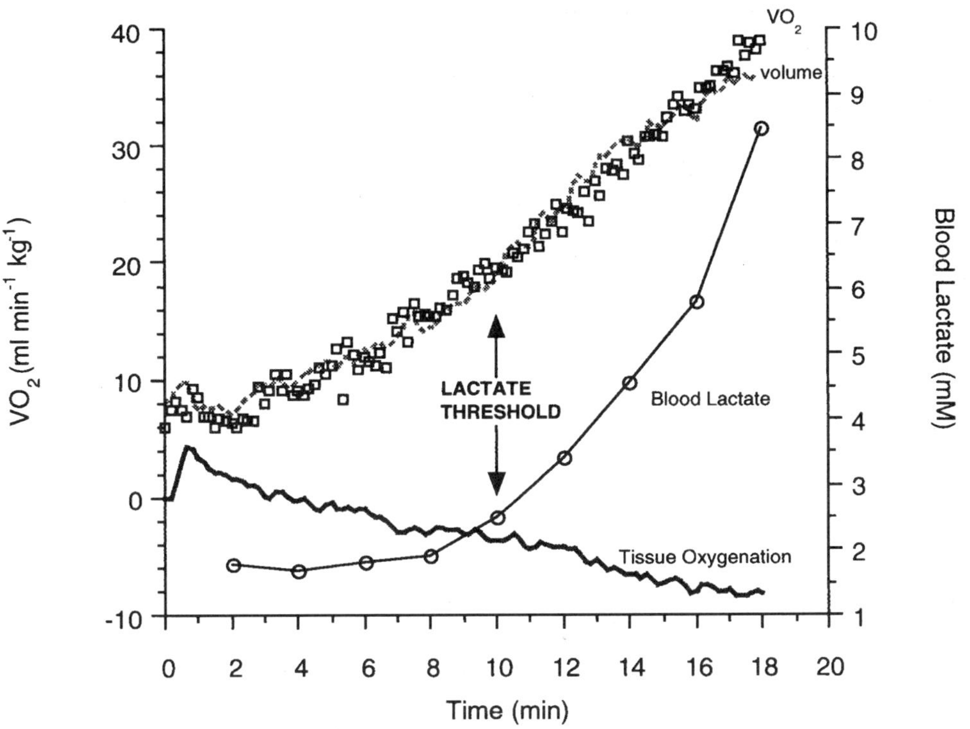

Figure 1. Changes in blood volume, blood oxygenation, whole body oxygen consumption and blood lactate during incremental cycling exercise. Average of 6 studies in a 34-year-old male subject (± SE for the blood lactate measurements). Cycling protocol started at time 0 as described in methods. The NIRS data is scaled arbitrarily (in particular the blood volume change is offset to illustrate its close coincidence with the VO_2 in this individual. Blood lactate (o), VO_2 (□), blood volume (---), muscle oxygenation (—).

Not all subjects showed such clear trends in the NIRS traces. It was impossible to use the $O_2 = 0$ method in 3 of the 14 subjects (21% failure rate); 2 of the subjects never deoxygenated below the baseline and 1 of them showed no initial oxygenation phase (i.e. there was a gradual desaturation from the start of the exercise). In 2 of the 14 subjects no identifiable change in the rate of the deoxygenation trace could be observed (i.e. the slope did not increase) resulting in a failure rate of 14%.

Figure 2 shows there is no correlation between the NIRS $O_2 = 0$ method and the true LT as we obtain a regression line of $y = 0.35x + 27.37$ with a correlation of $r = -0.22$, which is not significant ($n = 11$).

Figure 3 demonstrates a linear relationship between the NIRS slope change method and the true LT with a regression line of $y = 1.05x - 1.95$ and a correlation of $r = 0.75$, significant at $p < 0.01$ ($n = 12$). The slope is not significantly different from 1, suggesting that the two techniques are measuring the same parameter. However, a Bland-Altman plot, Figure 4, showed the mean difference in VO_2 between these two methods was 0.68 ml/kg/min with a SD of 3.39 ml/kg/min. This reveals, quite a large possible error between the two techniques i.e. in 5% of the population the difference between the methods could be as large as 6.5 ml/kg/min.

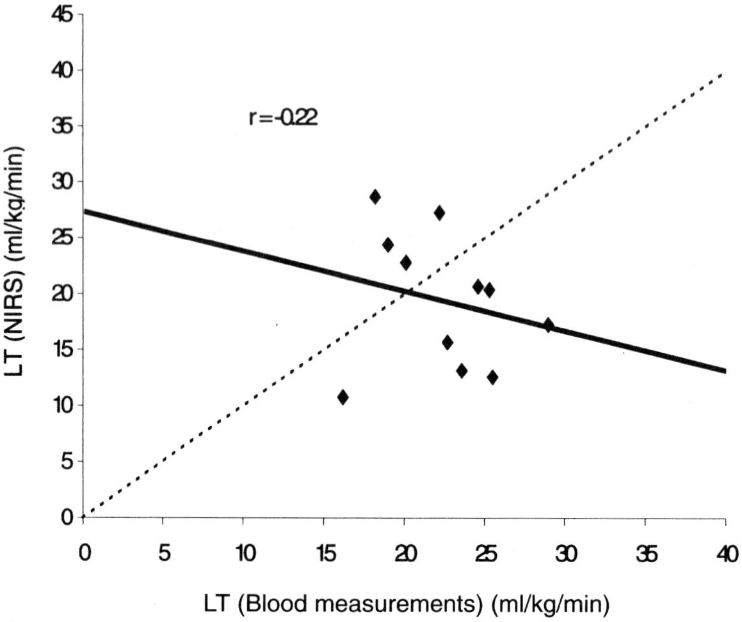

Figure 2. Correlation between blood lactate and NIRS ($O_2 = 0$) measurements of LT. Linear regression line (---) and line of unity (—) illustrated on graph.

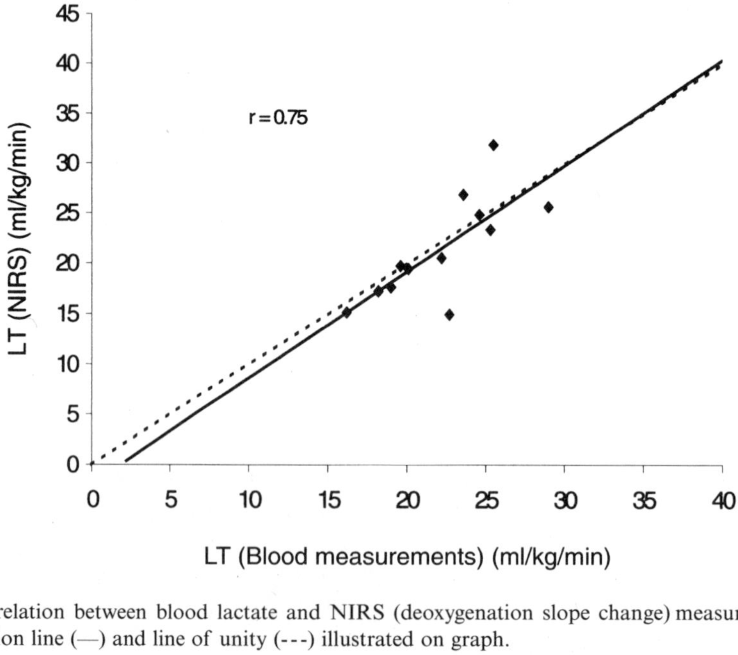

Figure 3. Correlation between blood lactate and NIRS (deoxygenation slope change) measurements of LT. Linear regression line (—) and line of unity (---) illustrated on graph.

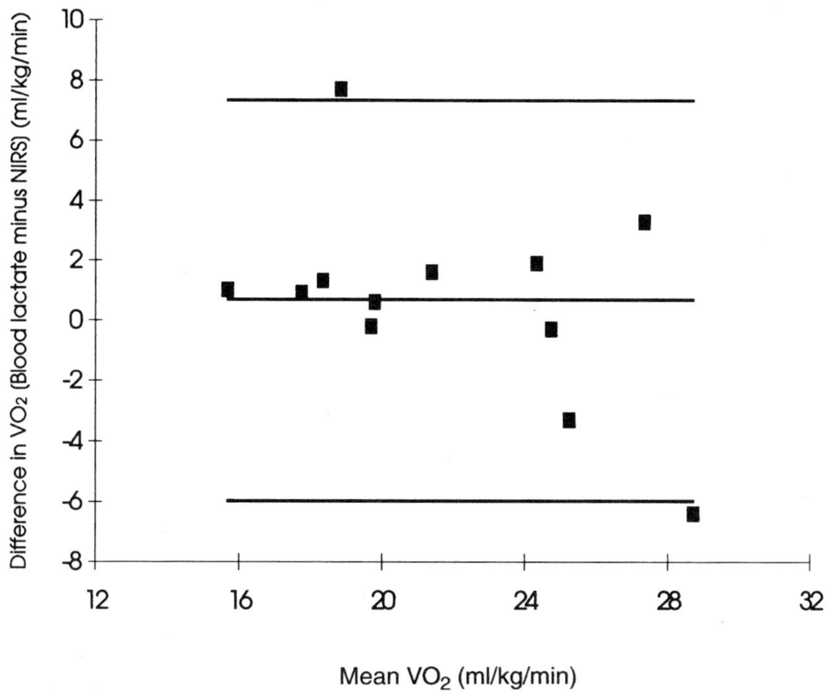

Figure 4. Bland Altman plot comparing blood lactate and NIRS (deoxygenation slope change) measurements of LT. The difference between the two methods is plotted against the average of them for each subject. The mean difference and the ±1.96 SD lines are illustrated.

4. DISCUSSION

This study has shown that in most subjects a predictable pattern occurs in the changes of muscle oxygenation and blood volume during an incremental exercise test in adults. In this our work broadly agrees with other studies (Bhambhani et al., 1997; Belardinelli et al., 1995). Despite using similar incremental cycling protocols, these workers developed different criteria for measuring the LT threshold by NIRS. On the basis of the results from this study, it appears that the point where muscle oxygenation crosses the baseline correlates very poorly with the lactate threshold. We have no immediate explanation for the differences between our studies and those of Bhambhani et al. (1997). Although they used an earlier version of the dual wavelength spectrometer described in this study, as the measuring technique is similar this should not have caused the large difference between the studies. We are currently using a multiwavelength NIR spectrometer to measure absolute oxy- and deoxyhaemo(myo)globin concentration changes in an attempt to resolve this discrepancy. However, we would caution against the use of a method that has no physiological basis i.e. *a priori* there is no reason why when the muscle oxygenation returns to the baseline level there should be a rise in blood lactate.

However, on the basis of the results in this study, the deoxygenation rate change may offer a practical non-invasive method for estimating the lactate threshold of adult athletes in the field, as it is easy, portable and relatively cheap test to administer. However, at this stage it should not be looked upon as a replacement method for determining the

LT as direct blood sampling and indirect gas exchange methods appear to be more generally applicable than this NIRS method (Wasserman et al., 1983). For example in this study we were not able to identify a slope change in all the subjects. The greater than 10% failure rate calls into question the use of NIRS as a general predictor of the LT. Indeed Bhambhani et al. (1997) mention that the deoxygenation rate change occurred in only 50% of their subjects studied. An analysis of the Bland-Altman plot also reveals that, even in those subjects where a deoxygenation rate change could be measured, there were several where the agreement with the blood lactate measurement was poor. These outliers suggest that the method, whilst useful in many individuals, may not be applicable to all subjects.

If the increased rate of deoxygenation corresponds to the LT, what is the physiological basis underpinning this? One intriguing possibility is that this is the point where the blood pH falls and the Bohr effect causes the further release of oxygen from haemoglobin. If true, this would mean that there was a physiological mechanism underpinning the correlation between the NIRS and the invasive blood lactate measurements of the anaerobic threshold, giving one confidence that the method might be applicable to a wide range of individuals of differing fitness levels.

To conclude, the results of this study show that in healthy adults the NIRS slope change method could in certain individuals be used to detect the lactate threshold, which corroborates the findings of Belardinelli et al. (1995). However, arbitrary points in the deoxygenation profile such as when the muscle oxygenation level returns to zero as proposed by Bhambhani et al. (1997) do not appear to be effective indicators of lactate threshold. However, it still remains to be seen whether NIRS will prove to be a more effective measurement of the lactate threshold than other promising non-invasive techniques (e.g. peak heart rate, EMG-activity or salivary alpha-amylase).

REFERENCES

Beaver WL, Wasserman K, and Whipp BJ (1986) A new method for detecting anaerobic threshold by gas exchange. J. Appl. Phys. 60:2020–2027.

Belardinelli R, Barstow TJ, Porszasz J, and Wasserman K (1995) Changes in skeletal muscle oxygenation during incremental exercise measured with near infrared spectroscopy. Eur. J. Appl. Phys. 70:487–492.

Bhambhani YN, Buckley SM, and Susaki T (1997) Detection of ventilatory threshold using near infrared spectroscopy in men and women. Med. Sci. Sports Exerc. 29:402–409.

Bland JM and Altman DG (1986) Statistical methods for assessing agreement between two methods of clinical measurement. Lancet 8:301–311.

Chance B, Dait MT, Zang C, Hamaoka T, and Hagerman T (1992) Recovery from exercise-induced desaturation in the quadriceps muscles of elite competitive rowers. Am. J. Physiol. 262:C766–C775.

Ferrari M, Binzoni T, and Quaresima V (1997) Oxidative metabolism in muscle. Phil. Trans. R. Soc. Lon. B 352:677–683.

Rundell KW, Nioka S, and Chance B (1997) Hemoglobin/myoglobin desaturation during speed skating. Med. Sci. Sports Exerc. 29:248–258.

Wasserman K, Whipp BJ, Koyal SN, and Beaver WL (1983) Anaerobic threshold and respiratory gas exchange during exercise. J. Appl. Phys. 35:236–243.

Weltman A (1995) The blood lactate response to exercise. (Current issues in exercise science: monograph no. 4) Human Kinetics, Champaign, Il, USA.

Wilson JR, Mancini DM, McCully K, Ferraro N, Lanoce V, and Chance B (1989) Noninvasive detection of skeletal muscle underperfusion with near-infrared spectroscopy in patients with heart failure. Circulation 80:1668–1674.

35

MYOGLOBIN AND O_2 CONSUMPTION IN EXERCISING HUMAN GASTROCNEMIUS MUSCLE

Thomas Jue,[1] Tuan Khanh Tran,[2] Paul Mole,[1] Youngran Chung,[1] Napapon Sailasuta,[2] Ralph Hurd,[2] Ulrike Kreutzer,[1] and Shinya Kuno[3]

[1]Department of Biological Chemistry
UC Davis, Davis, California 95616-8635, USA
[2]GE Medical Systems, Inc.
47697 Westinghouse Dr., Fremont, California
[3]Tsukuba University
Center for Tsukuba Advance Research Alliance (TARA)
Institute of Health and Sports Medicine
Tsukuba-Shi 305, Japan

1. INTRODUCTION

The regulation of oxygen transport to the mitochondria in exercising muscle is a key issue in biology, since oxygen demand can increase dramatically from the resting state. As the oxygen consumption (VO_2) increases, a coordinate set of controls must enhance the oxygen delivery from the lung to the cell. As VO_2 approaches its maximum rate (VO_{2max}), convection, diffusion or enzymatic activity must also become limiting. Identifying the rate- determining step is then a central issue (Sutton, 1992).

In particular many researchers have postulated a limiting step in O_2 transport from the capillary to the mitochondria (Hoppeler and Lindstedt, 1985; Stainsby, 1989; Wagner, 1995; Roca et al., 1989; Hogan et al., 1993). However, assessing the O_2 gradient from the capillary to the cell during exercise has posed many technical obstacles. Although near infra red spectroscopy (NIRS) methods can measure the oxygenated and deoxygenated states of Mb and Hb, the discrimination between Mb and Hb is not always clear (Jobsis, 1977). Some recent experiments suggest that the NIRS signals originate predominantly from Hb and reflect then only the vascular pO_2 (Wilson et al., 1989; Seiyama et al., 1988).

^1H NMR has presented an alternative approach to follow tissue oxygenation. The cellular pO_2 is reflected in the Val E11 signal of oxy Mb and the proximal histidyl $N_\delta H$

Oxygen Transport to Tissue XXI, edited by Eke and Delpy.
Kluwer Academic / Plenum Publishers, New York, 1999.

signal of deoxy Mb (Kreutzer et al., 1992; Kreutzer and Jue, 1991). In blood perfused tissue, the corresponding signals from Hb should also appear in the same spectral region (Kreutzer and Jue, 1991). Yet so far, studies have not detected any Hb signal (Richardson et al., 1995; Wang et al., 1992). Some researchers have now asserted that the erythrocyte Hb signal is perhaps NMR invisible, which is at odds with many other studies (Ho and Russu, 1981; Fetler et al., 1995; Wang et al., 1997; Lindstrom and Koenig, 1974).

Nevertheless if the NIRS and NMR techniques can discriminate the vascular and intracellular pO_2, then a unique research tool will emerge to help investigate VO_2 regulation in exercising skeletal muscle. We have undertaken a study to examine the NMR deoxy Mb signal in human skeletal muscle and to form a relationship between the NMR and the NIRS data. When blood flow to the human gastrocnemius muscle is reduced, the NMR spectra exhibit clearly the deoxy Mb signal at 78 ppm, which gradually rises with O_2 desaturation. Upfield at 73 ppm is a signal assigned to the β-deoxy-Hb His F8 $N_\delta H$ (Ho and Russu, 1981; Fetler et al., 1995; Wang et al., 1997). The kinetics of Mb and Hb desaturation are different, but the Mb and NIRS kinetics match closely. Even after Mb has attained a steady state level, the high energy phosphate levels, as reflected in the ^{31}P spectra, still remain unperturbed.

During dynamic exercise, as work increases, so also does the deoxy Mb signal intensity. A linear relationship is formed between VO_2 and deoxy Mb signal intensity. Concomitantly, the PCr level drops, while Pi concentration rises. The experimental results indicate that NMR can distinguish the proximal histidyl $N_\delta H$ signals of deoxy Mb/Hb, that the NIRS signals in these experiments are monitoring Mb, instead of Hb, desaturation, and that Mb desaturates progressively as work intensity increases. The study sets then a framework for the continuing study of VO_2 regulation in skeletal muscle.

2. METHODS

2.1. NMR

NMR measurements were performed on a 1 m bore diameter GE Signa scanner at 1.5 T. ^1H (63.86 MHz) NMR signal acquisition utilized a body coil transmit/surface coil (5 inch diameter) receiver configuration. Magnetic field shimming used a three-point Dixon method to improve the field homogeneity, yielding a water linewidth about 40 Hz (Schneider and Glover, 1991). A selective excitation pulse sequence was optimized to excite the deoxy-Mb His-F8 $N_\delta H$ signal, as confirmed by numerical simulation. Each data block comprised of 200 transients or 45 s signal averaging time.

^{31}P (25.85 MHz) signal acquisition utilized a conforming flexible coil, which wrapped around the leg. A 50 mm slice was selected and then excited with a self-refocused 45° radio frequency pulse. Each ^{31}P NMR spectrum consisted of 50 transients and required a total acquisition time of 140 s.

2.2. Near Infra Red Spectroscopy (NIRS)

NIRS measurements were made using a continuous-light source dual-wavelength spectrometer with a pair of colored LED as light sources and a photo diode detector (HEO100, Omron, Japan) (46). This system used two wavelengths on either side of the oxy/deoxy-Hb/Mb isobestic point. All wavelengths aside from 760 and 850 nm were filtered out. The probe was wrapped around the leg muscle with a Velcro strap and was engineered for a photon depth penetration of 2–3 cm. Because oxygen ligated heme groups have a greater absorbance at 850 nm than at 760 nm, while the corresponding deoxy heme groups have greater absorbance at 760 nm than at 850, the difference signal between 760 and 850 nm can reflect the changes in HbO_2/MbO_2 saturation on a relative scale (Mancini et al., 1994; Seiyama et al., 1988; Jobsis, 1977). The relative scale was calibrated against the difference signal observed when the muscle was at rest and at steady state after 15 min thigh occlusion at 240 mm Hg, which corresponds to 0% and 100% Mb/Hb deoxygenation, respectively.

2.3. Exercise and Cuffing Protocol

Healthy male volunteers (weight 120–150 lbs; age 20–27) were placed supine inside the magnet of a General Electric Signa 1.5 T scanner. The subject's calf muscle was positioned on top of a 5-inch diameter ^{1}H receiving coil and strapped down with Velcro. A ^{31}P conforming flexible coil was placed around the calf and the ^{1}H receiving coil. A pressure cuff connected to a manual air pump was wrapped around the subject's thigh just above the knee. Within a few strokes the cuff inflation reached the final pressure of 240 mm Hg.

The ergometer used for plantar flexion was constructed of wood. It consisted of a three-sided box with dimensions of 25.4 cm (W) × 25.4 cm (H) × 91.4 cm (L) with a foot pedal on an axle at one end and a moveable backplate at the other end of the box. Rubber tubing (1.3 cm dia. × 34.3 cm L) with a Hooke's constant of 31.12 N/cm was attached to the back plate and the axle of the foot pedal. Resistance to plantar flexion was adjusted by varying the number of tubes and/or by stretching the tubes to increase the distance between the axle and back plate. Mechanical work of plantar flexion involved moving the pedal against a specified resistance through an arc of 3.8 cm. The power output can be incremented by stops for forward and reverse movements with plantar flexion and relaxation. In this study, power was incremented by varying the contraction frequency from 30 to 70 contractions/min (0.5–1.17 Hz) with the resistance and plantar flexion arc held constant.

2.4. Steady State and Peak VO_2 for Plantar Flexion Exercise

Each subject performed 3 to 5 intensities of plantar flexion exercise of the dominant leg, each for 3 minutes with a 5 minute recovery period after each bout. Energy expenditure was determined continuously throughout the test by indirect calorimetry. The following sequence was followed. After resting for 10 minutes in a supine position, the subject breathed for 5 minutes through a mouthpiece and tubing connected to a SensorMedics metabolic cart (Model 2900) for breath-by-breath measurements of resting oxygen consumption (VO_2) and carbon dioxide production (VCO_2). Then the subject performed a series of 3 to 5 exercise bouts at progressively higher intensities by varying the frequency from 30 to 70 contractions/min (0.5 to 1.17 Hz) on the foot ergometer, with the resistance held constant. Each bout lasted 3 minutes and was followed by 5 minutes of resting recovery. The calculated mean value during the last 30 seconds of each bout was used to characterize the VO_2, VCO_2 and respiratory exchange ratio (RER) for the bout.

Following a rest period of 10 to 15 minutes, the subject's peak VO_2 was determined by holding the resistance constant and progressively increasing the contraction frequency each minute until the subject could no longer maintain the required cadence. VO_2 and

VCO_2 were determined throughout the test as described above. VO_2 was averaged over each 15 s interval. The highest VO_2 was designated as the subject's peak VO_2.

2.5. Statistical Analysis

All data are reported as standard error of the mean, ± SEM. Statistical significance in the Student's t test is ascribed for $P < 0.05$.

3. RESULTS AND DISCUSSION

Figure 1 shows the time course of the ^1H NMR spectra upon exercise and cuff pressure to reduce blood flow to the gastrocnemius muscle. During plantar flexion exercise of the gastrocnemius muscle at approximately 1 Hz, the deoxy Mb signal at 78 ppm emerges. The reference spectrum before exercise shows no signal between 50–100 ppm (Figure 1a). With exercise the Mb signal emerges at 78 ppm (Figure 1b–1h). Its signal

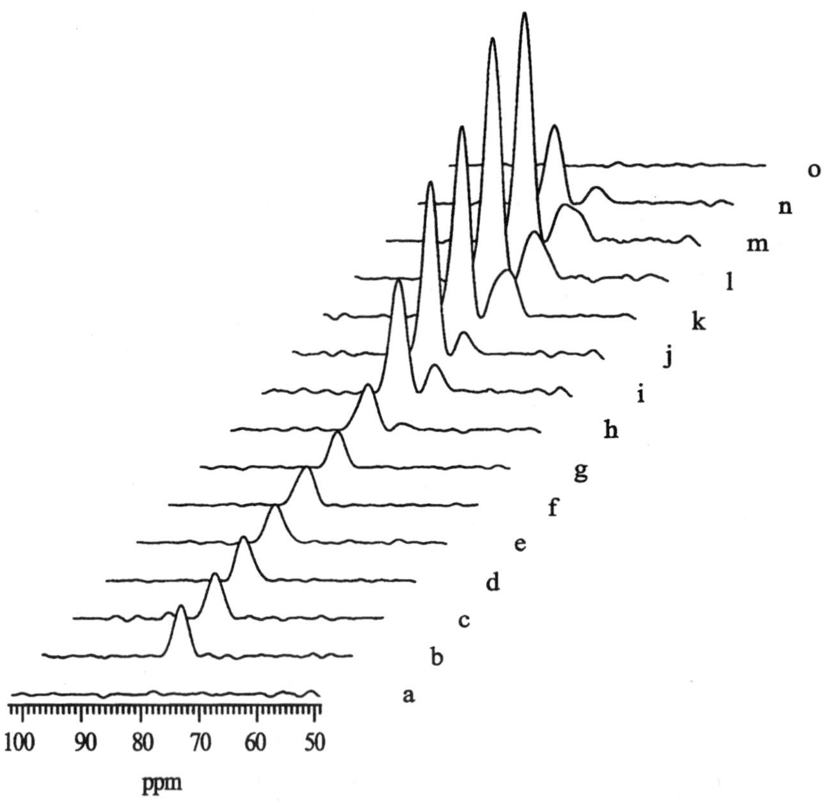

Figure 1. ^1H NMR spectra from human gastrocnemius muscle during exercise and subsequent cuff occlusion: The specific spectra correspond to a) control, b–h) plantar flexion exercise at 1 Hz, i–m) pressure cuffing at 180 mm Hg, and n–o) postexercise. During exercise, the His F8 $N_{\delta}H$ signals of both deoxy-Mb and deoxy-Hb are visible and rapidly reach a steady state. Upon pressure cuffing the leg at 240 mm Hg after the exercise, the deoxy-Mb and deoxy Hb signal intensities rise sharply at 78 and 73 ppm, respectively. These signals disappear after pressure cuff release.

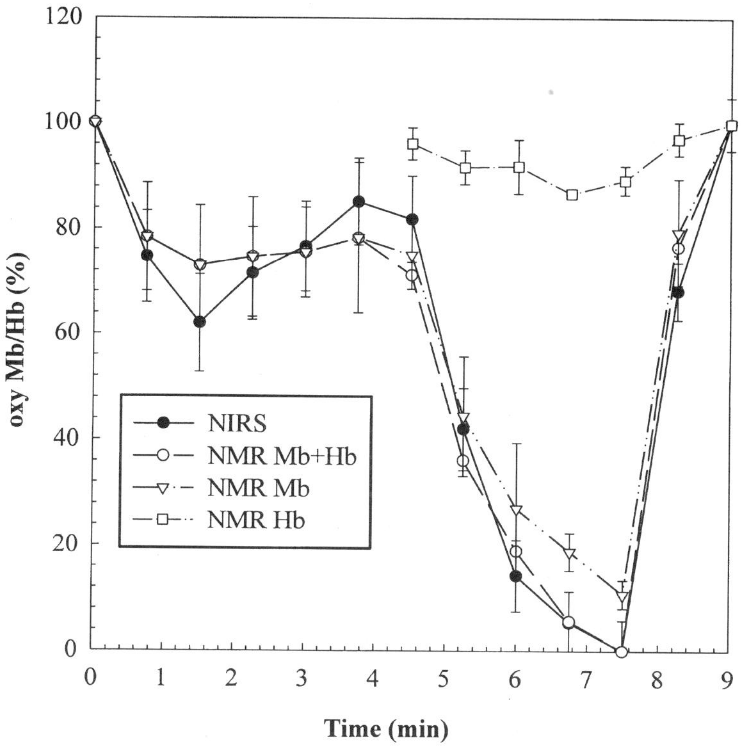

Figure 2. Graph of the relative NMR and NIRS signal intensity change as a function of exercise and pressure cuffing: With exercise the Mb oxygen saturation (▽) decreases to a steady state within 90s. The Hb signal (□) is below detectable limits. Immediately after exercise, pressure cuffing produces a signal of deoxy Hb and an increase in the deoxy Mb signal intensity. The Hb signal is visible and reaches a steady state level much faster than the Mb. The Hb signal represents a small contribution to the overall Mb/Hb signal intensity (○). The corresponding the NIRS data (●) match closely the observed Mb curve. Once the cuff pressure is released, both the deoxy Mb and deoxy Hb signals disappear rapidly.

intensity rises rapidly to a steady state level with exercise but increases even further at the onset of pressure cuff inflation (Figure 1i–m). The maximal Mb signal intensity (Figure 4m) is 4 times greater than the initial signal observed in Figure 1b. A signal assigned to the β-deoxy-Hb signal appears at 73 ppm after pressure cuffing the leg. Upon reperfusion both signals disappear within 90s (Figure 1n–o). The corresponding ^{31}P spectra reveal that PCr drops only during the exercise phase.

Figure 2 plots the deoxy Mb and Hb signals and the NIRS signal as a function of experiment time. During exercise the deoxy Mb signal is detectable but not the corresponding deoxy Hb signal. Within 45s of exercise at 1 Hz, the signal of deoxy Mb is visible and rapidly reaches a steady state level. The deoxy Hb signal intensity is not distinguishable during the exercise phase. Immediately after the exercise, the pressure cuffing induces a signal intensity increase in the deoxy Mb signal. The Hb signal now becomes evident. However, HbO_2 desaturates and reaches a steady state more rapidly than Mb. The respective NIRS and Mb rate constants during pressure cuffing are identically $0.34 \pm 0.020\,min^{-1}$. Moreover, the Hb contribution to the overall signal is insignificant.

During the exercise protocol, whole body oxygen consumption increases linearly with plantar flexion power attaining a peak value of 38 ± 6 ml O_2/min/100 g. At 70 repetition per min (rpm) the power output is about 15.0 watts and can be maintained for about 3 minutes. Mb desaturation increases in proportion to rpm increments and O_2 consumption attaining 48 ± 5% desaturation at the highest exercise intensity. As VO_2 increases, the PCr level falls, reaching 32 ± 2% of its resting value at the most intense exercise level. Muscle ATP concentration is unperturbed.

These findings, taken together, suggest that MbO_2 saturation decreases in response to respiration and energy needs and appears to be in contrast to the earlier finding (Richardson et al., 1995). The decreasing Mb saturation as a function of increasing work output suggests that the cellular pO_2 can modulate O_2 conductance from Hb to mitochondria and that the NIRS measurement is monitoring Mb desaturation.

REFERENCES

Fetler BK, Simplaceanu V, and Ho C (1995) 1H NMR investigation of the oxygenation of hemoglobin in intact human red blood cells. Biophys J 68:681–693.

Ho C and Russu I (1981) Proton nuclear magnetic resonance investigation of hemoglobins. In: Methods in Enzymology. Antonini E, Rossi-Bernardi L, and Chiancone E, eds. Academic Press, New York. pp. 275–312.

Hogan MC, Bebout DE, and Wagner PD (1993) Effect of blood flow reduction on maximal O2 uptake in canine gastrocnemius muscle in situ. J Appl Physiol 74:1742–1747.

Hoppeler H and Lindstedt SL (1985) Malleability of skeletal muscle in overcoming limitations: structural elements. J Exp Biol 115:355–364.

Jobsis FF (1977) Noninvasive, infrared monitoring of cerebral and myocardial oxygen sufficiency and circulatory parameters. Science 198:1264–1267.

Kreutzer U and Jue T (1991) 1H nuclear magnetic resonance deoxymyoglobin signal as indicator of intracellular oxygenation in myocardium. Am J Physiol 30:H2091–H2097.

Kreutzer U, Wang DS, and Jue T (1992) Observing the ^{1}H NMR signal of the myoglobin Val-E11 in myocardium: An index of cellular oxygenation. Proc Natl Acad Sci USA 89:4731–4733.

Lindstrom TR and Koenig SH (1974) Magnetic field dependent water proton spin lattice relaxatin rates of hemoglobin solutions and whole blood. J Magn Reson 15:344–353.

Mancini DM, Wilson JR, Bolinger L, Li H, Kendrick K, Chance B, and Leigh JS (1994) *In Vivo* Magnetic Resonance Spectroscopy Measurement of Deoxymyoglobin During Exercise in Patients With Heart Failure—Demonstration of Abnormal Muscle Metabolism Despite Adequate Oxygenation. Circulation 90:500–508.

Richardson RS, Noyszewski EA, Kendrick KF, Leigh JS, and Wagner PD (1995) Myoglobin O_2 Desaturation during Exercise. Evidence of Limited O_2 Transport. J Clin Invest 96:1916–1926.

Roca J, Hogan MC, Story D, Bebout DE, Haab P, Gonzalez R, Ueno O, and Wagner PD (1989) Evidence for tissue diffusion limitation of VO_2 max in normal humans. J Appl Physiol 67:291–299.

Schneider E and Glover GH (1991) Rapid *in vivo* proton shimming. Magn Reson Med 18:335–347.

Seiyama A, Hazeki O, and Tamura M (1988) Noninvasive quantitative analysis of blood oxygenation in the rat skeletal muscle. J Biochem 103:419–424.

Stainsby WN (1989) Oxidation/reduction state of cytochrome oxidase during repetitive contractions. J Appl Physiol 67:2158–2162.

Sutton JR (1992) VO_{2max}—new concepts on an old theme. Med Sci Sports Exerc 24:26–29.

Wagner PD (1995) Muscle O_2 transport and O_2 dependent control of metabolism. Med Sci Sports Exerc 27:47–53.

Wang D, Kreutzer U, Chung Y, and Jue T (1997) Myoglobin and hemoglobin rotational diffusion in the cell. Biophys J 73:2764–2770.

Wang Z, Wang DJ, Noyszewski EA, Bogdan AR, Haselgrove JC, Reddy R, Zimmerman RA, and Leigh JS (1992) Sensitivity of in vivo MRS of the N-delta proton in proximal histidine of deoxy myoglobin. Magn Reson Med 27:362–367.

Wilson JR, Mancini DM, McCully K, Feraro N, Lanoce V, and Chance B (1989) Noninvasive detection of skeletal muscle underperfusion with near-infrared spectroscopy in patients with heart failure. Circulation 80:1668–1674.

OXYGEN TRANSPORT TO MUSCULAR TISSUE UNDER ADAPTATION TO HYPOXIC HYPOXIA

I. N. Mankovskaya[1] and K. G. Lyabakh[2]

[1]Department of Hypoxia
Bogomolets Institute of Physiology
Kiev 252024, Ukraine
[2]Department of Information Technology in Medicine and Biology
Institute of Cybernetics
Kiev 252022, Ukraine

1. INTRODUCTION

The distribution of pO_2 values in mammalian skeletal muscle under adaptation to moderate altitude represents an area of research that has received little attention up till now. However, there are many data regarding the change in individual determinants of muscle pO_2 such as intercapillary distance, muscle fiber size, myoglobin concentration, volume of mitochondria, and oxidative enzymes activity as a result of exposure to prolonged hypoxic hypoxia (Reynafarje et al., 1966; Banchero, 1982; Cerretelli et al., 1984; Longmuir and Betts, 1987; Ward et al., 1989; Bae et al., 1998). These changes are consistent with the assumption that oxygen transport to muscle cells is improving in the course of adaptation to moderate altitude (Ward et al., 1989). To verify this assumption, we analyzed the distribution of pO_2 values in gastrocnemius muscle of rats, which were studied in Terskol near Mt Elbrus at 3100 m above sea level. An analysis of pO_2 distribution in skeletal muscle was performed using an original approach to mathematical modeling of O_2 transport separately in "white" and "red" regions of the muscle (Lyabakh and Shoshenko, 1991). As input values for this model served parameters of oxygen transport by blood, muscle blood flow, capillary bed structure, muscle fiber composition, and oxygen consumption rate. Also investigated were some factors which were found to influence muscle pO_2 but were unavailable for our mathematical model—the blood-tissue barrier thickness and mitochondria location in myocytes (Honig et al., 1984). Such study is of value for better understanding of regulatory mechanisms of pO_2 distribution in muscular tissue.

Oxygen Transport to Tissue XXI, edited by Eke and Delpy.
Kluwer Academic / Plenum Publishers, New York, 1999.

2. METHODS

Experiments were carried out on 60 adult male Wistar rats weighing 180–220 g in Kiev at sea level (control) and in North Caucasus, close to Mt Elbrus, at 3100 m. A comparison was made between the effects of one-week adaptation and the effects of 4-month adaptation to this altitude. The animals were anesthetized with intraperitoneal chloralose (0.05 g per kg body weight) augmented with urethane (0.5 g per kg body weight). Blood flow F and O_2 consumption (VO_2) were measured in normally perfused resting gastrocnemius muscles (Honig et al., 1971). An O_2 analyzer (Radiometer, Denmark) was used to determine pO_{2a} and pO_{2v}. The O_2 saturation (SO_2) of arterial and venous blood was measured spectrophotometrically (Philippov et al., 1984). The O_2 content (CO_2) in blood was estimated from SO_2 and oxygen blood capacity (Cmax O_2). For measurement of fiber and capillary structural parameters, serial frozen sections 12 to 16 μm were cut across the external layer of lateral gastrocnemius and the mid-belly of medial gastrocnemius. They were stained for (a) myoglobin to identify the type of muscle fibers (James, 1971), and (b) alkaline phosphatase to identify capillaries (Romanul and Bannister, 1962). To evaluate the number of open capillaries, india ink perfusion of gastrocnemius muscle was performed. Complete details of this procedure were described by (Gray and Renkin, 1978). In muscles perfused with india ink, a third section adjacent to the alkaline phosphatase section was counterstained with eosin as a background for counting filled capillaries. Photomicrographs of matched fields from these three serial sections were made at 100× total magnification (field areas 1.0 mm^2). Capillary length was measured using the longitudinal sections cut parallel to the muscle fibers (Plyley et al., 1976). Using the histochemical reaction for myoglobin, three fiber types were identified: large, uniformly light cells (white), the darkest small fibers (red) and a heterogeneous group of fibers with a wide range of darkening (intermediate). Muscle fibers of each type were counted in two field areas at 100× (0.8 to 1.0 mm^2). The fields contained 100 to 400 fibers. Myoglobin content in the skeletal muscle was estimated biochemically by the method of (Reynafarje, 1963). Electron microscopic investigations were performed using an electron microscope JEM-100CX (Japan). The thickness of the blood-tissue barrier (BTB) in rat gastrocnemius muscle was measured with a morphometric set MOR-AMO-3 (Austria).

All experimental results are reported as the mean ± S.E.M. with significance determined by the paired and unpaired Student's t-test; p values of <0.05 were considered significant for difference.

The pO_2 distribution in gastrocnemius muscle has been investigated by means of the mathematical computer model of O_2 diffusion-reaction in skeletal muscle (Lyabakh and Shoshenko, 1991). The model describes blood flow in capillaries and three-dimensional oxygen diffusion-reaction in microcirculatory unit MCU. Processes were considered on the assumption of the second-order kinetics of O_2 consumption. pO_2 and VO_2 distributions in MCU for steady state were computed. Calculations were based on the finite-difference method. To calculate the pO_2 histogram in gastrocnemius muscle, we proceeded from the following assumptions. Muscle tissue can be represented as a composition of red, intermediate and white fibers. Red and intermediate fibers are characterized by a high oxidative metabolic rate and intensive F. White fibers, in contrast, have a low oxidative metabolic rate and a slow F. Transcapillary O_2 exchange plays a key role in oxygen transport to red and intermediate fibers. It has been proposed that in white fibers arterioles may also supply neighbouring tissue with oxygen by direct diffusion—extracapillary, or transarteriolar O_2 exchange (Ye et al., 1995). That is why oxygen transport to different fiber areas was calculated by different means. pO_2 histograms in red and intermediate fibers were calculated on the usual basis of the experimental parameters of oxygen demand qO_2 and O_2 delivery (pO_2, SO_2, pH, $CmaxO_2$, $(a\text{-}v)O_2$, VO_2, parameters of capillary bed, F). pO_2 histograms in white fibers were calculated by applying the new approach. In the first place, we calculated pO_2 histogram in tissue fields supplied by transarteriolar oxygen flow as described before. Simultaneously, pO2 of arteriolar effluent blood was calculated in order to put it in the model and to compute pO_2 histogram in tissue supplied by transcapillary O_2 flux. We supposed that the amounts of tissue supplied by transarteriolar and transcapillary O_2 fluxes are in the ratio 1 : 1. This ratio and proportional fiber composition of the entire muscle permit to obtain the summarized pO_2 histograms in mixed gastrocnemius muscle (Lyabakh, this volume).

3. RESULTS

At the early stage of adaptation to hypoxia there was a distinct decrease in the parameters of oxygen transport by blood (pO_2, SO_2) as compared to control. At the same time, Hb concentration and hematocrit value were found to be higher than in control (Table 1). It was found that after a 4-month versus a 1-week adaptation period there was an increase in hemoglobin level and O_2 blood capacity. The hematocrit value also rose. There was also an increase in pO_2 of the arterial and venous blood and in regional (muscle) arterio-venous oxygen difference $(a\text{-}v)O_2$ (Table1). The shifts of arterial and venous pH were insignificant. However, a 1-week stay at the altitude resulted in the shift of arterial pH to the side of alkalosis compared to control. Muscle blood flow, F, decreased after 1-week stay on mountains compared to control. Even after 4-month stay at the altitude VO_2 and F did not reach the control value. However, after 4-month versus 1-week adaptation there was an increase in VO_2.

We have studied fiber composition and capillary/fiber ratio in the external layer and in the middle of the belly of lateral and medial gastrocnemii, respectively. They repre-

Table 1. Oxygen transport parameters in the dynamics of rat adaptation to moderate altitude

Parameters	Altitude a.s.l., exposure		
	Kiev control (n = 12)	3100 m 7 days (n = 16)	3100 m 4 months (n = 16)
pO_{2a}, Torr	92.5 ± 3.2	66.1 ± 1.7*	76.5 ± 2.6*
pO_{2v}, Torr	40.0 ± 2.1	32.8 ± 0.3*	43.5 ± 1.4
C_a O_2, ml/l	155.0 ± 6.3	154.0 ± 2.2	184.0 ± 0.9*
C_v O_2, ml/l	103.0 ± 0.3	96.0 ± 5.4	113.6 ± 1.0*
Sa O_2, %	86.0 ± 2.2	78.9 ± 0.9*	84.3 ± 1.9
Sv O_2, %	57.0 ± 0.9	48.9 ± 0.3*	52.1 ± 0.9*
VO_2, ml/min/100 g	1.27 ± 0.04	0.92 ± 0.04*	1.21 ± 0.02*
F, ml/min/100 g	24.4 ± 2.1	15.8 ± 1.2*	17.4 ± 1.8*
pHa	7.40 ± 0.010	7.46 ± 0.010*	7.42 ± 0.014
pHv	7.34 ± 0.051	7.31 ± 0.010	7.35 ± 0.080
Hb, g/l	131.5 ± 1.7	145.0 ± 1.9*	166.5 ± 3.6*
(a-v) O_2, ml/l	52.2 ± 4.8	57.6 ± 4.2	70.0 ± 5.4*
Hematocrit, %	43.5 ± 0.5	48.2 ± 0.3*	55.0 ± 0.9*
C max O_2, ml/l	180.1 ± 12.2	197.0 ± 8.5	226.2 ± 14.0*

The number of animals is given in the brackets. * $P < 0.05$ vs control.

sented arbitrarily "white" and "red" muscles according to histochemical classification of individual muscle fibers of the rat (Stein and Padykula, 1962). We have demonstrated on the basis of myoglobin distribution that both medial and lateral gastrocnemii contained fibers of all three types—red, white and intermediate. The lateral gastrocnemius contained a substantial number of white fibers (82.0 ± 2.2%), the proportion of intermediate and red fibers was 15.5 ± 0.9 and 2.5 ± 0.8%, respectively. The medial gastrocnemius showed the widest range of variation of composition internally (within a mid-belly cross-section), some areas having as few as 40% of red and intermediate fibers and some as many as 82% (mean 66%). After a 4-month adaptation to moderate altitude there was an increase in myoglobin content in the medial gastrocnemius: from 2.39 ± 0.15 (control) to 3.22 ± 0.10 mg/g tissue, $p < 0.001$; in the lateral gastrocnemius the change in myoglobin concentration was insignificant. It was also found that the proportional fiber composition based on myoglobin distribution did not change in the course of adaptation and remained the same as in control. Myoglobin accumulation was observed in the intermediate fibers of the medial gastrocnemius with initially slight or moderate partial darkening.

Marked changes were observed in the structural and diffusion parameters of the capillary system of m. gastrocnemius after 4 months of adaptation to moderate altitude compared to control (Table 2). These changes included an increase in the total capillary density and in the open capillary density as well as in the muscle capillary diameters, capillary/fiber and capillary surface/fiber surface ratios. There was also a decrease in capillary length and fiber size. The latter was expressed in a decrease of fiber diameter accompanied by an increase of fiber density. We found a decrease in intercapillary distances (Table 2) that provides evidence for improved O_2 diffusion in the muscle. After a prolonged stay on mountains we registered an intense branching of capillaries at the venous end with a formation of the microloops and micro-networks, an increased amount of intercapillary anastomoses. These findings can be regarded as some adaptive arrangements aimed to equalize O_2 diffusion conditions along the capillary. All changes of capillary bed parameters were expressed as a rule more markedly in "red" muscle regions than in "white" ones.

According to our electron microscopic investigations, the thickness of the BTB in the skeletal muscle constitutes a very sensitive indicator of tissue hypoxia. It was shown that a significant increase in the thickness of the BTB is characteristic for the 1-week term of adaptation to hypoxic hypoxia (0.92 ± 0.05 vs. 0.35 ± 0.02 μm in control, $p < 0.001$). Such increase occurred mainly at the expense of the endothelium hydration and a drastic expansion of the pericapillary space. Following a 4-month adaptation period, the BTB thickness was found to be almost the same as that of control (0.33 ± 0.03 vs. 0.35 ± 0.02 μm, $p > 0.05$). We also noted that mitochondria are most frequently located near capillaries and in the subsarcolemmal space in the myocytes of the adapted rats compared to the unadapted ones.

Our estimates of pO_2 distribution in the skeletal muscle have shown that the summarized control pO_2 histogram is characterized by the fact that the lowest pO_2 values approach 4–6 Torr and the highest ones-58–60 Torr and that the mode lies within a pO_2 range of 10–20 Torr (Figure 1a).

This pO_2 histogram represents a sum of pO_2 histograms obtained by calculations of pO_2 distribution in red and intermediate (Figure 2.1) and white fibers (Figure 2.2) under normoxic conditions. To calculate the latter pO_2 histograms, F and qO_2 were assumed to be 8–10 times greater in red and intermediate than white fibers of the rat muscle: F–30 and 4 ml/min/100 g, qO_2–2 and 0.2 ml/min/100 g, respectively (Lyabakh and

Table 2. Parameters of capillary bed in rat gastrocnemius muscle under adaptation to moderate altitude

Parameters	Altitude a.s.l., exposure	
	Kiev control (n = 7)	3100 m 4 months (n = 7)
Capillary density, cap./mm^2		
Gastrocnemius lat.	900 ± 14	1152 ± 32*
Gastrocnemius. med.	1768 ± 54	2627 ± 28*
Open capillaries, cap./ mm^2		
Gastrocnemius lat.	316 ± 8	415 ± 22*
Gastrocnemius med.	800 ± 28	998 ± 30*
Capillary length, μm		
Gastrocnemius lat.	740 ± 23	698 ± 31
Gastrocnemius med.	720 ± 18	655 ± 15*
Capillary diameter, μm		
Gastrocnemius lat.	5.8 ± 0.3	6.6 ± 0.2*
Gastrocnemius med.	6.2 ± 0.1	6.1 ± 0.2
Fiber density, fiber/mm^2		
Gastrocnemius lat.	885 ± 13	927 ± 10*
Gastrocnemius med.	1040 ± 34	1160 ± 34*
Fiber diameter, μm		
Gastrocnemius lat.	52.0 ± 1.1	48.6 ± 1.0*
Gastrocnemius med.	38.2 ± 1.4	32.0 ± 1.5*
Capillary/fiber ratio		
Gastrocnemius lat.	1.01 ± 0.05	1.24 ± 0.04
Gastrocnemius med.	1.70 ± 0.09	2.26 ± 0.04*
Capillary surface/fiber surface, %**		
Gastrocnemius lat.	11.1	16.7
Gastrocnemius med.	27.2	43.0
Intercapillary distance, μm		
Gastrocnemius lat.	56.2 ± 1.2	48.4 ± 1.5*
Gastrocnemius med.	36.6 ± 0.9	30.4 ± 1.0*

Muscles were taken bilaterally, two fields were counted for each muscle.The number of fields counted 28. The number of animals is given in the brackets. * $P < 0.05$ vs control; ** (Capillary density/fiber density) × (capillary radius/fiber radius) × 100.

Shoshenko, 1991). Our calculations revealed that pO2 histogram in white fibers shifted to the left and situated mainly between 4 to 30 Torr. At the same time, pO_2 histogram in red and intermediate fibers showed the prevalence of the higher pO_2 values, and was situated mainly between 16 to 44 Torr. Thus, the higher pO_2 values in summarized pO_2 histogram depend on the contribution of "red" region oxygen regime into oxygenation regime of the whole muscle.

On the early terms of adaptation the summarized pO_2 histogram shifted to the left and situated within a range of 2–30 Torr (Figure 1b). Such shift is characteristic of both "red" and "white" muscle regions but it was more pronounced in red and intermediate fibers than in white ones compared to control pO_2 histograms (Figure 3).

On the later terms of adaptation the summarized pO_2 histogram shows a distinct shift to the right as compared to pO_2 histogram achieved at 1-week adaptation period,

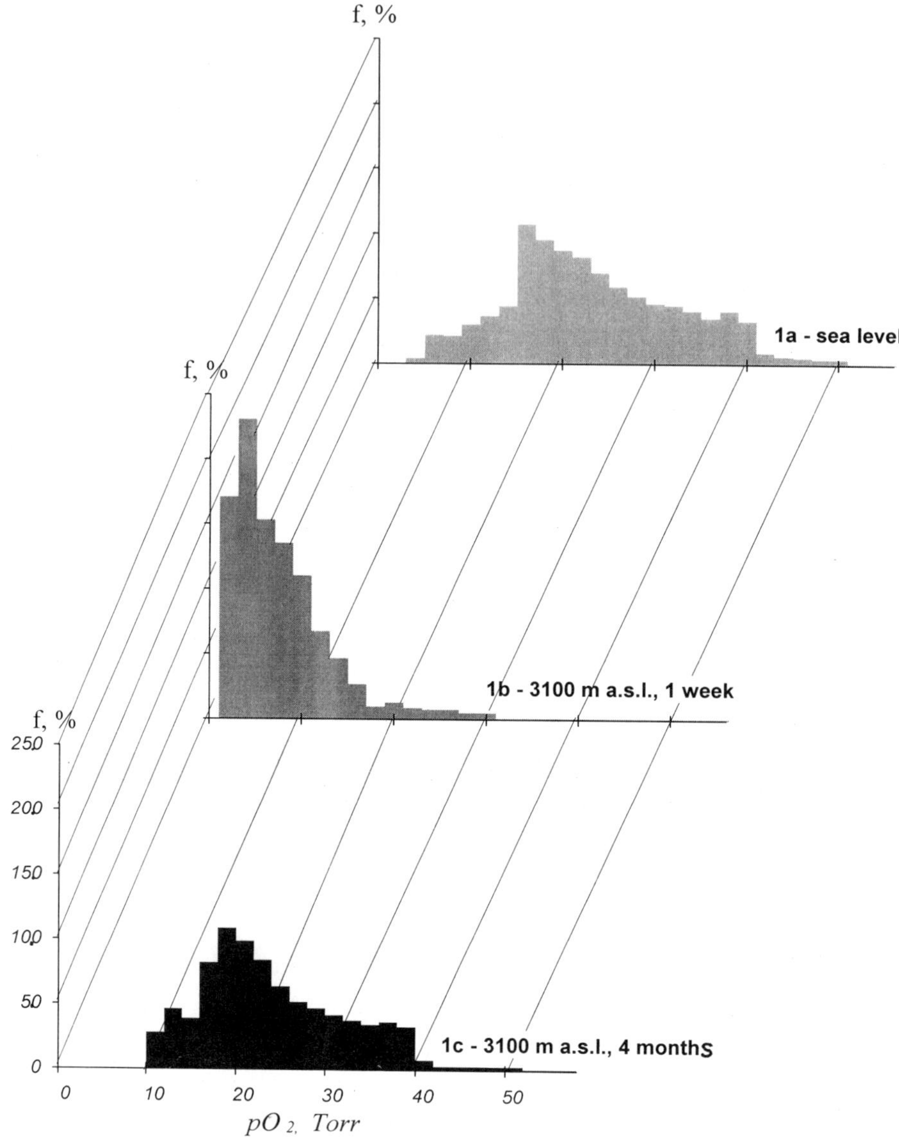

Figure 1. Predicted summarized pO_2 histograms of skeletal muscle in the dynamics of rat adaptation to moderate altitude of 3100 m.

and had a very similar configuration to the control one (Figure 1c). The components of this pO_2 histogram also tend to have the same configuration of pO_2 profile as in control (Figure 4).

We may conclude that our estimates of pO_2 distribution in the skeletal muscle have shown an improvement of oxygen transport to tissue which took place in the process of adaptation to hypoxic hypoxia. Computer calculations demonstrated that the most important factors contributing to an improvement of O_2 transport in skeletal muscle during adaptation to hypoxia are an increase in O_2 binding properties of the blood and rearrangements of the capillary network geometry.

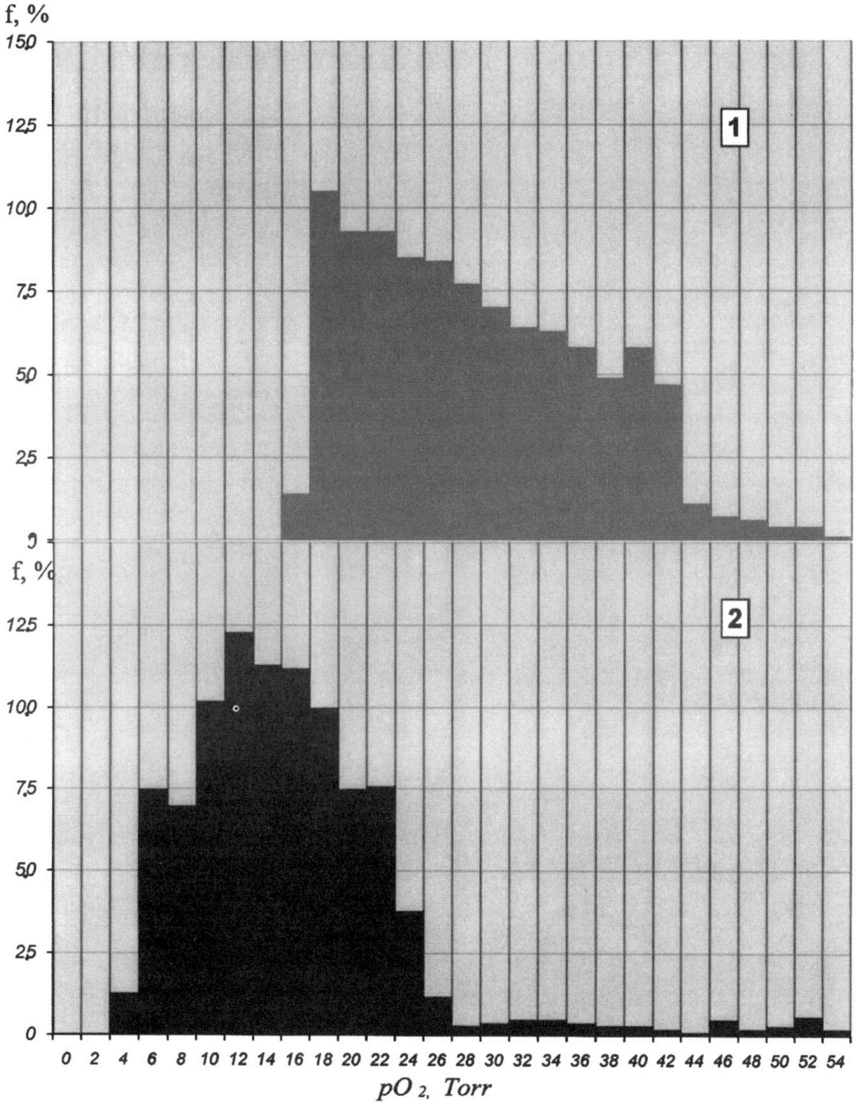

Figure 2. Predicted pO_2 histogram of red, intermediate and white muscle fibers under mormoxic conditions. 1—red and intermediate fibers, 2—white fibers.

4. DISCUSSION

It was noted that after-effects of a short-stay at moderate altitude are similar to those registered under acute hypoxic hypoxia of the same severity (simulated altitude in a barochamber or 14% O_2 in the inspired air) (Ward et al., 1989). It has been found that due to a long-term stay under reduced pO_2 conditions on mountains there was an increase in O_2 tension, O_2 saturation and O_2 content of arterial blood as well as in O_2 blood capacity, regional (muscle) arterio-venous O_2 difference and hematocrit as compared to a short-term stay. All these changes can be treated as some adaptive mechanisms aimed to increase O_2 supply to muscular tissue (Ward et al., 1989). Our findings are in line with

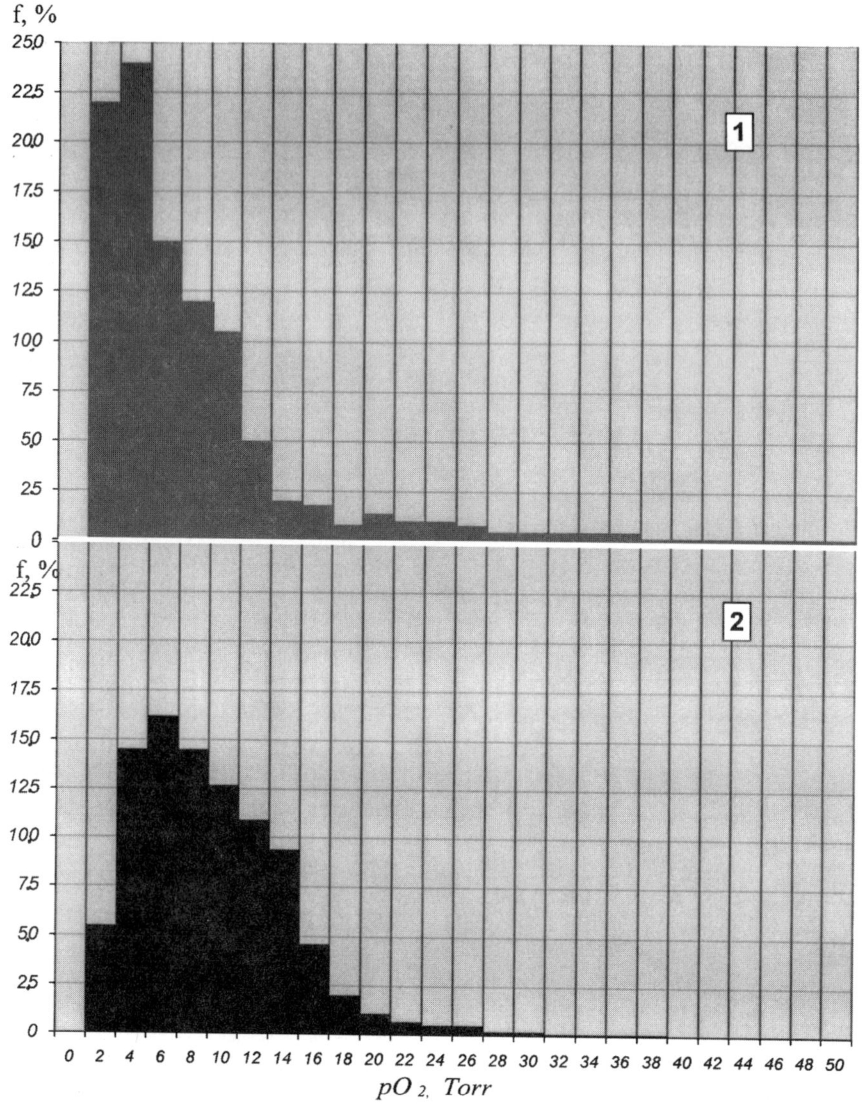

Figure 3. Predicted pO_2 histogram of red, intermediate and white muscle fibers on 1—week stay at the altitude 3100 m. 1—red and intermediate fibers, 2—white fibers.

those of many other investigators. There was usually a left shift of the oxygen dissociation curve (ODC) during chronic hypoxia due to respiratory alkalosis (Snyder et al., 1982). We found a shift of arterial pH to the side of alkalosis only after a short-term stay at the altitude of 3100 m a.s.l. It can be responsible for a shift of the ODC to the left and was accompanied by polycythemia. That is why the loading ability of blood can be enhanced, on the one hand, and blood viscosity can be increased, on the other. The latter may limit the muscle blood flow as revealed in our experiments. As adaptation terms increased, the muscle blood flow increased too but did not reach the level registered at sea level. We found a rise in F and hematocrit value after a 4-month stay on mountains compared to those registered on a short-term stay despite a rise of blood viscosity

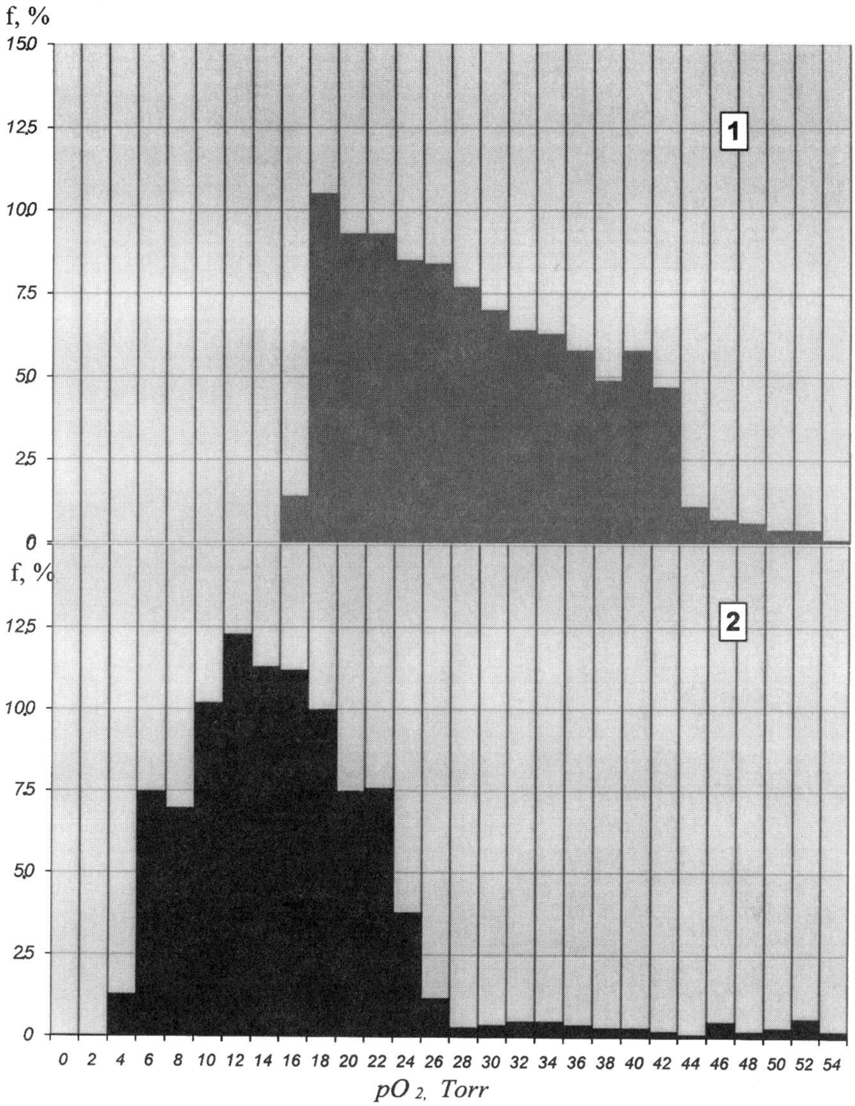

Figure 4. Predicted pO_2 histogram of red, intermediate and white muscle fibers on 4—month stay at the altiyude 3100 m. 1—red and intermediate fibers, 2—white fibers.

resulted from the increased hematocrit (Winslow et al., 1985). So, increasing blood viscosity cannot be the only reason for the limitation of F. The other reason may be the reduction of stroke volume caused by reduced cardiac filling or impaired myocardal contractility after a few weeks exposure to high altitude (Alexander et al., 1967). But, in our opinion, the most probable explanation could be the redistribution of organ blood flow with reducing muscle F in the rats after a short-term stay at moderate and high altitudes (Kasalicky et al., 1977). In this context it should be noted that the muscle diffusion pathlength may become longer than in control at the initial stage of adaptation to hypoxia. This fact was taken into consideration in our calculations of pO_2 histogram in the muscle (d = 80 mcm).

We have shown a marked increase of muscle capillarity as a result of adaptation to chronic hypoxic hypoxia. There was an increase in the number of capillaries per mm^2 of cross-sectional area in the muscle fibers. These data are consistent with the results of early studies which showed increased capillarity of the skeletal muscle in experimental animals exposed to low barometric pressures over several weeks (Tenney and Ou, 1970; Banchero, 1982). We found a reduction in size of the muscle fibers after a 4-month period of adaptation to moderate altitude. So, although capillary density rose in the muscle with exposure to middle altitude, this could not be caused by the formation of new capillaries but by atrophy of the muscle fibers (Banchero, 1982). However, our data about a decrease of mean capillary length in the muscle of adapted rats may reflect the formation of capillaries *de novo*. Established increase in capillary diameter, capillary/fiber and capillary surface/fiber surface ratios, and the number of intercapillary anastomoses also reflect intensified capillarity of the muscles and indicate that increased fraction of the surface area of each fiber is in contact with blood. The most important evidence indicating enhanced O_2 transport to muscle cells was found to be a decrease of capillary-tissue diffusion distances. Our experiments showed that adaptive changes in capillarity were more pronounced in "red" region of the muscle than in "white" one. All the above changes aimed to improve the conditions for oxygen transport from the capillary blood to the muscle cell membrane. To reach the latter, O_2 must diffuse through a gap of plasma between the red cells and the endothelium, the endothelium itself, and the extracellular space, i.e., through the BTB, or a "resistance layer". It was supposed that the resistance layer is a functional (diffusional) barrier rather than an anatomical one (Honig et al., 1984). The muscle BTB thickness was found to increase on a short-term stay at 3100 m and to reduce in the process of adaptation to this altitude. This fact is interesting in view of current concepts of tissue O_2 transport (Ivanov, 1993). As the resistance layer around the red cells acts to minimize the release of O_2, one can speculate that the extension of the muscle BTB could be an unfavourable factor for adequate O_2 delivery to the muscle cell membrane. On the contrary, a decrease of the BTB thickness in the muscle of rats on a long-term stay in the mountains may improve conditions for O_2 transport compared to those registered on a short-term stay in the mountains.

After a 4-month adaptation to moderate altitude, we registered an increase in myoglobin content in skeletal muscle playing no doubt a significant role in facilitation of O_2 diffusion from the capillary blood to mitochondria that is consistent with reported estimates (Kreuzer and Hoofd, 1984). Our data demonstrated the aggregation of mitochondria around the muscle capillaries of the adapted rats. It was shown that when mitochondria are clustered around the capillaries the Krogh cylinder radius shrinks from the anatomical value to that of the cluster (Mainwood and Rakusan, 1982; Honig et al., 1984). Thus, in our case the capillary pO_2 required for an adequate oxygen supply to the muscle of adapted animals may be somewhat lower compared to that which might have existed in the homogenous situation.

Our modeling of oxygen transport to muscle tissue properly reflected peculiarities of oxygen consumption rate, vascular bed structure and blood flow velocity in the different types of the muscle fibers. Of primary importance was modeling transarteriolar O_2 diffusion in the white fibers (Ivanov, 1993). This approach made it possible to analyze the contribution of each factor mentioned above in O_2 supply and O_2 distribution in the mixed skeletal muscle. The observation by many researchers revealed that pO_2 values, measured in *in vivo* muscle tissue, tend to be lower than those predicted by mathematical models, particularly in the capillary entrance regions. If our modeling is realistic, it explains how under normoxic conditions the calculated pO_2 histogram becomes very similar to that obtained experimentally (Harrison et al., 1984; Kessler et al., 1984). We have demonstrated that the contribution of "red" region oxygen regime into oxygenation of the whole muscle under normoxic and hypoxic conditions was more significant than the contribution of "white" region oxygen regime. Our calculations showed that pO_2 histogram in skeletal muscle under prolonged hypoxic hypoxia returned to its normal configuration despite the decreased values of pO_{2a} and SaO_2. This fact confirmed the assumption about improvement of the conditions for oxygen transport to the muscle cells in the course of adaptation to moderate altitude. The most important variables underlying this improvement were found to be the increased capillarity of red and intermediate muscle fibers and an increase in O_2 binding properties of the blood. The role of myoglobin concentration, the thickness of the BTB, and mitochondria distribution in the muscle cells is clearly important too and demonstrates the need for a better understanding of their changes in the course of adaptation to hypoxic hypoxia by means of mathematical modeling.

REFERENCES

Alexander J, Hartley L, Modelsky M, and Grover R (1967) Reduction of stroke volume during exercise in man following ascent to 3100 m altitude. J. Appl. Physiol 23:849–857.

Bae S, Kim C, Kim J, Yung Y, and Toshinai K (1998) Altitude training in juveniles—study of muscle oxygenation. In: The 3rd World Congress on Mountain Medicine and High Altitude Physiology and 18th Japanese Symposium on Mountain Medicine. Matsumoto, Japan, p. 183.

Banchero N (1982) Long-term adaptation of skeletal muscle capillarity. Physiologist 25:385–389.

Cerretelli P, Marconi C, Deriaz O, and Giesendanner D (1984) After effects of chronic hypoxia on cardiac output and muscle blood flow at rest and exercise. Eur. J. Appl. Physiol. 53:92–96.

Gray S and Renkin E (1978) Microvascular supply in relation to fiber metabolic type in mixed skeletal muscle of rabbits. Microvasc. Res. 16:406–425.

Harrison D, Hoper J, Gunther H, Vogel H, Frank K, Brunner M, Ellerman R, and Kessler M (1984) Microcirculation and PO_2 in Skeletal Muscle during respiratory Hypoxia and Stimulation. In: Oxygen Transport to Tissue-V, Adv. Exper. Med. Biol., 169 (Lubbers D, Acker H, Leniger-Follert E, and Goldstick T, Eds) Plenum Press, New York-London, pp. 477–485.

Honig C, Frierson J, and Nelson C (1971) O_2 transport and VO_2 in resting muscle: Significance for tissue-capillary exchange. Amer. J. Physiol. 220:357–363.

Honig C, Gayeski T, Federspiel W, Clark A, and Clark P (1984) Muscle O2 gradients from hemoglobin to cytochrome: new concepts, new complexities. In: Oxygen Transport to Tissue-V, Adv. Exper. Med. Biol., 169 (Lubbers D, Acker H, Leniger-Follert E, and Goldstick T, Eds) Plenum Press, New York-London, pp. 23–38.

Ivanov K (1993) The principles of energetics in organism: Theoretical and practical aspects. V.2. Biological oxidation and its providing with oxygen (Glebovsky V, Ed) Nauka, St.-Petersburg, 269 pp.

James N (1971) The histochemical demonstration of myoglobin in muscle spindles. Histochem. J. 3:333–338.

Kasalicky J, Ressl J, and Urbanova D (1977) Relative organ blood flow in rats exposed to intermittent high altitude hypoxia. Pflugers Arch. Ges. Physiol. 368:111–115.

Kessler M, Hoper J, Harrison D, Skolasinska K, Klovekorn W, Sebening F, Volkholz H, Beier J, Kernbach C, Rettig V, and Richter H (1984) Tissue O_2 supply under normal and pathological conditions. In: Oxygen Transport to Tissue-V, Adv. Exper. Med. Biol., 169 (Lubbers D, Acker H, Leniger-Follert E, and Goldstick T, Eds) Plenum Press, New York-London, pp. 69–80.

Kreuzer F and Hoofd L (1984) Facilitated Diffusion of Oxygen: Possible Significance in Blood and Muscle. In: Oxygen Transport to Tissue-V, Adv. Exper. Med. Biol., 169 (Lubbers D, Acker H, Leniger-Follert E, and Goldstick T, Eds) Plenum Press, New York-London, pp. 3–21.

Longmuir I and Betts W (1987) Tissue acclimation to altitude. Fed. Proc. 46:794.

Lyabakh K (1998) Mathematical modeling of oxygen transport in skeletal muscle during exercise: hypoxia and VO_2 max (this volume).

Lyabakh K and Shoshenko C (1991) PO_2 in rat mixed skeletal muscle with low value of O_2 diffusivity. Sechenov Physiol. J. 77:35–39.

Mainwood G and Rakusan K (1982) A model for intracellular energy transport. Can. J. Physiol. Pharmacol. 60:98–102.

Philippov M, Mankovskaya I, and Zaboluyev V (1984) The set for determining oxygemoglobin dissociation curves. Physiol J. (Ukraine) 30:754–756.

Plyley M, Sutherland G, and Groom A (1976) Geometry of the capillary network in skeletal muscle. Microvasc. Res. 11:161–173.

Reynafarje B (1963) Simplified method for the determination of myoglobin. J. Lab. Clin. Med. 61:138–145.

Reynafarje B, Narticoreva E, Guillen J, and Arrarte A (1966) Pigments and enzymes of human skeletal muscle at sea level and at high altitudes. Arch. Inst. Biol. Andina 1:170–178.

Romanul F and Bannister R (1962) Localized areas of high alkaline phosphatase activity in the terminal arterial tree. J. Cell Biol. 15:73–84.

Snyder L, Born S, and Lechner A (1982) Blood oxygen affinity in high- and low-altitude populations of the deer mouse. Respir. Physiol. 48:89–105.

Stein J and Padycula H (1962) Histochemical classification of individual skeletal muscle fiber of the rat. Amer. J. Anat. 110:103–124.

Tenney S and Ou L (1970) Physiological evidence for increased tissue capillarity in rats acclimatized to high altitude. Respir. Physiol 8:137–150.

Ward M, Milledge J, and West J (1989) High Altitude Medicine and Physiology. Chapman a. Hall Medical, London, 515 pp.

Winslow R, Monge C, Brown E, Klein H, Sarnquist F, Winslow N, and Mc Kneally S (1985) Effects of hemodilution on O_2 transport in high altitude polycythemia. J. Appl. Physiol. 59:1495–1502.

Ye G, Buerk D, and Jaron D (1995) Arteriolar contribution to microcirculatory CO_2/O_2 exchange. Microvasc. Res. 50:338–359.

INCOMPLETE OXYGEN EXTRACTION IN EXERCISING MUSCLE

Johannes Piiper

Max Planck Institute for Experimental Medicine
D-37075 Göttingen, Germany

INTRODUCTION

With increasing motor stimulation or exercise intensity of skeletal muscle its O_2 consumption and blood flow increase. As O_2 extraction increases, O_2 content and PO_2 in muscle venous outflow decrease, but never reach zero (e.g. Mercker et al., 1949).In this report, the various factors that have been considered to contribute to this "apparent inefficiency" of O_2 transfer will be analyzed in theory and on the basis of experimental data.

DIFFUSION LIMITATION

The most obvious mechanism of limited O_2 utilization in heavily exercising muscle is diffusion limitation, meaning that due to a finite resistance to O_2 diffusion, PO_2 in efferent venous blood cannot fall below a finite minimum value even when PO_2 in muscle tissue approaches zero (Mercker et al., 1949; Stainsby and Otis, 1964).

The local O_2 transfer results from tissue O_2 requirement, O_2 supply by blood flow and conditions for diffusion which for a simplifying model, are given by the effective surface area, diffusion distance and diffusivity and solubility of O_2 in the tissue (Fick's law). The anatomical substrate for blood-tissue O_2 transfer is complex and heterogeneous, comprising an irregularly branching arterial and venous vascular system ultimately leading to capillaries; the various tissue elements, of irregular geometry, offering resistance to O_2 diffusion (from red cell interior to the mitochondria); the irregularly shaped and distributed mitochondria where O_2 is consumed. Based on morpho-metry and physical material properties, the resistances to O_2 diffusion have been estimated for muscle (e.g. Groebe, 1995; Roy and Popel, 1996).

Oxygen Transport to Tissue XXI, edited by Eke and Delpy.
Kluwer Academic / Plenum Publishers, New York, 1999.

Another approach is to use physiological measurements of O_2 uptake, blood flow and blood gases for calculations on functional analog models (Piiper, 1992). In the Krogh cylinder model, a central blood capillary provides O_2, by radial diffusion, to surrounding tissue assumed homogeneous with respect to O_2 requirement and diffusivity. At maximal O_2 uptake (VO_2 max) there is good reason to assume PO_2 in the periphery of the venous end of the tissue cylinder to approach zero. Thus the PO_2 difference effective for diffusion is, all along the capillary, equal to venous effluent PO_2, PvO_2, and the diffusive O_2 conductance or O_2 diffusing capacity, DO_2, is obtained as maximum O_2 uptake (VO_2max) divided by PvO_2.

$$DO_2 = VO_2 \max/PvO_2 \tag{1}$$

A further simplification of the model assumes homogeneity of tissue with respect to PO_2, and zero tissue PO_2 at VO_2max. The PO_2 difference for diffusion decreases from the arterial PO_2 to venous PO_2. The mean effective capillary PO_2, PO_2(mean), can be obtained by an appropriate averaging procedure, e.g. Bohr integration. The O_2 diffusing capacity is

$$DO_2 = VO_2(\max)/PO_2(\text{mean}) \tag{2}$$

Wagner and associates have studied the relationship between VO_2(max), PvO_2 and PO_2(mean) in extensive studies on stimulated dog gastrocnemius (Hogan et al., 1988, 1989, 1990; Kurdak et al., 1996; Richardson et al., 1998) and exercising human quadriceps muscle (Roca et al., 1989; Richardson et al., 1993, 1995, 1998; Richardson and Saltin, 1998). It was shown that in conditions of VO_2max, changes of arterial PO_2, CO loading of hemoglobin, displacements of blood O_2 dissociation curve or changes of blood flow brought about changes in VO_2max proportional to changes in PO_2(mean) (or PvO_2) in accordance with the diffusion limitation model, i.e. DO_2 remaining close to constant. Thus at VO_2max muscle O_2 consumption could be considered as limited by diffusion (in addition to blood flow).

These simple models, describing capillary blood/muscle O_2 transfer by a single DO_2 value certainly are an oversimplification in many respects. Thus, there is now ample experimental evidence for extracapillary, in particular precapillary, O_2 transfer (Duling and Berne, 1970; Kerger et al., 1995; Intaglietta et al., 1996; Kobayashi and Takizawa, 1996). For modeling of O_2 transport, this means longitudinal unequal distribution of O_2 diffusion resistances. Furthermore, the dissociation reaction of oxyhemoglobin may exert a limiting effect. In the above-described approach by "overall" DO_2 this reaction is either assumed to be so rapid as not to be limiting, or to constitute another serial element of O_2 transfer resistance of constant magnitude i.e. to give rise to a PO_2 difference proportional to VO_2.

UNEQUAL DISTRIBUTION OF BLOOD FLOW

In a model without diffusion resistance but with unequal blood flow (with respect to O_2 requirement) the overall extraction of O_2 from blood may be considerably reduced compared to a homogeneous model, due to incomplete O_2 utilization by overperfused tissue. In such a case, a finite O_2 diffusing capacity (DO_2) would be calculated from the finite venous PO_2, whereas the real DO_2 is infinite. In a more general case of both finite diffusing capacity and unequal distribution of blood flow, DO_2 would be underestimated if perfusion heterogeneity is not taken into account (Piiper and Haab, 1991).

In all experimental studies using methods capable of detecting blood flow heterogeneity, a high degree of perfusion inhomogeneity has been found, both in resting and in stimulated or exercising skeletal muscle. The methods comprise measurement of local red cell flow, O_2 saturation, hematocrit, plasma PO_2 and other variables in microvessels (e.g. Damon and Duling, 1985; Sarelius, 1986; Duling and Damon, 1987; Ellsworth et al., 1988; Tyml, 1991; Torres Filho and Intaglietta, 1993; Ellis et al., 1994; Torres Filho et al., 1996; Zheng et al., 1996), distribution of embolizing microspheres (Cerretelli et al., 1984; Pendergast et al., 1985; Piiper et al., 1985; Marconi et al., 1988; Iversen et al., 1989, 1992; Iversen and Nicolaysen, 1990, 1991; Kurdak et al., 1996); washout of inert gases (Sparks and Mohrman, 1977; Cerretelli et al., 1984; Piiper and Meyer, 1984; Gronlund et al., 1989; Piiper and Haab, 1991) reviewed by Piiper (1990, 1992).

The unexpectedly large scatter in blood flow distribution, particularly that in stimulated muscle, was found to be difficult to reconcile with the relatively high blood O_2 utilization (Piiper et al., 1985).The hypothesis of locally oscillating blood flow distribution was not in conformity with the results of experiments in which microspheres were continuously infused during increasing time intervals, but the extent of unequal distribtuion remained unchanged (Iversen and Nicolaysen, 1990).

COMBINED AND SEPARATE EFFECTS OF DIFFUSION LIMITATION AND PERFUSION HETEROGENEITY

It is probable that both diffusion limitation and perfusion heterogeneity affect O_2 availability in exercising muscle, but it is difficult to quantitatively assess the role of each mechanism.

The situation is comparable to that in lungs where the factors contributing to the inefficiency of alveolar gas exchange are studied. By the multiple inert gas elimination technique (MIGET) the ventilation/perfusion heterogeneity and shunt can be determined and the alveolar-arterial PO_2 difference ($AaDO_2$) resulting therefrom can be estimated (Wagner et al., 1974). The difference between the measured and the calculated $AaDO_2$ is attributed to diffusion limitation and therefrom a value for the pulmonary diffusing capacity (DLO_2) can be estimated. In man in normal conditions most part of $AaDO_2$ is found to be explainable by unequal ventilation to perfusion distribution and shunt. Thus a straightforward calculation of pulmonary DO_2 from $AaDO_2$ would lead to a marked underestimate and to an overestimation of the role of diffusion limitation in pulmonary O_2 uptake.

To use a similar approach in the case of muscle, not only the unequal distribution of perfusion but also that of O_2 uptake (VO_2) should be known. If the local unequal distribution of blood flow and VO_2 were identical, the muscle would be homogeneous with respect to O_2 supply. Simultaneous measurements of blood flow distribution (by embolizing microspheres) and of oxidative metabolism (by deoxyglucose decay) showed both to be unequally distributed and, most important, little correlated with each other (Iversen and Nicolaysen, 1990, 1991). Thus blood flow distribution only slightly reflected the inhomogeneity of oxidative metabolism.

DIFFUSION SHUNT

Diffusion shunt for O_2 results from diffusion of O_2 from arterial to adjacent venous vessels. It leads to a decreased arterial O_2 supply to tissue, by reduction of precapillary blood O_2 content, and to reduced O_2 extraction, by elevation of O_2 content in effluent venous blood. It is generally believed that a diffusion shunt could not be important in exercising muscle due to high blood flow (e.g. Roth and Wade, 1986; Piiper, 1988). But some recent experiments on perfused skin and muscle have shown large effects (Kobayashi et al., 1991; Intaglietta et al., 1996), thus its role in exercising muscle needs reconsideration.

The effect of diffusion shunt for O_2 can easily be calculated for a simplified model consisting of a capillary bed supplying O_2 to tissue, with O_2 consumption VO_2(cons), and of an adjacent arterial and venous vessel constituting a counter-current system, with diffusive arterio-venous conductance DO_2(av), blood flow, Q, with linear blood O_2 equilibrium curve, of slope β (= effective solubility). The ratio of shunted O_2 flux, VO_2(sh), and consumed O_2, VO_2(cons), is

$$\frac{VO_2(\mathrm{sh})}{VO_2(\mathrm{cons})} = \frac{DO_2(\mathrm{av})}{Q \cdot \beta} \qquad (3)$$

Thus, as expected, increasing DO_2(av) increases VO_2(sh), and increasing Q (or β) diminishes VO_2(sh) relative to O_2 consumed. In heavy exercise, not only Q is high, but due to increased O_2 utilization PvO_2 is low leading to increase of effective β, and thus the diffusion shunt of O_2 in relation to O_2 consumption should be decreased.

For blood O_2 extraction EO_2 (=arterio-venous O_2 content difference/arterial O_2 content) for VO_2max conditions (assuming tissue PO_2 zero and no blood/tissue diffusion resistance) one obtains for this model:

$$EO_2 = \frac{1}{1 + DO_2(\mathrm{av})/(Q \cdot \beta)} \qquad (4)$$

Eq.(4) shows that high DO_2(av) reduces EO_2 by arterio-venous O_2 diffusion. But a high Q (and/or β) exerts an opposite effect on diffusion shunt, increasing EO_2.

The faster washout of less diffusible inert gas in resting and stimulated dog gastrocnemius muscle suggested operation of diffusion shunt (Piiper and Meyer, 1984). Since for O_2 the effective solubility β is much higher, due to chemical bonding, less diffusion shunt is expected for O_2. Furthermore, in practice any attempts to evaluate the role of diffusion shunt from multiple inert gases as washout is complicated by blood flow inhomogeneity and by possible tissue-blood diffusion limitation.

CONCLUSION

The experimentally observed changes of VO_2max proportional to changes in mean capillary (or venous) PO_2 are in accordance with blood/tissue diffusion limitation. But the considerable perfusion heterogeneity and shunt revealed by various methods puts in question the estimation of O_2 diffusing capacity on the basis of a homogenous model. The potential role of arterio-venous O_2 shunting in limiting O_2 extraction by muscle in heavy exercise should not be disregarded. An assessment of the role of each mechanism and of their combined effects in heavily exercising muscle awaits future studies.

REFERENCES

Cerretelli P, Marconi C, Pendergast D, Meyer M, Heisler N, and Piiper J (1984) Blood flow in exercising muscles by xenon clearance and by microsphere trapping. J Appl Physiol 65:24–30.

Damon DH and Duling BR (1985) Evidence that capillary perfusion heterogeneity is not controlled in striated muscle. Am J Physiol 249 (Heart Circ Physiol 18):H386–H392.

Duling BR and Berne RM (1970) Longitudinal gradients in periarteriolar oxygen tension. A possible mechanism for the participation of oxygen in the local regulation of blood flow. Circ Res 27:669–678.

Duling BR and Damon DH (1987) An examination of the measurement of flow heterogeneity in striated muscle. Circ Res 60:1–13.

Ellis CG, Wrigley SM, and Groom AD (1994) Heterogeneity of red blood cell perfusion in capillary network supplied by a single arteriole in resting skeletal muscle. Circ Res 75:357–368.

Ellsworth ML, Popel AM, and Pittman RN (1988) Assessment and impact of heterogeneities of convective oxygen transport parameters in capillaries of striated muscle. Microvasc Res 35:341–362.

Groebe K (1995) An easy-to-use model for O_2 supply to red muscle. Validity of assumptions, sensitivity to errors in data. Biophys J 68:1246–1269.

Gronlund J, Malvin GM, and Hlastala MP (1989) Estimation of blood flow distribution in skeletal muscle from inert gas washout. J Appl Physiol 66:1942–1955.

Hogan MC, Roca J, Wagner PD, and West JB (1988) Limitation of maximal O_2 uptake and performance by acute hypoxia in dog muscle in situ. J Appl Physiol 65:815–821.

Hogan MC, Roca J, West JB, and Wagner PD (1989) Dissociation of maximal O_2 uptake from O_2 delivery in canine gastrocnemius in situ. J Appl Physiol 66:1219–1226.

Hogan MC, Bebout DE, Gray AG, Wagner PD, West FB, and Haab PE (1990) Muscle maximal O_2 uptake at constant O_2 delivery with and without CO in the blood. J Appl Physiol 69:830–836.

Intaglietta M, Johnson PC, and Winslow RM (1996) Microvascular and tissue oxygen distribution. Cardiovasc Res 32:632–643.

Iversen PO and Nicolaysen G (1990) The distribution of blood flow and glucose uptake within single skeletal muscles in the awake rabbit. Acta Physiol Scand 14:373–381.

Iversen PO and Nicolaysen G (1991) Local blood flow and glucose uptake within resting and exercising rabbit skeletal muscle. Am J Physiol 260:H1795–H1801.

Iversen PO, Standa M, and Nicolaysen G (1989) Marked regional heterogeneity in blood flow within skeletal muscle at rest and during exercise hyperemia in the rabbit. Acta Physiol Scand 136:17–28.

Iversen PO, Flatebo T, and Nicolaysen G (1992) Uneven perfusion within single cat muscles: nitric oxide and citric synthase play no role. Respir Physiol 89:329–339.

Kerger H, Torres Filho IP, Rivas M, Winslow RM, and Intaglietta M (1995) Systemic and subcutaneous microvascular oxygen tension in conscious Syrian golden hamsters. Am J Physiol 268 (Heart Circ Physiol 37):H802–H810.

Kobayashi H and Takizawa N (1996) Oxygen saturation and pH changes in cremaster microvessels of the rat. Am J Physiol 270 (Heart Circ Physiol 39):H1453–H1461.

Kurdak SS, Grassi B, Wagner PD, and Hogan MC (1996) Blood flow distribution in working in situ canine muscle during blood flow reduction. J Appl Physiol 80:1978–1983.

Marconi C, Heisler N, Meyer M, Weitz H, Pendergast DR, Cerretelli P, and Piiper J (1988) Blood flow distribution and its temporal variability in stimulated dog gastrocnemius muscle. Respir Physiol 74:1–14.

Mercker H, Ochwadt B, and Schoedel W (1949) Der Einfluss der Erregungsfrequenz und der Belastung auf Durchblutung und Sauerstoffaufnahme des Muskels. Pflügers Arch 251:73–82.

Pendergast DR, Krasney JA, Ellis A, McDonald B, Marconi C, and Cerretelli P (1985) Cardiac output and muscle blood flow in exercising dogs. Respir Physiol 61:317–326.

Piiper J (1988) Role of diffusion shunt in transfer of inert gases and O_2 in muscle. In: Oxygen Transport to Tissue X. Adv Exp Med Biol 222 (Mochizuki M, Honig CR, Koyama T, Goldstick TK, Bruley DF Eds) Plenum Press, New York and London, pp. 55–61.

Piiper J (1990) Unequal distribution of blood flow in exercising muscle of the dog. Respir Physiol 80:129–136.

Piiper J (1992) Modeling of oxygen transport to skeletal muscle: blood flow distribution, shunt, and diffusion. In: Oxygen Transport to Tissue XIII. Adv Exp Med Biol 316 (Goldstick TK, McCabe M, Maguire DJ Eds) Plenum Press, New York, pp. 3–10.

Piiper J and Haab P (1991) Oxygen supply and uptake in tissue models with unequal distribution of blood flow and shunt. Respir Physiol 84:261–271.

Piiper J and Meyer M (1984) Diffusion-perfusion relations in skeletal muscle: model and experimental evidence from inert gas washout. In: Oxygen Transport to Tissue V. Adv Exp Med Biol 169 (Lübbers DW, Acker H, Lehniger-Follert E, Goldstick TK Eds) Plenum Press, New York and London, pp. 457–466.

Piiper J, Pendergast DR, Marconi C, Meyer M, Heisler H, and Cerretelli P (1985) Blood flow distribution in dog gastrocnemius muscle at rest and during stimulation. J Appl Physiol 64:241–251.

Richardson RL and Saltin B (1998) Human muscle blood flow and metabolism studied in the isolated quadriceps muscles. Med Sci Sports Exerc 30:28–33.

Richardson RS, Poole DC, Knight DR, Kurdak SS, Hogan MC, Grassi B, Johnson EC, Kendrick KF, Erickson BK, and Wagner PD (1993) High muscle blood flow in man: is maximal O_2 extraction compromised? J Appl Physiol 75:1911–1916.

Richardson RS, Knight DR, Poole DC, Kurdak SS, Hogan MC, Grassi B, and Wagner PD (1995) Determinants of maximal exercise VO_2 during single leg knee-extension exercise in humans. Am J Physiol 268 (Heart Circ Physiol 37):H1453–H1461.

Richardson RS, Tagore K, Haseler LJ, Jordan M, and Wagner PD (1998) Increased VO_2 max with right-shifted Hb-O_2 dissociation curve at a constant O_2 delivery in dog muscle in situ. J Appl Physiol 84:995–1002.

Roca J, Hogan MC, Story D, Bebout DE, Haab P, Gonzalez R, Ueno O, and Wagner PD (1989) Evidence for tissue diffusion limitation of VO_2 max in normal humans. J Appl Physiol 67:291–299.

Roth AC and Wade K (1986) The effects of transmural transport in the microcirculation: a two gas species model. Microvasc Res 32:64–83.

Roy TK and Popel AS (1996) Theoretical predictions of end-capillary PO_2 in muscles of athletic and nonathletic animals at VO_2max. Am J Physiol 271 (Heart Circ Physiol 40):H721–H737.

Sarelius IH (1986) Cell flow path influences transit time through striated muscle capillaries. Am J Physiol 250 (Heart Circ Physiol 19):H899–H907.

Sparks HV and Mohrman DE (1977) Heterogeneity of flow as an explanation of the multiexponential washout of inert gas from skeletal muscle. Microvasc Res 13:181–184.

Stainsby WN and Otis AG (1964) Blood flow, oxygen tension, oxygen uptake, and oxygen transport in skeletal muscle. Am J Physiol 206:858–866.

Torres Filho IP and Intaglietta M (1993) Microvessel PO_2 measurements by phosphorescence decay method. Am J Physiol 265 (Heart Circ Physiol 34):H1434–H1438.

Torres Filho IP, Kerger H, and Intaglietta M (1996) PO_2 measurements in arteriolar networks. Microvasc Res 51:202–212.

Tyml K (1991) Heterogeneity of microvascular flow in rat skeletal muscle is reduced by contraction and by hemodilution. Int J Microcirc Clin Exp 10:75–86.

Wagner PD, Saltzman HA, and West JB (1974) Measurement of continuous distributions of ventilation-perfusion ratios: theory. J Appl Physiol 36:588–599.

Zheng L, Golub AS, and Pittman RN (1996) Determination of PO_2 and its heterogeneity in single capillaries. Am J Physiol 271 (Heart Circ Physiol 40):H365–H372.

38

QUANTITATIVE MEASUREMENT OF OXYGEN CONSUMPTION AND FOREARM BLOOD FLOW IN PATIENTS WITH MITOCHONDRIAL MYOPATHIES

M. C. P. Van Beekvelt,[1,2] W. N. J. M. Colier,[2] R. A. Wevers,[1] and B. G. M. Van Engelen[1]

[1]Neuromuscular Centre Nijmegen
Department of Neurology
University Hospital Nijmegen
[2]Department of Physiology
Faculty of Medical Sciences
University of Nijmegen
P.O. Box 9101, 6500 HB Nijmegen
The Netherlands

ABSTRACT

Five patients with chronic progressive external ophthalmoplegia (CPEO) and 27 healthy controls were examined by near-infrared spectroscopy (NIRS) for the non-invasive and direct quantitative measurement of muscle oxygen consumption during rest as well as during static isometric handgrip exercise at 10% of their maximum voluntary contraction. In patients with CPEO, we found a significantly decreased oxygen consumption during exercise, but more remarkably already during rest. Our results suggest that NIRS is able to discriminate between CPEO patients and healthy controls, which makes NIRS a promising tool in the diagnostic work-up of patients suspected of a mitochondrial myopathy.

1. INTRODUCTION

Mitochondrial myopathies are neuromuscular disorders concerning a defect in the mitochondria. Since mitochondria provide about 90 percent of the energy that is needed

Oxygen Transport to Tissue XXI, edited by Eke and Delpy.
Kluwer Academic / Plenum Publishers, New York, 1999.

for the functionality of the body (Wallace, 1997), a defect will irrevocably lead to reduced energy production. Mitochondrial defects have, therefore, a major impact on the body, especially in high-energy tissues such as the brain or skeletal muscle.

Mitochondrial myopathies are phenotypically and genetically heterogeneous disorders sharing a characteristic exhibition of exercise intolerance, undue fatigue, and lactic acidosis during low- to moderate-intensity work (Taivassalo et al., 1998). Although mitochondrial myopathies are rare, they may mimic a large number of other disorders and, as a consequence, they are frequently considered in differential diagnosis but are rarely confirmed (Poulton, 1998).

Diagnostic confirmation of mitochondrial myopathies requires definite investigations like molecular genetic techniques, muscle histology and phosphorus magnetic resonance spectroscopy (^{31}P-MRS). However, molecular genetics identify only known defects, while sensitivity of muscle histology is limited, and routine use of ^{31}P-MRS is restricted by its limited availability and its time-consuming analysis (Argov et al., 1997).

So far, mitochondrial myopathies have been tested almost exclusively during exercise, since clinical symptoms are relatively minor during rest but become prominent during exercise (Taivassalo et al., 1998). Although exercise tests provide a valuable tool to study mitochondrial myopathies, they can be potentially hazardous in these patients, due to associated cardiac conduction abnormalities (e.g. in CPEO) (Bindoff et al., 1997; Poulton, 1998). Apart from that, there are patients who are incapable to perform exercise because they are severely affected or simply too young.

Despite these limitations of exercise tests, resting data of patients with mitochondrial myopathies are scarce. This is probably due to the current opinion that the level of oxygen consumption during rest is normal in most metabolic myopathies (Haller et al., 1990). However, direct evaluation of oxygen uptake and oxygen delivery in muscle has only recently become available with near-infrared spectroscopy (NIRS). NIRS, a direct, non-invasive method of tissue oximetry, has been used to study working muscle in patients with mitochondrial myopathies (Bank and Chance, 1994; Abe et al., 1997; Gellerich et al., 1998) but no studies have directly examined oxygen uptake and oxygen delivery in resting muscle.

The aim of the present study was, therefore, to investigate whether NIRS could discriminate between patients with chronic progressive external ophthalmoplegia (CPEO) and healthy controls; in rest as well as during exercise.

2. METHODS

2.1. Subjects

We studied a homogeneous group of 5 patients with chronic progressive external ophthalmoplegia (CPEO), according to the diagnostic criteria described by Emery (Bindoff et al., 1997), and compared them with 27 healthy controls. The physical characteristics of all subjects are shown in Table 1. The clinical characteristics of the patients with CPEO are shown in Table 2. The study was approved by the Faculty Ethics Committee and all subjects gave their written informed consent. The control subjects were free of medication except for one subject who used a cholesterol-decreasing medicine. Three of the patients used vitamin preparations while the other two patients had no medication. All subjects were right handed apart from 1 patient and 2 controls.

Table 1. Physical characteristics of patients with sporadic chronic progressive external ophthalmoplegia and controls

	Sex M/F		Length cm	Weight kg	Age** yrs	SFT** mm	MVC** Kg
Controls	17 M	Avg	178.9	70.0	28.8	2.3	28.6
	10 F	Sd	10.9	11.1	7.7	0.9	6.2
Patients	2 M	Avg	172.4	76.4	41.0	6.2	16.3
	3 F	Sd	6.3	17.1	14.6	1.9	6.1

M = male; F = female. SFT = Skinfold thickness measured on top of the flexor superficialis muscle and in between the two NIRS optodes using a Holtain skinfold caliper (Holtain Ltd., Crymmych, UK). MVC = maximum voluntary contraction. ** $P < 0.01$.

2.2. Procedures

The subjects were laid in a comfortable supine position, 15–20 min prior to the test. The right hand rested on a handgrip dynamometer with the upper arm at heart level and the forearm in an upward angle of 30° to avoid venous pooling of the blood. The arm was supported at the wrist and above the elbow so that there was no contact between forearm and dynamometer and circulation in the forearm was completely unrestricted. A pneumatic arm cuff was placed just above the elbow and was used to apply venous occlusion (50 mmHg) or arterial occlusion (260 mmHg). A wrist cuff, inflated to 260 mmHg, was used to exclude circulation changes from the hand. Near-infrared spectroscopy (NIRS) was used for the quantitative measurement of oxygen consumption and forearm blood flow (Van Beekvelt et al., 1998).

2.3. Near-Infrared Spectroscopy (NIRS)

NIRS is a non-invasive, continuous and direct method to determine oxygen consumption and hemodynamics in tissue. It can differentiate between oxyhemoglobin (O_2Hb) and deoxyhemoglobin (HHb) due to oxygenation-dependent absorption changes in the muscle. NIRS measurements were obtained with a continuous wave near-infrared spectrophotometer (OXYMON, University of Nijmegen) that generates light at 905, 850 and 770 nm (Van der Sluijs et al., 1998). The absorption changes in O_2Hb, HHb and total hemoglobin (tHb) were converted into concentration changes using the algorithm

Table 2. Clinical characteristics of patients with chronic progressive external ophthalmoplegia

Sex[a]/age (yrs)	Disease duration (yrs)	Fatigue[b]	Proximal muscle strength (MRC[c])	Modified Rankin scale[d]	Biopsy RRF[e]	Serum lactate[f] (μmol/L)
M/54	28	–	4/5	2	+	1520
F/54	27	++	3/5	3	+	2660
F/42	22	+++	3/5	3	++	844
F/36	4	–	5/5	2	++	1288
M/19	3	+	5/5	2	++	1586

[a] M = male; F = female.
[b] – = not present; + = mild; ++ = moderate; +++ = severe.
[c] MRC = Medical Research Council.
[d] Modified Rankin scale decribed by Van Swieten et al.[11]
[e] RRF = ragged red fibers; + = present; ++ = abundant.
[f] Normal values: <1700 μmol/L.

described by Livera et al. (1991). A pathlength factor of 4.0 was used to calculate absolute concentration changes (Ferrari et al., 1992; Duncan et al., 1996). The sample time was 0.1 s and all data were displayed real-time and stored on disk for analysis. The optical fibers were placed on top of the m. flexor digitorum superficialis with an inter-optode distance of 35 mm.

2.4. NIRS Measurements

2.4.1. Forearm Blood Flow. Forearm blood flow (FBF) was calculated from NIRS data by evaluating the rate of increase in [tHb] during venous occlusion (De Blasi et al., 1997). Concentration changes of tHb were expressed in μM/s and converted to milliliters blood per minute per 100 milliliters tissue ($ml{\cdot}min^{-1} \cdot 100\,ml^{-1}$). A linear increase within the first seconds of the occlusion was considered for the calculation of FBF.

2.4.2. O_2 Consumption. Oxygen consumption ($\dot{V}O_2$) was calculated from NIRS data by evaluating the rate of decrease in [Hb_{diff}] (De Blasi et al., 1997) during arterial occlusion. $\dot{V}O_2$ is expressed in milliliters O_2 per minute per 100 gram of tissue ($mlO_2{\cdot}min^{-1} \cdot 100\,g^{-1}$) using $1.04\,kg{\cdot}l^{-1}$ for muscle density (Vierordt, 1906) and assuming STPD conditions.

2.5. Protocol

FBF and $\dot{V}O_2$ were measured in rest and during exercise, which was performed at a workload that was equal to 10% of the subject's maximum voluntary contraction (MVC). The experiment started with a 5 min rest period. The wrist cuff was inflated after 4 min and followed, 1 minute later, by a venous occlusion of 20 s. After one minute of recovery an arterial occlusion was applied and lasted 30–45 s depending on the time to get a linear response of the NIRS variables. After 5 min recovery, the subject was asked to perform static isometric handgrip exercise at 10% MVC. Together with the start of exercise, the wrist cuff was inflated. After reaching the steady state (1 min), a venous occlusion was applied while exercise was maintained. The exercise was ended after 20 s occlusion and the wrist cuff released. After another 5 min of recovery, when NIRS had returned to pre-exercise levels, a second exercise session was performed but this time ending with an arterial occlusion instead of the venous occlusion. The experiment ended with 5 min recovery measurement.

2.6. Statistics

The CPEO patients were compared with healthy controls using a Student's t-test. A paired t-test was used for differences between rest and exercise within each group. $P < 0.05$ was considered statistically significant. All results are reported as mean ± SD.

3. RESULTS

We found a significantly lower oxygen consumption during rest ($P = 0.02$) (Figure 1A) in combination with a significantly higher blood flow ($P = 0.002$) (Figure 1B) in the CPEO group (respectively $0.063 \pm 0.027\,mlO_2{\cdot}min^{-1} \cdot 100\,g^{-1}$ and $1.344 \pm 0.264\,ml{\cdot}min^{-1} \cdot 100\,ml^{-1}$) as compared to the control group (respectively $0.106 \pm 0.006\,mlO_2{\cdot}min^{-1} \cdot 100\,g^{-1}$ and $0.719 \pm 0.063\,ml{\cdot}min^{-1} \cdot 100\,ml^{-1}$). This difference between CPEO and controls, con-

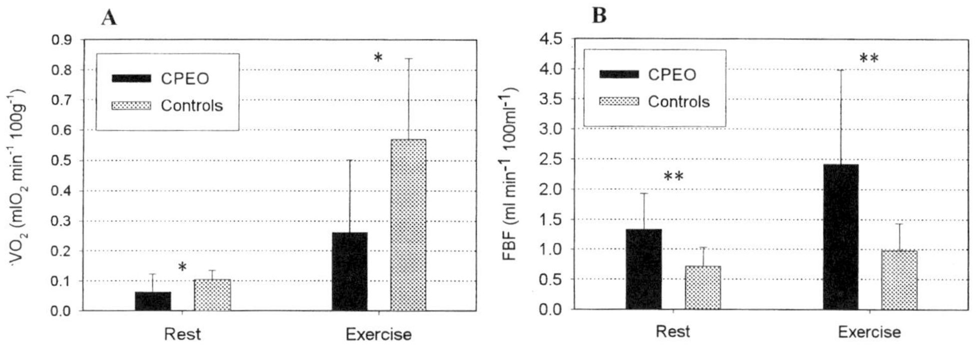

Figure 1. Average and standard deviation for near-infrared spectroscopy values of A) muscle oxygen consumption ($\dot{V}O_2$) and B) forearm blood flow (FBF) in patients with chronic progressive external ophthalmoplegia (CPEO) and healthy control subjects during rest and 10% static isometric handgrip exercise.

cerning the low oxygen consumption in combination with high blood flow, was even more pronounced during low-intensity exercise at 10% MVC (see Figure). During exercise, oxygen consumption was $0.262\,mlO_2{\cdot}min^{-1}\cdot 100\,g^{-1}$ ($\pm$ 0.107) in the CPEO group as compared to $0.570\,mlO_2{\cdot}min^{-1}\cdot 100\,g^{-1}$ ($\pm$ 0.052) in the control group ($P = 0.03$) (Figure 1A). Forearm blood flow in the CPEO group was $2.422\,ml{\cdot}min^{-1}\cdot 100\,ml^{-1}$ ($\pm$ 0.702) as compared to $0.984\,ml{\cdot}min^{-1}\cdot 100\,ml^{-1}$ ($\pm$ 0.090) in the control group ($P = 0.003$) (Figure 1B).

4. DISCUSSION

Although until recently direct evaluation of oxygen uptake and oxygen delivery in muscle was impossible, it is assumed that the level of oxygen consumption during rest is normal in most metabolic myopathies (Haller and Bertocci, 1990). In the present study, we used NIRS for direct examination of muscle oxygen consumption in patients with CPEO and were able to show, compared to controls, a decreased oxygen consumption in combination with an increased forearm blood flow that was not only present during exercise but also during rest.

Our NIRS data on oxygen consumption during exercise are compatible with previous NIRS studies (Bank and Chance, 1994; Abe et al., 1997) and support the hypothesis of a hyperkinetic circulatory response to exercise that is found in mitochondrial myopathies (Haller, 1994). This hyperkinetic response refers to a mismatch between oxygen delivery and oxygen uptake due to an exaggerated increase in oxygen transport relative to the increase in oxygen consumption (Haller, 1994). This mismatch in the transition from rest to exercise is perfectly seen in our CPEO data as the relatively minor increase in oxygen consumption in combination with a major increase in blood flow (Figure 1).

Furthermore, our NIRS data indicate that this mismatch between oxygen delivery and oxygen consumption is also present during rest (Figure 1). Oxygen consumption in the CPEO group is decreased compared to the control group, and forearm blood flow seems already exaggerated during rest relative to the low rate of oxygen consumption. Our resting data support the findings of several ^{31}P-MRS studies that found resting abnormalities in 35–85% of their patients (Arnold et al., 1985; Argov et al., 1987; Matthews et al., 1991) but does not correspond with the current opinion that the level of oxygen consumption is normal in most metabolic myopathies (Haller and Bertocci, 1990). However, reconsidering of this opinion seems recommendable based on the findings of this study as well as the abnormalities that are found with ^{31}P-MRS. In addition, it has to be taken into account that, apart from the present data, no studies have directly examined oxygen uptake and oxygen delivery in resting muscle before.

The conclusion of this study is that near infrared spectroscopy is able to discriminate between patients with CPEO, a form of a mitochondrial myopathy, and healthy controls. Although these preliminary data discriminate between groups and not between the individual patients, they suggest that NIRS, as a direct, non-invasive and relatively cheap device, can serve in the diagnostic work-up of patients suspected of a mitochondrial myopathy.

ACKNOWLEDGMENTS

The authors would like to thank Jos Evers and Marjo van de Ven for their assistance. This research was supported in part by EU contract BMH4-CT96.1658.

REFERENCES

Abe K, Matsuo Y, Kadekawa J, et al. (1997) Measurement of tissue oxygen consumption in patients with mitochondrial myopathy by noninvasive tissue oximetry. Neurology 49:837–841.

Argov Z, Bank WJ, Maris J, et al. (1987) Bioenergetic heterogeneity of human mitochondrial myopathies: phosphorus magnetic resonance spectroscopy study. Neurology 37:257–262.

Argov Z, De Stefano N, Taivassalo T, et al. (1997) Abnormal oxidative metabolism in exercise intolerance of undetermined origin. Neuromuscular Disord 7:99–104.

Arnold DL, Taylor DJ, and Radda GK (1985) Investigation of human mitochondrial myopathies by phosphorus magnetic resonance spectroscopy. Ann Neurol 18:189–196.

Bank W and Chance B (1994) An oxidative defect in metabolic myopathies: diagnosis by noninvasive tissue oximetry. Ann Neurol 36:830–837.

Bindoff L, Brown G, and Poulton J (1997) Mitochondrial myopathies. In: Diagnostic criteria for neuromuscular disorders. (Emery AEH, Ed.) Royal Society of Medicine Press Limited, London, 85–90.

De Blasi RA, Almenrader N, Aurisicchio P, and Ferrari M (1997) Comparison of two methods of measuring forearm oxygen consumption (VO2) by near infrared spectroscopy. J Biomed Optics 2:171–173.

Duncan A, Meek JH, and Clemence M, et al. (1996) Measurement of cranial optical path length as a function of age using phase resolved near infrared spectroscopy. Pediatr Res 39:889–894.

Ferrari M, Wei Q, Carraresi L, et al. (1992) Time-resolved spectroscopy of the human forearm. J Photochem Photobiol B 16:141–153.

Gellerich F, Müller T, Nioka S, et al. (1998) NIR-Spectroscopy investigation of m. vastus lateralis in patients with mitochondrial myopathies as detected by respirometric investigation of mitochondrial function in skinned fibers. SPIE 3194:121–132.

Haller RG (1994) Oxygen utilization and delivery in metabolic myopathies [editorial; comment]. Ann Neurol 36:811–813.

Haller RG and Bertocci LA (1990) Exercise evaluation of metabolic myopathies. In: Myology: Basic and Clinical. (Engel AG, Franzini-Armstrong C, Eds.) McGraw-Hill., Inc. New York, 807–821.

Livera LN, Spencer SA, Thorniley MS, et al. (1991) Effects of hypoxaemia and bradycardia on neonatal cerebral haemodynamics. Arch Dis Child 66:376–380.

Matthews PM, Allaire C, Shoubridge EA, et al. (1991) In vivo muscle magnetic resonance spectroscopy in the clinical investigation of mitochondrial disease. Neurology 41:114–120.

Poulton J (1998) Mitochondrial myopathies and related disorders. In: Neuromuscular disorders: clinical and molecular genetics. (Emery AEH, Ed.) John Wiley, Chichester, 203–213.

Taivassalo T, De Stefano N, Argov Z, et al. (1998) Effects of aerobic training in patients with mitochondrial myopathies. Neurology 50:1055–1060.

Van Beekvelt MCP, Colier WNJM, Van Engelen BGM, et al. (1998) Validation of measurement protocols to assess oxygen consumption and blood flow in the human forearm by near infrared spectroscopy. SPIE 3194:133–144.

Van der Sluijs MC, Colier WNJM, Houston RJF, and Oeseburg B (1998) A new and highly sensitive continuous wave near infrared spectrophotometer with multiple detectors. SPIE 3194:63–72.

Vierordt H (1906) In: Anatomische, Physiologische und Physikalische Daten und Tabellen. Anonymous.

Wallace DC (1997) Mitochondrial DNA in aging and disease. Sci Am 277:40–47.

39

CRITICAL CAPILLARY PO_2 AND THE ROLE OF LACTATE PRODUCTION IN OXYHEMOGLOBIN DISSOCIATION DURING EXERCISE

Karlman Wasserman

Division of Respiratory and Critical Care Physiology and Medicine
Harbor-UCLA Medical Center
UCLA School of Medicine
1000 W. Carson, Torrance, California 90509

1. INTRODUCTION

Lactate increase during exercise takes place above a certain metabolic rate varying with fitness and disease (Wasserman, 1994). The metabolic rate ($\dot{V}O_2$) above which lactate starts to increase is thought to be that at which energy (ATP) from anaerobic glycolysis supplements the energy generated from aerobic metabolism. The onset of the anaerobic production of ATP should therefore also be the onset of net lactate production by working muscle during exercise.

During the transit of blood through the capillary bed of muscle, PO_2 decreases depending on the ratio of blood flow to metabolic rate during exercise (Figure 1). While isolated mitochondria can respire and rephosphorylate ADP to ATP at a PO_2 of 1 mm Hg or less (Chance et al., 1962), the capillary PO_2 must be appreciably greater than 1 mm Hg to provide the pressure for O_2 to diffuse from the red cell to the sarcoplasm for muscle mitochondrial respiration. Wittenberg and Wittenberg (1989) estimated this pressure to be 15–20 mm Hg (Wittenberg and Wittenberg, 1989). It was termed the "critical" capillary PO_2 because it represents the lowest capillary PO_2 that allows the muscle mitochondria to receive the O_2 required to perform exercise, aerobically. The major factors determining the PO_2 difference between red cell and sarcoplasm are the resistances to O_2 diffusion by the red cell membrane, plasma, capillary endothelium, interstitial space and sarcolemma (Wittenberg and Wittenberg, 1989). By measuring oxymyoglobin saturation in the dog gracilis muscle during a moderate level of exercise, Gayeski and Honig (1988) estimated the PO_2 in the sarcoplasm to be about 5 mm Hg. The P50 for oxymyoglobin in

Oxygen Transport to Tissue XXI, edited by Eke and Delpy.
Kluwer Academic / Plenum Publishers, New York, 1999.

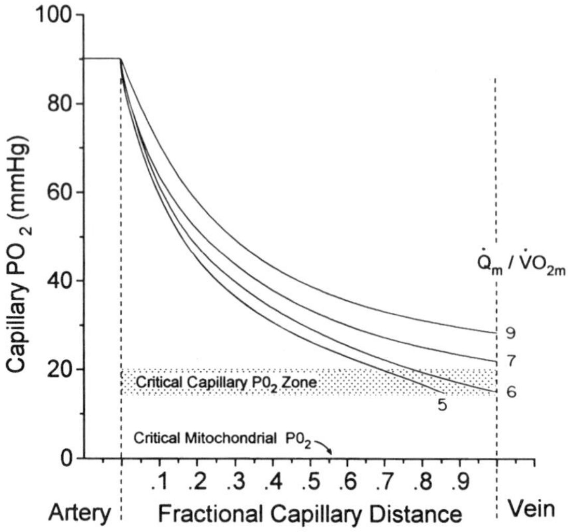

Figure 1. Model of muscle capillary O_2 partial pressure (Po_2) as blood travels from artery to vein. The model assumes hemoglobin concentration of 15 g/dl, arterial Po_2 of 90 mm Hg and a linear O_2 consumption along the capillary. The rate of fall of capillary Po_2 depends on the muscle blood flow (Qm)/muscle $\dot{V}o_2$ ($\dot{V}o_2$m) ratio. The curves include a Bohr effect due to a respiratory CO_2 production. The capillary Po_2 is heterogeneous along the capillary bed even with a homogenous Qm/$\dot{V}o_2$m. The end-capillary Po_2 cannot decrease below the critical capillary Po_2. Any muscle unit with a theoretical Qm/$\dot{V}o_2$m less than 6 will have increased anaerobic metabolism and lactate production. See text for application of model. (From Wasserman, 1994).

man is 6 mm Hg (Roughton, 1964). Because myoglobin would be less than $^1/_2$ saturated, oxymyoglobin can serve as a limited O_2 store for mitochondrial respiration during the period that capillary PO_2 is decreasing and $\dot{V}o_2$ is increasing at the start of exercise.

The importance of O_2 transport in both the exercise $\dot{V}o_2$ at which lactate starts increasing and the pattern of increase in $\dot{V}o_2$ above the lactate threshold is exemplified by acutely increasing the carboxyhemoglobin content of blood and thereby reducing O_2 transport. Increased carboxyhemoglobin not only reduces the $\dot{V}o_2$ at which lactate starts to increase (lactate threshold, *LT*) (Vogel and Gleser, 1972), but it also slows the rate of rise in $\dot{V}o_2$ as work rate increases, but only above the *LT* (Koike et al., 1990).

To determine the critical capillary Po_2 and its relationship to the lactic acidosis of exercise, normal subjects (Stringer et al., 1994) and patients with heart disease (Koike et al., 1994) were studied during exercise while femoral vein Po_2, Pco_2, pH and lactate were measured. When O_2 becomes diffusion limited during exercise the critical capillary PO_2 is reached (capillary PO_2 does not decrease despite increasing $\dot{V}o_2$). During exercise, the end capillary PO_2 is the lowest level to which the PO_2 falls in the capillary bed. As long as there is no diffusion limitation, the end-capillary PO_2 should continue to decrease as $\dot{V}o_2$ increases. If the capillary PO_2 decreased to its lowest limit (critical capillary PO_2) before the blood reached the end of the capillary, then no further O_2 will be consumed from that capillary, except through the "Bohr Effect" induced by the exercise lactic acidosis. The latter occurs in those myocytes undergoing anaerobic glycolysis. In this paper, the decrease in femoral vein PO_2, as the measure of end capillary PO_2, is analyzed in normal and heart failure subjects during constant and increasing levels of exercise. Using Fick's law of diffusion, the pattern of muscle-capillary recruitment is analyzed during exercise in normal subjects and patients with heart failure.

2. METHODS

A total of 10 normal subjects and 10 patients with stable chronic heart failure were studied during cycle ergometry. Five of the normal subjects did increasing work rate exercise protocols in ramp pattern to $\dot{V}O_2$ max. Five of the normal subjects did two constant work rate protocols, one at 40% of $\dot{V}O_2$ max and one at 85% of $\dot{V}O_2$ max. Each of the 10 patients did 3 work rates, one increasing work rate protocol in ramp pattern to $\dot{V}O_2$ max and two constant work rate protocols, one below the lactic acidosis threshold (*LAT*) and one above the *LAT*.

Each subject had a femoral vein catheter inserted 2 cm below the inguinal ligament to 6 cm cephalad. In addition, the normal subjects had a brachial catheter in place for simultaneous arterial and femoral vein measurements. In the normal subjects, sampling rate from the femoral vein was every 30 s for the incremental exercise studies. Sampling was every 5 s for the first two minutes and then every 30 s for the remaining four minutes of the constant work rate exercise tests. Sampling from the brachial artery was at 30 s intervals for the entire test. Blood sampling was every 30 s for the patients for each protocol.

Gas exchange was measured breath-by-breath using a Fleisch-type No. 3 pneumotachograph for measurement of respired flow and an MGA-1100 mass spectrometer (Perkin-Elmer, Pomona CA) for continuous measurement of gas concentrations in the normal subjects as previously described (Stringer et al., 1994). An Aeromonitor AE-280 gas exchange system (Minato Medical Science, Japan) was used for measuring gas exchange in the patients (Koike et al., 1994). PO_2, PCO_2 and pH were measured by electrode technique (Instrumentation Laboratories, Lexington, MA). Lactate was measured in each blood sample with a Yellow Springs Instrument lactate/glucose analyzer (model 2300, Yellow Springs, OH). Oxyhemoglobin saturation, hemoglobin concentration and carboxyhemoglobin were measured with an Instrumentation Laboratories Co-Oximeter.

3. RESULTS

3.1. Femoral Venous PO_2 Changes during Exercise in Normal Subjects

To determine the critical capillary PO_2 in normal subjects, 10 healthy adults were studied with femoral vein and arterial catheters, five during progressively increasing and five during two levels of constant work rate leg cycling exercise, one below and one above the *LAT*. During progressively increasing work rate in normal subjects, femoral vein blood PO_2 reached a "floor" or lowest value in the middle of the subjects' work capacities and before lactate concentration started to increase (Stringer et al., 1994). To confirm and investigate further the hypothesis that femoral vein PO_2 reached a "floor" value (critically low value) by the time the *LAT* was reached, normal subjects performed two constant work rate exercise tests for 6 minutes, one at a moderate work rate (40% of $\dot{V}O_2$ max, avg = 113 W, $\dot{V}O_2$ = 1.76 L/min) and one at a heavy work rate (85% of $\dot{V}O_2$ max, avg = 265 W, $\dot{V}O_2$ = 3.36 L/min). These tests were done with a high sampling density of femoral vein blood (every 5 s during the first two minutes and then every 30 s for four minutes) to achieve precise description of the kinetics and magnitude of change. The results of these studies are shown in Figure 2. The femoral vein PO_2 reached the same "floor" value at 30 to 60 s after the start of exercise in both moderate and heavy work intensities (Figure 2A). This value remained the same for the remainder of the exercise

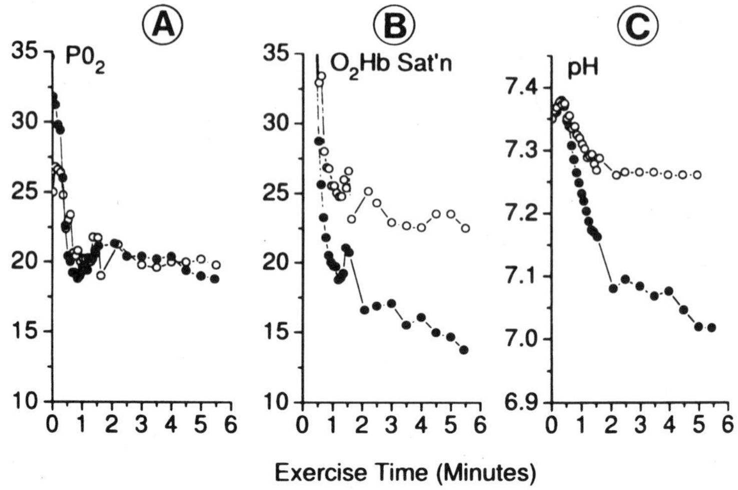

Figure 2. Femoral venous PO_2 (A), oxyhemoglobin (O_2Hb) saturation (B) and pH (C) as related to time of exercise for two constant work rate tests, one below (open circles) and one above (solid circles) the lactate acidosis threshold (*LAT*). The data are the average of 5 subjects. The below and above *LAT* work rates averaged 113 and 265 watts, respectively. Note that O_2Hb saturation is lower during the higher intensity exercise despite identical PO_2 values (modified from Stringer et al., 1994).

period, despite the large difference in work rate and $\dot{V}O_2$ (Figure 2A). In contrast to PO_2, oxyhemoglobin saturation (O_2Hb) continued to decrease past the time when the femoral vein (end-capillary) PO_2 became constant (Figure 2B). A striking difference can be seen in the pH changes for the moderate and heavy exercise (Figure 2C).

3.2. Oxyhemoglobin Dissociation above the Lactic Acidosis Threshold

For the 40% of maximum $\dot{V}O_2$ work rate (selected to be below the *LAT*), O_2Hb dissociation decreased rapidly for the first minute and then more slowly for the next one to two minutes before reaching a constant value. The change after one minute was similar to the pattern of pH decrease. For the work rate at 85% of maximum (above the *LAT*), the O_2Hb saturation decreased to a much greater degree and continued for the entire 6 minutes of exercise. The femoral vein O_2Hb desaturation not accounted for by the PO_2 decrease could be completely accounted for by the decrease in pH (Figure 2C). To illustrate this in a more graphical way, femoral vein O_2Hb saturation was plotted against the independently measured femoral vein PO_2 values for the moderate and heavy work rates (Figure 3), with pH isopleths overlaid on these data. This plot shows that the decrease in O_2Hb saturation that could not be accounted for by a decrease in PO_2 could be accounted for by the decrease in pH (Bohr Effect) (Figure 3). Thus the decrease in O_2Hb saturation below 25% was completely accounted for by acidification of the capillary blood.

3.3. Buffering the Exercise-Induced Lactic Acidosis

The net increase in lactic acid production during exercise must be buffered by intracellular HCO_3^-. This reaction releases CO_2 from the muscle over that expected from aerobic metabolism. Since this takes place above the *LAT*, end-capillary PCO_2 increases

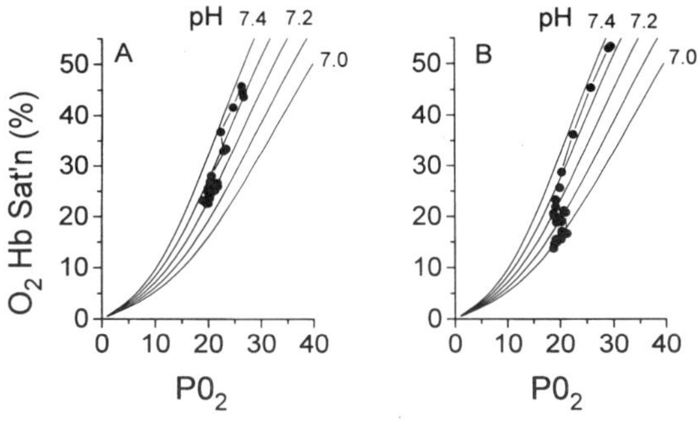

Figure 3. Changing femoral vein oxyhemoglobin saturation (O_2Hb Sat'n; see Figure 2B) as a function of femoral vein Po_2 (Figure 2A) for the 6-min constant work rate exercise tests shown in Figure 2. Superimposed are the lower part of oxyhemoglobin dissociation curves for pH values of 7.0–7.4, calculated from equations reported by Severinghaus (1979). Left panel: average data for below *LAT* exercise. Right panel data for above *LAT* exercise. Start of exercise is where O_2Hb saturation is highest. Femoral vein oxyhemoglobin saturation progressively decreased as exercise continued, as shown in Figure 2. O_2Hb saturation fell on pH isopleths in agreement with measured pH (modified from Stringer et al., 1994).

without a further fall in Po_2 (Figure 4). Simultaneously, intracellular HCO_3^- and therefore extracellular and femoral vein HCO_3^- decrease. Both the decrease in femoral vein HCO_3^- and the increase in PCO_2 serve to acidify the capillary blood of the muscle cells producing lactate (Wasserman, 1994). The lactic acidosis of exercise thereby facilitates O_2Hb dissociation. It is, therefore, an essential mechanism for achieving maximal O_2 extraction while simultaneously maintaining the partial pressure gradient needed to allow O_2 to diffuse from red cell to mitochondria at a rate appropriate for heavy exercise. Thus,

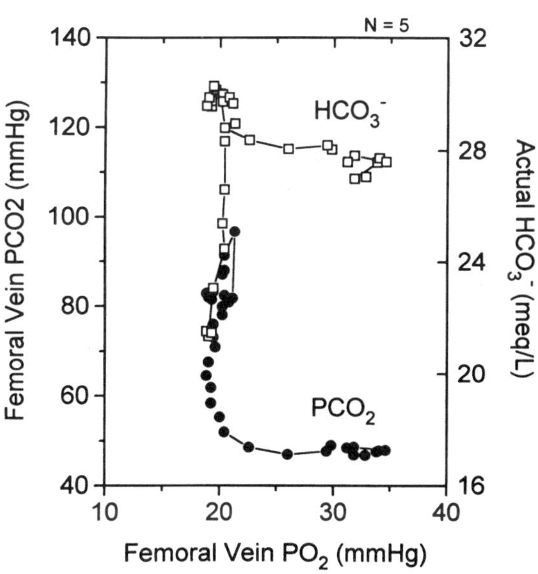

Figure 4. Femoral vein (end-capillary) Pco_2 and HCO_3^- as a function of end-capillary Po_2 during constant work rate heavy exercise. Values are the average for the 5 normal subjects whose data are shown in Figure 2. Values at the start of exercise are at the right and move leftward as $\dot{V}o_2$ increases. (From Wasserman and Stringer, 1996).

for work rates demanding more O_2 than that required at work rates just below the *LAT*, the further extraction of O_2 from O_2Hb is H^+ concentration dependent.

3.4. The Critical Capillary PO_2 in Normal Subjects

As would be predicted from the critical capillary PO_2 concept, femoral vein and therefore end-capillary PO_2 decreased to its lowest value before lactate started to increase (Figure 5). This was true for all subjects, whether performing increasing work rate exercise to maximum or heavy constant work rate exercise. When end-capillary blood reached a "floor" or "critical" value, lactate concentration increased consistent with the model shown in Figure 1.

3.5. Femoral Venous PO_2 Changes during Exercise in Stable Chronic Heart Failure

This patient study taken from the report by (Koike et al., 1994), consisted of 10 subjects with stable chronic heart failure. Each patient performed three exercise protocols, a progressively increasing work rate test in ramp pattern, a moderate intensity exercise test (80% of *LAT*) and a heavy intensity exercise test (the *LAT* plus 50% of the difference between the *LAT* and $\dot{V}O_2max$). Femoral vein PO_2 fell to a low value in the middle and not at the highest $\dot{V}O_2$ of the subjects' aerobic capacities (Figure 6). In 5 of the subjects, the PO_2 (and O_2Hb saturation) actually increased as VO_2 increased (Figure 6, right column). While this finding seems paradoxical, its reproducibility in a given subject, as seen in Figure 6, attests to the reliability of the phenomenon. The remaining

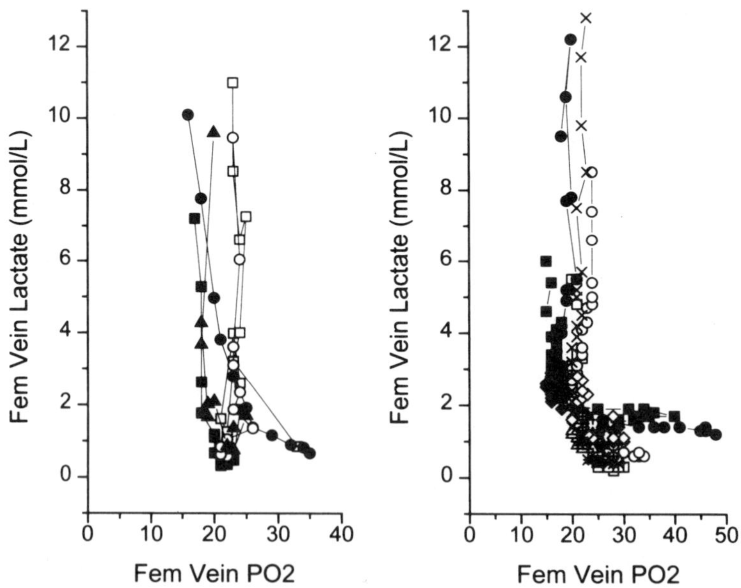

Figure 5. Femoral vein lactate as function of femoral vein Po_2 for incremental (ramp) exercise in 5 normal subjects (left panel) and 10 constant work rate exercise tests (five below and five above the *LT*) in 5 normal subjects (right panel). The highest Po_2 values are where exercise starts. Different symbols represent different subjects (modified from Wasserman and Stringer, 1996).

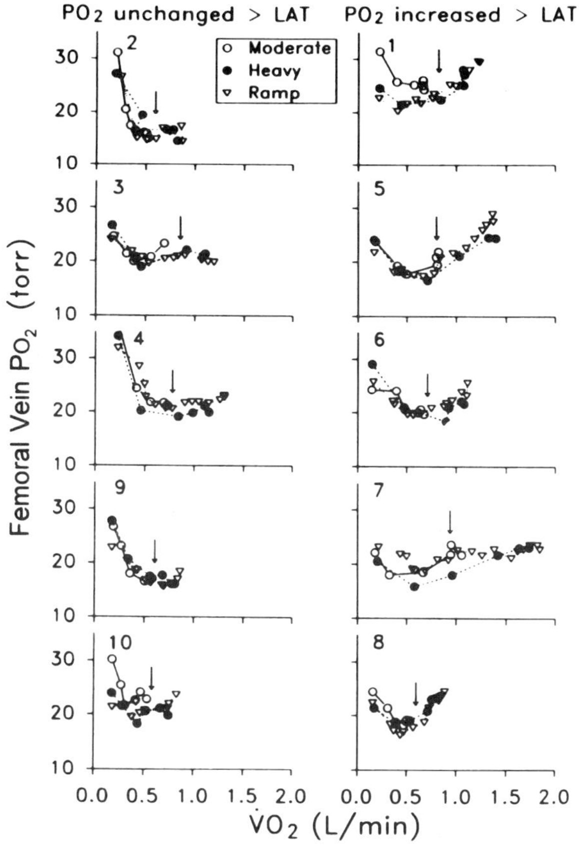

Figure 6. Femoral vein oxygen partial pressure (PO_2) plotted as a function of oxygen uptake ($\dot{V}O_2$) during two constant work-rate tests of moderate and heavy intensity and an incremental (ramp) exercise test for 10 patients with chronic heart failure. Numbers correspond to patients described in the original report. Femoral vein PO_2 rapidly decreased toward a minimal value with increasing $\dot{V}O_2$. After the femoral vein PO_2 reached its nadir, it increased in five patients despite increasing $\dot{V}O_2$ (right column), but was unchanged for the other five patients (left column). Arrows show the lactic acidosis threshold (*LAT*) determined non-invasively by the v-slope method during the incremental exercise test (from Koike et al., 1994).

5 subjects decreased their femoral vein PO_2 to a constant level (Figure 6, left column), similar to that of the normal subjects (Figure 5). The failure for the femoral vein PO_2 to decrease further, despite an increase in $\dot{V}O_2$, indicates that the end-capillary PO_2 had reached its lowest value and that O_2 had become diffusion-limited (critical capillary had been reached).

3.6. The Critical Capillary PO_2 in Patients with Stable Chronic Heart Failure

To estimate the femoral vein PO_2 at which femoral vein lactate concentration started to increase, the femoral vein blood lactate concentration was plotted against its respective PO_2 value for each exercise protocol for each patient. As for the normal subjects, lactate concentration did not increase until the lowest PO_2 had been reached in the femoral vein (Figure 7). This suggests that the lactate concentration increase occurred

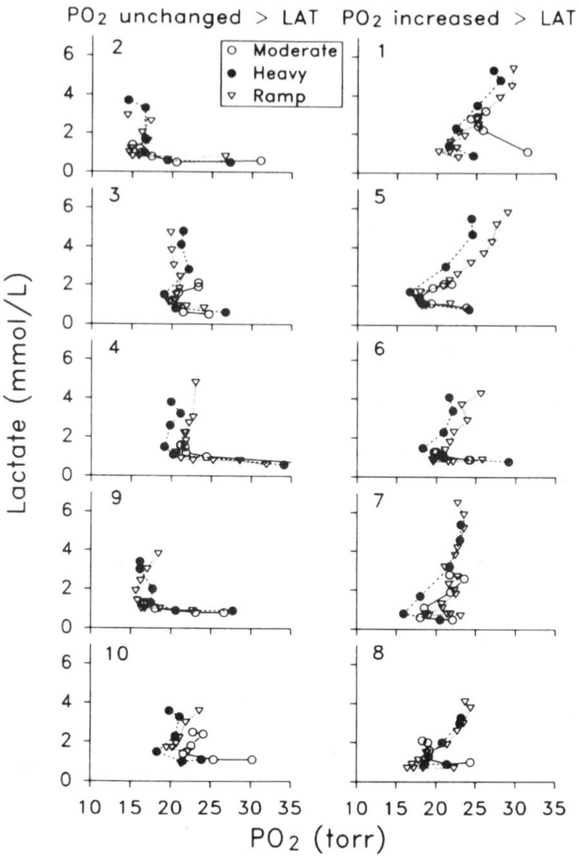

Figure 7. Relationship between femoral vein lactate and oxygen partial pressure (PO_2) for each exercise test for each patient shown in Figure 6. Each patient performed three tests, a progressively increasing work-rate test in "ramp" pattern, and moderate and heavy intensity constant work rate tests. The highest PO_2 values are where exercise starts. Femoral vein lactate increased after femoral vein PO_2 reached its lowest value. After reaching the minimum value, PO_2 was unchanged in five and increased in five patients. From Koike et al., 1994.

after the critical capillary PO_2 was reached. The changes in femoral vein and therefore critical capillary PO_2 was highly reproducible in the three protocols in each of the 10 patients (Figure 7).

3.6.1. Comparison of the Critical Capillary PO_2 in Normal Subjects and Patients with Stable Chronic Heart Failure. As was previously reported, the critical capillary Po_2 ranged between 16 and 22 mm Hg for the normal subjects and 14–20 mm Hg for the patient group (Figure 8). Critical capillary PO_2 was positively correlated with peak Vo_2/kg in the patient group and negatively correlated in the normal subjects. From Fick's Law of Diffusion,* the finding of a direct proportionality between critical capil-

* $VO_2 = D \times A/L \times P(\text{cap-mito})O_2$, where VO_2 = the rate of O_2 molecular diffusion per unit time into myocyte, D = rate of O_2 diffusion/distance/partial pressure gradient in the tissue fluid, P(cap-mito) = the partial pressure gradient of O_2 between the capillary and mitochondria, A = diffusion surface area and L = the path length for O_2 diffusion.

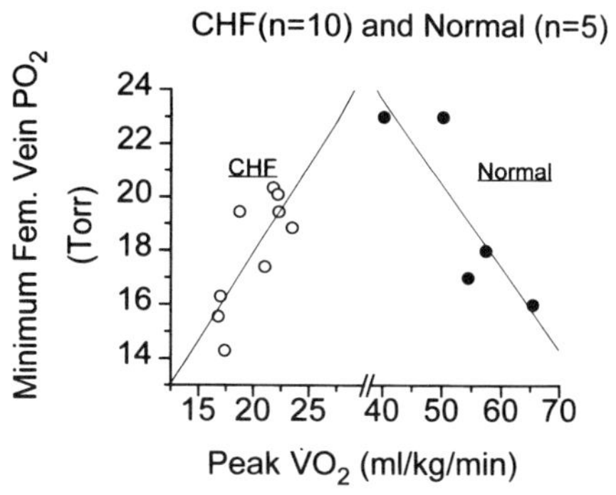

Figure 8. Lowest femoral vein PO_2 (critical capillary PO_2) versus peak $\dot{V}O_2$ in patients with chronic heart failure (CHF) and fit normal subjects. Linear regression lines were calculated for each group (from Wasserman and Stringer, 1996). For the patients, r = 0.70 (P = 0.01). For the normal subjects, r = 0.86 (P = 0.6). Data taken from Stringer et al., 1994 and Koike et al., 1994.

lary PO_2 and maximal $\dot{V}O_2$ is consistent with a relatively fixed capillary bed. In contrast, peak $\dot{V}O_2$ increased as critical capillary PO_2 decreased in normal fit subjects (Figure 8). From Fick's law of diffusion, this is possible only if there is an increase in capillary diffusion area and/or shorter diffusion path for O_2, i.e. increased capillarity. Obviously a correlation with only 5 subjects in the normal group must be interpreted cautiously. The different direction of the regressions for the patients and the physically fit subjects, however, suggests that different factors determine the maximal aerobic capacity, in the two groups.

3.6.2. The Pattern of Capillary Recruitment. The pattern of recruitment of capillary bed during exercise in our normal subjects and patients with heart failure are summarized in Figure 9. The slope of the relationship between $\dot{V}O_2$ and the PO_2 difference

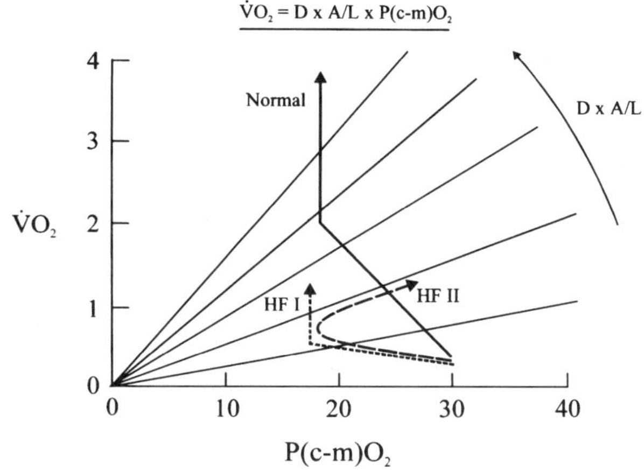

Figure 9. Pattern of change in $\dot{V}O_2$ as a function of partial pressure difference between the end-capillary PO_2 and myocyte or mitochondrial PO_2 in normal subjects and heart failure patients. The slopes radiating from the origin are the product of the diffusion coefficient for O_2 (D) and the capillary surface area (A) for diffusion divided by the diffusion distance (L) (from Fick's law of diffusion). As $\dot{V}O_2$ increases, femoral vein and end-capillary PO_2 reach floor or critical values, assuming the mitochondrial PO_2 is sustained at about 1 mm Hg. The normal subjects and patients have a similar minimal PO_2 but at different $\dot{V}O_2$ values. The femoral vein PO_2 changes in two different ways in heart failure patients. It either remains essentially constant, as in normal subjects (HF I), or increases (HF II). The arrow heads indicate the direction of change.

between capillary and mitochondrion is the product of the diffusion coefficient (D) and capillary surface area/diffusion distance (A/L) or the capillarity of the muscle, as defined by the equation in footnote 1. The finding that at the start of exercise, $\dot{V}O_2$ increases as PO_2 decreases, changing in the opposite direction to that expected if O_2 were diffusion limited, indicates that O_2 transfer to the mitochondria is not diffusion limited and that the increase in O_2 take up by the myocyte was achieved by an increase in $D \times A/L$ (non-diffusion limited capillary recruitment) and/or decrease in myocyte (mitochondrial) PO_2, thereby increasing $P(c\text{-}m)O_2$. When the critical capillary PO_2 or lactate threshold is reached, O_2 transfer becomes diffusion limited and capillary recruitment accounts for the increase in O_2 transport into cells (diffusion-limited capillary recruitment). It is unlikely that once the end capillary PO_2 has reached its lowest value, that myocyte PO_2 decreases further to increase $P(c\text{-}m)O_2$.

The heart failure patient population showed two different patterns of $\dot{V}O_2$-femoral vein PO_2 change (HF I and HF II). In both patterns, femoral vein PO_2 initially decreased as VO_2 increased showing early non-diffusion limited capillary recruitment. However this was short-lasting and the critical capillary PO_2, while similar to that of the normal subjects, took place at a lower $\dot{V}O_2$ than normal. Thereafter, the HF I group increased VO_2 by increasing $D \times A/L$, although this recruitment was much less than that of the normal subject. The HF II group increased their $\dot{V}O_2$ at work rates above the lactate threshold proportional to an increase in femoral vein (end-capillary) PO_2. Thus the relationship follows Fick's law of diffusion, revealing that $D \times A/L$ or muscle capillarity did not increase despite increasing $\dot{V}O_2$ in this patient group.

4. DISCUSSION

4.1. The Supply-Demand Balance of Exercise

O_2 is extracted from the capillary blood as it flows past the actively contracting muscles because O_2 is required for the aerobic production of ATP. To assure that an adequate capillary PO_2 is available to maintain O_2 diffusion to all myocytes involved in the exercise, a muscle blood flow increase of 6 liters/minute would be needed for an increase of 1 liter/minute muscle O_2 consumption (blood flow/metabolic rate ratio = 6) (Figure 1). This assumes a hemoglobin concentration of 15 gm/dl and an alveolar PO_2 adequate to saturate the arterial O_2Hb to at least 95% and a sink for capillary diffusion in the myocyte of about 1 mm Hg. This would leave about 1/6 of the arterial O_2 inflow into the capillary bed remaining at the venous end of the capillary bed. A blood flow/O_2 consumption ratio of 5:1 would cause obligatory anaerobiosis and lactic acidosis in the myocytes perfused by the venous capillary blood because 5 liters of blood contain only 1 liter of O_2. It is impossible to extract all of the O_2 out of the blood because of the diffusion pressure needed for O_2 in the blood to reach the mitochondria.

4.2. Muscle Capillary PO_2

The capillary PO_2 must be heterogeneous, even if there were a single "ideal" blood flow/metabolic rate ratio in the muscle, because the blood enters the muscle with a PO_2 = 90 mm Hg (in a normal subject at sea level) and leaves the capillary bed at a PO_2 which is approximately equal to that of the femoral vein PO_2 (15 to 20 mm Hg). The "critical" capillary PO_2 would be the lowest PO_2 to which the end-capillary PO_2 could fall. That the critical capillary PO_2 was reached should be evidenced by the failure of end capillary or femoral vein PO_2 to decrease further despite increasing work rate ($\dot{V}O_2$).

4.3. Critical Capillary PO_2 Exercise Lactic Acidosis and Oxyhemoglobin Dissociation

A lactic acidosis secondary to cellular hypoxia would be expected only if the critical capillary PO_2 was reached before the end of the muscle capillary network. When increasing work rate, the critical capillary PO_2 is reached before $\dot{V}O_2$ max is reached, capillary PO_2 no longer decreasing despite increasing work rate. However as work rate increased, lactate increased without a further fall in femoral vein PO_2. This is explained by the critical capillary PO_2 having been reached before the end of the muscle capillary. The lower the $Qm/\dot{V}O_2m$ ratio, the closer to the arterial end of the capillary that the critical capillary PO_2 would be reached and the greater the magnitude of anaerobic glycolysis.

4.4. Comparison of Critical Capillary PO_2 in Normal and Heart Failure Subjects

Femoral vein (end-capillary) Po_2 reached a critical (floor or lowest) value in the middle of the subjects' work capacities and before lactate started to increase in both the normal subjects and the patients. The critical capillary Po_2 ranged from 16 to 22 mm Hg for the normal subjects and 14 to 20 mm Hg for the patients (Figure 8). Femoral vein lactate did not start to increase (lactate threshold) until after the lowest femoral vein Po_2 had been reached. In contrast to Po_2, O_2Hb saturation continued to decrease above the lactate threshold. The mechanism for the further decrease in O_2Hb saturation above the lactate threshold was accounted for by the acidosis (Bohr effect) which became progressively more significant as $\dot{V}o_2$ increased. Since the increase in intracellular lactic acid concentration is buffered by intracellular HCO_3^-, both the extra CO_2 generated from intracellular HCO_3^- and the consumption of extracellular HCO_3^- by the cells serve to acidify the capillary blood of the muscles producing lactate (Figure 4).

The paradoxical finding of an increase in femoral vein PO_2 as work rate, $\dot{V}o_2$ and lactate increased in 5 of the 10 heart failure patients needs explanation. If more capillary bed was not recruited as $\dot{V}o_2$ increased, more flow must traverse the same capillary bed to increase the $\dot{V}o_2$, the latter achieved by increasing capillary PO_2 from increased blood flow and velocity through the fixed capillary bed. Alternatively, more inhomogeneity in the $Qm/\dot{V}o_2m$ ratios (Figure 1) as work of the exercising muscles increased could explain the paradoxical finding. The observation that the lowest femoral vein PO_2 does not occur at maximum exercise but rather in the mid-range of work capacity should caution investigators and model experts against assuming that the lowest femoral vein PO_2 is first reached at maximum work rate, or that the maximum work rate provides the lowest capillary PO_2 and O_2Hb. Saturation. It also should be noted that the decrease in capillary PO_2 would not be continuous to the end of the capillary if the capillary blood flow/$\dot{V}o_2m$ ratio was less than 6. Also, a single measurement of femoral vein PO_2 has limited meaning with respect to determining the critical capillary PO_2.

4.5. The Pattern of Femoral Vein PO_2 Change as $\dot{V}O_2$ Increases during Exercise in Normal and Stable Chronic Heart Failure Subjects

One of the pressing questions in heart failure research is whether the exercise limitation in heart failure patients is due to the failure to transport O_2 or a bioenergetic defect in the muscle cells resulting from a critical reduction in aerobic enzymes (Sullivan et al., 1991 and Drexler et al., 1992). We observed that patients with heart failure do not have abnormally elevated femoral vein PO_2 levels when lactate starts to increase and their estimated critical capillary Po_2 is in the same narrow range as that for normal subjects. This is consistent with the concept that decreased utilization of O_2 in heart failure patients is not a defect in mitochondrial function. However, a peripheral factor, such as the failure to recruit capillary bed as work rate increases, could contribute to reduced O_2 extraction at peak exercise and a reduced peak $\dot{V}o_2$.

5. SUMMARY

During exercise, O_2 transport from capillary to myocyte is not diffusion-limited until end-capillary PO_2 reaches its lowest value (critical capillary PO_2). Critical capillary PO_2 is reached at the mid-level of the subject's work capacity, not at $\dot{V}o_2$ max. Femoral vein lactate does not increase until the critical capillary PO_2 is reached. The increase in lactate is essential once the critical capillary PO_2 is reached because the increase in H^+ that accompanies the increase in lactate, through the "Bohr Effect", becomes the mechanism for oxyhemoglobin dissociation. Both normal subjects and patients with chronic stable heart failure have a critical capillary PO_2 between 14 and 22 mm Hg. The pattern of femoral vein (end capillary) PO_2 change as $\dot{V}o_2$ increases, describes the $\dot{V}o_2$ at which O_2 becomes diffusion limited and the extent to which new muscle capillary bed is recruited as $\dot{V}o_2$ increases above the lactate threshold.

REFERENCES

Chance, B, G Mauriello, and XM Aubert. ADP arrival at muscle mitochondria following a twitch. In: *Muscle As A Tissue*, edited by K Rodahl and SM Horvath. New York: McGraw-Hill, 1962, pp. 128–145.

Drexler, H, V Reide, T Munzel, H Konig, E Funke, and H Just. Alteration of Skeletal Muscle in Chronic Heart Failure. *Circulation* 85:1751–1759, 1992.

Gayeski, TEJ and CR Honig. Intracellular PO2 in long axis of individual fibers in working dog gracilis muscle. *Am.J.Physiol.* 254:H1179–H1186, 1988.

Koike, A, K Wasserman, K Taniguichi, M Hiroe, and F Marumo. Critical capillary oxygen partial pressure and lactate threshold in patients with cardiovascular disease. *J. Am. Coll. Cardiol.* 23:1644–1650, 1994.

Koike, A, D Weiler-Ravell, DK McKenzie, S Zanconato, and K Wasserman. Evidence that the metabolic acidosis threshold is the anaerobic threshold. *J. Appl. Physiol.* 68:2521–2526, 1990.

Roughton, FJW. Transport of oxygen and carbon dioxide. In: *Handbook of Physiology*, 1964, pp. 767–825.

Severinghaus, JW. Simple, accurate equations for human blood O_2 dissociation computations. *J. Appl. Physiol.* 46:599–603, 1979.

Stringer, W, K Wasserman, R Casaburi, J Porszasz, K Maehara, and W French. Lactic acidosis as a facilitator of oxyhemoglobin dissociation during exercise. *J. Appl. Physiol.* 76:1462–1467, 1994.

Sullivan, MJ, HJ Green, and FR Cobb. Altered skeletal muscle metabolic response to exercise in chronic heart failure: relation to skeletal muscle aerobic enzyme activity. *Circulation* 84:1597–1607, 1991.

Vogel, JA and MA Gleser. Effect of carbon monoxide on oxygen transport during exercise. *J. Appl. Physiol.* 32:234–239, 1972.

Wasserman, K. Coupling of external to cellular respiration during exercise: the wisdom of the body revisited. *Am. J. Physiol.* 266:E519–E539, 1994.

Wasserman, K and W Stringer. Critical capillary PO_2, net lactate production, and oxyhemoglobin dissociation: Effects on exercise gas exchange. In: *Exercise Gas Exchange in Heart Disease*, edited by K Wasserman. Armonk, NY: Futura Publishing Company, Inc., 1998, pp. 157–181.

Wittenberg, BA and JB Wittenberg. Transport of oxygen in muscle. *Ann. Rev. Physiol.* 51:857–878, 1989.

NADH FLUORIMETRY TO PREDICT ISCHEMIC INJURY IN TRANSPLANT KIDNEYS

J. M. C. C. Coremans, M. van Aken, H. A. Bruining, and G. J. Puppels

Department of General Surgery
University Hospital Rotterdam
dr Molewaterplein 40, 3015 GD Rotterdam
The Netherlands

1. INTRODUCTION

The probability of a successful kidney transplantation decreases as the warm ischemic time in the donor prior to nephrectomy increases. Based on this statistical background a large number of potential donor kidneys, namely those that can be harvested from non-heart beating donors, is discarded. It is expected that the pool of donor organs can be considerably enlarged when an objective test will become available which can evaluate graft viability before transplantation.

A possible pathophysiological feature involved in ischemic injury in transplant kidneys is mitochondrial dysfunction caused by desintegration of oxidative metabolic pathways (Rouslin, 1983; Rouslin and Ranagathan, 1983; Hardy et al., 1991; See et al., 1992; Nohl et al., 1993; Yano et al., 1993; Thorniley et al., 1994; Gonzalez-Flecha and Boveris, 1995; Cairns et al., 1997; Steinlechner-Maran et al., 1997). Oxidative phosphorylative activity is reflected by the kinetics of mitochondrial NADH oxidation. Since UV-light induced blue fluorescence from NADH in kidney is predominantly of mitochondrial origin, NADH fluorimetry can be used to monitor non-invasively the state of kidney cortex metabolism (Chance et al., 1962; Franke et al., 1976; Balaban, 1979).

Changes in steady state levels of NADH fluorescence have been associated with kidney and liver graft viability after hypothermic storage and transplantation (Thorniley et al., 1994; 1995). Upon reperfusion of ischemic tissue slower rates of NADH fluorescence decrease have been observed during cardiac allograft rejection and after ischemia/reperfusion damage in transplanted kidneys (Duboc et al., 1990; Thorniley, 1994). In isolated perfused rat livers, hepatic energy charge and decelerated NADH fluorescence kinetics correlated closely with survival rates after transplantation (Okamura, 1992).

Oxygen Transport to Tissue XXI, edited by Eke and Delpy.
Kluwer Academic / Plenum Publishers, New York, 1999.

The aim of the present study was to investigate whether NADH fluorimetry may be useful to evaluate the viability of preserved kidneys *prior* to transplantation. We have evaluated NADH oxidation kinetics in fluorescence images of preserved rat kidneys in relation to the extent of ischemia. Measurements were made with three groups of isolated perfused kidneys with minimal, moderate or severe ischemic damage.

2. METHODS

The experiments described in this paper were performed according to experimental protocols, which were approved by the Internal Animal Care and Use Committee of the Erasmus University Rotterdam.

2.1. Surgical Procedures

Male WagRij rats weighing 275–325 g were allowed free access to rat chow and acidified tap water. The animals were anesthetized by ether inhalation. A Zeiss OPMI-6 operating microscope was used for all microsurgical procedures. The rat was placed in supine position on a heated table to prevent hypothermia during preparation. A midline abdominal incision was made and intestines were retracted to the right side to expose the left kidney. The renal artery and vein were dissected down to the aorta and vena cava. Encountered branches were ligated. The urether was cut in its middle third. Heparin (15 IU/kg body weight, Leo Pharmaceutical Products B.V., Weesp, The Netherlands) was administered intraveneously and a flank incision was made in the kidney capsule. Immediately after exsanguination of the donor animal a perfusion canula was inserted through the aorta in the renal artery. The renal vein was cut in order to drain the blood and washout solution. The kidney was flushed *in situ* with maximally 10 ml of isotonic citrate solution (Na_3.citrate 78 mM, K_3.citrate 84 mM, $MgSO_4$ 40 mM, mannitol 100 mM, pH 7.1). This citrate solution was shown to be effective in preserving renal function in a rat kidney transplant model (Howden et al., 1983).

2.2. Experimental Groups

To simulate progressive ischemic injury flushed kidneys were decapsulated, excised and immediately transferred to a perfusion set-up (***direct***), stored for 1 h at room temperature (***RT***) or stored for 1 h at 37°C (***37°C***) ($3 \times n = 4$).

2.3. Experimental Protocol

Perfusion was performed at constant pressure (±90 mm Hg) at room temperature with citrate solution supplemented with ureum 4.0 mM, glucose 5.0 mM, $CaCl_2$ 2.0 mM, cysteine 0.5 mM, glycine 2.3 mM, Na.pyruvate 2.0 mM, Na.acetate 1.22 mM, Na.glutaminate 2.106 mM, Na.proprionate 0.208 mM, inosine 1.0 mM, alanine 5.0 mM, α-ketaglutaric acid 1.15 mM, ascorbate 2 mg/l, Na.lactate 0.32 ml/l, cholinecloride 1.0 g/l, pH 7.05. The medium was passed through a 1.2 μm filter (Schleicher and Schuell, Dassel, Germany) shortly before perfusion. Both nitrogen (N_2) and oxygen (O_2) equilibrated perfusates were used. After connection to the perfusion system, kidneys were allowed to stabilize to perfusion conditions for 10 min. Perfusion flow was monitored just before the arterial cannula with an in-line flow probe (T106; Transonic Systems, New York).

Perfusion pressure was measured by a pressure transducer (Hewlett-Packard 8805B corner amplifier). Pressure was adjusted by the pressure of the equilibrating gas.

Two protocols were performed to evaluate a possible reoxygenation dependent effect on NADH-oxidation kinetics of the perfused kidney cortex. In *protocol I*, the preserved hypoxic kidney was made completely anoxic with N_2-equilibrated perfusates. Following the 10 min stabilisation phase of perfusion with N_2-equilibrated perfusate, kidneys were perfused with O_2-equilibrated perfusate for 5 min. In *protocol II*, the sequence of aerobic and anaerobic perfusates was reversed.

Alanine, ascorbate, cholinecloride, cysteine, glucose, inosine, α-ketaglutaric acid, mannitol, Na.glutaminate, Na.proprionate, ureum were obtained from Sigma Chemical Company (St. Louis, MO). $CaCl_2$, glycine, Na.acetate, Na_3.citrate, K_3.citrate, Na.lactate, Na.pyruvate and $MgSO_4$ were obtained from Merck (Darmstadt, Germany).

2.4. NADH-Fluorimetry of Perfused Kidneys

During the experimental procedure surface NADH fluorescence images of kidneys were recorded with an image intensified CCD video camera equipped with a 105 mm macro lens (Figure 1A). 365 nm light from a 200 W Xenon Mercury lamp, was selected by means of an UG-1 color glass filter. This UV-excitation light was directed onto the cortex by a dichroic mirror (DM) which transmitted the fluorescence emitted by the kidney. A 470 nm narrow band interference filter was used to select NADH fluorescence from the tissue autofluorescence (Coremans et al., 1993; Ince et al., 1993).

Analysis of the metabolic state of tissue by means of NADH fluorimetry may be disturbed by tissue movements and by hemodynamic and oximetric effects due to changes in the absorption of the UV excitation light (Coremans et al., 1997). In the isolated perfused kidney metabolism-dependent changes in absorption of UV excitation light are

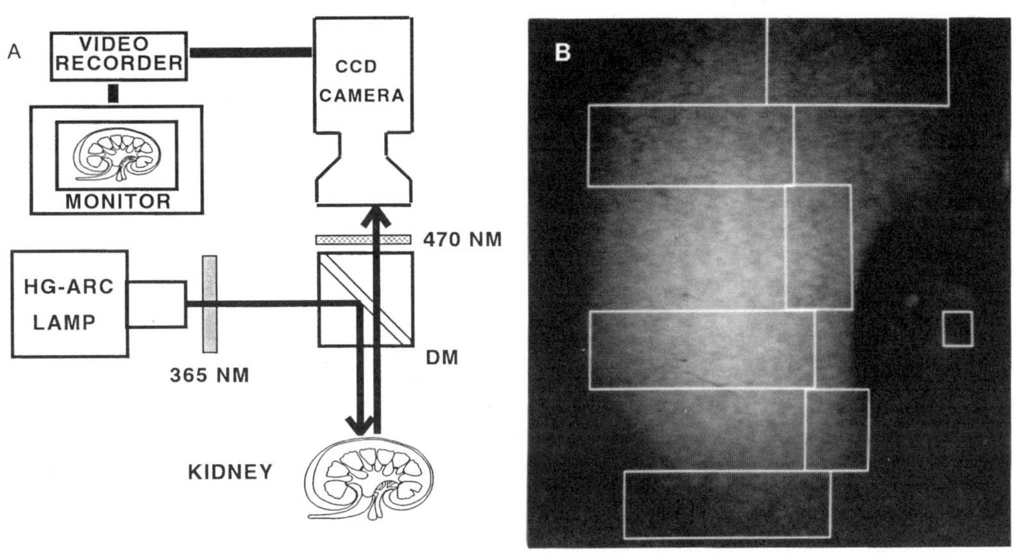

Figure 1. A. Schematic view of the NADH videofluorimeter, **B.** NADH fluorescence image of a perfused normoxic kidney showing 6 surface areas selected for fluorescence intensity analysis. The grey circle near the hilus of the kidney is a piece of fluorescence calibration glass.

small, so that uncorrected NADH fluorescence measurements are sufficient for these experiments (Franke et al., 1976).

NADH fluorescence images (Figure 1A) were computer analysed off-line. To obtain an impression of possible heterogeneity in kidney cortex metabolism, the surface was divided in 6 rectangular areas alternating along an imaginary line, connecting the poles of the kidney (Figure 1B).

2.5. NADH Oxidation Parameters

In isolated kidneys perfused with deoxygenated perfusate the mitochondrial NADH pool is reduced and a relatively high NADH fluorescence intensity is observed. Upon reoxygenation with O_2-saturated perfusate oxidative phosphorylative activity results in NADH oxidation. This is reflected by a decrease in NADH fluorescence intensity. After an initial rapid decline in B fluorescence intensity, a nearly steady state is reached after 1.5 min.

NADH oxidation kinetics were evaluated by two parameters (Figure 2A,B):

1. $\Delta F_{NADH/NAD^+}$: the difference in steady state NADH fluorescence intensities of reduced kidney perfused with N_2-equilibrated perfusate and after 1.75 min reoxygenation with O_2-equilibrated perfusate. $\Delta F_{NADH/NAD^+}$ is expressed in terms of percentage of the fluorescence intensity just before reoxygenation.

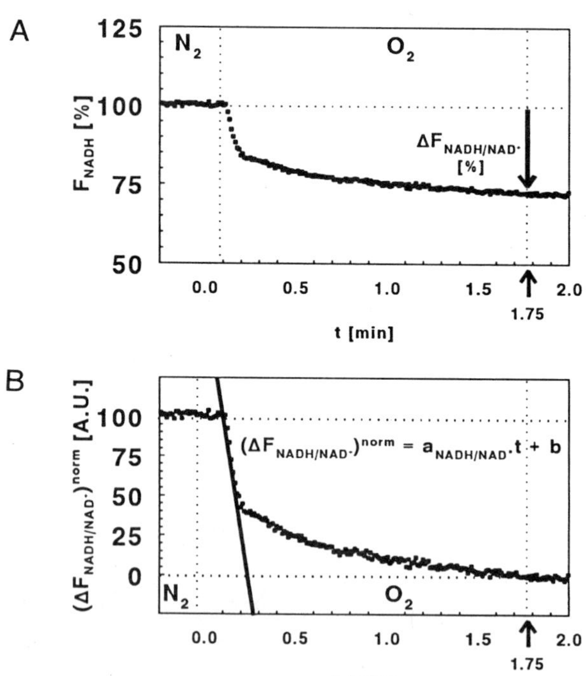

Figure 2. NADH oxidation kinetics in perfused rat kidney indicating **A.** $\Delta F_{NADH/NAD^+}$, the relative difference in NADH fluorescence intensity levels of the reduced and oxidized steady states. **B.** a_{NADH/NAD^+}, the slope of the tangent to the initial rapid NADH fluorescence intensity decrease.

2. $(\Delta F_{NADH/NAD^+})^{norm} = a_{NADH/NAD^+} . t + b$: the slope of the tangent of the initial rapid NADH fluorescence intensity decrease. a_{NADH/NAD^+} is given in [%.min^{-1}]. In this case the NADH fluorescence difference between reduced and oxidised kidney is normalised.

2.6. Statistics

The values presented in the tables are means ± SD. For statistical analysis the non-parametric Mann-Whitney U—Wilcoxon Rank Sum W tests for paired data were used. The accepted two-tailed probability for a significant difference between means was $p < 0.05$.

3. RESULTS

An example of surface NADH fluorescence intensity changes induced by an O_2 challenge of a kidney in reduced state is shown in Figure 3. After a short lag phase an abrupt decrease in NADH fluorescence intensity reflects the oxidation from NADH to NAD^+. After 5 min perfusion with O_2-equilibrated perfusate, perfusion with N_2-equilibrated perfusate is resumed and the $NADH/NAD^+$ redox equilibrium returns to the reduced state as witnessed by increasing NADH fluorescence intensity.

The three groups of kidneys with progressive ischemic injury displayed different NADH oxidation kinetics (Figure 4, Table I and II). In both experimental protocols the slope, **a**, of the tangent to the initial rapid NADH fluorescence intensity decrease was steepest for minimally damaged (*direct*) kidneys and smallest for kidneys stored for 1 h at *37°C*. On average the velocity of NADH oxidation of *RT* kidneys was somewhere in between. It should be noted, however, that one of the *RT* kidneys perfused according to protocol I showed NADH oxidation kinetics similar to *direct* kidneys (Table I). The slope of another *RT* kidney was in the range of those in the *37°C* group.

The relative difference in NADH fluorescence intensity levels of the reduced and oxidized steady states (ΔF_{NADH}) of the *37°C* kidneys was significantly larger than that of *direct* and *RT* kidneys.

There was no sizeable effect of an additional reoxygenation step on NADH oxidation parameters in protocol II. The kinetic parameters of *direct* and *37°C* kidneys perfused according to protocol I and II did not differ.

Frequently, intensity differences were observed in surface NADH fluorescence images. Apart from heterogeneity in NADH fluorescence intensity, regional differences in NADH oxidation kinetics were observed. To some extent this heterogeneity in kidney

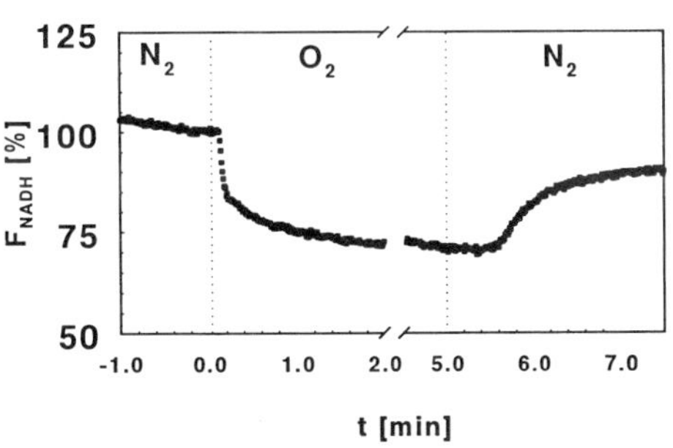

Figure 3. NADH fluorescence intensity changes in perfused rat kidney when N_2-saturated perfusate is replaced with O_2-saturated perfusate for 5 min.

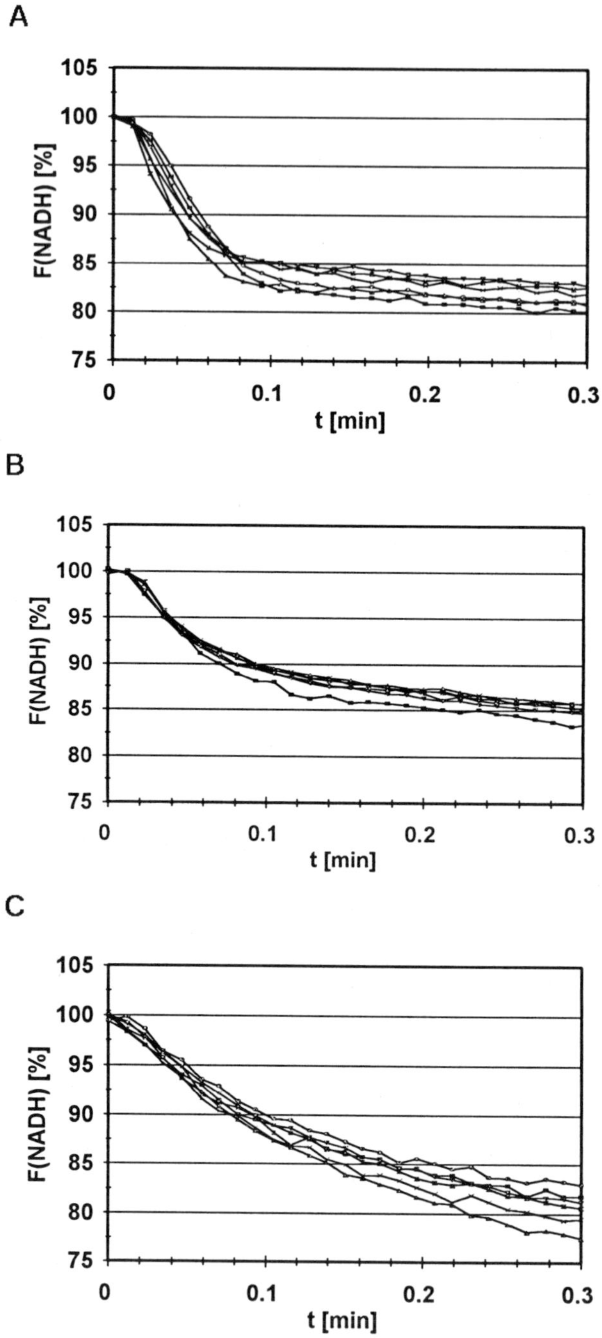

Figure 4. NADH oxidation kinetics in perfused rat kidneys with progressive ischemic injury. **A.** ***direct*** kidney, **B.** ***RT*** kidney and **C.** ***37°C*** kidney. The different symbols correspond to the average NADH fluorescence intensities of 6 surface areas of a single kidney.

cortex metabolism is reflected in the standard deviation of the kinetic parameters a and ΔF_{NADH}.

The initial rate of NADH oxidation did not correlate with the rate of O_2 delivery to the perfused kidney as monitored by overall perfusion flow. Perfusion flow did not

Table I. Effect of progressive ischemia injury on NADH oxidation kinetics of rat kidneys perfused according to *protocol I* (means ± sd)

Group	a_{NADH/NAD^+} [%.min^{-1}]	Group average a_{NADH/NAD^+} [%.min^{-1}]	$\Delta F_{NADH/NAD^+}$ [%]	Group average $\Delta F_{NADH/NAD^+}$ [%]
direct	−678.13 ± 119.64		−27.93 ± 1.51	
	−959.84 ± 153.19		−25.75 ± 1.10	
	−800.56 ± 49.34		−24.35 ± 0.54	
	−881.92 ± 41.64	**−830.12 ± 106.27**	−20.23 ± 1.30	**−24.56 ± 2.81**
RT	−612.62 ± 62.58		−23.98 ± 0.94	
	−870.21 ± 145.22		−22.83 ± 1.57	
	−554.59 ± 63.06		−23.28 ± 1.13	
	−688.20 ± 37.85	**−681.40 ± 118.86**	−21.53 ± 1.00	**−22.90 ± 0.89**
37°C	−571.38 ± 32.38		−41.07 ± 3.73	
	−425.52 ± 18.76		−26.92 ± 2.16	
	−446.83 ± 23.50		−34.97 ± 1.65	
	−453.92 ± 21.81	**−474.41 ± 56.95*†**	−36.53 ± 2.38	**−34.87 ± 5.11*†**

* $p < 0.05$; two tailed Mann-Whitney test relative to values of *direct* kidneys.
† $p < 0.05$; two tailed Mann-Whitney test relative to values of *RT* kidneys.

differ significantly between groups. For *direct*, *RT* and *37°C* kidneys perfusion flow was, 12.59 ± 2.74l.min^{-1}, 15.12 ± 2.57 ml.min^{-1} and 12.68 ± 2.66 ml.min^{-1}, respectively.

4. DISCUSSION

Mitochondrial NADH oxidation kinetics recorded by surface NADH videofluorimetry were compared for a model of progressive ischemic injury in rat kidney transplants. In this model the quality of the grafts during harvesting and storage, is expected to decrease in kidneys from, subsequently, the *direct*, *RT* and *37°C* groups. It is clear from Figure 4 and Tables I and II that the rate of NADH oxidation decreases with decreasing graft quality ($a_{direct} > a_{RT} > a_{37°C}$). This finding supports the view that oxidative phos-

Table II. Effect of progressive ischemic injury on NADH oxidation kinetics of rat kidneys perfused according to protocol II (means ± sd)

Group	a_{NADH/NAD^+} [%.min^{-1}]	Group average a_{NADH/NAD^+} [%.min^{-1}]	$\Delta F_{NADH/NAD^+}$ [%]	Group average $\Delta F_{NADH/NAD^+}$ [%]
direct	−785.49 ± 123.79		−25.18 ± 0.95	
	−805.88 ± 78.67		−25.83 ± 0.96	
	−880.67 ± 61.11		−27.70 ± 1.30	
	−709.13 ± 94.45	**−795.29 ± 61.08**	−23.67 ± 2.21	**−25.60 ± 1.45**
RT	−574.38 ± 50.25		−25.87 ± 1.41	
	−542.77 ± 42.23		−20.99 ± 1.42	
	−507.54 ± 29.07		−22.97 ± 1.88	
	−489.05 ± 43.96	**−528.44 ± 32.80***	−23.58 ± 2.90	**−23.35 ± 1.74**
37°C	−535.91 ± 53.21		−28.45 ± 1.59	
	−373.53 ± 36.60		−27.43 ± 2.19	
	−314.57 ± 29.62		−29.88 ± 1.52	
	−322.09 ± 20.09	**−386.53 ± 89.18***	−29.63 ± 1.80	**−28.85 ± 0.98*†**

* $p < 0.05$; two tailed Mann-Whitney test relative to values of *direct* kidneys.
† $p < 0.05$; two tailed Mann-Whitney test relative to values of *RT* kidneys.

phorylation is impaired by ischemic damage to mitochondria (Mittnacht and Farber, 1981; Saris and Eriksson, 1995; Ferrari, 1996), or by damage to one or more complexes (in particular complex I) of the respiratory chain (Rouslin, 1983; Rouslin and Ranagathan, 1983; Hardy et al., 1991; Thorniley et al., 1994; Gonzalez-Flecha and Boveris, 1995; Steinlechner-Maran et al., 1997; Cairns et al., 1997). An alternative explanation for the observed differences in NADH oxidation kinetics could be changes in regional renal perfusion induced by ischemic injury (Lennon et al., 1991). Overall perfusate flow, however, was equal for the three experimental groups.

The larger difference in NADH fluorescence intensity levels of the reduced and oxidized steady states (ΔF_{NADH}) of the *37°C* kidneys compared to the ΔF_{NADH} of *direct* and *RT* kidneys (Tables I and II), may reflect ischemia induced uncoupling of oxidative phosphorylation which leads to a more oxidized steady state under aerobic conditions (Franke et al., 1976; Saris and Eriksson, 1995). On the other hand the smaller ΔF_{NADH} of *direct* kidneys could be an indication of O_2 diffusion dependence of mitochondrial NADH oxidation in metabolically active isolated perfused kidney (Franke et al., 1976).

The presented results indicate that NADH oxidation kinetics monitored by surface NADH videofluorimetry can be useful in grading ischemic injury in transplant kidneys. The fact that NADH fluorescence measurements only provide information about the respiratory activity of proximal tubules which are located in the superficial layers of the outer cortex forms a limitation of this technique. However, the metabolically highly active proximal tubules are especially vulnerable to ischemic injury and changes in tubular reabsorption capacity directly influence renal function (Bahlmann et al., 1967).

Research is now focused on the question whether NADH oxidation kinetics of perfused donor kidneys can be used to predict their post-transplantation function. A first series of experiments with a rat model indicates a good correlation between NADH oxidation kinetics of the isolated perfused donor kidney and its post-transplantation function (Coremans et al., manuscript in preparation). Hence this non-invasive method for evaluation of the viability of preserved kidneys prior to transplantation may potentiate the use of kidneys from non-heart-beating donors, thus enlarging the number of kidneys that become available for transplantation by twenty percent (Kootstra et al., 1991).

ACKNOWLEDGMENTS

This study was supported in part by the Dutch Kidney Foundation (grant nr. C95.1433).

REFERENCES

Bahlmann J, Giebisch G, Ochwadt B, and Schoppe W (1967) Micropuncture study of isolated perfused rat kidney. Am. J. Physiol. 212:77–82.

Balaban RS (1979) The coupling of aerobic metabolism to active ion transport in the kidney (PhD dissertation), Durham, NC: Duke University.

Cairns CB, Ferroggiaro AA, Walther JM, Harken AH, and Banerjee A (1997) Post-ischemic administration of succinate reverses the impairment of oxidative phosphorylation after cardiac ischemia and reperfusion injury. Circ. 96(9 Suppl.):II-260–265.

Chance B, Cohen P, Jöbsis FF, and Schoener B (1962) Intracellular oxidation-reduction states *in vivo*. Science 137:499–508.

Coremans JMCC, Ince C, and Bruining HA (1993) NADH fluorimetry and diffuse reflectance spectroscopy on rat heart. In: Medical Optical Tomography: Functional imaging and monitoring. (Möller G, Chance

B, Alfano R, Arridge S, Beuthan J, Gratton E, Kaschke M, Masters B, Svanberg S, van der Zee P, Eds) SPIE Optical Engineering Press, Washington USA. Vol. IS 11: pp. 589–617.

Coremans JMCC, Ince C, Bruining HA, and Puppels GJ (1997) (Semi-)Quantitative analysis of reduced nicotinamide adenine dinucleotide fluorescence images of blood perfused rat heart. Biophys. J. 72:1849–1860.

Duboc D, Abastado P, Muffat-Joly M, Perrier P, Toussaint M, Marsac C, Francois D, Lavergne T, Pocidalo J-J, Guerin F, and Carpentier A (1990) Evidence of mitochondrial impairment during cardiac allograft rejection. Transplantation 50(5):751–755.

Ferrari R (1996) The role of mitochondria in ischemic heart disease. J. Cardiovasc. Pharmac. 28(Suppl. 1):S1–S10.

Franke H, Barlow CH, and Chance B (1976) Oxygen delivery in perfused kidney: NADH fluorescence and renal functional state. Am. J. Physiol. 231(4):1082–1089.

Gonzalez-Flecha B and Boveris A (1995) Mitochondrial sites of hydrogen peroxide production in reperfused rat kidney cortex. Biochim. Biophys. Acta 1243:361–366.

Hardy L, Clark JB, Darley-Usmar VM, Smith DR, and Stone D (1991) Reoxygenation-dependent decrease in mitochondrial NADH:CoQ reductase (Complex I) activity in the hypoxic/reoxygenated rat heart. Biochem. J. 274:133–137.

Howden B, Rae D, Jablonski P, Marshall VC, and Tange J (1983) Studies of renal preservation using a rat kidney transplant model. Transplantation 35(4):311–314.

Ince C, Ashruf JF, Avontuur JAM, Wieringa PA, Spaan JAE, and Bruining HA (1993) Heterogeneity of the hypoxic state in the rat heart is determined at capillary level. Am. J. Physiol. 264:H294–H301.

Kono Y, Ozawa K, Tanaka J, Ukikusa M, Takeda H, and Tobe T (1982) Significance of mitochondrial enhancement in restoring hepatic energy charge after revascularisation of isolated ischemic liver. Transplantation 33:150–155.

Kootstra G, Wijnen R, Van Hooff JP, and Van der Linden CJ (1991) Twenty percent more kidneys through a Non-heart beating program. Transplant. Proc. 23:910–911.

Lennon GM, Ryan PC, Gaffney EF, and Fitzpatrick JM (1991) Changes in regional renal perfusion following ischemia/reperfusion injury to the rat kidney. Urol. Res. 19:259–264.

Mittnacht Jr S and Farber JL (1981) Reversal of ischemic mitochondrial dysfunction. J. Biol. Chem. 256:3199–3206.

Nohl H, Koltover V, and Stolze K (1993) Ischemia/reperfusion impairs mitochondrial energy conservation and triggers $O_2^{\cdot -}$ release as a byproduct of respiration. Free Radic. Res. Commun. 18(3):127–137.

Okamura R, Tanaka A, Uyama S, and Ozawa K (1992) Low-temperature fluoreometric technique for evaluating the viability of rat liver grafts after simple cold storage. Transplant Int. 5:165–169.

Rouslin W (1983) Mitochondrial complexes I, II, III, IV and V in myocardial ischemia and autolysis. Am. J. Physiol. 244:H743–H748.

Rouslin W and Ranagathan S (1983) Impaired function of mitochondrial electron transfer complex I in canine myocardial ischemia. Loss of flavin mononucleotide. J. Mol. Cell. Cardiol. 15:537–542.

Saris NEL and Eriksson O (1995) Mitochondrial dysfunction in ischemia-reperfusion. Acta Anaesthesiol. Scand. 39: Supplementum 107:171–176.

See YP, Weisel RD, Mickle DAG, Teoh KH, Wilson GJ, Tumiati LC, Mohabeer MK, Madonik MM, Axford-Gatley RA, and Salter DR (1992) Prolonged hypothermic cardiac storage for transplantation. The effects on myocardial metabolism and mitochondrial function. J. Thorac. Cardiovasc. Surg. 104(3):817–824.

Steinlechner-Maran R, Eberl T, Kunc M, Schröcksnadel H, Margreiter R, and Gnaiger E (1997) Respiratory defect as an early event in preservation-reoxygenation injury of endothelial cells. Transplantation 63:136–142.

Thorniley MS, Lane NJ, Manek S, and Green CJ (1994) Non-invasive measurement of respiratory chain dysfunction following hypothermic renal storage and transplantation. Kidney Int. 45:1489–1496.

Thorniley MS, Simpkin S, Fuller B, Jenabzadeh MZ, and Green CJ (1995) Monitoring of surface mitochondrial NADH levels as an indication of ischemia during liver isograft transplantation. Hepatology 21(6):1602–1609.

Yano H, Takenaka H, Onitsuka T, Koga Y, and Hamada M (1993) Cardioplegic effect of University of Wisconsin solution on hypothermic ischemia of rat myocardium assessed by mitochondrial oxidative phosphorylation. J. Thorac. Cardiovasc. Surg.:106, 502–510.

EFFECT OF HYDROXYL RADICAL SCAVENGERS IN RENAL CORTICAL CELLS

G. Gronow, M. Moussavian, and M. Málυusz

Department of Physiology
Olshausenstr. 40, University of Kiel
D-24098 Kiel, Germany

1. INTRODUCTION

In renal cortical cells the reversibility of posthypoxic cellular function may be limited by their high demand for metabolic energy. Irreversible structural and functional damage may be detected after 15–20 min of normothermic oxygen deprivation (Gronow et al., 1984, 1990). Furthermore, cellular defense mechanisms may fail in the reoxygenation period due to the exhaustion of radical scavengers and antioxidants. Adenosine, the hypoxic break-down product of ATP will provide xanthine/hypoxanthine for an increased accumulation of superoxide radicals via xanthinoxidase (Paller and Neumann, 1991; Zager et al., 1995; Sogabe et al., 1996). Finally, posthypoxic cellular recovery may be limited in the kidney by the formation of reactive oxygen species (ROS). Among these the univalent reduction of hydrogen peroxide leads to the formation of the hydoxyl radical ($OH^{\cdot}$), a highly reactive biological oxidant.

Concurrently, benzoate may be present in substantial concentration in renal tissue and the urine. It derives either from food uptake and/or benzoyl-glycine (hippurate, Hip). Hydrolysis of Hip liberates cytoprotective glycine (Weinberg et al., 1989; Gronow et al., 1990, 1994) and the aromatic compound benzoate (BA). The latter may trap cytotoxic $OH^{\cdot}$-radicals similar to the known OH- scavenger salicylate (SA) via hydroxylation to hydroxy- (HB) and dihydroxybenzoates (DHB). Thus, the availability of benzoate may contribute to the posthypoxic recovery of renal cortical cells. In the following experiments we tested the liberation of hydroxylated benzoate metabolites in the postischemic rat kidney *in situ* as well as the effects of different hydroxyl radical scavengers and of deferoxamine on posthypoxic recovery of tubular cells from rat renal cortex.

Oxygen Transport to Tissue XXI, edited by Eke and Delpy.
Kluwer Academic / Plenum Publishers, New York, 1999.

2. METHODS

2.1. *In Situ* Experiments

Hydroxylation of benzoate was studied under pentobarbital anesthesia (60–8 mg/kg) in Hannover Wistar rats (BW 300–350 g) *in situ*. Hippurate (benzoyl-glycine) was infused intravenously at a rate of 0.5–1.0 $\mu mol \times g^{-1} \times h^{-1}$. Both kidneys were exposed by midsection and the ureters cannulated separately. Urine and blood samples served as probes for HPLC-analysis of hippurate (Hip), benzoate (BA), 2,5- and 2,3-dihydroxybenzoate (2,5- and 2,3-DHB), as well as the sum of 3- plus 4-hydroxybenzoate (3- plus 4-HB). Preceeding control periods yielded no significant differences in the excretion rates of benzoate metabolites of both kidneys. The right kidney was 30 min clamped, released and 60 min reperfused. Urine samples of both kidneys were analyzed for hippurate, benzoate and OH-metabolites (Mályusz et al., 1994). To compensate for variations in urine flow rates the excretion of benzoate metabolites was calculated as percent of excreted hippurate (% Hip).

2.2. ITS-Experiments

Isolated tubular segments of rat kidney cortex (ITS) were prepared as described elsewhere (Gronow et al., 1984, 1994). All tubules were suspended in Krebs-Ringer bicarbonate solution at 37°C in the presence of 5 mM glycine, 10 mM lactate, 2 mM butyrate and 0.5% (W:V) bovine albumine. ITS were 30 min gassed with 95% N_2:5% CO_2 at a $PO_2 < 1$ mm Hg and subsequently reoxygenated for 60 min with 95% O_2:5% CO_2. Effects of benzoate (BA), salicylate (SA), mannitol (MA), and deferoxamine (DFO) were compared after the 60 min reoxygenation period. Untreated reoxygenated tubules from the same preparation served as controls (=100%). Disruptive cellular swelling was indicated by the loss of the marker enzyme lactate dehydrogenase (LDH), and reversibility of posthypoxic cellular function was judged by intracellular potassium accumulation (K^+) as well as by the formation of glucose (gluconeogenesis, GNG). Assays of enzyme activity, glucose content of the incubation medium as well as of intracellular K^+ were performed according to standard procedures and previously described methods (Gronow et al., 1990, 1994). A *P*-value of 0.05 or less in Student's *t*-Test for paired observations was regarded to indicate a statistical significant difference.

3. RESULTS

3.1 Benzoate Hydroxylation *in Situ*

Plasma hippurate levels were maintained at about 1–2 $\mu mol \times g^{-1}$ in anesthetized rats using intravenous infusion of hippurate (Figure 1, upper panel. Each couple of columns represents left and right kidney. Fractional urinary excretion (% Hip) of 2,5- and 2,3-dihydroxybenzoate (2,5- and 2,3-DHB) as well as the sum of 3- plus 4-hydroxybenzoate (3- plus 4-HB) was not significantly different in 2 preceeding control periods (left side in Figure 1). During 30 min occlusion of the right renal artery there was no change in the urine of the untreated left kidney. After clamp release and reperfusion, the excretion of 2,5- and 2,3-DHB was not markedly changed in either the post-ischemic or the untreated kidney (Figure 1, middle). In contrast, excretion of 3-HB plus 4-HB rose

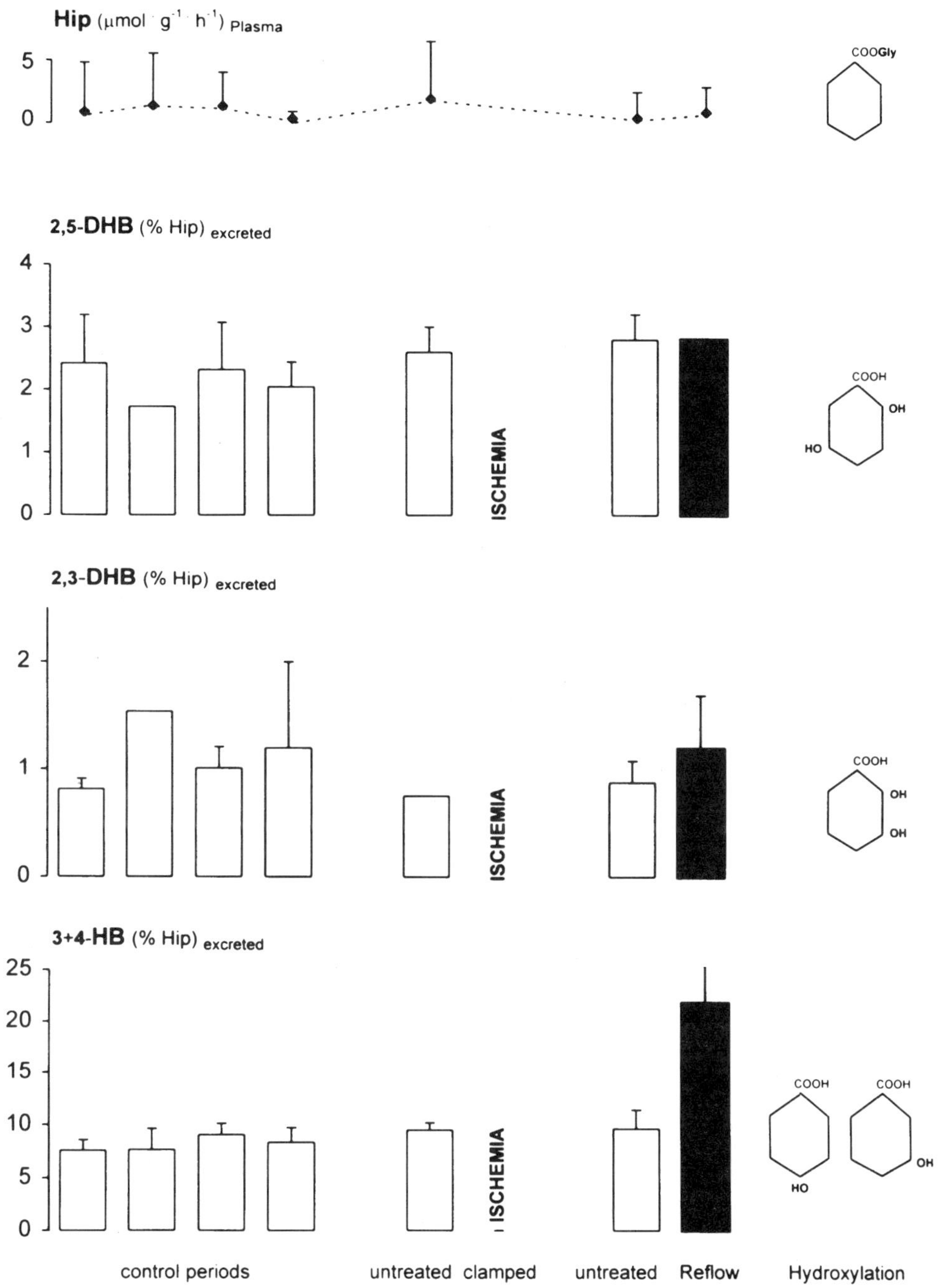

Figure 1. Hydroxyl radical-trapping in reperfused kidney *in situ.*

in the reperfused kidney by more than 100% with respect to the untreated control kidney (dark column). Addition of benzoate to the perfusate further stimulated the postischemic release of 3-HB plus 4-HB, whereas equimolar salicylate (SA) had no such an effect (not shown).

3.2. Comparison of OH· Radical Scavengers *in Vitro*

In reoxygenated isolated tubular segments benzoate significantly reduced LDH leakage at concentrations of 0.01–0.1 mM by about 28–55% (Figure 2). At 10 mM, however, benzoate became cytotoxic and induced LDH losses up to 160%. Salicylate was not as cytoprotective as benzoate. It exerted only at 0.1 mM a beneficial effect (52% of control), was ineffective at 1 mM and stimulated LDH-leakage at 10 mM up to 273% of control. Mannitol had no significant effect on enzyme losses with the exception of 10 mM (58% of control). Deferoxamine (DFO) was most effective among all tested substances, at 0.01–10 mM LDH-leakage fell by 34–78% (Figure 2).

Active cell function in the reoxygenation period, as indicated by intracellular K^+-accumulation (K^+, Figure 3) improved in the presence of 0.01–1 mM benzoate by 117–128%. 10 mM benzoate reduced the beneficial effect completely. All tested concentrations of salicylate exerted no effects on K^+-pumping activity with the exception of 10 mM, at this high concentration intracellular K^+ was reduced by 45%. Mannitol had no effect at low concentrations, but it significantly raised K^+ at 10 mM by 38%. Deferoxamine also increased K^+, at 1 mM up to 136%, and at 10 mM to 142% of untreated control. Gluconeogenesis (GNG) reflects availability of metabolic energy during reoxygenation. GNG was improved by 0.1 mM (161%) and by 1 mM benzoate (151%, Figure 4). 10 mM benzoate, however, reduced GNG significantly by 47%. Salicylate had no marked effect on GNG at low concentration (0.001–1 mM) and inhibited glucose formation, 28% at 1 mM, and 69% at 10 mM. Mannitol significantly supported GNG at concentrations >0.1 mM, Gluconeogenesis increased to 203% at 1 mM, and to 254% at 10 mM. Deferoxamine had no marked effects at low concentration (0.001–0.01 mM) but stimulated glucose formation in the higher range (0.1–10 mM) up to 137–204% of control.

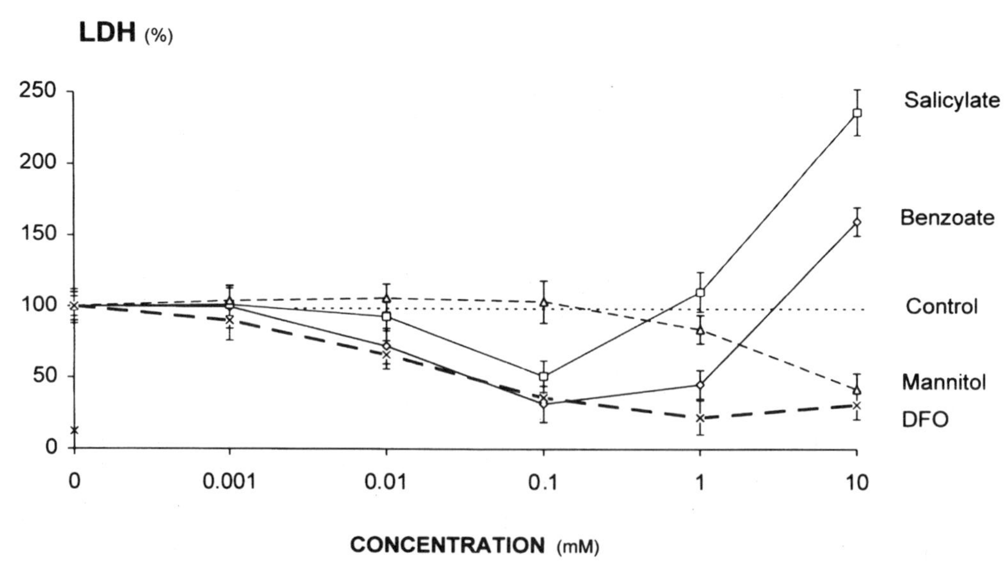

Figure 2. Cellular integrity of reoxygenated renal tubules.

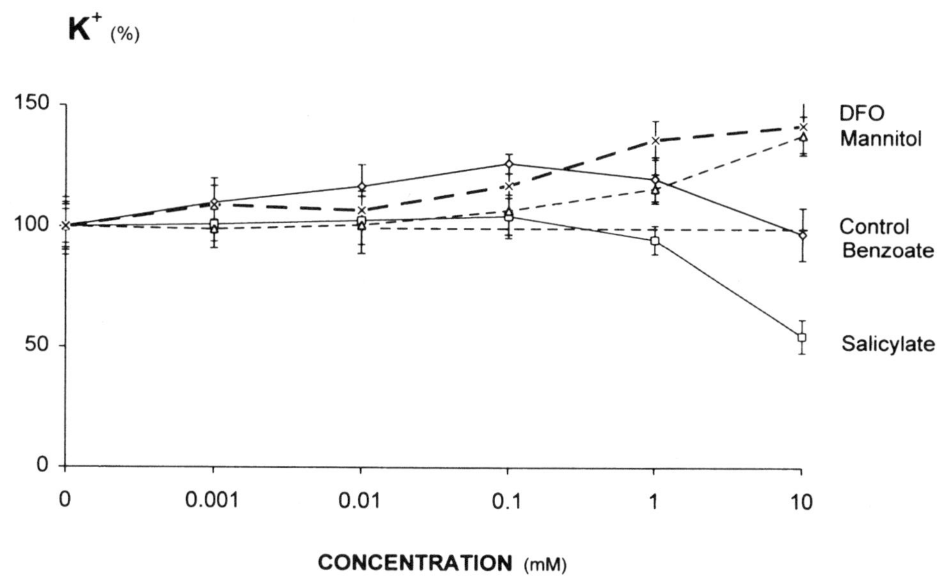

Figure 3. Cellular function in reoxygenated renal tubules.

4. DISCUSSION

Oxygen free radicals are generated in mammalian cells at the inner membranes of mitochondria, at the endoplasmatic reticulum, and in the cytosol. Under physiological conditions and at normal oxygen tension, oxygen radicals are detoxified by protective enzymes (SOD, Catalase, XOD) as well as by scavengers (α-tocopherol, GSH, polyalcohols, thiols, Chance et al., 1979). During reperfusion of the postischemic kidney, however, tissue activity of protecting enzymes and/or the capacity of oxygen radicals scavenging cell constituents may become insufficient, and free oxygen radicals may oxidize cellular macromolecules and membranes. Ionized iron which is present in biological material may than catalyze the Fenton reaction, i.e. the univalent reduction of hydrogen peroxide to $OH^{\cdot}$, the hydroxyl radical. $OH^{\cdot}$ may rapidly interact with surrounding macromolecules and extract electrons from unsaturated membrane lipids, aromatic amino acids or nucleic acids, for example. Inactivation of enzymes, cross-linking of DNA and destabilizing of cellular membranes may be the result of this forced oxidation (Sogabe et al., 1996).

Addition of $OH^{\cdot}$-radical scavengers (0.01–0.1 mM) to the incubation medium of reoxygenated renal cortical tubules significantly improved cellular integrity, which was indicated by a reduced loss of the cytosolic marker enzyme LDH, and supported cellular function, measured by means of intracellular K^+-accumulation and lactate gluconeogenesis (GNG). Thus, despite the presence of cytoprotective 5 mM glycine in all incubation media a lack of $OH^{\cdot}$ radical scavengers restricted recovery of cellular function in our experiments. Similar observations were made by Sogabe et al. (1996) who observed in rabbit kidney tubules a reduced cytoprotection by glycine during oxidant stress.

Benzoate (BA) may be present in renal tissue and thus contribute by $OH^{\cdot}$-trapping to the posthypoxic recovery of renal cortical cells. As demonstrated *in situ* (Figure 1), the preferred site for OH·-trapping was position 3 or 4 of the BA molecule. Binding of highly reactive OH· to BA probably also enhanced *in vitro* the recovery of reoxygenated tubular cells, as indicated in the presence of 0.01–1 mM BA by a significant reduction of LDH-loss (Figure 2) and a marked support of active cellular function K^+ and GNG (Figure 3 and 4). In the presence of salicylate (SA), however, no significant increase of 3-HB and 4-HB-hydroxylation could be observed during reperfusion *in situ.* Accordingly, cytoprotection in reoxygenated tubular cells was much smaller with equimolar SA than with BA. Hydroxylation of SA appears to be a more unspecific mechanism mediated by the P_{450}-system (Zager et al., 1995).

In contrast to the hydroxylation of benzoate, no reaction products of mannitol with hydroxyl ions could be estimated. Thus, no discrimination was possible between its unspecific osmotic effect and/or an expected OH·-scavanging activity (Gronow et al., 1989). At 10 mM, for example, mannitol was superior to equimolar benzoate or salicylate. This effect, however, may have been a combination of (osmotic) counteracting hypoxic cell swelling, OH·-scavenging activity, and/or cytotoxicity induced by high-dose benzoate or salicylate. At lower concentration (0.01–0.1 mM), however, equimolar benzoate was superior to mannitol, it significantly supported active cell function (K^+, GNG, Figures 3, 4) and suppressed LDH-leakage (Figure 2). Addition of equimolar deferoxamine (DFO) was even more effective than benzoate. Obviously, benzoate may trap hydroxyl radicals *after* they have been generated whereas DFO appears to prevent the generation of hydroxyl radicals by chelating catalytic iron ions in Fenton's reaction.

It is concluded that in reoxygenated, glycine-protected kidney cells benzoate may substantially contribute to cellular resistance against oxidant stress. Cytoprotection appears to be based on trapping hydroxyl radicals at position 3 or 4 of the benzoate molecule. Thus, the formation of 3- or 4-hydroxybenzoate appears to be a quantitative measure of oxidant stress in reoxygenated renal tissue.

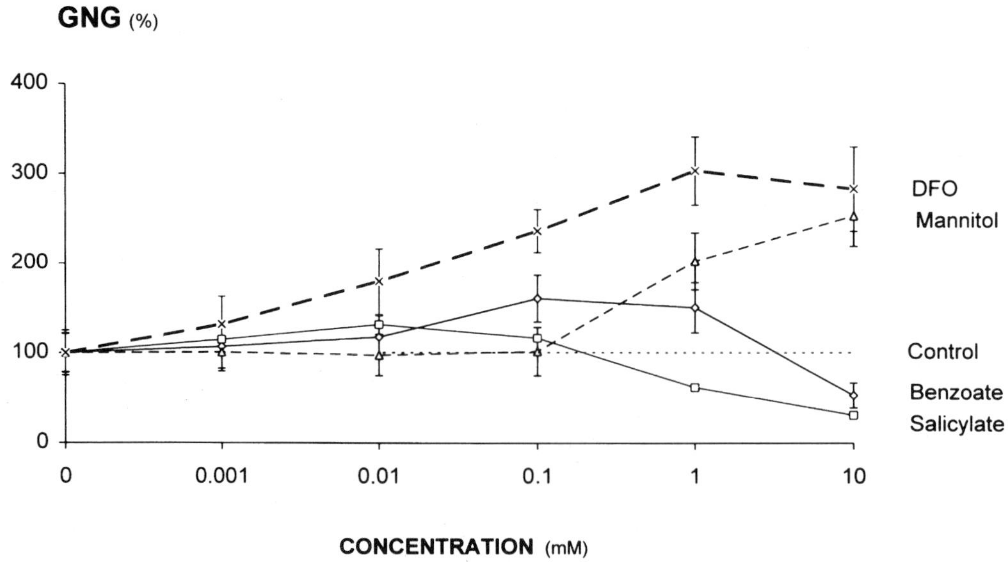

Figure 4. Cellular function in reoxygenated renal tubules.

ACKNOWLEDGMENTS

This work was supported by a grant of the Hensel Foundation, University of Kiel, Germany. The technical assistance of R. Lingg and M. Arnold is gratefully acknowledged.

REFERENCES

Chance B, Sies H, and Boveries A (1979) Hydroperoxide metabolism in mammalian organs. Physiol Rev 59:527–605.

Gronow G, Meya F, and Weiss Ch (1984) Studies on the ability of kidney cells to recover after periods of anoxia. Adv Exp Med Biol 169:589–595.

Gronow G, Prechel P, and Klause N (1989) Cytoprotective effect of isotonic mannitol at low oxygen tension. Adv Exp Med Biol 248:755–762.

Gronow G, Klause N, and Mályusz M (1990) Support of hypoxic renal cell volume regulation by glycine. Adv Exp Med Biol 277:705–712.

Gronow G, Klause N, and Mályusz M (1994) Restriction of hypoxic membrane defect by glycine improves mitochondrial and cellular function in reoxygenated renal tubules. Adv Exp Med Biol 361:585–589.

Mályusz M, Hackl A, Wrigge P, Lange M, Mályusz T, Sick H, and Gronow G (1994) Ammonia production from hippurate by the rat kidney in vitro. Renal Physiology 17:307–315.

Paller MS and Neumann TV (1991) Reactive oxygen species and rat renal epithelial cells during hypoxia and reoxygenation. Kidney Int 40:1041–1049.

Sogabe K, Roeser NF, Venkatachalam MA, and Weinberg JM (1996) Differential cytoprotection by glycine against oxidant damage to proximal tubule cells. Kidney Int 30:843–834.

Weinberg JM, Davis JA, Abarzua M, and Kiani T (1989) Relationship between cell ATP and glutathione content and protection by glycine against hypoxic tubular cell injury. J Lab Clin Med 113:612–623.

Zager RA, Burkhardt KM, Conrad DS, and Gmur DJ (1995) Iron, heme oxigenase, and glutathione: Effects on myoglobinuric proximal tubular injury. Kidney Int 48:1624–1634.

42

EFFECT OF FUROSEMIDE OR MANNITOL BEFORE INJECTION OF A NON-IONIC CONTRAST MEDIUM ON INTRARENAL OXYGEN TENSION

Per Liss,[1] Anders Nygren,[1] Hans R. Ulfendahl,[2] and Uno Erikson[1]

[1]Department of Diagnostic Radiology
University Hospital, Uppsala
[2]Department of Physiology and Medical Biophysics
University of Uppsala
Uppsala, Sweden

INTRODUCTION

Haemodynamic factors may play a role in the development of acute renal failure following administration of contrast media (CM). Today, one of the main the theories for this complication is hypoxia in the renal medulla (Brezis and Rosen, 1995), usually associated with an earlier impairment of renal function. In a recent study we found that low-osmolar and iso-osmolar CM caused a decrease in oxygen tension (pO_2) in the inner strip of the outer medulla (Liss et al., 1998). Since pretreatment with diuretics has been suggested and is also used as a prophylactic measure against contrast medium-induced acute renal failure (CM-ARF) (Levinsky and Bernard, 1988; Better et al., 1997), this study was undertaken to ascertain whether pretreatment with furosemide or mannitol could prevent the decrease in oxygen tension in the outer medulla that follows injection of CM.

MATERIAL AND METHODS

Animals

The studies were performed on 18 male Lewis-DA rats with an average body weight (BW) of 310 (±11) g. Anesthesia was induced with Inactin® (Byk-Gulden, Konstanz,

Oxygen Transport to Tissue XXI, edited by Eke and Delpy.
Kluwer Academic / Plenum Publishers, New York, 1999.

Germany), given intraperitoneally at a dose of 120 mg/kg BW. Tracheotomy was performed and each rat was placed on a servo-controlled heating pad to maintain body temperature at 37.5°C. The animals were divided into three groups: (1) furosemide followed by CM (n = 6); (2) mannitol followed by CM (n = 6), and (3) Ringer followed by Ringer (n = 6).

Surgical Procedures

A polyethylene catheter was placed in the left femoral artery for blood sampling and monitoring of blood pressure. The left femoral vein was catheterized for infusion of Ringer's solution (0.5 ml/h per 100 g BW). The right femoral vein was catheterized for infusion of CM. The left kidney was exposed by a left sub-costal flank incision and immobilised in a plastic cup. The kidney was embedded in pieces of cotton wool soaked in Ringer's solution, the surface covered with liquid paraffin. During the experiment the temperature of the kidney was monitored with a thermocouple probe.

Oxygen Microelectrodes

The oxygen microelectrodes used for the intrarenal measurements were a modified Clark type (Clark, 1956) used in earlier studies for measurements in the kidney (Liss, et al., 1997; Liss et al., 1998). They have an outer tip diameter of 2–8 μm and an inner hole diameter of 1–2 μm. The electrodes were polarized at –0.800 V, which gave a linear response between the oxygen tension and the electrode current. The electrodes were calibrated in water at 37°C bubbled with N_2 or air before and after each experiment. The drift of the micro-electrodes used was less than 0.5% per hour.

Experimental Protocol

When the surgical procedure were completed, the animal was allowed to recover for 20 min. Subsequently, after a 20 min control period, the experimental period began with an intravenous injection of furosemide (0.5 mg/ml) (NM Pharma AB, Stockholm) at a dose of 1 ml/kg BW or of mannitol (900 mOsm) at a dose of 2 ml/kg BW, injection of both furosemide or mannitol taking 30 sec. Ten minutes after the furosemide (or mannitol) injection, the non-ionic X-ray contrast medium iopromide (370 mg iodine/ml) (774 mOsm) (Schering, Berlin, Germany) was slowly injected intravenously taking 20 sec, at a dose of 600 mg iodine/kg BW. Ringer's solution in equivalent volumes was used as control. Throughout the experiment the oxygen tension in the renal cortex and outer medulla and the blood pressure, body temperature and kidney temperature were recorded continuously with a MacLab Instrument (AD Instruments, Hastings, England) connected to a Macintosh Power-PC 6100. The pO_2 was measured at a depth of 1.0 mm in the cortex and at an average depth of 3.9 mm (range 3.5–4.5) in the outer medulla. At the end of each experiment, the micro-electrodes were replaced by empty outer microelectrode casings of the same shape and size, placed on the same micro-manipulator and inserted to the same depth. After injection of a small amount of Indian ink, the outer casings were removed and the kidney was sectioned in order to verify the measurement sites.

Statistical Evaluation

Values are expressed as means ± SEM. The statistical significance of the data was tested by using a multivariate analysis of variance (Manova) model, comparing results obtained during the control period with those obtained after drug administration (repeated measurements). A p-value of <0.05 was regarded as significant in all analyses.

RESULTS

PO_2 – Mannitol + Iopromide

Ten minutes after the mannitol injection the pO_2 in the medulla had decreased from a control level of 32 ± 2 mm Hg to 28 ± 4. The addition of CM caused a further decrease in PO_2 to 24 ± 5 mm Hg. This reduction in pO_2 after CM was significant (Figure 1). No significant changes were observed in the cortex.

PO_2 – Furosemide + Iopromide

Ten minutes after the furosemide injection the pO_2 in the medulla had increased significantly, from a control value of 32 ± 2 to 44 ± 4 mm Hg. After injection of CM, pO_2 decreased to 37 ± 3 mm Hg during the 20 minutes the pO_2 was monitored. This decrease, from 44 ± 4 to 37 ± 3 mm Hg was significant (Figure 1). In the cortex, pO_2 decreased

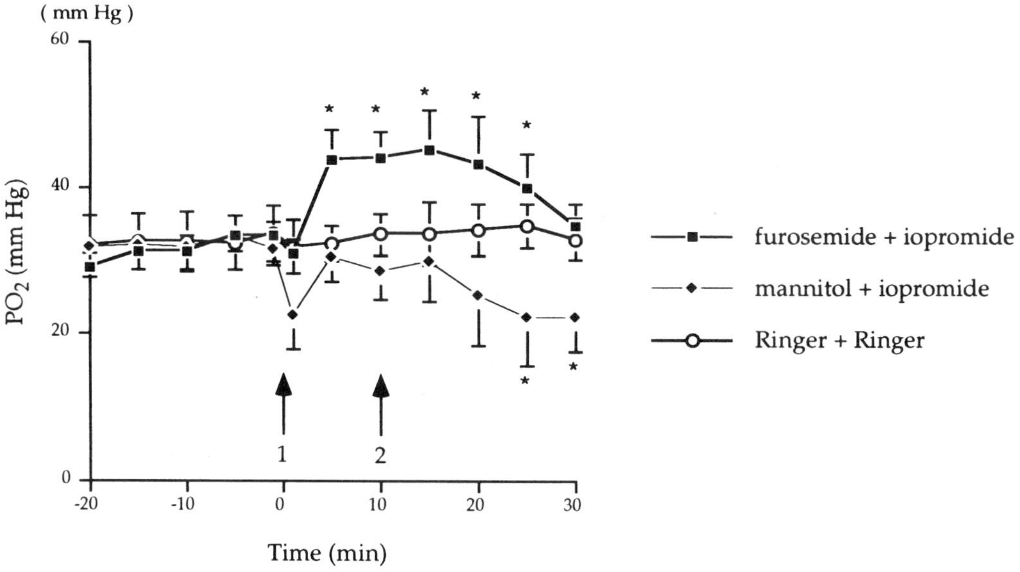

Figure 1. PO_2 in renal outer medulla (mm Hg), before injection of test substances (control value) after injection of a diuretic (mannitol n = 6; furosemide n = 6) and after injection of both diuretic and contrast medium (CM; iopromide). Ringer's (n = 6) used as control. * indicates a significant difference ($p < 0.05$) compared with the control period.

significantly from a control value of 47 ± 4 to 35 ± 4 mm Hg. After injection of CM, pO_2 increased to 42 ± 6 mm Hg (No figure).

DISCUSSION

The incidence of X-ray contrast medium-induced acute renal failure (CM-ARF) is reported to be less than 2% (Coggins and Fang, 1988), but the risk of this condition can increase up to 12% (Swartz et al., 1978) in patients with risk factors, e.g. impaired renal function, diabetes, or dehydration. Although there have been no published studies demonstrating a conclusively beneficial role for diuretics as prevention of CM-ARF (Rudnick et al., 1991; Kellum, 1998), mannitol and furosemide have long been used in clinical practice for this purpose (Levinsky and Bernard, 1988).

Today renal ischaemia, especially in the outer medulla, is suspected as the cause of contrast medium-induced ARF (Brezis and Rosen, 1995). In a recent study we found that iopromide induced a decrease in pO_2 in the outer medulla (Liss et al., 1998). In the present study we investigated the changes in oxygen tension in rat kidney after injection of either of these two diuretics prior to intravenous injection of the non-ionic, low-osmolar X-ray contrast medium iopromide. We found that pretreatment with mannitol (or furosemide) did not prevent the decrease in pO_2 in the outer medulla seen after injection of CM.

The hyperosmolar diuretic mannitol has been widely used to prevent and restore renal function (Better et al., 1997). One reason for this might be that an osmotic diuretic will almost invariably induce diuresis, even in conditions with a relatively severely impaired renal function. There are reports suggesting beneficial effects of mannitol on such renal functional parameters as serum creatinine. However, in the clinical setting there are several case reports claiming mannitol-induced ARF (Goldwasser and Fotino, 1984). Experimental studies have also shown impaired renal function after hyperosmolar mannitol injection (Nygren, 1989; Hellberg et al., 1990; Heyman et al., 1991).

Like hyperosmolar contrast media, mannitol is an osmotic diuretic, which is freely filtered by the glomeruli and poorly absorbed by the renal tubule. High mannitol concentration restricts the reabsorption of water and sodium from the tubule. This osmotic effect increases the amount of sodium in the tubule reaching the thick ascending limb (mTal) of the loop of Henle, in the outer renal medulla, where an active, oxygen-consuming resorption of sodium takes place. The increase in sodium in the tubule increases the sodium reabsorption in mTal (Pollock et al., 1991) resulting in a decrease in pO_2 in the outer medulla (Ullrich and Pehling, 1958; Aukland and Krog, 1961; Brezis et al., 1994). Since the pO_2 in the medulla is normally lower than that in the cortex (Aukland and Krog, 1960) it is believed that, since the medulla is on the verge of hypoxia under a normal workload, an increase in oxygen consumption may result in ischaemia in this region.

As a loop diuretic, furosemide exerts its effect by inhibiting the Na^+-K^+-$2Cl^-$ co-transporter (Greger, 1981) in mTal, whereby it will inhibit the active oxygen-consuming reabsorption of sodium (Burg et al., 1973) resulting in an increased pO_2 in the outer medulla. As about 80% of the oxygen consumption in the kidney is due to sodium resorption (Deetjen and Kramer, 1961; Lassen et al., 1961), a sodium reabsorption inhibitor will preserve the medullary pO_2.

Bayati et al. (1990) found that when given before induction of renal ischaemia in rats, furosemide improved the renal function as assessed one month after the injury. Also

Heyman et al. (1989) reported that injection of furosemide prior to a CM (iothalamate) could attenuate the severity of ARF and reduce mTal necrosis. Contradicting these findings of a preventive effect of furosemide on renal function, are experimental studies on glycerol-induced ARF, showing no prophylactic effect (Ufferman et al., 1975) or even exacerbation by this diuretic (Dubrow and Flamenbaum, 1988). Clinical investigators (Weinstein et al., 1992; Salomon et al., 1994) have also reported an increase in serum creatinine after injection of furosemide prior to injection of a CM. Weinstein et al. (1992) found that 8 patients given furosemide (1.5 mg/kg BW) before injection of CM showed a significant increase in serum creatinine, while in the control group of 10 patients, no change was observed. However, the authors also found that body weight in the furosemide/CM group decreased significantly, indicating a dehydrated state, whereas no such change was observed in the control group. In the study by Salomon et al. (1994) dehydration could be ruled out as all patients had a positive fluid balance.

In earlier studies by Nuutinen and Tuononen (1976) and by Brezis et al. (1994), injection of furosemide caused a sustained increase in pO_2. In the first study, pO_2 in the outer medulla was significantly increased during 40 min after injection. In the present study we found that even when furosemide increased the oxygen tension in the outer medulla for a short period, it did not prevent the decrease in pO_2 following injection of CM. If hypoxia in the renal medulla contributes to CM-ARF, it may be speculated that a low dose of furosemide—low enough not to cause dehydration may still have a protective effect. Brezis et al. (1994), injected mannitol after CM and found a further decrease in outer medullary pO_2. In the same study, injection of CM and mannitol followed by a furosemide injection increased pO_2. In our study we found that even though furosemide caused an increase in pO_2, injection of CM after furosemide significantly reduced pO_2 in the outer medulla. This decrease in pO_2 was not below the control value prior to injection of furosemide, though in this setting pO2 values were only monitored for 20 min after the injection of CM and pO_2 was declining. The important fact is that the furosemide injection did not prevent the decrease in pO_2 after injection of CM, which may be of clinical importance in patients with preceding renal impairment.

The results of this study indicate that use of diuretics does not prevent the decrease in pO_2 following injection of CM. In the clinical setting there is no evidence of any beneficial effect of diuretics in preventing CM-ARF. It is our opinion that furosemide and mannitol should not be injected before CM. Hydration with saline alone (Salomon et al., 1994) seems today to be the best prophylactic treatment.

SUMMARY

Oxygen tension (pO_2) in rat renal cortex and outer medulla was studied after an intravenous injection of mannitol or furosemide, followed 10 minutes later by an intravenous injection of the non-ionic X-ray contrast medium (CM) iopromide (370 mg iodine/ml). Ten minutes after mannitol injection, before injection of CM, pO_2 in the medulla had decreased from a control level of 32 ± 3 to 28 ± 4 mm Hg. The addition of CM caused a further decrease, to 24 ± 5 mm Hg, which was a significant reduction. Ten minutes after furosemide injection the pO_2 in the medulla had increased significantly, from a control value of 32 ± 2 to 44 ± 4 mm Hg. Injection of CM caused a significant decrease in pO_2 to $37 \pm 3^*$ mm. Ringer's solution (n = 6) caused no changes. We conclude that pretreatment with mannitol or furosemide does not prevent the CM-induced decrease in pO_2 in the outer medulla.

ACKNOWLEDGMENTS

This study was supported by the Gunvor and Josef Anér Foundation and the Marcus and Amalia Wallenberg Foundation.

REFERENCES

Aukland K and Krog J. Influence of various factors on urine oxygen tension in the dog. Acta Physiol Scand. 52:350–365. (1961).

Aukland K and Krog J. Renal oxygen tension. Nature. 188:671. (1960).

Bayati A, Nygren K, Källskog Ö, and Wolgast M. The effect of loop diuretics on the long-term outcome of post-ischemic acute renal failure in the rat. Acta Physiol Scand. 13:271–279. (1990).

Better O, Rubinstein I, Winaver JM, and Knochel JP. Mannitol therapy revisited (1940–1997). Kidney Int. 52:886–894. (1997).

Brezis M, Agmon Y, and Epstein F. Determinants of intrarenal oxygenation I. Effects of diuretics. Am J Physiol. 267:F1059–F1062. (1994).

Brezis M and Rosen S. Hypoxia of the renal medulla—its implication for disease. New Engl J Med. 332:647–655. (1995).

Burg M, Stoner J, Cardinal J, and Green N. Furosemide effect on isolated perfused tubules. Am J Physiol. 225:119–124. (1973).

Clark LC. Monitor and control of blood and tissue oxygen tensions. Trans Am Soc Art Int Org. 2:41–57. (1956).

Coggins CH and Fang LS. Acute renal failure associated with antibiotics, anesthetic agents, and radiographic contrast agents. In: Brenner BM, Lazareus JM, eds. Acute renal failure. Churchill Livingstone, 1988; 319–352.

Deetjen P and Kramer K. Sodium reabsorption and oxygen utilization of the kidney. Pflügers Arch. 273:636–650. (1961).

Dubrow A and Flamenbaum W. Acute renal failure associated with myoglobinuria and hemoglobinuria. In: Brenner BM, Lazareus JM, eds. Acute renal failure. Churchill Livingstone, 1988; 279–293.

Goldwasser P and Fotino S. Acute renal failure following massive mannitol infusion: appropriate response of tubuloglomerular feedback? Arch Intern Med. 144:2214–2216. (1984).

Greger R. Chloride reabsorption in the rabbit cortical thick ascending limb of the loop of Henle. Pflügers Arch. 390:38–43. (1981).

Hellberg O, Nygren A, Hansell P, and Fasching A. Post-ischaemic administration of hyperosmolar mannitol enhances erythrocyte trapping in outer medullary vasculature in the rat kidney. Renal Physiol Biochem. 13:328–332. (1990).

Heyman SN, Brezis M, Epstein FH, Spokes K, Silva P, and Rosen S. Early renal medullary hypoxic injury from radiocontrast and indomethacin. Kidney Int. 40:632–642. (1991).

Heyman SN, Brezis M, Greenfeld Z, and Rosen S. Protective role of furosemide and saline in radio-contrast-induced acute renal failure in the rat. Am J Kidney Dis. 14:377–385. (1989).

Kellum JA. Use of diuretics in the acute care setting. Kidney Int. Suppl 66:S67–S70. (1998).

Lassen NA, Munck O, and Thaysen JH. Oxygen consumption and sodium reabsorption in the kidney. Acta Pysiol Scand. 51:371–384. (1961).

Levinsky NG and Bernard DB. Mannitol and loop diuretics in acute renal failure. In: Brenner BM, Lazareus JM, eds. Acute renal failure. Churchill Livingstone, 1988; 841–856.

Liss P, Nygren A, Erikson U, and Ulfendahl HR. Injection of low and iso-osmolar contrast medium decreases oxygen tension in the outer renal medulla. Kidney Int. 53:698–702. (1998).

Liss P, Nygren A, Revsbech NP, and Ulfendahl H. Intrarenal oxygen tension measured by a modified Clark electrode at normal and low blood pressure and after injection of x-ray contrast media. Pflügers Arch. 434:705–711. (1997).

Nuutinen LS and Tuononen S. The effect of furosemide on renal blood flow and renal tissue oxygen tension in dogs. Ann Chir Gynaecol. 65:272–276. (1976).

Pollock CA, Lawrence JR, and Field MJ. Tubular sodium handling and tubuloglomerular feedback in experimental diabetes mellitus. Am J Physiol. 260:F946–F952. (1991).

Rudnick MR, Goldfarb S, and Murphy MJ. Mannitol and other prophylactic regimens in contrast media-induced acute renal failure. Coron Art Dis. 2:1047–1052. (1991).

Salomon R, Werner C, Mann D, D'Elia J, and Silva P. Effects of saline, mannitol and furosemide on acute renal function induced by radiocontrast agents. New Engl J Med. 331:1416–1420. (1994).

Swartz RD, Rubin JE, Leeming BW, and Silva P. Renal failure following major angiography. Am J Med. 65:31–37. (1978).

Ufferman RC, Jaenike JR, Freeman RB, and Pabico RC. Effects of furosemide on low-dose mercuric chloride acute renal failure in the rat. Kidney Int. 8:362–367. (1975).

Ullrich KJ and Pehling G. Active sodium transport and oxygen consumption in the outer medulla of the kidney. Pflügers Arch. 267:207–217. (1958).

Weinstein J-M, Heyman S, and Brezis M. Potential deleterious effect of furosemide in radiocontrast nephropathy. Nephron. 62:413–415. (1992).

IN VITRO PLACENTAL PERFUSION

Practical Parameters and Procedural Performance

D. J. Maguire,[1] G. R. Cannell,[2] R. S. Addison,[2] and B. Dawkins[3]

[1]School of Biomolecular and Biomedical Science
Griffith University, Nathan, Q4111
Australia
[2]Department of Medicine
The University of Queensland
Royal Brisbane Hospital
Herston, Australia
[3]Australian Red Cross Blood Service (Qld)
480 Queen St., Brisbane, Australia

1. INTRODUCTION

The placenta is a relatively short-lived tissue that is of absolute importance to the generation of future humans and indeed to the survival of all species of mammals. During its existence, it is responsible for a number of complex functions, which are normal features of separate organs and tissues within the body. Such functions include gas exchange, solute exchange, detoxification, primary endocrine functions, secondary endocrine functions and hormone degradation. In order to study the detailed biochemistry and physiology of full term placentae, *in-vitro* perfusion systems for this tissue were reported as long ago as 1922 (Schmitt), although the simplification of the procedure to incorporate the use of single placental lobules was not reported until much later (Panigel, 1962). The technique has been further refined since those early reports and has since been used to answer a wide range questions about normal and abnormal placental metabolism. In addition the technique has been applied to fields as diverse as toxicology, endocrinology and pharmacology, in order to determine the transfer kinetics of a range of xenobiotics. *In-vitro* placental perfusion using full-term human placentae has now been performed by our group in a number of locations for a period of years. As a result of a number of enquiries from other groups interested in establishing the procedure, it seemed timely to report our recent "success rate" with the procedure and to review the

Oxygen Transport to Tissue XXI, edited by Eke and Delpy.
Kluwer Academic / Plenum Publishers, New York, 1999.

major factors which require attention to ensure reasonable performance of this demanding procedure. We report here an analysis of the outcomes of ninety such perfusions attempted over the last two-year period (mid 1996–mid 1998). We also proffer a technique, which may be found valuable in attempts to normalise results for such parameters as oxygen consumption between perfusions as well as affording the prospect of earlier prediction of perfusion instability.

2. METHODS

2.1. Perfusion Technique

The dual recirculating perfusion system used was that reported by Addison et al. (1991), an adaptation of the method of Miller et al. (1985). Term human placentae were obtained, usually at caesarian section from healthy women with no significant drug history, and transported immediately in cold, oxygenated perfusate to an adjacent laboratory. Perfusions were established within a maximum period of twenty minutes of delivery, after carefully selecting a placental lobule with a well-defined foetal artery and vein pair. Maternal perfusion was established by insertion of two cannulae approximately 0.8 cm through the decidual plate. The volume of perfusate in the maternal and foetal circuits was 200 mL and 100 mL, respectively. The perfusate was tissue culture medium M199, obtained from Difco (Detroit, MI, U.S.A.) and supplemented with heparin (25 IU/mL), gentamicin (100 mg/L), glucose (2 g/L), sodium bicarbonate (2.9 g/L) and dextran (foetal 29 g/L; maternal 7.5 g/L). Perfusion was maintained in a dual recirculating mode with maternal and foetal flow rates of 25 and 3 mL/min, respectively. Tissue temperature was maintained at 37°C.

2.2. Perfusion Performance Criteria

Perfusions that leaked more than 12 mL per hour after connecting the perfusion circuit were abandoned immediately (failed attempts). If successful perfusion was established, maternal and foetal perfusate samples were collected at regular intervals throughout the period of perfusion and monitored for pH, pO2, pCO2, and glucose and, in some cases, lactate. Perfusions were continued up to a maximum of ten hours (successful perfusions), unless foetal arterial pressure exceeded 40 mmHg (abandoned perfusions).

2.3. Analyte Measurements

The oxygen content of perfusate delivery and return lines was assessed by collecting discrete samples throughout the period of the perfusion and assaying those samples in a blood gas analyser (ABL300, Radiometer, Copenhagen, Denmark). From these measurements, oxygen delivery, oxygen transfer and oxygen consumption by the placental tissue were derived. Glucose concentrations in perfusates were also determined by discrete sampling and analysis using a hexokinase method on an automated analyser (ABL625, Radiometer, Copenhagen, Denmark). Lactate measurements were similarly made on discrete samples using a lactate dehydrogenase spectrophotometric kit method on an automated analyser (ABL625, Radiometer, Copenhagen, Denmark).

2.4. Theoretical Calculations

The following derivation was developed to enable future estimation of the volume of fluid within the foetal vascular system in perfusions in which a bolus of any substance of known concentration had been added to the foetal circulation. Since the total volume (Vt) of the foetal perfusion circuit can be considered to consist of a known, measured volume (Vk) of perfusate and an unknown volume of foetal vascular circulation (Vf), then;

$$Vf = Vt - Vk \tag{1}$$

Since the volume (Vb) and initial concentration ([bi]) of the bolus added are known, then its expected concentration ([be]) after dilution into perfusate can be calculated as;

$$[be] = [bi] \times Vb/Vk \tag{2}$$

A similar equation relates the actual final concentration (bf) of the bolus, measured after dilution into the total perfusate present;

$$[bf] = [bi] \times Vb/Vt \tag{3}$$

By substituting equations 2 and 3 into equation 1, a new relationship is obtained which enables the calculation of the foetal vascular volume;

$$Vf = \{[bi] \times Vb/[bf]\} - \{[bi] \times Vk/[bf]\} \tag{4}$$

2.5. Empirical Calculations

Oxygen consumption rates, glucose utilisation rates and lactate production rates were calculated as described by Cannell et al. (1987). In an attempt to derive a better indicator of perfusion stability than the criteria used here (perfusate pressure), the raw pO_2 data was manipulated in a variety of ways and results compared between successful and unsuccessful perfusions. From a number of calculations attempted, the relationship;

$$\text{quotient} = \text{mod}(\text{Fven} - \text{Fart})/\text{mod}(\text{Mven} - \text{Mart}) \tag{5}$$

[where Fven is the foetal venous perfusate pO_2, Fart is the foetal arterial perfusate pO_2, Mven is the maternal venous perfusate pO_2, and Mart is the maternal arterial perfusate pO_2] provided a rapidly calculable potential empirical discriminator between success and failure of future perfusion.

3. RESULTS

In Table 1, the raw data for each category of experiment is listed. Selected categories from this table are plotted in Figure 1 as a fraction of total attempts for that period of the investigation. These same categories are broken down on an annual basis in Figure 1. It can be seen that there is a difference between perfusion success on an

Table 1. Raw data for each category of experiment

Year	Total attempts	Failed attempts	Perfusions abandoned	Perfusions maintained
1996	31	9	18	4
1997	33	8	16	9
1998	28	10	11	7

annual basis, being markedly different from the following two periods. This is particularly marked in terms of the fraction of perfusions which were established but failed to be maintained.

Application of the empirical equation;

$$\text{quotient} = \text{mod}(\text{Fven} - \text{Fart})/\text{mod}(\text{Mven} - \text{Mart}) \tag{5}$$

to the raw pO2 values of perfusate lines, measured at thirty minutes after establishing perfusion, to a series of successful and unsuccessful perfusions in the present work resulted in values shown in Table 2.

Table 2. Results obtained from retrospective application of equation 6

Perfusion category	Mean "quotient" (n = 6)	S.D.
Maintained	0.67	0.12
Failed	0.37	0.18

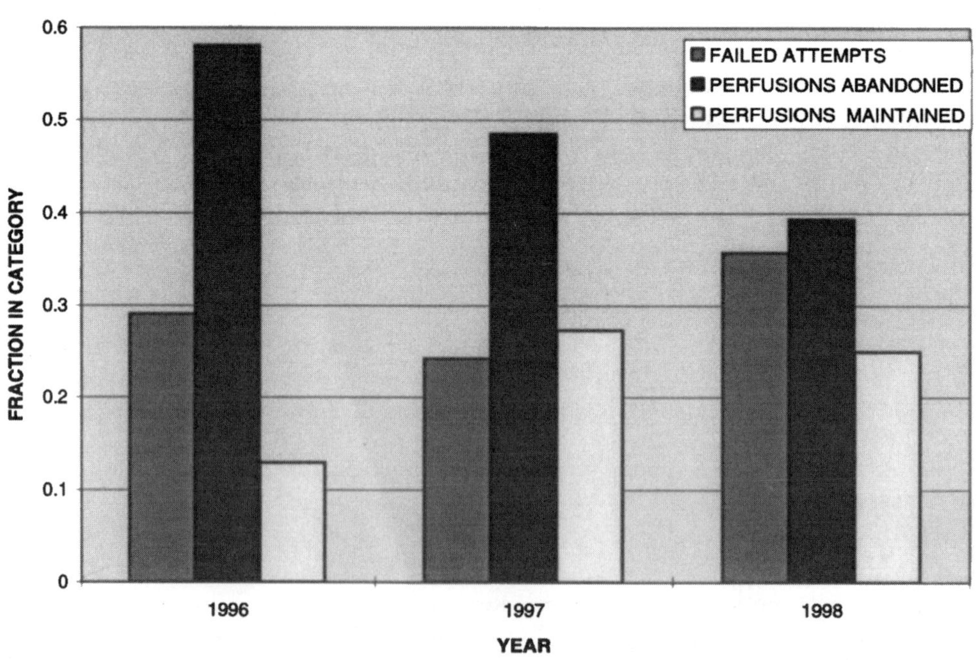

Figure 1. Perfusion success on an annual basis.

5. DISCUSSION

A total of 27 (30% of the total) attempts fell into the category of failed attempts (Table 1). To some extent the size of this category is dependent upon the degree to which the operator is prepared to persevere with specimens which might otherwise be abandoned immediately after inspection. Using the maintenance of pressure below 40 mmHg as the criteria for success, a further total of 45 perfusions (50% of the total) in the current series were abandoned at varying times after establishment of perfusion. The remaining perfusions, representing 20% of total perfusions attempted, were maintained for 5 hours or longer. This represents approximately 30% of perfusions initially successfully established. While this figure may appear low, it is dependent upon the condition of the initial material available and upon a variety of mechanical and biological factors operating during the perfusion process. These include the maintenance of integrity within perfusion lines, the maintenance of sterility within maternal and foetal perfusion lines, maintenance of anatomical integrity of, and between, maternal and foetal compartments and continued metabolic function within the placental tissue. Of particular concern within this series was the sudden development of high perfusion pressures in a number of perfusions at about one hour after the perfusion circuit had been established. Meticulous care with the quality of water used for preparation of perfusate fluid minimised but did not eliminate this problem. Although it was assumed that the cause of this problem is probably bacterial growth, it would be unlikely that there is sufficient time for bacterial division *per se* to provide sufficient material to physically block the capillary network of the placental lobule under investigation. If the problem is a result of bacterial contamination, then it is likely that the raised pressures reflect a metabolic response of the placental (foetal) vasculature to some bacterial product. It can be seen that there is a difference between perfusion success on an annual basis, which might be interpreted as the improvement in the technical skill of a new operator becoming familiarised with the idiosyncracies of the system. Finally, other parameters apart from those previously reported for this procedure (Maguire et al., 1992) were re-assessed at varying stages of successful and unsuccessful perfusions in an attempt to better define predictors of ultimate success of this extremely useful physiological model. While a number of parameters exist to compare transfer of a range of normal and introduced compounds, no satisfactory predictor of ultimate perfusion success exists other than the one previously cited here, namely the perfusion pressure at one hour. Lobule weight has previously been used to standardise parameters such as oxygen consumption between individual perfusions. However, variability in the amount of non-perfused tissue present in different lobule preparations introduces unacceptable variability into the derived parameter. To improve future inter-experimental comparisons, an attempt was made to identify a more satisfactory way to express oxygen consumption, rather than as placental lobule weight-related consumption rates. Foetal circulation volume should be more indicative of the amount of active placental tissue within a given lobule; it would certainly be expected to relate more closely to the surface area for transfer of metabolites across the placental barrier than a measurement such as total lobule weight, which cannot exclude non-respiring tissue. By considering the dilution of a bolus of known volume and concentration, the equation;

$$\mathrm{Vf} = \{[\mathrm{bi}] \times \mathrm{Vb}/[\mathrm{bf}]\} - \{[\mathrm{bi}] \times \mathrm{Vk}/[\mathrm{bf}]\} \tag{4}$$

was developed to derive the volume of the foetal vascular circulation. This value is advocated as a potentially better parameter to use than parameters such as lobule weight when attempting to "normalise" values for oxygen consumption rates obtained between different perfusion experiments. A similar equation could also be derived for the estimation of the maternal perfusion sink. It is important that the substance added in such an approach is one that does not bind significantly to perfusion lines or to the placental tissue. Using an empirical approach, application of the empirical equation;

$$\text{quotient} = \text{mod}(\text{Fven} - \text{Fart})/\text{mod}(\text{Mven} - \text{Mart}) \tag{5}$$

to the raw pO2 values obtained at thirty minutes after establishing perfusions to a series of successful and failed perfusions in the present work resulted in values of 0.67 +/− 0.23 for successful perfusions and 0.37 +/− 0.36 for unsuccessful perfusions. If the discrimination point is set at the lower limit of the normal range for this empirically derived quotient (0.44), then some perfusions likely to fail will be included, but all of the potentially successful perfusions will also be maintained. The ability to make a decision as to whether to continue or abandon perfusion after thirty minutes provides a considerable improvement over existing discriminators.

REFERENCES

Addison, RS, Maguire, DJ, Mortimer, RH, and Cannell, GR, 1991, Metabolism of Prednisolone by the Isolated Perfused Human Placental Lobule, *J. Steroid Biochem. Molec. Biol.*, 39:83–90.

Cannell, GR, Kluck, RM, Hooper, WD, and Dickinson, RG, 1987, Calculation of Viability Parameters for the Isolated Perfused Human Placental Preparation, *Intelligent Instruments and Computers*, 5:109–113.

Maguire, DJ, Addison, RS, Harvey, TJ, Mortimer, RH, and Cannell, GR, 1992, A Comparison of Parameters used to Standardise Results from In-Vitro Perfusions of Human Placentae, *Adv. exp. Biol. Med.*, 316:457–462.

Maguire, DJ, Voroteliak, Cowley, D, and Cannell, GR, 1992, Human Placental Oxygen Metabolism, *Adv. exp. Biol. Med.*, 316:457–462.

Miller, RK, Weir, PJ, Maulik, D, and Di Sant'agnese, PA, 1985, Human Placenta in vitro. Characterization during 12 Hours of Dual Perfusion, *Contr. Gynec. Obstet.*, 13:77–84.

Panigel, M, 1962, Placental Perfusion Experiments, *Am. J. Obstet. Gynecol.*, 84:1664–1683.

Schmitt, W, 1922, Untersuchungen zur Physiologie Plazentargefabe, *Z. Biol.*, 75:19–78.

DYNAMIC MEASUREMENTS OF OXYGEN TENSION IN ISLETS OF LANGERHANS

Henrik Ortsäter,[1] Per Liss,[2] Per-Eric Lund,[2] Karl E. O. Åkerman,[2] and Peter Bergsten[1]

[1]Department of Medical Cell Biology and
[2]Department of Physiology
Uppsala university, Uppsala, Sweden

1. INTRODUCTION

Oxidative metabolism plays a key role in glucose-induced insulin release (Ashcroft et al., 1973). With the development of a sensitive assay for insulin it was possible to characterize the dynamics of glucose-induced insulin release from the individual isolated islet of Langerhans (Bergsten and Hellman, 1993). To evaluate the role of the oxidative metabolism for the kinetics of insulin release a method for determining oxidative metabolism with a sensitivity comparable to that available for insulin is crucial. Glucose oxidation and oxygen consumption have been used to monitor glucose-induced metabolic changes (Ashcroft et al., 1973; Hutton and Malaisse, 1980; Longo et al., 1991). These studies indicate that insulin release and cellular metabolism, oxidation in particular, are tightly linked. However, the previously used techniques lack the sensitivity required to monitor the dynamics of metabolism from individual islets. Thus analysis of the coupling or synchronization between metabolic changes and secretion is hampered. To overcome this problem we have developed a new sensitive technique to monitor oxygen tension, which allows time-resolved measurements from individual islets, and characterized the responses in pO_2 following exposure to glycolytic and mitochondrial substrates.

2. METHODS

2.1. Materials

Reagents of analytical grade and deionized water were used. Collagenase, HEPES and bovine serum albumin (fraction V) were obtained from Boehringer Mannheim

Oxygen Transport to Tissue XXI, edited by Eke and Delpy.
Kluwer Academic / Plenum Publishers, New York, 1999.

GmbH (Mannheim, Germany). Poly-L-lysine (P-5899) was supplied by Sigma Chemical (St Louis, MO). Fetal calf serum (FCS) was provided by Gibco Ltd (Paisley, UK).

2.2. Preparation and Culture of Islets

Islets containing more than 90% β-cells were collagenase isolated from *ob/ob*-mice (Hellman, 1965) and cultured overnight in the presence of 5.5 mM glucose in RPMI 1640 medium supplemented with 10% FCS. Subsequent experimental handling was performed in a medium containing (in mM) NaCl 125, KCl 5.9, $MgCl_2$ 1.2, $CaCl_2$ 1.28, HEPES 25 and glucose 3, as well as bovine serum albumin (0.5 mg/ml) titrated to pH 7.4 with NaOH. During perifusion of individual islets either the glucose concentration was increased to 11 mM, or 5 mM α-keto isocaproic acid (KIC) was added to the medium.

2.3. Measurements of Oxygen Tension

Individual islets were placed in a perifusion chamber as previously described (Bergsten et al., 1994). The chamber was placed on a thermostated (37°C) stage of an inverted microscope (Nikon TMS, Tokyo, Japan) and perifused at a rate of 150–200 µl/min. Oxygen tension (pO_2) in the isolated islets was measured by modified Clark microelectrodes (Liss et al., 1997). The tip of the microelectrode was positioned inside or at different distances from the islet with the aid of a micromanipulator. When the microelectrodes penetrated into the islets the tip was positioned at a depth of 5–10 µm below the capsule of the islet. The electrodes are polarized at –0.8 V, which gives a linear response between the oxygen tension and the electrode current. The electrodes were calibrated in water saturated with NaS_2O_5 or air at 37°C before and after the experiments. The drift of the microelectrodes was less than 0.5% per hour.

2.4. Statistical Analysis

Values are expressed as means ± SEM. The statistical significance of the data was tested by using a multivariate analysis of variance (Manova) model, comparing results obtained during the control period with those obtained after either increase in the glucose concentration or addition of KIC (repeated measures). For comparison of data between groups of animals, a two tail Student's *t*-test for unpaired samples was used.

3. RESULTS

The oxygen tension (pO_2) was measured in individual islets of Langerhans by placing the modified Clark-electrode at different distances from islets perifused in the presence of 3 mM glucose. When the electrode was placed at distances greater than 50 µm pO_2 was 149 mmHg (Figure 1; n = 11). This value corresponded to the pO_2 in ambient air. A drop in pO_2 to 125 ± 8 mmHg was observed when the electrode was placed approximately 10 µm outside the islet, whereas 106 ± 9 mmHg was observed when the electrode was positioned inside the islet. With the electrode still positioned in the islet the glucose concentration was increased to 11 mM, which caused a prompt and significant ($p < 0.001$) drop in pO_2 (Figure 2; n = 6). The decrease in pO_2 started 17 ± 5 s after introduction of 11 mM glucose and was concluded 230 ± 30 s later after which pO_2 stabilized at 69 ± 6 mmHg. Islet perifused in the presence of 3 mM glucose were also stimulated

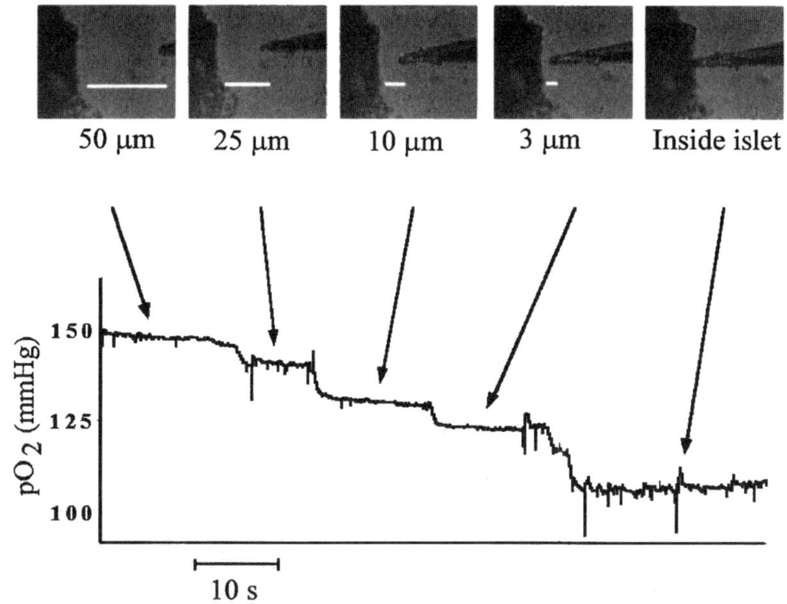

Figure 1. Recording of oxygen tension from an individual isolated islet in response to stimulatory glucose concentration. Islets were cultured overnight in the presence of 5.5 mM glucose. Individual islets were transferred to a microscope chamber and perifused in the presence of 3 mM glucose. A Clark microelectrode was positioned in the islet. The arrow indicates when the glucose concentration was increased to 11 mM. Representative experiment 6.

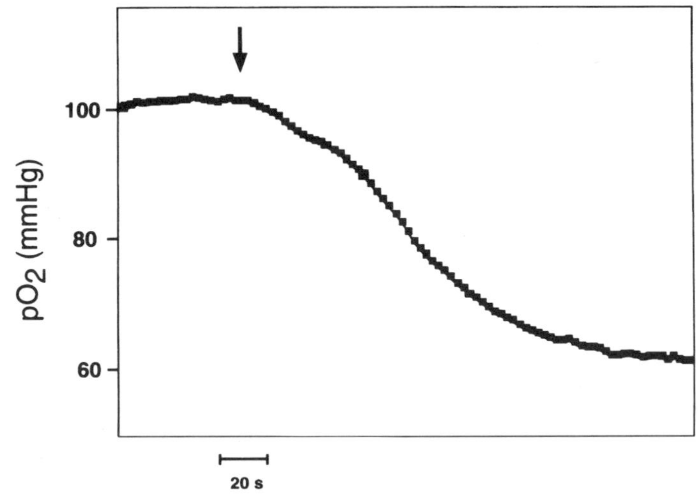

Figure 2. Recording of oxygen tension from an individual isolated islet in response to α-keto-isocaproic acid (KIC). Islets were cultured overnight in the presence of 5.5 mM glucose. Individual islets were transferred to a microscope chamber and perifused in the presence of 3 mM glucose. A Clark microelectrode was positioned in the islet. The arrow indicates when 5 mM KIC was added to the perifusion medium. Representative experiment 5.

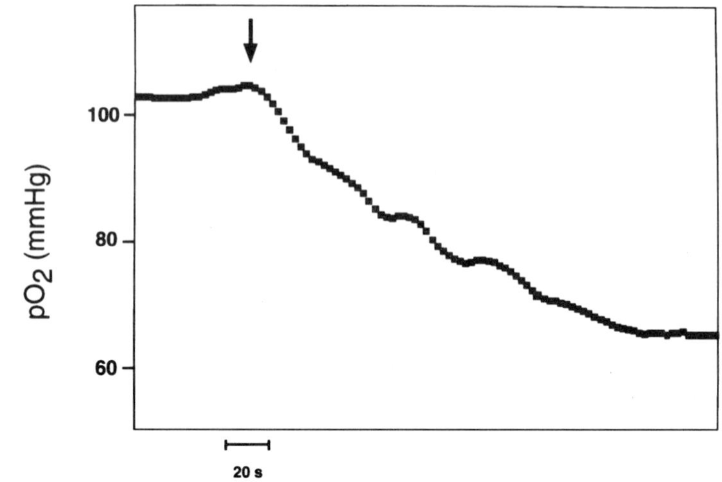

Figure 3. Recording of oxygen tension at different distances from an individual isolated islet. Islets were cultured overnight in the presence of 5.5 mM glucose. Individual islets were transferred to a microscope chamber and perifused in the presence of 3 mM glucose. The inserts show the distances of the tip of the Clark microelectrode from the islet. Representative experiment 11.

with 5 mM α-keto-isocaproic acid (KIC), which caused a similarly prompt and significant ($P < 0.001$) decrease in pO_2 measured with the electrode inside the islet (Figure 3; $n = 5$). The decline in pO_2 started 12 ± 3 s after introduction of KIC and lasted 290 ± 30 s, reducing the initial pO_2 level to 69 ± 5 mmHg.

4. DISCUSSION

Oxygen measurements of the islet of Langerhans have previously required groups of islets (Hutton and Malaisse, 1980; Longo et al., 1991; Conget et al., 1997) or large piscine islet organs (Schrezenmeir et al., 1992). With the development of a modified Clark electrode with a guard cathode and a very small (2–5 μm) tip diameter (Liss et al., 1997), the sensitivity of pO_2 measurements increased to such an extent that changes in the oxygen tension in a small number of cells could be detected.

In the present study we record changes in pO_2 in the vicinity of an isolated islet. So far this has only been attempted in the much larger piscine islet organs known as Brockmann bodies (Schrezenmeier et al., 1992). Our finding of a considerable pO_2 gradient outside individual islets relied upon the sensitive electrode. Although no measurements of pO_2 next to the cell during stimulation were carried out it can be anticipated that the observed sharp decline in pO_2 recorded inside the islet will also influence pO_2 in the media next to the islet. This would give a possibility to monitor changes in pO_2 outside the islet and possibly from single cells, which would open a new way of studying cellular metabolism.

Alterations in pO_2 caused by changes in the concentration of glucose or addition of α-keto-isocaproic acid were recorded with the electrode positioned inside the islet and is a continuation of our first measurements of oxygen tension inside individual islets of Langerhans, where we demonstrated a correlation between oscillations in insulin release and oxygen tension (Ortsäter et al., 1998). The pO_2 kinetics were similar for the two secretagogues. The rapid onset of the secretagogue-induced decrease in pO_2, which has also been observed when batches of islets were used (Hutton and Malaisse, 1980), supports the idea that increase in metabolism is an initial event triggering and preceding insulin release (Civelek et al., 1997). The subsequent increase in glucose-induced insulin release is critically dependent on adequate oxygen supply (Malaisse et al., 1988). Although glucose-induced increase in insulin release from the isolated islet is observed over at least 20 min (Bergsten and Hellman, 1993), there was a stabilization of pO_2 in the present study already after 4–5 min after introducing a stimulatory glucose concentration.

Non-insulin dependent diabetes mellitus is connected to impaired β-cell metabolism. Although the causes of the metabolic disease are largely unknown, recent findings point towards the importance of the mitochondria. Indeed, in a subpopulation of diabetics point mutations of the mitochondrial genome have been found (van den Ouweland et al., 1995). With the new, sensitive oxygen electrode it will be possible to monitor pO_2 changes in cells from normal as well as diabetic individuals and correlate these changes to simultaneous measurements of insulin release. The kinetics in pO_2 may serve as an index for the metabolic responsiveness of a given cell. In this way the new modified Clark microelectrode may give us a unique possibility to study the role of mitochondrial dysfunction in disease (Luft, 1995).

ACKNOWLEDGMENTS

The study was supported by grants from the Swedish Medical Research Council (12X-11203), the Swedish Diabetes Association, the Novo Nordisk Foundation, the Family Ernfors Foundation, the Gunvor and Josef Anér Foundation and the Marcus and Amalia Wallenberg Foundation.

REFERENCES

Ashcroft SJ, Weerasinghe LC, and Randle PJ. Interrelationship of islet metabolism, adenosine triphosphate content and insulin release. *Biochem J* 132:223–231, 1973.

Bergsten P and Hellman B. Glucose-induced amplitude regulation of pulsatile insulin secretion from individual pancreatic islets. *Diabetes* 42:670–674, 1993.

Bergsten P, Grapengiesser E, Gylfe E, Tengholm A, and Hellman B. Synchronous oscillations of cytoplasmic Ca^{2+} and insulin release in glucose-stimulated pancreatic islets. *J Biol Chem* 269:8749–8753, 1994.

Civelek VN, Deeney JT, Fusonie GE, Corkey BE, and Tornheim K. Oscillations in oxygen consumption by permeabilized clonal pancreatic beta-cells (HIT) incubated in an oscillatory glycolyzing muscle extract: roles of free Ca^{2+}, substrates, and the ATP/ADP ratio. *Diabetes* 46:51–56, 1997.

Conget I, Barrientos A, Manzanares JM, Casademont J, Vinas O, Barcelo J, Nunes V, Gomis R, and Cardellach F. Respiratory chain activity and mitochondrial DNA content of non-purified and purified pancreatic islet cells. *Metabolism* 46:984–987, 1997.

Hellman B. Studies in obese-hyperglycemic mice. *Ann NY Acad Sci* 131:541–558, 1965.

Hutton JC and Malaisse WJ. Dynamics of O_2 consumption in rat pancreatic islets. *Diabetologia* 18:395–405, 1980.

Liss P, Nygren A, Revsbech NP, and Ulfendahl HR. Intrarenal oxygen tension measured by a modified Clark electrode at normal and low blood pressure and after injection of x-ray contrast media. *Pflügers Arch* 434:705–711, 1997.

Longo EA, Tornheim K, Deeney JT, Varnum BA, Tillotson D, Prentki M, and Corkey BE. Oscillations in cytosolic Ca^{2+}, oxygen consumption, and insulin secretion in glucose-stimulated rat pancreatic islets. *J Biol Chem* 266:9314–9319, 1991.

Luft R. The development of mitochondrial medicine. *Biochim Biophys Acta* 1271:1–6, 1995.

Malaisse WJ, Rasschaert J, Zahner D, and Sener A. Hexose metabolism in pancreatic islets: the Pasteur effect. *Diabetes Res* 7:53–58, 1988.

Ortsäter H, Liss P, Lund PE, Åkerman K, and Bergsten P. Synchronous oscillations in oxygen tension and insulin release of individual mouse islets. *Diabetologia* 41:(Suppl 1)23A, 1998.

Schrezenmeier J, Gerö L, Laue C, Kirchgessner J, Müller A, Hüls A, Passmann R, Hahn HJ, Kunz L, Mueller-Klieser W, and Altman JJ. The role of oxygen supply in islet transplantation. *Transplant Proc* 24:2925–2929, 1992.

Van den Ouweland JM, Lemkes HH, Gerbits KD, and Maassen JA. Maternally inherited diabetes and deafness (MIDD): a distinct subtype of diabetes associated with a mitochondrial tRNA (Leu)(UUR) gene point mutation. *Muscle Nerve* 3:S124–S130, 1995.

INTERACTION BETWEEN NITRIC OXIDE AND THROMBOXANE A$_2$ IN THE REGULATION OF THE RESTING CEREBROVASCULAR TONE

Zoltán Benyó,[2] Christoph Görlach,[1] and Michael Wahl[1]

[1]Department of Physiology
Ludwig-Maximilians University
Pettenkoferstr. 12, D-80336 Munich
Germany
[2]Experimental Research Department—2nd Institute of Physiology
Semmelweis University of Medicine
Üllői út 78/A, H-1082 Budapest
Hungary

1. INTRODUCTION

Nitric oxide (NO) plays an important role in the regulation of the cerebral blood flow (CBF) (for review see: Faraci and Heistad, 1998). Inhibition of the L-arginine—NO pathway results in cerebral vasoconstriction and reduction of the CBF, indicating that the basal release of NO provides a resting relaxant tone in the cerebrovascular bed (for review see: Iadecola et al., 1994). This effect of basal NO production has generally been regarded as a direct relaxant action on the smooth muscle. The possibility that NO may regulate the resting cerebrovascular tone by modulation of the prostanoid synthesis has not been investigated to date.

NO reportedly stimulates the release of prostaglandins (Salvemini, 1997). A similar interaction in the regulation of the resting cerebrovascular tone is very unlikely, because inhibition of prostanoid synthesis by indomethacin does not mimic the contractile effect of NO synthase (NOS) blockade. L-arginine analogue NOS inhibitors have been shown recently to increase the release of thromboxane A$_2$ under in vivo and in vitro experimental conditions, suggesting an inhibitory effect of endogenous NO on thromboxane production (Stadler et al., 1993; Hellyer et al., 1997). A similar mechanism may also be involved in the maintenance of the NO-related resting cerebrovascular dilatory tone. The cere-

Oxygen Transport to Tissue XXI, edited by Eke and Delpy.
Kluwer Academic / Plenum Publishers, New York, 1999.

brovascular constriction that develops after inhibition of the NO synthesis would then be a consequence of disinhibition of thromboxane release. To test our hypothesis we investigated whether blockade of the biosynthesis or action of thromboxane A_2 attenuates the contractile effect of the NOS inhibitor N^G-nitro-L-arginine (L-NA) in isolated rat middle cerebral arteries (MCAs).

Selective inhibition of the neuronal isoform of the constitutive NOS (nNOS) reduces blood flow and increases the vascular resistance in the brain, indicating that basal NO release by nNOS contributes to the maintenance of the resting CBF (Kovách et al., 1994; Kelly et al., 1995; Wang et al., 1995). The cerebrovascular effect of selective nNOS blockade under in vitro experimental conditions has not been investigated before. Since nNOS is present both in the cerebrovascular endothelium and in cerebral perivascular nerves (Bredt et al., 1990; Nozaki et al., 1993; Gorelova et al., 1996), in the present study we investigated its possible involvement in the maintenance of the resting cerebrovascular tone in vitro by using the specific nNOS inhibitor 7-nitro indazole monosodium salt (7-NINA, Silva et al., 1995). Furthermore, we tested whether inhibition of prostanoid synthesis alters the cerebrovascular effect of 7-NINA. A part of these results has been published recently (Benyó et al., 1998a).

2. METHODS

Adult male Wistar rats were exsanguinated rapidly under deep ether anaesthesia. Ring segments of the MCA were prepared for measurement of isometric force as described previously (Görlach and Wahl, 1996). Special care was taken to preserve the endothelium during preparation of the artery. The segments were transferred into 5-ml organ baths filled with a modified Krebs solution of the following composition (mM): NaCl, 119; KCl, 4.6; NaH_2PO_4, 1.2; $CaCl_2{\cdot}2H_2O$, 1.5; $MgCl_2{\cdot}6H_2O$, 1.2; $NaHCO_3$, 15; glucose, 10. The bath solution was bubbled continuously with a humidified gas mixture (90% O_2/10% CO_2). The MCA segments were mounted on two L-shaped tungsten wires (50 µm diameter), one of which was fixed to the bath and the second connected to a force transducer for measuring isometric force. The vessels were allowed a 90-min equilibration period, during which the resting tension was adjusted to 1.5–2 mN. Thereafter each segment was exposed to a 124-mM K^+ Krebs solution to elicit a reference contraction. Following a 60-min resting period the functional integrity of the endothelium was tested by evaluation of the bradykinin (0.01–10 µM, Calbiochem)-induced, endothelium-dependent, NO-mediated (Görlach and Wahl, 1996) vasodilation after precontraction with 100 µM UTP (Serva). Segments that did not exhibit at least 20% relaxation of the precontraction were excluded from the study. Thereafter the experiments were continued with one of three different protocols. With the first protocol we tested the selectivity of 7-NINA for nNOS, because Zagvazdin and colleagues (1996) have reported some non-isoform-selective effects of 7-nitro indazole (7-NI) in vivo. Therefore, in the first group of the MCA segments we evaluated again the dilator effect of bradykinin in the presence of 7-NINA (100 µM, Calbiochem) or L-NA (100 µM, Sigma). In the second group of the MCA segments we determined the effect of L-NA or 7-NINA on the resting vascular tension in the absence or presence of indomethacin (5 µM, Sigma) to investigate a possible interaction between NO and prostanoids. Since indomethacin inhibited the effect of L-NA, in the third group of MCA segments we investigated the involvement of thromboxane A_2 in the mediation of the NO blockade-induced vasoconstriction. Therefore, the effect of L-NA on the resting tone of the vessel rings was determined in the presence of the thromboxane A_2 synthase inhibitor ridogrel ((E)-5-[[[3-pyridinil[3-(trifluoromethyl)phenyl]methylen]amino]oxy] pentanoic acid, 10 μM, Janssen), the thromboxane receptor antagonist ICI 192605 (4-(Z)-6-(2-o-chlorophenyl-4-o-hydroxyphenyl-1,3-dioxan-cis-5-yl)hexenoic acid, 100 μM, Tocris), or their vehicles. Indomethacin and ridogrel were dissolved in 2% Na_2CO_3 and 400 mM NaOH respectively and diluted with saline. ICI 192605 was dissolved in DMSO (Aldrich). Ridogrel reportedly has both thromboxane A_2 synthase inhibitor activity and antagonistic activity at the common receptor for thromboxane A_2 and its precursor prostaglandin H_2 (De Clerck et al., 1989). However, in the present experiments—according to the study of Descombes et al. (1993)—ridogrel is considered to act as a thromboxane A_2 synthase inhibitor, because the concentration of ridogrel selected (10 μM) is lower than its IC_{50} value at thromboxane receptors (De Clerck et al., 1989). Nevertheless, we can not exclude the possibility that ridogrel also antagonises the action of prostaglandin H_2, and therefore findings in the present study relating to the release of thromboxane A_2 also refer to prostaglandin H_2.

Results are presented as mean ± SEM, the number of MCA segments studied is expressed by n. The bradykinin-induced relaxation was calculated as percent decrease of precontraction due to UTP. All other values are expressed as percentage of the reference contraction induced by 124 mM K^+ Krebs solution. Results were analysed statistically using Student's t-test and a P value less than 0.05 was considered significant.

3. RESULTS

L-NA inhibited the vasodilator effect of bradykinin almost completely indicating an effective inhibition of the endothelial NO production (Figure 1). In contrast, incubation of the vessels rings with 7-NINA did not influence the bradykinin-induced dilation (Figure 1).

L-NA induced vasoconstriction that was not changed after incubation with the vehicle of indomethacin (Figure 2). Indomethacin reduced the resting tension weakly (Figure 3) and significantly attenuated the contractile effect of L-NA (Figure 2). In con-

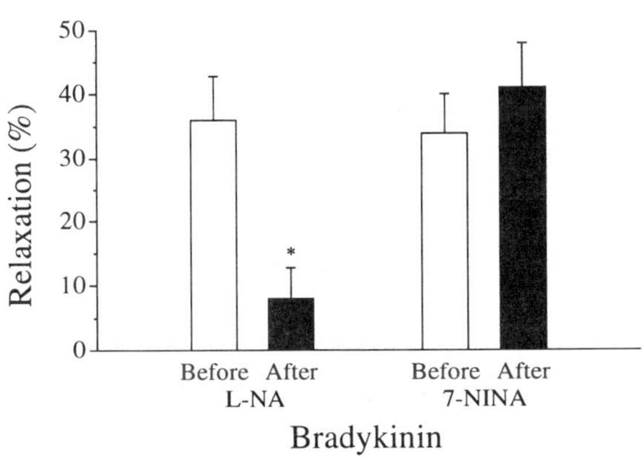

Figure 1. The relaxant effect of 1 μM bradykinin in isolated rat middle cerebral artery segments before (open bars) and after (solid bars) the administration of 100 μM N^G-nitro-L-arginine (L-NA, n = 10) or 100 μM 7-nitro indazole monosodium salt (7-NINA, n = 9). The values (mean ± SEM) are expressed as percentage of the precontraction induced by 100 μM UTP. *$P < 0.05$ versus "Before".

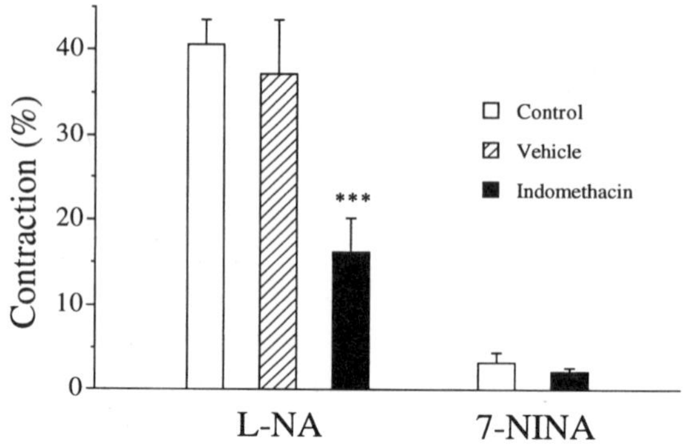

Figure 2. The effect of 100 μM N^G-nitro-L-arginine (L-NA) or 100 μM 7-nitro indazole monosodium salt (7-NINA) on the resting tone of isolated rat middle cerebral artery segments (n = 6–21) in the absence (open bars) or presence of 5 μM indomethacin (solid bars) or its vehicle (hatched bar). The values (mean ± SEM) are expressed as percentage of the reference contraction induced by 124 mM K^+ Krebs solution. ***$P < 0.001$ versus "Control".

trast, indomethacin did not influence the effect of 7-NINA on the resting tension (Figure 2).

The vehicles for ridogrel or ICI 192605 had no significant effect on the resting tone of the MCA segments (Figure 3). In contrast, a weak, but significant, reduction of the vascular tension was observed after administration of ridogrel or ICI 192605 (Figure 3). The contractile effect of L-NA was not altered in vehicle-treated segments when compared with untreated controls. Treatment with ridogrel or ICI 192605, however, markedly attenuated the L-NA-induced contractions (Figure 4).

4. DISCUSSION

Our observations demonstrate the involvement of thromboxane A_2 in the mediation of the L-NA-induced contraction of the rat MCA. These results indicate that

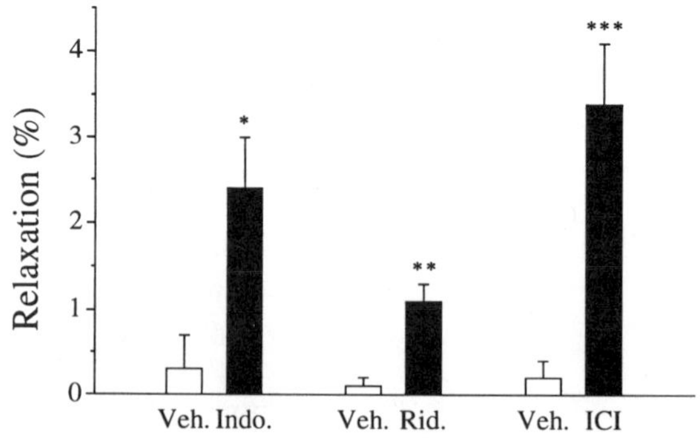

Figure 3. The effect of 5 μM indomethacin (Indo.), 10 μM ridogrel (Rid.), 100 μM ICI 192605 (ICI), or their vehicles (Veh.) on the resting tone of isolated rat middle cerebral artery segments (n = 11–27). The values (mean ± SEM) are expressed as percentage of the reference contraction induced by 124 mM K^+ Krebs solution. *$P < 0.05$, **$P < 0.01$, ***$P < 0.001$ versus the corresponding "Veh.".

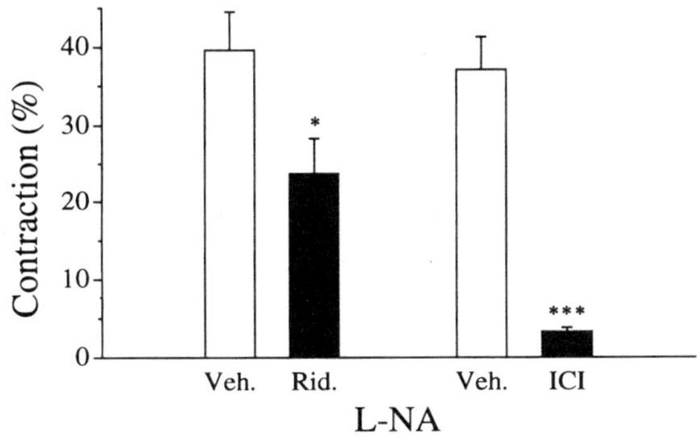

Figure 4. The effect of 100 μM N^G-nitro-L-arginine (L-NA) on the resting tone of isolated rat middle cerebral artery segments (n = 11–22) in the presence of 10 μM ridogrel (Rid.), 100 μM ICI 192605 (ICI), or their vehicles (Veh.). The values (mean ± SEM) are expressed as percentage of the reference contraction induced by 124 mM K^+ Krebs solution. *$P < 0.05$, ***$P < 0.001$ versus the corresponding "Veh.".

endogenous NO may maintain the resting cerebrovascular tone by regulation (suppression) of the synthesis and/or release of thromboxane A_2. Maintenance of a resting vasodilatory tone, however, may not be the only physiological function of NO mediated by suppression of thromboxane A_2. For instance, a similar mechanism may also be involved in the antiaggregatory effect of NO.

The vasoconstrictor effect of NOS inhibitors limits their potential therapeutic use in cerebral ischemia and brain trauma (for review see: Iadecola, 1997). Our present observation that this effect is mediated by thromboxane A_2 suggests that the cerebroprotective action of NO blockade in these pathophysiological states may be enhanced by simultaneous administration of thromboxane synthase inhibitors or receptor antagonists.

Previous studies have demonstrated constitutive thromboxane release from cerebrovascular endothelial cells (Spatz et al., 1997). The present finding that inhibition of the synthesis or action of thromboxane A_2 significantly reduces the resting tone of the MCA, together with our recent observation that ICI 192605 also relaxes rat basilar arteries (Benyó et al., 1998b), demonstrate that basal thromboxane release maintains a weak constrictor tone in the cerebral vasculature. Since NO may decrease the reactivity of thromboxane receptors (Wang et al., 1998) it is possible that attenuation of the resting thromboxane A_2-induced constrictor tone may also contribute to the dilator effect of NO. Furthermore, similar to the effect of endogenous NO, inhibition of the thromboxane-related resting constrictor tone may also contribute to the cerebrovascular action of nitrovasodilators.

Light and electron microscopic immunohistologic studies indicate that cerebrovascular endothelial cells contain both the endothelial (eNOS) and the neuronal (nNOS) isoforms of the constitutive NOS and that nNOS is also present in cerebral perivascular nerves (Bredt et al., 1990; Nozaki et al., 1993; Gorelova et al., 1996). In vivo inhibition of the nNOS with 7-NI results in a significant reduction of the CBF, indicating that nNOS-derived NO is an important factor in the maintenance of the resting perfusion of the brain (Kovách et al., 1994; Kelly et al., 1995; Wang et al., 1995). Since 7-NINA failed to induce a significant vasoconstriction in the present study cerebrovascular nNOS would not appear to be involved in the process of NO generation under resting conditions. Therefore, inhibition of the NO production by neurons appears to be responsible for the CBF-decreasing effect of 7-NI in vivo. Furthermore, 7-NINA did not influence the vasodilator effect of bradykinin, indicating that this substance stimulates only the eNOS in the rat MCA. Our recent studies, however, indicate that cerebrovascular nNOS may have an important physiological role in the mediation of the acetylcholine-induced vasorelaxation (Benyó et al., 1997).

In conclusion, our results indicate that the basal NO production suppresses thromboxane A_2 in rat middle cerebral arteries and that this interaction plays an important role in the maintenance of the resting cerebrovascular tone. Cerebrovascular nNOS is not involved in the process of NO generation under resting conditions. Further studies are needed to investigate the molecular mechanisms involved in the interaction between NO and thromboxane A_2.

ACKNOWLEDGMENTS

The authors express their gratitude to Dr. John Davis for critically reading the manuscript and to Josephine Specker for technical assistance. This work was supported by grants from the German BMBF, the Hungarian OTKA (F 013241, W 015186, T 017790) and ETT (518/1996) and the DAAD-MÖB Scientist Exchange Program. Dr. Benyó was supported by a Hungarian National Eötvös-Fellowship. Ridogrel was a generous gift from the Janssen Research Foundation.

REFERENCES

Benyó Z, Görlach C, and Wahl M (1997) Neuronal nitric oxide synthase is involved in agonist-induced but not in basal nitric oxide release in isolated rat basilar arteries. J Cereb Blood Flow Metab 17 (Suppl 1):S345.

Benyó Z, Görlach C, and Wahl M (1998a) Involvement of thromboxane A_2 in the mediation of the contractile effect induced by inhibition of nitric oxide synthesis in isolated rat middle cerebral arteries. J Cereb Blood Flow Metab 17:616–618.

Benyó Z, Görlach C, and Wahl M (1998b) Role of nitric oxide and thromboxane in the maintenance of cerebrovascular tone. Kidney Int 54 (Suppl 67):S218–S220.

Bredt DS, Hwang PM, and Snyder SH (1990) Localization of nitric oxide synthase indicating a neural role for nitric oxide. Nature 347:768–770.

De Clerck F, Beetens J, de Chaffoy de Courcelles D, Freyne E, and Janssen PAJ (1989) R 68 070: Thromboxane A_2 synthase inhibition and thromboxane A_2/prostaglandin endoperoxide receptor blockade combined in one molecule—I. Biochemical profile in vitro. Thromb Haemostas 61:35–42.

Descombes J-J, Devys M, Laubie M, and Verbeuren TJ (1993) Endothelial thromboxane production plays a role in the contraction caused by 5-hydroxytryptamine in rat basilar arteries. Eur J Pharmacol 243:193–199.

Faraci F and Heistad DD (1998) Regulation of the cerebral circulation: Role of endothelium and potassium channels. Physiol Rev 78:53–97.

Gorelova E, Loesch A, Bodin P, Chadwick L, Hamlyn PJ, and Burnstock G (1996) Localisation of immunoreactive factor VIII, nitric oxide synthase, substance P, endothelin-1 and 5-hydroxytriptamine in human postmortem middle cerebral artery. J Anat 188:97–107.

Görlach C and Wahl M (1996) Bradykinin dilates rat middle cerebral artery and its large branches via endothelial B_2 receptors and release of nitric oxide. Peptides 17:1373–1378.

Hellyer PW, Johnson LW, and Olson NC (1997) Effect of N^G-nitro-L-arginine-methyl-ester on cardiopulmonary function and biosynthesis of cyclooxygenase products during porcine endotoxemia. Crit Care Med 25:1051–1058.

Iadecola C (1997) Bright and dark sides of nitric oxide in ischemic brain injury. Trends Neurosci 20:132–139.

Iadecola C, Pelligrino DA, Moskowitz MA, and Lassen NA (1994) Nitric oxide synthase inhibition and cerebrovascular regulation. J Cereb Blood Flow Metab 14:175–192.

Kelly PAT, Ritchie IM, and Arbuthnott GW (1995) Inhibition of neuronal nitric oxide synthase by 7-niroindazole: effects upon local cerebral blood flow and glucose use in the rat. J Cereb Blood Flow Metab 15:766–773.

Kovách AGB, Lohinai Z, Marczis J, Balla I, Dawson TM, and Snyder SH (1994) The effect of hemorrhagic hypotension and retransfusion and 7-nitro-indazole on rCBF, NOS catalytic activity, and cortical NO content in the cat. Ann NY Acad Sci 738:348–368.

Nozaki K, Moskowitz MA, Maynard KI, Koketsu N, Dawson TM, Bredt DS, and Snyder SH (1993) Possible origins and distribution of immunoreactive nitric oxide synthase-containing nerve fibers in cerebral arteries. J Cereb Blood Flow Metab 13:70–79.

Salvemini D (1997) Regulation of cyclooxygenase enzymes by nitric oxide. Cell Mol Life Sci 53:576–582.

Silva MT, Rose S, Hindmarsh JG, Aislaitner G, Gorrod JW, Moore PK, Jenner P, and Marsden CD (1995) Increased striatal dopamine efflux in vivo following inhibition of cerebral nitric oxide synthase by the novel monosodium salt of 7-nitro indazole. Br J Pharmacol 114:257–258.

Spatz M, Kawai N, Merkel N, Bembry J, and McCarron RM (1997) Functional properties of cultured endothelial cells derived from large microvessels of human brain. Am J Physiol 272:C231–C239.

Stadler J, Harbrecht BG, Di Silvio M, Curran RD, Jordan ML, Simmons RL, and Billiar TR (1993) Endogenous nitric oxide inhibits the synthesis of cyclooxygenase products and interleukin-6 by rat Kupffer cells. J Leukoc Biol 53:165–172.

Wang G-R, Zhu Y, Halushka PV, Lincoln TM, and Mendelsohn ME (1998) Mechanism of platelet inhibition by nitric oxide: *In vivo* phosphorylation of thromboxane receptor by cyclic GMP-dependent protein kinase. Proc Natl Acad Sci USA 95:4888–4893.

Wang Q, Pelligrino DA, Baughman VL, Koenig HM, Albrecht RF (1995) The role of neuronal nitric oxide synthase in regulation of cerebral blood flow in normocapnia and hypercapnia in rats. J Cereb Blood Flow Metab 15:774–778.

Zagvazdin Y, Sancesario G, Wang YX, Share L, Fitzgerald ME, and Reiner A (1996) Evidence from its cardiovascular effects that 7-nitroindazole may inhibit endothelial nitric oxide synthase in vivo. Eur J Pharmacol 303:61–69.

ROLE OF NITRIC OXIDE IN THE CONTROL OF MITOCHONDRIAL FUNCTION

Paul R. Forfia,* Thomas H. Hintze, Michael S. Wolin, and Gabor Kaley

Department of Physiology
New York Medical College
Valhalla, New York 10595, USA

Nitric oxide (NO) has been a focus of intense scientific investigation over the past years. A vast array of physiologic as well as pathophysiologic roles for NO have been described in pulmonary, immunologic, neuronal, gastrointestinal, and reproductive systems.[21] Undoubtedly however, the preponderance of nitric oxide research has focused on the role of NO in the cardiovascular system, both in normal and pathophysiologic states. Nitric oxide is involved in the control of mean arterial blood pressure, vascular tone and regional blood flow in virtually all vascular beds.[1,34] In addition, NO prevents platelet aggregation and is an inhibitor of vascular smooth muscle proliferation. Thus, NO is believed to contribute significantly to the maintenance of vascular homeostasis. Despite ample evidence in support of the action of NO on vascular smooth muscle and the local abluminal environment, the role of NO in the control of tissue oxygen consumption has received relatively little attention. Thus, this chapter will discuss the role of NO in the control of tissue respiration, as well as possible physiologic and pathophysiologic implications linking endogenous NO production and oxygen utilization in mammalian systems.

SOURCES OF NITRIC OXIDE

In order to discuss NO in the control of cellular respiration, a brief review of the basic chemical and physical properties of NO is warranted. NO is free radical gas that is the end product of a five electron oxidation of one of the two guanidino nitrogen

* Paul R. Forfia is a third year medical student at New York Medical College.

Oxygen Transport to Tissue XXI, edited by Eke and Delpy.
Kluwer Academic / Plenum Publishers, New York, 1999.

groups attached to the semi-essential amino acid L-arginine.[19] This reaction is catalyzed by the enzyme nitric oxide synthase, having two constitutive, calcium dependent isoforms (ecNOS and nNOS), and an inducible, calcium independent isoform (iNOS). Endothelial NOS (ecNOS) activation occurs secondary to shear stress on the endothelial cell membrane exerted by the flowing blood, which leads to a deformation of cytoskeletal elements, opening membrane ion channels and an increase in calcium flux into the cytosol. NO is released from the endothelial cell in a radial fashion, and freely diffuses through adjacent cells into the surrounding tissue. However, due to its highly reactive nature, the diffusion distance of NO is limited by potent NO scavenging factors, such as hemoglobin and superoxide radical, present in the blood and surrounding tissues. Thus, due to its lipid permeability, the balance between NO production and NO inactivation in the local tissue milieu determines the functional range of NO. Under optimal conditions, the effective diffusion distance of NO has been estimated at 70 μm, which well exceeds the thickness of the microvascular wall, allowing NO access to the surrounding tissue.[17,35]

REGULATION OF METABOLISM BY NITRIC OXIDE

In 1979, Granger et al. demonstrated that an unknown substance derived from lipopolysaccharide-activated macrophages could potently inhibit mitochondrial respiration in a variety of neoplastic cell lines.[11] This macrophage-derived product, shown to be L-arginine dependent, was eventually identified as nitric oxide.[13,31] Numerous sites within the mitochondria were shown to be potential targets for the inhibitory actions of NO, including the TCA cycle enzyme aconitase, as well as complex I, complex II, and complex IV (cytochrome oxidase) of the electron transport chain.[2,5,12] Thus, in addition to providing a mechanism involved in macrophage-mediated cytotoxicity, these studies elucidated a role for NO in the control of mitochondrial respiration. However, because these effects were produced with supraphysiologic levels of NO (derived from the inducible isoform of NOS), the implication of these results was limited to high NO states, such as inflammation and septic shock. It remained unclear as to whether an inhibitory action on mitochondria could be demonstrated at substantially lower, more physiologic levels of NO. More recent studies have shown that inhibition of basal NO production *in vivo* leads to an elevation in canine skeletal muscle oxygen consumption, despite a fall in muscle blood flow.[18,28] Thus, NO released tonically from the vascular endothelium appears to modulate tissue oxygen extraction. Similarly, it was shown that blockade of endogenous NO production *in vivo* leads to a significant increase in myocardial oxygen consumption, suggesting that NO has a tonic inhibitory action on oxygen uptake in the working heart. These results were confirmed by *in vitro* studies showing that canine skeletal and cardiac muscle oxygen uptake could be suppressed by exogenous and endogenous sources of NO.[28,37,39] Similar NO-related mechanisms are suggested by *in vitro* studies of both primate and human cardiac tissue.[9,16] Moreover, it was demonstrated that microvessels isolated from these tissue preparations were a significant source of NO when stimulated by NO releasing agents. Thus, a physiologic linkage may exist between the microvascular production of NO and the control of mitochondrial respiration in adjacent parenchymal cells (Figure 1).

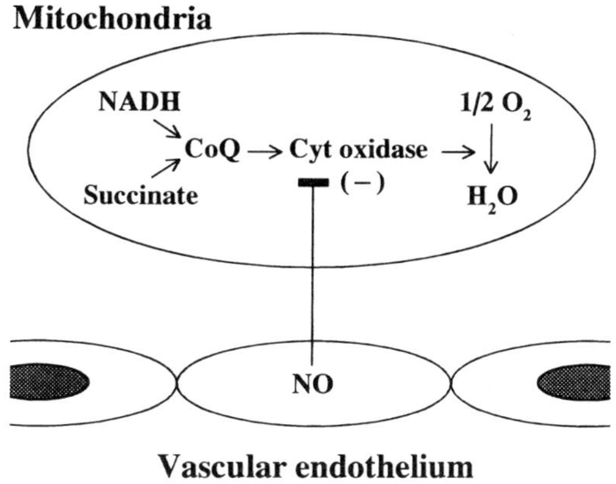

Figure 1. Nitric oxide synthesized in endothelial cells diffuses to parenchymal cells and inhibits mitochrodrial cytochrome oxidase, reducing tissue respiration and consequently oxygen consumption.

SITES OF REGULATION OF MITOCHONDRIAL FUNCTION BY NITRIC OXIDE

Potential sites of mitochondrial inhibition by nitric oxide, include aconitase, complex I, and complex II, however inhibition of these enzymes has only been observed in high NO states. In contrast, significant inhibition of cytochrome oxidase has been reported at nanomolar concentrations of NO. Considering that tissue levels of NO have been estimated at or below the nanomolar range it appears that cytochrome oxidase is the most likely target for inhibition at physiologic concentrations of nitric oxide. Indeed, Brown et al. estimated the K_i of NO for cytochrome oxidase to be between 60–270 nM.[2] This highly sensitive interaction can be explained partly by the physicochemical interaction between NO and cytochrome oxidase. Interestingly, this interaction has long been known, as biochemists have been using NO as a probe for over 25 years to evaluate the oxygen binding site of cytochrome oxidase.[38] Nitric oxide is a paramagnetic molecule, which avidly binds to transition metals, such as those found in the catalytic iron-sulfur or copper and heme centers of mitochrondrial enzymes, such as cytochrome oxidase. Access to these transition metal centers is limited by the hydrophobic domains that not only surround these catalytic sites but also permit enzyme insertion into the mitrochondrial membrane. However, it is presumed that due to its highly lipophilic nature, NO is able to diffuse through these hydrophobic membrane-embedded domains to access the heme and copper centers of cytochrome oxidase and cause a reversible inhibition of this enzyme. NO is also able to remove Fe from the Fe-S complex causing disruption of the catalytic center, by what appears to be an irreversible impairment of enzyme function and a reduction in cellular oxygen consumption. Interestingly, while this interaction is thought to be irreversible, a repletion of ferrous iron and reducing agents have been shown to restore enzyme function and normal cellular oxygen consumption.[5] Thus, NO-mediated inhibition of oxygen utilization may be modulated by the redox state and concentration of free iron in the mitochondrial compartment. Nevertheless, these interesting properties of NO do not explain the more sensitive, selective binding of NO to cytochrome oxidase as compared to similarly structured mitochrondrial enzymes. In an attempt to explain this phenomenon, a growing number of investigators have proposed that NO may actually be a remnant substrate for the enzyme cytochrome oxidase, reflecting the evolution of mitochondria in eukaryotic cells. In fact, there is a preponderance of biochemical, spectroscopic and genetic evidence that a significant homology exists between the electron accepting domain Cu_a of mitochondrial cytochrome oxidase, and that of bacterial nitrate reductase.[20,26,36] Early in evolution, reduced gases such as NO contributed significantly to the composition of the atmosphere.[15] Thus, NO played a central role in bacterial bioenergetics and the reduction of NO to N_2O appears to be a reaction that can be catalyzed by cytochrome oxidase.[3] Considering this, it is safe to assume that denitrification predates aerobic respiration and that the denitrification enzymes evolved before the first oxidases. Thus, it is likely that cytochrome oxidase originates from an exchange of genetic material between genes encoding the primitive oxidase and nitrate reductase.[24] Further support for this hypothesis is provided by studies showing that cytochrome oxidase possesses the ability to reduce NO to N_2O.[4] Therefore, the homology between cytochrome oxidase and nitrate reductase may provide the explanation for the selectivity of NO for cytochrome oxidase as well as provide insight into the origin of aerobic respiration.

NITRIC OXIDE AS A CONTROLLER OF CYTOCHROME OXIDASE

Using various spectroscopic techniques, investigators have demonstrated that NO is capable of binding to the oxygen-binding site of cytochrome oxidase.[4,7] Thus, it is plausible that NO and O_2 may compete for the same binding site within cytochrome oxidase. In fact, several studies have shown that the level of NO-mediated mitochondrial inhibition is inversely proportional to the concentration of oxygen in the system.[2,10,25,32] These findings have two important implications. Firstly, competitive inhibition of cytochrome oxidase by nitric oxide could serve to increase the Km for oxygen at cytochrome oxidase, and thus provide an explanation for the observation that the apparent Km for O_2 in tissues is consistently higher than expected from values *in vitro*.[14] Secondly, and more importantly, assuming normal electron flux through the respiratory chain, the rate of mitochondrial respiration may be dependent on the nitric oxide to oxygen ratio ($NO:O_2$) within the tissue. Normal tissue levels of NO and O_2 have been estimated at 100–500 nM and 30 μM respectively. These levels of NO and O_2 have been demonstrated to cause approximately half-maximal inhibition of mitochondrial respiratory rate.[2,25] Thus, at a physiologic $NO:O_2$, there is an inhibition of cellular respiration, suggesting that under normal physiologic conditions NO serves as a brake or tonic inhibitor of mitochondrial respiration. This would serve to increase the Km of cytochrome oxidase for oxygen, thus narrowing the effective diffusion gradient for oxygen between the interstitium and mitochondria. Recently, this effect of NO has been reported to significantly reduce ATP production, suggesting that NO serves to prevent both oxygen wasting and excessive ATP production at baseline tissue metabolic activity[10] Presumably, a constant $NO:O_2$ could be maintained at physiologic states of increased or decreased oxygen demand, as a rise or fall in oxygen delivery and blood flow would be accompanied by the same directional change in vascular endothelial shear stress and NO production. Thus, the amount of NO and O_2 delivered to the mitochondria may be matched proportionally over a wide range of metabolic states, providing an additional control of oxygen utilization and cellular energy production. Hence, any disruption in the $NO:O_2$ could presumably lead to pathologic alterations in cellular oxygen utilization. Such an alteration could occur in conditions of both increased and decreased NO production.

SIGNIFICANCE OF CONTROL OF OXYGEN CONSUMPTION BY NITRIC OXIDE

In addition to the well known alterations in hemodynamic function, it has also been shown that NO production is essentially absent in decompensated heart failure. This has been demonstrated by a reduction in mRNA expression for NO synthase, plasma levels of NO metabolites and tissue NO production in animals with end-stage heart failure.[16,30] It is also noteworthy that isolated failing cardiac muscle looses the bradykinin or acetylcholine-elicited inhibition of respiration.[37] Under such conditions, the $NO:O_2$ may be profoundly reduced, leading to disinhibition of the mitochondria and an increase in the level of O_2 consumption relative to the level of tissue metabolism. Such a phenomenon has been supported by *in vivo* studies showing that in the absence of NO, the amount of oxygen consumed across the coronary circulation is significantly higher than in normal animals at comparable levels of cardiac work. In other words, there is a reduction in cardiac efficiency. This increase in oxygen utilization creates an increased error signal at the level of the mitochondria that must be corrected by an appropriate match in O_2 delivery. However, as the deficit between oxygen demand and oxygen delivery increases, so does the likelihood for a mismatch between tissue perfusion and O_2 utilization, thus increasing the probability of ischemia. This is especially important when vascular reserve is compromised, as is often the case in disease settings in which NO production is decreased. In line with this hypothesis, a restoration of the $NO:O_2$ would restore physiologic mitochondrial function and may provide an explanation for the beneficial effects of NO-releasing agents such as organic nitrates and angiotensin converting enzyme inhibitors in patients with end stage heart failure.[6,23,27] In contrast, conditions such as septic shock represent a state of NO overproduction. Under such conditions, a markedly increased $NO:O_2$ leads to excessive inhibition of cellular respiration that may lead to a pathologic reduction in cellular energy stores resulting in cell dysfunction and death. Indeed, both *in vivo* and *in vitro* studies have demonstrated significant elevations in NO-mediated mitochondrial inhibition in animals with septic shock when compared to normal animals.[8,29] This effect, coupled with a reduction in tissue oxygen tension secondary to reduced tissue perfusion serves to further increase the $NO:O_2$ and exacerbate the suppression in cellular respiration. Until recently, clinicians have attempted to correct the pathology of sepsis by increasing tissue oxygenation through ventilatory and cardiovascular support. However, despite aggressive supportive therapy, 60% of patients progress to multiple organ failure and death. A novel therapy of septic shock may involve supportive therapy coupled with selective inhibition of inducible NOS derived NO. This would serve to normalize the $NO:O_2$ without perturbation of ecNOS function which serves to maintain baseline vascular tone and tissue blood flow (Figure 2).

SUMMARY AND CONCLUSIONS

In summary, NO is capable of decreasing mitochondrial respiration in a variety of mammalian tissues. This effect is mediated primarily via binding of NO to the O_2 binding site of cytochrome oxidase. This highly sensitive interaction presumably reflects a remnant homology between cytochrome oxidase and bacterial nitrate reductase. This effect has been demonstrated at physiologic levels of NO, highlighting the role for NO in the tonic control of cellular respiration. As this inhibition is dependent upon the levels

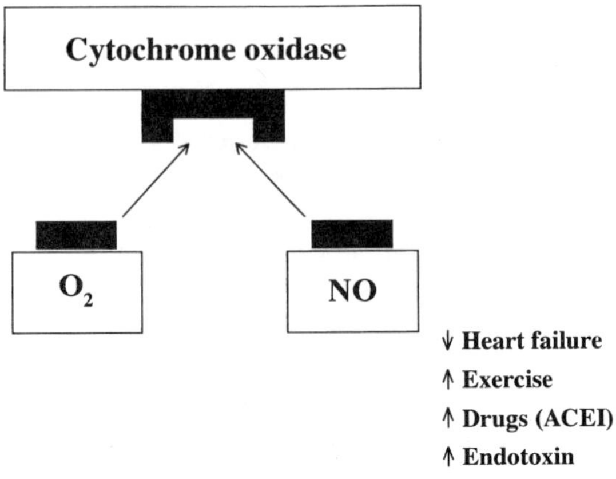

Figure 2. Nitric oxide is competing with oxygen for its binding site on cytochrome oxidase. Endothelial synthesis of nitric oxide is reduced in heart failure resulting in increased cellular respiration. In contrast, nitric oxide production is enhanced in exercise and after administration of drugs such as ACE inhibitors, tending to reduce oxygen consumption. Endotoxin administration or septic shock cause an overproduction of nitric oxide and a reduction in tissue oxygenation.

of NO and O_2 in the tissue, various states of NO production and oxygen supply dictate the ultimate respiratory rate of the mitochondria. Furthermore, deviation from a physiologic NO : O_2 may lead to an exacerbation of pathologic states, such as congestive heart failure and septic shock. Thus, NO may play a crucial role in the control of cellular respiration, providing an additional mechanism of action for this biologically diverse molecule that is distinct yet inseparable from its dilator effect on blood vessels.

ACKNOWLEDGMENT

Supported by NIH PO-1 HL43023, HL 50142, HL 53053, HL 31069.

REFERENCES

1. Bevan, JA, G Kaley, and GM Rubanyi. Flow dependent regulation of vascular function PP 163–177, Oxford University Press, 1995.
2. Brown, GC and CE Cooper. Nanomolar concentrations of nitric oxide reversibly inhibit synaptosomal respiration by competing with oxygen at cytochrome oxidase. *FEBS Letters* 356:295–298, 1994.
3. Brudwig, GW, TH Stevens, and SI Chan. Reactions of nitric oxide with cytochrome c oxidase. *Biochemistry* 19:5275–5285, 1980.
4. Cox, CD, WJ Payne, and DV Dervartanian. Electron paramagnetic resonance studies on the nature on hemoproteins in nitrite and nitric oxide reduction. *Biochemica. et Biophysica ACTA*. 253:290–294, 1971.
5. Drapier, JC and JB Hibbs, Jr. Murine cytotoxic activated macrophages inhibit aconitase in tumor cells; inhibition involves the iron-sulfur prosthetic group and is reversible. *J. Clin. Invest.* 78:790–797, 1986.
6. Elkayam, U, J Amin, A Mehra, J Vasquez, L Weber, and SH Rahimtoola. A prospective, randomized double-blind, crossover study to compare the efficacy and safety of chronic nifedipine therapy with that of isosorbide dinitrate and their combination in the treatment of chronic congestive heart failure. *Circulation* 82:1954–1961, 1990.
7. Farrar, JA, AJ Thomson, MR Cheesman, DM Dooley, and WG Zumft. A model of the copper centres of nitrous oxide reductase. *FEBS* 294(12):11–15, 1991.

8. Forfia, PR, X Zhang, F Ochoa, M Ochoa, X Xu, RD Bernstein, PB Sehgal, NR Ferreri, and TH Hintze. Relationship between plasma NOx and cardiac and vascular dysfunction after LPS induction in anesthetized dogs. *Am. J. Physiol.* 274 (Heart Circ. Physiol. 43):H193–H201, 1998.
9. Forfia, PR, X Zhang, D Knight, MS Wolin, and TH Hintze. The role of nitric oxide in the modulation of myocardial oxygen consumption in the non-human primate; an alternative mechanism of action for a calcium channel blocker. Submitted, 1998.
10. Giulivi, C. Functional implications of nitric oxide produced by mitochondria in mitochondrial metabolism. *Biochem. J.* 332:673–699, 1998.
11. Granger, DL, RR Taintor, JL Cook, and JB Hibbs, Jr. Injury of neoplastic cells by murine macrophages leads to inhibition of mitochondrial respiration. *J. Clin. Invest.* 65:357–370, 1980.
12. Granger, DL and A Lehninger. Sites of inhibition of mitochondrial electron transport in macrophage-injured neoplastic cells. *J. Cell Biol.* 95:527–535, 1982.
13. Hibbs, JB, Z Vavrin, and RR Taintor. L-arginine is reqired for expression of the activated macrophage effector molecule mechanism causing selective metabolic inhibition in target cells. *J. Immunol.* 138:550–565, 1987.
14. Jones, DP. Intracellular diffusion gradients of O_2 and ATP. *Am. J. Physiol.* 250:C663–C675, 1986.
15. Kasting, JF. Earth's early atmosphere. *Science* 259:920–926, 1993.
16. Kichuk, MR, N Seyedi, X Zhang, CC Marboe, RE Michler, LJ Addonizio, G Kaley, A Nasjletti, and TH Hintze. Regulation of nitric oxide production in human coronary microvessels and the contribution of local kinin formation. *Circulatioin* 94:44–51, 1996.
17. Kiechle, FL and T Malinski. Nitric oxide; biochemistry, pathophysiology, and detection. *Am J. Clin. Pathol.* 100:567–575, 1993.
18. King, CE, MJ Melinyshyn, JD Mewburn, SE Curtis, MJ Winn, SM Cain, and CK Chapler. Canine hindlimb blood flow and O2 uptake after inhibition of EDRF/NO synthesis. *J. Appl. Physiol.* 76:1166–1171, 1994.
19. Knowles, RG and S Moncada. Nitric oxide as a signal in blood vessels. *TIBS* 17:399–402, 1992.
20. Malmstrom, BG and R Aasa. The nature of the Cu_a center in cytochrome oxidase. *FEBS Letters* 325:49–52, 1993.
21. Moncada, S, MJ Palmer, and EA Higgs. Nitric oxide: physiology, pathophysiology, and pharmacology. *Pharmacol. Rev.* 43:109–142, 1991.
22. Okada, S, Y Takehara, M Yabaki, T Yoshioka, T Yasuda, M Inoue, and K Utsumi. Nitric oxide, a physiological modulator of mitochondrial function. *Physiol. Chem. Phys. and Med. NMR* 28:69–82, 1996.
23. Packer, ML, PD Kessler, and WH Lee. Calcium channel blockade in the management of severe chronic heart failure: a bridge too far. *Circulation*, 75 (suppl V):V56–V64, 1987.
24. Saraste, M and J Castresana. Cytochrome oxidase evolved by tinkering with denitrification enzymes. *FEBS Letters* 341:1–4, 1994.
25. Schweizer, M and C Richter. Nitric oxide potently and reversibly deenergizes mitochondria at low oxygen tensions. *Biochem. and Biophys. Res. Commun.* 204(1):169–175, 1994.
26. Scott, RA, WG Zumft, CL Coyle, and DM Dooley. *Pseudomonas stutzeri* N_2O reductase contains Cua-type sites. *Proc. Natl. Acad. Sci. USA.* 86:4082–4086, 1989.
27. Sharpe, DN, J Murphy, R Coxan, and SF Hannan. Enalipril in patients with chronic heart failure: a placebo controlled, randomized, double-blind study. *Circulation* 70:271, 1984.
28. Shen, WQ, TH Hintze, and MS Wolin. Nitric oxide; an important signaling mechanism between vascular endothelium and parenchymal cells in the regulation of oxygen consumption. *Circulation* 92:1086–1095, 1995.
29. Simonson, SG, K Welty-Wolf, Y-C Huang, JA Griebel, MS Caplan, PJ Fracica, and CA Piantadosi. Altered mitochondrial redox responses in gram negative septic shock in primates. *Circulatory Shock* 43:34–43, 1994.
30. Smith, CJ, D Sun, C Hoegler, BS Roth, X Zhang, G Zhao, X-B Xu, Y Kobari, K Pritchard Jr, WC Sessa, and TH Hintze. Reduced gene expression of vascular endothelial NO synthase and cyclooxygenase-1 in heart failure. *Circ. Res.* 78:58–64, 1996.
31. Stuehr, DJ and C Nathan. Nitric oxide; a macrophage product responsible for cytostasis and respiratory inhibition in tumor target cells. *J. Exp. Med.* 169:1543–1555, 1989.
32. Takehara, Y, H Nakamara, Y Inai, M Yabuki, K Hamazaki, T Yoshioka, M Inoue, AA Horton, and K Utsumi. Oxygen dependent reversible inhibition of mitochondrial respiration by nitric oxide. *Cell Structure and Function* 21:251–258, 1996.
33. Torres, J, V Darley-Usmar, and MT Wilson. Inhibition of cytochrome c oxidase in turnover by nitric oxide: mechanism and implications for control of respiration. *Biochem. J.* 312:169–173, 1995.

34. Ulmans, JG and R Levi. Nitric oxide in the regulation of blood flow and arterial pressure. *Annu. Rev. Physiol.* 57:771–790, 1995.
35. Vaughn, MW, L Kuo, and JC Liao. Effective diffusion distance of nitric oxide in the microcirculation. *Am. J. Physiol.* 274:H1705–H1714, 1998.
36. Vollack, KU, J Xie, E Hartig, U Romling, and WG Zumft. Localization of denitrification genes in the chromosomal map of *Pseudomonas aeruginosa. Microbiology* 144(2):441–448, 1998.
37. Xie, YW, W Shen, G Zhao, X Xu, MS Wolin, TH Hintze. Role of endothelium-derived nitric oxide in the modulation of canine myocardial mitochondrial respiration in vitro. Circ Res. 79:381–387, 1996.
38. Yonetani, T, JE Erman, JS Leigh, and GH Reed. Electromagnetic properties of hemoproteins. *J. of Biol. Chem.* 247(8):2447–2455, 1972.
39. Zhang, X, YW Xie, A Nasjletti, X Xu, MS Wolin, and TH Hintze. ACE inhibitors promote nitric oxide accumulation to modulate myocardial oxygen consumption. *Circulation* 95:176–182, 1997.

EASY QUANTIFICATION OF THE RESPIRATORY AND METABOLIC IMPACT OF BLOOD O_2-CO_2 EXCHANGE INTERACTIONS IN CRITICAL ILLNESS

Ivo Giovannini, Carlo Chiarla, Giuseppe Boldrini, Gennaro Nuzzo, Felice Giuliante, Gianfranco Lemmo, and Marco Castagneto

Geriatric Surgery and CNR Center for the Pathophysiology of Shock Catholic University School of Medicine
Rome, Italy

INTRODUCTION

When blood flows through the lungs, O_2 uptake causes unbinding of chemically bound CO_2, and contributes by this mechanism to CO_2 release from blood into pulmonary alveoli. This O_2-linked alveolar CO_2 discharge is thus a consequence of the Haldane effect (HE), a relatively well-known phenomenon (Christiansen et al., 1914). However, its magnitude and the relative respiratory implications are not recognized or quantified commonly, due to the lack of easily suitable methods. Our study has been carried out to perform such an assessment over a large group of patients, based on a previously validated model for blood CO_2 exchange and CO_2-O_2 interactions (Giovannini et al., 1993), and to develop new relationships which may ease quantification of the phenomenon in theoretical, experimental and clinical settings.

METHODS

By processing cardiorespiratory data from 247 measurements performed in 91 patients with sepsis, the necessary variables were generated by the model. Patient age was 53 ± 18 yr (mean $\pm$ SD), body weight 64.3 ± 12.8 kg and body surface area $1.72 \pm 0.19\,m^2$ (Du Bois and Du Bois, 1916). Causes of sepsis were peritonitis in 73 cases, retroperitoneal infection in 11, severe cholangitis in 6 and gangrenous fasciitis in 1 case. The diag-

Oxygen Transport to Tissue XXI, edited by Eke and Delpy.
Kluwer Academic / Plenum Publishers, New York, 1999.

nosis of sepsis was based on previously described criteria (Giovannini et al., 1989). The sepsis severity score, obtained at the time of each measurement (Elebute and Stoner, 1983), ranged from less than 10 to more than 20, with a normal Gaussian distribution and a mean ± SD of 15.0 ± 4.0. This patient population provided a continuous distribution of conditions, from relatively compensated to desperately ill states (Table 1), which was suitable for the assessment of CO_2-O_2 interactions over a wide range of cardiorespiratory and metabolic abnormalities.

Measurements, based mainly on simultaneous arterial and mixed venous blood gas analyses (all performed for clinical purposes), included arterial and mixed venous O_2 tension (PaO_2 and PvO_2, mmHg), and saturation (SaO_2 and SvO_2, fraction), CO_2 tension ($PaCO_2$ and $PvCO_2$, mmHg), plasma pH (pHa, pHv), hemoglobin concentration (Hb, g/dl) and other variables. Arterio-venous O_2 concentration difference (a-vDO_2, ml/dl) was calculated from the sum of hemoglobin-bound O_2 concentration difference (Hb-O_2, ml/dl $= 1.36 \cdot Hb \cdot [SaO_2\text{-}SvO_2]$) plus dissolved O_2 difference (D-O_2, ml/dl $= 0.003 \cdot [PaO_2\text{-}PvO_2]$). Based on the model (Giovannini et al., 1993), a stepwise iteration was carried out starting from $PaCO_2$ and from the corresponding blood CO_2 concentration ($CaCO_2$), by decreasing progressively CO_2 tension. At each step, the decrease in CO_2 concentration from the initial value (DCO_2), and the change in CO_2 concentration expected theoretically from reducing O_2 saturation to the mixed venous value (DCO_{2V}), were determined. The necessary calculations were completed by quantifying changes in plasma pH from blood desaturation at any given CO_2 tension, by linear interpolation of published experimental data (Kelman, 1968). The procedure was ended when DCO_2 equated DCO_{2V}, that is, when CO_2 tension decreased to the value which was necessary to maintain a blood

Table 1. Patient data and regressions (see text for symbols)

Pulmonary venous admixture*, fraction	0.19 ± 0.12	(0.04–0.68)
Wasted ventilation*, fraction	0.44 ± 0.13	(0.23–0.79)
Cardiac index*, L/min	4.4 ± 2.3	(1.8–13.7)
PaO_2, mmHg	73.1 ± 17.6	(32.3–154.0)
$PaCO_2$, mmHg	33.1 ± 8.2	(14.0–64.0)
pHa	7.446 ± 0.081	(6.951–7.680)
PvO_2, mmHg	35.6 ± 6.7	(18.3–62.1)
$PvCO_2$, mmHg	38.7 ± 9.0	(18.0–72.0)
pHv	7.414 ± 0.089	(6.860–7.621)
Hb, g/dl	10.4 ± 2.1	(4.5–17.2)
$PaCO_{2NH}$, mmHg	30.9 ± 8.1	(11.9–61.5)
$DPaCO_2$	2.2 ± 0.8	(0.4–5.2)
VA_H, %	6.8 ± 2.6	(1.1–15.9)

A. $PaCO_2 = 2.404 + 0.916\,(PvCO_2) - 2.021\,(v\text{-}aDCO_2) + 0.68\,(Hb\text{-}O_2)$
$r^2 = 0.99$, $p << 0.001$

B. $PaCO_2 = 2.428 + 0.914\,(PvCO_2) - 2.071\,(v\text{-}aDCO_2) + 2.720\,(v\text{-}aDCO_{2H})$
$r^2 = 0.99$, $p << 0.001$

C. $DPaCO_2 = 0.469\,(Hb\text{-}O_2)^{0.968} \cdot 1.014^{PaCO2}/1.026^{Hb}$
$r^2 = 0.99$ $p << 0.001$

D. $(PvCO_2\text{-}PaCO_2)/DPaCO_2 = 0.051 + 1.041\,(v\text{-}aDCO_{2NH}/v\text{-}aDCO_{2H})$
$r^2 = 0.99$ $p << 0.001$

E. $v\text{-}aDCO_{2NH}/v\text{-}aDCO_{2H} = -1.221 + 4.026\,(RQ)$
$r^2 = 0.99$ $p << 0.001$

* Determined as described by Comroe et al. (1962).

CO_2 concentration equal to $CaCO_2$ at an O_2 saturation equal to SvO_2. This corresponded to the value of $PaCO_2$ which should have been expected theoretically in the absence of O_2-linked CO_2 release, that is, in the absence of HE ($PaCO_{2NH}$, mmHg). Thus, the $PaCO_2$—$PaCO_{2NH}$ difference ($DPaCO_2$, mmHg) was the sparing allowed by the HE on the need to decrease CO_2 tension; while the related sparing on alveolar ventilation (VA_H, %) was equal to 100 (1-$PaCO_{2NH}/PaCO_2$) since, for a constant CO_2 production, the $PaCO_{2NH}/PaCO_2$ ratio corresponds to the ratio between the actual value of alveolar ventilation and the value expected in the absence of HE. The model yielded additionally the venoarterial CO_2 concentration difference (v-$aDCO_2$, ml/dl), a separate quantification of the component related to HE (v-$aDCO_{2H}$, ml/dl) and of that related to the increase in CO_2 tension from $PaCO_2$ to $PvCO_2$ at constant O_2 saturation (v-$aDCO_{2NH}$, ml/dl), and the respiratory quotient (RQ) as the ratio between v-$aDCO_2$ and a-vDO_2. In spite of the patient illness, which was in some cases extremely severe, attention was paid to perform all measurements in a condition of "steady state" (that is, in the absence of acute changes of cardiorespiratory patterns) to avoid distortion of the results from transients in blood and body CO_2 stores. Data from a small sub-set of measurements ($n = 20$) in acutely hypoventilating subjects were reported in the Discussion, but not included in the main study. Statistical analysis was peformed by using regression analysis by the least square technique, analysis of residuals, skewness and kurtosis control, and a "simplest best fit" procedure, allowing for selection of the simplest possible regressions yielding the best control of variability, based on Mallow's Cp criteria (Seber, 1977).

RESULTS

Indirect evidence of the modulation exerted by the HE on $PaCO_2$ was provided in regression A (Table 1). This showed that, for any given $PvCO_2$, an expected decrease in $PaCO_2$ of 2.02 mmHg per ml/dl blood CO_2 unloading was buffered by a simultaneous and opposite effect related to O_2 loading and increasing Hb-O_2. Regression B reconfirmed dependency of this buffering on the HE-mediated CO_2 unloading in the lungs. The effects of Hb-O_2 and v-$aDCO_{2H}$ on $PaCO_2$ were highly significant ($p < 0.001$ for both), however with wide confidence limits for regression coefficients, as the likely consequence of indirect linear estimation of nonlinear phenomena. To improve this quantification, the sparing allowed by hemoglobin-O_2 saturation on the need to decrease $PaCO_2$ ($DPaCO_2$, mmHg) was calculated iteratively by our model, thus obtaining the $PaCO_{2NH}$, $DPaCO_2$ and VA_H values shown in Table 1. Main pathophysiologic correlates of $DPaCO_2$ were assessed in regression C, which is suitable also to assess correlates of VA_H, by considering that $VA_H = DPaCO_2/PaCO_2$. HE accounted, as a mean, for 31% of total CO_2 exchange (mean $vaDCO_{2H}$/v-$aDCO_2 = 0.31$, range 0.20–0.40). The modulation exerted by HE on $DPaCO_2$ was assessed further by quantifying the relationships between $DPaCO_2$, the venoarterial CO_2 tension difference ($PvCO_2$-$PaCO_2$), the HE-mediated and HE-independent CO_2 release components (v-$aDCO_{2H}$, v-$aDCO_{2NH}$), and the respiratory quotient (RQ), which is the metabolic factor regulating the magnitude of O_2 exchange relative to CO_2 exchange, and thus the magnitude of O_2-CO_2 exchange interactions (regressions D and E).

DISCUSSION

The interaction between O_2 and CO_2 exchange in lung capillaries involves unbinding and discharge of CO_2 from blood as a consequence of O_2 binding to hemoglobin

(Haldane effect), sparing in part the need to decrease $PaCO_2$ and thus to increase alveolar ventilation. This effect has long been known, and its relevance in physiological and clinical situations has been emphasized, however without the availability of simple methods for its quantification. An easily suitable tool for this purpose has been provided by our work, by calculating $DPaCO_2$ and VA_H and by assessing the main pathophysiologic determinants setting their magnitude. Indeed, regression C in Table 1 provides an accurate quantification of the dependency of $DPaCO_2$ on hemoglobin-O_2 binding (which is the phenomenon generating the HE) and of the exponential effect of hypercapnia on $DPaCO_2$ (which is related to the shape of the blood CO_2 equilibration curve, causing a steeper relationship between changes in $PaCO_2$ and $CaCO_2$ in hypercapnia). Maximum $DPaCO_2$ in our cases was 5.2 mmHg, however much larger values may be predicted from regression C in extreme hypercapnia, and are also expected on the basis of previously published experiments (Joffe and Poulton, 1921, Stadie and O'Brien, 1937). From our quantifications, a proportionally smaller effect on VA_H may be estimated: this depends on the hyperbolic relationship binding $PaCO_2$ to alveolar ventilation, for a given CO_2 production, so that in hypercapnia small changes in ventilation are associated with large changes in $PaCO_2$ (and vice-versa). In reality, however, the full impact of these phenomena on the work of breathing is also magnified hyperbolically by increasing wasted ventilation, and even further exponentially by increasing total ventilation (Giovannini et al., 1989).

Regression C allows a deeper insight within the pathophysiologic interactions affecting $DPaCO_2$ by considering that Hb-O_2 is practically equivalent to CO_2 production/(cardiac output $\times$ RQ), and that $PaCO_2$ is proportional to CO_2 production/(cardiac output $\times$ ventilation:perfusion ratio). Substitution of these quantities in the regression allows one to emphasize the increasing relevance of HE in hypodynamic states, in states with low RQ, with low ventilation and ventilation:perfusion ratio, or when these conditions are combined. In hypodynamic states, this is related obviously to the larger O_2 uptake and O_2-linked CO_2 discharge per unit pulmonary blood flow, and in hypoventilation to the larger effect on $PaCO_2$. More in particular, however, it is the ratio between the HE-independent and the HE-mediated components of CO_2 exchange (v-$aDCO_{2NH}$/v-$aDCO_{2H}$) which is affected in low-RQ states, as shown by the other fits.

From regression E it may also be estimated that the v-$aDCO_{2NH}$/v-$aDCO_{2H}$ ratio decreases down to 0 when RQ decreases to about 0.3. Although this is only an extrapolation outside the distribution of measurements, and although in reality such a condition is supposed to take place transiently (in unsteady state) this is an interesting finding. It reflects the decreasing contribution of the non-Haldane component, compared to the Haldane component, in maintaining CO_2 exchange at low RQ, up to an RQ of 0.3 when CO_2 exchange is maintained only by the HE. A parallel consideration may be extended to decreasing alveolar ventilation, as implicated better in regression D.

In severe hypoventilation with extreme reduction of the measured RQ, the latter reflects CO_2 retention with increasing body CO_2 stores, and not the truly metabolic respiratory quotient (and it is defined more properly "respiratory exchange ratio"). However, the dynamics of O_2-CO_2 exchange interactions, although unsteady and evolving, are not affected in their fundamental nature, at least in the initial stages. In effect, when additional data from patients with acute hypoventilation were included in our study, these fell outside the main distribution, but the relationships predicted by regressions D and E in Table 1 remained basically unchanged. These considerations on the role of HE in maintaining CO_2 release from pulmonary capillaries into the alveoli apply to diseased states (acute hypoventilation, or respiratory arrest), to non-pathologic conditions (apnoea during diving, apneic oxygenation) (Mertzlufft and Brandt, 1989, Hicks and White, 1992) and also to the maintenance of CO_2 exchange in pulmonary alveoli with very low ventilation:perfusion ratios (Klocke, 1987). As mentioned, in unsteady state much more complex dynamics become involved and should be taken into account (Mithoefer, 1959, Nahas and Verosky, 1966), while our approach has been limited to simpler interactions, ignoring also the impact of ventilation:perfusion maldistribution. Nevertheless, the results are consistent with those obtained by more complicated procedures (Grant, 1982).

In conclusion, this study provides a simplified method for assessing the contribution of HE to pulmonary CO_2 release and to the modulation of $PaCO_2$ and alveolar ventilation. The results suggest an increasing quantitative relevance of HE in low-blood-flow and low-ventilation states, which should be magnified further in extreme hypercapnia, breath holding, and respiratory arrest. These results are important to understand mechanisms of endurance in physiologic states such as voluntary apnoea, and mechanisms of adaptation and survival in extreme illness.

ACKNOWLEDGMENTS

The authors acknowledge the kind assistance offered in the development of this work by Fabio Cesari, Antonio Marchetti, Emilio Meneschincheri and Vincenzo Tracuzzi from the Centro Elaborazione Dati (CED) and Infocenter, Catholic University, Rome.

REFERENCES

Christiansen J, Douglas CG, and Haldane GS (1914) The absorption and dissociation of carbon dioxide by human blood. J Physiol 48:244–271.

Comroe JH jr, Forster RE, Du Bois AB, Briscoe WA, and Carlsen E (1962) In: The Lung, Chicago, Year Book.

Du Bois D and Du Bois EF (1916) A formula to estimate the approximate surface area if height and weight be known. Arch Int Med 17:863–871.

Elebute EA and Stoner HB (1983) The grading of sepsis. Br J Surg 70:29–31.

Giovannini I, Chiarla C, Boldrini G, and Castagneto M (1989) Impact of fat and glucose administration on metabolic and respiratory interactions in sepsis. JPEN 13:141–146.

Giovannini I, Chiarla C, Boldrini G, and Castagneto M (1993) Calculation of venoarterial CO_2 concentration difference. J Appl Physiol 74:959–964.

Grant BJB (1982) Influence of Bohr-Haldane effect on steady-state gas exchange. J Appl Physiol 52:1330–1337.

Hicks JW and White FN (1992) Pulmonary gas exchange during intermittent ventilation in the American alligator. Respir Physiol 88:23–36.

Joffe J and Poulton EP (1921) The partition of CO_2 between plasma and corpuscles in oxygenated and reduced blood. J Physiol 54:129–151.

Kelman GR (1968) Nomogram for calculation of buffer line shift due to reduction of haemoglobin at constant PCO_2. Clin Chim Acta 22:277–280.

Klocke RA (1987) Carbon dioxide transport. In: Handbook of Physiology, The Respiratory System, Vol. 4 (Farhi LE, Marsh Tenney S, Eds), American Physiological Society, Bethesda MD, pp. 173–197.

Mertzlufft F and Brandt L (1989) An in vivo model for the proof of the Christiansen-Douglas-Haldane effect. In: Oxygen Transport to Tissue, XI (Rakusan K, Biro GP, Goldstick TK, Turek Z, Eds), Plenum Press, New York, pp. 397–405.

Mithoefer JC (1959) Mechanism of pulmonary gas exchange and CO_2 transport during breath holding. J Appl Physiol 14:706–710.

Nahas GG and Verosky M (1966) The storage of CO_2 during apneic oxygenation. Ann NY Acad Sci 133:134–141.

Seber GAF (1977) In: Linear Regression Analysis, New York Wiley.

Stadie WC and O'Brien H (1937) The carbamate equilibrium. II. The equilibrium of oxyhemoglobin and reduced hemoglobin. J Biol Chem 117:439–470.

48

ENHANCEMENT EFFECT OF METHYL α-D-GLUCOSIDE FOR INHIBITORY EFFECTS OF ANTIOXIDANTS ON ADP/Fe^{2+}-INDUCED LIPID PEROXIDATION IN RAT LIVER MITOCHONDRIA

Hitoshi Hori,[1] Masaki Ishibashi,[1] Saharuddin B. Mohamad,[1]
Hideko Nagasawa,[1] Yoshihiro Uto,[1] Hiroyuki Sakamaki,[1] Ning Pan,[1]
Kazuto Ohkura,[1] and Sansei Nishibe[2]

[1]Department of Biological Science and Technology
Faculty of Engineering
The University of Tokushima
Tokushima, 770-8506
Japan
[2]Faculty of Pharmaceutical Sciences
Health Sciences University of Hokkaido
Hokkaido, 061-0293
Japan

INTRODUCTION

Overproduction of oxygen free radical is closely related to aging, cancer, inflammation, ischemia/reperfusion injury, and arteriosclerosis (Kerr et al., 1996). For protection against these free radical related diseases, many kinds of traditional Kampo medicine and traditional Chinese medicine, which are all crude drugs, have been used clinically (Otsuka, 1988). The data on their pharmacological mechanisms increased, but these data are not enough to be evaluated scientifically (Takahashi et al., 1993, Akamatsu et al., 1998). We found acteoside (ACT) and its derivatives, isolated from oriental medicine *Phacellantus tubiflorus*, as potent inhibitors of lipid peroxidation in mitochondria (Pan and Hori, 1994, Pan and Hori, 1996). We recently found that methyl α-D-glucoside (Me-Glc; Figure 1) enhanced the anti-peroxidative activity of acteoside in rat liver mitochondria (Sakamaki et al., 1996). For the investigation of the importance of non-antioxidant glycosides contained commonly in Kampo medicines, we present here the enhancement effect of the

Oxygen Transport to Tissue XXI, edited by Eke and Delpy.
Kluwer Academic / Plenum Publishers, New York, 1999.

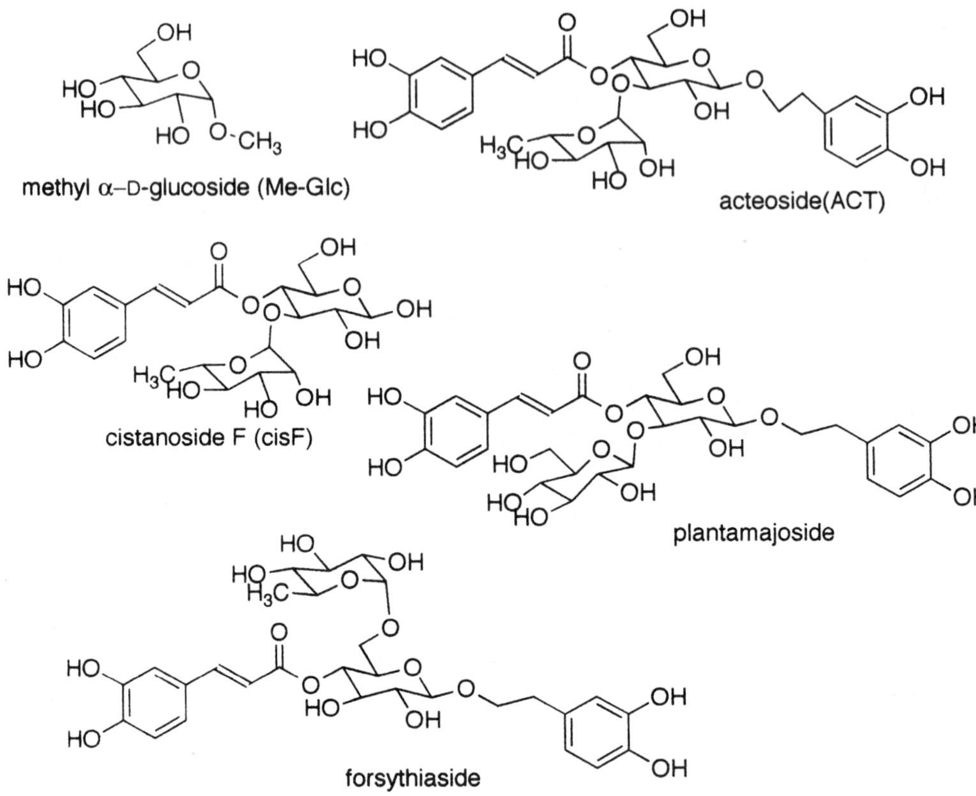

Figure 1. Chemical structures of methyl α-D-glucoside and phenylpropanoid glycosdies.

simplest glycoside Me-Glc for inhibitory effects of a variety of antioxidants, including ACT and α-tocopherol, in iron-induced mitochondrial lipid peroxidation reaction.

MATERIALS AND METHODS

Chemicals

Methyl α-D-glycoside (Me-Glc), 1, 1-diphenyl-2-picrylhydrazyl (DPPH), 2-(*N*-morpholino)ethanesulfonic acid monohydrate (MES) and other reagents were purchased from Wako Pure Chemical Industries, Ltd, Osaka. Phenylpropanoids such as acteoside and cistanoside F were isolated from *Phacellanthus tubiflorus* (Pan and Hori 1994, Pan and Hori, 1996). Other phenylpropanoids such as plantamajoside and forsythiaside were isolated from *Plantago asiatica* (Ravn et al., 1992) and *Forsythia suspensa* (Nishibe et al., 1982), respectively. Chemical structures of methyl α-D-glycoside and phenylpropanoid glycosides are shown in Figure 1.

Preparation of Rat Liver Mitochondria (RLM)

RLM were isolated from male Wistar rats (250–300 g) by homogenization followed by differentiated centrifugation in ice-cold medium (pH 7.4) (Myers and Slater, 1957,

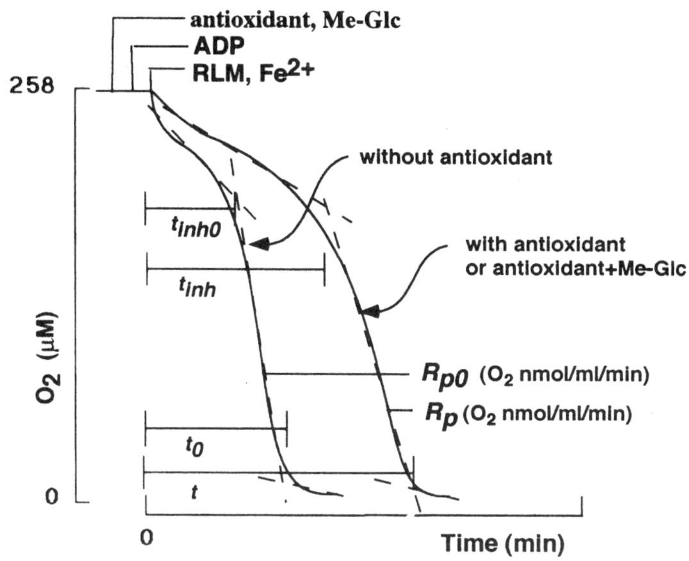

Figure 2. The experimental chart for the calculation of the percent inhibition on lipid peroxidation in rat liver mitochondria.

Lash and Sall, 1993). Fresh and freeze-thawed RLM were used for this study. The mitochondrial protein content was determined by the biuret method using bovine serum albumin as a standard (Gornall et al., 1949).

Assay of Lipid Peroxidation

Mitochondrial lipid peroxidation reactions were initiated by ferrous sulfate and ADP, and their oxygen consumption rates were measured with a Clark-type oxygen electrode. The rapid oxygen consumption rate (*Rp*) in mitochondrial lipid peroxidation, its Percent inhibition with drugs were calculated as described previously (Pan and Hori, 1994). Briefly, lipid peroxidation in mitochondria was measured by assuming an oxygen concentration of 516 natom O/mL (258 μM O_2)/mg protein in the experimental medium at 25°C. The percent inhibition with drug on mitochondrial lipid peroxidation was calculated using Eq.1, where parameters were defined as shown in Figure 2.

$$Inhibition(\%) = \left(1 - \frac{R_p t}{t_{inh}} k\right) \times 100\% \quad k = \frac{t_{inh0}}{R_{p0} t_0}$$

R_p = Rate of lipid peroxidation (nmol O/min)
R_{p0} = Rate of lipid peroxidation of control (nmol O/min)
t = total time (min)
t_0 = total time of control (min)
t_{inh} = induction time (min)
t_{inh0} = induction time of control (min) (1)

The mean effective concentration of antioxidant tested required for its 50% inhibition was defined as IC_{50} in the mitochondrial lipid peroxidation reaction.

Free Radical Scavenging Activity

The interaction of test compound with free radical species was measured by using DPPH at 517 nm according to the method of Blois (1958) with some modifications. To 2.5 mL of 100 μM DPPH solution, containing 20 μM MES, pH 5.5, was added 5 μL of different concentrations of antioxidants tested. The change in optical absorbance at 517 nm was measured every 5 min for 40 min. The mean effective concentration of antioxidants tested required to decrease the absorption by 0.20 in the reaction with DPPH after 30 min ($IC_{0.20}$) was calculated from the absorbance concentration curve.

RESULTS AND DISCUSSION

The Inhibition of Antioxidants on Mitochondrial Lipid Peroxidation

Some Kampo medicines have immunostimulating activities partially related to its antioxidant and radical scavenging activities. First we checked antioxidant activities of acteoside, cistanoside F, plantamajoside, and forsythiaside isolated as active constituents from Kampo medicines and α-tocopherol. As RLM used here, we mainly used freeze-thawed RLM instead of fresh RLM, because of no influences of ferrous-iron [Fe(II)] to the energy transducing system in the ferrous-iron induced mitochondrial lipid peroxidation (Shivaswamy et al., 1993). Generally, in the ADP/Fe(II)-induced mitochondrial lipid peroxidation reaction without inhibitors, oxygen started to be consumed slowly (initiation: evaluated by the induction time, t_{inh}, in Eq. 1), followed by rapid consumption (propagation: evaluated by the rate of lipid peroxidation, R_p, in Eq. 1). With inhibitors such as acteoside or cistanoside F, the initiation time and total reaction time were extended to yield less amount of oxygen consumption than that of control. Plantamajoside and forsythiaside also inhibited weakly during both the initiation and progression periods. Among the tested antioxidants, α-tocopherol was the best inhibitor of the ADP/Fe(II)-induced lipid peroxidation reaction in freeze-thawed rat liver mitochondria. The order of their inhibitory activities (IC_{50}) was: α-tocopherol (3.0 μM) > cistanoside F (13.6 μM) > acteoside (14.7 μM) > forsythiaside (29.0 μM) > plantamajoside (29.1 μM).

Effects of Methyl α-D-Glucoside (Me-Glc) on Inhibitory Activities of Antioxidants in Mitochondrial Lipid Peroxidation

Acteoside and other antioxidants used here are chemically classified to be phenylpropanoid glycosides. Methyl α-D-glycoside (Me-Glc), the simplest and unnatural glycoside, was used as the simplest one of non-antioxidant glycosides contained commonly in Kampo medicine. Influence of Me-Glc against the effects of acteoside on oxygen consumption due to ADP/Fe(II)-induced lipid peroxidation was examined in fresh RLM and freeze-thawed RLM. Previously, we reported that in fresh RLM, addition of Me-Glc (20 μM) enhanced the anti-lipid peroxidation activity or inhibitory activity of ACT (20 μM) by 10% from 37% to 47% (Sakamaki et al., 1996). In freeze-thawed RLM, however, Me-Glc with the same concentration ratio as that in fresh RLM enhanced its inhibitory activity by 14.1% (from 42% to 56.1%) as shown in Figure 3 (a). Me-Glc (20 μM) alone showed no inhibitory activity in either fresh RLM

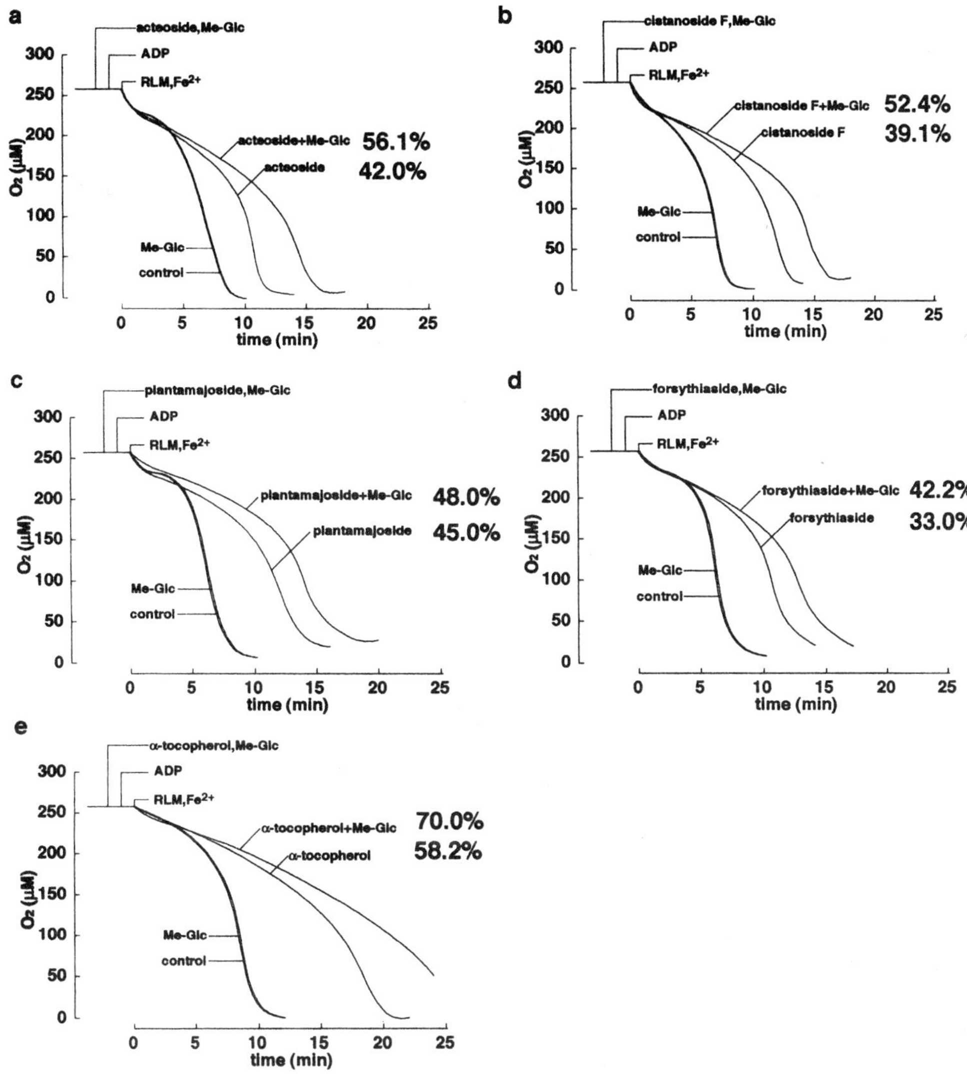

Figure 3. Influence of Me-Glc for inhibitory effects (%) of phenlypropanoid glycosides (a–d) and α-tocopherol (e) on oxygen consumption due to ADP/Fe(II)-induced lipid peroxidation in freeze-thawed RLM.

or freeze-thawed RLM. This enhancement effect of Me-Glc was dose-dependent. Glucose itself also showed no enhancement effect. From these preliminary data of Me-Glc on the inhibitory activities of antioxidants in mitochondrial lipid peroxidation, we mainly examined the enhancement effect of Me-Glc to antioxidants on the freeze-thawed RLM. Both concentration of Me-Glc and antioxidants tested were established appropriately at the concentration close to the IC_{50} of each antioxidant. The enhancement effects of Me-Glc were also found for other antioxidants. Thus Me-Glc enhanced the inhibitory activities of cistanoside F (20 μM), plantamajoside (30 μM), and forsythiaside (30 μM) by 13.3% (from 39.1% to 52.4%) at 20 μM Me-Glc, 3.0% (from 45.0% to 48.0%) at 30 μM Me-Glc, and 9.2% (from 33.0% to 42.2%) at 30 μM Me-Glc, respectively. For the generality of its enhancement effect, we checked the effect of Me-Glc to α-

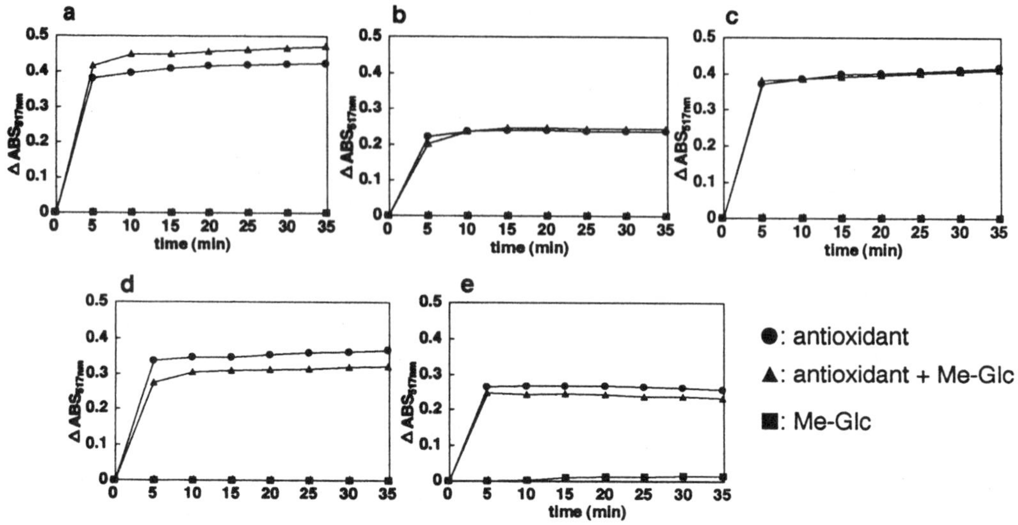

Figure 4. Influence of Me-Glc (10 μM) for the radical scavenging activities of antioxidants (10 μM), aceteoside (a), cistanoside F (b), plantamajoside (c), forsythiaside (d), and α-tochopherol (e).

tocophenrol as a typical antioxidant or membrane-lipid-peroxidation inhibitor. As a result, Me-Glc also enhanced the inhibitory activity of 4 μM α-tocopherol by 12.5% (from 58.2% to 70.7%). Me-Glc enhanced the inhibitory activity (66%) of 4 μM α-tocopherol dose-dependently.

Influence of Me-Glc on the Radical Scavenging Effects of Antioxidants

To understand the enhancement effect of Me-Glc on the inhibition of antioxidants in mitochondrial lipid peroxidation, we examined whether Me-Glc had a free radical scavenging activity. First, we examined the radical-scavenging potencies ($IC_{0.20}$) of antioxidants and Me-Glc. Every antioxidant used here showed radical-scavenging activities, the potencies were ranked as follows: acteoside ($IC_{0.20}$ = 5.0 μM) > plantamajoside ($IC_{0.20}$ = 5.4 μM) > cistanoside F ($IC_{0.20}$ = 7.3 μM) > forsythiaside ($IC_{0.20}$ = 8.3 μM) > α-tocopherol ($IC_{0.20}$ = 11.9 μM). Me-Glc alone did not have effects. Next we examined the effect of Me-Glc on radical-scavenging activities of antioxidants. Figure 4 showed that 10 μM Me-Glc had no effect on radical-scavenging activities of 10 μM antioxidants. From these data, the enhancement effect of Me-Glc does not depend on a free-radical mechanism.

CONCLUSION

We found that the simple glycoside Me-Glc enhanced the inhibitory activities of antioxidants on the mitochondrial membrane lipid peroxidation. Our results will become an initial step in pharmacological study of large amounts of glycosides as common components in crude drugs as Kampo medicines. Me-Glc may also be an effective additive to enhance an antioxidant effect in western medicines as well as oriental medicines such as Kamp medicines.

ACKNOWLEDGMENT

This work was supported in part by a Grant-in-Aid for Scientific Research (No. 05671872 to H. Hori) from the Ministry of Education, Science, Sports and Culture, Japan.

REFERENCES

Akamatsu H, Asada Y, and Horio T (1998) Inhibitory effect of shofu-san, a Japanese kampo medicine, on neutrophil functions in vitro. *Am. J. Chin. Med.* 26:57–64.

Blois MS (1958) Antioxidant determination by the use of a stable free radical, *Nature* 181:1199–1200.

Gornall AG, Bardawill CJ, and David MM (1949) Determination of serum proteins by means of the biuret reaction. *J. Biol. Chem.* 177:751–766.

Kerr ME, Bender CM, and Monti EJ (1996) An introduction to oxygen free radicals. *Heart Lung* 25:200–209.

Lash LH and Sall JM (1993) Mitochondrial Isolation from Liver and Kidney: Strategy, Techniques, and Criteria for Purity, In: Methods in Toxicology, Vol.2: Mitochondrial Dysfunction (Lash, LH, Jones DP, Eds), Academic Press, Inc., San Diego.

Mellors A and Tappel AL (1966) The inhibition of mitochondrial peroxidation by ubiquinone and ubiquinol. *J. Biol. Chem.* 241:4353–4356.

Myers DK and Slater EC (1957) The enzymatic hydrolysis of adenosie triphosphate by liver mitochondria I. Activities at different pH values. *Biochem. J.* 67:558–572.

Nishibe S, Okabe K, Tsukamoto H, Sakushima A, and Hisada S (1982) The structure of forsythiaside isolated from *Forsythia suspensa*. *Chem. Pharm. Bull.* 30:1048–1050.

Otsuka Y (1988) Pharmacotherapy in orienetal medicine. In: Recent Advances in the Pharmacology of KAMPO (JAPANESE HERBAL) MEDICINES (Hosoya E and Yamamura Y), Excerpta Medica, Tokyo.

Pan N and Hori H (1994) The interaction of acteoside with mitochondrial lipid peroxidation as an ischemia/reperfusion injury model. *Adv. Exp. Med. Biol.* 361:319–325.

Pan N and Hori H (1996) Antioxidant action of acteoside and its analogs on lipid peroxidation. *Redox Report* 2:149–154.

Ravn H, Nishibe S, Sasahara M, and Xuebo L (1992) Phenolic compounds from *Plantago asiatica. Phytochemistry* 29:2627–3631.

Sakamaki H, Hori H, Pan N, Namiki A, and Hikawa S (1996) Enhancement effect of a simple glycoside against the inhibitory activity of anti-lipid peroxidative compound, *Lipid Peroxide Res.*, 20:63.

Takahashi S, Yoshikawa T, Naito Y, Minamiyama Y, Tanigawa T, and Kondo M (1993) Antioxidant properties of antiulcer Kampo medicines. *Free Radic. Res. Commun.* 19 Suppl 1:S101–S108.

A PATHOGENESIS-BASED CLASSIFICATION SCHEME FOR MITOCHONDRIAL DISORDERS

D. J. Maguire

School of Biomolecular and Biomedical Science
Griffith University
Nathan, Q4111, Australia

1. INTRODUCTION

Mitochondria are the small energy "powerhouses" of animal cells and, during dark periods, plant cells. The adenosine triphosphate (ATP) synthesised by these organelles is used to provide most of the energy for cellular metabolism. In recent years, it has become accepted that they have probably evolved from bacterial strain(s) that developed endosymbiotic relationships with the host cells. Among the most compelling evidence for this theory of endosymbiosis is the existence, within mitochondria, of a mitochondrial genome, the translation of which involves unique codons, each with its own tRNA species. The two rRNA molecules encoded by mitochondrial genes show more structural similarity to the equivalent bacterial entities than to the nuclear-encoded ones. It is now believed that during subsequent evolution following the initial endosymbiotic event, the mitochondrial genome has become increasingly conservative of space until, in humans, the total length of coded information is a mere 16,569 nucleotide base pairs. This reduction in encrypting ability has been achieved by the transfer of most of the original bacterial DNA into the host eucaryotic cell's nucleus and also by the use of devices such as overlapping genes. The remaining extra-nuclear coding information provides for the synthesis of only 13 protein subunits, all directly involved in the process of electron transport-linked ATP synthesis. Unlike other sub-cellular organelles, mitochondria are therefore a combination of nuclear-encoded and mitochondrially-encoded translation products. This situation is represented in a highly simplistic fashion in Figure 1. By comparison, the encryption of those gene products which do not undergo translation, such as the 22 tRNA species required to satisfy the minimum demand for translation ability in addition to the two unique mitochondrial rRNA species, have been retained within the mitochondrial genome. This separated storage and processing of encryption data has a

Oxygen Transport to Tissue XXI, edited by Eke and Delpy.
Kluwer Academic / Plenum Publishers, New York, 1999.

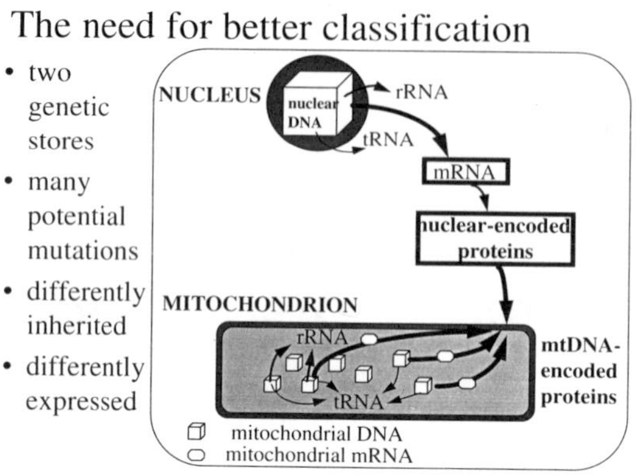

Figure 1. A simplistic representation of mitochondrial protein origins and the need for a clearer classification of the mitochondrial disorders.

number of implications for human disease expression and for the genetic analyses of those diseases. There are some disorders of mitochondrial metabolism which have been known for a considerable time to be clearly encrypted by errors in genes exclusive to the nuclear storage pool. There is also clear evidence for the existence of disorders in which the mitochondrial gene pool of an individual is uniformly defective; the disorder is homoplasmic in its expression. In relatively recent times, there have been an increasing number of reports of disorders in which mutations are present in only a portion of the mitochondrial population of an individual (heteroplasmy). Some previously recognised disorders fall clearly into one or other of these categories. However, a number of other previously described disorders of diverse symptoms and modes of expression now appear to comprise cases which are completely different in their genetic origin. Among this group, one of the best examples is Leigh's syndrome, characterised by a range of genetic defects (Table 1), including errors in mitochondrial mRNA, tRNA or rRNA (Schoffner, 1996). Those forms of the disorder generally are expressed via ATP synthetase malfunction. Nevertheless, it appears that the most common form of this disorder involves nuclear encryption instead (Wijburg and Barth, 1992), in terms of errors resulting in a defective pyruvate dehydrogenase complex (PDHC). On the basis of that heterogenous genetics alone, there would appear to be a basis for redefinition of some mitochondrial disorders. To avoid confusion among the medical profession and among

Table 1. The complex range of aetiologies of Leigh's syndrome

Leigh's syndrome - variable symptoms and variety of genetic defects suggest it is not a single disease entity

- **NUCLEAR ENCODED FORMS**
  - **PDHC - pyruvate dehydrogenase complex**
  - **COX 1 (Cytochrome C oxidase) - complex IV**
  - **NADH dehydrogenase - complex I**
- **MITOCHONDRIALLY ENCODED FORM**
  - **ATP synthetase**
    - **ATPase 6;**
      - **T8993G (leu to arg) (Uziel et al, 1997), T8993C (leu to pro)**
      - **others at 3243, 8344, 8593**

allied health workers, it would also seem timely to provide a clear definition of all the existing mitochondrial disorders via a system of classification based upon the aetiology, rather than the symptoms, of the disorders. In the present report an attempt has been made to use that approach upon a limited, but representative number of mitochondrial disorders.

2. HUMAN MITOCHONDRIAL DISORDERS

The use of the word "mitopathy" has become accepted practice as a descriptor for the general class of disorders of mitochondrial metabolism. It is an appropriate term insofar as it describes the underlying feature which is ultimately expressed in symptoms of the class, but it also has the added advantage of brevity when compared to the original descriptor ("mitochondrial disorder"). This brevity has added usefulness because there is often a need for the addition of modifying decriptors to provide a clear meaning to particular disorders. For example, in order to distinguish between those mitopathies of nuclear genome origin and those of mitochondrial gene origin, the descriptors, "nuclear mitopathy" and "mitochondrial mitopathy", respectively, might be recommended.

2.1. Nuclear Mitopathies (Mitochondrial Matrix Disorders)

Mitochondrial involvement in human disease has been recognised at one level for many years—a number of genetically well-defined disorders are known to be expressed via disturbances to metabolic pathways which are, at least partly, located within mitochondria. These disorders are inherited in a Mendelian pattern. Disorders of urea cycle enzymes located within the mitochondria are one such example. Such disorders arise as a result of errors in coding which have originated in nuclear genes as opposed to genes situated within mitochondria, so it would seem appropriate to call disorders of this general type "nuclear mitopathies" (Figure 2).

The disorders specifically characterised by episodes of hyperammonemia, such as the urea cycle enzymes might more appropriately be called "(nuclear) mitazopathies", or "(nuclear) mitammopathies", to distinguish them from disorders of lipid metabolism for example. This latter group would arguably be called "(nuclear) mitolipopathies".

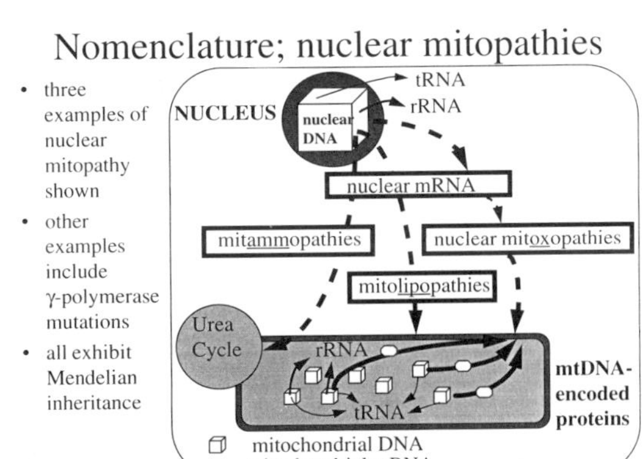

Figure 2. A simple representation of nuclear mitopathies.

2.1.1. Nuclear Mitazopathies (or Mitammopathies), Of the five recognised disorders of urea cycle, the best representatives for this discussion are ornithine transcarbamylase deficiency (OTC), and carbamoyl phosphate synthetase deficiency. Even within OTC, distinctions are made between early onset and late onset forms of the disorder. The causes of the difference between the two OTC forms appears to be the milder enzymic defect of the latter and the dependence upon a protein challenge for its expression. Both disorders are associated with nuclear gene defects and are characterised by the inability of the respective mutant gene products to function correctly within the mitochondrial matrix.

2.1.2. Nuclear Mitolipopathies. Medium chain acyl decarboxylase deficiency (MCAD) is one example of such a disorder, though by no means the only one. This is claimed to be the most common disorder of mitochondrial fatty acid oxidation (Ruitenbeck et al., 1992). The disorder is clearly associated with a nuclear gene, 90% of cases reported being attributable to an A,G mutation at position 985 of the MCAD gene (Bross, 1995).

2.2. Disorders of Oxygen Metabolism (Mitochondrial Electron Transport Chain Disorders)

The full medical significance of the extranuclear information stored in mitochondria has only relatively recently began to be appreciated as an increasing number of disorders which were previously believed to have "loose genetic links" have now been associated directly with mitochondrial DNA (mtDNA) variants. Almost all of the early discovery of mitochondrial genome disorders have been associated with relatively rare defects. Unfortunately, most were also introduced as "named" diseases, often long before the existence of the mitochondrial genome was suspected. The range of disorders that are potentially classifiable as disorders attributable to mutations in the mitochondrial genome span from obscure diseases expressed early in development to well-established disorders generally associated with the aged; from diseases of neuromuscular involvement to diseases of endocrinological aetiology; from disorders with clear genetic patterns to diseases which require additional environmental factors to initiate expression. Needless to say, wide recognition of the importance of mtDNA within the medical fraternity and amongst biomedical researchers has not yet been achieved. Much of the problem seems to arise as a result of the inability of workers inculcated with the concepts of Mendelian inheritance to accept some of the notions involved in mitochondrial genetics, including such concepts as purely maternal inheritance, homoplasmic versus heteroplasmic expression, and disease threshold. For the purpose of clarity of meaning and interpretation, it seems appropriate to separate these disorders from the nuclear-encoded group alluded to above by using the term "mitochondrial mitopathies" (literally "sick mitochondria caused by (mutated) mitochondrial genes") for this group. A general feature of such disorders is the malfunction of one or more of the complexes of the electron transport chain. The widespread effect upon many aspects of intermediary glucose metabolism is one consequence of the disruption of oxygen metabolism accompanying mutations expressed at the electron transport chain complexes. Symptoms such as lactic acidosis are therefore a feature of these mitochondrial mitopathies.

2.2.1. Mitochondrial Mitoxopathies. The presence within the cell of two separate coding sources contributing subunits to the formation of electron transport chain com-

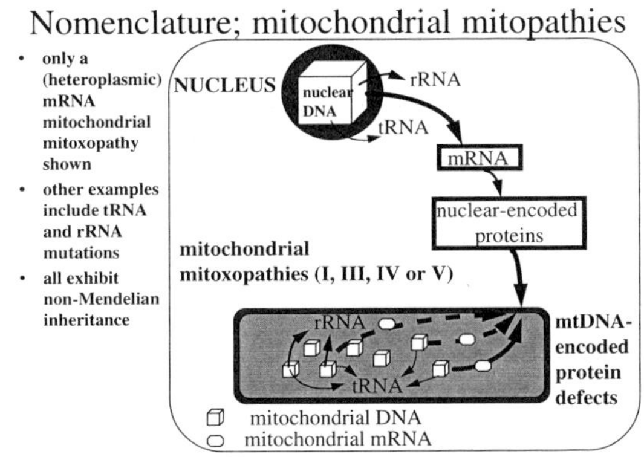

Figure 3. A simple representation of mitochondrial mitoxopathies.

plexes provides the rationale for a natural division of mitopathies into two further groups. In the first group are those mitopathies that are caused by mutations in the translated genes encoded within mitochondria. It would seem appropriate to call this group of disorders "mitochondrial mitoxopathies" (Figure 3).

Genetic features of this group of disorders include:

1 non-mendelian inheritance patterns
2 the potential for genetic drift within the lifetime of the individual
3 in the case of heteroplasmic mutation distribution, the existence of a level of mutation burden, a "threshold" above which disease expression occurs
4 tissue mosaicism in the case of heteroplasmic mitochondrial mitopathies

2.3. Further Subdivisions within the Mitochondrial Mitoxopathies

A case is made for further subdividing this group so as to reflect their origins more specifically. The fact that there are a group of mitochondrial genes which code for non-translated products predicates the need for a separate classification.

2.3.1. Mitochondrial rRNA Mitoxopathies. The existence of mutations within the sequence of mitochondrial DNA encoding the two mitochondrial rRNA genes had not been suspected to be clinically significant until the report of work by Hutchin et al. (1993). That report was the first to associate aminoglycoside-induced deafness syndrome to such a mutation although since then there have been reports of additional mutations on 12SrRNA associated with this syndrome. There have also been a number of reports of cases in which mitochondrial rRNA mutations have been associated with deafness in the absence of aminoglycoside exposure. The aminoglycoside-induced deafness syndrome described by Hutchin's group might therefore be a representative of a separate group which could be called the "iatrogenic mitochondrial rRNA mitoxopathies", Hutchin's work sounds a warning that there may be other, as yet undescribed, disorders which would fall into the category of "environmental mitochondrial mitoxopathies". The existence of mutations within the sequence of mitochondrial DNA encoding the two mitochondrial rRNA genes had not been suspected but clearly present an experimental model which might be extended to a range of xenobiotics in future studies.

2.3.2. Mitochondrial tRNA Mitoxopathies. Various reports have linked mutations in mitochondrial tRNA species to the expression of disorders of the electron transport chain. It is difficult at this time to accurately assess real significance of all of these reports against a background of conflicting evidence. However, the mitochondrial tRNA mutations reviewed by Schoffner (1996) are representative of this class of mutation. The defective tRNA presumably leads to inappropriate translation, thereby producing the same effect as would occur if multiple open reading frame mutations were present.

2.4. Nuclear Mitoxopathies

While there are a number of recognised disorders which can definitely be classified as mitochondrial mitopathies, there is great potential for the discovery of the members of a second class of mitopathies which is caused by mutations in the translated genes encoded within the nucleus. It would seem appropriate to call these disorders "nuclear mitoxopathies". Nuclear genes are involved in all facets of mitochondrial gene expression, including DNA synthesis (the mitochondrial DNA polymerase is encoded in the nucleus) and translation (all of the protein subunits of the mitochondrial ribosomes are encoded in nuclear genes). In addition, but with the express exception of complex II of the electron transport chain, all of the electron transport chain complexes contain some subunits which are mitochondrially encoded and some which are nuclear-encoded Therefore, one could expect to observe nuclear mitoxopathies which mimic each of the mitochondrial mitoxopathies in terms of symptoms. In addition, there should also be a number of nuclear mitopathies resulting from the disruption to mitochondrial DNA synthesis. By comparison to mitochondrial mitoxopathies, however, nuclear mitoxopathies would have Mendelian patterns of expression and would not be expected to exhibit any genetic drift within the lifetime of the individual.

3. DISCUSSION

The development of a classification scheme based upon pathogenesis will make it easier for health professionals as well as allied health workers to better understand their client's disorders. Clear delineation of the origin of a particular individual's disorder will also ensure their access to existing and emerging therapies as they becomes available. A clearly defined classification system will also help to ensure the access of individuals and their families to important health-related services such as appropriate genetic counselling. In the present work, an attempt has been made to classify the known mitopathies on a pathogenetic basis and to draw attention to apparent anomalies in the literature.

REFERENCES

Bross, P, Jespersen, C, Jensen, TG, Andresen, BS, Kristensen, MJ, Winter, V, Nandy, A, Krautle, F, Ghisla, S, Bolund, L, Kim, JP, and Gregersen, N. Effects of two Mutations Detected in Medium Chain Acyl-CoA Dehydrogenase (MCAD)-deficient Patients on Folding, Oligomer Assembly and Stability of MCAD enzyme. J. Biol. Chem. 270:10284–10290 (1995).

Hutchin, T, Haworth, I, Higashi, K, Fischel-Ghodsian, N, Stoneking, M, Saha, N, Arnos, C, and Cortopassi, G. A Molecular Basis for Human Hypersensitivity to Aminoglycoside Antibiotics. Nucleic Acids Res. 21:4174–4179, (1993).

Ruitenbeck, W, Poels, PJE, Turnbull, DM, Garvaglia, B, Chalmers, RA, Taylor, RW, and Gabreels, FJM. Rhabdomyolosis and Acute Encephalopathy in Late Onset Medium Chain Acyl-CoA dehydrogenase Deficiency. J. Neurol Neurosurg Psychiatry, 58:209–214 (1992).

Schoffner, JM. Maternal Inheritance and the Evaluation of Oxidative Phosphorylation Diseases. Lancet, 348:1283–1287, (1996).

Wijburg, FA and Barth, BG. Leigh Syndrome Associated with a Deficiency of the PDHC. Neuropediatrics, 23:147–152, (1992).

50

A POSSIBLE NEW mtrRNA MUTATION SITE FOR AMINOGLYCOSIDE-INDUCED DEAFNESS SYNDROME

D. J. Maguire, A. J. Sun, and W. Rosner

School of Biomolecular and Biomedical Science
Griffith University
Nathan, Q4111, Australia

1. INTRODUCTION

The aminoglycoside group of antibiotics include some of the most powerful agents in the arsenal available for use against gram-negative bacterial infections and include such compounds as gentamycin, tobramycin, amikacin, netilmicin, kanamycin, streptomycin and neomycin. Their primary intracellular site of action has been shown to be the 30S ribosomal subunit, consisting of 21 proteins and a single 16S molecule of RNA, the 16SrRNA molecule. Bacterial resistance to aminoglycosides has been associated with mutations in both the protein subunits and the 16SrRNA. Gentamycin has been advocated as an antibiotic for aggressive treatment of lung infections in Cystic Fibrosis (CF) patients. This same antibiotic is widely applied in a range of other situations. The primary intracellular site of the antimicrobial action of aminoglycosides is the bacterial 30S ribosomal subunit, consisting of 21 proteins and a single 16S RNA molecule. One unfortunate but relatively rare side-effect of these antibiotics is their ototoxicity; deafness is induced in approximately 1 in 10,000 individuals treated with these antibiotics. Another reported side-effect of the aminoglycosides, nephrotoxicity, may be induced by a similar mechanism to that suggested for ototoxicity. There is now considerable evidence that the aminoglycoside-induced deafness is of mitochondrial origin. The human mitochondrial 12S subunit of ribosomal RNA (rRNA) transcribed from mitochondrial DNA (mtDNA) shares more homology with bacterial 16S rRNA than with the human cytosolic 18S rRNA. As reported by Hutchin et al. (1993) and Prezant et al. (1993), the mtDNA of a group susceptible individuals was shown to contain a mutation, corresponding to nucleotide position 1555 of the 12S rRNA sequence. The mutation was shown to exhibit homoplasmic expression—all copies of the mitochondrial gene in the tissues examined contained the mutation. It was proposed that

Oxygen Transport to Tissue XXI, edited by Eke and Delpy.
Kluwer Academic / Plenum Publishers, New York, 1999.

the slight change in the 12SRNA sequence resulting from this mutation was sufficient to alter the nucleic acid conformation to a more "bacterial-like" structure. Since then there have been a number of reports of other mutated 12S rRNA sites which confer susceptibility to aminoglycoside-induced deafness (Table 1) along with one report of a 16S rRNA mutation associated with this syndrome. It is believed that these mutated structures fail to bind appropriately to the (nuclear-encoded) ribosomal protein components of the mitochondrial translation process. It has been proposed that this leads to the formation of stable mismatched aminoacyl-tRNAs in the 70S ribosome, thereby causing misreading of the (mitochondrially-transcribed) mRNA during protein synthesis. It is assumed that the products of this aberrant mitochondrial protein synthesis are unable to form appropriate

Table 1. Comparison of the predicted stem-loop structures and thermodynamic characteristics for wild type and mutant rRNA at some of the positions previously recorded and the one reported here

Mutation Position (Sequence analysed)	Wild Type Energy	Wild Type Shape	Mutant Energy	Mutant Shape
663 A→G (648-667)	-5.48 KJ	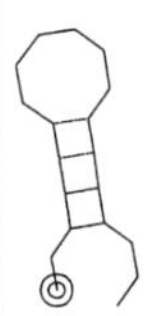	-4.44 KJ	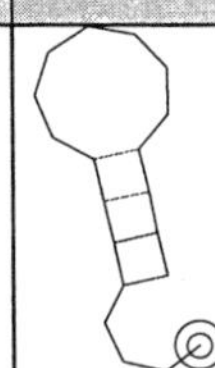
709 G→ A (690-820)	-66.07 KJ	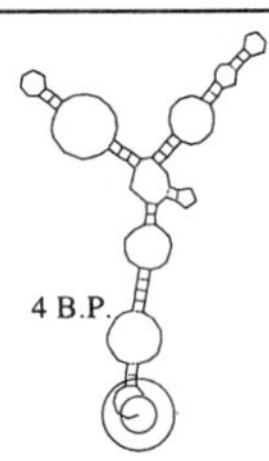	-58.95 KJ	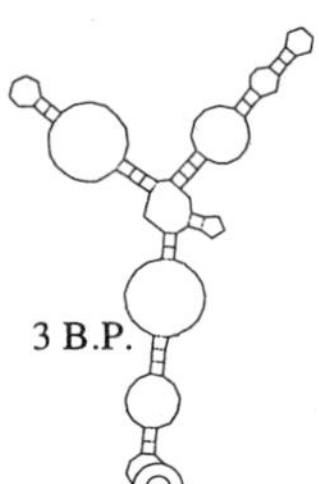
752 C→ T (690-820)	-66.07 KJ	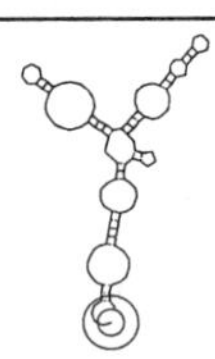	-66.90KJ	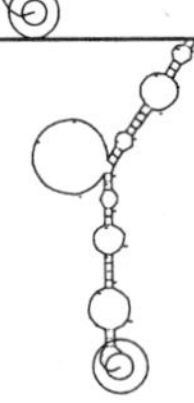
1555 A→ G (1485 - 1566)	9.50 KJ	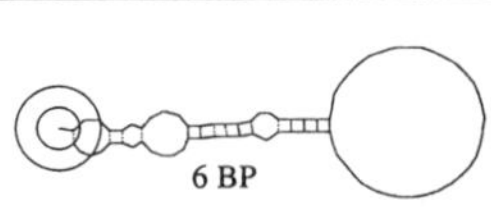	-2.01 KJ	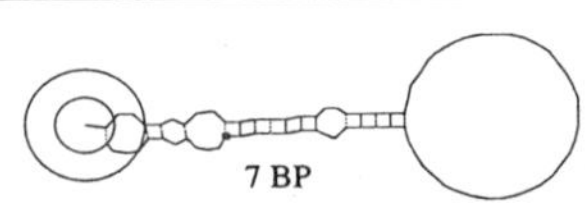
1445 G→A	-19.76	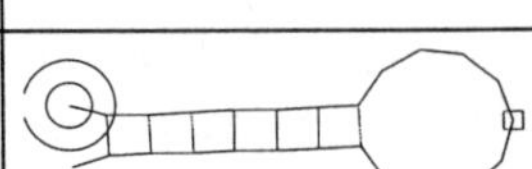	-19.76	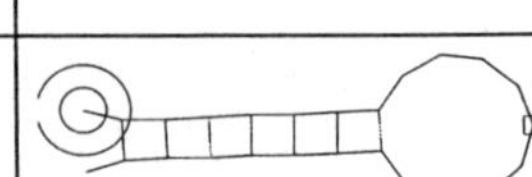
1382 A→ C	-7.16	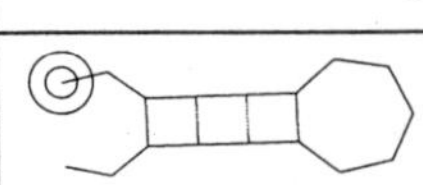	-2.18	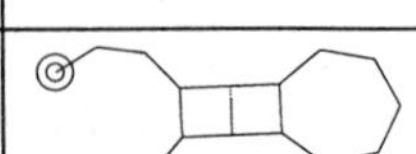

electron transport chain complexes. We report a similar example of gentamycin-induced deafness in which no evidence for any of the previously reported mutations could be found. In this instance, a nucleotide (G at position 1445 of 12S rRNA) was found to have undergone mutation to A. The only other change from the reported sequence of 12S rRNA was at position 1438 (A to G) which was also present in our control samples and reflects an error in the published Cambridge sequence.

2. METHODS

2.1. DNA Samples

DNA was extracted from human blood samples which had been collected in EDTA as the anticoagulant. DNA samples were from one individual whose deafness had appeared after treatment with gentamycin and two control individuals who had shown no adverse reactions to gentamycin treatment. DNA extraction involved the use of a standard commercial extraction kit (Micromix 200 Extraction Kit (Biotech)).

2.2. Primer Design

Forward (F) and reverse (R) oligonuclootide primers followed the design of Bacino et al. (1995). Each was 20 bases long and primed the amplification of a 293 bp amplicon. The sequences of these oligonucleotides were;

F (1261–1280) 5′CCG CCA TCT TCA GCA AAC CC 3′, and
R (1654–1673) 5′AGC TCA GAG CGG TCA AGT TA 3′

2.3. Primer Synthesis and Preparation

Primers, synthesised at Southern Cross University (Lismore, Australia), were received as desalted pellets. These pellets were dissolved in 200 microlitres of distilled water and their concentrations determined spectrophotometrically using a 1/50 dilution of this stock solution.

2.4. PCR (Polymerase Chain Reaction) Amplification

The mitochondrial DNA segment containing base numbers 1261 to 1673 was amplified using the primers described above in a thermal cycler (Hybaid Omnigene). The amplification mixture consisted of genomic DNA (100–200 ng), forward and reverse primers (each at 20 pmol), KCl (50 mM), $MgCl_2$, (1.5 mM), dNTPs (each at 200 μM), Taq DNA polymerase (2 units) and Tris-HCl, pH 8.3 in a total volume of 50 μL. Amplification conditions included a first cycle of 5 minutes at 95°C, followed by thirty cycles, each consisting of 30 seconds at 95°C, a minute at 54°C and 1.5 minutes at 65°C. The final primer extension step involved incubation for 5 minutes at 65°C.

2.5. Analysis and Purification of Amplimers

2 uL of gel loading buffer was added to 10 uL of each amplimer solution and loaded onto a 1.5% agarose gel containing 4 uL of ethidium bromide (10 mg/mL).

Electrophoresis was conducted at 100 V for 1 to 2 hours in TBE (Tris-Borate buffer, 0.01 M, pH 8.0, containing 1 mM EDTA). Amplimers were sized against standard DNA markers to confirm that the appropriate product was present. A phenol/chloroform method was then used to extract the PCR product from the amplification mixture. 250 µL Tris-EDTA buffer (10 mM Tris-HCl, 1 mM EDTA), pH 8.0, was added to 50 µL of PCR product, followed by 400 µL of TE-saturated phenol and 400 µL of chloroform/isoamyl alcohol (24:1). After careful mixing of the two phases to extract DNA, the phases were separated by centrifugation at 14,000 g for 5 minutes. The top (aqueous) phase was removed and after a further extraction using chloroform, added to 40 µL of 3 M sodium acetate, pH 5.2, and 1 mL of 95% ethanol then mixed and cooled to −20°C for one hour. Precipitated DNA was collected by centrifugation at 4°C for fifteen minutes and washed with 70% ethanol before being dried in a vacuum dessicator for 20 minutes. Pellets were redissolved in 50 µL of Tris-EDTA buffer.

2.6. Cycle Sequencing

The cycle sequencing method employed was the "Terminator Ready" cycle sequencing kit. The cycle sequencing mixture consisted of 8.0 µL of terminator ready reaction mix, 100 ng of amplimer, 3.2 pmol of primer in a final volume of 50 µL. The mixture was overlayed with one drop of light mineral oil and placed in a thermal cycler (Hybaid Omnigene). The conditions used for cycle sequencing involved thermal ramping to 96°C for 30 seconds, 15 seconds at 50°C and 4 minutes at 60°C. Products were purified by ethanol precipitation, dried briefly in a vacuum dessicator and analysed on a DNA Sequencer (Applied Biosystems).

2.7. Stem-Loop Structure Modeling

The sequences in the region of each of the previously reported aminoglycoside susceptibility mutations and the wild type sequences were compared with the published sequence of bovine mitochondrial 12S rRNA, human mitochondrial 12SrRNA and E. coli 12SrRNA. Stem-loop modelling was performed on those sequences for which there was clear homology with these sequences, using the program 'RNAdraw' (Matzura and Wennborg, 1996).

3. RESULTS

3.1. Amplification of 12S rRNA

An electrophoretic separation of the amplimers obtained from PCR reactions involving varying amounts of wild type and mutant mtDNA are shown in Figure 1. As can be seen, DNA amounts between 100 and 200 ng were suitable for production of amplimers in the expected size range.

3.2. Sequencing of 12S rRNA Amplimers

In Figure 2, the results obtained from the sequencing reactions performed on a normal and an affected individual are shown. It can be seen that the 1438 position is a G (guanosine) in each case, whereas the G to A (adenosine) variation at position 1445

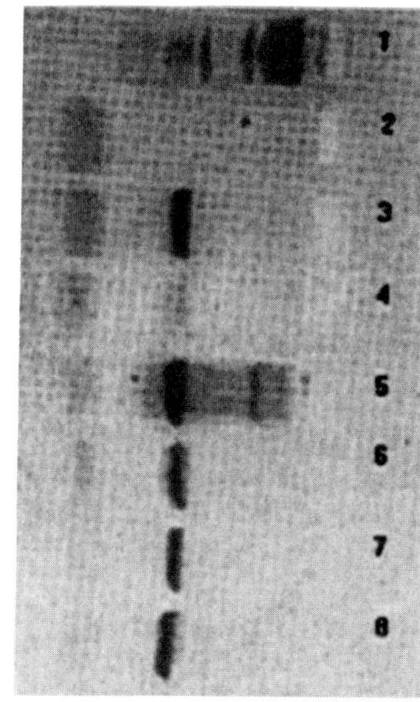

Figure 1. Electrophoretic separation of the amplimers obtained from PCR reactions involving varying amounts of wild type and mutant DNA. Lane 1, DNA molecular weight markers; Lanes 3 & 5, normal individual's amplified DNA at 100 ng and 200 ng DNA, respectively; Lanes 6–8, affected individual's amplified DNA at 100 ng, 150 ng and 200 ng, respectively.

occurs only in the affected individual. The 1445 variant appears to be expressed in a homoplasmic mode, as evidenced by the single base representation at that position. Another control individual tested also showed the 1438 A to G variant and was also free of the 1445 variant observed in the affected individual.

3.3. Stem-Loop Structure Modeling

In Table 1, the recorded aminoglycoside susceptibility mutations are listed. Where it was possible to perform stem-loop modeling, the structures derived by that operation are shown alongside the wild type structures in the region of the mutation position. Thermodynamic characteristics for each structure is also shown. It can be seen that in

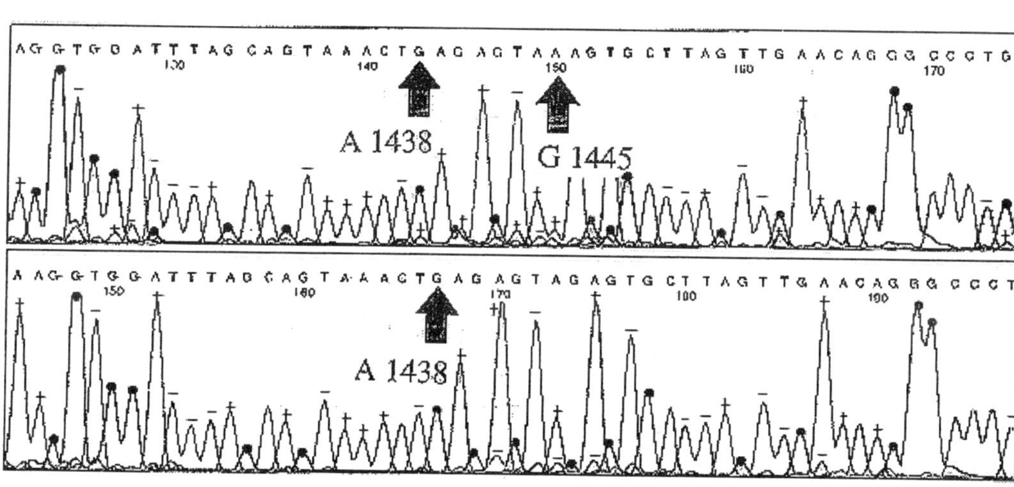

Figure 2. mtDNA sequences obtained on an unaffected (lower trace) and an affected individual (upper trace).

all cases where the analysis could be performed, the pairs of predicted structures (mutant and wild type) have similar stability, except for the originally reported 1555 mutation and the 1382 mutation. Previously reported mutations which are not shown in this table include those at positions 827, 961, 1005, 1048, 1107, 1243, and 1520.

4. DISCUSSION

Aminoglycoside-induced deafness has been linked to a homoplasmic mutation (1555 A to G) in the 3′ end of the small ribosomal gene of the human mitochondrion. It has been suggested that this mutation creates a more bacterial-like rRNA (Hutchin, 1993; Prezant, 1993; Bacino, 1995). Since the discovery of that mutation, other point mutations in the same gene have been linked to susceptibility to this affliction. In the present work, yet another mutation site is indicated on the basis of amplification and sequencing of a region of the mitochondrial rRNA gene. Another position which is at variance with the published Cambridge sequence of human mtDNA occurs in both affected and control subjects. It has been suggested that this represents an error in the original Cambridge sequence (Prezant et al., 1993).

4.1. Stem-Loop Structure Modeling

A structure for bovine 12S mitochondrial rRNA has been proposed, as well as for human mitochondria and for bacteria (*E.coli*), (Gutell, 1994). Previously, comparisons have been made between the region surrounding the 1555 mutation and the partially homologous region of the 3′ end of a bacterial (*E.Coli*) 16S rRNA (Hutchin et al., 1993; Prezant et al., 1993). An attempt was made in the present work to compare stem-loop structures in the region of each of the reported aminoglycoside susceptibility mutations with the normal sequences, using the corresponding structures in the bovine and human 12S mitochondrial rRNA as models for the mutated structure. Using this "localised structure prediction model", a number of model stem-loops with considerable similarity to the wild-type structures. Furthermore, it was possible to show that there were clear changes in the shape of some such structures when the reported mutations were introduced. Finally, the pairs of predicted structures (mutant and wild type) were shown to have similar stability, as detemined by thermodynamic criteria. The initial step in the process of localised modeling was not successful for the mutations previously recorded at positions 827, 961, 1005, 1048, 1107, 1243, and 1520, but was successful for the one reported here. An analysis of the structural implications of these other mutations reported will await the development of a more complete analysis of the human mitochondrial 12S rRNA and its interaction with ribosomal subunits and amino-acyl tRNA species. In any event, it appears that the mutation reported here and a number of other mutations do not alter the structure of stem loops directly involved in the binding of aminoacyl-tRNA molecules. It is also unlikely that they affect the spatial position of such stem-loops by either lengthening or shortening the distance betwen interacting elements of the 12SrRNA molecule. Such a mechanism has been proposed for one of the reported aminoglycoside susceptibility mutations (12SrRNA, 961T–>C, Bacino et al., 1995). An alternative proposal is that there are other regions of the rRNA which are important in forming relationships with either the aminoacyl-tRNA or protein subunits which contribute to the binding of aminoacyl-tRNA. The mutation reported here and at least one previously reported mutation lie on an "exposed loop" region which may interact with protein subunits. The possibility exists therefore, that an allosteric interaction may induce a change in the stability of binding of aminoacyl-tRNA.

4.2. Metabolic and Diagnostic Implications

The precise mechanism by which cell death is induced in aminoglycoside susceptible individuals is unknown. It is assumed that the sudden, and presumably complete, inhibition of appropriate mitochondrial translation processes would result in grossly impaired oxygen metabolism. In the absence of experimental evidence to support that assumption, other mechanisms might also be proposed. One such mechanism might involve the production of a toxic translation product which exerts its action at the mitochondrial level or, following transport from the mitochondria, at some other cellular site. To explain the accute sensitivity of the cells involved in ototoxicity, there are several possible explanations. One possible mechanism is that cells which are not sensitive are inaccessible to the antibiotic. Another possibility is that the antibiotics may be concentrated by some mechanism within the susceptible cells. The involvement of mitochondrial genes in human disease processes is becoming increasingly accepted, although there is still resistance in some quarters to wide acceptance of that involvement, especially for those mutations which do not exhibit homoplasmic expression. The level of incidence of this new mutation in a general population has not yet been investigated, but should present no great problems for routine testing of close relations provided it exhibits homoplasmic expression in those family members. Obviously, only the maternal inheritance line would need to be investigated before administration of aminoglycosides.

REFERENCES

Bacino, C, Prezant, TR, Bu, XD, Fournier, P, and Fischel-Ghodsian, N, "Susceptibility Mutations in the Mitochondrial Small Ribosomal RNA Gene in Aminoglycoside Induced Deafness" *Pharmacogenetics*, 5:165–172 (1995).

Gutell, R, "Collection of Small Subunit (16S- and 16S-like) Ribosomal RNA Structures" *Nucleic Acid Res.* 22:3502–3507 (1994).

Hutchin, T, Haworth, I, Higashi, K, Fischel-Ghodsian, N, Stoneking, M, Saha, N, Arnos, C, and Cortopassi, G. "A Molecular Basis for Human Hypersensitivity to Aminoglycoside Antibiotics" *Nucleic Acid Res.* 21:4174–4179 (1993).

Matzura, O and Wennborg, A. "RNAdraw: an Integrated Program for RNA Secondary Structure Calculation and Analysis under 32-Bit Microsoft Windows" *Computer App. Biosci.* 12:247–249 (1996).

Prezant, TR, Agapian, JV, Bohlman, MC, Bu, XD, Oztas, S, Qiu, W, Arnos, KS, Cortopassi, GA, Jaber, L, Rotter, JI, Shohat, M, and Fischel-Ghodsian, N. "Mitochondrial Ribosomal RNA Mutation Associated with both Antibiotic-Induced and Non-Syndromic Deafness. *Nature Genet.* 4:289–294 (1993).

51

EFFECTS OF DIFFERENT VASODILATORS ON HUMAN INTERNAL MAMMARY ARTERY *IN VITRO*

M. Mahmoudian,[1] M. Shafiei,[1] H. Mirkhani,[1] and Gh. R. Omrani[2]

[1]Department of Pharmacology
[2]Department of Cardiovascular Surgery
Iran Univ. Med. Sci., P.O. Box 14155-6183
Tehran, Iran

1. INTRODUCTION

In coronary artery bypass grafting (CABG) surgery, arterial grafts have been used with increased frequency. They are expected to be superior to vein grafts because of their long-term patency. The arterial grafts that have been used include internal mammary artery (IMA), gastroepiploic artery, inferior epigastric artery and radial artery (He et al., 1997). However, the use of arterial grafts has brought up the question of hypoperfusion that may occur due to the small diameter of the arterial grafts, the spastic characteristic, or both (He et al., 1995; He and Yang, 1995). Perioperative infarction, due to restriction of flow caused by spasm of the arterial grafts which was resistant to pretreatment with papaverine has been reported (Meldrum-Hanna and Ross, 1986) and the incidence of cardiac related deaths in patients with IMA grafts is significantly greater than in patients with saphenous vein grafts (Molenaar et al., 1988). Recent studies have also demonstrated that blood flow through arterial grafts may be inadequate for maximal exercise (Kawasuji et al., 1993) and hence causes a hypoperfusion syndrome (Loop and Thomas, 1993). The cause for arterial graft vasoconstriction is still unknown. Studies on the pharmacology of human IMA, as the most frequent used arterial graft, may provide an understanding of the mechanism of spasm and suggest a rational approach for its reversal. Vasodilator drugs may affect vascular tissues either directly by acting on smooth muscle cells or indirectly via endothelial cells. The present study was designed to investigate the direct vasorelaxant effects of nifedipine (NFD), a calcium channel blocker, isosorbide dinitrate (ISDN), a nitrovasodilator, isoproterenol (ISPT), a non-selective-, and BRL 37344, a β_3-selective-adrenoceptor agonist.

Oxygen Transport to Tissue XXI, edited by Eke and Delpy.
Kluwer Academic / Plenum Publishers, New York, 1999.

2. METHODS

2.1. Preparation and Sampling of Tissues

Tissues were obtained from patients undergoing coronary bypass surgery. The protocol was approved by the hospital ethical committee. All the patients studied were diagnosed with ischaemic heart disease and their clinical characteristics including drug therapy are listed in Table 1. During the operation, the discarded distal end of the IMA was carefully removed and transported to the laboratory in a sealed flask at 4°C in oxygenated Krebs buffer (composition mM: NaCl 118, KCl 4.8, Ca Cl_2 2.5, KH_2PO_4 1.2, $MgSO_4$ 1.2, $NaHCO_3$ 25, glucose 10).

2.2. Organ Bath Studies

In the laboratory (20–25 min following surgical removal) isolated blood vessels were carefully cleared from surrounding tissues and cut into rings of 3 mm length. The endothelium was removed mechanically by gentle rubbing of the intimal surface of the segments with a metal rod for 30–50 s. Rings were placed over two parallel wires and suspended in the organ bath containing Krebs buffer maintained at 37°C and bubbled with a mixture of 95% O2 and 5% CO_2. The rings were stretched to a tension of 2.5–3 g and allowed to equilibrate for 2.5 h. Tissues were washed at 15 min intervals. At the beginning of the experiments, the absence of endothelium was functionally assessed using acetylcholine (10^{-6} M). Isometric contractions were recorded on a Beckman polygraph. To generate the concentration-response curves of the vasodilators, the tissues were submaximally contracted with a single concentration of a contractile agent and then cumulative concentrations of relaxant compounds were added into the organ bath. As will be discussed later, with regard to the mechanisms involved in vasorelaxation and in order to obtain maximal relaxant effects, KCl (40 mM) and phenylephrine (PE) (10^{-5} M) were used as the contractile agents, the former for NFD and the latter for ISDN and ISPT. Relaxant responses were expressed as the percentage of the precontraction induced by either PE or KCl. When the antagonist, i.e. propranolol, was used, it was added to the bath and tissue was allowed to equilibrate for 30 min before the agonist concentration-response curve was repeated.

Table 1. Clinical characteristics of 26 patients undergoing internal mammary artery graft surgery

Clinical characteristics	n
Age (years)	
56 ± 7.67	26
Sex	
Male	22
Female	4
Drug therapy	
β-adrenoceptor antagonists	25
nitrates	24
Ca^{2+} channel blockers	17
ACE inhibitors	6
digoxin	2

2.3. Drugs

NFD was a gift from Tolid-darou Pharmaceutical Co. DL-isoproterenol hydrochloride and DL-propranolol hydrochloride (Sigma Chemical Co.), BRL 37344 (RR + SS)(4-[2-[2-(3-chlorophenyl)-2-hydroxyethyl]phenoxyl]acetic acid (Tocris Cooksen Ltd.), and ISDN was also a gift from Arya Pharmaceutical Co.

In each experiment, NFD was dissolved in di-methyl sulfoxide (DMSO) to make 1 mM solution. This was then further diluted with distilled water. The prepared solutions were protected from light. The dilutions of acetylcholine, BRL 37344, propranolol, ISPT and ISDN were made using saline, except for the highest concentration of ISDN (10^{-4} M) which was prepared in a mixture of DMSO and saline (1:5). PE and KCl were made in Krebs buffer.

2.4. Data Analysis

Concentration-response curves were generated by plotting the percentage of the initial tension versus the negative logarithm of concentration of vasodilator. EC50 values (effective concentration of vasodilator required for half maximal relaxation) were calculated from the concentration-response curves and pEC50 value (-log EC50) used for comparison of drug potency. pEC50 values and maximal relaxations were compared by analysis of variance and Student's t test. In all cases a p value of <0.05 was taken as statistically significant.

3. RESULTS

Figure 1 shows traces from three experiments using PE and KCl precontracted rings in which responses to ISPT, ISDN and NFD, respectively, were established. Concentration-response curves for vasodilator-induced relaxation are shown in Figure 2. The pEC50 values are listed in Table 2 and the order of vasodilator potency to relax human IMA pre-contraction was as follows: NFD > ISDN > ISPT. Maximal relaxation of vasodilators are also shown in Table 2. The order of maximal relaxation was as follows: ISDN > NFD > ISPT.

In the majority of patients the relaxant effects of ISPT (10^{-8} M-10^{-4} M) was completely inhibited by propranolol (2×10^{-7} M). However, in some patients the antagonism was found to be relative. The percent of maximal inhibition between these two groups was significantly ($p < 0.05$) different (Figure 3). BRL 37344 (10^{-6} M-10^{-4} M) relaxed PE-induced contracted rings of some, but not all, of our patients (Figure 4).

Table 2. Relaxant effects (mean ± S.D.) of vasodilators on phenylephrine- and potassium-induced contractions of human IMAs

Spasmogen	Vasodilator	Maximum (% relaxation)	pEC50
KCl (40 mM)	nifedipine	69.13 ± 15.32	7.67 ± 0.12
Phenylephrine (10 μM)	isoproterenol	49.19 ± 23.84	5.82 ± 0.22
Phenylephrine (10 μM)	isosorbide dinitrate	98.80 ± 1.68	6.26 ± 0.24

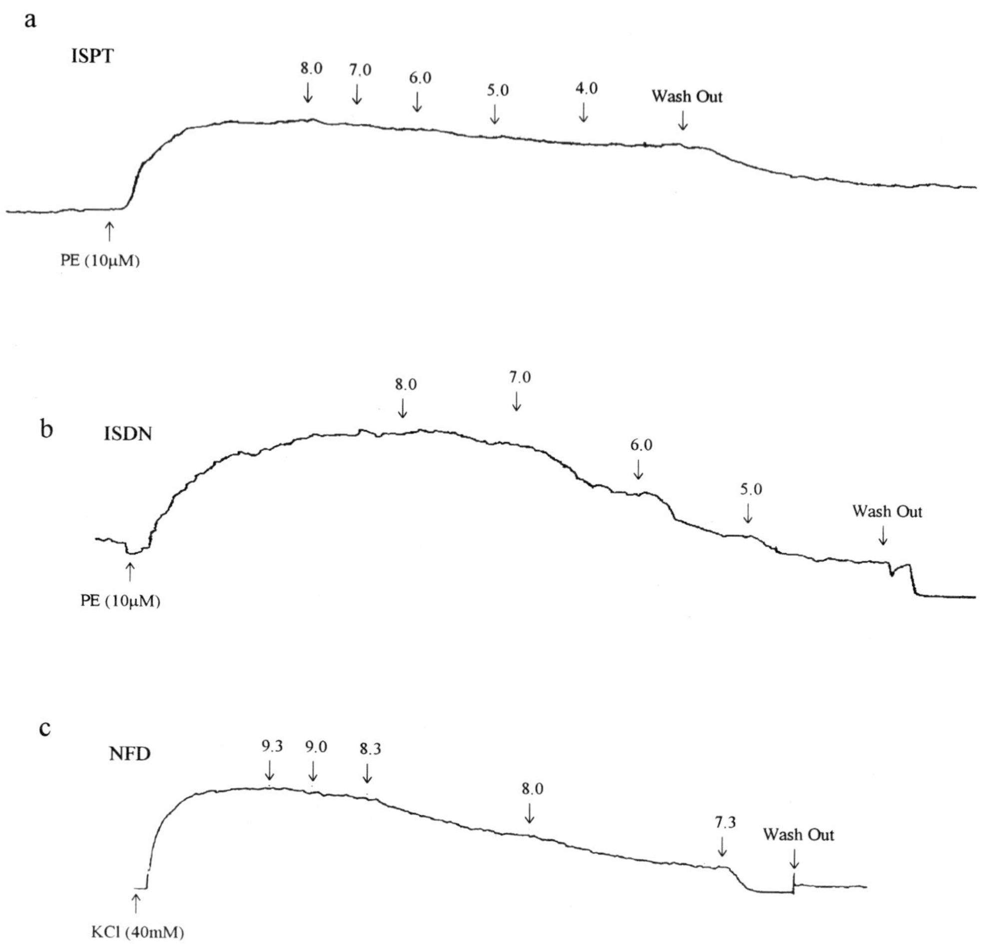

Figure 1. Traces from three experiments in which responses to a) isoproterenol (ISPT) and b) isosorbide dinitrates (ISDN) in phenylephrine (PE) precontracted rings and c) nifedipine (NFD) in KCl precontracted rings of human internal mammary artery without endothelium were established. Concentration of drugs are expressed as minus logarithm of concentrations.

4. DISCUSSION

The IMA is currently the preferred conduit for myocardial revascularization (Loop et al., 1986; Tector et al., 1986). The management of vasospasm improves the immediate and long-term flow through the IMA after the operation. This study investigated the effectiveness of various vasodilators to relax pre-contracted human IMA. To produce the contraction of IMA, two contractile agents were used. Potassium produces contraction of vascular smooth muscle by direct membrane depolarization, which results in an influx of extracellular calcium across the cell membrane through voltage-operated calcium channels (Loutzenhiser and Epstein, 1984; Hogestatt, 1984). On the other hand, PE causes contraction of vascular muscle by stimulating α-adrenergic receptors, which cause a release of intracellular calcium (Loutzenhiser and Epstein, 1984; Hogestatt, 1984) and influx of extracellular calcium through receptor-operated channels (Haeusler, 1978;

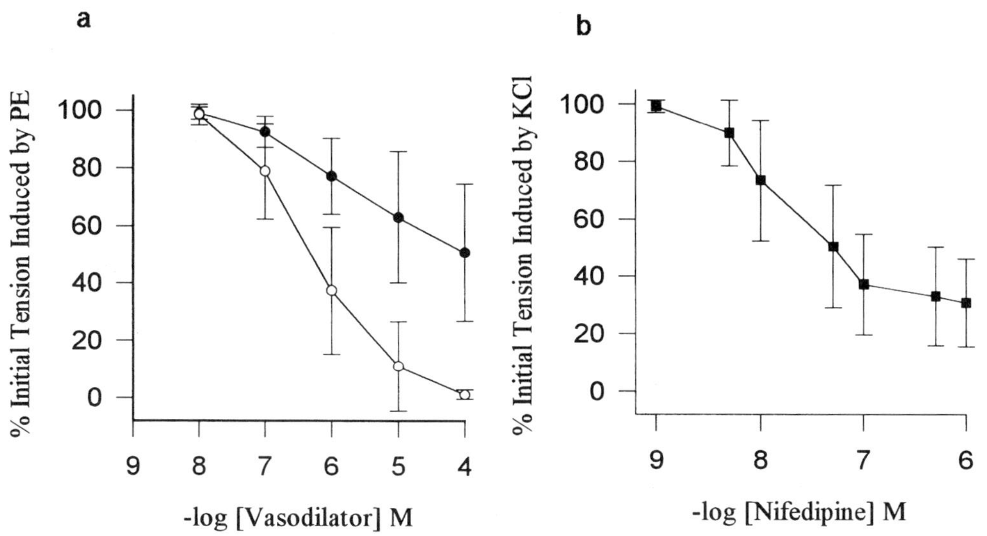

Figure 2. Mean concentration-response (% initial tension) curves for a) isoproterenol (ISPT, ●, $n = 8$) and isosorbide dinitrates (ISDN, ○, $n = 5$) in phenylephrine (10 µM) precontracted rings and b) nifedipine (NFD, ■, $n = 8$) in KCl (40 mM) precontracted rings of endothelium-denuded human internal mammary artery. Each point represents mean ± S.D.

Hogestatt, 1984). Calcium entry blockers act selectively on the voltage-dependent channels. The potency and maximal response of nifedipine has been shown to be greater for inhibition of potassium-induced contraction than α-agonist induced contraction (Jett et al., 1992). The mechanism of nitrate-induced relaxation is not voltage-dependent (Schmaar and Sparks, 1972). Nitrovasodilators act by releasing nitric oxide in the muscle cell which stimulates guanylate cyclase to raise cGMP (Rapoport and Murad, 1983). These steps lead to smooth muscle relaxation (Rapoport et al., 1983; Ignarro and Kadowitz, 1985). β-Adrenergic agonists also act by increasing adenylate cyclase activity and causing an increase in cAMP (Triner et al., 1971) which results in the removal of intracellular calcium from the cell. Thus, nitrovasodilators and β-adrenergic agonists

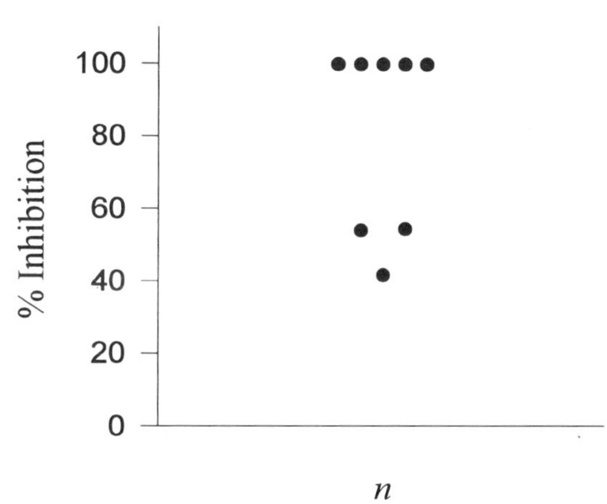

Figure 3. Inter-individual variation observed in inhibition induced by propranolol (2×10^{-7} M) against isoproterenol-induced relaxation in phenylephrine (10 µM) precontracted IMA rings without endothelium.

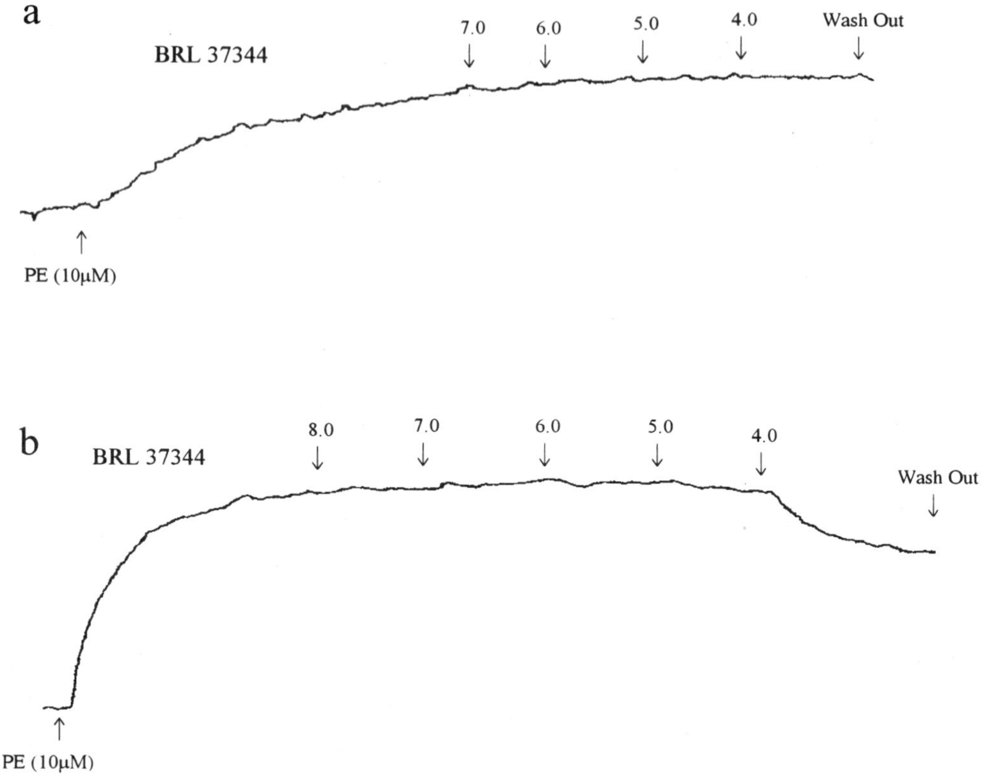

Figure 4. Traces from two experiments in which responses to BRL 37344 in phenylephrine (PE) (10 μM) precontracted rings of endothelium-denuded human internal mammary artery belonging to two patients were established. Concentration of drugs are expressed as the minus logarithm of concentrations.

appear to be more effective vasodilators against α_1-adrenoceptor agonists (Jett et al., 1992). Therefore, in the present study, the relaxant effect of NFD was investigated in the tissues precontracted by KCl, while those of ISPT and ISDN in tissues precontracted by PE.

Every effort was made to reduce or avoid the influence of previous drug therapies. For instance, the equilibration period was 2.5 h and the bathing solution was changed every 15 min. (Ahlner et al., 1986) have indicated that 90 min equilibration was sufficient for human tissue experiments to give a stable and reproducible contraction. Therefore, we were nearly confident that in our experiments the influence of the previous drug therapy was minimum.

The relaxation studies (Table 2) demonstrated that ISDN induced a maximal or near-maximal relaxation in PE-mediated contraction and the results showed the least inter-individual variations (varied from 96.5% to 100%). NFD and ISPT could partially relax the human IMA but a large extent of inter-individual differences were observed (maximal responses for ISPT varied from 22.5% to 100% and for NFD from 45% to 88%). In addition, there are differences between maximal responses obtained in the present study and the other reports. For example, the maximal responses to ISPT varied from 35 ± 9% (Jett et al., 1992) to 80.52 ± 10.5% (Molenaar et al., 1988) in α-agonist-mediated contractions, and those to NFD were equivalent to 90–92% in KCl-mediated contractions (Jett et al., 1992; Cooper and Locke, 1994). Whether these differences are due to the inter-ethnic variations or the protocol used in the above experiments needs further studies.

In the present study, ISPT produced concentration-dependent relaxant responses in endothelium-denuded rings of IMA. It has been previously reported that the potency and maximal effect of ISPT were not changed by removal of endothelium in IMA, and the β-adrenoceptors on the smooth muscle is probably responsible for the ISPT-induced relaxation (Molenaar et al., 1988). The relaxant response to ISPT was completely antagonized by propranolol (2×10^{-7} M). Propranolol has high affinity only for classic β-adrenoceptors (β_1, β_2). This concentration of propranolol, i.e. 2×10^{-7} M, has been shown to block β_1- and/or β_2-adrenoceptors, but does not affect atypical β-adrenoceptors significantly (Kaumann and Molenaar, 1996). Thus it appears that β-adrenoceptors in the group of IMAs which showed complete inhibition by propranolol, are of classic subtype(s). But in those with relative antagonism, the involvement of atypical β-adrenoceptors is probable. It is also interesting to note that high concentration (10^{-4} M) of BRL 37344, a β_3-selective-agonist, produced vasorelaxation in some patients. The study is being continued to characterize β-adrenoceptor subtype(s) on the smooth muscle cells of human IMA.

5. CONCLUSION

Our results indicate that the potency of isosorbide dinitrate to relax phenylephrine-induced contraction is less ($p < 0.05$) than that of nitroglycerin in the same situation (present study, data not shown; He and Yang, 1997) and takes longer time ($p < 0.05$) to reach its peak response (15 min), its effectiveness makes it a suitable vasodilator for long-term use. Nifedipine is a potent and relatively effective vasodilator but it is difficult to predict its efficacy because of the variation observed between individuals. IMA undergoes β-adrenoceptor-mediated vasorelaxation, but the maximal response to β-adrenergic agonists and receptor subtype(s) involved, appear to show inter-individual variability.

REFERENCES

Ahlner J, Andersson RGG, Axelsson KL, Bergdahl B, Dahlstrom U, and Rydell EL (1986) The relaxant effect of glyceryl trinitrate on isolated human peripheral vein and its relation to cyclic GMP metabolism. *Acta Pharmac Tox* **58**:129–136.

Cooper GJ and Locke TJ (1994) The in vitro response of human internal mammary artery to vasodilators. *Cardiovasc Surg* **107**:1155–1157.

Haeusler G (1978) Relationship between noradrenaline-induced depolarization and contraction in smooth muscle. *Blood Vessels* **15**:46–54.

He G-W and Yang C-Q (1995) Comparison among arterial grafts and coronary artery. An attempt at functional classification. *J Thorac Cardiovasc Surg* **109**:707–715.

He G-W, Yang C-Q, and Starr A (1995) Overview of the nature of vasoconstriction in arterial grafts for coronary surgery. *Ann Thorac Surg* **59**:676–683.

He G-W and Yang C-Q (1997) Comparison of nitroprusside and nitroglycerin in inhibition of angiotensin II and other vasoconstrictor-mediated contraction in human coronary bypass conduits. *Br J Clin Pharmacol* **44**:361–367.

Hogestatt ED (1984) Characterization of two different calcium entering pathways in small mesenteric arteries from rat. *Acta Physiol Scand* **122**:483–495.

Ignarro LJ and Kadowitz PJ (1985) The pharmacological and physiological role of cyclic GMP in vascular smooth muscle relaxation. *Ann Rev Pharmacol Toxicol* **25**:171–191.

Jett GK, Gyton RA, Hatcher CR, and Able PW (1992) Inhibition of human internal mammary artery contractions. *J Thorac Cardiovasc Surg* **104**:977–982.

Kaumann AJ and Molenaar P (1996) Differences between the third cardiac β-adrenoceptor and the colonic β_3-adrenoceptor in the rat. *Br J Pharmacol* **118**:2085–2098.

Kawasuji M, Tedoriya T, Takemura H, Sakakibara N, Taki J, and Watanabe Y (1993) Flow capacities of arterial grafts for coronary artery bypass grafting. *Ann Thorac Surg* **56**:957–962.

Loop FD, Lytle BW, and Cosgrove DM, et al. (1986) Influence of the internal mammary artery graft on 10-year survival and other cardiac events. *N Engl J Med* **314**:1–6.

Loop FD and Thomas JD (1993) Hypoperfusion after arterial bypass grafting. *Ann Thorac Surg* **56**:812–813.

Loutzenhiser R and Epstein M (1984). Activation mechanisms of human renal artery: effects of KCl, norepinephrine, and nifedipine upon tension development and ^{45}Ca influx. *Eur J Pharmacol* **106**:47–52.

Meldrum-Hanna WG and Ross DN (1986) The problem of the internal mammary artery. *Texas Heart Institute J* **13**:171–172.

Molenaar P, Malta E, Jones CR, Buxton BF, and Summers RJ (1988) Autoradiographic localization and function of β-adrenoceptors on the human internal mammary artery and saphenous vein. *Br J Pharmacol* **95**:225–233.

Rapoport RM and Murad F (1983) Endothelium-dependent and nitrovasodilator-induced relaxation of vascular smooth muscle: role of cyclic GMP. *J Cyclic Nucl Prot Phosphoryl Res* **9**:281–296.

Rapoport RM, Draznin MB, and Murad F (1983) Endothelium-dependent relaxation in rat aorta may be mediated through cyclic GMP-dependent protein phosphorylation. *Nature* **306**:174–176.

Schmaar RL and Sparks HV (1972) Response of large and small arteries to nitroglycerin, $NaNO_2$, and adenosine. *Am J Physiol* **223**:223–228.

Tector AJ, Schmahl TM, and Canino VR (1986) Expanding the use of the internal mammary artery to improve patency in coronary artery bypass grafting. *J Thorac Cardiovasc Surg* **91**:9–16.

Triner L, Nahas GG, Overweg NIA, Verosky M, Habif DV, and Ngai SH (1971) Cyclic AMP and smooth muscle function. *Ann NY Acad Sci* **185**:458–476.

RED BLOOD CELL DEGRADATION PRODUCTS MODIFY RESPONSIVENESS OF ERYTHROPOIETIN GENE TO HYPOXIA

E. Nečas,[1] K. Šulc,[1] J.H. Živný,[1] V. Sýkora,[1] and P. Ponka[2]

[1]Department of Pathophysiology
First Medical Faculty
Charles University
Prague, Czech Republic
[2]Lady Davis Institute for Medical Research
Jewish General Hospital and Department of Physiology
McGill University
Montreal, Quebec, Canada

1. INTRODUCTION

The oxygen sensors monitor oxygen tension in the organism and trigger appropriate feedback responses when the tension declines. Peripheral chemoreceptors in the carotid and aortic bodies, and erythropoietin (Epo) producing cells in the kidney are the most notoriously known oxygen sensors.

Various pathophysiological conditions which cause tissue hypoxia, such as low oxygen tension in the arterial blood (p_aO_2), anemia, increased affinity of blood to oxygen, and carbon monoxide (CO) intoxication stimulate Epo production. All these conditions have in common decreased oxygen tension at the venous capillary ends. However, except in the case of low p_aO_2, the oxygen tension is not affected at the arterial capillary ends. Epo stimulates red blood cell production and oxygen carrying capacity of the blood. This diminishes the longitudinal gradient of oxygen tension in the capillary by improving oxygen tension at the venous capillary ends. Thus, the Epo-based oxygen sensor is primarily concerned with monitoring and regulating the oxygen tension along the capillary particularly at its venous ends. In contrast, the peripheral chemoreceptors are primarily concerned with oxygen tension in the arterial blood (Figure 1).

Hypoxia activates the Epo gene in fibroblastoid-like peritubular cells in the renal cortex (Koury et al., 1988; Le Hir et al., 1991; Bachmann et al., 1993). Epo gene activa-

Oxygen Transport to Tissue XXI, edited by Eke and Delpy.
Kluwer Academic / Plenum Publishers, New York, 1999.

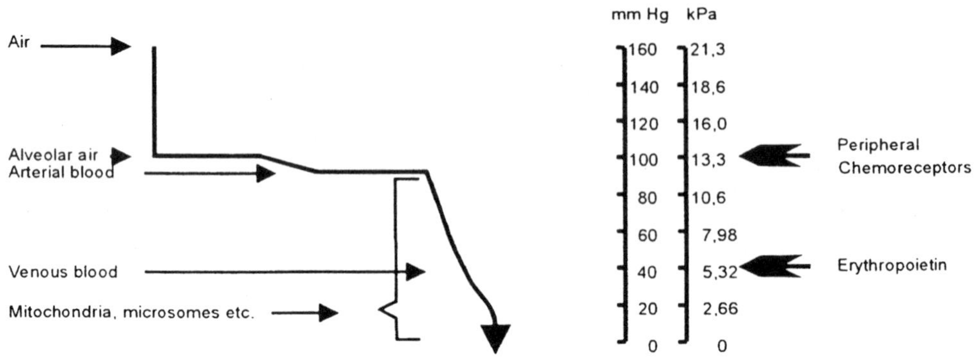

Figure 1. Peripheral chemoreceptors and of Epo production in monitoring oxygen delivery to tissues.

tion is mediated by hypoxia sensitive transcription factor HIF-1 (reviewed recently by Semenza, 1998). The Epo response is mediated by oxygen tension ("oxygen concentration") because in anemic experimental animals it was abrogated by normobaric or hyperbaric oxygenation (Nečas and Neuwirt, 1969a,b). On the other hand, Epo production was not stimulated by inhibitors of oxidative phosphorylation including cyanide (Nečas and Neuwirt, 1972; Nečas and Thorling, 1972; Tan et al., 1991). This indicates that metabolic consumption of the oxygen delivered to tissues is not involved in the control of Epo production.

The control of Epo production by a feedback mechanism utilizing oxygen tension is confronted by two paradoxes. First, the clinical and experimental conditions regarded as "compensated anemia" or "compensated hemolysis" correspond to intensive red blood cell production in the presence of a low degree anemia. Very mild tissue hypoxia appears to activate Epo production under these conditions (Nečas and Neuwirt, 1969c). Second, high degree polycythemia loses its beneficial effect on oxygen delivery because of the high blood viscosity (Thorling and Erslev, 1968). Consequently, Epo production and red blood cell production could resume creating a circulus vitiosus leading to the death of organism. However, excessive polycythemia significantly inhibits Epo and red blood cell production (Živný et al., 1970). It was also demonstrated that polycythemic animals produce less erythropoietin compared to animals with normal or low hematocrits when challenged by hypoxia (Nečas et al., 1970; Něas et al., 1972).

Oxygen sensing leading to Epo production could be mediated by a heme protein (Goldberg et al., 1988; Gleadle et al., 1995; Ho and Bunn, 1996; Huang et al., 1997). Therefore, we have studied effects of heme availability on Epo gene responses to tissue hypoxia. In the present experiments, heme synthesis in mice exposed to acute hypoxia was either inhibited or heme was administered. Epo mRNA in the kidney and plasma Epo levels were subsequently measured.

2. METHODS

2.1. Mice

Eight- to 11-week-old male Balb/c and C3H/HeN mice were used.

2.2. Hypoxia Stimulation

Mice were exposed to a low oxygen tension atmosphere generated by flushing the hypoxia chamber with nitrogen. Oxygen tension was decreased and maintained between 11.9% and 6.3%. Alternatively, carbon monoxide (CO) was added to the air in concentrations from 0.05% to 0.3%. The exposure time was from one to five hours.

2.3. Epo Plasma Levels

Epo was measured by RIA in plasma samples from heparinized blood collected immediately after exposure to hypoxia.

2.4. Epo mRNA

The guanidinium isothiocyanate method (Chomczynski and Sacchi, 1987) was used to isolate total RNA from both kidneys collected into the liquid nitrogen immediately after exposure to hypoxia. RNA samples were denatured by heating at 95°C for 2 min in an RNA loading buffer containing 50% (v/v) deionized formamide, 6.5% (v/v) formaldehyde, 5% (w/v) bromophenol blue and 5% (v/v) glycerol in 1 mM MOPS buffer. Twenty or 30 µg RNA samples were loaded onto a 1% agarose gel containing 2% (v/v) formaldehyde and electrophoresis was performed at 4°C. RNA was stained with ethidium bromide, visualized by UV light and photographed to confirm equal loading of RNA according to the 28S and 18S bands of ribosomal RNA. The RNA was then transferred to a nitrocelullose or nylon membrane and hybridized to ^{32}P labeled probes (10^9 cpm per µg DNA). Two genomic probes for the murine Epo gene, two Pst I fragments, MSEP1.1 and MSEP1.0, subcloned into the plasmid pUC19 and encompassing exons 2 through 4 and part of 5, were provided by Dr. E. Goldwasser (Beru et al., 1986). Other probes were those for rat heme oxygenase (HO), the 883 base-pair EcoRI/Hind III fragment of pRH01, a plasmid containing full-length cDNA for rat HO provided by Dr. S. Shibahara (Shibahara et al., 1985), and for the murine β-actin provided by Dr. L. Wall. The membranes were prehybridized with salmon sperm DNA and subsequently hybridized with the probes at 42°C in hybridization solutions (50% formamide, 0.1% polyvinylpyrrolidone, 0.1% Ficoll 400, 0.1% BSA, 0.1% SDS, 5 × SSPE (1 × SSPE is 0.15 M NaCl, 0.01 M NaH_2PO_4 and 0.001 M EDTA, sodium salt), 10% (w/v) dextran sulphate and 100 µg/ml of denatured herring sperm DNA. After washing the membranes, the autoradiography was performed using Kodak X-ray film.

2.5. Drugs

Bovine hemin (Eastman), Sn-protoporphyrin-IX (Sn-PP) and protoporphyrin-IX (PP-IX) (Porphyrin Products, Utah, USA) were dissolved in 0.01 N NaOH and diluted with plasma from normal mice). Sn-PP and PP-IX were handled under subdued light. The drugs were administered to mice either subcutaneously (s.c.) or intravenously (i.v.) in 0.1 to 0.2 ml volumes. 1,4-dioxoheptanoic acid (succinylacetone, SA) was purchased from Sigma.

2.6. Red Blood Cell Hemolyzate

Red blood cells were separated from murine blood by centrifugation and washed with saline. Packed red blood cells were frozen to −80°C and subsequently warmed up to 50°C in three repetitive cycles.

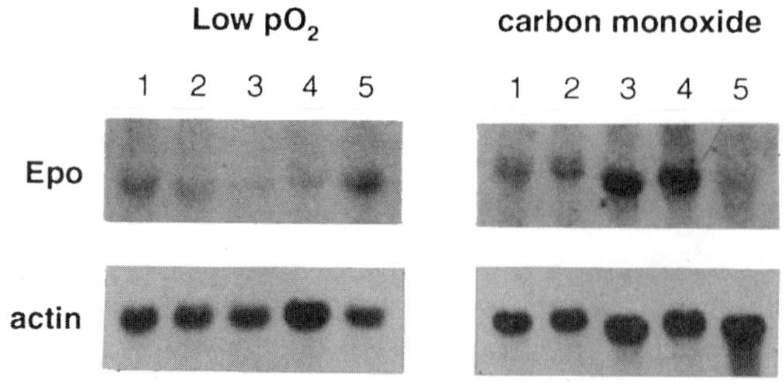

Figure 2. Epo mRNA levels in the kidney of Balb/c mice exposed for four hours to either low oxygen: lanes 2 to 5 show mice exposed to oxygen concentrations of 11.9% (4.500 m altitude), 9.7% (6.000 m altitude), 7.9% (7.500 m altitude) or 6.3% (9.000 m altitude), or CO in concentration of 0.05%, 0.10%, 0.20% or 0.30% (lanes 2 to 5). Lanes 1 show results from control mice.

2.7. Heat Damaged Red Blood Cells

Packed murine red blood cells, after being washed three times in saline, were heated for 60 min to 42°C, 46°C or 50°C. Control red blood cells were maintained at 4°C.

2.8. Heme Synthesis in the Spleen

Mice were injected with 1 μCi of ^{59}Fe i.v. five hours before sacrifice. ^{59}Fe-heme radioactivity in spleens was determined as described by Borová et al. (1973).

3. RESULTS

3.1. Epo mRNA and Epo after Hypoxia

Exposure of mice to either decreased oxygen concentration or to the CO-containing air increased Epo mRNA in the kidney (Figure 2). Balb/c and C3H/HeN mice responded similarly to both kinds of stimuli (Figure 3).

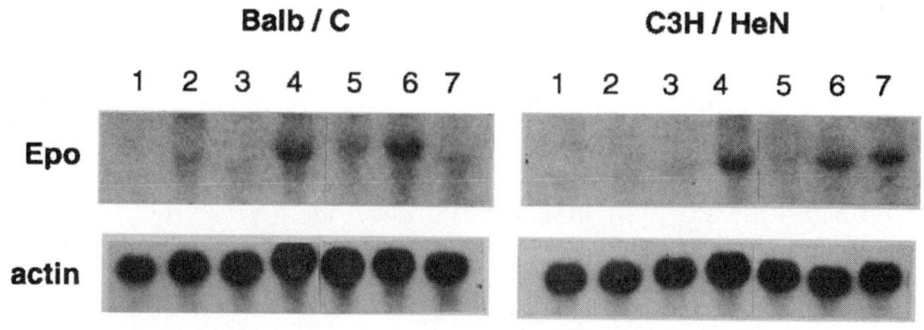

Figure 3. Epo mRNA in Balb/c and C3H/HeN mice exposed for four hours to either low oxygen (mice in lanes 2 to 4 were exposed to 11.1%, 8.4% or 6.3% oxygen), or to 0.05%, 0.10% or 0.20% of CO (lanes 5 and 7). Lane 1 show results from control mice.

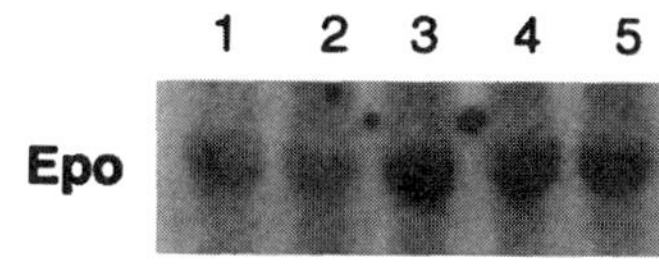

Figure 4. Epo mRNA in kidneys of Balb/c mice exposed to 0.10% CO for two, three, four and five hours (lanes 2 to 5).

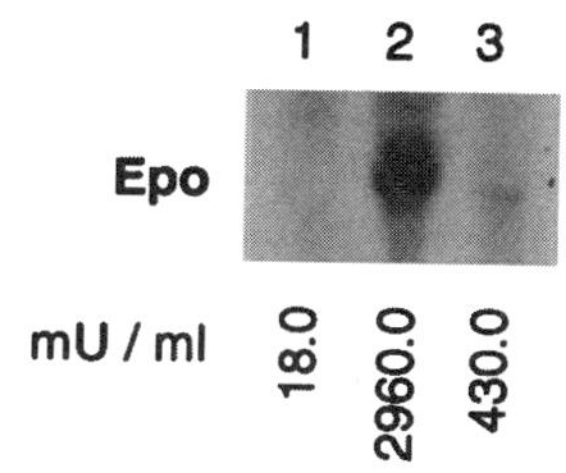

Figure 5. Epo mRNA and Epo in C3H/HeN mice exposed for four hours to either 0.20% CO (lane 2) or to 6.9% oxygen (8.500 m altitude) (lane 3). Lane 1 is a control mouse.

Increased Epo mRNA was demonstrated in the kidney after three to five hours of CO exposure (Figure 4).

Exposure to CO induced Epo mRNA more consistently than exposure to low oxygen, which was reflected by higher Epo plasma levels in mice exposed to CO (Figure 5).

3.2. Inhibition of Heme Synthesis

Heme synthesis was inhibited by i.v. administration of 1,4-dioxoheptanoic acid (succinylacetone; SA) by 70 to 80% shortly after SA.

SA administered to mice, one or two hours before exposure to CO, diminished response of Epo mRNA and Epo (Figures 6 and 7).

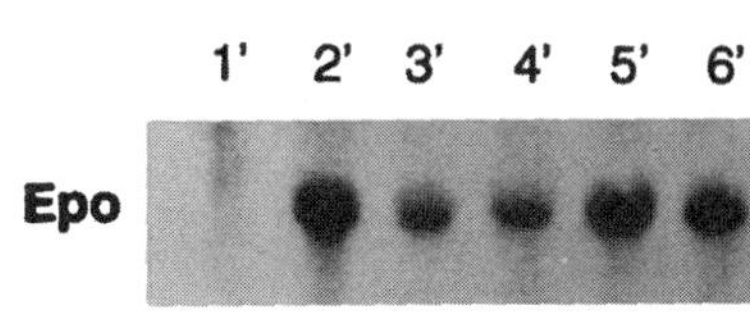

Figure 6. Effect of 40 mg of SA, administered (i.v.) 5, 6, 8 or 22 hours before sacrifice to C3H/HeN mice (lanes 3 to 6) on Epo mRNA response to four hours exposure to 0.10% CO. The mouse in lane 2 was exposed to CO but not treated with SA, and the mouse in lane 1 was untreated.

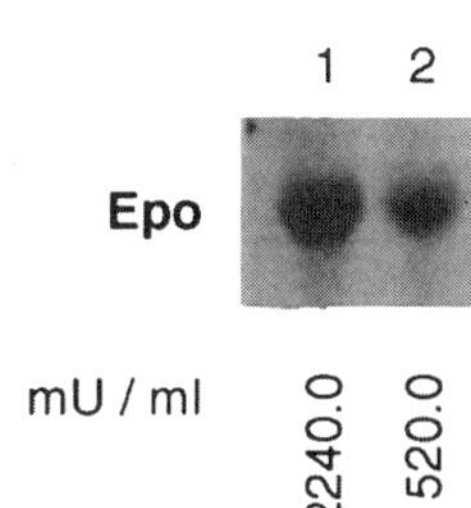

Figure 7. Epo mRNA and plasma Epo levels in C3H/HeN mice exposed to 0.15% CO for four hours. The mouse in lane 2 was given 40 mg of SA (i.v.) six hours before sacrifice.

Table 1. Effects of 40 mg of SA on heme synthesis in the spleen

Time after SA	Cpm of ^{59}Fe-heme per 100 mg of the spleen	Inhibition
Without SA	45,783	0%
	42,609	
	38,927	
	51,713	
5	8,679	81%
6	13,623	70%
8	30,501	32%
22	23,637	47%

3.3. Hemin and Sn-Protoporphyrin Administration

The administration of 6.5 mg of hemin before exposure to CO abolished the increases in Epo mRNA and in Epo, and increased heme oxygenase (HO) mRNA in the kidney (Figure 8).

Results presented in Figures 9 and 10 confirm suppressive effects of hemin on Epo mRNA and demonstrate its dose-dependence.

In contrast to hemin, Sn-PP did not inhibit the response of Epo mRNA or plasma Epo to CO exposure (Figure 11, lane 5). Administration of Sn-PP together with hemin potentiated suppression by hemin alone (Figure 11, lanes 6 to 8 as compared to lanes 2 to 4).

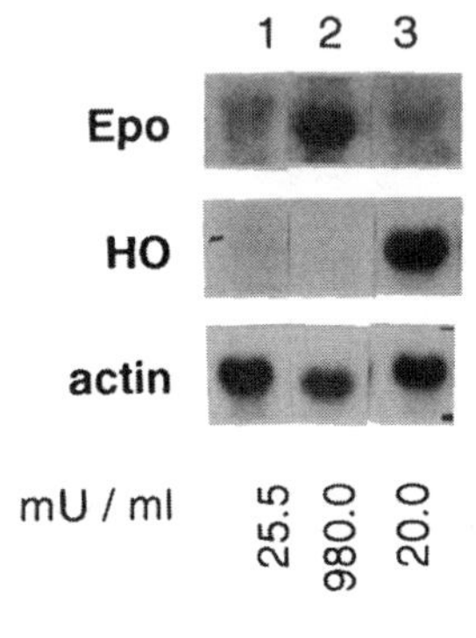

Figure 8. Epo mRNA, HO mRNA and Epo in plasma in C3H/HeN mice. Mice in lanes 2 and 3 were exposed to 0.10% CO for four hours. The mouse in lane 3 was given 6.5 mg of hemin five hours before sacrifice.

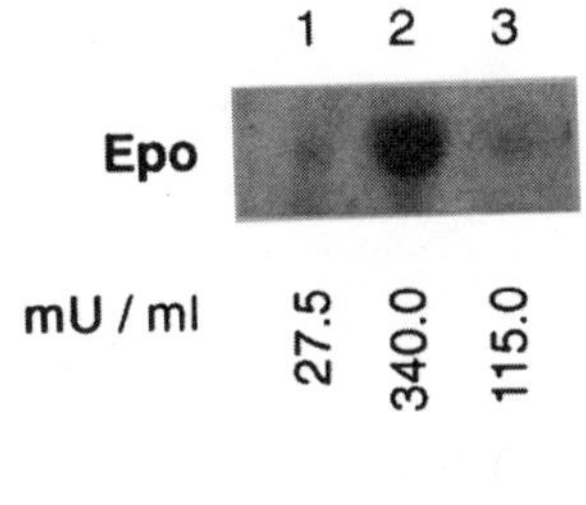

Figure 9. Effect of 2.5 mg of hemin given to C3H/HeN mouse five hours before sacrifice (lane 3). Mice in lanes 2 and 3 were exposed to 0.15% of CO for four hours.

Table 2. Distribution of ^{59}Fe from pre-labeled, heat damaged red blood cells injected i.v. to mice

	Percentage of radioactivity from 59Fe labeled red cells determined 2.5 hr after red blood cell transfusion			
Temperature degrees C	Blood	Spleen	Left kidney	Right kidney
4	78	0.8	1.3	1.2
42	79	1.3	1.4	1.3
46	55	11.4	1.0	1.1
50	12	18.7	6.2	6.3

3.4. Effect of Hemolyzate from Red Blood Cells and of Heat-Damaged Red Blood Cells

Administration of hemolyzate from red blood cells or of heat damaged red blood cells diminished the responses of Epo mRNA and Epo to CO exposure (Figures 12 and 13).

Red blood cells heated to 46°C or 50°C were rapidly cleared from the circulation as shown in Table 2.

4. DISCUSSION

Identification and cloning of the erythropoietin gene (Jacobs et al., 1985) raised new interest in control of Epo production by hypoxia. Jelkmann (1992) and Bunn and Poyton (1996) reviewed this research.

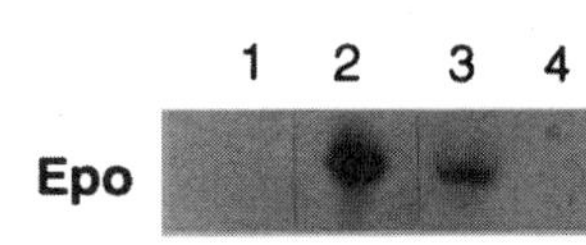

Figure 10. Effect of 1.2 mg (lane 3) and 6.0 mg (lane 4) of hemin on Epo mRNA in C3H/HeN mice exposed to 0.15% CO for four hours. Hemin was administered five hours before sacrifice. The mouse in lane 1 was untreated.

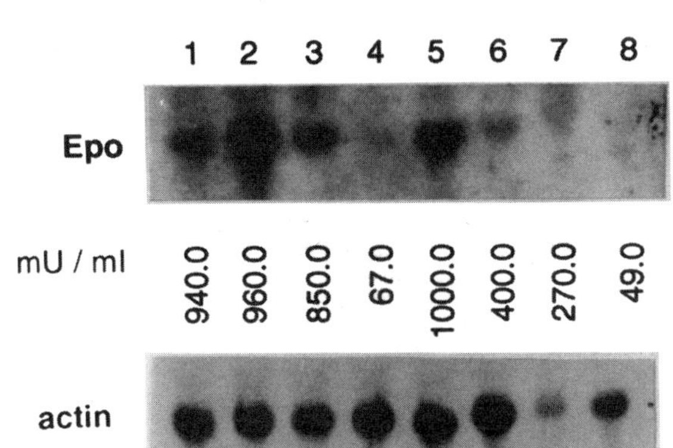

Figure 11. Effects of hemin, Sn-PP, and of hemin plus Sn-PP. Epo mRNA and Epo were measured in C3H/HeN mice exposed to 0.15% of CO for 3 hours before sacrifice. Three mg of Sn-PP were administered (s.c.) to mice in lanes 5 to 8, four and half-hours before sacrifice. Hemin was administered by i.v. route in doses of 0.1 mg (lanes 2 and 6), 0.5 mg (lanes 3 and 7), 1.0 mg (lanes 4 and 8), 3.5 hours before sacrifice.

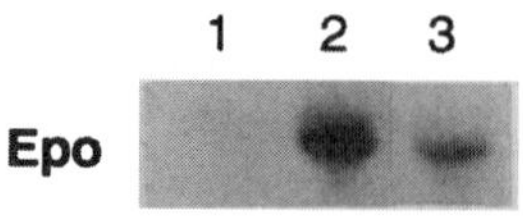

Figure 12. Epo mRNA in C3H/HeN mice exposed to 0.15% CO for three hours (lanes 2 and 3). The mouse in lane 3 was pretreated with 0.4 ml of hemolyzed red blood cells administered both intravenously (one third) and subcutaneously (two thirds) 3.5 hours before sacrifice. Lane 1 is from an untreated mouse.

Cellular mechanisms that allow recognition of oxygen tension ("concentration of oxygen") are ubiquitous (Maxwell et al., 1993). Indirect evidence suggests that a heme containing molecule, capable of oxygen binding, is involved in the initial step of oxygen sensing (Goldberg et al., 1988; Ho and Bunn, 1996). Deoxygenation of this hypothetical molecule, or replacement of its Fe^{2+} by Co^{2+}, leads to the transcription factor HIF-1 (Hypoxia Inducible Factor) in the nucleus, which can then bind to a hypoxia responsive element (HRE) in the 3′ untranscribed part of the Epo gene, acting as transcriptional enhancer (Semenza et al., 1991). HIF-1 is a heterodimer consisting of subunits HIF-1α and HIF-1β (identical to ARNT- the aryl hydrocarbon nuclear translocator). HIF-1α is ubiquinated and quickly degraded when oxygen tension is elevated. On the other hand, hypoxia triggers the production of HIF-1α, increases its stability and leads to its activation. The current knowledge regarding HIF-1 has been recently reviewed by Semenza (1998). Another regulation may involve the stability of Epo mRNA with relation to oxygen status (Goldberg et al., 1991).

Heme, an Fe^{2+} containing protoporphyrin-IX, is involved in many cellular processes and is putatively an essential molecule for cellular oxygen sensing. Inhibition of heme synthesis by succinylacetone (SA) administration diminished Epo mRNA and Epo production in mice rendered hypoxic (Figures 6 and 7). However, this result can be only hypothetically ascribed to the lack of heme for an oxygen sensing heme protein. If this were the case, our experimental set up would indicate a relatively rapid turnover of the heme protein.

Administration of exogeneous heme (hemin) to mice rendered hypoxic also diminished or abrogated increases in Epo mRNA and Epo in plasma (Figures 8 to 11). Hemin administration induces the gene for the heme oxygenase-1 enzyme (HO) that degrades heme to biliverdin, releasing Fe^{2+}. We demonstrated an induction of HO mRNA by hemin (Figure 8) in our experiments. However, suppression of Epo responses to hypoxia does not seem to be caused by degradation of heme. First, the activity of heme oxygenase increases with delay after an increase in HO mRNA (Sardana and Kappas, 1987), and thus may not be significant in the present experiments, which lasted approximately

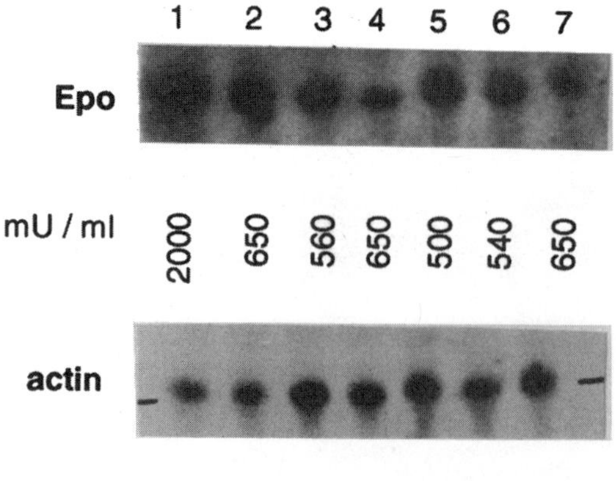

Figure 13. Epo mRNA and Epo plasma levels from Balb/c mice exposed to 0.15% CO for four hours before sacrifice. Mice were given (i.v.) either 0.2 ml of heat damaged red blood cells (46°C—lanes 2 and 5, 50°C—lanes 3 and 6) or of hemolyzate (lanes 4 and 7), either 20 hours (lanes 2 to 4) or 46 hours (lanes 5 to 7) before sacrifice.

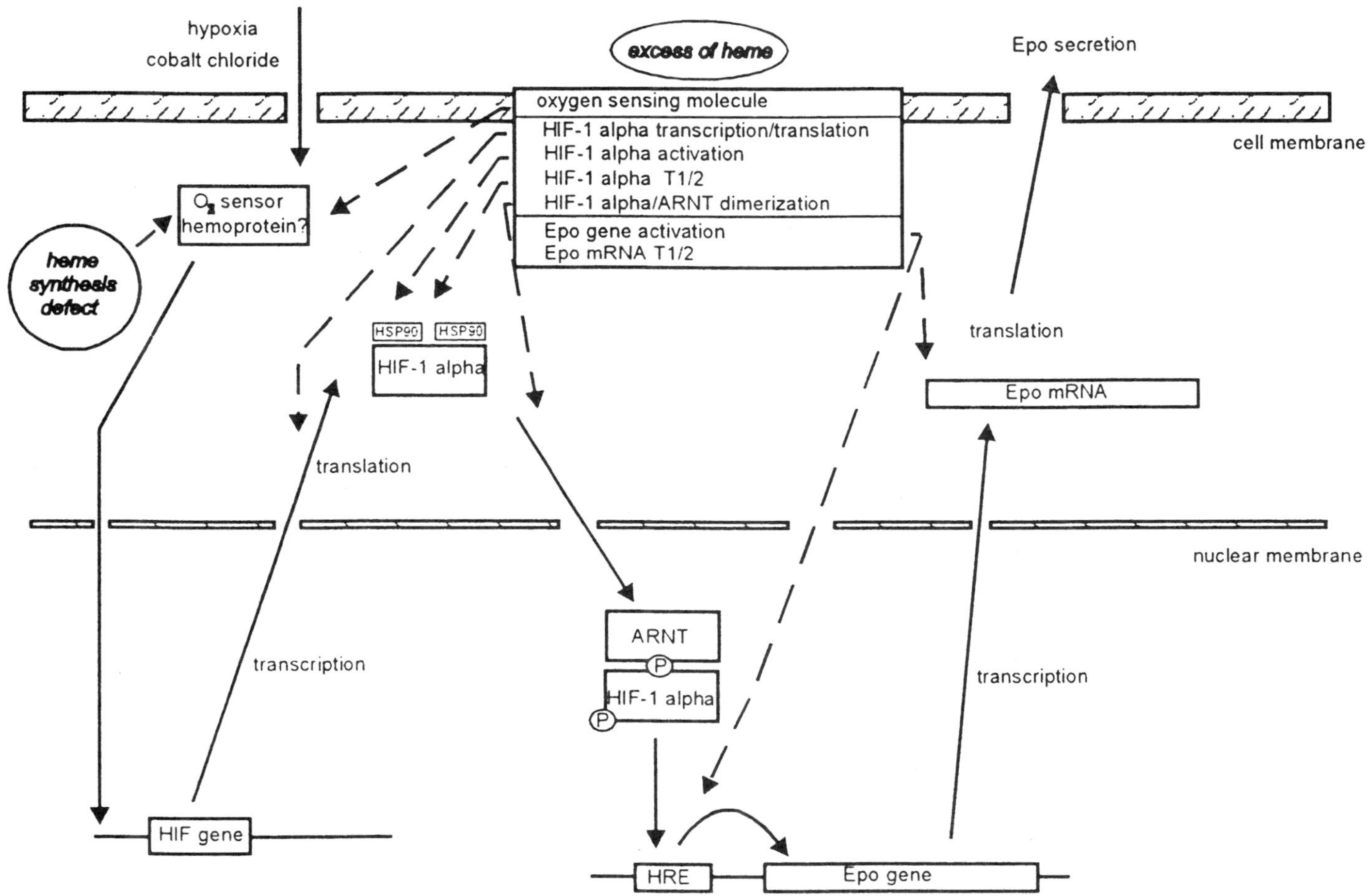

Figure 14. Control of Epo secretion by hypoxia or cobalt chloride induction (full arrows) and possible interference of lack or excess of heme (dashed arrows).

5 hours. Second, and more importantly, inhibition of heme oxygenase by Sn-PP did not abrogate the effect of hemin on Epo induction and even potentiated this hemin action (Figure 11). Wang and Semenza (1993) did not demonstrate any effect of 40 μM/L of hemin on HIF-l and Epo in hypoxic Hep3B cells. This dose is comparable to low doses of 0.1 and 0.5 mg of hemin injected to mice in the present experiments.

Our results suggest that the control of Epo production by oxygen tension requires heme supply in particular concentrations; both its decrease and increase interfere with the control of Epo production making it less sensitive to hypoxia. Some possible sites of heme action are depicted in Figure 14. It must be noted, however, that mechanisms of regulation of Epo secretion are certainly more complex than is shown in Figure 14 (see e.g. Huang et al., 1997; Fandrey et al., 1994). Other sites of hemin action are therefore possible.

Because degradation of red blood cells can be a natural source of heme, we have also tested effects of red blood cell hemolyzate and of heat damaged red blood cells on Epo mRNA and circulating Epo in mice rendered hypoxic. There was a long lasting decrease in responsiveness to hypoxia after these treatments (Figures 12 and 13). It was previously suggested that there might be a physiological safegard mechanism against excessive polycythemia based on desensitization of Epo responsiveness to hypoxia (Nečas et al., 1972). This mechanism could use red blood cell degradation products to diminish stimulation of Epo gene. Polycythemia naturally increases the absolute amount of degraded red blood cells. This could counterbalance negative influence of excessive polycythemia on tissue oxygen supply (Thorling and Erslev, 1968) and prevent undesired Epo gene expression.

ACKNOWLEDGMENT

This work was supported by grant 96/305/0619 from the Grant Agency of the Czech Republic and VS97–108 from the Ministry of Education of the Czech Republic.

REFERENCES

Bachmann S, Le Hir M, and Eckardt KU (1993) Co-localization of erythropoietin mRNA and ecto-5′-nucleotidase immunoreactivity in peritubular cells of rat renal cortex indicates that fibroblasts produce erythropoietin. J Biochem Cytochem 41:335–341.

Beru N, McDonald J, Lacombe C, and Goldwasser E (1986) Expression of the erythropoietin gene. Mol Cell Biol 6:2571–2575.

Borová J, Ponka P, and Neuwirt J (1973) Study of intracellular iron distribution in rabbit reticulocytes with normal and inhibited heme synthesis. Biochim Biophys Acta 320:143–156.

Bunn HF and Poyton RO (1996) Oxygen sensing and molecular adaptation to hypoxia. Physiol Rev 76:839–885.

Chomczynski P and Sacchi N (1987) Single-step method of RNA isolation by acid quanidinium thiocyanate-phenol-chlorofor extraction. Anal Biochem 162:156–159.

Fandrey J, Frede S, and Jelkmann W (1994) Role of nitrogen peroxide in hypoxia-induced erythropoietin production. Biochem J 303:507–510.

Goldberg MA, Dunning PS, and Bunn HF (1988) Regulation of the erythropoietin gene: evidence that oxygen sensor is a heme protein. Science 242:1412–1415.

Goldberg MA, Gaut CC, and Bunn HF (1991) Erythropoietin mRNA levels are governed by both the rate of gene transcription and posttranscriptional events. Blood 77:271–277.

Ho VT and Bunn HF (1996) Effect of transition metals on the expression of the erythropoietin gene: further evidence that the oxygen sensor is a heme protein. Biochem Biophys Res Commun 223:175–180.

Huang LO, Ho V, Arany Z, Krainc D, Galson D, Tendler D, Livingston DM, and Bunn HF (1997) Erythropoietin gene regulation depends on heme-dependent oxygen sensing and assembly of interacting transcription factors. Kidney Int 51:548–552.

Jacobs K, Shoemaker CB, Rudersdorf R, Neill SD, Kaufman RJ, Mufson A, Seehra J, Jones SS, Hewick R, Fritsch EF, Kawakita M, Shimizu T, and Miyake T (1985) Isolation and characterization of genomic and cDNA clones of human erythropoietin. Nature (London) 179:633–636.

Jelkmann W (1992) Erythropoietin: structure, control of production and function. Physiol Rev 72:449–489.

Koury ST, Bondurant MC, and Koury MJ (1988) Localization of erythropoietin synthesizing cells in murine kidneys by in situ hybridization. Blood 71:524–527.

Le Hir M, Eckardt KU, Kaissling B, Koury ST, and Kurtz A (1991) Structure-function correlations in erythropoietin formation and oxygen sensing in the kidney. Klin Wschr 69:567–575.

Maxwell PH, Pugh CW, and Ratcliffe J (1993) Inducible operation of the EPO gene 3′ enhancer in multiple cell lines: Evidence for a widespread oxygen sensing mechanism. Proc Natl Acad Sci USA 90:2423–2427.

Nečas E and Neuwirt J (1969a) Response of erythropoiesis to blood loss in hyperbaric air. Am J Physiol 216:800–803.

Nečas E and Neuwirt J (1969b) Lack of erythropoietin in plasma of anemic rats exposed to hyperbaric oxygen. Life Sci 8:1221–1228.

Nečas E and Neuwirt J (1969c) The role of oxygen supply in the regulation of erythropoiesis in compensated anaemia. Scand J Haematol 6:179–185.

Nečas E and Neuwirt J (1970) Feedback regulation by red cell mass of the sensitivity of the erythropoietin-producing organ to hypoxia. Blood 36:754–763.

Nečas E and Neuwirt J (1972) Effects of inhibitors of energy metabolism on erythropoietin production. J Lab Clin Med 79:388–396.

Nečas E and Thorling EB (1972) Unresponsiveness of erythropoietin-producing cells to cyanide. Am J Physiol 222:1187–1190.

Nečas E, Živný J, and Neuwirt J (1972) Effect of polycythemia on erythropoietin production in the hypoxic rat. Am J Physiol 223:809–811.

Sardana MK and Kappas A (1987) Dual control mechanism for heme oxygenase: Tin (IV)-protoporphyrin potently inhibits enzyme activity while markedly increasing content of enzyme protein in liver. Proc Natl Acad Sci USA 84:2464–2468.

Semenza GL (1998) Hypoxia-inducible factor 1 and the molecular physiology of oxygen homeostasis. J Lab Clin Med 131:207–214.

Semenza GL, Nejfelt MK, Chi SM, and Antonarkis SE (1987) Hypoxia-inducible nuclear factors bind to an enhancer element located 3′ to the human erythropoietin gene. Proc Natl Acad Sci USA 88:5680–5684.

Shibahara S, Muller R, Taguchi H, and Yoshida T (1985) Cloning and expression of cDNA of the rat heme oxygenase gene. Proc Natl Acad Sci USA 82:7865–7869.

Tan CC and Ratcliffe PJ (1991) Effect of inhibitors of oxidative phosphorylation on erythropoietin mRNA in isolated perfused rat kidneys. Am J Physiol 261:F982–987.

Thorling EB and Erslev AJ (1968) The “tissue” tension of oxygen and its relation to hematocrit and erythropoiesis. Blood 31:332–343.

Wang GL and Semenza GL (1993) Desferrioxamine induces erythropoietin gene expression and hypoxia-inducible factor 1 DNA-binding activity: Implications for models of hypoxia signal transduction. Blood 82:3610–3615.

Živný J, Neuwirt J, and Trávníček T (1970) Erythropoietin bioassay in polycythemic mice: comparison of three various methods. Endocrinol Exp 4:143–150.

53

PRINCIPLES UNDERLYING VASCULAR ADAPTATION/ANGIOGENESIS

Karel Rakusan

Department of Cellular and Molecular Medicine
University of Ottawa, Ottawa, ON K1H 8M5, Canada

1. INTRODUCTION

The first step in the analysis of any functional change is the proper description of the structures in which a given function takes place. Similarly, morphometric data are a necessary prerequisite for satisfactory modelling of these functional changes. As a matter of fact, the validity of any model is limited by the quality of its entry data. In this presentation, we will concentrate on the topology of the coronary microvascular bed and its changes due to formation of new channels (angiogenesis).

The terminal vascular bed is composed of arterioles, the capillary net and venules. The degree of constriction of arterioles will ultimately regulate the capillary blood flow, while the surface of the microvascular bed and its geometrical distribution will provide the structural basis for diffusion. Proper evaluation of the geometry of the microvascular bed will vary from tissue to tissue, but the basic morphometric principles will remain the same. As global indicators of capillary supply one can mention capillary numerical density and length density per tissue cross-section (N/mm^2 or mm/mm^2) or per volume tissue (N/mm^3 or mm/mm^3) and volume density (mm^3/mm^3). From length density per tissue volume one can also estimate the total length of microvessels per total organ or its parts. Finally, the capillary surface density (mm^2/mm^3) may be also derived.

At the local level one can measure on cross-section the size of microvessels, fibre-capillary ratio, intercapillary distances and the size of the tissue—regions supplied by each individual capillary, i.e., capillary domains. The variability of the domain size yields an estimate of heterogeneity of capillary spacing. On longitudinal sections one can measure the minimal capillary length, defined as the shortest contiguous pathway from terminal arteriole to collecting venule, capillary set length, which includes all the capillaries connected to the same arteriole venule and the length of capillary segments (the distance between two ensuing branch points). Branching of the larger vessels may be the subject of fractal analysis.

Oxygen Transport to Tissue XXI, edited by Eke and Delpy.
Kluwer Academic / Plenum Publishers, New York, 1999.

Measurements from both cross- and longitudinal sections may be used for calculation of the capillary supply unit, defined as the volume of a three-dimensional section of tissue supplied by a capillary segment, and which can be calculated as the product of the capillary domain area and capillary segment length. For more details consult the previous descriptions (Weibel, 1979; Rakusan et al., 1993; Rakusan, et al., 1994).

Presence of the microvascular structures does not imply that all of them are perfused continuously. The degree of possible capillary recruitment varies from organ to organ. For instance, adaptive response of the microcirculation to changes in local metabolic rate is considerable in organs like gut and skeletal muscle (both resistance-vessel responses and exchange responses, such as in capillary recruitment, Renkin 1984). Observations on the coronary capillary bed *in vivo* are restricted to the most superficial layers of the myocardium. Reported data vary but it seems that capillary reserve in hypoxic hearts is relatively small. In our experiments, it was found to be around 11% and it decreased even more in hearts of larger and older animals, as well as in cardiac hypertrophy (Rakusan, 1987). One has also to distinguish between perfusion with plasma only and perfusion with blood varying in the proportion of red blood cells (local hematocrit).

Another important aspect influencing oxygen supply is the direction of blood flow in neighbouring microvessels. This is a relatively easy task to evaluate in certain tissues which yield to direct observations in vivo, but the contractile nature of cardiac activity presents formidable difficulties for such analysis in the heart muscle. For this reason, we developed a method for estimation of microvascular flow vectors based on sequential *in vivo* infusion of two or more differently colored microspheres. We found a very high percentage of concurrent blood flow between adjacent capillaries in both normal and hypertrophic rat heart: 87% in midmyocardium and 68% in subendomyocardium. (Cicutti and Rakusan, 1992). Interestingly, under conditions of acute hypoxia, the percentage of countercurrent flow between neighbouring capillaries increased (Cicutti and Rakusan, 1994).

2. ANGIOGENESIS

We have made a distinction between the microvascular bed as a morphological structure and the actually perfused microvascular bed. Nevertheless, one may expect that in situations requiring maximal perfusion as in severe hypoxia, all the existing structures will be utilized. At the same time, we should be aware that microvascular structures are not static either. Restructuring of the microvascular bed due to appearance and disappearance of new vessel segments is probably a constant feature even under normal conditions and it is more pronounced under pathological conditions. These changes are surprisingly fast. Hudlicka and Tyler (1986) reviewed the older literature concerning the rate of angiogenesis in various experimental situations. Observations on capillary growth in regenerating tadpole tails revealed that capillary sprouts met and connected within 2–4 hours from the time when they were noticed, opened within 4–6 hours and flow was established within 10 hours. In the case of wound healing, one may notice indication of growth 24 hours after injury, capillaries subsequently appearing within a few days, and finally vascularization completed within 6–8 days. In regenerating minced muscles, vascular growth started 2–3 days after implantation and by 9 days vascularization was completed.

Angiogenesis is a complex process, which involves all morphological components of the vascular wall as well as of the surrounding tissue. The three main stimuli for angiogenesis are: 1) mechanical factors, 2) energy imbalances as in hypoxia, and 3) inflammatory processes (see Figure 1). Among mechanical factors one should consider an increased red blood cell-endothelial cell interaction, as in polycythemia, an increased wall tension as a result of an increased capillary pressure and an increased shear stress associated with an increased velocity of flow. Obviously, these mechanisms may be interrelated. Ultimately, angiogenesis is a resultant of a delicate balance of angiogenic growth stimulators and inhibitors. A survey of the literature would reveal more than fifty putative factors potentially involved in angiogenesis. The most prominent stimulators are probably vascular endothelial growth factor, which is specific for endothelial cells, and stimulators from the fibroblast growth factor family, which are involved in the whole tissue response. The main inhibitors are probably thrombospondin, angiostatin and endostatin. (For more details see Hudlicka et al., 1992; Folkman, 1995; Rakusan, 1995; Hudlicka and Brown, 1996.)

As an example of well documented angiogenic processes one may focus on coronary angiogenesis. Morphometric analysis of the mammalian heart muscle has revealed a substantial degree of angiogenesis in the early stages of the postnatal development. Olivetti and coworkers (1980) reported that during the first postnatal days in the rat myocardium, capillaries grow two to three times more rapidly than does the myocardial mass. According to our own estimates, close to half of all the capillaries found in adult rat left ventricle are formed during the relatively short of 3–4 week postnatal period. These estimates were based on capillary density, myocyte-to-capillary ratio and the total number of myocytes in the rat left ventricle (Rakusan and Turek, 1985). Considerable angiogenesis in the early postnatal development process is not limited to coronary capillaries only but involves more complex structures such as arterioles. In our recent study on arteriolar growth and maturation in the postnatal rat heart, we found the proportion of immature, new arterioles to be 22% at day 1, which decreased to about half of these values within a week and by postnatal day 28 there were essentially no new immature arterioles detected (Heron and Rakusan, 1998).

In contrast to many situations with documented coronary angiogenesis, our knowledge of possible disappearance of coronary capillaries is rather limited. We have found an evidence for capillary involution in rat hearts undergoing atrophy following heterotopic isotransplantation (Rakusan et al., 1997). In this case, decreased hemodynamic load leads to a reduction of cardiac mass in transplanted hearts which is associated with an involution of coronary capillaries. Thus, three weeks after transplantation of normal

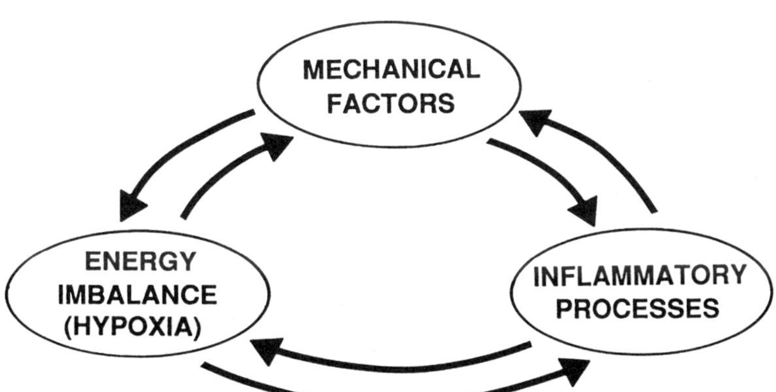

Figure 1. Major factors leading to the initiation of angiogenesis. From Rakusan 1995.

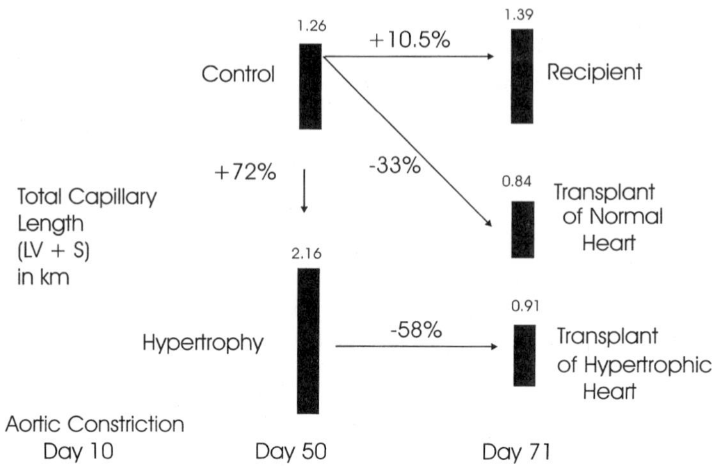

Figure 2. Capillary growth in cardiac hypertrophy and their involution in transplanted rat hearts.

hearts the aggregate capillary length decreased by one third. Decrease in the aggregate capillary length of the originally hypertrophic transplanted hearts was even more pronounced: 52% in male and 64% in female hearts (Figure 2).

Based on data from Rakusan et al., 1997. Total capillary length in the left ventricle and septum of control and hypertrophic rat hearts on postnatal day 50 as compared with recipient hearts, transplants of normal heart and transplants of hypertrophic heart three weeks later. Data for male and female hearts are combined.

3. CONCLUSIONS

Remarkable plasticity of the microvascular bed has been demonstrated by our studies of the microvascular bed primarily in rat hearts. Similar results, however, may be found in various organs of diverse species. For instance, twenty sessions of intermittent normobaric hypoxia influences vascularization of human myometrium. Hypoxia training increased the percentage of cross-sectional area of myometrial tissue occupied by vessels by almost half (from 7.9% of the area to 11.8%). The increase was due to both increases in vessel size and vessel density (Rakusan et al., 1998).

In conclusion: we have stressed the importance of the geometry of the microvascular bed for optimal oxygen supply based on various studies. It is not sufficient simply to know the geometrical features of the microvascular bed but also its perfusion at any given moment. Moreover, the microvascular bed is remarkably plastic and it can adjust to the functional environment within a relatively short time-frame. Knowledge of the factors regulating initiation, extent and cessation of angiogenic processes is essential.

ACKNOWLEDGMENTS

These studies were supported by the Medical Research Council of Canada and Ontario Heart and Stroke Foundation.

REFERENCES

Cicutti N and Rakusan K (1992) Microvascular flow vectors in normal and hypertrophic myocardium as determined by the method of colored microspheres. Microvasc. Res. 43:267–275.

Cicutti N and Rakusan K (1994) The effect of hypoxia on capillary flow direction in the isolated perfused rat heart. Can. J. Cardiol. 10:367–373.

Folkman J (1995) Clinical applications of research on angiogenesis. N. Engl. J. Med. 333:1757–1763.

Heron MI and Rakusan K (1998) Arteriolar growth in the postnatal rat heart. J. Mol. Cell. Cardiol. 30:A162.

Hudlicka O and Tyler KR (1986) Angiogenesis. The growth of the vascular system. Academic Press. London.

Hudlicka O, Brown M, and Egginton S (1992) Angiogenesis in skeletal and cardiac muscle. Physiol. Rev. 72:369–417.

Hudlicka O and Brown M (1996) Postnatal growth of the heart and its blood vessels. J. Vasc. Res. 33:266–287.

Olivetti G, Anversa P, and Loud AV (1980) Morphometric study of early postnatal development in the left and right ventricular myocardium of the rat. Circ. Res. 46:503–512.

Rakusan K and Turek Z (1985) Protamine inhibits capillary formation in growing rat hearts. Circ. Res. 57:393–398.

Rakusan K (1987) Microcirculation in the stressed heart, In: The Stressed Heart. Legato MJ, Ed. Martinus Nijhoff Publishing, Boston, pp. 107–124.

Rakusan K, Batra S, Veprek P, and Turek Z (1993) Capillary density, distribution, and length parameters related to oxygen supply in myocardial hypertrophy and atrophy. In: Idiopathic Dilated Cardiomyopathy. (Figulla HR, Kandoff R, McManus B, Eds.) Springer-Verlag, Berlin, pp. 155–164.

Rakusan K, Batra S, and Heron MI (1994) A new approach for quantitative evaluation of coronary capillaries in longitudinal sections, In: "Oxygen Transport to Tissue—XVI". (Hogan MC, Mathieu-Costello O, Poole DC, Wagner PD, Eds) Plenum Press, New York, pp. 407–416.

Rakusan K (1995) Coronary angiogenesis. From morphometry to molecular biology and back. Ann. N.Y. Acad. Sci. 752:257–266.

Rakusan K, Heron MI, Kolar F, and Korecky B (1997) Transplantation-induced atrophy of normal and hypertrophic rat hearts: effect on cardiac myocytes and capillaries. J. Mol. Cell Cardiol. 29:1045–1054.

Rakusan K, Ehrenburg IV, Gulyaeva NV, and Tkatchouk EN (1998) The effect of intermittent normobaric hypoxia on tissue restructuring. Hypoxia Med. 6:59–60.

Renkin EM (1984) Control of microcirculation and blood-tissue exchange. In: Handbook of Physiology, Section 2. The Cardiovascular System, Volume IV., Microcirculation, Part 2, (Renkin EM, Michel CC, Eds) American Physiological Society, Bethesda, MD, pp. 627–688.

Weibel ER (1979) Stereological Methods, Vol. 1. Practical Methods for Biological Morphometry. Academic Press: London.

54

SOME ASPECTS CONCERNING THE ANTIOXIDANT CAPACITY OF VENOUS BLOOD IN LOWER LIMBS VARICOSE VEINS

Ioana-Raluca Siska, Jecu Avram, Carmen Tatu, Carmen Bunu, Francisc Schneider, and Rosana Maties

Department of Physiology
*1st Department of Surgery
University of Medicine and Pharmacy
Timisoara - 1900, Romania

1. INTRODUCTION

Varicose veins are a very common pathology, but the etiology is not well understood. Epidemiological studies have shown that, besides hereditary and hormonal factors, prolonged standing position may contribute to the development of this disease. The relation between the standing position and the typical physiopathological modifications of the varicose veins is still unclear (Beaglehole, 1986; Michiels et al., 1994).

Michiels et al. explained the alterations observed in the pathological vein as consequences of hypoxia. They showed that hypoxia, for example resulting from venous stasis, induces the activation of endothelial cells which then release inflammatory mediators able to activate neutrophils (which generate significant quantities of $O_2^{\bullet -}$ and H_2O_2 extracellularly) and to induce their infiltration, as well as growth factors for smooth muscle cells. These processes could represent the beginning of a cascade of events eventually leading to structural and functional modifications of the venous wall similar to that one observed in the varicose vein wall (Michiels et al., 1994; Michiels et al., 1995; Michiels et al., 1996; Michiels et al., 1997).

It has been firmly established that oxidant products of activated neutrophils are critically involved in the killing of endothelial cells. The continuous exposure to blood components including prooxidant agents makes the venous wall susceptible to oxidative stress and free radical-mediated reactions. Under pathological conditions, an excessive amount of macromolecules are broken down by free radicals and proteases, the varicose saphenous vein losing as much as 60% of the collagen in its wall (Robert, 1992).

Oxygen Transport to Tissue XXI, edited by Eke and Delpy.
Kluwer Academic / Plenum Publishers, New York, 1999.

Thus the adequacy of the defense systems against free radicals is critical for the susceptibility of the vessel wall to oxidative damage. Among the systems capable to protect the cells against oxidative injury are superoxide-dismutase (SOD) and reduced glutathione (GSH).

Superoxide dismutase (SOD) is a scavenger for the superoxide radical ($O_2^{\bullet-}$), accelerating its conversion to hydrogen peroxide (H_2O_2). SOD works in collaboration with H_2O_2-removing enzymes, such as catalase and glutathione-peroxidases (Fulbert and Cals, 1992; Halliwell et al., 1992; Halliwell, 1996).

$$2O_2^{\bullet-} + 2H^+ \xrightarrow{sod} O_2 + H_2O_2$$

Reduced glutathione (GSH) is a cysteine-containing tripeptide with reducing properties which play an important role in cellular protection against oxidative damage. During oxidative stress, GSH can scavenge various free radicals directly, as well as being a cofactor for enzymes, such as glutathione peroxidases (GPx). GSH presents synergistic interactions with other components of the antioxidant defense system, including superoxide dismutase (Mezzetti et al., 1990; Fulbert and Cals, 1992; Halliwell et al., 1992; Halliwell, 1996; Gerard-Monnier and Chaudiere, 1996).

$$2GSH + 2R\bullet \rightarrow GSSG + 2RH$$
$$2GSH + H_2O_2 \xrightarrow{gpx} 2H_2O + GSSG$$

2. MATERIAL AND METHODS

Before and after 30 minutes of venous stasis by the standing position, blood samples were drawn from the distal portion of the saphenous vein in 21 patients with primary varicose veins (14 females and 7 males, mean age = 50.4 years). Superoxide dismutase (SOD) blood level, reduced glutathione (GSH) plasma concentration, and total leukocyte count were determined.

Statistical analysis was performed using Student's T test.

2.1. SOD Determination (G. Szegli, Cantacuzino Institute, Bucharest, Romania—unpublished)

2.1.1. Principle. The $O_2^{\bullet-}$ generating system is the redox couple formed by methylene blue and tetramethylenediamine (TMEA), as indicator being used the Nitro blue tetrazolium (NBT). When $O_2^{\bullet-}$ is produced, NBT is reduced to formazan, a colored compound with a maximum rate of absorption at 560 nm. The rate of absorption is directly proportional to $O_2^{\bullet-}$ concentration. SOD in the reaction mixture induces $O_2^{\bullet-}$ consumption, reducing the formation of formazan.

2.1.2. Reagents.

Phosphate buffer (KH_2PO_4 + $Na_2HPO_4 \cdot 12H_2O$) 0.05 M, pH = 7.8
Tetramethylenediamine (TMEA) 1.16 ml/10 ml distilled water
Methylene blue 0.003 g/50 ml distilled water
Nitro blue tetrazolium (NBT) 0.26 g/40 ml distilled water

2.1.3. Technique. The reaction mixture for one sample contains:

Phosphate buffer	2.70 ml
Tetramethylenediamine (TMEA)	0.05 ml
Methylene blue	0.05 ml
Nitro blue tetrazolium (NBT)	0.20 ml

Heparinized blood was used. Before determination, SOD was extracted as following: 1 ml blood + 0.25 ml chloroform + 0.5 ml ethyl alcohol; the mixture is centrifuged, just the supernatant being used. The sample contains 0.1–0.3 ml supernatant + 3 ml reaction mixture. The blank contains the same mixture, without the supernatant. The tubes were uniformly illuminated for 15 minutes with a neon lamp of 17 W, in a closed space, without other light sources. The extinction of both samples and blanks was determined spectrophotometrically at 560 nm, compared to the phosphate buffer.

One SOD unit is defined as the amount of enzyme which reduces with 50% the formazan in the blank. SOD is expressed in U/ml probe, its calculation depending on the dilution of the sample; if the extinction of the sample (Es) is less than that of the blank (Eb), the calculation formula is:

$$[(\text{Eb} - \text{Es})/\text{Eb} \times 2 \times 1]/\text{sample volume} \times \text{sample dilution}$$

The formula is used as long as the enzyme reduces the formazan of the blank to 40–60%.

2.2. GSH was Determined using a Modified Form of Ellman's Method (Albini 1980)

2.2.1. Principle. Dithiobisnitrobenzoic acid (DTNB) was added to plasma, at pH = 8; the reaction of DTNB with the -SH groups generates a colored compound with a maximum rate of absorption at 420 nm.

2.2.2. Reagents. Phosphate buffer pH = 8.0 (0.3683 g NaH_2PO_4 + 16.9749 g $Na_2HPO_4 \cdot 12\ H_2O$ + 4.675 g NaCl, in 1000 ml distilled water).

5,5′ dithiobis-(2-nitrobenzoic acid) (DTNB) 40%, solved in the phosphate buffer.

2.2.3. Technique.

The sample contains: 0.15 ml plasma + 3.5 ml phosphate buffer + 0.5 ml DTNB
The blank contains: 0.15 ml plasma + 4 ml buffer
The optical density is measured spectrophotometrically after 15 minutes at 420 nm.
Calculation (μM/l): absorption × 2020
Normal values: 370–500 ± 65 μM/l

All chemicals were purchased from Sigma Chemical Co. (St Louis, MO, USA)

3. RESULTS

After 30 minutes in the standing position SOD fell significantly ($p < 0.0001$) (Figure 1), GSH level also presented a significant increase ($p = 0.0002$) (Figure 2) and the total leukocyte number rose by about 1.2% ($p = 0.01$) (Figure 3). The statistical analysis is presented in Table I.

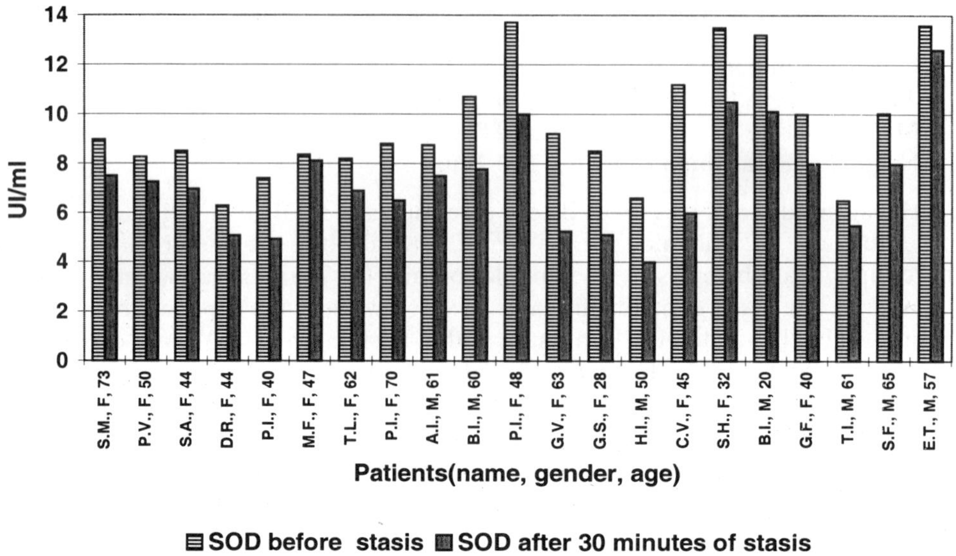

Figure 1. SOD concentrations in the venous blood before and after 30 minutes of venous stasis (standing position).

Other investigators (Flagg et al., 1993; Bolzan et al., 1997) reported the existence of a large variability of SOD, GSH and GSH-related enzymes ascribed, at least in part, to the sex and age of the individuals. Our study also showed that SOD and GSH in the human blood exhibit a wide interindividual variability, but there were no statistically significant differences related to sex and age ($p > 0.05$).

4. DISCUSSION

Similar results, e.g. a significant decrease of superoxide dismutase blood concentrations and an increase of total leukocyte count after venous stasis in patients with varicose veins and with venous hypertension were reported by Ciuffetti et al. (1994). They also reported that no significant modification was observed at any stage of the study in healthy controls.

We consider that our data are significant due to the following considerations:

1) SOD decrease could represent a negative factor for the endothelial cell lining the vessel wall: during the following supine position, oxygen free radicals (espe-

Table I. SOD, GSH blood levels and leukocyte number before and after 30 minutes of venous stasis

Statistical analysis	SOD before stasis (UI/ml)	SOD after stasis (UI/ml)	GSH before stasis (μM/I)	GSH after stasis (μM/I)	Leukocyte number before stasis (no/mm^3)	Leukocyte number after stasis (no/mm^3)
Average	9.534	7.314	339.161	389.576	5176.19	5235.714
Standard deviation	2.335	2.137	80.962	73.484	860.758	825.097
p value	$p < 0.0001$		0.0002		0.010	

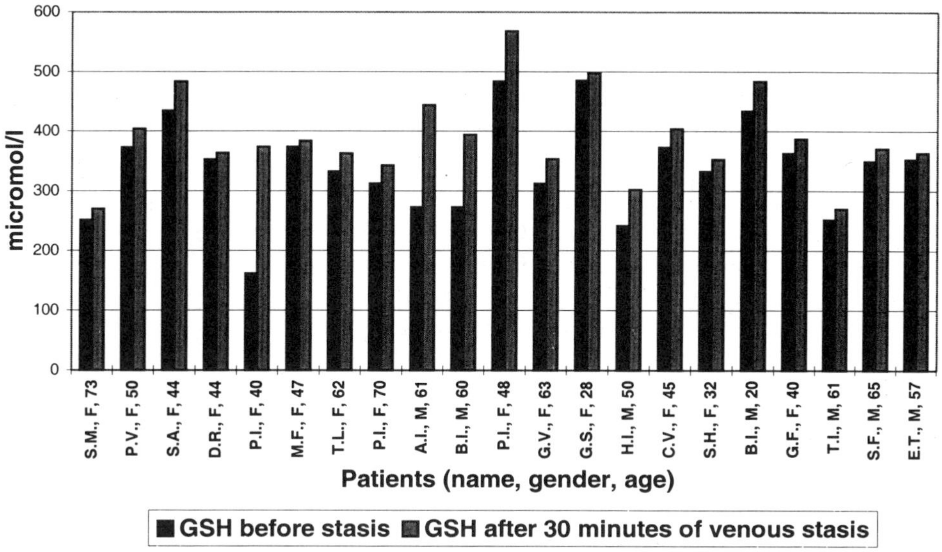

Figure 2. GSH variations in the venous blood before and after 30 minutes of venous stasis (standing position).

cially superoxide) production will rise with damaging effects in the endothelium. The importance of the decline of SOD activity is underlined by the findings of Mickle et al. (1990), who demonstrated that SOD and catalase seem to be more effective as antioxidants for the saphenous vein endothelial cells than other water-soluble antioxidants (such as ascorbic acid or Trolox, a water-soluble analogue of alpha-tocopherol).

Other data suggesting the significance of SOD determinations are those of Scott et al. (1990). They described that in the supine position blood from varicose veins had a significantly higher oxygen tension than that from normal veins. After 30 minutes in the standing position, the PO_2 was markedly reduced: 18% in normal veins and 35% in varicose veins. In all subjects there was a higher

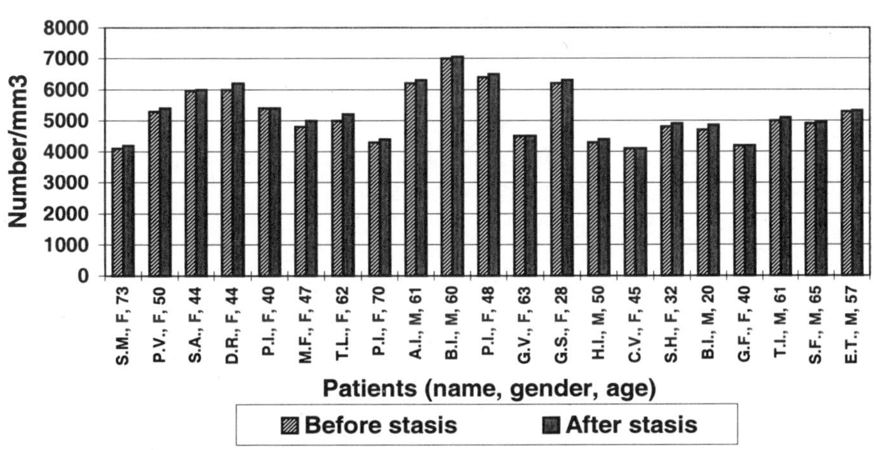

Figure 3. The number of leukocytes in the blood of the varicose veins before and after 30 minutes of venous stasis (standing position).

oxygen tension in leg vein blood when lying than while standing. This could be important taking into account the experimental research of Li et al. (1992), who investigated the antioxidant enzyme activities (superoxide dismutase, catalase and glutathione peroxidase) in human saphenous vein endothelial cells cultured at atmospheric (pO_2 of 150 mmHg) or low (pO_2 of 40 mmHg) oxygen tensions. The cells cultured at the low oxygen tension had significantly lower antioxidant activities than the cells cultured at the high oxygen tension, suggesting that postcapillary venous endothelial cells should have lower antioxidant enzyme activities than the precapillary arterial ones.

Therefore, in varicose veins both plasmatic and intracellular levels of SOD seem to decrease during hypoxic conditions. This observation could be important because hydrogen peroxide also modulates the venous tone: lower concentrations of H_2O_2 induce contractions of the human saphenous vein preparations, while higher doses are followed by venodilation (unpublished data).

2) The importance of GSH study relies on two facts:
   - GSH is an important scavenger of H_2O_2 and may be involved in the modulation of H_2O_2-induced leukocyte rolling within postcapillary venules. It has been demonstrated that endothelial cells, similar to leukocytes and erythrocytes, contain large quantities of glutathione and catalase, which scavenge H_2O_2 and protect cells from oxidant-induced damage in vitro (Harlan et al., 1984). The glutathione redox cycle is sufficient to detoxify normal amounts of intracellular oxidants in catalase-deficient cells, but hydrogen peroxide in non-toxic concentrations may overcome this single system and thus contribute to in vitro cellular damage. Furthermore, depletion of glutathione activity induced an increase of leukocyte rolling in rat mesenteric venules superfused with 10 mM H_2O_2 (Johnston et al., 1996).
   - recent studies confirmed that GSH modulates vasomotor reactivity in human coronary arteries (Cheung and Schultz, 1997; Kugyiama et al., 1998), but no data concerning the influence of GSH on venous smooth muscle reactivity are available.

   Our data are the first to show a very significant augmentation of GSH plasmatic level after 30 minutes of venous stasis in patients with lower limb varicosities. It was not the aim of our study to investigate the precise mechanisms which may explain this modification, but this could be due to a decrease of intracellular GSH stores by leakage, reducing the intracellular antioxidative capacity and increasing the risk for oxidative reperfusion injury. This modification may also be related to an impaired function of cell antioxidant enzymes in response to hypoxia/hyperoxia.

   Because the metabolic turnover of glutathione represents one of the main cytoprotective systems against the peroxide attack, further studies are required to clarify these aspects (amplitude and duration of GSH intracellular depletion, GSH resynthesis, activity of GSH-related enzymes).

3) The accumulation of white cells in the venous blood during stasis agrees with the findings of Moyses et al. 1987, Coleridge-Smith et al. 1988, Thomas et al. 1988, Johnston et al. 1993, Ciuffetti et al. 1994, Dormandy 1995, Iwao et al. 1995, Saharay et al. 1996, Saharay et al. 1997. This process could be the first step leading to localized leukocyte activation in the microcirculation, e.g. their subsequent adhesion, migration into tissues and release of toxic metabolites

such as free radicals. This process may overcome the natural scavenging mechanisms, some of them being also impaired, and generate the tissue damage.

5. CONCLUSIONS

These results suggest impairments in leukocyte rheology and local alterations of blood antioxidant systems. Altogether and because the inflammation is frequently persistent, these processes could eventually lead to the alterations of the venous wall observed in varicose veins.

REFERENCES

Albini A (1980) Standardizatione di un metodo di dosaggio dei gruppi tiolici -SH nel sero umano a seguito di alcune considerationi cinetiche, Boll Soc It Biol Sper LVI:1892–1898.

Beaglehole R (1986) Epidemiology of varicose veins. World J Surg 10:898–902.

Bolzan AD, Bianchi MS, and Bianchi NO (1997) Superoxide dismutase, catalase and glutathione peroxidase activities in human blood: influence of sex, age and cigarette smoking. Clin Biochem 30:449–454.

Bradbury AW, Murie JA, and Ruckley CV (1993) Role of the leucocyte in the pathogenesis of vascular disease. Br J Surg 80:1503–1512.

Cheung P-Y and Schultz R (1997) Glutathione causes coronary vasodilation via a nitric-oxide and soluble guanylate cyclase-dependent mechanism. Am J Physiol 273, H1231–H1238.

Ciuffetti G, Mannarino E, Paltriccia R, Malagigi V, Sergi F, Paulisch P, Pasqualini L, and Lupattelli G (1994) Leucocyte activity in chronic venous insufficiency. Int Angiol 13(4):312–316.

Coleridge-Smith PD, Thomas P, Scurr JH, and Dormandy JA (1988) Causes of venous ulceration: a new hypothesis. Br Med J 296:1726–1727.

Dormandy JA (1995) Influence des éléments figures et du flux sanguin sur l'endothélium veineux. Phlébologie 48, special issue:137–140.

Flagg EW, Coates RJ, Jones DP, Eley JW, Gunter EW, Jackson B, and Greenberg RS (1993) Plasma total glutathione in humans and its association with demographic and health-related factors. Br J Nutr 703:797–808.

Fulbert JC and Cals M-J (1992) Les radicaux libres en biologie clinique: origine, role pathogene et moyens de defense. Path Biol 40, 1:66–77.

Gerard-Monnier D and Chaudiere J (1996) Métabolisme et fonction antioxydante du glutathion. Path Biol 44, 1:77–85.

Halliwell B, Gutteridge JMC, and Cross CE (1992) Free radicals, antioxidants, and human disease: Where are now?. J Lab Clin Med 119, 6:598–620.

Halliwell B (1996) Mechanisms involved in the generation of free radicals. Path Biol 44, 1:6–13.

Harlan JM, Levine JD, Callahan KS, Schwartz BR, and Harker LA (1984) Glutathione redox cycle protects cultured endothelial cells against lysis by extracellularly generated hydrogen peroxide. J Clin Invest 73:706–713.

Iwao T, Toyonaga A, Shigemori H, Sumino M, Oho K, and Tanikawa K (1995) Microcirculation in venous disorders: the role of the white blood cells. Int J Microcirc Clin Exp 90:3–8.

Johnston B, Kanwar S, and Kubes P (1996) Hydrogen peroxide induces leukocyte rolling: modulation by endogenous antioxidant mechanisms including NO. Am J Physiol 40, H614–H621.

Kugyiama K, Ohgushi M, Motoyama T, Hirashima O, Soejima H, Misumi K, Yoshimura M, Ogawa H, Sugiyama S, and Yasue H (1998) Intracoronary infusion of reduced glutathione improves endothelial vasomotor response to acetylcholine in human coronary circulation. Circulation 97:2299–2301.

Li RK, Shaikh N, Weisel R, Tumiati L, Wu TW, and Mickle D (1992) Cultured vascular endothelial cell susceptibility to extracellularly generated oxidant injury. J Mol Cell Cardiol 24(6):595–604.

Mezzetti A, Di Ilio C, Calafiore AM, Aceto A, Marzio L, Frederici G, and Cuccurullo F (1990) Glutathione peroxidase, glutathione reductase and glutathione transferase activities in the human artery, vein and heart. J Mol Cell Cardiol 98:935–938.

Michiels C, Arnould T, and Remacle J (1994) Role de l'hypoxie et des cellules endotheliales dans le developpement des veines variqueuses. Med Sci 10:845–853.

Michiels C, Arnould T, Janssens D, Bajou K, Geron I, and Remacle J (1995) Interactions entre les cellules musculaires lisses apres activation par l'hypoxie. Une etiologie possible de la maladie veineuse. Phlebologie 48:141–149.

Michiels C, Arnould T, Janssens D, Bajou K, Geron I, and Remacle J (1996) Interactions between endothelial cells and smooth muscle cells after their activation by hypoxia. A possible etiology for venous disease. Int Angiol 15:124–130.

Michiels C, Arnould T, Thibaut-Vercruyssen R, Bouaziz N, Janssens D, and Remacle J (1997) Perfused human saphenous veins for the study of the origin of varicose veins: role of the endothelium and of hypoxia. Int Angiol 16:134–141.

Mickle DA, Li RK, Weisel RD, Tumiati LC, and Wu TW (1990) Water-soluble antioxidant specificity against free radical injury using cultured human ventricular myocytes and fibroblasts and saphenous vein endothelial cells. J Mol Cell Cardiol 24:1297–1304.

Moyses C, Cederholm-William SA, and Michel CC (1987) Haemoconcentration and accumulation of white cells in the feet during venous stasis. Int J Microcirc Clin Exp 5:311–320.

Robert A (1992) Extracellular matrix and vascular pathologies. Role of proteases and oxygenated free radicals. Ann Cardiol Angiol 24:A18–A23.

Saharay M, Addison IE, Shields DA, Scurr JH, and Coleridge-Smith PD (1996) Leukocyte migration in the leg in response to experimental venous hypertension. J Vasc Surg 24:725–731.

Saharay M, Shields DA, Porter JB, Scurr JH, and Coleridge-Smith PD (1997) Leukocyte activity in the microcirculation of the leg in patients with chronic venous disease. J Vasc Surg 25:265–273.

Scott HJ, Cheatle TR, McMullin GM, Coleridge-Smith PD, and Scurr JH (1990) Reappraisal of the oxygenation of blood in varicose veins. Br J Surg 77:934–936.

Thomas PR, Nash GB, and Dormandy JA (1988) White cell accumulation in dependent legs of patients with venous hypertension: a possible mechanism for trophic changes in the skin. Br Med J 27:1693–1695.

ASSESSMENT OF TISSUE OXYGENATION IN A CARDIOVASCULAR ICU WITH RADIOMETER'S ABL-625, SIGGAARD ANDERSEN'S OSA AND ASKIT'S NHMS

Gábor Vereczkey

Cardiovascular Surgery Clinic
Internal Medical and Cardiovascular Centre
Haynal Imre University of Health Sciences
Budapest, H-1135, Hungary

1. INTRODUCTION

In an Intensive Care Unit it is crucial to have a reliable method to monitor and evaluate total body tissue oxygenation.[1,2] Although in physiology studies it is relatively simple to determine arterial and mixed venous blood gas values and measure cardiac output with an aortic flow-meter, it raises ethical and principle problems in human clinical practice.

1.1. Oxygen Consumption

1.1.1. The Principle of Oxygen Consumption. As a principle, O_2 consumption is an important indicator of a patient's general condition. To gain information about VO_2, the difference between arterial ($ctO_2(a)$) and mixed venous oxygen content ($ctO2(\bar{v})$) gives the extracted volume of O_2 from one litre of blood. If this $ctO_2(a - \bar{v})$ is multiplied by the actual cardiac output, the result is the total body oxygen consumption in a minute.

$$VO_2 = ctO_2(a-\bar{v}) * CO$$

1.1.2. The Simple But Clear Picture. In biology lessons at the elementary school everything was simple and clear regarding oxygen consumption. In the human arterial blood there is 20 ml/dl (200 ml/L) oxygen, while in venous blood there is only 15 ml/dl (150 ml/L). If one heart contraction ejects 70 ml of blood into the aorta, 70 heartbeats a

Oxygen Transport to Tissue XXI, edited by Eke and Delpy.
Kluwer Academic / Plenum Publishers, New York, 1999.

minute pumps 70×70 ml = 4900 ml blood into the circulation. This approximately 5 L of blood contains 5×200 ml = 1000 ml O_2 in the arterial side, and 5×150 ml = 750 ml O_2 in the venous side. The difference between these two is the well known 250 ml, that is utilised in a minute by the body.

1.1.3. Traditional Blood Gas Interpretation in Intensive Care Patients. Following our university studies, in clinical practice the interpretation of blood gases is an important issue in the intensive care unit. To assess the condition of a patient from arterial and mixed venous pO_2, pCO_2, pH, sO_2, HCO_3^-, BE, together with all the additionally available confusing parameters like $avDO_2$, $avpO_2$, $AapO_2$, $pO_2(a - v)$, $sO_2(a - v)$, $pO_2(A - a)$, pO2(a/A), pO2(a)/pO2(v) seems difficult and not clear at all.

1.2. The Problem of Traditional Blood Gas Analysers

Traditional blood gas analysers measure pH, pCO_2, and pO_2 only. All other parameters are calculated, and therefore can be misleading. After measuring arterial pO_2 with a classic blood gas analyser sO_2 is calculated from an ideal oxygen dissociation curve that is built in the program of the analyser.

1.2.1. The "Ideal" Oxygen Dissociation Curve. Most blood gas analysers use a "built in" ideal oxygen dissociation curve to calculate the likely sO_2 from the measured pO_2. Some devices use conductivity as a principle to determine the haematocrit value and then calculate haemoglobin values from this, but only a few determine some of the dyshaemoglobins to correct the "ideal" oxygen saturation.

None of these methods give information about the position of the ODC that is crucial in all forms of respiratory and metabolic alkalosis and acidosis.

1.2.2. Millilitres of Oxygen Meet Grams of Haemoglobin. Traditionally 1 g haemoglobin can bind 1.39 ml O_2. When haemoglobin values are given in grams and oxygen content in millilitres, the fraction of oxygenated haemoglobin (the real saturation) is hard to determine at a glance.

1.2.3. Partial Pressures of Gases. When measuring partial gas pressures inside the body the generally used units are either mmHg or kPa, while the ambient pressure is usually indicated in mbars or in atm. Life is often made complicated for no other reason than tradition.

1.3. RADIOMETER's ABL-625 Blood Gas Analyser

The ABL-625 besides measuring the classic blood gas parameters, like pH, pCO2 and pO2 also measures tHb, cO_2Hb, cHHb, cCOHb, cMetHb, cSHb, cFHb, and calculates FO_2Hb, FHHb, FCOHb, FMetHb, FSHb, FFHb. Electrolyte (K^+, Na^+, Ca^{++}, Cl^-), glucose and lactate levels are also measured.

1.3.1. The Traditional Use of the RADIOMETER's ABL-625 Blood Gas Analyser. As both oxygen and haemoglobin is measured in mmol, fraction of oxygenated haemoglobin (FO_2Hb) and total haemoglobin (tHb) yields oxygen content (ctO_2). If oxygen content is measured from both arterial ($ctO_2(a)$) and mixed venous blood ($ctO_2(\bar{v})$) the difference gives the arterial—mixed venous oxygen content difference ($ctO_2(a - \bar{v})$), that when multiplied with cardiac output yields tissue oxygen consumption.

1.3.2. The "Deep Picture" of the RADIOMETER's ABL-625 Blood Gas Analyser. The ABL-625 with Siggaard Andersen's built in Oxygen Saturation Algorithm (OSA) is used to measure the arterial pH(a), $pCO_2(a)$, $pO_2(a)$, and tHb(a) (including $FO_2Hb(a)$, FHHb(a), FCOHb(a), FMetHb(a), FSHb(a), FFHb(a)), $ctO_2(a)$, p50(a), and to determine the new arterial parameters of oxygen extraction tension (p_x), extractable oxygen content (c_x) and the cardiac output compensation factor (Q_x). There is no need for central venous or pulmonary artery catheters to determine whether oxygen availability to the tissues is acceptable or not.

1.4. Cardiac Output Measurement

Obviously the "Traditional Use" still requires invasive techniques and rather complicated cardiac output measurements to determine oxygen consumption. This approach is often regarded as the gold standard, mainly due to the thought that the more invasive the techniques are, the more reliable information we gain.

1.4.1. Cardiac Output Measurement in Physiology Studies. In physiology studies cardiac output (CO) is usually determined with an invasive cardiac output measurement technique, like dye dilution, thermodilution and with the use of an aortic-flowmeter. They require either the insertion of a pulmonary artery catheter or opening of the chest to apply two flowmeter rings around the aorta.

1.4.2. Cardiac Output Measurement in the Intensive Care Unit. Similarly, in current intensive care practice it is usual to use pulmonary artery catheters. The aim of these devices is to determine mixed venous blood gas parameters, like $pO_2(\bar{v})$, $sO_2(\bar{v})$, $pCO_2(\bar{v})$, $pH(\bar{v})$.

An arterial line is inserted to monitor blood pressure that is displayed as a curve with minute updated numeric figures of systolic, diastolic and mean arterial pressures and trends of these figures available.

A central venous line is inserted via the jugular, subclavian or femoral veins to administer infusions and iv. medication. This line is often used to draw central venous blood samples to determine central venous blood gas values like $sO_2(cv)$, $pO_2(cv)$, $pCO_2(cv)$ that are obviously not equal to mixed venous values taken from the pulmonary artery $sO_2(\bar{v})$, $pO_2(\bar{v})$, $pCO_2(\bar{v})$.

To monitor cardiac output it has been almost obligatory in the past two decades to introduce a pulmonary artery catheter. The aim of this Swan-Ganz catheter is to determine cardiac output by the thermodilution technique, to measure pulmonary pressures and with the inflation of a balloon in a pulmonary end artery to estimate pulmonary capillary wedge pressure to gain information about left atrial and ventricular filling pressures.

Other invasive techniques have also been used like dye-dilution and continuous cardiac output monitoring using heaters in the catheter itself.

1.4.3. Non-Invasive Monitoring. Automatic blood pressure monitors are available to determine systolic, diastolic and mean arterial blood pressures in cases where invasive arterial monitoring is not required anymore. Trends of these 1 to 60 minute readings are also available.

Non-invasive cardiac output measurement techniques like measuring temperature drop along the extremities can be used to estimate peripheral vascular resistance and with blood pressure readings cardiac output could be determined.

Other non-invasive techniques using the Doppler principle seem to be reliable, but rather cumbersome to use them continuously.

The introduction of oesophageal Doppler heads seemed to have solved the problem of long term continuous monitoring, but the position of the head can change depending on patient position.

Contour analysis based on pulse wave form processing gives information on the local flow and pressure conditions of the actual extremity only.

Metabolic computers measuring O_2 consumption and CO_2 production from inspired and exhaled gases made a breakthrough in both cardiac output and oxygen consumption measurements.

Impedance technology as a principle has proved impressive, but early impedance cardiographs were rather unreliable due to insufficiencies in the measurement principles and software analysis.

2. METHODS

The combined use of RADIOMETER's ABL 625 Blood Gas Analyser, that applies the new blood gas parameters from Siggaard Andersen and ASKIT's NHMS yields a reliable and reproducible semi-invasive haemodynamic monitoring technique that provides trend records as well. Only an indwelling arterial catheter is required to take the blood samples and the rest is completely non-invasive.

2.1. RADIOMETER's ABL-625

This complex haemoxymeter, blood gas-, electrolyte-, glucose- and lactate analyser,[11] together with Siggaard Andersen's built in Oxygen Saturation Algorithm (OSA) is used to determine the arterial pH, pCO_2, pO_2 (classic blood gas parameters), ctHb, ceHb, FO_2Hb, FHHb, FCOHb, FMetHb, FSHb, FFHb, ctO_2, p50 (Haemoxymeter parameters) and p_x, c_x and Q_x values (Siggaard Andersen's newly introduced Oxygen Saturation Algorithm parameters) and K^+, Na^+, Ca^{++}, Cl^- (electrolytes), glucose and lactate levels. Due to the new parameters of p_x, c_x and Q_x, that are printed immediately with the blood gas results, there is no need for central venous or pulmonary artery catheterisation to determine whether oxygen availability to the tissues is acceptable.

The simple but ingenious solution to solve the problems of traditional blood gas analysers came from Siggaard Andersen after more than 10,000 Scandinavian blood gas samples were examined to find correlations between sO_2 and pO_2 versus ctO_2 and pO_2. In this study pO_2 and sO_2 had no correlation with ctO_2.

2.1.1. Oxygen Content versus Oxygen Saturation. With the help of a haemoxymeter that measures tHb, eHb, O_2Hb, HHb, COHb, MetHb, SHb and that can be corrected for FHb, one can easily tell the fraction of oxygenated hemoglobin (FO_2Hb), that is a real saturation parameter versus sO_2 where COHb, MetHb, SHb and FHb are ignored.

2.1.2. O_2 Molecules Meet Haemoglobin Molecules. As one haemoglobin molecule binds one oxygen molecule and if both oxygen and haemoglobin are measured in mmol/L, saturation and fraction of oxygenated haemoglobin can be plotted on the same "y" axis.

2.1.3. Partial Pressures of Gases. If pressures both inside and outside the body are indicated in kPa, calculations are made simple. The ambient pressure approximately is 100 kPa at sea level, which means that fractions of inspired oxygen given in % (FiO_2) also tells us the inspired O_2 pressure in kPa.

2.1.4. The "y" Axis. Plotting ctO_2 and ctHb instead of sO_2 on the "y" axis changed the principal approach to blood gas analysis. Now both pO_2 and ctO_2 together with ctHb are measured and p50 and the position of the ODC curve is determined, versus the measurement of pO_2 only and calculation of sO_2 from an ideal ODC curve in other blood gas machines. It means that the actual position of the ODC is determined for each and every blood sample. Haemoglobin (ctHb), effective haemoglobin (ceHb) and oxygen content (ctO_2) are indicated on the "y" axis against $pO_2(a)$ on the "x" axis.

2.1.5. Interpretation of the Clear Picture. Now the blood gas status of a patient can be easily interpreted. The difference between arterial ($ctO_2(a)$) and venous oxygen content ($ctO_2(\overline{v})$) tells the consumed oxygen per litre of blood. If cardiac output is measured, oxygen consumption (VO_2) can be determined.

2.2. The OSA and the "Deep Picture"

Siggaard Andersen's Oxygen Saturation Algorithm (OSA) program introduced three new arterial parameters, the oxygen extraction tension (p_x), the extractable oxygen content (c_x) and the cardiac output compensation factor (Q_x) into blood gas analysis. He also introduced the term of oxygen availability where the association-dissociation relation between oxygen and haemoglobin is also indicated, versus the classical term of oxygen delivery (DO_2) where only the amount of bound haemoglobin is indicated. This new concept of blood gas analysis is called the "Deep Picture".

2.2.1. Measuring ctO_2, ($ctHb = FO_2Hb + FHHb + FCOHb + FMetHb + FSHb + FFHb$). It is not just the question of plotting different parameters, but measuring versus calculating that makes the principal difference.

If oxygen saturation is calculated from arterial pO_2 the result can be acceptable only in ideal situations which is definitely not the case in the intensive care unit.

2.2.2. ceHb: Effective Haemoglobin Concentration. Effective haemoglobin concentration is the amount of haemoglobin that can participate in O2 association and dissociation. ceHb = ctHb − (FCOHb + FMetHb + FSHb + FFHb)

2.2.3. p50 and the Position of the ODC. If all haemoglobin and dyshaemoglobin components are determined with a photometer and plotted on the "y" axis versus the pO_2 readings on the "x" axis, and these measurements are also performed at half saturation, p50 then the position of the actual patient's actual ODC curve is determined.

2.2.4. The Real Arterio-Venous Oxygen Content Difference ($ctO_2(a-\bar{v})$). If ctO_2 and ctHb (including FO_2Hb + FHHb + FCOHb + FMetHb + FSHb + FFHb) are measured in both arterial and mixed venous blood, the difference of the ctO2(a) and $ctO_2(\bar{v})$ will give the real $ctO_2(a-\bar{v})$ versus the calculated value from tHb measurements and saturation calculations.

2.2.5. Arterial Oxygen Availability. This term makes clear the difference between the "Traditional Blood Gas Analysis" and the "Deep Picture Concept". The available arterial oxygen is the amount of oxygen that is really available for the tissues, in other words this is the amount of oxygen that can dissociate from haemoglobin. This is defined by p_x, from pO_2, cteHb and the position of the ODC.

2.2.6. p_x or $pO_2(x)$: Arterial Oxygen Extraction Tension. In principle p_x is the calculated pO_2 in arterial blood after the extraction of 2.3 mmol O_2 from 1 L arterial blood. These calculations are based on the determined position of the ODC, that makes p_x comparable to $pO_2(\bar{v})$ (mixed venous pO_2) in normal oxygen consumption.

2.2.7. c_x or $ctO_2(x)$: Extractable Arterial Oxygen Content. In principle c_x is the calculated amount of oxygen that can be extracted from arterial blood at normal p_x value (5.1 kPa), that is comparable to normal $pO_2(\bar{v})$.

2.2.8. Q_x: Cardiac Output Compensation Factor. The combined ratio between the actual p_x and the normal p_x of 5.1 kPa and the actual c_x and the normal c_x of 2.3 mmol/L gives Q_x. This value will inform us about the extent by which we have to improve cardiac output to achieve acceptable tissue oxygenation with the given respiratory and metabolic conditions.

2.2.9. Lactate. Following diminished peripheral perfusion of tissues due to increase in systemic vascular resistance, anaerobic processes take place that result in lactate production. Following open heart surgery in extra-corporeal circulation and hypothermia these lactate readings can be extremely high.

2.2.10. Interpreting the "Deep Picture" with the new parameters: p_x, c_x, Q_x. One can question why these c_x and p_x values are comparable to the real $ctO_2(a-\bar{v})$ and $pO_2(\bar{v})$. They are definitely not equivalent to these real parameters. But if Q_x is also involved one can improve cardiac output, to the extent as determined by Q_x, to achieve the goal of a normal balance between oxygen supply and demand. To make it simple, if Q_x is 1.0, it also means that p_x and c_x are equivalent to $pO_2(\bar{v})$ and $ctO_2(a-\bar{v})$. If Q_x is 1.8, it means that we have to improve cardiac output by 1.8 times the actual value to achieve a condition where p_x and c_x will be equivalent to $pO_2(\bar{v})$ and $ctO_2(a-\bar{v})$.

If one static parameter alone is to be chosen, that describes oxygen availability to the tissues at a given moment, it should be p_x. If p_x is in the normal range, we can be assured that the patient's global tissue oxygenation is acceptable. On the other hand, if dynamic information is requested to follow-up improvement or deterioration in the patient's condition one has to monitor and trend some other parameters as well. If p_x is in the normal range (that is individually defined for the actual patient by the OSA program and printed in the blood gas report) the modified VO_2 measurements are $VO_2 = c_x * CO$. If p_x is not in the normal range, then cardiac output has to be corrected according to the Q_x value. For the safety of the patient, lactate measurements are always there to check if anaerobic processes are present. In other words the "Deep Picture" gives information about oxygen availability to the tissues without the use of pulmonary artery catheters and mixed venous blood sample data.

2.3. ASKIT's NHMS

2.3.1. The Principle of the Non-Invasive Haemodynamic Monitor System. ASKIT's NHMS, a Non-invasive Haemodynamic Monitor System uses parallel ECG (electrocardiogram), ICG (impedance cardiogram), PCG (phono-cardiogram), ABPM (automatic blood pressure measurements) and a unique software that calculates beat to beat values and displays all three curves with the blood pressure readings that can easily be called back continuously, together with the first derivative of the ICG, the dz/dt curve. Though in principle this device is a non-invasive haemodynamic monitor it is also capable of accepting and displaying continuous blood pressure readings from invasive blood pressure monitors.[13,14,15]

2.3.2. The Electro-Mechanical Components of Systole. Thanks to the exact timing of the electro-mechanical components with the help of the combined analysis of ECG, ICG and PCG, the PEP (pre-ejection period = from the beginning of the Q-wave to the second component of the first heart sound), the VET (ventricular ejection time = from the second component of the first heart sound to the first component of the second heart sound) and the QS2 (complete electro-mechanical systole = PEP + VET) can be continuously recorded and displayed.

2.3.3. The First Derivative Curve (dz/dt) of the Impedance Cardiogram. The dz/dt curve represents flow of blood in the chest, from the inlet of the inferior and superior cava veins to the outlet of the great vessels of the aortic arch and the diaphragmatic outlet of the aorta. It gives information about the changes in thoracic blood content. The shape of this derivative curve is comparable to aortic and carotid artery flow patterns. From the height and the steepness of the individual cycles of the dz/dt curve, stroke volume is determined on a beat to beat basis. Cardiac output is determined from stroke volume and heart rate. Furthermore the shape of the dz/dt curve contains all the information regarding flow and contractility, in the same way as the information in an ECG curve is much more than the measurable heights and distances in it.

2.3.4. "Long Term Monitoring" with ASKIT's NHMS. Following open heart surgery, patients are monitored continuously by ASKIT's NHMS "long-term monitoring" program. This Non-invasive Haemodynamic Monitoring System (NHMS), based on combined impedance cardiographic (ICG), electrocardiographic (ECG), phonocardiographic (PCG) and automatic blood pressure monitoring (ABPM), is used for continuous, "long-term monitoring" and measurement of postoperative stroke volume (SV), cardiac output (CO), systolic-, diastolic- and mean arterial pressures (PS, PD, MABP), systemic vascular resistance (SVR), pre-ejection period (PEP), ventricular ejection time (VET), electro-mechanical systolic time (QS2) and all their indexes, rate pressure product (RPP), diastolic and systolic peak dz/dt ratio (D/S), ECG R peak and dz/dt_{max} index (RZ) and acceleration index (ACI).

2.3.5. Displayed Curves of ECG, ICG, PCG. ECG, ICG, PCG are continuously displayed on the screen.

2.3.6. Numeric Display of Z0, CO, SVR and PEP. Four, alarm limited, continuously updated numeric parameters are displayed on the screen. In our practice Z0 (basic impedance), CO (cardiac output), SVR (systemic vascular resistance) and PEP (pre-ejection period) are displayed on the screen and continuously updated according to the last 9 cardiac cycles. (Any 4 parameters can be displayed, but must be defined before the start of "Long-Term Monitoring").

2.3.7. Parameter Trends. Trends of any 4 parameters can be viewed at a time either in numeric or in graphic form. Change of trend selection can be made even during the "long-term monitoring" process.

2.3.8. The xxx.wks File. The measured data are all available in trends and if saved in xxx.wks format they are ready for further analysis in Excel or Works Spreadsheets.

3. RESULTS

A male patient was monitored for 16 hours following coronary bypass surgery. From ASKIT's NHMS the saved xxx.wks files of this patient were imported to the Spreadsheet of MS.Works 3.0 and these were further imported to Powerpoint in MS.Office 97 to create diagrams for the examination of his 16 hour trends of different parameters. Non-invasive blood pressure measurements were automatically set to 15 minutes. All related parameters were also updated according to these readings. The monitored parameters are grouped according to their function and values that are comparable in one chart. The blood gas parameters were obtained from the ABL-625 and the measured and calculated parameters were included in the MS.Works 3.0 Spreadsheet according to the sampling times of the arterial blood that was marked with the "event key" of the ASKIT's NHMS.

3.1. Figure 1: Basic Impedance and Blood Pressure and Heart Rate

This diagram shows the trends of Z0, MABP, PS, PD, HR and SV.

3.1.1. Z0 (Basic Impedance). Basic impedance has not changed significantly which means that the patient's overall fluid balance was normal.

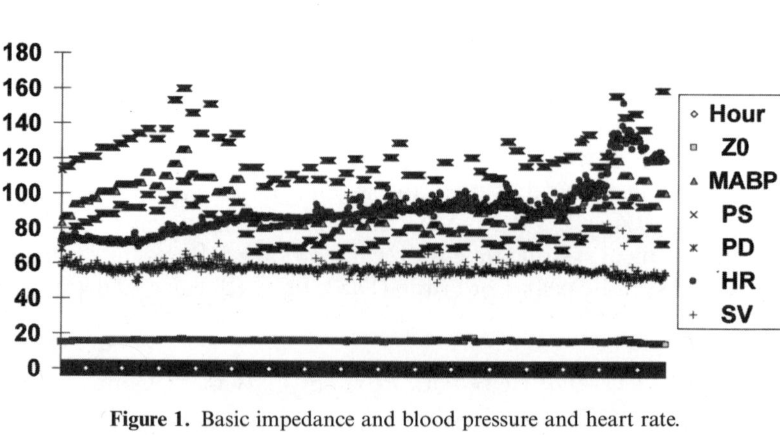

Figure 1. Basic impedance and blood pressure and heart rate.

3.1.2. MABP (Mean Arterial Blood Pressure). There is a peak in mean arterial blood pressure due to an increase in systemic vascular resistance that is predominantly caused by reduced body temperature following the relatively short re-perfusion period following ECC. As this patient was given combined midazolam and fentanyl anaesthesia this early rise in blood pressure also falls in the range of recovering from midazolam. Following the first peak there is a gradual rise within the next 8 hours with a second peak just before the end of monitoring. This later falls into the range of full recovery from anaesthesia including the elimination of fentanyl and extubation.

3.1.3. PS (Systolic Blood Pressure). Generally the same applies as for MABP.

3.1.4. PD (Diastolic Blood Pressure). Generally the same applies as for MABP.

3.1.5. HR (Heart Rate). There is a continuous gradual increase in heart rate from 75/min to 90/min with a sharp increase at the end of monitoring, that falls within the period of extubation.

3.1.6. SV (Stroke Volume). Stroke volume shows a gradual decrease from 60 ml to 53 ml.

3.2. Figure 2: Contractility and Pre-Ejection Period

This diagram shows the trends of HR, SV, LVSERI and PEP.

3.2.1. LVSERI (Left Ventricular Systolic Ejection Rate Index). This is a contractility parameter that is derived from a per se three dimensional parameter, so it describes the improvement of contractility during the course of this early postoperative recovery. Though SV is reduced from 60 ml to 53 ml, the contractility of the myocardium is gradually improving.

3.2.2. PEP (Pre-Ejection Period). The electric component of the electromechanical systole shortens to a small extent from 120 msec to 105 msec. It is worth looking at the period of increased heart rate and blood pressure readings during the 15th hour when due to full recovery from anaesthesia and the last minutes with the endotracheal tube in, the adrenaline outflow reduced PEP to around 80 ms.

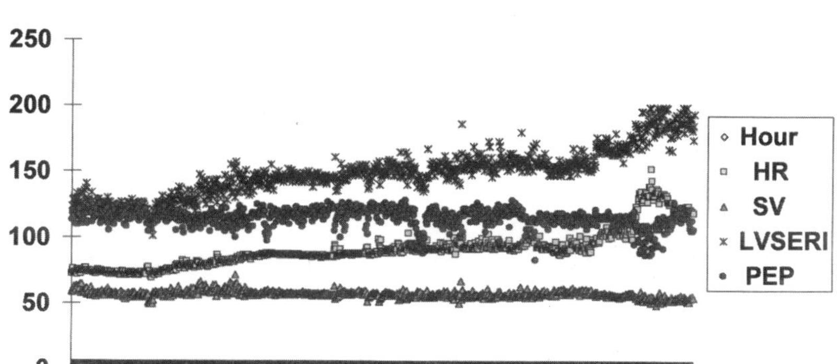

Figure 2. Contractility and pre-ejection period.

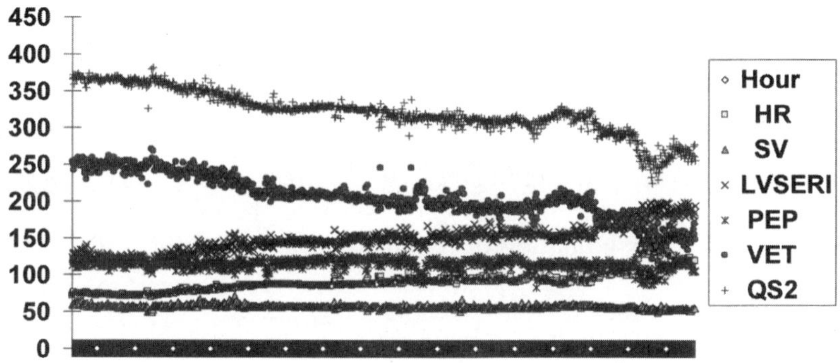

Figure 3. Contractility and electro-mechanical components.

3.3. Figure 3: Contractility and Electro-Mechanical Components

This diagram shows the trends of HR, SV, LVSERI, PEP, VET and QS2.

SV, PEP, VET and QS2 gradually decrease while LVSERI and HR gradually increase, in a way that are consistent with the improvement of myocardial function, due to both revascularisation and recovery from anaesthesia.

3.4. Figure 4: Contractility and Myocardial Oxygenation

This diagram shows the trends of P/V, A2/A1, D/S and ACI.

3.4.1. P/V (PEP/VET). A minimal increase from 0.45 to 0.5 can be observed that is due to a reduction of VET to a larger extent than PEP. This also means that the risk of arrhythmia is relatively low, in spite of the fact that PEP in itself also diminished.

3.4.2. A2/A1 (Regurgitation or Dyskinesis Index). This index if bigger than 0.85 suggests that there is either regurgitation (due to either aortic or mitral valve incompetence) or dyskinesis of the heart. It is worth looking at the 3 and 4 hour period of the 3rd, 4th , 5th and 8th, 9th, 10th, 11th hours when there were dyskinetic periods due to an increase in SVR, MABP, PS, PD and the start of full recovery from anaesthesia when adrenaline levels started to rise. These periods are parallel with the increase in D/S value that suggests myocardial oxygen supply insufficiency.

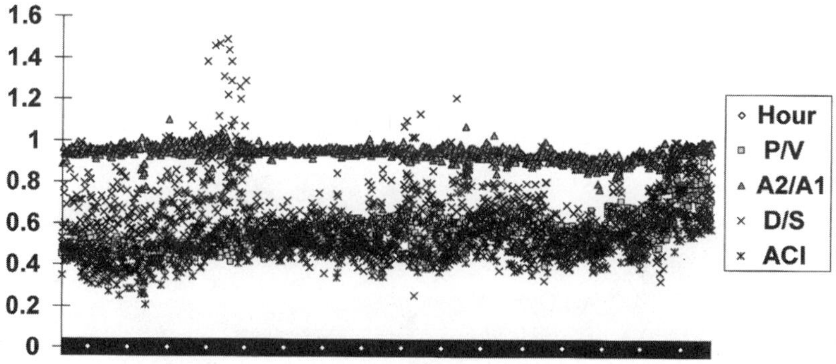

Figure 4. Contractility and myocardial oxygenation.

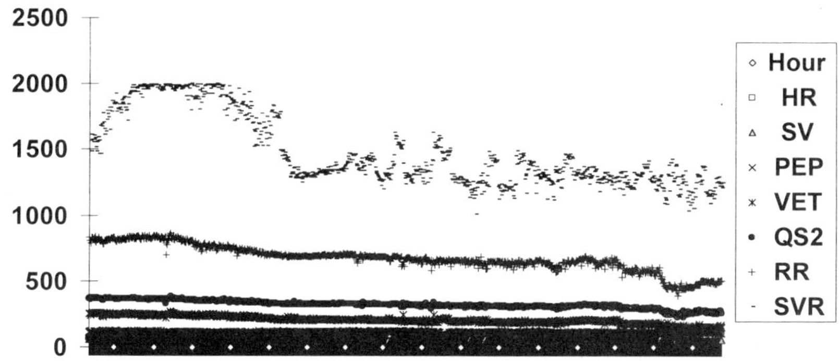

Figure 5. Contractility and systemic vascular resistance.

3.4.3. D/S (Diastolic Peak / Systolic Peak of the dz/dt Curve). If there is an increase in the height of the diastolic component of the dz/dt curve it means that the end diastolic filling pressure is high. It leads to dyskinesis that is an earlier sign of myocardial oxygen supply insufficiency than ST-segment changes in an ECG, than the patient complaints about angina, or than any kinetic changes in echocardiography.

3.4.4. ACI (Acceleration Index). Acceleration is a three dimensional parameter in impedance cardiography that together with LVSERI informs about myocardial function. Values of ACI decrease when there is an increase in A2/A1 or D/S.

3.5. Figure 5: Contractility and Systemic Vascular Resistance

This diagram shows the trends of HR, SV, PEP, VET, QS2, RR and SVR.

3.5.1. SVR (Systemic Vascular Resistance). There is a sharp increase in SVR due to a body temperature drop following the relatively short re-perfusion period after the ECC at 28°C.

3.5.2. RR (R-R distance in the ECG). With the increase in heart rate there is a parallel decrease in RR.

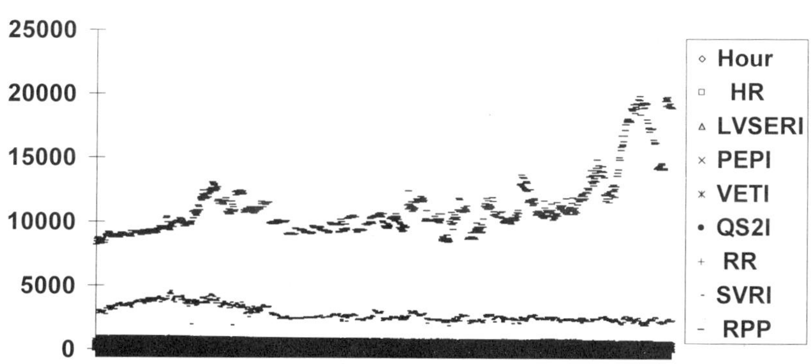

Figure 6. Contractility and rate pressure product and systemic vascular resistance.

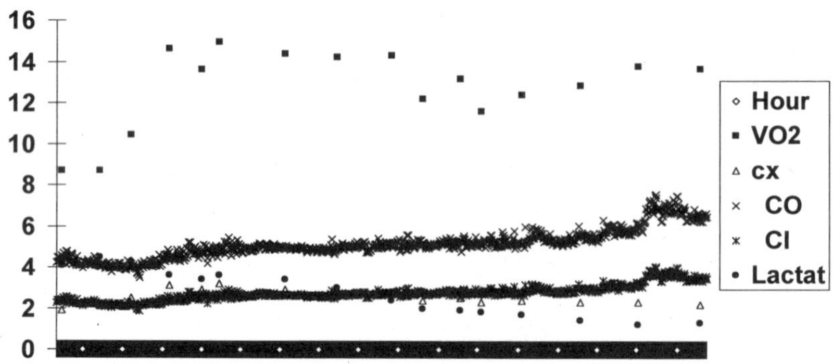

Figure 7. Cardiac output and oxygen consumption and lactate.

3.6. Figure 6: Contractility and Rate Pressure Product and Systemic Vascular Resistance

3.6.1. SVRI (Systemic Vascular Resistance Index (Easier to Compare with RPP)). There is a peak of 4000 dyn.sec.cm^{-5} that is parallel with the first peak in RPP increase. It is important that at the second peak of RPP there is no increase in SVR and SVRI.

3.6.2. RPP (Rate Pressure Product (HR × PS)). There is a parallel peak of RPP with SVR and SVRI increase, caused by a dominant increase in systolic blood pressure (PS), while there is no increase in SVR and SVRI at the second RPP peak. This later RPP increase is predominantly caused by an increase in HR.

3.7. Figure 7: Cardiac Output and Oxygen Consumption and Lactate

In this chart a gradual increase in cardiac output (CO) and cardiac index (CI) is seen with a low VO_2 at the beginning of the early postoperative period. This low VO_2 increases with opening of the peripheral vascular beds and remains high due to an oxygen deficit caused by an increase in peripheral vascular resistance in the early postoperative period that also resulted in anaerobic metabolism and lactate production. By the time of extubation VO_2 almost came back to normal, but adrenaline outflow caused a second increase in oxygen consumption. Lactate levels were high during the first hours, and they were back to normal after the oxygen deficit was settled.

4. DISCUSSION

Following open heart surgery, there is a critical 24–48 hour period when thorough monitoring is required to maintain a satisfactory oxygen supply for both the heart and all other tissues of the body. To properly control this situation it is important to be continuously aware of all the haemodynamic changes (including SV, HR, CO, SVR, PEP, VET, ACI) and the globally available oxygen for the tissues[3,4,5,6] (that is determined not only by the classic $pO_2(a)$, and $sO_2(a)$ figures, but the ceHb, $ctO_2(a)$ and the position of the ODC[12] as defined by the p50). Calculating with all these parameters in mind is time consuming and requires an accurate mathematical approach.[7,8,9,10] The RADIOMETER ABL-625 together with the above parameter's automatically displays Siggaard Andersen's oxygen extraction tension or p_x value (an all-in-one indicator of pO_2, ctO_2 and p50 of the ODC[4]), that with the extractable oxygen or c_x value will determine the compensation factor Q_x, that indicates what haemodynamic changes are required to achieve satisfactory tissue oxygenation at the given respiratory condition. Based on this Qx value, inodilator therapy (dopamine, dobutamine, dopexamine, adrenaline, noradrenaline or PDE-inhibitors) is used to improve blood flow and MAP to optimise afterload for the heart and perfusion for coronary, cerebral, renal, mesenteric and other vascular beds. With the continuously displayed PEP, VET, SV, CO, MAP and SVR the infusion speed of the syringe pumps containing inodilator agents can be optimised. If further increase in a given inodilator agent or a combination of them will not improve CO and px, but reduces the PEP, it is not worth further increasing the speed of the pump as the $\alpha 1$, $\beta 1$, $\beta 2$ and δ receptor stimulation of these agents will only provoke arrhythmia and myocardium oxygen supply deficiency. Instead of risking the patients myocardium and general condition with trial and error, the treating physician can decide immediately at the bedside to stop further increasing the dose of inotropic and vasodilator agents.

4.1. Conclusions

The aim of this concerted respiratory and inodilator therapy, continuously monitored by the ABL-625 and the NHMS is to achieve a p_x in the reference range and a Q_x close to 1.0 with low lactate levels. If these goals are met and the patient's clinical condition is satisfactory, we can be satisfied with the postoperative treatment of our patient.

ACKNOWLEDGMENTS

I would like to say thanks to those who provided me with excellent material and critical comments to complete this paper: Dr. Ivar Gothgen, Head of Intensive Care Unit, University Hospital, Copenhagen; Denmark, Ms. Inge Birkemose Møller, Radiometer International, Copenhagen, Denmark; Mr. János Hegyi, WIP-Radiometer, Hungary; Mr. Imre Jakobicz, General Manager, ASKIT, Hungary; Mr. János Lakatos, Technical Manager, ASKIT, Hungary.

Blood Gas Parameters (in order of appearance in the text):

VO_2	Oxygen consumption
$ctO_2(a)$	Arterial oxygen content
$ctO2(\bar{v})$	Mixed venous oxygen content
$ctO_2(a - \bar{v})$	Arterial-mixed venous oxygen content difference
pO_2	Partial pressure of oxygen
pCO_2	Partial pressure of carbon dioxide
pH	Negative logarithm of hydrogen ion concentration
sO_2	Oxygen saturation
HCO_3^-,	Bicarbonate
BE	Base excess
$avDO_2$	Arterio-venous oxygen difference (in haemodynamic calculations)
$avpO_2$	Arterio-venous oxygen tension difference
$pO_2(a - v)$	Arterio-venous oxygen tension difference
$sO_2(a - v)$	Arterio-venous oxygen saturation difference
$pO_2(A - a)$	Alveolo-arterial oxygen tension difference

$AapO_2$	Alveolo arterial oxygen tension difference
pO2(a/A)	Ratio of arterial and alveolar oxygen tension
ctHb	Concentration of total haemoglobin
ceHb	Concentration of effective haemoglobin
cO_2Hb	Concentration of oxygenated haemoglobin
cHHb	Concentration of reduced haemoglobin
cCOHb	Concentration of carboxy-haemoglobin
cMetHb	Concentration of met-haemoglobin
cSHb	Concentration of sulf-haemoglobin
cFHb)	Concentration of foetal haemoglobin
FO_2Hb	Fraction of oxygenated haemoglobin
FHHb	Fraction of reduced haemoglobin
FCOHb	Fraction of carboxy-haemoglobin
FMetHb	Fraction of met-haemoglobin
FSHb	Fraction of sulf-haemoglobin
FFHb)	Fraction of foetal haemoglobin
ctO_2	Oxygen content
p50	Oxygen tension at half saturation
$p_x = (pO_2(x))$	Oxygen extraction tension in arterial blood
$c_x = (ctO_2(x))$	Extractable oxygen content from arterial blood
Q_x	Cardiac output compensation factor
$sO_2(cv)$	Central venous oxygen saturation
$pO_2(cv)$	Central venous oxygen tension
$pCO_2(cv)$	Central venous carbon-dioxide tension
$sO_2(\bar{v})$	Mixed venous (pulmonary artery) oxygen saturation
$pO_2(\bar{v})$	Mixed venous (pulmonary artery) oxygen tension
$pCO_2(\bar{v})$	Mixed venous (pulmonary artery) carbon-dioxide tension

Non-Invasive Haemodynamic Parameters (in order of appearance in the text):

NHMS	Non-invasive Haemodynamic Monitoring System
ICG	impedance cardiograph
ECG	electrocardiograph
PCG	phonocardiograph
ABPM	automatic blood pressure monitor
SV	stroke volume
CO	cardiac output
CI	Cardiac Index
PS	Systolic Arterial Blood Pressure
PD	Diastolic Arterial Blood Pressure
MABP	Mean Arterial Blood Pressure
SVR	Systemic Vascular Resistance
PEP	Pre-Ejection Period
VET	Ventricular Ejection Time
QS2	Electro-Mechanical Systole Time
RPP	Rate Pressure Product (HR * PS)
D/S	Ratio of Diastolic Peak dz/dt and Systolic Peak dz/dt
RZ	ECG R peak and dz/dt_{max} index
ACI	Acceleration Index

Z0	Basic Impedance
LVSERI	Left Ventricular Systolic Ejection Rate Index
P/V	Ratio of electric and mechanical systole (PEP/VET)
A2/A1	Regurgitation or Dyskinesis Index

REFERENCES

1. Luce JM. Hemodynamic and respiratory monitoring in critical care medicine. In: Kelley WN, ed. *Textbook of Internal Medicine.* Philadelphia: J.B. Lippincott Company, 1989:2016–2021.
2. Rackov EC, Astiz ME, and Weil MH. Cellular oxygen metabolism during sepsis and shock. *JAMA* 1988;259:1989–1993.
3. Russell JA and Phang PT. The oxygen delivery/consumption controversy, Approaches to management of the critically ill. *Am J Respir Crit Care Med* 1994;149:533–537.
4. Pinsky MR. Beyond global oxygen supply-demand relations: in search of measures of dysoxia. *Intensive Care Med* 1994;20:1–3.
5. Samsel RW, Schumacker PT. Pathological supply dependence of oxygen utilization. In: Principles of Critical Care. Hall JB, Schmidt GA, and Wood LDH (Eds.). New York, McGraw-Hill, 1992:667–678.
6. Schumacker PT, Cain SM. The concept of a critical oxygen delivery. Intensive Care Med 1987;13: 223–629.
7. Siggaard-Andersen O, Gøthgen IH, Wimberley PD, and Fogh-Andersen N. The oxygen status of the arterial blood revised: relevant oxygen parameters for monitoring the arterial oxygen availability. *Scand J Clin Lab Invest* 1990; 50, Suppl 203:17–28.
8. Siggaard-Andersen O and Siggaard Andersen M. The oxygen status algorithm: a computer program for calculating and displaying pH and blood gas data. *Scand J Clin Lab Invest* 1990; 50, Suppl 203:29–45.
9. Siggaard-Andersen O, Wimberley PD, Fogh-Andersen N, and Gøthgen IH. Arterial oxygen status determined with routine pH/blood gas equipment and multiwavelength hemoximetry; reference values, precision and accuracy. *Scand J Clin Lab Invest* 1990; 50, Suppl 203:57–66.
10. Siggaard-Andersen O, Gøthgen IH, Fogh-Andersen N, and Larsen VH. Oxygen status of arterial and mixed venous blood. *Crit Care Med* 1995;23:1284–1293.
11. Vincent J-L, Dufaye P, Berre J, Leeman M, Degaute J-P, and Kahn RJ. Serial lactate determinations during circulatory shock. *Crit Care Med* 1983;11:449–451.
12. Willis N, Clapman MCC, and Mapleson WW. Additional blood-gas variables for the rational control of oxygen therapy, with allowance for shifts of the oxygen dissotiation curve. *Br J Anaesth* 1989;59: 1160–1170.
13. Kubicek WG, Karnegis JM, Patterson RP, Witsoe DA, and Mattson RH. Development and evaluation of an impedance cardiac output system. *Aerospace Med* 1966;37:1208–1212.
14. Gömöry A, Horváth S, Thöring J, Fésüs L, Asbóth R, and Naszlady A. Keringési parameterek non invazív meghatározása és követése az impedancia kardiográfia módszerével. *Orvosi Hetilap* 1988;129:2083–2089.
15. Horváth I, Mezey B, Juricskay I, Simon A, and Jávor T. Data for the clinical evaluation of impedance cardiography measurements. *Cardiologia Hungarica Suppl.* 1993:29–32.

56

NOREPINEPHRINE VS NITRIC OXIDE SYNTHASE INHIBITION IN ENDOTOXEMIA

Klaus Wagner,[1] Reiner Schäfer,[1] Angela Roth-Isigkeit,[2] Karl-F. Klotz,[1] Hans-Jürgen Friedrich,[2] and Peter Schmucker[2]

[1]Department of Anesthesiology
[2]Institut für medizinische Biometrie und Statistik
Medical University Lübeck,
Germany

1. INTRODUCTION

The mortality rate in sepsis reemains high despite many therapeutic efforts. Initial treatment of sepsis, in addition to surgical treatment of the underlying cause, is aimed to prevent circulatory shock leading to multiple organ failure, including liver failure. Therefore, vasopressor agents such as norepinephrine (NE) are frequently administered to restore vascular tone and improve tissue perfusion (Vincent et al., 1991).

Some findings in sepsis and endotoxemia, like hypotension and vasodilation, are thought to be mainly due to increased nitric oxide (NO) production (Evans et al., 1993; Schneider et al., 1993; Lorente et al., 1993) controlled by the enzyme NO synthase. Inhibition of NO synthase leads to vasoconstriction and a rise in arterial pressure in normal as well as endotoxemic humans and animals (Kilbourne et al., 1990; Petros et al., 1994).

Although the administration of NO synthase inhibitors in endotoxemia appears to be the logical consequence of this physiological and pathophysiological knowledge, data on the benefit of this procedure were inconsistent with regard to cardiac output (Klabunde et al., 1991) and survival rate (Wright et al., 1992). In addition, in previous experiments we showed that NO synthase inhibition led to a significant hypoxia in skeletal muscle (Pohl et al., 1993). A large dose of NO synthase inhibitor which completely blocked NO synthase, led to circulatory shock with adverse effects on outcome, while this was rarely seen when reduced amounts of NO synthase inhibitor were administered.

Therefore, a prerequisite of this study was to identify a dose of NO synthase inhibitor that would reverse some changes of hyperdynamic endotoxemia and restore

Oxygen Transport to Tissue XXI, edited by Eke and Delpy.
Kluwer Academic / Plenum Publishers, New York, 1999.

(approximate) normal values without circulatory shock. It was considered undesirable to measure effects of circulatory collapse caused by large amounts of NO synthase inhibitor since these findings would have no counterpart in clinical practice.

Since the liver is frequently affected in multiorgan failure in sepsis, we were particularly interested to investigate the influence of conventional circulatory support with NE administration in comparison to a NO synthase inhibitor on liver circulation and liver oxygen metabolism in sepsis.

Therefore the aim of this experimental study was to measure the effect on liver blood flow, liver oxygen metabolism and liver tissue pO_2 of NE compared with NO synthase inhibition with L-NAME in hyperdynamic endotoxemia.

2. MATERIAL AND METHODS

The porcine hyperdynamic endotoxemia model used in these studies was modified from the one first described by Breslow (Breslow et al., 1987) Female pigs of landrace breed (weight 19–25 kg) were anesthetised with intramuscular ketamine (10 mg/kg bw), benzodiazepine (1 mg/kg), and 0.5 mg atropine. The animals were tracheotomized and anesthesia was maintained with i.v. thiopentone (100–200 mg/hr), fentanyl (0.2–0.5 mg/hr), and pancuronium (5 mg/hr). The animals were normoventilated with an EVA ventilator (Dräger, Lübeck, Germany) and the FIO_2 adjusted to keep an arterial hemoglobin saturation between 97% and 99%. Fluid loss was replaced by normal saline and hydroxy-ethyl-starch infusion to maintain normal values of central venous pressure and pulmonary capillary wedge pressure. All animals were in the supine horizontal position.

A cannula was inserted into the femoral artery, the right jugular internal vein was exposed and a Swan-Ganz pulmonary artery catheter advanced to wedge position. A median laparotomy was performed, the common hepatic artery and portal vein were carefully isolated, preserving the hilar nerve plexus. An appropriately sized flow probe was attached to the portal vein. The common hepatic artery was exposed and all branches bypassing the liver parenchyma were carefully ligated. A second flow probe was attached to the common hepatic artery. The abdominal superior vena cava was punctured in cranio ventral to dorso caudal direction and a plastic cannula advanced into a hepatic vein. When all surgical procedures were completed, the abdominal wall was reapproximated and towels were placed on the surface to minimize heat loss. The position of the cannula in the hepatic vein was verified in all cases by autopsy.

Blood flow of the portal vein was measured with an electromagnetic flow meter (Narcomatic RT 500, Narco Bio-Systems, Housten, Texas, USA). Blood flow of the hepatic artery was measured with an ultrasonic flow meter (Burton and Gorewit, 1984, Transonic System Inc, Ithaca, USA). Tissue pO_2 was measured with a computer controlled polarographic pO_2 needle probe (pO_2 Histograph, Eppendorf-Netheler-Hinz GmbH, Hamburg) (Vaupel, 1991). The tip diameter of the pO_2 probe was 250 μm, the polarisation voltage was −700 mV (against a Ag/AgCl calomel electrode). Single pO_2 values were measured in a step-wise fashion, where the needle probe was rapidly advanced 1.0 cm and withdrawn for 0.3 cm as to reduce tissue compression at the tip of the pO_2 probe (Schramm et al., 1992; Wagner et al., 1992). Thereafter a data collection time of 1.2 seconds ensued. This sequence was repeated until 8 to 12 single pO_2 values were measured in a continuous path through the parenchyma, thereafter the pO_2 probe was withdrawn and repositioned to avoid traversing previously measured tissue. In this way 80 to

100 single pO_2 values were recorded and plotted as a pO_2 histogram. Tissue temperature was determined with a thermocouple (Philips, type K, tip diameter 260 μm). The pO_2 probes were calibrated before and after a pO_2 histogram was measured. pO_2 data were corrected for changes in pO_2 sensitivity, temperature and probe drift. Blood samples were analysed for hemoglobin content and O_2 saturation with an OSM3 hemoximeter set to animal mode swine (Radiometer Copenhagen). Values of hepatic oxygen delivery (DO_2) and hepatic oxygen consumption (VO_2) were calculated according to the formula of Nöldge et al. (1992). Standard monitoring equipment and pressure transducer systems (Sirecust, Siemens, Germany) were used for measurement of systolic, diastolic blood pressure and mean arterial pressure (MAP), pulmonary artery pressure (PAP), pulmonary capillary wedge pressure (PCWP), central venous pressure (CVP). Cardiac output was measured according to the thermodilution principle and performed in triplicate for each measurement block. From these primary data, cardiac index (CI) and systemic vascular resistance (SVR) were calculated by the monitoring device.

Preliminary experiments were performed to establish the dose of NE and L-NAME. In these preliminary experiments, animals (n = 3) received 0.025 μg/kg/min LPS (strain O127:B8; Sigma, Germany) for 20 minutes to test the effects on systemic hemodynamics without supportive NE administration. All three animals developed profound hypotension leading to septic shock and death within 60 minutes. Thus, the possibility of investigating a group of untreated endotoxemic animals with a MAP between 80 and 100 mm Hg was precluded. In a set of animals (n = 5) L-nitro arginine methyl ester (L-NAME, Serva Germany) was given immediately after LPS administration was stopped instead of the temporary NE infusion. Several different doses and administration regimens of L-NAME (up to 30 mg/kg) given as a bolus injection or continuous infusion (10 mg/kg/min) were unsuccessfully tested, since all animals died in circulatory shock. Only temporary circulatory support with NE administration kept the animals alive.

LPS administration was stopped (cut-off criteria for LPS) when systolic pulmonary artery pressure lay above 60 mm Hg or MAP fell below 80 mm Hg, whichever came first.

2.1. Experimental Protocol

The study comprised three groups. The NE group (n = 7), the L-NAME group (n = 5) and a control group (n = 7). In all animals, 30 minutes were allowed for hemodynamic stabilization after the surgical preparation. Baseline data were recorded at time 0 (T0). Thereafter an infusion of LPS (0.05 μg/kg/min) was started in the NE group and the L-NAME group, the control group received no LPS. LPS was stopped when cut-off criteria for LPS infusion were reached. The animals of the NE group and the L-NAME group were primarily stabilised by fluid administration and temporarily supported with NE infusion (0.06 ± 0.05 μg/kg/min, Table 2). The second measurement block (T20) was 20 minutes after T0. In the NE group treatment was continued with NE and the dose adjusted to keep MAP between 70 and 100 mm Hg. In the L-NAME group L-NAME was infused at a rate of 1 mg/kg/min for 4 minutes and the NE infusion was stopped. The control group received no NE. Further measurement blocks were at t = 40, t = 60, t = 90 and t = 120 minutes.

For each measurement block the following set of data was obtained: blood samples were drawn from the femoral artery, pulmonary artery, portal, and hepatic vein. Each of these samples was analysed for Hb content and Hb saturation, lactate, GOT/AST. A liver pO_2 histogram was recorded from 80 to 100 single pO_2 values and liver temperature was measured. Flow readings of the hepatic artery and the portal vein were taken. The circulatory parameters described above were read and CI averaged from three measurements.

2.2. Statistics

To test for differences within each group the Friedman-test was used as a global test followed by the Wilcoxon-Wilcox test if the p-value of the global test indicated a significant difference. The tests for differences between the groups were the H-test of Kruskal-Wallis (global) and the Nemenyi-test as the detailed one. Differences were considered significant at $p < 0.05$.

3. RESULTS

In the NE group at t = 120 minutes SVR was significantly decreased and CI was significantly increased. This inverse relation of the two parameters, characteristic of a hyperdynamic state in septicemia, was first observed at t = 40 and remained unchanged throughout the experiment. Circulatory changes in the L-NAME group were similar at t = 40 minutes and t = 60 minutes (data not shown) indicating a hyperdynamic state. During the later course of the experiment SVR gradually increased and CI declined in the L-NAME group, findings consistent with the effect of NO synthase inhibition (Table 1).

Hepatic artery flow (Figure 1) decreased at 20 minutes. Just prior to this measurement the infusion of LPS had been stopped. Thereafter, hepatic artery flow increased by approximately 100% and was significantly increased at t = 40 and t = 60 minutes in the NE group and the L-NAME group. In the NE-group the increased flow was maintained (with a minor decrease towards the end of the experiment) while administration of L-NAME reduced hepatic artery flow to baseline levels at the end of the experiment. In the control group hepatic artery flow remained constant during the the entire study period.

Results of portal vein flow are shown in Table 1. Values remained at baseline level without hyperemic flow increase. Given the large contribution of portal vein to total hepatic blood flow, total hepatic blood flow did not change significantly over the course of the experiment, although at t = 120 minutes it tended to decrease in the NE group.

Hepatic O_2 extraction ratio (VO_2/DO_2, Figure 2) of the L-NAME group increased between measurements at t = 0 and t = 20 minutes due to the LPS infusion. Afterwards

Table 1. CI cardiac index, MAP mean arterial pressure, SVR systemic vascular resistance, PVF portal vein flow, Lactate Serum lactate. THBF Total hepatic blood flow

	Control t = 0	Control t = 120	NE t = 0	NE t = 120	L-NAME t = 0	L-NAME t = 120
CI [ml/kg/min]	138 ± 26	#137 ± 6	#148 ± 23	212 ± 63	165 ± 12	#110 ± 16
MAP [mm Hg]	106 ± 12	93 ± 12	113 ± 15	89 ± 28	128 ± 10	112 ± 27
SVR [dyn*sec*cm^{-4}]	2851 ± 873	*2412 ± 435	*2792 ± 758	1470 ± 522	2469 ± 279	*3070 ± 1437
Lactate [mmol/l]	0.78 ± 0.26	0.63 ± 0.15	0.92 ± 0.19	§2.04 ± 1.46	0.92 ± 0.43	§1.50 ± 0.76
PVF [ml/min]	1040 ± 170	1045 ± 120	1184 ± 190	750 ± 110	1200 ± 175	1040 ± 290
THBF [ml/min]	1202 ± 463	1223 ± 287	1719 ± 279	1316 ± 411	1418 ± 515	1281 ± 698

All data are given as mean values ±SD. # $p < 0.05$ compared to CI of the NE group at t = 120. * $p < 0.05$ compared to SVR of the NE group at t = 120. § $p < 0.05$ compared to lactate of the control group at t = 120.

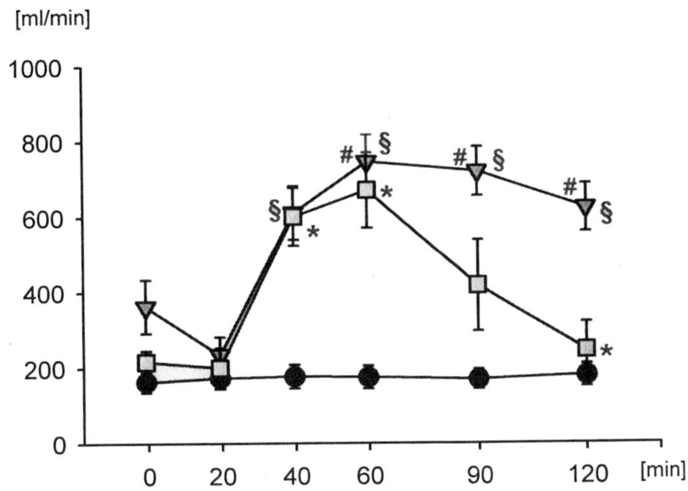

Figure 1. Hepatic artery flow [●] control group; [▽]NE group, [•]L-NAME group #p < 0.05 NE group t = 0 vs. T = 60, t = 90, t = 120. *p < 0.05 L-NAME group vs. control group for t = 40, t = 60, t = 120. §p < 0.005 NE group vs. control group for t = 40, t = 60, t = 90, t = 120.

in the NE group hepatic oxygen delivery did increase significantly ($p < 0.01$) between measurements at t = 20 minutes and at t = 60 and t = 90 minutes. A comparable tendency was found for the L-NAME group for measurements at t = 20 minutes compared to t = 60 minutes and the hepatic O_2 extraction ratio declined. For the L-NAME group at t = 120 minutes hepatic DO_2 was significantly reduced ($p < 0.05$) and the hepatic O_2 extraction ratio significantly increased compared to NE and controls.

Data on arterial blood lactate levels are given in Table 1. In both septicemic groups blood lactate levels were significantly elevated without differences between the LPS treated groups.

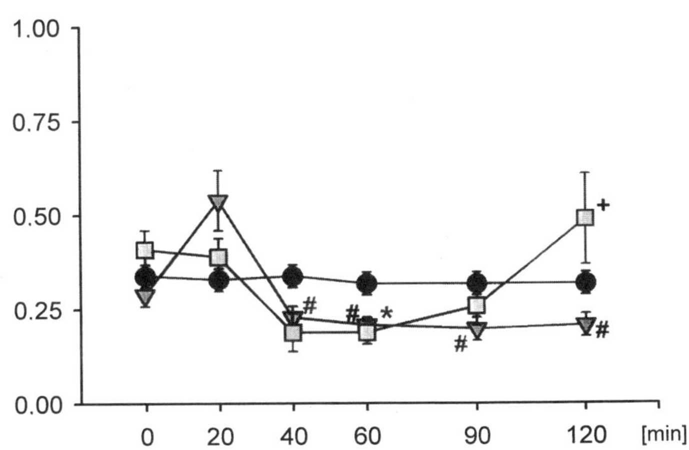

Figure 2. VO_2/DO_2 liver [●] control group; [▽]NE group, [•]L-NAME group #p < 0.05 NE group t = 20 vs. T = 40, t = 60, t = 90, t = 120, t = 0 vs. t = 90. *p < 0.05 L-NAME group vs. control group for t = 60. +p = 0.07 L-NAME group vs. NE group for t = 120.

Table 2. Liver tissue pO_2 of the control group, the NE group and the L-NAME group

	t = 0	t = 20	t = 40	t = 60	t = 90	t = 120
Control group	36.5 ± 12.7	37.7 ± 9.0	36.5 ± 10.1	33.8 ± 4.9	32.6 ± 6.8	29.5 ± 4.3
NE group	35.7 ± 11.3	40.3 ± 12.3	38.2 ± 13.6	37.7 ± 13.9	33.2 ± 15.5	33.6 ± 11.0
L-NAME group	30.7 ± 4.6	35.2 ± 12.6	37.3 ± 9.1	*49.2 ± 6.8	39.8 ± 10.2	37.6 ± 9.1

pO_2 values for all measuring points were normally distributed without significant changes of shape or curtosis between the groups or over time (mean ± SD).
*$p < 0.05$ compared to the control group at t = 60.

Despite the marked changes in hepatic artery flow, O_2 delivery and O_2 extraction ratio, mean liver tissue pO_2 (Table 2) remained stable over the entire study period without significant differences between the groups (except a change of significance between the control and the L-NAME group at t = 60 minutes).

4. DISCUSSION

The key findings of this experimental study were:

1. Hepatic artery blood flow and not portal vein flow were increased in hyperdynamic endotoxemia with NE stabilization, while administration of the NO synthase inhibitor L-NAME led to a marked reduction of hepatic artery flow without affecting portal vein flow.
2. Hepatic oxygen consumption remained constant with NO synthase inhibition and was slightly reduced in the NE treated group.
3. Liver tissue mean pO_2 values were within normal limits in both treatment groups.

4.1. Critique of the Model

In preliminary experiments a suitable dose and administration regimen of LPS was determined. A regimen of 0.025 µg/kg/min LPS (given for approximately 20 minutes) was chosen for two main reasons. First, in all animals a hyperdynamic circulatory state developed. Second, the circulatory deterioration of the animals during LPS administration could safely be halted when the LPS infusion was stopped and the immediate and temporary infusion low dose of NE was instituted.

A different set of experiments was carried out in which the dose of L-NAME was titrated to reduce the increased CI of hyperdynamic endotoxemia to approximately baseline values. The amount of L-NAME given was 1 mg/kg/min over 4 minutes. This dose was small in comparison to those reported in the literature, which frequently lay between 25 and 300 mg/kg.

The effect of NE administration in hyperdynamic endotoxemia had been described by Breslow et al. (1987). When NE was given in doses of 3.0 ± 1.6 µg/kg/min to increase MAP levels similar to those as in this study, a small but not statistically significant increase in CI was measured without redistribution of organ blood flow away from the liver. In addition, the markedly reduced SVR of the hyperdynamic state was unchanged during NE administration.

We did not investigate the effect of NE, but review of the similarities between the experimental design of this study and the investigations cited above, together with the findings of this study that SVR was markedly reduced in the NE group and the NE dose was small (0.05–0.64 μg/kg/min) compared to Breslow, it seemed justified to deduce that NE administration had no decisive effect on the hyperdynamic state and the liver blood flow in the present study.

The cannulation of a hepatic vein was performed without puncture of hepatic parenchyma and, therefore, without local disturbance of intrahepatic microcirculation, which could have affected the measurement of hepatic tissue pO_2 distribution. Correct position of the catheter tip in a hepatic vein was verified at postmortem examination.

4.2. Hepatic Circulation

In this model of hyperdynamic endotoxemia a marked statistically significant increase of hepatic artery flow was measured and portal vein flow remained at baseline levels over the whole experimental period. Given the comparatively small contribution of hepatic artery flow (app. 17%) to the total hepatic blood flow, total hepatic blood flow did not show significant changes over the experimental period.

With regard to hepatic artery flow our findings are in agreement with those of Lang et al. (1984) and Ottosson et al. (1991) who found a 160% increase in a rat peritonitis model of septicemia. Wang et al. (1991) found no change in hepatic artery flow but measured a significant rise in portal vein flow, while Ottoson et al. (1991) measured a significant reduction of portal vein flow; both authors used a rat model of hyperdynamic septicemia.

Hepatic oxygen delivery was significantly increased in the NE group and remained elevated apart from a minor decline at the end of the experiment. Our data confirm those of Fink et al. (1989), who found an increased mesentic DO_2 using a porcine hyperdynamic septicemia model. In contrast are data of Arvidsson et al. (1991), Hutardo et al. (1992), and Antonson et al. (1993) who found a reduced mesenteric and hepatic oxygen delivery, respectively. It should be noted, however, that the latter investigations were carried out in normo- or hypodynamic septicemia models.

A tendency of a slight decrease of hepatic VO_2 was found in this investigation in the NE group but these differences did not reach statistical significance. Data from the literature on hepatic VO_2 in septicemia are diverse. Arvidsson et al. (1991) and Antonson et al. (1993) found in a hypodynamic porcine septicemia model an increased hepatic and mesenteric VO_2, respectively, while Dahn et al. (1990) found in human normodynamic septicemia an increased mesenteric VO_2. Fink et al. (1989) noted that mesenteric VO_2 was unchanged in porcine hyperdynamic septicemia. Similar results were found by Nöldge et al. (1993), who found unchanged hepatic VO_2 during the first 3 hr of a 4 hr experiment, and a terminal increase in liver VO_2.

4.3. Liver Tissue pO2

Liver tissue pO_2 in this study was measured with O_2 needle probes measuring pO_2 in the depth of liver parenchyma. Possible artifacts due to compression caused by the 250 mm probe seem to be successfully compensated by the rapid stepwise electrode movement combined with a 30% withdrawal step at the end of the forward motion (Schramm et al., 1992; Wagner et al., 1992). In contrast, most previous investigations of liver tissue pO_2 were carried out either with surface pO_2 probes or with silicon tubes. In this investigation, hyperdnamic septicemia at no time led to a significant decrease of mean liver pO_2, or to a left shift, or to a microcirculatory maldistribution of the liver pO_2 values.

Our data confirm the results of Nöldge et al. (1993), who measured a normal liver tissue pO_2 frequency distribution throughout the experiment with a mean tissue pO_2 between 38 mm Hg and 60 mm Hg. Most other investigations frequently showed a deterioration of tissue oxygenation in septicemia or hemorrhagic shock. Fry et al. (1979), using a rat sepsis model with cecal ligation and puncture, noted a markedly decreased liver oxygenation from a mean pO_2 of 24.5 mm Hg to 1.5 mm Hg 6 hours after cecal ligation and puncture, but cardiac output was not measured. Apparently without supportive fluid administration and from the reported LD100 at 8 hours it can be assumed that a hypodynamic state developed in this animal model. Dahn et al. (1990) derived liver pO_2 in humans in hyperdynamic septicemia from the Krogh tissue cylinder model and arterial pO_2, their calculated mean liver pO_2 values were reduced to 16.2–25.8 mm Hg from their calculated normal values of 30.0 to 33.5 mm Hg. Mäkisalo et al. (1988) measured mean liver pO_2 with a silicon tube in porcine hemorrhagic shock and found a reduction of mean liver pO_2 from 32.4 ± 1. to 16.1 ± 0.4 mm Hg. Kessler et al. (1976) found similar data during hemorrhagic shock in dogs. Apart from the study of Dahn et al. (1990) all data of low mean liver tissue pO_2 were measured in hypodynamic circulation which most likely accounts for these differences in findings.

Induction of endotoxemia in this investigation led to an increase of blood lactate levels similar to those reported in the literature (Mäkisalo et al., 1988). Raised blood lactate levels were commonly interpreted as the result of tissue hypoxia (Hutardo et al., 1992). But Curtis (Curtis and Cain, 1992) showed that in septicemia lactate degradation is reduced thus possibly accounting for the increase of blood lactate levels.

4.4. NO Synthase Inhibition

The NO synthase inhibitor used in this study was L-NAME given at a dose of 4 mg/kg over 4 minutes. The effects of L-NAME administration in this porcine model of hyperdynamic endotoxemia were an increase in SVR, a decrease in CI and a stabilization of MAP after the infusion of NE was stopped.

There is evidence that NO is generated from two distinct forms of NO synthase: Whereas a basal amount of NO—essential for the physiological regulation of microcirculation—is released from a constitutive, Ca^{2++} dependent form of NO synthase (Moncada et al., 1991), a Ca^{2++} independent form of NO synthase can be induced by endotoxin and cytokines in vascular smooth muscle cells, hepatocytes and endothelial cells (Killburn et al., 1990).

The NO synthase inhibitor L-NAME, like many other inhibitors (L-NMA, L-NAC) does not discriminate between the two forms. Therefore the effect of the chosen dose of L-NAME (no circulatory shock, achievement of normalization of CI, SVR and MAP) was not due to a selective, unique inhibition of inducible NO synthase with the constitutive NO synthase left intact, but presumably a partial overall inhibition of constitutive NO synthase and inducible NO synthase alike but to a lesser degree. This degree of NO synthase inhibition was empirically elaborated in preliminary experiments since we could not measure activity of constitutive NO synthase or inducible NO synthase. In addition it is very likely that the "right" dose of an unspecific NO synthase inhibitor is at least species specific, since Meyer (Meyer et al., 1994) infused L-NAME in a dose of 25 mg/kg in a sheep sepsis model with the same correction of sepsis induced circulatory changes compared to 4 mg/kg/min over 4 minutes in this study.

In this investigation in a porcine model of hyperdynamic endotoxemia the effect of NO synthase inhibition on hepatic circulation was mainly limited to the hepatic artery resulting in a more than two-fold reduction of hepatic arterial flow compared to the hyperdynamic state. It was noteworthy that the reduction of hepatic artery flow did not fall below the baseline value of the L-NAME group. In contrast, no effect of NO synthase inhibition on portal vein flow was seen. A similar pattern had been reported by Meyer (Meyer et al., 1994) measuring total liver blood flow. After administration of endotoxin total liver blood flow tended to increase by 150–200%; a bolus injection of L-NAME (25 mg/kg) led to a reduction of total liver blood flow to just below (–9%) baseline levels. In an animal model of cirrhosis of the liver Pizcueta (Pizcueta et al., 1992) found a statistically significant reduction in portal vein flow after NO synthase inhibition.

The measured effect after L-NAME administration can not be attributed to the substance itself, since previous experiments in rats with the biologically inactive enantiomere D-NAME showed that D-NAME had no effect on MAP, skeletal muscle blood flow and skeletal muscle pO_2 distribution (Pohl et al., 1993).

The onset of circulatory effects after NO synthase inhibition with L-NAME (1 mg/kg/min for 4 minutes) was gradual, delayed and did not prevent the development of a hyperdynamic circulation with an increased hepatic DO_2 comparable to the NE group, but this hyperdynamic circulation was only transient and at the end of the experiment hepatic oxygen delivery had returned to baseline values. In experiments on skeletal muscle oxygenation the administration of 30 mg/kg L-NAME over 10 minutes led to an immediate and sustained increase of MAP and a reduction of skeletal muscle blood flow. Administration of the NO synthase inhibitor in this study had no statistically significant effect on hepatic VO_2. During the course of the experiment hepatic VO_2 tended to decrease but had reached baseline levels at the end of the experiment. Measuring whole animal VO_2 before and after NO synthase inhibition with L-NAME Meyer (Meyer et al., 1994) found no significant changes.

Given the relative contribution of portal venous (app. 85%) and hepatic arterial (app. 15%) blood to total liver blood flow the observed changes here were not statistically significant but total hepatic blood flow tended to decrease at the end of the experiment. With the hepatic VO_2 remaining constant this led to a marked rise of the hepatic oxygen extraction ratio and a significant decrease of the liver vein saturation at $t = 120$ minutes. Under these circumstances one might have expected to measure a reduced or low mean liver pO_2. But data measured revealed a physiological mean pO_2 of the liver. This was found not only at $t = 120$ minutes. At $t = 40$ minutes, when a hyperdynamic state with increased hepatic DO_2 and a reduced oxygen extraction ratio had developed mean liver tissue pO_2 was not increased compared to the value measured at $t = 120$. Tissue pO_2, amongst other factors (Kessler et al., 1976), is mainly influenced by the oxygen consumption of the cells and the oxygen supply delivered via the microcirculation. The fact that we measured a normal mean liver pO_2 indicates first that liver VO_2 was not supply dependent despite the reduction of hepatic artery blood flow, second that the ability to regulate oxygen availability to the liver cells was not compromised.

In conclusion, in a hyperdynamic endotoxemia model stabilized with NE administration we found hepatic artery flow and hepatic oxygen delivery increased while liver VO_2 was unchanged and mean liver tissue pO_2 was in the normal range. These findings indicate, that NE administration in hyperdynamic septicemia does not compromise liver oxygen supply. When NE was substituted for the NO synthase inhibitor L-NAME the hyperdynamic circulation was returned to normal. Hepatic artery flow and not portal vein flow were decreased, and a reduced hepatic oxygen delivery led to a marked rise in hepatic oxygen extraction ratio, while mean liver tissue pO_2 remained within the physiological range. These data lend support to the notion, that the administration of

NO synthase inhibitors might be an effective therapy of hyperdynamic circulation in septicemia.

REFERENCES

Antonsson Jan B, Kari Kutilla, Juha Niinikoski, and Ulf Hagrund: Subcaneous and gut tissue perfusion and oxygenation changes as related to oxygen transport in experimental peritonitis. Circ Shock 1993;41:261–267.

Arvidsson D, Rasmussen I, Almqvist P, Niklasson F, and Haglung U: Splanchnic oxygen consumption in septic and hemorrhagic shock. Surgery 1991;109:190–197.

Breslow MJ, Miller CF, Parker SD, Walmanet AT, and Traystman RJ: Effect of vasopressors on organ blood flow during endotoxin shock in pigs. Am J Physiol 1987;252:H291–H300.

Burton RG and Gorewit RC: Ultrasonic flowmeter uses wide-beam transit-time technique. Med Elec 1984;34:68–73.

Curtis SE and Cain SM: Regional and systemic oxygen delivery/uptake relations and lactate flux in hyperdynamic, endotoxin-treated dogs. Am Rev Respir Dis 1992;145:348–354.

Dahn MS, Wilson RF, Lange MP, Stone A, and Jacobs LA: Hepatic parenchymal oxygen tension following injury and sepsis. Arch Surg 1990;125:441–443.

Evans TE, Carpenter A, Kinderman H, and Cohen J: Evidence of increased nitric oxide production in patients with the sepsis sydrome. Circulatory Shock 1993;41:77–81.

Fink MP, Rothschild HR, Deniz YF, Wang H, Lee PC, and Cohn SM: Systemic and mesenteric O2-metabolism in endotoxic pigs: effect of ibuprofen and meclofenamate. J Appl Physiol 1989;67:1950–1957.

Hurtado FJ, Gutierrez AM, Silva N, Fernandez E, Khan AE, and Gutierrez G: Role of tissue hypoxia as the mechanism of lactic acidosis during E. coli endotoxemia. J Appl Physiol 1992;72:1895–1901.

Kessler M, Höper J, and Krumme BA: Monitoring of tissue perfusion and cellular function. Anesthesiology 1976;45:184–197.

Kilbourn RG, Jubran A, Gross SS, Griffith OW, Levi R, Adams J, and Lodato RF: Reversal of endotoxin-mediated shock by B-methyl-l-arginine, an inhibitor of nitric oxide synthesis. Biochem Biophys Res Commun 1990;172:1132–1138.

Klabunde RE and Ritger RC: N^G-monomethyl-L-arginine (NMA) restores arterial blood pressure but reduces cardic output in a canine model of endotoxic shock. Biochem Biophys Res Commun 1991;178:1135–1140.

Lang Ch, Bagby GJ, Ferguson JL, and Spitzer JJ: Cardiac output and redistribution of organ blood flow in hypermetabolic sepaia. Am J Physiol 1984;246:331–337.

Lorente J, Landin L, Renes E, de Pablo R, Jorge P, Rodena E, and Liste D: Role of nitric oxide in the hemodynamic changes of sepsis, Crit Care Med 1993;21:759–767.

Mäkisalo HJ, Soini HO, Lalla MLT, and Höckerstedt KAV: Subcutaneous and liver tissue oxygen tension in hemorrhagic shock: an experimental study with whole blood and two colloids. Crit Care Med 1988;16:857–861.

Meadows D, Edwards JD, Wilkins RG, and Nightingale P: Reversal of intractable septic shock with norepinephrine therapy, Crit Care Med 1988;16:124–132.

Meyer J, Hinder F, Stothert Jr. J, Traber LD, Herndon DN, Flynn JT, and Traber DL: Increased organ blood flow in chronic endotoxemia is reversed by nitric oxide synthase inhibition. J Appl Physiol 1994;76:2785–2793.

Moncada S, Palmer RMJ, and Higgs EA: Nitric oxide: physiology, pathophysiology and pharmacology. Pharmacol Rev 1991;43:109–142.

Nöldge GFE, Priebe H-J, and Geiger K: Splanchnic hemodynamics and oxygen supply during acute normovolemic hemodilution alone and with isoflurane-induced hypotension in the anaesthetized pig. Anesth Analg 1992;75:660–674.

Nöldge GFE, Pannen BHJ, Armbruster K, and Gelger KK: Endotoxemia induces more profound impairment of small intestinal than of hepatic oxygenation in pigs. Anaesthesiology 1993;79:A253.

Nöldge GFE, Armbruster K, Pannen BHJ, Müller W, and Gelger KK: Changes in intestinal mucosal pH during endotoxemia are not necessarily reflected in respective changes in splanchnic metabolic and acid-base status. Anaesthesiology 1993;79:A761.

Ottosson J, Dawidson I, Brandberg A, Idvall J, and Sandor Z: Cardiac output and organ blood flow in experimental septic shock: effect of treatment with antibiotics, corticosteroids, and fluid infusion, Circulatory Shock 1991;35:14–24.

Petros AJ, Lamb G, Leone A, Moncada S, Bennett D, and Vallance P: Effect of a nitric oxide synthase inhibitor in patients with septic shock: a randomised study. Cardiovascular Research 1994;28:34–39

Pizcueta P, Piqué JM, Fernandez M, Bosch J, Rodés J, Whittle BJR, and Moncada S: Modulation of the hyperdynamic circulation of cirrhotic rats by nitric oxide inhibition. Gastroenterology 1992;103:1909–1915.

Pohl U, Wagner K, and De Witt C: Endothelium-derived nitric oxide in the control of tissue perfusion and oxygen supply: physiological and pathophysiological implications. Europ Heart J 1993;14(S I):93–98.

Schneider F, Lutun P, Vouchot A, Bilbault P, and Tempe JD: Plasma cyclic guanosine 3′-5′ monophosphate concentrations and low vascular resistance in human septic shock. Intensive Care Med 1993;19:99–104.

Schramm U, Wagner K, Bossen W, and Weiss C: PO_2 Messung mit einem neuen Sauerstoffsensor: Morphologische Veränderungen im Bereich des Stichkanals. Anat Anz 1992;174:78–81.

Vaupel P, Schlenger K, Knoop C, and Höckel M: Oxygenation of human tumors:evaluation of tissue oxygen distribution in breast cancers by computerized O_2 tension measurements. Cancer Res 1991; 51:3316–3322.

Vincent JL: Diagnostic and medical management/supportive care of patients with gram-negative bacteremia and septic shock. Infect Dis Clin North Am 1991;5:807–816.

Wagner K, Bossen W, and Schramm U: Tissue alterations by the penetration of a pO_2 sensing needle. Adv Exp Med Biol 1992;317:639–644.

Wang P, Ba ZF, and Chaudry IH: Increase of hepatic blood flow during early sepsis is due to increased portal blood flow. Am J Physiol 1991;261:R1507–R1512.

Wright CE, Rees DD, and Moncada S: Protective and pathological roles of nitric oxide in endotoxin shock. Cardiovasc Res 1992;26:48–57.

57

CHANGES IN MICROVASCULAR BLOOD FLOW PATTERNS AND OXYGEN SATURATION IN BREAST SKIN DURING RADIOTHERAPY

D. K. Harrison,[1] E. M. Harrison,[2] D. J. Newton,[3] and P. M. Windsor[4]

[1]Regional Medical Physics Department
Durham Unit, Dryburn Hospital
Durham, DH1 5TW, United Kingdom
[2]Vascular Laboratory
[3]Department of Medicine
[4]Department of Radiotherapy and Oncology
Ninewells Hospital and Medical School
Dundee DD1 9SY, United Kingdom

INTRODUCTION

Acute skin reactions, ranging from mild erythema to moist desquamation, can be seen in patients receiving standard fractionated radiotherapy to the breast for conservation therapy of breast carcinoma. The main erythematous reaction becomes maximal around the 15th day of treatment and lasts for 20–30 days after the treatment is completed. If a severe reaction occurs the breast may become swollen, oedematous and tender, gradually resolving over the next 24 months (Mendelson, 1993). Late radiation reactions can occur from 3 months onwards in breast tissue whereby scarring, fibrosis and calcification develop, with increasing incidence from 6 to 36 months (Mendelson, 1993). As yet, clinical studies have been unable to predict the late effects of irradiation from the severity of the early skin erythema reaction. However, both early and late reactions to radiation therapy can cause great discomfort to patients and severely affect their quality of life.

In previous studies lightguide spectrophotometry, laser Doppler imaging (LDI) and digital thermographic imaging (DTI) have been used to study oxygen supply and blood flow in inflammatory reactions induced experimentally in forearm skin (Harrison et al., 1993). More recently, the technique of LDI has been demonstrated as a promising tool for assessing skin perfusion in the breast (Seifalian et al., 1995). The present study is an attempt to combine these techniques to investigate the mechanisms involved in the early

Oxygen Transport to Tissue XXI, edited by Eke and Delpy.
Kluwer Academic / Plenum Publishers, New York, 1999.

radiation reaction in the breast at selected time points during and after a standard radiotherapy technique. A better understanding of the processes will facilitate the testing of interventions to lessen or prevent the occurrence of severe skin reactions.

This is a preliminary report of results from the study, concentrating mainly on changes in skin oxygen saturation (SO_2).

METHODS

An EMPHO II micro-lightguide spectrophotometer (BGT, Überlingen, Germany, Frank et al., 1989) was used to measure the oxygen saturation (SO_2) profiles in the skin along a central line of the treated breast from lateral to medial and proximal to distal within the marked treatment area. Histograms, consisting of 100 measurement points, were also recorded centrally at the medial and lateral extremities of the treatment area. LDI of each breast was performed using an LDI 1.0 instrument (Moor Instruments, Axminster, UK) with a 633 nm laser (Essex & Burns, 1991). Similarly, DTI was carried out using an Agema 880 series digital system.

Radiation treatment consisted of a dose of 45 Gray administered in 20 fractions over 4 weeks. Measurements were carried out prior to the first fraction, after 5, 10 and 20 fractions and, whenever possible, after 3 months. LDI and DTI images were recorded from both the treated and untreated breasts throughout. However, because of the total time involved for the patient, LS measurements were made on the treated breast and only carried out on the untreated breast once, prior to the start of treatment.

RESULTS

Measurements were carried out on 8 patients, of whom it was possible to follow up 6 at 3 months. Figure 1 shows laser Doppler images from the treated breast of a patient prior to treatment (control) and after 20 fractions of radiation.

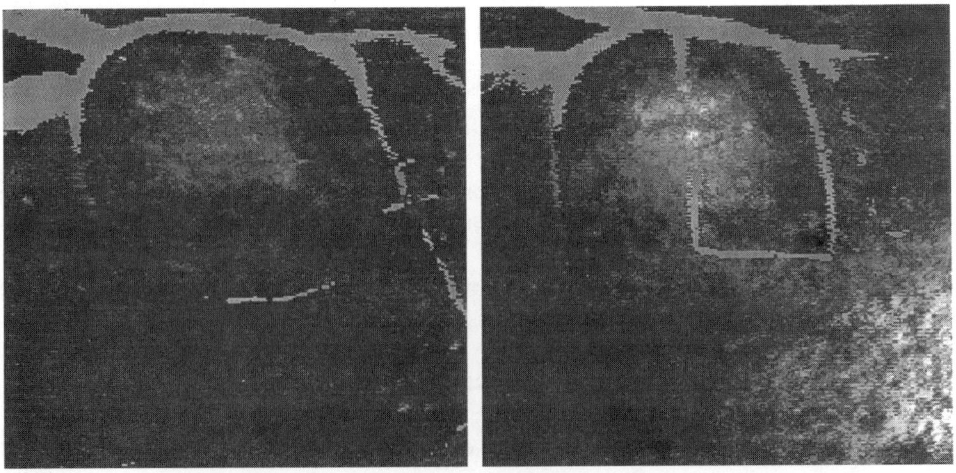

Figure 1. Grey-scale Laser Doppler images of flux (perfusion units) (left) prior to treatment and (right) after 20 fractions of radiation.

It can be seen that there is an overall increase in the laser Doppler flux over the areolar region and, in particular, the medial aspect of the breast and the adjacent skin at the sternum. Although not clearly seen in the grey scale image, the perfusion demonstrates a marked degree of heterogeneity within the areas of increased perfusion. The thermographic images, although not shown here, reveal a similar pattern.

SO_2 profiles from the proximal to the distal edges of the treatment area across the breast of the same patient for all stages of treatment are shown in Figure 2. It can be seen that there is an increase in SO_2 across the breast at each stage of treatment which, in this case, is starting to resolve at 3 months. There is a less marked increase in the areolar area. The SO_2 histograms recorded at the medial and lateral sites on the same patient are shown in Figure 3. The changes are similar to those indicated by Figure 2, but the changes at the medial site are generally greater and the histogram remains shifted to the right even at 3 months.

The mean changes in SO_2, together with the standard deviation (SD) in the skin at the medial and lateral sites are summarised in Figure 4. In addition, the mean and SD of the SO_2 values in the areolar area, taken from the lateral to medial and proximal to lateral profile measurements, are included for all stages. The populations of SO_2 values were found to be normally distributed and Student's t test was applied to investigate significance. There was no significant difference between the SO_2 values at the medial and lateral sites prior to treatment or after 20 fractions. Apart from these two stages, the differences between the three areas were all significant ($p < 0.001$). The changes in SO_2 were all significant compared with the control values ($p < 0.001$). The mean SO_2 values were also significantly different between each stage of treatment and the next at the $p < 0.001$ level with the exception of the means for the 5th and 10th fractions in the areolar area and at the lateral site which were not significant.

DISCUSSION

The results presented here are preliminary and the full study has yet to be completed. Furthermore, for reasons of space and reproduction it is not possible to present

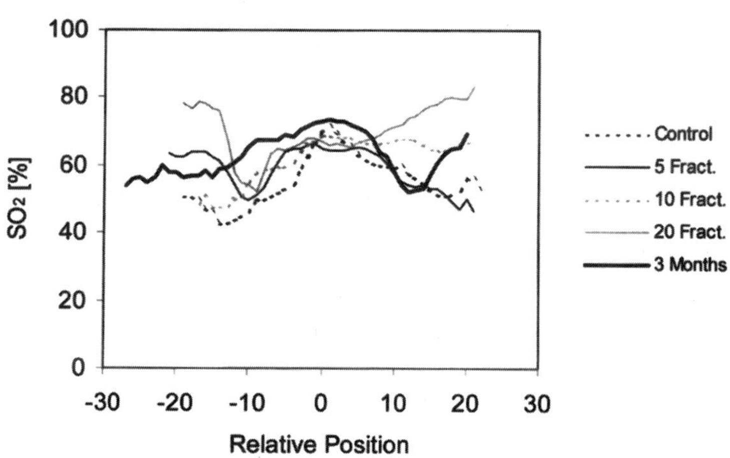

Figure 2. SO_2 profiles across the breast, measured from the proximal to the distal edge of the treatment area, at all stages of treatment. The profiles are aligned relative to the position (0) of the nipple, SO_2 values are recorded at 3–4 mm intervals.

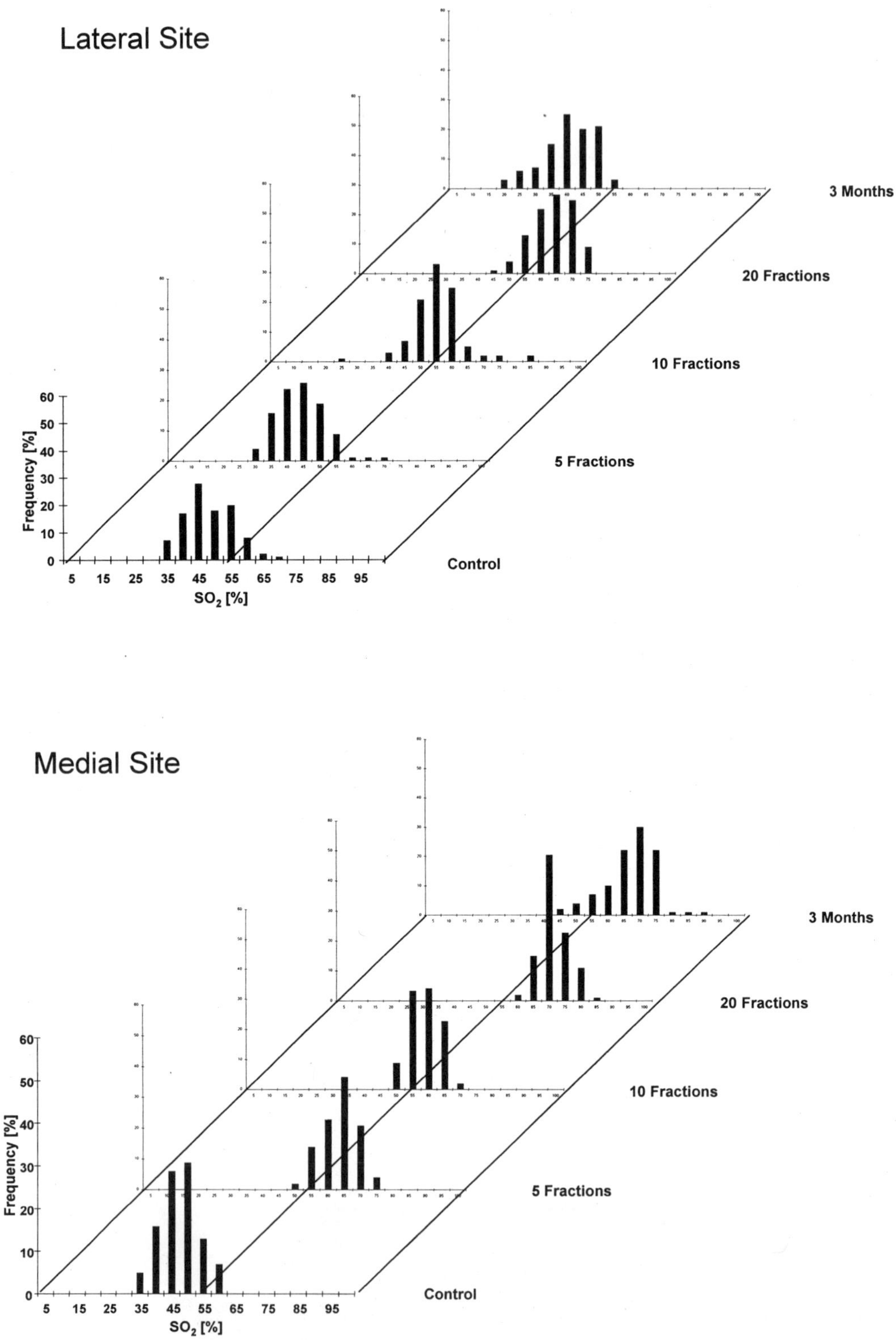

Figure 3. SO_2 histograms at the lateral and medial sites at all stages of treatment.

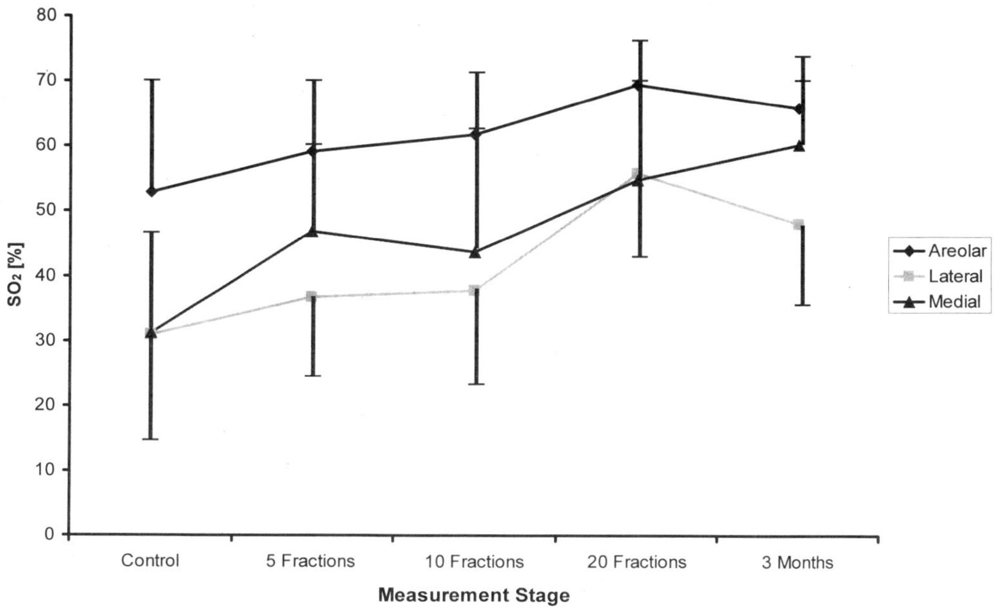

Figure 4. Mean and SD of all SO_2 values measured at the medial site, the lateral site and in the areolar area at all stages of treatment.

the full range of thermographic and laser Doppler images. The laser Doppler images presented are, therefore, from a patient who represented the "mean response" of the group and they can be compared with the corresponding SO_2 profiles and histograms.

However, the measured changes in SO_2 are highly significant statistically even at this stage, so it is clear that these results are likely to be representative of the final outcome. It is interesting to note the differences between the lateral and medial sites with the SO_2 at the medial site continuing to increase at the 3-month stage. This is consistent with the fact that the most severe skin reactions tend to occur medially and may reflect continuing erythema.

In a recent study using spectrophotometry and conventional single point laser Doppler flowmetry to investigate erythema (Simonen et al., 1998), the authors suggested that inflammatory responses may play a role in the mediation of the erythematous response to radiation in human skin. The mean changes in SO_2 and shifts in the SO_2 histograms measured here are similar to those measured during the course of the tuberculin reaction in skin (Harrison et al., 1992; Newton et al., 1994). The changes in laser Doppler flow patterns and thermographic images also show a similarity to those observed in the tuberculin reaction (Harrison et al., 1993). In particular, the heterogeneity in perfusion patterns in the areas of increased flow are characteristic of an inflammatory response. Such changes are often accompanied by a decrease in tissue pO_2, as measured transcutaneously, possibly due to the increased diffusion distances induced by oedema, increased oxygen consumption by infiltrating white cells and decreased capillary transit times for the erythrocytes (Harrison et al., 1994; Newton et al., 1996). Not all of these factors may be present in the acute reaction to radiation. There is certainly no evidence of tissue hypoxia from the SO_2 histograms. However, this in itself may be significant in conditions of inflammation and further studies should include measurement of pO_2 in the skin.

There were no severe acute skin reactions amongst this group. It is therefore not possible to determine at this stage which of the three methods or which area of measurement may be most predictive of a severe reaction occurring, although it is likely that a combination of the techniques will prove to be most sensitive. However, the current results demonstrate that the technique of tissue spectrophotometry alone is sufficiently sensitive to detect significant changes in skin SO_2, in all of the areas of the breast investigated, after only 5 fractions of the radiation dose.

In addition to studying further patients, it is intended to follow up those already investigated in order to determine whether there are any predictive features of the results which may be indicative of any late radiation reactions that may occur.

ACKNOWLEDGMENTS

The authors are most grateful to Professor JJF Belch and Dr Faisel Khan, Department of Medicine, University of Dundee, for use of the laser Doppler imaging equipment. The willing participation of the patients in the study is also gratefully acknowledged.

REFERENCES

Essex TJH and Burns PO (1991) A laser Doppler scanner for imaging blood flow in skin. J Biomed Eng 13:189–194.

Frank KH, Kessler M, Appelbaum K, and Dümmler W (1989) The Erlangen micro-lightguide spectrophotometer (EMPHO I). Phys Med Biol 34:1883–1990.

Harrison DK, Evans SD, Abbot NC, Beck JS, and McCollum PT (1992) Spectrophotometric measurements of haemoglobin saturation and concentration in skin during the tuberculin reaction in normal human subjects. Clin Phys Physiol Meas 13:349–363.

Harrison DK, Abbot NC, Swanson Beck J, and McCollum PT (1993) A preliminary assessment of laser Doppler perfusion imaging using the tuberculin reaction as a model. Physiol Meas 14:241–252.

Harrison DK, Abbott NC, Carnochan FMT, Beck JS, James PB, and McCollum PT (1994) Protective regulation of oxygen uptake as a result of reduced oxygen extraction during chronic inflammation. Adv Exp Med Biol 345:789–795.

Mendelson EB (1993) Radiation changes in the breast. Sem Roentgenol 28:344–362.

Newton DJ, Harrison DK, Delaney CJ, Beck JS, and McCollum PT (1994) Comparison of macro- and micro-lightguide spectrophotometric measurements of microvascular haemoglobin oxygenation in the tuberculin reaction in normal human skin. Physiol Meas 15:115–128.

Newton DJ, Harrison DK, and McCollum PT (1996) Oxygen extraction rates in inflamed human skin using the tuberculin reaction as a model. Int J Microcirc Clin Exp 16:118–123.

Seifalian AM, Chaloupka K, and Parbhoo SP (1995) Laser Doppler perfusion imaging—a new technique for measuring breast skin blood flow. Int J Microcirc Clin Exp 15:125–130.

Simonen P, Hamilton C, Ferguson S, Ostwald P, O'Brien M, O'Brien P, Back M, and Denham J (1998) Do inflammatory processes contribute to radiation induced erythema observed in the skin of humans? Radiother Oncol 46:73–82.

IMPACT OF THE ANTIMETASTATIC DRUG BATIMASTAT ON TUMOR GROWTH AND PO_2 MEASURED BY EPR OXIMETRY IN A MURINE MAMMARY ADENOCARCINOMA

Philip E. James, Julia A. O'Hara, Stalina Grinberg, Tomasz Panz, and Harold M. Swartz

EPR Center for the Study of Viable Biological Systems
Radiology Department 7785 Vail Building
Dartmouth Medical School
Hanover, New Hampshire 03755

1. INTRODUCTION

Patients treated for early breast cancer have only a 10–15% probability of local relapse (Fowble, 1991), but the likelihood of eventual recurrence with distant metastases can be much higher depending on the original stage of the cancer. It is from this systemic disease that patients die (Willner et al., 1997). Treatment of malignant tumors has relied heavily on the use of cytotoxic agents, targeted at the increased metabolic and proliferative activity of the cancer cell. An alternative strategy has been to develop low toxicity therapies that can be administered continuously over a period of several years to stabilize malignant disease, and it is in this category that treatment with matrix metalloproteinase inhibitors (MMPIs) lie.

Matrix metalloproteinase enzymes (MMPs) are capable of degrading components of the extracellular matrix (Powell and Matrisian, 1996). They play a primary role in tissue remodeling, such as during wound healing (Agren et al., 1998) or embryogenesis (Reponen et al., 1995), and participate in passage of tumor cells across basement membranes (Chambers and Matrisian, 1997). There is evidence to suggest that MMPs are responsible for breakdown of extracellular matrix required for new blood vessel ingrowth supplying the tumor cells with oxygen and nutrients. MMP activity is normally held in check by the latency of the secreted pro-enzymes, native inhibitors such as alpha-2-macroglobulin (a general plasma proteinase inhibitor), and tissue inhibitors of matrix metalloproteinases (TIMP-1 and TIMP-2).

Oxygen Transport to Tissue XXI, edited by Eke and Delpy.
Kluwer Academic / Plenum Publishers, New York, 1999.

BB-94 (British Biotechnology 94) or Batimastat ([4-(N-hydroxyamino)-2R-isobutyl-3S-(thiopen-2-ythiomethyl)-succinyl]-L-phenylalanine-N-methylamide), is a synthetic, low molecular weight inhibitor of matrix metalloproteinases. It has been shown to inhibit angiogenesis, tumor growth and metastasis in animals (Davies et al., 1993a). It has been successful in a phase I dose escalation trial for patients with malignant ascites at the ICRF Medical Oncology Unit (Western General Hospital) in Edinburgh U.K. The drug is administered intraperitoneally, and in a recent report (Beattie et al., 1994) seven patients tolerated the drug well and showed no significant toxicity, and only one patient re-accumulated ascites within the 28-day study period following treatment. BB-94 administered to human ovarian carcinoma xenografts growing in nude mice caused resolution of ascitic disease and increased survival 5–6 fold. BB-94 treated tumors were avascular, and contained necrotic tumor cells in a highly developed stroma, encapsulated by a dense capsule (Davies et al., 1993b). Marimastat is a close derivative of BB94 which can be administered orally. This drug's effectiveness following chemotherapy in treating metastatic breast cancer is currently under investigation in patient volunteers across the US.

The oxygenation of a tumor is thought to be a pivotal factor that influences growth, malignant potential, and sensitivity to non-surgical treatment modalities (e.g. irradiation or certain anticancer drugs (Brizel et al., 1997, Hockel et al., 1996), the most recognized being the radiosensitizing potential of oxygen—in general approximately two to three times higher radiation doses are needed to kill anoxic cells when compared with well-oxygenated cells (Hall, 1988). The clinical importance of tumor oxygen in relation to angiogenesis and/or metastasis however, remains uncertain, since valid methods for the routine measurements of intratumoral oxygen tensions in tumors until recently have been lacking (Stone et al., 1993).

In vivo electron paramagnetic resonance oximetry (EPR) using particulate tumor-implanted paramagnetic materials can be used to monitor pO_2 in tumors in a method that is sensitive and that can be repeated non-invasively as frequently as necessary or desirable (Goda et al., 1996, Goda et al., 1997, Goda et al., 1995, O'Hara et al., 1998, O'Hara et al., 1997, O'Hara et al., 1995). Therefore we have used *in vivo* EPR using gloxy (James et al., 1997b) to monitor pO_2 in murine mammary adenocarcinomas during the course of growth and after treatment with BB-94. This report describes how we tested the hypothesis that treatment of tumors with MMPIs at an early stage in their development will prevent growth and invasion of the primary tumor into surrounding tissue. If hypoxia predicts for malignant progression of tumors, then tumors with higher metastatic potential, such as our untreated MTG-B tumors might be expected to have lower tumor pO_2 than those that have diminished metastatic potential as a consequence of treatment with BB94. The relation between tumor pO_2 and tumor volume, and the ability of BB-94 to produce higher pO_2 values compared to untreated tumors of the same volume, was also assessed.

2. METHODS

2.1. Tumor Model

The MTG-B tumor is a murine mammary adenocarcinoma which was maintained via serial passage in five to seven week-old female C3H/HeJ mice (The Jackson Lab, ME). This tumor was initially obtained from a spontaneous C3H mouse mammary tumor originating in the glandular epithelium of a female C3H mouse (Clifton et al., 1966, Clifton and Draper, 1963). It has been utilized in our laboratory for many years for studies on drug and radiation sensitivity (Douple et al., 1987, Jones et al., 1989). Tumors were transplanted by removal of tumor tissue aseptically from donor mice, mincing the tumor in minimum essential media (MEM) without serum and pressing the minced tumor particles through a 70 μm cytosieve (Falcon) and centrifuging at 60 g for 5 minutes. The pellet of tumor cells was resuspended in fresh MEM and 0.05 ml (10^6 cells) tumor cell suspension was injected s.c. into the right flank of recipient mice as previously described (Bacic et al., 1993, O'Hara et al., 1995). All animal experiments were conducted under protocols approved by the Dartmouth Institutional Animal Use and Care Committee.

Injection of Gloxy:

A suspension of gloxy pieces (the oxygen-reporting, paramagnetic material) was prepared by grinding with an agate ball and pestle on a mechanical grinder (Wifflebug, Crescent Dental Manufacturing Co., IL). The powder was suspended in saline and filtered through a 20 μm pore diameter filter. Material greater than 20 μm was made up in saline (50 mg/ml) and 25 μl injected directly into each tumor. We have previously used gloxy extensively in conjunction with EPR oximetry to measure the pO_2 in several tumor models and in normal tissue (Jackson et al., 1997, James et al., 1997a, James et al., 1997b). Its high spin density and spectral sensitivity to oxygen (particularly at lower pO_2) make it particularly suitable for measuring the pO_2 in tumors.

2.2. BB-94 Treatment

A solution of BB-94 (10 mg/ml) was made up in 0.01% Tween 80 in sterile saline just before use. It was solubilized by placing in an ultrasonic water bath for 10 min. 100 μl was injected i.p. per 20 g mouse body weight (to achieve 50 mg/kg) on treatment days specified below. Controls were injected with vehicle only on the same schedule as treated animals.

Treatment Groups:

Two experiments were performed differing only in the timing of BB-94 treatment, the number of animals per group, and the overall length of the observation period.

2.2.1. Intense Treatment Regimen

1. CONTROL (Vehicle only)—12 mice. Mice were injected daily. Gloxy was injected on day 7 following tumor inoculation. 2 mice were sacrificed on each of days 10, 12, 14, 16, 18, 20, and tumor tissue dissected and fixed for histological analysis.
2. TREATED (BB-94 starting on 5th day post inoculation)—12 mice. Mice were injected with BB-94 daily and gloxy injected as for the control mice above.

pO_2/volume: All animals had tumor volumes measured by calipers daily and pO_2 measured by EPR every second day. Tumor volumes were calculated from measurement of 3 orthogonal diameters, such that Tumor Volume = $(\pi/6)(d_1{*}d_2{*}d_3)$.

2.2.2. Less Intense Treatment Regimen. A total of 32 animals were used for the pO_2/volume study and all had tumor volumes measured by calipers daily and pO_2 measured by EPR every second or fourth day on a schedule that allowed for 8 from each group to be measured at each session. All animals were sacrificed at the end of the 32 day study. Gloxy was injected into tumors when they reached approx. 6 mm diameter.

1. CONTROL (Vehicle Only)—16 mice.
2. TREATED (BB-94 starting 5th day post inoculation)—16 mice. Mice were injected with BB-94 every 2nd day (in contrast to the short term study, in which mice were injected every day).

2.3. Histology Study

Gloxy was injected into a further group of 36 animals as for the pO_2 group above. Tumor pO_2 and volume was measured on the day the animals were sacrificed (in general, 3 animals from each of control and treated histology groups were sacrificed every 4th day). Tumors were dissected and fixed in formalin, then prepared in paraffin blocks prior to slicing, standard staining with hematoxylin and eosin, and histological assessment. Three to four slices (separated by 0.5 mm) from each tumor were analyzed and the necrotic fraction calculated by measuring the ratio of the areas of viable: necrotic tissue in each slice. Digitized color images of the tumor slices were prepared by scanning photographic slides of microscopic slides. Areas containing intact cells stained with hematoxylin (viable) and areas containing eosinophilic acellular material (necrotic) were outlined. The relative areas were measured on Gray scale images using NIH Image software. A point count method was used to estimate the number of capillaries. Briefly, the number of vessels falling on intersections of a 10 by 10 grid were counted. The grid area was calibrated using a stage micrometer.

1. CONTROL (Vehicle only)—12 mice.
2. TREATED (BB-94 starting 5th day post tumor inoculation)—24 mice.

All studies were blind, in that animals were assigned to treatment groups at the time of tumor inoculation and were pre-assigned into groups to be measured at particular time points.

2.4. Electron Paramagnetic Resonance (EPR) *in Vivo* EPR Oximetry

EPR spectra of gloxy in tumors were obtained using a 1.2 GHz ("L-band") EPR spectrometer constructed in the EPR Center at Dartmouth. Typical settings for the spectrometer were: magnetic field, 425 Gauss; modulation frequency, 27 kHz; modulation amplitude, 0.2 Gauss; incident microwave power, 15 mW. The mice were not anesthetized for EPR measurements, but were restrained in a plastic holder with a hole through which the tumors were exposed. The detecting EPR coil was positioned over the tumor and five spectra were recorded per animal (30 seconds per spectrum, total measurement time was less than 10 min., including positioning of the animal and spectrometer adjustment). Spectral line widths of gloxy were measured and the pO_2 values calculated from a calibration curve of pO_2 *vs.* line width in the presence of tumor cells using a Bruker, ER 220D-SRC EPR spectrometer (9.6 GHz, X-band) and a commercial oxygen analyzer (Sensor Medics Co., Model OM-11, Anaheim CA). The calibration procedure was the same as described previously, including verification of its validity for 1.2 GHz studies (James et al., 1997b, James et al., 1996).

2.5. Statistics

Values given are mean ± SEM. Differences were statistically significant at a level of $p < 0.05$.

3. RESULTS

Average tumor pO_2 was similar in both control and treated groups initially, but with time and tumor growth, these values differed significantly (indicated by an asterisk in Figure 1(a) and (b)). For these figures, data from all animals in either control or treatment groups measured on a given day were averaged.

The two different treatment regimens appeared to have characteristic changes in pO_2 with time, the overall trend being increased tumor pO_2 with BB-94 treatment. Daily treatment with BB-94 (short-term study) resulted in tumors having higher average pO_2 compared to controls, with a relatively small decrease in average pO_2 over 23 days (6.3+/–0.7 mm Hg to 5.9+/–0.6 mm Hg). Control tumors decreased from 6.0+/–0.6 mm Hg to an average pO_2

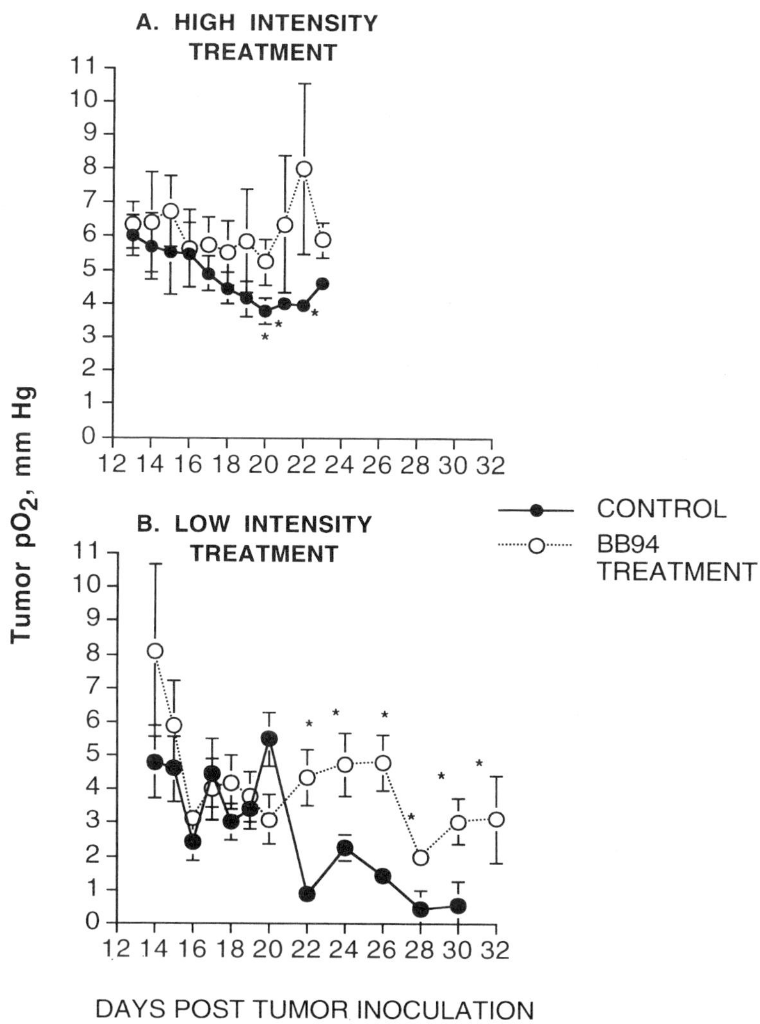

Figure 1. Changes in tumor pO_2 with time following tumor cell inoculation. Each data point represents the average from all tumors measured in either control or treated groups on a given day, n = 8–10.*—indicate where means of the BB-94 treated tumors were significantly different from controls. In this figure (and throughout), (A) shows data from the high intensity treatment study, and (B) data from the low-intensity treatment study.

of 4.3+/–0.2 mm Hg over the same time period. Less intense treatment (given over a longer period of time) caused a greater decrease in pO_2 from 8.1+/–2.2 mm Hg to 4.7+/–1.0 mm Hg over 23 days (control values were 4.8+/–1.0 to 2.2+/–0.3, respectively). At 32 days post tumor inoculation, treated tumors had an average pO_2 of 3.2+/–1.4 mm Hg (compared to 0.5+/–0.7 mm Hg in controls). Tumor pO_2 was highest in animals given intense (daily injections) BB-94 treatment, and values were statistically different from controls as of day 18 post tumor inoculation (compared to day 22 in the low intensity tumor treatment).

A high intensity treatment regimen also resulted in delayed appearance of palpable nodules, and reduced tumor burden up to 22 days post tumor inoculation (Figure 2(a)); tumor volume increased from 52.0+/–13.3 mm^3 to 511.3+/–152.9 mm^3 in treated

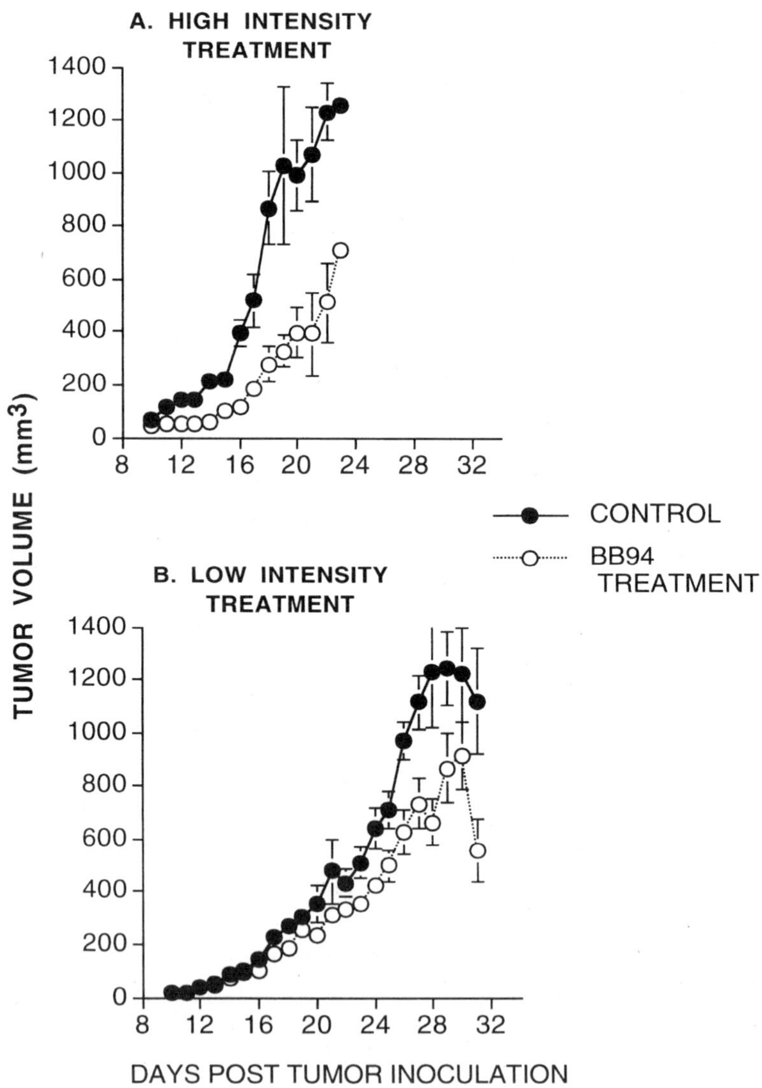

Figure 2. Changes in tumor volume with time following tumor inoculation. In the High Intensity experiment (A) there was a significant difference in tumor volume between the control and treated groups by the 11th day post tumor inoculation ($p < 0.05$). There was no difference between the groups in the low intensity treatment experiment (B). Error bars are SEM.

animals (compared to an initial volume of 215+/–19.6 mm^3 increasing to 1257+/–150.0 mm^3 in control tumors). Long term treatment given less frequently also resulted in an overall decrease in tumor volume, but this was apparent only at later stages in tumor growth (treated = 492.7+/–302.0 mm^3; control = 1238.2+/–246.9 mm^3 on day 32; Figure 2(b)).

From a comparison of these same data plotted as tumor pO_2 versus tumor volume (Figure 3 (a) and (b); averages of a group on a given day) it appears that for a given tumor volume, the corresponding tumor pO_2 is slightly higher for treated tumors (this was particularly pronounced at larger tumor volumes).

The necrotic fraction (calculated from % necrotic tissue) was larger in control tumors compared to those treated with BB-94 on a high intensity regimen (data not shown). The extent of necrosis in tumors given low intensity BB-94 treatment was not significantly different from control tumors.

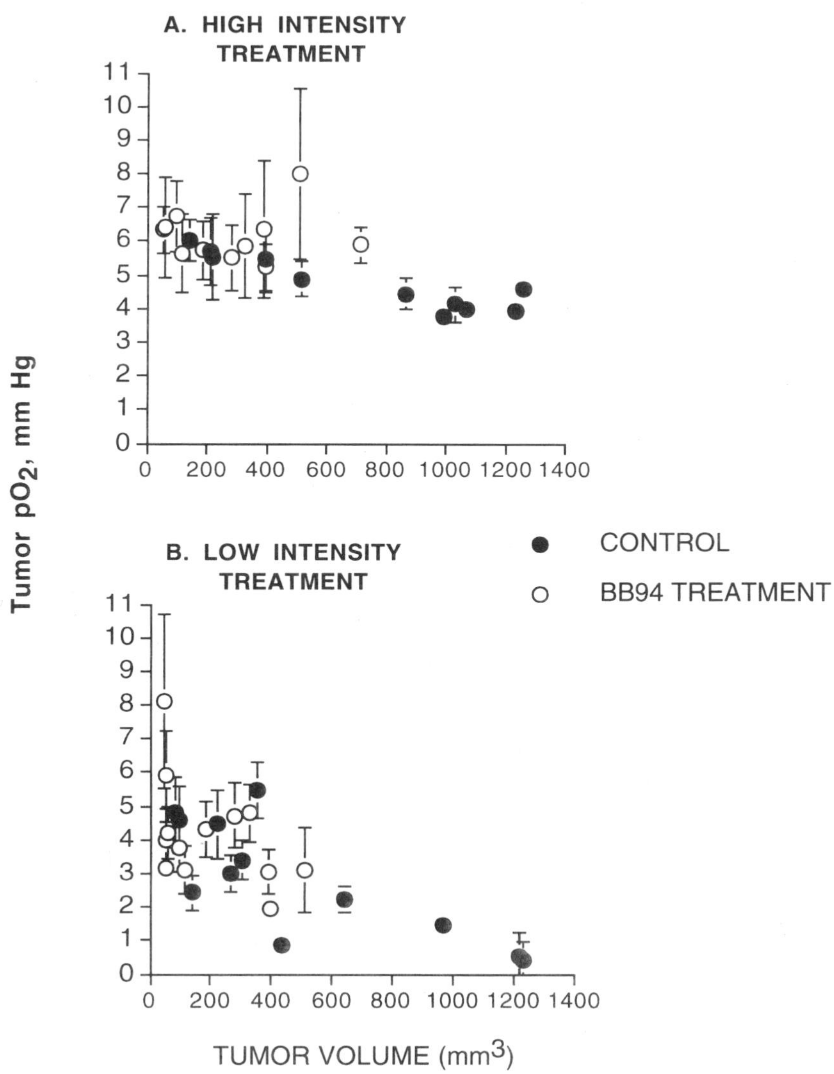

Figure 3. Same data as in Figures 1 and 2 plotted as a function of tumor pO_2 versus volume.

We also selected tumor slices for histochemical staining with an antibody specific for matrix metalloproteinase II. We failed to observe a difference in specific stain binding between control and treated tumor slices (data not shown).

The number of capillaries present in tumors given BB-94 was, on average, lower than that in control tumors. This became more significant at later time points in tumor growth. The average value for control tumors was 36, compared to 21 (+/–6) vessels per 0.1 mm^2 in BB-94 treated tumors.

4. DISCUSSION

These results clearly demonstrated that BB-94 treatment caused decreased tumor volume, increased tumor pO_2, and decreased tumor burden. The frequency of BB-94 treatment had an effect on the magnitude and time needed to observe these differences—an intense treatment regimen (daily injections) resulted in a significant difference in tumor pO_2 between control and treated tumors as early as day 18 post tumor inoculation.

The potentially most important effect of the drug was to change the pO_2 from what it would have been in an untreated tumor of the same volume. The volume was a most important issue because in general smaller tumors tend to have higher average pO_2, so any drug that decreases the growth rate of a tumor, if compared to untreated tumors, would tend to increase the pO_2 at the same time point after tumor inoculation. Our data suggest that for a given tumor volume, the corresponding tumor pO2 was slightly higher for treated tumors (perhaps only at larger tumor volumes). Tumor pO_2 is thought to be a highly important factor governing the success of radiotherapy and post-irradiation tumor reoxygenation may contribute to a tumor's response to subsequent doses of radiation (Withers, 1987). We have recently shown that EPR measurements of tumor pO_2 can be used to time split dose irradiation of a murine tumor (O'Hara et al., 1998) for an improved effect on tumor regrowth delay. The data in this study have implications for both understanding the mechanisms of this class of drugs and for providing a rationale for the use of combined therapy. An example of the latter would be to use the timing of the increased pO_2 as a guide as to when to use radiation therapy to get the maximum combined effectiveness with the drug.

MMPs are known to play an important role in tumor progression and angiogenesis (MacDougall and Matrisian, 1995). Ingrowth of blood vessels, necessary to supply the increasing demand for oxygen and other nutrients, is undoubtedly influenced by inhibition of MMPs (Thorgeirsson et al., 1993). BB-94 has previously been shown to affect tumor physiology/anatomy (Beattie et al., 1994)—causing smaller, avascular tumors composed of a high density of extracellular matrix enclosed in a dense capsule. Our results especially as shown in Figure 2A, further support these findings, and the higher pO_2 values measured in BB-94—treated tumors could have been the result of decreased number of cells in a given tumor volume, giving rise to a higher average pO_2. The pO_2 measured is a balance between oxygen delivery (blood supply) and oxygen consumption. If BB-94 was active in preventing blood vessel growth, then an overall decrease in tumor pO_2 might have been expected. We tested the influence of BB-94 on oxygen consumption of cells (both CHO and J774 macrophages) and found no effect. We cannot, however, rule out the possibility of a direct effect on tumor cell oxygen consumption *in vivo*. If the actions of the drug are related to changes in growth factors, then these data may provide important insights as to the amount and timing of the effects on the pO_2. The data from both the high and low-intensity studies showed a difference between average tumor pO_2

following BB-94 treatment compared to control tumors. It is interesting, however, that this difference was observed even in the case of low-intensity treatment, where tumor volume and necrotic fraction were not significantly different.

Our failure to observe a difference in the amount of MMP between control and treated tumors may not be surprising, since BB-94 inhibits proteinase enzyme action and is not thought to suspend or affect production of MMPs.

These data were conclusive in showing that BB-94 treatment influenced tumor growth and tumor pO_2 in MTG-B tumors in mice, and suggest the potential use of BB-94 to decrease tumor burden and increase tumor pO_2 so as to improve the effectiveness of radiotherapy.

ACKNOWLEDGMENTS

This work was supported in part by British Biotech Pharmaceuticals Limited, Oxford, UK, Program Project Grant P01 GM51630 (H.M.S.), Core Grant CA23108 of the Norris Cotton Comprehensive Cancer Center and utilized the facilities and personnel of the Dartmouth Medical School EPR Center for the Study of Viable Systems supported by NIH Grant P41 RR11602. The authors wish to thank Dr. Elizabeth Bone of British Biotech for input on experimental plans and helpful discussions.

REFERENCES

Agren MS, Jorgensen LN, Andersen M, Viljanto J, Gottrup F (1998) Matrix metalloproteinase 9 level predicts optimal collagen deposition during early wound repair in humans. British Journal of Surgery 85(1):68–71.

Bacic G, Liu KJ, O'Hara JA, Harris RD, Szybinski K, Goda F, Swartz HM (1993) Oxygen tension in a murine tumor: a combined EPR and MRI study. Magn Reson Med 30(5):568–572.

Beattie G, Young HA, Smyth JF (1994) Phase I study of intra-peritoneal metalloproteinase inhibitor BB94 in patients with malignant ascites {Abstr 015}. Ann Oncol 5:72.

Brizel DM, Sibley GS, Prosnitz LR, Scher RL, Dewhirst MW (1997) Tumor hypoxia adversely affects the prognosis of carcinoma of the head and neck. International Journal of Radiation Oncology, Biology, Physics 38(2):285–289.

Chambers AF, Matrisian LM (1997) Changing views of the role of matrix metalloproteinases in metastasis. Journal of the National Cancer Institute 89(17):1260–1270.

Clifton KH, Briggs RC, Stone HB (1966) Quantitative radiosensitivity studies of solid carcinomas *in vivo*: methodology and effect of anoxia. J Natl Canc Inst 36:965–974.

Clifton KH, Draper NR (1963) Survival curves of solid transplantable tumors irradiated *in vivo*: a method of determination and statistical evaluation; comparison of cell-survival and ^{32}P uptake in DNA. Int J Radiat Biol 7:515–535.

Davies B, Brown PD, East N, Crimmin MJ, Balkwill FR (1993a) A synthetic matrix metalloproteinase inhibitor decreases tumor burden and prolongs survival of mice bearing human ovarian carcinoma xenografts [published erratum appears in Cancer Res 1993 Aug 1;53(15):3652]. Cancer Res 53(9):2087–2091.

Davies B, Waxman J, Wasan H, Abel P, Williams G, Krausz T, Neal D, Thomas D, Hanby A, Balkwill F (1993b) Levels of matrix metalloproteases in bladder cancer correlate with tumor grade and invasion. Cancer Res 53(22):5365–5369.

Douple E, O'Hara J, Jones E (1987) Paraplatin enhancement of radiation therapy in a murine tumor (MTG-B). In Anticancer Drug Research (eds. Lapis, K and Eckhardt, S) Akademiai Kiado, Budapest, pp. 71–80.

Fowble B (1991) Local-regional treatment options for early invasive breast cancer. In Breast Cancer Treatment (eds. Fowble, B, Goodman, RL, Glick, JH and Rosato, EF) Mosby, St Louis, p. 43.

Goda F, Bacic G, O'Hara JA, Gallez B, Swartz HM, Dunn JF (1996) The relationship between partial pressure of oxygen and perfusion in two murine tumors after x-ray irradiation: a combined gadopentetate dimeglumine dynamic magnetic resonance imaging and in vivo electron paramagnetic resonance study. Cancer Res 56(14):3344–3349.

Goda F, O'Hara JA, Liu KJ, Rhodes ES, Dunn JF, Swartz HM (1997) Comparisons of measurements of pO_2 in tissue in vivo by EPR oximetry and micro-electrodes. In Oxygen Transport to Tissue (eds. Nemoto, EM and LaManna, JC) Plenum Press, New York, pp. 543–549.

Goda F, O'Hara JA, Rhodes ES, Liu KJ, Dunn JF, Bacic G, Swartz HM (1995) Changes of oxygen tension in experimental tumors after a single dose of X-ray irradiation. Cancer Research 55(11):2249–52.

Hall EJ (1988) Radiobiology for the Radiologist. JB Lippincott, Philadelphia.

Hockel M, Schlenger K, Aral B, Mitze M, Schaffer U, Vaupel P (1996) Association between tumor hypoxia and malignant progression in advanced cancer of the uterine cervix. Cancer Res 56(19):4509–4515.

Jackson SK, James PE, Swartz HM (1997) Effect of lipopolysaccharide on intra-phagosomal oxygen concentration as measured by EPR oximetry. Advances in Experimental Medicine & Biology 428:613–620.

James PE, Goda F, Grinberg OY, Szybinski KG, Swartz HM (1997a) Intrarenal pO2 measured by EPR oximetry and the effects of bacterial endotoxin. Advances in Experimental Medicine & Biology 411:557–568.

James PE, Grinberg OY, Goda F, O'Hara JA, Swartz HM (1997b) Gloxy: An oxygen-sensitive coal for accurate measurement of low oxygen tensions in biological systems. Magn Reson Med 38(1):48–58.

James PE, O'Hara JA, Grinberg OY, Panz T, Swartz HM (1996) Intratumoral pO_2 measured using a new oxygen sensitive paramagnetic material, Gloxy. In Oxygen Transport to Tissue (eds. Harrison, DK and Delpy, DT) Plenum Press, New York, pp. 97–106.

Jones E, Lyons B, Douple E, Filimonov A, Dain B (1989) Response of a brachytherapy model using ^{125}I in a murine tumor system. Radiat Res 118:112–130.

MacDougall JR, Matrisian LM (1995) Contributions of tumor and stromal matrix metalloproteinases to tumor progression, invasion and metastasis. Cancer & Metastasis Reviews 14(4):351–362.

O'Hara JA, Goda F, Demidenko E, Swartz HM (1998) Effect on regrowth delay in a murine tumor of scheduling split dose radiation based on direct pO_2 measurements by EPR oximetry. Radiation Research 150:549–556.

O'Hara JA, Goda F, Dunn JF, Swartz HM (1997) Potential for EPR oximetry to guide treatment planning for tumors. Advances in Experimental Medicine & Biology 411:233–242.

O'Hara JA, Goda F, Liu KJ, Bacic G, Hoopes PJ, Swartz HM (1995) The pO_2 in a murine tumor after irradiation: an in vivo electron paramagnetic resonance oximetry study. Radiat Res 144(2):222–229.

Powell WC, Matrisian LM (1996) Complex roles of matrix metalloproteinases in tumor progression. Current Topics in Microbiology & Immunology 213(Pt 1):1–21.

Reponen P, Leivo I, Sahlberg C, Apte SS, Olsen BR, Thesleff I, Tryggvason K (1995) 92-kDa type IV collagenase and TIMP-3, but not 72-kDa type IV collagenase or TIMP-1 or TIMP-2, are highly expressed during mouse embryo implantation. Developmental Dynamics 202(4):388–396.

Stone HB, Brown JM, Phillips TL, Sutherland RM (1993) Oxygen in human tumors: correlations between methods of measurement and response to therapy. Summary of a workshop held November 19–20, 1992, at the National Cancer Institute, Bethesda, Maryland. Radiat Res 136(3):422–434.

Thorgeirsson UP, Lindsay CK, Cottam DW, Gomez DE (1993) Tumor invasion, proteolysis, and angiogenesis. Journal of Neuro-Oncology 18(2):89–103.

Willner J, Kiricuta IC, Kolbl O (1997) Locoregional recurrence of breast cancer following mastectomy: always a fatal event? Results of univariate and multivariate analysis. International Journal of Radiation Oncology, Biology, Physics 37(4):853–863.

Withers HR (1987) Biologic basis of radiation therapy. In Principles and Practice of Radiation Oncology (eds. Perez, CA and Brady, LW) J B Lippincott Co, Philadelphia, PA, pp. 68–98.

BLOOD PERFUSION OF SUBCUTANEOUS TUMOURS IN MICE FOLLOWING THE APPLICATION OF LOW-LEVEL DIRECT ELECTRIC CURRENT

T. Jarm,[1] Y. A. B. D. Wickramasinghe,[2] M. Deakin,[2] M. Čemažar,[3] J. Elder,[2] P. Rolfe,[2] G. Serša,[3] and D. Miklavčič[1]

[1]Faculty of Electrical Engineering
University of Ljubljana
Tržaška 25, SI-1000 Ljubljana
Slovenia
[2]School of Postgraduate Medicine, Keele University
Thornburrow Drive, Hartshill, Stoke-on-Trent, ST4 7QB
United Kingdom
[3]Department of Tumour Biology
Institute of Oncology
Zaloška 2, SI-1000 Ljubljana
Slovenia

ABSTRACT

Electrotherapy with low-level direct electric current has been proved to be an effective local treatment of solid tumours. In the presented study an attempt was made to evaluate the effect of a single treatment with electrotherapy on blood perfusion of solid subcutaneous fibrosarcoma Sa-1 tumours in A/J mice. The tissue-staining method with Patent blue-violet dye, the rubidium extraction technique, and the noninvasive near-infrared spectroscopy method were used for this purpose. Results of all methods indicate that perfusion and subsequently oxygenation of tumours were reduced due to application of electrotherapy.

1. INTRODUCTION

It has been shown in several studies by various authors that electrical current can be used for treatment of solid malignancies. Antitumour effectiveness of electrotherapy

Oxygen Transport to Tissue XXI, edited by Eke and Delpy.
Kluwer Academic / Plenum Publishers, New York, 1999.

has been demonstrated on various experimental tumour models as well as in clinical trials (Humphrey and Seal, 1959; David et al., 1985; Marino et al., 1986; Heiberg et al., 1991; Griffin et al., 1994; Plesničar et al., 1994; Xin, 1994). In all reported studies electrotherapy was used as a local treatment and as such it can also be used as a suitable adjuvant treatment with other therapies in order to potentiate their localised effectiveness (Serša et al., 1992; Serša et al., 1993b). Many attempts have been made to try explain effectiveness of electrotherapy alone or of its combined use with other therapies. The understanding of the mechanisms of antitumour action is of importance for optimising existing and developing new treatment strategies. One can hardly expect that one single mechanism would be responsible for demonstrated antitumour effectiveness because wide ranges of different tumours and various treatment modalities were used in reported studies. In the present study we concentrate on one particular experimental tumour model which has been extensively used in the past in studies of the effects of direct current electrotherapy (DC-ET) alone and of its use in combination with certain anticancer drugs (Serša et al., 1992, 1993a, 1993b). Growth of murine fibrosarcoma Sa-1 tumours in A/J mice is temporarily retarded by application of a single DC-ET and tumours can be permanently eradicated when DC-ET is applied together with tumour necrosis factor TNF-α at doses which are not effective when TNF-α alone is administered (Serša et al., 1994). Growth delay of treated tumours with respect to control tumours exhibits dose dependence on amplitude and duration of applied current (Serša et al., 1993a). The effects of different electrode materials and configurations on tumour growth have been studied as well as the effects of DC-ET on tissue pH and temperature, on production of tissue necrosis, and on electrode material deposition (Miklavčič et al., 1993, 1994). None of the above effects were found to be responsible for the demonstrated tumour growth retardation in the case when electrodes were placed outside the tumour with the tumour itself being located between the electrodes.

One of the proposed mechanisms of antitumour action of DC-ET for electrode configurations where tumour is not penetrated by the stimulation electrodes has been the effect of direct current on the vessels supplying the tumour at the site of electrode insertion (Griffin et al., 1995). Solid tumours in general are known to be hypoxic in comparison to normal tissues. Poor supply of oxygen and nutrients to tumours is a result of structurally and functionally abnormal microcirculation of blood in tumours. Deficient perfusion and oxygenation are not only limiting factors for tumour growth but can also result in resistance of tumours to radiotherapy and to treatment with various chemotherapeutics (Vaupel, 1979; Jain, 1988). Therefore it has been suggested that altering the blood supply to tumours by means of DC-ET might lead to eradication of tumour mass (Griffin et al., 1995).

In the presented work we assessed the effects of DC-ET on parameters of tumour perfusion by means of three methods: invasively by the tissue staining method using Patent blue-violet biological dye (PBV) and by the rubidium 86 extraction technique (^{86}Rb), and non-invasively using the optical method of near-infrared spectroscopy (NIRS).

2. METHODS

2.1. Tumour Model

Healthy female and male mice of immunocompetent A/J strain (8–12 weeks of age) were used in the experiments. Only female animals were used for the near-infrared spectroscopy part of the study and animals of both sexes were used in experiments involving staining of tissue with Patent blue-violet (PBV) and rubidium extraction technique. Animals were purchased from Rudjer Bosković Institute, Zagreb, Croatia. Mice were kept in standard animal colonies and fed *ad libitum*. Solid subcutaneous tumours were initiated dorsolaterally in the right flank of animals with injection of saline suspension of about $5 \cdot 10^5$ viable fibrosarcoma Sa-1 cells. Cells were obtained from ascitic form of Sa-1 tumours and their viability was assessed with standard trypane blue staining method. The experiments were performed 7 to 10 days after inoculation when tumours reached volume of about $50\,mm^3$. Tumour volume was calculated from three mutually orthogonal diameters of the tumour (a, b, c) measured with calliper and using the formula $V = \pi abc/6$. The experiments were conducted in accordance with the European Council directives and the Slovenian and the British legislation concerning animal welfare.

2.2. Electrotherapy (DC-ET)

A single treatment with DC-ET consisted of one-hour treatment with a current of 0.6 mA amplitude. An adjustable current generator was used to provide regulated direct current which was delivered via a pair of subcutaneously inserted needle electrodes (Pt/Ir 90/10% alloy, diameter 1 mm, length 2 cm, rounded tip). Electrodes were inserted in parallel on two opposite sides of the tumour at least 5 mm away from its edge, with the anode (positive electrode) always on the caudal side of the tumour. Thus the tumour was not in physical contact with the electrodes (Figure 1). Animals in the control groups were treated in exactly the same way as animals subjected to DC-ET except that no current was applied. Throughout the experiments animals were lightly physically restrained and anaesthetised using the inhalation agent enflurane which was delivered in a mixture of oxygen and nitrous oxide. Body temperature was maintained at physiological values by a regulated heating pad on which the animals were placed.

2.3. Tissue Staining with Patent Blue-Violet Dye (PBV)

Saline solution of biological dye Patent blue-violet (Byk Gulden, Switzerland) was injected into the tail vein of animals (dose 0.2 ml, concentration 1.25%). After the dye had been left to distribute evenly through the tissues for 2 minutes, animals were euthanised and tumours were removed from the skin and cut in half along their largest diameter. The percentage of stained area of tumour cross-section (perfused) as opposed to non-stained area (non-perfused) was visually estimated by two persons. An average of both estimations was used as an indicator of tumour perfusion. Experimental groups of 6 to 9 animals were formed with respect to the predefined time interval between the start of DC-ET and the moment at which animals were injected with the dye. Corresponding control groups were also formed (Szabo, 1985; Sagar et al., 1991).

2.4. Rubidium Extraction Technique

Relative tissue perfusion (RTP) was assessed by the ^{86}Rb extraction technique at predefined time intervals after DC-ET. For each time interval a separate group of animals was used with 4 to 6 animals per group. Corresponding control groups were also formed. For injection, a 50 mCi/ml solution of $^{86}RbCl$ (Amersham PLC, Little Chalfont, Bucks., UK) was used. At predefined intervals the 0.1 ml aliquots of solution were injected into the tail vein of animals. Animals were euthanised one minute after injection. It has been experimentally verified that the amount of ^{86}Rb extracted by tissues one minute after injection was well within the plateau region required for validity of the method

(Sapirstein, 1958; Zanelli and Fowler, 1974). The whole tumour, thigh muscle sample from the left hind leg, and the tail from the same animal were removed and placed in preweighed glass vials. Muscle samples from the leg were used for assessment of the effects of electrotherapy applied locally to the tumour in the right flank on perfusion of other tissues not directly exposed to electric current. All samples were weighed and their radioactivity was measured using a gamma counter (Institute Jožef Stefan, Ljubljana, Slovenia). In each measurement sequence the activity of three 0.1 ml aliquots of injection solution and the background activity were also measured.

Radioactivity measured in the tumour and the muscle sample was corrected for the background activity and the dose radioactivity was corrected for the residual activity in the tail. All animals with more than 10% of the injected dose retained in the tail due to improper injection were excluded from further evaluation. The following equation was applied to calculate the RTP of tissues expressed as a percentage of dose extracted in the tissue sample normalised per sample weight:

$$RTP(\%/\mathrm{g}) = 100 \cdot \frac{(activity_{SAMPLE} - activity_{BACKGROUND})/weight_{SAMPLE}}{activity_{DOSE} - activity_{TAIL}}$$

2.5. Near-Infrared Spectroscopy

Two NIRO2-X2 instruments (Keele University, UK) were used in the study for simultaneous monitoring of concentration changes of oxygenated and deoxygenated haemoglobin (HbO and Hb) in tumour and normal reference tissue of the same animal. Pulsed laser near-infrared light was delivered to the tissue and collected from the tissue by two optical light guides (optodes). Concentration changes were calculated from changes in attenuation of light in tissue at three wavelengths (904 nm, 845 nm, 775 nm) using a modified Beer-Lambert law (Wickramasinghe et al., 1992). Two additional parameters were calculated from measured HbO and Hb changes; change of total haemoglobin concentration Hb-tot defined as HbO+Hb which is correlated with changes of blood volume in the tissue under observation, and change of differential haemoglobin concentration Hb-dif defined as HbO-Hb. Monitoring of normal tissue (thigh muscles of left hind leg as opposed to tumour growing in the right flank) served for evaluation of general stability of the animal under anaesthesia and for assessing systemic responses to DC-ET. Before measurement all hair was removed from the skin covering the tumour and the muscle. Due to the small dimensions of the animals and tumours it was impossible to attach optodes to the site of measurement itself (Figure 1). Therefore both pairs of optodes were held fixed to desired positions by special holders.

Monitoring of NIRS parameters was started approximately 20 minutes after induction of anaesthesia. DC-ET was started only when stable NIRS signals from both tumour and muscle tissue had been recorded for at least 20 minutes. NIRS monitoring was maintained until approximately 60 minutes after the end of DC-ET and for a corresponding period in control animals.

3. RESULTS

3.1. Tissue Staining with PBV Dye

We used biological dye Patent blue-violet (PBV) instead of Evans blue which is otherwise a more widely used dye in perfusion studies. The reason for this is that the

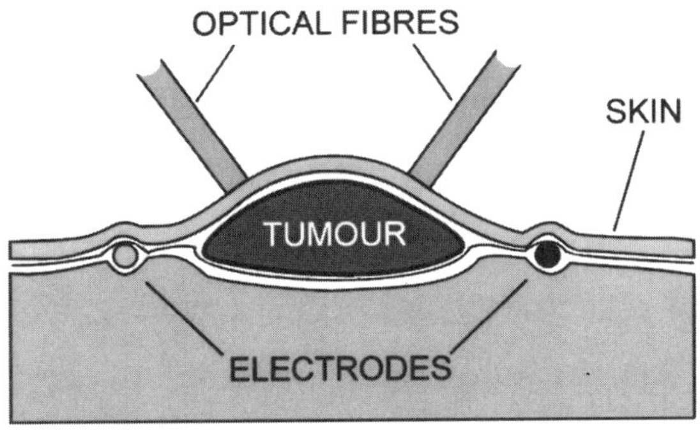

Figure 1. Position of electrodes for DC-ET and optical guides for NIRS measurements in tumours.

colour contrast between stained and non-stained areas of tissue was far better with PBV. We verified in a separate pilot study that both dyes produce virtually the same results for control tumours and for tumours treated with DC-ET (data not shown).

Results of staining the tumours with PBV at different intervals during and after ET are presented in Figure 2. The degree of staining of control tumours did not change during the period of observation and was about 90% of cross-section area. Five minutes after the onset of DC-ET tumours were still equally stained as the controls. Percentage of stained area then rapidly declined to minimum level of about 15% which was reached in 15–30 minutes. This reduced level was maintained for nearly 24 hours when some minor increase in perfusion could already be observed. Three days after DC-ET treat-

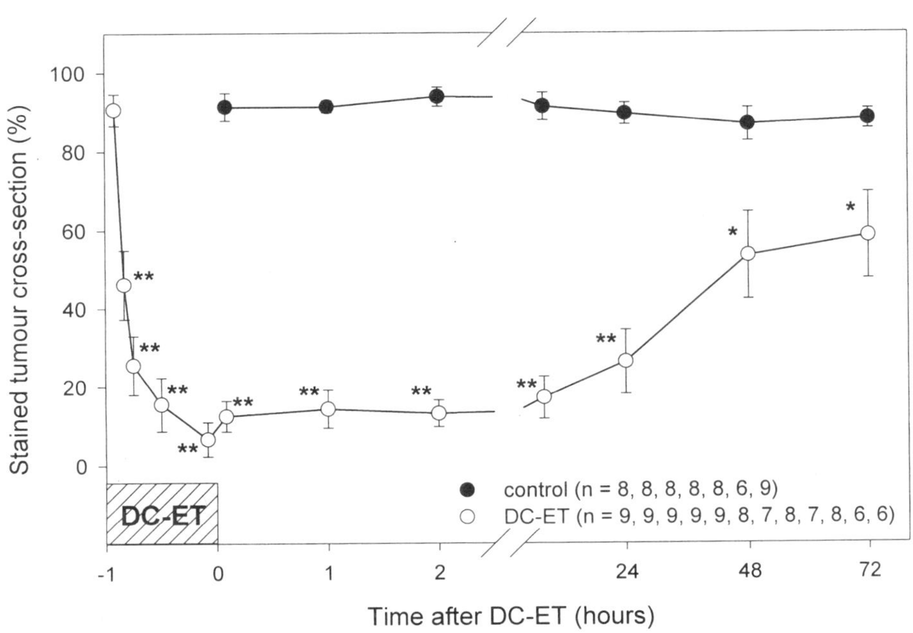

Figure 2. Perfusion of tumours at different time intervals during and after DC-ET assessed as percent of stained area of tumour cross-section. Mean values and standard error bars are shown (*P < 0.05, **P < 0.001; Student t test).

ment the percentage of stained area reached about 60% which was still substantially less than in the corresponding control tumours. It should be noted that the results from individual tumours at different time intervals were very homogeneous in all control groups. However, scattering of individual data was increased for DC-ET-treated tumours especially during dynamic changing of perfusion level, i.e. during "deperfusion" in the first half hour of DC-ET and during "reperfusion" 24 to 72 hours after DC-ET.

3.2. Rubidium Extraction Technique

Results of measuring the relative tissue perfusion (RTP) of tumours at different intervals during and after DC-ET are presented in Figure 3.

As soon as 15 minutes after the start of application of electrotherapy, the RTP of the tumours had already reached the minimum level of about 1.5%/g which was more than a 50% reduction with respect to control tumours where RTP was about 3.5%/g. A statistically significant decrease in RTP was maintained for two days except for the group of tumours 24 hours after treatment. The mean value of these tumours was surprisingly high but this was due to three tumours which were apparently not successfully treated. Three days after treatment a partial reperfusion of treated tumours was demonstrated. Control tumours on their own exhibited a gradual decrease in perfusion with increasing size and age of tumours as expected. As with the staining technique, the scattering of results of individual animals was higher for DC-ET than for control animals.

No significant change due to the application of DC-ET to tumours was observed in thigh muscles which were not directly exposed to electrical treatment. Perfusion of muscles also did not change significantly during the observation period of three days.

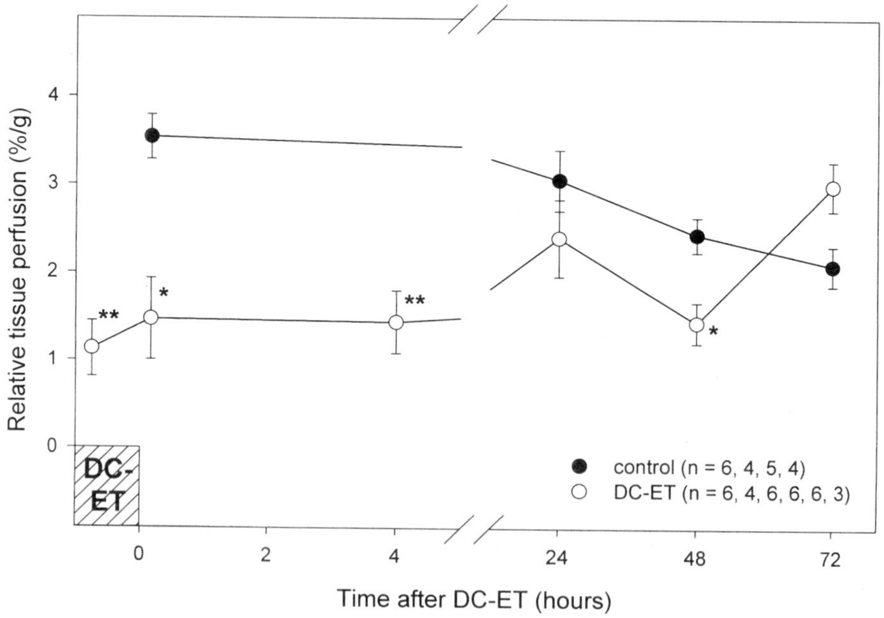

Figure 3. Relative tissue perfusion of tumours at different time intervals during and after DC-ET assessed by the rubidium extraction technique. Mean values and standard error bars are shown (*P < 0.05, **P < 0.001; Student t test).

The mean RTP values of muscles in all treated and control animals were within the 3–4%/g range (data not shown).

3.3. Near-Infrared Spectroscopy

Only those animals were included in evaluation for which stable baseline values of measured parameters prior to therapy were established in both tumour and normal tissue. All cases where measurements became invalid during monitoring due to disruption of optical contact between the optodes and tissue were also excluded from further evaluation. Total rejection rate was about 30%. Thus 5 and 6 animals remained in DC-ET and control group, respectively.

There was a large variability in magnitudes of response of the measured parameters to DC-ET between individual tumours. Nevertheless, common typical features were observed in the majority of tumours.

Figure 4 presents data recorded from two typical tumours from DC-ET and control groups. Onset of DC-ET resulted in mild transient excitation of the animals despite anaesthesia. After initial excitation HbO decreased rapidly to a level lower than before DC-ET. This level was maintained for a period of about 30 minutes in most animals but then HbO started to increase gradually. Immediately following the excitation Hb started to increase thus partially matching the rapid decrease of HbO. After the minimum HbO level had been reached Hb continued to increase at a lesser but steady rate. Hb-tot was initially decreased due to the rapid decrease of HbO but then it started to increase due to the increasing Hb and later also due to the increasing HbO. Hb-dif decreased in all treated tumours without exception and in most tumours it continued to decrease during the whole period of observation. In tumours of control animals and also in muscles of

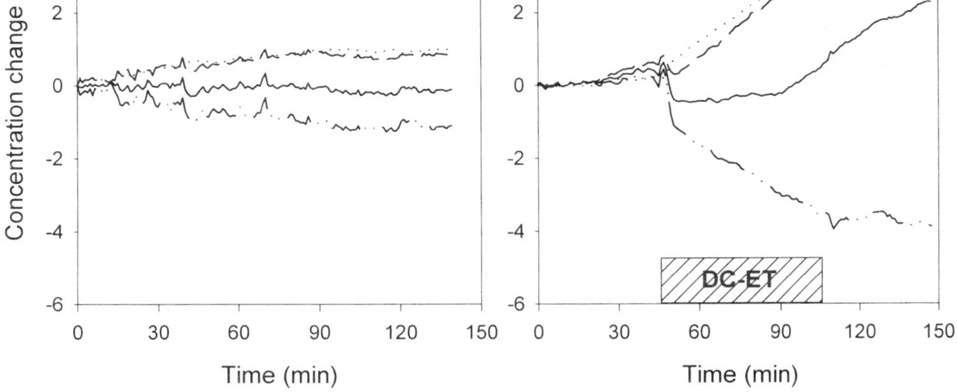

Figure 4. Typical NIRS measurements of individual response of control tumour (a) and of tumour treated with DC-ET (b).

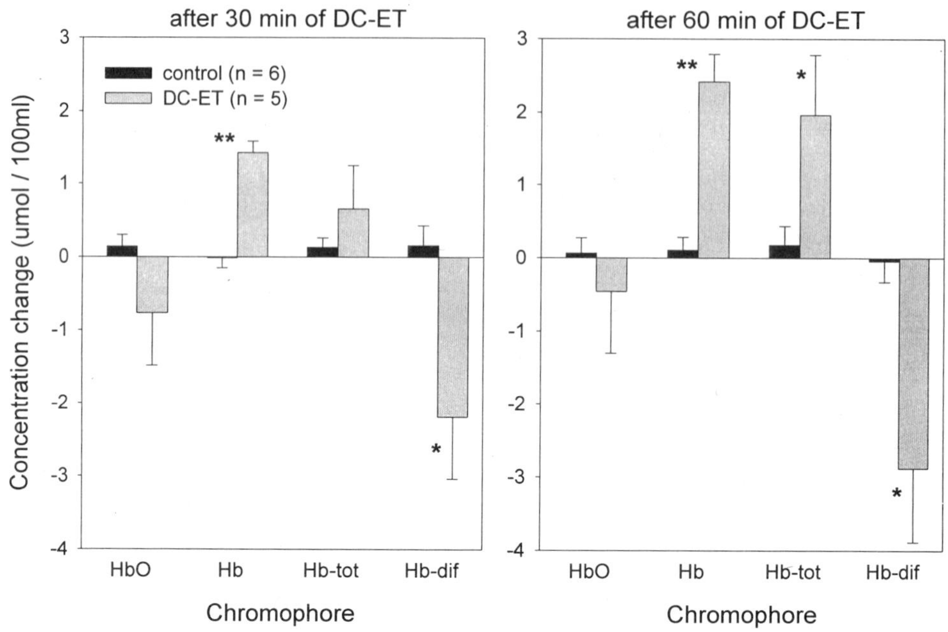

Figure 5. Comparison of changes in concentration of NIRS parameters (HbO, Hb, Hb-tot, Hb-dif) in control tumours and tumours treated with DC-ET. Mean values and standard error bars are shown (*P < 0.05, **P < 0.001; Student t test).

DC-ET-treated animals there were no significant changes of measured parameters. Average changes of measured tumour parameters in both experimental groups at two characteristic time intervals with respect to the moment of start of DC-ET are presented in Figure 5.

4. DISCUSSION AND CONCLUSIONS

The perfusion data obtained with the tissue staining method and by the rubidium extraction technique clearly demonstrate that blood supply to the tumours was severely decreased due to application of DC-ET. Although the direct histological evidence is not available at the moment, we suggest that the most probable reason for this was damaged blood vessels and subsequent vascular occlusion at the site of insertion of the electrodes especially since the skin between the electrodes and the tumour was also not stained in most cases and the skin where the electrodes were inserted appeared macroscopically damaged. Extreme pH changes as the result of electrolysis in the immediate vicinity of the electrodes demonstrated by (Miklavčič et al., 1993) could explain this effect. In the samples taken during the "deperfusion" period (first 30 minutes after the start of ET) we observed that non-perfused areas in tumours occurred first in the part of tumour that was closer to the cathode. This is in accordance with the results of previous studies where it was demonstrated that post-treatment necrosis at the site of insertion of the electrodes was more extensive at the cathode than at the anode in both tumour and normal tissue. Rapid decrease in HbO and Hb-tot after the start of DC-ET as demonstrated by NIRS in the majority of treated tumours correlates well with the perfusion data. The dynamics of the decrease in PBV staining and HbO are very similar. Maximum decrease in HbO and minimum level of tissue staining are reached on average in about 15 minutes and in 15–30 minutes respectively after onset of DC-ET. Faster decrease of HbO could be explained by increased extraction of oxygen from the blood (deoxygenation of haemoglobin) which would occur in case of decreased blood flow through tumour. This hypothesis is supported by the fact that Hb was increasing more rapidly during decreasing of HbO than at the latter stage after the minimum HbO level had been reached which is an indicator of increased rate of deoxygenation of haemoglobin. The slow but steady increasing of Hb and Hb-tot observed at latter times indicates that blood volume was increasing. This could be explained if venous outflow from the tumour was more severely impaired than arterial inflow, for which however, no direct evidence is available at the moment. This would mean that fresh blood was still able to enter some parts of the tumour in reduced quantities but it was not able to leave the tumour. This would lead to gradual accumulation of blood (increasing of Hb-tot) in the tumour. We observed substantial quantities of spilled blood within some of tumours treated with DC-ET and in their immediate vicinity. This effect was absent in control tumours. We think that the large Hb-tot increase at the end of NIRS monitoring accounts for this blood pooling. The moment when HbO started to increase in some tumours (but not in all) probably indicated when leaking from vessels in the tumours began. It is however possible that the recorded NIRS signals also contained contributions from the tissue surrounding the tumours due to the small size of tumours.

The reperfusion dynamics in the days following the DC-ET treatment as demonstrated by tissue staining and rubidium extraction (Figures 2, 3) were in good agreement with known dynamics of regrowth of Sa-1 tumours after DC-ET (Serša et al., 1993a). On average in the first days following DC-ET the size of these tumours was reduced. Then 3 to 4 days after treatment tumours started growing again. Tissue staining with PBV and rubidium extraction revealed that 3 days after treatment partial (PBV) or total (rubidium) reperfusion occurred.

We have previously observed that not all tumours are equally responsive to DC-ET resulting in wider ranges of values of all measured parameters for DC-ET-treated tumours in comparison to control tumours. In the light of the effect of DC-ET on tumour-feeding vessels the main possible factor besides the biological differences between tumours would be the relative proximity of these vessels to electrodes during treatment and the ratio of affected and unaffected supplying vessels.

Measurements taken from normal muscle tissue in case of NIRS and rubidium extraction technique confirmed our expectations that the effects of DC-ET were indeed localised in the area between the electrodes and that DC-ET did not produce significant perfusion changes in other tissues.

We conclude that perfusion and consequently oxygenation of subcutaneous Sa-1 fibrosarcoma tumours in A/J mice were significantly decreased due to electrotherapy. The results obtained by means of all three methods are mutually supportive and are in good agreement with known growth dynamics of tumours after treatment. In this particular tumour model the effect on tumour blood supply is most probably the primary mechanism of antitumour action of DC-ET.

ACKNOWLEDGMENT

The studies were supported by the Ministry of Science and Technology of the Republic of Slovenia and by the EC Concerted Action Project on Near Infrared

Spectrophotometry and Imaging of Biological Tissue (funded by Commission of the European Communities, Directorate-General for Science, Research and Development, International Scientific Cooperation, Brussels, Belgium).

REFERENCES

David SL, Absolom DR, Smith CR, Gams J, and Herbert MA (1985) Effect of low level direct current on in vivo tumor growth in hamsters. Cancer Res. 45:5626–5631.

Griffin DT, Dodd NJF, Moore JV, Pullan BR, and Taylor TV (1994) The effects of low-level direct current therapy on a preclinical mammary carcinoma: tumour regression and systemic biochemical sequelae. Br. J. Cancer 69:875–878.

Griffin DT, Dodd NJF, Zhao S, Pullan BR, and Moore JV (1995) Low-level direct electrical current therapy for hepatic metastases. I. Preclinical studies on normal liver. Br. J. Cancer 72:31–34.

Heiberg E, Nalesnik WJ, and Janney C (1991) Effects of varying potential and electrolytic dosage in direct current treatment of tumors. Acta Radiol. 32:174–177.

Humphrey CE and Seal EH (1959) Biophysical approach towards tumor regression in mice. Science 130:388–390.

Jain RK (1988) Determinants of tumor blood flow: a review. Cancer Res. 48:2641–2658.

Marino AA, Morris D, and Arnold T (1986) Electrical treatment of Lewis lung carcinoma in mice. J. Surg. Res. 41:198–201.

Miklavčič D, Serša G, Kryžanowski M, Novaković S, Bobanović F, Golouh R, and Vodovnik L (1993) Tumor treatment by direct electric current—tumor temperature and pH, electrode material and configuration. Bioelectroch. Bioenerg. 30:209–220.

Miklavčič D, Fajgelj A, and Serša G (1994) Tumor treatment by direct electric current: electrode material deposition. Bioelectroch. Bioenerg. 35:93–97.

Plesničar A, Serša G, Vodovnik L, Jančar J, Zaletel-Kragelj L, and Plesničar S (1994) Electric treatment of human melanoma skin lesions with low level direct electric current: an assesment of clinical experience following a preliminary study in five patients. Eur. J. Surg. Suppl. 574:45–49.

Sagar KB, Pelc LR, Rhyne TL, Howard J, and Warltier DC (1991) Estimation of myocardial infarct size with ultrasonic tissue characterization. Circulation 83:1419–1428.

Sapirstein LA (1958) Regional blood flow by fractional distribution of indicators. Am. J. Physiol. 193:161–168.

Serša G, Miklavčič D, Batista U, Novaković S, Bobanović F, and Vodovnik L (1992) Anti-tumor effect of electrotherapy alone or in combination with interleukin-2 in mice with sarcoma and melanoma tumors. Anti-Cancer Drugs 3:253–260.

Serša G and Miklavčič D (1993a) The feasibility of low level direct current electrotherapy for regional cancer treatment. Reg. Cancer Treat. 1:31–35.

Serša G, Novaković S, and Miklavčič D (1993b) Potentiation of bleomycin antitumor effectiveness by electrotherapy. Cancer Lett. 69:81–84.

Serša G, Golouh R, and Miklavčič D (1994) Anti-tumor effect of tumor necrosis factor combined with electrotherapy on mouse sarcoma. Anti-Cancer Drugs 5:69–74.

Szabo G (1985) New steps in the intra-arterial chemotherapy of head and neck tumors. Oncology 42:217–223.

Vaupel P (1979) Oxygen supply to malignant tumors. In Tumor Blood Circulation: Angiogenesis, Vascular Morphology and Blood Flow of Experimental and Human Tumors (Peterson H-I, Ed.) CRC Press, Boca Raton, pp. 143–168.

Wickramasinghe YABD, Livera LN, Spencer SA, Rolfe P, and Thorniley MS (1992) Plethysmographic validation of near infrared spectroscopic monitoring of cerebral blood volume. Arch. Dis. Child. 67:407–411.

Xin Y-L (1994) Advances in the treatment of malignant tumours by electrochemical therapy. Eur. J. Surg. Suppl 574:31–36.

Zanelli GD and Fowler JF (1974) The measurement of blood perfusion in experimental tumors by uptake of ^{86}Rb. Cancer Res. 34:1451–1456.

EFFECT OF BREATHING CARBOGEN ON THE OXYGEN TENSION OF MURINE AND HUMAN TUMOURS MEASURED USING AN EPPENDORF pO_2 HISTOGRAPH

Andrew I. Minchinton,[1] Karen H. Fryer,[1] Peter Lim,[2] and Christina Aquino-Parsons[2]

[1]BC Cancer Research Centre and
[2]BC Cancer Agency
600 West 10th Avenue, Vancouver
British Columbia V5Z 1L3, Canada

INTRODUCTION

For decades the presence of hypoxic cells within solid tumours has been regarded as an impediment to the efficacy of radiotherapy (Thomlinson and Gray, 1955). Several treatment modalities have been developed with the aim of overcoming the problem posed by hypoxic cells though, so far, they have had minimal clinical impact (Hall, 1994). Most efforts have centred around either supplying more oxygen to tumour tissue, with the aim of increasing the oxygen tension in the hypoxic cells and thus increasing their sensitivity to radiation or administering drugs which selectively increase the sensitivity of hypoxic cells to radiation (Dische, 1991). More recently drugs have been developed which exhibit selective toxicity to cells at lower than normal oxygen tension (Brown, 1993). The aim of treatment with these "bioreductive cytotoxins" is to kill the hypoxic cells whilst conventional radiation / chemotherapy eradicates the more oxygenated tumour cells.

Much interest is now focused on the measurement of tumour oxygen tension and correlation of this with outcome after therapy. To date, the most extensive study has been performed on cervix tumours (Höckel et al., 1993; Höckel et al., 1996) where a positive correlation has been observed between mean tumour pO_2 and overall survival or local response. These studies have been facilitated by the development of a convenient needle based oxygen sensor (Vaupel et al., 1991). Other studies using the same measurement technique on other tumour sites has also shown a positive correlation between tumour oxygenation measured using an Eppendorf pO_2 histograph and therapeutic outcome

Oxygen Transport to Tissue XXI, edited by Eke and Delpy.
Kluwer Academic / Plenum Publishers, New York, 1999.

(Nordsmark et al., 1996; Brizel, 1996). These observations suggest that increasing tumour oxygen tension can improve treatment outcome.

Soon after hypoxic tumour cells were postulated as being responsible for tumour radioresistance (Thomlinson and Gray, 1955) breathing gas mixtures with increased oxygen content from that of air (21%) was employed with the aim of improving tumour oxygenation (Churchill-Davidson et al., 1957). Since that time several clinical trials have combined radiation therapy with the breathing of gases of increased oxygen content, primarily hyperbaric oxygen (Cade et al., 1978; Dische, 1979; Dische et al., 1983; Fletcher et al., 1977; Henk, 1986; Watson et al., 1978). The conclusions from these studies regarding improvement to therapy are equivocal.

Interest in using carbogen breathing to improve tumour radiation sensitivity (particularly in combination with the administration of nicotinamide) has recently intensified (Kjellen et al., 1991; Rojas, 1991). Here we report some preliminary results from a study examining the change in tumour pO_2 after breathing carbogen (95% O_2, 5% CO_2) in patients with cervix cancer and in implanted tumours in mice measured using an Eppendorf pO_2 histograph.

The most important question that this study addresses concerns the relationship between oxygen tension measured using the Eppendorf pO_2 sensor and the presence of hypoxic cells. Carbogen breathing has been shown to increase mean and median tissue pO_2 values with a concurrent decrease in the proportion of measurements registering oxygen tension less than 5 or 2.5 mmHg (Falk et al., 1992). However since there is no direct method of assessing hypoxic cells, the decrease in the presence of the population of hypoxic cells within the tumour is inferred. Horsman has demonstrated in C3H mammary carcinoma that pO_2 measured using an Eppendorf pO_2 sensor correlates with the hypoxic fraction measured using radiobiological methods in mice that were subjected to various treatments that altered tumour oxygenation and radiosensitivity (Horsman et al., 1993). However, when different types of tumours were surveyed no correlation was found between hypoxia measured radiobiologically and oxygenation status measured using the Eppendorf pO_2 sensor (Horsman et al., 1995). These studies suggest that the relationship between pO_2 measured using Eppendorf sensors and the presence of radioresistant hypoxic cells is complex.

MATERIALS AND METHODS

Measurements of pO_2 were made using an Eppendorf pO_2 histograph (Hamburg, Germany). In both the human and animal studies the needle sensor was advanced into tissue stepwise using the standard format employing a 1 mm advance followed by a 0.3 mm pull-back. No anaesthesia was used. The cervical tumours were assessed using 4 parallel tracks at the 12, 3, 6 and 9 o'clock positions relative to the cervical canal similar to the method described previously (Höckel et al., 1991). Patients breathed carbogen at a flow rate of between 8–15 l/min via a Bains anaesthetic circuit using a Morgan mouthpiece and a nose clip. After inhaling carbogen for 4 minutes a second set of measurements using 4 parallel tracks in approximately the same position as previously were obtained whilst the patient was continuing to breath carbogen. Fifteen patients with FIGO stage Ib1 to IIIb epithelial cervical cancers were assessed.

The murine tumours used were SCCVII implanted subcutaneously on the backs of female C3H/hen mice and used when they reached approximately 8 mm in diameter. Mice were gently restrained in a lead jig whilst measurements were made and carbogen was breathed by the mice by supplying the front of the lead box with 1 litre of gas per minute. Mice and humans breathed carbogen for approximately 4 minutes prior to pO_2 measurements being made. Ten mice breathing air and ten breathing carbogen were also examined. The measurement procedure lasted 2–5 minutes in humans and animals.

RESULTS

SCCVII tumours of air breathing mice in this study exhibited consistent pO_2 values as a function of distance along the track through the tumour. Several tracks were sampled by inserting the needle through a puncture until it was just subcutaneous and then advancing the needle in different directions through the tumour. In this way 24 to 40 measurements were made in each tumour. Typical examples of these profiles of pO_2 as a function of distance are shown in Figure 1. Figure 1a and 1b shows tumour pO_2 profiles from air breathing mice and Figures 1c and 1d show tumour pO_2 profiles in mice which have breathed carbogen for approximately 4 minutes prior to the commencement of mea-

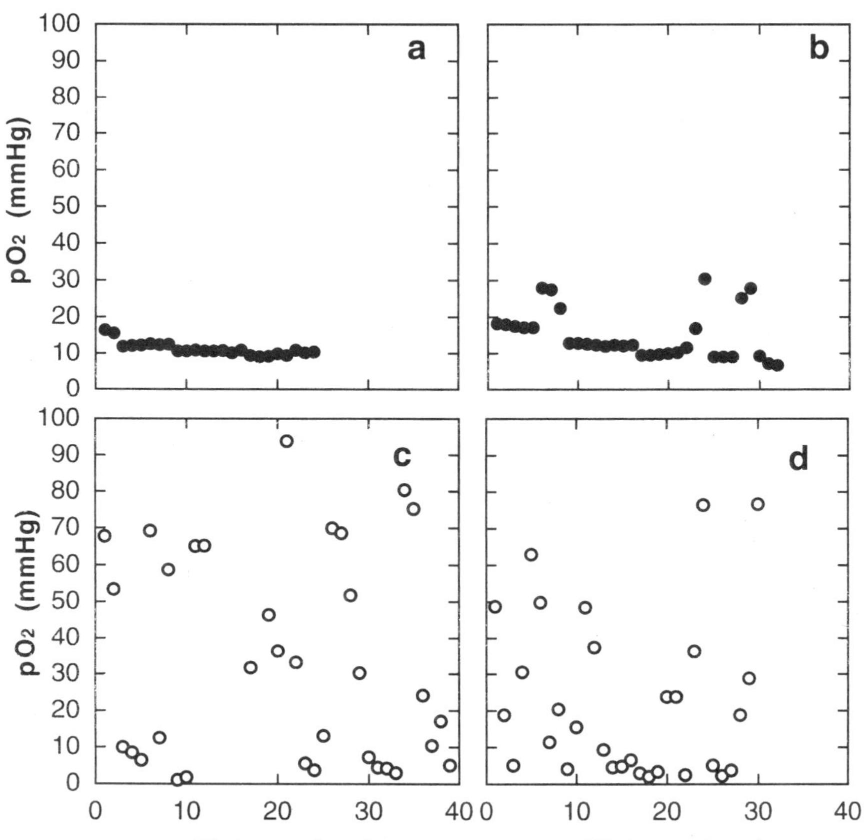

Figure 1. Tumour pO_2 as a function of accumulated distance through tumours (usually 4 tracks through a tumour approximately 8 mm in diameter were made. Panels a & b show the profile of pO_2 values seen in two typical SCCVII tumours in air-breathing mice. Panels c & d show the pO_2 values seen in the same tumours after the mice breathed carbogen for 4 minutes. Each point represents an individual pO_2 measurement.

surements. Figure 2a and 2b shows the histograms of pO_2 values obtained in 10 animals examined breathing air and carbogen respectively. Individual tumour pO_2 profiles in air breathing animals were homogeneous, as has been noted previously (Minchinton, 1995), but the tumour pO_2 profiles of carbogen breathing animals were much more variable. Many values were similar to those from air-breathing animals, but superimposed on this were many values of much higher, but no lower, pO_2 values. The histograms of tumours of carbogen breathing animals showed that the mean pO_2 values at least doubled compared to air breathing animals.

Human tumour pO_2 profiles were more heterogeneous when patients were breathing either room air or carbogen. Figure 3a shows the pO_2 histogram of patients breathing air and figure 3b shows the tumour pO_2 values from the same patients after breathing carbogen. There is a dramatic increase in measured pO_2 values when patients breath carbogen, amounting to an increase of over 10-fold in mean pO_2 from 8.3 to 118 mmHg. The median pO_2 increases similarly. The proportion of measurements less than 2.5 mmHg and less than 5 mmHg also decrease dramatically from 34% and 62% to 4.8 and 6.7% respectively. This increase in measured pO_2 is much greater than seen in murine tumours (Figure 2).

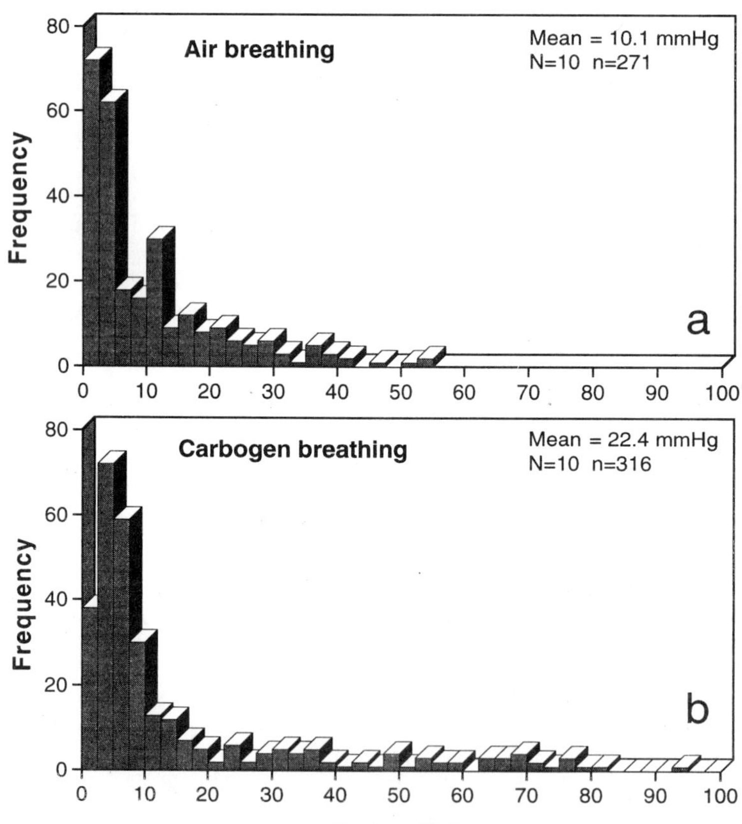

Figure 2. Frequency histograms of SCCVII tumour pO_2 in mice breathing air (panel a) and after breathing carbogen for 4 minutes (panel b). Details the same as described in Figure 1.

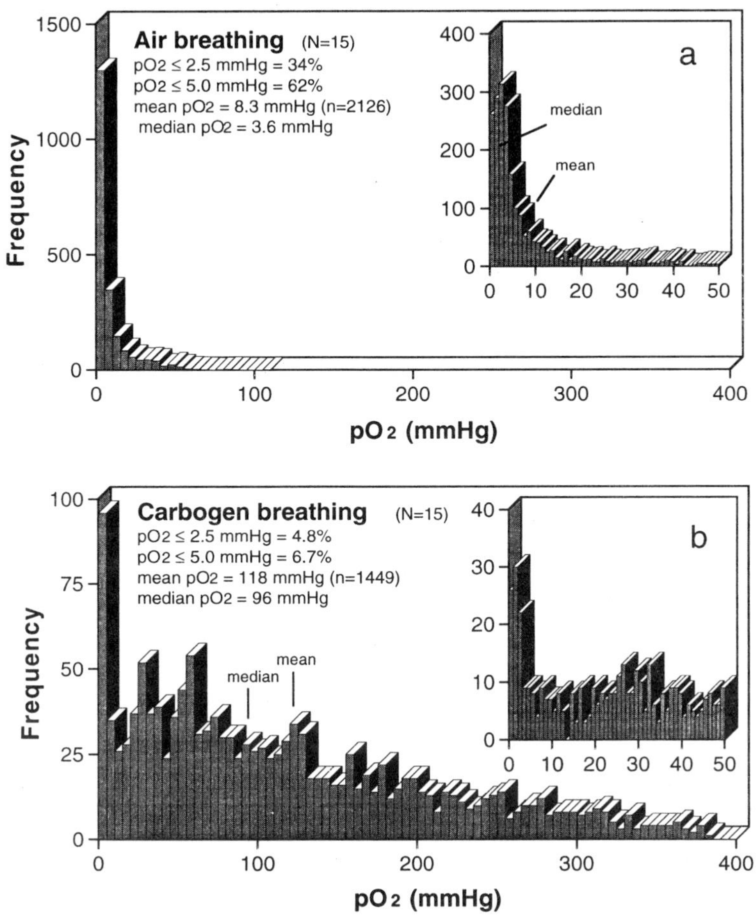

Figure 3. Frequency histograms of cervix tumour pO_2 in 15 patients breathing air (panel a) and after breathing carbogen for 4 minutes (panel b). Inset graphs show the low pO_2 range (0–15 mmHg) classified by 1.0 mmHg intervals. The main graph shows pO_2 range from 0–400 mmHg classified by 5 mmHg intervals.

DISCUSSION

One of the most important questions relating to the use of carbogen as a means of decreasing tumour hypoxia in patients undergoing radiation therapy is whether the promising effects seen in animal model tumours is translated to human malignancies. Several studies have show that carbogen breathing can increase tumour radiation sensitivity presumably via a decrease in the hypoxic cell population within the tumour (Rojas, 1991). A method of directly quantifying the oxygenation of tumours is therefore desirable and the use of Eppendorf pO_2 histography is compelling because of its ease of use in accessible tumours.

Our results show that carbogen increases the measured pO_2 in both animal and human tumours, but that the magnitude of the increase seen in cervix tumours in patients is considerably larger than observed in murine SCCVII tumours. The increase in pO_2 measured in cervix tumours is larger than would be expected on the basis of the increased oxygen carrying capacity of carbogen saturated blood (Churchill-Davidson et al., 1957).

One explanation that might account for the large increase in tumour oxygen tension measured by the Eppendorf pO_2 sensor in human cervix tumours is that carbogen breathing may induce an increase in tumour blood supply and thus enhance oxygen delivery via tumour blood vessel dilation (Honess and Bleehan, 1995) and possibly other mechanisms. Another explanation is that the needle sensor is influenced by the high pO_2 of blood that is exposed to the sensor on the needle via trauma and haemorrhage that occurs as the needle tracks through tissue.

In conclusion, tumour oxygenation, as measured using an Eppendorf pO_2 sensor is dramatically increased in cervical tumours after breathing carbogen for a few minutes. This increase in tumour oxygenation is greater than that observed in murine SCCVII tumours after the same exposure to carbogen and employing an identical measurement technique. Further studies of the relationship between pO_2 measured using oxygen sensors and the presence of hypoxic cells is warranted, particularly in human malignancies.

ACKNOWLEDGMENTS

This research is supported by the National Cancer Institute of Canada and the Department of Radiation Oncology, British Columbia Cancer Agency. The advice and assistance of Alex Green with regard to the breathing of carbogen is also gratefully acknowledged.

REFERENCES

Brown, JM (1993) SR 4233 (Tirapazamine): a new anticancer drug exploiting hypoxia in solid tumours. *Br. J. Cancer*, **67**:1163–1170.

Cade, IS, McEwen, JB, Dische, S, Saunders, MI, Watson, ER, Halnan, KE, Wiernik, G, Perrins, DJD, and Sutherland, I (1978) Hyperbaric oxygen and radiotherapy: a MRC trial in ca. Bladder. *Br. J. Radiol.*, **51**:876–878.

Churchill-Davidson, I, Sanger, C, and Thomlinson, RH (1957) Oxygenation in radiotherapy II. Clinical application. *Br. J. Radiol*, **30**:406–422.

Dische, S (1979) Hyperbaric oxygen: the MRC trials and their clinical significance. In Br. J. Radiol. (Vol. 51, pp. 888–894).

Dische, S (1991) A review of hypoxic cell radiosensitization. *Int. J. Radiat. Oncol. Biol. Phys.*, **20**:147–152.

Dische, S, Anderson, PJ, Sealy, R, and Watson, ER (1983) Carcinoma of the cervix-anaemia, radiotherapy and hyperbaric oxygen supports the importance of hypoxia in radiotherapy. *Br. J. Radiol.*, **56**:251–255.

Falk, SJ, Ward, R, and Bleehen, NM (1992) The influence of carbogen breathing on tumour tissue oxygenation in man evaluated by computerised pO_2 histography. *Br. J. Cancer*, **66**:919–924.

Fletcher, GH, Lindberg, RD, Caderao, JB, and Wharton, JT (1977) Hyperbaric oxygen as a radiotherapeutic adjuvant in advanced cancer of the uterine cervix—Preliminary results of a randomized trial. *Cancer*, **39**:617–623.

Hall, EJ (1994) *Radiobiology for radiologists* (4th ed.). Philadelphia: Harper and Row.

Henk, JM (1986) Late results of a trial of hyperbaric oxygen and radiotherapy in head and neck cancer: A rationale for hypoxic cell radiosensitizers. *Int. J. Radiat. Oncol. Biol. Phys.*, **12**:1339–1341.

Höckel, M, Knoop, C, Schlenger, K, Vorndran, B, Baußmann, E, Mitze, M, Knapstein, PG, and Vaupel, P (1993) Intratumoral pO_2 predicts the survival in advanced cancer of the uterine cervix: updated analysis of an open prospective clinical study. *Radiother. Oncol.*, **26**:45–50.

Höckel, M, Schlenger, K, Aral, B, Mitze, M, Schaffer, U, and Vaupel, P (1996) Association between tumor hypoxia and malignant progression in advanced cancer of the uterine cervix. *Cancer Research*, **56**: 4509–4515.

Höckel, M, Schlenger, K, Knoop, C, and Vaupel, P (1991) Oxygenation of Carcinomas of the Uterine Cervix: Evaluation by Computerized O_2 Tension Measurements. *Cancer Res*, **51**:6098–6102.

Honess, DJ and Bleehan, NM (1995) Perfusion changes in the RIF-1 tumour and normal tissues after carbogen breathing and nicotinamide, individually and combined. *Br. J. Cancer*, **71**:1175–1180.

Horsman, MR, Khalil, AA, Nordsmark, M, Grau, C, and Overgaard, J (1993) Relationship between radiobiological hypoxia and direct estimates of tumour oxygenation in a mouse tumour model. *Radiotherapy and Oncology*, **28**:69–71.

Horsman, MR, Khalil, AA, Nordsmark, M, Siemann, DW, Hill, SA, Lynch, EM, Chaplin, DJ, Stern, S, Thomas, CD, Guichard, M, Grau, C, and Overgaard, J (1995) The use of oxygen electrodes to predict radiobiological hypoxia in tumours. *Tumour Oxygenation*, 49–57.

Kjellen, E, Joiner, MC, Collier, JM, Johns, H, and Rojas, A (1991) A Therapeutic Benefit from Combining Normobaric Carbogen or Oxygen with Nicotinamide in Fractionated X-Ray Treatments. *Radiother Oncol*, **22**, 81–91.

Minchinton, AI (1995) Differences in the tumour oxygen tension in mice and man. In PW Vaupel, DK Kelleher, and Günderoth (Eds.), *Tumour Oxygenation* (pp. 211–218). Stuttgart: Gustav Fischer Verlag.

Nordsmark, M, Overgaard, M, and Overgaard, J (1996) Pretreatment oxygenation predicts radiation response in advanced squamous cell carcinoma of the head and neck. *Radiotherapy and Oncology*, **41**:31–39.

Rojas, A (1991). Radiosensitization with Normobaric Oxygen and Carbogen. *Radiother. Oncol.*, **20**:65–70.

Thomlinson, RH and Gray, LH (1955) The histological structure of some human lung cancers and the possible implications for radiotherapy. *Br. J. Cancer*, **9**:539–549.

Vaupel, P, Schlenger, K, Knoop, C, and Höckel, M (1991) Oxygenation of human tumors: Evaluation of tissue oxygen distribution in breast cancers by computerized O_2 tension measurements. *Cancer Res.*, **51**: 3316–3322.

Watson, ER, Halnan, KE, Dische, S, Saundres, MI, Cade, IS, McEwan, JB, Wierneik, G, Perrins, DJD, and Sutherland, I (1978) Hyperbaric oxygen and radiotherapy. A MRC trial in carcinoma of the cervix. *Br. J. Radiol.*, **51**:879–887.

OXYGEN CONSUMPTION AND ITS TRANSPORT IN TISSUES DURING PHOTODYNAMIC THERAPY OF TUMORS AND WHOLE BLOOD

A. A. Stratonnikov, A. J. Douplik, and V. B. Loschenov

General Physics Institute
38 Vavilov Street, Moscow 117942
Russia

1. INTRODUCTION

The analysis of microcirculation is of great importance for tissue differential diagnostics and understanding of oxygen transport. Being the main component of cell metabolism the availability of molecular oxygen in tissues is an important factor for normal cell function. But from the other side sometimes the presence of oxygen in sufficient concentrations in tissues may be necessary when one wants to destroy abnormal cells (tumors). It is understood now that availability of molecular oxygen in tissues plays a very important role during photodynamic therapy (PDT).

Photodynamic therapy is an effective method to treat cancer and other nononcological lesions by means of light action on photosensitizers in tissue. It is assumed that the destroying effect is mainly due to the formation of singlet oxygen resulting from the interaction of light excited photosensitizer with molecular oxygen (triplet ground state). So the destroying effect will be proportional to the rate of singlet oxygen formation which in turn depends on light intensity, photosensitizer concentration and molecular oxygen availability in tissues. The present work deals with the investigation of blood oxygen saturation (SO_2) in microcirculatory vessels of tissues in vivo and blood samples during light irradiation in the PDT process.

During PDT the special drug (photosensitizer) is injected to the patient. After some time period required for the photosensitizer to be selectively accumulated in tumor the tumor spot is irradiated by light with appropriate wavelength. The light is absorbed by the photosensitizer resulting in the formation of reactive species destroying the tumor cells. It is considered that singlet oxygen is the most important component that oxidizes

Oxygen Transport to Tissue XXI, edited by Eke and Delpy.
Kluwer Academic / Plenum Publishers, New York, 1999.

different cell structures. So the availability of molecular oxygen in tissues during light irradiation of photosensitized tissues is the important factor of a successful therapeutic effect. The oxygen consumption rate may serve as a dose metric for PDT efficiency.

2. OXYGEN TRANSPORT TO TISSUES AND PHOTODYNAMIC THERAPY

The evaluation of blood oxygen saturation in capillary vessels by means of absorption spectra analysis of tissues in the visible wavelength range has been considered earlier (Frank K. H. et al., 1989; Harrison D. K. et al., 1992, 1993; Adler J. et al., 1996; Kessler M. et al., 1996). The value characterizing the blood oxygen saturation degree SO_2 is determined by:

$$SO_2 = C_{HbO2}/(C_{HbO2} + C_{Hb}), \tag{1}$$

where C_{HbO2} and C_{Hb} are the concentrations of hemoglobin associated and free of oxygen, respectively. Oxygen transport to tissues is based on supply units having one shareable arteriole and venule connected by capillaries (Adler J. et al., 1996; Kessler M. et al., 1996). The number of capillaries in one supply unit may be as many as one hundred. The diameter of capillaries is very small (7–8 µm) allowing erythrocytes to pass through them one by one. It is while passing through the capillary that blood transfers nutrients and oxygen to cells and takes away the products of metabolism. Capillaries form a very dense network in the tissue so that the distance from the cell to the nearest capillary does not exceed 0.5 mm.

At the arterial end of the capillary (Figure 1) the oxygen saturation (SO_2) is close to its maximum value and in normal conditions comes to 96–98%. It is this value which can be measured by the serial pulse oximeter. During their movement through the capillary from arterial to venous end, the erythrocytes give out oxygen which dissociates from hemoglobin and diffuses through the capillary walls to the cells where it is consumed in

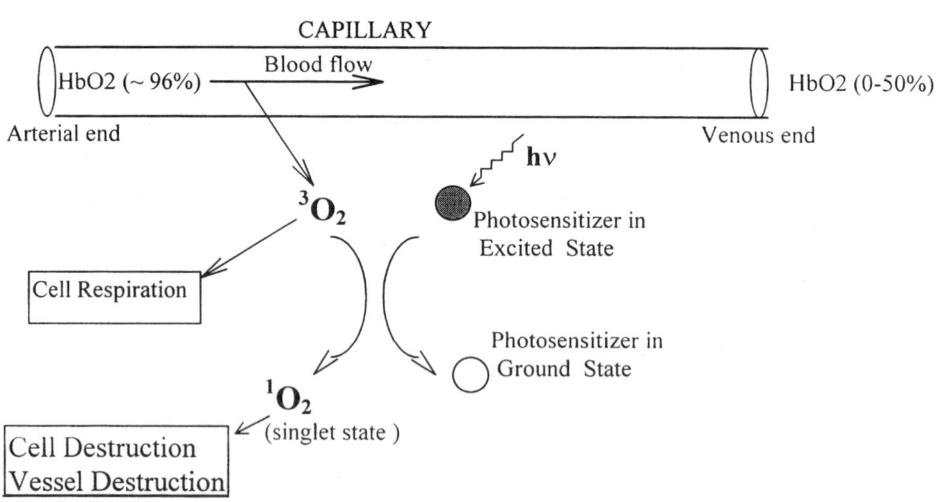

Figure 1. Oxygen transport and consumption in tissues during photodynamic therapy.

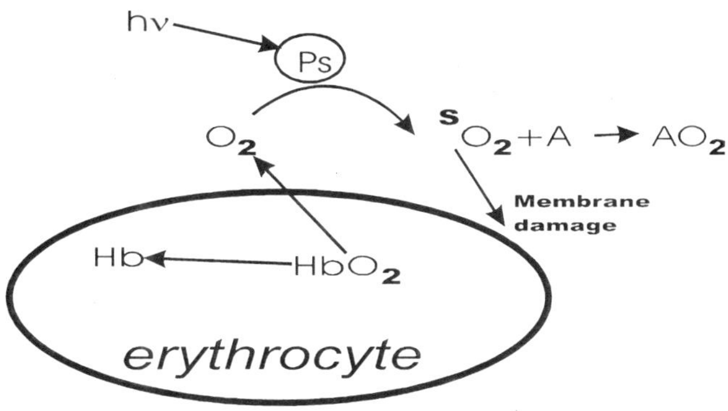

Figure 2. Oxygen consumption during light irradiation of whole blood incubated with photosensitizer.

the process of cell metabolism. So the oxygen saturation falls along the capillary and at the venous end may amount to 50% in the normal state, 20%–50% (reserve range), or 0–20% (critical range) (Adler J. et al., 1996). The regulatory mechanisms of oxygen supply are based on the arterial venous anastomoses which redistribute the blood flow if the critical range is achieved.

The normal state oxygen saturation, averaged for many capillaries, is in the range 65–75% (mean value between arterial and venous end). In the process of light irradiation during PDT the consumption rate of oxygen may be essentially increased due to its interaction with excited photosensitizer resulting in transferring oxygen from the triplet to the singlet state and subsequent tissue oxidation (see Figure 1) (Foster T. H. et al., 1991, 1993; Zilberstein J. et al., 1997). This additional oxygen consumption may drastically suppress the blood oxygen saturation in capillary vessels (Stratonnikov A. A. et al., 1997). Moreover the capillaries may also be destroyed as a result of PDT. It should be especially pronounced for photosensitizers acting through a "vessel mechanism". The predominating capillary destruction should also be observed if light irradiation is applied during a short time interval after photosensitizer injection when a significant amount of photosensitizer still circulates in the blood. Capillary thrombosis and capillary hemorrhage will also result in the blood oxygen saturation decreasing. This fact can be used for evaluation of the efficiency of vessel destruction in the PDT session.

Blood oxygen saturation changes during light irradiation of blood samples incubated with photosensitizer are easier to interpret (Figure 2). In the case of a blood sample there is no oxygen transport by vessels and we should observe only the decrease of oxygen saturation during light irradiation (Douplik A. J. et al., 1996). The diffusion distance for an oxygen molecule bound to hemoglobin to its consumption site is very small in the case of a blood sample and the equilibrium time between blood oxygen saturation (SO_2) and oxygen consumption rate is very short. So oxygen saturation changes in a blood sample should stop immediately as soon as the light is switched off.

3. MATERIALS AND METHODS

3.1. Experimental Setup

For the measurements of absorption spectra of tissues in vivo during PDT we applied a fiber optic spectrometer LESA-5 (Biospec, Russia). The experimental setup is

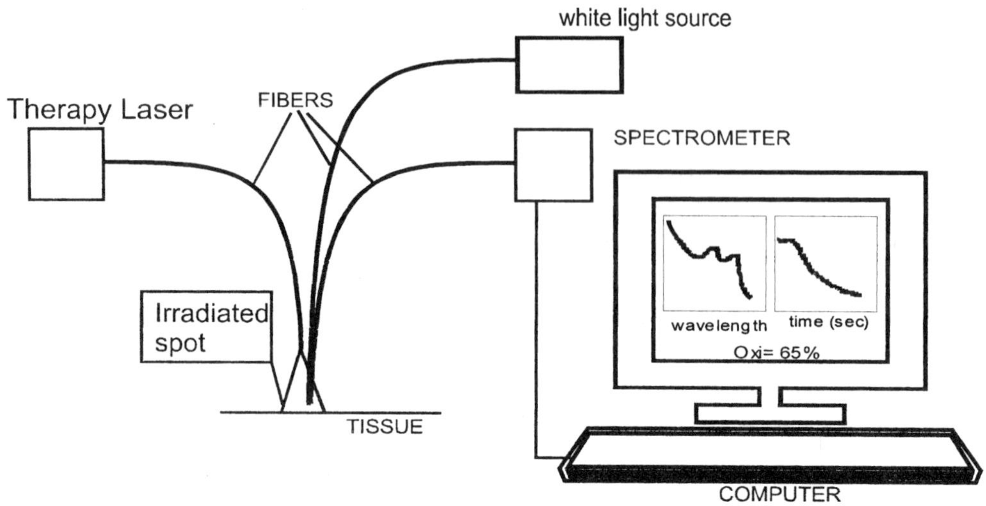

Figure 3. The experimental setup for real time monitoring of hemoglobin oxygen saturation in microcirculatory vessels in vivo during photodynamic therapy.

shown in Figure 3. The light from a halogen lamp was delivered to the tissue by means of a quartz fiber. For blood sample measurements the white light delivery fiber and receiving spectrometer fibers were on opposite sides of the sample, and we measured the diffuse transmitted light. For in vivo measurements these fibers were on the same side of the tissue and we measured the diffuse reflected light (as shown in Figure 3). After passing through the tissue the light was collected by the receiving fiber. The output end of the receiving fiber was the input slit of the spectrometer. The spectrometer was controlled by a personal computer. A special program operating in the MS Windows environment allows one to acquire and display spectra in real time. The desired spectra can be saved for following sophisticated spectral analysis.

The blood oxygen saturation is calculated from the tissue absorption spectra in real time and the time dependence is displayed on the computer screen. For SO_2 estimation (see Eq. 1) the complete spectral range from 520 nm to 590 nm where the spectral differences between Hb and HbO_2 are more pronounced has been used. The relative error in evaluating blood oxygen saturation was about 3% while the absolute error was less than 10%. The sampling size of the measurement is determined by the distance between delivering and receiving fibers and is estimated to be 1.5 mm. This sampling size averages the contributions from many capillary units.

We used a solid state laser with doubled frequency (672 nm) corresponding to the absorption maximum of the photosensitizer used in this work. A special adjustment was provided so that the oxygen saturation measurement would be performed in the center of an irradiated spot.

3.2. Blood Sample Preparation

The blood from one healthy volunteer was used in these investigations. The initial levels of blood oxygen saturation were in the range from $SO_2 = 80$–85%. The photosensitizer was added to the blood one hour before sample preparation. The blood drop was placed between two glass slides. The thickness of the blood layer was controlled with the use of a special film and was in the range 110–130 μm.

3.3. Animal Preparation

The 10 C57Bl/DBA (BDF_1) breeding females bearing Ehrlich tumor were used in the current work. All mice were 10–12 weeks old, and their body weight was 20–22 g. The photosensitizer was injected intravenously at a dose of 4 mg/kg. Two PDT sessions were applied 6 hours and 28 hours after photosensitizer injection.

3.4. Photosensitizer

We used Photosence photosensitizer produced by NIOPIC, Russia. It is a mixture of sulphanated aluminum phthalocyanines with different sulphanation degree (30% tetra-sulphanated, 50% tri-sulphanated and 20% di-sulphanated), absorption maximum 675 nm, extinction coefficient is 10^5 liter/mole*cm. The photosensitizer was dissolved in a physiological saline solution for intravenous injection.

4. RESULTS AND DISCUSSION

The typical absorption spectra (without correction for scattering) of mouse tumor during photodynamic therapy at different time intervals (30, 40, 50 and 60 seconds) after the beginning of laser irradiation with 100 mW/cm^2 power density are shown in Figure 4. The spectral changes are due to the transformation of oxygenated hemoglobin (HbO_2) which has a two-humped absorption spectrum to deoxygenated hemoglobin (Hb) in the microcirculatory blood vessels. The corresponding oxygen saturation calculated from these spectra is also shown on the plot. The decrease in oxygen saturation is due to the additional oxygen consumption during the PDT process and due to the possible vessel thrombosis as a result of the PDT process.

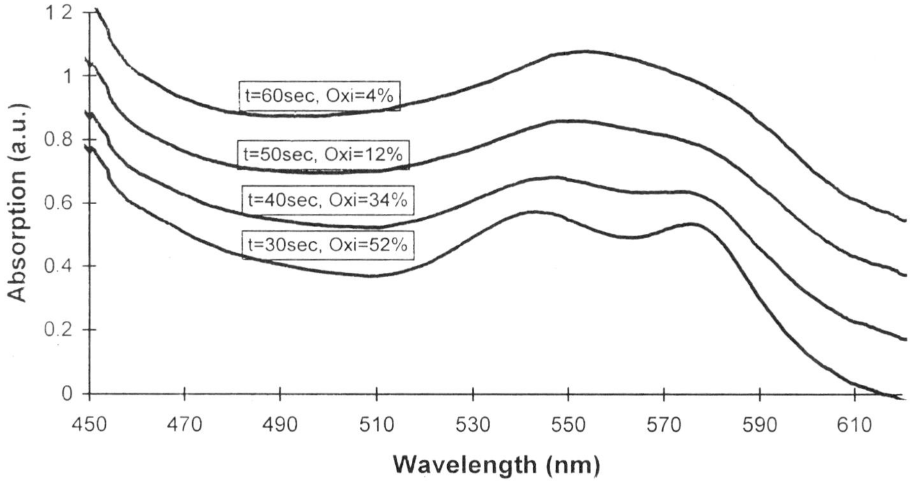

Figure 4. The typical absorption spectra of mouse tumor during photodynamic therapy at different time intervals (30, 40, 50 and 60 seconds) after beginning of laser irradiation with 100 mW/cm^2 power density.

Figure 5 demonstrates the SO_2 time dependence of mouse tumor during first (a) and second (b) PDT sessions with multiple laser irradiation. The direct correlation between laser switching on/off and SO_2 is clearly observed. The difference in SO_2 recovery rate after switching off the laser is significant between the first and second sessions. It may be due to the fact that during the first session 6h after photosensitizer injection while its concentration in blood is relatively high, the preferential capillary destruction is

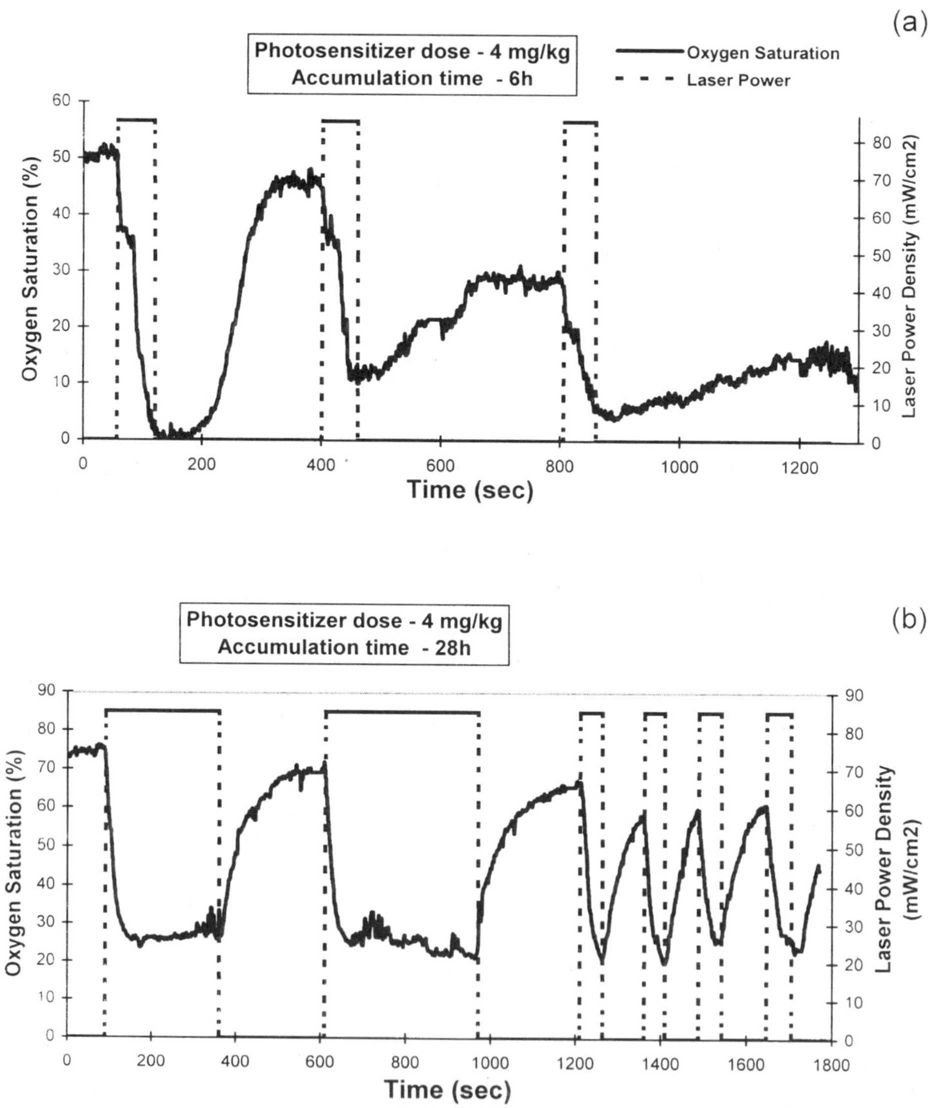

Figure 5. The influence of laser irradiation (85 mW/cm^2, 670 nm) on hemoglobin oxygen saturation in microcirculatory blood vessels of mice during photodynamic therapy with Photosence in injection dose 4 mg/kg. The time dependence of laser power is shown by dotted line. (a) First PDT session (photosensitizer accumulation time 6 hours). The irreversible vessel damage is observed. (b) Second PDT session (photosensitizer accumulation time 28 hours). The oxygen saturation is restored reversibly after switching off the laser.

obtained during laser irradiation. From Figure 5a we see that after the third laser irradiation the SO_2 is stabilized around 25% and is not restored to its previous value.

During a second session, 28 h after photosensitizer injection (with the same laser power density 85 mW/cm^2) we see quite a different SO_2 response to laser irradiation. First, the SO_2 does not fall to zero during laser irradiation but is stabilized at the 25% level not depending on laser irradiation time. Second, the quick recovery of SO_2 to its initial values is observed after switching off the laser. These facts point out that there is no irreversible capillary destruction during the second session and the SO_2 decrease in this case results mainly from the increased oxygen consumption rate due to PDT. The characteristic saw-like curve shown in the range of 1200–1800 sec on Figure 5 (b) is not changed after switching the laser on and off 20 times.

Figure 6 shows the dependence of the SO_2 stabilization value on laser power density. We see that increasing the laser power density three times up to 250 mW/cm^2 does not induce a fall of SO_2 to zero but stabilizes it at the level of 15%. After switching the laser off we observe the SO_2 recovery to its previous values. This proves that dynamic equilibrium is achieved between the oxygen delivery rate by blood and the oxygen consumption rate due to PDT.

Figure 7 demonstrates the influence of pure oxygen breathing on SO_2 values during PDT. Pure oxygen flow was delivered to the mouse. The time of switching oxygen flow on and off is shown in the figure by arrows. We see that this increases both the initial SO_2 value (before switching on laser) and stabilizing SO_2 value during laser irradiation by an amount of 12–15%.

Figure 8 presents the blood oxygen saturation dynamics in blood samples incubated with different photosensitizer concentrations. The lowest concentration (1 μM) is the value observed inside the tissues during PDT and the highest concentration (20 μM) is the value observed in blood after systemic administration. As a control we used a blood sample without photosensitizer (0). The strict correlation between the oxygen consumption rate and photosensitizer concentration incubated in the blood sample proves that

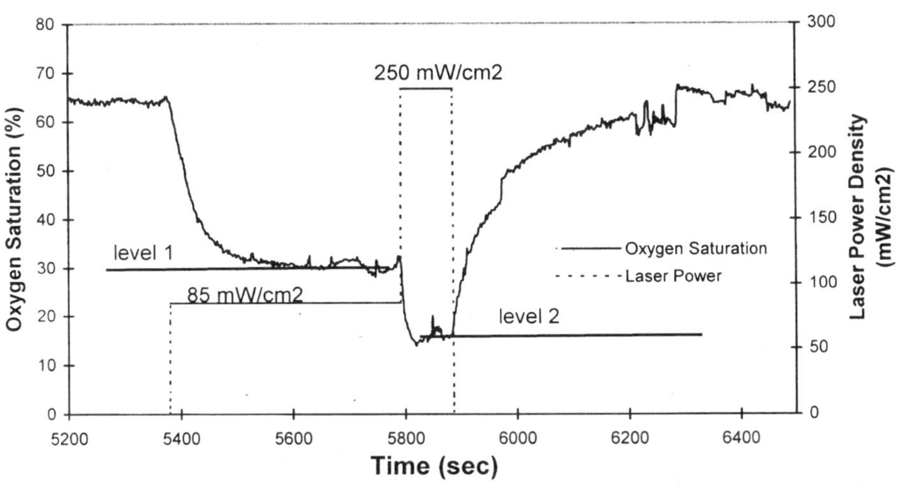

Figure 6. The influence of laser irradiation power density on oxygen saturation stabilization level during photodynamic therapy with Photosence (sulphanated Al phthalocyanine) in injection dose 4 mg/kg (second session, accumulation time 28 h). The dynamic equlibrium between oxygen saturation and laser power density is achieved.

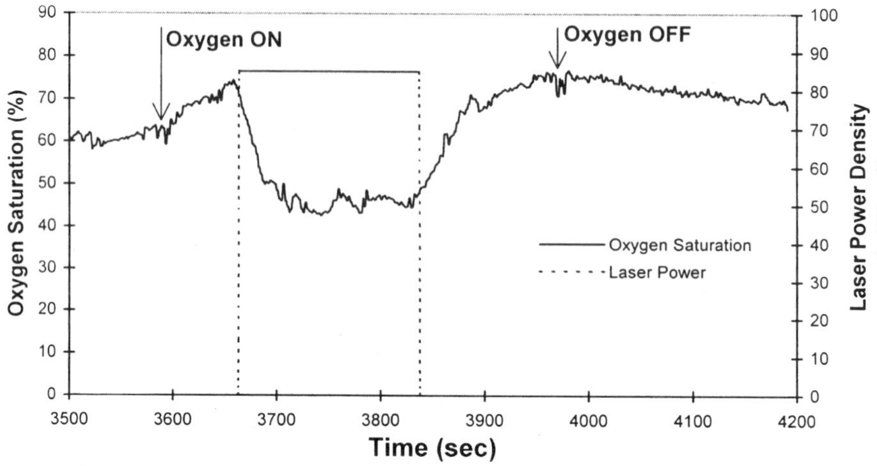

Figure 7. The influence of pure oxygen breathing on hemoglobin oxygen saturation in microcirculatory blood vessels of mouse during photodynamic therapy with Photosence in injection dose 4 mg/kg (second session, accumulation time 28 hours). The arrows show the beginning and end of pure oxygen supply. The dynamic equilibrium is achieved at higher oxygen saturation values.

oxygen is consumed in blood is a result of PDT reactions due to the interaction of light with photosensitizer.

5. CONCLUSIONS

As the availability of molecular oxygen in tissues in sufficient quantity is the necessary condition for a successful PDT result, blood oxygen saturation control may be an

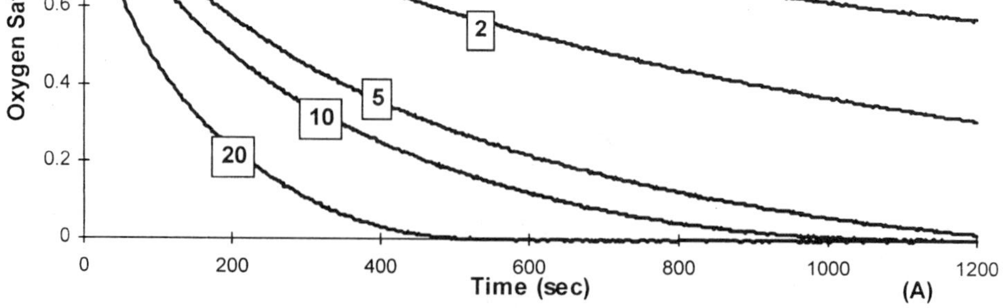

Figure 8. Blood oxygen saturation dynamics in blood samples incubated with different photosensitizer concentrations (Photosence) during PDT with laser power densities at 100 mW/cm^2.

effective monitoring tool. The analysis of SO_2 kinetic dependencies from injection dose, accumulation time and also from the laser power density will help one to take into account the changes in blood supply during PDT and optimize the session to attain the most effective results. It may serve also as a tool to control damage to the vessel network as a result of PDT.

The model of light irradiation of photosensitized tissues in vivo and the resulting changes in blood oxygen saturation may be also very useful for the microcirculation investigations. In this model one can manipulate the oxygen consumption rate in tissue simply by changing the incident light power density and observing the blood oxygen saturation response to it.

REFERENCES

Adler J, Kuchenreuther S, Doblanzki W, and Georgieff M (1996) "The Erlangen Microlightguide photometer—a new way of monitoring oxygen supply of tissue", SPIE 2628:349–355.

Douplik AJ, Loschenov VB, and Stratonnikov AA (1996) "Spectra changes of whole blood with Sulphanated Aluminum Phtalocyanine photosensitizer during photodynamic therapy in vitro" SPIE 2923:44–48.

Foster TH, Murant RS, Bryant RG, Knox RS, Gibson SL, and Hilf R (1991) "Oxygen Consumption and Diffusion Effects in Photodynamic Therapy", Radiation Research 126:296–303.

Foster TH, Hartley DF, Nichols MG, and Hilf R (1993) "Fluence Rate Effects in Photodynamic Therapy of Multicell Tumor Spheroids", Cancer Research 53:1249–1254.

Frank KH, Kessler M, Appelbaum K, and Dummler W (1989) "The Erlangen micro-lightduide spectrophotometer EMPHO I", Phys. Med. Biol. 34:1883–1900.

Harrison DK, Evans SD, Abbot NC, Swanson J Beck, and McCollum PT (1992) "Spectrophotometric measurements of haemoglobin saturation and concentration in skin during the tuberculin reaction in normal human subjects", Clin. Phys. Physiol. Meas. 13:349–363.

Harrison DK, Delaney C, Brown L, Newton D, and McColum PT (1993) "Clinical applications of lightguide diffuse reflectance spectrophotometry in vascular disease", SPIE 2084:195–203.

Kessler M, Krug A, and Hoper J (1996) "Non-invasive topographical investigation of functional parameters in the human skin", SPIE 2679:227–235.

Stratonnikov AA, Douplik AY, Klimov DV, Loschenov VB, Meerovich GA, Mizin SV, Fomina G, Kazachkina NI, Yakubovskya RI, and Budenok YV (1997) "The Absorption Spectroscopy as a Tool to Control Blood Oxygen Saturation during Photodynamic Therapy", SPIE 3191:58–66.

Zilberstein J, Bromberg A, Frantz A, Rosenbach-Belkin V, Kritzmann A, Pfefermann R, Salomon Y, and Scherz A (1997) "Light-dependent Oxygen Consumption in Bacteriochlorophyll-Serine—treated Melanoma Tumors: On-line Determination Using a Tissue-inserted Oxygen Microsensor", Photochem. & Photobiol. 65(6):1012–1019.

62

CAN TUMOR OXYGENATION BE IMPROVED BY REDUCING CELLULAR OXYGEN CONSUMPTION?*

Oliver Thews, Debra K. Kelleher, Monika Hummel, and Peter Vaupel

Institute of Physiology and Pathophysiology
University of Mainz, D-55099 Mainz
Germany

1. INTRODUCTION

Tumor hypoxia, which can be found in many experimental and human tumors is an important factor influencing the therapeutic efficacy of standard radiotherapy, O_2-dependent chemotherapy and photodynamic therapy (Hall, 1994) and might be responsible for the development of aggressive tumor cell subpopulations (Graeber et al., 1996). Since this oxygen deficiency results from a disparity between O_2 supply to the tumor tissue and the oxygen demand of the cells, several attempts have been undertaken to improve tumor oxygenation primarily by increasing the arterial oxygen supply. The O_2 supply to the tumor cells can be improved by (a) increasing the arterial O_2 content (by breathing hyperoxic gases under normobaric or hyperbaric conditions, or by correcting tumor anemia); (b) improving tumor perfusion or increasing its uniformity; (c) enhancing oxygen release from blood by shifting the O_2-hemoglobin dissociation curve. However, theoretical studies by Secomb et al. (1995) using a comprehensive model of the tumor microcirculation (Secomb et al., 1993) postulated that tumor oxygenation could be improved much more effectively by reducing the oxygen consumption of the tumor cells. A reduction of 30% has been predicted to be necessary if tumor hypoxia is to be completely eradicated.

One possible way of reducing cellular O_2 consumption is to cool the tumor cells below 30°C thus reducing the cellular respiration rate by approx. 30% (Vaupel, 1990). However, under *in vivo* conditions cooling the tumor tissue will result in a reduction of O_2 supply caused by decreasing tumor perfusion due to vasoconstriction. Pharmaceutical

* Supported by the Dr.med.h.c. Erwin Braun Foundation, Basel (Grant 5.5.6).

agents have also been tested for their ability to reduce cellular respiration (Biaglow et al., 1984; Biaglow et al., 1986; Thews et al., 1996; Vaupel et al., 1972; Vaupel and Mueller-Klieser, 1986). Agents which inhibit cell proliferation do not reduce the oxygen consumption significantly, thus demonstrating that the high proliferation rate of fast growing tumors is not the main reason for the high O_2 demand of tumors (Thews et al., 1996). However, lovastatin, an inhibitor of cholesterol biosynthesis, decreases the O_2 demand by 33% (Thews et al., 1996) indicating that other biochemical pathways might be relevant for the oxygen consumption of tumors. Unfortunately, lovastatin is not suitable for therapeutic application *in vivo* because it is not possible to obtain the necessary tissue concentration without toxic side effects.

In the present study the effect of another pharmacological agent, m-iodobenzylguanidine (MIBG) on the cellular O_2 consumption and on the oxygenation of solid tumors *in vivo* has been investigated. This agent is known to inhibit the mitochondrial respiratory chain at site I shifting the energy metabolism from oxidative glucose utilization to glycolysis resulting in an acidification of the tumor tissue (Jaehde et al., 1992; Kuin et al., 1994; Loesberg et al., 1990).

2. MATERIALS AND METHODS

2.1. Animals and Tumors

Male Sprague-Dawley rats (Charles River Wiga, Sulzfeld, Germany; body weight 200 to 320 g) housed in our animal care facility were used for all experiments in this study. Animals received standard diet and acidified water *ad libitum* before and prior to the investigation. All experimentation had previously been approved by the regional animal ethics committee and was conducted according to German federal law.

The DS-sarcoma of the rat was used either as a hemorrhagic ascites or as solid tumors. The hemorrhagic ascites was induced by i.p. injection of 1 ml DS-sarcoma cell suspension (approx. 10^4 cell/µl). After 5–6 days, when the animals had developed adequate ascites (>10 ml), animals were randomly assigned to one of the experimental groups. To induce solid growing DS-sarcomas, 0.4 ml of DS-ascites (approx. 10^4 tumor cells/µl) was injected subcutaneously into the hind food dorsum. Tumors grew as flat, spherical segments and replaced the subcutis and corium completely. Tumor volumes were determined by measuring the three orthogonal diameters of the tumor and using an ellipsoid approximation with the formula: $V = d_1 \times d_2 \times d_3 \times \pi/6$. Tumors reached the target volume of 1.0–1.7 ml approx. 6 to 8 days after inoculation.

2.2. Drugs

Meta-iodobenzyguanidine (MIBG, Sigma-Aldrich Chemie, Deisenhofen, Germany) was dissolved in phosphate buffered saline (PBS). For measurements in the hemorrhagic ascites, a MIBG solution was used at a concentration of 2 mg/ml. Animals received a dose of 20 mg/kg body weight 2 h prior to the measurements. For experiments in solid tumors, MIBG was dissolved in PBS at a concentration of 2.5 mg/ml and animals treated at a dose of 25 mg/kg or 100 mg/kg body weight 4 h prior to measurements. Animals receiving only PBS served as controls.

2.3. Measurement of the Cellular O_2 Consumption Rate

The O_2 consumption rate of an ascites sample was analyzed using a photometric technique (Mueller-Klieser et al., 1986). In brief, ascites (2 ml) was equilibrated in a thermostated tonometer with water saturated carbogen (95% O_2, 5% CO_2) for at least 15 min. Thereafter, 50 µl of this sample were aspirated in a gas-tight syringe (Hamilton, Reno NV, USA). At precise time intervals of 5 min, 10 µl aliquots were injected from the syringe into an elution chamber where cells were lysed and all oxygen contained in the injected volume was washed out by an O_2-free nitrogen gas flow. The O_2 concentration in this carrier gas was analyzed by an O_2-sensitive dye in a continuously recording photometer (Zeiss, Oberkochen, Germany). By integrating the photometric reading, the O_2 content of the injected aliquot was determined. Since the O_2 content in the gas-tight syringe decreases due to respiration by the cells, the consumption rate in the sample can be calculated from the time intervals between the aliquot injections and the decrease in the O_2 content of the aliquots. The O_2 consumption rate of the sample was normalized to the number of cells in the sample, determined by automatic cell counting in a flow cytometer EPICS Profile II (Coulter, Hialeah FL, USA). Prior to and after each measurement series the system was calibrated by injecting 5 µl of room air containing a defined O_2 content.

2.4. Measurement of the Tumor Oxygenation

The distribution of the tumor oxygen tensions (pO_2) was measured polarographically using steel-shafted microelectrodes (outer diameter: 300 µm) and pO_2 histography (Eppendorf-Netheler-Hinz, Hamburg, Germany; for more details of this method see Vaupel et al., 1991). Animals were anesthetized with sodium pentobarbital (40 mg/kg, i.p., Nembutal®, Ceva, Paris, France), and catheters placed into the carotid artery and the jugular vein. The rats laid supine on a heated operating pad. Rectal temperature was maintained at 37.5–38.5°C and animals breathed room air spontaneously. Mean arterial blood pressure (MABP) was continually monitored through the connection of the arterial catheter to a Statham pressure transducer (type P 23 ID, Gould, Oxnard CA, USA). The Ag/AgCl reference electrode was placed in the lower abdomen between the skin and the underlying musculature. For tumor pO_2 measurement, a small incision was made into the skin overlying the tumor and the O_2-sensitive electrode was advanced to a depth of approx. 1 mm. The electrode was then automatically moved through the tissue in preset steps of 1.0 mm immediately followed by a fast electrode retraction of 0.3 mm so that an effective step length of 0.7 mm resulted. Approximately 80 pO_2 values were obtained from each tumor. The oxygenation status of each tumor was described by the mean and median pO_2 and the fraction of pO_2 values $\leq$2.5 mmHg and $\leq$5 mmHg. Additionally, arterial blood gas analysis was performed immediately before and after tumor tissue pO_2 measurements using a pH/blood gas analyzer (Type 178; Ciba Corning, Fernwald, Germany) to ensure that animals had arterial blood gases within the physiological range during the measuring period.

2.5 Determination of Global Metabolite Concentrations

Tumors and skeletal muscle samples were taken from anesthetized animals, rapidly frozen in liquid N_2, ground to a fine powder and freeze-dried. Global glucose and lactate concentration were assayed enzymatically using standard enzymatic test kits (kits #245158 and #256773, Boehringer-Mannheim, Mannheim, Germany). ATP, ADP, and AMP concentrations were measured using HPLC techniques. Here, aliquots of 2–3 mg of freeze-dried tissue samples were extracted with 1 ml 0.6 M perchloric acid, centrifugated, and neutralized with 2 M KOH. The isocratic separation of nucleotides was performed with a reversed phase column (Eurospecial Oligo 100, C18, 5 μm, 250 × 4 mm) and a Eurospecial Oligo 100 guard cartridge system (5 × 4 mm; Knauer, Berlin, Germany) at 254 nm. The mobile phase consisted of 50 mM $NH_4H_2PO_4$, 15% (v/v) methanol, and 38 mM tetrabutylammonium hydroxide, adjusted to pH 3.85. The flow rate was 1 ml/min and the sample size 20 μl.

2.6. Statistical Analysis

Results are expressed as means ± standard error of the mean (SEM). Differences between the groups were assessed by two-tailed Wilcoxon test for unpaired samples. The significance level was set at $\alpha = 5\%$ for all comparisons.

3. RESULTS

Treating DS-sarcoma ascites *in vivo* with MIBG at a dose of 20 mg/kg b.w. reduced the cellular O_2 consumption significantly (Figure 1). In control animals, the oxygen consumption was $36.7 \pm 2.0 \cdot 10^{-18}$ mole · s^{-1} per cell whereas in MIBG-treated animals the O_2 consumption was reduced by approx. 43% down to $20.9 \pm 1.1 \cdot 10^{-18}$ mole · s^{-1} per cell.

Treating solid growing DS-sarcomas *in vivo* with MIBG induced a shift of energy metabolism from oxidative glucose utilization to aerobic glycolysis. This was indicated by a significant increase in the lactate concentration in tumors from 11 to approx. 15 mmol/l (Figure 2) and in muscle tissue from 1.1 to 1.7 mmol/l. In parallel, ATP levels in tumors (Figure 2) as well as in skeletal muscle (data not shown) decreased by approx. 40%. At the same time, AMP concentrations increased significantly with MIBG treatment maybe indicating a reduced ATP formation by aerobic glycolysis. MIBG treatment also leads to a distinctly higher glucose concentration in tumors (and in the skeletal muscle; data not shown) than in control animals (Figure 2).

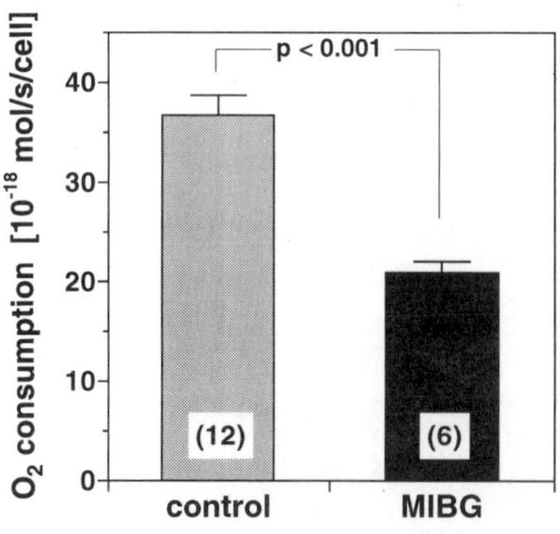

Figure 1. Oxygen consumption rate of DS-ascites cells treated in vivo with MIBG (number of animals in parentheses).

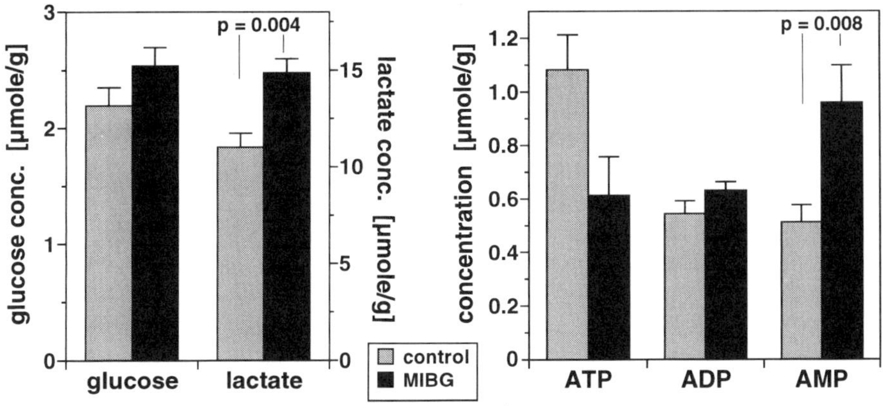

Figure 2. Metabolic and bioenergetic parameters of solid DS-sarcomas in animals treated with MIBG (each column represents a minimum of 10 tumors).

The DS-sarcoma is known to be a rather hypoxic tumor entity. In untreated control tumors the median pO_2 was 2.5 mmHg and the fraction of hypoxic pO_2 measurements ≤2.5 mmHg was approx. 50%. Although the metabolic parameters were substantially different following MIBG treatment as compared to the control group, the tumor oxygenation status was approximately the same in both groups (Figure 3). Both showed a median pO_2 of ≈2 mmHg and a fraction of pO_2 values ≤2.5 mmHg of about 50%. When MIBG was used at a much higher concentration of 100 mg/kg b.w., the median pO_2 in the MIBG group was approx. the same (≈2 mmHg) however, the fraction of hypoxic pO_2 values was distinctly higher (62%). In these experiments the mean arterial blood pressure in the MIBG group was significantly lower than in control animals, and thus the depression of the blood pressure might account for the worsening of the tumor oxygenation.

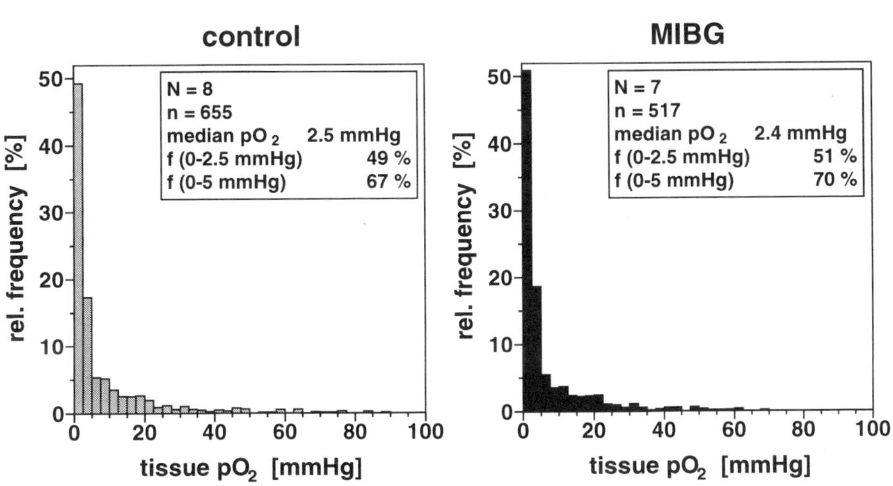

Figure 3. pO_2 distribution of solid DS-sarcomas in control animals and in animals treated with MIBG (N: number of tumors, n: number of measured pO_2 values).

4. DISCUSSION

In DS-ascites m-iodobenzylguanidine at a dose of 20 mg/kg b.w. results in a pronounced inhibition of cellular respiration (Figure 1). Although in solid DS-sarcomas the O_2 consumption rate was not measured, the bioenergetic parameters indicate a MIBG-induced inhibition of glucose oxidation. In the MIBG group, the tumor lactate concentration was significantly higher indicating an increase in glycolytic rate with lower ATP and higher AMP levels suggesting a reduced ATP yield. These data are in good accordance with others showing that MIBG treatment results in a reduction of the extracellular pH and a higher glucose uptake by the tumor tissue (Jaehde et al., 1992; Kuin et al., 1994).

From the intensified glycolysis, a lower tissue glucose concentration might have been expected in the MIBG group. However, the glucose levels in tumor and muscle were found to be distinctly higher. This might be the result of an MIBG-induced increase in blood glucose concentration which has been recently demonstrated by Biaglow et al. (data presented at the 10th International Conference on Chemical Modifiers of Cancer Treatment, Clearwater, Florida, USA, 1998).

From the theoretical considerations (Secomb et al., 1995), a reduction of cellular oxygen consumption rate of more than 30% should be sufficient to improve tumor tissue oxygenation substantially. However, no differences in the O_2 status were found between the MIBG-treated animals and the control group. Since the mean arterial blood pressure (MABP) and the arterial blood gases were monitored and found to be in the normal range in both groups, side effects of MIBG (at a dose of 20 mg/kg b.w.) causing a reduction in the arterial O_2 supply to the tumor can be excluded. At higher MIBG concentrations (100 mg/kg b.w.) however, a decrease of the MABP was seen resulting in a reduction of the O_2 supply thus accounting for the worsening of the oxygenation status.

Several other possible mechanisms of reducing cellular O_2 consumption of tumors have been scrutinized over the last years:

(i) blocking cholesterol biosynthesis by lovastatin (Thews et al., 1996),
(ii) calcium channel blockers (Biaglow et al., 1986; Vaupel and Mueller-Klieser, 1986),
(iii) narcotics, e.g., pentobarbital (Biaglow et al., 1984),
(iv) acute hyperglycemia (Crabtree-effect) (Vaupel et al., 1972).

However, all of these measures are unsuitable for a reduction of respiration rate in solid tumors *in vivo* in order to improve the tumor oxygenation and in turn to increase the radiosensitivity of the tissue. The inhibition of cellular respiration by an acute hyperglycemia (Crabtree-effect) is only a transient effect which disappears almost completely 10 min after the start of a glucose infusion (Vaupel et al., 1972). Calcium channel blockers (Biaglow et al., 1986; Vaupel and Mueller-Klieser, 1986) as well as pentobarbital (Biaglow et al., 1984) show only minor effects on the O_2 consumption rate at concentrations tolerable *in vivo*. Only at higher, unphysiological concentrations *in vitro* is the impairment of the respiration rate more pronounced (Biaglow et al., 1984; Biaglow et al., 1986; Vaupel and Mueller-Klieser, 1986). Lovastatin is suitable to reduce the oxygen consumption in DS-ascites cells *in vivo* by more than 30% (Thews et al., 1996). However, in solid tumors it is not possible to reach a sufficient concentration of lovastatin to inhibit cell proliferation and to reduce the oxygen demand. This agent is therefore not suitable for investigating the impact of oxygen consumption on the oxygenation of solid tumors.

Tissue hypothermia is another means of decreasing cellular O_2 consumption effectively (Vaupel, 1990). In *in vitro* experiments the O_2 demand can be reduced by more than 80% by cooling tumor cells down to 15°C (Vaupel, 1990) which should be sufficient to improve the oxygenation of solid tumors. First studies on the effect of hypothermia on tumor oxygenation supported this theory since it could be demonstrated that cooling the tumor from 35°C down to 15°C reduced the fraction of hypoxic pO_2 values (≤2.5 mmHg) significantly from 18 to 8% (Figure 4) (Kelleher et al., 1998). However, the improvement in tumor oxygenation is smaller than expected from the reduction of the respiration rate due to

(i) constriction of vessels feeding the tumor,
(ii) an increase in blood viscosity, both yielding a reduced arterial blood supply to the tumor (Kelleher et al., 1998), and
(iii) a left shift of the HbO_2-dissociation curve resulting in a hampered O_2 release from erythrocytes to the tissue.

These three mechanisms reduce the O_2 supply to the tumor cells and thereby diminish the effect of hypothermia on tumor oxygenation. In particular, the hypothermia-induced vasoconstriction seen in these experiments resulted in a severe reduction of tumor perfusion by more than 60% (Kelleher et al., 1998) thus counteracting the beneficial effect (in terms of tumor oxygenation) of reducing the O_2 demand of the tumor.

The question arises, why in the hypothermia experiments at a tumor temperature of 15°C a reduction of tumor hypoxia was found (Figure 4) despite the occurrence of a severe vasoconstriction (Kelleher et al., 1998) whereas in the present study no effect of reducing tumor O_2 consumption on the oxygenation was found. Results from the *in vitro* experiments showed that lowering the temperature down to 15°C reduces the O_2 consumption of tumor cells by approx. 84% (Vaupel, 1990). From these data it seems to be likely that a more pronounced reduction in O_2 consumption than the postulated value of 30% (Secomb et al., 1995) is necessary to improve tumor oxygenation (a reduction of cellular respiration by approx. 64% by cooling the tumor down to 25°C also had no effect on tumor hypoxia; Kelleher et al., 1998). Whether such pronounced reductions in O_2 consumption can be achieved using pharmacological agents is questionable.

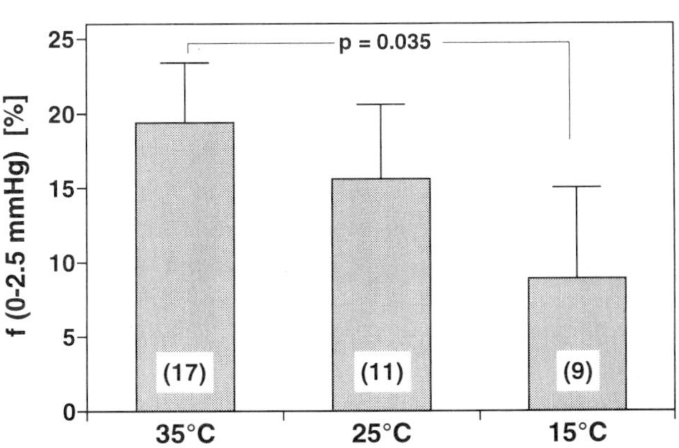

Figure 4. Tumor hypoxia of DS-sarcomas expressed by the fraction of hypoxic pO_2 measurements ≤2.5 mmHg in control tumors and during hypothermic treatment (25°C and 15°C, number of animals in parentheses).

5. CONCLUSIONS

The present study suggests that a reduction of the O_2 consumption rate of approx. 40% as obtained by treating tumors with MIBG, a blocker of the mitochondrial respiration chain, is not sufficient to improve tumor oxygenation. From the data obtained during hypothermia treatment of the tumors, it might be possible through a further reduction of the O_2 consumption rate by more than 80% to markedly reduce tumor hypoxia. However, the decreased tumor perfusion resulting from a hypothermia-induced vasoconstriction may limit the suitability of hypothermia as a means of improving radiosensitivity in solid tumors.

REFERENCES

Biaglow JE, Suit HD, Durand RE, and Dosoretz DE (1984). On the mechanism for enhancement of tumor radiation to hyperbaric oxygen in sodium pentobarbital anesthetized rodents. Adv. Exp. Med. Biol. 180:301–310.

Biaglow JE, Varnes ME, Jacobson B, and Suit HD (1986). Effect of calcium channel blocking drugs on tumor cells oxygen utilization. Adv. Exp. Med. Biol. 200:583–589.

Graeber TG, Osmanian C, Jacks T, Housman DE, Koch CJ, Lowe SW, and Giaccia AJ (1996). Hypoxia-mediated selection of cells with diminished apoptotic potential in solid tumors. Nature 379:88–91.

Hall EJ (1994). Radiobiology for the radiologist (4th ed.). Philadelphia: J.B. Lippincott.

Jaehde E, Volk T, Atema A, Smets LA, Gluesenkamp KH, and Rajewsky MF (1992). pH in human tumor xenografts and transplanted rat tumors: effect of insulin, inorganic phosphate, and m-iodobenzylguanidine. Cancer Res. 52:6209–6215.

Kelleher DK, Nauth C, Thews O, Krueger W, and Vaupel P (1998). Localized hypothermia: Impact on oxygenation, microregional perfusion, metabolic and bio-energetic status of subcutaneous rat tumors. Br. J. Cancer 78:56–61.

Kuin A, Smets L, Volk T, Paans A, Adams G, Atema A, Jaehde E, Maas A, Rajewsky MF, Visser G, and Wood P (1994). Reduction of intratumoral pH by the mitochondrial inhibitor m-iodobenzylguanidine and moderate hyperglycemia. Cancer Res. 54:3785–3792.

Loesberg C, Van Rooij H, Nooijen WJ, Meijer AJ, and Smets LA (1990). Impaired mitochondrial respiration and stimulated glycolysis by m-iodobenzylguanidine (MIBG). Int. J. Cancer 46:276–281.

Mueller-Klieser W, Zander R, and Vaupel P (1986). A new photometric method for oxygen consumption measurements in cell suspensions. J. Appl. Physiol. 61:449–455.

Secomb TW, Hsu R, Dewhirst MW, Klitzman B, and Gross JF (1993). Analysis of oxygen transport to tumor tissue by microvascular networks. Int. J. Radiat. Oncol. Biol. Phys. 25:481–489.

Secomb TW, Hsu R, Ong ET, Gross JF, and Dewhirst MW (1995). Analysis of the effects of oxygen supply and demand on hypoxic fraction in tumors. Acta Oncol. 34:313–316.

Thews O, Kelleher DK, and Vaupel P (1996). In vivo oxygen consumption rate of DS sarcoma cells on inhibition of DNA synthesis. Cancer Res. 56:2009–2012.

Vaupel P (1990). Pathophysiological mechanisms of hyperthermia in cancer therapy. In: Biological Basis of Oncologic Thermotherapy (M. Gautherie ed.). Springer, Berlin, pp. 73–134.

Vaupel P, Guenther H, and Grote J (1972). Einfluss einer Hyperglykaemie auf die Sauerstoff- und Glucoseaufnahme von Implantationstumoren (DS-Carcinosarkom) in vivo. Z. Krebsforsch. 77:17–25.

Vaupel P and Mueller-Klieser W (1986). Verapamil inhibits the respiration rate of cancer cells. Adv. Exp. Med. Biol. 200:645–648.

Vaupel P, Schlenger K, Knoop C, and Hoeckel M (1991). Oxygenation of human tumors: Evaluation of tissue oxygen distribution in breast cancers by computerized O_2 tension measurements. Cancer Res. 51:3316–3322.

PREDICTIVE POWER OF THE TUMOR OXYGENATION STATUS*

P. Vaupel[1] and M. Hoeckel[2]

[1]Institute of Physiology and Pathophysiology
University of Mainz
D-55099 Mainz
[2]Department of Obstetrics and Gynecology
University of Leipzig
D-04103 Leipzig, Germany

1. INTRODUCTION

In most solid tumors the tissue pO_2 status is poorer than in normal tissue at the site of tumor growth (for reviews see Vaupel et al., 1996, 1998; Vaupel and Hoeckel, 1998). Peculiarities of tumor tissue oxygenation can mainly be attributed to characteristic structural and functional abnormalities of the tumor microcirculation (perfusion-limited O_2 delivery), to a deterioration of the diffusion-geometry (diffusion-limited O_2 delivery), and—in some cases—to a reduced O_2 carrying capacity of the blood due to tumor-associated anemia.

Systematic pO_2 data acquisition from accessible tumors in conscious patients became possible 13 years ago when we started to use a computerized histography system with mechanically stable O_2 electrodes with a 17 μm sensitive gold cathode in a stainless steel shaft of approximately 250 μm (Kallinowski et al., 1990). This novel technique (Eppendorf, Hamburg, Germany) allowed, for the first time, quick and reliable polarographic tumor pO_2 readings in the routine clinical setting. Applying this device in patients we performed systematic studies on breast carcinomas and locally advanced cancers of the uterine cervix. In the meantime, several other groups have carried out similar studies on these tumor entities while others have extended this type of measurement to soft tissue sarcomas, primary and metastatic tumors of the head and neck, rectal carcinomas, and glioblastomas (see Table 1).

*Supported by the Deutsche Krebshilfe (Grant 70-1920-Va 2).

Oxygen Transport to Tissue XXI, edited by Eke and Delpy.
Kluwer Academic / Plenum Publishers, New York, 1999.

Table 1. Systematic assessment of pretherapeutic oxygenation status of human malignancies by computerized pO_2 histography (selected references)

Tumor entity	Authors
Breast carcinomas	Vaupel et al. (1991), Runkel et al. (1994), Vaupel (1997), Hohenberger et al. (1998)
Cervical cancers	Hoeckel et al. (1991), Brizel et al. (1995), Vaupel (1996), Lyng et al. (1996), Sundfor et al. (1997), Wong et al. (1997), Fyles et al. (1998a)
Rectal cancers	Kallinowski & Buhr (1995), Mattern et al. (1996)
Soft tissue sarcomas	Brizel et al. (1994, 1996), Nordsmark et al. (1994, 1996a, 1997), Hohenberger and Dragon (1995)
Head and neck tumors	Lartigau et al. (1993), Nordsmark et al. (1994, 1996b), Terris et al. (1994), Strnad et al. (1997), Lyng et al. (1997), Vaupel (1997), Becker et al. (1998)
Glioblastomas	Rampling et al. (1994)

Applying the pO_2 histography system in patients with locally advanced cancer of the uterine cervix, we initiated a controlled prospective trial to evaluate the clinical relevance of these tumors' oxygenation status. The measurements were performed according to the principles of systematic random sampling to take into account intratumoral heterogeneity. After having presented preliminary data from the first analysis demonstrating the powerful predictive value of tumor oxygenation on local control and overall survival for patients treated with curative intent there is a growing body of evidence unequivocally supporting our earlier data on cervix cancer. In addition, similar findings have been obtained for soft tissue sarcomas and for squamous cell sarcomas of the head and neck.

2. OXYGENATION STATUS AS A PREDICTIVE PARAMETER IN LOCALLY ADVANCED CANCER OF THE UTERINE CERVIX

The study of tumor oxygenation in more than 120 patients with locally advanced cervical cancer has demonstrated that tumor oxygenation is independent of various patient and tumor characteristics including hemoglobin level, FIGO stage, clinical tumor size, histopathology, grade of malignancy, patient age, menopausal status and parity. In our first report published, 33 patients receiving standard radiotherapy with or without chemotherapy were analyzed for treatment outcome (Hoeckel et al., 1993). In these cervical tumors, tissue O_2 tension measurements revealed median pO_2 values ≤ 10 mmHg in approx. 50% of the patients. After a median follow-up of 2 years, Kaplan-Meier life table analysis showed significantly shorter survival and recurrence-free survival for patients with treated hypoxic tumors. According to the Cox proportional hazards model, the hypoxic pO_2 fraction (≤ 5 mmHg) was the most powerful single predictor of survival and recurrence-free survival. The results of this earlier study were consistent with the notion that radiobiologically hypoxic tumors are less radiosensitive. Already at this early stage, we pointed out that mechanisms of treatment failure other than the effect of hypoxia on the radiation response could not be excluded.

By the end of 1996, 105 patients with advanced cancers of the uterine cervix had entered the study (Hoeckel et al., 1996b). Histopathological examination of the surgical specimens following radical tumor resection in 50% of the patients showed that hypoxic tumors exhibited larger tumor extensions and more frequent (occult) parametrial spread, as well as lymphvascular space involvement, compared to well-oxygenated tumors of similar clinical stage and size. All patients were analyzed for treatment outcome after a median observation period of 3 years. Again, Cox regression analysis revealed tumor oxygenation as the strongest independent prognostic factor followed by FIGO stage. The potential of tumor oxygenation in predicting clinical outcome increased with tumor sizes (Hoeckel et al., 1996b).

Five year overall as well as disease-free survival probabilities calculated for patients who underwent standard primary treatment for cure were again significantly lower for hypoxic tumors than for well-oxygenated malignancies of similar clinical stages and sizes (see Figure 1). Most interesting was the finding that the disadvantage in outcome for hypoxic tumors was independent of the mode of primary treatment (radiation or radical surgery). This latter finding led to the conclusion that there may be an association between tumor hypoxia and malignant progression in advanced cancer of the uterine cervix (Hoeckel et al., 1996b; Hoeckel and Vaupel, 1998).

In a more recent study by Fyles et al. (1998b), the predictive power of oxygenation with regard to tumor control and survival was re-assessed. There was again clear evidence that hypoxia and tumor size are significant adverse predictors of disease-free survival. Furthermore, a high risk group of patients with bulky hypoxic tumors could be defined. Hypoxic tumors also appeared to have a greater propensity to relapse in the pelvis and para-aortic lymph nodes as compared to well-oxygenated tumors.

In clinical studies on locally advanced cervix cancers, Haensgen et al. (1997) and Schorcht et al. (1997) have confirmed our earlier data. They have again unequivocally shown that tumor hypoxia strongly correlates with poor local control and overall survival.

Assuming that hypoxic cervical cancers exhibit a higher glycolytic rate than normoxic tumors, hypoxia and a poor overall survival should be associated with high lactate levels. This notion has been experimentally supported by Schwickert et al. (1995). They have shown a significant correlation of high lactate levels in the primary tumors with incidence of distant metastasis.

In a cohort of patients with locoregional relapses in the pelvis, measurements revealed pO_2 values significantly lower than the median pO_2 values of primary lesions of comparable sizes (Hoeckel et al., 1998). Median tumor pO_2 values <4 mmHg indicated a

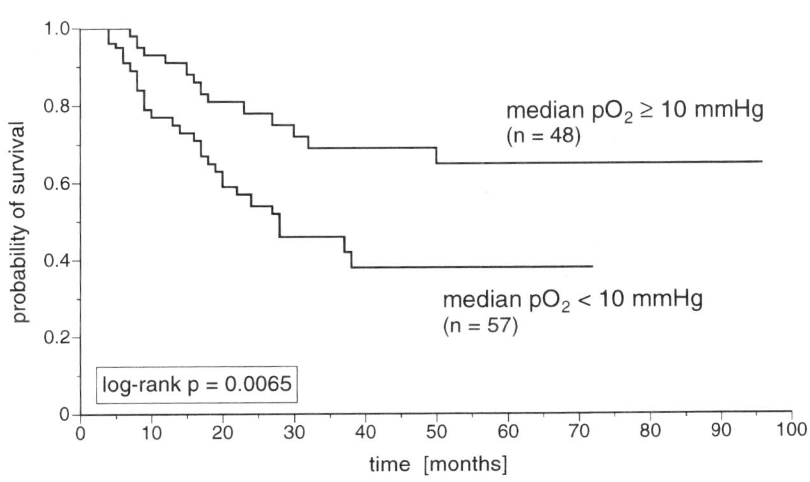

Figure 1. Overall survival probability for patients with cervical cancer treated with curative intent, stratified for tumor oxygenation (adapted from Hoeckel et al., 1996a).

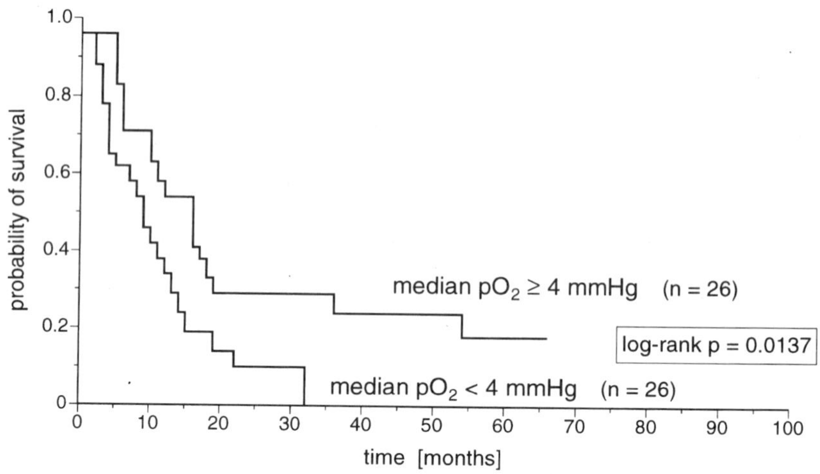

Figure 2. Survival of patients with recurrent cancer of the uterine cervix stratified for median pO_2 (adapted from Hoeckel et al., 1998).

significantly shorter median survival time as compared to median tumor pO_2 values ≥4 mmHg (9 months vs. 16 months, log-rank p – 0.01, see Figure 2).

3. PREDICTIVE POTENTIAL OF THE OXYGENATION STATUS OF SOFT TISSUE SARCOMAS

Tumor oxygenation status was assessed prior to treatment in 25 patients with primary soft tissue sarcomas by Nordsmark et al. (1998). They found that patients with the more hypoxic tumors were associated with a poorer disease specific survival and overall survival at 4 years observation time. In a study by the same group on the relationship between tumor oxygenation and cell proliferation, the fastest growing tumor cells were found in the worst oxygenated soft tissue sarcomas, independent of clinical and histopathological parameters (Nordsmark et al., 1996a).

In another study by Brizel et al. (1996) on high-grade soft tissue sarcomas which underwent preoperative irradiation and localized hyperthermia the 18-month actuarial disease-free survival was 70% for patients with well-oxygenated tumors, but only 35% for those with hypoxic malignancies.

4. PREDICTIVE VALUE OF THE OXYGENATION STATUS OF CARCINOMAS OF THE HEAD AND NECK

In a study on the correlation between the tissue oxygenation status and radiation response in advanced metastases of squamous cell carcinomas of head and neck (SCCHN), Nordsmark et al. (1996b) have shown that the most hypoxic tumors had significantly lower loco-regional tumor control than well-oxygenated nodes. Cox multiple regression analysis revealed the fraction of pO_2 values less than 2.5 mmHg to be the strongest independent variable in predicting radiation response when using tumor control in the site of pO_2 assessment as treatment endpoint.

From a study by Brizel et al. (1997) there is also a clear indication that tumor hypoxia adversely affects the prognosis of patients suffering from primary and metastatic SCCHN. In a most recent study on 43 patients, tumor hypoxia adversely affected locoregional control, disease-free survival and overall survival (Brizel et al., 1998).

As was the case with cervix cancers, there was a significant correlation of high lactate levels (corresponding to tumor hypoxia) in head and neck cancers with the incidence of metastasis (Walenta et al., 1997).

5. SUMMARY

We have previously demonstrated with a prospective study in primary cancer of the uterine cervix (clinical size >2 cm) that tumor hypoxia not only indicates decreased radiocurability but is generally associated with malignant progression of the disease. This finding also holds true for other tumor entities (soft tissue sarcomas, squamous cell carcinomas of the head and neck region), for lymph node metastases of head and neck lesions, and for locoregional recurrences of cervical cancers. All data available so far support our thesis that in cervical cancers (and in other solid tumors as well), tumor hypoxia and clinical aggressiveness in terms of resistance to therapy and tumor dissemination, are interrelated.

REFERENCES

Becker A, Haensgen G, Richter C, and Dunst J (1998) Comparison of oxygenation status of primary tumours, their neck node metastases and normal tissue in squamous cell carcinoma of the head and neck. Strahlenther Onkol, in press.

Brizel DM, Rosner G, Harrelson J, Prosnitz LR, and Dewhirst MW (1994) Pretreatment oxygenation profiles of human soft tissue sarcomas. Int J Radiat Oncol Biol Phys 30:635–642.

Brizel DM, Rosner GL, Prosnitz LR, and Dewhirst MW (1995) Patterns and variability of tumor oxygenation in human soft tissue sarcomas, cervical carcinomas, and lymph node metastases. Int J Radiat Oncol Biol Phys 32:1121–1125.

Brizel DM, Scully SP, Harrelson JM, Layfield LJ, Dodge RK, Charles HC, Samuslski TV, Prosnitz LR, and Dewhirst MW (1996) Radiation therapy and hyperthermia improve the oxygenation of human soft tissue sarcomas. Cancer Res 56:5347–5350.

Brizel DM, Sibley GS, Prosnitz LR, Scher RL, and Dewhirst MW (1997) Tumor hypoxia adversely affects the prognosis of carcinoma of the head and neck. Int J Radiat Oncol Biol Phys 38:285–289.

Brizel DM, Scher RL, and Dewhirst MW (1998) Oxygenation of head and neck cancer: Impact on treatment outcome and changes during radiotherapy. 10th International Conference on Chemical Modifiers of Cancer Treatment, Clearwater, FL.

Fyles AW, Milosevic M, Wong R, Kavanagh M-C, Pintilie M, Sun A, Chapman W, Levin W, Manchul L, Keane TJ, and Hill RP (1998a) Oxygenation predicts radiation response and survival in patients with cervix cancer. Radiother Oncol 48:149–156.

Fyles AW, Milosevic M, Wong R, Kavanagh M-C, Pintilie M, Chapman W, Levin W, Manchul L, Keane TJ, and Hill RP (1998b) Oxygenation in cervix cancer predicts radiation response and survival. 10th International Conference on Chemical Modifiers of Cancer Treatment, Clearwater, FL.

Haensgen G, Becker A, and Dunst J (1997) Untersuchungen zum Verhaeltnis des Sauerstoffpartialdruckes unter der Strahlentherapie am Beispiel des inoperablen Uteruszervixkarzinoms. Strahlenther Onkol 173:591.

Hoeckel M and Vaupel P (1998) The prognostic significance of hypoxia in cervical cancer: A radiobiological or tumor biological phenomenon? In: Medical Radiology—Diagnostic imaging and radiation oncology.

Blood perfusion and microenvironment of human tumors (Molls M, Vaupel P, Eds) Springer, Berlin, Heidelberg, New York, pp. 73–79.

Hoeckel M, Schlenger K, Knoop C, and Vaupel P (1991) Oxygenation of carcinomas of the uterine cervix: Evaluation of computerized O_2 tension measurements. Cancer Res 51:6098–6102.

Hoeckel M, Knoop C, Schlenger K, Vorndran B, Mitze M, Knapstein PG, and Vaupel P (1993) Intratumoral pO_2 predicts survival in advanced cancer of the uterine cervix. Radiother Oncol 26:45–50.

Hoeckel M, Schlenger K, Mitze M, Schaeffer U, and Vaupel P (1996a) Hypoxia and radiation response in human tumors. Sem Radiat Oncol 6:3–9.

Hoeckel M, Schlenger K, Aral B, Mitze M, Schaeffer U, and Vaupel P (1996b) Association between tumor hypoxia and malignant progression in advanced cancer of the uterine cervix. Cancer Res 56:4509–4515.

Hoeckel M, Schlenger K, Hoeckel S, Aral B, Schaeffer U, and Vaupel P (1998) Tumor hypoxia in pelvic recurrences of cervical cancer. Int J Cancer 97:365–369.

Hohenberger P and Dragon S (1995) In situ oxygen partial pressure measurements in human soft tissue sarcomas. In: Tumor Oxygenation (Vaupel PW, Kelleher DK, Guenderoth M, Eds) Gustav Fischer, Stuttgart, Jena, New York, pp. 327–333.

Hohenberger P, Felgner C, Haensch W, and Schlag PM (1998) Tumor oxygenation correlates with molecular growth determinants in breast cancer. Breast Cancer Res Treatment 48:97–106.

Kallinowski F and Buhr HJ (1995) Tissue oxygenation of primary, metastatic and xenografted rectal cancers. In: Tumor Oxygenation (Vaupel PW, Kelleher DK, Guenderoth M, Eds) Gustav Fischer, Stuttgart, Jena, New York, pp. 205–209.

Kallinowski F, Zander R, Hoeckel M, and Vaupel P (1990) Tumor tissue oxygenation as evaluated by computerized pO_2-histography. Int J Radiat Oncol Biol Phys 19:953–962.

Lartigau E, Le Ridant AM, Lambin P, Weeger P, Martin L, Sigal R, Lusinchi A, Luboinski B, Eschwege F, and Guichard M (1993) Oxygenation of head and neck tumors. Cancer 71:2319–2325.

Lyng H, Sundfor K, Trope G, and Rofstad EK (1996) Oxygen tension and vascular density in human cervix carcinoma. Brit J Cancer 74:1559–1563.

Lyng H, Sundfor K, Tanum G, and Rofstad EK (1997) Oxygen tension in primary tumours of the uterine cervix and lymph node metastases of the head and neck. Adv Exp Med Biol 428:55–60.

Mattern J, Kallinowski F, Herfarth C, and Volm M (1996) Association of resistance-related protein expression with poor vascularization and low levels of oxygen in human rectal cancer. Int J Cancer 67:20–23.

Nordsmark M, Bentzen SM, and Overgaard J (1994) Measurement of human tumour oxygenation status by a polarographic needle electrode. Acta Oncol 33:383–389.

Nordsmark M, Hoyer M, Keller J, Nielsen OS, Jensen OM, and Overgaard J (1996a) The relationship between tumor oxygenation and cell proliferation in human soft tissue sarcomas. Int J Radiat Oncol Biol Phys 35:701–708.

Nordsmark M, Overgaard M, and Overgaard J (1996b) Pretreatment oxygenation predicts radiation response in advanced squamous cell carcinoma of the head and neck. Radiother Oncol 41:31–39.

Nordsmark M, Keller J, Nielsen OS, Lundorf E, and Overgaard J (1997) Tumour oxygenation assessed by polarographic needle electrodes and bioenergetic status measured by ^{31}P magnetic resonance spectroscopy in human soft tissue tumours. Acta Oncol 36:565–571.

Nordsmark M, Keller J, Nielsen OS, Jensen OM, and Overgaard J (1998) Tumor hypoxia is associated with poor prognosis in patients with soft tissue sarcomas. 10th International Conference on Chemical Modifiers of Cancer Treatment, Clearwater, FL.

Rampling R, Cruickshank G, Lewis AD, Fitzsimmons SA, and Workman P (1994) Direct measurement of pO_2 distribution and bioreductive enzymes in human malignant brain tumours. Int J Radiat Oncol Biol Phys 29:427–432.

Runkel S, Wischnik A, Teubner J, Kaven E, Gaa J, and Melchert F (1994) Oxygenation of mammary tumors as evaluated by ultrasound-guided computerized-pO_2-histography. Adv Exp Med Biol 345:451–458.

Schorcht J, Neumann H, Eberhardt HJ, and Neumann U (1997) Untersuchungen zur Oxygenierung inoperabler gynaekologischer Tumoren. Strahlenther Onkol 173:589.

Schwickert G, Walenta S, Sundfor K, Rofstad EK, and Mueller-Klieser W (1995) Correlation of high lactate levels in human cervical cancer with incidence of metastasis. Cancer Res 55:4757–4759.

Strnad V, Keilholz L, Kirschner M, Meyer M, and Sauer R (1997) Sauerstoffdruckverteilung in Lymphknotenmetastasen und die Veraenderungen waehrend akuter respiratorischer Hypoxie. Strahlenther Onkol 173:267–271.

Sundfor K, Lyng H, Kongsgard U, Tropé C, and Rofstad EK (1997) Polarographic measurement of pO_2 in cervix carcinoma. Gynecol Oncol 64:230–236.

Terris DJ and Dunphy EP (1994) Oxygen tension measurements of head and neck cancers. Arch Otolaryngol Head Neck Surg 120:283–287.

Vaupel P (1996) Oxygen transport in tumors: Characteristics and clinical implications. Adv Exp Med Biol 388:341–351.

Vaupel P (1997) Blood flow and O_2-status of head and neck carcinomas. Adv Exp Med Biol 428:89–95.

Vaupel P and Hoeckel M (1998) Oxygenation of human tumors. In: Medical Radiology—Diagnostic Imaging and Radiation Oncology. Blood Perfusion and Microenvironment of Human Tumors—Implications for Clinical Radiooncology (Molls M, Vaupel P, Eds) Springer, Berlin, Heidelberg, New York, pp. 63–72.

Vaupel P, Schlenger K, Knoop C, and Hoeckel M (1991) Oxygenation of human tumors: Evaluation of tissue oxygen distribution in breast cancers by computerized O_2 tension measurements. Cancer Res 51:3316–3322.

Vaupel P, Thews O, and Hoeckel M (1996) Tumor oxygenation: Characterization and clinical implications. In: rhErythropoietin in Cancer Supportive Treatment (Smyth JF, Boogaerts MA, Ehmer BR-M, Eds) Marcel Dekker, New York, Basel, pp. 205–239.

Vaupel P, Thews O, Kelleher DK, and Hoeckel M (1998) Current status of knowledge and critical issues in tumor oxygenation: Results from 25 years research in tumor pathophysiology. Adv Exp Med Biol, in press.

Walenta S, Salameh A, Lyng H, Evensen JF, Mitze M, Rofstad EK, and Mueller-Klieser W (1997) Correlation of high lactate levels in head and neck tumors with incidence of metastasis. Amer J Pathol 150:409–415.

Wong RKW, Fyles A, Milosevic M, Pintilie M, and Hill RP (1997) Heterogeneity of polarographic oxygen tension measurements in cervix cancer: an evaluation of within and between tumor variability, probe position, and track depth. Int J Radiat Oncol Biol Phys 39:405–412.

64

THE CARDIOME PROJECT

An Integrated View of Cardiac Metabolism and Regional Mechanical Function

James B. Bassingthwaighte, Hong Qian, and Zheng Li

University of Washington
Seattle, Washington 98195-7962

INTRODUCTION

In this presentation the first goal is to give an idea of the broader aspects of the Physiome Project, its origins and its current state. The second goal is to give an example of the kind of approaches that will support the Physiome Project by contributing a component that is in the process of being developed into a functioning model and a database entry point: a piece of the Cardiome Project.

THE PHYSIOME PROJECT

The PHYSIOME PROJECT is a loosely integrated multi-centric program to design, develop, implement, test and document, archive and disseminate qualitative and quantitative information, e.g., databases and models, of the functional behavior of organelles, cells, tissues, organs, and organisms. It is a successor to the Genome Project. The focus of the Human Physiome Project is on the human organism, its physiology and pathophysiology, to provide eventually full working models of physiological systems that integrate the observations from many laboratories into quantitative, self-consistent, comprehensive descriptions.

The goal is to provide functional descriptions of human biological systems in health and disease to the community of scientists, physicians, teachers, and manufacturers. A major feature of the project is the databasing of laboratory and clinical observations on all organisms for retrieval and evaluation. A network of Physiome Centers would comprise an adaptable international resource for integrated physiological systems, structured for accessibility via Internet, for the immense databases of information on methods, data, and models.

Oxygen Transport to Tissue XXI, edited by Eke and Delpy.
Kluwer Academic / Plenum Publishers, New York, 1999.

There is a growing effort to raise consciousness about integrative biology and to provide a setting for the scientific results from research laboratories concerned with genomics, molecular and cell biology, biology in general, and clinical medicine. For those with a primary interest in therapeutics, the Physiome Project can provide a framework for determining the effects of pharmaceutical or genetic interventions on target molecules and functional systems through an in depth, comprehensive understanding of biology at the level of the cell, the tissue, the organ and the organism. The Project will need to be supported by the development of databases that follow upon those developed from the Genome and the Proteome.

In addition to the structural and static information in the databases, a salient goal of the Physiome Project is to integrate dynamic information and modeling, which may be considered to include the protein folding, DNA/RNA conformational dynamics, protein-DNA and other protein-protein interactions, solute-ligand competition for binding sites on enzymes and receptors, and antigen-antibody reactions. Integrations only at static, structural levels (at the level of molecules, organelles, cells, and broader), may not always be sufficient as the bases for understanding human biology and physiology. Many important biological and physiological phenomena have to be explained by a non-dynamic perspective at the molecular as well as the systemic level.

The Physiome Project combines engineering applications with biology, informatics and databasing, and large scale systems analysis. It is potentially a vehicle for discovery and invention, and is a good subject for society-sponsored conferences which would provide an open forum for scientific presentations and advancement of plans, much as in the style of the Gordon Research Conferences in the US. The topics covered would be both on the logistics of specific aspects of the Physiome Project and on its strategies and scientific advances. Government agencies are demonstrating clear interest (particularly NIH, NSF, NASA, and DOE).

In parallel, we need meetings with officials from government agencies to discuss funding mechanisms. One that I feel is important is to obtain better national funding for those agencies which are appropriate to supporting a "set of networked research resource facilities" to take on specific aspects of the program in various countries.

Meetings in 1997 and 1998 that were pertinent to the Physiome Project are summarized or linked at the Physiome Website (www.physiome.org).

THE CARDIOME PROJECT

The Cardiome Project aims to develop a visualizable, interactive model of the physiologically normal working heart. It is a part of the Physiome Project. The Physiome Project is the all-encompassing project, endorsed at an international conference in St. Petersburg in July 1997, "On Designing the Physiome Project," by the International Union of Physiological Sciences, and the project is targeted now by units within NSF, NIH, and DOE. *The Cardiome*, our specific target, is one of the eight fields selected at the St. Petersburg meeting to demonstrate the practicality and utility of the approach.

The Cardiome is the description, in quantitative, testable form, of the functioning of the normal heart and its responses to intervention. It will prove helpful in integrating results from many years' efforts into a comprehensive understandable scheme. These efforts have spanned the fields of transport within blood vessels, the distributions of regional coronary blood flows, permeation processes through capillary and cell walls, mediated cell membrane transport, diffusion, electrophysiology, metabolism of the prime substrates (fatty acid and glucose), and studies of metabolism of the purine nucleosides and nucleotides (mainly adenosine and ATP).

The overall Cardiome Project is a large-scale multidisciplinary, multiuniversity effort. It involves physicians, computer scientists, pharmacologists, electrophysiologists, and biomedical engineers working together to develop a model of the human heart. A model heart put together through the collaborative efforts of Hunter, Smaill, Noble, Winslow, McCulloch and others provides fully quantitative descriptions of the electrical currents and ion pump, of the spread of electrical activity across a finite element representation of the fibre directions in the myocardium, and the cardiac contraction itself. We at the University of Washington are expanding this to account for the spatial distribution of the coronary arteries, the regional myocardial blood flows, the uptake and metabolism of glucose, fatty acids and oxygen used for the energy to form ATP, which is in turn used to fuel the work of contraction and ion pumping. In order to link the metabolism with the finite element representations of the mechanics of contraction we will collaborate with Peter Hunter at Auckland University, New Zealand, Andrew McCulloch at UC San Diego and Theo Arts, Maastricht University, Maastricht, The Netherlands. The ionic currents in the Hunter-McCulloch model use the equations either of Denis Noble (Oxford University, UK) or of Yoram Rudy (Case Western Reserve University). Rai Winslow (Johns Hopkins University) used these currents to drive the spread of excitation throughout the three-dimensional cardiac structure.

To model the chemical-mechanical conversion of high energy phosphate for contraction we have begun a collaboration with Amir Landesberg and Samuel Sideman, at Technion University, Haifa, and with Martin Kushmerick, Michael Regnier and Albert Gordon at the University of Washington. The fatty acid metabolism studies are a continuation of work in collaboration with Ger van der Vusse, Rob Reneman, and Jan Glatz at Maastricht University and James Caldwell at the University of Washington. The glucose studies are a continuation of a collaboration with Claudio Cobelli and Paolo Vicini at The University of Padua which, like most of the others above, is formalized through their participation in the activities of our Simulation Resource Facility at the University of Washington.

DESIGN PHASE

In order to provide a demonstration of the utility of modeling we would like to have a functional version of the Cardiome developed within three years to the level where it is useful for predicting responses to blockage of coronary arteries, to pharmaceutical agents used for blocking channels or pumps (e.g., digitalis in cardiac failure), and to metabolic stimuli or exercise. It will require many additional steps of subsequent development to account for intracellular regulatory processes, so our planned initial result can only be regarded as a primitive beginning.

The goal is to develop an integrated synthesis of cardiac metabolic, electrical, and mechanical events into a visualizable form that can be seen by the *public* to represent cardiac mechanics in health and disease and can be used by *investigators* as a stepping stone for developing insight into the complex metabolic and biophysical interrelationships that are the basis for cardiac function. Within this context we define four linked modules which, along with two modules already well developed by others for the electrophysiology and the mechanics, describe cardiac function. The six modules are as diagrammed in Figure 1.

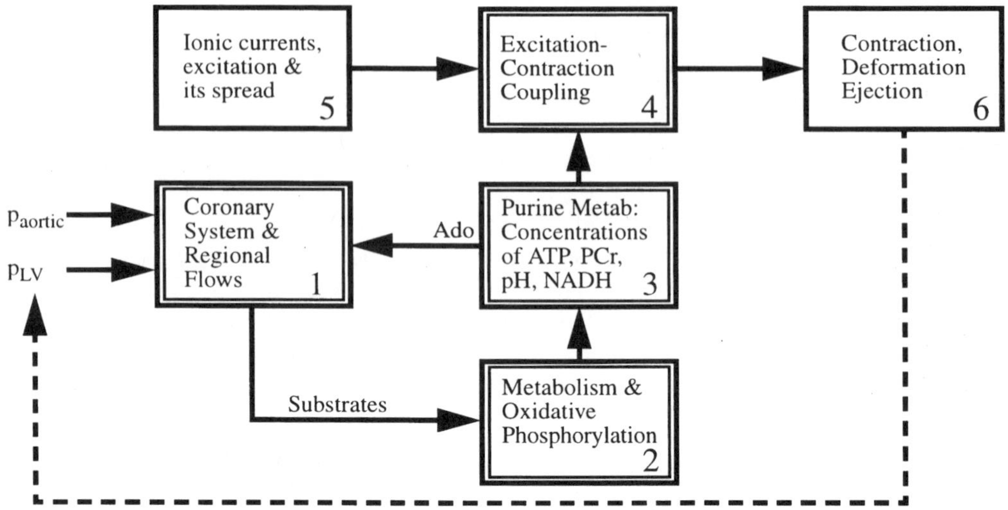

Figure 1. Overall diagram of the Cardiome Model components.

1. Coronary artery anatomic localization and **regional myocardial flows**, providing substrate and oxygen delivery,
2. **Metabolism of the substrate for energy metabolism**, fatty acid and glucose, the tricarboxylic acid cycle, and **oxidative phosphorylation**,
3. **Purine nucleoside and purine nucleotide metabolism**, describing the formation of ATP and the regulation of its degradation to adenosine in endothelial cells and myocytes, and its effects on coronary vascular resistance,
4. **Excitation-contraction coupling (ECC)**, calcium release and reuptake, and the relationships between these and the strength and extent of sarcomere shortening.
5. **The electrical currents and their spread across the myocardium**: these are described by either the Luo-Rudy model (Luo and Rudy, 1994a, 1994b; Shaw and Rudy, 1995; Rudy and Shaw, 1997), or the Noble-DiFrancesco model (DiFrancesco and Noble, 1985; Ch'en et al., 1997).
6. **The three-dimensional finite element representation of the contracting heart** has been brought to an advanced stage by Peter Hunter and his group at University of Auckland, New Zealand (Hunter et al., 1988; Hunter and Arts, 1997) and Andrew McCulloch and his group at San Diego. (See McCulloch et al., 1998.) It is into their three-dimensional finite element reconstructions of the mechanically contracting heart that we will put our components of the physiological responses.

The major inputs to the model will be: aortic blood pressure, hematocrit, pO_2, pCO_2, heart rate, concentrations of palmitate, lactate, acetate, glucose, pyruvate, pH, Na^+, Ca^{2+}, K^+, Cl^-, PO_4, HCO_3, Mg^{2+}. To take this through the first stage, gross simplifications must be made, so we will temporarily omit incorporating: (1) mechanico-electrical feedback, (2) systolic compression of intramyocardial coronary arteries and veins, (3) regulation of intracellular enzymes by secondary processes, (4) vascular and tissue remodeling, (5) protein metabolism, (6) systemic influences on total body vascular resistance, (7) changes in cardiac pool sizes of glycogen and di- and triphosphoglycerides, and many other features. Any list of omissions should be nearly infinite; these are merely the more obvious ones. These, and others found to be critical, will be the subject of later modifications.

The overall schema (Figure 1) is centered around the fluxes of ATP, ADP/Pi, and H^+ in Module 3, linking the other modules in the regulation in the functioning heart. Module 4 for the excitation-contraction coupling now exists only as a purely empirical relationship without adaptability or regulatory control, or any link to ATP. The other modules, for coronary flow distribution and substrate and oxygen delivery to tissue, for substrate utilization and oxidative phosphorylation, and for purine metabolism are new undertakings. We will focus initially on the single sarcomere level modules and their interactions with one another. Periodically, we will incorporate each module into the integrated whole organ model for testing there. Each stage remains archived through our Source Code Control System to preserve all previous versions. As information from experiments is gathered, or inconsistencies recognized from the behavior of the whole organ model, improvements will be made in the sarcomere modules. To give users prompt responses we must code for computational efficiency. Periodic reviews to improve speed will be essential in a large long-range project like this.

MODULE 1. CORONARY SYSTEM AND REGIONAL FLOW DISTRIBUTIONS

Intramyocardial flow distributions are remarkably heterogeneous (Yipintsoi et al., 1973; Bassingthwaighte et al., 1990), showing a range of 6 to 10-fold in normal hearts in awake or anesthetized animals. This is explained by the fractal self-similar branching of the coronary arteries (Bassingthwaighte et al., 1989). Anatomic data from van Bavel et al. (1992) and Kassab et al. (1993) serve as a more realistic basis for network reconstructions than we achieved earlier (van Beek et al., 1989). Network reconstructions developed by Daniel Beard exhibit appropriate pressure distributions, regional flow heterogeneities and fractal spatial correlations in flows, and the same power law form ($\sim 1/t^3$) for the washout time course as is observed (Bassingthwaighte and Beard, 1995).

The next stages concern the delivery of substrates and oxygen. The mechanism for glucose delivery to myocytes is through the interendothelial clefts, and from the interstitium via a transporter across the cell membrane (Kuikka et al., 1986). Fatty acid transport is more complex because it binds tightly to albumin; it does not traverse the clefts, but must be stripped off the albumin before being translocated (reviewed by van der Vusse et al., 1999, submitted). The inhibition of tracer palmitate permeability by high albumin concentration was explained by Bassingthwaighte, Noodleman et al. (1989) in a model form for the fatty acid translocation. In a further refinement, Caldwell et al. (1994) found in exercising dogs that the regional density of fatty acid transporters in the heart is proportional to the regional flows.

The non-invasive measurement of regional myocardial blood flow (MBF) and oxygen consumption (MRO_2) using positron emission tomography (PET) that we have developed in the past two years contributes directly to the Cardiome project, and leads to an important simplification. The most direct measure of oxidative tissue metabolism is the conversion rate of oxygen to water via mitochondrial respiration. A sequence of cardiac images obtained at two-second intervals after single-breath inhalation of ^{15}O-oxygen (Li et al., 1997) gives, in each region of interest, the time-activity curves of the sum of the ^{15}O-oxygen and its product, ^{15}O-water. Their relative amounts can be distinguished by kinetic modeling (Figure 2) because the mean transit time for water produced in the tissue is much longer than that for untransformed oxygen, most of which is in the form of oxyhemoglobin within the red blood cells. Consequently, the extraction of ^{15}O-oxygen can be estimated and used to calculate the steady-state oxygen consumption within each ROI. The model (Figure 2) accounts for intravascular convection, penetration of capillary and parenchymal cell barriers, the metabolism to ^{15}O-water in parenchymal cells, and ^{15}O-water transport into the venous effluent. Nonlinear binding of oxygen to hemoglobin in erythrocytes and to myoglobin in myocytes is explicitly incorporated in the model. This model is more accurate and much easier to use than the linear model by Deussen and Bassingthwaighte (1996) because the local volumes of distribution for oxygen are calculated automatically from oxygen supply and demand at steady state.

With this PET method we were able to study the correlation between regional myocardial blood flow and oxygen consumption *in vivo*. Our results from a limited number of experiments showed that the regional oxygen consumptions were approximately proportional to the regional flows as measured by the microsphere technique (Yipintsoi et al., 1973; Bassingthwaighte, Malone et al., 1990) (Figure 3). These results support the generality that flows, transport conductance for substrate (Caldwell et al., 1994), and oxygen consumption are regionally matched. This strong evidence for proportional regional requirements will simplify the overall modeling project if we find that there is regional matching with ATP usage.

The methods for using models to determine the parameters of transport and metabolism are discussed by Bassingthwaighte and Goresky (1984), by Kuikka et al. (1986), and in a book slated for publication next month (Bassingthwaighte, Goresky and

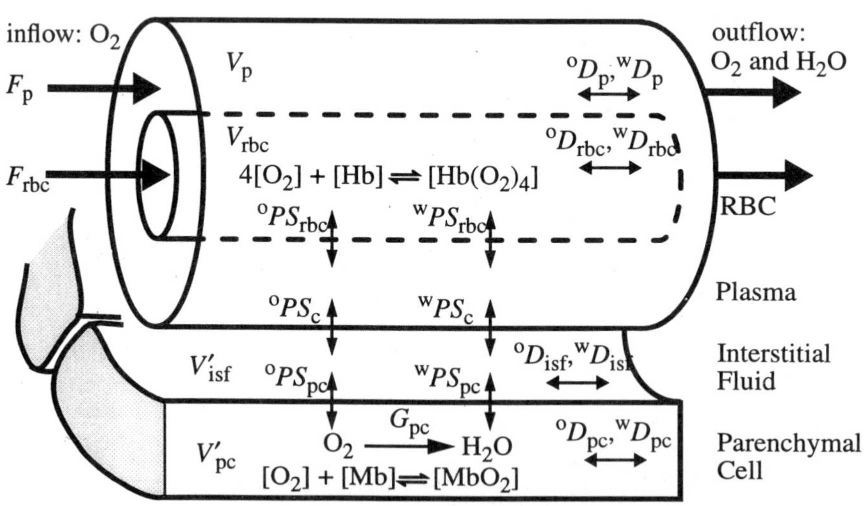

Figure 2. Dual capillary-tissue model for ^{15}O-oxygen transport and metabolism into ^{15}O-water. The four concentric regions are red blood cells (rbc), plasma (p), interstitial fluid (isf) and parenchymal cells (pc). The symbols used are F for flow, ml min^{-1} g^{-1}; V for anatomical volumes and V' for volumes of distribution, ml g^{-1}; PS for permeability-surface area product, ml min^{-1} g^{-1}; G for the first-order consumption rate, ml min^{-1} g^{-1}; and D for the axial dispersion coefficient, cm^2 s^{-1}. The superscripted "o" and "w" denote oxygen and water, respectively.

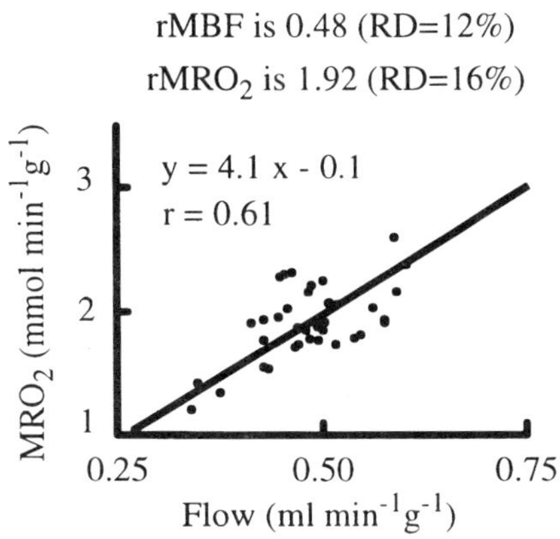

Figure 3. Correlation between blood flow (rMBF) and oxygen consumption ($rMRO_2$) in regions of interest (0.9 ± 0.3 g) of the LV myocardium of a closed-chest dog.

Linehan, 1998). Optimized parameter adjustment can be accomplished manually under XSIM, a graphical user-friendly interface developed by Richard King et al. in our group, or by using automated optimization, e.g., by a steepest descent technique modified from that of Bronikowski et al. (1980) with our sensitivity function based approach (Chan et al., 1993). As a standard technique for evaluating the error and bias in estimates, we use Monte Carlo assessment of the effects of adding noise to form "pseudo data" from model solutions.

A satisfactory first definition for regional substrate delivery therefore provides for the fluxes by flow and transmembrane transport of oxygen, the substrates, and the metabolites, including lactate and CO_2. The oxygen model diagrammed in Figure 2 is of the same general form as for these other solutes, all of which can be metabolized, but the others are more complex because they need specialized transporters, e.g., for fatty acid and for adenosine.

MODULE 2. METABOLISM AND OXIDATIVE PHOSPHORYLATION IN MITOCHONDRIA

Module 2, "Metabolism & Oxidative Phosphorylation," contains the detailed enzymatic reaction network for **Glycolysis**, **TCA cycle**, **oxidative phosphorylation**, and **acyl CoA mobilization** from fatty acids and pyruvate. General overviews of cardiac metabolism are given by Randle and Tubbs (1979) and Bassingthwaighte (1991). Mitochondria, which comprise one-third of myocyte volume, are well dispersed among and closely apposed to bundles of myofilaments.

The heart uses substrates principally for energy production. The main substrate is fatty acid, supporting about two-thirds of the caloric needs; the range of estimation is very wide, from 35% to 95%, which reflects the ability of the myocyte to shift fuels under different circumstances. The remainder of energy comes from glucose, lactate and amino acids. Glucose also provides pyruvate and lactate needed to maintain the constituents of the TCA cycle. Amino acids maintain intracellular proteins and enzymes and support transport reactions, exemplified by the role of carnitine in the transfer of fatty acids across the mitochondrial membrane. The overall processes of substrate metabolism include: delivery by flow into capillaries; transport across the capillary walls and across the sarcolemma of myocyte; diffusion and reaction in the cytosol; transport across the mitochondrial inner membrane; intramitochondrial reactions; breakdown of 2-carbon fragments into CO_2 and water through the TCA cycle; the reactions of the electron transport chain; and the production of ATP from ADP, coupled to the electron transfer chain and to other reactions in the glycolytic series. For fatty acids, Chatham et al. (1995) provide data on TCA fluxes of 2-carbon fragments and the shifts between oxidative and anaerobic metabolism.

For fatty acid metabolism, the thiokinase reaction requires one ATP-to-AMP reaction per free fatty acid. β-oxidation produces one molecule of $FADH_2$ and one molecule of NADH per 2-carbon fragment (total eight 2-carbon fragments from palmitate), and yields 35 molecules of ATP (one NADH yields 3 ATP, one $FADH_2$ yields 2 ATP). Each acetate molecule gives a total of 12 molecules of ATP in TCA and respiratory chain. So the total ATP generation per palmitate is $8 \times 12 + 5 \times 7 - 2 = 129$ molecules of ATP.

In glucose metabolism, for every molecule of glucose, 4 molecules of ATP are produced in anaerobic (2 ATP) and aerobic (2 ATP) glycolysis, and 32 molecules of ATP are produced in TCA and respiratory chain by oxidation of NADH and $FADH_2$ (for every molecule of glucose, 6 molecules of NADH and 2 molecules of $FADH_2$ are formed in the TCA cycle; 2 molecules of NADH are formed during aerobic glycolysis and 2 molecules of NADH are formed during pyruvate oxidation). To relate ATP generation to oxygen utilization, for glucose, 6ATP per O_2; for palmitate, 5.6ATP per O_2. So the oxygen cost per ATP is similar with fatty acid versus glucose during aerobic metabolism. Module 2 can therefore in its most elementary form for the normal physiological state be represented by the stoichiometric conversion of substrate and oxygen to ATP, with proportional usage the same everywhere.

Through the supply/demand imbalance the cardiac contractile performance influences and indirectly governs the blood flow and oxygen delivery in normal physiological states. The rate of carbon flux through TCA cycle is reduced when the O_2 supply is low, inducing a switch from aerobic fatty acid and glucose metabolism to glycolysis and lactate production. This shift toward anaerobic metabolism means there is zero ATP from palmitate (actually a net loss due to diacylglycerophosphate production) and only 2 ATP per glucose, in glycolysis, and at the expense of H^+ production and whose metabolism must be included (Achs and Garfinkel, 1977). The apportionment between glucose and fatty acid as substrate is governed by their availability and by pO_2, pH, and NADH levels.

To develop this part of the model we need values for the various biochemical constants for the enzymes; Evgeni Selkov has developed an enzyme database for the metabolic pathways (Selkov et al., 1996, 1997). This module is complex, but simplified versions have been worked out, for example by Magnus and Keizer (1997). Theirs has the explicit advantage of incorporating the influences of Ca^{2+} on the transformations within the mitochondrion. Selkov has such a model himself.

MODULE 3. ATP AND ADENOSINE METABOLISM AND RELEASE

The "Purine Metabolism" module contains the kinetic reactions for myocardial phosphoenergetics (Figure 1). Purine, as both ATP and adenosine, is an important biochemical regulator for the cardiac muscle mechanics. This key to understanding the energetics was extensively studied experimentally and computationally by the late Professor Keith Kroll (Kroll and Stepp, 1996; Kroll et al., 1997; Kroll and Bassingthwaighte, 1998). Based on experimental measurements on rabbit heart and subsequent comparison with the simulation from this kinetic model, he showed that an energy imbalance arises in the underperfused heart, and that the intracellular free energy loss is minimized by losing adenosine and AMP from the cell. Further study (Gustafson and Kroll, 1998) suggests also that the enzyme 5′-nucleotidase is down-regulated when PCr and ATP decrease.

The descriptive element, the imbalance, Δr_{ATP}, must be replaced by the appropriate reactions for formation and degradation of ATP. The former are in the "Oxidative Phosphorylation" module and the latter in the modules for "Contraction" and for the ion pumps, included in "Ionic currents" and "ECC". All modules are tested for adherence to mass balance; for this module the test is to set the membrane permeabilities to zero for all forms of creatine, phosphate, and purine base while running model solution for conditions that degrade ATP and PCr and to see whether or not their totals remain constant; such constancy is a minimal requirement, without which the model is invalid. There is never a proof of validity, but when many tests against observed data are satisfied one gains confidence in such a model as a working hypothesis.

The computer simulation program we developed for this uses mouse clicks to bring out deeper features. A click on a bottom "Cr K'ase" pops up a new window to show the underlying mechanics for the enzyme.

MODULE 4. EXCITATION-CONTRACTION COUPLING

Our earliest model of excitation-contraction coupling (Bassingthwaighte and Reuter, 1972) included our kinetics of the slow inward calcium current reported formally in the Beeler-Reuter (1977) model of the action potential. It was timely as a step in integrated modeling and it incorporated the concepts of that time on how calcium cycling in the myocyte served as the key factor in the governance: trigger Ca^{2+} current, calcium release from the SR, calcium-troponin binding, contractile filament shortening and reuptake of calcium by the sarcoplasmic reticulum, SR. A more modern view is provided for example by Bers (1991) in an excellent monograph. Module 4 requires information from the Ionic Currents (Module 5). Initially we will not use the detailed time course of the currents, but to speed computation will use the total net quantities of each ion transferred during the action potential to determine the concentrations within the cytosol. These are the quantities that need to be removed by the pumps and exchanges during the remainder of the cardiac cycle during any steady state. These will determine the ATP usage for ionic balance.

The ATP used for cross-bridge shortening is influenced by the tension and the time course of the shortening velocity in the sarcomere (Landesberg et al., 1994a,b; 1996a,b; Regnier et al., 1995, 1997; Blei et al., 1993; McFarland et al., 1994; Wiseman et al., 1995; Chase et al., 1995). Because local Ca^{2+} and early systolic pre-ejection phase shortening cause a shift from the strongly to the weakly bound form of ATP-myosin, this reduces the amount of ATP required during a contraction. Consequently our modeling needs to take this level of detail into account when it comes to examining the relationship between local flows, local metabolism, and contractile function, especially in situations such as prolonged ventricular pacing where remodeling occurs (Arts et al., 1994, 1995), thinning the ventricular wall at sites of early activation and causing hypertrophy in late activated regions. Since there is potential benefit in using pacing as an inhibitor of outflow tract obstruction in hypertrophic subaortic stenosis, for example, testing this approach in a model of an intact heart may be one way to use the modeling in defining therapy. Since one can obtain individual cardiac structural descriptions from MR and regional physiological information from PET and ultrasound, this could evolve to a stage, as it has for repairing knees, where the modeling is key to the therapeutic intervention.

The myosin ATP form and Ca^{2+} influence the patterns of deformation through the ventricle; the ventricular structure has been precisely detailed in two dog hearts (LeGrice et al., 1995, 1997) with fiber directions over the whole ventricle, and fibre sheets and bundles between which the vessels are positioned. These have been integrated into whole heart mechanical models (Module 6) by our collaborators (McCulloch et al., 1989, 1991, 1998; Hunter et al., 1988; Hunter and Arts, 1997) and it is with these model structures and the Ionic Current Module 5, that we will integrate our Modules 1 to 4.

SUMMARY

The goal, to develop a functioning three-dimensional computational model of the excitation, metabolism and contraction of the heart within three years, is one of the beginnings for the Cardiome Project. Our first stage will not be likely to provide highly accurate prediction of physiological behavior in general, but will be focussed so that it is adequate for at least three specific purposes: response to regional flow reduction, response to heart rate changes, and response to increased metabolic drive. We would like to make the model visualizable by three-dimensional viewing, with cross-sectional and transparency viewing approaches, illustrate the fiber directions, the arteries, the deformation with contraction and images of regional functions such as oxygen consumption, pre-ejection strain, or lactate concentration. The display techniques developed by Hunter et al. and by McCulloch et al. would be excellent for such demonstration and teaching purposes, and should be attractive enough for public display.

The Physiome Project is underway now, with growing government and private support. Now we are going from the era of molecular biology, led by the Genome Project, into a new era of integrative biology. The goal is to understand biology so deeply and so broadly that predictions about interventions can be made. Methods of experimentation and of diagnosis are critical to acquiring the data, and therefore in making the prediction, and so all aspects of our Society's efforts and interests are relevant to undertaking this grand challenge.

REFERENCES

Achs, MJ and Garfinkel, D (1977) Computer simulation of energy metabolism in anoxic perfused rat heart. *Am J Physiol* 232 (*Regulatory Integrative Comp. Physiol.* 1):R164–R174.

Arts, T, Prinzen, FW, Snoeckx, LHEH, Rijcken, JM, and Reneman, RS (1994) Adaptation of cardiac structure by mechanical feedback in the environment of the cell: A model study. *Biophys J* 66:953–961.

Arts, T, Prinzen, FW, Snoeckx, LHEH, and Reneman, RS (1995) "A model approach to the adaptation of cardiac structure by mechanical feedback in the environment of the cell." In: *Molecular and Subcellular Cardiology: Effects of Structure and Function.* (S Sideman and R Beyar, Eds.) Plenum. New York. pp. 217–228.

Bassingthwaighte, JB and Reuter, H (1972) Calcium reversal potential in cardiac muscle. *Biophys J* 12:214a.

Bassingthwaighte, JB and Goresky, CA (1984) "Modeling in the analysis of solute and water exchange in the microvasculature." In: *Handbook of Physiology. Sect. 2, The Cardiovascular System. Vol IV, The Microcirculation.* (EM Renkin and CC Michel, Eds.) Am. Physiol. Soc. Bethesda, MD. pp. 549–626.

Bassingthwaighte, JB, Noodleman, L, van der Vusse, GJ, and Glatz, JFC (1989) Modeling of palmitate transport in the heart. *Mol Cell Biochem* 88:51–58.

Bassingthwaighte, JB, Malone, MA, Moffett, TC, King, RB, Chan, IS, Link, JM, and Krohn, KA (1990) Molecular and particulate depositions for regional myocardial flows in sheep. *Circ Res* 66:1328–1344.

Bassingthwaighte, JB (1991) "The myocardial cell." In: *Cardiology: Fundamentals and Practice, 2nd Ed.* (ER Giuliani, V Fuster, BJ Gersh, MD McGoon, and DC McGoon, Eds.) Mosby-Year Book Inc. St. Louis, MO. pp. 113–149.

Bassingthwaighte, JB, R Friesner, B Honig, CF Starmer, and VZ Marmarelis. *Modeling and Simulation.* Bethesda MD: NIH/NCRR Workshop on Technologies for the Future: Biomedical Computing for Visualization, Modeling and Decision Support, April 12–13, 1991, pp. 1–15.

Bassingthwaighte, JB (1992) Fractal vascular growth patterns. *Acta Stereol* 11 (Suppl. 1):305–319.

Bassingthwaighte, JB (1995) "Toward modeling the human physionome." In: *Molecular and Subcellular Cardiology: Effects on Structure and Function.* (S Sidema and R Beyar, Eds.) Plenum. New York. pp. 331–339.

Bassingthwaighte, JB and Beard, DA (1995) Fractal ^{15}O-water washout from the heart. *Circ Res* 77:1212–1221.

Bassingthwaighte, JB, Goresky, CA, and Linehan, JH editors. (1998) *Whole organ approaches to cellular metabolism. Capillary permeation, cellular uptake and product formation.* Springer Verlag. New York.

Beeler, GW and Reuter, H (1977) Reconstruction of the action potential of ventricular myocardial fibres. *J Physiol Lond* 268:177–210.

Bers, DM (1991) *Excitation—Contraction Coupling and Cardiac Contractile Force.* Kluwer Academic Publishers. London.

Blei, ML, Conley, KE, and Kushmerick, MJ (1993) Separate measures of ATP utilization and recovery in human skeletal muscle. *J Physiol* 465:203–222.

Bronikowski, TA, Linehan, JH, and Dawson, CA (1980) A mathematical analysis of the influence of perfusion heterogeneity on indicator extraction. *Math Biosci* 52:27–51.

Caldwell, JH, Martin, GV, Raymond, GM, and Bassingthwaighte, JB (1994) Regional myocardial flow and capillary permeability-surface area products are nearly proportional. *Am J Physiol* 267 (*Heart Circ. Physiol.* 36):H654–H666.

Chan, IS, Goldstein, AA, and Bassingthwaighte, JB (1993) SENSOP: A derivative-free solver for non-linear least squares with sensitivity scaling. *Ann Biomed Eng* 21:621–631.

Chase, PB and Kushmerick, MJ (1995) Effect of physiological ADP concentration on contraction of single skinned fibers from rabbit fast and slow muscles. *Am J Physiol* 268:C480–C489.

Chatham, JC, Forder, JR, Glickson, JD, and Chance, EM (1995) Calculation of absolute metabolic flix and the elucidation of the pathways of glutamate labeling in perfused rat heart by [C-13] NMR spectroscopy and nonlinear least squares analysis. *J Biol Chem* 270:7999–8008.

Ch'en, F, Clarke, K, Vaughan-Jones, R, and Noble, D (1997) "Modeling of internal pH, ion concentration, and bioenergetic changes during myocardial ischemia." In: *Analytical and Quantitative Cardiology.* (S Sideman and R Beyar, Eds.) Plenum. New York. pp. 281–290.

Deussen, A and Bassingthwaighte, JB (1996) Modeling [^{15}O]oxygen tracer data for estimating oxygen consumption. *Am J Physiol* 270 (*Heart Circ. Physiol.* 39):H1115–H1130.

DiFrancesco, D and Noble, D (1985) A model of cardiac electrical activity incorporating ionic pumps and concentration changes. *Philos Trans R Soc Lond B Biol Sci* 307:353–398.

Glass, L, Hunter, P, and McCulloch, A (1991) *Theory of Heart: Biomechanics, Biophysics, and Nonlinear Dynamics of Cardiac Function.* Springer-Verlag. New York.

Gustafson, LA and Kroll, K (1998) Downregulation of 5'nucleotidase in rabbit heart during coronary underperfusion. *Am J Physiol* 274 (*Heart Circ. Physiol.* 42):H529–H538.

Hunter, P and Arts, T (1997) "Tissue remodeling with micro-structurally based material laws." In: *Analytical and Quantitative Cardiology.* (S Sideman and R Beyar, Eds.) Plenum. New York. pp. 215–225.

Hunter, PJ, McCulloch, AD, Nielsen, PMF, and Smaill, BH (1988) A finite element model of passive ventricular mechanics. *ASME BED* 9:387–397.

Kassab, GS, Rider, CA, Tang, NJ, and Fung, YB (1993) Morphometry of pig coronary arterial trees. *Am J Physiol* 265 (*Heart Circ. Physiol.* 34):H350–H365.

Kroll, K and Stepp, DW (1996) Adenosine kinetics in the canine coronary circulation. *Am J Physiol* 270 (*Heart Circ. Physiol.* 39):H1469–H1483.

Kroll, K, Kinzie, DJ, and Gustafson, LA (1997) Open system kinetics of myocardial phosphoenergetics during coronary underperfusion. *Am J Physiol* 272 (*Heart Circ. Physiol.* 41):H2563–H2576.

Kroll, K and Bassingthwaighte, JB (1998) "Ch. 11. Role of capillary endothelial cells in transport and metabolism of adenosine in the heart: an example of the impact of endothelial cells on measures of metabolism." In: *Whole Organ Approaches to Cellular Metabolism.* (JB Bassingthwaighte, CA Goresky, and JH Linehan, Eds.) Springer Verlag. New York. pp. 261–275.

Kuikka, J, Levin, M, and Bassingthwaighte, JB (1986) Multiple tracer dilution estimates of D- and 2-deoxy-D-glucose uptake by the heart. *Am J Physiol* 250 (*Heart Circ. Physiol.* 19):H29–H42.

Landesberg, A and Sideman, S (1994) Mechanical regulation of cardiac muscle by coupling calcium kinetics with cross-bridge cycling: a dynamic model. *Am J Physiol* 267 (*Heart Circ. Physio.* 36):H779–H795.

Landesberg, A and Sideman, S (1994) Coupling calcium binding to troponin C and cross-bridge cycling in skinned cardiac cells. *Am J Physiol* 266 (*Heart Circ. Physio.* 35):H1260–H1271.

Landesberg, A, Markhasin, VS, Beyar, R, and Sideman, S (1996) Effect of cellular inhomogeneity on cardiac tissue mechanics based on intracellular control mechanisms. *Am J Physiol* 270 (*Heart Circ. Physio.* 39):H1101–H1114.

Landesberg, A (1996) End-systolic pressure-volume relationship and intracellular control of contraction. *Am J Physiol* 270 (*Heart Circ. Physio.* 39):H338–H349.

LeGrice, IJ, Smaill, BH, Chai, LZ, Edgar, SG, Gavin, JB, and Hunter, PJ (1995) Laminar structure of the heart: Ventricular myocyte arrangement and connective tissue architecture in the dog. *Am J Physiol* 269 (*Heart Circ. Physiol.* 38):H571–H582.

LeGrice, IJ, Hunter, PJ, and Smaill, BH (1997) Laminar structure of the heart: a mathematical model. *Am J Physiol* 272:H2466–H2476.

Li, Z, Yipintsoi, T, and Bassingthwaighte, JB (1997) Nonlinear model for capillary-tissue oxygen transport and metabolism. *Ann Biomed Eng* 25:604–619.

Luo, CH and Rudy, Y (1994) A dynamic model of the cardiac ventricular action potential: II: Afterdepolarizations, triggered activity, and potentiation. *Circ Res* 74:1097–1113.

Luo, C and Rudy, Y (1994) A dynamic model of the cardiac ventricular action potential I. Simulations of ionic currents and concentration changes. *Circ Res* 74:1071–1096.

McCulloch, AD, Smaill, BH, and Hunter, PJ (1989) Regional left ventricular epicardial deformation in the passive dog heart. *Circ Res* 64:721–733.

McCulloch, AD and Omens, JH (1991) Non-homogeneous analysis of three-dimensional transmural finite deformation in canine ventricular myocardium. *J Biomech* 24:539–548.

McCulloch, A, Bassingthwaighte, JB, Hunter, P, and Noble, D (1998) *Computational Biology of the Heart: From Structure to Function Progress in Biophysics and Molecular Biology Volume 69, 151–572.* Elsevier/Pergamon.

McFarland, EW, Kushmerick, MJ, and Moerland, TS (1994) Activity of creatine kinase in a contracting mammalian muscle of uniform fiber type. *Biophys J* 67:1912–1924.

Magnus, G and Keizer, J (1997) Minimal modeling of beta-cell mitochondrial Ca$sup 2 + $ handling. *Am J Physiol* 273 (Cell. Physiol. 42):C717–C733.

Marmarelis, VZ, Bassingthwaighte, JB, D'Argenio, DZ, and Foster, DM (1994) Overview of NIH-funded biomedical modeling and simulation resources. *Proc Int Fed Automat Control "Modeling and Control in Biomedical Systems"*:14–15.

Noble, D (1995) The development of mathematical models of the heart. *Chaos, Solitons and Fractals* 5:321–333.

Randle, PJ and Tubbs, PK (1979) "Carbohydrate and fatty acid metabolism." In: *Handbook of Physiology, Section 2, The Cardiovascular System.* (RM Berne and N Sperelakis, Eds.) Am. Physiol. Soc. Bethesda, MD. pp. 805–844.

Regnier, M, Morris, C, and Homsher, E (1995) Regulation of the cross-bridge transition from a weakly to strongly bound state in skinned rabbit muscle fibers. *Am J Physiol* 269 (Cell. Physiol. 38):C1532–C1539.

Regnier, M, Martyn, DA, and Chase, PB (1998) Calcium regulation of tension redevelopment kinetics with 2-deoxy-ATP or low [ATP] in rabbit skeletal muscle. *Biophys J* 74:2005–2015.

Rudy, Y and Shaw, RM (1997) "Cardiac excitation: an interactive process of ion channels and gap junctions." In: *Analytical and Quantitative Cardiology.* (S Sideman and R Beyar, Eds.) Plenum. New York. pp. 269–279.

Selkov, E, Basmanova, S, Gaasterland, T, Goryanin, I, Gretchkin, Y, Maltsev, N, Nenashev, V, Overbeek, R, Panyushkina, E, Pronevitch, L, Selkov, E jr, and Yunus, I (1996) The metabolic pathway collection from EMP: the enzymes and metabolic pathways database. *Nucleic Acids Res* 24:26–28.

Selkov, E, Galimova, M, Goryanin, I, Gretchkin, Y, Ivanova, N, Komarov, Y, Maltsev, N, Mikhailova, N, Nenashev, V, Overbeek, R, Panyushkina, E, Pronevitch, L, and Selkov, E jr (1997) The metabolic pathway collection: an update. *Nucleic Acids Res* 25:37–38.

Shaw, RM and Rudy, Y (1995) The vulnerable window for unidirectional block in cardiac tissue: characterization and dependence on membrane excitability and cellular coupling. *J Cardiovasc Electrophysiol* 6:115–131.

van Bavel, E and Spaan, JA (1992) Branching patterns in the porcine coronary arterial tree. Estimation of flow heterogeneity. *Circ Res* 71:1200–1212.

van Beek, JHGM, Roger, SA, and Bassingthwaighte, JB (1989) Regional myocardial flow heterogeneity explained with fractal networks. *Am J Physiol* 257 (*Heart Circ. Physiol.* 26):H1670–H1680.

van Bilsen, M and Chien, KR (1993) Growth and hypertrophy of the heart: towards an understanding of cardiac specific and inducible gene expression. *Cardiovasc Res* 27:1140–1149.

van der Vusse, GJ, Little, SE, Reneman, RS, and Bassingthwaighte, JB (1999) Transfer of fatty acids across myocardial endothelium. *Am J Physiol* (*Heart Circ. Physiol.* in press).

Wiseman, RW and Kushmerick, MJ (1995) Creatine kinase equilibration follows solution thermodynamics in skeletal muscle. *J Biol Chem* 270:12428–12438.

Yipintsoi, T, Dobbs, WA jr., Scanlon, PD, Knopp, TJ, and Bassingthwaighte, JB (1973) Regional distribution of diffusible tracers and carbonized microspheres in the left ventricle of isolated dog hearts. *Circ Res* 33:573–587.

COMPUTATIONAL MODELING OF OXYGEN TRANSPORT FROM COMPLEX CAPILLARY NETWORKS

Relation to the Microcirculation Physiome

D. Goldman and A. S. Popel

Department of Biomedical Engineering
Johns Hopkins University School of Medicine
Baltimore, Maryland 21205

1. INTRODUCTION

Microvascular networks, even in skeletal muscle, are highly complex structures, and their function in oxygen transport remains to be addressed completely. In order to do this, a combination of detailed experiments and biophysically-detailed modeling is needed. In the past ten years, there have been several computational studies of the role of non-uniform geometry in oxygen transport in skeletal muscle. Models that allow a significant degree of geometric realism include those of Groebe (1990), Secomb and Hsu (1994), and Hoofd and Turek (1996). To continue investigating the role of physiologically realistic complexity, we have developed a computational model that is general enough to directly utilize detailed geometric, biophysical, and hemodynamic data in oxygen transport calculations.

Currently, as described below, the model is based mainly on experimental data on the hamster cheek-pouch retractor muscle. Since no complete three-dimensional data are available on the structure of specific capillary networks in this muscle, the model also includes a geometric component. In addition, the model contains a realistic simulation of two-phase (plasma/erythrocyte) blood flow in the network.

The computational model we have developed is sufficiently general to allow its application to the study of transport in other skeletal muscles and other tissues. What is needed now is the detailed anatomical and physiological data on which these calculations could be based. One of the principal goals of the microcirculation physiome project is the construction of databases containing such information. Therefore, this project will

Oxygen Transport to Tissue XXI, edited by Eke and Delpy.
Kluwer Academic / Plenum Publishers, New York, 1999.

have a significant impact on the modeling and understanding of oxygen transport in the microcirculation.

In what follows, we first present the computational model developed, then discuss new results obtained with the model. We also discuss the relation of this work to the microcirculation physiome project.

2. METHODS

The results described in this paper were obtained using a computational model that was constructed using relevant transport equations and physiological data. Most of the anatomical and hemodynamic data was from experiments on the hamster cheek-pouch retractor muscle. In the work described, there were three different parts to the model used to obtain oxygen distribution results. The principal and most general part of the model was developed to accurately simulate the distribution of oxygen in a block of tissue, given a capillary network geometry and blood flow distribution. The second part, also very general, was designed to accurately describe both total blood flow and hematocrit distributions through a given capillary network. The third and most tissue-specific part of the model was designed to capture certain known anatomical features, such as capillary diameter, capillary number density, and tortuousity. All three parts of the model were used to obtain each of the oxygen distributions presented below.

2.1. Transport Model

The computational model of oxygen transport used is based on the transport equations described in Popel (1989). In the tissue, which is taken to be a continuous, homogeneous medium, oxygen partial pressure $P(x,y,z,t)$ evolves according to the following reaction-diffusion equation:

$$\frac{\partial P}{\partial t}=\left[1+\frac{c_{Mb}}{\alpha}\frac{dS_{Mb}}{dP}\right]^{-1}\left\{D\nabla^2 P-\frac{1}{\alpha}M(P)+\frac{1}{\alpha}D_{Mb}c_{Mb}\nabla\cdot\left(\frac{dS_{Mb}}{dP}\nabla P\right)\right\} \quad (1)$$

where D is the diffusion coefficient for oxygen, α is the solubility, and $M(P)$ is the rate of consumption. Facilitated transport in the tissue is included using the myoglobin concentration c_{Mb}, diffusion coefficient D_{Mb}, and oxygen saturation $S_{Mb}(P) = P/(P + P_{50,Mb})$. The particulate nature of blood (Federspiel and Popel, 1986; Groebe and Thews, 1989) was accounted for by considering a continuous mixture of plasma and erythrocytes and using a mass transfer coefficient for the intracapillary resistance. At each axial location θ along any given capillary, blood flow was described by the hematocrit and average velocity. The transport of oxygen is described by the mass balance equation for the blood oxyhemoglobin saturation $S(\xi,t)$:

$$\frac{\partial S}{\partial t}=-u\frac{\partial S}{\partial \xi}-\frac{1}{\pi R}\left[C+\alpha_b\frac{dP_b}{dS}\right]^{-1}\oint \mathrm{j}\cdot d\theta \quad (2)$$

where u is the mean blood velocity, R is the capillary radius, j is the outward oxygen flux at (ξ,θ), and C is the oxygen-binding capacity of blood, which is directly proportional to the local hematocrit. P_b is the intracapillary oxygen tension (PO_2), and α_b is the solubility of oxygen in plasma. The capillary-tissue oxygen flux is given by:

$$j = \kappa(P_b - P_w) \tag{3}$$

where κ is the mass transfer coefficient and P_w is the tissue PO_2 at the capillary surface. Although κ is known to vary with hematocrit (Federspiel and Popel 1986), for the calculations presented it was fixed at 7.96×10^{-6} ml $O_2/cm^2/s/torr$. The boundary condition at the capillary-tissue interface is given by:

$$-D\alpha \frac{\partial P_w}{\partial n} = j \tag{4}$$

where n is the unit vector normal to the capillary surface and j is given by Eq. (3).

Equations (1)–(4) represent a basic mathematical model of oxygen transport. We now specify Michaelis-Menten consumption kinetics, $M = M_0\, P/(P + P_{cr})$, and the Hill equation for oxyhemoglobin saturation $S(P) = P^n/(P^n + P_{50}{}^n)$. The model is completed by specifying no-flux conditions at each outer tissue boundary. Alternatively, periodic boundary conditions may be specified in one or more directions. Parameters for the oxygen transport model were those used in the study of Ellsworth et al. (1988), except where other parameters were needed or more reliable values have become available (see Roy and Popel, 1996;). For example, the value $D_{Mb} = 1.73 \times 10^{-7}\,cm^2/s$ was taken from Juergens et al. (1994).

To solve the above problem numerically for the tissue and capillary oxygen distributions, it is necessary to discretize these equations. This is done using the finite-difference method to approximate spatial derivatives, with central differencing used in Eq. (1), upwinding used in Eq. (2), and third-order schemes used for boundary conditions on P. The tissue domain is covered with a uniform grid and the capillaries are divided into short straight cylindrical segments, each of which has a given hematocrit and mean blood velocity, and oxyhemoglobin saturation (SO_2) which is determined during the transport calculation. Because capillary segments may be arbitrarily oriented, the interpolation between tissue grid points and capillary surface/grid line intersections requires some care. Also, since no uniformity is assumed in the directions of blood flow and oxygen transport, the three-dimensional steady-state oxygen distribution cannot be obtained by performing successive 2-D steady-state calculations, as was possible in the work of Ellsworth et al. (1988). Therefore, the time-dependent equations are solved, using an explicit, first-order scheme, to evolve the full, 3-D capillary/tissue oxygen distribution from its initial condition to a steady state. In addition, allowing capillaries to be curvilinear and to intersect makes it necessary to properly handle non-parallel segment intersections and vessel bifurcations.

We have developed an accurate, efficient, and flexible code for solving the general, 3-D problem of oxygen transport and consumption described by Eqs. 1–4. Several criteria were used to validate the simulation code. Once capillary SO_2 and tissue PO_2 were found to converge to steady states, the results were checked for several different network configurations and parameter sets. Refinement of the tissue and capillary discretizations showed the expected decrease in both local solution errors and global error measures, such as the difference between total capillary surface oxygen flux and tissue oxygen consumption. Results for a single straight capillary were also compared directly to the analytical result (Krogh, 1929) which can be obtained for this case when consumption is constant everywhere and diffusion parallel to the capillary is neglected.

For the results presented, a minimum of $56 \times 56 \times 200$ tissue points were used. Each run required approximately 50 Mb of computer memory and 1 day of CPU time on a

Silicon Graphics O2 workstation. Increased network complexity, as seen in brain for example, should not significantly affect computational requirements, and transport calculations for networks (or tissue volumes) several times larger than the present ones are possible. In addition, increasing the efficiency of the computational model would allow much larger networks to be considered.

2.2. Flow Model

The computational model of blood flow used is based on the formulation of Pries et al. (1990, 1994) and incorporates most of the important physiological phenomena observed in microvessels. These include the Fahraeus-Lindqvist effect (viscosity dependence on vessel size and discharge hematocrit), the Fahraeus effect (dependence of discharge hematocrit on tube diameter and hematocrit), and phase separation at divergent bifurcations. The relevant rheological equations were rescaled from the original rat data to the appropriate form for hamster blood, using the volume of the hamster erythrocyte. A numerical code was then developed which could solve for the blood flow rates and hematocrits in a very general network, given input hematocrits and either pressure or flow boundary conditions. The results of the flow computations, the hematocrit and volumetric flow rate in each capillary segment, along with the inlet SO_2's, are used as the basis for our oxygen transport calculations. Although steady-state blood flow and oxygen transport are our main focus, our code has the capability to compute time-dependent blood flows. Only steady-state flows with pressure boundary conditions were used in the calculations presented.

2.3. Network Model

The geometric model of skeletal-muscle capillary networks that we have developed begins with an idealization of the observed muscle-fiber structure. Parallel, cylindrical fibers of constant diameter are packed hexagonally, and the accompanying parallel capillaries are placed in the spaces between the fibers. This can be done uniformly, but here a certain degree of randomness is added, as described below. The construction is such that both the capillary density and capillary-to-fiber number, on average, agree with existing data for hamster cheek-pouch retractor muscle. Specifically, we used the experimental results of Ellsworth et al. (1988), Bennett et al. (1991), and Dong (1997). With our basic geometric model, we are able to calculate oxygen transport in a block of muscle tissue from networks of straight parallel capillaries, parallel capillaries with anastomotic cross-connections, and capillaries with tortuousity, which is observed in skeletal muscle (particularly during contraction; see Mathieu-Costello, 1987 and Mathieu-Costello et al., 1989).

3. RESULTS

Using the computational model described above, we have constructed a number of capillary networks and examined their blood flow and oxygen transport properties. For all cases discussed here the overall blood flow was unidirectional (concurrent) and the length of the tissue block was 400 μm. The tissue cross-section was approximately square with an average area of 7820 μm^2 for the cases discussed, and the number of parallel capillaries was 11. The blood flow (5 times resting, actual values given below) and oxygen consumption (10 times resting, 1.57×10^{-3} ml O_2/ml/s) rates used for oxygen transport calculations correspond to working conditions. We now present results for several different types of capillary networks, beginning with those containing no tortuousity or anastomoses.

3.1. Fiber-Structured

These capillary networks (F1–3) were constructed using the geometric model described above, namely the capillaries were placed in the spaces between hexagonally packed muscle fibers. This was done randomly, subject to the total number of capillaries and number of capillaries per fiber (5.25, on average) being fixed. Since the blood pressures were uniform across the inlets and outlets, and the inlet hematocrits were also uniform, the flow rate (585 μm/s) and hematocrit (0.43) were the same for all capillary segments.

3.2. Random

These networks (R1–3) were constructed by placing the capillaries randomly in the x-y plane, without considering the role of muscle fibers. However, the capillaries were constrained to be at least two capillary diameters apart, and the total number of capillaries and number of capillaries per fiber were held constant. All the capillaries in these networks had the same flow velocity and hematocrit as those in the fiber-structured networks.

3.3. Fiber-Structured with Anastomoses

These networks (FA1–3) were identical to the above fiber-structured ones (F1–3), except that cross-connections were placed between some of the parallel capillaries. This was done randomly, but the endpoints of each cross-connection were constrained to be on capillaries feeding the same muscle fiber and to be between 12.5 μm and 50 μm apart in the z-direction. FA1–2 had 11 anastomoses, while FA3 had 22. Although the pressures were uniform across the inlets and outlets, and the inlet hematocrits were also uniform, the presence of cross-connections produced significant variations in both flow rates and hematocrits. This can be seen in the histograms of discharge hematocrit and mean blood flow velocity, shown in Figure 1.

3.4. Fiber-Structured with Tortuousity

To study the effect of tortuousity, we have compared networks having straight capillaries with structurally similar ones in which each of the capillaries has 19% tortuousity, in accordance with experimental data (Mathieu-Costello, 1987). The hematocrits and velocities are again uniform across the vessels, but since the overall capillary length is increased the actual flow velocity decreases to 498 μm/s.

4. DISCUSSION

We now examine the various simulation results described above, beginning with a comparison of the tissue oxygen distributions produced by the fiber-structured (F1–3)

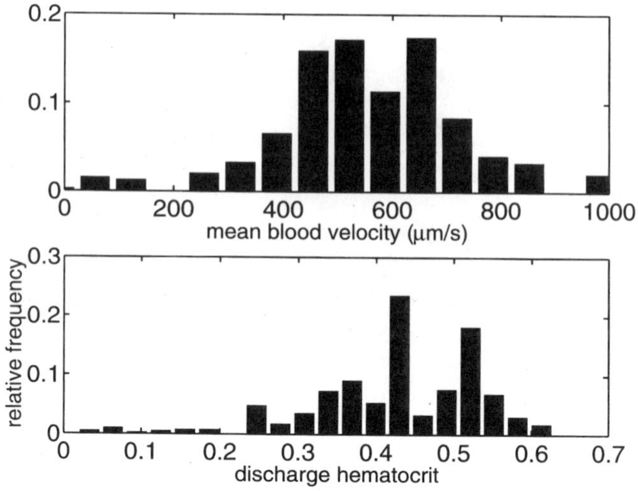

Figure 1. Normalized histograms of mean blood velocity and discharge hematocrit for the anastomosed network FA3.

and random (R1–3) networks. In general, the geometric structure produced by considering the muscle fibers results in a more uniform distribution of capillaries, and hence of oxygen. Figure 2 shows the combined tissue PO_2 frequency distributions for these two types of networks. It can be seen that oxygen is more evenly distributed for the fiber-structured cases, with the mean PO_2 being 23.2 ± 5.5 torr (mean $\pm$ SD), versus 21.7 ± 6.9 torr for the random cases.

We next compare the fiber-structured cases with (FA1–3) and without (F1–3) anastomoses. As an example of the geometries studied, Figure 3 shows network FA3, which

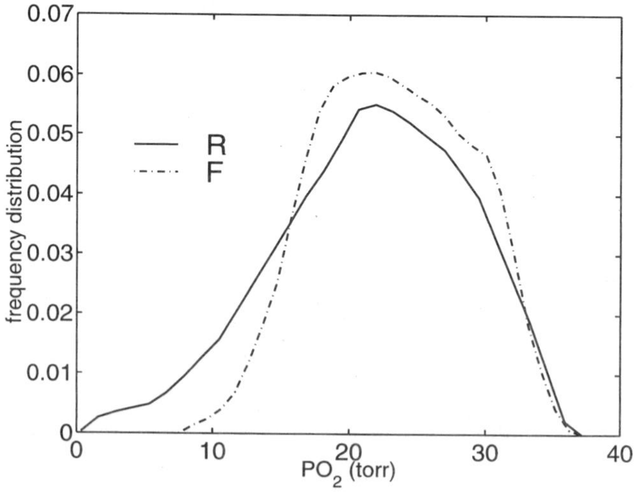

Figure 2. Tissue PO_2 frequency distributions for two types of networks studied, random (R) and fiber-structured (F).

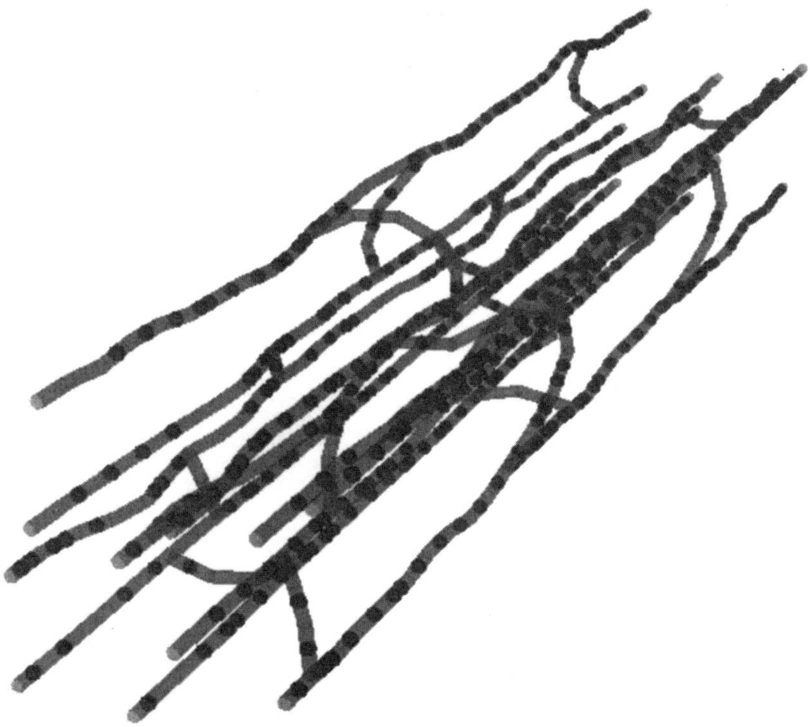

Figure 3. Geometry of capillary network FA3. For visualization, erythrocytes are shown with average density proportional to the local hematocrit.

produced the flow distributions in Figure 1. The effect on blood flow of anastomoses can be seen in Figure 4, which shows blood velocity histograms for networks having 11 and 22 anastomoses. Six parallel networks were used to obtain these results, F1–3 plus three additional similar ones that were generated. It can be seen that with more cross-connections the flow becomes more heterogeneous, the coefficient of variation (CV = SD/mean) being 0.47 for networks with 22 anastomoses, versus 0.31 for those with 11. This increase in heterogeneity, which is also seen for the hematocrit, could be partly due to the uniformity of pressure drops across the parallel vessels. In reality, the role of anastomoses might be to smooth out highly non-uniform flow patterns or enable circumvention of blocked vessels, however these possibilities remain to be investigated. The mean tissue PO_2 in the networks studied with anastomoses was 22.8 ± 5.8, versus 23.2 ± 5.5 in those without. This small change in the PO_2 distributions is due to the relatively small coefficient of variation for blood flow in the anastomosed networks and the uniformity of inlet SO_2. This result is consistent with those of Popel et al. (1986), who did not see a significant change in the CV of tissue PO_2 until the CV of erythrocyte flux was 0.6 and the CV of inlet PO_2 was 0.2.

We are currently examining the effect of capillary tortuousity on oxygen transport. Networks containing tortuous capillaries have produced increased oxygen distribution heterogeneity, compared to networks of straight capillaries, under fixed pressure drop conditions. Fixed flow boundary conditions are being studied.

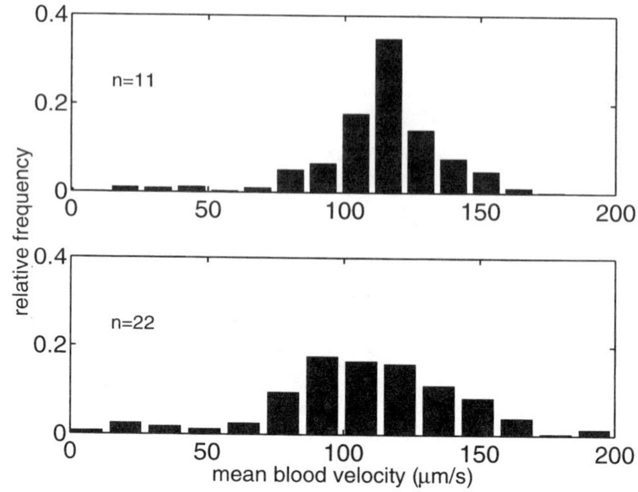

Figure 4. Normalized histograms of the mean resting blood velocity in networks having 11 and 22 anastomoses.

The results described in this paper were obtained using simulated capillary networks, constructed by matching the basic structure and geometric parameters of experimental networks. The retractor muscle was chosen for this study because of the large amount of available data. However, it was necessary to use a geometric model for the exact network structure because reconstruction data from experiments were not available. The microcirculation physiome project, by combining experimental results from different laboratories and making them publicly available, should allow biophysically-detailed computational studies to be performed. For example, blood flow and oxygen transport calculations could be done using reconstructed experimental networks, giving complete, accurate results of direct physiological relevance. This could be done for a variety of tissues, allowing potentially revealing comparisons of flow and transport characteristics. For example, it would be of interest to compare results for different skeletal muscles or for the same muscle in animals of different ages.

ACKNOWLEDGMENTS

This work was supported by NIH grants HL18292 and HL09595.

REFERENCES

Bennett, RAO, Pittman, RN, and Sullivan, SM (1991) Capillary spatial pattern and muscle fiber geometry in three hamster striated muscles. Am. J. Physiol. 260:H579–H585.

Dong, M (1997) Influence of aging on oxygen transport in the microcirculation of skeletal muscle. Ph.D. thesis, Virginia Commonwealth University, Richmond, VA.

Ellsworth, ML, Popel, AS, and Pittman, RN (1988) Assessment and impact of heterogeneities of convective oxygen transport parameters in capillaries of striated muscle: experimental and theoretical. Microvasc. Res. 35:341–362.

Federspiel, WJ and Popel, AS (1986) A theoretical analysis of the effect of the particulate nature of blood on oxygen release in capillaries. Microvasc. Res. 32:164–189.

Groebe, K and Thews, G (1989) Effects of red cell spacing and red cell movement upon oxygen release under conditions of maximally working skeletal muscle. Adv. Exp. Med. Biol. 248:175–185.

Groebe, K (1990) A versatile model of steady state O_2 supply to tissue. Application to skeletal muscle. Biophys. J. 57:485–498.

Hoofd, L and Turek, Z (1996) Realistic modelling of capillary spacing in dog gracilis muscle greatly influences the heterogeneity of calculated tissue oxygen pressures. Adv. Exp. Med. Biol. 388:333–340.

Juergens, KD, Peters, T, and Gros, G (1994) Diffusivity of myoglobin in intact skeletal muscle cells. Proc. Natl. Acad. Sci. USA 91:3829–3833.

Krogh, A (1929) The anatomy and physiology of capillaries. Yale Univ. Press, New Haven, CT.

Mathieu-Costello, O (1987) Capillary tortuousity and degree of contraction or extension of skeletal muscles. Microvasc. Res. 33:98–117.

Mathieu-Costello, O, Hoppeler, H, and Weibel, ER (1989) Capillary tortuousity in skeletal muscles of mammals depends on muscle contraction. J. Appl. Physiol. 66:1436–1442.

Popel, AS, Charny, CK, and Dvinsky, AS (1986). Effect of heterogeneous oxygen delivery on the oxygen distribution in skeletal muscle. Math. Biosci. 81:91–113.

Popel, AS (1989) Theory of oxygen transport to tissue. Crit. Rev. Biomed. Eng. 17:257–321.

Pries, AR, Secomb, TW, Gaehtgens, P, and Gross, JF (1990) Blood flow in microvascular networks: experiments and simulation. Circ. Res. 67:826–834.

Pries, AR, Secomb, TW, Gessner, T, Sperandio, MB, Gross, JF, and Gaehtgens, P (1994) Resistance to blood flow in microvessels in vivo. Circ. Res. 75:904–915.

Roy, TK and Popel, AS (1996) Theoretical predictions of end-capillary PO_2 in muscles of athletic and non-athletic animals at VO_{2max}. Am. J. Physiol. 271:H721–H737.

Secomb, TW and Hsu, R (1994) Simulation of O_2 transport in skeletal muscle: diffusive exchange between arterioles and capillaries. Am. J. Physiol. 267:H1214–H1221.

REGULATION OF LOCAL PERFUSION

Levels of Control and Contributing Factors

Karlfried Groebe

Institut für Physiologie und Pathophysiologie
Johannes Gutenberg-Universität Mainz
Duesbergweg 6, D-55099 Mainz, Germany

1. INTRODUCTION

The present paper is going to report some basic information on the regulation of local perfusion that is important for modeling the involved systems. Moreover, approaches to and problems in doing so are discussed.

Control mechanisms in vascular systems exhibit extreme variabilities in a number of aspects:

- Vascular control occurs on different *time scales*, *e.g.*, within days to months for vascular growth, revascularization, vascular hypertrophy, or changes in gene expression. On the other hand, responses to acute changes in pressure or energy demand take place within seconds to minutes.
- In different organs or functional states, there are different *primary purposes* or *control principles* underlying vascular control, *e.g.*, blood volume control and body homeostasis in the kidney, optimal exchange of respiratory gases in the lungs, or adequate ATP production by energy metabolism in muscle. In inflammation, generally speaking, the integrity of the organism is the primary goal governing perfusion control.
- Finally, vascular control involves a number of *central* and *local mechanisms*:
  - autonomic nervous control *via* changes in sympathetic tone,
  - humoral control by circulating epinephrine or angiotensin, by locally released prostacyclin or kinins, *etc.*,
  - local metabolic control, which will be dealt with below, and
  - highly organ specific mechanisms like the initial vasodilation at the onset of muscular work which is thought to be caused by extrinsic or intrinsic vasodilator nerves or by K^+-release from depolarized muscle cells.

Oxygen Transport to Tissue XXI, edited by Eke and Delpy.
Kluwer Academic / Plenum Publishers, New York, 1999.

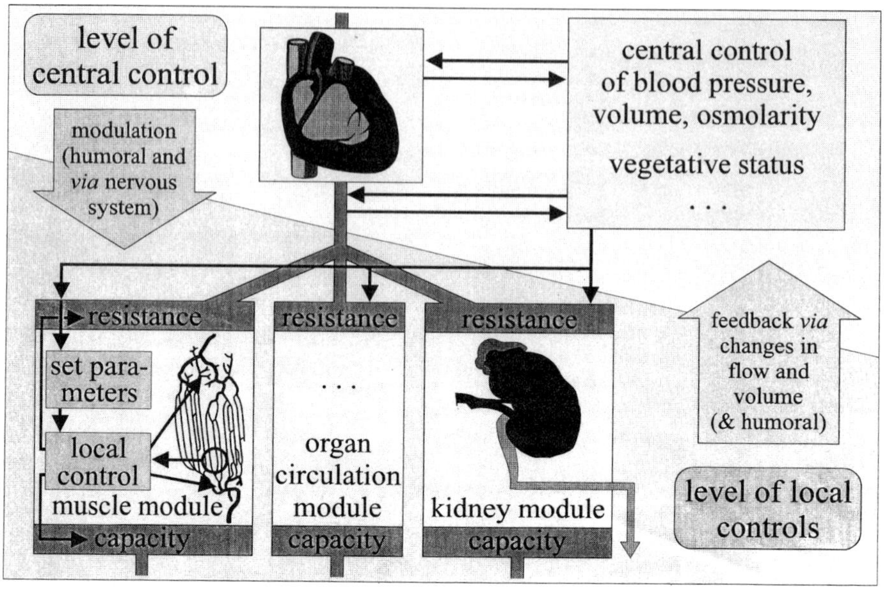

Figure 1. Interaction of central (top) and local (bottom) mechanisms in the control of circulation. For details see text.

Central and local mechanisms cooperate in a fairly complex fashion[15] (see Figure 1). Even though the exact mechanisms of their interaction are still unclear, it appears that local control largely operates as a stand-alone system which is modulated in its function by central mechanisms. The latter seem to play a role in blood pressure regulation, coordinate the demands for perfusion in different organ systems, and adjust organ perfusions to the vegetative status of the organism.

2. PRINCIPLES GOVERNING LOCAL CONTROL AND PROBLEMS IN UNDERSTANDING

As a consequence of the outlined composition of the vascular control system, a modular approach to modeling central/local interactions should be adequate, for which one requires profound understanding of local control and knowledge of the mechanisms that mediate the modulating actions by central control systems. This is illustrated in Figure 1 which shows the level of organ controls in its lower part. Local controllers collect information from the supplied organ (in particular, about tissue metabolic status) and convert it to changes in vascular conductance (and capacity). Controller operation is governed by a number of parameters (*e.g.*, vessel diameters at zero pressure and flow) which may be reset by central mechanisms. Central controls are shown in the upper part of Figure 1 and include blood pressure, volume, and osmolarity regulations, vegetative status of the organism, and other factors. While central modulation of local controls is achieved *via* humoral and nervous systems, changes in organ blood flow and volume are fed back to and integrated within the central controller.

Whereas for central control of circulation a number of modeling approaches exist, the function of local microvascular controllers is much less clear. Their major tasks are *(i)* providing sufficient blood flow for supplying nutrients, most importantly O_2, and for removing metabolic waste products; *(ii)* maintaining local water balance[15] (*i.e.*, the amount of water filtrated and reabsorbed by the vasculature must roughly be the same). *(iii)* Local controllers should work economically (*i.e.*, organ blood flows should not exceed the ones actually required in order to avoid overloading the cardiovascular system). *(iv)* All of these tasks need to be performed at variable arterial blood pressure levels.

Figure 2 shows a schematic drawing of the microcirculation consisting of arteriolar tree, capillary bed, and venular tree. Arterial, venous, and mean capillary pressures are P_a, P_v, and $\bar{P}_c$, respectively, and perfusion rate is $\dot{Q}$. The shaded area indicates the region of exchange vessels in which transvascular fluid movement takes place. Effectors of microvascular control are pre- and postcapillary resistance R_a and R_v of which changes in R_a are most pronounced. Changes in capillary resistance R_c are largely consequences of altered R_a. The figure illustrates how the controlled variables, perfusion rate $\dot{Q}$ and net fluid movement J_v, depend on the distribution of flow resistance in the network (see insets): $\dot{Q}$ is inversely proportional to total resistance which is the sum of precapillary, capillary, and postcapillary resistance. J_v, on the other hand, most importantly depends on the permeability-surface area product L_pS and on average capillary hydrostatic pressure $\bar{P}_c$ which, in a first approximation, is a function of the ratio pre- over postcapillary resistance. Thus for separately controlling the variables $\dot{Q}$ and J_v, it must be possible to vary pre- and postcapillary resistance independently.

The physiological mechanisms by which active changes in pre- and postcapillary resistance can be brought about are *(i)* myogenic vasoconstriction in arterioles upon increasing transvascular pressure[9] that attenuates changes in perfusion pressure; *(ii)* dependence of arteriolar and venular diameters on the concentrations of vasoactive

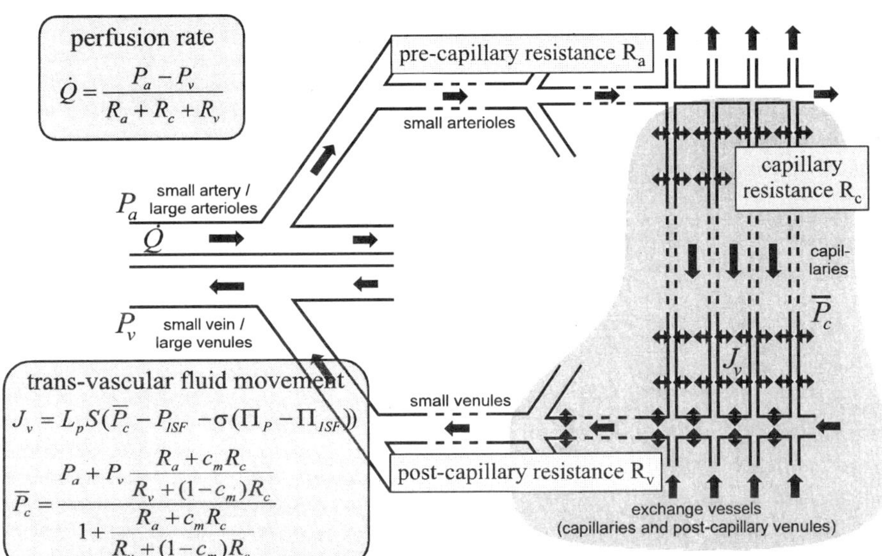

Figure 2. Schematics of skeletal muscle microcirculation consisting of arteriolar tree, capillary bed, and venular tree. Shaded: region of exchange vessels (capillaries and small venules). $\dot{Q}$: perfusion rate. P_a, P_v: mean arterial and central venous pressures. $\bar{P}_c$: mean pressure in exchange vessels. Insets: dependence of perfusion rate $\dot{Q}$ and net transvascular fluid movement J_v on resistances (for details see text and[15]).

substances,[15,17] a number of which are released by the tissue when oxygen consumption rate increases or tissue energetic status turns critical; *(iii)* shear-dependent[11] and conducted[16] vasodilation which represent amplifiers of increases in luminal blood flow.

At this point, some remarks on modeling these elements of microvascular control seem to be adequate: It has been proposed that myogenic response is based on a regulator operation[9] in which arteriolar wall tensile stress or wall tension $T = r \cdot P$ is the regulated variable (Figure 3): A pressure change dP causes a distortion in wall tension $r \cdot dP$ to which the regulator responds with a change in vessel radius dr that partly restores wall tension. Regulator action is governed by its gain G_T that needs to be positive and less than 1 for stable operation. Solving this differential equation results in the *P-r* relation shown on the right in which r^* and P^* are initial pressure and radius. This equation exhibits a singularity at zero pressure which, of course, is not realistic. However, in the above it has not been considered that at lowest pressures arterioles distend passively. Assuming some threshold pressure at which vessel behaviour turns from passive to active, one finds the pressure-diameter relation shown at bottom, right of Figure 3. This is not too much different from the relation measured by Davis[3] (solid line in left bottom panel). Improved solutions are obtained if passive vessel characteristics are included in the differential equation.

In a similar fashion, shear-dependent vasodilation can be modeled in which wall shear stress or wall shear rate $\dot{r} \approx \dot{Q}/pi/r^3$ is the regulated variable (Figure 4). Again, a distortion of wall shear rate by a change in blood flow $d\dot{Q}$ is partly compensated for by a corresponding change in vessel radius dr. The differential equation is slightly more complex, but the result is similarly simple. Resulting flow-diameter relations are in reasonably good agreement with experimental data[12] (bottom panels).

Vasoactive metabolites are produced in or consumed by tissue, *e.g.*, adenosine, osmolarity, carbon dioxide, hydrogen ions, or oxygen. Modeling their action poses most severe problems:

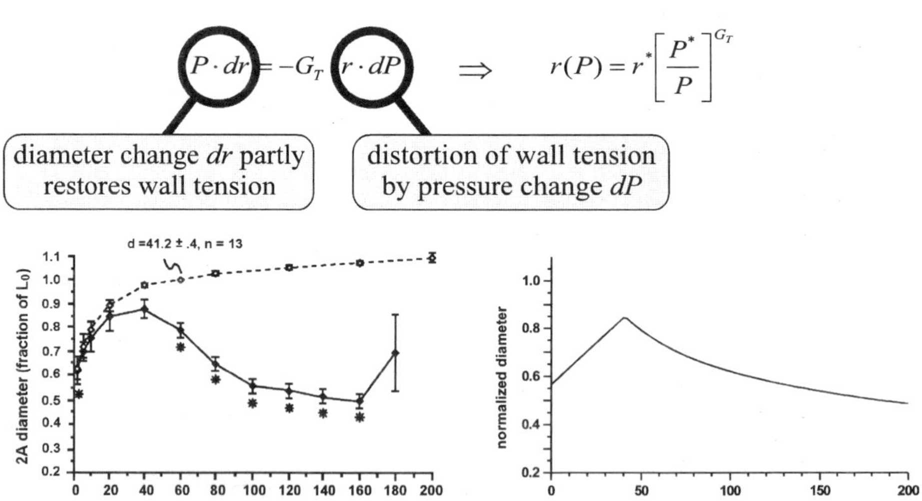

Figure 3. Modeling myogenic response as a regulator operation in which arteriolar wall tensile stress or wall tension $T = r \cdot P$ is the regulated variable. Bottom panels: relations between transmural pressure (abscissa) and normalized vessel diameter (ordinate) as measured by Davis[3] (left) and as resulting from the model (right). For more details see text.

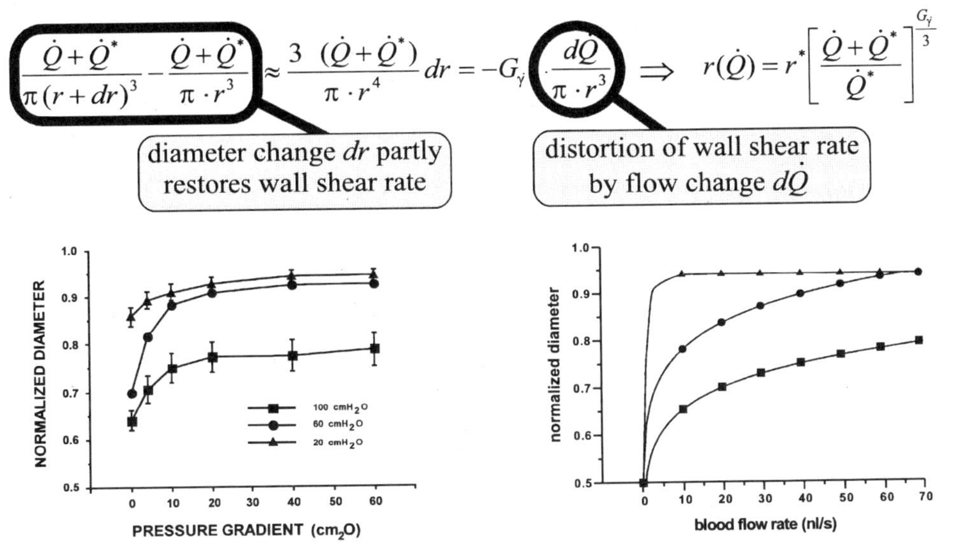

$$\frac{\dot{Q}+\dot{Q}^*}{\pi(r+dr)^3}-\frac{\dot{Q}+\dot{Q}^*}{\pi\cdot r^3}\approx\frac{3\ (\dot{Q}+\dot{Q}^*)}{\pi\cdot r^4}dr=-G_{\dot{\gamma}}\cdot\frac{d\dot{Q}}{\pi\cdot r^3}\Rightarrow r(\dot{Q})=r^*\left[\frac{\dot{Q}+\dot{Q}^*}{\dot{Q}^*}\right]^{\frac{G_{\dot{\gamma}}}{3}}$$

Figure 4. Modeling shear-dependent vasodilation as a regulator operation in which arteriolar wall shear stress or wall shear rate $\dot{\gamma} \approx \dot{Q}/pi/r^3$ is the regulated variable. Bottom panels: relations between indicators of $\dot{\gamma}$ (*i.e.* longitudinal pressure gradient or flow rate; abscissa) and normalized vessel diameter (ordinate) as measured by Kuo et al.[12] (left) and as resulting from the model (right).

- Production rates as functions of $\dot{V}_{O_2}$ and/or tissue energetic status, as well as
- dose-response curves for vasoactive metabolite action on vascular beds are unknown.
- The mode of vasoactive metabolite action (which ultimately needs to occur at the arterioles) is unclear. Current hypotheses include:
  - diffusion of metabolites from tissue to arterioles,
  - conduction of a vasodilator signal in the vessel wall,[18]
  - metabolite diffusion from venules to paired arterioles.[8]

The problem in understanding the mode of vasoactive metabolite action is related to the fact that tissue nutrient concentrations and, most notably, oxygen partial pressure fall along the capillary and with distance from the capillary. Consequently, the worst supplied tissue regions—that have been termed "lethal corners"—are located next to the venous capillary ending. These are the first tissue sites to generate a metabolic signal when their energetic status turns critical. This is the signal which a tightly coupled metabolic perfusion control needs to react to by dilating the corresponding feeding arterioles and thus restoring adequate supply.

Geometrical relations are shown in a detail of the vasculature of cat tenuissimus muscle (Figure 5) illustrating some of the difficulties in understanding local metabolic perfusion control (*c.f.*[14]). There is a distance of several hundred μm between lethal corners (shaded) and feeding arterioles which prohibits tissue diffusion of vasoactive metabolites. Upstream conduction of dilator signals along the capillary wall appears to be effective in controlling the feeding arteriole only, if signals are generated within the proximal one third of the capillary[18]—which excludes the lethal corners. Diffusion of vasodilator substances from venules to their accompanying arterioles is the best supported explanation of all the ones presently proposed. Yet, it is not satisfactory either for a number of

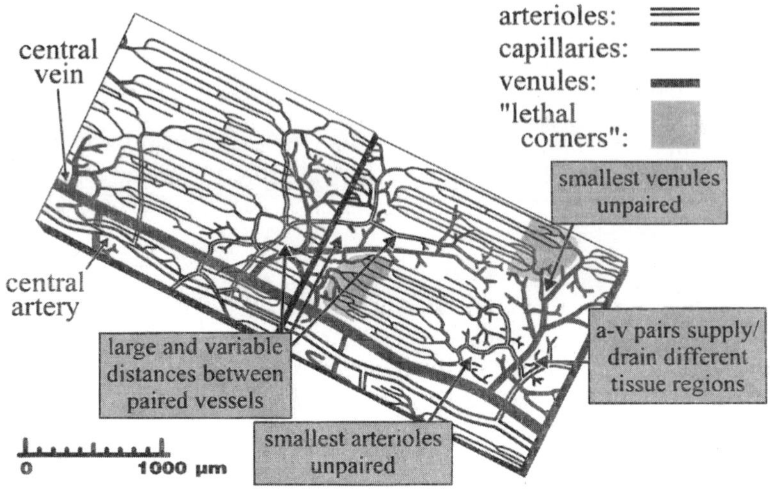

Figure 5. Detail of the microvasculature in cat tenuissimus muscle (redrawn from[5]). This figure illustrates the difficulties in understanding local metabolic perfusion control. For details see text.

reasons: venular-arteriolar (v-a) distances are very large and highly variable (some $10\,\mu m$ up to $1\,mm$); the smallest branches of the vascular tree are not arranged as v-a pairs at all; and in the larger vessel pairs, an arteriole often supplies a tissue microregion different from the one drained by its accompanying venule. These observations all strongly challenge the suitability of venular-arteriolar diffusion for establishing precise spatial and temporal adjustment of local blood flow to metabolism.

Beyond these difficulties, there are further problems that complicate understanding of how local microvascular control systems may function: It is unclear how the individual physiological mechanisms of myogenic response, shear-dependent and conducted vasodilation, as well as vasoactive substance effects might be coordinated to result in specific responses of the system to changes in $\dot{V}_{O_2}$, arterial blood O_2 content, *etc.* Finally, local microvascular controllers need to regulate not only perfusion rate but also transvascular fluid exchange (or, equivalently, hydrostatic pressure in the capillary bed). As has been discussed, this may be achieved by coordinated adjustment of pre- and postcapillary resistance. To date, nothing is known about how these tasks may be integrated.

3. A MODELING APPROACH

Before going into further details of this approach, it is outlined briefly how this technical problem of local blood flow and blood pressure control might be tackled from an engineering point of view. Probably one would

- position a sensor for indicators of tissue energetic status released to the blood downstream of the lethal corner in order to record the unabridged local signal,
- place the effector next to the sensor and control flow by changing venous flow path conductance, and

- build a servo-mechanism into the arteriolar tree that keeps capillary pressure low and (almost) constant at varying systemic pressures, and that adjusts it as needed for maintaining the blood flow preset by the venular flow controller.

Thus, an engineer would solve the problem of integrating the requirements of pressure and perfusion controls by constructing two largely independent mechanisms.

There is evidence that nature has applied the very same blueprint since all of the components constituting the above control system actually do exist in the microcirculation: Venules are reactive to vasoactive substances[2] and thus may represent combined sensors of tissue energetic status and effectors in flow control. Arterioles show myogenic response[9] and flow-induced dilation[11] which together may be viewed as a servo-mechanism for control of capillary hydrostatic pressure during variations of blood flow rate or arterial pressure. These mechanisms have been included in a mathematical model of the microvascular system in cat tenuissimus muscle.[7] Symmetrical vascular trees were assumed and flow was governed by Hagen-Poiseuille's law in which apparent viscosity η depends on vessel diameter and blood flow velocity. Furthermore passive distensibility in capillaries and venules was considered. All of the morphometrical and functional data required to perform the actual calculations were taken from experimental studies from the literature (among others;[1,3,10,11,13,19] for details see[7]) and were specific to cat *tenuissimus muscle* whenever available. The validity of the model was ascertained by comparing measured resting pressures and velocities in the various vascular generations of *m. tenuissimus*[6,19] to computed pressures and velocities which are in reasonable agreement.

Figure 6, left panels display calculated transmural pressures (top) and vessel diameters (bottom) as functions of position along the microvasculature for resting perfusion $\dot{Q}_{rest}$ and arterial pressures P_a ranging from 56 to 200 $mmHg$. Compared to standard conditions (solid, $P_a = 105\,mmHg$), a $95\,mmHg$ increase in P_a raises pressure at the capillary origin by no more than $6\,mmHg$ which demonstrates the system's capability of capillary pressure autoregulation. In the right panels, same quantities are shown for an arterial pressure of $105\,mmHg$ and perfusion rates ranging from 0.5 to 8 times the resting perfusion $\dot{Q}_{rest}$. It turns out that an increase in perfusion above $\dot{Q}_{rest}$ *reduces* pressure drops in arterioles, and pressure at the capillary origin rises from 38 to $54\,mmHg$ at $8 \cdot \dot{Q}_{rest}$. This is an effect of flow-induced vasodilation which is partly counteracted by myogenic response. In working muscles, a rise in capillary perfusion pressure with blood flow rate is essential for achieving highest performance levels at which capillary resistance may become limiting to perfusion. These data show that blood flow autoregulation by metabolite-mediated active changes in venular diameters in combination with arteriolar control of capillary pressure by myogenic response and shear-induced dilation should be potent enough to play a dominant role in the adjustment of perfusion to tissue metabolic needs.

The present model is intended to propose a plausible and realistic concept of a stable closed feedback loop for metabolic perfusion regulation which as such exists independently of central controls. Individual components of this feedback loop may, however, be modulated in their actions by a variety of factors. In this way, the operation of the controller could be adapted to extraordinary conditions, its regulatory range could be shifted, *etc.* Modulation may be achieved by modifying "set points" of the local mechanisms, *e.g.*,

- unstressed vessel diameters at zero flow,
- parameters of pressure-diameter relations,
- parameters of vascular blood flow-diameter relations, or
- parameters of dose-response curves of vasodilator action on vessel diameters.

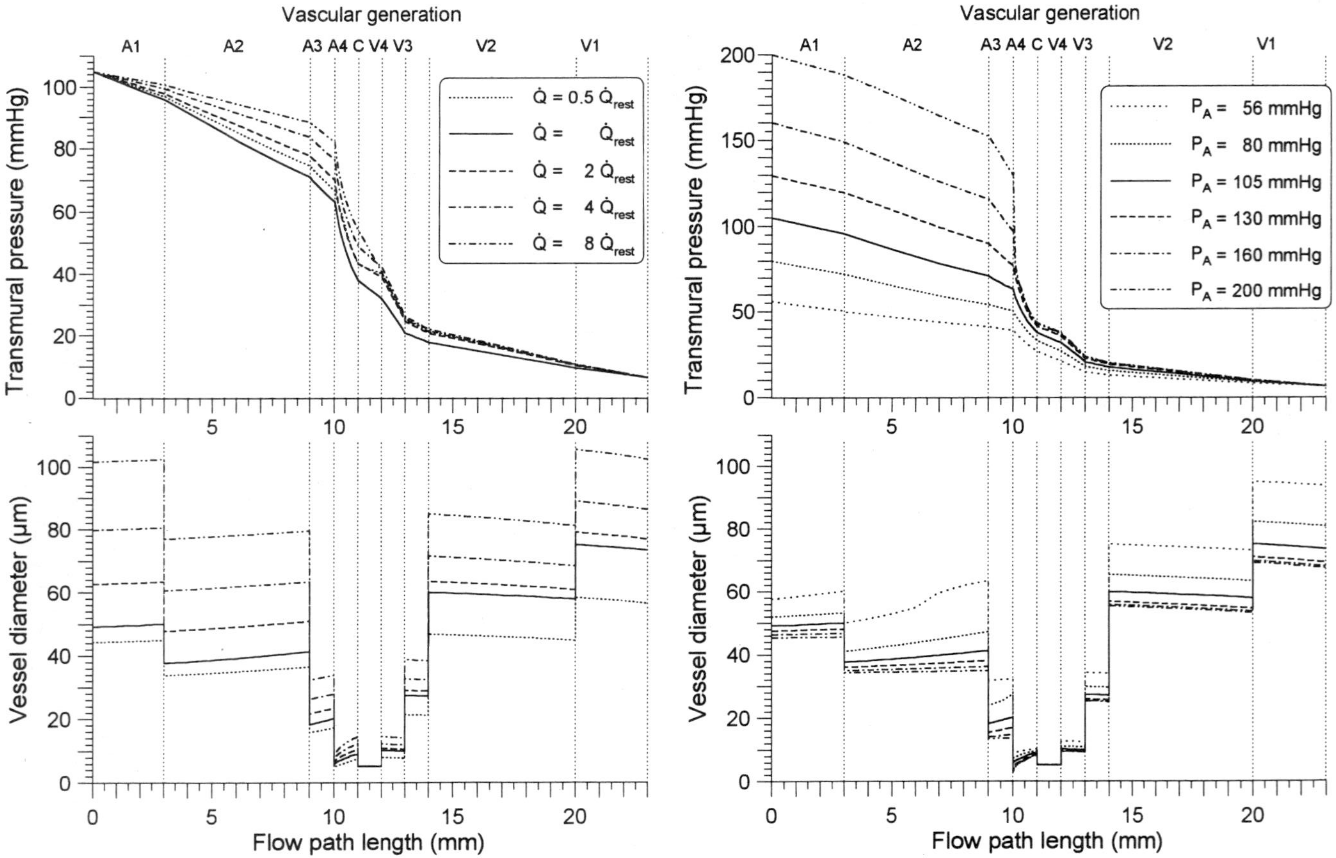

Figure 6. Calculated transmural pressures (top) and vessel diameters (bottom) as functions of position along the microvasculature for: *(i)* arterial pressures P_A ranging from 56 to $200\,mmHg$ (see in-figure legends) at resting perfusion $\dot{Q} = \dot{Q}_{rest}$ (left panels); *(ii)* arterial pressure $P_A = 105\,mmHg$ and blood flow rates $\dot{Q}$ ranging from $0.5 \cdot \dot{Q}_{rest}$ to $8 \cdot \dot{Q}_{rest}$ (which about equals blood flow capacity of cat tenuissimus muscle;[4] right panels). Vertical dotted lines delineate arteriolar generations A1 to A4, capillaries C, and venular generations V4 to V1.

4. PROSPECTS FOR FURTHER DEVELOPMENT AND CONCLUSION

In future, more sophisticated approaches to modeling vascular control systems, further combined experimental/theoretical efforts should be taken to clarify the mechanisms of local microvascular controllers and their central modulation. Secondly, vasodilator release curves (and time courses) as functions of $\dot{V}_{O_2}$ and tissue energetic status (or tissue P_{O_2}) as well as joint vasodilator dose-response curves and time courses of vasodilator action for different microvessels are required. Thirdly, further developments of vascular control models should account for non-symmetrically branching vascular systems, consider vasodilator effects on microvessels quantitatively, and include transient system behaviour (*e.g.*, at changing $\dot{V}_{O_2}$). On this basis, local microvascular control models should be combined with models of O_2 supply and consumption. In a next step, central modification of local controllers should be added, and composite models of central and local control of the circulation should be tackled. Finally, also long-term control mechanisms need to be considered.

In conclusion, a framework for modeling short-term control of vascular dynamical systems has been outlined in which local metabolic control systems are designed to function independently and central control mechanisms serve as modulators of local controllers. The principal determinants of tissue perfusion and water balance (pre- and post-capillary flow resistances) and their physiological correlates have been identified, approaches to modeling the elements involved in local microvascular control have been suggested, and problems in understanding the coordination of local control mechanisms have been discussed. A novel conceptualization of the integration of microvascular control mechanisms has been proposed according to which blood flow is regulated by metabolite-mediated active diameter changes in the venules and arterioles serve as controllers of capillary hydrostatic pressure *via* myogenic response and shear-induced vasodilation. This view resolves several shortcomings in current understanding by being based on an unambiguous and strictly local closed feedback loop for metabolic perfusion control, offering an explanation to how blood flow and hydrostatic pressure control in vascular beds may be coordinated, and suggesting that the blood flow itself represents the transporting medium for signals transferred from lethal corners to arterioles.

REFERENCES

1. PL Altman and DS Dittmer, "Biology Data Book", Federation of American Societies for Experimental Biology, Bethesda, 1972.
2. DN Damon and BR Duling, Venular reactivity in the hamster cheek pouch and cremaster muscle, *Microvasc.Res.* 31:379–383 (1986).
3. MJ Davis, Myogenic response gradient in an arteriolar network, *Am.J.Physiol.* 264:H2168–H2179 (1993).
4. E Eriksson and B Lisander, Change in precapillary resistance in skeletal muscle vessels studied by intravital microscopy, *Acta Physiol.Scand.* 86:295–305 (1972).
5. E Eriksson and M Myrhage, Microvascular dimensions and blood flow in skeletal muscle, *Acta Physiol.Scand.* 86:211–222 (1972).
6. K Fronek and BW Zweifach, Microvascular pressure distribution in skeletal muscle and the effect of vasodilation, *Am.J.Physiol.* 228:791–796 (1975).
7. K Groebe, Precapillary servo control of blood pressure and postcapillary adjustment of flow to tissue metabolic status: A new paradigm for local perfusion regulation, *Circulation* 94:1876–1885 (1996).

8. RL Hester, Venular arteriolar diffusion of adenosine in hamster cremaster microcirculation, *Am.J.Physiol.* 258:H1918–H1924 (1990).
9. PC Johnson, The myogenic response, **in**: "Handbook of Physiology, Sect. 2: The Cardiovascular System, Vol. IV: Microcirculation", EM Renkin, CC Michel, eds., pp. 409–442, American Physiological Society, Bethesda, 1984.
10. MF Kiani and AG Hudetz, A semi-empirical model of apparent blood viscosity as a function of vessel diameter and discharge hematocrit, *Biorheology* 28:65–73 (1991).
11. A Koller and G Kaley, Endothelial regulation of wall shear stress and blood flow in skeletal muscle microcirculation, *Am.J.Physiol.* 260:H862–H868 (1991).
12. L Kuo, WM Chilian, and MJ Davis, Interaction of pressure and flow induced responses in porcine coronary resistance vessels, *Am.J.Physiol.* 261:H1706–H1715 (1991).
13. HH Lipowsky, Network hemodynamics and the shear rate dependency of blood viscosity, **in**: "Microvascular Networks: Experimental and Theoretical Studies", AS Popel, PC Johnson, eds., pp. 182–196, Karger, Basel, 1986.
14. R Myrhage and E Eriksson, Vascular arrangements in hindlimb muscles of the cat, *J.Anat.* 131:1–17 (1980).
15. EM Renkin, Control of microcirculation and blood-tissue exchange, **in**: "Handbook of Physiology, Sect. 2: The Cardiovascular System, Vol. IV: Microcirculation", EM Renkin, CC Michel, eds., pp. 627–687, American Physiological Society, Bethesda, 1984.
16. SS Segal and BR Duling, Flow control among microvessels coordinated by intercellular conduction, *Science* 234:868–870 (1986).
17. JT Shepherd, Circulation to skeletal muscle, **in**: "Handbook of Physiology, Sect. 2: The Cardiovascular System, Vol. III: Peripheral Circulation and Organ Blood Flow", JT Shepherd, FM Abboud, eds., pp. 319–370, American Physiological Society, Bethesda, 1983.
18. HS Song and K Tyml, Evidence for sensing and integration of biological signals by the capillary network, *Am.J.Physiol.* 265:H1235–H1242 (1993).
19. BJ Zweifach and HH Lipowsky, Pressure-flow relations in blood and lymph microcirculation, **in**: "Handbook of Physiology, Sect. 2: The Cardiovascular System, Vol. IV: Microcirculation", EM Renkin, CC Michel, eds., pp. 251–307, American Physiological Society, Bethesda, 1984.

MODELING IN THE EVALUATION/ CALIBRATION OF MODERN OXYGEN MEASURING TECHNIQUES

L. Hoofd

237 Department of Physiology
Faculty of Medical Sciences
University of Nijmegen
PO Box 9101, 6500 HB Nijmegen
The Netherlands

1. INTRODUCTION

A variety of techniques are available for measuring oxygen or oxygen-related signals in biological media (Swartz et al., 1998). With these techniques, a signal is measured that is somehow related to oxygen, and the O_2 is derived from this measured signal. The signal can be quite close to O_2, e.g., mass spectrometry of oxygen, or only indirectly related. Examples of indirect signal are analysis of fluorescence lifetime or of magnetic signals such as EPR, that are influenced by O_2.

In any case, the instrument has to be considered—modelled—for a correct interpretation of its output, and the medium for a correct interpretation of the measurement. For the instrument, this means that the relation between signal and O_2 has to be established. For the medium where the instrument is used, it has to be analyzed whether and in how far the measured value of O_2 reflects the O_2 status of the medium.

A classical example is the polarographic O_2 electrode. It measures electric current. Modeling the instrument reveals that this current is related to "the" oxygen partial pressure pO_2 just outside the tip of the electrode. When applied in a gas mixture, such as air, that value will be the pO_2 of the gas mixture. When applied in blood, an additional gradient outside the electrode leads to underestimation of the blood pO_2. This is shown experimentally by the flow dependency of the polarographic electrode; the measured blood pO_2 varies according to the blood flow velocity. When applied in tissue, questions arise about the implication of the measured pO_2. In particular, transcutaneous O_2 electrodes need a modeling of the environment for any interpretation of their measurements.

Oxygen Transport to Tissue XXI, edited by Eke and Delpy.
Kluwer Academic / Plenum Publishers, New York, 1999.

2. METHODS OF MODELING

Modeling is a much more general term than is mostly realised. Any measurement assumes certain properties of the measured medium so that the quantity of interest can be determined; in fact, this is elementary modeling. The O_2 electrode would not measure anything if the O_2 could not freely move to the electrode tip. This transport is considered to be by—read, is modelled as—free diffusion. It can be modelled in a very elementary and simple way. On the other hand, models can be made very complicated, up to the limits of computer capacity. Here, four types of modeling will be discerned:

1. Elementary modeling, for understanding a process;
2. Modeling of the measuring device; this can often be quite elementary too;
3. Modeling of the adjacent environment of the measuring device; to test the sensitivity to external interference;
4. Modeling of the measured entity in order to understand what actually is measured.

These four types will be elucidated for the transport and measurement of oxygen.

2.1. Elementary Modeling

It is generally assumed that oxygen is distributed in tissue and model systems by diffusion. For the time scale of this distribution, we have to solve the time-dependent diffusion equation:

$$\frac{\partial p}{\partial t} = D\nabla^2 p \tag{1}$$

where p is oxygen partial pressure, t is time, D is the diffusion coefficient, in $m^2 \cdot sec^{-1}$, and ∇ is the gradient operator. In a 1-dimensional model, $\nabla = \partial/\partial x$ and the solution is a classical one, for a sudden change in p from 0 to p_0 at $x = 0$, $t = 0$:

$$\frac{p}{p_0} = 1 - \text{erf}\left[\frac{x}{2\sqrt{Dt}}\right] \tag{2}$$

where the integrated probability function or error function erf(z) starts from zero at $z = 0$ and approaches the limit value 1 for $z \to \infty$ quickly; in fact, this limit is almost reached, within 95%, for $z \approx 2$. This allows estimation of the time t needed for the O_2 to diffuse over a distance x by equating $x/(2\sqrt{Dt}) = 2$; for a typical D of $2 \times 10^{-9}\,m^2/sec$ we have:

$x =$	10 µm	100 µm	1 mm	1 cm
$t \approx$	3 msec	0.3 sec	30 sec	50 min

which gives an instant idea of the time scales of diffusion.

For the penetration of O_2 into tissue, we need a steady state description coupled to oxygen consumption:

$$\begin{aligned} -\vec{\nabla}\cdot\vec{J} &= O_2 \text{ disappearance rate} \\ \vec{J} &= -\mathcal{P}\vec{\nabla}p \end{aligned} \tag{3}$$

where $\vec{J}$ is oxygen flux, in $\text{mol} \cdot \text{m}^{-2} \cdot \text{sec}^{-1}$, and $\mathcal{P}$ is classically defined as the oxygen permeability, in $\text{mol} \cdot \text{m}^{-1} \cdot \text{kPa}^{-1} \cdot \text{sec}^{-1}$. There is much confusion about this latter quantity. It is totally different from the diffusion coefficient D though linked to it through the oxygen solubility α by $\mathcal{P} = D\alpha$. Without going in detail, it must be remarked that terms like permeability and diffusion constant are used inconsistently and often improperly; therefore, it is imperative that they be correctly and neatly defined in modeling.

When O_2 disappears only through the tissue's oxygen consumption $\dot{M}$, in $\text{mol} \cdot \text{m}^{-3} \cdot \text{sec}^{-1}$, elementary homogeneous modeling leads to solutions, in different dimensions:

$$p = \frac{\dot{M}}{2\mathcal{P}}\{x^2 - 2dx + d^2\} \qquad 1\text{D}$$

$$p = \frac{\dot{M}}{4\mathcal{P}}\left\{r^2 - 2d^2 \ln\left[\frac{r}{d}\right] - d^2\right\} \qquad 2\text{D}$$

$$p = \frac{M}{6\mathcal{P}}\left\{r^2 + \frac{2d^3}{r} - 3d^2\right\} \qquad 3\text{D} \qquad (4)$$

where x, r are the (radial) coordinates, respectively, and d is the penetration distance in terms of that coordinate. Examples are shown in Figure 1; for a border pO_2 of 10 kPa (75 mmHg) at $r = 2.5\,\mu\text{m}$ and a typical rat heart $\dot{M}/\mathcal{P}$ of $0{,}2\,\text{kPa} \cdot \mu\text{m}^{-2}$, the calculated penetration distances d are:

1D: 34 μm   2D: 24 μm   3D: 17 μm

So, the oxygen can penetrate from outside into such a tissue slab (1D) up to a distance of about 32 μm (from 2.5 to 34 μm), and from a spherical source (3D) up to about 14 μm.

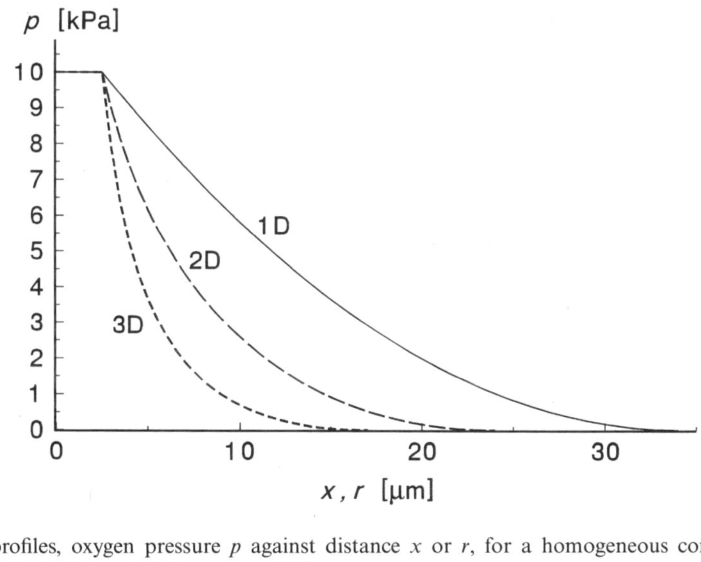

Figure 1. pO_2 profiles, oxygen pressure p against distance x or r, for a homogeneous consuming medium modelled in 1 (1D), 2 (2D) and 3 dimensions (3D).

Note, that O_2 can be involved in more reactions such as reversible binding to myoglobin (Mb). Then, corresponding reaction terms have to be added to the O_2 disappearance term (Kreuzer and Hoofd, 1987).

The 1-dimensional case would give an estimate of how thick a tissue layer might be in order to still obtain the O_2 from outside; the 2-dimensional case is the classical Krogh cylinder of O_2 supply around a capillary; the 3-dimensional case gives an idea of O_2 release around an erythrocyte. It can already be seen from such very simple modeling, that steep gradients in pO_2 are possible in tissue, as found by later more sophisticated modeling (Groebe, 1990; Bos et al., 1995).

2.2. Modeling of Measuring Device

Often, a measuring device is modelled quite simply. Again, the O_2 electrode is a good example. O_2 transport is modelled as 1-dimensional from outside to the cathode, e.g., a flat cathode and parallel layering. The O_2 flux to the cathode is:

$$J = \frac{\mathcal{P}}{L}(p_0 - p_1) \tag{5}$$

where p_0 and p_1 are the outside and cathode pO_2, respectively, and L is the pathway length of the oxygen. At the cathode, all the oxygen is ionized according to the reaction $O_2 + 4e^- \rightarrow 2O^{2-}$, transforming the oxygen flux J into an electric current I actually measured. Consequently, no pO_2 is present at the cathode, $p_1 = 0$, and $I \propto J \propto p_0$, the current reflects the outside O_2 pressure.

2.3. Modeling of Device Environment

Following the electrode example above, it is conceivable that the O_2 path will have to include part of the environment; in other words, O_2 has to travel not only inside but also towards the electrode tip. This can be complicated to model, but we might start from the simple 3-dimensional solution as applied above now in reverse direction, diffusion not from a source but to a sink—note, that the electrode consumes O_2:

$$p = p_\infty - \frac{\dot{n}}{4\pi\mathcal{P}r} \tag{6}$$

where p and p_∞ are the pO_2s at distance r and "far away", respectively, and $\dot{n}$ is the oxygen consumption of the electrode, in $\text{mol}\cdot\text{sec}^{-1}$. The latter is found from the electrode current or from the product JA where A is the cathode surface area. As an example, assuming the same $\mathcal{P}$ inside and outside the electrode and $A \approx \pi r^2$, the last two formulas could be combined into a startling result:

$$p \approx p_\infty - \frac{r}{4L} p_0 \tag{7}$$

which indicates that the pressure drop outside the electrode might easily be of the same order as inside, if the electrode is not designed correctly so as to minimise the apparent r/L.

2.4. Modeling of the Medium

An O_2 electrode measures a pO_2 of 4 kPa in tissue. What does it mean? Some would say, nothing, and even that might not be untrue. For a correct interpretation of the measured value, modeling can be very helpful. This is particularly so because both measurements and calculations have shown that tissue pO_2 values can vary considerably even over short distances (Honig et al., 1984; Federspiel, 1986; Groebe, 1990; Hoofd and Kreuzer, 1997). The field of tissue pO_2 modeling, however, is too broad to be more than touched here; for a review, see Popel (1989); and also Groebe (1996).

3. RELATION BETWEEN MODEL AND MEASURING TECHNIQUE

In the above treatment, the O_2 electrode was dealt with extensively. It measures electric current, but that signal is quite a direct measurement of oxygen partial pressure provided it is sure that no electrons are produced by, or disappear through, other processes. Being one of the oldest measuring techniques, it is still commonly and widely used.

Most close to a direct measurement of O_2 is mass spectrometry. Here, molecules are counted of the same molecular weight, so that also oxygen isotopes can be discerned, such as ^{18}O in tracer experiments. Radioactive measurements, ^{15}O, are also quite direct. All the other known techniques are indirect (Swartz et al., 1998). These measure a signal different from O_2 but influenced by O_2 and infer either the partial pressure pO_2 or the concentration cO_2 from this modified signal. It is not often made clear which of the two is inferred; this might be important since $cO_2 = \alpha O_2 pO_2$ and αO_2 can be different for different media, e.g., different tissue fractions (Egginton and Sidell, 1989; Hoofd and Egginton, 1997). While the laws of diffusion require the pO_2 distribution to be continuous, this consequently cannot hold for the cO_2 which might "jump" from one value to the other when crossing any boundary. It makes averaged values of cO_2 even more unreliable than averaged pO_2 to draw conclusions from.

Possibly the first indirect method for determining O_2 was measurement of saturation sO_2 of haemoglobin (Hb) or myoglobin (Mb) in blood and tissue. From the sO_2, the pO_2 is derived through the equilibrium relationship, the saturation curve, mostly determined in a similar but different medium. Before that, the sO_2 is inferred from absorption of visible and/or infrared light. Apart from the experimental problems, the light measurement is over a light spot, a certain area of tissue—where steep pO_2 gradients might be present. So, no extreme values of pO_2 will be seen but only averaged values. However, it was shown recently from correct modeling of Mb in skeletal muscle tissue, that these averaged values do not reflect the average pO_2 (Hoofd, 1998). There are two reasons for that. The first reason is the nonlinear saturation curve, the hyperbola $sO_2 = pO_2/(p_{50} + pO_2)$. For a p_{50} of 0.7 kPa, close saturations of 90% and 80% correspond with pO_2s of 6.3 and 2.8 kPa, respectively. The average of 85% corresponds with 3.97 kPa, underestimating the average pO_2 which is $(6.3 + 2.8)/2 = 4.55$ kPa. The second reason is the skew pO_2 distribution, ample low pO_2s but high pO_2s scattered over a broad range. If 60% is the result of 90% once and 50% thrice, it corresponds with an average pO_2 of 2.1 kPa, whereas from a non-skew distribution such as 80% and 40% 1.63 kPa would be derived, again an underestimation. Note, that it is unlikely that high pO_2s will be found with this method—in accordance with the published results (Honig et al., 1984).

The problem of nonlinearity lurks in every method of measuring. With phosphorescence quenching (Rumsey et al., 1988), the cO_2 is inversely related to the phosphorescence lifetime, and the same problems can arise as pointed out for the myoglobin measurements. This was nicely analyzed by Pittman (1998).

The steep gradients can present another problem, demonstrated for Hb. This species is present in the erythrocytes where pO_2 might be smooth so the average pO_2 can be inferred from sO_2. At the same time, the gradients around the erythrocytes can be steep so that the latter value reflects neither the plasma pO_2 nor the effective pO_2 that drives diffusion into the tissue. A pO_2 measured in the adjacent tissue, even with high spatial resolution up to the capillary, would never show values close to the erythrocyte pO_2. This apparent pO_2 gap, as deduced from theoretical models, was termed Extraction Pressure (Hoofd et al., 1990; Hoofd and Bos, 1998). When found experimentally, without the correct modeling background it might lead to totally erroneous conclusions, such as the presence of a diffusion barrier at the capillary wall. Already without measurements, such a pO_2 gap as predicted by several authors has been falsely understood as a physical diffusion barrier. It is a functional gap and the misinterpretation shows how important correct modeling can be for understanding a system.

An additional problem with photometric measurements arises when there is no discernment between Hb and Mb, as with near-infrared (NIRS) measurements (Jöbsis, 1977) in muscle. This can be elucidated by the following example of a calculation for rat heart. For such tissue, local pO_2s can be calculated using the multicapillary model (Hoofd, 1995a; 1995b). It allows calculation not only of pO_2 distributions but also for cO_2Mb; for the data see (Hoofd et al., 1990). From the corresponding blood data the average capillary cO_2Hb is found. Adding estimates about arteriolar and venular amounts "seen" by the NIRS (both taken as 20% of the blood total) the total amount of O_2Hb can be calculated that is detected. It is shown in the left grey bar of Figure 2. On top of it is the solid bar for O_2Mb; NIRS only detects the sum of both. The next column shows the situation for a heart flow supply of 25% less than the former case. The difference is shown in the right column; it is this difference signal that is the result of a NIRS measurement. 21% of this signal is due to O_2Mb. NIRS has not (yet) been applied to rat heart, but for

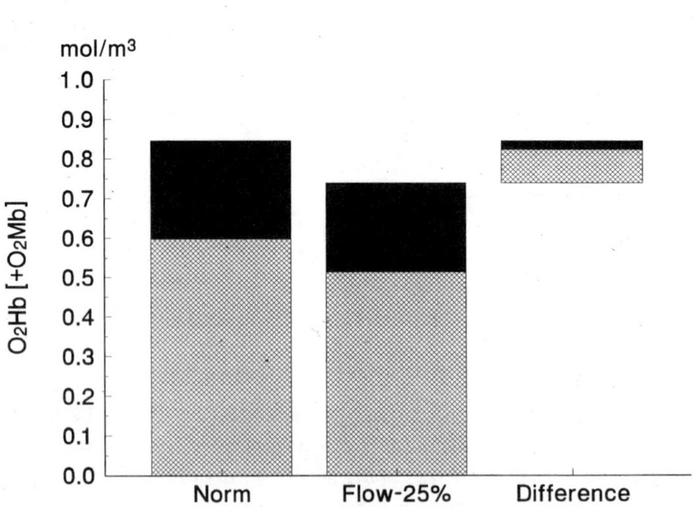

Figure 2. Calculated amounts of O_2Hb (grey bars) and O_2Mb (solid bars) as measured by NIRS for normal working rat heart with adequate blood supply (Norm) and for 25% decreased supply (Flow-25%); the differential signal (Difference) is seen to contain a considerable O_2Mb component.

this rat heart situation all the necessary data were available and it clearly shows that one should be cautious when interpreting NIRS measurements in muscle.

With electron paramagnetic resonance (EPR) measurements (also called electron spin resonance, ESR) the O_2 is inferred from its influence on magnetic phenomena in certain materials (Povich, 1975; Swartz et al., 1998). These can be inserted in a tissue in the form of tiny crystals, small in our perception (~100 μm) but large with respect to tissue pO_2 gradients as mentioned above. Thus, a first estimate of its influence might come from one-dimensional modeling as done above:

$$p = p_c + \frac{\dot{M}}{2P} x^2 \tag{8}$$

where p_c is the crystal surface pO_2 and x is the distance from the surface. When the O_2 source is at distance d and it normally supplies O_2 up to a distance R the surrounding pO_2 without crystal would be:

$$p = p_c + \frac{\dot{M}}{2\mathcal{P}} \{x^2 + 2(x-d)(R-d)\} \tag{9}$$

and we can calculate the pO_2 at the "crystal location" substituting $x = 0$. This pO_2 can be either higher ($d > R$) or lower ($d < R$) than p_c so that the detected pO_2 is not predicted to be biased towards either lower or higher values. Whether it reflects the "tissue pO_2" or not will have to be found from better modeling.

Magnetic resonance (NMR or MRI) measurements can also be influenced by O_2. In particular, ^{19}F in perfluorochemicals can be used to measure O_2 indirectly, through relaxation rates (Nöth et al., 1995). Here, to the existing problems of evaluating peaks and shifts, the problem of nonuniform O_2 distribution in fluosol-blood mixtures has to be added. So far, nobody seems to have paid attention to this problem.

4. DISCUSSION

It can be concluded that modeling plays an important role in various stages of the measuring and evaluation process.

- Evaluation of measurements should be according to an underlying model of the measuring technique, i.e., the measuring device should be modelled adequately.
- Interpretation of measurements should be along a corresponding (tissue) model.

As shown above, even simple modeling can be helpful. Often, the simple models help to find out in what respects the modeling should go into more detail, leading to an improved model. If a model is applied for more than just a rough estimate, in particular for prediction, the following constraint should be obeyed.

- Any simplification in a model should be verified or validated; if that is not possible, it should be clearly stated in what respect the model is simplified.

See Groebe (1996) for a more extensive list of constraints to modeling.

In many fields of science, there is a natural, balanced interaction between experimental and modeling work, leading to improvement of both the modeling and the experiments. It is unfortunate that this is much less pronounced in the oxygen measurement field, where there is often an imbalance or theoretical work is barely appraised or is lacking completely. Consequently, correct interpretation of the measurements is hampered and is often done in an actually (semi-)qualitative manner. The misunderstanding about capillary wall diffusion resistance is a good example.

This paper is not concerned with experimental or technical problems nor with other functions of modeling, like understanding a system. Also in these fields, interactions between the two can play a role. Examples are, the evaluation of a conversion matrix from NIRS light absorption to Hb saturation, and understanding how to test whether a tissue oxygenation is improved or not.

REFERENCES

Bos C, Hoofd L, and Oostendorp T (1995) Mathematical model of erythrocytes as point-like sources. Math Biosci 125:165–189.

Egginton S and Sidell BD (1989) Thermal acclimation induces adaptive changes in subcellular structure of fish skeletal muscle. Am J Physiol 256:R1–R9.

Federspiel WJ (1986) A model study of intracellular oxygen gradients in a myoglobin-containing skeletal muscle fiber. Biophys J 49:857–868.

Groebe K (1990) A versatile model of steady state O_2 supply to tissue; application to skeletal muscle. Biophys J 57:485–498.

Groebe K (1996) Practical applications of models of oxygen supply, diffusion, and consumption. Past, perspectives, and problems. In: Oxygen Transport to Tissue XVII, Adv Exp Med Biol 388 (Ince C, Kesecioglu J, Telci L, and Akpir K, Eds) Plenum Press, New York, pp. 161–175.

Honig CR, Gayeski TEJ, Federspiel W, Clark A Jr, and Clark P (1984) Muscle O_2 gradients from hemoglobin to cytochrome: new concepts, new complexities. In: Oxygen Transport to Tissue V, Adv Exp Med Biol 169 (Lübbers DW, Acker H, Leniger-Follert E, and Goldstick TK, Eds) Plenum Press, New York, pp. 23–38.

Hoofd L (1995a) Calculation of oxygen pressures in tissue with anisotropic capillary orientation I: Two-dimensional analytical solution for arbitrary capillary characteristics. Math Biosci 129:1–23.

Hoofd L (1995b) Calculation of oxygen pressures in tissue with anisotropic capillary orientation II: Coupling of two-dimensional planes. Math Biosci 129:25–39.

Hoofd L (1998) A first estimate of the effect of catchment area on pO_2 histograms derived from O_2Mb saturation measurements. In: Oxygen Transport to Tissue XX, Adv Exp Med Biol 454 (Hudetz AG and Bruley DF, Eds) Plenum Press, New York, pp. 391–397.

Hoofd L and Bos C (1998) Extraction pressures calculated for rat heart and dog skeletal muscle and application in models of tissue oxygenation. In: Oxygen Transport to Tissue XIX, Adv Exp Med Biol 428 (Harrison DK and Delpy DT, Eds) Plenum Press, New York, pp. 679–685.

Hoofd L and Egginton S (1997) The possible role of intracellular lipid in determining oxygen delivery to fish skeletal muscle. Respir Physiol 107:191–202.

Hoofd L and Kreuzer F (1997) Oxygen transfer from blood to mitochondria. In: The Lung: Scientific Foundations, second edition (Crystal RG and West JB, Eds) Lippincott—Raven Publishers, Philadelphia, pp. 1913–1923.

Hoofd L, Olders J, and Turek Z (1990) Oxygen pressures calculated in a tissue volume with parallel capillaries. In: Oxygen Transport to Tissue XII, Adv Exp Med Biol 277 (Piiper J, Goldstick TK, and Meyer M, Eds) Plenum Press, New York, pp. 21–29.

Jöbsis FF (1977) Noninvasive infrared monitoring of cerebral and myocardial oxygen sufficiency and circulatory parameters. Science 198:1264.

Kreuzer F and Hoofd L (1987) Facilitated diffusion of oxygen and carbon dioxide. In: Handbook of Physiology. The Respiratory System—IV (Fahri LE and Tenney SM, Eds) Amer Physiol Soc, Betheshda, Maryland, pp. 89–111.

Nöth U, Morrissey S, Deichmann R, Adolf H, Schwarzbauer C, Lutz J, and Haase A (1995) *in vivo* measurement of partial oxygen pressure in large vessels and in the reticuloendothelial system using fast 19F-MRI. Magn Reson Med 34:738–745.

Pittman RN, Golub AS, Popel AS, and Zheng L (1998) Interpretation of phosphorescence quenching measurements made in the presence of oxygen gradients. In: Oxygen Transport to Tissue XX, Adv Exp Med Biol 454 (Hudetz AG and Bruley DF, Eds) Plenum Press, New York, pp. 375–383.

Popel AS (1989) Theory of oxygen transport to tissue. Crit Rev Biomed Eng 17:257–321.

Povich MJ (1975) Measurement of dissolved oxygen concentrations and diffusion coefficients by electron spin resonance. Anal Chem 47:346–347.

Rumsey WL, Vanderkooi JM, and Wilson DF (1988) Imaging of phosphorescence: A novel method for measuring the distribution of oxygen in perfused tissue. Science 241:1649–1651.

Swartz HM, Dunn J, Grinberg O, O'Hara J, and Walczak T (1998) What does EPR oximetry with solid particles measure—and how does this relate to other measures of pO_2? In: Oxygen Transport to Tissue XIX, Adv Exp Med Biol 428 (Harrison DK and Delpy DT, Eds) Plenum Press, New York, pp. 663–670.

MATHEMATICAL MODELING OF OXYGEN TRANSPORT TO SKELETAL MUSCLE DURING EXERCISE

Hypoxia and VO_2max

K. G. Lyabakh

Department of Information Technology in Medicine and Biology
Glushkov Institute of Cybernetics
Kiev 252022, Ukraine

1. INTRODUCTION

Diffusion of oxygen to tissue has been well studied, mainly by means of mathematical models. The effect of blood flow on oxygen supply to the muscle has been widely explored for almost a century since the first studies of Krogh. Nevertheless, a number of problems of hard work, such as muscle hypoxia under load as a function of local conditions, are not clearly understood (Saltin and Rowell, 1980; Ivanov, 1993; Convertino, 1996). We reasoned to examine tissue hypoxia by means of a model. From many oxygen modes of a contracting muscle, we have chosen the mode of maximum oxygen consumption with possible tissue hypoxia. The present study is intended to analyze the influence of local conditions—the apparent Km, muscle blood flow and intercapillary distance on oxygen mode of a hard-working muscle. The aim of the investigation was to find the effect of Km on the parameters of hypoxia zone in the muscle fiber, to consider the notion of maximum oxygen consumption of a muscle (VO_2max), and to investigate quantitatively the values of VO_2max as a function of muscle blood flow and intercapillary distance.

2. METHODS

A computer model of oxygen delivery to and consumption in a contracting muscle was used to describe a steady-state of capillary blood flow and three-dimensional oxygen

Oxygen Transport to Tissue XXI, edited by Eke and Delpy.
Kluwer Academic / Plenum Publishers, New York, 1999.

diffusion-reaction in the microcirculatory unit (MCU). The configuration of MCU was taken from (Grunewald and Sowa, 1977).

Basic assumptions include:

Repeating Grunewald capillary-tissue microcirculatory units
Oxygen transport in capillary by convection
Oxygen transport in tissue by diffusion
Capillary wall permeability is infinite
Second-order kinetics of O_2 consumption in tissue (Bagdasarova and Haskin, 1977)
Homogeneous tissue
Homogeneous blood
Steady-state
Bohr effect ignored

List of symbols

pO_{2t}	oxygen partial pressure in tissue
pO_{2a}	oxygen partial pressure in arterial blood
pO_{2v}	oxygen partial pressure in venous blood
pO_{2c}	oxygen partial pressure in capillary
qO_2	oxygen demand
VO_2	oxygen consumption rate
VO_2max	limit of oxygen consumption rate
Dt	velocity of O_2-debt production
C_a	oxygen concentration in arterial blood
C_v	oxygen concentration in venous blood
C_c	oxygen concentration in capillary
Ca max	maximal oxygen concentration in arterial blood
d	intercapillary distance
v	linear blood velocity in capillary
F	muscle blood flow
sr	area of capillary cross-section
L	length of capillary
D	O_2-diffusivity in tissue
α	O_2-solubility in blood and tissue
K_m	apparent value of Michaelis-Menten constant
R	volume of MCU
r	surface of MCU
e	normal to r

System of equations and boundary conditions are presented for tissue in Eq. 1 and for any capillary in Eqs. 1–6.

$$\frac{\partial^2 pO_{2t}}{\partial x^2}+\frac{\partial^2 pO_{2t}}{\partial y^2}+\frac{\partial^2 pO_{2t}}{\partial z^2}=\frac{qO_2 \cdot pO_{2t}^2}{\alpha D(pO_{2t}^2+K_m^2)}; \tag{1}$$

$$\frac{\partial C_{Ci}}{\partial z}=\frac{n\cdot AVD\cdot\left(\frac{\partial pO_2}{\partial x}+\frac{\partial pO_2}{\partial y}\right)\frac{1}{v_i}}{\sum_{i=1}^{n}\frac{1}{v_i}\cdot\int_0^L\left(\frac{\partial pO_2}{\partial x}+\frac{\partial pO_2}{\partial y}\right)\partial z}; \tag{2}$$

$$pO_{2C} = \frac{-\ln\left(1-\sqrt{\dfrac{C_C}{C_a\max}}\right)}{(0.4492\cdot pH - 0.10093\cdot pH^2 + 0.0066815\cdot pH^3 - 0.454)}; \tag{3}$$

[Grodins et al. 1967]

$$AVD = \frac{qO_2}{F}\cdot\int_R\left(\frac{pO_{2t}^2}{pO_{2t}^2+K_m^2}\right)dR; \tag{4}$$

$$C_c(0) = C_a; \tag{5}$$

$$\frac{\partial pO_{2t}}{\partial e_{e\perp r}} = 0; \tag{6}$$

To solve the equations, the finite difference technique was used. The solution of the equation system allowed to get pO_2 and VO_2 distribution in MCU. In the general case it covers the arbitrary values of v, the arbitrary distribution of parallel capillaries and their arterial and venous ends inside MCU. To investigate tissue hypoxia we used a biochemical criterion of hypoxia—the state when oxygen consumption rate $VO_2(x,y,z)$ becomes less than oxygen demand qO_2 because of the lack of oxygen at (x,y,z). That provides an opportunity to get a graphical image of tissue hypoxia. The calculated value of ratio $S(x,y,z) = VO_2(x,y,z)/qO_2$ describes the degree of hypoxia at (x,y,z). The surfaces $VO_2(x,y,z)$ = constant set up a field of VO_2 provided that $pO_2 < pO_2crit$. We refer to a mode as anoxia if $VO_2 < 0.1\ qO_2$ and hypoxia if $0.1\ qO_2 \le VO_2 \le 0.9\ qO_2$. This field represents an image of hypoxia in MCU. It is an area where oxygen debt is produced. The rate of oxygen debt production $Dt = qO_2\ (1 - s)$ where s is average value of S. For simplicity, we show the MCU cross-section which passes through the center of MCU and a capillary, thus obtaining a flat picture of hypoxia zone (Figure 1).

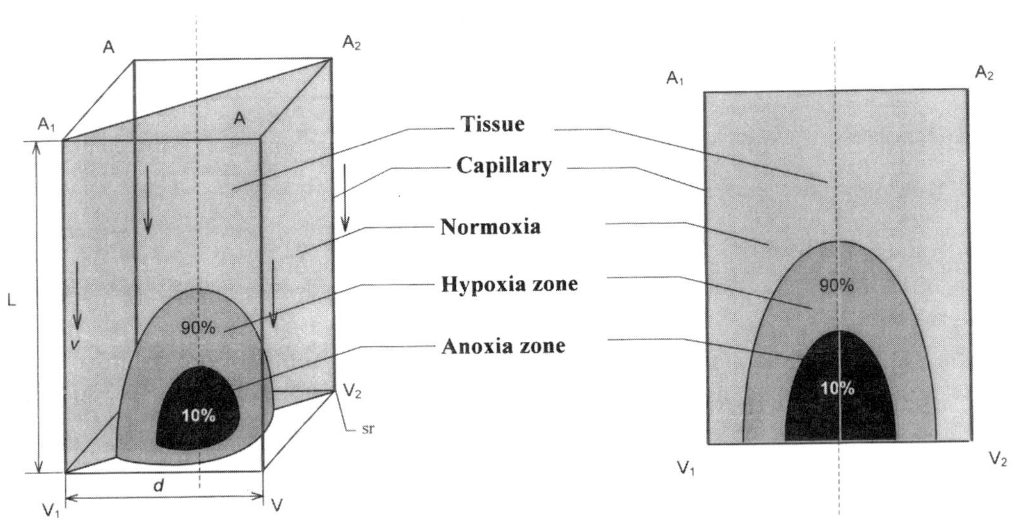

Figure 1. Microcirculatory unit (MCU) with hypoxia and anoxia zones inside. On left: three-dimensional image of MCU, on right the MCU cross-section which passes through the center of MCU and a capillary. % values represent (VO2/ qO2) × 100.

We rendered the simplest case: concurrent capillary blood flow with equal values of v, n = 4. This provides a symmetry distribution of VO_2-lines. In order to estimate the influence of Km on hypoxia zone in MCU during hard work, the computation of the muscle fiber VO_2 with different Km (1 Torr–5 Torr) was carried out. The influence of microcirculation on tissue oxygenation of a hard working muscle was simulated by means of qO_2 increasing and changing blood flow F with different values of the intercapillary distances d. The plateau of $VO_2(qO_2,F,d)$ was reached through increasing qO_2. The value of VO_2 at the plateau was taken as VO_2max. In the computational experiment, the combinations of parameters providing maximum oxygen consumption have been found.

The morphological and physical values of a model were chosen to correspond with the normal conditions of the muscle: $sr = 16 \times 10^{-8}\,cm^2$, $D = 1.3 \times 10^{-5}\ cm^2/s$, $\alpha = 2.8 \times 10^{-5}\,Torr^{-1}$, $L = 9 \times 10^{-4}\,cm$.

3. RESULTS AND DISCUSSION

The nonzero kinetics of oxygen consumption were considered many years ago but the relative effect of Km on hypoxia zone had not been investigated and had not been quantified. The results of modeling of the second-order kinetics of oxygen consumption with different values of Km are presented in this paper. The values of Km vary greatly. Km for mitochondria depends on its metabolic state, local physical and chemical environment. Km can reach values equal to small fractions of one Torr (Chance, 1965; Starlinger and Lubbers, 1973). In accordance with (Buerk and Longmuir, 1977; Wilson et al., 1979; Kreuzer and Hoofd, 1982), the order of magnitude of apparent Km can reach one Torr and greater, e.g., 8 Torr (Bagdasarova and Haskin, 1977). Thus, one has to distinguish between a true mitochondrial Km and an apparent Km in the tissue. That is why we reverted to the problem of the influence of Km value on working muscle oxygen mode, on hypoxia zone in particularly. We investigated the effect of the Km value on the area of hypoxia zone and on VO_2-gradients within this zone. Despite Km estimates in literature ranging from 0.01 Torr to over 5 Torr we used the model values not less than 1 Torr to guarantee the accuracy of computations 0.5 Torr.

The experimental conditions were as follows: qO_2 = 8.57 ml/min/100 g, F = 42 ml/min/100 g, d = 40 mcm, Ca = 0.17, Cmax = 0.20, pH = 7.15 for several values of Km, within the interval (1–5 Torr). (Table 1 and Figure 2).

Figure 2 presents the comparison of hypoxia images in a working muscle for different values of apparent Km. The results of modeling show that during exercise O_2-debt is produced in spite of normoxic arterial blood flow oxygenation. Hypoxia zone is located in the center of the muscle fiber. Actually hypoxia zone does not touch the blood vessels. From the picture of hypoxia in MCU it is evident that the shape and size of hypoxia zone essentially depend on the value of Km. The more the value of Km the more the

Table 1. Effect of value of apparent Km on parameters of oxygen mode in the muscle MCU during exercise

Km, Torr	1	3	5
s, %	77	74	71
$(a\text{-}v)_{O2}$	0.155	0.148	0.142
pO_{2V}, Torr	10	12	14
pO_{2t}, Torr	9	11	13
Dt, ml/min/100 g	1.97	2.23	2.48

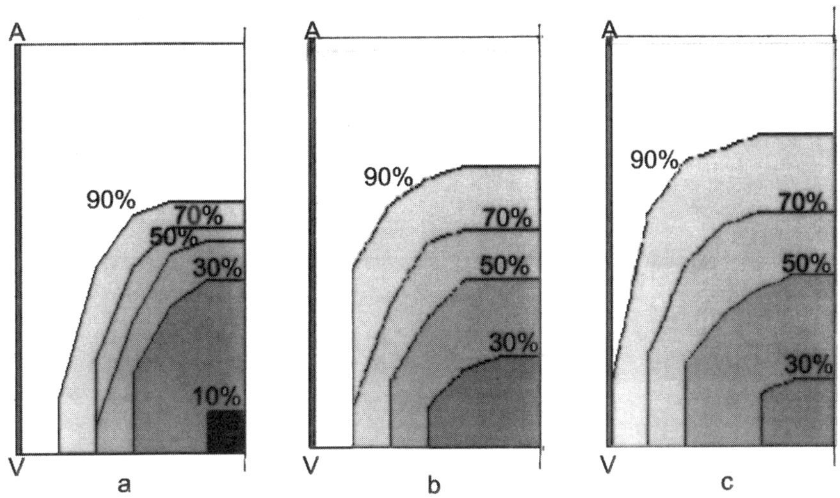

Figure 2. The effect of value of apparent Km on hypoxia image in rectangular MCU cross-section of working muscle. x axis—O_2 path length in MCU, y axis—the capillary length. a – Km = 1 Torr, b – Km = 3 Torr, c – Km = 5 Torr. % values represent (VO2/ qO2) × 100.

hypoxic area. It is situated in the center of MCU. The decrease of Km affects not only the area of hypoxic zone, shrinking its boundaries, but increases the gradients of VO_2 inside the hypoxia zone toward the lethal corner. Hypoxia zone partially transforms into anoxia zone. The area of anoxia is dependent on the value of Km. The less Km the more the anoxic zone and the more the average value of oxygen consumption rate. Extrapolation of this tendency assumes that the largest oxygen uptake of the whole MCU would occur at zero value of tissue pO_2 at the lethal corner. The value of Km affects not only the parameters of hypoxia zone but pO_2 in blood and tissue (Table 1). If the value of Km increases O_2 extraction from blood rate drops, $(a\text{-}v)_{O2}$ or AVD are lowered and mean value of tissue pO_2 becomes greater.

These results indicate the great significance of the value of Km for hypoxia modeling. The latter seems to be impossible without taking into consideration Km.

To examine the relationship between aerobic muscle capacity and microcirculation we gave a working definition of VO_2max as value of oxygen consumption that the 10% increase of qO_2 accompanies a change of $VO_2(qO_2)$ not greater than 1%. (Lyabakh,1997) The VO_2max mode in a mathematical model was reproduced as follows. The muscle was supposed to be supplied with blood flow F. Under a constant capillarity and blood pH = 7.15, which corresponds to the shift of the oxygen dissociation curve to the right, the muscle function grows with an increasing intensity. The latter fact is presented in the model with the growth of the input value, oxygen demand.

We have to perform some calculations to find the level of Vo_2max for the maximum acceptable blood flow with limit values of the parameters of oxygen supply to the tissue.

The relationship $VO_2(qO_2)$ (Table 2) demonstrates the existence of VO_2max, its quantitative magnitude and the other characteristics of oxygenation in working muscle. The same is shown in Figure 3, the value of VO_2max as a fixed point of $VO_2(qO_2)$.

Three portions can be distinguished on this curve. The first portion ranges from 0 to the point Vt, the second one is located between the points Vt and Vmax (the beginning of the plateau of the curve), the third portion contains all values of oxygen demand

Table 2. Oxygen mode parameters in the muscle at different levels of its activity F = 80 ml/min/100 g, d = 40 mcm

qO_2, ml/min/100 g	8.57	11.76	12.85	14.28	17.14	20.07
$(a\text{-}v)_{O2}$,	0.102	0.115	0.119	0.122	0.125	0.127
VO_2, ml/min/100 g	7.59	8.94	9.00	9.02	9.18	9.40
s, %	88	76	70	63	54	47
Dt, ml/min/100 g	0.96	2.82	3.85	5.26	7.96	10.67
pO_{2t}, Torr	18	12	10	9	7	6
pO_{2V}, Torr	24	21	20	19	18	17

to the right of Vmax. In the first portion, linear growth of oxygen consumption rate can be observed. In the second range, oxygen consumption rate lagged behind oxygen demand, but VO_2 increased with increased load. In the third portion, the load still increases, but the possibilities for the muscle fiber to increase its oxygen consumption rate are exhausted and the oxygen consumption rate remains constant. In theory, it slowly increases with the growth of oxygen demand, but this increase could be neglected. The third point is at the value of VO_2max.

To investigate VO_2max as a function of F and d we changed qO_2 from 6 ml/min/100 g to 16 ml/min/100 g for different values of F within the interval of 60 ml/min/100 g < F < 120 ml/min/100 g and kept all other parameters constant. Parameter d was within the interval 40–70 mcm. In any dependence, VO_2max (qO_2,F,d) was found. The results of experiments are shown in Figure 4. The computed values of VO_2max (F,d) extrapolated to the whole body muscle (mass 35 kg) indicate that under chosen conditions VO_2max does not exceed 70% qO_2.

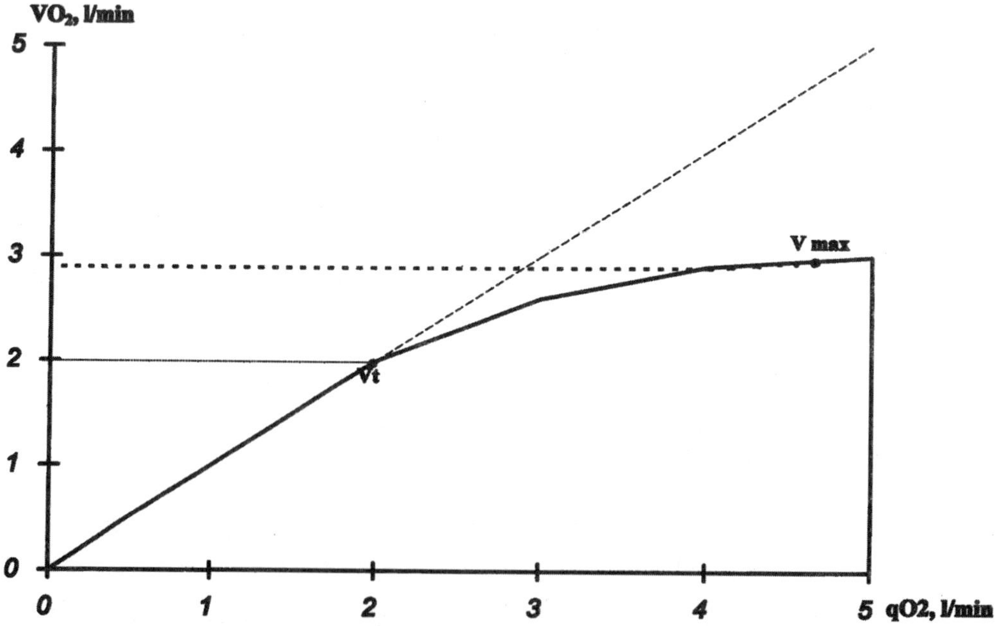

Figure 3. The rate of muscle oxygen consumption VO_2 plotted against oxygen demand qO_2 in the muscles of mass 35 kg (F = 32 l/min, d = 50 mcm, Km = 4 Torr).

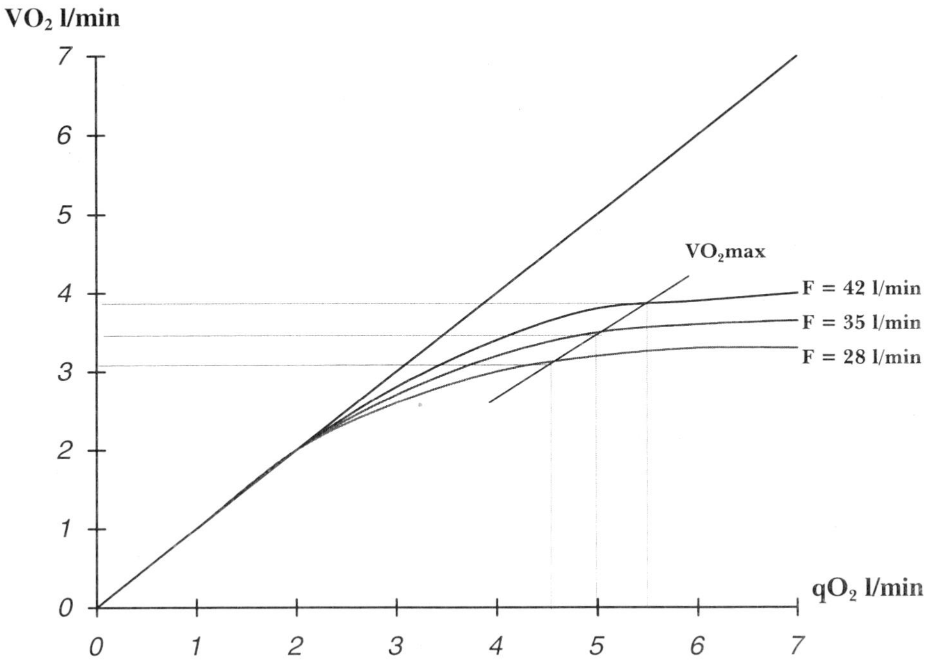

Figure 4. VO_2-dependencies on oxygen demand qO_2 for different values of muscle blood flow F in the muscles of mass 35 kg (d = 40 mcm, Km = 4 Torr).

It is seen that aerobic work can increase by increased muscle blood flow. But an increase in VO_2max may take place without a rise of F if diffusion pathlength becomes shorter. The family of curves VO_2max (F,d) (Figure 5) demonstrates that VO_2max can increase not only due to the growth of muscle blood flow but due to the increase of number of opened capillaries. If so the sensitivity of VO_2max to F becomes higher too. On the other hand, the increase of F contributes to the rise of sensitivity VO_2max to the changes of intercapillary distance d.

It is known that during adaptation to physical activity cardio-respiratory system improves its capability that provides the rise of the value of VO_2max of the whole organism. Our results showed quantitatively the role of local muscle factors in formation of VO_2max of the muscle. This may in turn have an effect on the well known value—VO_2max of a whole organism. It follows that adaptation of organism to hard muscle work is handled by the response of cardio-respiratory system as well as local changes in the muscle microcirculation. It can be seen from our computation that if F reaches high values, muscle adaptation to prolonged exercise may take place just through the growth of new capillaries. The structural parameters of muscle tissue and its vascular bed play a very significant role in VO_2max formation. They can extend the limits of muscle adaptation and enhance the aerobic capacity of skeletal muscle.

4. CONCLUSION

1. Hypoxia takes place in skeletal muscle during heavy exercises under normoxic saturation of arterial blood. The spatial parameters of hypoxia depend not only

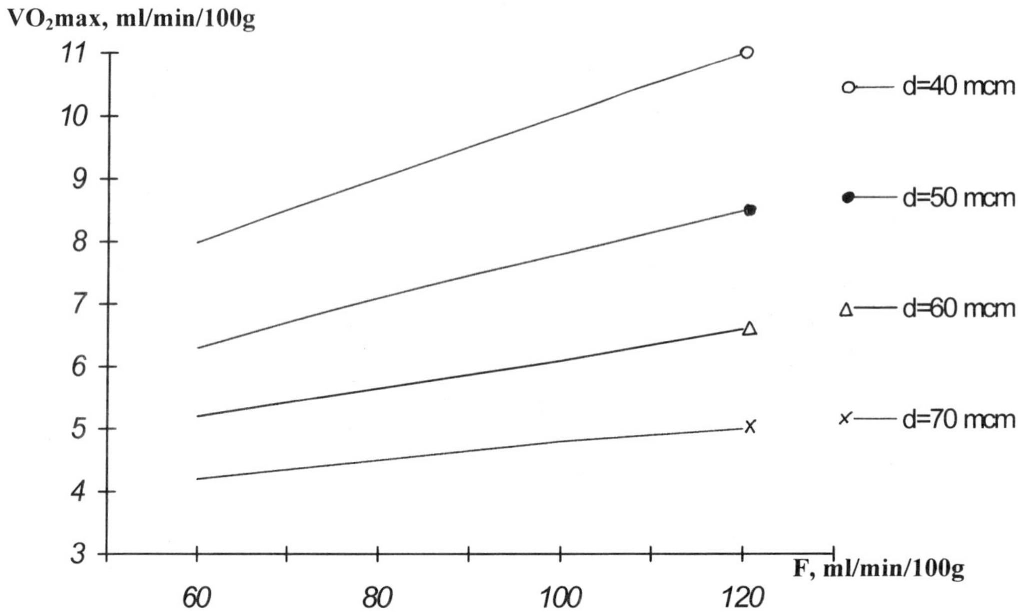

Figure 5. The effect of muscle capillary blood flow on the VO_2max for different values of intercapillary d (Km = 4 Torr).

on the metabolic rate and blood flow but on the value of apparent Km. A decrease of Km provides the partial transformation of hypoxia zone into anoxia zone.

2. VO_2max in the skeletal muscle does not exceed 70% oxygen demand.
3. The sensitivity of VO_2max to the variation of blood flow F and intercapillary distance d depends on the values of F and d: the more F and the less d, the more the sensitivity VO_2max to their variations.

REFERENCES

Bagdasarova T and Haskin V (1977) The effect of adaptation to hypoxia on the dependence of tissue breathing of temperature and oxygen pressure in vitro. Sechenov Physiol. J. 63:1598–1609.

Buerk D and Longmuir I (1977) Evidence for nonclassical respiratory activity from oxygen gradient measurements in tissue slices. Microvasc. Res. 13:345–353.

Buerk D and Saidel G (1978) A comparison of two nonclassical models for oxygen consumption in brain and liver tissue. In: Oxygen Transport to Tissue-III (Silver I, Erechinska M, Bicher H, Eds) Adv. Med. Biol, Plenum Press, New York-London, pp. 225–232.

Chance B (1965) Reaction of oxygen with the respiratory chain in cells and tissues, J. Gen. Physiol. 49:163–188.

Convertino V (1996) Exercise and adaptations to microgravity environments. In: Handbook of physiology. Environmental physiology. Oxford University Press, New-York, pp. 815–843.

Grodins F, Buel J, and Bart A (1967) Mathematical analysis and digital simulation of the respiratory control system. J. Appl. Physiol. 22:260–276.

Grunewald W and Sowa W (1977) Capillary structures and O_2 supply to tissue. Rev. Physiol. Biochem. Pharmacol. 77:140–209.

Ivanov K (1993) The principles of energetics in organism: Theoretical and practical aspects. V.2. Biological oxidation and its providing with oxygen (Glebovsky V, Ed) Nauka, St.-Petersburg, pp. 212–243.

Kreuzer F and Hoofd L (1984) Facilitated Diffusion of Oxygen: Possible Significance in Blood and Muscle. In: Oxygen Transport to Tissue-V, Adv. Exper. Med. Biol., 169 (Lubbers D, Acker H, Leniger-Follert E, Goldstick T, Eds) Plenum Press, New York-London, pp. 3–21.

Lyabakh K (1997) Quantitative analysis of determinants of aerobic metabolism threshold and maximal oxygen consumption in working muscle. The proceeding of the modern olympic sport. International scientific congress (May 16–19) (Boloban, Kochur, Levitsky, Eds) Ukrainian University of Physical Education and Sport. Kyiv, p. 146.

Saltin B and Rowell L (1980) Functional adaptations to physical activity and inactivity. Fed. Proc. 39:1506–1513.

Starlinger H and Lubbers D (1973) Polarographic measurements of the oxygen pressure performed simultaneously with optical measurements of the redox state of the respiratory chain in suspensions of mitochondria under steady-state conditions at low oxygen tensions. Pflugers Arch. 341:15–22.

PARADOXICAL RELATION OF MITOCHONDRIAL PO_2 TO $\dot{V}o_2$

John W. Severinghaus

Department of Anesthesia
University of California Medical School
San Francisco, Calofornia 94143-0542

1. OXYGEN PHYSICAL CHEMISTRY

The study of oxygen is at the base of all physiology, and its mysteries are among the most important for our understanding of our bodies. My perspective of oxygen reactions is colored by a background in physics and physical chemistry, particularly in the understanding of polarographic oxygen electrodes, and the 35 year effort in my laboratory to understand the impact of high altitude on climbers and sojourners. I thus may espouse different schemes than are common among biochemists. Let's begin with a perplexing idea: Oxygen consumption is maximum when cytochrome metal surface Po_2 is zero. This idea helps understand oxybiology of muscle, carotid body, some mucosa, and other tissues. Consider an analogy to oxygen polarographic electrodes.

2. OXYGEN ELECTROCHEMISTRY

2.1. Oxygen and Electrons

O_2 reacts with energetic electrons at wet metal surfaces, becoming a free radical with one electron, or hydrogen peroxide with two electrons, or 4 hydroxyl ions with 4 electrons.

2.2. Oxygen Electrodes

Polarographic oxygen electrodes resemble reduced mitochondria. Both consume O_2 by providing electrons. At the usual operating voltage, about –0.6v, an electrode (Figure 1) consumes all the oxygen reaching the metallic surface. The relationship

Oxygen Transport to Tissue XXI, edited by Eke and Delpy.
Kluwer Academic / Plenum Publishers, New York, 1999.

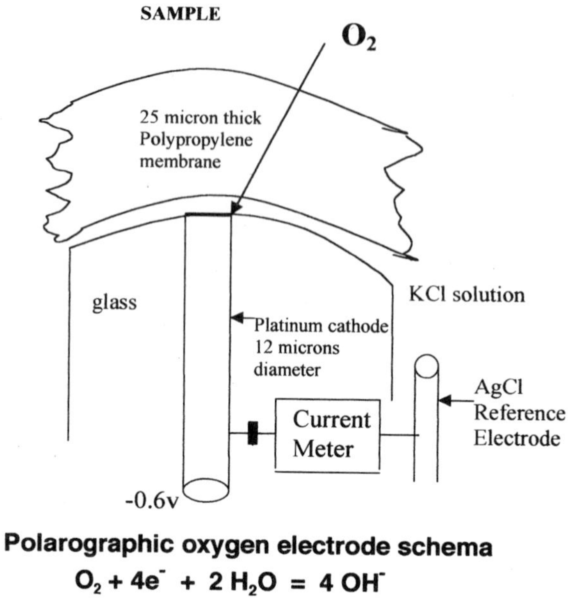

Figure 1. Schema of a polarographic Clark type oxygen electrode as a model of cytochrome oxygen dynamics. Oxygen diffuses through a limiting material (which in-vivo includes tissues between capillary lumen and the cytochrome). An electrode is operated at –0.6v in order to consume all the oxygen which reaches its surface, but this occurs in life only at VO_{2max} or at the critical threshold of oxygen delivery in shock or hypoxia.

between cathode potential and current is a polarogram (Figure 2). Current at a constant voltage is a linear function of oxygen supply to the cathode. To measure Po_2 in solutions or gases a semipermeable membrane separates metal from sample, providing a diffusion barrier. If the permeability of this membrane is sufficiently low, almost all the diffusion gradient occurs within the membrane, so the current of the electrode becomes a linear function of the Po_2 on the outside surface of the membrane, hence in the sample. In

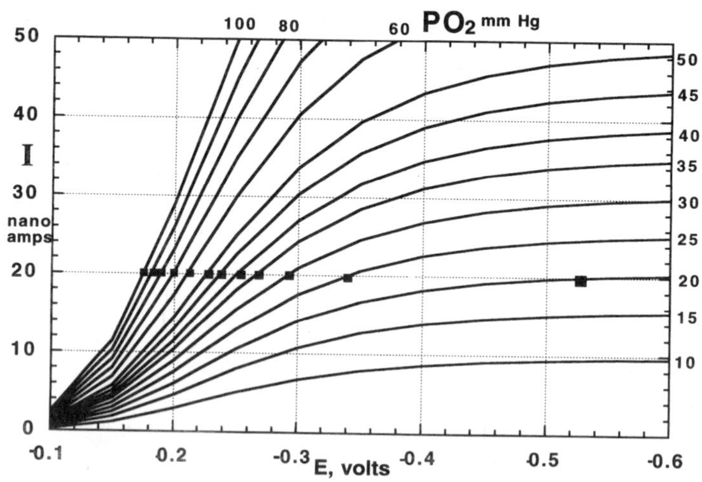

Figure 2. Schema of oxygen polarograms at varying sample Po_2 values. At a fixed voltage, E, current I is a linear function of Po_2. In vivo, the potential varies on the cytochrome metal because the electron flux or oxidative metabolic rate is determined by ATP consumption, not by Po_2. The series of solid squares at I = 20 nanoamps show the rise of negative potential required in either an electrode or a cytochrome when current is held constant as Po_2 falls. In this example, a Po_2 of 20 mm Hg is the critical threshold at which all the oxygen arriving at the cathode (or aa_3) is reduced.

liquids, to accomplish this without great reduction of permeability, the cathode metal area is made so small that almost all of the diffusion gradient is within the membrane. In today's blood gas electrodes, cathode diameter is about 10 microns. The membrane material is usually polypropylene, 12–25 microns thick.

2.3. Making O_2 Consumption Independent of Po_2

In life, O_2 consumption is held independent of Po_2 (above a critical threshold) by the Pasteur effect, in which consumption of ATP releases ADP which activates both glycolysis and mitochondrial metabolism. To maintain a constant metabolic rate, the mass law indicates that the product of electron pressure (electrical potential, or redox state) and oxygen flux must be constant. As oxygen supply decreases, redox potential rises (more negative). The process of constant oxygen consumption with varying oxygen pressure can be simulated in an oxygen electrode by varying the voltage to maintain a constant current and oxygen consumption. The resulting relationship of E, potential, to PO_2 is shown in Figure 2 at the I = 20 nanoamp line. As PO_2 falls, the cathode potential must be increased negatively to maintain constant current, reaching a critical level at 20 mm Hg in Figure 2. The relationship of PO_2 to E is replotted from Figure 2 in Figure 3, illustrating the hyperbolic rise of potential as oxygen tension falls.

2.4. Mitochondria in Hypoxia

Mitochondria undergo a similar steep rise in potential as oxygen availability falls toward the critical threshold, the lowest oxygen supply at which oxygen consumption can be maintained constant. In dilute stirred solutions with excess substrate, stimulated mitochondria maintain a constant O_2 consumption down to nearly zero, measured as $PO_2 = 0.1$ mm Hg (Naqui et al., 1986). At this level, they are almost fully reduced, which means their electrons are at a strongly negative potential, ready to react with any O_2 molecule, like the platinum cathode of an O_2 electrode. As with an O_2 electrode, at a constant metabolic rate, mitochondrial electron potential varies with Po_2. In progressive hypoxia, as the number of

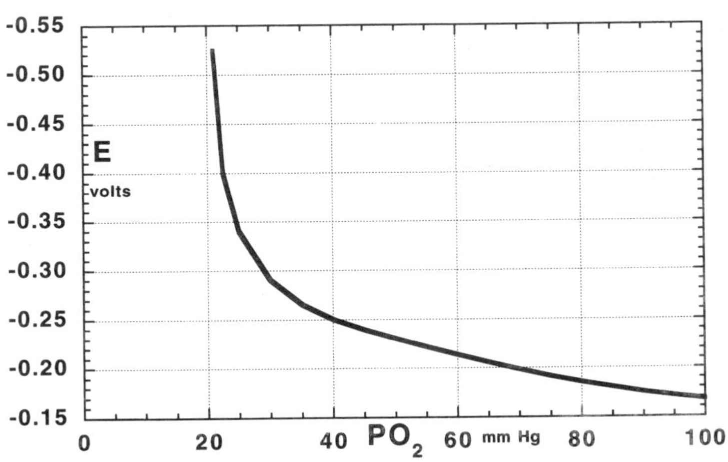

Figure 3. Data from Figure 2 at 20 nanoamps are replotted to illustrate the hyperbolic rise of potential needed at the cathode (or aa_3) at constant current (or oxidative metabolic rate) as Po_2 falls.

O_2 molecules at or near the cytochrome surface falls, in order for O_2 consumption to continue to equal the flux of electrons (i.e. reducing equivalents, or metabolites), with fewer oxygen molecules available, the potential charge on the electrons must rise (this is the mass law). We may think of this as an accumulation of electrons in hypoxia which we term a redox reduction. The relationship of the reduction (i.e. increasingly negative potential) to the supply of oxygen is hyperbolic similar to that shown for an electrode in Figure 3.

2.5. Effect of Hypoxia on Substrate Concentrations

This cytochrome reduction causes the concentrations of each of the metabolic intermediates in the citric acid cycle to increase because each step in the citric acid cycle has an electrical potential drop dependent on its flux. When rate is constant, a rise of negative potential at the terminal aa_3 cytochrome, or change in redox caused by a change in oxygen, raises the absolute potential at each preceeding step, and thus drives up the concentration of each metabolic intermediate. This causes the observed rise of pyruvate (and thus of lactate) as a function of work.

2.6. Lactate and Pyruvate

Lactate rises in proportion to pyruvate (and further as accumulation of H^+ ions drives pyruvate to lactate). Some of this lactate diffuses out to blood as long as blood concentration of La is lower than tissue concentration. When it becomes easily measurable it is called the anaerobic threshold. It does not imply a reduction in VO_2. O_2 consumption is still equal to ATP production and energy demand, rising in direct proportion during exercise with increasing work rate (showing that work is actually still essentially fully aerobic) (Severinghaus, 1994). The ATP generated by increasing lactate generation and export is a small fraction of total energy production until work rate exceeds oxygen supply (at the critical threshold).

2.7. Critical Threshold

Now consider what happens when capillary Po_2 falls below the critical threshold (Figure 4). This threshold varies from less than 10 mm Hg in trained athlete muscle to about 30 mm Hg in gut. In brain it is about 20 mm Hg. At this critical threshold the Po_2 at the cytochrome surface may be assumed to have reached zero, all the delivered O_2 being consumed on arrival. Like an O_2 electrode, which always operates with zero Po_2 at the metal surface, below the critical threshold, O_2 consumption becomes proportional to the Po_2 at the membrane's external surface, which in the life analogy is the capillary endothelial surface. At the fully reduced cytochrome surface, there are sufficient electrons available such than no O_2 molecules reaching the metal atom remain unreduced, leaving none to diffuse away from the site, and this by definition is a Po_2 of 0 at the cytochrome in the region below $Po_2 = 20$ in Figure 4 (Severinghaus, 1994).

3. MUSCLES

3.1. Paradoxes

In skeletal muscle, we thus have the paradox in which the cytochrome Po_2 is zero at VO_{2max}. The foregoing electrode analogy helps us to understand how hypoxia can have

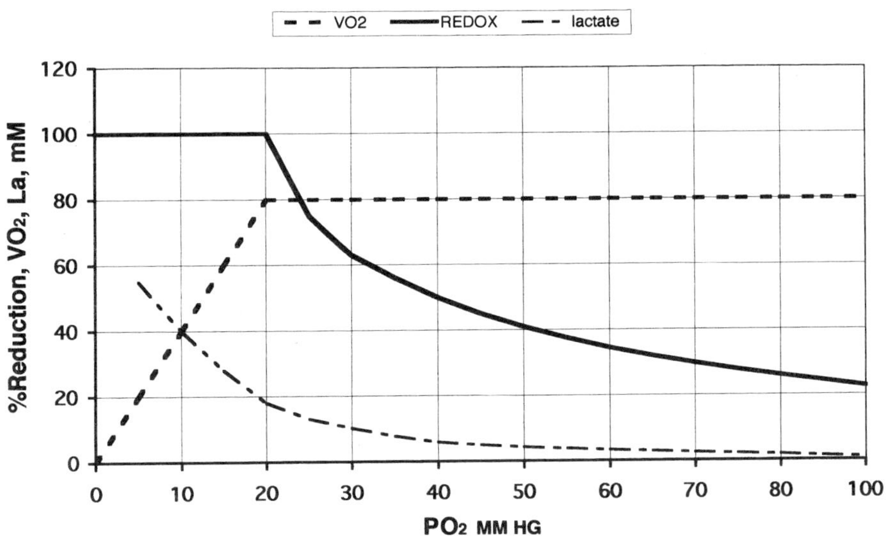

Figure 4. A schema of the effects of falling Po_2 on oxygen consumption, cytochrome aa_3 reduction and lactate concentration (arbitrary units). Substrate concentrations in the citric acid cycle and of pyruvate and lactate rise before oxygen consumption falls as needed to maintain constant oxygen consumption as reduction increases (i.e. as aa_3 negative potential rises).

such profound biochemical effects without a decrease in O_2 consumption, and why muscles generate lactate at high work rate even when O_2 consumption and aerobic ATP production continue to match work rate.

3.2. Limits

The physiologic limit of exercise in a trained subject is oxygen supply. VO_{2max} falls in direct proportion to muscle or leg venous Po_2 in subjects breathing low O_2 during exercise (Wagner et al., 1990). Studies of capillary Po_2 and O_2 consumption are consistent with the hypothesis that Po_2 at the cytochromes falls to zero at VO_{2max}, from which a generalization is appropriate: Steady state exercise is limited by O_2 delivery, peaking at the point at which all the O_2 which can be extracted from passing blood is being consumed. Work can be increased by increased flow, as seen with one arm or one leg exercise where cardiac output does not limit flow. The evidence is clearest in skeletal muscle because a well trained working muscle has a remarkably homogeneous circulation, so its venous blood is a good index of end-capillary blood, permitting estimation of mean capillary Po_2.

3.3. Function of Anaerobic Metabolism

If anaerobic metabolism contributes little to ATP generation, what does it do? The more important function of the lactic acid is to acidify blood and raise Po_2 (Wasserman, 1994). Muscle venous Po_2 in exercise falls to about 15 mm Hg at about 60% of VO_{2max}, but falls no further at higher work rate due to lactic acidification of blood shifting the O_2 dissociation curve rightward, i.e. raising blood Po_2 at given saturation (Wasserman, 1994). Muscle venous blood saturation can fall below 10%.

3.4. Zero Po_2 Hypothesis

This insight of zero Po_2 at VO_{2max} permitted modelling of tissue Po_2 in muscle as a function of the work, plotted as a fraction of VO_{2max} (Severinghaus, 1994). The model (Figure 5) helps understand how the lactic acid increases VO_2 max by about 20% before Po_2 reaches zero.

4. CAROTID BODY

4.1. Questions

Now I want to consider how the zero Po_2 hypothesis may apply to the carotid body and help understand how it works. There are several unanswered questions. What is the

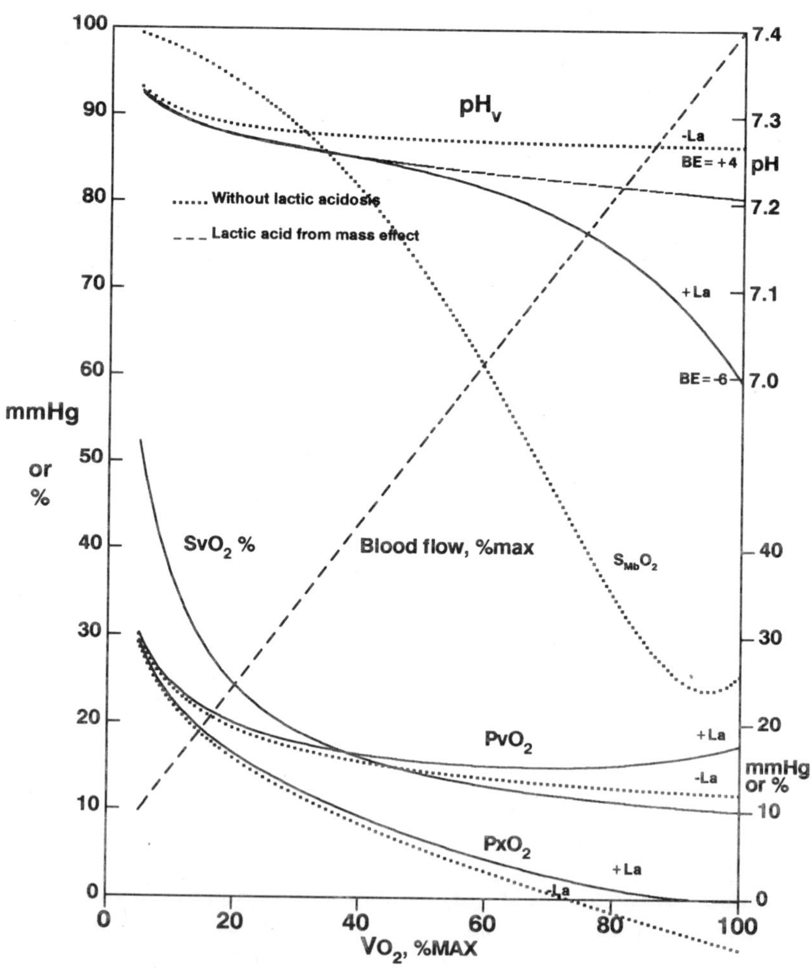

Figure 5. A model of skeletal muscle blood flow, oxygen consumption, venous Po_2, PxO_2 (at cytochtome aa_3), venous pH and myoglobin saturation as a function of work rate as % of VO_{2max}. The role of lactic acid generation in facilitating oxygen delivery is shown as dotted lines compared with solid lines where no acid excretion occurs and thus no Bohr shift of the oxyhemoglobin dissociation curve raises blood Po_2.

normal Po_2 at the sensing site in the neurone? How can the carotid body sense arterial Po_2, not oxygen content of blood or blood saturation, considering its very high metabolic rate? How can the response to hypoxia be generated as a nearly linear function of hemoglobin desaturation?

4.2. Effect of High Flow

Blood flows through sinusoids at high velocity, rather than through capillaries (McDonald, 1981). It is therefore reasonable to postulate that the very high O_2 consumption is mostly derived from a thin well mixed film of plasma near the endothelial surfaces, making it independent of the presence of hemoglobin in the blood, despite the very high metabolic rate of the sensory cells. This lack of an effect of blood O_2 content was thought to require separation of plasma from red cells by streaming. Such mechanism is not needed at high flow rate. Carotid bodies are stimulated by hypotension and reduced flow (McDonald, 1981). The stimulus due to locally injected vasodilators such as papaverine (Nims et al., 1953) suggest that shunts open which reduce flow velocity past the critical sensory endothelial surfaces.

4.3. Tissue Po_2

The very high metabolic rate of the mammalian carotid body generates a diffusion gradient within the tissue (Acker and Lübbers, 1977). A long-standing target of much research has been to determine the Po_2 in the sensory mechanism. Microelectrode studies in the carotid body have reported both very low and almost arterial values. The current value (Lahiri et al., 1994) is a mean oxygen pressure in the microvessels of the carotid body of about 50–60 torr at normoxia at which level the cellular oxygen pressures are only 20 to 30 torr. Let me assume that the normal carotid body has a gradient due to its own metabolic rate of about 30 mm Hg between blood and cytochrome. While this tissue value is similar to brain and other tissues, the reduction of Po_2 from a normal of 100 torr to 50–60 in the microvessels suggests significant blood Hb reduction. Because of the high flow, it may be possible that the endothelial surface Po_2 falls considerably lower than 50–60 due to high metabolic rate, and the observed value represents the later equilibrium as O_2Hb unloads after passing the site of consumption.

4.4. Hyperbolic Response

The carotid body's response plot is hyperbolic, with a low Po_2 asymptote of about 30 mm Hg (Figure 6). Like other highly active tissues (e.g. brain) O_2 consumption should be independent of Po_2 down to some critical threshold Po_2, below which consumption falls as a linear function proportional to Po_2. In carotid body, at the critical PaO_2 threshold of about 30 mm Hg the Po_2 at the cytochrome aa_3 may reach zero, all the delivered O_2 being consumed on arrival. With falling Pao_2, the gradient remains about 30 mm Hg until PaO_2 reaches 30, and cytochrome aa_3 reaches zero. O_2 gradients must be steep immediately around the cytochrome aa_3 metal electron donors in mitochondrial sites of high consumption. The output of the carotid body, nerve impulses, reaches a maximum when aa_3 is fully reduced. Below this level, O_2 consumption falls in proportion to PaO_2. In the actual heterogeneous situation, some of the aa_3 metal reaches zero Po_2 early, so the transition is not abrupt.

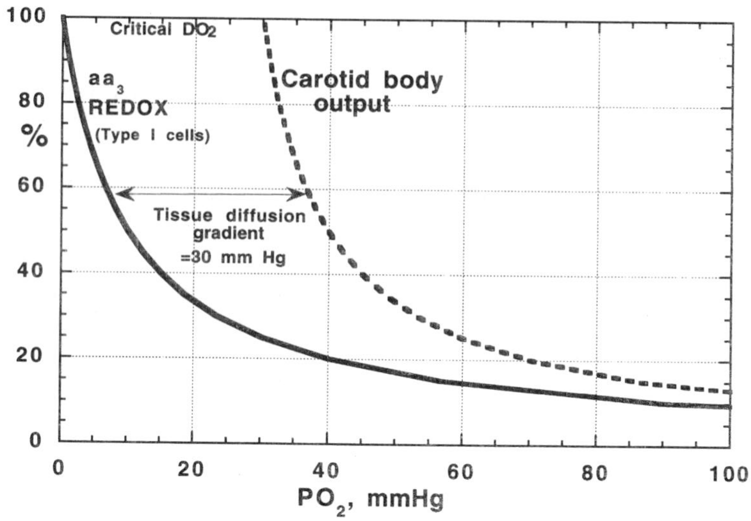

Figure 6. A hypothetic explanation of the hyperbolic relationship found between Po_2 and neural output of the carotid body. If the type I cells have a constant high metabolic rate, the diffusion gradient between endothelial surface and aa_3 is constant. As Po_2 falls, aa_3 becomes reduced, with a hyperbolic rise of its potential as shown in Figure 3, until it reaches full reduction. When PaO2 = 30 mm Hg, neural output reaches a maximum and oxygen consumption begins to fall.

4.5. Blood Oxygen Dissociation Curve

Why is the carotid body response, and the resulting ventilatory response to isocapnic hypoxia a hyperbolic function of PaO_2? The shape of the carotid body response, in this hypothesis, is hyperbolic because of the hyperbolic relationship of cytochrome redox potential to its oxygen supply as illustrated in Figure 3. Because the upper portion of the blood Hb O_2 dissociation curve is also hyperbolic, down to about 70%, the carotid body neural output and the resulting ventilatory response happen to be linear functions of SaO_2. Perhaps it will be possible to determine, using its absorption spectrum whether aa_3 in type I cells becomes fully reduced when PaO_2 reaches 30 torr.

5. BRAIN EFFECTS OF HYPOXIA (AMS, HACE)

5.1. Angiogenesis and HACE

HACE is not due to either high cerebral blood flow or lack of ATP. Some mechanism must account for the petechial hemorrhages seen in the retina. A new candidate is the process of angiogenesis (Severinghaus, 1995). As tissues grow or become hypoxic, the cells attract macrophages which express mRNA for VEGF (vascular endothelial growth factor), followed within 3–12 hours by VEGF protein, also known as VPF, (vascular permeability factor), a protein cytokine which dissolves basement membranes of capillaries, permitting the endothelial junctions to break, plasma and red cells to leak out, and later, with help of other angiogenesis factors, facilitating growth of capillary tubules into hypoxic (growing or infarcted) tissue. This first stage of angiogenesis is blocked by dexamethazone, which is used both to slow cancer tissue growth, and to prevent or treat acute mountain sickness and high altitude cerebral edema. This supports the concept that HACE is due to this VEGF attack during the first step of angiogenesis.

5.2. Brain Injury at High Altitude

Climbers who have strong hypoxic ventilatory responses breathe more deeply at altitude and thus are better oxygenated, but they have been suggested to suffer more permanent (tested at 6 mo) loss of certain subtle brain functions (number memory, finger tapping speed, etc). The hypothetic mechanism is hypocapnic cerebral vasoconstriction, not hypoxemia itself (Hornbein et al., 1989).

6. SURFACE TISSUE SENSITIVITY TO HYPOXIA

6.1. Deep Cellular Anoxia at Altitude

The observation of common nosebleeding during the first week at altitudes of 3–5 Km ft is unexplained, but commonly assumed to be due to dryness (although bleeding is not common with sojournes in desert areas). Like skin, it is probable that surface layers of nasal mucosa obtain their metabolic oxygen from the ambient air rather than from blood. There should be a vulnerable layer of deeper cells which are normally marginally supplied with oxygen, but which at altitude are too deep to get oxygen through the living more superficial layers. This deeper layer, becoming anoxic, may invoke an angiogenesis response in which VEGF causes local mucosal bleeding. To test this, we obtained nasal washings on two subjects during 4 days at 3810 m altitude, and showed that VEGF concentration rose on the first day in both subjects. It rapidly returned to normal in one, who experienced no nose bleeding, but remained elevated in the other who did have bleeding. We conclude that nose bleeds may result from angiogenic VEGF released in deeper layers of nasal mucosa at altitude.

REFERENCES

Acker H and Lübbers DW (1977) Relationship between local flow, tissue Po_2 and total flow of the cat carotid body. In Chemoreception in the Carotid Body (H Acker et al., Eds) Springer-Verlag, New York, pp. 271–276.

Hornbein TF, Townes BD, Schoene RB, Sutton JR, and Houston CS (1989) The cost to the central nervous system of climbing to extremely high altitude. N Engl J Med 321:1714–1719.

Lahiri S, Wilson DF, Iturriaga R, and Rumsey WL (1994). Microvascular PO_2 regulation and chemoreception in the cat carotid body. Advances in Experimental Medicine and Biology 345:129–135.

McDonald, DM (1981) Peripheral Chemoreceptors, In: Regulation of Respiration Part I (Hornbein TF, Ed). Lung Biology in Health and Disease, Vol 17, (Claude Lenfant, Ed). Dekker, New York, pp. 105–319.

Naqui, A, Chance B, and Cadenas E (1986) Reactive oxygen intermediates in biochemistry. Annu Rev Biochem 55:137–166.

Nims RG, Severinghaus JW, and Comroe JH Jr (1953) J Pharm Exp Therap 109:58–61.

Severinghaus JW, Bradley AF Jr (1958) Electrodes for blood Po_2 and Pco_2 determination. J Appl Physiol 13:515–520.

Severinghaus JW (1994) Exercise O_2 transport model assuming zero cytochrome Po_2 at VO_{2max}. J Appl Physiol 77:671–678.

Severinghaus JW (1995) Hypothetic roles of osmotic pressure and angiogenesis in high altitude cerebral edema. J Appl Physiol 79:375–379.

Wagner PD, Roca J, Hogan MC, Poole DC, Bebout DE, and Haab P (1990) Experimental support for the theory of diffusion limitation of maximum oxygen uptake. Adv Exp Med Biol 227:825–833.

Wasserman K (1994) Coupling of external to cellular respiration during exercise: the wisdom of the body revisited. Am J Physiol 266:E519–E539.

SENSITIVITY OF PROTEIN C IMMUNOSENSOR WITH AND WITHOUT HUMAN SERUM ALBUMIN

H. I. Balcer,[1] J. O. Spiker,[2] and K. A. Kang[3]

[1]Department of Biochemistry and Molecular Biology
[2]Department of Biological Sciences
[3]Department of Chemical and Biochemical Engineering
University of Maryland Baltimore County
Baltimore, Maryland 21250

1. INTRODUCTION

Protein C (PC) acts as an anticoagulant and antithrombotic in the hemostasis pathway. The PC deficiency can result in thrombosis and other clinical complications due to the blockage of oxygen transport. The proposed PC immunosensor is designed to diagnose a PC deficient patient prior to the onset of thrombosis and also to monitor a patient's PC level in blood plasma. The current work focuses on the effect that non-competitive inhibitors have on the immunosensor's sensitivity compared to pure PC samples in buffer solution. Human serum albumin (HSA) was chosen as the first testing protein because it represents approximately 55% of the proteins in human blood plasma.

1.1. Protein C

PC is a vitamin K dependent (VKD) zymogen produced in the liver that is critical in the regulation of hemostasis. This role is accomplished by functioning as an anticoagulant and antithrombotic. Additionally, PC assists in the regulation of thrombin complex production in blood. PC has a normal physiological concentration of 3 to 5 μg/mL (Marlar et al., 1990).

The PC deficiency can result from a variety of conditions including PC gene disorders, liver disease, and acquired risk factors including pregnancy, aging, and physical trauma (Marlar et al., 1990). Individuals with the homozygous deficiency have 0–20% of

Oxygen Transport to Tissue XXI, edited by Eke and Delpy.
Kluwer Academic / Plenum Publishers, New York, 1999.

the normal PC concentration in their blood, while heterozygous individuals have PC concentrations 40–60% of normal concentration. PC deficiency causes a thromboembolic episode, meaning that a blood clot forms and blocks blood circulation (Paleyanda et al., 1991).

When a thrombus is formed as a result of PC deficiency, the affected cells are not supplied with oxygen. The loss of oxygen can be caused by a direct blockage of the blood flow through the formation of a thrombus or by a piece of a thrombus breaking off, flowing through the blood stream, and blocking blood flow in another part of the body (Seligsohn, 1987).

1.2. Human Serum Albumin

HSA accounts for 55% of the total protein in blood plasma, with a physiological concentration of 40–50 mg/mL. HSA's primary function is to maintain osmotic pressure between blood vessels and tissue. HSA accomplishes this function through its colloidal nature, which prevents tissue from absorbing fluids as blood is circulated (Britannica Online, 1998). In this study, HSA is used to model the system after physiological conditions to determine if it affects the sensitivity of the PC biosensor. The study is necessary because blood plasma contains many other proteins. By observing the effects that HSA has on the system, it can be determined if the PC immunosensor will be effective in a hospital setting. Other immunosensor research indicates that the use of plasma negatively affects signal intensity by decreasing the signal intensity compared to pure samples (Rowe et al., 1997).

1.3. Need for Real Time Biosensor

The present method for measuring PC concentration, an enzyme linked immunosorbent assay (ELISA), is inadequate for immediate patient care. An ELISA may take as long as 6 hours to complete, requires a skilled technician to complete, and is expensive. The PC immunosensor under development requires as little as fifteen minutes to complete an assay, is simple to operate, and reusable to reduce cost. Additionally, the biosensor is portable, allowing easier access in the hospital bedside setting.

1.4. PC Biosensor

The PC immunosensor consists of three main components: the laser source and detector, the computer, and the sensor in the sample chamber (Figure 1).

The immunosensor operates on the principal of an immunoassay, similar to an ELISA. For the biosensor, the primary anti-PC monoclonal antibody (1° mAb) is bound to the surface of an optical fiber via the use of avidin-biotin binding. When the sample is injected into the sample chamber the PC binds to the 1° mAb via a strong antibody-antigen reaction. The fiber is washed with buffer solutions to remove all unbound molecules. The secondary anti-PC monoclonal antibody (2° mAb), tagged with a fluorophore which is specific to another epitope on PC, is then injected into the sample chamber to react with PC on the fiber. A laser beam is applied to the fiber to excite the fluorophores and the fluorescence is measured by the fluorometer. The amount of fluorescence is correlated to the concentration of PC [Spiker et al., 1997, Spiker et al., 1998].

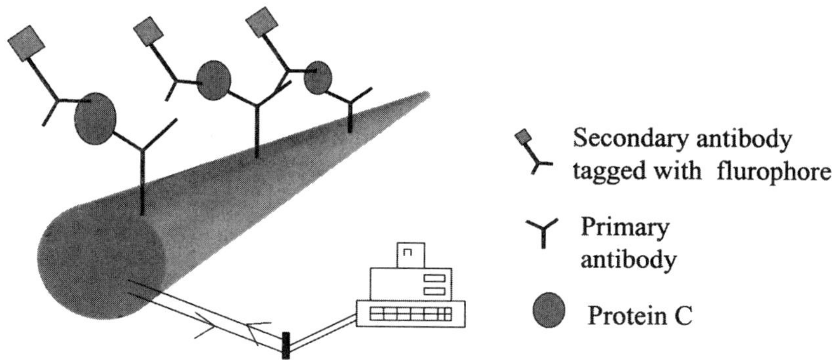

Figure 1. Diagram of the PC immunosensor. The 1° mAb is immobilized on the quartz optical fiber. The PC in the sample is bound to the 1° mAb by an antigen-antibody reaction. The 2° mAb in the solution recognizes the PC, which is bound to the 1° mAb. The laser beam passes through the fiber and the PC concentration is quantified from the signal intensity.

1.5. Fluorometer

The Analyte 2000 (Research International; Woodinville, WA) is a laser source and detector that uses a power pack running on 110 V, and a RS 232 link cable for attaching to a communications port on the computer. The multifiber harness consists of an output cable and an input cable that join at a "ST" type female fiber connector which attaches to the optical fiber used in the test chamber. The laser source is generated by a laser diode, with a peak power 5.0 mW. The signal is determined by the use of a filter that discriminates the excitation signal at 635 nm from the emission signals at 667 nm. The software package is customizable, capable of utilizing one to four fibers at the same time, allows calibration of the fibers, and adjustment of the signal intensity range to be graphed by the program. The signal received by the fluorometer is displayed in picoamperes (pA) (Yung, 1995).

2. METHODS

2.1. Monoclonal Antibody Preparations

The 1° mAb (MAB7D7, MW = 150,000) and the 2° mAb (MAB8861, MW = 150,000) are donated by the American Red Cross (Rockville, Maryland). The 1° mAb is linked to biotin (MW = 575.7; Sigma; St. Louis, MO) according to Savage, et al. (1992). The 2° mAb is linked to Cy5 (Research Organics; Cleveland, Ohio) according to the manufacturer's specifications and the linked antibodies are gel filtrated using Biogel P10 gel (Bio-Rad Laboratories; Hercules, CA) to separate the linked antibodies from the free Cy5 or biotin. The concentration of the linked mAb's is determined at 280 nm.

2.2. Optical Fiber Preparation

The quartz optical fibers were donated from Research International. The fiber is constructed of seven smaller fibers, one core fiber carrying the excitation signal and six surrounding fibers returning the emission signal to the fluorometer (Yung, 1995). The fibers, 25 cm long with a 1 mm nylon cladding, are stripped of 12.5 cm of cladding from the end. The fibers are then dipped in hydrofluoric acid to taper them for greater photon signal return. The fibers are silanated and linked to crosslinker added according to Bhatia, et al. (1989). The fibers are then placed in a phosphate buffered solution containing avidin (MW = 53,000; Sigma) at a concentration of 0.05 mg/mL for twenty-four hours. During this time avidin reacts with the crosslinker at the terminal amino acid's NH_2 group to covalently bond the avidin to the fiber. The fiber is taken from the PBS solution and carefully placed in a sample through the T-connectors and glued at both ends. The sample chamber is a 200 µL disposable micro pipet (Fisher Scientific; Pittsburgh, PA) with plastic T-connectors attached at both ends. Sample input is accomplished through syringe injection at one end with the output solution exiting via an exhaust port at the other T-connector. A fiber in a chamber is reacted with the biotinylated 1° mAb at a concentration of 65 µg/mL. The biotin, which is bound to the 1° mAb, binds with the avidin at specific sites on the protein. The reason for using avidin-biotin binding of the 1° mAb is that it prevents distortion of the three-dimensional structure of the antibody, which may occur when the 1° mAb is directly bound to the fiber (Narang et al., 1997). The 1° mAb is incubated for at least 12 hours to allow sufficient binding time to the fiber. The chamber is washed out with phosphate buffered saline solution with 0.03% azide and 0.1% Tween 20 (PBST; Sigma).

2.3. Sample Preparation

PC samples (MW = 62,000) are stored in a –70°C, and prepared for testing on the day of an experiment. The samples consist of PC, donated by the American Red Cross, in PBST. The PC is from a stock source of purified PC and is diluted for sample preparation. The HSA (MW = 66,250; Sigma) is added into the samples at various concentrations.

2.4. Measurements

All fibers are washed with PBST buffer prior to the experiments. The PC sample is injected into the sample chamber and incubated for ten minutes. The fiber is then washed out with PBST to remove any unbound biomolecules and a baseline signal is taken. The 2° mAb, at a concentration of 5µg/mL is then injected and allowed to bind to the PC for five minutes. The system is washed to remove unbound 2° mAb, and then fluorescent intensity is measured. The fluorescent intensity for each PC sample is correlated with PC concentration.

3. RESULTS AND DISCUSSION

For the experiments, the PC concentration range is 0.2 µg/mL to 2.0 µg/mL, the range of PC concentrations of heterozygous deficient patients. Previous research using pure samples detected PC concentrations accurately in this range (Spiker et al., 1998). Optical readings of pure PC samples were compared to samples containing 50 µg/mL, 1 mg/mL, and 5 mg/mL of HSA. The experimental procedure was to inject samples at increasing PC concentrations.

The results from the 50 µg/mL HSA (Figure 2) samples indicate that the sensor is capable of detecting PC concentration within the range of interest, with the HSA having a slight effect on the system's sensitivity. In the figure, the samples containing pure PC

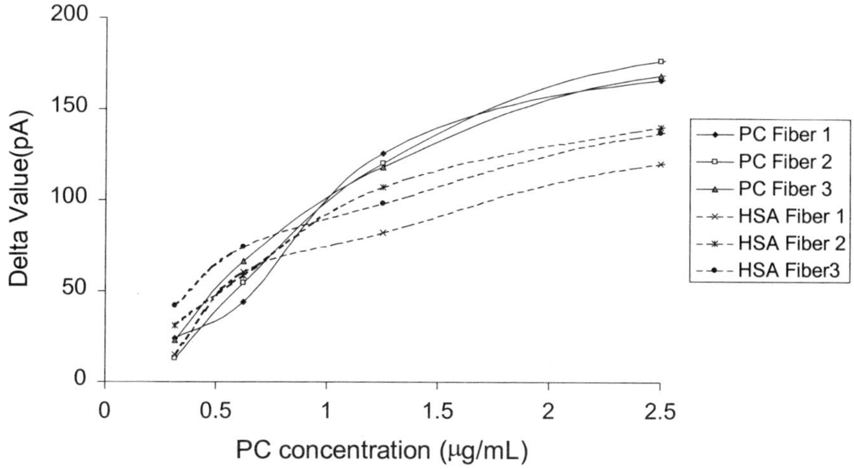

Figure 2. Comparison of optical intensities of pure PC samples to those of samples containing 50 μg/mL HSA. The signal intensities of the samples with HSA are similar to those of the pure samples, especially at lower concentrations. The data indicates that HSA at this concentration does not significantly affect the immunosensor's sensitivity.

are drawn with solid lines, and the samples containing both PC and HSA are drawn with dashed lines. The semblance of the results can be observed by the similar values between the signals at the lower PC concentrations. However, PC samples of 2 μg/mL with HSA show 15% decrease in the optical signal intensity compared to the pure PC samples. This indicates that HSA at these concentrations affects signal intensity.

PC samples containing 1 mg/mL of HSA also indicate that the immunosensor is capable of discriminating between PC concentrations (Figure 3). The signals of the samples at the lower concentrations of PC are similar to those of the pure PC samples. At the higher PC concentrations, 2 μg/mL and greater, a decrease in the signal intensities of 20–30% of the samples containing PC and HAS is observed, again suggesting that some type of interference by HSA is occurring at these levels.

When samples of PC containing 5 mg/mL of HSA are tested, the signal depression occurs across the entire range of PC concentrations (Figure 4). The signal is depressed approximately 40% at all the concentrations of PC. However, the curve shape, which is indicative of the ability of the immunosensor to differentiate between PC concentrations, suggests that in spite of the depression in the signal intensity, the immunosensor should be capable of discriminating between different PC concentrations at this level.

There are plausible reasons why the depression in the signal intensities was observed. One reason may be that the HSA could be acting as a diffusional barrier for PC binding to the 1° mAb. HSA would serve as a barrier by blocking the shortest path of the PC to the 1° mAb. With some of the paths blocked to the 1° mAb, less PC binds to the fiber during the incubation period because its rate of diffusion has been decreased. Less PC bound to the fiber allows less 2° mAb to be bound, and as a result, the signal intensity is decreased. Another possible reason may be that the HSA is non-specifically binding to the 1° mAb instead of being washed out of the sample chamber. This coating would hinder the binding of the PC to the 1° mAb and reduce signal intensity since the 2° mAb could bind to less PC. In the near future, it is intended to determine the cause of the signal loss at the higher HSA and PC concentrations.

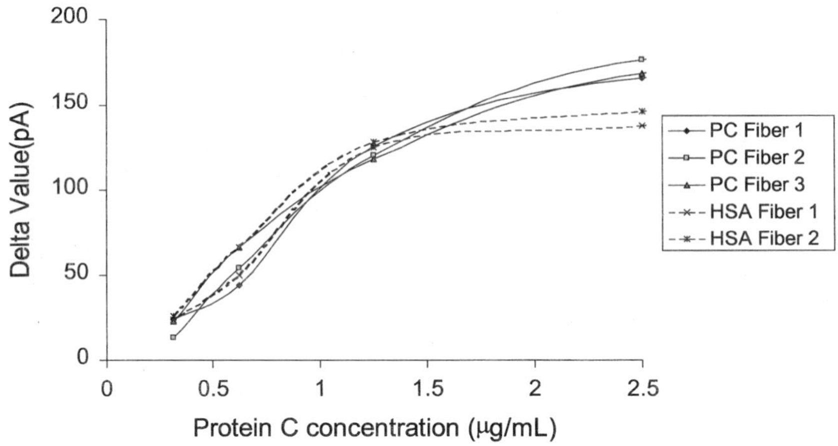

Figure 3. Comparison of optical intensities of pure PC samples to those of samples containing 1 mg/mL HSA. The signal intensities are similar to those of the pure samples at low PC concentrations, but there is a decrease in signal intensity at higher levels. This data indicates that HSA may interfere with the binding of PC at this higher concentration.

4. CONCLUSIONS

The experimental results suggest that the biosensor can measure PC concentrations in a system that has a preliminary physiological modeling. The depression of the optical signal values in the samples containing higher concentrations of HSA suggests that the HSA may be affecting the signal intensity. This effect may be the result of a diffusional barrier effect or a coating of the fiber with HSA may be causing the decreased signal intensity. This data implies that when blood plasma samples are tested the signal inten-

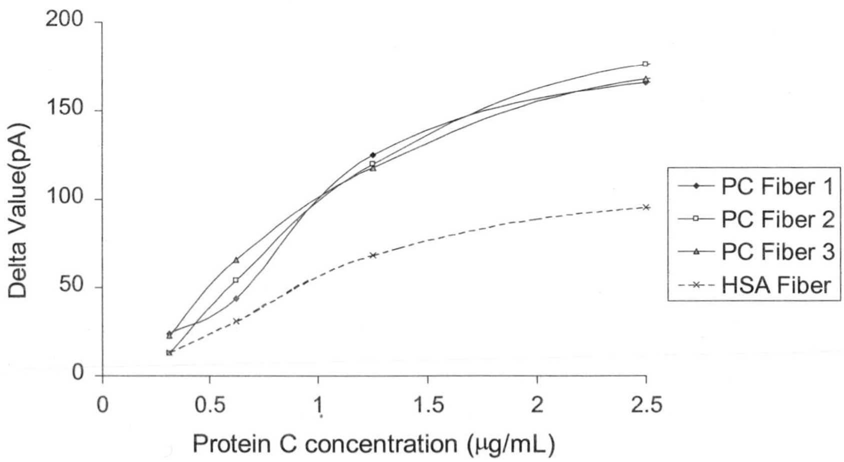

Figure 4. Comparison of optical intensities of pure PC samples to those of samples containing 5 mg/mL HSA. The signal intensities are depressed at all levels compared to the pure samples. This data indicates that HSA at this concentration may be interfering with the PC antibody-antigen reaction.

sities should be less than those of the samples containing only PC. The ability to use the immunosensor without altering the experimental protocols during the experiments will allow its continued use as a real time device.

5. FUTURE WORK

Samples containing HSA at its physiological concentration, 50 mg/mL, will be tested. The concentration of Tween 20, a detergent, will also be increased from 0.1% to 1–2% in order to reduce possible non-specific binding. A recirculation pump may be included in the system in order to increase the chance of the PC binding to the 1° mAb. Later the system will be tested using whole plasma samples. In the future, the immunosensor will also be automated to allow easier operation.

ACKNOWLEDGMENTS

We would like to thank the American Red Cross of Rockville, Maryland, The Whitaker Foundation, and the National Science Foundation (grant BES-9733207), for their financial and material support. We would also like to thank the Naval Research Laboratory in Washington, D.C. and Research International Inc. for their technical assistance.

REFERENCES

Bhatia S, Shriver-Lake L, Prior K, Geoger J, Calvert J, Bredehorst R, and Ligler F (1989) "Use of Thiol-Terminal and Hetrobifunctinoal Crosslinkers for Immobilization of Antibodies on Silica Surfaces." *Anal Biochem*. 178:408–413.

"serum albumin." *Britannica Online.* www.eb.com:180/cgi-bin/g?DocF=micro/538/40.html [Accessed 10 August 1998].

Marlar R and Adcock D (1990) "Protein C Replacement Therapy as a Treatment Modality for Homozygous Protein C Deficiency." In: *Protein C and Related Anti-Coagulants.* (Bruley D and Drohan W Eds) Gulf Publishing Company, Houston, pp. 165–178.

Narnag U, Anderson G, Ligler F, and Buans J (1997) "Fiber optic-based biosensor for Ricin." *Biosen and Bioelec* 12:937–945.

Paleyanda R, et al. (1991) "The Expression of Therapeutic Proteins in Transgenic Animals." In: *Recombinant Technology in Hemostasis and Thrombosis*. (Hoyer L and Drohan W Eds) Plenum Press, New York.

Rowe C, Bolitho J, Jane A, Bundesen P, Rylatt D, Eisenberg P, and Ligler F (1997) "Rapid Detection of D-dimer Using a Fiber Optic Biosensor," *Throm and Haemost* 79:94–98.

Savage D, Mattson G, Desai S, Nielander G, Morgensen S, and Conklin E (1992) *Avidin-Biotin Chemistry: A Handbook.* Pierce, Rockford.

Seligsohn U (1987) *Report at the ISTH Subcommittee Meeting*, Brussels, July.

Spiker J, Kang K, Drohan W, and Bruley DF (1997) "Protein C Detection Via Fluorophore Mediated Immuno-Optical Biosensor." ISOTT XIX, *Advances in Experimental Medicine and Biology.* 428:621–627.

Spiker J, Kang K, Drohan W, and Bruley D (1998) "Preliminary Study of Biosensor Optimization for the Detection of Protein C." ISOTT XX, *Advances in Experimental Medicine and Biology,* Plenum, New York, 1998.

Yung C (1995) *Analyte 2000 Fiber Optic Multianalyte Sensor System Instruction Manual*, Research International Inc., Woodinville, Washington.

71

SIMULTANEOUS MEASUREMENTS OF MECHANICAL PRESSURE AND OXYGEN PRESSURE DURING PUNCTURING OF SEMISOLID POLYMERIC NETWORKS BY POLAROGRAPHIC NEEDLE ELECTRODES

H. Baumgärtl, W. Zimelka, and D. W. Lübbers

Max-Planck-Institut für Molekulare Physiologie
Rheinlanddamm 201, 44139 Dortmund, Germany

1. INTRODUCTION

Many investigations exist about the electrochemical properties of polarographic pO_2 needle electrodes and the technical difficulties when using these electrodes in biological objects, but only a few are dealing with the influence of mechanical pressure on the measuring signal during the insertion of the electrode into a solid or semisolid medium as, for example, biological tissue or some polymeric matrices of heterogeneous structure. These investigations, however, demonstrate clearly that the measuring signal can be much influenced by puncturing technique, geometry of the electrode and mechanical properties of the medium (Schuchardt and Lösse, 1973; Baumgärtl et al., 1974; Schäfer and Höper, 1976a, 1976b; Klinowski et al., 1982; Freund, 1983; Himmelberg, 1983; Baumgärtl et al., 1987; Beunink et al., 1989; Schramm et al., 1990; Wagner et al., 1992; Müller et al., 1994; Baumgärtl et al., 1997). In such media the measured polarographic signal often does not mirror the "true" oxygen pressure, but is determined by artefacts. According to our knowledge systematic investigations of the influence of mechanical pressure on the polarographic signal have not been performed up to now. Since the analysis of the applied mechanical pressure may help to explain the artefacts, in this study mechanical pressure and polarographic signal were measured simultaneously during the insertion of the electrode into solidified agar layers. The experiments showed that in dependency of the consistency of the medium a specific interaction exists between the exerted mechanical pressure and the actual measured signal. The results allowed us to work out a special technique of stepwise puncturing. By properly choosing path-length

Oxygen Transport to Tissue XXI, edited by Eke and Delpy.
Kluwer Academic / Plenum Publishers, New York, 1999.

of the step, velocity of penetration during the step and frequency of the steps it was possible to minimize the artefacts.

2. MATERIALS AND METHODS

2.1. Preparation of Agar Layers

Solutions of 0.2%, 1.0%, 2% and 5% of Agar were prepared using as solvents distilled water, Theorell-buffer, physiological saline or 1% KCl solution. The necessary Agar concentrations (Agar, highly purified, Merck art.1613) were transferred to the boiling solvent and continuously stirred until a clear solution was obtained without streaks. The solution was then filled in Petri dishes so that a ca. 15–20 mm thick layer was formed. After cooling down and complete temperature equalization (ca. 22°C) the solidified layer was covered with a thin film of the particular solvent.

To study the structure of the different Agar preparations (1%, 2% and 5%) ca. 1 cm^3 blocks were cut. These blocks were frozen at –40°C on the object stage of the freezing microtome Frigocut Mod.2700 (Reichert-Jung, 69226 Nusslach, Germany) using "Tissue Freezing Medium" (Electron Microscopy Science, Washington, USA). Because of the different consistencies of the Agar preparations different cutting temperatures and slice thicknesses had to be applied. 5% Agar was cut at –2°C to –5°C (slice thickness 10–30 μm) and 1% or 2% Agar at –20°C (slice thickness 15–25 μm). Some of the slices were stained by Hematoxilin or Toluidin Blue O (Fa. Merck, Darmstadt, Germany). The slices were placed on a slide and analysed by light microscopy (Axiophot, Fa. Zeiss, Germany). For photographs PAN-F-Film was used.

2.2. Polarographic pO_2 Needle Electrodes

The pO_2 measurements were performed using the membrane-covered microcoaxial needle electrode according to Baumgärtl and Lübbers (1983). In this type of electrode the cathode is formed by an electrochemically etched platinum wire which is fused in a capillary of a properly selected glass. At the tip of the electrode there is a recess the length of which is adapted to the size of the diameter of the platinum cathode. A thin layer of Ta-Pt-Ag/AgCl sputtered on the outer surface of the glass capillary forms the reference electrode and serves simultaneously as electric shielding. This construction has the advantage that measuring electrode (cathode) and reference electrode (anode) can be covered by the same membrane to separate the measuring circuit from the medium to be measured (Clark principle). A double membrane of collodion and polystyrene was used.

The electrodes applied for puncturing were photographed before sputtering and membranization. The photographs were used to quantify their geometry by determining the diameter at several distances from the tip. The geometrical data are used to estimate the forces applied during puncturing. 9 electrodes were used in the experiments. The mean diameter of the tip of the 9 electrodes used in our experiments was ca. 3 ± 2 μm. It increased at a distance of 3000 μm to 247 ± 54 μm corresponding to a $\tan\alpha$ of $3.2° \pm 1.3°$ over this length. The mean diameter of the platinum cathode within the recess was 4.5 ± 1.8 μm and the relation "length of the recess to diameter of the cathode" 2.1 ± 0.4. To calculate volume and surface of the shaft it is assumed that the shaft resembles a cone.

The electrodes were calibrated in the different solvents. The solvents were equilibrated with 5 different gas mixtures (0–25% oxygen in nitrogen) produced from highly purified oxygen and nitrogen by a precision gas mixing pump (Fa. Wösthoff, Bochum, Germany). The mean reduction current in air-equilibrated buffer solution (Theorell-buffer, pH = 7.2, t = 21.5°C) amounted to 0.11 nA/145 Torr. The corresponding "zero" current (nitrogen) was 0.0018 nA, i.e. 61 times smaller than the just mentioned reduction current. The response time t_{90} of the electrode varied between 25 and 250 ms dependent on the length of the recess and the properties of the double membrane. The "stirring" effect—measured as signal difference between a medium stirred by gas bubbles and without bubbling—was 1.2%. Temperature sensitivity was ca. 2.4%/°C between 18–25°C. The optimum polarization voltage varied in different media: in Agar layers prepared with saline or KCl-solutions the optimum voltage range was 600–750 mV, with Theorell-buffer it changed to 700–850 mV. During a measuring period of 8.8 hours the drift of the polarographic signal amounted to only 0.4%/hour. Before the experiments the Agar layer was covered with a thin layer of the solvent and equilibrated with air at room temperature. To compare the different experiments the pO_2 signal measured without pressure directly on the surface of the Agar layer was taken as 100%. Because of the smaller oxygen conductivity of the Agar layer this signal was about 1.5 times smaller than the signal in buffer solution.

The signal was recorded on a 3 channel graphic recorder (Servogor 330, Fa Servogor, Nürnberg, Germany). The response time of the whole system was between 0.5–1.0 s. The values were also registered on a computer and represented by a graphic program.

2.3. Measurement of Mechanical Pressure

The mechanical pressure during the insertion of the needle electrode was monitored by fixing the Petri dish (diameter ca. 50 mm) with the agar layer on the scale of an electronic balance (Sartorius Handy H110, Fa. Sartorius, Göttingen, Germany) the signal of which was continuously recorded on the graphic recorder. The weight signal corresponds to the forces developed during the puncture (Himmelberg, 1983). The total force, F_t, is composed of three parts:

1. The force F_p which is exerted by the tip to overcome the resistance of the medium. The corresponding pressure at the tip, p, is given by $p = F_p \cdot (\pi r^2)$ with r = radius of the tip.
2. The force of the wedge-shaped shaft of the electrode, $F_n \cdot \sin\alpha$
3. The frictional force $F_n \cdot \mu \cdot \cos\alpha$ with μ = coefficient of friction.

$$F_t = F_p + F_n \cdot \sin\alpha + F_n \cdot \mu \cdot \cos\alpha \qquad (1)$$

with F_n = force in the direction of the puncture, α = angle of the shaft of the electrode.

2.4. Puncturing Technique

The insertion of the electrode was carried out by a puncturing apparatus (Nanostepper, Bachhofer Reutlingen, Germany) which allowed short and quick perpendicular steps. A single step has a length of 0.125 µm and lasts 20 µs. The frequency of the steps can be varied between 1 and 25 kHz; so e.g. a distance of 100 µm corresponds to 800 single steps. They last 0.8 s at a frequency of 1 kHz and only 0.032 s at 25 kHz.

Dependent on the step frequency the speed of insertion varies between 125 μm/s (1 kHz) and 3125 μm (25 kHz). In our experiments pathlengths of 10, 25 and 100 μm with low (1 kHz) and high frequencies (25 kHz) were used. To position the electrode the Nanostepper was mounted on a micromanipulator (Fa. Bachofer, Reutlingen, Germany). The punctures were microscopically observed.

3. RESULTS

3.1. Structure of Solidified Agar and Changes of Mechanical Pressure during Puncturing

Microscopic investigations showed that during solidification 0.2% Agar formed only small single polymeric fibers separated by relatively large distances. Fibers forming polymeric networks are obtained with larger concentrations. In 1% and 2% Agar a continuous network of polymeric fibers was formed. Figure 1 shows, as an example, the microscopic structure of 2% Agar (H_2O) and the damage caused by the electrode. The polymeric structure varies somewhat from preparation to preparation and is also influenced by the solvent. In 5% Agar a fibrous structure could not be anymore observed, but thick, amorphic aggregations appeared. Table 1 gives the balance output during puncturing layers of different Agar concentrations. The balance output, for example, amounts at a depth of 1000 μm to 5 ± 3 mg in 0.2%, to 38 ± 20 mg in 1%, to 79 ± 32 mg in 2% and to 256 ± 105 mg in 5% Agar. All measured values have a relatively large standard deviation demonstrating the variability of the actual polymeric structure. Furthermore the output increases with the depth of insertion: for example in 2% Agar from 79 ± 32 mg at 1000 μm to 149 ± 71 mg at 2000 μm and 209 ± 95 mg at 3000 μm. This should be expected since diameters and volume of the shaft of the electrode increase with increased distances from the tip.

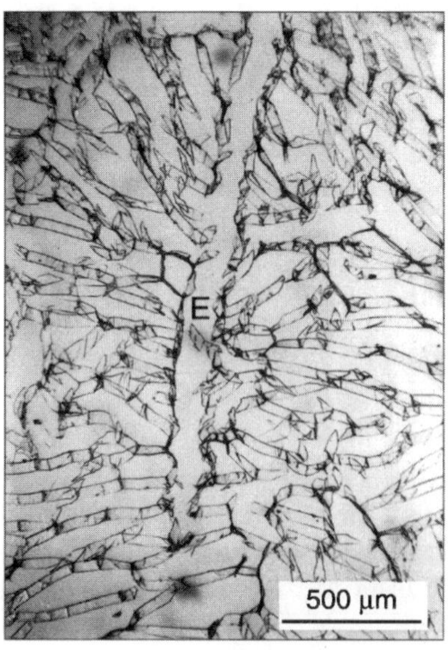

Figure 1. Fibrous structure of solidified 2% Agar (H_2O). Frozen section (−10°C). Mean thickness of the fibers 35 μm. E: puncturing channel.

Table 1. Balance output during insertion of the electrode into Agar of different concentrations depending on the puncturing depth. Puncturing technique: step length 1×100 μm, step frequency 25 kHz, 10 s interval

Depth of puncture μm	0.2% (n=10)	1% (n=7)	2% (n=8)	5% (n=10)
100	1 ± 0.8	4 ± 3	6 ± 3	10 ± 2
500	3.5 ± 2.5	26 ± 18	40 ± 16	125 ± 63
1000	5 ± 3	38 ± 20	79 ± 32	256 ± 105
1500	7.5 ± 5.5	47 ± 13	118 ± 53	420 ± 146
2000	9.5 ± 8	65 ± 27	149 ± 71	524 ± 223
2500	14 ± 12	83 ± 35	172 ± 88	624 ± 70
3000	17 ± 14	108 ± 48	209 ± 95	743 ± 53
3500	19.5 ± 15	153 ± 62	326 ± 140	902 ± 209
4000	24 ± 19	206 ± 70	401 ± 216	1185 ± 426

3.2. The Behaviour of the Weight Signal and the Simultaneously Measured pO_2 Signal during Puncturing Agar Preparations of Different Concentrations

Figure 2 shows a part of an original record of puncturing a 2% Agar layer (Theorell buffer) by the described membrane covered needle electrode starting at a depth of 0.89 mm. The Agar layer was covered by air saturated water. Puncturing was carried out

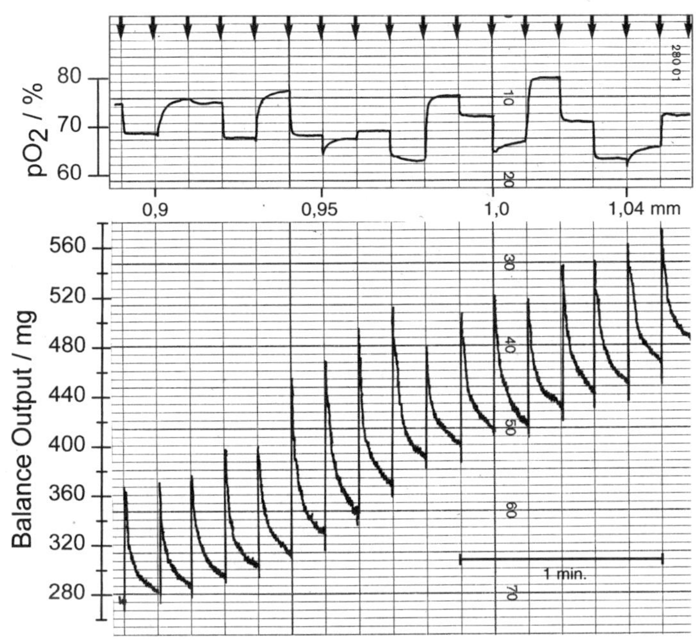

Figure 2. Part of an original record of puncturing solidified 2% Agar. Upper trace: pO_2 sinal, lower trace: balance output. Puncturing technique: step length 10 μm, step frequency 25 kHz (i.e. 80 steps of 0.125 μm in 0.003 s) without pressure release, 10 s interval.

by steps of a length of 10 μm (25 kHz) followed by a stop of 10 s. The balance signal (lower trace) increases steeply during insertion and decreases more slowly during the stop. The amplitudes vary between 71.3 and 119.4 mg (mean value 91.5 ± 15.4 mg). The lowest values increase because of the increasing electrode volume, in this case from 275.0 mg to 487.5 mg. The pO_2 signal (upper trace) shows quite a variable picture: it may decrease, increase or remain similar. Quick or slower changes with distinctly different amplitudes are observed. This shows that the insertion of the electrode can change the polymeric structure of Agar in a very variable way producing different local diffusion conditions for oxygen.

In the following examples it will be shown in which way increasing Agar-Agar concentrations influence the puncturing force and the resulting pO_2 signal. To demonstrate the behaviour of the signals over longer distances in the following figures the pO_2 values measured before the insertion are directly connected omitting the other values. The amplitude of the corresponding balance signal is given as a perpendicular line.

In 1% Agar-Agar (saline) during stepwise puncturing (step length 25 μm, step frequency 25 kHz without pressure release, 10 s interval) characteristic pO_2 peaks are formed in about similar distances (Figure 3). With increasing depth the pO_2 signal decreases steadily reaching 82% at 1000 μm and 41% at 3000 μm of its initial value. The balance output increases from 20.1 mg at 1000 μm to 98.6 at 3000 μm. This increase follows about the data given by the geometry of the electrode (Figure 3 upper part).

In 2% Agar-Agar (Theorell buffer) a larger pressure increase was found. The mean pO_2 signal decreased only from 80% to 70% with a mean value of 77.2 ± 9.4%, but the variability was large; for example during the first 4 punctures the actual pO_2 signal decreased to 46% of its initial value of 145 Torr (= 100%). During further insertion (up to 1100 μm) 32 distinct pO_2 peaks of variable amplitudes in distances of ca 34 μm were formed. Withdrawing the electrode gave a homogeneous pO_2 signal of 94% without any peaks.

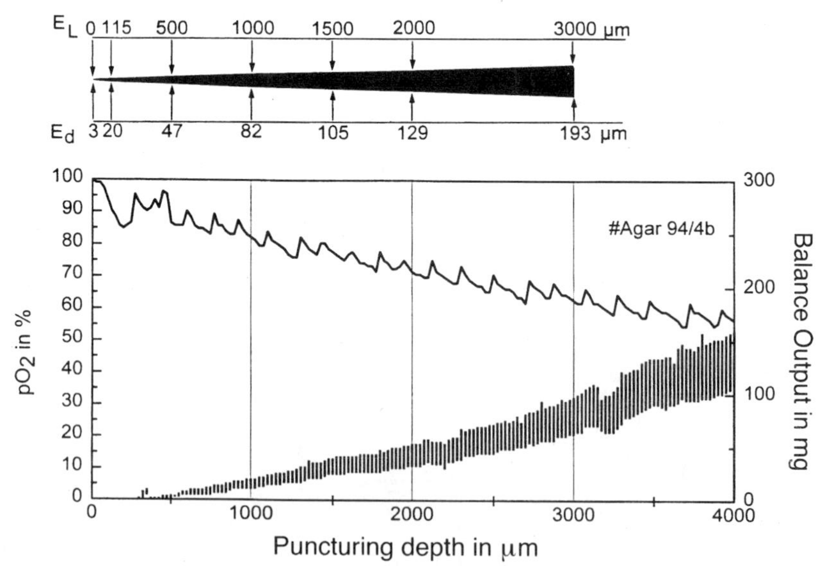

Figure 3. Stepwise insertion of a pO_2 microelectrode into a solidified layer of 1% Agar. Upper part. Geometry of the electrode (E_L length in μm, E_D diameter in μm). Lower part: upper trace = polarographic pO_2 signal, lower trace = balance output.

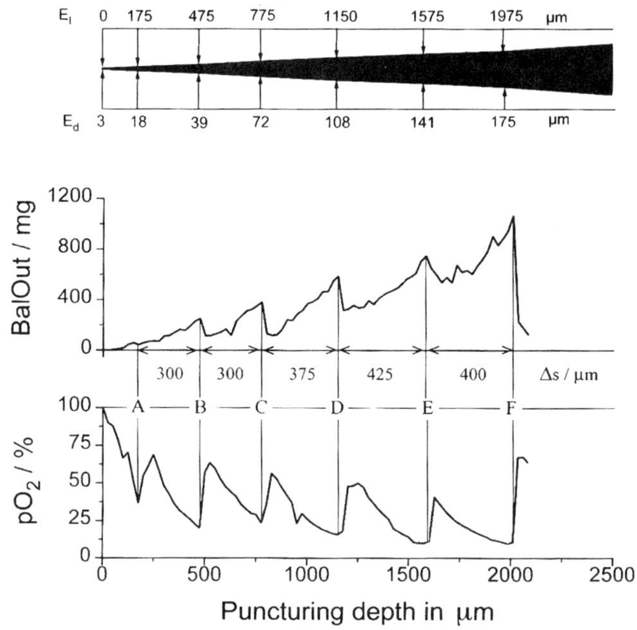

Figure 4. Stepwise insertion of a pO_2 microelectrode into a solidified layer of 5% Agar. Upper part: geometry of the electrode (E_l length in µm, E_d diameter in µm). Lower part: upper trace = balance output., lower trace = polarographic pO_2 signal.

Figure 4 shows the puncturing of a 5% Agar layer (saline). During insertion (step length 25 µm, step frequency 1 kHz without pressure release, 20 s interval) balance output increases steadily (5 mg/10 µm) with small variations and the pO_2 signal decreases, but suddenly at a depth of 475 µm a steep decrease of the balance output occurs (252 mg to 113 mg). During this decrease a large increase of the pO_2 signal from 20% to 64% of its initial value is observed. During further insertion these changes repeat (B–F) in similar distances (ca. 300 to 425 µm, mean value 360 ± 51 µm); the mean value of the pressure decrease of 221 ± 51 mg (B–F) is accompanied by a mean increase of the pO_2 signal of 35 ± 5% (51 ± 7 Torr).

By using almost the same puncturing technique (step length 25 µm, step frequency 1 kHz without pressure release, 10 s interval) in a 0.2% Agar layer (saline) quite different pressure and pO_2 traces are obtained (Figure 5). During the first 31 steps (775 µm) balance output remains close to zero with small constant amplitudes, then it increases stepwise to a higher,but rather small value (0 to 4 mg); the amplitudes remain similar and constant. The stepwise change repeats at 1400 µm, at 2725 µm and at 3700 µm with a similar increase of the balance signal. The pO_2 signal starts to react at a depth of 350 µm and decreases almost continuously reaching 95% at 1000 µm and 75% at 4000 µm; steps in the traces as in Figures 2–4 were not observed.

3.3. Influence of Puncturing Technique

To investigate the influence of the puncturing technique solidified 2% Agar (H_2O) was punctured by 2 × 100 µm steps forward (25 kHz) immediately followed by a 100 µm step backward to release the pressure and by an interval of 10 s (Figure 6). These 3 steps

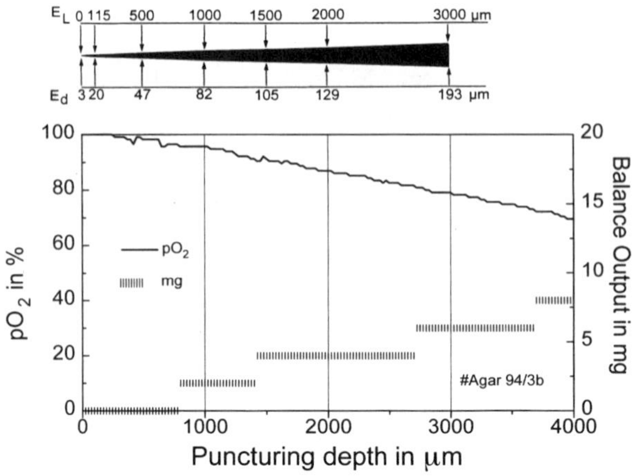

Figure 5. Stepwise insertion of a pO_2 microelectrode into a solidified layer of 0.2% Agar. Upper part: geometry of the electrode (E_L length in µm, E_D diameter in µm). Lower part: upper trace = pO_2 signal, lower trace = balance output.

are carried out in ca. 1 s. The trace of the balance output demonstrates that during the large steps weight increases distinctly, but that after pressure release weight becomes over a long distance always zero. The changes are so short that the influence on the pO_2 is minimized: there are only short changes during the insertion which are canceled during the pressure release (Figure 6 insert). There is still a decrease of the pO_2 signal, but no peaks are formed. During the withdrawal of the electrode the 100% value was always recorded in the puncturing channel. It was found that the best results were obtained by using relatively large 2 (or more) steps forwards with high insertion speed immediately followed by a step backward to release the pressure.

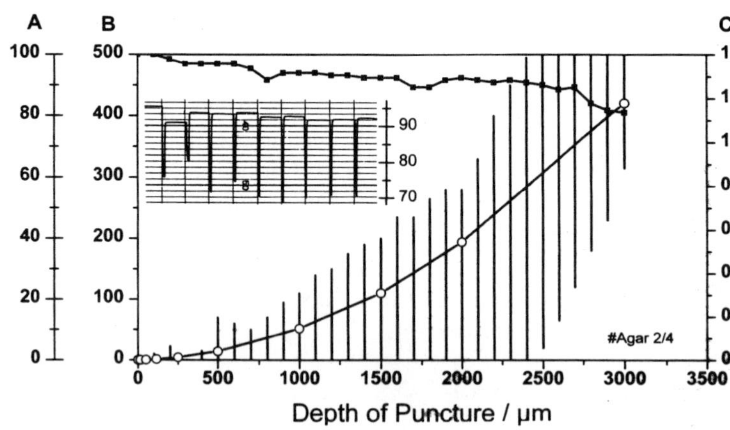

Figure 6. Stepwise puncturing of a solidified layer of 2% Agar with pressure release using a pO_2 microelectrode. Upper trace: polarographic pO_2 signal (scale A). Perpendicular lines: amplitudes of the balance output (scale B). Lower trace: changes of the surface of the electrode tip given in mm^2 (scale C). Insert see text.

4. DISCUSSION

Already in 1973 Schuchardt and Lösse reported that during puncturing of a 3% air equilibrated Agar layer with a relatively large, glass insulated, recessed platinum electrode (50 μm tip dameter) using steps of 500 μm variable pO_2 signals were observed, mostly a signal decrease, although the pO_2 in the air saturated Agar was constant. The variability of the signal could be reduced by inserting the electrode by smaller steps of 10 μm. Similar observations were reported by Klinowski et al., (1982). They found that during insertion of a recessed needle electrode into an air saturated, gelled Agar layer (0.9% NaCl) the pO_2 signal decreased in an almost exponentional way if the Agar concentration was larger than 0.5%. The steepness of the decrease increased with increasing Agar concentrations. Crawford and Cole (1985) showed that such artefacts could be minimized by using needle electrodes with a proper, membrane-filled recess producing a small reduction current. The possible reason for such concentration dependent artefacts and their variability was found by Baumgärtl et al. in 1974. They observed that during continuous puncturing in concentrated Agar solutions undissolved, microscopically visible particles were accumulating in front of the tip of the needle electrode (tip size 1–3 μm). During accumulation the pO_2 signal decreased sharply, but when the tip penetrated the accumulated particles, the pO_2 signal increased suddenly to the initial recorded pO_2 of the solution. It is well documented that in such solutions filamentous structures exist (see Rees and Welsh, 1977; Bezrukov, 1979; Martinsen et al., 1989). The accumulation depended on the concentration of Agar and the length of the step; it could be strongly reduced by stepwise puncturing. In solidified polymeric structures and in biological material, however, even the small tip of microelectrodes can form polymeric barriers during continuous as well as stepwise puncturing and thereby falsify the pO_2 measurements by changing the oxygen conductivity of the medium (Grunewald, 1970; Baumgärtl et al., 1997). Our experiments show that there is a strong coupling between pressure measured by the balance output and the pO_2 signal. During insertion the Agar layer was compressed and polymeric material was accumulated in front of the tip of the electrode. Therefore, the actual pressure necessary to penetrate the Agar layer increased during the insertion, but, unfortunately, the actual response of the polarographic signal could not be predicted. These reactions can be explained by the varying structures of the Agar gel, but they are also influenced, as already mentioned, by the puncturing technique (see Baumgärtl et al., 1997). Our investigations showed that in layers of solidified Agar the best results were obtained

1. with relatively large steps,
2. with high insertion speed and
3. with 2 (or more) steps forward immediately followed by a step backward to release the pressure.

Since the optimal technique depends on the structure of the medium to be measured the puncturing technique must always be adapted to the actual measuring situation.

REFERENCES

Baumgärtl, H: Systematic investigations of needle electrode properties in polarographic measurements of local tissue pO_2. In: AM Ehrly, J Hauss, and R Huch (eds.): Clinical Oxygen Pressure Measurement. Springer-Verlag Heidelberg, New York, London, Paris, Tokyo (1987) pp. 17–42.

Baumgärtl, H, W Grunewald, and DW Lübbers: Polarographic determination of the oxygen partial pressure field by Pt microelectrodes using the O_2 field in front of a Pt macroelectrode as a model. Pflüg. Arch. 347:49–61 (1974).

Baumgärtl, H and DW Lübbers: Microcoaxial needle sensor for polarographic measurement of local O_2 pressure in the cellular range of living tissue. Its construction and properties. In: E Gnaiger and H Forstner (eds.): Polarographic Oxygen Sensors, Springer-Verlag, Berlin/Heidelberg (1983) pp. 37–65.

Baumgärtl, H, W Zimelka, and DW Lübbers: Effects of puncturing on the measurement of local oxygen pressure using polarographic microelectrodes. Adv. Exp. Med. Biol. 411:527–541 (1997).

Beunink, J, H Baumgärtl, W Zimelka, and H-J Rehm: Determination of oxygen gradients in single Ca-alginate beads by means of oxygen-microelectrodes. Experientia 45:1041–1047 (1989).

Bezrukov, MG: Die Bildung der Raumstruktur von Proteingelen. Angew. Chem. 91:634–646 (1979).

Crawford, DW and MA Cole: Performance evaluation of recessed miocrocarthodes: criteria for tissue pO_2 measurement. J. Appl. Physiol. 58(4):1400–1405 (1985).

Freund, W: Morphologische Veränderungen bei Messungen mit Einstichelektroden in der Rattenniere, Dissertation Univ. Kiel (1983).

Himmelberg, H: Mechanisches Verhalten von Mikrostichelektroden zur Sauerstoffdruck-messung in Muskelgewebe. Dissertation Univ. Münster (1983).

Grunewald, W: Diffusionsfehler und Eigenverbrauch der Pt-Elektrode bei pO_2-Messungen im steady state. Plüg. Arch. 322:24–44 (1970).

Klinowski, J, SW Korsner, and CP Winlove: Problems associated with the micropolarographic measurement of the arterial wall pO_2. Cardiovasc. Res. 16:448–456 (1982).

Martinsen, G, G Skjak-Braek, and O Smidsrod: Alginate as Immobilization Material: I. Correlation between Chemical and Physical Properties of Alginate Gel Beads. Biotechnol. Bioeng. 33:79–89 (1989).

Müller, W, A Winnefeld, O Kohls, T Scheper, W Zimelka, and H Baumgärtl: Real and pseudo oxygen gradients in Ca-alginate beads monitored during polarographic pO_2 measurements using Pt-needle microelectrodes. Biotechnol. Bioeng. 44:617–625 (1994).

Rees, DA and EJ Welsh: Sekundär- und Tertiärstruktur von Polysacchariden in Lösungen und in Gelen. Angew. Chem. 89:228–239 (1977).

Schäfer, D and J Höper: The influence of glass needle electrodes on rat liver cells and tissue. Sixth European Congress on Electron Microscopy, Jerusalem 1976a, pp. 304–305 .

Schäfer, D and J. Höper: Alterations in rat liver cells and tissue caused by needle electrodes. In: M Kessler, LC Clark Jr, DW Lübbers, I Silver, and W Simon (eds.): Ion and Enzyme Electrodes in Biology and Medicine. Urban and Schwarzenberg, München (1976b) pp. 217–222.

Schramm, U, W Fleckenstein, and C Weber: Morphological Assessment of Skeletal Muscular Injury Caused by pO_2 Measurements with Hypodermic Needle Probes. In: AM Ehrly et al. (eds): Clinical Oxygen Pressure Measurement. Blackwell Ueberreuter Wissenschaft Berlin (1990) pp. 38–50.

Schuchhardt, S and B Lösse: Methodological problems when measuring with pO_2-needle electrodes in semi-solid media. In: M Kessler, DF Bruley, LC Clark, DW Lübbers, IA Silver, and J Strauss (eds.): Oxygen Supply. Urban & Schwarzenberg, München (1973) pp. 108–109.

Wagner, K, W Bossen, and U Schramm: Tissue alterations by the penetration of a pO_2 sensing needle probe. Adv. Exp. Med. Biol. 317:639–644 (1992).

A NEW COMBINED DEEP-BODY-TEMPERATURE/NIRS PROBE FOR NON-INVASIVE METABOLIC MEASUREMENTS ON HUMAN SKELETAL MUSCLE

T. Binzoni,[1] R. Springett,[2] J. C. P. Dalton,[3] and D. Delpy[2]

[1]Departments of Physiology and Radiology
University of Geneva
Switzerland
[2]Department of Medical Physics and Bioengineering
University College London
United Kingdom
[3]Deep Body Thermometers Ltd.
Little Eversden, Cambridge
United Kingdom

1. INTRODUCTION

Temperature modulates tissue metabolism through thermodynamic phenomena, e.g. changes in enzymatic activities, pH variations, etc. In humans, temperature changes may be induced by stressful environmental conditions (e.g. immersion in cold water, etc.) or, artificially, during therapeutic procedures (extracorporeal circulation, cancer treatment using hyperthermia, etc.). Moreover, temperature distribution is not uniform in the body. In a forearm immersed in cold water at 20°C, it is possible to observe radial gradient values reaching 2°C/cm (Ducharme et al., 1991). Thus, the measurement of total tissue metabolism often results from the contribution of several subsets of tissue exposed to different temperatures, making the interpretation of the data sometimes difficult. In this context, the aim of the present study was to develop a non-in-vasive probe allowing one to simultaneously: 1) establish in a human skeletal muscle a region of uniform temperature; 2) measure the temperature; 3) measure the oxidative metabolism in the corresponding region.

Oxygen Transport to Tissue XXI, edited by Eke and Delpy.
Kluwer Academic / Plenum Publishers, New York, 1999.

2. METHODS

The probe was built using a modified Deep Body Temperature (DBT) sensor (Deep Body Thermometers Ltd., Little Eversden, Cambridge, UK). This sensor (shown in Figure 1), is attached to the surface of the tissue, and contains two temperature sensors and a heating element via which it is possible to establish zero heat flux through the underlying tissue. When this zero heat flux is established, the surface and the tissue core temperature are equal (Solman et al., 1973). The probe was modified to include two slots in the silicone rubber body moulding into which two optical fibres with end mounted prisms could be mounted (fibre separation 3 cm). These fibres were attached to a white light source (Oriel 77501, Stratford CT, USA) and a CCD based spectrophotometer (Cope et al., 1989) from which it was possible to measure the near infrared (NIR) spectrum of the tissue under the probe. An appropriate heat insulator was installed on the back of the DBT probe to permit its use when both it and the tissue were immersed in cold water.

This combined DBT/NIR probe enabled a uniform temperature region of approximately 3 × 3 × 1.5 (depth) cm to be established and changes in oxyhaemoglobin [HbO_2] and deoxyhaemoglobin [Hb] concentration in the tissue measured using an optical pathlength derived from the tissue water spectrum measured at the start of the study and assuming a tissue water content of 62% (Matcher et al., 1993). Preliminary measurements were performed with the probe fixed on the forearm of a subject (over the hand flexors). Muscle temperature was changed using a thermostated bath, and [Hb] and [HbO_2] changes were followed with a time resolution of 2 s i.e., 1.5 s for the acquisition of one

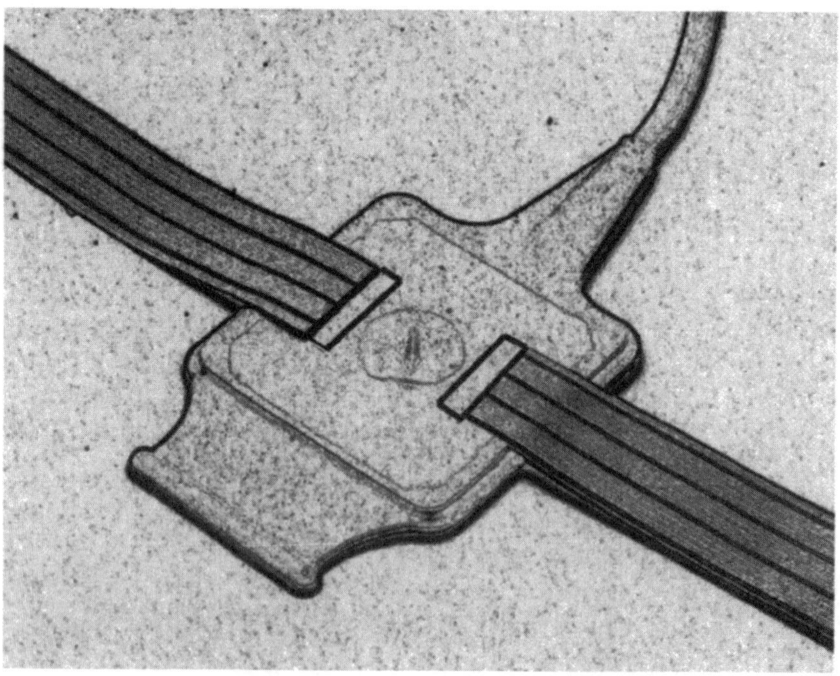

Figure 1. A schematic diagram of the combined DBT/NIRS probe.

spectrum and 0.5 s delay. O_2 consumption was calculated by measuring the slope of [Hb] changes, during 3 minutes of arterial occlusion (Cheatle et al., 1991, De Blasi et al., 1993). Measurements were made at 26.0, 28.2, 30.5, 33.3, 34.8, 35.9, 37.0 and 37.8°C.

3. RESULTS

The 8 curves appearing in Figure 2a each represent the mean of 10 baseline spectra (sample intensity = I), acquired before the start of the ischaemia, at the different temperatures defined in the METHODS section. The curves are shifted toward the bottom of the graph with decreasing temperature. For a given wavelength, $-\log_{10}(I/I_{air})$ varies linearly ($p < 0.01$) as a function of temperature when fitted to an expression of the form $-\log_{10}(I/I_{air}) = aT + b$. Figure 2b shows the correlation coefficient (r) of the linear regressions over the wavelength range studied.

Figure 3 shows the slope of the regression lines. Note that the slope increases with wavelength and exhibits an abrupt change near 840 nm.

Figure 4 shows the intramuscular oxygen consumption of the forearm at rest as a function of temperature, derived from NIRS measurements made during the arterial occlusion.

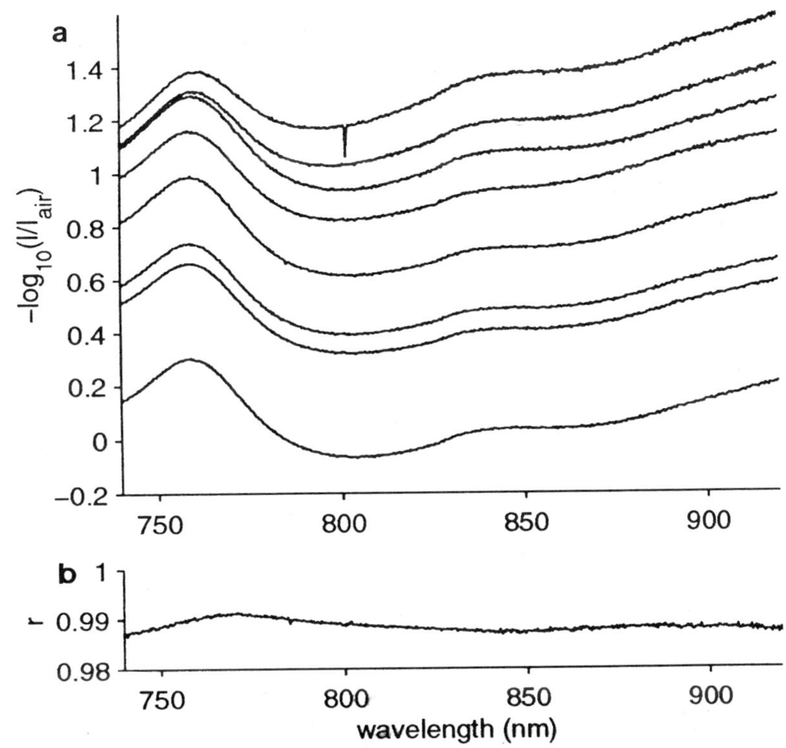

Figure 2. a) Baseline spectra at T = 26.0 (bottom curve), 28.2, 30.5, 33.3, 34.8, 35.9, 37.0 and 37.8 (top curve) °C. b) Correlation coefficient (r) as a function of the wavelength of the linear regression: $-\log_{10}(I/I_{ref}) = aT + b$.

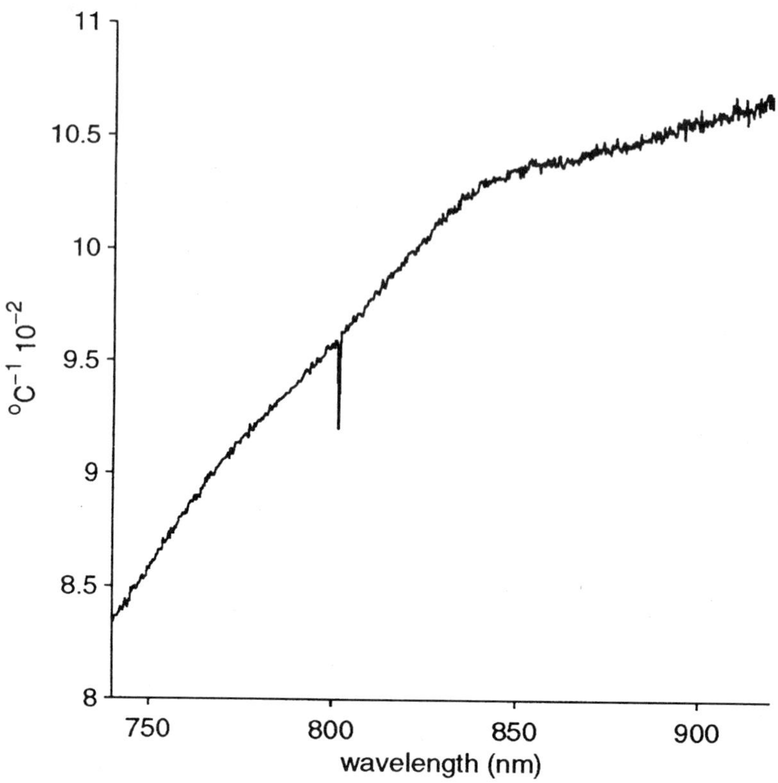

Figure 3. Slope (a) of the linear regression $-\log_{10}(I/I_{ref}) = aT + b$ as a function of the wavelength (Note that the "spike" around 810 nm is an artifact due to a cosmic ray hitting a CCD pixel).

4. DISCUSSION

If it is assumed that [Hb] and [HbO_2] remain reasonably constant during intramuscular temperature changes, then the results reported in Figure 2 allow one to conclude that $-\log_{10}(I/I_{air})$ shifts by an offset proportional to the intramuscular temperature and that the slope itself also depends on the wavelength. This means that the attenuation spectra expressed as $-\log_{10}(I_T/I_T(0))$, where $I_T(0)$ is a reference spectrum acquired at the same temperature (T) as I_T, will eliminate the effect of the offset, i.e.:

$$\begin{aligned} -\log_{10}(I_{T1}/I_{T1}(0)) &= \log_{10}(I_{T1}(0)) - \log_{10}(I_{T1}) \\ &= \{\log(I_{T1}(0)) + \text{offset}\} - \{\log_{10}(I_{T2}) + \text{offset}\} \\ &= -\log_{10}(I_{T_2}/I_{T2}(0)) \end{aligned}$$

where I_{T2} and $I_{T2}(0)$ are measure at any reference temperature T2. From the practical point of view, this allows us to compare [Hb] and [HbO_2] *changes* at different temperatures, i.e. O_2 consumption can reasonably be compared. Of course, the hypothesis that the [Hb] and [HbO_2] remain constant is not completely true but, fortunately, this problem only seems to become noticeable in the Hb region (~760 nm) when the muscle reaches the critical temperature of ~36°C. (In Figure 2a, the third curve from the top, at 35.9°C, is not parallel to the others). In fact, it has been demonstrated that during the present pro-

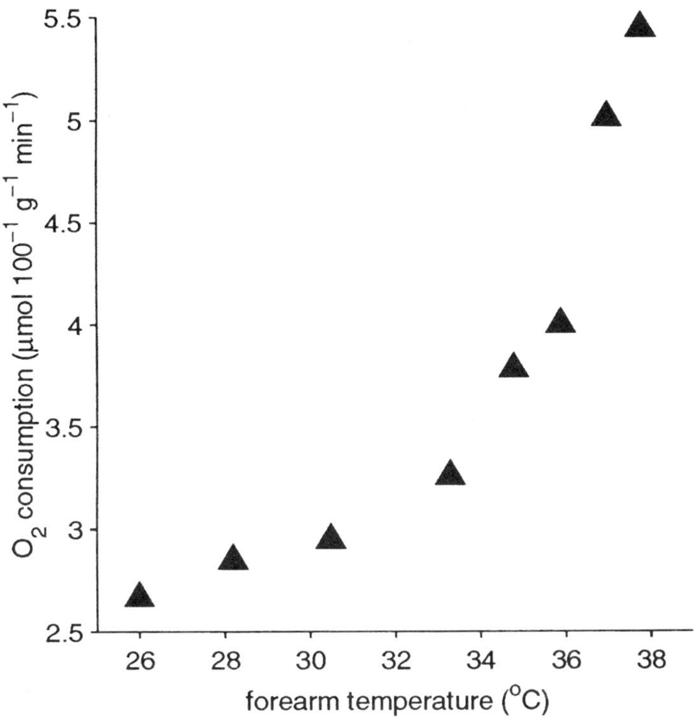

Figure 4. Forearm oxygen consumption as a function of temperature.

tocol there is a variation in blood perfusion, heat flux and tissue blood volume when the temperature exceeds ~36°C (Binzoni et al., 1995). Thus, it is likely that the observed offset is due to changes in the scattering coefficient of the tissue induced by the temperature variations. The present results are in agreement with analogous *in vitro* studies in tissue (Kelly et al., 1995). It must be noted that scattering associated changes are not observed *in vitro* when studying the spectra of pure Hb, HbO_2 or H_2O (Kelly et al., 1995, Sfareni et al., 1997). However, there are small temperature dependent changes in the absorption spectra of these compounds, and the precision in the calculation of [Hb] and $[HbO_2]$ changes can be improved by using temperature corrected extinction coefficients.

The above results allow us to calculate and to compare intramuscular O_2 consumption values measured at different temperatures if care is taken to acquire $I_T(0)$ at the same temperature as I_T. The results presented in Figure 4 seem to be in agreement with early findings on animals (Cossin and Bowler 1987). Moreover, we have previously demonstrated (Binzoni et al., 1998) that during the present protocol, the skeletal muscle metabolism remains aerobic even during the three minute ischaemic period so the O_2 consumption data should represent the total energy consumption of the tissue. It will be necessary to perform further experiments to define the exact mathematical function describing these data. It is however noteworthy, that a variation of 2°C, i.e. of the order of the possible intramuscular temperature gradients, are sufficient to produce measurable changes in oxygen consumption. This point shows the importance of clamping the temperature at a fixed value, by means of the DBT/NIRS probe, when measuring precise O_2 consumption values. To our knowledge, these are the first data for resting human intramuscular O_2 consumption as a function of a well defined temperature.

The results shown in Figure 3 are not essential for the calculation of the O_2 consumption but it is interesting to notice the abrupt change in the slope of the regression near ~840 nm. It is difficult to explain the change in slope in this region in terms of variations in [Hb], [HbO_2] or tissue water content, since changes in their absorption spectra as a function of temperature are too "smooth" in this wavelength range (Sfareni et al., 1997), and the wavelength dependence of pathlength is approximately linear (Matcher et al., 1997). It would thus be interesting to use this type of analysis to study the behaviour of the scattering coefficients of the skeletal muscle in this region when changing temperature.

5. CONCLUSIONS

The combined DBT/NIRS probe should enable new studies of the biological effects of temperature on humans. The probe should allow one to take "temperature biopsies" in a defined muscle region in order to investigate muscle metabolism and changes in optical properties. The non-invasive character of the measurements is ideal for the follow-up of time dependent processes. Further studies are now needed to improve the algorithm for [Hb] and [HbO_2] calculation in the presence of large temperature changes.

ACKNOWLEDGMENTS

We thank the Swiss National Science Foundation (#31-47075.96) and Hamamatsu Photonics KK for the financial support.

REFERENCES

Binzoni T, Hiltbrand E, Kayser B, Ferretti G, and Terrier F (1995) Human intramuscular temperature and heat flow transients at rest. J. Appl. Physiol. 79(5):1736–1743.

Binzoni T, Quaresima Q, Barattelli G, Hiltbrand E, Gürke L, Terrier F, Cerretelli P, and Ferrari M (1998) Energy metabolism and interstitial fluid displacement in human gastrocnemius during ischaemic preconditioning. J. Appl. Physiol. (in press)

Cheatle TR, Potter LA, Cope M, Delpy DT, Coleridge-Smith PD, and Scurr JH (1991) Near infrared spectroscopy in peripheral vascular disease. Br. J. Surg. 78:405–408.

Cope M, Delpy DT, Wray S, Wyatt JS, and Reynolds EOR (1989) A CCD spectrometer to quantitate the concentration of chromophores in living tissue utilising the absorption peak of water at 975 nm. Adv. in Exp. Med. & Biol., 248:33–40.

Cossin AR and Bowler K (1987) Temperature biology of animals. Chapman and Hall, London, New York.

De Blasi RA, Cope M, Elwell CE, Safoue F, and Ferrari M (1993) Noninvasive measurement of human forearm oxygen consumption by near infrared spectroscopy. Eur. J. Appl. Physiol. Occup. Physiol. 67:20–25.

Ducharme MB, VanHelder WP, and Radomski MW (1991) Tissue temperature profile in the human forearm during thermal stress at thermal stability. J. Appl. Physiol. 71(5):1973–1978.

Kelly JJ, Kelly KA, and Barlow CH (1995) Tissue temperature by near-infrared spectroscopy. SPIE 2389:818–828.

Matcher SJ, Cope M, and Delpy DT (1993) Use of the water absorption spectrum to quantify tissue chromophore concentration changes in near infrared spectroscopy. Phys. Med. Biol., 39:177–196.

Matcher SJ, Cope M, and Delpy DT (1997) In vivo measurements of the wavelength dependence of tissue scattering coefficients between 760 nm and 900 nm measured using time resolved spectroscopy. Appl. Opt. 36:386–396.

Sfareni R, Boffi AJ, Quaresima V, and Ferrari M (1997) Near infrared absorption spectra of human deoxy- and oxyhaemoglobin in the temperature range 20–40°C. Biochim. Biophys. Acta. 1340:165–169.

Solman AJ and Dalton JCP (1973) New thermometers for deep tissue temperature. Biomedical Engng. 8:432–435.

LASER-ASSISTED OPTICAL ROTATIONAL RED CELL ANALYZER (LORCA) IN CLINICAL PRACTICE

Hemorheological Kinetics and Tissue Oxygenation

G. Cicco,[1] A. J. van der Kleij,[2] G. D. Stingi,[1] M. S. Tarallo,[1] and A. Pirrelli[1]

[1]Department of Clinical Methodology and
Medico-Surgical Technologies Internal Medicine and Hypertension
Lab. of Hemorheology and
Oxygen Transport to Tissue
University of Bari, Italy
[2]Department of Surgery/Hyperbaric Medicine
Academic Medical Center, University of Amsterdam
The Netherlands

INTRODUCTION

Relevant hemodynamic changes in arterial hypertension (i.e. vasoconstriction induced by an organic arterial wall thickening or a functionally induced vasospasm) can lead to increased peripheral resistance. In the microvasculature of hypertensive subjects it is often possible to detect decreased peripheral perfusion together with hemorheological alterations (such as the presence of red blood cell aggregates, decreased erythrocyte deformability). These hemorheological changes can raise blood pressure (increasing the resistance of the blood flow in the microvasculature) and may finally induce a reduction in peripheral tissue oxygenation. These conditions are negatively influenced by lipoidoproteinosis, diabetes and smoking, which very important risk factors in hypertensives for acute myocardial infarction, transient ischemic attack, ictus, and peripheral occlusive arterial disease (Cesarone, 1992; Cicco, 1993, 1998). It has been shown that a good pharmacological control of the blood pressure and the lipoidoproteinosis influence the hemorheological patterns in the microvasculature, improving peripheral perfusion and tissue oxygenation (Cicco, 1994, 1995, 1997).

Oxygen Transport to Tissue XXI, edited by Eke and Delpy.
Kluwer Academic / Plenum Publishers, New York, 1999.

Table 1. Characteristics of the study population (14 males, 11 females)

*Age range 47–53 years *Non-smokers *Family history of hypertension *Without respiratory or haematic pathology *With lipoidoproteinosis (Stage 2b; Fredrickson Classification)	*without insulin resistance *without diabetes *family history of lipoidoproteinosis *with peripheral occlusive arterial disease (stage II type; Leriche-Fontaine Classification)
*Obesity—body Mass Index (BMI)	*M : 28 ± 2.1 F : 30 ± 3.9
*Systolic Blood Pressure (SBP)	*172 ± 9 mmHg
*Diastolic Blood Pressure (DBP)	*95 ± 8 mmHg
*Mean Blood Pressure (MBP)	*120 ± 5 mmHg
*Heart Rate (HR)	*85 ± 5 mmHg

AIM OF THE STUDY

The aim of this study was to evaluate the effect of an antihypertensive medication combined with a hypocholesterolemic drug and a seasonable diet (to modulate hypertriglyceridemia) on peripheral tissue oxygenation. In addition, hemorheological parameters were assessed with a new instrument, the Laser-assisted Optical Rotational Red Cell Analyzer (LORCA) as described by Hardeman et al. (1994) and a microscopical index: the Erythrocyte Morphologic Index (EMI) described by Zipursky et al. (1983) and introduced in Italy by Forconi et al. (1994, 1997, 1998).

MATERIALS AND METHODS

We studied 25 patients (14 male and 11 female aged 50 ± 3 years) (Table 1) characterized by mild hypertension stage (World Health Organization Classification), hypercholesterolemia, hypertriglyceridemia, (lipoidoproteinosis type 2b; Fredrickson Classification) (Fredrickson, 1978) (Table 2) with obesity (Body Mass Index BMI = 29 ± 3), without insulin resistance, with peripheral occlusive arterial disease (POAD) Stage II type a (Leriche-Fontaine Classification) (Table 3). All patients were non-smokers, without respiratory or hematological pathologies and without diabetes. The patients were evaluated after 60 days of oral treatment with amlodipine (10 mg OD), pravastatine (20 mg OD) and a low calorie diet (1400 Kcal). At the start of the study, every 15 days and at the end of the study period we measured systolic and diastolic blood pressure (SBP, DBP), mean blood pressure (MBP) and heart rate (HR), total cholesterolemia, triglyc-

Table 2. Fredrickson classification for hyperlipoproteinemia

	Increase in		
Type	Lipids	Lipoproteins	Cardiovascular risk straddlehead
1	Triglycerides (TG)	Chylomicrons	Low
2a	Cholesterol (CH) and TG	LDL-CH	Very high
2b	CH-TG	LDL-CH VLDL-CH	Very High
3	CH-TG	Abnormal mean lipoprotein	Very High
4	TG	VLDL-CH	Maybe high
5	TG	VLDL-CH Chylomicrons	Maybe low

Table 3. Leriche-Fontaine Peripheral Occlusive Arterial Disease Classification

Stage I	Patients with various undefined symptoms, paresthesias, heaviness in legs, and uncertain objectivity
Stage II	"Claudicatio Intermittens"
	a: pain free time in legs whilst walking more than 100 m
	b: pain free time in legs whilst walking less than 100 m
Stage III	Patients with spontaneous or nocturnal leg pain reduced by sitting or movement
Stage IV	Presence of trophic lesions.

eridemia. Two transcutaneous oximeters (Microgas 7650 Kontron Instruments with the Combi sensor) were used to evaluat the transcutaneous oxygen partial pressure ($TcpO_2$) and the transcutaneous carbon dioxide partial pressure ($TcpCO_2$) (van der Kleij, 1996, 1998, 1998). Moreover, peripheral oxygen saturation percentage (%Sat O_2) was measured using a pulse oximeter at the second finger of the right hand (7845 Kontron Instruments Pulse Oximeter) and oxygen concentration (ctO_2) was also measured. To establish the level of vascular disease the Regional Perfusion Index (RPI = $TcpO_2$ foot/$TcpO_2$ in the subclavicular standard area) (Hauser,1983) and the Doppler guided Windsor Index (WI) were used. In order to study erythrocyte deformability and the hemorheological kinetics under dynamic conditions we measured the elongation index (EI). The venous fresh blood samples suspended in polyvinyl pyrrolidone (PVP) were examined under different shear stresses: 0.30–0.53–0.95–1.69–3.00–5.34–9.48–16.86–30.05 Pascals (Pa). We considered only those values detected at 30 Pascals at a temperature of 37°C. These examinations were done with a new computerized laser rays (III a) instrument (Mechatronics, Hoorn, The Netherlands) called the Laser Assisted Optical Rotational Red Cell Analyzer (LORCA) (Figure 1) (Hardeman et al., 1994, 1995, 1998).

Using laser rays the instrument evaluates red blood cell deformability, under several shear stresses (from 0.30 to 30 Pa). Using fresh venous blood samples (without PVP) the device also evaluates red blood cell aggregation, detecting the Aggregation Index, the $t^1/_2$

Figure 1. Laser-assisted Optical Rotational Red Cell Analyzer (LORCA).

Table 4. Hemodynamic parameters measured during study

Days	0	15	30	45	60
SBP (mmHg)	172 ± 9	160 ± 6	156 ± 5	153 ± 7	152 ± 8*
DBP (mmHg)	95 ± 8	88 ± 6	86 ± 5	84 ± 5	82 ± 9*
MBP (mmHg)	120 ± 5	112 ± 3	109 ± 2	107 ± 4	105 ± 6*
HR (beats/min)	85 ± 5	86 ± 4	86 ± 5	87 ± 2	88 ± 6*

* $p < 0.01$ Values are means ± standard deviation.

of the aggregation in seconds, and the shear stress threshold to prevent red blood cell aggregation (Hardeman, 1994).

Moreover, to evaluate red blood cell deformability under static conditions the Erythrocyte Morphologic Index (EMI) was calculated according to the method of Zipursky (Zipursky, 1983) and Forconi (Forconi, 1994, 1997. 1998). We put a fresh venous blood sample in E.D.T.A. then immediately add 50 μl of whole blood to 0.5 ml of PO_4 buffer, 0.125 M. Immediately after mixing, we add 0.5 ml of glutaraldehyde at 0.3% in PO_4 buffer 0.125 M. After a 5 minute resting period to allow red blood cell fixing, the fixed erythrocytes are studied using an optical microscope with an objective immersed in glycerol. Then, the outline of the red blood cell (RBC) is considered to have a bowl-shape or discocyte form according to the Bessis standard (Bessis, 1978). The bowl-shaped RBC indicates a higher level RBC deformability, the discocyte shape indicates less RBC deformability. The normal value is greater than 1 (bowl-shaped are more numerous than discocytes). All measurements were made under standard conditions. Statistical analysis was done using the Student's t-test for paired values.

RESULTS

After 60 days of oral treatment with amlodipine (a calcium antagonist) 10 mg OD, pravastatine 20 mg OD, and a low calorie diet (1400 Kcal) there was a normalization of the blood pressure values (SBP % ch = −12 $p < 0.01$, DBP % ch = −14 $p < 0.01$, MBP % ch = −13 $p < 0.01$) (Table 4). Lipid parameters were also normalized (total cholesterolemia % ch = −15 $p < 0.01$, triglyceridemia % ch = −16 $p < 0.01$) (Table 5). The Elongation Index (EI) detected with the LORCA under dynamic conditions, showed a significant increase (% ch = +5 $p < 0.01$), associated with an improvement in red blood cell deformability. The Erythrocyte Morphologic Index (EMI) calculated after microscopic examination of the blood sample, was also increased (% ch = +18 $p < 0.01$) related to an increase in bowl-shaped RBC indicating an increase in red blood cell deformability. The Aggregation Index (AI), detected with LORCA was decreased (% ch = −14 $p < 0.01$) related with an increase of the aggregation half time (expressed in seconds) (% ch = +35 $p < 0.01$) indi-

Table 5. Data on lipidic parameters

Days	0	15	30	45	60
Total cholesterol (mmol/l)	6.8 ± 0.7	6.6 ± 0.6	6.4 ± 0.6	6.1 ± 0.6	5.8 ± 0.7*
Triglycerides (mmol/l)	2.7 ± 0.2	2.6 ± 0.1	2.5 ± 0.2	2.4 ± 0.2	2.2 ± 0.2*

* $p < 0.01$ Values are mean ± standard deviation.

Table 6. Data on hemorheological parameters

Days	0	15	30	45	60
Elongation index	0.55 ± 0.02	0.56 ± 0.02	0.57 ± 0.03	0.57 ± 0.02	0.58 ± 0.02*
Erythrocyte morphologic index	0.75 ± 0.05	0.84 ± 0.05	0.82 ± 0.06	0.84 ± 0.03	0.89 ± 0.06*
Aggregation index	70.0 ± 6.00	67.0 ± 4.00	65.0 ± 3.00	63.0 ± 2.00	60.0 ± 5.00*
t½sec.	1.70 ± 0.30	2.00 ± 0.20	2.10 ± 0.40	2.20 ± 0.30	2.30 ± 0.40*

* $p < 0.01$ Values are mean ± standard deviation.

cating normalization of the time of red blood cell aggregation (Table 6). There was also an improvement of $TcpO_2$ measured at the standard subclavicular area (% ch = +7 $p < 0.02$) the $TcpO_2$ measured on the back of the foot (% ch = +21 $p < 0.03$).

The oxygen saturation percentage measured by pulse oximetry was also slightly improved (% ch = +1 $p < 0.02$). The oxygen concentration increased (% ch = + +6 $p < 0.01$) (Table 7). The peripheral perfusion indices were also improved by the end of the study. The Regional Perfusion Index increased (% ch = +11 $p < 0.01$) and the Ankle—Brachial Pressure Index (Windsor index = WI) also improved (% ch = +14 $p < 0.01$) (Table 8).

DISCUSSION

It is known (Forconi, 1989) that study of erythrocyte morphology reveals alterations which underline a worsening of the hemorheological condition and finally leads to decreased transport of oxygen to tissues. The results of this study demonstrated an improvement of red blood cell deformability and tissue oxygenation (especially at the back of the foot), and normalization of lipidic parameters. Moreover, determination of the peripheral perfusion indices (such as the Regional Perfusion Index and the Windsor Index) showed improvement in the microvasculature blood, emphasizing the clinical value of use of these perfusion indices.

Table 7. Data on tissue oxygenation parameters

Days	0	15	30	45	60
$TcpO_2$ chest (mmHg)	67 ± 4	68 ± 3	69 ± 4	71 ± 3	72 ± 5*
$TcpO_2$ foot (mmHg)	47 ± 2	49 ± 3	51 ± 2	55 ± 3	57 ± 3*
$TcpCO_2$ (mmHg)	35 ± 0.6	34 ± 0.7	35 ± 0.8	34 ± 0.7	34 ± 0.8*
Sat O_2 (%)	95 ± 1.17	95 ± 1.09	96 ± 1.10	96 ± 1.4	96 ± 1.57*
ctO_2 (mmol/l)	7.3 ± 0.6	7.4 ± 0.7	7.4 ± 0.6	7.6 ± 0.3	7.8 ± 0.5*

* $p < 0.01$ Values are mean ± standard deviation.

Table 8. Data on Windsor Index and regional Perfusion Index

Days	0	15	30	45	60
Windsor index	0.67 ± 0.04	0.71 ± 0.02	0.72 ± 0.03	0.76 ± 0.02	0.77 ± 0.05*
Regional perfusion index	0.70 ± 0.04	0.72 ± 0.03	0.74 ± 0.04	0.77 ± 0.03	0.78 ± 0.05*

* $p < 0.01$ Values are mean ± standard deviation.

Obviously, these results are related to a normalization of blood pressure and lipidic parameters. Improved rheological conditions can be beneficial for the microcirculation and lead to better peripheral perfusion and oxygen transport to the cell. In this study, red blood cell deformability and aggregability was assessed using the LORCA and the EMI and these improved rheological parameters were associated with improved tissue oxygenation. In our opinion, it may be concluded that use of the LORCA device enables clinical evaluation of hemorheological parameters. Finally, it may be concluded that use of amlodipine (10 mg OD, oral) and pravastatine (20 mg OD, oral) combined with a low calorie diet (1400 Kcal) is beneficial for hypertensives with lipoidoproteinosis.

REFERENCES

Bessis M (1978) Reintegrazione degli studi di sangue. Piccin ed. Padova.

Cesarone MR, Laurora G, and Belcaro GU (1992) Microcirculation in systemic hypertension. Angiology; 43(11):899–903.

Cicco G and Pirrelli A (1993) Emoreologia clinica ed ipertensione arteriosa. Min. Cardioangiol.; 41:6:213–223.

Cicco G, Dolce E, Mancini M, and Pirrelli A (1994) Arteriosclerosis and hemorheology in hypertension: organ damage. Med. J. Surg. Med.; 3:131–138.

Cicco G, Dolce E, Vicenti P, Gigante G, and Pirrelli A (1995) Peripheral oxygen release evaluation using 3 transcutaneous oximeters contemporarily on patients with hypertension and lipoidoproteinosis. Clin. Hemorheology; (15):3:P25 539.

Cicco G, Vicenti P, Pirrelli A, and van der Kleij AJ (1997) Peripheral perfusion and tissue oxygenation improvement induced by antihypertensive medication combined with lipoidoproteinosis treatment. In: Adv. in Exp. Med. Biol. 411:261–266.

Cicco G and Pirrelli A (1998) Hemorheology and tissue oxygenation transport during hypertension. Clin. Hemorheology and Microcirculation IOS Press, Amsterdam; 19; 1:61–63.

Cicco G (1998) Red blood cell deformability; blood viscosity and tissue oxygenation in hypertension. 2nd Int. Conf. Hemorheology and Tissue Oxygenation in Hypertension Abstract Book L3, 27–28.

Forconi S, Guerrini M, Cappelli R, Sarri P, and Furesi L (1989) Rapporti fra endotelio e cellule circolanti a livello microcircolatorio influenza su coagulazione, fibrinolisi, mobilità della parete vascolare ed emoreologia. In: Tesi M. Microcircolazione 1989, Monduzzi Editore Bologna 145–151.

Forconi S, Turchetti V, De Matteis C, Leoncini F, Pierolini F, Tribalzini L, Pastorino G, and Altavilla R (1994) Behaviour of blood viscosity, intra-erythrocytic calcium, total erythrocyte morphology in ischaemic diseases. Proceed. XVII Eur. Conf. Microcicul. Roma 68.

Forconi S (1997) Relationship between erythrocyte morphology and deformability—Boletim de Societade Portuguesa de Emorreologia e Microcirculaçao, X Eur. Conf. on Clin. Hemorheology L2.

Forconi S (1998) Morphological approach to red cell deformability in a clinical environment. 2nd Int. Conf. Hemorheology and Tissue Oxygenation in Hypertension Abstract Book L28, 79.

Fredrickson DS, Goldstein JL, and Brown JS (1978) The Familial Hyperlipoproteinemias. In: The metabolic basis of inherited disease. Stanbury JB; Wyngaarden JB, and Fredrickson DS eds. New York, McGraw Hill 604–655.

Hardeman MR, Goedhart PT, Dobbe JGG, and Lettinga KR (1994) Laser-Assisted Optical Rotational Red Cell Analyzer (LORCA) I: a new instrument for measurement of various structural hemorheological parameters. Clin. Hemorheology 14:605–618.

Hardeman MR, Goedhart PT, and Schut NH (1994) Laser. Assisted Optical Rotational Red Cell analyzer (LORCA) II: red blood cell deformability; elongation index versus cell transit analyzer. Clin. Hemorheology 14:619–630.

Hardeman MR and Goedhart PT (1994) Laser-Assisted Optical Rotational Red Cell Analyzer (LORCA): Red Cell Aggregometry. In: Hemorheology & Erythrocyte Aggregation Vol. 4 J.F. Stoltz ed., Editions Médicales Internationales Paris.

Hardeman MR and Goedhart PT (1995) Laser Assisted Optical Rotational Red Cell Analyzer (LORCA): "viscoscan" of RBC deformability and estimation of RBC internal viscosity. Biorheology, 32:357.

Hardeman MR (1998) Clinical hemorheology placed in to hypertension: introduction and implementation of new methodology. Clin. Hemorheology and Microcirculation, IOS Press, Amsterdam 19; 1:50–54.

Hardeman MR (1998) LORCA, a new versatile hemorheological instrument. 2nd Int. Conf. Hemorheology and tissue oxygenation in Hypertension Abstract Book C27, 77, 78.

Hauser CJ and Shoemaker W (1983) Use of a transcutaneous pO_2 Regional Perfusion Index to quantify tissue perfusion in peripheral vascular disease. Am. J. Surg. 197:337–343.

Kleij van der AJ and Bakker DJ (1996) Oximetry. In: "Handbook of hyperbaric medicine" G Oriani, A Marroni, and F Wattel (Ed. Springer-Verlag Italy Milano) 670–675.

Kleij van der AJ (1998) Which method can be used to evaluate the tissue oxygenation? Clin. Hemorheology and Microcirculation, IOS Press, Amsterdam 19; 1:55–60.

Kleij van der AJ (1998) Invasive and non-invasive pO_2 measurements in Clinical Practice—2nd Int. Conf. Hemorheology and tissue oxygenation in hypertension. Abstract Book C23, 69–70.

Zipursky A, Brown E, Palko J, and Brown E (1983) The erythrocyte differential count in newborn infants. Am. J. of Ped. Haemat. Oncol. 45–51.

SEPARATION OF HUMAN PROTEIN C FROM COMPONENTS OF TRANSGENIC MILK USING IMMOBILIZED METAL AFFINITY CHROMATOGRAPHY

A. S. Dave and D. F. Bruley*

University of Maryland Baltimore County
Department of Chemical and Biochemical Engineering
ECS 101
1000 Hilltop Circle
Baltimore, Maryland 21250, USA

1. INTRODUCTION

Human Protein C (PC) is an important Vitamin K-dependent anticoagulant and antithrombotic in the human bloodstream. Deficiency in any of the Vitamin K-dependent proteins will disturb hemostasis (Bruley, 1990). A deficiency of PC may cause unnecessary, dangerous blood clots to occur in various parts of the body. A common location of a blood clot due to PC deficiency is in a deep vein in the leg, thereby congesting the blood flow within the vein of the leg. The blood may reach the heart via an alternate route if a particular vein has a clot. However, the congestion will eventually do its damage by disturbing the overall blood flow, similar to how a traffic jam disturbs the overall flow of traffic, despite the fact that alternate routes are available. Due to a lack of oxygen and nutrient transport to the tissue surrounding the clot, the tissue could die, potentially resulting in the amputation of the leg, or even death in some cases (Seligsohn, 1987). It is also possible that blood clots due to PC deficiency could occur in other parts of the body. A blood clot in the brain may deprive it of necessary oxygen, possibly resulting in major brain damage, if not death. A blood clot near the heart may lead to a heart attack, which will most probably result in death.

* To whom correspondence should be directed

Oxygen Transport to Tissue XXI, edited by Eke and Delpy.
Kluwer Academic / Plenum Publishers, New York, 1999.

Currently, there are pharmaceuticals available on the market which act as anticoagulants and are used to treat PC deficient patients. Two major types of drugs used are heparin and Coumadin (MSN, 1980). The disadvantage to using heparin and Coumadin is that both drugs have dangerous side effects. For example, a major side effect resulting from their use is excessive bleeding, which may result in a stroke. Coumadin cannot be used by pregnant women, even though anticoagulant deficiencies often occur in conjunction with pregnancy, because it can pass into the fetus and cause major damage, and can also transfer into a newborn through the milk of the mother. Coumadin has caused birth defects such as blindness and mental handicap. In general, Coumadin can damage any organ and cause Coumadin-induced skin necrosis in patients if not monitored carefully (PDR, 1996). Due to the need for anticoagulants and the dangerous side effects of currently used pharmaceuticals for anticoagulation treatment, it is necessary to search for a safer anticoagulant. Literature has shown that it will be safe to use reasonable doses of PC because it does not cause the bleeding complications and skin necrosis attributed to heparin and Coumadin, and research is being conducted to find a way to produce large amounts of PC at low cost.

1.1. The Function of Protein C in the Human Body

PC normally circulates in human blood plasma in its inactive form. When a cut occurs, thromboplastin is released into the bloodstream (MSN, 1980). Thromboplastin then cleaves prothrombin to produce thrombin, which then cleavesfibrinogen to form fibrin. Fibrin is made up of dense, intricate threads that trap blood cells. Finally, fibrin becomes insoluble and causes the blood to clot. PC is activated when thrombin binds to a protein called thrombomodulin to form a thrombin-thrombomodulin complex. While thrombin can activate PC by itself, the formation of the thrombin-thrombomodulin complex expedites the activation of PC 20,000 times (Esmon, 1989). Once PC has been activated, it binds with a cofactor called Protein S and forms a complex (Henkens, 1995). Together, these proteins deactivate Factor Va and Factor VIIIa (Mather, 1996). Because Factor Va has been deactivated, it cannot react with prothrombin and Factor Xa. The end result is that thrombin and fibrin cannot be produced, and the blood can no longer clot. In addition to its anticoagulant capabilities, PC also binds with PAI-1 (plasminogen activator inhibitor), thus preventing the PAI-1 from deactivating the tPA (tissue plasminogen activator). The tPA can then activate plasminogen, which in turn activates plasmin. The plasmin is able to break down blood clots (Bruley, 1990).

1.2. Transgenic Animal Milk as a Source for Protein C

Because of the high cost of producing PC from human plasma, genetic engineering of animals is being currently researched as an alternative method to produce PC. The most commonly used method to produce transgenic animals is by microinjection of DNA into an embryo, and then the embryo is transplanted into a host animal that produces offspring which may be transgenic (Jaenisch, 1988). The DNA contains the desired hybrid gene, which in this case is the gene that expresses PC, and a promoter. After the protein/promoter has been microinjected into the embryo, the resulting animal might be able to produce milk in its mammary gland which contains the desired protein (Jaenisch, 1988). Furthermore, the gene for the desired protein can be passed on to the offspring when the animal reproduces. A group of researchers from Virginia Polytechnic Institute

(VPI) and the American Red Cross (ARC) used swine as transgenic animals for PC production. There were several reasons why swine were used. First, swine have already been used to produce pharmaceuticals used to treat humans without causing disease (Velander, 1997). Second, swine have a short gestation period (4 months). As a result, they reproduce often, and can have up to 12 offspring at a time. Finally, 300 liters of milk can be produced by swine annually (Avellar, 1997).

The hybrid gene for PC was microinjected into the embryo of a pig. It was discovered that the resulting offspring was able to express PC in its milk at concentrations as high as 1 mg/ml (Velander, 1992), which is 250 times the concentration of PC in blood plasma. Because of the high concentration of PC, it may be advantageous to research PC purification from transgenic swine milk.

A process has been developed for obtaining 95% pure PC from transgenic swine milk, as outlined in Figure 1 (Drohan, 1994).

The current research focuses on the third step of the process, which is the separation of PC from proteins called caseins. Caseins are proteins found in the milk that aggregate to form casein micelles when they bind with calcium ions. These micelles can trap PC and prevent it from binding to the chromatography column during the separation process. Nearly 40% of the PC may be associated with the micelles. In an attempt to avoid this problem, a polyethylene glycol pellet is dissolved in phosphate buffered saline, and dissolved in EDTA. This strips calcium from the micelles, causing them to break up, and causes PC to be released. After the casein micelles are broken, polyethylene glycol is dissolved in tris buffered saline, and this solution is used to dissolve PC and precipitate a majority of caseins and other milk proteins (Drohan, 1994). In this step of the process, a significant amount of PC is lost because much of it may still be trapped between broken casein micelles. The PC recovery in this step of the process is only fifty-four percent (54%).

Step	Protein C Recovery
1. Fat removed from milk	100%
2. Precipitation by polyethylene glycol in phosphate buffered saline	80%
3. Casein micelles broken up by EDTA	43%
4. Viral inactivation by solvent/ detergent mixture	43%
5. Barium chloride precipitation	30%
6. Anion exchange chromatography	24%

Figure 1. Original process to purify PC from transgenic swine milk as developed by researchers at the American Red Cross and Virginia Polytechnic Institute.

1.3. Immobilized Metal Affinity Chromatography

Currently at the University of Maryland Baltimore County, our research group is conducting research on using immobilized metal affinity chromatography (IMAC) as an alternative method for separating PC from Cohn Fraction IV-1 because it is the most common source. The ultimate goal of this research is to see if IMAC can be used in the process that separates PC from transgenic swine milk. Specifically, this research focuses on applying IMAC in the step in which caseins are separated from the protein mixture. The two major objectives in this research are to investigate whether PC can bind to and be eluted from an IMAC column in large amounts, and if IMAC can remove the casein from the protein mixture.

In IMAC, the stationary phase is a metal ion which captures proteins flowing through the column. Proteins have a special group of molecules on their surface known as histidine units, which have a special affinity for the metal ion in the IMAC column (Winzerling, 1992). The greater the number of histidine units on the surface of the protein molecule, the higher the probability that the metal ion in the IMAC column will bind the protein.

The following mechanism describes immobilized metal affinity chromatography (IMAC) (Winzerling, 1992):

Gel matrix-----Spacer arm----Chelator-----Metal ion

The gel used for IMAC is usually an ion exchange matrix made up of agarose or silica. A spacer arm is used to connect the gel with a chelating agent. The purpose of the spacer arm is not only to help the chelating agent bind with the gel, but also to provide space so that the metal ion, which binds to the chelator, can adsorb the protein. The chelating agent binds with a metal ion by means of a coordinate interaction. The outer shell electrons on the metal ion interact with the chelating agent (Winzerling, 1992). Once the metal binds with the chelating agent and forms the metal-chelate complex, it becomes immobilized and can then bind protein, as shown by the mechanism below:

$$\text{Ch-Metal} + \text{Protein} \Leftrightarrow \text{Ch-Metal---Protein}$$

The most common chelating agent used is iminodiacetic acid (IDA) because it is the most available chelating agent on the market. Other examples of possible chelators are nitrilotriacetic acid (NTA) and triscarboxymethylethylenediamine (TED)(Winzerling, 1992).

The choice of the metal ion used in IMAC is determined by the strength of the affinity of the metal to the protein. Copper has experimentally been found to have the highest affinity for many proteins. Copper only needs one histidine unit to bind to a protein, whereas other metals, such as nickel, zinc and cobalt require two histidine units (Tadepalli, 1997). From the results of former students in this research group, it was determined experimentally that copper was the metal ion that gave the best recovery of PC in IMAC.

There are several advantages of using IMAC as opposed to alternative methods. For example, IMAC is relatively inexpensive as compared to immunoaffinity chromatography, the method currently used to purify PC from Cohn Fraction IV-1 (Winzerling, 1992). Moreover, IMAC has a high loading capacity for proteins (12 mg protein/ml of gel) (Dalton, 1996). Finally, there is a minimal amount of ligand leakage

Component	Number of Histidines
Protein C	12
Alpha$_{s1}$-Casein	11
Alpha$_{s2}$-Casein	6
Kappa-Casein	3
Beta-Casein	5
Alpha-lactalbumin	3
Serum albumin	18
Beta-lactoglobulin	3

Figure 2. Histidine content of various proteins in transgenic milk.

from the column, as opposed to immunoaffinity chromatography, in which antibody leakage has been an expensive problem (Dalton, 1996).

To determine if IMAC may be useful in a process to separate PC from transgenic swine milk, it is important to know the histidine content of the various proteins in the milk. Figure 2 shows the number of histidine units on the milk protein molecules from our recent survey (Dalton, 1996).

According to the figure, serum albumin, PC and α_{s1}-Casein are the proteins with the highest number of histidine units as compared to the other proteins in the milk. Only α_{s1}-Casein has a similar number of histidine units as PC among all of the caseins. Because PC has a relatively large number of histidine units as compared to caseins in general, it appears that IMAC can be used to separate PC from caseins.

2. METHODS

2.1. Source Material

The PC used in this research was supplied to us by the American Red Cross in the form of a PC/albumin mixture. The ratio of the albumin to PC in this mixture is 10 to 1. Because of this concentration difference and albumin having more histidine units than PC, it was necessary to remove the albumin as much as possible from the mixture in order to prevent the interference of albumin when applying IMAC to bind PC. Most of the albumin (60–90%) was removed via ion-exchange chromatography. Caseins were then added at an amount approximately 25 times the amount of PC in the mixture.

2.2. IMAC

In this experiment, a small chromatography column (BioRad; Hercules, CA), about 10 cm in height, and 0.8 cm in diameter, was packed with Chelating Fast Flow (IDA) gel (Pharmacia; Uppsala, Sweden), to a height of 2 cm. The volume of the gel within the column was 1 ml. Because PC has a strong affinity for copper, it may strip some of the copper from the gel as it flows through the system. This phenomenon, known as metal-ion transfer (MIT), may cause copper to leak from the IMAC column if copper is fully loaded on to the gel. Therefore, it is better to load the column at less than the full copper capacity. MIT may not be prevented by loading less than the maximum capacity of copper, but the chance of metal leakage is decreased since there is now extra space in the gel for stripped copper to rebind. Therefore, the column was arbitrarily loaded with copper at 80% of the capacity. The copper solution was composed of 100 mM $CuSO_4 \cdot 5 H_2O$, 100 mM sodium acetate, and 0.5 M NaCl, at a pH of 3.9. After the copper was loaded into the gel, a peristaltic pump system (Pharmacia; Uppsala, Sweden) was used to pump equilibrium buffer through the column. The equilibrium buffer was a solution that was 20 mM sodium phosphate and 0–0.5 M NaCl at a pH of 7.0. The flow rate of the buffer was constant at 23 ml/hr. Once the liquid was pumped into the column, it flowed through the column and went through a UV detector (Pharmacia; Uppsala, Sweden). The UV detector read the protein at a wavelength of 280 nm, and the output was recorded on a chart recorder (Goerz; Austria). The equilibrium/wash buffer was used to establish a baseline on the chart recorder so that a peak could be recorded when the protein was eluted out of the column. After the baseline was established, 1 ml of a PC/albumin/casein mixture was loaded into the column, and 30 minutes were allowed for the mixture to bind to the copper in the column. After 30 minutes, the equilibrium buffer was pumped through the column again to wash out any protein that did not bind to the copper. The liquid flowed through the UV-detector and an output peak was recorded to confirm the presence of unbound protein. An elution buffer (15 mM imidazole, 20 mM sodium phosphate, 0.5 M NaCl) was then pumped through the column. Again, the flow rate was constant at 23 ml/hr. The elution buffer desorbed the protein mixture from the column and flowed through the UV detector. The output was then recorded as a chromatogram. The wash fallthrough and eluant were each collected in different tubes.

2.3. ELISA

After collecting samples, an enzyme-linked immunoadsorbent assay (ELISA) was performed to detect the concentration of PC in the collected samples. An ELISA can indirectly measure the absorbance of PC in a sample. The absorbance is measured at 490 nm. Several standards of known PC concentration are used to establish a mathematical relationship between absorbance and the concentration of PC, and this relationship can then be used to determine the concentration of PC in the samples collected.

2.4. UV Spectrophotometer

A UV-VIS Scanning Spectrophotometer (Shimadzu) was used to measure total protein concentration in the samples. Protein readings were taken at 280–320 nm. A combination of the ELISA results and total protein readings yielded an approximation of how much casein was removed via IMAC, as well as how much PC was recovered.

3. RESULTS

Higher concentrations of sodium chloride are used in equilibrium conditions to reduce non-specific binding (Winzerling, 1992). In order to verify this concept experimentally, two different sodium chloride concentrations (0 and 0.5 M) were used in the equilibrium buffer. Six experiments were run for 0 M sodium chloride, and thirteen experiments for 0.5 M sodium chloride case, as shown in the Figures 3(a) and (b). More experiments were run using 0.5 M sodium chloride because these experiments were spread out

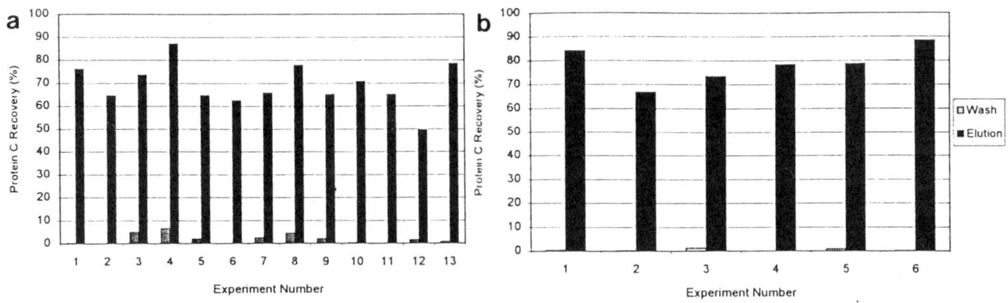

Figure 3. The recovery of Protein C from the IMAC column when the sodium chloride in the equilibrium buffer was (a) 0 M, and (b) 0.5 M.

over a period of three months, whereas experiments have been run using 0 M sodium chloride over a period of about one month.

In the case where no sodium chloride was included in the equilibrium buffer, the average recovery (+/– standard deviation) of PC in the wash was 0.52 +/– 0.49%. The average recovery of PC (+/– standard deviation) in the elution was 78.40 +/– 7.63%. In the case where 0.5 M sodium chloride was included in the equilibrium buffer, the average recovery of PC (+/– standard deviation) in the wash was 2.14 +/– 2.16% and the average recovery (+/– standard deviation) in the elution was 69.41 +/– 9.42%. The standard deviation of PC recovery in the wash fallthrough was high because the ELISA results have a higher error at lower PC concentrations.

Figures 4(a) and (b) show a graph of the approximate amount of casein removed via IMAC in the cases of no salt in the equilibrium buffer and 0.5 M salt in the equilibrium buffer.

In the figures, the values of the percentage casein removed were dependent on the amount of albumin still present in the eluant. The exact percentage of casein removed could not be determined because the methods for determining the albumin concentration were unavailable, but a decent approximation could be made by giving a minimum and maximum percentage casein removed, based on the assumptions of zero and 100% albumin recovery in the eluant. The minimum percentage of casein removed was calculated under the assumption that 100% of the albumin was recovered in the eluant, whereas the maximum percentage was calculated based on the assumption that none of the albumin was recovered. The actual percentage of casein removed is somewhere within

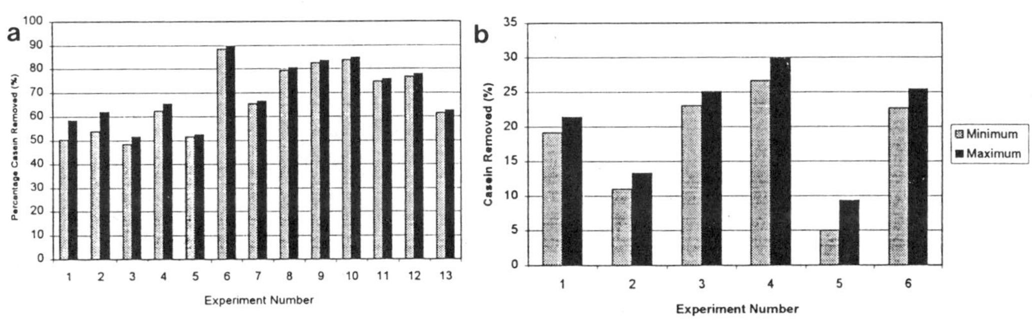

Figure 4. The approximate percentage of casein removed from the PC/albumin/casein mixture when the sodium chloride in the equilibrium buffer was (a) 0 M, and (b) 0.5 M.

this range of values. For the case where no sodium chloride was included in the equilibrium buffer, the average amount of casein removed (+/– standard deviation) was 17.92–20.76 +/– 8.25–7.90%, where the average minimum amount of casein removed (+/– standard deviation) was 17.92 +/– 8.25% and the average maximum removed (+/– standard deviation) was 20.76 +/– 7.90%. For the case in which 0.5 M sodium chloride was included in the equilibrium buffer, the average amount of casein removed (+/– standard deviation) was 67.65–70.11 +/– 14.10–12.67%, where the average minimum amount of casein removed (+/– standard deviation) was 67.65 +/– 14.10% and the average maximum removed (+/– standard deviation) was 70.11 +/– 12.67%.

4. DISCUSSION

For the case in which no sodium chloride was included in the equilibrium buffer, the recovery of PC was very high (78%), whereas the PC recovery was not as high (69%) when 0.5 M sodium chloride was included in the buffer. This is consistent with theory in the sense that more protein binding is expected to occur in IMAC when there is no sodium chloride used in the equilibrium conditions. However, a major consequence of not using sodium chloride in this situation is that, according to theory, not only will more PC bind, but other proteins (in this case casein and albumin) will also bind. The experimental results confirmed this hypothesis. As shown in Figures 4(a) and (b), less caseins are removed in the wash fallthrough when no sodium chloride is used in equilibrium conditions, which seems to indicate that more caseins bound to the copper in the column in the absence of sodium chloride. Figure 4(b) shows that more caseins were removed when 0.5 M sodium chloride was present in the equilibrium buffer, implying that less casein bound to the column. The general goal of this research is to maximize the amount of PC recovered from the IMAC column as well as the amount of casein removed from the column. Finding the condition in which the highest possible recovery of PC is combined with the lowest recovery of casein is an optimization problem in which more conditions will have to be varied besides sodium chloride concentration in the equilibrium buffer.

5. CONCLUSIONS

From this preliminary study of PC/casein separation via IMAC, it can be concluded that IMAC can recover high amounts of PC and remove a significant amount of casein from the PC/albumin/casein mixture. For obtaining a higher purity of PC, it is better to use 0.5 M sodium chloride in equilibrium conditions, as opposed to 0 M sodium chloride, because the higher sodium chloride concentration appears to reduce non-specific binding. The goal of this project is to obtain PC at higher recoveries using IMAC than the 54% PC recovery achieved by the precipitation of caseins in the original process. The results of this study seem to show that this goal is attainable, using either 0 or 0.5 M sodium chloride in the equilibrium buffer, but more casein is removed when 0.5 M sodium chloride is used.

6. FUTURE STUDIES

Finding the conditions in which IMAC can recover a maximum amount of PC and remove a maximum amount of casein is an optimization problem in which several parameters must be varied. In this study, sodium chloride concentration in the equilibrium buffer was varied. In the future, even higher concentrations of sodium chloride (1 M) will be tested. Furthermore, parameters such as imidazole and sodium chloride concentration in the elution buffer, and the amount of copper loaded in the column will be varied to see how the PC/casein separation is affected. Other future studies include running activity assays on the PC to see if the biological activity of the PC is maintained throughout the IMAC process. Quantitative studies for determining the amount of copper in the eluant must also be developed to ensure there is no copper present. Copper in the eluant could be a potential threat to patients to whom the PC is being administered. The ultimate goal of this project is to incorporate IMAC in a process that purifies PC from transgenic animal milk to increase yield and reduce cost.

ACKNOWLEDGMENTS

We would like to thank the American Red Cross and The Whitaker Foundation for their support on this project.

REFERENCES

Dalton J. (1996) *Separation of Human Protein C using Immobilized Metal Affinity Chromatography. M.S. Thesis.*

Drohan WN, et al. (1994) "A Scalable Method for the Purification of Recombinant Human Protein C from the Milk of Transgenic Swine," *Advances in Bioprocess Engineering*, 501–507.

Esmon CT, (1989) "The Roles of Protein C and Thrombomodulin in the Regulation of Blood Coagulation." *J Biol Chem* **264** (9): 4743.

Henkens CM, et al., (1995) "Plasma Levels of Protein S, Protein C, and Factor X: Effects of Sex, Hormonal State and Age," *Thrombosis and Haemostasis* **74**.

Jaenisch R (1988) "Transgenic Animals." *Science* **240**:1468.

Mather T, et al. (1996) "The 2.8 A Crystal Structure of Gla-Domainless Activated Protein C." *The EMBO Journal.* **15**(24), 6822.

Medical-Surgical Nursing, 2nd Ed. W.B. Saunders Co., Phil. PA, (1980).

Physicians Desk Reference, 50th Ed. Medical Economics Data Prod. Co., Montvale NJ (1996).

Protein C and Related Anti-Coagulants (1990) Ed: Duane F Bruley and William N Drohan. Gulf Publishing Company; New York.

Seligsohn U (1987) *Report at the ISTH Subcommittee Meeting*, Brussels, July.

Tadepalli S (1997) *Feasibility Study of the Binding and Elution of Human Protein C from Cohn Fraction IV-1 using Immobilized Metal Affinity Chromatography. M.S. Thesis.*

Velander WH, et al. (1997) "Transgenic Livestock as Drug Factories." *Scientific American*, 70.

Velander WH, et al. (1992) "High Level Expression of a Heterologous Protein in the Milk of Transgenic Swine using the cDNA Encoding Human Protein C," *Proc Natl Acad Sci.* **89**:12003–12007.

Winzerling JJ, Berna P, and Porath J (1992) "How to Use Immobilized Metal Ion Affinity Chromatography," *METHODS: A Companion to Methods in Enzymology* **4**:4–13.

ANALYSIS OF PHOSPHORESCENCE DECAY FOR NONUNIFORM OXYGEN TENSION

Consequences of Finite Excitation Flash Duration

Aleksander S. Golub,[1] Aleksander S. Popel,[2] Lei Zheng,[1] and Roland N. Pittman[1]

[1]Department of Physiology
Medical College of Virginia Campus
Virginia Commonwealth University
Richmond, Virginia
[2]Department of Biomedical Engineering
Johns Hopkins University
School of Medicine
Baltimore, Maryland

1. INTRODUCTION

The time-resolved analysis of phosphorescence decays using the Stern-Volmer equation has proved to be a reliable means of determining the concentration of oxygen, $[O_2]$, or oxygen tension, PO_2, under conditions in which oxygen is distributed homogeneously (Vanderkooi et al., 1987; Rumsey et al., 1990; Shonat et al., 1995 and 1997; Lo et al., 1996; Sinaasappel and Ince, 1996; Sinaasappel, 1997; Buerk, 1998). Methods developed to analyze phosphorescence decays under conditions where the PO_2 is nonuniform have been based on the assumption that the excitation is produced by a brief, momentary flash that can be represented by a Dirac δ-function at time $t = 0$ (Vinogradov and Wilson, 1994, Golub et al., 1997). This condition is an idealization, however, since a real excitation flash generally has a finite duration that could have a complex time course (Edgerton, 1979; Marshak, 1984). For a finite duration flash, some excited phosphor molecules are decaying, while others are still being excited so that rates of excitation and deactivation of phosphor molecules are dependent on the oxygen concentration. This poses problems for the analysis of the decay, such as the choice of time point at which to begin analysis of the emission curve and the factors that must

Oxygen Transport to Tissue XXI, edited by Eke and Delpy.
Kluwer Academic / Plenum Publishers, New York, 1999.

be considered to optimize the analysis. We will examine the consequences of a finite excitation flash on the analysis of phosphorescence decay in cases where the PO_2 is nonuniform.

The analysis of phosphorescence quenching data is based on the Stern-Volmer equation, written in the form:

$$\tau_0/\tau = 1 + (k_q/k_0)P \quad (1)$$

where τ_0 and τ are the phosphorescence lifetimes corresponding to PO_2's of 0 and P, respectively, and k_q and k_0 ($= 1/\tau_0$) are constants (Vanderkooi et al., 1987; Buerk, 1993). Oxygen is the primary quencher in biological systems and Eq. 1 forms the basis of calibrations relating $[O_2]$ or PO_2 to lifetime. The accuracy of Eq. 1, based on lifetimes, has been demonstrated for many phosphorescent dioxygen probes (Vanderkooi et al., 1987; Vinogradov and Wilson, 1995; Vinogradov et al., 1996, Lo et al., 1996).

A number of publications give the following equation as an alternative expression of the Stern-Volmer equation in terms of initial phosphorescence intensities (Torres Filho and Intaglietta, 1993, Lo et al., 1996, Sinaasappel and Ince, 1996, Sinaasappel, 1997):

$$I(0)/I(P) = \tau_0/\tau = 1 + (k_q/k_0)P \quad (2)$$

where I(0) and I(P) are the phosphorescence light intensities for PO_2's of 0 and P under conditions of steady-state excitation, and the other symbols were defined after Eq. 1. Application of Eq. 2 carries the implicit assumption that the excitation flash lasts long enough to yield a steady state concentration of the excited phosphor molecules (Alberty, 1983). Since the typical excitation flash does not satisfy this steady state condition, one cannot expect the ratio of the phosphorescence intensities given by Eq. 2 generally to be valid.

For the more general case of a nonuniform PO_2, as would be expected for oxygen in the microcirculation, the emission intensities arising from regions of an excited volume having a range of PO_2 are different in magnitude, and this consequence must be taken into account during analysis of the observed phosphorescence emission curve (Buerk, 1993).

The determination of $[O_2]$ or PO_2 by the phosphorescence method is based on fitting the phosphorescence intensity data with a variety of fitting algorithms based on specific models (Vinogradov and Wilson, 1994; Golub et al., 1997). Finding the appropriate model to describe the phosphorescence signal is crucial for the accurate recovery of key parameters (e.g., mean PO_2 and its standard deviation).

2. METHODS

2.1. Actual Excitation Flash of Xenon Lamp

Traditionally the phosphorescence pulse method has employed a xenon flash lamp as an excitation source. This light pulse is not infinitesimally short, but rather has a finite duration and a complicated waveform. Its amplitude and duration depend upon the voltage, capacitance, resistance and inductance of the discharge circuit (Edgerton, 1979; Marshak, 1984). A short time after the high voltage trigger, the light output reaches a maximum value and then decays (see curve labelled "Flash" in Figure 2).

To analyze, with mathematical models, the effect of flash duration and fitting delay (definition see below) on the phosphorescence decay curve for a typical excitation flash, the time profile of the flash lamp can be approximated by the following empirical function:

$$L(t) = L \cdot t \cdot [\exp(-t/\tau_2) - \exp(-t/\tau_1)] \tag{3}$$

where L is a light intensity scale factor, and τ_1 and τ_2 are the parameters that describe the rising and falling phases, respectively, of the flash. The excitation flash duration is generally defined as the elapsed time between where the rising and falling parts of the pulse equal one third of the maximum pulse amplitude (Edgerton, 197; Marshak, 1984).

2.2. Mathematical Modeling

The full phosphorescence waveform obtained from a region of uniform PO_2 is usually expressed as a convolution of the exponential impulse response and the excitation intensity time course or lamp function, L(u) (Ware et al., 1973; Ware, 1991):

$$Y(t) = \int_0^t A \exp[-(t-u)/\tau]\, L(u)\, du \tag{4}$$

where Y(t) is the normalized phosphorescence waveform function, A is a normalization constant, t is time, u is the flash time and τ is the phosphorescence decay lifetime.

A convolution model is valid if the system under consideration is stationary and linear, implying that the excitation light intensity must be small enough to yield a linear response in concentration of the excited phosphor. We can use the convolution model (Eq. 4) to simulate continuous excitation expressed by a step function with constant intensity of absorbed light, L, and duration t. At the end of the period of excitation, the intensity of phosphorescence having lifetime τ will be:

$$I(\tau,t) = \int_0^t \exp[-(t-u)/\tau]\, L\, du = \tau\, L\, [1 - \exp(-t/\tau)] \tag{5}$$

The ratio of phosphorescence intensities can be obtained from Eq. 5 as:

$$I_0/I = (\tau_0/\tau)\cdot[1 - \exp(-t/\tau_0)]/[1 - \exp(-t/\tau)] \tag{6}$$

where I_0 and τ_0 refer to intensity and lifetime, respectively, in the absence of oxygen.

The disadvantage of the convolution model is that simulation of the system response to an external stimulus does not provide insight into the physical basis of the process under consideration. Appropriate modeling cannot be made using the convolution approach, because of the possible nonlinear conditions that exist in tissue volumes excited by long duration and high intensity light concentrated into a microscopic region. Under these circumstances, the increase of excited phosphor concentration is not proportional to the excitation light intensity, but the process can be modeled so that the change of excited phosphor concentration in a uniform medium is described by the following differential equation:

$$dN/dt = k_x\, L(t)(N_T - N) - (k_0 + k_q\, P)\, N \tag{7}$$

where N is the concentration of excited phosphor, N_T is the total phosphor concentration, k_x is a rate coefficient of excitation and k_0, k_q and P are the traditional Stern-Volmer parameters (see Eq. 1). The first term on the right hand side describes the photoexcitation process, where the rate of increase in the excited phosphor concentration is proportional to the excitation light intensity, L(t), and the concentration of phosphor in the ground state, (N_T-N). The second term describes deactivation by phosphorescent emission and collisional quenching. We will express $[O_2]$ in micromolar and note that an oxygen concentration (saline at 37°C) of 100 µM is approximately equivalent to 63 mmHg. The connection between µM and mmHg can be calculated using Henry's law and the solubility of oxygen in the medium.

The normalized form of Eq. 7 is:

$$dn/dt = k_x\, L(t)\cdot(1-n) - (k_0 + k_q\, P)\, n \qquad (8)$$

where $n = N/N_T$ is the fraction of phosphor in the excited form. Since the phosphorescence intensity is proportional to the absolute or relative concentration of the excited form of phosphor, the function n(t) has the same time course as the corresponding phosphorescence emission curve.

3. RESULTS

3.1. Results of Simulation

Results of the modeling of the phosphorescence intensity after excitation by a rectangular light pulse show that the Stern-Volmer equation for I_0/I is accurate only for infinitely long excitation. Phosphorescence probes used for measurements of oxygen have lifetimes of several hundred microseconds, while the excitation flash duration usually is less then 30 microseconds. Thus, the intensity ratio must be very different from that given by Eq. 2. The plot in Figure 1, based on the calculation of Eq. 6, demonstrates the dependence of I_0/I on τ_0/τ for durations of the excitation light pulse from 0.1 to 30 µs. In this idealized example, $\tau_0 = 600\,\mu s$ and τ_0/τ varies from 1 to 10. The differences in I_0/I among samples having different τ diminishes as the excitation pulse duration is shortened, and approaches unity for very short light pulses.

Simulation of the phosphorescence response with Eq. 8 for different durations of excitation confirms the results given by the simplified convolution model. Equation 8 can be solved numerically for an arbitrary lamp function, L(t), yielding the dimensionless concentration of the excited form. In this model, flash and photochemical processes have the same time parameter. We used the Runge-Kutta method (Mathcad 6.0 PLUS, MathSoft; Cambridge, MA) to obtain solutions of Eq. 8 for the parameter values of phosphor listed above at a concentration of 1 mM. Values for k_x and the scale factor for L(t) were chosen to produce a preset concentration of the excited form that would yield a pseudolinear or nonlinear response function, Y(t).

As a practical example, the behavior of Pd-meso-tetra (4-carboxyphenyl) porphine, in the presence of molecular oxygen, has been simulated using the following values: $k_0 = 17.4\cdot10^{-4}\,\mu s^{-1}$ and $k_q = 2.06\cdot10^{-4}\,\mu s^{-1}\,\mu M^{-1}$ (Sinaasappel and Ince, 1996).

Phosphorescence time courses in response to different excitation functions for variable oxygen concentrations (0–100 µM) were calculated according to Eq. 8. The range 0–100 µM was chosen since it corresponds to the normal physiological range of oxygen

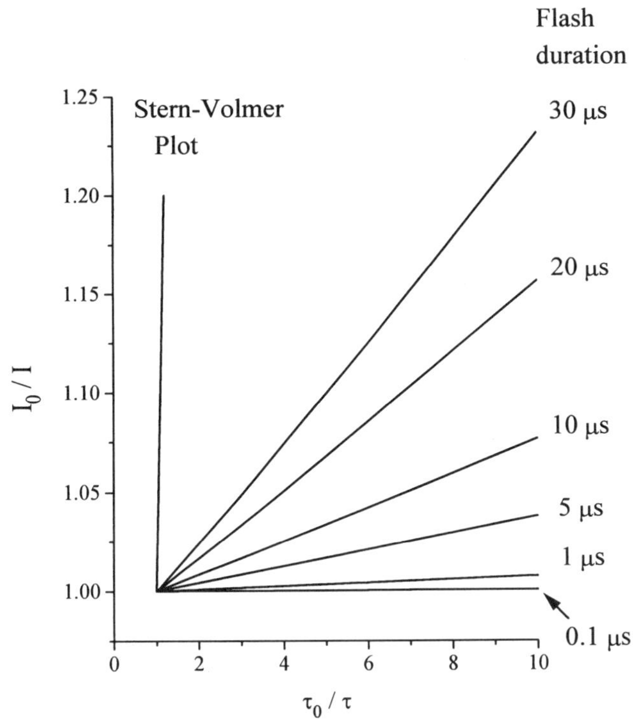

Figure 1. The dependence of the ratio of intensities, I_0/I, on τ_0/τ for different durations of the rectangular excitation light pulse of a constant intensity obtained from the convolution model (Eq. 6). In this example $\tau_0 = 600\,\mu s$ and τ varies from 60 to $600\,\mu s$. The identity $I_0/I \equiv \tau_0/\tau$ (Stern-Volmer ratio) is only valid for very long excitation times which significantly exceed the flash duration of a short arc xenon lamp typically applied in the method. Shortening the flash duration from 30 to $0.1\,\mu s$ causes the convergence of I_0/I to unity, so that the phosphorescence intensity for different lifetimes remains the same.

concentration in the microcirculation. The non-linear least-squares Marquardt Levenberg fitting algorithm was used for data analysis (Origin 3.5, Microcal Software Inc., Northampton, MA).

For the purpose of simulating the excitation process, the time profiles for the step function $L(t) = L$, for $t > 0$, were calculated to estimate the time necessary to reach a steady state concentration of the triplet form for different oxygen concentrations. In order to simulate the pulse excitation, the flash waveform (xenon flash lamp 35-S Chadwick-Helmuth; 600 V, 4 J) was described empirically by Eq. 3 with $\tau_1 = 6\,\mu s$, $\tau_2 = 8\,\mu s$ (flash duration was $30\,\mu s$). The fitting delay (i.e., the initial part of the phosphorescence emission curve usually skipped in the signal analysis; see further explanation below) was taken as $60\,\mu s$ to determine differences in initial intensities for different quencher concentrations. For comparison, the response for a light pulse of the same time profile, but $1\,\mu s$ duration, has also been simulated and analyzed using a fitting delay of $3\,\mu s$. To ensure the linear response, the excitation light intensity in these simulations was chosen so that no more than 1% of the phosphor was excited.

The results show that the time required to reach a steady state depends on the oxygen concentration and generally is about 6τ, where τ is the phosphorescence lifetime at the given $[O_2]$. For our example of Pd-meso-tetra (4-carboxyphenyl) porphine and oxygen, 6τ is more than 3 milliseconds for the lowest oxygen concentrations. The ratio of initial intensities obeys the Stern-Volmer equation in the form of Eq. 2 only after a steady state has been achieved. The difference of initial intensities between samples with different $[O_2]$ diminishes with the shortening of the excitation period. Unfortunately, the concentration of excited phosphor also decreases, with the consequence that the intensity of the excitation light must be increased in order to obtain a detectable phosphorescence light emission.

The results of the simulation of phosphorescence intensity caused by the excitation flash with 30 μs duration are shown in Figure 2, which displays the relationship of the time courses between excitation light intensity and phosphorescence intensity obtained from two identical compartments having oxygen concentrations of 0 and 100 μM. The phosphorescence peak is delayed relative to the flash peak longer for the lower oxygen concentration component and the ratio of peak intensity amplitudes obtained from the two compartments is 1.5.

In practice, data points obtained during the rising part of the phosphorescence emission curve and the early part of its decay are contaminated with the excitation light.

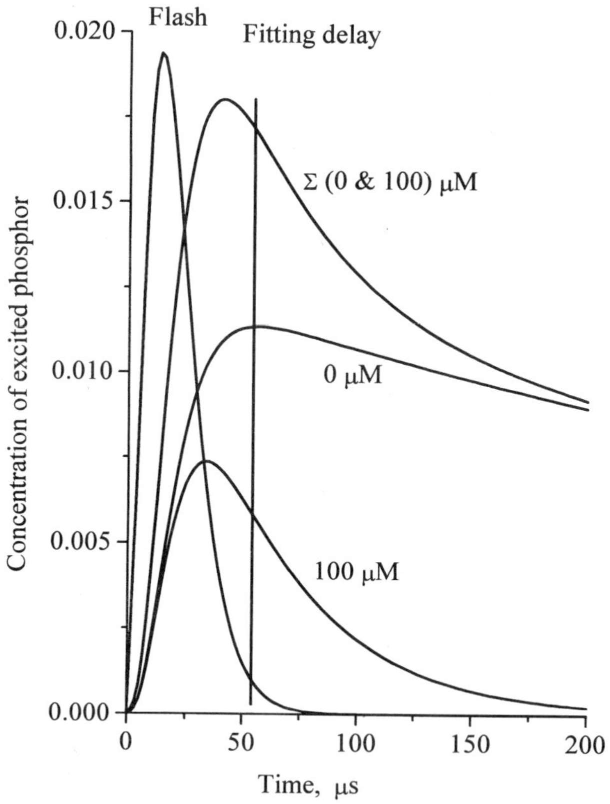

Figure 2. Phosphorescence curves caused by an excitation flash with a duration of 30 μs. The fraction of phosphor molecules in the excited state, n, is linearly related to the light emission intensity. The fitting delay was taken as 60 μs to determine differences in initial intensities for different quencher concentrations. The plot displays the relationship of the time courses of excitation light intensity and phosphorescence intensity obtained from two identical compartments having oxygen concentrations of 0 and 100 μM. Thus, if the ratio of peak intensity amplitudes obtained from these two compartments is 1.5, the amplitude ratio for a 60 μs fitting delay will be 2.2. Therefore, analysis of a mixed signal (dashed line) will yield a significant underestimation of the mean quencher concentration in this simple two-compartment system.

These points are usually skipped, so that the analysis begins at a time t_0, after the initiation of excitation when the signal begins to decrease (Torres Filho and Intaglietta, 1993; Vinogradov and Wilson, 1994; Kerger et al., 1995; Shonat et al., 1995 and 1997; Sinaasappel and Ince, 1996; Vinogradov et al., 1996; Zheng et al., 1996). The results of Figure 2 illustrate this important concept of the "fitting delay," a time (t_0) that corresponds to the interval between the start of excitation ($t = 0$) and the beginning of the monotonic decay. This approach helps to exclude the flash artifact and excitation process from the analyzed curve. For the purpose of analysis of the decay curve, the first data point after the fitting delay corresponds to a shifted initial time (new $t = 0$).

Does use of the fitting delay cause a problem in terms of the magnitude of the initial intensities in a heterogeneous system? Consider two compartments that are each half the size of the reference volume. Since the decay of the mixed curve is dominated by the decay of the slowest component, the first point in the analysis of the mixed curve ($t = t_0$) has an even smaller contribution from the compartment with a high oxygen concentration (see Figure 2). Thus, if the ratio of peak intensities $I(0\,\mu M)/I(100\,\mu M)$ in our example is 1.5, the amplitude ratio at the time of the fitting delay ($t_0 = 60\,\mu s$) will be 2.2. Therefore, recovery analysis of a mixed signal yields a significant underestimation of the mean oxygen concentration in the reference volume, due to the dominant contribution of the component with lower oxygen concentration.

This example could be applied to a mixed phosphorescence signal obtained from a micro- or macroscopic volume of a living organ containing arterioles with high PO_2 and adjacent tissue with low PO_2 (i.e., phosphor is distributed in both the intravascular and extravascular spaces). The results of our simulations show that the parameters returned by a monoexponential analysis could substantially underestimate mean oxygen concentration.

In order to deal with heterogeneity in the initial intensities, we can define the function $G(P) = I(0,t_0)/I(P,t_0)$, analogous to Eq. 2, but specific for any excitation flash function and generally different from τ_0/τ until the excitation illumination lasts long enough to produce a steady state. Since all components of the mixed phosphorescence curve obtained from a heterogeneous system have the same fitting delay, we can also write this function as $G(P) = I(0)/I(P)$, which indicates how much the initial phosphorescence intensities (at $t = 0$ for the decay curve, after shifting the time base by the fitting delay) for PO_2 P are lower than the initial intensity for an oxygen-free medium ($P = 0$). This function can be measured in calibration experiments for the same parameters of excitation light as used for the measurement procedure.

We will use the value of G(100) for $[O_2] = 100\,\mu M$ as a reference for a simple comparison of functions G(P) for different flash durations. Results of this simulation reveal that a brief flash with a $1\,\mu s$ duration has a relatively uniform initial intensity, with G(100) = 1.06, whereas for a flash of $30\,\mu s$ duration, G(100) = 2.18. Thus, the shorter the flash pulse, the weaker is the dependence of initial intensities on oxygen concentration. When the excitation illumination lasts as long as 6τ, the ratio of initial intensities G(P) approaches the relationship predicted by the Stern-Volmer equation; in this example G(100) = 13.3.

3.2. Correction for Nonuniformity of Initial Intensities

In a heterogeneous system the initial intensities originating from elementary volumes having different quencher concentrations are different at the initial point of the phosphorescence decay curve (i.e., at the fitting delay). The contributions of the various regions of a nonuniform system to the total phosphorescence output depend on the energy and duration of the excitation light and on the fitting delay. All these parameters must be set to the same fixed values for calibration and for experimental measurements to ensure compatibility. The correction function G(P) (for first point after the fitting delay), is obtained from the calibration procedure, and can help to establish the volume fraction of each quencher concentration on the basis of its contribution to the overall phosphorescence light intensity.

It is convenient to analyze of the phosphorescence decay curves obtained from a nonuniform system using two independent variables. First, the distribution of the reciprocal lifetimes is recovered in terms of weight coefficients. Second, these weight factors are corrected using G(P) to yield the volume distribution of oxygen concentration.

3.3. Fitting Function for Distribution of Reciprocal Lifetimes, f(λ)

Application of the phosphorescence quenching technique for measuring dioxygen in living tissue at the microscopic scale faces additional problems arising from the existence of local gradients in oxygen concentration. The nonuniformity of initial intensities from elementary volumes having different PO_2 distorts the shape of the volume distribution of oxygen tension, v(P). The undistorted information that can be revealed directly from analysis of the phosphorescence decay in a heterogeneous volume is the distribution of phosphorescence lifetimes, τ, or reciprocal lifetimes, $\lambda = 1/\tau$. One can consider f(λ) to be the spectrum, or weight distribution, of reciprocal lifetimes into which the time course of the phosphorescence emission curve can be decomposed. The motivation for using λ instead of P as a primary variable comes from the connection between the reciprocal lifetime, $\lambda = 1/\tau$, and the PO_2, P. These two variables are linearly related by the lifetime version of the Stern-Volmer equation (Eq. 1):

$$\lambda = 1/\tau = k_0 + k_q P \quad \text{or} \quad P = (\lambda - k_0)/k_q \tag{9}$$

The problem of recovering the volume distribution of oxygen tension, v(P), can be solved in three sequential steps: (1) finding the fitting function, Y(t), of the normalized phosphorescence decay curve in terms of the continuous distribution of reciprocal lifetimes, f(λ); (2) converting the distribution f(λ) into the volume distribution of reciprocal lifetimes v(λ) using the correction function for initial intensities; and (3) obtaining the volume distribution v(P) from v(λ). A correction function G(P) for fixed flash parameters and fitting delay must be established in calibration experiments and then rescaled as G(λ). Using the substitution of variables in Eq. 9, G(P) can be converted to G(λ), then v(λ) to v(P), the desired volume distribution of quencher. The function G(λ) indicates the ratio of initial intensity at λ_0 relative to that at λ:

$$G(\lambda) = I(\lambda_0)/I(\lambda) \tag{10}$$

The function f(λ) as a fraction of total light emitted for any λ is proportional to the corresponding volume fraction v(λ) divided by G(λ):

$$f(\lambda) = a \cdot v(\lambda)/G(\lambda) \tag{11}$$

where the constant α is a proportionality coefficient.

If the function $f(\lambda)$ is obtained by fitting the phosphorescence decay curve, and the correction function, $G(\lambda)$, is known from calibration experiments, then we can rewrite Eq. 11 by collecting the known functions of λ on the right hand side:

$$\alpha \cdot v(\lambda) = f(\lambda) \cdot G(\lambda) \tag{12}$$

Since $v(\lambda)$ is defined as a normalized distribution of volume fractions, we have

$$\alpha = \int_{\lambda_0}^{\infty} \alpha \cdot v(\lambda)\, d\lambda = \int_{\lambda_0}^{\infty} f(\lambda) \cdot G(\lambda)\, d\lambda \tag{13}$$

Substituting this expression for α into Eq. 11 and solving for $v(\lambda)$, we have

$$v(\lambda) = f(\lambda) \cdot G(\lambda)/\alpha = f(\lambda)\, G(\lambda) \Big/ \int_{\lambda_0}^{\infty} f(\lambda) \cdot G(\lambda)\, d\lambda \tag{14}$$

Finally $v(\lambda)$ can be converted to $v(P)$ using the substitution in Eq. 9.

In order to construct fitting functions that model the continuous distribution of λ having a minimal number of fitting parameters, we used the convolution of simple functions of statistical distributions with exponential decay. To obtain the pulse response function $y(t)$ for any distribution of λ we use the Laplace transform to express the normalized decay curve as:

$$y(t) = I(\lambda, t)/I(\lambda, 0) = \int_{0}^{\infty} \exp(-\lambda t) \cdot f(\lambda) d\lambda \tag{15}$$

This distribution function satisfies the following normalization condition:

$$\int_{\lambda_0}^{\infty} f(\lambda) d(\lambda) = 1 \tag{16}$$

Furthermore, the first two central moments of $f(\lambda)$ can be written as

$$m = \int_{0}^{\infty} \lambda \cdot f(\lambda) d(\lambda); \qquad \sigma^2 = \int_{0}^{\infty} (\lambda - m)^2 \cdot f(\lambda) d(\lambda) \tag{17}$$

Statistical models of a continuous distribution can be used to evaluate the moments of $f(\lambda)$, by fitting the phosphorescence decay to specific distributions of PO_2. For instance, the Rectangular distribution is defined as:

$$f(\lambda) = 1/2\delta \qquad \text{for } \lambda_{min} \leq \lambda \leq \lambda_{max} \tag{18}$$

where $\delta = (\lambda_{max} - \lambda_{min})/2$. The first two central moments are:

$$m = (\lambda_{min} + \lambda_{max})/2 \quad \text{and} \quad \sigma^2 = \delta \tag{19}$$

The associated fitting equation that describes the phosphorescence decay is, according to the integration of Eq. 15 from $m - \delta$ to $m + \delta$:

$$y_R(t) = \exp(-mt) \cdot \sinh(\delta t)/\delta t \tag{20}$$

4. DISCUSSION

The Stern-Volmer equation for phosphorescence intensities has been formulated with the assumption of equilibrium between phosphor excitation and deactivation. The duration of the excitation light pulse necessary to achieve this equilibrium must exceed six phosphorescence lifetimes. For shorter excitation periods, the Stern-Volmer ratio of phosphorescence intensities is not valid, but the ratio of lifetimes remains correct.

A typical excitation flash of finite duration is not long enough to produce steady state excitation and not short enough to provide equal excitation of all compartments having different oxygen concentrations. In the case of a heterogeneous distribution of oxygen concentration, a typical excitation flash will produce different initial intensities (which do not obey the Stern-Volmer equation) and different times of peak emission intensity from compartments having different oxygen concentration. Thus, the overall signal from a heterogeneous system at the moment of the fitting delay ($t = t_0$) is composed of contributions with different weight factors. This unknown difference of initial intensities coming from different fractions of the reference volume significantly complicates the task of recovery of the volume distribution of oxygen in strongly heterogeneous systems like a microvessel together with the adjacent interstitium and parenchymal cells.

Using a fitting delay in the analysis of a phosphorescence decay curve helps to avoid the flash artifact. However, this procedure significantly decreases the fraction of the fastest decaying components used in the analysis of the mixed signal (i.e., highest PO_2). Furthermore, it leads to an inappropriate weighting of the various components of the signal and an inaccurate recovery of the mean value of PO_2. Moreover, since the time course of the fastest decaying components will be almost completed by the time of the fitting delay, the accuracy of measurements of the faster components (i.e., higher PO_2) will be less than those of the slower ones. The domination of the amplitude of the decay signal by the low PO_2 component in a two-compartment system (e.g., blood vessel/surrounding tissue) can yield a large systematic error weighted heavily toward the lower PO_2 component of the system.

The parameters of a real excitation flash involve a trade-off between high flash intensity and short flash duration. On the one hand, a high flash intensity is desirable in order to produce a high signal-to-noise ratio. On other hand, a short flash duration is desirable to produce equal excitation of all compartments in a heterogeneous system and to enable use of a short fitting delay. This means that the excitation light pulse must be short, not only in the usual sense (i.e., width of pulse at 1/3 of maximum light intensity), but the afterglow of the flash lamp in the excitation part of the spectrum should also be as short as possible.

Comparison of the models for two excitation lamp functions expressed as exponential pulse and step functions, shows that I_0/I increases with the time of illumination, finally reaching the Stern-Volmer steady state limit. The difference of initial phosphorescence intensities between compartments having different quencher concentrations can be avoided by shortening the flash, but this leads to a decrease in phosphor excitation. Also, this effect cannot be evaluated and accounted for using the intensity ratio of the Stern-Volmer equation (Eq. 2), because the flash duration is not long enough to produce a steady state for compartments having near zero oxygen concentration. However, the quantitative analysis of the processes generating a decay signal from a multicompartmental volume can be accomplished using the function G(P) to correct the volume fractions on the basis of weight coefficients obtained from the deconvolution procedure. The correction function can be determined for any fitting delay and flash energy by calibrating the phosphorimeter for the same conditions as those under which the measurements are performed.

ACKNOWLEDGMENTS

This work was supported in part by grant HL-18292 from the National Heart, Lung and Blood Institute and by a Grant-in-Aid from the American Heart Association, Virginia Affiliate, Inc.

REFERENCES

Alberty R (1983) Physical Chemistry. New York, NY: John Wiley and Sons, Inc.

Buerk D (1993) Biosensors (Theory and Applications). Lancaster, PA: Technomic Pub. Co., Inc.

Buerk D, Tsai A, Intaglietta M, and Johnson P (1998) Comparing tissue PO_2 measurement by recessed microelectrode and phosphorescence quenching. In: Oxygen Transport to Tissue XX, (Hudetz A, Bruley D, Eds) Plenum Press, New York, pp. 367–374.

Edgerton H (1979) Electronic Flash, Strobe. Cambridge, MA: The MIT Press.

Gardner D, Gardner J, Laush G, and Menke W (1959) Methods for analysis of multicomponent exponential decay curves. J Chem Phys 31:978–986.

Golub A, Popel A, Zheng L, and Pittman R (1997) Analysis of phosphorescence in heterogeneous systems using distribution of quencher concentration. Biophys J 73:452–465.

Kerger H, Torres Filho I, Rivas M, Winslow R, and Intaglietta M (1995) Systemic and subcutaneous microvascular oxygen tension in conscious Syrian golden hamsters. Am J Physiol 268:H802–H810.

Lo L-W, Koch C, and Wilson D (1996) Calibration of oxygen-dependent quenching of the phosphorescence of Pd-meso-tetra (4-carboxyphenyl) porphine: a phosphor with general application for measuring oxygen concentration in biological systems. Anal Biochem 236:153–160.

Marshak I (1984) Pulsed Light Sources. Consultants Bureau, New York.

Rumsey W, Schlosser C, Nuutinen E, Robiolio M, and Wilson D (1990) Cellular energetics and the oxygen dependence of respiration in cardiac myocytes isolated from adult rat. J Biol Chem 265:15392–15399.

Shonat R, Richmond K, and Johnson P (1995) Phosphorescence quenching and the microcirculation: An automated, multipoint oxygen tension measuring instrument. Rev Sci Instr 66:5075–5084.

Shonat R, Wachman E, Niu W, Koretsky A, and Farkas D (1997) Near-simultaneous hemoglobin saturation and oxygen tension maps in mouse brain using an AOTF microscope. Biophys J 73:1223–1231.

Sinaasappel M and Ince C (1996) Calibration of Pd-porphyrine phosphorescence for oxygen concentration measurements in vivo. J Appl Physiol 81:2297–2303.

Sinaasappel M (1997) Microvascular pO_2 in Models of Shock. Doctoral Thesis. University of Amsterdam. Amsterdam, The Netherlands.

Torres Filho I and Intaglietta M (1993) Microvessel pO_2 measurements by phosphorescence decay method. Am J Physiol 265:H1434–H1438.

Vanderkooi J, Maniara G, Green T, and Wilson D (1987) An optical method for measurement of dioxygen concentration based upon quenching of phosphorescence. J Biol Chem 252:5476–5482.

Vinogradov S and Wilson D (1994) Phosphorescence lifetime analysis with a quadratic programming algorithm for determining quencher distribution in heterogeneous systems. Biophys J 67:2048–2059.

Vinogradov S and Wilson D (1995) Metallobenzoporphyrins. New phosphorescent probes for oxygen measurements. J Chem Soc Perkin Trans 2:103–111.

Vinogradov S, Lo L-W, Jenkins W, Evans S, Koch C, and Wilson D (1996) Noninvasive imaging of the distribution in oxygen in tissue in vivo using near-infrared phosphors. Biophys J 70:1609–1617.

Ware W (1991) Recovery of fluorescence lifetime distributions in heterogeneous systems. In: Photochemistry in Organized and Constrained Media. (Ramamurthy V, Ed) VCH Pub, New York, pp. 563–602.

Ware W, Doemeny L, and Nemzek T (1973) Deconvolution of fluorescence and phosphorescence decay curves. A least square method. J Chem Phys 77:2038–2048.

Zheng L, Golub A, and Pittman R (1996) Determination of PO_2 and its heterogeneity in single capillaries. Am J Physiol 271:H365–H372.

A THREE WAVELENGTH STUDY ON ARTERIAL OCCLUSION IN SKIN TISSUE

R. Graaff,[1] A. Talsma,[1] F. Folkerts,[2] J. van der Zee,[1] and W. Ament[1]

[1]Department of Biomedical Technology/Artificial Organs
University of Groningen
Bloemsingel 10, 9712 KZ Groningen
The Netherlands
[2]Technical Research Department
University Hospital Groningen
The Netherlands

1. INTRODUCTION

Near infrared spectroscopy (NIRS) is a non-invasive optical method to monitor changes in the oxygenation and concentration of hemoglobin and myoglobin in biological tissue. Moreover, information on changes in the oxygenation of cytochrome aa_3 can be obtained. A large number of studies have been performed during the last decades to develop the method (*e.g.*, Jöbsis, 1977, Chance et al., 1988, Delpy et al., 1988). Apart from the large number of studies to apply this technique to the monitoring of brain oxygenation, several studies have also been performed on other applications such as skeletal muscle oxygenation (*e.g.*, Hampson and Piantadosi, 1988) and on monitoring flaps in plastic surgery (Irwin et al., 1995).

Values for the mentioned variables are obtained from the measurement of the changes in the light intensity at several wavelengths after light propagation through the tissue. The observed changes are mainly caused by the changes of the volume fraction of blood and by the changes in the oxygenation of blood and myoglobin in the sample volume under the sensor. Cytochrome aa_3 has a concentration in tissue that is at least an order of magnitude smaller than that of hemoglobin (Colier, 1995, Elwell, 1995). Therefore, we neglect contributions by changes in the oxygenation of cytochrome aa_3. Furthermore, we neglect differences between the absorption spectra of hemoglobin and myoglobin (Hampson and Piantadosi, 1988). Therefore, only two wavelengths may be sufficent for some cases, even for the case of muscle studies.

Oxygen Transport to Tissue XXI, edited by Eke and Delpy.
Kluwer Academic / Plenum Publishers, New York, 1999.

In the past, we have performed studies on reflectance pulse oximetry (RPO), a non-invasive optical method to monitor the arterial oxygen saturation. We developed a measuring system for RPO based on a red and an infrared light emitting diode (660 nm and 940 nm LEDs). This system has the option to measure at several distances from the light sources simultaneously, with the LEDs and detectors integrated into one sensor (Graaff et al., 1993*a*). In a later stage, we showed that this system could also be used for NIRS, as it is capable to determine changes in tissue oxygenation and tissue blood content simultaneously (Talsma et al., 1995). Since that time, we modified the measuring system. A third wavelength was added (730 nm), as described in the present paper.

Besides the instrumentation, also the theoretical basis of NIRS is related to that of pulse oximetry. The signals in methods as RPO and NIRS have often been explained with modifications of the Lambert-Beer model, that are based on several assumptions. One of these assumptions is that the photon path lengths between light source and detector are constant. In some models, it has been assumed that these photon path lengths are equal for all wavelengths used. However, previous studies on light scattering within tissue and blood have shown that absorption and scattering properties are influencing light path lengths, which makes the signal interpretation more complex. For that reason, we based the calculations of changes in oxygenation and blood volume fraction from the measured signals on Monte Carlo simulations of light propagation in tissue.

The purpose of the present study is to contribute to the theoretical basis of NIRS. For that aim, it is investigated whether the measurements at the third wavelength (730 nm) could be explained by the calculated changes in blood volume and tissue oxygenation, which were obtained from the signals of the 660 nm and 940 nm LEDs only. This study has been based on measurements on the forearm of human volunteers during total occlusion of the upper arm, using an inflatable cuff.

2. METHODS

2.1. Instrumentation

For the three wavelengths study we modified our reflectance pulse oximeter instrumentation that was used in previous studies (Graaff et al., 1993*a*,*b*, and 1996, Dassel et al., 1992, 1995, and 1996). The reflectance pulse oximeter system was developed for the simultaneous measurement of the ratio between the red and infrared intensity changes on the skin surface at several distances between light sources and detectors.

A similar probe with a diameter of 18 mm was used as in our previous studies, as shown in Figure 1. The sensor consists of three LEDs [CR 10 1M (660 nm), CR 10 IR-C (940 nm), and CR 10 IR-K (730 nm), Elcos, Pfaffenhofen, Germany]. Furthermore, three photodiodes have been integrated into the probe (BPX 90, 1.65 by 3.05 mm, Siemens, München, Germany). In the experiments described below, only the signals from two photodiodes have been used, that have their centers at distances of 7.0 and 9.5 mm from the LEDs. The probe has been subdivided into two optically isolated parts in order to prevent the detection of photons that have only travelled within the probe. The bottom and internal walls of the probe housing as well as the sidewalls of the photodiodes have been painted black in order to avoid photons from re-entering the tissue by means of reflectance within the probe. Both probe compartments have been filled with clear epoxy resin (EPOTEK 301).

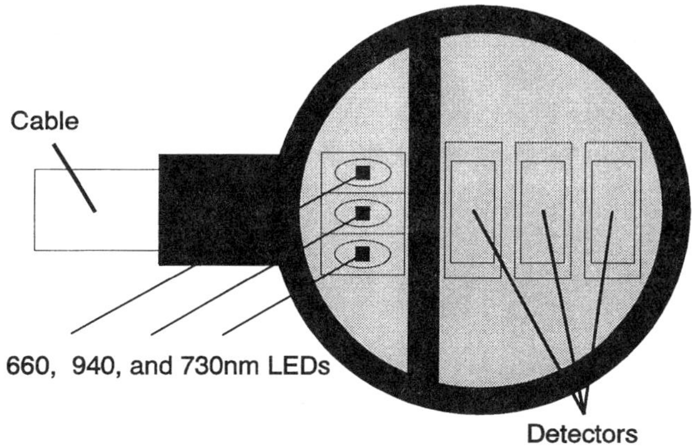

Figure 1. Multi-site sensor.

A block diagram of the electronic circuits is shown in Figure 2. The LEDs are activated quasi-continuously during periods of 20 μs, with a repetition rate of 1 kHz. This process has been programmed in an EPROM. The LED drivers are activated with TTL pulses, which have been isolated optically. These LED drivers are situated in an interface box in the vicinity of the patient. The detected intensities from the three detectors are preamplified at that location as well. Electrical isolation of the signals was obtained with isolation amplifiers. The signals, which contain information on the intensity changes for each of the three LEDs, were demultiplexed with sample-hold circuits. New electronic circuits were developed for demultiplexing three signals for each detector. After subtraction of background illumination these circuits take samples during the last part of each of the LED-pulses. Timing of these processes could be realized with the same EPROM that activates the LED-drivers. Subsequently, the signals were amplified logarithmically, and converted to 12 bit digital values, which were stored in a personal computer. Further processing was performed with a personal computer, on-line as well as off-line.

2.2. Occlusion Experiment

Measurements were obtained on four male volunteers at the inner side of the right forearm, approximately 8 cm from the elbow, with the sensor attached to the skin with a

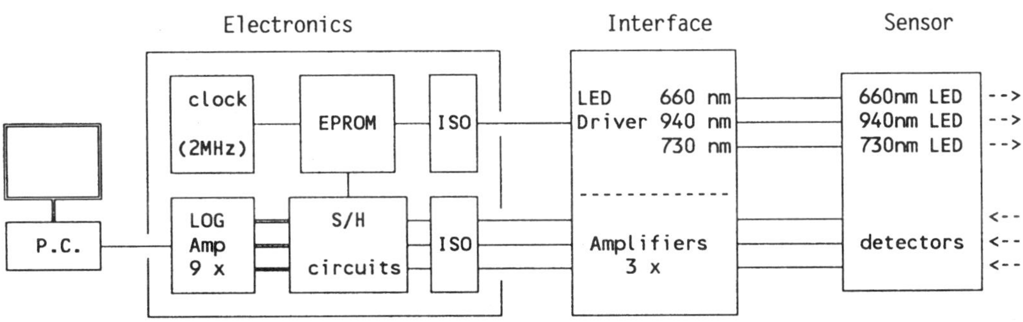

Figure 2. Block diagram of the NIRS instrument.

transparent, double-sided adhesive ring. The subjects were sitting in a chair with their right lower arm resting on a table. The changes in the intensity were measured during 10 minutes. After one minute of registration of the baseline conditions, a cuff at the upper arm was inflated to obtain a total occlusion during five minutes. After that period, the cuff was released and four minutes of the recovery were registrated. Intensities obtained from all three LEDs and at both detectors were recorded each second during the whole period of measurement.

2.3. Simulations of Light Propagation

Monte Carlo simulations of light propagation in a homogeneous semi-infinite medium were used to relate the changes in light intensity to changes in absorption and scattering. The simulations were performed for both rectangular detectors, located at 7.0 and 9.5 mm from the light source. For the tissue medium we used a refractive index $n_{refr} = 1.4$ against air (Bolin et al., 1989).

To investigate the influence of the chosen phase function, we used two different scattering profiles: the isotropic scattering phase function as well as forwardly scattering Henyey-Greenstein scattering, with $g = 0.875$, where g is average cosine of the scattering angle, or the asymmetry factor of the phase function (Henyey and Greenstein, 1941). Furthermore, to investigate the influence of the internal reflection at the tissue interface caused by the refractive index mismatch, simulations were not only obtained for $n_{refr} = 1.4$, but also for the matched boundary condition, using $n_{refr} = 1.0$. Finally, numerical results for each detector were also obtained with diffusion theory for both mentioned boundary conditions.

Values of the captured photon fraction as a function of the absorption coefficient, μ_a, and the reduced scattering coefficient, μ_s', for each set of data were obtained by using the *condensed* Monte Carlo technique (Graaff et al., 1993*c*). This technique allows to use the Monte Carlo results for the semi-infinite medium again to obtain results for other absorption and scattering coefficients.

A fitted equation was obtained for each set of data to describe the numerical results. Figure 3 shows the Monte Carlo results and the fitted equation for the natural logarithm of the fraction of emitted photons that is captured by the detector at 7 mm, both for the case of the isotropic scattering function.

The changes in the intensity I are proportional to the changes in the captured photon fraction in a given situation. The derivatives $d\ln(I)/d\mu_a$ and $d\ln(I)/d\mu_s'$ in a given situation have therefore the same values as the corresponding derivatives of the logarithm of the photon fraction. The corresponding values can therefore be obtained by partial differentiation of the fitted expression for ln(*Photon Fraction*). These derivatives are a function of both the absorption and reduced scattering coefficients.

2.4. Theoretical Model to Interpret the Measurements

We assumed that changes in the measured intensities for each LED are caused by changes in the blood volume fraction, denoted by f_b, and the mean fraction of hemoglobin and myoglobin molecules in the tissue under the probe that are saturated with oxygen, $S_{O2,tis}$. Therefore, we applied the following equation to the intensity changes with the 660 nm LED and the 940 nm LED:

$$d\ln[I(\lambda)] = dI(\lambda)/I(\lambda) = X_1(\lambda)df_b + X_2(\lambda)dS_{O2,tis}. \quad (1)$$

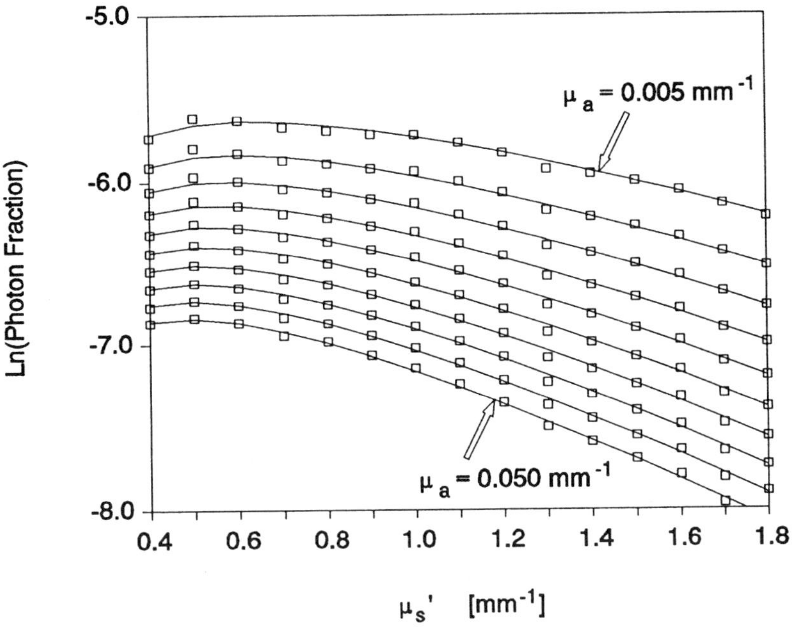

Figure 3. Results of a condensed Monte Carlo simulation. The logarithm of the photon fraction, captured by the rectangular sensor at 7 mm from the light source, is given as a function of the absorption and reduced scattering coefficient of a semi-infinite medium; $n_{refr} = 1.4$. The solid line describes the fitted equation.

For simplicity, we thereby assumed that hemoglobin and myoglobin have the same molar absorptivity. Using the measured changes of the intensity from the 660 nm and 940 nm LEDs, the changes in f_b and $S_{O2,tis}$ can then be obtained for a given sensor. Herein, the values X_1 and X_2 are the partial derivates with respect to changes in f_b and $S_{O2,tis}$, respectively, keeping the other variables constant. Initial values for f_b and $S_{O2,tis}$ have to be chosen in this model, as well as scattering coefficients for the tissue.

The changes in blood volume and oxygen saturation cause changes in the absorption and scattering properties. Therefore, the values for X_1 and X_2 for each sensor and at each wavelength could be based on numerical results for light propagation in a semi-infinite turbid medium, using diffusion theory or Monte Carlo simulations, as described above (Talsma et al., 1995).

3. RESULTS

3.1. Calculated Changes in Blood Volume and Tissue Saturation at Both Detectors

From the measured intensity differences we calculated the changes in blood volume and tissue oxygen saturation for each detector separately. We assumed an initial tissue oxygen saturation of 70%, and an initial tissue blood volume of 5%, based on the tissue optical model. Results have been worked out using the intensity changes from the 660 nm and 940 nm LEDs, for both detectors separately.

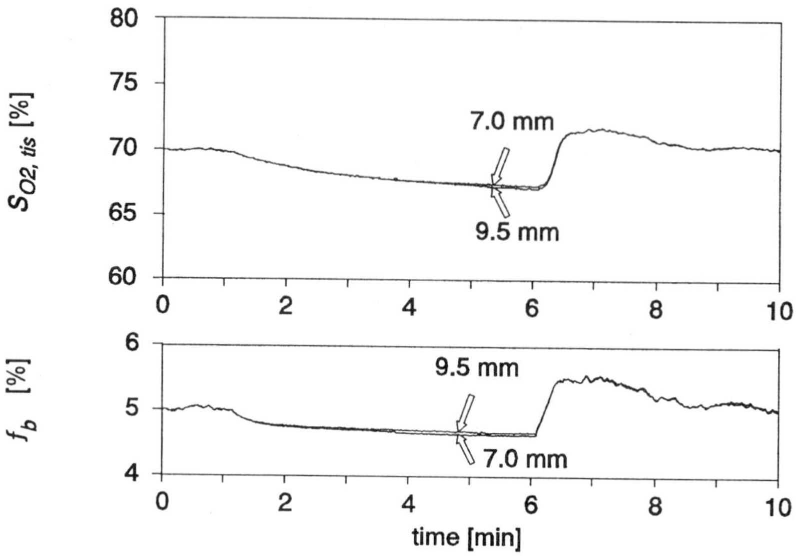

Figure 4. Example of calculated changes in $S_{O2,tis}$ and f_b, based on measurements during an occlusion experiment at detectors at 7 and 9.5 mm. Assumed values at $t = 0$: $S_{O2,tis} = 70\%$ and $f_b = 5\%$. Subject #4.

The results show that the calculated changes in blood volume and tissue saturation as obtained for the detector at 7.0 mm were in some subjects practically the same as the results obtained for the detector at 9.5 mm, as shown in Figure 4.

In other subjects the results from both detectors showed more differences (Figure 5). We suppose that these differences are related to an increasing influence of optical properties of subdermal tissues, especially at the farthest detector. This was consistent

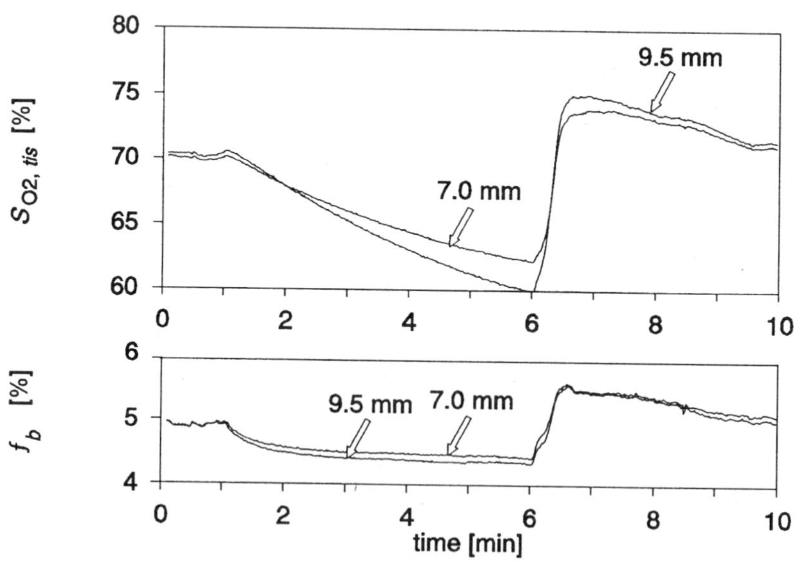

Figure 5. Example of calculated changes in $S_{O2,tis}$ and f_b, based on measurements during an occlusion experiment at detectors at 7 and 9.5 mm. Assumed values at $t = 0$: $S_{O2,tis} = 70\%$ and $f_b = 5\%$. Subject #3.

with the skin thickness: the differences were largest in subject #3, who had the thinnest skin (Figure 5), and smallest in subject #4, who had the thickest skin of the four volunteers (Figure 4).

3.2. Discrepancies between the Measured and Calculated Intensities with a Third LED

With the calculated changes in blood volume and tissue saturation, based on the measured signals from the 660 nm and 940 nm LEDs only, it is possible to calculate the changes in intensity at other wavelengths and compare them with measured intensity changes. We used this method to validate the two-wavelength model by comparing calculated intensity variations at the wavelength of the third LED with the measured intensity variations as obtained with that LED. It appeared that the slope of the measured intensity decrease during occlusion was always steeper than that of the calculated intensity, also for the person with the thickest skin, as shown in Figure 6.

The assumptions with respect to the optical properties of the tissue below the sensor, as given in Table 1, may have influenced the calculated results. Therefore, we have varied the tissue optical properties in the calculation procedure to see whether they may explain the discrepancies as observed with the third wavelength.

To eliminate the influence of optical properties of deeper layers in the tissue, we selected for this analysis only the measured data on which Figures 4 and 6 are based. The close agreement between the results in Figure 4 at both detectors indicates that the results at both detectors can be considered to be determined by changes in the same tissue layer. This is probably only the skin, and not partly the muscles of the arm just beneath the skin.

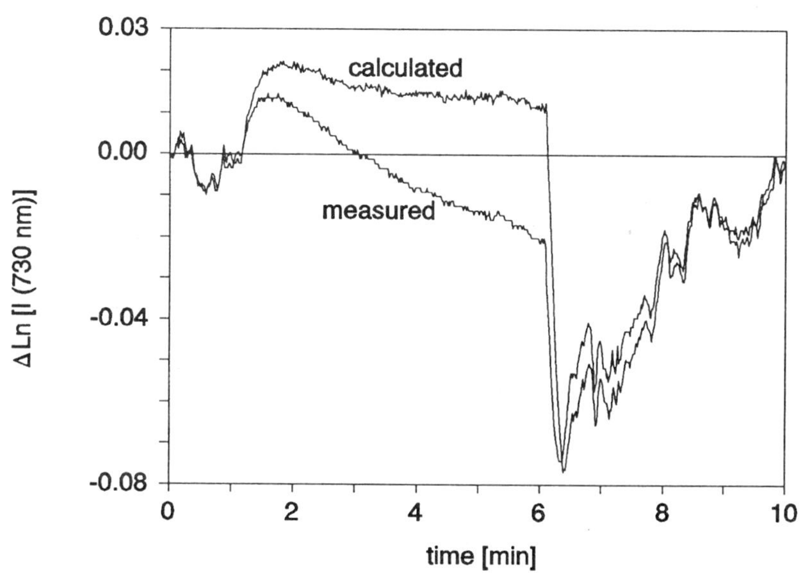

Figure 6. Measured and calculated changes in the intensity for the 730 nm LED at 9.5 mm. The calculations are based on the values of $S_{O2,tis}$ and f_b for subject #4, as given in Figure 4, that were obtained from measurements with the 660 and 940 nm LEDs.

Table 1. The difference in Ln(I) between $t = 3$ and $t = 6$ minutes during total occlusion of the arm for the 730 nm LED and the detector at 9.5 mm. Data as in Figure 6

	Ln[$I(t = 6)$] − Ln[$I(t = 3)$]
Measured difference	−0.0214
Calculated difference	
Blood and Myoglobin:	
initial blood volume fraction f_b = 5% at $t = 0$.	
initial oxygen saturation $S_{O2,tis}$ = 70% at $t = 0$.	
myoglobin concentration $c_{Mb,tot}$ = 0.0 mmol/l.	−0.0036
Blood Free Tissue Scattering:	
$\mu_{s,tis}' = 0.83\,mm^{-1}$ at 760 nm.	
Wavelength dependence factor: $(\lambda/760)^{-1.8}$ ($\beta = 0.1$).	
Phase function: diffusion theory.	
Tissue refractive index: 1.4.	

The basic tissue optical properties in the calculation that have to be varied to eliminate the disagreement are the absorption and reduced scattering coefficients of the medium. In the present model, the absorption coefficients are influenced by the blood volume fraction, f_b, the tissue oxygen saturation, $S_{O2,tis}$, and the total myoglobin concentration, $c_{Mb,tot}$, and by water absorption, that plays a role in the signals from the 940 nm LED. The reduced scattering coefficients of the tissue consist of scattering contributions by blood and blood-free tissue, and can be described by the reduced scattering coefficients of the tissue at a given wavelength, μ_s', and its wavelength dependence (Graaff, 1993*b*). The wavelength dependence of the forwardly scattering part of μ_s' is assumed to be proportional to $\lambda^{-0.6}$ (Talsma et al., 1995), whereas the remaining part is assumed to be caused by Rayleigh scattering, which is proportional to λ^{-4}. The wavelength dependence of μ_s' has also been varied in an attempt to eliminate the disagreement between the measured and calculated results in Figure 6. This influence has been determined by variation of β, the fraction of Rayleigh scatterers at 633 nm (Jacques et al., 1987). Furthermore, the influence of the tissue optical model itself was determined: the assumptions of diffusion theory versus the much more reliable results obtained by the Monte Carlo simulation of light propagation, the choice of the phase function for each scattering event, and the influence of refractive index mismatch at the tissue surface.

To search for an explanation of the discrepancies, we have varied the initial values of the optical properties at $t = 0$, and we compared the calculated changes in ln(I) for the 730 nm LED between $t = 3$ and $t = 6$ minutes. The properties that were changed and the effect on the calculated changes in ln(I) for the 730 nm LED are given in Table 2. The results make clear that it is not possible to explain the measured difference of Table 1 by varying any of these tissue optical properties.

4. DISCUSSION

The calculations in Table 2 make clear that the measured results cannot be described with the present model using changes in blood volume and tissue oxygen saturation in a homogeneous medium. Even for the condition where the calculated changes in blood volume fraction and tissue oxygen saturation for different detectors are practically the

Table 2. Study on the influence of variation of the optical properties on the results of Table 1. The correction term for the calculated difference in Ln(I) is given as a function of the variation of the indicated property. The other properties for each case have the values as given in Table 1

	Correction term for calculated ΔLn[I]
Initial blood volume fraction, f_b at $t = 0$: from 5% to 1%	0.0050
Myoglobin concentration, $c_{Mb,tot}$: from 0.0 to 0.1 mmol/l.	−0.0003
Initial oxygen saturation, $S_{O2,tis}$ at $t = 0$: from 70% to 50%	0.0010
Tissue scattering, $\mu_{s,tis}'$ at 760 nm: from 0.83 to 0.60 mm^{-1}	−0.0005
Wavelength dependence factor for tissue scattering:	−0.0017
from $(\lambda/760)^{-1.8}$ to $(\lambda/760)^{-0.6}$ (from $\beta = 0.1$ to $\beta = 0.0$)	
Phase function:	
from diffusion theory to Monte Carlo simulation with:	−0.0017
isotropic scattering	−0.0017
Henyey-Greenstein scattering, $g = 0.875$	
Refractive index, from $n_{refr} = 1.4$ to $n_{refr} = 1.0$	0.0009

same, no agreement could be obtained by varying the tissue optical properties. To our opinion, the deviations in Figure 6 are too large to be explained by effects of cytochrome aa_3 or differences between hemoglobin and myoglobin absorption spectra. Therefore, we expect that other phenomena have to be taken into account. The disagreement in Figure 6 may be due to the fact that our numerical results are based on the assumption that the optical properties of the biological tissue are the same at each position under the sensor. We do not expect that this is caused by tissue layers with different properties, because of the agreement between the calculated changes in $S_{O2,tis}$ and f_b in Figure 4. We expect that this effect is caused by blood vessels, because blood vessels make it impossible for blood cells to be distributed homogeneously over the tissue. Therefore, we want to focus in a future study on the influence of blood vessels on NIRS, which cause an inhomogeneous distribution of blood within the tissue.

ACKNOWLEDGMENTS

The authors thank W. Kaan and S. ten Have (University Hospital Groningen), and the students G.C. de Jong, J. Toolsema, and R. Kolkman for support during the improvement of the equipment.

REFERENCES

Bolin FP, Preuss LE, Taylor RC, and Ference RJ (1989) Refractive index of some mammalian tissues using a fiber optic cladding method. Appl Opt 28:2297–2303.

Chance B, Nikoa S, Kent J, McCully K, Fountain M, Greenfeld R, and Holtom G (1988) Time-resolved spectroscopy of haemoglobin and myoglobin in resting and ischemic muscle. Anal Biochem 174:698–707.

Colier WNJM (1995) Near infrared spectroscopy: toy or tool. Dissertation, University of Nijmegen, The Netherlands.

Dassel ACM, Graaff R, Aarnoudse JG, Elstrodt JM, Heida P, Koelink MH, Mul FFM de, and Greve J (1992) Reflectance Pulse Oximetry in Fetal Lambs. Pediatr Res 31:266–269.

Dassel ACM, Graaff R, Sikkema M, Meijer A, Zijlstra WG, and Aarnoudse JG (1995) Reflectance pulse oximetry at the forehead improves by pressure on the probe. J Clin Monit 11:237–244.

Dassel ACM, Graaff R, Meijer A, Zijlstra WG, Aarnoudse JG (1996) Reflectance pulse oximetry at the forehead of newborns: the influence of varying pressure on the probe. J Clin Monit 12:421–428.

Delpy DT, Cope M, van der Zee P, Arridge S, Wray S, and Wyatt J (1988) Estimation of optical pathlength through tissue from direct time of flight measurement. Phys Med Biol 33:1433–1442.

Elwell CE (1995) A practical users guide to near infrared spectroscopy. Hamamatsu Photonics KK, pp. 1–25.

Graaff R, Dassel ACM, Koelink MH, Aarnoudse JG, Mul FFM de, Zijlstra WG, and Greve J (1993*a*) Condensed Monte Carlo simulations applied to reflectance pulse oximetry. In: Photon Migration and Imaging in random media and tissues, Proc SPIE 1888 (Chance B and Alfano RR, Eds), SPIE, Bellingham, pp. 201–212.

Graaff R, Dassel ACM, Koelink MH, Mul FFM de, Aarnoudse JG, and Zijlstra WG (1993*b*) Optical properties of human dermis *in vitro* and *in vivo*. Appl Opt 32:435–447.

Graaff R, Koelink MH, Mul FFM de, Zijlstra WG, Dassel ACM, and Aarnoudse JG (1993*c*) Condensed Monte Carlo simulations for the description of light transport. Appl Opt 32:426–434.

Graaff R, Dassel ACM, Zijlstra WG, Mul FFM de, and Aarnoudse JG (1996) How tissue optics influences reflectance pulse oximetry. In: Oxygen Transport to Tissue XVII (Ince C, Kesecioglu J, Telci L, and Akpir K, Eds.), Plenum Press, New York, pp. 117–132.

Hampson NB and Piantadosi CA (1988) Near infrared monitoring of human skeletal muscle oxygenation during forearm ischemia. J Appl Physiol 64:2449–2457.

Henyey LG and Greenstein JL (1941) Diffuse radiation in the galaxy. Astrophys J 93:70–83.

Irwin MS, Thorniley MS, Doré CJ, and Green CJ (1995) Near infra-red spectroscopy: a non-invasive monitor of perfusion and oxygenation within the microcirculation of limbs and flaps. Br J Plast Surg 48:14–22.

Jacques SL, Alter CA, and SA Prahl (1987) Angular dependence of He-Ne laser light scattering by human dermis. Lasers Life Sci 1:309–333.

Jöbsis FF (1977) Non-invasive infrared monitoring of central and myocardial oxygen sufficiency and circulatory parameters. Science 198:1264–1267.

Talsma A, Graaff R, Mul FFM de, Ament W, and Greve J (1995) A tissue-optical model to determine changes in saturation and blood volume during ischemia. In: Optical Biopsies, Proc SPIE 2627 (Cubeddu R, Mordon SR, and Svanberg K, Eds) SPIE, Bellingham, pp. 186–195.

HETEROGENEITY DEPTH LOCALIZATION USING NEAR INFRARED TIME RESOLVED SPECTROSCOPY (NIR-TRS)

M. W. Hemelt, B. Yang, and K. A. Kang

Department of Chemical and Biochemical Engineering
University of Maryland Baltimore County (UMBC)
1000 Hilltop Circle, ECS 101
Baltimore, Maryland 21250

1. INTRODUCTION

Time Resolved Spectroscopy (TRS) introduces a sharp pulse of light into a system, and an attenuated broadened output is measured and analyzed. Previous studies (Kang et al., 1994) have shown that TRS is useful for both homogeneous system characterization, and two-dimensional heterogeneity localization. Recently, frequency response analysis theory has been used to reduce Near Infrared (NIR) TRS spectra for more efficient characterization (Kang et al., 1994; Kang et al., 1996). The advantage of using frequency response analysis is that five engineering parameters, used to identify and characterize the system, can be obtained from a single spectrum. The use of frequency response analysis also allows for the calculation of two of these parameters, the magnitude ratio (MR) and the phase shift (ϕ), over a range of modulation frequencies from one spectrum. The MR is defined as the ratio of the output to the input photon counts at a particular modulation frequency, and is useful for two-dimensional heterogeneity localization. The output spectrum lags behind the input by a phase shift (ϕ), which is also a function of the modulated frequency and provides useful information on the heterogeneity depth. Another parameter analyzed in this study is the time constant (τ), defined as the response time of the system to an input disturbance, and it is independent of modulation frequency.

This study intends to locate physiological heterogeneities in three dimensions using system parameters from the frequency response analysis of NIR-TRS spectra. The depth localization was mainly accomplished through the use of the ϕ and τ values.

Oxygen Transport to Tissue XXI, edited by Eke and Delpy.
Kluwer Academic / Plenum Publishers, New York, 1999.

2. METHODS

Computer simulations of various systems were performed to observe the phenomenon clearly without noise and error that may occur during experiments. The system of study is composed of a continuum with an absorber in it. The model size was specified to be $20 \times 20 \times 20$ cm with absorbing boundaries. Since, the goal of the simulation was to simulate a leg, the absorption and scattering coefficients of the surrounding medium were matched to the leg and were 0.1 and 6.0 cm^{-1}, respectively. An absorber ($1.0 \times 1.0 \times 1.0$ cm) was placed at a depth of 1.5 cm (model surface to the heterogeneity center) in the model [Figure 1]. Reflectance measurements were used. The computer simulation involves solving a mathematical model based on the photon diffusion equation (Patterson et al., 1989) using the B-W-K numerical technique (Williford et al., 1980; Bruley, 1980; Bruley, 1995; Kang and Bruley, 1997). The surface of the model was scanned at a constant source-detector (S-D) separation in 0.5 cm increments. The obtained TRS spectra were then transferred into the frequency domain using a computer code developed by Bruley (Bruley, 1974). To analyze the results systematically, a Δ value is used. The Δ value is defined as the obtained value at each S-D position subtracted by a reference value measured far away from the absorber.

3. RESULTS AND DISCUSSION

3.1. Effect of S-D Separation on Photon Penetration Depth

To study the relationship between the S-D separation and the heterogeneity depth, the model was scanned in the horizontal direction in 0.5 cm intervals for S-D separations of 2.0, 3.0, and 4.0 cm. For simplicity, only the values scanned along the straight line immediately above the absorber are illustrated.

3.1.1. Phase Shift. The ϕ is negatively proportional to the mean photon pathlength in the media. Since the photon path changes when a heterogeneity passes through it, the ϕ is a useful tool for localizing the heterogeneity depth. Figure 2 demonstrates the ϕ values

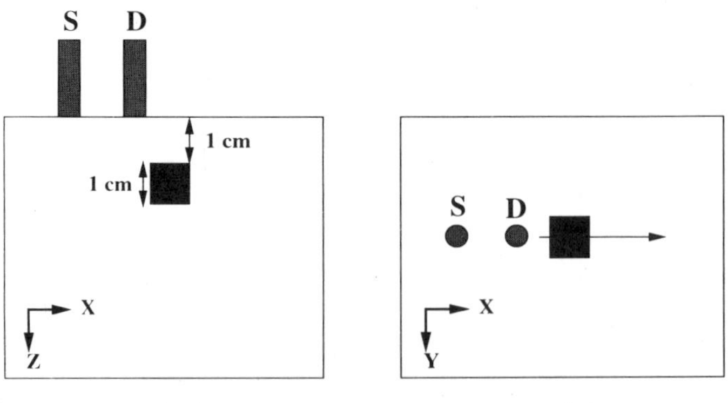

Figure 1. A schematic diagram of the $20 \times 20 \times 20$ cm model with an absorber located at a surface to center depth of 1.5 cm as seen from (a) the side view, and (b) the top view. The absorption and scattering properties of the surrounding medium were 0.1 cm^{-1} and 6.0 cm^{-1}, respectively.

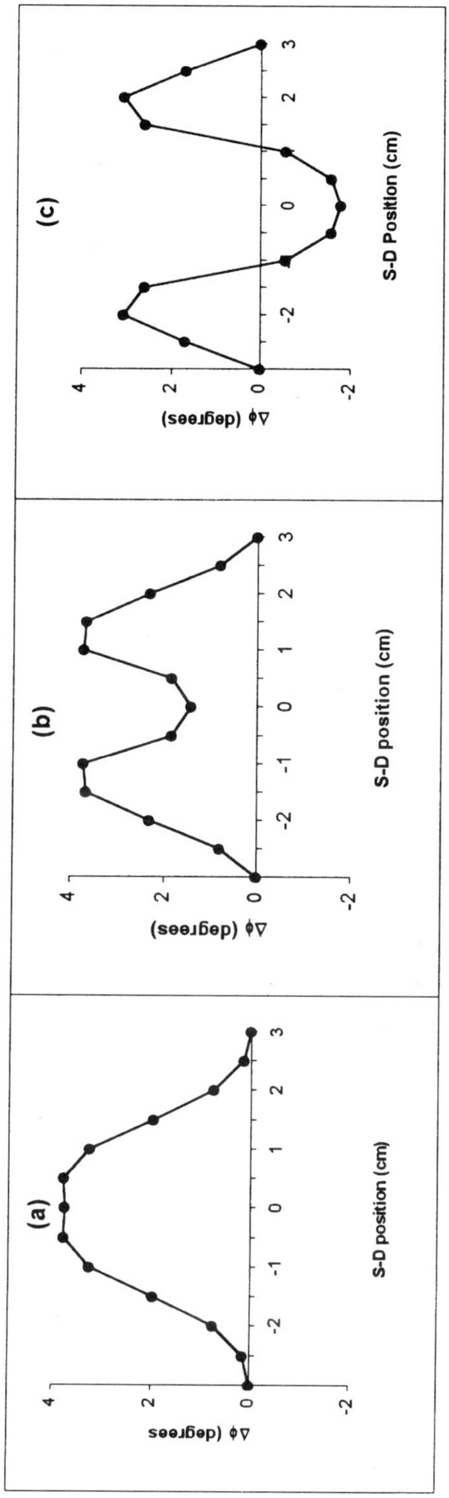

Figure 2. One dimensional profiles of the $\Delta\phi$ values vs. S-D position at an arbitrarily chosen modulation frequency of 100 MHz for (a) 2.0 cm, (b) 3.0 cm, and (c) 4.0 cm S-D separation. As the S-D separation increased, an "M" formation is obtained. The absorber is located at position 0.

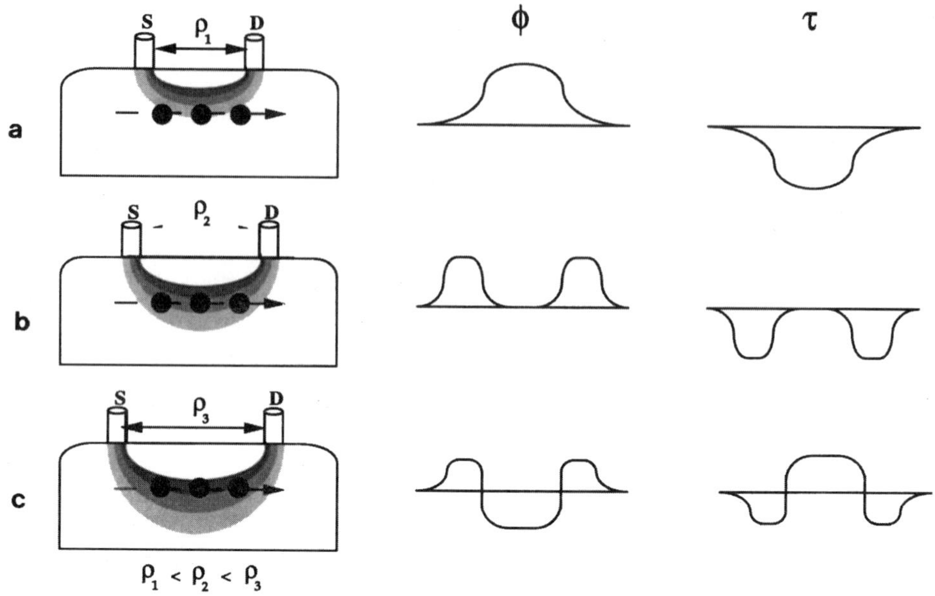

Figure 3. A schematic diagram of the effect of S-D separation on the photon penetration depth. This explains the formation of the "M" shape in the ϕ plots, and the "W" shape in the τ plots.

at an arbitrarily chosen modulation frequency of 100 MHz for S-D separations of 2.0, 3.0, and 4.0 cm.

At a small S-D separation (2.0 cm), the photons do not penetrate deep in the media, and therefore, only the longer photon path is absorbed by the absorber [Figure 3 (a); Barnett et al., 1997; Hemelt et al., 1997]. As a result of the longer path being absorbed by the absorber, the mean photon path decreases, and the ϕ values increase near the absorber [position 0; Figure 2 (a)]. However, when the model is scanned at larger S-D separations a "M" profile is obtained [3.0 cm; Figure 2 (b)]. This shape can be accounted for by the fact that, during the scanning procedure, the absorber intersects various positions of the photon path [Figure 3 (b)].

In Figure 3 (b), it is seen that the heterogeneity first intersects the long path causing an increase in ϕ values. As the S-D is moved to the next position, the heterogeneity absorbs the mean path, and the ϕ values return to the baseline. When the S-D unit is once again moved, the heterogeneity intersects the long path causing an increase in ϕ values. When these effects are combined, the "M" formation is obtained. When the S-D separation increases, the photons penetrate deeper, and the heterogeneity in this case intersects the shorter path and causes the ϕ values to decrease below the baseline value at the heterogeneity position [Figure 3 (c)]. The results of the 4.0 cm S-D separation scan show this trend [Figure 2 (c)].

To actually determine the heterogeneity depth from the information obtained above, a correlation needs to be made between the S-D separation and the heterogeneity depth. The "ρ-to-depth" ratio is defined for this purpose. It is a ratio of the S-D separation at which the ϕ values return to the baseline (absorber intersects the mean path), to the depth of the heterogeneity. In this particular case, the "ρ-to-depth" ratio for the ϕ values at 100 MHz is 2:1 (3.0 cm : 1.5 cm). Since the ϕ values are frequency dependent, the "ρ-to-depth" ratio changes according to the chosen frequency. Therefore, the "ρ-to-depth" ratio must be determined for various frequency values as well as various optical properties of the continuum.

3.1.2. Time Constant. The time constant (τ), is the response time of the system to a disturbance input (independent of modulation frequency). Therefore, the heterogeneity depth can also be correlated to the profile of the S-D distance versus the τ values. Figure 4 shows the $\Delta\tau$ values at various S-D separations. As seen in Figure 4 (a), the $\Delta\tau$ values for a 2.0 cm S-D separation decrease over the position of the heterogeneity, meaning the system response time becomes faster, because the longer photon paths are being absorbed [Figure 3 (a)]. When the S-D separation is increased to 3.0 cm, the $\Delta\tau$ value at position 0 [Figure 4 (b)] meets the baseline value. This is similar to the ϕ profile in which the heterogeneity intersects the mean photon path causing little change in the $\Delta\tau$ values [Figure 3 (b)]. At larger S-D separations [4.0 cm; Figure 4 (c)], the $\Delta\tau$ values increase at position 0, but decrease on both sides. This trend appears because when the S-D position is directly over the absorber, it absorbs the shorter photon paths resulting in a larger response time. As the S-D is moved away from the absorber, it absorbs the longer photon paths, causing a decrease in the $\Delta\tau$ values [Figure 3 (c)].

A "ρ-to-depth" ratio can also be determined for the τ values. In this case (absorption and scattering coefficients, 0.1 and 6.0 cm^{-1}), the "ρ-to-depth" ratio is 2:1 (3.0 cm: 1.5 cm). Since the τ is independent of modulation frequency, it can be also used as a reference measurement for the ϕ values, which are frequency dependent.

3.2. Effect of Modulation Frequency on Photon Penetration Depth

3.2.1. Phase Shift. In order to determine how the modulation frequency affects the photon penetration depth in a scattering medium, the ϕ values at various modulation frequencies (100 MHz, 584 MHz, 1 GHz, 2 GHz) were studied for a S-D separation of 4.0 cm. The frequencies were chosen to give one steady state value (100 MHz), one value around the break frequency (584 MHz), and two frequencies equally spaced after the break frequency (1 GHz, 2 GHz). To normalize the $\Delta\phi$ values, they were multiplied by a ratio of the chosen frequency to 100 MHz. For example, if the maximum change for 100 MHz is 10.0, the maximum change for 1 GHz would be 100, ten times higher because there are ten times the number of waves for the same time interval. Therefore, the ϕ values at 1 GHz would be multiplied by 0.1.

Figure 5 displays the ϕ values for 4.0 cm S-D separation scan. The "M" shape shown in previous studies is displayed at lower modulation frequencies [Figure 5 (a), (b)], which indicates the absorber intersects both the short, long, and mean photon paths during scanning. However, at greater frequencies, the ϕ values begin to flatten, meaning that only longer photon paths are absorbed [Figure 5 (c), (d)]. Since the heterogeneity is held at a constant depth, this trend means that, as the modulation frequency increases, the penetration depth of the photons in the media decreases. This effect allows the heterogeneity depth to be determined in three dimensions using a single S-D separation only by changing the modulation frequency. Since a wide range of photon penetration depth into the media for one S-D separation can be obtained, this approach saves time for measurements and data reduction.

3.3. Modulation Frequency Effect on Photon Penetration Depth

To confirm the results obtained in the previous studies, where the absorber position was held constant and the S-D position was moved across the medium, a new

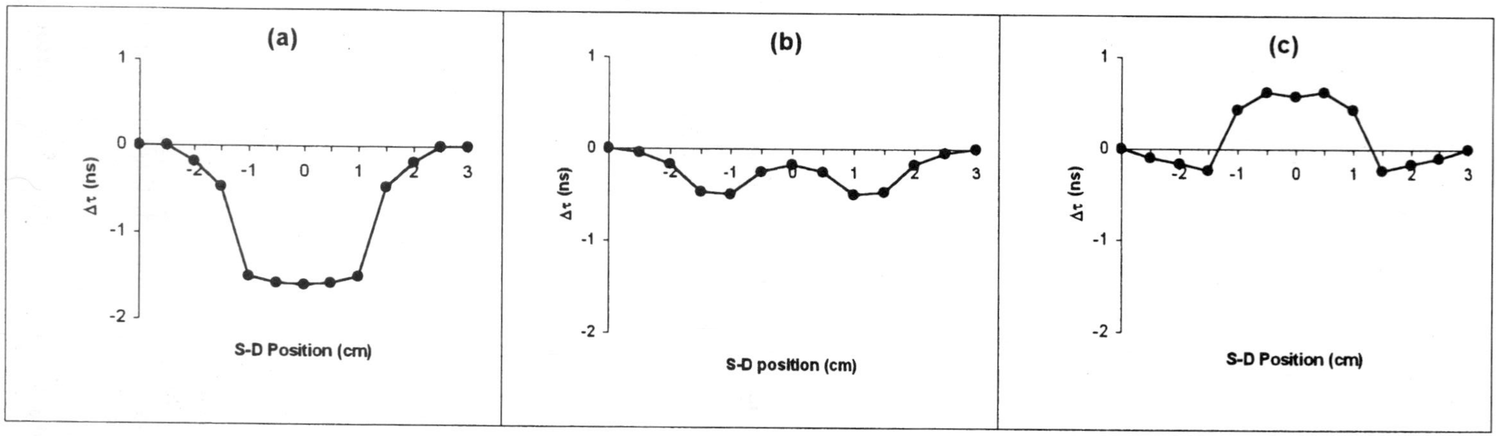

Figure 4. One dimensional profiles of the $\Delta\tau$ values vs. S-D position for S-D separations of (a) 2.0 cm, (b) 3.0 cm, and (c) 4.0 cm. As the S-D separation is increased a "W" formation is obtained. The absorber is located at position 0.

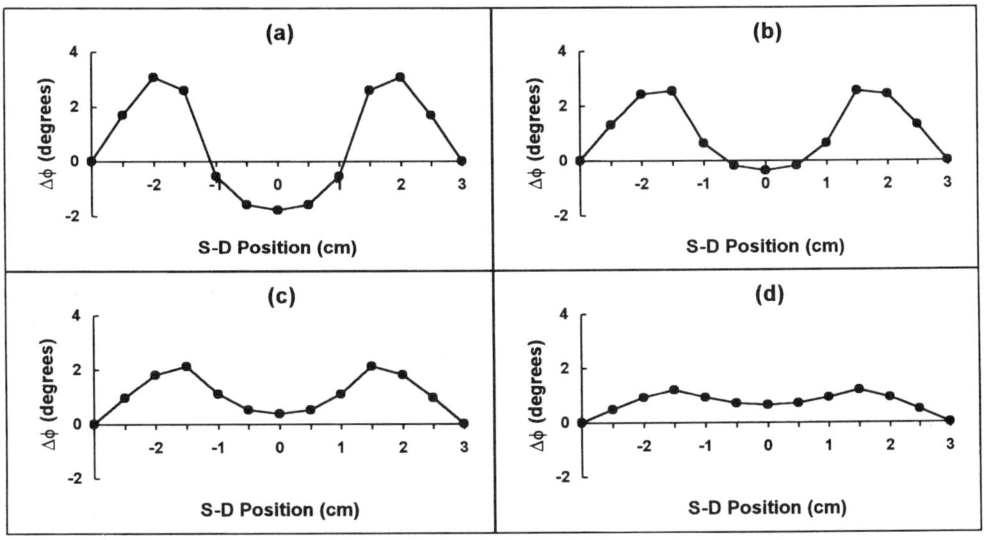

Figure 5. One dimensional profiles of the $\Delta\phi$ values at a 4.0 cm S-D separation for modulation frequencies of (a) 100 MHz, (b) 584 MHz, (c) 1 GHz, and (d) 2 GHz. At lower modulation frequencies negative values are obtained at positions near to the absorber. However, at higher modulation frequencies, the ϕ values begin to flatten at the position of the heterogeneity and only the longer photon paths are being absorbed. The absorber is located at position 0.

approach was introduced. In order to observe the effect of the absorber location on the photon path, the position of the absorber was changed on a plane (both horizontally and vertically). The changes in MR due to the presence of the absorber at various locations are studied. The S-D unit was placed at the center of the surface of the model with a constant S-D distance of 3.0 cm. A cubic absorber (0.5 cm × 0.5 cm × 0.5 cm) was moved on the plane parallel to the S-D unit (x–z plane) at 0.5 cm intervals [Figure 6].

The modulation frequencies studied were 20 MHz, 0.2 GHz, 1 GHz and 2 GHz.

Figure 7 is the contour plots of the MR values of the x–z plane at various frequencies The shape of the photon path is nearly the same for all four selected frequencies.

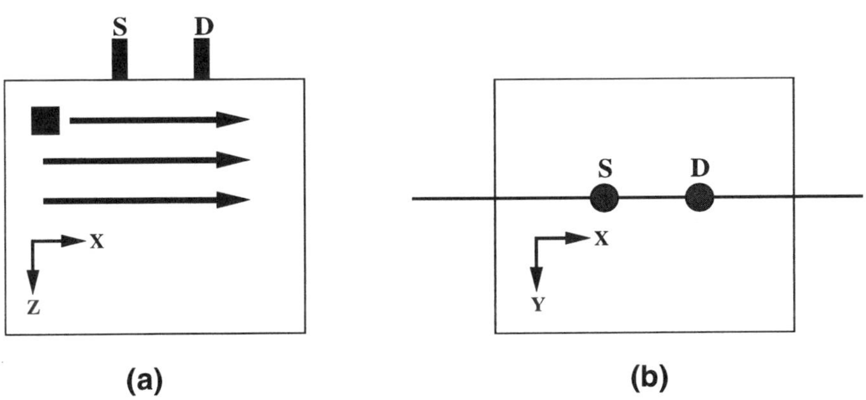

Figure 6. Schematic diagram of a model used where the S-D was positioned on the surface with a 3.0 cm separation as seen from (a) the side view, and (b) the top view. The absorber was moved in 0.5 cm increments on the x–z plane.

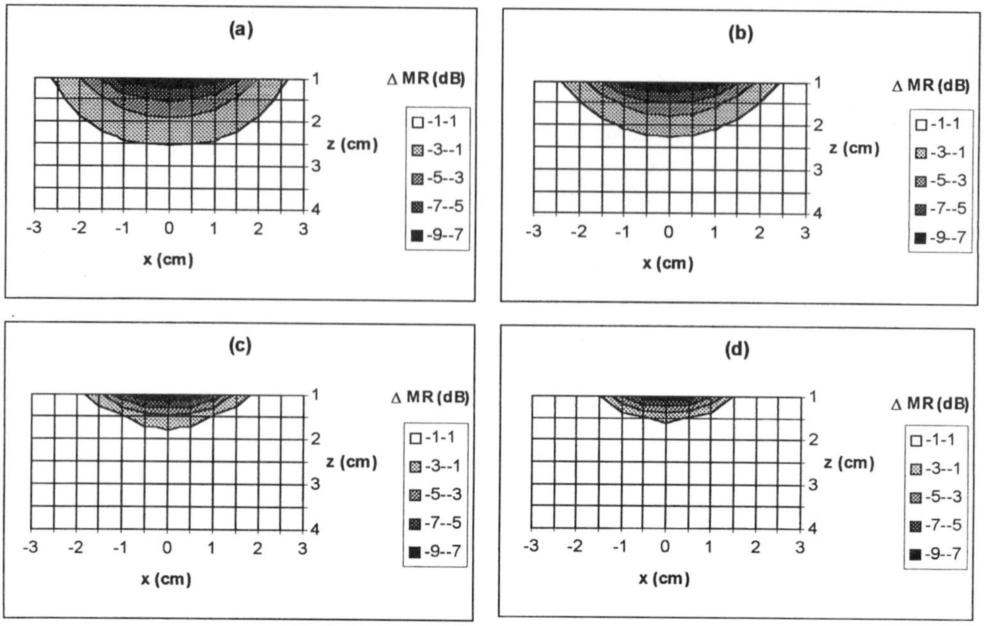

Figure 7. Contour plots of ΔMR values from an x–z plane (z = 1.0 to 4.0 cm) for a S-D separation of 3.0 cm at modulation frequencies of (a) 20 MHz, (b) 0.2 GHz, (c) 1 GHz, and (d) 2 GHz.

The banana shaped photon paths in the medium were observed [Figure 7]. There is a continuous decrease in the MR value when the absorber position was moved close to the S-D unit. Since the MR is based on the amount of photons absorbed, a greater decrease in MR means that more photons are absorbed. This indicates that more photons are distributed near the S-D position. For the analysis, the photon penetration depth was defined as the distance between S-D surface and the deepest point in z direction with magnitude ratio change 1 dB or more. The penetration depth changes with frequency from 2.6 cm to 2.3 cm, 1.8 cm, and 1.6 cm, respectively. By correlating the values from the legend, it is seen that there is less photon penetration as the modulation frequency increases.

4. CONCLUSIONS

Time Resolved Spectroscopy can be an effective, non-invasive, inexpensive way of detecting a heterogeneity in a scattering media, such as human tissue. From the study results, it was found that, once the range of photon penetration depth at a specific S-D separation is estimated, multi-frequency analysis would allow for localizing a heterogeneity in three dimensions from a single S-D distance scan. This approach would be more efficient than taking measurements at multiple S-D separations to find the heterogeneity depth, due to the less measuring time and data reduction.

ACKNOWLEDGMENTS

This research is supported by The Whitaker Foundation, Biomedical Research grant. The authors would also like to thank Dr. Duane Bruley for his advice, and use of the computer reduction code.

REFERENCES

Barnett JT, Hemelt MW, Kang KA, and Bruley DF (1997) Near Infrared Time Resolved Spectroscopy for the detection of Deep Vein Thrombosis within the Human Leg. In: ISOTT XIX, Advances in Experimental Medicine and Biology. (Harrison D, Delpy D, Eds) Plenum, New York, pp. 31–43.

Bruley DF (1995) Probabalistic Solutions and Models: Oxygen Transport in the Brain Microcirculation. Math Microcirc Phen 134–157.

Bruley DF (1995) Empirical Modeling for Oxygen Transport Processes and Related Physiological and Bioprocess Systems. In: ISOTT XVII, Advances in Experimental Medicine and Biology. (Ince C, Kesecioglu J, Telci L, Akpir K, Eds) Plenum, New York, pp. 109–116.

Bruley, DF (1974) Pulse reduction code for process identification.

Hemelt MW, Barnett JT, Bruley DF, and Kang KA (1997) The Application of NIR-TRS and Frequency Response Analysis for Deep Vein Thrombosis Detection. Biotech Prog 13(5):640–648.

Kang KA, Bruley DF, Kitai M, and Chance B (1996) System Parameter Analysis of NIR-TRS Spectra from Homogeneous Media with and without an Absorbing Boundary and Heterogeneous Media with a Single Absorber. In: ISOTT XVII, Advances in Experimental Medicine and Biology. Plenum, New York, pp. 399–413.

Kang KA, Bruley DF, Londono JM, and Chance B (1994) Highly Scattering Optical System Identification via Frequency Response Analysis of Optical NIR-TRS Spectra. Ann Biomed Eng 22:240–252.

Kang KA and Bruley DF (1997) The B-W-K Technique Applied to Problems of Hypothermia. AIChE Symposium Series 93(314):174–184.

Patterson MS, Chance B, and Wilson BC (1989) Time Resolved Reflectance and Transmittance for the Non-invasive Measurement of Tissue Optical Properties. Appl Optics 28(12):2331–2336.

Williford Jr. C, Bruley D, and Artigue R (1980) Probabilistic Modeling of Oxygen Transport in Brain Tissue. Neurol Res, 2(2):153–170.

78

TOWARD ABSOLUTE QUANTITATION OF BOLD FUNCTIONAL MRI

Ikuhiro Kida,[1,3] Fahmeed Hyder,[2] Richard P. Kennan,[2] and Kevin L. Behar[1]

[1]Department of Neurology
[2]Department of Diagnostic Radiology, Yale University
New Haven, Connecticut 06520
[3]Biophysics, Research Institute for Electronic Science
Hokkaido University
Sapporo 060, Japan

INTRODUCTION

Blood-oxygenation level dependent (BOLD) image contrast in magnetic resonance imaging (MRI) has been widely used in the field of functional imaging to interpret changes in focal brain activity in response to stimuli. The BOLD image contrast relies on activation-induced changes in the magnetic properties of blood (Ogawa et al., 1990; 1993a; 1993b; Shulman et al., 1993; Kennan et al., 1994). The presence of paramagnetic deoxyhemoglobin in the microvasculature creates a magnetic susceptibility difference between the vessels and the surrounding tissue, thus producing a microscopic magnetic field gradient. The microscopic field gradients affect the value of R2* (i.e., apparent transverse relaxation rate of tissue water) which can be mapped by a gradient-echo MRI sequence. The equilibrium between deoxy- and oxyhemoglobin can be shifted by altering the blood oxygenation, and since deoxy- and oxyhemoglobin are para- and diamagnetic, respectively, BOLD image contrast can be created, whereby hemoglobin acts as an endogenous MRI contrast agent. A BOLD functional MRI sequence measures the changes in R2* upon activation. The quantitative change in R2* is determined by parameters which influence the microscopic field gradient such as the geometry of vessels, static magnetic field strength, and the concentration of deoxyhemoglobin within the vessels (Ogawa et al., 1993a; 1993b; Kennan et al., 1994). The concentration of deoxyhemoglobin or the local blood oxygenation fraction (Y) in the microvasculature is determined by the regional values of cerebral metabolic rates of oxygen consumption

Oxygen Transport to Tissue XXI, edited by Eke and Delpy.
Kluwer Academic / Plenum Publishers, New York, 1999.

(CMR_{O2}), cerebral blood flow (CBF), and volume (CBV) (Ogawa et al., 1993a; 1993b; Kennan et al., 1994). While geometry and morphology of microvessels are important for the basal BOLD image contrast at a particular magnetic field strength, it is only the change in concentration of deoxyhemoglobin, due to a short-lived and/or transient physiological perturbation, that is important for being able to quantitate the BOLD signal change for functional MRI (see Appendix).

A current limitation of the BOLD functional MRI method is that the relationship between the BOLD signal change and the underlying physiological parameters of interest (i.e., CBF and CMR_{O2}) is not known quantitatively. Therefore, the magnitude of a change in the BOLD signal, or $\Delta R2^*$, may not be directly proportional to the magnitude of alteration in cortical activity. The interpretation of previous studies investigating the relationship between R2* and Y have been limited due to the absence of corresponding CBF and CMR_{O2} measurements (Ogawa et al., 1992; 1993a). Furthermore, other studies have examined the changes in R2* with respect to extreme physiological perturbations, e.g., hypoxia (Prielmeier et al., 1994; Kida et al., 1996). While these studies were informative in revealing the trends of the physiological variables which affect the BOLD signal, a quantitative understanding of BOLD signal with respect to CBF, CBV, or CMR_{O2} is lacking. More recently, (Kennan et al., 1997; 1998) have quantitatively examined the relation of the BOLD signal with CBV in rat cortex at a lower magnetic field strength of 2 Tesla. These studies in conjunction with the rat functional studies of (Hyder et al., 1996; 1997) have revealed that while CBV has some influence on the BOLD signal, the predominant parameters are CBF and CMR_{O2}. Absolute quantitation of the BOLD signal can be achieved by the calibration of R2* with multi-parametric physiological data (i.e., CBF, CMR_{O2}, CBV) acquired under various levels of brain activity.

In this paper, absolute values of R2* and CBF were measured under different levels of cortical activity in 30-day-old rat cortex at a magnetic field strength of 7 Tesla in order to calibrate the BOLD signal. Absolute CBF was measured by an inversion recovery MRI method (Schwarzbauer et al., 1996) with echo planar imaging (EPI). Maximum and minimum estimates of R2* (see Appendix) at 7 Tesla were determined from multi-echo EPI spin-echo and gradient-echo data (Bandettini et al., 1994). Additional CMR_{O2} measurements (Hyder et al., 1996) under the same conditions allowed estimation of the changes in CBV during these physiological perturbations by comparison of observed and predicted values of CMR_{O2}. The combined multi-parametric physiological data indicates that for the range of brain activity studied under these conditions, the calibrated BOLD signal at 7 Tesla is affected predominantly by changes in CBF and CMR_{O2}.

MATERIALS AND METHODS

Male, Sprague-Dawley 30–32 day-old rats (n = 20; 65–90 g; fasted overnight) were tracheotomized and artificially ventilated under 0.7–1.2% halothane with 70/30 N_2O/O_2. A femoral artery was cannulated for continuous blood pressure monitoring and periodic sampling of blood Pco_2, Po_2, and pH. A femoral vein was cannulated for nicotine infusion. In some cases (n = 8), an additional femoral vein was cannulated for [1-^{13}C] glucose infusion for the CMR_{O2} measurements. Intraperitoneal lines were inserted for administration of some anesthetic and paralyzing agents. The scalp was retracted and the dura removed. After the surgery, anesthesia was maintained throughout the experiment with morphine sulfate (initial dose 50 mg/kg; supplemental 30 mg/kg/30 min, i.p.) and the rat was paralyzed with D-tubocurarine chloride (initial dose 0.5 mg/kg; supplemental 0.25

mg/kg/30 min, i.p.). The rat was placed prone in a cradle and covered with a hot water pad to maintain body temperature (37°C). The head was secured using a bite bar and foam cushions were placed to minimize gross head movement. Three different levels of brain activity were studied: (A) baseline condition of morphine sulfate only, (B) lower activity condition of morphine sulfate with sodium pentobarbital (initial dose 30 mg/kg; supplemental 5 mg/kg/30 min, i.p.), and (C) higher activity condition of morphine sulfate with nicotine hydrogen tartrate (maintenance dose of 15 μl/kg/min, i.v.). Physiological variables were maintained near normal values throughout the measurements (P_{CO2} 35–45 mmHg, $P_{O2} \geq 100$ mmHg). Mean arterial blood pressures maintained for the three conditions were as follows: 100 ± 10 mmHg (morphine), 130 ± 5 mmHg (nicotine), and 70 ± 8 mmHg (pentobarbital). These are within the range of autoregulation in the rat cortex (Hernandez et al., 1978). Absolute CBF, multi-echo EPI, spin-echo and gradient-echo MRI, and CMR_{O2} measurements were made under conditions A, B, and C above. In some cases, CMR_{O2} was determined from C-4 glutamate turnover using the proton observe carbon edit (POCE) method (Hyder et al., 1996, 1997).

All *in vivo* experiments were performed on a highly modified 7 Tesla Bruker Biospec horizontal bore spectrometer. High resolution Fast Low Angle Shot anatomical images (image matrix = 128 × 128; in-plane resolution = 141 × 141 μm^2; slice-thickness = 1000 μm; echo time = 15 ms) were obtained to determine the slice positions in the sensorimotor area of the rat cortex. A region of 7 × 2 × 5 mm^3 volume was locally shimmed (Gruetter et al., 1993) to obtain half height line-widths of the water signal of 13–18 Hz. The EPI data were acquired with sequential sampling (Hyder et al., 1995) using the following parameters: image matrix = 32 × 32; in-plane resolution = 560 × 500 μm^2; slice thickness = 1000 μm. Gradient-echo and spin-echo EPI images were obtained with five echo times (15, 20, 25, 30 and 40 ms for gradient-echo and 28, 38, 48, 58 and 68 ms for spin-echo) using long repetition times. A single exponential fit to the multi-echo EPI data (Bandettini et al., 1994) from the gradient-echo and spin-echo experiments provided R2*(obs) and R2(obs) maps, respectively (see Appendix). The contribution of macroscopic magnetic field inhomogeneity to R2*(obs) within the volume (R2*(ΔB_0), see Appendix) was estimated by a double-DANTE tagging MRI sequence quantitatively (Mosher and Smith, 1990). In brief, a radio-frequency mesh (where each grid was 16 × 16 pixels, and each line in the mesh was 2 pixels wide) is displayed as low intensity pixels over the entire field of view. For a completely homogeneous magnetic field, adjacent vertical and horizontal lines are parallel. Therefore, based on the imaging parameters used and deviation of parallel lines in certain locations within the field of view, the macroscopic field distortions can be determined (see Mosher and Smith, 1990). The imaging parameters were as follows: image matrix = 128 × 128; in-plane resolution = 141 × 141 μm^2; slice-thickness = 1000 μm; echo time = 30 ms; DANTE pulse spacing = 3.33 ms; number of DANTE pulses = 8; gradient strength for DANTE tagging = 150 Hz/mm. The CBF maps were obtained using a spin-echo EPI sequence which followed a slice-selective and a non-slice-selective inversion recovery technique as previously described (Schwarzbauer et al., 1996). The inversion recovery times ranged from 0.5 to 1.8 s. R1 maps describing the longitudinal relaxation rate of tissue water were obtained using the slice-selective and non-slice-selective inversion recovery data (i.e., R1s and R1n). The calculation of CBF (Schwarzbauer et al., 1996) required an estimate of the relaxation rate of arterial blood (R1b). Ultra-high resolution spin-echo MRI (image matrix = 256 × 256; in-plane resolution = 72 × 72 μm^2; slice-thickness = 250 μm) inversion recovery data were used to obtain R1b under condition A (0.50 ± 0.03 1/s; n = 3), and assumed to be the same under conditions B and C. The blood partition coefficient (λ) was assumed to be 0.95 ml/g. CBF

Table 1. Absolute cerebral blood flow (ml/g/min), the observed and estimated transverse relaxation times (1/s) and the cerebral metabolic rates of oxygen consumption (μmol/g/min) in rat brain under different levels of cortical activity

				R2*(Y)		
Anesthesia	CBF	R2*(obs)	R2(obs)	Maximum	Minimum	CMR_{O2}
Pentobarbital	0.93 ± 0.43	36.3 ± 2.1	22.8 ± 1.9	29.4 ± 4.0	6.6 ± 4.0	1.7 ± 0.3
Morphine	1.24 ± 0.33	31.8 ± 2.0	22.4 ± 2.9	25.1 ± 5.2	2.7 ± 5.2	4.1 ± 0.2
Nicotine	1.60 ± 0.22	29.4 ± 1.5	21.8 ± 2.1	22.8 ± 4.5	1.0 ± 4.5	–

Values represent means ± S.D.

was calculated as $\lambda(R1s\text{-}R1n)/(1 + \varepsilon)$, where ε is a small correction factor which includes the R1b term (see Schwarzbauer et al., 1996 for details).

RESULTS AND DISCUSSION

Table 1 summarizes the measured values of CBF (from inversion recovery technique), R2* (from gradient-echo), R2 (from spin-echo), and CMR_{O2} (from POCE data). For increased and decreased levels of cortical activity, CBF was elevated to 1.60 ± 0.19 ml/g/min and reduced to 0.93 ± 0.43 ml/g/min, respectively, from a baseline value of 1.24 ± 0.32 ml/g/min. The altered values of CBF from baseline are in good agreement with previous autoradiographic results in adult rats (1.44 ml/g/min under halothane/nitrous oxide anesthesia with a dose of nicotine, (Grunwald et al., 1991); 0.70 ml/g/min with sodium pentobarbital (Otsuka et al., 1991). The values of CBF and CMR_{O2} determined here for 30-day-old rats are in good agreement with our previous results obtained using identical methods and conditions in adult rats (Hyder et al., 1998a). The changes in R2*(obs) and R2(obs) with respect to changes in CBF (or CMR_{O2}) were negatively correlated (see Table 1).

Based on the relationship between R2*(obs) and R2(obs), maximum and minimum values of R2*(Y), which is the relaxation rate of tissue water due to blood oxygenation, can be estimated (see Appendix). Figure 1 shows the estimated values of R2*(Y) plotted

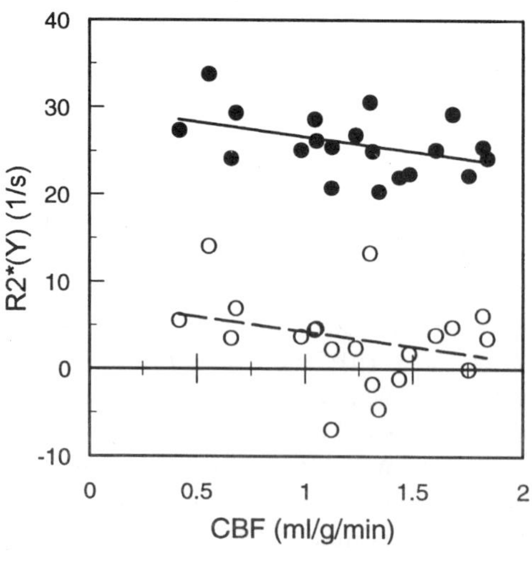

Figure 1. The estimated values of R2*(Y) plotted against the absolute CBF under three levels of cortical activity. (●) maximum (○) minimum estimation. Data represent morphine (n = 10), pentobarbital (n = 6), and nicotine (n = 4).

against CBF under different levels of cortical activity. This is the first quantitative demonstration of a linear dependence of R2*(Y) with CBF, which corresponds to a positive relationship between BOLD signal and CBF. R2*(Y) can be defined as

$$R2^*(Y) = C\nu_{max}(1 - Y)b \tag{1}$$

where ν_{max}, is the magnetic field dependent deoxyhemoglobin susceptibility frequency shift, b is the blood volume fraction in a voxel, and C is a BOLD signal proportionality constant (Ogawa et al., 1993a; 1993b; Kennan et al., 1994; Buxton and Frank 1997; Hyder et al., 1998b). Based on Fick's principle and Eq. (1), it can be shown that

$$\Delta CMR_{O2}/CMR_{O2} = (1 + \Delta CBF/CBF) \times (1 + \Delta R2^*(Y)/R2^*(Y))/(1 + \Delta b/b) - 1 \tag{2}$$

where the terms without and with the "Δ" indicates basal and relative differences, respectively, and $\Delta b/b$ is equivalent to ΔCBV/CBV. Therefore, based on a range of values for ΔCBV/CBV under different levels of cortical activity, R2*(Y) and CBF can be used to predict values of $\Delta CMR_{O2}/CMR_{O2}$. Figure 2 shows the best agreement that can be obtained between the predicted and measured values of $\Delta CMR_{O2}/CMR_{O2}$. By comparison of the predicted and measured values of $\Delta CMR_{O2}/CMR_{O2}$, we can find the range of ΔCBV/CBV that would be in agreement with the current data. These range of changes for CBF and CMR_{O2} can be modeled with a 2–10% change in CBV (see Eq. (2)). Based on the recent results of (Kennan et al. 1997; 1998) in rat cortex, small changes in CBV are expected within this physiological range. Changes in CBV for similar physiological perturbations using a different method (Wei et al., 1993; Chen et al., 1995) are in agreement with our predictions.

The predicted slope of the curve in Figure 2 (i.e., $(\Delta CMR_{O2}/CMR_{O2})/(\Delta CBF/CBF) = \Psi = 0.8$) is also in good agreement with our results in adult rats (Hyder et al., 1998a). Based on our recent proposed model of cerebral oxygen delivery (Hyder et al., 1998b), a linear relationship is expected between changes in CBF and CMR_{O2} within the range of autoregulation (where Ψ is close to 1). However, this observed trend in our data and

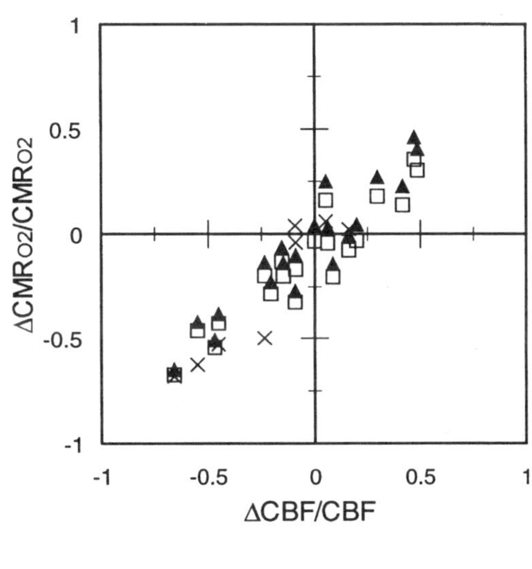

Figure 2. Relationship between fractional CBF and CMR_{02} changes under three levels of neuronal activity. Predicted values of $\Delta CMR_{02}/CMR_{02}$ were calculated using by Eq. (2) from $\Delta R2^*(Y)/R2^*(Y)$ and ΔCBF/CBF with assumption $\Delta b/b$ ((▲) 0.02, (□) 0.10). Measured values of of $\Delta CMR_{02}/CMR_{02}$, as determined by POCE measurements (X), were performed under conditions of morphine (n = 4) and pentobarbital (n = 4) only. The uncertainty in the measured CMR_{02} data allow for a possible change in CBV of ≤10%.

model is in contrast to another model of cerebral oxygen delivery which predicts a highly non-linear relationship between the changes in CBF and CMR_{O2} (Buxton and Frank, 1997). Our finding of a linear relationship (Figure 2) indicates that demand and delivery of oxygen is very tightly regulated within the autoregulation range.

Our current observation of a tight correlation between changes in CBF and CMR_{O2} with global anesthesia is in excellent agreement with physiological changes during sensory activation. In a rat forepaw model, we and others (Ueki et al., 1988; Hyder et al., 1997) have shown that the significant increases in CBF and CMR_{O2} are matched, along with concomitant but small increases in BOLD signal (Gyngell et al., 1996; Hyder et al., 1997) and CBV (Kennan et al., 1998). These observations of tightly matched changes between CBF and CMR_{O2} in conjunction with a positive BOLD signal in the rat cortex are in contrast to the available human data. To date, the predicted (Ogawa et al., 1993b; Menon et al., 1994; Kim and Ugurbil, 1997) and observed (Fox and Raichle, 1986; Fox et al., 1988) changes in CMR_{O2} are insignificant in comparison to the changes in CBF during functional activation. Future human studies with calibrated R2*(Y) in a human system would provide a different approach to the measurement of CMR_{O2} *in vivo* at high spatial resolution.

CONCLUSION

Absolute measurements of relaxation rates and multi-parametric physiological data (R2*(obs), R2(obs), CBF and CMR_{O2}) acquired under a range of cortical activity were used to calibrate the BOLD signal at 7 Tesla in 30-day-old rats. A linear relationship was observed between the BOLD signal and CBF. The dominant parameters which modulate the BOLD signal are CBF and CMR_{O2}. The linear relationship observed between CBF and CMR_{O2} indicates a tightly regulated system of cerebral oxygen delivery for the young rat. Future MRI studies of CBV under the same conditions will allow the possibility of a high resolution CMR_{O2} based images.

ACKNOWLEDGMENTS

We thank Bei Wang for help with the animal preparations and Scott McKintyre, Peter Brown and Terry Nixon for technical support. This work was supported by the Japan Society for the Promotion of Science and grants from the National Institutes of Child Health and Development PO1 HD32573 and the National Science Foundation DBI-9730892.

REFERENCES

Bandettini PA, Wong EC, Jesmanowicz A, Hinks RS, and Hyde JS (1994) Spin-echo and gradient echo EPI of human brain activation using BOLD contrast: a comparative study at 1.5 T. *NMR in Biomed* 7:12–20.

Buxton RB and Frank LR (1997) A model for the coupling between cerebral blood flow and oxygen metabolism during neural stimulation. *J Cereb Blood Flow Metab* 17:64–72.

Chen JL, Wei L, Bereczki D, Hans FJ, Otsuka T, Acuff V, Ghersi-Egea JF, Patlak C, and Fenstermacher JD (1995) Nicotine raises the influx of permeable solutes across the rat blood-brain barrier with little or no capillary recruitment. *J Cereb Blood Flow Metab* 15:687–698.

Fox PT and Raichle ME (1986) Focal physiological uncoupling of cerebral blood flow and oxidative metabolism during somatosensory stimulation in human subjects. *Proc Natl Acad Sci USA* 83:1140–1144.

Fox PT, Raichle ME, Mintun MA, and Dence C (1988) Nonoxidative glucose consumption during focal physiologic neural activity. *Science* 241:462–464.

Gruetter R (1993) Automatic, localized in vivo adjustment of all first- and second order shim coils. *Magn Reson Med* 29:804–811.

Grunwald F, Schrock H, and Kuschinsky W (1991) The influence of nicotine on local cerebral blood flow in rats. *Neurosci Lett* 124:108–110.

Gyngell ML, Bock C, Schmitz B, Hoehn-Berlage M, and Hossmann KA (1996) Variation of functional MRI signal in response to frequency of somatosensory stimulation in a-chloralose anesthetized rats. *Magn Reson Med* 36:13–15.

Hernandez MJ, Brennan RW, and Bowman GS (1978) Cerebral blood flow autoregulation in the rat. *Stroke* 9:150–155.

Hyder F, Rothman DL, and Blamire AM (1995) Image reconstruction of sequentially sampled echo-planar data. *Magn Reson Imaging* 13:97–103.

Hyder F, Chase JR, Behar KL, Mason GF, Siddeek M, Rothman DL, and Shulman RG (1996) Increased tricarboxylic acid cycle flux in rat brain during forepaw stimulation detected with ^{1}H [^{13}C] NMR. *Proc Natl Acad Sci USA* 93:7612–7617.

Hyder F, Rothman DL, Mason GF, Rangarajan A, Behar KL, and Shulman RG (1997) Oxidative glucose metabolism in rat brain during single forepaw stimulation: a spatially localized ^{1}H[^{13}C] nuclear magnetic resonance study. *J Cereb Blood Flow Metab* 17:1040–1047.

Hyder F, Kennan RP, Sibson NR, Mason GF, Behar KL, Rothman DL, and Shulman RG (1998a) Cerebral oxygen delivery in vivo: NMR measurements of CBF and CMR_{O2} at different levels of brain activity. in *"Proc. of ISMRM, 6th Scientific Meeting, 1998"* p. 1160.

Hyder F, Shulman RG, and Rothman DL (1998b) A model for the regulation of cerebral oxygen delivery. *J Appl Physiol* 85:554–564.

Kennan RP, Zhong J, and Gore JC (1994) Intravascular susceptibility contrast mechanisms in tissue. *Magn Reson Med* 31:9–21.

Kennan RP, Scanley BE, and Gore JC (1997) Physiologic basis for BOLD MR signal changes due to hypoxia/hyperoxia: separation of blood volume and magnetic susceptibility effects. *Magn Reson Med* 37:953–956.

Kennan RP, Scanley BE, and Gore JC (1998) Physiological basis for BOLD MR signal changes due to neuronal stimulation: separation of blood volume and magnetic susceptibility effects. *Magn Reson Med* (in press).

Kida I, Yamamoto T, and Tamura M (1996) Interpretation of BOLD MRI signals in rat brain using simultaneously measured near-infrared spectrophotometric information. *NMR in Biomed* 9:333–338.

Kim SG and, Ugurbil K (1997) Comparison of blood oxygenation and cerebral blood flow effects in fMRI: estimation of relative oxygen consumption change. *Magn Reson Med* 38:59–65.

Menon RS, Ogawa S, Tank DW, and Ugurbil K (1993) 4 Tesla gradient recalled echo characteristics of photic stimulation-induced signal changes in the human primary visual cortex. *Magn Reson Med* 30:380–386.

Mosher TJ and Smith MB (1990) A DANTE tagging sequence for the evaluation of translational sample motion. *Magn Reson Med* 15:334–339.

Ogawa S, LEE TM, Sha A, Nayak S, and Glynn P (1990) Oxygenation-sensitive contrast in magnetic resonance imaging of rodent brain at high magnetic fields. *Magn Reson Med* 14:68–78.

Ogawa S, Tank DW, Menon RS, Ellermann JM, Kim SG, Merkle H, and Ugurbil K (1992) Intrinsic signal changes accompanying sensory stimulation: Functional brain mapping with magnetic resonance imaging. *Proc Natl Acad Sci USA* 89:5951–5955.

Ogawa S, Menon RS, Tank DW, Kim SG, Merkle H, Ellermann JM, and Ugurbil K (1993a) Functional brain mapping by blood oxygenation level-dependent contrast magnetic resonance imaging: a comparison of signal characteristics with a biophysical model. *Biophys J* 64:803–812.

Ogawa S, LEE TM, and Barrere B (1993b) The sensitivity of magnetic resonance image signal of a rat brain to changes in the cerebral venous blood oxygenation. *Magn Reson Med* 29:205–210.

Otsuka T, Wei L, Acuff VR, Shimizu A, Pettigrew KD, Patlak, CS, and Fenstermarcher JD (1991) Variation in local cerebral blood flow response to high-dose pentobarbital sodium in the rat. *Am J Physiol* 261:H110–H120.

Prielmeier F, Nagatomo Y and Frahm J(1994) Cerebral blood oxygenation in rat brain during hypoxic hypoxia. Quantitative MRI of effective transverse relaxation rates. *Magn Reson Med* 31:678–681.

Schwarzbauer C, Morrissey SP, and Haase A (1996) Quantitative magnetic resonance imaging of perfusion using magnetic labeling of water proton spins within the detection slice. *Magn Reson Med* 35:540–546.

Shulman RG, Blamire AM, Rothman DL, and McCarthy G (1993) Nuclear magnetic resonance imaging and spectroscopy of human brain function. *Proc Natl Acad Sci USA* 90:3127–3133.

Ueki M, Linn F, and Hossmann KA (1988) Functional activation of cerebral blood flow and metabolism and after global ischemia of rat brain. *J Cereb Blood Flow Metab* 8:486–494.
Wei L, Ostuka T, Acuff V, Bereczki D, Pettigrew K, Patlak C, and Fenstermacher JD (1993) The velocities of red cell and plasma flows through parenchymal microvessels of rat brain are decreased by pentobarbital. *J Cereb Blood Flow Metab* 13:487–497.

APPENDIX

The observed MRI signals obtained using gradient-echo (GE) and spin echo (SE) pulse sequences are given by

$$S_{GE} = S_{0,GE}\exp(-TE_{GE}R2^*(obs)) \quad [1]$$

and

$$S_{SE} = S_{0,SE}\exp(-TE_{SE}R2^*(obs)) \quad [2]$$

respectively, where S_0 is the intrinsic signal of the system for the sequence, TE is the echo time which determines the weighting for the sequence, and R2*(obs) and R2(obs) are the observed values of relaxation rates of tissue water using the GE and SE pulse sequence, respectively. The value of R2*(obs) is influenced by the static magnetic field strength, local magnetic field inhomogeneity, geometry of local blood vessels, and blood oxygenation. Since geometry and morphology of microvessels do not affect the BOLD signal during a short-lived and/or transient physiological perturbation, the BOLD signal change for functional MRI is mainly dependent on the local concentration of deoxyhemoglobin. The value of R2*(obs) can be further expanded to

$$R2^*(obs) = R2^*(Y) + R2(other) + R2^*(\Delta B_0) \quad [3]$$

where $R2^*(\Delta B_0)$ is the macroscopic static magnetic field inhomogeneity, R2(other) is the relaxation rate of materials other than tissue water, and R2*(Y) is the relaxation rate of tissue water. The values of R2*(Y) dependent on the microscopic magnetic field gradients caused by the presence of deoxyhemoglobin,

$$R2^*(Y) = R2(Y) + R2'(Y) \quad [4]$$

where R2(Y) and R2′(Y) are the reversible and irreversible components of the relaxation rate of tissue water that are dependent on blood oxygenation, respectively. In contrast to R2*(obs) measured in the GE experiment, the value of R2(obs) measured in the SE experiment is given by

$$R2(obs) = R2'(Y) + R2(other) \quad [5]$$

because the extra π pulse in a SE pulse sequence refocuses most of the dephasing transverse magnetization that gives rise to the signal in the GE experiment. From the above equations, R2*(Y) is given by

$$R2^*(Y) = R2^*(obs) - R2(obs) + R2'(Y) - R2^*(\Delta B_0) \quad [6]$$

which is equivalent to

$$R2^*(Y) = R2^*(obs) - R2(other) - R2^*(\Delta B_0) \quad [7]$$

If $R_2*(\Delta B_0)$ is known, then the minimum and maximum values of $R2*(Y)$ are given by

$$R2^*_{\text{min}}(Y) = R2^*(\text{obs}) - R2(\text{obs}) - R2^*(\Delta B_0) \quad [8]$$

and

$$R2^*_{\text{min}}(Y) = R2^*(\text{obs}) - R2^*(\Delta B_0) \quad [9]$$

respectively. Using the double-DANTE tagging MRI method described in MATERIALS AND METHODS, the mean value of $R2*(\Delta B_0)$ was measured to be $6.7 \pm 0.8\ s^{-1}$ ($n = 20$).

MEASUREMENT OF OXYGEN DIFFUSION THROUGH CULTURED ENDOTHELIAL CELL MONOLAYERS

J. P. W. M. Lamers-Lemmers,[1] L. J. C. Hoofd,[1] I. Otte-Höller,[2]
R. M. W. de Waal,[2] and B. Oeseburg[1]

[1]Department of Physiology
Faculty of Medical Science
University of Nijmegen
PO Box 9101, 6500 HB Nijmegen
The Netherlands
[2]Department of Pathology
University Hospital St. Radboud, Nijmegen
The Netherlands

1. INTRODUCTION

Accurate values for the oxygen diffusion coefficient (DO_2), the oxygen solubility (αO_2), and/or the oxygen permeability ($PO_2 = DO_2 \times \alpha O_2$) of substances which contribute significantly to the diffusion resistance to oxygen in tissue are important for understanding oxygen diffusion from capillaries into tissue. One of the possible important resistances to the transport of oxygen in tissue is the capillary endothelial cell wall, because of the high flux of oxygen through this wall out of the capillary. About the importance of this resistance, e.g., Groebe (1996) estimates the capillary vicinity to account for about half of the total resistance to oxygen diffusion, in dog gracilis muscle.

The barrier formed by the endothelial cell monolayer alone is difficult to determine in vivo, because of the presence of perivascular cells and extracellular matrix. Endothelial cells can be propagated in culture, however, and form a continuous monolayer with intercellular junctions. Such endothelial monolayers have been used extensively to study permeability for different types of molecules. The major source of human endothelial cells is the umbilical cord from which primary cultures can easily be derived, and expanded for up to 10–15 passages. The present study describes direct measurement of the oxygen diffusion through human umbilical vein endothelial cell monolayers, in a diffusion chamber.

Oxygen Transport to Tissue XXI, edited by Eke and Delpy.
Kluwer Academic / Plenum Publishers, New York, 1999.

2. METHODS

MEM-100 silicone membranes (Membrane Products Corporation, Salt Lake City USA) were coated with fibronectin for 30 minutes at 37°C. Human umbilical vein endothelial cells (passage 6) were cultured in a monolayer on these silicone membranes for 1–2 days by adding the cells 1:1 in EMEM complete culture medium (EMEM complete culture medium: Eagle Minimum Essential Medium (Boehringer Ingelheim) + antibiotics + 10% human serum + 20% Newborn Calf Serum + glutamine + heparine + Endothelial Cell Growth Factor (ECGF)). The surface coverage of the monolayers was checked after each experiment using eosin staining. These model layers, composed of a silicon membrane and an endothelial cell monolayer, were washed twice with EMEM culture medium (EMEM culture medium: Eagle Minimum Essential Medium + antibiotics) and were then placed in the diffusion chamber between two adjacent gas chambers, which could be filled with gases of different composition. A thin layer of the EMEM culture medium was placed on top of the monolayer to keep the cells wet during the experiment. The model layer was thus composed of a silicone membrane (105 μm), an endothelial cell monolayer (approximately 1 μm), and a layer of EMEM culture medium (<1 μm) (Figure 1).

The oxygen permeability (PO_2) of the endothelial cell monolayers was determined using a steady-state method as described previously (Hoofd et al., 1986). Figure 3 shows the recordings of one of the steady-state measurements for the determination of PO_2 of the endothelial cell monolayer. The oxygen partial pressures in the top and bottom chambers are plotted against time. At the beginning of the experiment the gas chambers were flushed with gases of different composition and at t = 0 the valves of the gas inlets were closed. The values for dp/dt of both top and bottom chamber were determined as the initial increase or decrease in oxygen pressure, using a least squares fit method. The oxygen flux J, which is the amount of oxygen that has diffused through a layer with area A, is derived from these values, using the ideal gas law:

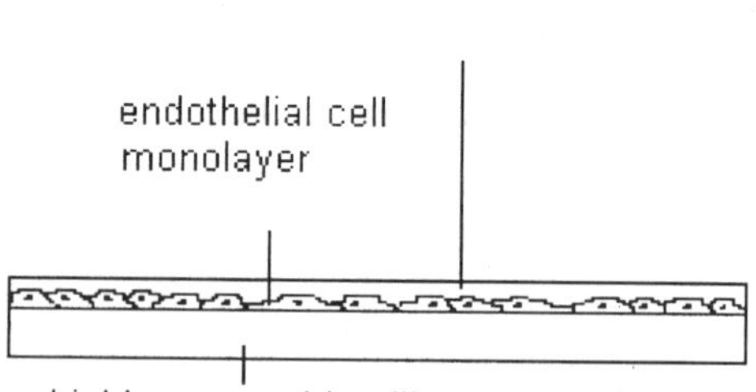

Figure 1. Model layer for determination of PO_2 of endothelial cell monolayer. The diffusion chamber that was used for the present study was based on those described previously (de Koning et al., 1981, Hoofd and Lamboo, 1985, Bouwer, 1987) (Figure 2A), extended with a special holder to place the model layer in the diffusion chamber. As a consequence the volumes of the top and bottom chamber were somewhat larger than the volumes reported in these earlier studies. The temperature of the diffusion chamber was controlled by a built-in circulating system connected to a water bath (Figure 2B) and was maintained at 25°C during the experiment to reduce oxygen consumption of the endothelial cells. The actual temperature was measured continuously during the measurement, using a temperature probe.

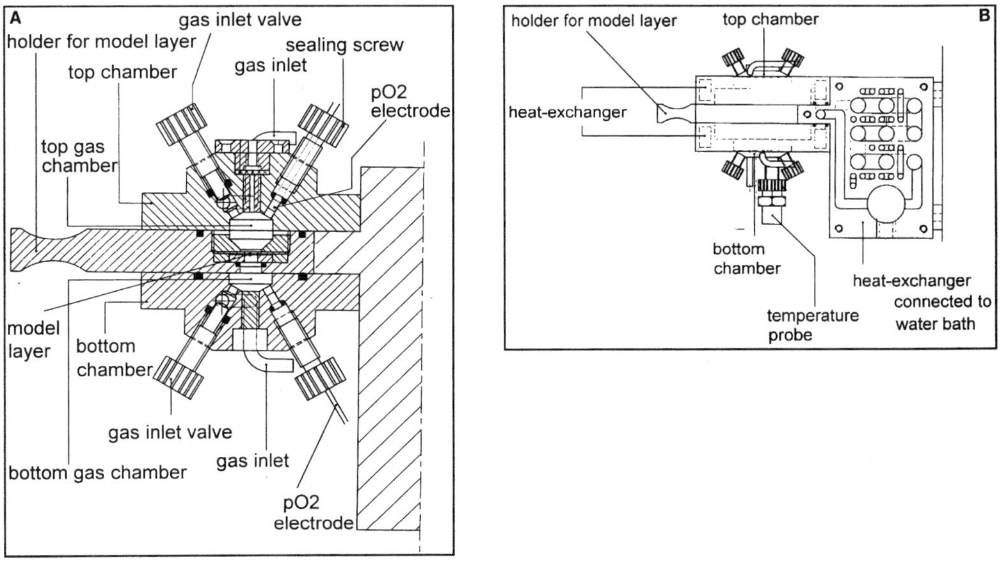

Figure 2. The diffusion chamber set-up. Left panel (A): diffusion chamber with special holder for the model layer; Right panel (B): built-in circulating system for temperature control of the diffusion chamber.

$$\frac{dn}{dt} = J \cdot A = \frac{d}{dt}\left(\frac{p \cdot V}{R \cdot T}\right) \tag{1}$$

(n = amount of oxygen in moles, t = time, A = area of the layer, p = oxygen partial pressure, V = chamber volume, R = gas constant, T = temperature).

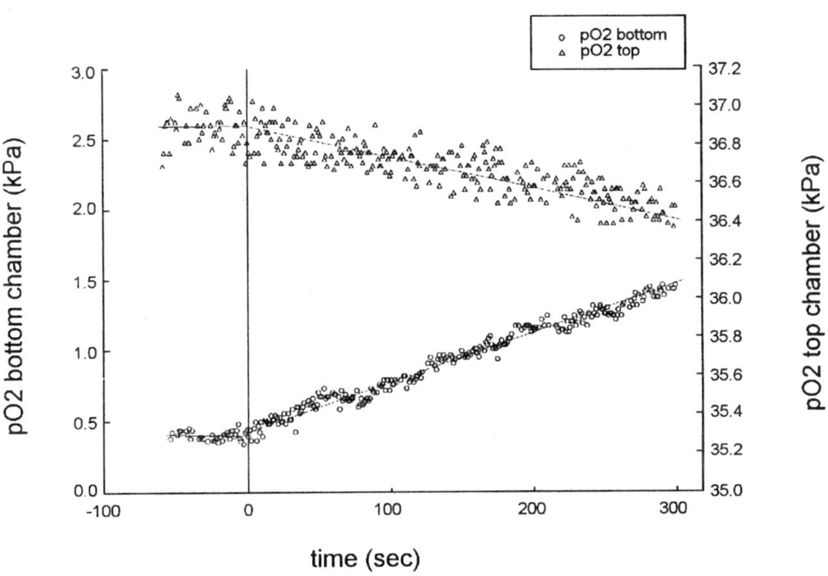

Figure 3. Recordings of a steady-state measurement for the determination of PO_2 of the endothelial cell monolayer.

The oxygen permeability PO_2 of the layer is then calculated from the oxygen flux:

$$J = P\frac{\Delta p}{L} \tag{2}$$

(Δp = oxygen pressure difference between top and bottom chamber, L = layer thickness) (For details, see Hoofd et al., 1986).

Using this steady state method, the set-up was checked by determining the oxygen permeabilities of Teflon membranes ($PO_2 = 4.46 \pm 0.13 \times 10^{-12}\,\text{mol·m}^{-1}\text{·sec}^{-1}\text{·kPa}^{-1}$), silicon membranes ($PO_2 = 1.72 \times 10^{-12}\,\text{mol·m}^{-1}\text{·sec}^{-1}\text{·kPa}^{-1}$), and water ($PO_2 = 3.27 \pm 0.12 \times 10^{-12}\,\text{mol·m}^{-1}\text{·sec}^{-1}\text{·kPa}^{-1}$) at 25*007C. As a further check, this set-up was also applied to determine the DO_2 of water ($DO_2 = 2.56 \pm 0.3 \times 10^{-9}\,\text{m}^2\text{·sec}^{-1}$) at 25°C, using a non-steady state method (van Dijk et al., 1996). All of these values were comparable with literature values.

3. RESULTS

The PO_2 of the endothelial cell monolayer was determined from the PO_2 of the total model layer by correcting for the PO_2 of silicone membrane and EMEM culture medium, determined in two separate experiments, and the layer thicknesses (L) of these layers:

$$\frac{L_{\text{endothelium}}}{PO_{2,\text{endothelium}}} = \frac{L_{\text{total model layer}}}{PO_{2,\text{total model layer}}} - \frac{L_{\text{silicone membrane}}}{PO_{2,\text{silicone membrane}}} - \frac{L_{\text{EMEM}}}{PO_{2,\text{EMEM}}} \tag{3}$$

PO_2 of the endothelial cell monolayer at 25,00 ± 0.005°C was determined to be $9 \times 10^{-12}\,\text{mol·m}^{-1}\text{·sec}^{-1}\text{·kPa}^{-1}$, which was comparable with the value of $9.15 \pm 0.38 \times 10^{-12}\,\text{mol·m}^{-1}\text{·sec}^{-1}\text{·kPa}^{-1}$ at 37°C reported by Liu et al., (1994). These authors already mentioned that this value falls in the expected range for tissue.

4. DISCUSSION

We confirmed that the endothelial resistance to oxygen is not of dominant importance in oxygen transport to tissue as already reported by Liu et al. (1994). It might not be necessary to take the endothelial resistance into account as a separate factor in theoretical models.

During this study, we also tried to determine the DO_2 of endothelial cells using a nonsteady state method (van Dijk et al., 1996). Using this method we determined the delay time in two separate experiments: one experiment with just the silicon membrane as a model layer and one experiment with the endothelial cell monolayer on top of the silicon membrane. The delay time of the silicone membrane was determined to be about 15 sec, which corresponds to a DO_2 of $0.1 \times 10^{-9}\,\text{m}^2\text{sec}^{-1}$. The difference in delay time between the two experiments would yield the delay time for just the endothelial cell monolayer, but no significant difference between the two values was found. An estimate of the delay time from the value for DO_2 for endothelial cells, as reported by Liu et al. (1996) is:

$$DO_2 = \frac{0.15 \cdot (L_{\text{model layer}})^2}{t_{\text{delay}}} = 8.73 \times 10^{-10}\ \text{m}^2/\text{sec}$$
$$\Rightarrow t_{\text{delay}} = 1.7 \times 10^4\ \text{sec} \tag{4}$$

This delay time is an order of magnitude shorter than that of the silicone membrane, so that the expected extra delay of the endothelial monolayer could not be determined in our set-up; the minimum delay time, which can be determined, is about 1 second. This set-up is thus useful for determination of PO_2, but not for the determination of DO_2 of endothelium. It can also be used for determination of PO_2 and DO_2 of other model layers.

In this study we used human umbilical vein endothelial cells. Although these cells have been used extensively in permeability studies (van Hinsbergh, 1997), it is obvious that the obtained results may be influenced by the cell type chosen. In addition, it has been described that endothelial permeability is a function of passage number. Future experiments may therefore explore the influence of these and other parameters.

ACKNOWLEDGMENTS

This research was supported by Grant number: 902-19-109 by NWO (Netherlands Organization for Scientific Research).

REFERENCES

Bouwer ST (1987) Experimental set-up. In: Facilitated oxygen diffusion through hemoglobin solutions. Measurement of diffusion and reaction parameters. (PhD Thesis) University of Nijmegen, pp. 46–48.

van Dijk AMC, Hoofd LJC, and Oeseburg B (1996) Diffusion coefficient of oxygen in various model layers as determined by analysis of time-dependent diffusion. Adv Exp Med Biol 388:327–331.

Groebe K (1996) Practical applications of models of oxygen supply, diffusion, and consumption. Past, Perspectives, and Problems. Adv Exp Med Biol 388:161–175.

van Hinsbergh VWM (1997) Endothelial permeability for macromolecules. Mechanistic aspects of pathophysiological modulation. Arterioscler Thromb Vasc Biol 17:1018–1023.

Hoofd L and Lamboo A (1985) Oxygen permeability of methemoglobin solutions soaked in millipore filters. Adv Exp Med Biol 191:565–570.

Hoofd L, de Koning J, Kreuzer F, and Lamboo A (1986) Determination of permeabilities or two gases from recording the partial pressure of one gas. Pflügers Arch 407:320–326.

de Koning J, Hoofd LJC, and Kreuzer F (1981) Oxygen transport and the function of myoglobin. Theoretical model and experiments in chicken gizzard smooth muscle. Pflügers Arch 389:211–217.

Liu CY, Eskin SG, and Hellums JD (1994) The oxygen permeability of cultured endothelial cell monolayers. Adv Exp Med Biol 345:723–730.

80

MONITORING CHANGES IN OXYGEN TENSION AND MICROCIRCULATION IN A FLAP WITH A MODIFIED OXYGEN MICROELECTRODE AND LASER-DOPPLER FLOWMETRY

Anders G. Liss[1] and Per Liss[2]

[1]Department of Plastic Surgery
[2]Department of Diagnostic Radiology
University of Uppsala
Uppsala, Sweden

1. INTRODUCTION

Free tissue transfer is an established reconstructive technique in plastic surgery, yet it still has an estimated 5–20% anastomotic failure rate.[1] Several ways to monitor blood perfusion in free cutaneous flaps have been described[2] but there is still a need for an ideal free tissue flap monitoring procedure. Tissue oxygen tension (pO_2) has been claimed to be the a reliable measure of tissue perfusion and viability of tissue flaps.[3,4] Without early detection and re-exploration, irreversible tissue damage occurs after 8–12 hours of ischaemia in a free flap, and even with re-exploration, a decrease in flap circulation can also occur due to the no-reflow phenomenon.[5] As venous occlusion seems to be more detrimental than arterial occlusion[6] it is clinically important not only to be able to execute accurate monitoring of the viability of the flap, but also to distinguish between venous stasis and arterial occlusion. The purpose of this study was to compare perfusion in a pedicled flap as monitored by a modified Clark electrode[7,8] with changes in blood flow (BF) as measured by laser-Doppler flowmetry, and to establish whether it is possible with these two procedures to determine if ischaemia in a flap is due to venous stasis or arterial occlusion.

Oxygen Transport to Tissue XXI, edited by Eke and Delpy.
Kluwer Academic / Plenum Publishers, New York, 1999.

2. MATERIAL AND METHODS

2.1. Animals

Male Sprague-Dawley rats with an average body weight (BW) of 308 ± 7 g were used. They were anaesthetized with Inactin® (Byk-Gulden, Konstanz, Germany), given intraperitoneally. Tracheostomy was performed and the rat was placed on servo-controlled heating pad to maintain the body temperature at 37.5°C.

2.2. Surgical Procedures

A catheter was introduced into the left femoral artery for blood sampling and continuous monitoring of blood pressure, and another in the left femoral vein for infusion of Ringer's solution (0.5 ml/h per 100 g BW). The right epigastric inferior artery and vein were dissected free and a pedicled groin fascio-cutaneous flap was raised.

2.3. Oxygen Microelectrode

The oxygen microelectrodes used in this study were of modified Clarke-type[9] constructed as described by Revsbech[8] and used in vivo as described in earlier studies.[7,10] They had a tip diameter of 2–8 μm and an inner hole diameter at the tip of 1–2 μm. The electrodes were polarized at −800 mV, an electrical potential at which a linear response is observed between oxygen tension and electrode current. The electrodes were calibrated in water at 37°C bubbled with N_2 gas and air before and after the experiments. The electrode current in N_2 gas was 10 ± 1.1 pA, and in air, 239 ± 28 pA (n = 12). The stirring effect in bubbling water was 1.3 ± 0.3%, the 90% response time was 2.6 ± 0.5 sec.

2.4. Laser-Doppler Flowmetry

The technique is based on detecting the velocity of the movement of erythrocytes through microvessels and yields a parameter (red blood cell flux) continuously and non-invasively. Assuming that hematocrit is constant during the measurement, the LDF-signal is proportional to the blood flow (BF) through the monitored tissue region. In this study the LDF-signal is presented in perfusion units, PU. The tissue under study was illuminated by a heliumneon laser light via an optical fibre, and the Doppler shift of light scattered by moving red blood cells was analysed (Periflux master PF 4001–2, PeriMed, Stockholm, Sweden).

2.5. Experimental Protocol

When the surgical procedure was completed, the animal was allowed to recover for 30 min. In 11 rats the oxygen tension and the laser-Doppler signal were then measured continuously in the flap. The electrode was inserted using micromanipulators (WPI, USA) to a depth of 1 mm from the skin surface into the tissue. The laser-Doppler probe registered changes in the surface tissue at a distance of 3–4 mm apart from the oxygen electrode. After a 30 minute control period, the artery or the vein of the flap was clamped. The decrease in pO_2 was slower than that in BF. The clamp was opened again when the pO_2 low level had stabilized (Figure 1).

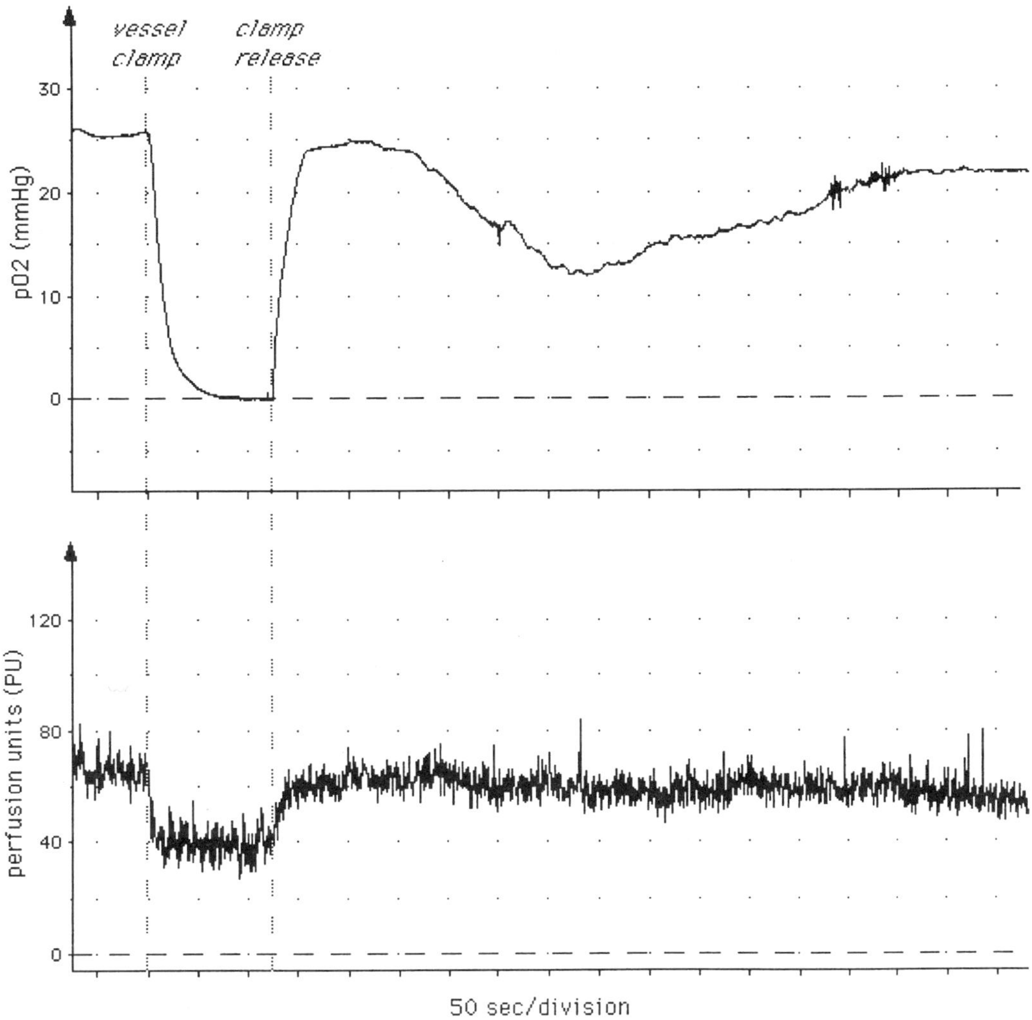

Figure 1. Typical record of pO_2 (*top*) and perfusion changes (*bottom*) before, during and following an arterial occlusion by a vessel clamp.

A value was considered stable after clamping or releasing of the clamp when the recording remained unchanged for 30 sec, at which point in time the clamp was released. The arterial clamp was opened after 380 ± 110 sec (6.3 ± 1.8 min) the venous clamp after 700 ± 150 sec (11.7 ± 2.5 min). The recovery of the flap's oxygen tension and BF was studied following the release of the clamp. A total of 16 arterial and 12 venous occlusions were performed. In 7 animals a primary arterial occlusion was performed. In 2 animals, primary vein clamping was performed. In another 2 animals no data were obtained from the first occlusion, due to technical problems. After reperfusion, the flap was allowed to recover for 30 min before the next clamping was performed. Repeated occlusions gave responses of a similar pattern in pO_2 and BF. Oxygen tension in the flap, laser-Doppler signal, blood pressure, body temperature and flap temperature were recorded continuously with a MacLab Instrument (AD Instruments, Hastings, England) connected to a Macintosh Power-PC 6100 computer.

2.6. Statistical Evaluation

Values are expressed as means ± SEM. The statistical significance of the data was tested by using a multivariate analysis of variance (Anova) model, comparing results obtained during the control period with those obtained after drug administration (repeated measures). For comparison of data between groups of animals, a two-tail Student's t-test for unpaired samples was applied. A p-value less than 0.05 was accepted as significant and was marked by the symbol * in all analyses.

3. RESULTS

3.1. Changes in pO_2 in the Pedicled Flap

The mean pO_2 in the flap before clamping of an artery or vein was 19.7 ± 1.8 mmHg and 17.4 ± 1.7 mmHg, respectively. After arterial or venous clamping, pO_2 decreased to 0.3 ± 0.1 mmHg and 1.2 ± 0.6 mmHg*, respectively. Reperfusion of the artery or vein resulted in a increase to 17.6 ± 1.2 mmHg and 11.6 ± 2.3 mmHg, respectively.

Clamping of the artery (n = 16) resulted in a decrease in pO_2 which reached a stable minimum after 3.2 ± 0.4 min. After opening the arterial clamp, the signal stabilized after 5.5 ± 1.4 min. In 8 of the 16 arterial vessel occlusions, releasing the clamp resulted in a steady rise in pO_2, reaching a final level that remained stable for 30 sec and was taken as 100%, while the other 8 arterial occlusions resulted in an overshoot. This overshoot was 138 ± 13% higher than the stable end-value and its duration was 3 ± 0.6 min. In 4 cases the overshoot was followed by a decline (rebound) in pO_2 (62 ± 17%). The duration of this decline was 6.3 ± 3.5 min.

Clamping of the vein (n = 12) resulted in a decrease in pO_2 which reached a stable minimum after 6.9 ± 1.1 min*. After opening of the venous clamp, a stable value was reached after 24.0 ± 8.3 min*. After release of the venous clamp, 2 out of 12 recordings showed an overshoot (108 ± 0% with a duration of 1.2 ± 0.6 min). In both of these 2 cases the overshoot in pO_2 was followed by a rebound (78 ± 2% with a duration of 8.1 ± 7.7 min).

3.2. Perfusion Changes in the Flap as Assessed by Laser-Doppler Flowmetry

Before clamping the artery or vein the mean BF was 117 ± 20 PU and 91 ± 13 PU, respectively. After clamping the vessels, the mean minimum levels were 52 ± 17 PU and 34 ± 6* PU, after reperfusion 100 ± 16 PU and 58 ± 12 PU, respectively.

Clamping of the artery (n = 16) resulted in a decrease in BF. The duration of this decrease was 0.4 ± 0.1 min. After release of the arterial clamp a stable end-value in BF was seen after 3.6 ± 0.8 min. In 7 of the 16 arterial occlusions releasing the clamp resulted in a steady increase in BF, while in 9 of the arterial reperfusions an overshoot in BF was seen, compared with the stable end-value (129 ± 8% with a duration of 2.2 ± 0.8 min). In 4 cases the overshoot was followed by a decline in BF prior to reaching a stable end-value (60 ± 15% with a duration of 5 ± 1.2 min).

Clamping of the vein resulted in a decrease in BF during 1.8 ± 0.4 min*. In 4 of 12 cases, opening of the clamp on the veins showed a higher BF value after the clamp was released, compared with the stable end-value (160 ± 35% with a duration of 2.6 ± 0.9 min). This overshoot was followed by a rebound in all these 4 cases (62 ± 8% with a duration of 6.0 ± 2.5 min).

4. DISCUSSION

The monitoring of a free tissue flap is of major importance, as early detection of impairment in microcirculatory flow in the flap is crucial for assessing the possibility of re-operating and re-establishing an adequate metabolic state in the flap. In the present study we have shown that, in a rat pedicle flap, it is possible to distinguish between a venous and an arterial flap stasis with a modified Clark electrode (Figure 2) and that a decrease in the pO_2 of a flap is a more specific and sensitive indicator of stasis in a vessel than is the laser-Doppler measurement (Figure 1).

In clinical practice, laser-Doppler flowmetry (LDF) is widely used, allowing continuous recording, and the measurements are easily obtained by means of a non-invasive cutaneous probe.[11] However, the recordings require careful interpretation[12] and do not meet the criterion of being a simple monitoring technique.

Tissue oxygen tension (pO_2) has been suggested to reflect perfusion.[3,4] However, pO_2 measurement with microelectrodes is an invasive procedure and as the size of the electrodes affect the measurement of local oxygen tension, the electrodes have to be fine if accurate and stable values are to be obtained and consequently are both extremely fragile and also complicated to handle in clinical practice. The microelectrode used in our study had a tip diameter of 2–8 µm.

In the tissue the oxygen tension at any given location may differ due to factors as distance to capillaries and to tissue regions having different oxygen consumption. A flap

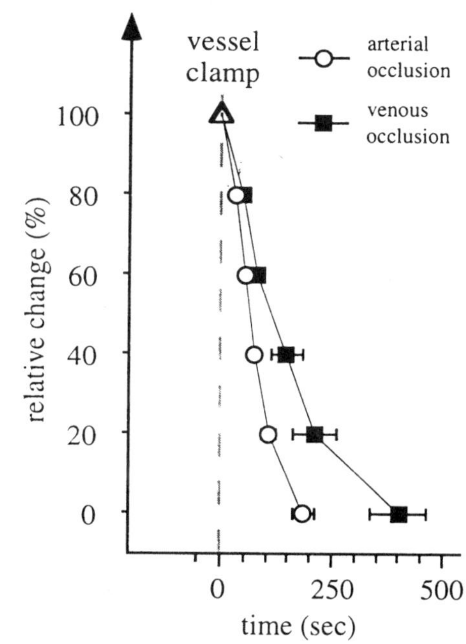

Figure 2. Relative change in pO_2 in pedicled fasciocutaneous flaps descriminating between an arterial and venous occlusion (mean ± 1 SEM).

may also have poor circulation in some parts, which further increase heterogeneity of the oxygen tensions. Measurements taken at several location in such heterogeneous tissue are required to represent the heterogeneous population. Once can reduce the impact of heterogeneity on the measurement by raising, as we did, a flap containing mainly subcutaneous fat tissue of relatively homogeneous distribution of microvasculature and oxygen sinks relative to the volume of the pO_2 and LDF measurements.

Hjortdal et al.[13] had a 90–120 sec 90% (t90) response with their Continucath® equipment and reported a decrease in pO_2[14] after occlusion to a stable value after 15 min. They found no difference between arterial and venous occlusions. In our study we found a significant difference in the time course of the decrease in pO_2 during arterial and venous occlusions. The reason for the faster response in this study is most likely due to the shorter reaction time of the electrode ($t90\% = 2.6 \pm 0.5$ sec).

Discriminating between arterial and venous occlusion in the clinical situation may be of importance since a venous stasis is more harmful to the tissue.[5] From the clinical point of view, venous congestion displays typical signs, like a change in the colour of the tissue from normal to a darker tone of redish blue. In our study, the appearance of this clinical sign in the tissue was not assessed. However, our experience from these experiments is that the decrease in BF and pO_2 precedes the manifestation of this clinical sign.

The significantly faster decrease in pO_2 after arterial occlusion, compared with venous occlusion, is probably due to a fairly constant O_2 consumption in the tissue, while during venous occlusion pooling of blood occures in the post-capillary vessels allowing for a while the inflow of arterial, saturated blood into the volume of pO_2 measurement thus delaying its decline to a minimum value. This effect could even be augmented if a larger size flap, that that of this study would be used. The laser-Doppler signal declines faster after arterial occlusion than after venous occlusion, which is in accord with the pO_2 measurement.

Reperfusion of an occluded artery is significantly faster than reperfusion of an occluded vein. Most likely, the pooling of blood in the tissue at a venous occlusion increases the resistance in the tissue and thus hinders the blood-flow and consequently slows the reperfusion. This can be even more exaggerated when there is microthrombosis in the capillary system or oedema in the tissue. The laser-Doppler signal increases after reperfusion of an occluded artery and vein, which is also in accord with the pO_2 measurement. Hjortdal et al.[6] reported that the pO_2 in tissue increased faster in arterial insufficiency flaps than in venous stasis flaps when the venous or arterial clamp was removed. The same group have also reported that, after venous occlusion the flaps became severely congested.[14] Our finding of a quicker increase in both pO_2 and blood flow after arterial occlusion compared with venous stasis, confirm these results.

Although the pO_2-electrode is fast, the shift in laser-Doppler signal is significantly faster. However, the interpretation of the decrease in laser-Doppler value is more difficult, as the signal does not fall to a zero value but rather to half of the starting level. Also, the baseline of the laser-Doppler recording varies over time as well as between different locations in a specific tissue, and this complicates the interpretation as to whether an occlusion of a vessel has occurred, or not.

ACKNOWLEDGMENTS

Financial support was received from the following funds: Gunvor and Josef Anér, Marcus and Amalia Wallenberg, Åke Wiberg and Thuréus.

REFERENCES

1. Shaw, WW, Discussion: Direct monitoring of microvascular anastomosis with the 20 MHz ultrasonic Doppler probe. An experimental and clinical study. By Swartz WM, Jones NF, Cherup L, Klein A, *Plastic Reconstr Surg*, 81:149–159. 1988.
2. Jones, BM, Monitors for cutaneous microcirculation. *Plastic Reconstr Surg*, 73:843–850. 1984.
3. Hjortdal, VE, Henriksen, TB, Kjølseth, D, Stender Hansen, E, Djurhuus, JC, and Gottrup, F, Tissue oxygen tension in myocutaneous flaps. *Eur J Surg*, 157:307–311. 1991.
4. Gottrup, F, Firmin, R, Hunt, TK, and Mathes, SJ, The dynamic properties of tissue oxygen in healing flaps. *Surgery*, 11:527–536. 1984.
5. May, JW, Chait, LA, O'Brien, BM, and Hurley, JV, The no-reflow phenomenon in experimental free flaps. *Plast Reconstr Surg*, 61:256–267. 1978.
6. Hjortdal, VE, Hauge, E, and Stender Hansen, E, Differential effects of venous stasis and arterial insufficiency on tissue oxygenation in myocutaneous islands flaps; an experimental study in pigs. *Plastic Reconstr Surg*, 89(3):521–529. 1992.
7. Liss, P, Nygren, A, Revsbech, NP, and Ulfendahl, H, Intrarenal oxygen tension measured by a modified Clark electrode at normal and low blood pressure and after injection of X-ray contrast media. *Pflügers Arch*, 434:705–711. 1997.
8. Revsbech, NP, An oxygen microsensor with a guard cathode. *Limnol Oceanogr*, 34(2):474–478. 1989.
9. Clark, LC, Monitor and control of blood and tissue oxygen tensions. *Trans Am Soc Art Int Org*, 2:41–57. 1956.
10. Liss, P, Nygren, A, Erikson, U, and Ulfendahl, HR, Injection of low and iso-osmolar contrast medium decreases oxygen tension in the outer renal medulla. *Kidney Int*, March. 1998.
11. Hedén, P, Hamilton, R, Arnander, C, and Jurell, G, Laser-Doppler surveillance of the circulation of free flaps and replanted digits. *Microsurgery*, 6:11–19. 1985.
12. Svensson, H, Holmberg, J, and Svedman, P, Interpreting laser-Doppler recordings from free flaps. *Scand J Plast Reconstr Surg Hand Surg*, 27(2):81–87. 1993.
13. Hjortdal, VE, Timmenga, EJ, Kjølseth, D, Henriksen, TB, Stender Hansen, E, Djurhuus, JC, et al., Continuous direct tissue oxygen tension measurement. *Ann Chir Gynaecol*, 80:8–13. 1991.
14. Hjortdal, VE, Awwad, AM, Gottrup, F, Kirkegaard, L, and Gellett, S, Tissue oxygen tension measurement for monitoring musculocutaneous and cutaneous flaps. *Scand J Plast Reconstr Surg Hand Surg*, 24:27–30. 1990.

A NEW MEASURING DEVICE FOR NON-INVASIVE DETERMINATION OF OXYGEN PARTIAL PRESSURE AND OXYGEN CONDUCTANCE OF THE SKIN AND OTHER TISSUES

K. Niehoff* and W. K. R. Barnikol

Bereich Klinische Physiologie
Private Universität Witten/Herdecke and GlukoMediTech AG
D-58455 Witten/Ruhr, Germany

SUMMARY

About fifty percent of the oxygen consumption of the skin is supplied by diffusion through the surface. This portion of the skin oxygen supply becomes of high importance in case of arterial occlusion. The oxygen permeation coefficient (P) of the upper layers and the oxygen pressure field within the skin determine the diffusive oxygen uptake from the outside. To our knowledge, the permeation coefficient (P) until now was only estimated by indirect methods of little practicability (Baumberger et al., 1951; Eberhard et al., 1978). An oxygen partial pressure of the skin is conventionally measured by modified CLARK type electrodes. A disadvantage of this so-called transcutaneous electrode is its oxygen consumption and the fixed coupling of the consumption with the oxygen pressure to be determined. Therefore the measurement always induces a systematic error (the so-called stirring effect) which depends, among other factors, on the oxygen availability of the skin under the electrode.

The new device combines a consumption-free oxygen partial pressure detector on the basis of luminescence quenching by oxygen with an independently working specific oxygen consumer realized by an active galvanic chain (silver-lead element). The chain permits setting any oxygen mass flow ($\dot{m}O_2$) in a certain range by varying the electrode current choosing different resistors within the electrical circuit. According to the

* This contribution contains an excerpt of the thesis of K. Niehoff.

Oxygen Transport to Tissue XXI, edited by Eke and Delpy.
Kluwer Academic / Plenum Publishers, New York, 1999.

diffusion law, the surface oxygen pressure (ePO_2) being measured is a linear function of the oxygen flow ($\dot{m}O_2$) directed to the cathode: $ePO_2 \equiv -(1/P)\cdot(\dot{m}O_2/A) + icPO_2$; A: area under the cathode. The intracutaneous oxygen partial pressure ($icPO_2$) is a virtual quantity defined by the equation given. Only by using an active electrode different oxygen mass flows can be set and so the oxygen conductance of the upper skin layers can be assessed.

First experiments on human skin in the gluteal region of an adult delivered an estimated value of the permeation coefficient (P): $2.2\cdot10^{-5}$ ml O_2 (STPD)/(atm·s·cm^2) at 42°C skin surface temperature; the intracutaneous partial pressure obtained was 5.5 kPa (41 mmHg) (STPD: "standard temperature pressure dry" conditions of the gas). At 42°C skin temperature no burning occurs.

The determined O_2-conductance is in the same range as estimated formerly (see above). The intracutaneous oxygen partial pressure determined seems to be a realistic value of the tissue at 42°C. By in vitro measurements with technical membranes the new device and procedure was validated giving precise values of the oxygen conductance. Hence the new method may be used for calibration of the oxygen flux optode (Holst et al., 1993). The O_2-permeation coefficient (P) could be an important parameter for evaluating dermatological applications (which attempt to increase P) in the treatment of local dermal oxygen deficiency. The intracutaneous oxygen partial pressure found is a reasonable value for a surface temperature of 42°. Because of the independence of the O_2-partial pressure determined and the O_2-consumption the new device exhibits no stirring effect and it provides more insight into the limitations of conventional transcutaneous oxygen measurement.

1. INTRODUCTION AND PROBLEM

The oxygen partial pressure of the skin is of fundamental importance as a global indicator for the regular function of the entire oxygen transporting system of the organism (Huch et al., 1975; Tremper et al., 1979). A normal oxygen pressure field in the skin requires an allover intact oxygen transporting system, including the lung and the cardiovascular system. Moreover the intracutaneous oxygen pressure field gives information about the oxygen supply to the skin itself, being altered in local hypoxic states (Lucas et al., 1988; Ewald, 1984). As about fifty percent of the dermal oxygen consumption is delivered by passive diffusion from the outside (Fitzgerald, 1957; Großmann et al., 1981) the factors determining this diffusive uptake, i.e. the oxygen permeation coefficient (P) of the upper skin layers and the three-dimensional intracutaneous oxygen pressure field, become of great importance in such local perfusion disturbances.

The three-dimensional cutaneous oxygen pressure field is practically not exactly determinable. An oxygen partial pressure (PO_2) of the skin is normally measured by modified CLARK type electrodes. This so-called transcutaneous electrode measures an epicutaneous oxygen partial pressure (ePO_2) and gives under certain conditions an average intracutaneous oxygen pressure of the upper dermis. A disadvantage of the polarographic CLARK type electrode is its oxygen consumption and the fixed coupling of the consumption with the oxygen pressure to be determined. Therefore the determination of the intracutaneous oxygen pressure always generates a systematic error (the so-called stirring effect). Under certain circumstances, the epicutaneous oxygen pressure (ePO_2) approximates the arterial oxygen pressure (aPO_2). This requires heating of the skin to a temperature of about 44°C, which leads to burns within a few hours.

To our knowledge, the oxygen permeation coefficient (P) of the skin until now may be estimated only roughly by indirect methods of little practicability. Baumberger and Goodfriend obtained a value (P) from the adjusting time of the oxygen partial pressure in a warm (45°C) buffer solution around a human finger. They measured the oxygen pressure polarographically (Baumberger et al., 1951). Eberhard and Severinghaus provided a CLARK type electrode with membranes of variable diffusive resistance for oxygen (Eberhard et al., 1978). On the basis of a reference diffusion resistor they were able to get indirectly an approximate value of P.

Here we present a new combined measuring device not only to assess an intracutaneous oxygen partial pressure, but also an oxygen conductance. The device is applicable to the surface of other tissues likewise.

2. PRINCIPLE OF THE NEW PROCEDURE

Figure 1 displays graphically the principle of the device via the cutaneous capillaries a certain oxygen average partial pressure ($icPO_2$) is provided. The galvanic oxygen sink of the measuring head generates an outwards directed oxygen flow ($\dot{m}O_2$), reversing the

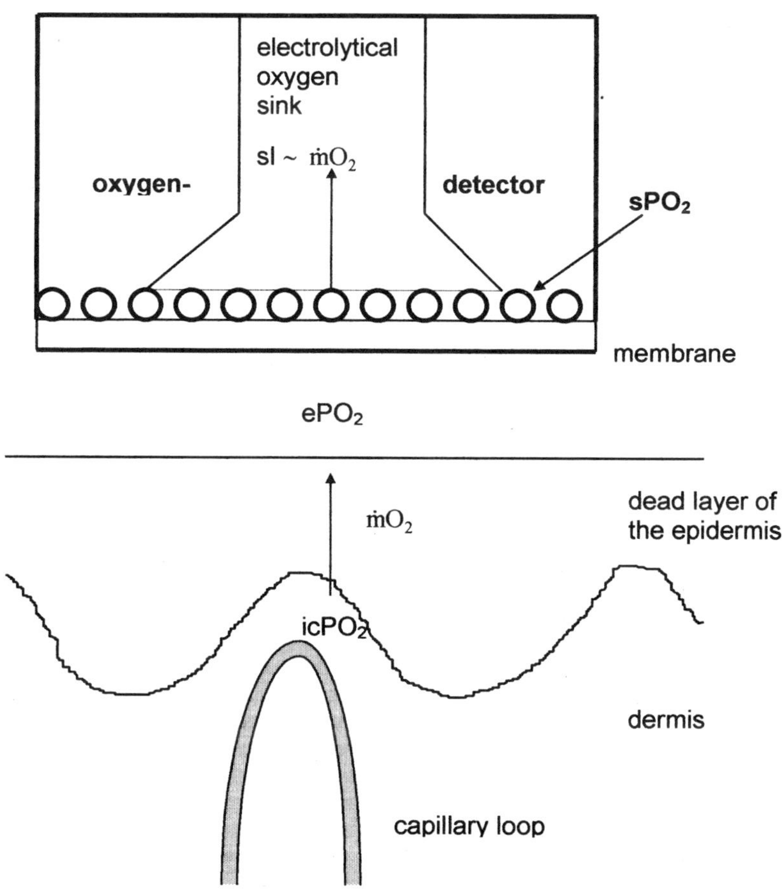

Figure 1. Graphical display of the device.

natural oxygen flow. Therewith a certain epicutaneous oxygen partial pressure (ePO_2) is set, which can be determined by the PO_2 sensor of the head (so-called dynamic mode of measuring). Deactivating the sink by interruption of its electrical circuit effects an equilibration of the oxygen partial pressures: $ePO_2 = icPO_2$ (so-called static mode).

In the dynamic (stationary) mode FICK's first law is valid:

$$\frac{\dot{m}O_2}{A} \equiv P(icPO_2 - ePO_2) \tag{1}$$

where A: area under the head, P: permeation coefficient for oxygen.

Equation (1) defines an average intracutaneous oxygen partial pressure. Rearranging the equation gives:

$$ePO_2 = icPO_2 - \frac{1}{A \cdot P} \cdot \dot{m}O_2 \tag{2}$$

Varying the oxygen flow ($\dot{m}O_2$) and measuring the corresponding epicutaneous oxygen pressure (ePO_2) gives the function ePO_2 ($\dot{m}O_2$).

From the slope of this function the permeation coefficient of oxygen (P) may be derived and from the intersection of the y-axis the intracutaneous oxygen partial pressure ($icPO_2$). This quantity may be also measured directly in the static mode. A coincidence of both values means an inner validation of the measuring procedure.

3. DEVICE AND PROCEDURE

The new device combines a consumption-free oxygen partial pressure detector on the basis of luminescence quenching by oxygen with an independently working specific oxygen consumer used as the oxygen sink.

We employed the oxygen detector presented by Burkhard and Barnikol (Barnikol et al., 1990; Burkhard, 1988; Burkhard et al., 1983), measuring the fluorescence-intensity of an excited dye. The intensity depends inversely on the oxygen concentration. Tetraphenylporphyrin (Fa. Ventron GmbH, Karlsruhe/D) adsorbed on the chromatographic material Porapak Q (Fa. Waters Assoc., Framingham, Massachusetts/USA) served as the fluorescence dye. The schematic experimental setup is shown in Figure 3. The "dye-carrier-balls" are fixed on a glass lens by double-sided adhesive transparent tape. The visible exciting light is given by a stabilized mercury lamp (Fa. Eppendorf,

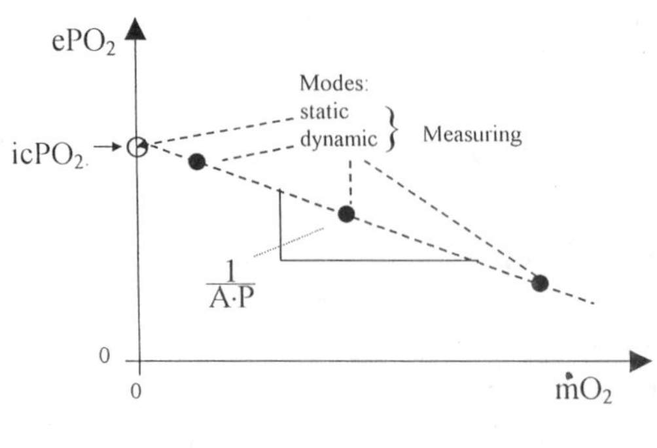

Figure 2. Principle of the graphical evaluation of the procedure.

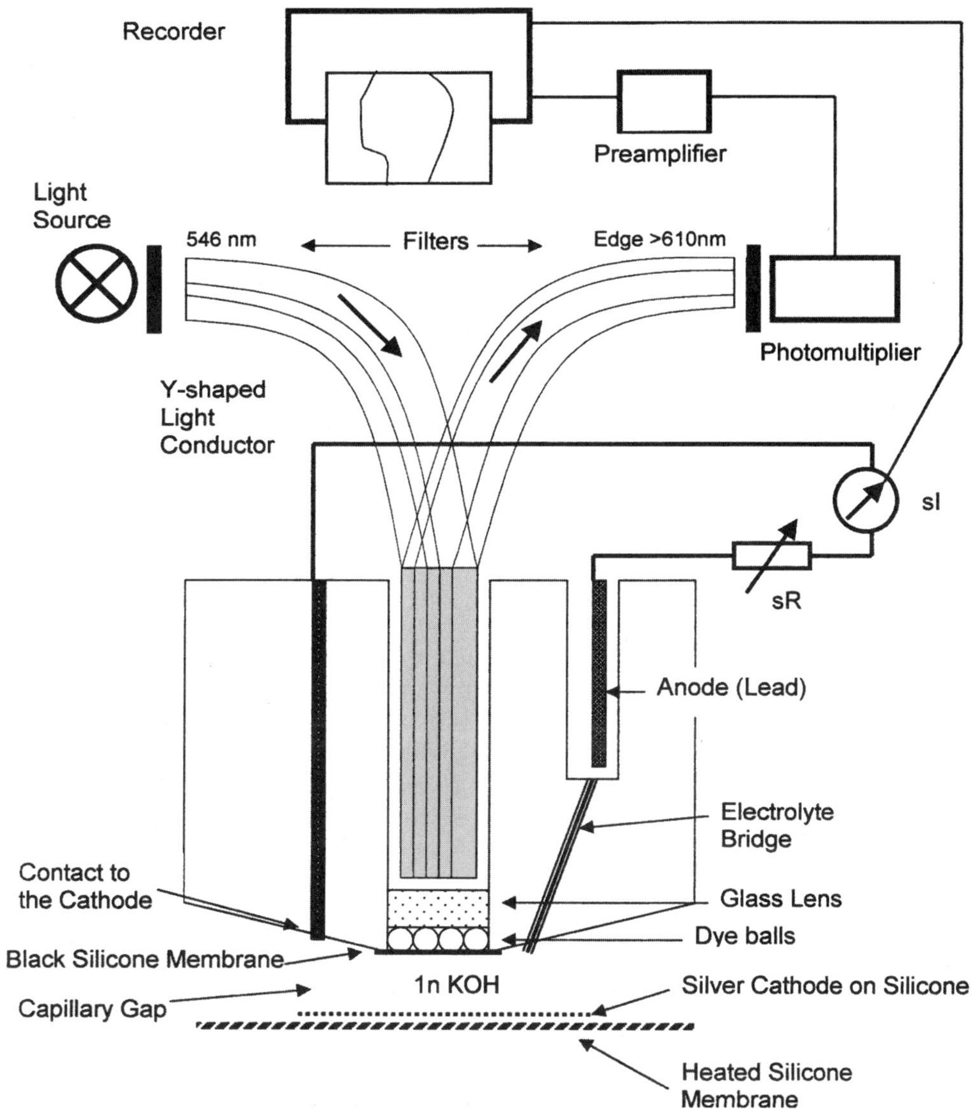

Figure 3. Display of the experimental setup.

Hamburg/D). After passing a green filter (546 nm, Fa. Schott Glaswerke, Mainz/D) the exciting light reaches the dye through one of the arms of a Y-shaped light conductor (Fa. Schölly Fiberoptik, Denzlingen/D). The fibres of the excitation and detection arm are mixed statistically in the common arm of the light conductor. As the fluorescence light has a higher wavelength than the exciting light, the latter can be eliminated by an optical edge (>610 nm). The intensity of the light is measured with a photomultiplier (type C 31034 A, Radio Corp. of America/USA). After amplifying the signal it is displayed by a recorder (Fa. Linseis, Selb/D). The fluorescence signal has to be calibrated. We determined the relative fluorescence intensity, F_{rel} (Burkhard, 1988):

$$F_{rel} = (F_{N_2} - F)/(F_{N_2} - F_{air}) \tag{3}$$

with

$$F_{rel} = PO_2/(a + b \cdot sPO_2) \quad (4)$$

a,b: constant parameters; F: fluorescence intensity measured; F_{N_2}: fluorescence intensity under N_2; F_{air}: fluorescence intensity under air, sPO_2: sensor oxygen partial pressure.

The fluorescence signal is temperature dependent which has to be regarded. When located on the skin, the oxygen detector measures the surface partial pressure of the cathode (sPO_2) which almost equals the epicutaneous oxygen pressure (ePO_2), because the oxygen diffusion resistance of the outer silicone membrane is low compared with the resistance of the upper skin layers: $sPO_2 \approx ePO_2$.

The independent oxygen consumer is realized by an active galvanic chain, which is constructed according to the active Syland® oxygen electrode (Syland Scientific GmbH, 1989). The chain (a silver-lead element in 1n KOH) reduces oxygen at the silver cathode surface without any external voltage applied according to the following reaction:

$$O_2 + 2H_2O + 4\,e^- = 4OH^-$$

Because of the absolute oxygen specificity the oxygen mass flow ($\dot{m}O_2$) can be calculated directly from the electrode current (sI) using the Faraday constant (**F**). The resistor (sR) in the electrode circuit is varied to regulate the oxygen flow ($\dot{m}O_2$) in a certain range (see Figure 3). We used resistors between 500 Ω and 80 kΩ. The current was measured by an amperemeter (Fa. Gossen GmbH, Erlangen/D).

The construction of the cathode has to fulfill several needs: It should have little influence on the response time of the device, its oxygen diffusion resistance should be low and the distribution of oxygen permeable areas and oxygen reducing areas should be as homogeneous as possible. These requirements are fulfilled by a layer of micro silver flakes (1–5 µm) on a self-adhesive silcone SilGel 612 (all silicones from Fa. Wacker Chemie, München/D).

The outer membrane may be heated by a resistance wire out of constantan embedded into the silicone (RTV ME 625). The temperature is controlled by a small (2 mm) thermistor located between skin and device. The diameter of the cathode membrane was 20 mm. The dye-carrier-balls are protected against the KOH-solution and against scattered light by a black silicone membrane (VP 7209 or E 43).

A measuring cycle is performed as follows: The oxygen detector is calibrated under N_2 and ambient air. It is located on the skin and fixed, so that it does not exert too much pressure on the skin surface. We checked the adequate pressure before by not impairing the local perfusion. The skin surface is heated to a temperature chosen. Now the PO_2 signal reaches a stable plateau; this is the first direct assessment of $icPO_2$, as the electrical circuit of the electrochemical chain is still interrupted (so-called static mode with $\dot{m}O_2 = 0$). Then the first phase of oxygen reduction starts by closing the electrical circuit and allowing a finite value of oxygen flow ($\dot{m}O_2$,1): measurement in the dynamic mode with $\dot{m}O_2 > 0$. It is finished, when a first stationary and stable oxygen pressure level is reached. This gives a first value for the epicutaneous oxygen partial pressure (ePO_2,1). Thereafter a greater oxygen mass flow ($\dot{m}O_2$,2) is set by choosing a smaller resistor (sR) and a second (lower) stationary and stable oxygen pressure level is achieved (ePO_2,2). The whole procedure may be repeated several times. In the end, for control, a second oxygen pressure level ($icPO_2$) in the static mode (sI = 0: interruption of the electrical curcuit) is measured. Thus it is possible to estimate the drift of the oxygen sensor. At last, the calibration of the oxygen detector with gases is repeated in order to regard any change of sensitivity.

4. RESULTS

Before medical application the new device and procedure was rectified and validated with the aid of in vitro measurements on technical membranes, especially equation 4. First experiments were done in the gluteal region of an adult person heating the human skin to 42°C surface temperature.

In Figures 4 and 5 two measuring cycles from the same site of the skin are displayed. The sensor oxygen pressure (sPO_2, with $ePO_2 \approx sPO_2$) is shown to be dependent on the electrode current (sI). The directly measured intracutaneous oxygen pressure as determined in the static mode ($icPO_2 = ePO_2$, with $\dot{m}O_2 = 0$) is marked by a big point on the y-axis. The average of this intracutaneous partial pressure was about 5.5 kPa (41 mmHg). If measuring in the stationary (i.e. dynamic) mode and assessing the linear diffusion equation in the first cycle (Figure 4) from different value pairs (sPO_2 and $\dot{m}O_2$) the extrapolated intracutaneous oxygen pressure ($icPO_2$) is clearly lower than the directly measured one. The extrapolations in Figure 4 show, that the higher the oxygen flow preset by the resistor (sR), the lower is the intracutaneous oxygen pressure ($icPO_2$) determined by extrapolation.

In the second cycle shown in Figure 5 the intracutaneous oxygen pressure determined in the static mode equals the value determined in the dynamic mode. From this measurement we calculated an oxygen permeation coefficient (P) of the upper skin layers as shown in Table 1.

During these measurements (at 42° Celsius) even after many hours of measurement no burns of the skin were observed.

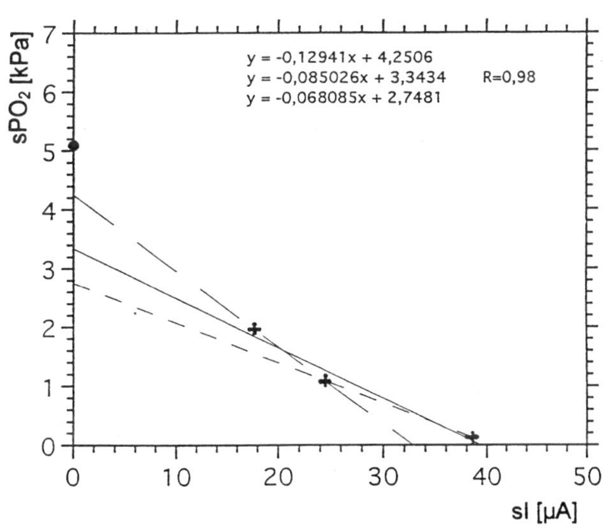

Figure 4. First measurement cycle on human gluteal skin (42°C surface temperature), sPO_2: sensor oxygen partial pressure; sI: sensor current; ●: static mode at $\dot{m}O_2 = 0$; +: dynamic mode with a finite transcutaneous oxygen current, the linear regression functions in each case of two respectively three points are given.

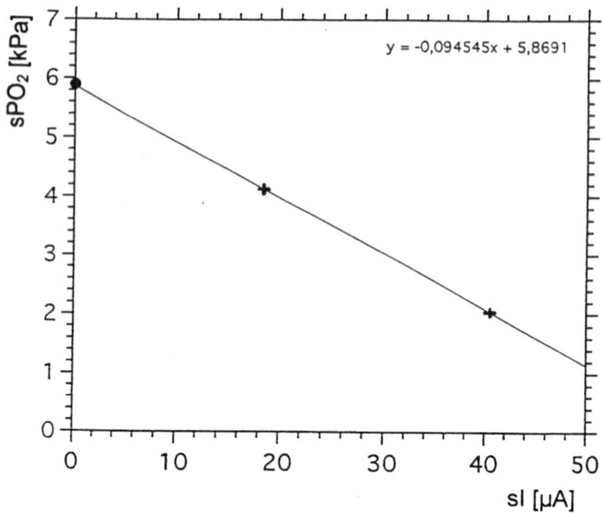

Figure 5. Second measurement cycle on human gluteal skin, denotations see Figure 4.

5. DISCUSSION

Before application to the human skin the new device presented here was evaluated with appropriate technical membranes, i.e. first it was shown, that the linear equation (2) holds for the setup and second it was shown that a preset oxygen partial pressure outside of the technical membrane, could be assessed correctly from dynamic (stationary) measurements by extrapolation (Niehoff, 1998).

In contrast, the directly (static mode) determined intracutaneous oxygen pressure ($icPO_2$) is clearly greater than the extrapolated one from the oxygen pressure levels obtained in the stationary mode (see Figure 4). Hence it may be concluded, that during these measurements the oxygen consumption of the electrode is in significant competition to the oxygen consumption of the skin. I.e.: the electrode is "robbing" oxygen from the skin. From this point of view the choice of small value of oxygen flow is favourable. On the other hand it has to be regarded that setting smaller oxygen flows means that the oxygen sensor has to fulfill higher needs with respect to stability and sensitivity because smaller oxygen partial pressure steps have to be determined.

But during thc second measuring cycle (Figure 5) the intracutaneous pressure ($icPO_2$) appears not to be decreased by the oxygen consumption of the electrode. The

Table 1. The oxygen permeation coefficient (P) of the upper skin layers from different authors (STPD: "standard temperature pressure dry" conditions of the gas)

Source	P [ml O_2 (STPD)/(atm·s·cm^2)]
Baumberger, Goodfriend, 1951	$2.5\cdot 10^{-5}$
Eberhard, Severinghaus, 1978	$1.5\cdot 10^{-5}$
This publication	$2.2\cdot 10^{-5}$

coincidence of the directly measured (static mode) value of $icPO_2$ and the extrapolated one (dynamic mode) means an inner validation of the analytical procedure. It can be concluded, that in this situation the oxygen consumption of the galvanic chain is low compared with the oxygen availability within the skin. Therefore from this measurement it is possible to calculate a correct oxygen permeation coefficient (P) (see Table 1). So the new device provides information about the limitations of conventional transcutaneous oxygen measurement in which the desired oxygen pressure signal and oxygen consumption of the device are strictly coupled. The conventional transcutaneous electrode gives by itself no information whether it is in significant competition to the skin oxygen consumption and may decrease the intracutaneous oxygen pressure level significantly beyond the stirring effect, which is a systematic constant error to low values. The decrease of the intracutaneous oxygen level by the electrode will be more severe with lower skin surface temperatures. This is one reason for conventional transcutaneous measurement at about 44°C.

The stirring effect of the conventional transcutaneous electrode is supposed to be about 6% (Severinghaus et al., 1978). The new method works without this error. This is one reason why we suppose that the value of the intracutaneous oxygen pressure (icPO2) determined here (41 mm Hg) at 42°C skin surface temperature is closer to the "real" physiologic value than that obtained by conventional transcutaneous measurement a 44°C surface temperature. This value is about 70 mm Hg and reflects in some way the arterial oxygen partial pressure paralyzing the microvascular system by heat. 41 mm Hg for the intracutaneous oxygen pressure seems to be a reasonable value and it should be emphasized, that it is assessed in the thick gluteal skin. A device equipped with a passive polarographic electrode does not allow to determine the parameters of the system equation (2), because it works with a fixed oxygen consumption at a given oxygen partial pressure.

The value of the oxygen permeation coefficient (P) of the upper skin layers determined here is in the same range as formerly estimated (Baumberger et al., 1951; Eberhard et al., 1978)—see Table 1. This coefficient (P) together with the oxygen pressure gradient in the upper skin determines the diffusive oxygen uptake from the outside, which is estimated to be about 50% of the physiological normal oxygen consumption of the skin (Fitzgerald, 1957; Großmann et al., 1981). Substances like fluocarbons are supposed to raise P and could thus increase the diffusive oxygen uptake of the skin. So the permeation coefficient (P) could be an important parameter in evaluating local dermatological applications in the treatment of skin oxygen deficiency (e.g. arterial occlusive disease).

The prototype of the device presented here is still too big for routine clinical use, as it has a diameter of 6 cm. A miniaturisation seems to be necessary. Another drawback is the bleaching of the dye, which leads to an unstable oxygen signal, when assessing the intensity of the fluorescence light. Therefore in future the fluorescence lifetime should be determined instead of the intensity, as assessing the lifetime gives a more stable oxygen signal because this time is independent from the absolute fluorescence intensity. (Barnikol et al., 1996; Vanderkooi et al., 1987).

REFERENCES

Barnikol WKR, Burkhard O, and Weiler N: Meßvorrichtung zur Bestimmung des Sauerstoffpartialdruckes, des Sauerstoffgehaltes und des Sauerstoff-Flusses in biologischen Systemen; Offenlegungsschrift am Deutschen Patentamt, Nr. DE 3900191 A1, Aktenzeichen P 3900191.1, Offenlegungstag: 12.7.1990.

Barnikol WKR, Burkhard O, Trübel H, Petzke F-H, Weiler N, and Gaertner Th: Ein neuartiges Verfahren der Sauerstoffdetektion für Medizin, Biologie, Umweltforschung und Biotechnik Biomedizinische Technik *41* (1996) 170–177.

Baumberger JP and Goodfriend RB: Determination of arterial oxygen tension in man by equilibration through intact skin; Fed Proc Am Physiol Soc 10 (1951) 10.

Burkhard O and Barnikol WKR: Ein schneller O_2-Detektor für die Analyse der Lungenfunktion; Prax Klin Pneumol 37 (1983) 805–807.

Burkhard O: Ein neuer Detektor zur Messung des Sauerstoffpartialdruckes in biologischen Systemen; thesis Mainz, Germany, 1988.

Eberhard P and Severinghaus JW: Measurement of heated skin-O_2-conductance and PO_2—sensor-induced O_2-gradient; Acta Anaesthesiol Scand Suppl 68 (1978) 1–3.

Ewald U: Evaluation of the transcutaneous oxygen method used at 37 degrees C for measurement of reactive hyperaemia in the skin; Clin Physiol 4 (1984) 413–423.

Fitzgerald LR: Cutaneous respiration in man; Physiol Rev 37 (1957) 325–336.

Großmann U, Huber J, Fricke K, and Lübbers DW: A new model for simulating the oxygen pressure field of skin; Adv Physiol Sci 25 (1981): Oxygen Transport to Tissue, 319–320; eds.: AGB Kovách, E Dorá, M Kessler, and JA Silver; Pergamon Press, Akadémiai Kiadó, Budapest. Personal information from U. Grossmann, Max-Planck-Institut für Systemphysiologie, Rheinlanddamm 201, Dortmund, Germany.

Holst GA, Lübbers DW, and Voges E: O_2-flux optode for medical application; SPIEJ: Advances in Fluorescence Sensing Technology 1885 (1993) 216–227.

Huch R, Lübbers DW, and Huch A: The transcutaneous measurements of oxygen and carbon dioxide tensions for the determination of arterial blood-gas values with control of local perfusion and peripheral perfusion pressure. Theoretical analysis and practical application. In: JP Payne and DW Hill (eds.): Oxygen Measurement in Biology and Medicine, Butterworth, London, 1975, 121–138.

Lucas A, Agache P, Risold JC, and Cuenot C: Effet chronique sur la pression partielle de l'oxygène transcutanée: Presse Med 17 (1988) 279.

Niehoff K: Ein neuartiger Meßkopf zur nichtinvasiven Bestimmung eines Sauerstoffpartialdruckes und einer Sauerstoffleitfähigkeit der Haut oder anderer Gewebe sowie technischer Membranen; Thesis, Witten/Ruhr, Germany, 1998.

Severinghaus JW and Thunstorm A: Problems of calibration and stabilization of $tcPO_2$ electrodes; Acta Anaesthesiol Scand Suppl 68 (1978) 68–72.

Syland Scientific GmbH: Firmeninformation (enterprise information) "Sauerstoff-Meßtechnologie", Syland Scientific GmbH; Tiergartenstr./Humboldtstr.2, Heppenheim, Germany, 1989.

Tremper KH, Waxman K, and Shoemaker WC: Effects of hypoxia and shock on transcutaneous PO_2 values in dogs; Crit Care Med 7 (1979) 526–531.

Vanderkooi JM, Maniara G, Green TJ, and Wilson DF: An optical method for measurement of dioxygen concentration based upon quenching of phosphorescence; J Biol Chem 262 (1987) 5476–5482.

OXYGEN TRANSPORT IN 10μm ARTIFICIAL CAPILLARIES

T. C. Page,[1] W. R. Light,[1] and J. David Hellums[2]

[1]Research and Development
Biopure Corporation
11 Hurley St.
Cambridge, Massachusetts 02141
[2]Cox Laboratory for Biomedical Engineering
Rice University, MS 142
6100 S. Main St.
Houston, Texas 77005-1892

ABSTRACT

In order to further define the influence of microvessel diameter on intraluminal oxygen transport a previously described *in vitro* artificial capillary system was modified from a vessel diameter of 25μm to 10μm. Oxygen uptake and release rates were measured for hemoglobin solutions and red blood cell (Rbc) suspensions of the same overall hemoglobin concentration (10g/dl). The modified apparatus was tested by comparing data for the hemoglobin solutions with predictive simulations from a validated mathematical model of oxygen transport. Preliminary data for oxygen uptake by Rbc suspensions flowing in 10μm diameter capillaries are presented. As observed previously oxygen uptake is faster in hemoglobin solutions than in Rbc suspensions.

1. INTRODUCTION

A limitation of theoretical models of oxygen transport has been the scarcity of experimental data from sufficiently controlled systems to critically test the underlying simplifying assumptions (Hellums et al., 1996). For this reason we previously developed an *in vitro* "artificial capillary" apparatus where all flow, geometric, and physical parameters are known. The apparatus was used to model intraluminal oxygen transport in the microcirculation (here defined to be vessels <100μm in diameter). Within this range

Oxygen Transport to Tissue XXI, edited by Eke and Delpy.
Kluwer Academic / Plenum Publishers, New York, 1999.

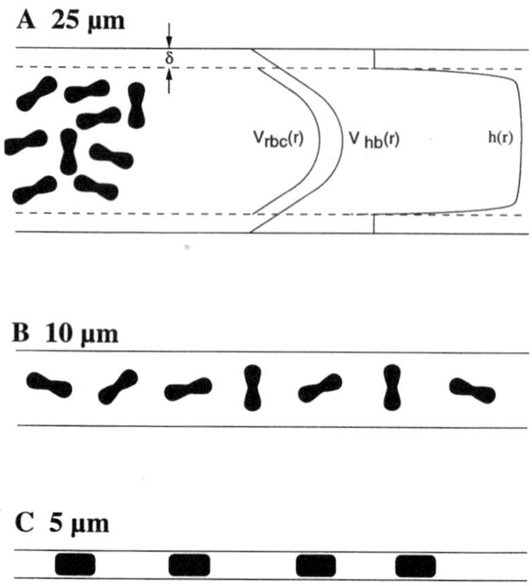

Figure 1. Schematic illustration of three distinct flow regimes in the microcirculation corresponding to arterioles, intermediate sized capillaries, and true capillaries.

there are a variety of vessels from arterioles to capillaries to venules that influence oxygen transport. We have simplified this complex *in vivo* distribution into three representative flow regimes which require different approaches to model oxygen transport processes in blood (Figure 1). The first regime is that of arteriolar-sized vessels in the range of 20 to 100 μm in diameter. This regime has been extensively studied in our laboratories (Boland et al., 1984; Page et al., 1998a). Predictive mathematical models have also been developed and published (Lemon et al., 1987; Nair et al., 1989; Page et al., 1998b). A second regime is that of vessels in the intermediate diameter range of 8 to 15 μm. In this intermediate regime red blood cells exhibit complex flow patterns and can travel in single file or in multiple layers. The third regime is that of the true capillaries, with diameters of less than 7 μm where erythrocytes deform and flow in single file. In the current work we describe modifications to the artificial capillary apparatus which allow modeling of the intermediate diameter range—specifically 10 μm diameter capillaries. Data on oxygen transport in hemoglobin solutions is presented to validate the apparatus. Preliminary oxygen uptake and release data for erythrocyte suspensions are also presented.

2. MATERIALS AND METHODS

2.1. Apparatus

The experimental system is illustrated schematically in Figure 2. The apparatus has been previously described in detail (Page et al., 1998). The system includes a "capillary" (either 25 μm or 10 μm diameter) embedded in a thin silicone rubber film. Silicone rubber was selected because of its high permeability to oxygen combined with sufficient optical clarity to permit spectroscopic determination of hemoglobin oxygen saturation. A silicone rubber film containing a single capillary is mounted on the stage of a microscope. The capillary is cannulated at each end and perfused with red blood cell suspensions or hemoglobin solutions. A dual wavelength microspectrophotometric method is used to determine the oxygen saturation of the sample at various axial positions along the

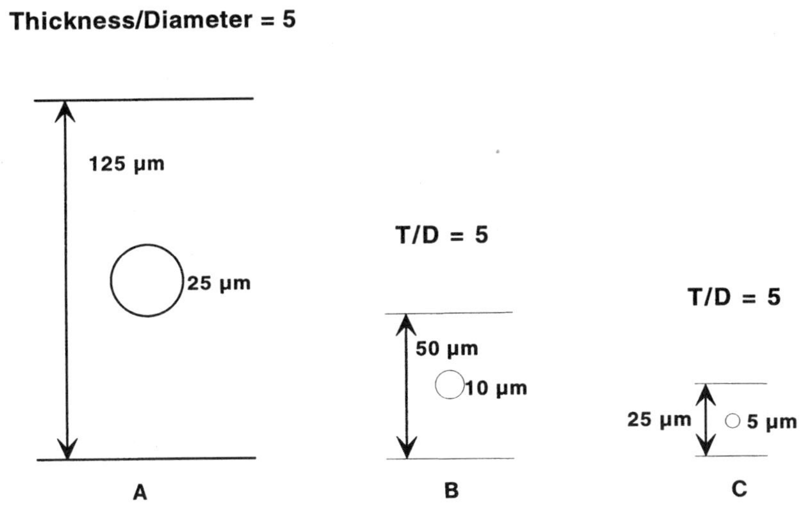

Figure 2. Schematic of Experimental Approach. Hemoglobin-containing samples (Rbc suspensions, hemoglobin solutions, mixtures, etc.) are passed through a 10 μm diameter lumen formed in a 50 μm thick silicone rubber film. Both oxygen uptake and release experiments are done by controlling the oxygen tension of the feed sample and the external suffusing gas. In this figure a cross-section of the capillary is shown. Both surfaces of the silicone film are exposed to humidified gas (air or nitrogen), and the sample uptakes or releases oxygen. Fractional oxygen saturation is measured spectrophotometrically.

capillary. The sample in the feed reservoir is equilibrated with a humidified gas of the desired oxygen tension and suffused with the same gas (often either air or nitrogen) while in the feed reservoir. The capillary film is suffused with a humidified gas of an independently selected oxygen tension. In a typical oxygen uptake experiment the feed reservoir contains a deoxygenated sample, and the gas suffusing the capillary film is air. In a typical oxygen release experiment the feed reservoir contains an oxygen-saturated sample and the gas suffusing the capillary film is nitrogen.

2.2. Preparation and Description of Samples

The specimens used in the experimental work discussed here all contained a total of 10 g/dl hemoglobin—whether in the form of erythrocytes or hemoglobin in solution. The erythrocytes were drawn from human subjects, washed and resuspended as previously described (Page et al., 1996). The suspended cells had a P_{50} of 27 ± 1 mmHg and a Hill coefficient of approximately 2.65. The hemoglobin solution was a purified bovine material prepared by Biopure Corporation with a P_{50} of 25 ± 1 mmHg and a Hill coefficient of approximately 2.65.

3. RESULTS

3.1. Hemoglobin Solution

In Figure 3 oxygen uptake and release data for Bhb at 10 g/dl are compared with predictions from the theoretical mathematical model. The curves on the figure are from a pre-

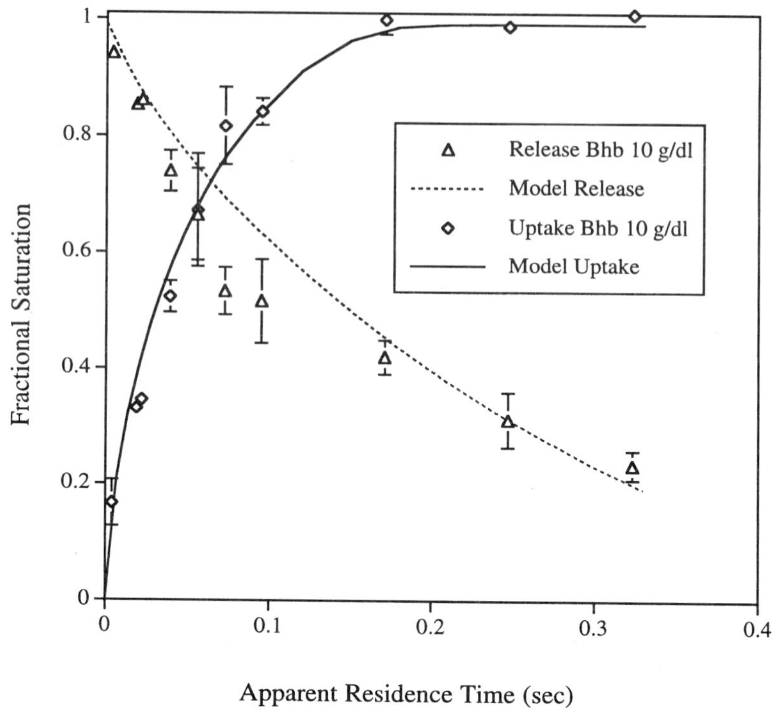

Figure 3. Hemoglobin solution oxygen uptake and release experiments for a 10 μm diameter capillary. Experimental determinations of hemoglobin oxygen saturation are presented as a function of residence time in the capillary. The error bars denote the standard deviation for replicate experiments. The curves are from predictive theoretical calculations. The good agreement between the experimental and theoretical values serves as a validation of the experimental apparatus.

viously published theoretical model which had been experimentally validated (Lemon et al., 1987) and the points are experimental measurements. This relatively simple model (as compared to red cells) can be used to validate the experimental system because modeling oxygen transport in hemoglobin solutions has a firm theoretical basis. There was good agreement between the experimental data points and the mathematical simulation.

3.2. Rbc Suspension

The first data for oxygen uptake and release by Rbc suspensions flowing in a 10 μm capillary is shown in Figure 4. Variability is larger for Rbc suspensions than for hemoglobin solutions due to the complex flow behavior of Rbc's in this flow regime and due to the sieving effect where light may pass through variable concentrations of hemoglobin and plasma gaps. Direct passage of light through plasma gaps reduces the signal to noise ratio. Though preliminary, this data gives good indication of the differences between Rbc suspension and hemoglobin solution behavior in 10 μm capillaries.

3.3. Oxygen Uptake Rbc versus Hemoglobin Solution in 10 μm Capillaries

Oxygen uptake data for hemoglobin solutions and Rbc suspensions in 10 μm capillaries are presented in Figure 5. Uptake data for the 10 μm capillary are shown as fitted

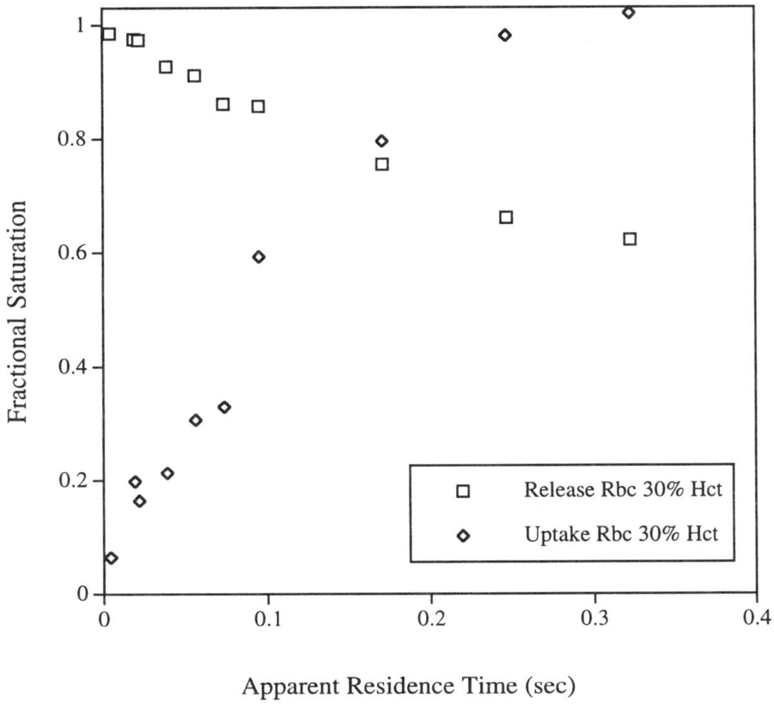

Figure 4. Oxygen Uptake and Release for Erythrocytes in 10 μm Diameter Capillary. In the figure fractional saturation is plotted as a function of apparent residence time in the capillary. Both uptake (diamonds) and release (squares) are shown. The error bars are plus/minus standard deviation. Error bars are larger for erythrocyte suspensions than for hemoglobin solutions due to sieving effect where light may pass through variable concentrations of hemoglobin including passing through no hemoglobin (plasma gaps). This direct passages reduces the signal to noise ratio. The uptake of oxygen may occur up to 4 times faster than in the 25 μm capillary.

curves. Oxygen uptake by hemoglobin solution is faster than by Rbc suspension in both flow regimes. Uptake in the 10 μm capillary samples is much faster than for the 25 μm samples (0.2 sec vs 0.8 sec for saturation in hemoglobin solutions, respectively).

4. DISCUSSION

The mathematical simulation of oxygen transport in hemoglobin solutions has a firm theoretical basis, making it useful to compare results of mathematical simulations with experimental results as a test of the experimental methodology. Experimental results for Bhb solutions of 10 g/dl agree well with mathematical predictions of both oxygen uptake and release.

Oxygen uptake by hemoglobin solution occurs more rapidly in smaller vessels. In the 10 μm capillary uptake of oxygen by Bhb to full saturation occurs in less than 0.2 sec. The corresponding time for full saturation in the 25 μm capillary is on the order of 4 times longer (Figure 5; Page et al., 1998). The surface area to volume ratio for the 10 μm capillary is 2.5 times larger than for the 25 μm capillary.

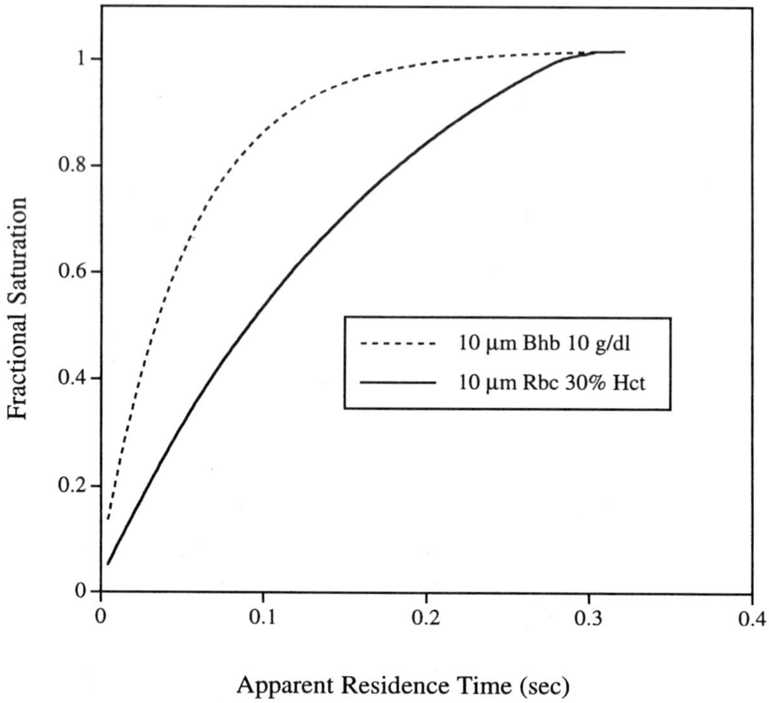

Figure 5. Comparison of oxygen uptake in 10 μm and 25 μm diameter capillaries. Fractional saturation is plotted versus residence time in the capillary (sec) for hemoglobin solutions and Rbc suspensions of the same overall hemoglobin concentration for both the 10 μm and 25 μm vessels. The arrows represent the difference in saturation between hemoglobin solutions and Rbc suspensions for a residence time which yields a fractional saturation of 0.8 in the hemoglobin solution. This distance is larger in the 10 μm capillary than in the 25 μm capillary. Data series for the 10 μm capillary are shown as fitted curves to reduce clutter in the plot.

The data for oxygen uptake and release by Rbc suspensions in 10 μm capillaries may be used to critically test mathematical models of oxygen transport. Only two theoretical studies have addressed oxygen transport in capillaries of the 10 μm range. Wang and Popel (1993) and Frank et al. (1997) utilized the "parachute" shape of erythrocytes flowing single file through axisymmetric and 2-D conduits. Similar approaches may be useful in modeling the oxygen transport measurements presented in this paper for Rbc suspensions, though the intraluminal flow parameters will have to be more complex.

It has been shown that hemoglobin solutions take up oxygen at a higher rate than red cell suspensions of the same overall hemoglobin concentration. The above data reconfirms this finding and extends it to 10 μm capillaries. Future work will continue to refine this 10 μm apparatus and extend it study of Rbc/hemoglobin solution mixtures.

REFERENCES

Boland, EJ, Unno, H, Olson, JS, and Hellums, JD (1984). An in-vitro method for simulation of oxygen transport in the microcirculation. *Adv. Exp. Med. Biol.*, **180**:251–259.

Boland, EJ, Nair, PK, Lemon, DD, Olson, JS, and Hellums, JD (1987). An in vitro capillary system for studies on microcirculatory O_2 transport. *J. Appl. Physiol.*, **62**:791–797.

Federspiel, WJ and Popel, AS (1986). A theoretical analysis of the effect of the particulate nature of blood on oxygen release in capillaries. *Microvasc. Res.*, **32**:164–189.

Frank, AO, Chuong, CJ, and Johnson, RL (1997). A finite-element model of oxygen diffusion in the pulmonary capillaries. *J. Appl. Physiol.*, **82**(6):2036–2044.

Gaehtgens, P, Duhrssen, C, and Albrecht, KH (1980). Motion, deformation, and interaction of blood cells and plasma during flow through narrow capillary tubes. *Blood Cells*, **6**:799–812.

Hellums, JD, Nair, PK, Huang, NS, and Oshima, N (1996). Simulation of intaluminal gas transport processes in the microcirculation. *Annals of Bomedical Engineering*, **24**:1–24.

Hellums, JD (1980). Deformation of blood cells in capillaries: A commentary. *Blood Cells*, **6**:815–817.

Lemon, DD, Nair, PK, Boland, EJ, Olson, JS, and Hellums, JD (1987). Physiological factors affecting O_2 transport by hemoglobin in an in vitro capillary system. *J. Appl. Physiol.* **62**:798–806.

Nair, PK, Hellums, JD, and Olson, JS (1989). Prediction of oxygen transport rates in blood flowing in large capillaries. *Microvasc Res.*, **38**:269–285.

Page, TC, McKay, CB, Light, WR, and Hellums, JD (1996). Experimental Simulation of oxygen transport in microvessels. *In* "Blood Substitutes: New Challenges" (Winslow, RM, Vandegriff, KD, and Intaglietta, M, Eds.), Birkhauser, Boston.

Page, TC, McKay, CB, Light, WR, and Hellums, JD (1998a). Oxygen transport by erythrocyte/hemoglobin solution mixtures in an in vitro capillary as a model of hemoglobin-based blood substitute performance, *Microvasc Res.*, **55**:54–64.

Page, TC, Light, WR, and Hellums, JD (1998b). Prediction Of Microcirculatory Oxygen Transport By Erythrocyte/Hemoglobin Solution Mixtures, *accepted*, *Microvasc Res.*

Page, Light, and Hellums. (1998) "Chapter 11: Experimental and mathematical simulation of oxygen transport by hemoglobin-based blood substitutes," in *Present and Future Perspectives of Blood Substitutes*, E. Tsuchida, editor, Elsevier, Lausanne, pp. 135–145.

Wang, CH and Popel, SA (1993). Effect of red blood cell shape on oxygen transport in capillaries. *Math. Biosci.* **116**:89–110.

Zarda, PR, Chien, S, and Slalak, R (1977). Interaction of viscous incompressible fluid with an elastic body. Symp. Fluid-Structure Interaction, ASME, New York, pp. 65–82.

A DUAL WAVELENGTH SPECTROPHOTOMETER FOR USE IN PLASTIC SURGERY

Comparison with a Hamamatsu NIRO-500 Instrument

J. A. Pickett,[1] M. S. Thorniley,[2] E. Balogun,[3] and D. P. Jones[4]

[1]Clinical Physics Group
St. Bartholomew's and the Royal London Hospitals NHS Trust
Whitechapel, London E1 2BL
United Kingdom
[2]Department of Instrumentation and Analytical Science
UMIST, PO Box 88, Manchester M60 1QD
United Kingdom
[3]Department of Surgical Research
Northwick Park Institute for Medical Research
Northwick Park Hospital
Harrow, Middlesex HA1 3UJ
United Kingdom
[4]Department of Medical Electronics and Physics
St. Bartholomew's & The Royal London Hospitals Medical School
Queen Mary & Westfield College
Charterhouse Square, London EC1M 6BQ
United Kingdom

1. INTRODUCTION

The use of free flaps is well established in plastic and reconstructive surgery. However, some 10% of free flaps fail in the first 48 postoperative hours. Of these flaps 83% can be saved with immediate surgical intervention (Furnas and Rosen, 1991). A monitor of tissue viability is therefore required which should be continuous, easily interpreted and should reliably indicate changes in flap perfusion, particularly venous ischaemia which has been shown to be the most detrimental to rate of survival and area

Oxygen Transport to Tissue XXI, edited by Eke and Delpy.
Kluwer Academic / Plenum, New York, 1999.

of survival of both skin and myocutaneous flaps (Thorniley et al., 1998). It should also, ideally, distinguish venous ischaemia from that caused by arterial or total occlusion. Recent studies have shown near infrared spectroscopy to satisfy many of these requirements as a monitor of flap perfusion (Irwin et al., 1995; Thorniley et al., 1998). However this technique, intended as it is for cerebral monitoring, is over-specified and more expensive than is necessary for our needs. The aim of this work is to compare our new device with NIRS using standard occlusion tests to establish whether this device can distinguish venous from total occlusions as effectively as a commercial NIRS device. Dual wavelength near-infrared instruments have been in use for some time in a variety of applications (Mancini et al., 1991; Chance et al., 1992; Chance, 1994). Our device differs in the use of an inexpensive, readily available dual light source from a pulse oximeter probe and signal processing by a "LabView virtual instrument" residing on a laptop computer. The resulting instrument is small, easily portable and can use a miniature, sterilisable probe suitable for clinical use.

2. METHODS

2.1. Apparatus Description

The new device employs an inexpensive dual wavelength light emitting diode (LED), operating at 660 and 925 nm and a photodiode detector (Pickett et al., 1996). These components were obtained from a neonatal pulse oximeter probe (Critikon "Oxyshuttle", Johnson and Johnson Medical Ltd., U.K.). The components were mounted in a plastic shell approximately 1 cm apart in a reflectance configuration with a black plastic shield between them (Figure 1). The LEDs are driven at 300 Hz with a mark : space ratio of 1 : 3. The LED current is continuously variable from 0 to 180 mA. The output from the photodiode is amplified and demultiplexed into red and infra-red

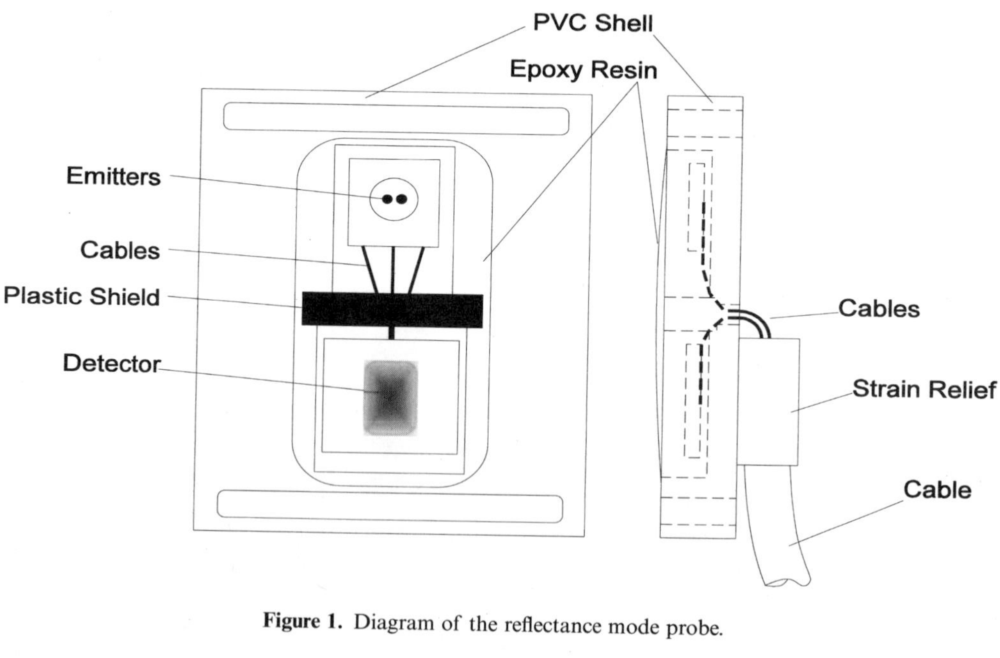

Figure 1. Diagram of the reflectance mode probe.

channels and the resulting signals low pass filtered (Pickett et al., 1997). These signals form the input to a 12-bit analogue-to-digital converter connected to a laptop computer.

A "LabView" program is used to calculate and display the measured concentration changes of oxygenated haemoglobin, HbO_2 and reduced haemoglobin, Hb. Two further parameters are derived from these concentrations, the total haemoglobin concentration change, $[HbT] = [HbO_2] + [Hb]$ and the haemoglobin oxygenation index, $[HbD] = [HbO_2] - [Hb]$.

The calculation is performed by first storing a reference value for each wavelength. Absorbance changes from this reference are then calculated. The changes in concentration of Hb and HbO_2 can then be calculated by solving the following simultaneous equations:

$$\Delta A_{660} = \Delta c_{Hb}\alpha_{Hb660}d + \Delta c_{HbO_2}\alpha_{HbO_2660}d$$
$$\Delta A_{925} = \Delta c_{Hb}\alpha_{Hb925}d + \Delta c_{HbO_2}\alpha_{HbO_2925}d$$

where

ΔA_λ = change in absorbance at wavelength λ
Δc_{Hb} = change in concertration of Hb
Δc_{HbO_2} = change in concentration of HbO_2
$\alpha_{Hb\lambda}$ = molar absorption coefficient of Hb at wavelength λ
$\alpha_{HbO_2\lambda}$ = molar absorption coefficient of HbO_2 at wavelength λ
d = Optical pathlength

This calculation assumes that all the observed changes in A are due to changes in the concentrations of Hb and HbO_2. This allows the terms for all other light losses to be subtracted out of the equations. The optical pathlength d is a product of the separation between emitters and detector and a differential pathlength factor (DPF). This DPF was assumed for these initial measurements to be equal for the red and infrared wavelengths; this assumption is unlikely to be valid but provided the pathlength for each wavelength remains constant for the duration of the measurement this should introduce a systematic error only. A DPF for adult human forearm of 4.16 was used in these measurements (Duncan et al., 1995) and was used in both the new device and the Hamamatsu NIRO-500. The values used for α are from published tables (Matcher et al., 1995).

2.2. Experimental Protocol

Measurements were made on twelve healthy volunteers in the age group 25 to 50 years. The new device and a Hamamatsu NIRO-500 with an interoptrode separation of 1 cm were used.

The NIRO-500 was used first and the position of the optrodes marked on the forearm. The new instrument was then used with the probe as close as possible to the position used by the NIRO-500 probes. Both probes were fixed to the forearm with adhesive tape and shielded from ambient light. A standard blood pressure cuff on the upper arm was used to produce venous (60 mm Hg) and total (180 mm Hg) occlusions. Measurements were made with the sensors attached to the left forearm before, during and following venous and total occlusions of two minutes duration. The changes were measured after the initial increase in blood volume, due to the relatively slow inflation of the blood pressure cuff, until just before release of the cuff. The points at which the

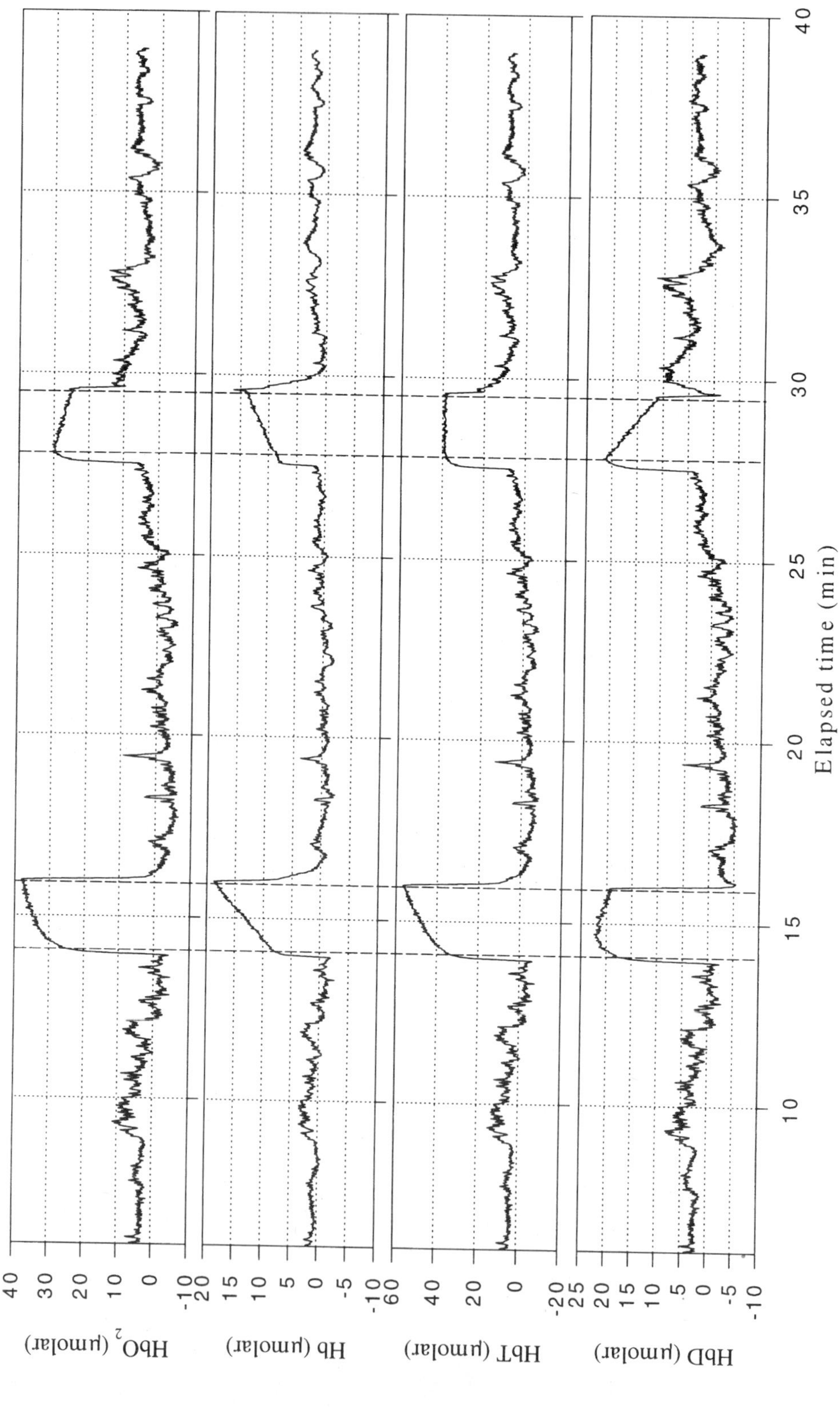

Figure 2. Typical output from the NIRO-500.

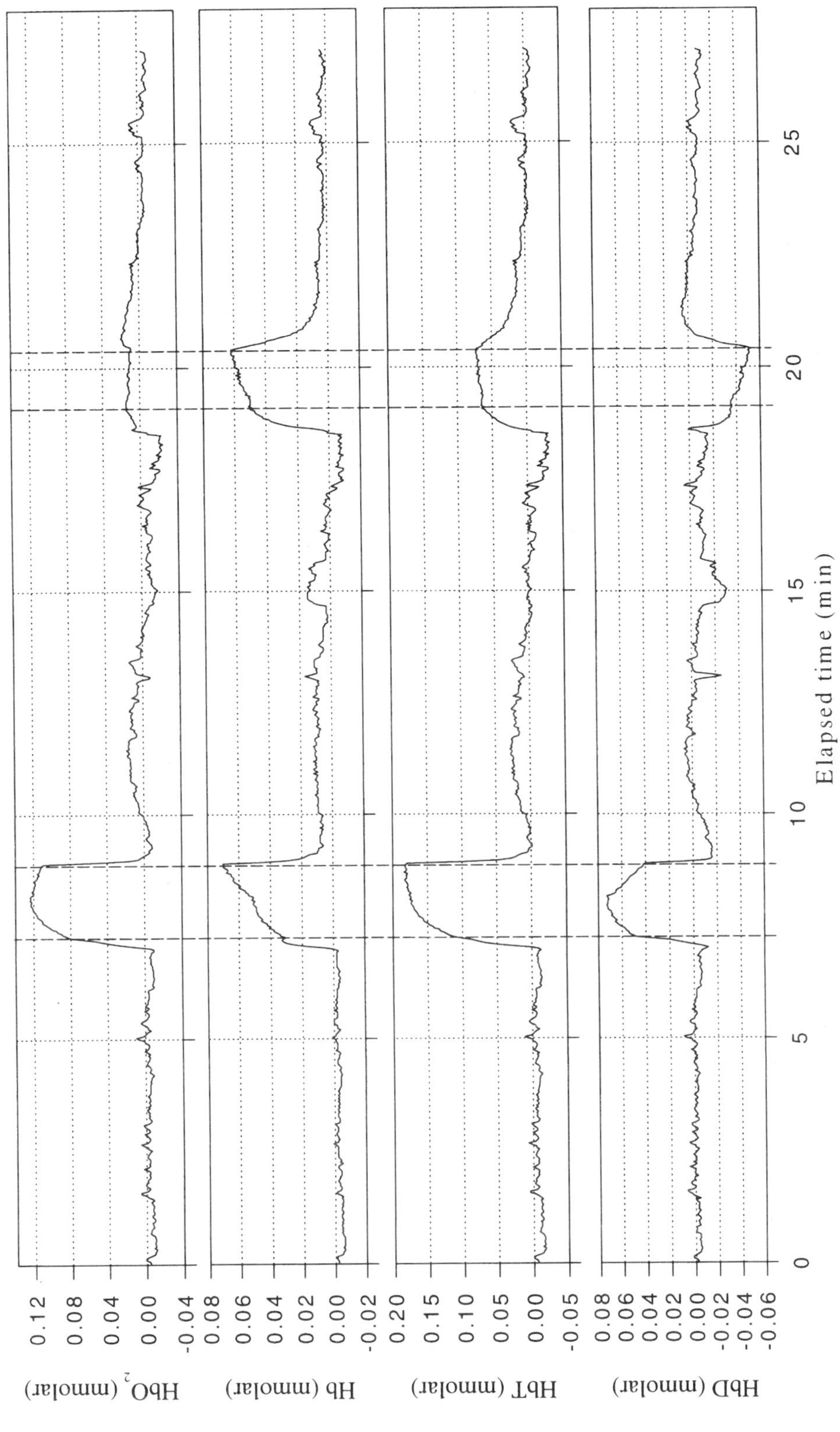

Figure 3. Typical output from the new device.

measurements were made are indicated by the dashed lines in Figures 2 and 3. Ten minutes baseline measurements were made before and after each occlusion. The measurements were repeated on each subject after a lengthy recovery period in excess of one hour.

2.3. Statistical Analysis

The results obtained for each subject were averaged to produce twelve data sets. Data are presented as mean ± SEM. Statistical significance between groups was tested by paired Student's t-test.

3. RESULTS

3.1. Occlusion Studies

Typical traces from the two instruments are shown in Figures 2 and 3. On each trace the first occlusion is a venous occlusion and is characterised by a large initial increase in blood volume consisting of an increase in both Hb and HbO_2, the increase in total haemoglobin in most cases being comprised in the majority by deoxygenated haemoglobin. As the occlusion progresses the rise in HbO_2 becomes less, in some cases reaching a plateau and beginning to fall.

The second occlusion (total) in each trace shows a smaller initial rise in blood volume (HbT). This then remains approximately constant for the rest of the duration of the occlusion. During this period an approximately linear increase in Hb is accompanied by a decrease in HbO_2.

In venous occlusions characteristic patterns of change were obtained using both devices, namely an increase in Hb and HbT, and a consistent fall in HbD. Total occlusions also gave a characteristic pattern, that is an insignificant change in HbT, a fall in HbO_2, a rise in Hb and a fall in HbD.

Table 1 shows the maximum changes in the parameters expressed as mean (SEM). Inspection of the average values reveals broad agreement between the two devices. Exact agreement would not be expected due to the use of different wavelengths by the two

Table 1. Mean (SEM) Maximum changes for venous and total occlusions using both devices

	Maximum change mean (SEM) μM	
	NIRO-500 n = 12	New device n=12
Venous occlusion		
HbO_2	7.5 (1.6)	2.5 (2.5)
Hb	13.5 (1.7)	18.7 (1.3)
HbT	21.1 (2.0)	21.3 (3.3)
HbD	−6.0 (2.5)	−16.2 (2.3)
Total occlusion		
HbO_2	−5.1 (2.3)	−12.3 (1.5)
Hb	9.5 (2.1)	15.4 (1.1)
HbT	4.5 (0.9)	3.2 (1.4)
HbD	−14.6 (4.3)	−27.7 (2.2)

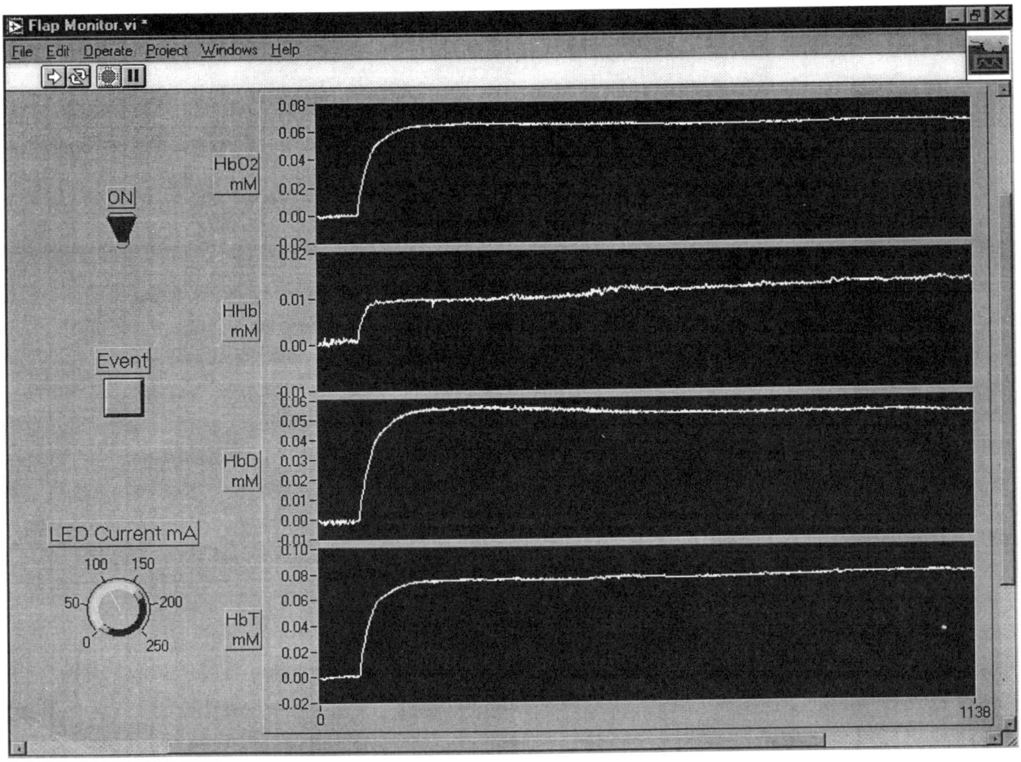

Figure 4. Front panel display of the new device showing reperfusion of a radial forearm flap.

devices. Pathlength differences between the two wavelengths used by the new device would be expected to introduce some error. However, the similarity of the sign and magnitudes measured by the two instruments is encouraging. A clear difference between venous and total occlusion is apparent from the table for HbO_2 and HbT using both devices. This is confirmed by a paired Student's "t" test which shows a significant difference between the two types of occlusion for HbO_2 and HbT ($p < 0.01$) using both devices.

3.2. Recent Ongoing Surgical Applications

The new device has been used to make measurements during surgery in several cases. In Figure 4 is shown a front panel display of the new device measuring changes in reperfusion of a human radial forearm flap. This measurement was made after the arm had been exsanguinated while the flap was raised. The reperfusion shown occurred when the tourniquet was removed but before division of the pedicle. The part of the recording shown covers a period of approximately thirty eight minutes.

4. DISCUSSION

The new device yielded repeatable characteristic patterns for the different types of occlusions in broad agreement with those measured by the NIRO-500. We are aware that the wavelengths used limit the performance of the system but we still obtained reliable measurements and furthermore using the new device were able to discriminate between venous and total ischaemia.

Near-infrared spectroscopy has been shown to be useful in monitoring flaps. Since the new device is also effective in discriminating between venous and total occlusions in the same way as current NIR instruments, then it also should be useful for assessing tissue viability in plastic surgery (Irwin et al., 1995; Thorniley et al., 1998). Whilst our device could not be considered as providing an equivalent measurement to the NIRS system, it may be as effective in the application for which it is intended. In addition, it has the advantages of being inexpensive, portable and has a sterilisable probe which can be applied directly to a flap during surgery.

ACKNOWLEDGMENTS

The authors would like to thank the Joint Research Board of the Royal Hospitals NHS Trust and the Royal Society (Dr Thorniley) for Financial Support. We would like to acknowledge the help of the Departments of Surgical Research and Clinical Physics. Pulse Oximeter probes were supplied by Johnson and Johnson Medical Ltd. UK.

REFERENCES

Chance B, Dait MT, Zhang C, Hamaoka T, and Hagerman F (1992) Recovery from exercise-induced desaturation in the quadriceps muscles of elite competitive rowers. Am J Physiol 262(3 Pt 1):C766–C775.

Chance B (1994) Non-invasive approaches to tissue bioenergetics. Biochem Soc Trans 22(4):983–987.

Duncan A, Meek JH, Clemence M, Elwell CE, Tyszczuk L, Cope M, and Delpy DT (1995) Optical pathlength measurements on adult head, calf and forearm and the head of the newborn infant using phase resolved optical spectroscopy. Phys Med Biol 40(2):295–304.

Furnas H and Rosen JM (1991) Monitoring in microvascular surgery. Ann Plast Surg 26(3):265–272.

Irwin MS, Thorniley MS, Dore C, and Green CJ (1995) Near infra-red spectroscopy: a non-invasive monitor of perfusion and oxygenation within the microcirculation. Brit J Plast Surg 48:14–22.

Mancini DM, Ferraro N, Nazzaro D, Chance B, and Wilson JR (1991) Respiratory muscle deoxygenation during exercise in patients with heart failure demonstrated with near-infrared spectroscopy. J Am Coll Cardiol 18(2):492–498.

Matcher S, Elwell CE, Cooper C, Cope M, and Delpy DT (1995) Performance Comparison of several published tissue near infra-red spectroscopy algorithms. Anal Bioch 227:54–68.

Pickett JA, Amoroso P, Nield DV, and Jones DP (1996) Optical Technique for continuous monitoring of flap viability in plastic surgery. In: "Bridging Disciplines for Biomedicine" Proc of the 18th Ann Int Conf of the IEEE Eng in Med Biol Society, Amsterdam, Netherlands. (IEEE Piscataway, NJ, USA) Abstract p46, CDROM paper #520.

Pickett JA, Amoroso P, Nield DV, and Jones DP (1997) Pulse oximetry and PPG measurements in plastic surgery. In: "Magnificent Milestones and Emerging Opportunities in Med Eng" Proc of the 19th Ann Int Conf of the IEEE Eng in Med and Biol Society, Chicago, Illinois, USA. (IEEE Piscataway, NJ, USA) CDROM paper pp. 2330–2332.

Thorniley MS, Sinclair JS, and Green CJ (1998) The use of Near Infra-red spectroscopy for assessing flap viability during reconstructive surgery. Brit J Plast Surg 51:218–226.

84

REUSABILITY STUDY OF FIBER OPTIC BASED PROTEIN C BIOSENSOR

James O. Spiker,[1] William N. Drohan,[3] and Kyung A. Kang[2]

[1]Department of Biological Sciences
[2]Department of Chemical and Biochemical Engineering
University of Maryland Baltimore County
Baltimore, Maryland 21250, USA
[3]American Red Cross, 15601 Crabbs Branch Way
Rockville, Maryland 20855

1. INTRODUCTION

Protein C (PC) is a vitamin K-dependent plasma zymogen produced in the liver. It is present at a concentration of 4.0 μg/ml in normal plasma and, importantly, acts as both a potent anticoagulant and antithrombotic. The major activator of PC is thrombin that is formed during blood coagulation. Activation occurs on the surface of endothelial cells by means of a complex with thrombin and thrombomodulin. The PC exhibits its anti-coagulating ability by cleaving activated co-factors V and VIII into inactive fragments. This proteolytic cleavage requires, among other things, protein S and Ca^{++}. The PC exhibits its antithrombotic nature by stimulating fibrinolysis through the indirect activation of plasminogen (Esmon et al., 1995, Reltsma, 1997).

Deficiency of PC was first discovered in cases of venous thrombosis. An insufficient PC supply leads to a disruption in hemostasis, a physiological state in which the concentration of coagulants are normally well balanced with that of anticoagulants. Clot formation resulting from low PC levels may prevent the transport of oxygen and nutrients throughout the circulatory system. If left untreated, the patient may experience major clotting complications or even death. Individuals having a heterozygous deficiency are particularly difficult to diagnose since they may be symptomatic or asymptomatic. Moreover, the diagnosis is complicated since most of the heterozygotes may still be producing as much as 50% of the typical PC levels. Homozygous PC deficiency is a serious condition, causing thrombotic complications which may lead to death after birth. In addition to gene disorders, other situations, such as pregnancy, liver disease, or severe injury, may also lead to diminishing levels of PC (Bruley and Drohan, 1990; Reltsma, 1997).

Oxygen Transport to Tissue XXI, edited by Eke and Delpy.
Kluwer Academic / Plenum Publishers, New York, 1999.

1.1. Need for an Efficient Protein C Assay

Due to the detrimental consequences of a deficiency, the monitoring of PC levels is extremely important. However, assay methods are complicated since there are more than 30 major plasma factors related to hemostasis, some of which are structurally homologous to PC. The conventional assay for acquiring PC concentrations is the Enzyme-Linked Immunosorbent Assay (ELISA). This assay is time consuming, expensive, and requires the need of a skilled technician to perform difficult procedures. Such a process is not practical for incorporation in a clinical setting. Therefore, it is desired to have a PC assay method that is easy to operate, highly sensitive and selective, and provides real-time measurement. To this end, an immuno-optical PC biosensor is currently in development at the University of Maryland Baltimore County (UMBC).

1.2. Protein C Biosensor

The PC biosensor makes use of a sandwich fluoro-immunoassay on an optical fiber [Figure 1]. A tapered quartz fiber is inserted into a glass capillary tube, and then monoclonal antibody against PC is immobilized on the fiber surface. When a sample is injected into this chamber, PC binds specifically to the antibody in a high affinity reaction. Next, the system is probed with a fluorophore tagged secondary antibody against PC. Following this, a laser diode emits excitation light through the fiber. The intensity of fluorescence generated by the tagged antibodies is then correlated with the PC concentration.

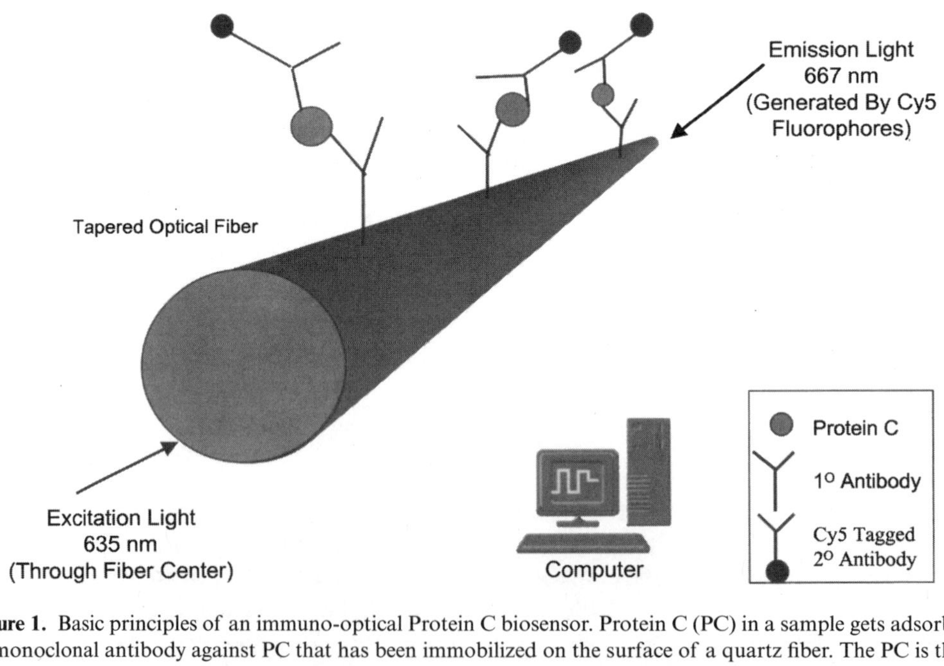

Figure 1. Basic principles of an immuno-optical Protein C biosensor. Protein C (PC) in a sample gets adsorbed to monoclonal antibody against PC that has been immobilized on the surface of a quartz fiber. The PC is then probed with a fluorophore tagged secondary antibody. A laser diode supplies light that excites the tagged antibody. The fluorescent signal is then detected and displayed on a computer, where it may be correlated with the concentration of PC in the sample.

The instrumentation and methods employed in the PC biosensor have been applied to the detection of other small biomolecules. Other groups have reported success with the detection of, for example, D-dimer and ricin (Narang, 1997; Rowe, 1998). In each case, detection sensitivities rivaling those of ELISA have been achieved. To date, the PC biosensor has been shown to detect PC levels near 0.1 µg/ml, in pure buffered samples (Spiker et al., 1997). The current study has been focused upon fiber reusability, as this directly affects cost effectiveness and the reproducibility of results: (1) Direct binding and the use of an avidin-biotin reaction to attach the primary antibody to the fiber; (2) Saturation of the primary antibody with antigen was studied; as well as (3) the feasibility of eluting captured antigen from the system. The research presented here is a continuation of the work contributed at the 23rd, 24th and 25th Annual ISOTT meetings (Spiker et al., 1996 and 1997; Kang et al., 1997).

2. METHODS

2.1. Protein Preparation

All PC (MW = 62,000) and anti-PC monoclonal antibodies (MW = 150,000) were provided by the American Red Cross (Rockville, Maryland). Antibodies were directed against two different PC surface epitopes. The fluorophore, Cy5 (Biological Detection Systems, Inc.; Pittsburgh, PA), was conjugated with the secondary antibody (MAb 8861) using the manufacturer's instructions. Ideally, a ratio of two to four fluorophore molecules per antibody was achieved. Higher ratios interfered with the three-dimensional shape of the antibody, thereby lowering its affinity for PC. Biotin (MW = 575.7) was conjugated with the primary antibody (MAb 7D7) according to Savage et al. (1992).

2.2. Fiber Preparation

Initially, all quartz fibers were chemically treated according to the methods described by Bhatia, et al. (1989). In the case of direct binding, the fibers were then stored in 65 µg/ml antibody in phosphate buffered saline (PBS) for 12 hours. In the case of avidin-biotin binding, the fibers were placed in a 50 µg/ml avidin (MW = 53,000) solution for 12 hours. They were then incubated for 12 hours in 65 µg/ml antibody-biotin in PBS. Both sets of fibers, direct binding or avidin-biotin, were introduced to a blocking buffer (2.0 mg/ml bovine serum albumin in PBS) for 30 minutes in order to occupy any areas of fiber surface not covered with primary antibody.

2.3. Fluorometer and Sample Chamber

The fluorometer, the Analyte 2000, and all optical fibers were obtained from Research International, Inc. (Woodinville, Washington). Following preparation, the fiber is placed inside a glass capillary tube (200 µl capacity) that is then flanked with two plastic "T" connectors and sealed with hot glue. Sample or buffer injections occur via syringe at the input port [Figure 2]. The fluorometer generates excitation light at 635 nm that causes the fluorophore to produce an emission signal at 667 nm. A dichroic long-pass filter interrogates the signal, then the system amplifies and converts the photon energy to electrical energy. Finally, the intensity of arriving photons is converted to a digital signal by a computer for display on a monitor.

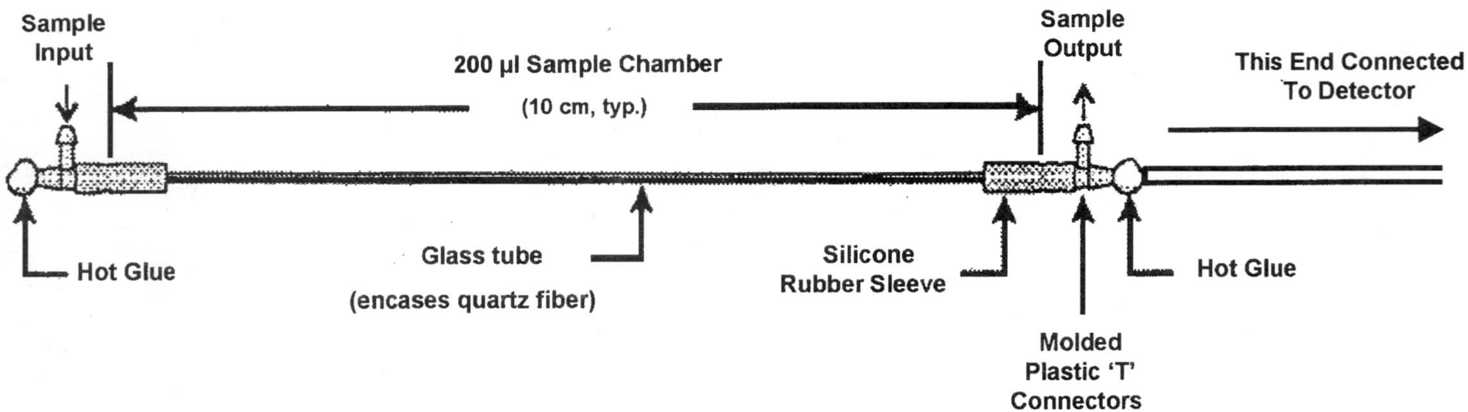

Figure 2. Diagram of sample chamber assembly. A quartz fiber (1 mm × 25 cm) is inserted into a glass tube flanked by two plastic "T" connectors. The connections are sealed with hot glue, creating a sample chamber size ~200 µl (diagram based on work performed by Research International).

2.4. General Procedure

A typical assay proceeds as follows:

Step 1. Sample incubation, 10 min.
Step 2. Wash buffer [1 ml of PBS, 0.01% Tween 20 (PBST)], 1 min.
Step 3. Cy5-Anti-PC (300 µl at 5 µg/ml in PBST), 5 min.
Step 4. Wash buffer (1 ml PBST), 1 min.

In order to determine the PC concentration, the fluorescent signal measured at the end of Step 2 is subtracted from the signal measured at the end of Step 4. The difference (ΔpA) represents the amount of tagged antibody specifically bound to PC (Spiker et al., 1997). Any non-specifically bound probe antibody is washed from the system during Step 4. Therefore, the ΔpA may be correlated with the concentration of PC by constructing a standardization curve. The measuring time is a maximum of 20 min, and may be compared to a typical ELISA (4~6 hr).

3. RESULTS AND DISCUSSION

3.1. Direct versus Avidin-Biotin Binding

During covalent immobilization, antibodies frequently denature or incur damaging changes to their three-dimensional structure. Such changes reduce the antibody's affinity for antigen. To address this problem, avidin-biotin immobilization has gained interest in recent years. Avidin is strongly specific and multivalent for biotin. Therefore, avidin that has been immobilized on the fiber surface will bind to biotin that is attached to the primary antibody. This technique is beneficial since it provides a firm yet specific indirect-covalent immobilization of the antibody (Narang et al., 1997). To test this approach, assay results were compared among six fibers, three utilizing direct binding and three utilizing avidin-biotin binding. A series of three different PC concentrations (0.3125, 1.25, and 2.5 µg/ml) were assayed on each fiber, and the mean values are shown in Figure 3. In all cases, fibers utilizing the avidin-biotin bridge demonstrated enhanced

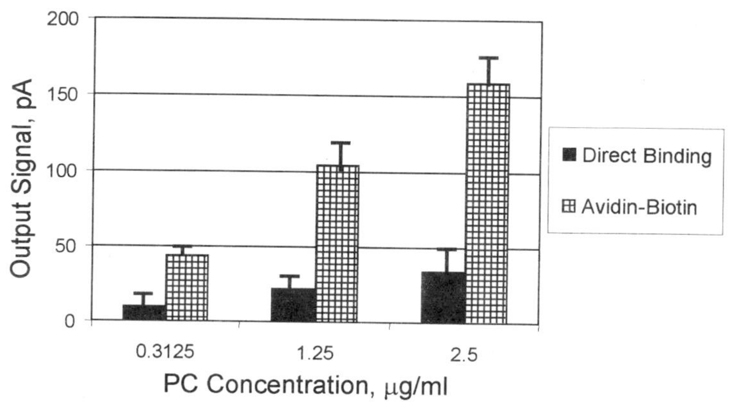

Figure 3. Comparison of antibody immobilization methods. Primary antibody was immobilized on groups of fibers using either directing binding or an avidin biotin bridge. PC samples were assayed on each group of fibers and, in all cases, enhanced signal intensity was observed with fibers utilizing the avidin-biotin bridge.

binding capabilities of the primary antibody. For example, at 2.5 µg/ml of PC, a 367.7% increase in signal intensity was achieved (34pA vs. 159pA). Based upon these results, it is suspected that the antibodies retain a greater portion of their antigen specificity during immobilization with the avidin-biotin bridge. Therefore, all subsequent experiments utilized the avidin-biotin method of immobilization.

3.2. Standard Curve

In order to generate a standard curve, various PC concentrations were measured in triplicate on separate fibers. The mean values obtained are shown in Figure 4. Our primary area of concern is PC concentrations in the range of 0.2–2.5 µg/ml, representing the PC range of heterozygous deficient patients. Using pure buffered PC, the biosensor is currently able to demonstrate the ability to measure PC concentrations in this area. This range shows a linear relationship on the inset plot, with an R^2 of 0.991 [Figure 4].

3.3. Saturation Study

To examine system reusability, it was desired to determine the number of consecutive assays that could be performed per fiber. After repeated runs on a single fiber, the primary antibody slowly becomes saturated with PC. Eventually, PC in a sample has no place to bind to the fiber, and further assays become useless. Therefore, the number of consecutive assays is suspected to be directly related to the saturation of the primary antibody with antigen. In order to determine the number of uses per fiber, assays were repeated (using samples of identical PC concentration) until a significant drop in sensitivity was observed. This drop was defined as an output signal that was 50% less than the mean value of all previous signals on that fiber. With repeated 1.0 µg/ml PC samples, and average of 4.5 runs were achieved per fiber. Using a more dilute PC concentration, 0.5 µg/ml, the onset of saturation was delayed. Here, an average of 6.3 assays were realized per fiber [Figure 5]. Generally, using dilute PC concentrations extends fiber lifetime

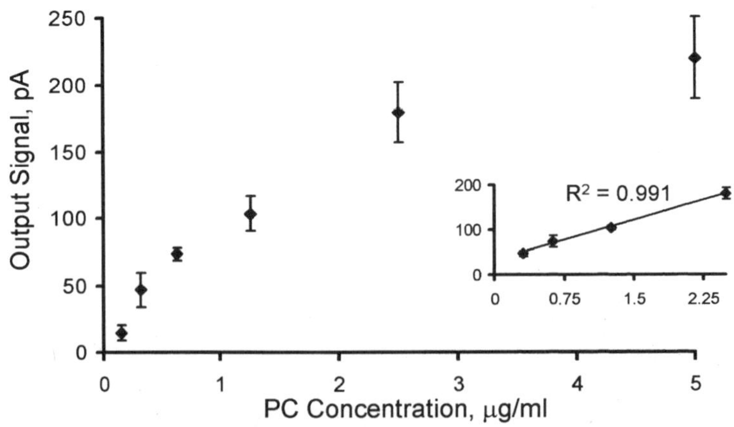

Figure 4. Standard curve. Various PC concentrations were assays in triplicate on different fibers, the mean values of which of plotted above. Our primary PC range of interest, 0.2–2.5 µg/ml, is graphed in the inset with a R^2 of 0.991.

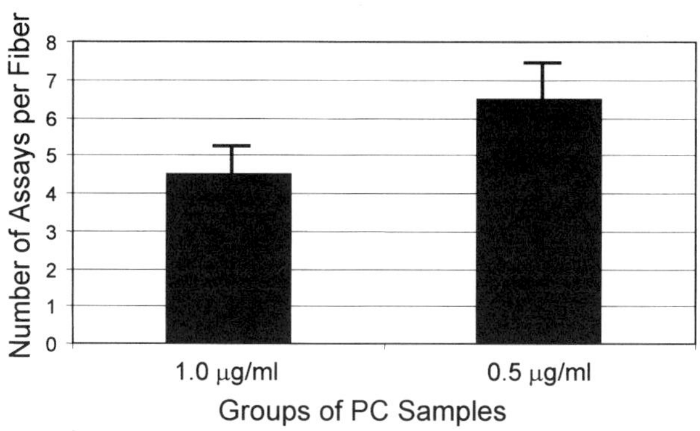

Figure 5. Primary antibody saturation study. The number of potential consecutive runs per fiber was determined. When using repeated samples of 1.0 and 0.5 μg/ml of PC, and average of 4.5 and 6.3 runs, respectively, per fiber were achieved. The number of repeated assays is suspected to be related to the saturation state of the primary antibody.

since it takes a larger number of samples to saturate active antibody sites on the fiber surface with antigen.

3.4. Regeneration Study

After demonstrating antibody saturation, we examined the potential to eliminate this condition with the inclusion of an elution step. If captured PC could successfully be eluted from the system, fiber lifetime would be extended. Experiments were designed to make use of the fact that the primary antibody is calcium dependent. In the presence of a calcium containing buffer, the configuration of the antibody is changed such that it releases bound PC. Assays were repeated using 1.0 μg/ml PC samples until a drop in sensitivity was achieved. A two-minute elution was performed following each assay. Three different elution conditions were tested: a 10 mM $CaCl_2$ solution in PBST at (1) pH 6.0, (2) 7.2, or (3) 9.4. These conditions were chosen based upon protocols from immunoaffinity purification of PC (Kang et al., 1992). When compared to fibers on which no elution buffer was used, the data suggests that the calcium buffers at pH 6.0 and 7.2 are not capable of eluting bound PC (average of 4.2 and 4.5 assays per fiber, respectively). Utilizing the calcium buffer at pH 9.0, a significant increase in fiber lifetime was achieved. In this situation, an average of 6.3 assays were achieved per fiber [Figure 6]. It is suspected that with this buffer, the conditions were adequate to allow the antibody to release a significant amount of bound PC. However, once the calcium was washed out of the system, the antibody was again primed to bind PC in the next assay.

4. CONCLUSION

The PC biosensor in development offers the ability to monitor PC levels in real-time. It is a viable alternative to other costly and time-consuming assays, such as an ELISA. The ability to rapidly diagnose PC deficiency, in the range of minutes, provides physicians with the opportunity to immediately provide their patients with treatment. Early treatment is crucial in order to avoid the thrombotic complications that may ensue

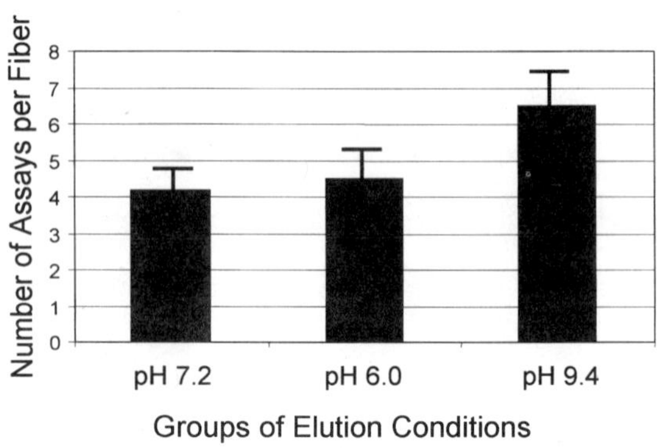

Figure 6. Fiber regeneration study. 1.0 µg/ml PC samples were repeatedly assayed on fibers utilizing a two-minute elution step after each assay. The elution condition was a 10 mM $CaCl_2$ solution in PBS at (1) pH 6.0, (2) 7.2, or (3) 9.4. Elution buffers at pH 6.0 and 7.2 achieved an average of 4.2 and 4.5 runs per fiber, respectively, and did not appear to extend fiber lifetime. Eluting with $CaCl_2$ at pH 9.4 did, however, extend fiber reusability by achieving an average of 6.3 runs per fiber.

as the result of a PC deficiency. In this paper, experimental results have demonstrated the advantages of employing an avidin-biotin reaction to immobilize antibody to the fiber. The reusability of the optical fibers has been illustrated, with the ability to reproduce a series of runs per fiber. Furthermore, preliminary results confirm the potential for employing an elution step in order to better regenerate the system. Eluting bound PC from the system serves to increase fiber lifetime and thereby makes the system easier to operate as well as increases cost efficiency.

5. FUTURE WORK

Conditions of the elution step need to be fully explored in order to yield an optimal level of fiber longevity. Adjusting the $CaCl_2$ concentration, pH, or even the duration of the elution step itself may lead to the maximal removal of PC from the system.

ACKNOWLEDGMENTS

This research is supported, in part, by The Whitaker Foundation, the American Red Cross, and the National Science Foundation (Grant BES-9733207). The authors would like to thank Research International and Drs. Fran Ligler, Chris Rowe, and Lisa Shriver-Lake at the Naval Research Laboratory of Washington D.C. for their technical support.

REFERENCES

Bhatia S, Shriver-Lake L, Prior K, Georger J, Calvert J, Bredehorst R, and Ligler F (1989) Use of thiol-terminal silanes and heterobifunctional crosslinkers for immobilization of antibodies on silica surfaces. Anal Biochem 178:408–413.

Bruley D and Drohan W (1990) Activation Cascade. In: Protein C and Related Anticoagulants—Advances in Applied Biotechnology Series. Gulf Publishing Company, Houston.

Esmon C and Schwarz HP (1995) An update on clinical and basic aspects of the Protein C anticoagulant pathway. Trends Cardiovasc Med 5:141–148.

Kang K, Anis N, Eldefrawi M, Drohan W and Bruley D (1997) Reusable, Real-Time, Immuno-Optical Protein C Biosensor. Adv Exp Med Biol 411:437–444.

Narang U, Anderson G, Ligler F, and Burans J (1997) Fiber optic-based biosensor for ricin. Biosens Bioelectron 9–10:937–945.

Reltsma P (1997) Protein C deficiency: from gene defects to disease. Thromb Haemost 78:344–350.

Rowe C, Bolitho J, Jane A, Bundesen P, Rylatt D, Eisenberg P, and Ligler F (1998) Rapid detection of D-dimer using a fiber optic biosensor. Thromb Haemost 79:94–98.

Spiker J, Kang K, Drohan W, and Bruley D (1996) Protein C Detection Via Fluorophore Mediated Immuno-Optical Biosensor. Adv Exp Med Biology 428:621–627.

Spiker J, Kang K, Drohan W, and Bruley D (in press) Preliminary study of biosensor optimization for the detection of Protein C.

85

OXYGEN AND HYDROGEN GAS TRANSPORT THROUGH LIVING C6 CELL SUSPENSIONS IS FASTER THAN THAT IN DEAD CELLS

M. Tomita,[1] N. Tanahashi,[1] M. Kobari,[1] M. Yokoyama,[1] M. Inoue,[1] I. Schiszler,[2] and Y. Fukuuchi[1]

[1]Department of Neurology
School of Medicine
Keio University
Tokyo 160, Japan
[2]JSPS Fellow from Semmelweis University
Budapest, Hungary

INTRODUCTION

Gas transport through the living cerebral cortex has been reported by Gotoh et al. (1961) to be faster than that in dead cortex. This enhancement was attributed by them to a stirring effect due to the vasomotor action of the microvasculature in the tissue whereby gas molecules could be transported to far distant areas. At the *ex vivo* level, facilitation of oxygen gas transport across removed living cortical tissues was observed by Longmuir and Bourke (1960), who obtained values which were approximately 3-fold higher than those through free water. MacDougall and McCabe (1967) also demonstrated a high but very wide range of values for the oxygen diffusion coefficient in removed tissues.

The present paper examines whether or not such enhanced transport of oxygen gas and hydrogen gas, a nonmetabolizing gas, occurs in a more simple system: dense living cell suspensions.

THEORY AND METHODS

The general expression for gas diffusion in an aqueous fluid is $dc/dt = -D.d^2c/dx^2$, where D = Fick's diffusion coefficient, c = concentration, t = time, and x = distance.

Oxygen Transport to Tissue XXI, edited by Eke and Delpy.
Kluwer Academic / Plenum Publishers, New York, 1999.

However, in tissue, diffusion is a very complicated process: in addition to the Brownian motion of molecules, it is affected by molecular interactions, the mean free path, collision frequency, water solubility and its variations in the biological fluid, and the biological stirring effect. In this paper, we broadly use the apparent diffusion coefficient "D" instead of D: "D" = D + ξ, where ξ denotes such above factors which enhance or suppress the diffusion process within the biological fluid. The value of the "D"/D ratio represents enhancement if it is >1, and suppression if it is <1.

For a cylinder of simple solution, like saline, in the recess of a polarographic oxygen platinum electrode (see Figure 1), the following equation was developed by Davies and Brink (1942) assuming that the diffusion is broadly one-directional along the tension gradient when gas is introduced from the orifice of the recess:

$$c(x,t) = c(2/\sqrt{\pi})\int_0^{x/2\sqrt{\text{"D"}t}} \exp(-y^2)dy \tag{1}$$

where c(x,t) is the concentration of gas at a distance x from the orifice at a time t after gas introduction, C the uniform gas concentration at the orifice, and y a variable of integration. When gas is supplied as a step function into the orifice of the recess which is filled with a solution, the function is transported effectively but in a smearing manner through the solution, and the response curve yielded by the electrode builds up after a short lag time. The response curve traces the above equation in the rising part, reaching a peak. However, sooner or later when the gas supply is discontinued, it begins to decline after the peak due to back-diffusion of gas to the orifice. This back-diffusion process may also be analyzed by the equation. Since x is the distance from the orifice to the platinum electrode and therefore fixed in the above equation, the height c(x,t) is now simply a function of the product of the two variables, "D" and the specified time t.

Based on these considerations, we measured the gas transit times ($T_{control}$ and T_{cell}) for the peak height or for the mean height through the control solution ("D"$_{control}$ = D) and a given suspension ("D"), respectively. The ratio $T_{cell}/T_{control}$ was thus regarded as the reciprocal of the "D"/D ratio, since $T_{cell} \times$ "D" must be equal to $T_{control} \times D$ in order to get the same height of c.

C6 cells (astrocytoma of rat) were inoculated into flasks containing 10% Dulbecco's modified Eagle's medium plus 10% Gibco newborn calf serum (DMEM). Immediately before the experiment, cells were harvested, and centrifuged. The sedimented cells were diluted with DMEM so that suspensions with a cytocrit (Cct) of 0% (n = 13), ca. 10–20% (n = 4), ca. 50% (n = 5), and ca. 80% (n = 1) were prepared. Additional cells were either subjected to asphyxia (n = 5) or rigidified with glutaraldehyde (n = 2). For the asphyxiation, the sedimented cells were placed in a glass syringe, left airtightly packed and stored for 24 h at room temperature. We confirmed that the PO_2 of the packed cells fell quickly to zero, which suggested that the cells had become asphyxiated through oxygen deprivation and the accumulation of carbon dioxide and acid metabolites that occurred in the suspensions confined within the syringe. Suspensions with living cells and those with dead cells adjusted to Cct = 50% were used for subsequent experiments.

To measure the transit times of oxygen gas and hydrogen gas through the suspensions, a modified platinum electrode (Gotoh et al., 1966) of the recessed type (Davies and Brink, 1942) fabricated by us was used. A diagrammatic illustration of the platinum electrode is shown in the upper part of Figure 1. Briefly, a 100-µm platinum wire was sealed in a glass rod of 3.0 mm in outer diameter, and only the cut surface was exposed at the tip of the rod end. The glass rod containing the platinum wire at its center was inserted

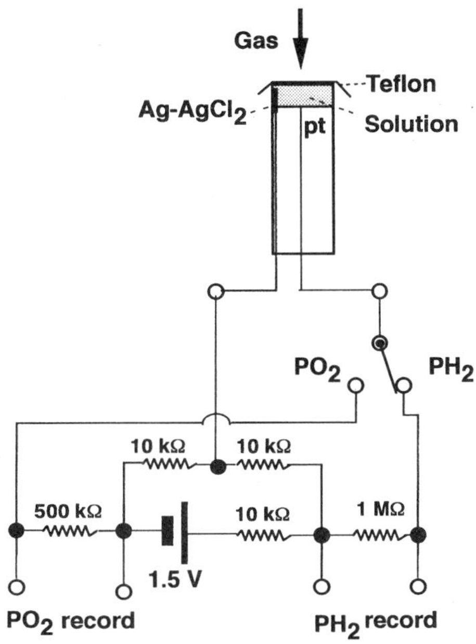

Figire 1. Diagrammatic sketch of the platinum electrode and electrical circuit for monitoring oxygen gas and hydrogen gas.

into a glass sheath of 4.5 mm in inner diameter together with an Ag-AgCl reference electrode, so that a recess of 1.3 mm in depth (the length was selected because Davies and Brink employed a length of 1.6 mm–0.6 mm) and 4.5 mm in diameter was formed in front of the sensitive platinum surface. The recess was filled with one or other of the living cell or dead cell suspensions and covered with a thin Teflon membrane, which was freely permeable to gases. With a parallel resistance of 1 MΩ and an applied potential of +0.5 V for hydrogen gas, or a resistance of 500 kΩ and an applied potential of –0.5 V for oxygen gas, the electrode became linearly sensitive to gas concentration (Gotoh et al., 1966). When 100% gas was blown towards the membrane, the gas was transported to the platinum surface through the layer of the cell suspension, building up a response curve (c(t)) after a short lag time (appearance time:AT). The speed of transport through the living cell suspensions was compared with that of the medium alone or of dead cells in terms of the peak time (PT), and mean transit time (MT) which was calculated from the formula $\Sigma t.c(t)/\Sigma c(t)$.

RESULTS

The response curves to the gases were found in general to be smooth, but with a rather rapid rise, single peak, and gradual descending part. The polarity was positive for hydrogen gas (Figure 2), and negative for oxygen gas (Figure 3). As shown in the middle part of Figure 2, the response curve to hydrogen gas was shortened with a 50% Cct cell suspension, and much more so with an 80% Cct suspension (bottom of Figure 2). A summary of the calculated values of AT, PT, and MT for hydrogen gas is given in the upper part of Table 1. Both PT and MT were decreased from those of DMEM (control),

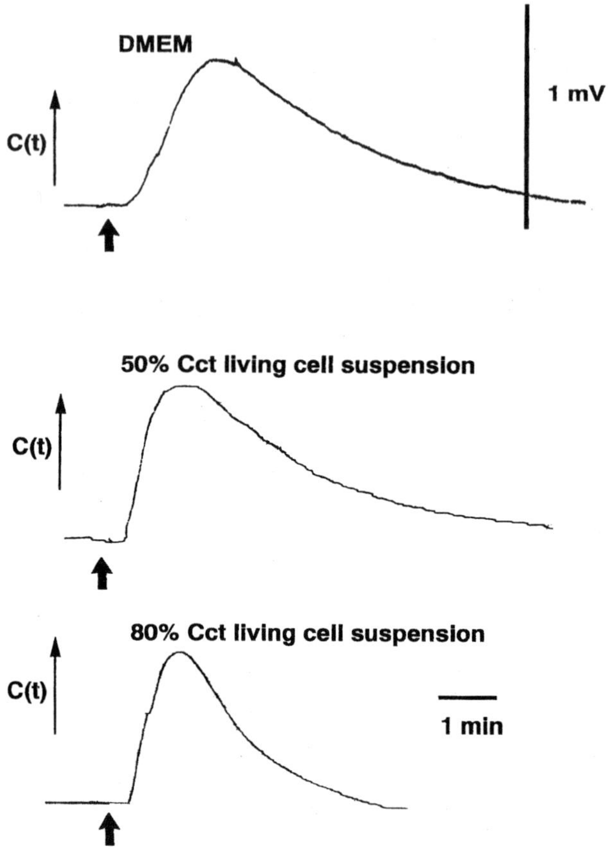

Figure 2. Response curves to 100% hydrogen gas blown for 20 s towards the electrode in a recess containing different suspensions of living cells: top, recess with DMEM (medium) alone; middle, a 50% Cct C6 cell suspension; bottom, an 80% Cct C6 cell suspensions. It should be noted that the response curves became shortened with the living cells, respectively. The time scale is indicated by the bar. Note the enhancement of gas diffusion with increase in density of the living cell suspension.

yielding an average increase in "D"/D ratio of 1.5. Although there was only one case, the ratio was high (3.7) in the 80% Cct cell suspension. On the other hand, PT and MT were both prolonged when the cells were killed either with glutaraldehyde or by asphyxiation. The ratio "D"/D declined to 0.8 in these dead cell suspensions.

Response curves to oxygen gas are shown with DMEM in the top part of Figure 3, with a 50% Cct cell suspension in the middle part and with an 80% Cct cell suspension at the bottom. Unlike those in the case of hydrogen gas, the curves with living cell suspensions were rather distorted due probably to oxygen consumption by the cells. The values of AT, PT, and MT for oxygen gas are given in the lower part of Table 1. The values were reduced, except for AT which became rather increased. The MT with 50% Cct cell suspensions (102.2 ± 38.1 s) was statistically significantly smaller than that for DMEM (229.4 ± 16.9 s, $p < 0.05$). The response curve with the 80% cell suspension was completely suppressed indicating that oxygen supplied to and diffusing into the living cell suspensions is all consumed by the cells and no oxygen reaches the electrode (bottom of

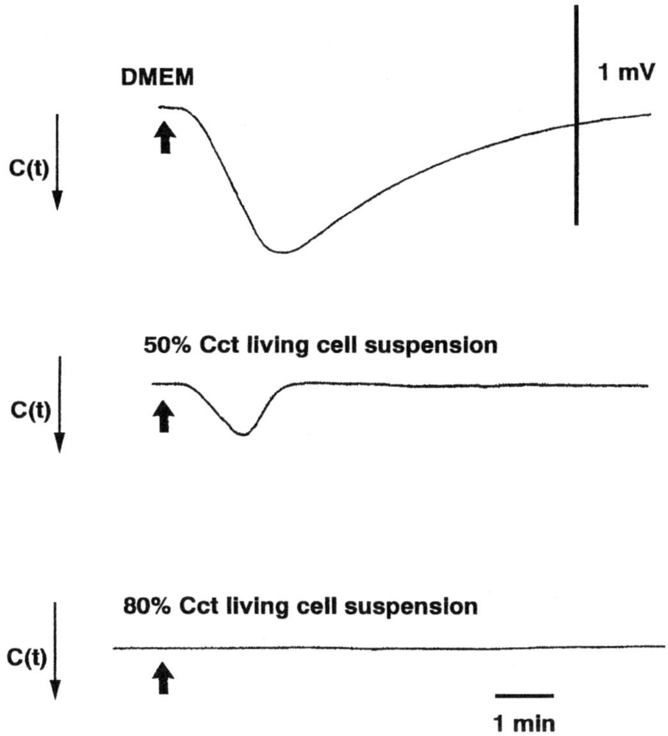

Figure 3. Response curves to 100% oxygen gas: top, DMEM; middle, 50% Cct cell suspension; bottom, 80% Cct cell suspension. See text for further explanation.

Figure 3). The "D"/D ratio for oxygen gas was calculated to be 1.9 for 50% Cct living cell suspensions.

As shown in Table 1, both PT and MT for oxygen gas in DMEM were longer than those for hydrogen gas. This could be due to the difference in Fick's diffusion coefficient D of the two gases: D for oxygen gas is $10.9 \times 10^{-4} cm^2 \cdot min^{-2}$ at 20°C, while that for hydrogen gas is $16.6 \times 10^{-4} cm^2 \cdot min^{-2}$ at 10°C (Altman and Dittmer, 1971). The ratio of D is broadly equivalent to the reciprocal of the ratio of the transit times for hydrogen gas/oxygen gas.

Table 1. Summary of results

No.	Gas	Material	n	AT	PT	MT	"D"/D ratio
1	H_2	DMEM	13	9.0 ± 0.4 s	116.0 ± 24.0 s	164.2 ± 31.2 s	
2	H_2	50% Cct cells	5	8.3 ± 0.4 s	65.0 ± 5.8 s*	126.4 ± 6.2 s	1.5
3	H_2	80% Cct cell	1	–	30.0 s	47.4 s	3.7
4	H_2	Glutaraldehyde	2	–	151.0 ± 8.2 s	195.8 ± 21.5 s	0.8
5	H_2	Asphyxiated	5		138.0 ± 6.0 s	192.8 ± 23.8 s	0.8
6	O_2	DMEM	5	14.3 ± 2.2 s	125.0 ± 21.2 s	229.4 ± 16.9 s	
7	O_2	50% Cct cells	5	16.3 ± 3.2 s	79.0 ± 9.2 s	102.2 ± 38.1 s*	1.9
8	O_2	80% Cct cells	2		no build up		

The "D"/D ratio was calculated from $(PT_{control}/PT_{cell} + MT_{control}/MT_{cell})/2$. * $p < 0.05$ compared with DMEM (control).

DISCUSSION

The above data yielding "D"/D ratio >1 indicate that the diffusion of both hydrogen gas and oxygen gas through C6 cell suspensions is enhanced. Such enhancement disappeared when the cells were killed with glutaraldehyde or by asphyxiation. In the living C6 cell, we have measured a membrane potential of >60 mV membrane potential employing an intracellular microelectrode with reference to an outside reference electrode (Tomita et al., 1997). The thermal energy level of the living cells is greatly elevated when compared to that of dead cells. Since the temperature rise is restricted by the environmental water, the energy produced is transformed into the kinetic energy of molecules. The value of the Boltzman constant (k) at the cell membrane was assumed by Fröhlich (1975) to the three times larger (3k) than that in fluid containing non-living materials.

The present findings for enhanced gas diffusion in living cell suspensions may have several practical significances, as follows. 1) Instantaneous equilibrium of gases between blood and tissue at the capillary level could be established in a more efficient way than has ever been envisaged. 2) Oxygen gas flus to an activated cell region is enhanced as if the gas were being sucked up by the cells. 3) Such enhancement and suppression depending on cell activity could provide a plausible explanation for the enigmatical changes in apparent diffusion coefficient of protons in ischemic brain tissue detected by MRI (Moseley et al., 1991). 4) Cell density-dependent enhancement of the apparent diffusion coefficient must be incorporated when constructing oxygen tension maps for the tissue.

SUMMARY

The diffusion of both oxygen gas and hydrogen gas through a 1.3-mm layer of cultured glioma C6 cell suspension was examined with a modified recessed-type platinum electrode. We found that diffusion was enhanced when the suspension contained living C6 cells, and that the observed enhancement disappeared after killing the cells by rigidification with glutaraldehyde or asphyxiation. Based on these findings, we conclude that oxygen gas and hydrogen gas are transported much faster in living cell suspensions than in dead cell suspensions. This enhancement could be ascribable to certain biological energy-consuming stirring mechanisms at the cell membrane and its vicinity.

ACKNOWLEDGMENT

The authors would like to thank Prof. Akimichi Kaneko, National Institute for Physiological Sciences, for his helpful advice and courtesy in allowing us to utilize his facilities at the National Institute for Physiological Sciences, Okazaki 444, Japan.

REFERENCES

Altman PL and Dittmer DS (1971) Diffusion coefficients of gases in water. In: Respiration and Circulation, Federation of American Societies for Experimental Biology, Maryland, p. 24.

Davies PW and Brink F (1942) Microelectrodes for measuring local oxygen tension in animal tissue. Rev Sci Instr 13:524–533.

Fröhlich H (1975) The extraordinary dielectric properties of biological materials and the action of enzymes. Proc Natl Acad Sci USA 72:4211–4215.

Gotoh F, Meyer JS, and Tomita M (1966) Hydrogen method for determining cerebral blood flow in man. Arch Neurol 15:549–559.

Gotoh F, Tazaki Y, and Meyer JS (1961) Transport of gases through brain and their extravascular vasomotor action. Exp Neurol 4:48–58.

Longmuir IS and Bourke A (1960) The measurement of the diffusion of oxygen through respiratory tissue. Biochem J 76:225–229.

MacDougall JDB and McCabe M (1967) Diffusion coefficient of oxygen through tissue. Nature 215:1173–1174.

Moseley ME, Wendland MF, and Kucharezyk J (1991) Magnetic resonance imaging of diffusion and perfusion. Top Magn Reson Imaging 3:50–67.

Tomita M, Fukuuchi Y, Tanahashi N, Kobari M, Takeda H, Takao M, Yokoyama M, and Ito D (1997) Glutamate-induced cultured astrocytic swelling, depolarization and ionic changes. XVIII Int Symp Cereb Blood Flow Metabol, Baltimore, June 1997; Abstract: J Cereb Blood Flow and Metab 17(Suppl):S730.

86

CEREBRAL BLOOD FLOW IN HUMANS MEASURED WITH NEAR INFRARED SPECTROSCOPY IS NOT REPRODUCIBLE

M. J. T. van de Ven,[1] W. N. J. M. Colier,[2] D. Walraven,[2] B. Oeseburg,[2] and H. Folgering[1]

[1]Department of Pulmonology Dekkerswald
University of Nijmegen
P.O. Box 9001, 6560 GB Groesbeek
The Netherlands
[2]Department of Physiology
Faculty of Medical Sciences
University of Nijmegen
P.O. Box 9101, 6500 HB Nijmegen
The Netherlands

1. INTRODUCTION

Theoretically, Near Infrared Spectroscopy (NIRS) can be used as a non-invasive and fast method to quantify cerebral blood flow (CBF) in neonates (Wyatt et al., 1990) and adults (Elwell et al., 1994). In contrast to current methods such as ^{133}Xe washout, positron emission tomography (PET) or magnetic resonance imaging (MRI), repeated measurements are possible at the bedside and the method is non-invasive: no ionizing radiation or injection of an intravascular tracer is needed. Therefore, the method seems very attractive for practical clinical use.

The principles of NIRS were first described in 1977 (Jöbsis, 1977) and are based on absorption of near infrared light by chromophores, mainly oxyhemoglobin [O_2Hb] and deoxyhemoglobin [HHb]. Changes of optical density of a perfused tissue are converted into concentration changes of [O_2Hb] and [HHb]. When combining the changes of concentrations with an absolute value of arterial saturation obtained by pulse-oximetry, absolute values of CBF and cerebral blood volume (CBV) can be calculated.

However, the reproducibility is subject of discussion (Fallon et al., 1993; Elwell et al., 1994; Newton et al., 1997).

Oxygen Transport to Tissue XXI, edited by Eke and Delpy.
Kluwer Academic / Plenum Publishers, New York, 1999.

Recently, Newton et al. (1997) validated NIRS measurements with other accepted techniques in dogs. They concluded that NIRS measurements in the reflectance mode using [O_2Hb] as an intravascular tracer, did not correlate with the microsphere technique nor with the venous outflow technique. They suggested that a faster sampling rate and better signal-to-noise (s/n) ratio might improve the accuracy of the measurements.

Another drawback of the method was the high rate of rejection of measurements after presetting criteria for reproducibility. In Fallon's study (1993), Elwell's study (1994), and in Newton's study (1997), 66% of all measurements could not be used as they did not fulfill preset criteria. The aim of our study was to evaluate the effect of improvements of NIRS instrumentation on the reproducibility of CBF in adults, using [O_2Hb] as an intravascular tracer. In contrast to other investigators, we used a highly sensitive continuous wave near infrared spectroscope (OXYMON, Physiology Department and Instrumentation Department, KU Nijmegen, The Netherlands) with faster sampling rates and a lower signal to noise (s/n) ratio (van der Sluijs et al., 1998). We focused on CBF measurements and the intra-individual variance on predefined well-controlled stable baseline values of arterial saturation (SaO_2), mean arterial blood pressure (MAP) and end-tidal CO_2 ($PetCO_2$), heart rate (HR) and respiratory rate (RR).

2. METHODS

2.1. NIRS

The OXYMON measures changes in optical density (OD) in the near infrared region at three different nominal wavelengths (λ): 905, 845 and 770 nm. We used a distance between transmitting and receiving fibers of 5.5 cm. This ensues deep enough penetration of light into the brain. A sampling rate of 10 Hz was used. A pathlength factor of 6.0 was used. The NIRS data were collected and displayed in real time. Data were stored on disk for off-line analysis and calculation of CBF.

2.2. CBF Measurements

We used the "oxygen swing method" for measuring the CBF. The method is based on the Fick's principle in which a rapidly induced change in arterial oxyhemoglobin concentration [O_2Hb] is used as an intravascular tracer. The rate of accumulation of the tracer is monitored with NIRS. Edwards (1988) described this CBF technique using NIRS in infants. A detailed description of the analysis technique and quantification of CBF in adults using NIRS was described by Elwell (1992, 1994).

CBF is given by the formula:

$$\mathrm{CBF(ml.100g^{-1}.min^{-1})} = \frac{K * \Delta OI}{2 * [Hb]_0 \int^t \Delta SaO_2 dt}$$

where K is a constant, which accounts for molecular weight of hemoglobin, cerebral tissue density and the cerebral-to-large vessel hematocrit ratio. [Hb] represents the total hemoglobin concentration in blood (mmol/l), acquired by a blood sample. The s/n ratio of the NIRS data can be improved by measuring the change in the difference between [O_2Hb] and [HHb], defined as the oxygenation index (OI). [tHb] is the sum of [O_2Hb] and [HHb]. If [tHb] is constant, the absolute changes in [O_2Hb] and [HHb] will be equal.

The difference between these changes will be approximately twice the amplitude of the changes of [O_2Hb]. The SaO_2 is measured by pulse-oximetry in the beat to beat mode, the OI with NIRS. When the rate of accumulation is measured within the mean transit time (MTT) of 6 to 10 s (Leenders et al., 1990) of the tracer, the venous outflow of the tracer will be zero. First, a slow gradual decrease in saturation is performed by lowering the FiO_2 to 11–12%. This results in a decrease of SaO_2 of about 10–15%. Next, after a stable baseline of at least 20 s at this level is reached, a bolus of pure oxygen is given in the inspiratory air during one breath.

2.3. Experimental Procedure

The subjects were in supine position during the whole experiment. Peripheral SaO_2 and HR were monitored with a pulse oximeter (N200, Nellcor Puritan Bennet, USA), with the sensor attached to the right-frontal forehead. The optodes of the NIRS were positioned on the left side of the forehead with the receiving probe 2 cm from the midline. The optodes were held in position by an elastic band around the head.

Arterial blood pressure was measured from the middle finger of the right hand non-invasively with a finger plethysmographic device (Finapres, Ohmeda, USA). Blood pressure values were monitored as mean arterial blood pressure (MAP) during each step change in the protocol.

The subjects breathed through a facemask, covering mouth and nose. Separated valves for in- and expiratory gasmixture were integrated in the mask. Changes in inspiratory gasmixture of oxygen, nitrogen and carbon dioxide were induced by means of a computer controlled massflow system (Bronckhorst, Hitec, The Netherlands). This system generates maximum flow rates for O_2 and nitrogen (N_2) of 50 l/min and for CO_2 of 2.5 l/min. The FiO2 was monitored continuously using an oxygen analyzer (OM-11, Beckmann Inc., USA). Fast changes in inspiratory gasmixture could be generated; the aimed changes were reached within one breath. This is of prime importance to prevent slurring of the input function. The expiratory part of the system monitored $PetCO_2$ and the respiratory rate (N1000, Nellcor Puritan Bennet, USA). The analogue data were linked directly to the NIRS computer for real time display and storage simultaneously with the NIRS data.

2.4. Protocols

2.4.1. Series 1. We started to measure seventeen healthy subjects, 9 male and 8 female, aged 21 to 54 years (mean 31 years). Four measurements of CBF in baseline, normocapnic conditions were performed. After the first two measurements, the entire equipment was disconnected from the subjects, to determine the effect of repositioning of the optodes on the reproducibility of the measurements. This was done to get informed on the variation in measurements, due to optode localisation and attachment. Baseline, steady state conditions were defined as stable values of SaO_2, MAP, $PetCO_2$, RR, HR, [tHb], [O_2Hb] and [HHb] for at least 20 s.

2.4.2. Series 2. After finishing the first experiments, alterations were made to the protocol. We emphasized more prolonged baseline steady states (30–60 sec). Ten healthy subjects, 5 males and 5 females, aged 21 to 54 years (mean 30) were investigated. After one try-out CBF procedure, we performed 4 repeated measurements in each subject during baseline normocapnic conditions without disconnection of the optodes. [Hb] was determined at the start of each series of experiments in each subject. Values of $PetCO_2$, HR, RR, MAP and CBF were calculated as means (SD). A paired t-test was used to compare the MAP data between baseline, normocapnic condition and just before and after the $[O_2Hb]$ bolus was given. The level of statistical significance was set at $p < 0.05$. To compare reproducibility within subjects, the coefficient of variation (CV) was calculated.

All volunteers gave informed consent. This study was approved by the ethical committee of the Department of Pulmonology Dekkerswald, University of Nijmegen.

3. RESULTS

Figure 1 shows the upstroke of arterial saturation (upper panel) and OI (lower panel) versus time. For CBF calculations, we took the maximal value of the slope of the OI curve, plotted against the change in SaO_2. In our algorithm, we could observe an unstable ΔtHb or ΔSaO_2, based on two measurements of 6 seconds, which were started at $t = -20$ s and at $t = -6$ s, before the takeoff upstroke point. To ensure that all of the CBF measurements took place within the mean cerebral transit time, an integration time of 3 sec was used.

3.1. Series 1

CBF was measured 4 times in each individual (Table 1). The ΔSaO_2, at which level a stable baseline was obtained, varied between 10 and 15% (mean 12%). Values of $PetCO_2$ varied little (mean SD of 0.37 kPa, range 0.07 to 0.80); the HR and RR varied by values

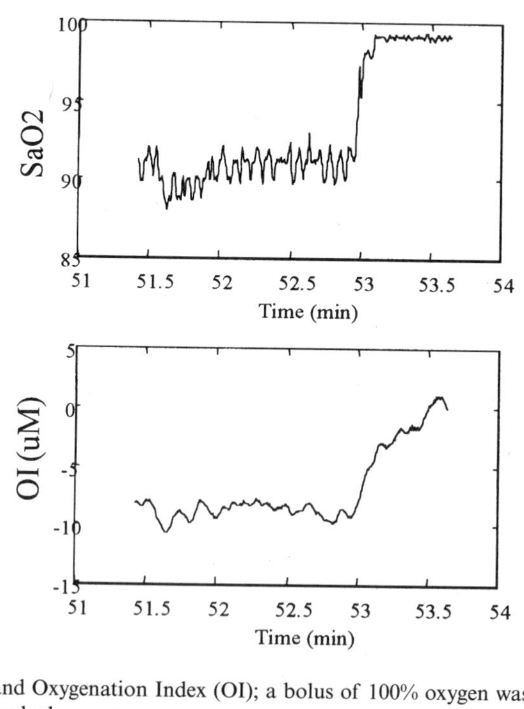

Figure 1. Signal of SaO_2 and Oxygenation Index (OI); a bolus of 100% oxygen was given after a stable baseline of desaturation was reached.

Table 1. Age and mean values (with SD) for cerebral blood flow (CBF), change in saturation (deltaSaO_2), heart rate, PetCO_2, and respiratory rate in 17 healthy subjects; each subject challenged 4 measurements: two duplicated measurements, disconnection of the optodes and repeated measurements

Subject	Age	CBF (ml.100g^{-1}.min^{-1}) Mean(SD)	delta SaO_2 (%) Mean(SD)	MAP (mm Hg) Mean(SD)	PetCO_2 (kPa) Mean(SD)	HR (min^{-1}) Mean(SD)	RR (min^{-1}) Mean(SD)
1	38	28(26)	11(3)	105(4)	4.47(0.25)	76(11)	14(2)
2	21	30(15)	10(1)	78(8)	4.12(0.26)	80(8)	10(2)
3	22	20(8)	10(2)	97(8)	4.55(0.26)	79(10)	12(1)
4	29	11(3)	11(5)	88(4)	5.43(0.07)	63(8)	11(1)
5	54	36(15)	11(7)	108(8)	4.95(0.28)	68(9)	8(5)
6	22	20(4)	12(4)	77(7)	4.27(0.18)	62(8)	19(2)
7	26	19(4)	13(5)	92(8)	4.26(0.33)	56(8)	14(2)
8	22	19(6)	11(2)	98(8)	4.22(0.42)	78(10)	15(2)
9	28	15(6)	14(7)	101(7)	5.14(0.41)	65(3)	9(2)
10	41	27(12)	11(1)	98(5)	4.40(0.56)	66(6)	12(3)
11	21	21(6)	15(1)	103(5)	4.46(0.68)	60(6)	16(2)
12	22	21(6)	14(2)	84(2)	4.24(0.27)	68(10)	11(3)
13	41	27(19)	11(2)	109(6)	4.26(0.61)	61(13)	16(1)
14	30	13(8)	10(1)	92(4)	4.30(0.80)	60(14)	17(2)
15	33	36(15)	11(1)	77(3)	4.65(0.11)	73(13)	13(2)
16	45	36(9)	11(3)	89(4)	4.72(0.33)	60(17)	8(3)
17	32	35(15)	10(2)	107(5)	4.83(0.29)	55(14)	9(2)
Mean	31(10)	26(9)	12(2)	94(6)	4.55(0.37)	66(8)	13(3)

of 8 min^{-1} and 3 min^{-1} respectively. The MAP within subjects, showed significant changes ($p < 0.01$) between the normocapnic baseline condition and just after the oxygen bolus was given. When a distinction was made between systolic and diastolic blood pressure, both systolic and diastolic blood pressure values were significant ($p < 0.05$ and $p < 0.01$, respectively). Individual values of CBF between 11 to 36 (mean 26) ml/min/100 g brain tissue were found.

The within subject CV ranged from 22.3 to 92.4% (mean 42.4%). Figure 2 shows the considerable variability of CBF values within the subjects in the first setup.

3.2. Series 2

The results are shown in Table 2. The ΔSaO_2 varied between 5 to 14% (mean 11%). Values of PetCO_2 varied with a mean SD of 0.26 kPa (range 0.14 to 1.39); the HR and RR varied by 9 min^{-1} and 4 min^{-1} respectively. The MAP within subjects showed significant changes ($p = 0.03$) between the normocapnic baseline condition and just before the oxygen bolus was given. When a distinction was made between systolic and diastolic blood pressure, only systolic blood pressure was significant ($p < 0.01$). Individual values of CBF between 12 to 37 (mean 20) ml/min/100 g brain tissue were found.

The within subject CV ranged from 31 to 52 (mean 22.6%). Figure 3 shows the variability of CBF values within the subjects during the second setup.

4. DISCUSSION

In this study the reproducibility of CBF measurements in human adults was investigated. Using a new and improved continuous wave NIRS instrument, we fulfilled some

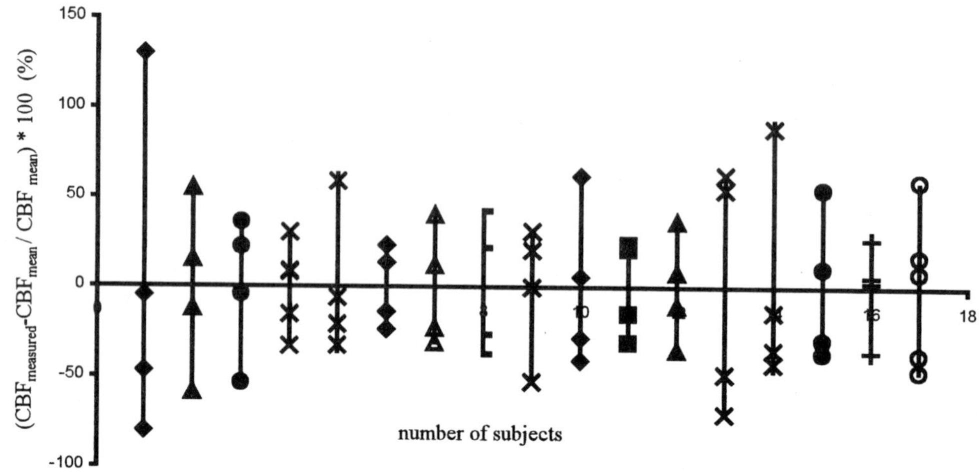

Figure 2. The within subject variability of CBF in the first series: measurements with disconnection of the optodes.

of the recommendations by other authors to improve accuracy and reliability (Elwell et al., 1994; Newton et al., 1997) of CBF measurements in adults. However, the results showed a wide intra-subject variation of the CBF measurements, in standardized experimental conditions.

4.1. Physiological Considerations

CBF is controlled by metabolic, neural and chemical stimuli. The autoregulation of CBF, with a lower and an upper blood pressure limit (Heistad et al., 1983), can be influenced by a number of vascular mechanisms (Strandgaard, 1992; Edvinsson, 1993).

Table 2. Age and mean values (with SD) for cerebral blood flow (CBF), change in saturation (deltaSaO_2), heart rate, $PetCO_2$, and respiratory rate in 10 healthy subjects; each subject 4 measurements without disconnection of the optodes

Subjects	Age	CBF ($ml.100g^{-1}.min^{-1}$) Mean(SD)	delta SaO_2 (%) Mean(SD)	MAP (mm Hg) Mean(SD)	$PetCO_2$ (kPa) Mean(SD)	HR (min^{-1}) Mean(SD)	RR (min^{-1}) Mean(SD)
1	21	27(4)	10(1)	75(5)	4.68(0.14)	80(5)	12(1)
2	22	15(5)	5(1)	91(3)	4.86(0.28)	55(3)	11(1)
3	29	16(1)	14(2)	83(8)	4.50(0.39)	76(7)	11(2)
4	54	25(4)	14(5)	86(6)	4.52(1.39)	72(6)	6(2)
5	44	37(12)	10(2)	100(9)	4.45(0.14)	73(5)	8(2)
6	21	20(2)	9(3)	71(5)	4.45(0.25)	65(3)	17(3)
7	30	12(6)	10(3)	85(3)	4.60(0.47)	50(5)	18(2)
8	22	16(2)	10(2)	94(4)	4.21(0.58)	66(5)	12(1)
9	21	20(3)	10(0)	92(4)	4.05(0.35)	72(5)	12(2)
10	41	14(5)	13(2)	95(7)	4.10(0.56)	65(4)	11(2)
Mean	31(12)	20(8)	11(3)	87(5)	4.44(0.26)	67(9)	12(4)

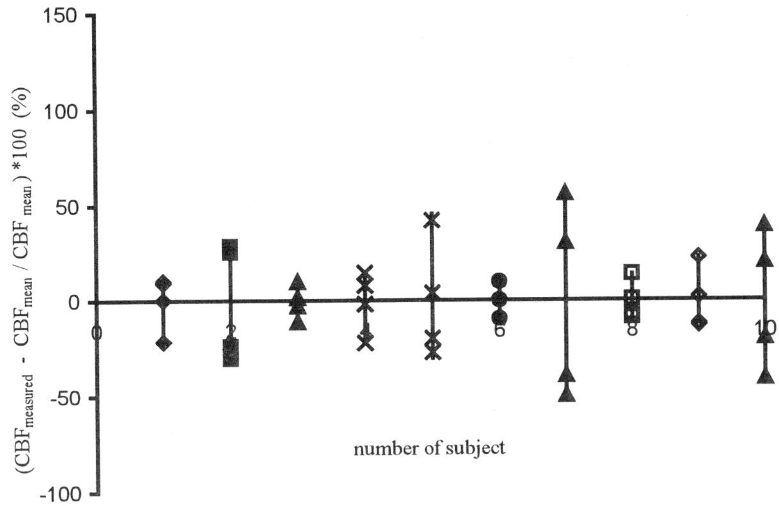

Figure 3. The within subject variability of CBF in the second series: four consecutive measurements without disconnection of instrumentation.

A 7% increase in flow per 10-mmHg increase in MAP in normal humans was described (Moyer et al., 1954). All subjects in our study were considered normotensive, although we found mean MAP values up to 109 mm Hg. It is well known that within-individual alterations in blood pressure occur and differences between home and clinic measurements are common (Ayman et al., 1940). Considering the results of Moyer, alterations of (mean) 6-mmHg (series 1) and 5-mmHg (series 2) during the whole experiment could lead to an increase in (mean) flow of approximately 3.5 to 4%. These small alterations are unlikely to be of relevant importance for the wide CBF variance.

In both series, the MAP showed significant changes within subjects, when mean blood pressure values were compared. In the first series, both systolic and diastolic blood pressure values were significantly different between normocapnic baseline condition and just after the oxygen bolus was given. However, these differences in MAP still fall well within the range of cerebral autoregulation. When a distinction was made between systolic and diastolic blood pressure in the second series, only systolic blood pressure was significant ($p < 0.01$). The latter can be influenced very easily by environmental factors (Ayman et al., 1940).

Alterations of heart rate values and respiratory rate values could have an influence on CBF variability. However, both variables showed no significant differences during the experiments and are therefore unlikely to be responsible for the CBF variability.

The main chemical stimuli of CBF are CO_2, pH and O_2. CO_2 is known as an important modulator of cerebral vascular resistance: hypocapnia induces cerebral vasoconstriction, whereas hypercapnia causes cerebral vasodilation. Concerning intrasubject alterations of PetCO2 during series 1 and 2, changes ranged from 0.07 to 0.8 kPa in series 1 and from 0.14 to 1.39 kPa in series 2. Another chemical stimulus, hypoxia, could result in vasodilation and therefore increased CBF (Heistad et al., 1983). We applied only a slight desaturation (mean ΔSaO_2 12 and 11%, series 1 and 2, respectively). Previous physiological studies showed that CBF is virtually constant when arterial oxygen tension varies between 5.7 and 12.3 kPa (Siesjö, 1978) and therefore no significant influence on CBF may be expected.

The calculated mean value of all CBF measurements from series 1 was lower than the mean value of all CBF measurements of series 2. This could be partly due to the lower $PetCO_2$ seen in series 2 (mean 4.44 kPa), compared to series1 (4.55 kPa). It is well known that hypocapnia leads to vasoconstriction with lowered CBF. As a broad rule of thumb, CBF increased by 46%/per kPa change in arterial PCO_2 during moderate hypercapnia when compared to the normocapnic value for blood flow (Olesen et al., 1971). Hyperventilation ($PaCO_2 < 3.4$ kPa) reduces flow by a maximum of 40–50% (Kety and Schmidt, 1948). CBF reactivity during moderate hyperventilation is not known in terms of percentage/kPa CO_2. Based on Olesen's values for CBF reactivity during moderate hypercapnia, only a (mean) 4.5% lowered value of CBF could account for the lower $PetCO_2$ in our experiments. This is not sufficient to explain the lowered (mean) CBF value of the second series.

4.2. Methodological Considerations

We used an advanced NIRS instrument, the Oxymon, with improved noise level, making rapid sampling (10 Hz) possible. The maximal sampling rate of the peripheral arterial SaO_2 signal, using a pulse oximeter, however is limited to a beat-to-beat modus. This is a weakness, which cannot be solved.

We measured with 3 s integration time; due to the high sampling rate of the NIRS instrument we collected a semi-continuous signal, in contrast to the SaO_2 signal of which only 2 or 3 values can be collected due to the heart beat. This will mainly explain the high CV in our group and among others (Newton et al., 1997; Elwell et al., 1994) and may explain why neonates with higher heart rates have better intrasubject CV's (Bucher et al., 1993).

The influence of the disconnection of the optodes could cause a lower reproducibility; the CV of the second series however, showed little improvement compared to the first series. Therefore small positioning errors are not of relevant importance for the measurements.

The mean CBF in both protocols were in the same range as found by Elwell et al. (1994) and Owen-Reece et al. (1996).

As shown by others (Harris et al., 1993; Lam et al., 1996), the interoptode distance should have a minimum distance of 5 cm for significant intracranial signal sampling. The suggestion that extracranial blood contents could disturb intracranial CBF measurements was investigated by Owen-Reece et al. (1996). He investigated cerebral blood volume in healthy subjects, before and after inflating a pneumatic tourniquet proximal to the measurement site. He saw no effect on NIRS measurement of cerebral hemodynamics when the scalp blood flow was clamped. This finding was in contrast with the results of Newton et al. (1997) in dogs and Lam et al. (1996) in patients during carotid endarterectomy, in which the external carotid artery was clamped for a full 2 minutes. In 50% of the patients, they saw an extracranial contribution to the NIRS signal. In our experiments, disturbances due to extracranial contribution are not of great importance in terms of reproducibility, since the applied changes in the experiments were equal on internal and external carotid circulation.

In conclusion, measurements of CBF with the NIRS are not reproducible despite improved faster sampling and lower noise level. Generally, the measured values of CBF were in the same range as measured by other investigators, using the NIRS method.

Considering methodological and physiological problems, the limited frequency of the pulse oximetry remains an important limitation to the use of O_2Hb as an indicator.

ACKNOWLEDGMENTS

This study is granted by the Dutch Asthma Foundation.

REFERENCES

Ayman D and Goldshine AD (1940) Blood pressure determinations by patients with essential hypertension. I. The difference between clinic and home readings before treatment. Am J Med Sci 200, 465.

Bucher HU, Edwards AD, Lipp AE, and Duc G (1993) Comparison between near infrared spectroscopy and ^{133}Xenon clearance for estimation of cerebral blood flow in critically ill preterm infants. Pediatr Res 33:56–60.

Edvinsson L, MacKenzie ET, and McCulloch J. Cerebral blood flow and metabolism (1993). Part 4. Fundamental responses of the cerebral circulation. 553–580.

Edwards AD, Wyatt JS, Richardson C, Delpy DT, Cope M, and Reynolds EO (1988) Cotside measurement of cerebral blood flow in ill newbron infants by near infrared spectroscopy. Lancet 2:770–771.

Elwell CE, Cope M, Edwards AD, Wyatt JS, Delpy D, and Reynolds EO (1992) Measurement of cerebral blood flow in adult humans using near-infrared spectroscopy: methodology and possible errors. Adv Exp Med Biol 317:235–245.

Elwell C, Cope M, Edwards D, Wyatt J, Delpy D, and Reynolds EO (1994) Quantification of adult cerebral hemodynamics by near infrared spectroscopy. J Appl Physiol 77:2753–2760.

Elwell CE, Cope M, Kirkby D, Owen-Reece H, Cooper CE, Reynolds EOR, and Delpy DT (1994) An automated system for the measurement of the response of cerebral blood volume and cerebral blood flow to changes in arterial carbon dioxide tension using near infrared spectroscopy. Adv Exp Med Biol 361:143–55.

Fallon P, Roberts I, Kirkham FJ, Elliott JM, Lloyd-Thomas A, Maynard R, and Edwards AD (1993) Cerebral hemodynamics during cardiopulmonary bypass in children using near-infrared spectroscopy. Ann Thorac Surg 56:1473–1477.

Harris DNF and Bailey SM (1993) Near infrared spectroscopy in adults. Anaesthesia 48:694–696.

Heistad RD and Kontos HA (1983) Cerebral circulation. In: Handbook of Physiology, section two: The cardiovascular system, volume III, part 1. (Shepherd JT, Abboud FM, Eds) Am Phys Soc, Bethesda, Maryland, pp. 137–182.

Jöbsis FF (1977) Noninvasive infrared monitoring of cerebral and myocardial oxygen sufficiency and circulatory parameters. Science 198:1264–1267.

Kety SS and Schmidt CF (1948) The effects of altered arterial tensions of carbon dioxide and oxygen on cerebral blood flow and cerebral oxygen consumption of normal young men. J Clin Invest 27:484–492.

Lam J, Smielewski P, al Rawi P, Griffiths P, Pickard JD, and Kirkpatrick PJ (1997) Internal and external carotid contribution to near infrared spectroscopy during carotid endarterectomy. Stroke 28(5):906–11.

Leenders KL, Perani D, Lammertsma AA, Heather JD, Buckingham P, Healy MJR, Gibbs JM, Wise RJS, Hatazawa J, Herold S, Beaney RP, Brooks DJ, Spinks T, Rhodes C, Frackowiak RSJ, and Jones T (1990) Cerebral blood flow, blood volume and oxygen utilisation. Normal values and effect of age. Brain 113:27–47.

Moyer JH, Morris G, and Snyder H (1954) A comparison of the cerebral hemodynamic response to aramine and norepinephrine in the normotensive and the hypotensive subject. Circulation 10:265–270.

Newton CRJC, Wilson DA, Gunnoe E, Wagner B, Cope M, and Traystman RJ (1997) Measurements of cerebral blood flow in dogs with near infrared spectroscopy in the reflectance mode is invalid. J of Cerebr Blood flow and Metab 17:695–703.

Olesen J, Paulson OB, and Lassen NA (1971) Regional cerebral blood flow in man determined by the initial slope of the clearance of intra-arterially injected ^{133}Xe. Stroke 2:519–540.

Owen-Reece H, Elwell CE, Harkness W, Goldstone J, Delpy DT, Wyatt JS, and Smith M (1996) Use of near infrared spectroscopy to estimate cerebral blood flow in conscious and anaesthetized adult subjects. Brit journ of Anaesth 76:43–48.

Siesjö BK (1978). Brain Energy Metabolism. John Wiley & Sons, New York.

Strandgaard S and Paulson OB (1992) Regulation of cerebral blood flow in health and disease. J of cardiovasc Pharmac 19(suppl.6):S89–S93.

Van der Sluijs MC, Colier WNJM, Houston RJF, and Oeseburg B (1997) A new and highly sensitive continuous wave near infrared spectrophotometer with multiple detectors SPIE 3194:63–72.

Wyatt JS, Cope M, and Delpy DT (1990) Quantitation of cerebral blood volume in human infants by near infrared spectroscopy. J Appl Physiol 68(3):1086–91.

EVALUATION OF CRITICAL OXYGEN SUPPLY IN THE BRAIN CORTEX

F. Wenzel,[1] K.-M. Scheufler,[2] and J. Grote[1]

[1]Department of Physiology
University of Bonn
Nußallee 11, 53115 Bonn, FRG
[2]Department of Neurosurgery
University of Freiburg
Freiburg, FRG

INTRODUCTION

Insufficiency of cerebrocortical oxygen supply is associated with typical changes of several biochemical and biophysical parameters. Among these the fall of tissue oxygen tension and a left shift of the oxygen tension distribution (Grote et al., 1996), the rapid increase of the lactate/pyruvate ratio (Siesjö 1978), the decrease in the ratio of PCr/P_i or ATP/P_i (Chance 1989) as well as the decline in the cytochrome oxidase redox state (Sylvia et al., 1985; Edwards et al., 1991) are most sensitive to characterize bioenergetic deficiency. Due to the heterogeneity of microcirculation, cellular activity and cellular metabolism (Leniger-Follert & Lübbers 1976; Feng et al., 1988; Lübbers et al., 1994; Grote et al., 1996) as well as due to differences in the spatial resolution of the applied methods, the interpretation of the derived data is complicated and frequently source of controversial discussions.

Since recently non-invasive techniques have been introduced to evaluate insufficient tissue oxygen supply by near-infrared spectroscopy or reflectance spectroscopy (Cooper et al., 1994; Chance et al., 1997; Steinberg et al., 1997) we investigated the regulation of local oxygen supply in the brain cortex of rabbits during arterial hypoxia and arterial hyperoxia by simultaneously recording tissue oxygen tension, redox state of cytochrome aa_3, capillary oxygen saturation and local cerebral blood flow using NIR reflectance spectroscopy together with conventional oxygen tension microelectrode and laser-Doppler-flow measurements. In addition, the experiments offered the opportunity of studying the limitations of the techniques applied under conditions of critical oxygen supply.

Oxygen Transport to Tissue XXI, edited by Eke and Delpy.
Kluwer Academic / Plenum Publishers, New York, 1999.

METHODS

Seven adult male New Zealand White Rabbits weighing 2–3 kg were studied under light general anaesthesia induced by ketaminehydrochloride (Ketavet, 40 mg/kg i.m. initially, 2 mg/kg i.v. every 25 min during the experiment) and midazolamehydrochloride (Dormicum, 0.3 mg/kg i.v. every 25 min). The animals were premedicated with xylazine-hydrochloride (Rompun 2%, 5 mg/kg i.m.). For intravenous drug administration an ear vein was cannulated. Following tracheotomy muscle relaxation was maintained by pancuronium bromide (Pancuronium Organon, 0.1 mg/kg i.v. every 25 min), and the rabbits were artificially ventilated (Mark 7 respirator, Bird, U.S.A.). Tidal volume and ventilation frequency were adjusted to provide normoxic normocapnia. Hyperoxic or hypoxic conditions were established by increasing or decreasing the inspiratory oxygen fraction. End-tidal carbon dioxide fraction was continuously measured (HP 78356A, U.S.A.). The right carotid artery was cannulated to record arterial blood pressure (CM120, Honeywell and Philips, U.S.A) and to collect blood samples for analysis of arterial oxygen tension (P_aO_2), arterial carbon dioxide tension (P_aCO_2) and arterial pH (ABL 330, Radiometer Copenhagen, D.K.). Brain surface temperature was maintained between 35 and 37°C with a heating lamp.

The heads of the rabbits were fixed in a stereotactic frame and a craniotomy was performed over the right parietal cortex. To permit tissue oxygen tension (P_tO_2) measurements with surface microelectrodes (Lübbers et al., 1969; Grote et al., 1981), the dura was stripped from the craniotomy site. The multiwire electrodes were calibrated at 37°C with saline solutions of different oxygen tension. To avoid impairment of blood flow in the measuring field the electrodes were mounted on a counterbalanced system. Local blood flow was determined using laser-Doppler-flowmetry (MBF 3D, Moor Instruments, G.B.). For determination of the capillary oxygen saturation (SO_2 cap) as well as of the oxidation level of cytochrome aa_3 (Cyt_{ox}/Cyt_{tot}) in the brain tissue near-infrared spectroscopy was applied (Multiscan OS30, NIOS, F.R.G.). The needle probe for the flow measurements as well as the photosensor of the spectroscope were attached to a micromanipulator and positioned under microscopic guidance approximately 0.5 mm above the cortical surface.

Arterial blood pressure was recorded continuously throughout the experiment while arterial blood gas tensions and pH were analyzed before and after each measuring period. All experiments were performed after a 30 min stabilization period following craniotomy. After onset of hyperoxic or hypoxic ventilation an equilibration period of 15 min was allowed. At the conclusion of the experiments normoxic conditions were reestablished, and all measurements were repeated.

Statistical comparison for differences of mean values was performed by unpaired Mann-Whitney-test. A P value < 0.05 was considered statistically significant. All data are presented as mean value ± SD.

RESULTS

The mean values for the respiratory gas tensions and pH of the arterial blood measured initially during normocapnic normoxia were $P_aO_2 = 90 \pm 12$ mmHg, $P_aCO_2 = 40 \pm 3$ mmHg and pH $= 7.23 \pm 0.08$. The corresponding values determined during hyperoxia were 157 ± 41 mmHg, 41 ± 2 mmHg and 7.23 ± 0.08. Stepwise lowering of the oxygen fraction in the inspiratory gas resulted in arterial oxygen tensions of 58 ± 12 mmHg (mild hypoxia) and 36 ± 4 mmHg (severe hypoxia). The simultaneously measured P_aCO_2 was found to be 36 ± 3 mmHg and 34 ± 5 mmHg, respectively. The mean arterial pH was 7.27 ± 0.07 during mild hypoxia and 7.04 ± 0.17 during severe hypoxia. In all experiments mean arterial blood pressure was within normal range (80–110 mmHg).

The P_tO_2 values measured in the superficial cell layers of the brain cortex under conditions of normocapnic normoxia showed the typical oxygen tension distribution of the brain tissue (Grote et al., 1982, 1996; Lübbers et al., 1994). The oxygen tensions ranged from very low values near 0 mmHg to values of about 85 mmHg. The mean tissue oxygen tension was found to be 32 ± 6 mmHg. As shown in Figure 1, normocapnic hyperoxia induced a significant elevation of the cortical oxygen tension with a shift of the frequency histogram to the right. The mean value was 44 ± 3 mmHg. Stepwise lowering of the arterial oxygen tension to 58 and 36 mmHg caused a pronounced shift of the P_tO_2 distribution to low values. The corresponding mean tissue oxygen tensions were 22 ± 2 and 9 ± 3 mmHg, respectively. Both values were found to be significantly below the normoxic control level.

The results of the corresponding oxygen saturation measurements in the capillary blood of the brain cortex are summarized in Figure 2. During normoxia the mean capillary oxygen saturation was found to be 46.5 ± 7.4%. Hyperoxia resulted in a slight increase to 49.7 ± 8.8%, mild hypoxia caused a slight decrease to 42.8 ± 3%. A significant SO_2 cap decrease could only be seen during severe arterial hypoxia. The respective mean value was 31.9 ± 2.3%.

While the capillary oxygen saturation showed less pronounced changes during arterial hypoxia or hyperoxia significant reactions to alterations of the arterial oxygen tension could be observed in the oxidation level of cytochrome aa_3 in the brain cortex, recorded

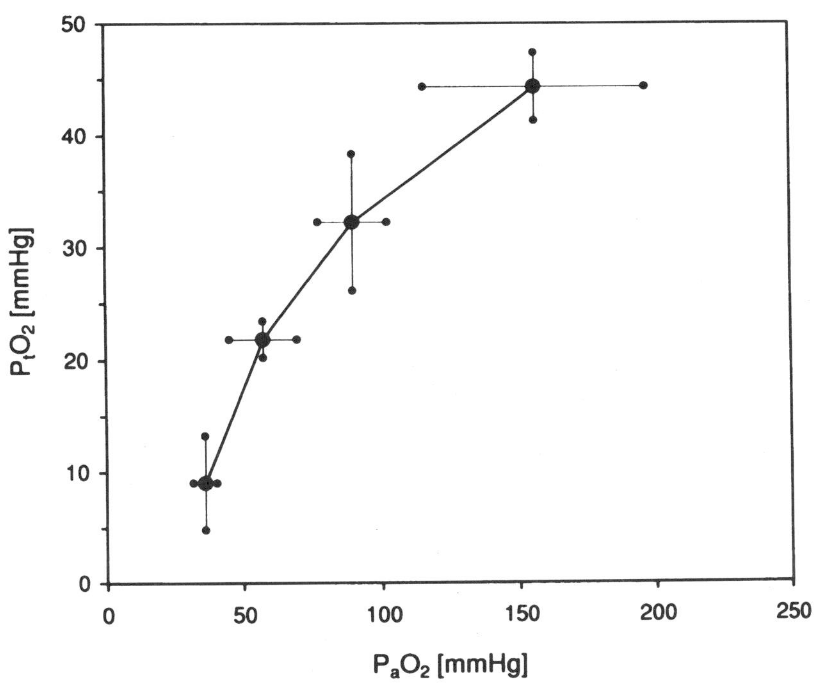

Figure 1. Mean tissue oxygen tension (P_tO_2) in the brain cortex of rabbits in response to changes of arterial oxygen tension (P_aO_2) during normocapnia and normal arterial blood pressure.

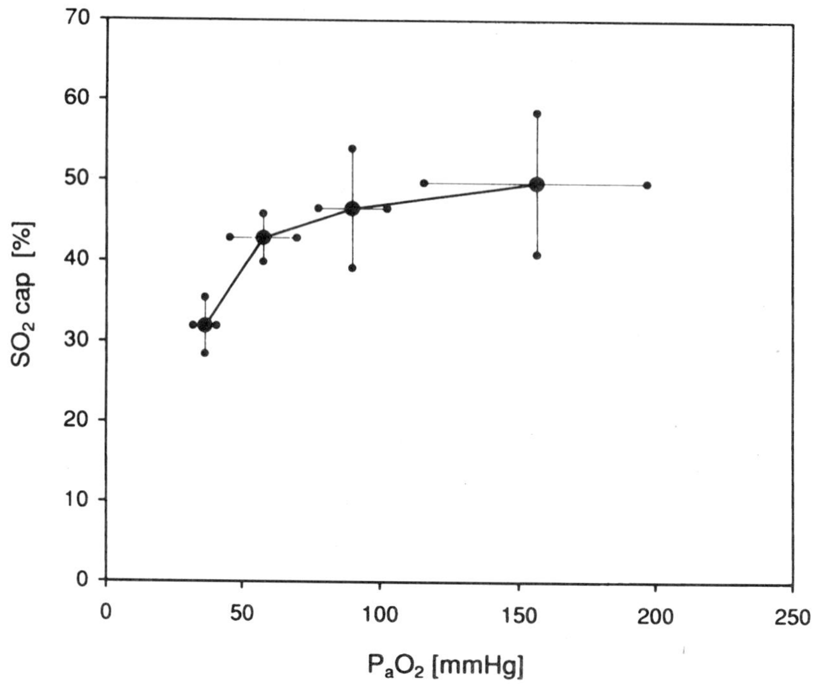

Figure 2. Capillary oxygen saturation (SO_2 cap) in the brain cortex of rabbits in response to changes of arterial oxygen tension (P_aO_2) during normocapnia and normal arterial blood pressure.

simultaneously. During normoxia a mean value for the ratio Cyt_{ox}/Cyt_{tot} of 1.6 ± 0.2 was determined. Arterial hyperoxia as well as mild or severe hypoxia induced significant changes of the redox state of the cytochrome oxidase (Figure 3). Under hyperoxic conditions the mean ratio of Cyt_{ox}/Cyt_{tot} was 1.8 ± 0.5. The respective mean values found during mild and severe arterial hypoxia were 1.3 ± 0.3 and 1.0 ± 0.1.

In addition to the significant fall in the tissue PO_2 as well as in the oxidation level of cytochrome aa_3, a pronounced increase in local cortical blood flow could be observed when arterial oxygen tensions below 60 mmHg were attained. As can be seen in Figure 4, which gives the data of a typical experiment, stepwise lowering of the arterial oxygen tension from a normal level (106 mmHg) to approximately 50 mmHg and to 30 mmHg caused vasodilation and a subsequent blood flow increase of about 100 and 250%, respectively. The mean values of the tissue PO_2 as well as of the redox state of cytochrome aa_3 determined in parallel showed a pronounced decrease. The corresponding oxygen saturation levels in the capillary blood were found to be 52%, 46% or 36%.

DISCUSSION

The oxygen tension distribution determined in the superficial cell layers of the brain cortex is in close agreement with those of deeper cortical regions up to at least 500 µm below the brain surface (Grote et al., 1996). The derived tissue oxygen tension histograms, therefore, reflect the microheterogeneity of oxygen delivery and oxygen consumption not only at the brain surface but also within the cortical laminae below. Under comparable conditions almost identical oxygen tension frequency distributions were determined in

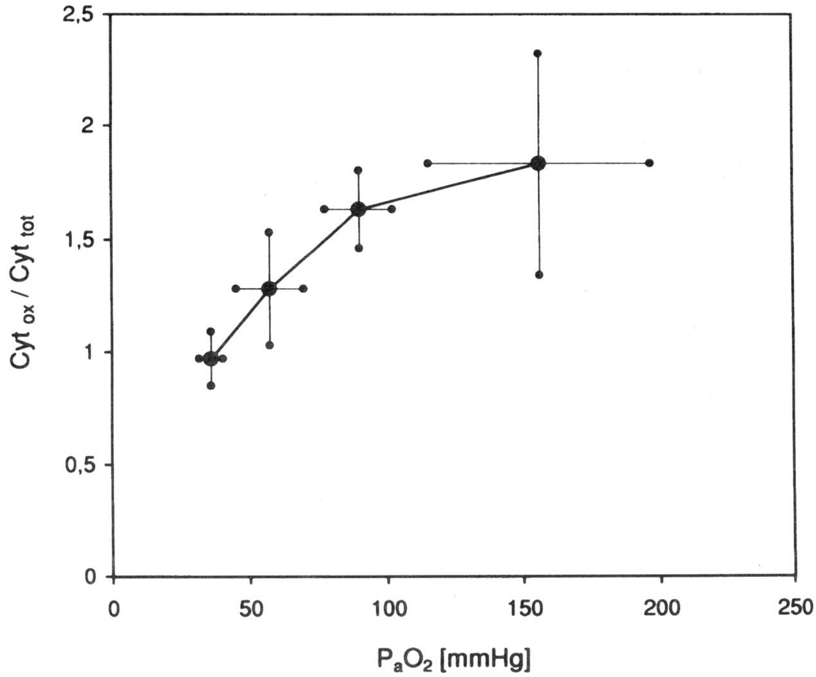

Figure 3. Redox state of the cytochrome oxidase (Cyt_{ox}/Cyt_{tot}) in the brain cortex of rabbits in response to changes of arterial oxygen tension (P_aO_2) during normocapnia and normal arterial blood pressure.

the cerebral cortex of different species using needle or surface microelectrodes (Grote et al., 1981, 1996, Lübbers et al., 1994).

The shift of the tissue oxygen tension distribution to high values as observed during normocapnic hyperoxia is caused by the increase in the oxygen delivery although a slight

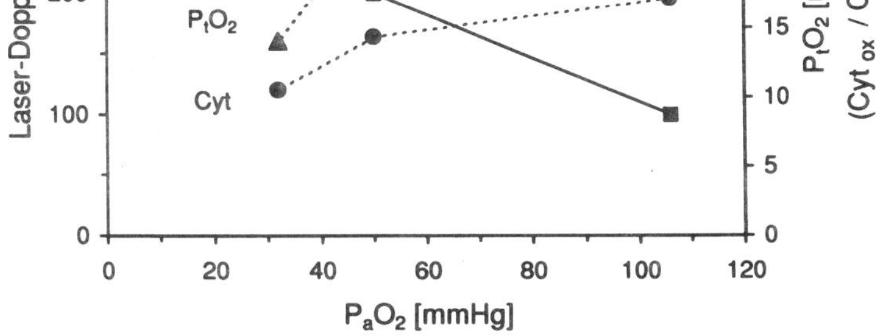

Figure 4. Effects of stepwise decrease in arterial oxygen tension (P_aO_2) on the oxygen tension (P_tO_2) and the redox state of the cytochrome oxidase (Cyt_{ox}/Cyt_{tot}) as well as on the local blood flow (laser-Doppler-flow), as determined during normocapnia and normal arterial blood pressure in the brain cortex of a rabbit.

decrease in local blood flow was present. The hypoxia induced left shift of the oxygen tension histograms and the increase in the number of tissue oxygen tension values between 0 and 5 mmHg describe the increasing restriction of cerebral oxygen supply at low arterial oxygen tensions which could no longer be compensated by the pronounced blood flow increase (Grote & Schubert 1982). First hypoxia-induced reactions in local cerebral blood flow were observed under conditions of mild arterial hypoxia with oxygen tensions between approximately 50 and 70 mmHg.

The effect of changes in the arterial oxygen tension on the tissue PO_2 of the brain cortex is comparable to that on the redox state of cytochrome aa_3. The findings agree with results of in vivo measurements of the cytochrome oxidase redox relationship in hypoxic rat brain, performed by Jöbsis-VanderVliet and his coworkers (Sylvia et al., 1985). In addition, our data indicate that the redox state of cytochrome aa_3, as determined by the applied NIR reflectance spectroscopy, correctly describes the oxygen supply conditions in the brain area under investigation.

The capillary oxygen saturation measurements performed simultaneously in the same region, however, resulted in less pronounced changes. Hyperoxia as well as mild hypoxia induced no significant increase or decrease in the oxygen saturation of the capillary blood. The results can be explained by the characteristics of the O_2 binding of hemoglobin and by the blood flow reactions during hyperoxia or hypoxia. As described by the oxygen dissociation curve of hemoglobin elevation of the arterial oxygen tension from normal levels to about 200 mmHg or lowering of the arterial oxygen tension up to 50–60 mmHg cause only slight changes in the oxygen saturation. The expected SO_2 increase during hyperoxia is in part compensated by the simultaneous blood flow restriction, while the hypoxia induced hyperemia reduces the fall in SO_2 following arterial oxygen tension decrease. Since at low oxygen tensions the blood SO_2-PO_2 relationship is given by the steep part of the oxygen dissociation curve a pronounced decrease in the capillary oxygen saturation has to be expected during severe arterial hypoxia. The oxygen saturation values measured in the capillary blood of the brain cortex at arterial oxygen tension levels of approximately 35 mmHg, indeed, show the assumed SO_2 drop which obviously could no longer be compensated by the blood flow increase and by the right shift of the oxygen dissociation curve caused by the decrease in blood pH.

The data indicate that capillary SO_2 is less sensitive to characterize brain tissue oxygen supply conditions and to detect bioenergetic deficiency of the brain tissue as compared to the results of tissue PO_2 measurements or of determinations of the redox state of cytochrome aa_3.

REFERENCES

Chance B (1989) What are the goals of magnetic resonance research?. NMR in Biomedicine 2:179–187.

Chance B, Mayevsky A, Guan B, and Zhang Y (1997) Hypoxia/ischemia triggers a light scattering event in rat brain. Adv Exp Med Biol 428:457–467.

Cooper C, Matcher S, Wyatt J, Cope M, Brown G, Nemoto E, and Delpy D (1994) Near infrared spectroscopy of the brain: relevance to cytochrome oxidase bioenergetics. Biochem Soc Trans 22:971–980.

Edwards A, Brown G, Cope M, Wyatt J, McCormick D, Roth S, Delpy D, and Reynolds E (1991) Quantification of changes in the concentration of cerebral oxidised cytochrome oxidase. J Appl Physiol 71:1907–1913.

Feng Z, Roberts E, Sick T, and Rosenthal M (1988) Depth profile of local oxygen tension and blood flow in rat cerebral cortex, white matter and hippocampus. Brain Res 445:280–288.

Grote J, Zimmer K, and Schubert R (1981) Effects of severe arterial hypocapnia on regional blood flow regulation, tissue PO_2 and metabolism in the brain cortex of cats. Pflügers Arch 391:195–199.

Grote J and Schubert R (1982) Regulation of cerebral perfusion and PO_2 in normal and edematous brain tissue. In: Oxygen transport to human tissues. (Loeppky J, Riedesel M, Eds) Elsevier/North Holland, Amsterdam, pp. 169–178.

Grote J, Laue O, Eiring P, and Wehler M (1996) Evaluation of brain tissue O_2 supply based on results of PO_2 measurements with needle and surface microelectrodes. J Autonom Nerv Syst 57:168–172.

Leniger-Follert E and Lübbers D (1976) Behavior of microflow and local PO_2 of the brain cortex during and after direct electrical stimulation. Pflügers Arch 366:39–44.

Lübbers D, Baumgärtl H, Fabel H, Kessler M, Kunze K, Riemann H, and Seiler D, Schuchardt S (1969) Principle and construction of various platinum electrodes. Progr Resp Res 3:136–146.

Lübbers D, Baumgärtl H, and Zimelka W (1994) Heterogeneity and stability of local PO_2 distribution within the brain tissue. Adv Exp Med Biol 345:567–574.

Siesjö B (1978) Brain energy metabolism. John Wiley and Sons, New York.

Steinberg F, Röhrborn H, Otto T, Scheufler K, and Streffer C (1997) NIR reflection measurements of hemoglobin and cytochrome aa_3 in healthy tissue and tumors. Adv Exp Med Biol 428:69–77.

Sylvia A, Piantadosi C, and Jöbsis-Vander-Vliet F (1985) Energy metabolism and in vivo cytochrome c oxidase redox relationships in hypoxic rat brain. Neurol Res 7:81–88.

88

COMPARISON OF CEREBRAL BLOOD VOLUME MEASURED BY NEAR INFRARED SPECTROSCOPY AND CONTRAST ENHANCED MAGNETIC RESONANCE IMAGING

M. Wolf,[1] O. Weber,[3] M. Keel,[1] X. Golay,[3] M. Scheidegger,[3]
H. U. Bucher,[1] S. Kollias,[2] P. Boesiger,[3] and O. Bänziger[1]

[1]Clinic for Neonatology
[2]Institute of Neuroradiology
University Hospital Zurich
CH-8091 Zurich
[3]Institute Biomed. Engineering
University and Swiss Federal Institute of Technology Zurich
CH-8092 Zurich

ABSTRACT

Cerebral blood volume (CBV) can be quantified by both near infrared spectroscopy (NIRS) and magnetic resonance imaging (MRI). The aim is to compare CBV results obtained by NIRS and MRI in adult patients.

10 adult patients, 6 females and 4 males, age median 24 (range 21 to 76) years, were included in this study. All needed a MRI investigation with contrast medium for clinical reasons.

The NIRS instrument, the Cerebral RedOx Monitor 2020 from Critikon, quantifies cerebral haemoglobin concentration using a sensor with two receiving channels at different distances. Geometrical detector arrangements of this type enable a ratio measurement to be achieved, which reduces the contribution of the skull and skin, thus allowing quantification. Cerebral haemoglobin concentration can be converted in CBV, as the haemoglobin concentration in the blood is known.

CBV can be quantified by MRI using an indicator dilution method. The method requires an injection of a paramagnetic contrast agent. The input function can be measured at the throat and thus perfusion images can be quantified.

CBV was measured by NIRS just before the patient entered the magnet and after

Oxygen Transport to Tissue XXI, edited by Eke and Delpy.
Kluwer Academic / Plenum Publishers, New York, 1999.

he had left it. The sensor for the NIRS measurement was applied to the patients front three times for 1 minute to each side, avoiding the sinuses. CBV was determined by contrast enhanced MRI between the NIRS measurements.

The mean CBV (NIRS) was 8.6 (SD 1.3) ml/100 g and CBV (MRI) was 7.1 (SD 2.5). The correlation between CBV (NIRS) and CBV (MRI) was Pearson's correlation coefficient −0.297 ($p = 0.204$) respectively Spearman's rho (nonparametric) −0.266 ($p = 0.257$). The CBV values obtained by NIRS and MRI, even though they are in the same range, do not correlate.

1. INTRODUCTION

Near infrared spectrophotometry (NIRS) has become a widely used method to investigate cerebral oxygenation by determining cerebral haemoglobin concentration (tHb in µmol/l) (Wolf et al., 1997), cerebral blood volume (CBV in ml/100 g) (Wyatt et al., 1990), cerebral blood flow (Skov et al., 199;, Bucher et al., 1993) and tissue oxygen saturation (Wolf et al., 1997). Several studies (Okada et al., 1995; Firbank et al., 1996; Owen-Reece et al., 1996) have raised doubts about its accuracy in adults, because the extracerebral layers of tissue (skin, skull and cerebro spinal fluid) will overwhelm the cerebral signal. However, none of the instruments has been compared to another method in human adults. The algorithm of the Cerebral RedOx Monitor 2020 from Critikon is designed to reduce the influence of the extracerebral layers of tissue. The best way to test its effectiveness is a comparison to another method in vivo. There is no gold standard for this purpose.

Magnetic resonance imaging (MRI) is a technique which is able to quantify CBV using a tracer without imposing any additional discomfort on patients undergoing a conventional contrast enhanced MRI for clinical reasons. The only adverse effect for the patient is a prolongation of the MRI study by a few minutes.

The aim of this study is to compare CBV results obtained by NIRS to MRI in adults.

2. PATIENTS

Inclusion criteria for the adult patients:

1. A contrast enhanced MRI is indicated for clinical reasons.
2. The MRI for clinical reasons should have an expected duration of less than 45 min.
3. The patient does not undergo a MR angiography.
4. The patient is conscious and responsible to be able to decide whether he or she wants to participate in this study.

Ten adult patients, 6 females and 4 males, age median 24 (range 21 to 76) years, were included in this study. The patient's haemoglobin concentration (cHb in g/100 ml) was determined for clinical reasons and was median 13.7 (9.6 to 15.4) g/dl.

3. METHODS

3.1. NIRS

CBV measured by NIRS will be called CBV (NIRS). The NIRS instrument, the Cerebral RedOx Monitor 2020 from Critikon, quantifies tHb using a sensor with two

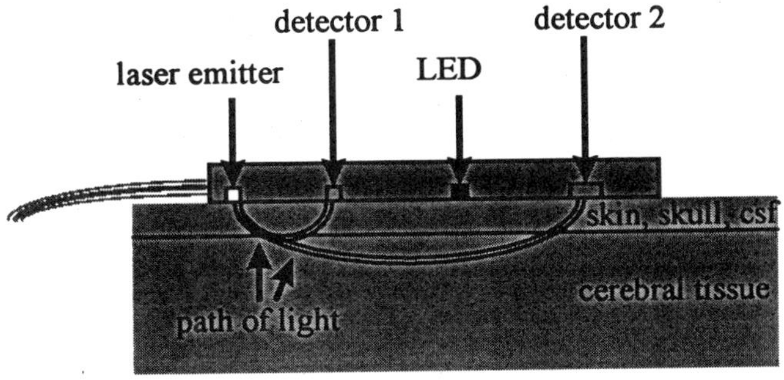

Figure 1. A schematic diagram of the sensor to measure absolute cerebral haemoglobin concentration by the geometrical principle. The system based on two detectors reduces the contribution of the extra-cerebral tissue. (LED = light emitting diode, csf = cerebro spinal fluid).

receiving channels at different distances. By taking a ratio between the signal of detector 2 and the signal of detector 1, the influence of the extracerebral tissue is reduced. An LED is placed exactly in the middle between the two detectors to check the light coupling at the two detectors. The combination of these two features enables to determine a baseline for the calculation of absolute oxy-, deoxy- and total haemoglobin concentrations. So far this baseline has been missing in NIRS and only relative concentration changes were quantified. The principle has been described in detail by (Wolf et al., 1997).

tHb can be converted in CBV by adjusting for the haemoglobin concentration in the blood: CBV = 0.89 tHb/cHb.

CBV (NIRS) was measured just before the patient entered the magnet and after he left it. The sensor for the NIRS measurement was applied to the patients forehead three times for 1 minute on each side, avoiding the sinuses. The position of the sensor was measured, such that the CBV value could be determined by MRI for the same region. The data were recorded with a sample rate of 1/s and a median was calculated over the one minute. All the data for each side were then averaged.

3.2. MRI

CBV measured by MRI (CBV (MRI)) can be quantified by an indicator dilution method (Rempp et al., 1994) requiring a bolus injection of paramagnetic (e.g., Gadolinium-based) contrast agent. The arterial input function can be measured in the carotid arteries at the throat and allows the quantification of the perfusion images.

To determine CBV (MRI), contrast enhanced images were acquired on a Philips Gyroscan ACS-NT 1.5 Tesla whole body system (Philips Medical Systems, Best, the Netherlands) between the NIRS measurements. Before, during, and after bolus injection of Gd-DTPA and subsequent flushing with physiologic saline solution, dynamic images were acquired with a fast gradient-echo planar imaging (FFE-EPI) sequence. During 2 minutes, 60 dynamic images of 6–8 slices covering part of the brain and of one slice through the throat were acquired in an interleaved manner. Relative CBV pixel values were determined based on the tracer kinetic theory. A semi-circular region of interest covering approximately the same brain tissue as examined in the NIRS experiment was defined interactively, on which the mean relative CBV was calculated. Absolute CBV (MRI) was then obtained by normalisation with the arterial input function as determined in the carotid arteries (Rempp et al., 1994).

3.3. Statistics

The following statistics were carried out using SPSS 7.5.1: The means and standard deviations were calculated for the CBV (NIRS) and CBV (MRI). The test retest variability (TRV in %) was calculated using analysis of variance. Differences between two paired parameters were determined using a paired sample t-test and paired sample Wilcoxon signed rank test. Correlations were analysed using Pearson's correlation coefficient and Spearman's rank correlation coefficient (rho).

3.4. Ethical Approval

This study was approved by the local ethical committee and written informed consent was obtained from each patient.

4. RESULTS

The mean CBV (NIRS) was 8.6 (SD 1.3) ml/100g and CBV (MRI) was 7.1 (SD 2.5). The difference between the two means was not statistically significant.

The TRV for CBV (NIRS) was 11.5%.

The mean CBV (NIRS) before the MRI was 8.5 (SD 1.4) ml/100g and after the MRI was 8.4 (SD 1.3) ml/100g. The mean difference between CBV (NIRS) before and CBV (NIRS) after the MRI was 0.065 (SD 0.42) ml/100g, which was not statistically significant. The correlation between the CBV (NIRS) before and the CBV (NIRS) after was 0.952 and highly significant: $p < 0.001$.

The mean CBV (NIRS) of the left hemisphere was 8.7 (SD 1.4) ml/100g and the right hemisphere was 8.5 (SD 1.3) ml/100g. The mean CBV (MRI) of the left hemisphere was 6.9 (SD 2.3) ml/100g and the right hemisphere was 7.2 (SD 2.7) ml/100g. The difference between left and right hemisphere was 0.16 (SD 0.91) ml/100g ($p > 0.05$) for CBV (NIRS) and −0.32 (SD 1.48) ml/100g ($p > 0.05$) for CBV (MRI). The CBV (NIRS) of the left and the right hemisphere correlated with a Pearson's coefficient of 0.773 ($p = 0.009$). For CBV (MRI) the correlation was 0.834 ($p = 0.003$).

The correlation between CBV (NIRS) and CBV (MRI) was −0.297 (Pearson) ($p = 0.204$) and −0.266 (Spearman) ($p = 0.257$), respectively. Thus the two methods were not correlated (Figure 2).

5. DISCUSSION

Despite similar mean values of CBV (NIRS) and CBV (MRI), a low TRV of 11.5% and reasonable correlations between CBV on left and right hemisphere, CBV (NIRS) and CBV (MRI) do not correlate. We looked carefully at the two outliers with the highest CBV (MRI), but could not find any particularities. If these outliers are taken out of consideration, we get a positive correlation (line of regression CBV (MRI) = 0.871*CBV (NIRS)−1.69, Pearson 0.57 $p = 0.14$, Spearman 0.41 $p = 0.32$, mean difference (Bland and Altman, 1986) = 2.9 ± 1.3 ml/100g). There are many possible reasons for the discrepancy between CBV (NIRS) and CBV (MRI)-values.

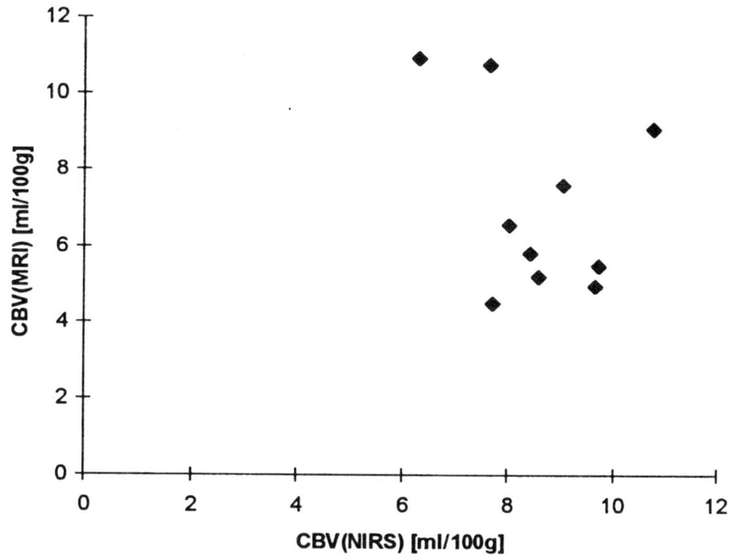

Figure 2. The cerebral blood volume measured by near infrared spectrophotometry (CBV (NIRS) in ml/100 g) against the value measured by magnetic resonance imaging (CBV (MRI) in ml/100 g).

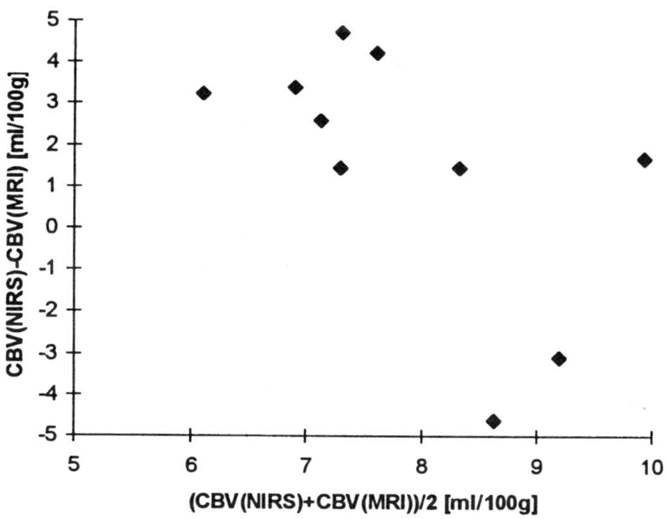

Figure 3. According to Bland & Altman (1986) this figure shows the difference between the cerebral blood volume measured by near infrared spectrophotometry (CBV (NIRS) in ml/100 g) and that measured by magnetic resonance imaging (CBV (MRI) in ml/100 g) against their mean. The mean difference equals to 1.5 (SD 3.1) ml/100 g. The two measurements, which show a difference below zero, correspond to the ones with the two highest CBV (MRI).

5.1. Possible Reasons for the Discrepancy between CBV (NIRS) and CBV (MRI)

The NIRS measurements have a test TRV of 11.5%. By averaging 6 samples per hemisphere, this variability is reduced to 4.7%.

The influence of the extracerebral tissue may be overwhelming the NIRS-signal from the brain (Okada et al., 1995; Firbank et al., 1996) despite the ratio measurement.

The kinetic tracer dilution theory applied for CBV (MRI) relies on an intact brain-blood-barrier. It might be that this condition was not fulfilled in the two patients with significantly higher perfusion rates than the other subjects.

Shifting the region of interest by few millimeters changed the mean CBV (MRI) value by up to ten percent. Reproducibility of the processing is therefore in the same range.

Even though the position of the NIRS-sensor on the head was carefully measured, it is difficult to estimate the region of tissue that is actually interrogated. The shape may be different to the one used for the MRI method, which may have a considerable effect.

The measurements could not be carried out exactly at the same time, because the NIRS instrument was not able to tolerate the strong magnetic field of the MRI-scanner. The size of this effect is small and can be estimated by the difference between the CBV (NIRS) measured prior and after the MRI scan. The CBV was increased by 0.7% (not significant). However, there could also be differences due to the not uniform mental activation of the patients.

6. CONCLUSION

The CBV values obtained by NIRS and MRI, even though they are in the same range, do not correlate. It is difficult to determine which factor/s cause this discrepancy.

ACKNOWLEDGMENTS

This work was funded by the Swiss federal office of education and science (BBW No 95.0701-1) in the framework of the EU cost shared project on near infrared spectroscopy and imaging and by the Department of Education of the County of Zurich.

REFERENCES

Bland JM and Altman DG (1986) Statistical methods for assessing agreement between two methods of clinical measurement. Lancet 2:307–310.

Bucher HU, Edwards AD, Lipp AE, and Duc G (1993) Comparison between near infrared spectroscopy and 133Xenon clearance for estimation of cerebral blood flow in critically ill preterm infants. Pediatr Res 33:56–60.

Firbank M, Arridge SR, Schweiger M, and Delpy DT (1996) An investigation of light transport through scattering bodies with non-scattering regions. Phys Med Biol 41:767–783.

Okada E, Firbank M, Schweiger M, Arridge SR, Cope M, and Delpy DT (1995) A theoretical and experimental investigation of the effect of sulci on light propagation in brain tissue. SPIE 2626:2–8.

Owen-Reece H, Elwell CE, Harkness W, Goldstone J, Delpy DT, Wyatt JS, and Smith M (1996) Use of near infrared spectroscopy to estimate cerebral blood flow in conscious and anaesthetized adult subjects. Brit J of Anaesthesia 76:43–48.

Rempp KA, Brix G, Wenz F, Becker CR, Gückel F, and Lorenz WJ (1994) Quantification of regional cerebral blood flow and volume with dynamic susceptibility contrast-enhanced MR Imaging. Radiology 193:637–641.

Skov L, Pryds O, and Greisen G (1991) Estimating cerebral blood flow in newborn infants: comparison of near infrared spectroscopy and 133Xe clearance. Pediatr Res 30:570–573.

Wolf M, Evans P, Bucher HU, Dietz V, Keel M, Strebel R, and von Siebenthal K (1997) The Measurement of Absolute Cerebral Haemoglobin Concentration in Adults and Neonates. Adv Exp Med Biol 428: 219–227.

Wyatt JS, Cope M, Delpy DT, Richardson CE, Edwards AD, Wray S, and Reynolds EO (1990) Quantitation of cerebral blood volume in human infants by near-infrared spectroscopy. J Appl Physiol 68:1086–1091.

89

BIOLUMINESCENCE FOR BIOLOGICAL SENSING IN LIVING MAMMALS

Weisheng Zhang, Pamela R. Contag, Ashima Madan,
David K. Stevenson, and Christopher H. Contag

Division of Neonatal and Developmental Medicine
Department of Pediatrics
Stanford University School of Medicine
Stanford, California 54305-5208

1. INTRODUCTION

Noninvasive *in vivo* assays are required for the study of biological processes that are dynamic involving intact organ systems and complex physiologic changes, such as tissue oxygenation. Optical methods that use external light sources in the near infrared (nir) have been developed for oxygenation determinations and have widespread application in medicine and biomedical research (Jobsis, 1977; Benaron et al., 1997). Further refinements in these methods are providing powerful tools for research; however, access to *in vivo* information relating to some physiologic changes can not be assessed using nir monitoring and imaging. To address this unmet need for additional real-time bioassays, we have developed an *in vivo* method utilizing bioluminescent reporters, or photoproteins, as indicators of biological functions (Contag et al., 1995). In contrast to other optical methods which use external sources of light, the photoproteins provide an internal source of light that can be monitored externally as an indicator of biological processes such as infection and gene expression (Contag et al., 1995, 1996, 1997).

1.1. Photoproteins

Luciferases are photoproteins that have been used for decades as biological labels for gene expression and for determination of ATP concentrations in cells, *ex vivo* and *in vitro* (Campbell, 1988; Hastings, 1996). Wavelengths of bioluminescent light have been described from 460 nm to 700 nm, although the most well characterized and most often used bioluminescent reporters are blue to yellow-green (490–560 nm, bacterial, jellyfish and firefly luciferases; Campbell, 1988; Hastings, 1996). For application of luciferases to

Oxygen Transport to Tissue XXI, edited by Eke and Delpy.
Kluwer Academic / Plenum Publishers, New York, 1999.

Table 1. Photoproteins that have been used as reporters in cells and tissues

Enzyme	Species of Origin	Emission Color and Spectral Peak	Unique Properties
Aequorin	*Aequorea victoria*	Blue (466 nm)	Ca^{2+} sensitive
Renilla luciferase	*Renilla reniformis*	Blue (470 nm)	
Firefly luciferases			Requires O_2 and ATP;
North American	*Photinus pyralis*	yellow-green (562 nm)	mutants with enhanced
Modified	*Photinus pyralis*	yellow-green (562 nm)	expression in mammalian
Red mutant[a]	*Photinus pyralis*	orange-red (610 nm)	cells are available
Japanese	*Luciola cruciata*	yellow-green (562 nm)	Red mutants identified
	Luciola mingrelica	yellow-green (582 nm)	substrate biosynthesis
Bacterial luciferases			characterized, low
Vibrio spp.	*Vibrio fischeri*	Blue (489)	temperature optima
	Vibrio harveyi	Blue (489)	full function at 37°C
Photorhabdus spp.	*Photorhabdus luminescens*	Blue (489)	

[a] Five amino acid substitutions in *luc* from *P. pyralis* (Contag and Eames unpublished data).
[b] One color variant has been mutated for orange-red emission.

the study of living animal models of human physiology, ultra-sensitive cameras are required to detect the few photons that escape the scattering and absorbing environment of mammalian tissues after being emitted internally. Fortunately, photonic detection technologies have advanced in the areas of night vision and quality control in the silicon chip industry, so that cameras with this capability are commercially available.

1.1.1. Bacterial Luciferases. For labeling bacterial cells, the *lux* operons from bioluminescent bacteria are ideal since these operons not only encode the luciferase genes (two genes for the heterodimeric enzymes, gene locus: *lux* and enzyme: Lux) but also encode the genes for the biosynthetic enzymes (fatty aldehyde synthases) for the substrate (decanal; Frackman et al., 1990; Szittner and Meighen, 1990; Xi et al., 1991; Meighen, 1993). Use of these operons obviates the need for an exogenous supply of substrate. Moreover, no deleterious effects of substrate biosynthesis on bacterial metabolism or virulence have been observed (Contag et al., 1995). Among the *lux* operons from bioluminescent bacteria, that from *Photorhabdus luminescens* (previously known as *Xenorhabdus luminescens*) appears to be ideally suited for use in mammalian animal models given that mammalian body temperatures lie within the temperature optimum for this enzyme (Frackman et al., 1990; Meighen, 1993). This is in contrast to the low temperature optima for beetle luciferases (Luc) and other characterized bacterial luciferases (those from *Vibrio spp.*).

1.1.2. Local Determinants of Photon Emission from Photoproteins. Photon emission from the luciferase reaction is dependent on each of components of the reaction: luciferase, substrate (luciferin), Mg^{++}, energy (ATP or NADPH), and oxygen. Constitutive expression of the photoprotein can be achieved such that enzyme levels are uniform over time and when cells are located in different tissue compartments. Since the bacterial luciferase from *P. luminescens* is encoded on an operon which also encodes enzymes for the biosynthesis of substrate (decanal), and all genes on the operon are expressed in nearly equimolar concentrations, substrate is not limiting in bacterial cells expressing this operon. The other components of the light emitting reaction (Mg^{++} and energy) are available, in excess, in living cells. Oxygen concentrations can be variable and

affect photon emission such that signals, on a per cell basis, reflect local oxygen concentrations.

Since there are aerobic and anaerobic compartments in the GI tract and Salmonella is an intestinal pathogen by virtue of its ability to move from the anaerobic lumen of the intestine to the oxygenated microenvironments of cell surfaces and cytoplasmic vacuoles, the bioluminescent signals from an O_2 dependent bacterial luciferase expressed in these bacterial cells is indicative of the position of the pathogen in this pathogenic process (Contag et al., 1995). In these studies a large fraction of Salmonella, that were expressing luciferase, were not bioluminescent due to the anaerobic environment within the lumen of the intestine. Injection of oxygen, in the form of air, into the intestinal lumen of infected animals resulted in a strong and immediate bioluminescent signal demonstrating that luciferase could be used as an *in vivo* indicator of oxygen levels.

1.2. Regulation of Gene Expression by Changes in Oxygen Concentration in Mammalian Tissues

Similar to labeling bacterial cells, mammalian cells can be labeled with luciferases from eukaryotic organisms. Thus, the study of mammalian gene expression and can be performed in vivo (Contag et al., 1996, 1997). A significant advantage of *in vivo* luciferase monitoring over conventional assays is that a given bioluminescent tag can be used in cell culture correlates of *in vivo* processes, then used in animals to test predictions made in these correlative assays, and finally used in *ex vivo* assays to confirm *in vivo* observations. Thus, luciferase is a biological tag that can streamline assay development from cell culture to living animal models and biochemical analyses.

The effects of shifts in oxygen homeostasis in mammalian tissues (hypo- or hyperoxia) can be observed at the level of gene expression and a number of genes have been identified that can be altered in their expression by transient changes in oxygen concentration (Table 2). Alterations in gene expression due to changing tissue oxygenation are mediated by transcription factors, primarily HIF-1, that bind defined responsive elements in the promoters of these regulated genes (Table 2). Linking the promoter elements of these genes to the luciferase coding sequence would enable external monitoring of the effects that changes in oxygen levels have on cells in living animals. In this case the luciferase enzyme serves as a reporter for gene expression and is not directly sensing the oxygen concentrations. Thus, light emission would be an indication of transcriptional changes due to transient fluctuations in tissue oxygenation. Monitoring would be performed hours after removal of the oxygen deprivation insult to the cells or tissues. A luciferase derived from a eukaryotic organism, such as the firefly, would be used in this scenario and the substrate supplied exogenously rather than made by the luciferase-expressing cells.

Mammalian expression vectors can be designed to respond to changes in tissue conditions and report, with light, when changes occur. This would include use of the oxygen sensitive genetic elements to monitor the effects of hyperoxia or hypoxia/ischemia. These vectors are propagated in *E. coli* and used to transfect mammalian cell lines in culture, or transduce primary mammalian cells. These cell cultures serve as correlative assays for testing responsiveness of the genetic constructs to the desired stimuli and may be predictive of the response in living animals. Once tested in cell culture, stable lines of labeled cells can be used for monitoring the biological response following adoptive transfer. Alternatively, the expression vectors may also be used to directly transfect cells *in vivo*,

Table 2. Genes sensitive to changes in tissue oxygenation

Protein and Gene	Signal and Response	Regulation[a]	Unique features[b]
Heme oxygenase 1 (hmox-1 or HO-1)	Induction with hyper- and hypoxia	HIF-1	General stress responsive promoter; HO protein is an antioxidant
Vegf	Induced by hypoxia	HIF-1	Angiogenesis
Erythropoetin (Epo)	Induced by hypoxia	HIF-1	Erythropoiesis
p53	Induced by hypoxia	Regulated at the post transcriptional level	Regulates ischemic apoptosis and is a tumor suppressor gene
Tyrosine hydroxylase	Induced by hypoxia	HIF-1, AP-1	Rate limiting enzyme in synthesis of catecholamines
Nitric oxide synthase 1 (NOS1)	Induced by hypoxia	HIF-1	Synthesizes nitric oxide which regulates vascular tone
Glucose trasporter 1 and glycolytic enzymes: aldolase A, enolase 1, lactate dehydrogenase A, phosphofructokinase L, phosphoglycerate kinase 1	Induced by hypoxia	HIF-1	Allows for increased anaerobic ATP synthesis
Transferrin	Induced by hypoxia	HIF-1	Delivers iron to the bone marrow for incorporation into hemoglobin

[a] The transcription factors that are thought to be involved in the oxygen dependent regulation are listed.
[b] The features that are related to tissue oxygenation are listed.

as in DNA-based therapeutics and vaccines, or transferred to embryos to create transgenic lines of mice that have "built-in" reporters. *In vivo* transfections and embryo injections can be monitored noninvasively and delivery methods optimized through external detection. In transgenic animals each of the animals' cells has the potential of responding with light to specific signals. To test each of these approaches for studying mammalian gene expression, we have linked the promoter from the inducible heme oxygenase (HO-1) gene to the coding sequence of a modified firefly luciferase, and studied its expression after stable integration into the chromosomes of cells in culture and as a transgene in mice.

Heme oxygenases are enzymes which cleave heme to form equal molar amounts of CO and biliverdin, which is subsequently converted to bilirubin by biliverdin reductase (Tenhunen et al., 1968): HO is the key catabolic enzyme in this pathway. Two distinct isoforms of heme oxygenase, HO-1 and HO-2, have been identified in human, rat, and mouse (Maines et al., 1988; Trakshel and Maines, 1989; Applegate, 1991). Although these two isoenzymes are biochemically and genetically distinct, substrate and cofactor requirements are similar (Trakshel and Maines, 1989). It has been shown that expression of HO-1, also known as Heat shock protein 32 (Hsp32), is dramatically induced by its substrate heme and by several forms of biological stress, for example, heavy metals, heat shock, UV radiation, and oxidative species (Reviewed by Maines, 1992). Modulation of HO-2 expression has been only observed in response to glucocorticoids (Weber et al., 1994).

The 5′-flanking regulatory sequence of HO-1 and the responsive *cis*-elements within this promoter region have been well documented (reviewed by Choi and Alam, 1996).

2. METHODS

2.1. Labeling Bacterial Cells

Well characterized vectors have been used for the expression of the *lux* operons in Gram negative organisms. The original pUC-based plasmid used to clone the *lux* operon from *P. luminescens* including the promoter region (vector designated pCGLS1; Frackman et al., 1990), is well suited for expressing luciferase from Gram-negative organisms, such as *E. coli* and *Salmonella*. Introduction of the *lux* encoding vector into cells of these bacterial species can be carried out using standard methods of bacterial cell transformation for Gram-negative organisms.

2.2. Transient Transfection of Mammalian Cells and Selection of Stable Integrations

Although a wide variety of methods have been described for transfection of cell lines in culture and cells *in vivo*, we have used cationic liposome mediated gene transfer for each of these applications, using procedures recommended by the manufacturer (Gibco-BRL, Gaithersberg, MD). A wide variety of mammalian expression vectors that express firefly luciferase have been described. Many of these mammalian promoters either contain cryptic bacterial promoters or read-though occurs from upstream promoters in bacterial cells such that *E. coli* containing an intact *luc* reading frame can be selected by growing the transformed cells on bacterial plates containing the substrate luciferin (5 mM, Molecular Probes, Analytical Luminescent Laboratories or Biosynth). Note, however, that even though this permits selection of transformed bacteria with intact *luc* reading frames, the intensity of bioluminescence in these bacterial cells does not predict levels of expression in mammalian cells.

For labeling mammalian cells in culture, DNA delivery was optimized with respect to DNA and Lipofectin concentrations for each cell line. For cell transfections DNA ranges of 0.5 to 5 μg and Lipofectin amounts of 5 to 20 μl (in 5 μl increments) with each DNA concentration were used in the optimization, as recommended by the manufacturer (Gibco-BRL). Optimal conditions were determined by monitoring luciferase expression in living cells by adding luciferin to the transiently expressing cells at 24 h post transfection and monitoring expression (light production) with an intensified charge coupled device (ICCD) camera (Hamamatsu model C2400-32) at 24 h intervals for 72 h. For transient expression assays in several cell lines, 2 μg of DNA and 6 μl of Lipofectin were optimal in 1 ml of medium.

Stable cell lines were generated using optimal conditions as determined in the transient expression assays. In order to facilitate selection of stable lines, the luciferase expression vector (pGL3) was cotransfected with a vector encoding a Zeocin resistance gene (pZeo, Invitrogen, San Diego, CA) at a ratio of either 10:1 or 5:1 (pZeo to the luciferase expression vector). After 24 h Zeocin (a bleomycin derivative, Invitrogen, San Diego, CA) was added to the wells at a predetermined optimal concentration between 100 and 500 μg/ml. Such optima were determined prior to transfections by testing the susceptibility of cells to Zeocin for each cell line. Drug concentrations of 5 μg/ml to 1000 μg/ml

(5, 10, 25, 50, 100, 250, 500 and 1000 μg/ml) were evaluated in triplicate for each cell line. The lowest drug concentration that resulted in significant killing of cultures in 3 days was used for selection of stable transfectants.

Optimal conditions for transfection were determined for NIH 3T3 and HeLa cells, and stable bioluminescent lines were generated by cotransfection with Luc expression vectors and pZeo in 1 ml culture volume for 3 h (in six well dishes). Cultures were expanded to 6 ml in a 100 mm dish, and after 24 h Zeocin was added to a final concentration of 500 μg/ml. 250 μg/ml of Zeocin has been used for the selection of stable human T cell lines derived from Jurkat and CEM. When colonies were visible, luciferin was added to the cultures at a concentration of 5 mM and cultures screened for bioluminescent colonies with the Hamamatsu ICCD camera. Bioluminescent colonies were selected by pipetting cells off the culture dish directly, or using a cloning ring, and cells were transferred to a 96 well dish. Cultures were expanded over a period of 15–21 d. Subsequently, samples were frozen in medium containing fetal Bovine serum, growth medium and dimethyl sulfoxide (in a ratio of 5:4:1)

2.3. Transgenic Mice

A transgenic line was generated in our laboratory where the transgene was the promoter from murine heme oxygenase 1 (HO-1; Alam et al., 1994; Lee et al., 1997) fused to modified version of the North American firefly luciferase gene (Promega, Madison WI). Modifications to this Luc coding sequence included optimization of mammalian codon usage and removal of the peroxisome targeting site for cytoplasmic expression. This transgenic mouse line is designated HO-luc Tg. All mice were maintained in the Research Animal Facility at Stanford University under strict adherence to institutional guidelines.

2.4. Imaging

For imaging, mice and rats were anesthetized and imaged as previously described (Contag et al., 1995, 1997) using pentobarbital (35–70 mg/kg body weight). Low doses (35 mg/kg) were used both with neonatal mice and rats, both of which are sensitive to the effects of pentobarbital. While under anesthetic animals were observed closely, when not in the dark box, and respiration monitored. Following the imaging animals were kept warm and allowed to recover.

An aqueous solution of the substrate, luciferin (15 milligrams pre milliliter; mg/ml), was injected into the peritoneal cavity 20 min. prior to imaging for systemic delivery at a dose of 150 mg/kilogram body weight; kg). At the time of imaging, anesthetized animals were placed in a light-tight chamber, and a grayscale body surface reference image was collected in low room light provided by leaving the chamber door slightly open. Subsequently, the door to the chamber was closed, to exclude the room light that would obscure the relatively dimmer luciferase bioluminescence. Photons emitted from luciferase within the animal, and then transmitted through the tissue, were collected and integrated for a period of time between 5 and 30 min. A pseudocolor image representing light intensity (generally blue representing least intense and red most intense light) was generated on an Argus 20 image processor (Hamamatsu, Japan); images were transferred using a plug-in module (Hamamatsu, Japan) for Adobe Photoshop (Adobe, Mountain View, CA) to a computer (Macintosh 8100/100, Apple Computer, Cupertino, CA). Grayscale reference images and pseudocolor images were superimposed using the image processing software, and saved as PICT files. Superimposed images were opened within another graphics software package (Canvas, Deneba, Miami FL) and composites made and annotations added. An intensified CCD camera (C2400-32, Hamamatsu), fitted with a 50 mm f 1.2 Nikkor lens (Nikon, Japan) and an image processor (Argus 20, Hamamatsu) was used. The capabilities of alternative low light imaging devices employing different technologies have been assessed (Contag et al., 1997).

3. RESULTS

3.1. Intestinal Pathogens in Aerobic and Anaerobic Environments

Luciferase reactions are oxygen dependent and thus when these enzymes are expressed constitutively in tissues the level of light emission can be a function of tissue oxygenation. The direct effects of *in vivo* O_2 levels on enzyme activity in living animals has been demonstrated in the Salmonella gastrointestinal (GI) infection model (Figure 1) (Contag et al., 1995). In this model, it was demonstrated that the technique was quantitative, and its use resulted in more information in less time than conventional assays.

The fact that Salmonella in the lumen of the intestine expressed luciferase, but were not bioluminescent, was demonstrated in mice with bioluminescent signals from cecum, and not the colon, that excreted bioluminescent Salmonella (data not shown). This was then tested and confirmed by injecting oxygen, in the form of air, into the intestinal lumen at positions above and below the luminescent cecum (Figure 1). Upon introduction of oxygen, bioluminescent signals were apparent in the regions where signals were previously absent. This indicated that the bioluminescent signal was indicative of an intimate or pathogenic association of Salmonella with the intestinal epithelial cells, either form of attachment or entry into mouse cells, where the bacterial cells had access to oxygen from the animal (Contag et al., 1995).

3.2. Gene Expression

In HO-luc transgenic mice, gene induction could be observed following treatment of mice with cadmium ions (Cd^{2+}). Injection of 5 mM $CdCl_2$ into the peritoneal cavity

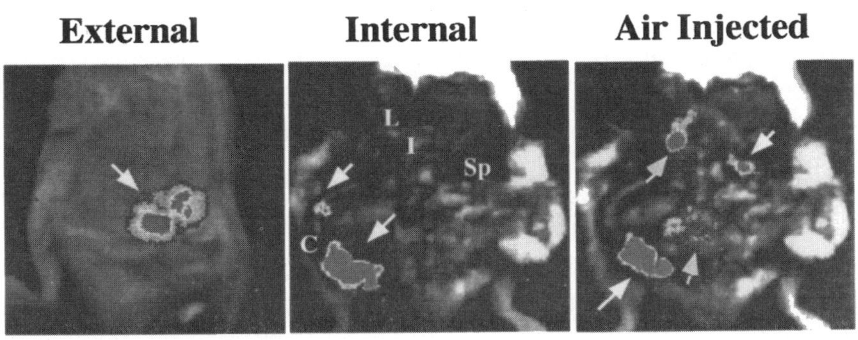

Figure 1. Oxygen dependence of luciferase reaction during a GI infection with Salmonella. Mice were infected with a virulent strain of *S. typhimurium* (strain SL1344) and imaged externally. When a focal signal (marked with arrows) was apparent, the animal was imaged again following laprotomy; a focal signal was again apparent. Oxygen was injected in the form of air at two places above and below the cecum and the animal imaged. Signals were apparent at the sites of injection (marked with arrows).

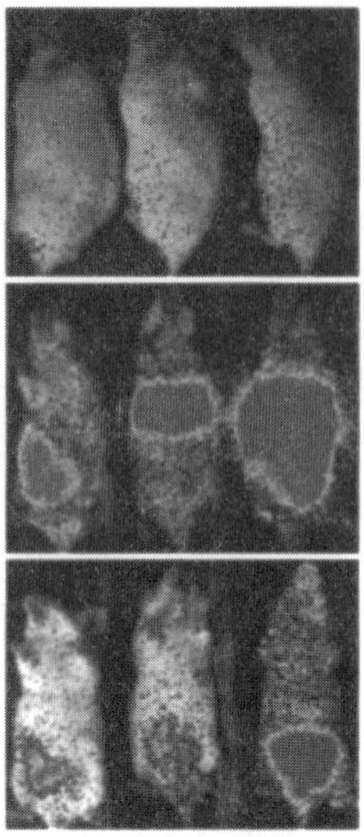

Figure 2. Spatiotemporal analysis of gene expression in living transgenic mice. Three mice containing the HO-luc transgene were treated with 10* μM $CdCl^{++}$ (intraperitoneal injection), and expression of the HO-1 promoter followed over time (shown in hours after treatment) using photon emission from firefly luciferase as an indicator. Peak expression was observed at 24 h followed by a gradual decrease.

resulted in significant increases in light production with peak expression at about 9 h post $CdCl_2$ treatment (Figure 2). This signal over the abdominal region diminished over the period of 24 to 96 h post treatment. Thus, transcriptional regulation of this gene could be monitored in vivo and the spatiotemporal changes monitored over time. The HO-1 promoter is sensitive to changes in local oxygen concentrations, both to hypoxia and hyperoxia, therefore this transgenic mouse line may serve as a good indicator strains for oxygen sensing in vivo.

4. DISCUSSION

The examples reviewed here demonstrate that external detection of *in vivo* gene expression is possible and quantitative, that oxygen availability in compartments of the GI tract affect the luciferase reaction, and that changes in these processes can be quantitatively assessed *in vivo*, suggesting that bioluminescence may provide a rapid real time method for monitoring tissue oxygenation noninvasively in living laboratory animals. Understanding biological processes in the context of the living mammal is particularly relevant to the study of tissue oxygenation. The *ex vivo* methods by which these events are typically monitored require removal of tissue and consequently the loss of contextual influences of the living animal. External detection of luciferase expression in the tissues of living experimental animals may be useful in assessing both the dynamic changes in tissue oxygenation and the physiological changes that occur after exposure to hyper- or hypoxic conditions in real time.

Noninvasive monitoring of light emitted from within a living mammal, where the light is constitutively expressed or reporting fluctuations in gene expression, is a platform technology that results in increased information about specific physiological processes and integrated, whole mammalian systems. The monitoring of biological light requires less time and fewer animals than conventional methods, thus reducing the cost of obtaining biologically relevant data in living animals. This is especially relevant to the costly processes of drug discovery where significant advances have been made in generating large numbers of potentially useful therapeutic compounds (e.g. combinatorial chemistries, genomics and high throughput screening). In the absence of significant advances in *in vivo* analyses, the bottleneck in understanding key processes and facilitating drug development is at the animal model step and the extraordinary developments in the earlier steps accentuates this bottleneck. *In vivo* luciferase monitoring addresses this limitation with broad applications in the areas of infectious disease (viral, bacterial and fungal), tumor progression and metastasis, gene therapy, mammalian development, and many other areas in which animal models are used as predictors for the human response to therapy.

ACKNOWLEDGMENTS

We thank Hamamatsu Corporation for technical support in use of their photonic detection systems, This work was funded in part through grants from NIH 1R01 HL58013, ONR contract N-00014-94-1024 and unrestricted gifts from the Mary L. Johnson Research Fund, The Providence Foundation, and the and Baxter Foundations.

REFERENCES

Alam, J, J Cai, and A Smith. 1994. Isolation and characterization of the mouse heme oxygenase-1 gene. Distal 5′ sequences are required for induction by heme or heavy metals. J. Biol. Chem. 269(2):1001–1009.

Applegate, LA, A Neol, G Vile, E Frenk, and RM Tyrrell. 1991. Two genes contribute to different extents to the heme oxygenase activity measured in cultured human skin fibroblasts and keratinocytes: implications for protection against oxidant stress. Photochem Photobiol. 61:285–291.

Benaron, D, W-F Cheong, and DK Stevenson. 1997. Tissue Optics. Science 276:2002–2003.

Campbell, AK. 1988. Chemiluminescence. Principles and Applications in Biology and Medicine. Ellis Horwood Ltd. and VCH Verlagsgesellschaft mbH, Chichester, England.

Choi, AMK and J Alam. 1996. Heme oxygenase-1: Function, regilation, and implication of a novel stress-inducibale protein in oxidant-induced lung injury. Am. J. Respir. Cell Mol. Biol. 15:9–19.

Contag, CH, SD Spilman, P Contag, M Oshiro, BF Eames, P Dennery, DK Stevenson, and DA Benaron. 1997. Visualizing gene expression in living mammals using a bioluminescent reporter. Photochem Photobiol. 66:523–531.

Contag, CH and PR Contag. 1996. Viewing disease progression through a bioluminescent window. Optics & Photonics News:22–23.

Contag, CH, PR Contag, JI Mullins, SD Spilman, DK Stevenson, and DA Benaron. 1995. Photonic detection of bacterial pathogens in living hosts. Mol. Microbiol. 18:593–603.

Contag, CH, PR Contag, SD Spilman, DK Stevenson, and DA Benaron. 1996. Photonic monitoring of infectious disease and gene regulation, pp. 220–224. *In* E Sevick-Muraca and D Benaron (ed.), OSA TOPS on Biomedical Optical Spectroscopy and Diagnostics, vol. 3. Optical Society of America.

Contag, PR, IN Olomu, DK Stevenson, and CH Contag. 1998. Bioluminescent reporters in living mammals. Nature Med. 4:245–247.

Frackman, S, M Anhalt, and KH Nealson. 1990. Cloning, organization, and expression of the bioluminescence genes of *Xenorhabdus luminescens*. J. Bact. 172:5767–5773.

Hastings, JW. 1996. Chemistries and colors of bioluminescent reactions: a review. Gene 173:5–11.

Jobsis, FF. 1977. Noninvasive, infrared monitoring of cerebral and myocardial oxygen sufficiency and circulatory parameters. Science 198:1264.

Lee PJ, BH Jiang, BY Chin, NV Iyer, J Alam, GL Semenza, and AM Choi. 1997. Hypoxia-inducible factor-1 mediates transcriptional activation of the heme oxygenase-1 gene in response to hypoxia. J. Biol. Chem. 272(9):5375–81.

Maines, MD. 1988. Heme oxygenase: Function, multiplicity, regulatory mechanisms and clinical applications. FASEB J. 2:2557.

Maines, MD. 1992. Heme oxygenase: Clinical application and functions, CRC Press, Inc., Boca Raton, FL.

Meighen, E. 1993. Bacterial bioluminescence: organization, regulation, and application of the *lux* genes. FASEB J. 7:1016–1022.

Sanderson, KE and BAD Stocker. 1987. *Salmonella typhimurium* strains used in genetic analysis, pp. 1220–1224. *In* FC e. i. c. Neidhardt and JL Ingraham and KB Low and B Magasanik and M Schaechter and E Umbarger (ed.), *Escherichia coli* and *Salmonella typhimurium*. Cellular and Molecular Biology, vol. 2. American Society of Microbiology, Washington DC.

Szittner, R and E Meighen. 1990. Nucleotide sequence, expression and properties of luciferase coded by *lux* genes from terrestrial bacterium. J. Biol. Chem. 265:16581–16587.

Tenhunen, R, HS Marver, and R Schmid. 1968. The enzymatic conversion of heme to bilirubin by micrisomal heme oxygenase. Pro. Natl. Acad. Sci. USA 61:748–755.

Trakshel, GM and MD Maines. 1989. Multiplicity of heme oxygenase isozymes. HO-1 and HO-2 are different molecular species in rat and rabbit. J. Biol. Chem. 264:1323–1328.

Weber, CM, BC Eke, and MD Maines. 1994. Corticosterone regulates heme oxygenase-2 and NO synthase transcription and protein expression in rat brain. J. Neurochem 63:953–962.

Xi, L, KW Cho, and SC Tu. 1991. Cloning and nucleotide sequences of *lux* genes and characterization of luciferase of *Xenorhabdus luminescens* from a human wound. J. Bact. 173:1399–1405

MONTE CARLO SIMULATION AND THE SPECTROSCOPIC ANALYSIS OF THE RESPONSE TO NEURAL STIMULATION IN BRAIN

Ying Zheng,[1] Bob Vuksanovic,[1] Yuqian Hou,[2] and John Mayhew[2]

[1]Department of Automatic Control and Systems Engineering
University of Sheffield
Mappin Street, Sheffield S1 3JD
United Kingdom
[2]Artificial Intelligence Vision Research Unit
and Department of Psychology
University of Sheffield
Western Bank, Sheffield S10 2TP
United Kingdom

1. INTRODUCTION

Optical imaging spectroscopy is a technique for investigating the metabolic effects of brain activity (Jobsis et al., 1977; Malonek and Grinvald, 1996). It exploits the fact that the brain contains chromophores and these affect the spectra of light remitted (transmitted or reflected) from the brain's surface. Spectrophotometry can be used to estimate changes in the relative proportions or volume fractions of these chromophores. The three most easily measured and common chromophores in brain are involved in the metabolic pathway satisfying the demands of active neurons. They are oxygenated and deoxygenated hemoglobin and the cytochromes, (particularly cytochrome oxidase) (Chance, 1991).

The issue that is central to all the methods of analysing in vivo optical spectroscopic data is the complication introduced by the effects of tissue scattering. There are grounds for challenging the adequacy of the optical spectroscopic analysis methodology used in many studies. The difficulty arises from the effects of tissue scattering and weaknesses in the algorithms which on the whole are unable to deal adequately with the non-linearities introduced by the variation of optical path length with wavelength.

Oxygen Transport to Tissue XXI, edited by Eke and Delpy.
Kluwer Academic / Plenum Publishers, New York, 1999.

This problem is well known (Cope et al., 1991; Lubbers and Wodick, 1969), however, especially in the visible wavelengths, we know of no algorithms suitable for the analysis of in vivo diffuse reflectance spectroscopic data.

To address this issue we describe below a Monte Carlo Simulation of light transport in tissue to simulate reflectance spectroscopy experiments using diffuse illumination over a physiologically plausible range of tissue absorption and scattering parameters. The results of the MCS were used to develop two functionals, one to characterise the wavelength dependent path lengths of remitted photons, the other to correct the measured absorption spectra. These functionals were then used as the basis of modification of the Beer-Lambert Law, we call path length scaling analysis (PLSA).

We know of no study that actually measured the distribution of path lengths from reflectance imaging in the visible ranges 500–640 nm, but our current MCS results suggest a range of mean path lengths from ~1 to ~3 mm over the wavelengths 505–645 nm. (the mean maximum depths are about a factor of 3 smaller). The path lengths at 620 nm are ~3 times longer than at 577 nm. Below 600 nm the characteristic shape of the haemoglobin absorption peaks can be detected. This form of wavelength-dependent relationship might be expected as the expression for the mean free path length (the mean distance between scattering and absorption 'interactions' in a turbid medium) is $1/(\mu_a + \mu_s)$ (Wilson and Jacques, 1990) where μ_a is the coefficient of absorption which is determined largely by the haemoglobin present in the tissue, and μ_s is the scattering coefficient, which is well approximated by a ramp decreasing with wavelength (van der Zee et al., 1991).

The principle ways MCSs have been used to interpret spectrographic (predominantly NIRS) data are: i) to invert the measurements by either using a form of precomputed look up table from which it is possible by interpolation to find the tissue absorption and scattering parameters that best fit the measurement data (Flock et al., 1989; van der Zee et al., 1991); or ii) to use the MCS to produce estimates of the mean path length at different wavelengths in the appropriate imaging geometry which can be then used in a multi-component spectral analysis algorithm using a variant of the Beer-Lambert law. This is to regard the tissue as a complex cuvette having a distribution of 'widths' (Lubbers and Wodick, 1969), and the MCS is used to estimate that distribution.

We describe below a spectrographic analysis algorithm called path length scaling analysis (PLSA) which takes into account the effects of light scattering by tissue to introduce a wavelength dependency on remitted photon path length. The PLSA spectrographic analysis algorithm i) models the wavelength dependency of the path lengths with a functional of the form $1/f(\mu_a, \mu_s)$, and ii) corrects the measured attenuation. We then use least squares to solve for the terms corresponding to changes in the concentration of the chromophores and changes in the path length parameters in an extended variation of the Beer-Lambert Law.

2. METHODS

The MCS of light transport through tissue tracks photon paths through a history of scattering events until the photon is either remitted or absorbed. The outcome of each simulation was determined by three parameters: the absorption coefficient μ_a, the scattering coefficient μ_s and the scattering anisotropy parameter g which determines the change in direction following a scattering event.

Our MCSs were carried out for g = 0.85 and over twelve wavelengths. At each wavelength, the absorption and scattering coefficients were calculated using:

$$\begin{aligned}\mu_a(\lambda) &= \frac{1}{\log_{10}(e)}[c_o\varepsilon_o(\lambda) + c_d\varepsilon_d(\lambda)] \qquad (\text{mm}^{-1}) \\ &= \frac{c_t}{0.4343}[s\varepsilon_o(\lambda) + (1-S)\varepsilon_d(\lambda)]\end{aligned} \tag{2.1}$$

$$\mu_s(\lambda) = -\frac{\lambda}{30} + \frac{110}{3} + \Delta\mu_s \qquad (\text{mm}^{-1}) \tag{2.2}$$

where c_o, c_d are the concentrations of oxy- and deoxygenated hemoglobin, respectively; $\varepsilon_o, \varepsilon_d$ are the extinction coefficients for oxy- and deoxygenated hemoglobin, respectively; $c_t = c_o + c_d$ is the concentration of hemoglobin (μM) (also denoted by Hbt), S is the oxygen saturation in tissue given by $S = c_o/c_t = c_o/(c_o + c_d)$, and $\Delta\mu_s$ is change or offset in μ_s. Note that we have modelled μ_s as a linear function of wavelength based on the data given by van der Zee (van der Zee et al., 1991). The change in scattering coefficient is assumed independent of wavelength and hence shifts the linear relationship by a constant term. Using MCS, we investigate how changes in hemoglobin concentration c_t, oxygen saturation S and scattering offset $\Delta\mu_s$ affect path length and absorption, both of which are modelled as functions of wavelength.

2.1. Path Length Modelling

2.1.1. Diffuse Illumination without Camera Optics. In this case, the MCSs were carried out over a circular region of 5 mm radius. For each simulation, the number of photons emitted into the tissue was 4 millions. Simulations were run for various parameter settings, and under each condition, the mean path length was calculated for each of the twelve wavelengths. It was found that path length was not constant over wavelengths. Moreover the inverse path length looked similar in shape to the specific extinction coefficients for HbO_2 and Hbr. Thus we used the following model for path length estimation:

$$\frac{1}{P(\lambda)} = \alpha(S\varepsilon_o(\lambda) + (1-S)\varepsilon_d(\lambda)) + \beta\mu_s(\lambda) + \gamma \tag{2.3}$$

The parameters α, β and γ can be estimated by the least squares algorithm, the results of which are shown in Figure 1. It can be seen that α, β and γ are dependent on Hbt, saturation and scattering, in particular Hbt. However the residues obtained indicate that eqn. (2.3) contains most information for path length modelling. Provided the total haemoglobin concentration (Hbt) does not change significantly during an experiment, the path length model thus obtained yields adequate results on simulated data, as will be seen in section 3 below.

2.1.2. Diffuse Illumination through a Spectrographic Slit with Camera Optics. To improve the accuracy of our simulation, additional camera optics with spectrographic slit were added to the existing MC model. The mean path lengths for (i) photons going through the slit, (ii) photons missing the slit, and (iii) photons being absorbed in tissue can then be calculated, shown in Figure 2. Clearly they are different from each other. As expected, a different set of parameters α, β and γ were obtained in each case.

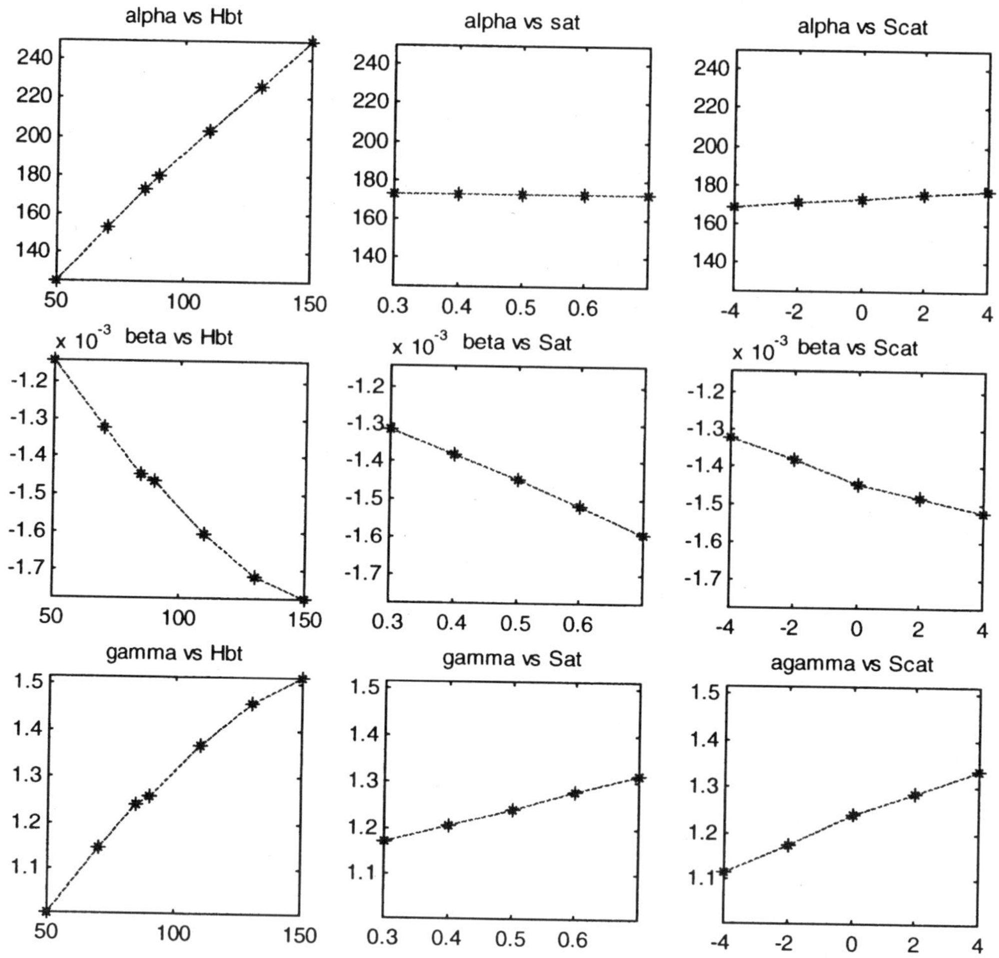

Figure 1. Least squares estimates of α, β and γ as Hbt, saturation and scattering vary.

2.2. Attenuation Modeling

The Beer Lambert Law states that the attenuation of transmitted collimated light through tissue (without scattering) is given by $A_B(\lambda) = P(\lambda)(c_o\varepsilon_o(\lambda) + c_d\varepsilon_d(\lambda))$. Our MCS showed that in general there was a difference between the measured attenuation $A_m(\lambda)$ obtained from the number of remitted photons and the attenuation $A_B(\lambda)$ obtained from the Beer-Lambert Law. This difference exists whether we use unity path length, the path length obtained from the diffuse illumination, or the three path lengths calculated when the camera optics were introduced. In order to relate the measured attenuation to the physiological parameters, the following absorption modelling strategies were investigated.

(a) Define the path length as

$$P_1(\lambda) = \frac{A_m}{c_o\varepsilon_o + c_d\varepsilon_d} = \frac{A_m}{A_{B1}}$$

where A_{B1} was the attenuation from the Beer-Lambert Law assuming unity path length. We then modelled the inverse path length $1/P_1(\lambda)$ using eqn. (2.3). However when exam-

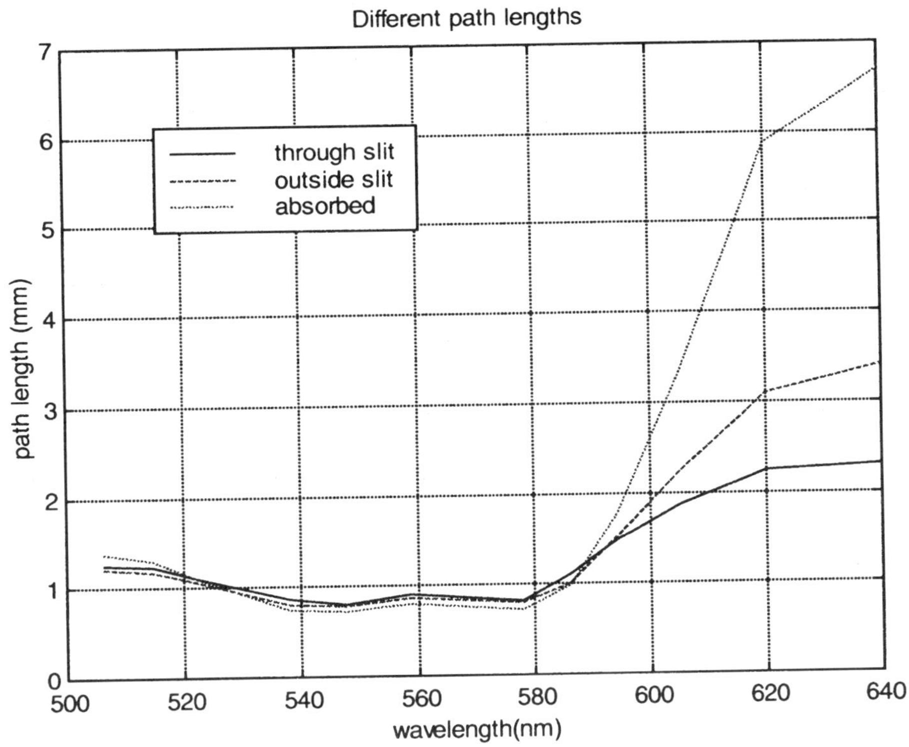

Figure 2. Mean path lengths for photons entering the slit (solid line), missing the slit (broken line) and absorbed in tissue (dotted line) for $c_t = 85$, $S = 0.5$, $\Delta\mu_s = 0$.

ining the residue it was unacceptably large for long wavelength range. This was due to modelling the inverse path length. Attempts were made to model path length instead. At this stage, the results have not led to any more satisfactory improvement.

(b) Modelling the path length by a composition of path lengths obtained using the camera optics and slit. In this case the path length was modelled as

$$P_2(\lambda) = \frac{A_m}{A_{B1}} = w_1 P_{abs}(\lambda) + w_2 P_r(\lambda)$$

where P_{abs} and P_r were the mean path lengths of the photons absorbed in the tissue and remitted through the slit, respectively. The two weightings w_1 and w_2 were obtained using the least squares algorithm. Again when examining the residue, the results were not satisfactory. Various combinations of path lengths were used without significantly improving the result.

(c) Model the measured attenuation and the attenuation from the Beer Lambert Law as

$$A_B = aA_m + b(\lambda) + c \tag{2.4}$$

where a, b and c were parameters to be estimated. Figure 3 shows a, b and c estimated using the least squares algorithm for the diffuse illumination case. The residue was considerably reduced. We therefore used eqn. (2.4) in the path length scaling

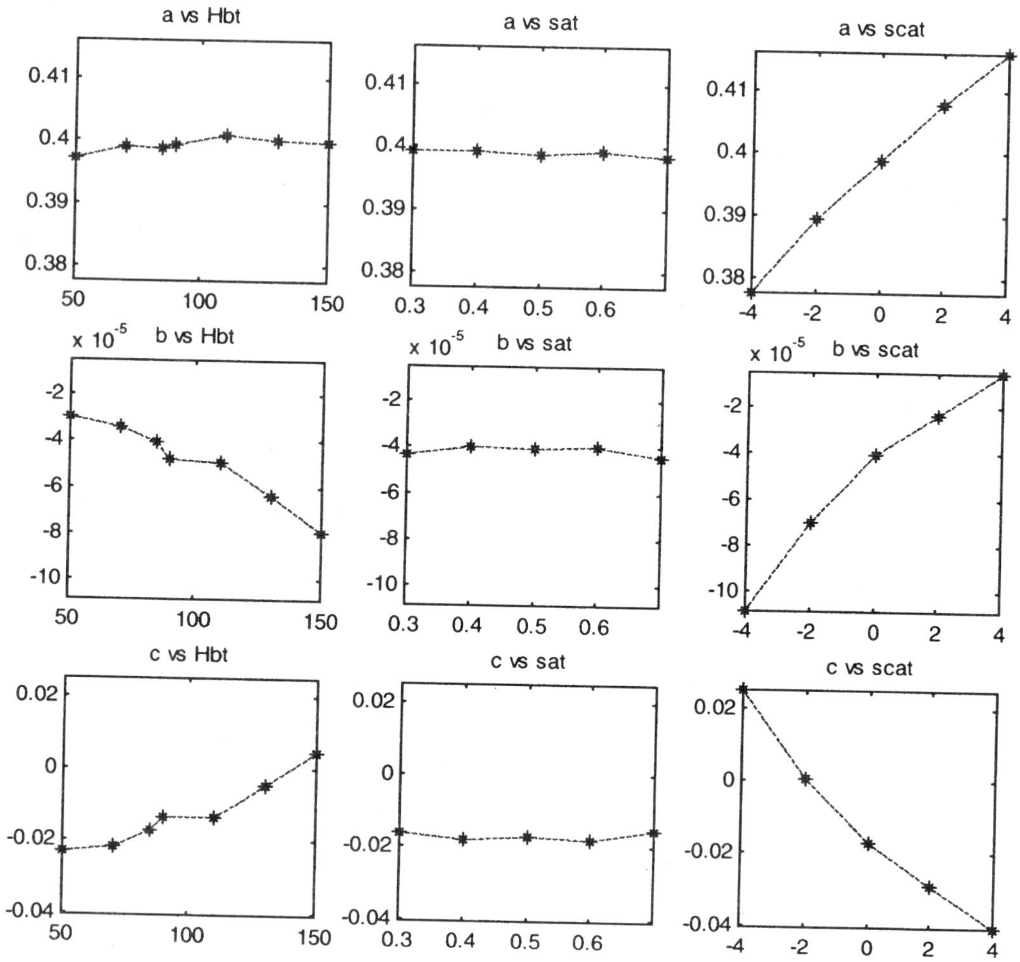

Figure 3. Least squares estimates of a, b and c as of Hbt, saturation and scattering vary.

analysis below to correct for the measured attenuation before applying the Beer-Lambert Law.

2.3. Estimating Changes in Haemoglobin Parameters

The objective of modelling path length and attenuation is to investigate small changes in the haemoglobin parameters. Specifically we are interested in estimating changes in c_o and c_d, as well as changes due to scattering, reflected by changes in β and γ. (For simplicity in this paper, we have neglected cytochrome in our model.) These changes can be estimated as follows. First, the measured absorption is corrected using (2.4), yielding

$$A_c(\lambda) = \hat{A}_B = aA_m(\lambda) + b\lambda + c$$

where $A_c(\lambda)$ is the corrected attenuation measurement which would satisfy the Beer Lambert Law. Hence from the Beer Lambert Law,

$$A_c(\lambda) = P(\lambda)(c_o\varepsilon_o(\lambda) + c_d\varepsilon_d(\lambda))$$
$$= \frac{c_o\varepsilon_o(\lambda) + c_d\varepsilon_{d(\lambda)}}{a(S\varepsilon_o(\lambda) + (1-S)\varepsilon_d(\lambda)) + \beta\lambda + \gamma} \tag{2.5}$$

It can be seen that A_c is a ratio, and hence is not affected by multiplying the numerator and denominator by an arbitrary scaling factor. It is thus necessary to keep one of the path length parameters (α, β or γ) constant. We have selected α because we find that variations in α over a wide range of values of the physiological parameters have relatively little effect on the path length once the baseline value for total blood concentaration (Hbt) has been set.

After a little algebra, the expression relating changes in the measurements of attenuation to changes in the concentrations of the chromophores and the path length parameters is:

$$\Delta A_c = A_c^t - A_c^0 \approx \frac{\partial A_c}{\partial c_0}\Delta c_0 + \frac{\partial A_c}{\partial c_d}\Delta c_d + \frac{\partial A_c}{\partial \beta}\Delta\beta + \frac{\partial A_c}{\partial \gamma}\Delta\gamma$$

where A^t_c is the (corrected) post stimulation attenuation at time t, and A^0_c is the (corrected) pre stimulation baseline attenuation. Using equation (2.5), we obtain the following.

$$\frac{\partial A_c}{\partial c_o} = P\left(\varepsilon_o - \frac{A_c^0 \alpha c_d(\varepsilon_o - \varepsilon_d)}{(c_o + c_d)^2}\right) \qquad \frac{\partial A_c}{\partial c_d} = P\left(\varepsilon_d - \frac{A_c^0 \alpha c_o(\varepsilon_o - \varepsilon_d)}{(c_o + c_d)^2}\right)$$

$$\frac{\partial A_c}{\partial \beta} = -P * A_c^0 * \lambda \qquad \frac{\partial A_c}{\partial \gamma} = -P * A_c^0$$

A series of MCS experiments were conducted with various changes in the total concentration of haemoglobin (Hbt) and saturation (S). Provided the changes are small (about less than 10% for Hbt and less than 20% for S), we can accurately recover changes in c_o and c_d. These changes are well within the range of normal physiological variation.

3. RESULTS

The path length and attenuation modelling described above were used to analyse the spectra of the haemodynamic responses to 1 s vibrissal stimulation in rat somatosensory cortex. (For details of the experimental procedures, see (Mayhew et al., 1998).) Briefly, six second time series of spatial spectroscopic data were collected at 30 Hz from somatosensory cortex of anaesthetised rats (8 subjects) through a thinned skull cranial window. Stimulation of the whisker for 1 s occurred during the third second. Pre processing and averaging over trials produced a spectral time series of 180 frames over the wavelength range 505–640 nm. The individual subject data was analysed using the path length scaling analysis (PLSA) algorithm.

The parameters used in the PLSA are:

$$c_t = 120\ \mu M,\ S = 0.5,\ \alpha = 210,\ \beta = -0.0017,\ \gamma = 1.4,\ a = 0.4,\ b = -0.00006,\ c = -0.009.$$

The results of the PLSA algorithm were compared with results obtained from the simple linear form of the Beer-Lambert Law which assumes no wavelength dependency of path length (i.e., unity path length). We refer to this method as linear multi-component analysis (LMCA). The MCS experiments show a very large change in path length between the wavelengths below 586 and above 600 nm, thus a test of the stability of the two algorithms is to compare the estimates of the haemoglobin changes using spectral data between the ranges 505 to 586 nm with those between 586 and 645 nm. The results of the analysis are shown in Figure 4.

Over the range 505–586 nm the hemoglobin results using both PLSA and LMCA methods are very similar (Figures 4a and c), i.e., there is an increase in oxygenated haemoglobin and a decrease in deoxygenated haemoglobin below baseline. However, over the range 586–645 nm, the result using PLSA (Figure 4b) is very similar to that over the low wavelength range, whereas the analysis using LMCA (Figure 4d) shows a pronounced biphasic change in the concentration of de-oxygenated hemoglobin which is inconsistent with the results obtained using LMCA for data in the low wavelength range.

These results can be explained as follows: in the region of high haemoglobin absorption (505–585 nm) the differences in the wavelength-dependent path length are relatively slight and the assumption of constant path length is 'almost correct'. In contrast over the full range of wavelengths (505–645 nm) the assumption of constant path length is

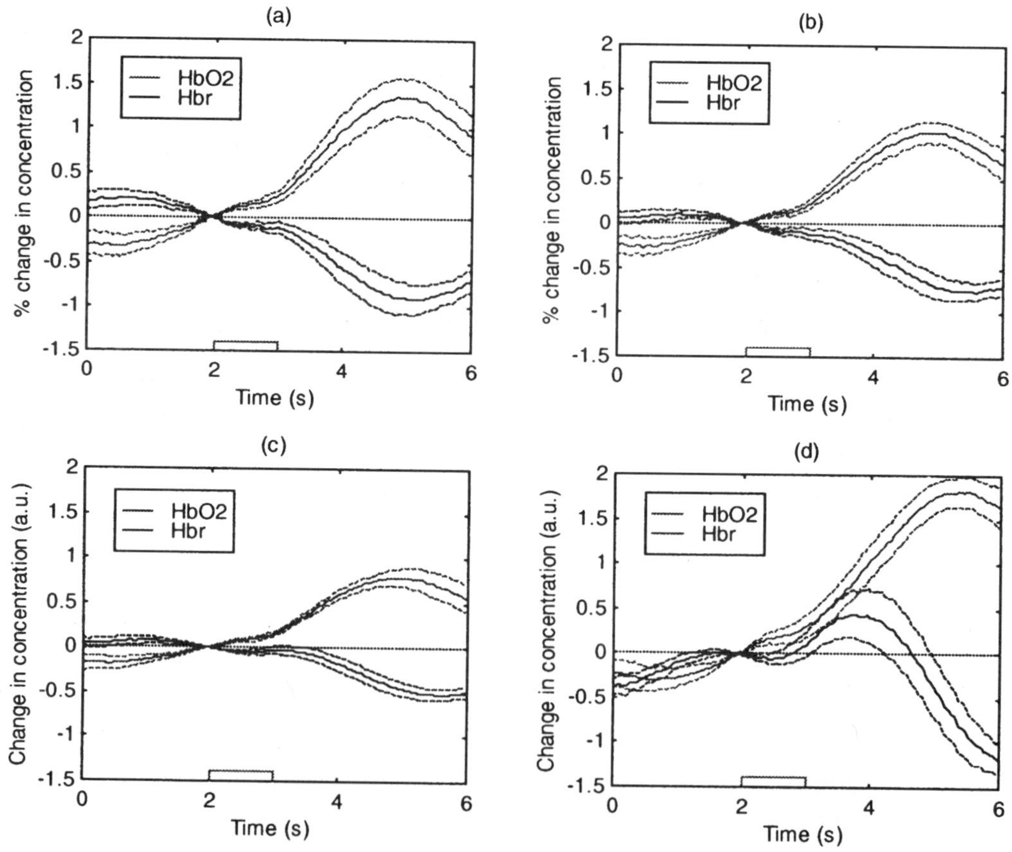

Figure 4. Time series of chromophore concentration changes. Solid lines are the temporal changes and dotted lines represent error bounds. Results of PLSA over the wavelength (a) 505–645 nm, (b) 586–645 nm. Results of LMCA over the wavelength (c) 505–645 nm, (d) 586–645 nm.

incorrect, with a factor of three change between 577 and 620 nm and the results of the LMCA in this region may be not only quantitatively incorrect but also qualitatively misleading. The PLSA shows no such pronounced qualitative changes dependent on which range of wavelength of data were used.

4. DISCUSSION

The conclusion from our MCS experiments is that in the visible wavelengths the volume of tissue sampled is very different at the different illumination wavelengths, and furthermore the depth of tissue sampled may easily differ by more than a factor of 2 or more at different wavelengths. This wavelength dependency of the volume of tissue sampled must be taken into account if the estimates of the changes in chromophore concentrations from the spectroscopic analysis of remitted illumination from brain are to be quantitatively accurate. If the simple linear form of the Beer-Lambert Law is used then there is the possibility that the estimates of the chromophore concentration changes will be in error.

Another important implication of the path length research presented here is that differences in the depths of tissue sampled could be important in studies to investigate the hemodynamic and metabolic responses to stimulation. Some of the differences in the results of mapping studies using different wavelength illumination (Frostig et al., 1990) might in part be due to the different depths of tissue sampled. However, in the preparation we report here, the similarity between the results of the PLSA analysis over the different ranges of wavelength suggest that the response to neural activation in rat somatosensory cortex is 'equally' distributed throughout the depth of tissue sampled.

ACKNOWLEDGMENTS

The authors gratefully acknowledge the contribution by Jason Berwick, Steve Askew, and Pete Coffey from the Neuro Imaging Research Group in Sheffield for their contribution to the collection and analysis of the physiological data we describe. Dr. Y. Hou was supported by EPSRC GR/K36171 and by Wellcome Trust project no. 050080/Z/97/Z. Dr. B. Vuksanovik was supported by EPSRC GR/L35386.

The physiological component of the work was supported using start up funding provided by UniSheff.

REFERENCES

Chance, B 1991. Optical Method. *Annu. Rev. Biophys. Chem.* **20**:1–28.

Cope, M, van der Zee, P, Essenpreis, M, Arridge, SR, and Delpy, DT 1991. Data analysis methods for near infrared spectroscopy of tissue: problems in determining the relative cytochrome aa3 concentration. *SPIE* **1431**:251–262.

Flock, ST, Wilson, BC, and Patterson, MS 1989. Monte Carlo modelling of light propagation in highly scattering tissues- II: Comparison with measurements in phantoms. *IEEE. Trans. Biom. Eng.* **36**:1169–1173.

Frostig, RD, Lieke, EE, Ts'o, D, and Grinvald, A 1990. Cortical functional architecture and local coupling between neuronal activity and the microcirculation revealed by in vivo high-resolution optical imaging of intrinsic signals. *Proc. Natn. Acad. Sci. U.S.A.* **87**:6082–6086.

Jobsis, FF, Keizer, JH, LaManna, JC, and Rosenthal, M 1977. Reflectance spectrophotometry of cytochrome aa3 in vivo. *J. Appl. Physiol.: Respirat. Eviron. Exercise Physiol.* **43(5)**:858–872.

Lubbers, DW and Wodick, R 1969. The examination of multicomponent systems in biological materials by means of a rapid scanning photometer. *Applied Optics* **8**:1055–1062.

Malonek, D and Grinvald, A 1996. Interactions between electrical activity and cortical microcirculation revealed by imaging spectroscopy: implications for functional brain mapping. *Science* **272**:551–554.

Mayhew, J, Zheng, Y, Hou, Y, Vuksanovic, B, Berwick, J, Askew, S, and Coffey, P 1998. Spectroscopic Analysis of Changes in Remitted Illumination: The Response to Increased Neural Activity in Brain. *AIVRU Memo 127 (submitted to Neuroimage)*.

van der Zee, P, Essenpreis, M, Delpy, DT, and Cope, M 1991. Accurate determination of the optical properties of biological tissues using a Monte Carlo inversion technique., ICO: Atmosphere, volume and surface scattering and prpagation, Florence Italy, pp. 125–128.

Wilson, BC and Jacques, SL 1990. Optical reflectance and transmittance of tissues: principles and applications. *IEEE. J. Quantum Mechanics.* **26**:2186–2199.

AUTHOR INDEX

SUBJECT INDEX

Participants Photograph (Taken on Trinity square in front of Matthins Church.)

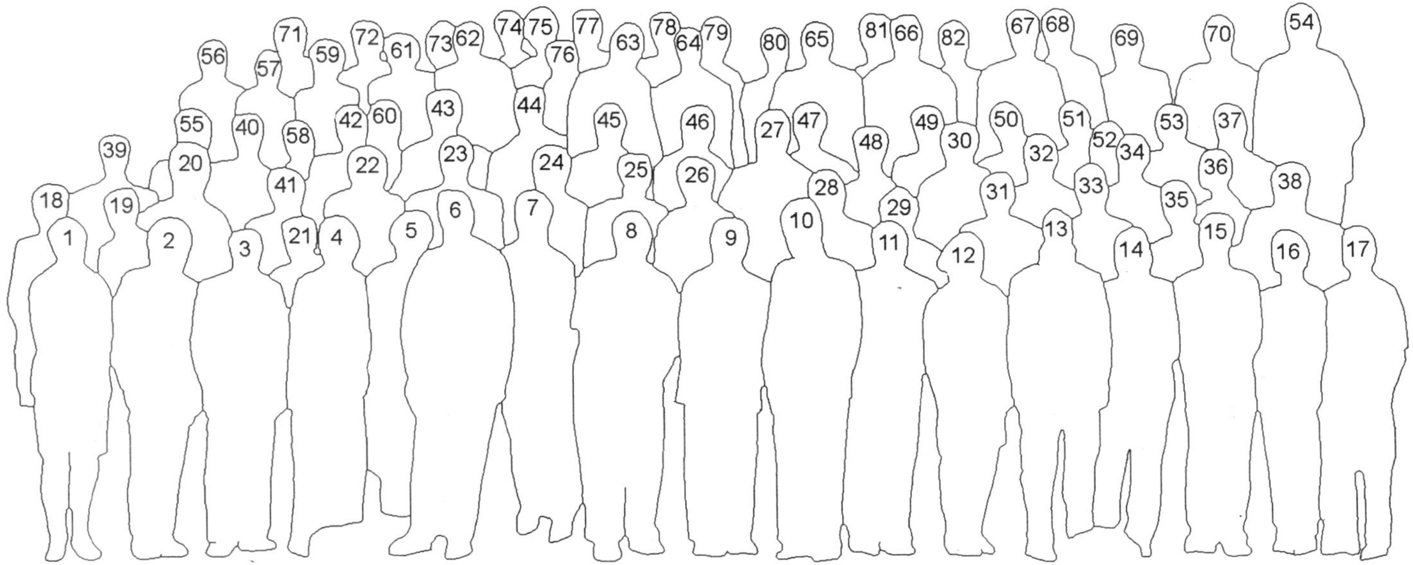

1 Thorniley, MS
2 Kida, I
3 Harrison, DK
4 Jue, T
5 Nioka, S
6 Wolf, M
7 Springett, R
8 Vojnovic, B
9 Chance, B
10 Eke, A
11 Martinez-Coll, A
12 Quaresima, V
13 Lemmers, JPWM
14 van Beekvelt, MCP
15 Colier, W
16 Hori, Mrs
17 Hori, H
18 Tomita, M
19 Tomita, Mrs
20 Graaff, R
21 Harrison, Mrs
22 Hyder, F
23 Stratonnikov, A
24 Thews, O
25 Swartz, HM
26 Vaupel, P
27 Elwell, CE
28 Sugioka, K
29 LaManna, JC
30 Popel, AS
31 Hoofd, L
32 Jarm, T
33 Oeseburg, B
34 Ferrari, M
35 van de Ven, M
36 Binzoni, T
37 Erdmann, W
38 Hudetz, AG
39 van Beek, HJHGM
40 van der Kleij, AJ
41 Delpy, DT
42 Faithfull, NS
43 Groebe, K
44 Cooper, C
45 Cicutti, NJ
46 Lovell, AT
47 Dunn, JF
48 Bruley, Mrs
49 Bruley, DF
50 Mayevsky, A
51 Schleinkofer, L
52 Erdmann, Mrs
53 Contag, CH
54 Maguire, D
55 Barnikol, WKR
56 Sakanoue, J
57 Kerr, ME
58 Keipert, P
59 Coremans, A
60 van Rossem, K
61 Lübbers, DW
62 Mayhew, JEW
63 Hamar, J
64 Lutz, J
65 Goldman, D
66 Balcer, HI
67 Benyó, Z
68 Dave, AS
69 Pittman, RN
70 Turek, Z
71 Barnett, N
72 Severinghaus, JW
73 Zatelli, R
74 Ortsäter, H
75 Liss, P
76 Kang, KA
77 Liss, AG
78 Bassingthwaighte, JB
79 Rakusan, K
80 Necas, E
81 Wilson, DF
82 Spiker, JO